THIRD EDITION

Transport Phenomena Fundamentals

CHEMICAL INDUSTRIES

A Series of Reference Books and Textbooks

Founding Editor

HEINZ HEINEMANN
Berkeley, California

Series Editor

JAMES G. SPEIGHT
CD & W, Inc.
Laramie, Wyoming

MOST RECENTLY PUBLISHED

Transport Phenomena Fundamentals, Third Edition, Joel Plawsky

Synthetics, Mineral Oils, and Bio-Based Lubricants: Chemistry and Technology, Second Edition, Leslie R. Rudnick

Modeling of Processes and Reactors for Upgrading of Heavy Petroleum, Jorge Ancheyta

Synthetics, Mineral Oils, and Bio-Based Lubricants: Chemistry and Technology, Second Edition, Leslie R. Rudnick

Fundamentals of Automatic Process Control, Uttam Ray Chaudhuri and Utpal Ray Chaudhuri

The Chemistry and Technology of Coal, Third Edition, James G. Speight

Practical Handbook on Biodiesel Production and Properties, Mushtaq Ahmad, Mir Ajab Khan, Muhammad Zafar, and Shazia Sultana

Introduction to Process Control, Second Edition, Jose A. Romagnoli and Ahmet Palazoglu

Fundamentals of Petroleum and Petrochemical Engineering, Uttam Ray Chaudhuri

Advances in Fluid Catalytic Cracking: Testing, Characterization, and Environmental Regulations, edited by Mario L. Occelli

Advances in Fischer-Tropsch Synthesis, Catalysts, and Catalysis, edited by Burton H. Davis and Mario L. Occelli

Transport Phenomena Fundamentals, Second Edition, Joel Plawsky

Asphaltenes: Chemical Transformation during Hydroprocessing of Heavy Oils, Jorge Ancheyta, Fernando Trejo, and Mohan Singh Rana

Chemical Reaction Engineering and Reactor Technology, Tapio O. Salmi, Jyri-Pekka Mikkola, and Johan P. Warna

Lubricant Additives: Chemistry and Applications, Second Edition, edited by Leslie R. Rudnick

Catalysis of Organic Reactions, edited by Michael L. Prunier

The Scientist or Engineer as an Expert Witness, James G. Speight

Process Chemistry of Petroleum Macromolecules, Irwin A. Wiehe

THIRD EDITION

Transport Phenomena Fundamentals

Joel L. Plawsky

CRC Press is an imprint of the
Taylor & Francis Group, an **informa** business

CRC Press
Taylor & Francis Group
6000 Broken Sound Parkway NW, Suite 300
Boca Raton, FL 33487-2742

Printed on acid-free paper
Version Date: 20131004

International Standard Book Number-13: 978-1-4665-5533-4 (Hardback)

Visit the Taylor & Francis Web site at
http://www.taylorandfrancis.com

and the CRC Press Web site at
http://www.crcpress.com

For my wife, Gail, who endured my total silence for way too long as I revised and corrected the text for this edition

Contents

PART II Multidimensional, Convective, and Radiative Transport

Preface

The primary purposes of this third edition were to

1. Eliminate, as much as possible, the number of typographical, cut-and-paste, and numerical errors that remained or cropped up during the writing of the second edition
2. Eliminate some old-fashioned/graphical approaches to working problems in favor of solutions that make use of more modern tools like MATLAB®, Maple®, or COMSOL®
3. Introduce some new concepts that may expand the appeal of the text beyond chemical engineering alone
4. Introduce new problems and reorganize the problems at the end of each chapter to make it easier for the instructor to use the text and for students to understand how to approach the problems

The objective of this book remains to streamline how transport phenomena is taught so that breadth and depth need not be pitted against one another to the point where a student's education in the subject is severely diluted. The basis for this is a unified treatment of heat, mass, and momentum transport based on a balance equation approach (In − Out + Gen = Acc). The text is used in a two-term transport phenomena sequence offered at Rensselaer Polytechnic Institute. Part I, Chapters 1 through 8, forms the first term in the transport sequence. It takes the student through the balance equation in the context of diffusive transport, be it momentum, energy, mass, or charge. Each chapter adds a term to the balance equation highlighting the effects of that addition on the physical behavior of the system and the underlying mathematical description. The students also become familiar with modeling efforts and developing mathematical expressions based on the analysis of a control volume, the derivation of the governing differential equations, and the solution of that equation with appropriate boundary conditions.

Part II, Chapters 9 through 15, forms the second term in the sequence and builds on the balance equation description of diffusive transport by introducing convective transport terms. The second term focuses on partial rather than ordinary differential equations. The Navier–Stokes and convective transport equations are derived from balance equations in both macroscopic and microscopic forms. Students are introduced to paring down the full, microscopic equations to model a situation and determine a solution to their problem. When that approach fails or is too detailed, they are introduced to macroscopic versions of those models that arise upon averaging over one or more coordinate directions. This leads to a discussion of the momentum, Bernoulli, energy, and species continuity equations and a brief description of how these equations are applied to heat exchangers, continuous contactors, and chemical reactors.

Following the derivation of the convective transport equations, the students are introduced to the three fundamental transport coefficients: the friction factor, the heat transfer coefficient, and the mass transfer coefficient. Boundary layer theory is used as the basis for deriving theoretical values for these coefficients. The book introduces boundary layers on a flat plate as the student's first line of attack and builds from there to handle flow over and within curved surfaces and finally turbulent flow. Appropriate correlations for the transport coefficients are also introduced. The last chapter of the text covers the basics of radiative heat transfer and introduces students to important concepts such as the blackbody, radiation shields, and enclosures.

This third edition of the text incorporates many changes to material covered in the first and second editions. At least 70 new homework problems have been introduced, and a number of the original problems have been revised. The problems are now organized at the back of each chapter so that they are easier to assign and the students have a better idea of what is being asked. The changes begin in Chapter 2 where the discussion on non-Newtonian fluid behavior has been updated and expanded. In Chapter 3, a section on how to describe the viscosity of a suspension has been added. A discussion of interfacial thermal resistance was added to Chapter 4 following the introduction of flux discontinuities in mass transfer. The two topics are intimately related, and as microelectronics and MEMS devices proliferate, interfacial thermal resistance becomes an increasingly important topic. The major change in Chapter 5 was to replace the transistor example with that of a diode. The diode is easier to understand and also to present in both the text and for the instructor. Major changes to Chapter 6 include removing references to Heisler charts for transient convective problems in favor of one-term approximations using the actual series solution. Modern tools such as MATLAB, Maple, or Mathematica make it very easy to evaluate, plot, and even invert the series so that the use of graphical representations is no longer needed. In the semi-infinite section, where moving boundaries are discussed, the topic of front propagation and traveling waves is introduced along with the related phenomenon of the Kirkendall effect in solids. A discussion of Turing patterns is introduced in Chapter 9 to show how diffusion, when coupled with reaction, does not always act to smear out concentration gradients. Lift concepts are reinforced in Chapter 10, especially in the problems. Chapter 13 introduces some changes in the discussion of the drag curves around single spheres and cylinders and incorporates some equation-based representations of the drag curves. Chapter 14 introduces a number of updated correlations for turbulent flows, including more recent theoretical correlations by Churchill and coworkers. In Chapter 15, changes include an updated discussion of blackbody radiation and the incorporation of the data regarding the fraction of radiation emitted within a given region to Appendix D. Finally, the numerical modules introduced as problems in the second edition have been removed from the text and placed in Appendix G. Since these are constantly changing, it seemed more appropriate to add them to an appendix and refer to a website.

I hope this book will be relatively straightforward for students to follow. The first nine chapters represent a form of "phonics" for transport phenomena. Close coupling of the transport phenomena and the repetition of model development and solution sequences are stressed, though not every mathematical step could be shown in every case. This provides for a form of pattern matching via analogy between the phenomena that should reinforce basic concepts. The level of mathematics is sufficient to solve interesting problems and most importantly to validate basic numerical solutions and so help the students understand when the software they will eventually rely on provides a useful answer and when it provides a physical solution. This validation exercise is perhaps the most important component of a student's education. It will hopefully force them to be perpetually skeptical, to reconnect with the physical world, and to build the deeper understanding that is the key to being a good engineer.

Joel L. Plawsky

Maplesoft
615 Kumpf Drive
Waterloo, ON Canada N2V 1K8
Toll Free (Canada & USA): 1-800-267-6583
Phone: +1-519-747-2373
Fax: +1-519-747-5284
Sales: info@maplesoft.com
www.maplesoft.com

COMSOL, Inc.
1 New England Executive Park
Suite 350
Burlington, MA 01803 USA
Tel: +1-781-273-3322
Fax: +1-781-273-6603
info@comsol.com
www.comsol.com

Mathematica
Wolfram Research
100 Trade Center Drive
Champaign, IL 61820-7237 USA
Phone: +1-217-398-0700
+1-800-WOLFRAM (965-3726, US and Canada only)
Fax: +1-217-398-0747
www.wolfram.com

Acknowledgments

Special thanks are due to Professor Susan Sharfstein and her transport students at the State University of New York's College of Nanoscale Science and Engineering. They exhibited extraordinary patience and perseverance, were the best technical editors one could hope to have, and have helped to sharpen up a number of technical discussions in the text.

I would like to thank my wife, Gail, for her assistance in writing the book and her help in making sure my explanations were clear and my grammar passable. Finally, I want to express my thanks to Barbara Glunn, Kari Budyk, Jill Jurgensen, and the staff at CRC Press and Taylor & Francis Group for all their assistance in making this book a reality and for giving me the opportunity to improve the approach and execution of that approach in this third edition.

Author

Joel L. Plawsky received his BS in chemical engineering from the University of Michigan and his MSCEP and ScD in chemical engineering from the Massachusetts Institute of Technology. After graduation, Joel worked for Corning Inc. in their research division before returning to academia at Rensselaer Polytechnic Institute. He is currently a professor of chemical engineering in the Howard P. Isermann Department of Chemical and Biological Engineering.

Joel was a NASA Faculty Fellow in 1999 and 2000 and has had four experiments fly in the microgravity environment of the Space Shuttle and the International Space Station. He was a visiting professor of chemical engineering at Delft University of Technology in 2002. He is a fellow of the American Institute of Chemical Engineers where he has served as the chairman of the Transport and Energy Processes Division and has received the Institute's Herbie Epstein Award. Joel serves on the editorial board of *Chemical Engineering Communications* and is the holder of four patents.

The cover art shows three recent examples from Joel's research. The top left figure is a simulation of the heat generation rate within the furnace portion of a chemical reactor used in the Claus process to transform hydrogen sulfide gas produced during petroleum refining into elemental sulfur. The work was supported by the New York State Pollution Prevention Institute. The image on the lower left is an interference pattern superimposed on a color enhanced reflectivity image of the inside surface of the CVB heat pipe experiment run on the International Space Station in 2010. The image allows one to determine the liquid film thickness on the walls of the object, the shape of the vapor-liquid interface, and the fluid mechanics and heat transfer responsible for the operation of the device. This work was supported by NASA. The image on the right is a Ge photonic crystal structure deposited on a Si wafer by a process known as oblique angle deposition. This work was supported by NSF as part of the Smart Lighting Engineering Research Center program.

Part I

Transport Fundamentals and 1-D Systems

1 Introductory Concepts

1.1 INTRODUCTION

Classical transport phenomena is concerned with the exchange of energy, mass, or momentum between systems of engineering interest and is usually taught in courses on fluid mechanics, heat transfer, and mass transfer. Though now ranking along with thermodynamics, mechanics, and electromagnetism as one of the cornerstones of engineering education, a course or text devoted exclusively to transport phenomena became popular only in the late 1950s and early 1960s [1–3]. Over the past several decades, the unifying transport concept has revolutionized the way engineering science is taught. Momentum, energy, and mass transport can now be grouped together due to their similar mathematical framework, the similar molecular mechanism through which the transport of these quantities may be visualized, and their importance to all engineering disciplines. The scope of transport phenomena encompasses all agents of physical change, and transport processes are fundamental to the evolution of the universe and to the success of all life on Earth. As engineers, we can use the concepts embodied in transport phenomena to analyze the workings of natural systems [4] and to mimic those systems for the betterment of the environment, society, and the human condition.

1.2 SCOPE OF TRANSPORT PHENOMENA

A true appreciation of the universe is impossible without an understanding of the concepts involved in transport phenomena. All change occurs through some kind of transport process. Momentum transport is responsible for much of the physical geography of the planet. As an example of this, erosion due to the flow of the Oxara River formed the canyon near Thingvellir in Iceland shown in Figure 1.1. The buttes, mesas, and terraces of the American southwest were also created by running water. In those cases, flash floods generated by intermittent rainfall eroded specific regions of the layered sediment forming the characteristic patterns. Running water due to the incessant rainfall of the Kaanapali coast of Kauai formed the features seen in Figure 1.2.

Similar stresses can be created by flowing gases. Wind erosion is responsible for shifting sand dunes in the great deserts of the world as shown in Figure 1.3 [5]. Water, however, is more powerful and, in its solid form, created the barrier islands of North Carolina, the fjords of Norway, and the Great Lakes. Figure 1.4 shows the mouth of the Vatnajokull glacier in Iceland. The icebergs in the photo look black due to the sediment they carved from the face of the mountain.

The transport of energy leads to momentum transport and causes the continents to move. The temperature difference between the Earth's core and the crust drives the flow of mantle fluid. Hotter mantle fluid near the core is less dense and rises, while cooler mantle fluid near the crust is denser and sinks. Figure 1.5 shows a computer simulation of mantle convection. The streamlines illustrating the flow path and arrows showing the direction of motion are indicated in the figure. Light areas represent higher temperatures and darker colors represent cooler temperatures.

The exchange of energy between the Sun and the Earth provides the energy to drive the weather. Couplings of heat and momentum transport in the atmosphere and in the oceans, similar to those occurring between the core and crust of the Earth, result in tornadoes, hurricanes, the major weather front systems of the midlatitudes, and the great ocean currents [6]. Figure 1.6 shows a cross section of a typical hurricane or tropical cyclone driven by these processes. The exchange of radiative energy between the Sun and the Earth also makes most life on this planet possible.

FIGURE 1.1 A canyon in Iceland near Thingvellir. The canyon was created by water-based erosion of the rock and soil—an example of momentum transport.

FIGURE 1.2 Erosion of a mountain on the Kaanapali coast of Kauai.

Beyond the inanimate examples mentioned earlier, all living systems exist by exploiting the processes of transport phenomena. At a fundamental cellular level, the synthesis of DNA and proteins involves sophisticated transport mechanisms for delivering energy and raw materials to the synthesizing organelles [7]. Similar situations exist for the transport of nutrients and waste materials across cell walls and membranes. Ions are transported between the neurons in the brains and bodies of animals to make movement, thought, and sensation possible [8].

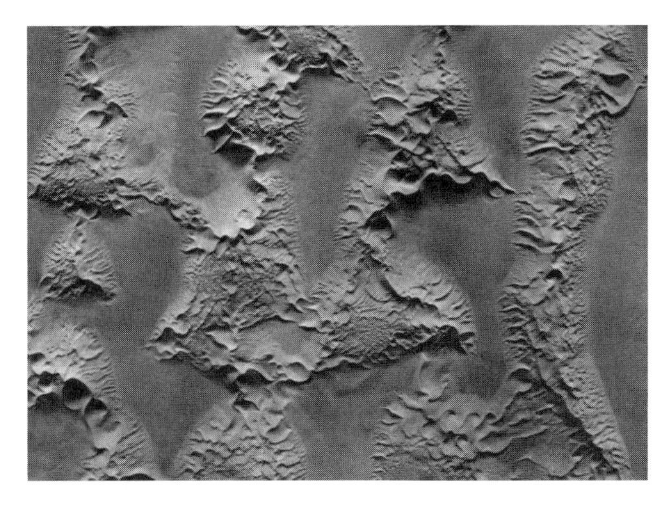

FIGURE 1.3 Sand dunes of the Libyan Desert seen from an altitude of 10 km. The peaks rise to a height of 180 m above the intervening flat ground. (Photo courtesy of Aero Service Corporation.)

FIGURE 1.4 An iceberg from the Vatnajokull glacier. The dark color is volcanic sediment of the glacier carved from the mountain in the background.

At a higher level, fluids are pumped through every part of a plant or animal to deliver nutrients and remove wastes. Oxygen and carbon dioxide are exchanged in the lungs of land animals, the gills of fish, and the stomata of plants. Figure 1.7 is a drawing of the bronchial system within the human respiratory tract [9]. The great surface area in these systems assists in the absorption and exchange of gases and represents an extreme example of an engineering concept humans exploit called extended surfaces.

Heat is exchanged between the bodies of plants or animals and the environment to keep biological processes operating at a necessary rate. At the level of the entire organism, we can see that nature has designed complete packages for exploiting transport phenomena in ways that make man-made systems crude by comparison. One need only see how a shark moves through the water or how a hawk or vulture rides the thermal updrafts for hours without beating its wings to see how expertly nature has designed its systems for exploiting momentum transport. Those outward manifestations are meager compared to the sophisticated transport mechanisms

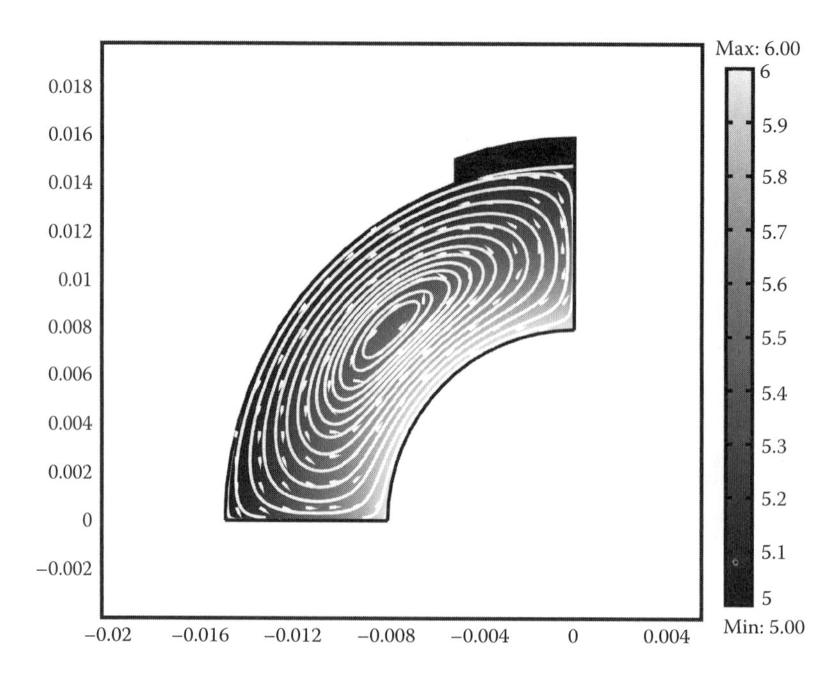

FIGURE 1.5 The plate tectonic mechanism for continent motion. The motion is driven by the coupling of heat and momentum transport between the core and crust of the Earth. This figure shows how cellular convection forced the motion of the early crust and current continents (simulation performed using Comsol®).

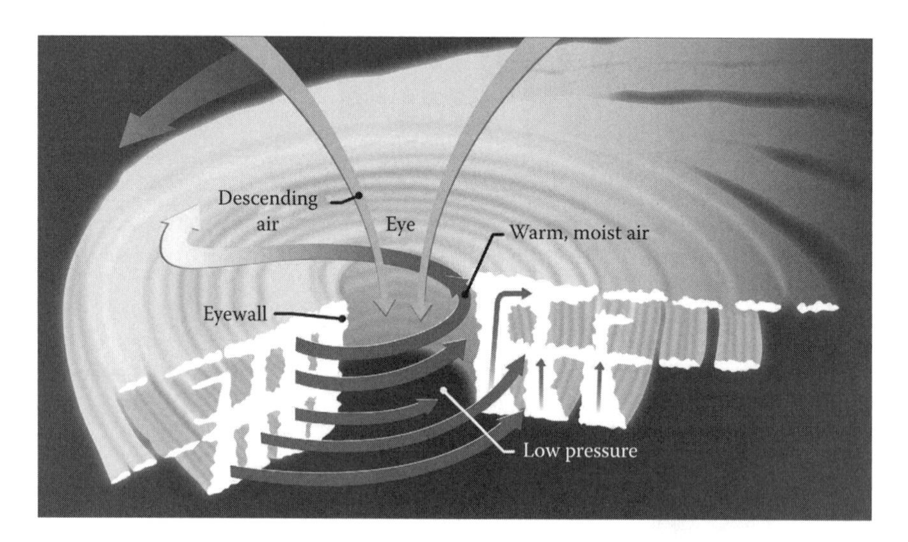

FIGURE 1.6 Cross section of a typical tropical cyclone. Winds and violent weather are driven by coupled heat, mass, and momentum transport. (From National Aeronautics and Space Administration, http://spaceplace.nasa.gov/hurricanes/hurricane_diagram_large.en.jpg.)

keeping the human body alive. As just one example, the energy conversion and transport mechanism, the Krebs cycle, taking ingested sugars and converting them into energy, carbon dioxide, and water, runs at an efficiency of 37% [7]. This makes the Krebs cycle at 37°C, one of the most efficient engines known. By contrast, power plants must operate at many hundreds of degrees to approach that kind of efficiency.

Though we rely on the exploitation of transport phenomena for the routine maintenance of our bodies, transport phenomena is of fundamental importance to the way we interact with our

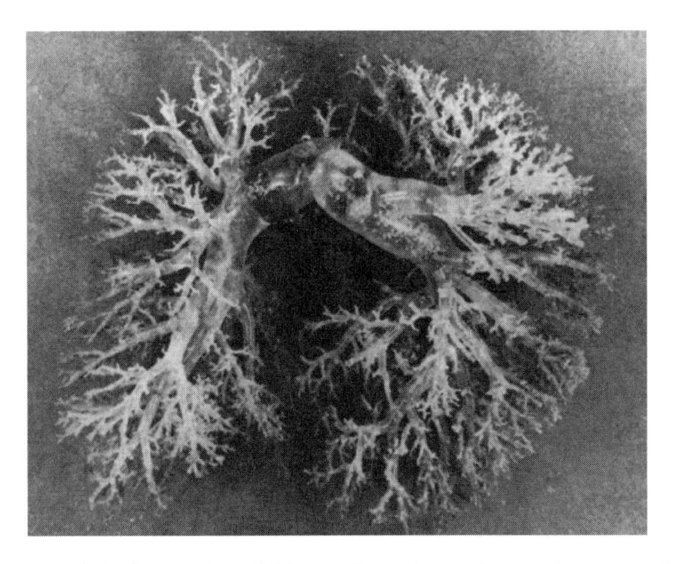

FIGURE 1.7 Diagram of the human bronchial tree from the trachea to the terminal bronchioli. (From Cumming, G. et al., The functional morphology of the pulmonary circulation, in Fishman, A.P. and Hect, H.H., eds., *The Pulmonary Circulation and Interstitial Space*, The University of Chicago Press, Chicago, IL, 1969. With permission.)

environment and has shaped our lives and culture. The discovery of fire and the use of clothing for insulation enabled us to expand our habitable environment to include nearly all places on Earth. The use of heat transfer to cook and preserve food enabled early humans to eat a wider variety of foods and prevent illnesses occurring from raw foods or those containing natural heat-labile toxins. Following heat transfer came the exploitation of momentum transport. Irrigation, as an example, enabled us to water our crops regardless of the weather and so provide a reliable source of food. The aqueducts of Rome (Figure 1.8) enabled the Romans to provide clean water to large cities and provided for sewage disposal as well.

The bubonic plague was conquered only when these principles of momentum transport (and sewage disposal) were rediscovered and exploited. Perhaps the most dramatic example of momentum

FIGURE 1.8 The Pont du Gard Bridge over the aqueducts of Nemausus. (From Hauck, G., *The Aqueduct of Nemausus*, McFarland & Co., Jefferson, NC, 1988.)

FIGURE 1.9 Sailboat in Sydney Harbor. Harnessing momentum transport allowed humans to colonize the globe.

transport exploitation was when humans first took to the sea and discovered that sails could be used to capture the wind and power their ships (Figure 1.9). In modern times, we have exploited momentum transport for flight. We have exploited heat transfer to build power plants (Figure 1.10), run our automobiles, and move materials from one location to another. We have exploited mass transfer to engineer drug delivery systems (patches, timed-release pills) and foodstuffs, to separate and manufacture chemicals and pharmaceuticals, to develop perfumes and scents, to produce medical devices (artificial kidneys and lung machines), and to synthesize the basic building blocks of life (DNA and proteins). We have exploited charge transfer to develop batteries, vacuum tubes, semiconductors, and the world of devices derived from these basic items. Finally, we have exploited the transport of photons to produce electromagnetic waves for radio, television, radar, lasers, optical communications, and the Internet.

FIGURE 1.10 Geothermal power plant outside Reykjavik, Iceland. The large cloud in the distance is a new well, while the foreground devices separate high-temperature steam from water allowing for electricity generation as well as providing hot water for heating and bathing to the people of Reykjavik.

Regardless of your engineering discipline, transport phenomena is fundamental to your field. The primary goal of this text will be to organize and present the elements of transport phenomena in a manner that will highlight the interrelationships between the phenomena and enable the scientist, engineer, mathematician, or humanist to recognize and use transport processes for the betterment of our society. Along the way, we will introduce some new computational techniques, forged on our exploitation of charge transport, that allow us to solve more realistic problems and probe the depths of the interrelationship between transport phenomena in ways that were not possible just a couple of decades before.

1.3 PRELIMINARY ASSUMPTIONS

In a text such as this, we can consider only a classical transport phenomena. We exclude a discussion of transport in relativistic systems and also neglect transport in systems displaying quantum mechanical effects. These are important areas at the forefront of research in transport phenomena yet are beyond the scope of the treatment here. We will consider both space and time to be continua making up a four-dimensional space. Time will always point in the positive direction for convenience though time reversal can be incorporated in any of the equations presented. The continuum approximation is a useful one whose chief advantage is that it enables us to use the language of differential and integral calculus to describe the physical phenomena occurring.

We must consider the following question: What constitutes the transport phenomena? Clearly, momentum, heat, and mass transfer are examples and yet limiting our definition to these examples is overly restrictive. How do we fold the motion of electrons in semiconductors into our concept of transport phenomena or the flow of photons through an optical fiber? Where does the flow of solid materials under pressure such as glaciers or creep in Teflon fit in? How do we describe the flow of particulate systems such as sand or cement?

In this century, theoretical and experimental physics have shown that all matter is composed of fundamental particles and that the four forces allowing for particle interaction are themselves mediated by particles (mass or massless). These forces and their mediating particles are listed in Table 1.1 [11,12]. All interactions between particles involve an exchange of particles. For example, the electromagnetic interaction involves the exchange of virtual photons. Virtual photons are undetectable as particles, and only their action is noticed. Imagine two charged particles separated by a finite distance. One particle emits a virtual photon. This photon carries some momentum with it and so the momentum of the emitting particle is changed. The virtual photon is captured by the receiving particle. This particle also has its momentum changed due to the exchange. This change in the momentum of the particles we perceive as a force between them. A similar situation exists for all the particle exchanges listed in Table 1.1. Each exchange changes the momentum of the exchanging particles. We perceive this change as a force between the particles.

TABLE 1.1
Physical Interactions

Name	Intrinsic Strength	Mediating Particle	Rest Mass	Spin	Range of Interaction	Sign
Strong	1	Gluon	0	1	Infinite $\propto 1/r$	Attractive
Electro-magnetic	0.01	Photon	0	1	Infinite $\propto 1/r$	Attractive or repulsive
Weak	10^{-14}	Intermediate boson W^+, W^-, Z°	$\approx 10^5 \frac{\text{MeV}}{c^2}$	1	$\approx 10^{-18}$ m	
Gravitational	10^{-40}	Graviton	0	2	Infinite $\propto 1/r$	Always attractive

Particle motion and exchange imply a transport process of some kind. We can expand our definition of transport phenomena to include particle exchanges fundamental to the four forces of nature and every other particle interaction derived from them. In a more abstract sense, we can view all exchanges of mediating particles and hence all particle interaction as an exchange of information. In this light, transport phenomena will encompass any situation that involves a net transfer of information between particles and hence an exchange of particles. These particles may be photons, gravitons, molecules, or planets depending on the scale of the system. The particles we generally consider in classical transport phenomena range in size from electrons in a semiconductor to individual atoms, molecules, and polymers. Moreover, as long as transport is occurring over a distance, large compared to the size of the particle, our continuum approximation will be valid. If the particles under consideration are galaxies, then the universe can be regarded as a continuum, and we can speak of galactic diffusion within a gravitational field in a matter similar to mass diffusion in a concentration field or heat diffusion in a temperature field.

Since our particles are relatively large, and spaced far apart, the two forces of interaction we normally deal with are electromagnetic and gravitational in nature. The mediating particles are the photon and the graviton. We will consider some aspects of the transport of photons since the electromagnetic force is so strong and since photon transport plays an important role in radiative energy exchange.

We stated previously that transport involves an exchange of information between particles. All the basic information about a particle is contained in a specification of its quantum state. Any time the particle exchanges information about its quantum state with another particle, the state of both particles changes and we have transport. Historically, we have pigeonholed this quantum state information into a number of categories: linear momentum, angular momentum, energy, charge, chemical composition, color, charm, strangeness, etc. These subdivisions arose quite naturally as we uncovered the fundamental laws of physics governing particle interaction. It is important to realize that quantum mechanics has given us a framework for unifying transport phenomena at a fundamental level and equating transport with an exchange of information about quantum states. The exchange is always mediated by one of the fundamental particles listed in Table 1.1.

The transport of linear and angular momentum we call fluid mechanics when dealing with liquids and gases and solid mechanics or dynamics when dealing with solids. The transport of energy we refer to as heat transfer, and the transport of chemical species makes up the field of mass transfer. We will see that charge transfer, when applied to ionic particles, can be considered as a variant of mass transfer. The transport of photons is called the field of electromagnetics in dealing with macroscopic phenomena and quantum electrodynamics when dealing with phenomena at the atomic level and below. The transport of gravitons is called celestial mechanics, gravitation, or general relativity. The transport of gluons is called quantum chromodynamics and this field describes the exchange of color among quarks. Finally, the exchange of information about chemical composition (fundamentally an electromagnetic interaction) pervades all of transport and manifests itself as differences in physical properties for momentum and energy transport and as chemical reactions in mass transport.

1.4 EQUILIBRIUM FOUNDATIONS

Fundamentally, transport phenomena describes the processes that take a system of particles from a nonequilibrium state toward an equilibrium state, from an equilibrium state toward a nonequilibrium state, or from one nonequilibrium state to another. The situation is shown schematically in Figure 1.11.

To describe a transport process, we must know three things: the state of the particles at their origin in space and time, the state of the particles at their destination, and the path the particles travel from their origin to their destination. The state of the particles is defined in terms of their thermodynamic state variables. Generally, for the purposes of this text, we use temperature, pressure, and n composition variables to describe the system, but there are many other combinations we can specify

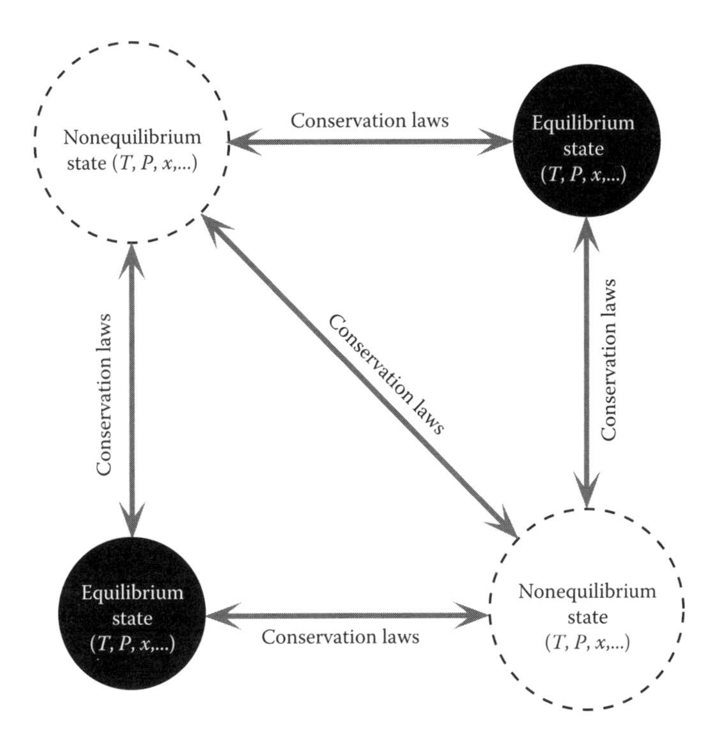

FIGURE 1.11 Starting points, destinations, and allowed paths for transport processes.

and other characteristics that may be important such as charge or surface area. The set of variables we use depends upon how many variables are required to uniquely define the state of our system.

As Figure 1.11 shows, we cannot travel between states at will. For instance, we cannot travel directly from one equilibrium state to another. We must pass through a nonequilibrium state(s) along the way. We must also travel along a particular path between states. The route and the travel time along the path are defined and controlled by fundamental conservation laws that apply to any quantity that is conserved within the system:

$$\begin{matrix}\text{Flow into} \\ \text{system}\end{matrix} - \begin{matrix}\text{Flow out} \\ \text{of system}\end{matrix} + \begin{matrix}\text{Generation} \\ \text{within system}\end{matrix} = \begin{matrix}\text{Accumulation} \\ \text{within system}\end{matrix} \tag{1.1}$$

Equilibrium states represent special fixed points. They attract particles from nonequilibrium states. To keep the particles in a nonequilibrium state or to take particles to another nonequilibrium state requires interference, generally in the form of expending energy or work. We must stop the particles along their route much as we stop a car for a traffic light by applying the brakes. A similar situation exists when we want to take particles from an equilibrium state to a nonequilibrium state. We must provide them with a kick in the form of energy or work to force them from equilibrium and get them to travel to the nonequilibrium state.

The equilibrium state is not a static situation. On the microscopic level, transport is always occurring within a system or collection of systems at equilibrium. The net effect of transport in an equilibrium situation is to leave the system undisturbed. For a transport process to be perceived, we must be able to measure a change in the state variables describing the system of particles of interest. In a way, it is similar to the exchange of information between a computer's memory and a disk drive. Transport involves writing and erasing information on a magnetic disk. If one writes and then immediately erases the same material, transport is occurring, but the net effect is to leave the disk and memory unchanged, an equilibrium state of sorts.

Clearly, to be able to discuss transport processes, we must first be able to define the equilibrium states. The remainder of this chapter will be devoted to a brief review of thermodynamics. The discussion is not meant to provide you with a detailed understanding. We cannot hope to do justice to the excellent background material that can be found in the references at the end of this chapter [13,14]. The purpose of the chapter is to jog your memory so that you can go back to your reference text(s) and review the concepts if necessary.

1.5 DEFINING EQUILIBRIUM

Consider the overall system shown in Figure 1.12. This system is composed of two subsystems that may interact with one another but that are isolated from the surroundings. Subsystem 1 is filled with a substance a and subsystem 2 is filled with substances a, b, and c that form a homogeneous solution. The phase of material in subsystem 1 may differ from the phase present in subsystem 2. The subsystems are separated by a rigid, weightless membrane that is permeable to species a, allows for heat to pass between the subsystems, and is free to move within the confines of the two subsystems.

For both subsystems to be in equilibrium with one another, the total entropy of the system must be maximized. We can use the conservation law, Equation 1.1, and apply it to the entropy. Since the system, as a whole, is isolated from the rest of the universe, we have no entropy flowing in or out. Similarly, we can have no entropy being generated. Any generation of entropy signifies that a change is taking place and so we would not have equilibrium. The only term left is the entropy accumulated, and here we see that this must also be zero:

$$\begin{matrix} \text{Entropy in} & - & \text{Entropy out} & + & \text{Entropy generated} & = & \text{Entropy accumulated} \\ 0 & - & 0 & + & 0 & = & \dfrac{dS}{dt}=\dfrac{dS_1}{dt}+\dfrac{dS_2}{dt} \end{matrix} \tag{1.2}$$

Entropy maximization

$$dS = dS_1 + dS_2 = 0 \tag{1.3}$$

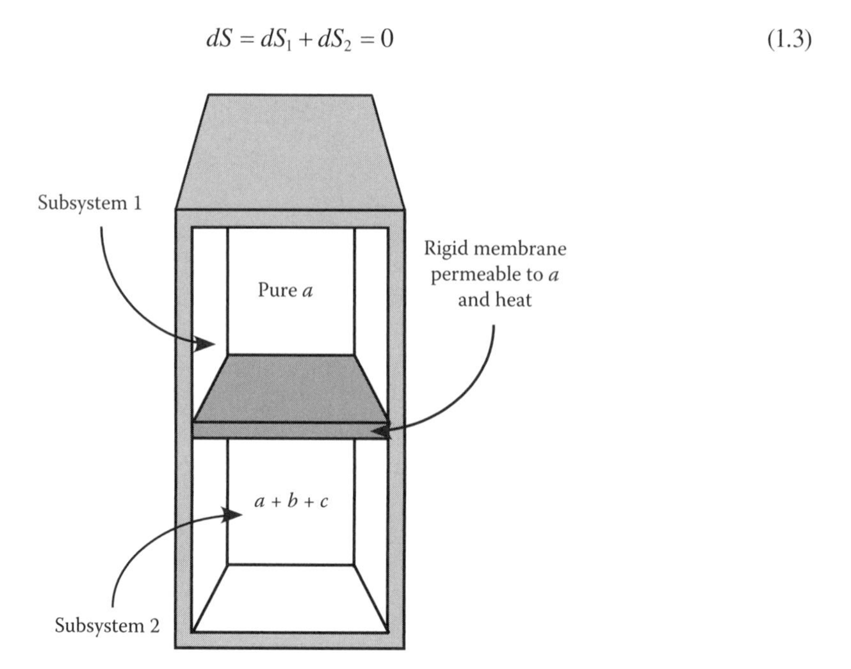

FIGURE 1.12 Equilibrium of subsystems exposed to various internal forces and stresses.

Equations 1.2 and 1.3 do not give us much information about the two subsystems. We must relate the entropy of each subsystem to the $n + 2$ state variables that define it uniquely. We can use the internal energy, volume, and composition of the system to define it. These variables are convenient for the isolated systems discussed here. Other variables are more appropriate for open systems (see [12–16]) that exchange mass with their surroundings. The change in entropy for each subsystem can be related to changes in energy, volume, and composition:

$$dS_1 = \frac{1}{T_1} dU_1 + \frac{P_1}{T_1} dV_1 - \frac{\mu_{a1}^c}{T_1} d\mathcal{N}_{a1} \tag{1.4}$$

$$dS_2 = \frac{1}{T_2} dU_2 + \frac{P_2}{T_2} dV_2 - \frac{\mu_{a2}^c}{T_2} d\mathcal{N}_{a2} - \frac{\mu_{b2}^c}{T_2} d\mathcal{N}_{b2} - \frac{\mu_{c2}^c}{T_2} d\mathcal{N}_{c2} \tag{1.5}$$

We can substitute directly into Equation 1.3, but we have some external constraints on Figure 1.12 that link the two subsystems and make the solution easier. Since the overall system is isolated from the surroundings, the total internal energy must be constant and the total volume must also be constant. We can have no flow of material into or out of the system, so the total number of moles of species a, species b, and species c will be constant:

$$dU = dU_1 + dU_2 = 0 \tag{1.6}$$

$$dV = dV_1 + dV_2 = 0 \tag{1.7}$$

$$d\mathcal{N}_a = d\mathcal{N}_{a1} + d\mathcal{N}_{a2} = 0 \tag{1.8}$$

$$d\mathcal{N}_{b2} = 0 \tag{1.9}$$

$$d\mathcal{N}_{c2} = 0 \tag{1.10}$$

Using Equations 1.6 through 1.10, we can write the requirement of equilibrium referring only to changes occurring in subsystem 1:

$$dS = \left(\frac{1}{T_1} - \frac{1}{T_2}\right) dU_1 + \left(\frac{P_1}{T_1} - \frac{P_2}{T_2}\right) dV_1 - \left(\frac{\mu_{a1}^c}{T_1} - \frac{\mu_{a2}^c}{T_2}\right) d\mathcal{N}_{a1} = 0 \tag{1.11}$$

The trivial solution for this equation would be if $dU_1 = dV_1 = d\mathcal{N}_{a1} = 0$. This implies no interaction between the two subsystems and we know that this is not the case. Thus, for dS to be 0, we must have each of the quantities in brackets equal to zero. The requirements for equilibrium now state that the temperatures of the subsystems, the pressures in the subsystems, and the chemical potentials of species a in the subsystems must be equivalent. We can say nothing about species b or c since they are restricted to subsystem 2 and cannot leave:

$$\frac{1}{T_1} = \frac{1}{T_2} \qquad T_1 = T_2 \tag{1.12}$$

$$\frac{P_1}{T_1} = \frac{P_2}{T_2} \qquad P_1 = P_2 \tag{1.13}$$

$$\frac{\mu_{a1}^c}{T_1} = \frac{\mu_{a2}^c}{T_2} \qquad \mu_{a1}^c = \mu_{a2}^c \tag{1.14}$$

The chemical potential of a species, μ^c, is a nebulous quantity. Generally, we try to relate it to something we can measure such as the concentration, mole fraction, or partial pressure. Often there exists a trivial relationship such as in an ideal mixture. We define the chemical potential for ideal gases as

$$\mu^c_{a,vap} = RT\ln\left(f^c_a\right) = RT\ln\left(\frac{P_a}{P}\right) = RT\ln\left(y_a\right) \tag{1.15}$$

where

f^c_a is the fugacity of component a
P_a is its partial pressure
y_a is its mole fraction

We define the chemical potential of liquids in terms of the activity, a. Here, in an ideal case, the activity, a_a, and mole fraction, x_a, are equivalent:

$$\mu^c_{a,liq} = RT\ln\left(a_a\right) = RT\ln\left(x_a\right) \tag{1.16}$$

We often have additional complications in our subsystems including chemical reactions and reversible work interactions (in addition to the PV work we have already taken into account). Let us deal with chemical reactions first and assume that in subsystem 2 a chemical reaction occurs taking ν_1 moles of a and reacting them with ν_2 moles of b to form ν_3 moles of c:

$$\nu_1 a + \nu_2 b \rightarrow \nu_3 c \qquad \text{or in algebraic form } \nu_3 c - \nu_1 a - \nu_2 b = 0 \tag{1.17}$$

Clearly, the number of moles of species a, b, and c is related at all times and only one may be varied independently. To simplify the accounting, we introduce a new variable, ξ, called the extent of reaction:

$$d\xi = \frac{d\mathcal{N}_j}{\nu_j} \qquad j = 1,2,3\ldots n \tag{1.18}$$

ξ is defined in such a way that it is the same for all species. All variations in the number of moles of the species, $d\mathcal{N}_j$, can now be expressed in terms of a change in the extent of reaction, $d\xi$. We incorporate the reaction into the equilibrium considerations by changing the constraints on the number of moles of each species. To account for the reaction, we formulate Equations 1.8 through 1.10 in terms of a change in the extent of reaction rather than a change in moles. In our case, we know the overall change in moles of species a is related to how much a has reacted to form c. This constraint is given by

$$d\mathcal{N}_a = d\mathcal{N}_{a1} + d\mathcal{N}_{a2} = -\nu_1 d\xi \tag{1.19}$$

Similarly, the changes in the number of moles of b and c are

$$d\mathcal{N}_b = d\mathcal{N}_{b1} + d\mathcal{N}_{b2} = -\nu_2 d\xi \tag{1.20}$$

$$d\mathcal{N}_c = d\mathcal{N}_{c2} = \nu_3 d\xi \tag{1.21}$$

Substituting Equations 1.6, 1.7, and 1.19 through 1.21 into the equilibrium condition, Equation 1.3, we find

$$dS = 0 = \left(\frac{1}{T_1} - \frac{1}{T_2}\right)dU_1 + \left(\frac{P_1}{T_1} - \frac{P_2}{T_2}\right)dV_1 - \left(\frac{\mu^c_{a1}}{T_1} - \frac{\mu^c_{a2}}{T_2}\right)d\mathcal{N}_{a1} + \left(\nu_1\frac{\mu^c_{a2}}{T_2} + \nu_2\frac{\mu^c_{b2}}{T_2} - \nu_3\frac{\mu^c_{c2}}{T_2}\right)d\xi \tag{1.22}$$

Again, it is immediately obvious that if we are not to have a trivial solution, that is, no reaction or system interaction, the terms in brackets must be equal to zero:

$$T_1 = T_2 \qquad P_1 = P_2 \qquad \mu_{a1}^c = \mu_{a2}^c \qquad \nu_1\mu_{a2}^c + \nu_2\mu_{b2}^c = \nu_3\mu_{c2}^c \tag{1.23}$$

The chemical reaction has added another equilibrium constraint linking the chemical potentials of a, b, and c in subsystem 2. Notice that we still require the chemical potentials of species a in subsystems 1 and 2 to be equal.

So far, we have said nothing about reversible work interactions between the two systems. Such work interactions may be considerable and may arise due to the subsystems being in potential fields, like gravity, or subject to stresses. Let us consider the case of Figure 1.12 but where we allow the two subsystems to interact in such a way that they may have reversible work interactions due to each other's electric field, magnetic field, or stress field. We will lump all these interactions into one term, W_{rev}. Details of how this term is decomposed can be found in [15]. To account for this work, we need to add one extra term to Equation 1.3:

$$dS = \frac{1}{T}dU + \frac{P}{T}dV - \sum_{i=1}^{n}\frac{\mu_i^c}{T}d\mathcal{N}_i + \frac{1}{T}dW_{rev} \tag{1.24}$$

We can proceed with our equilibrium analysis and write one equation for subsystem 1 and another for subsystem 2:

$$dS_1 = \frac{1}{T_1}dU_1 + \frac{P_1}{T_1}dV_1 - \frac{\mu_{a1}^c}{T_1}d\mathcal{N}_{a1} + \frac{1}{T_1}dW_{rev1} \tag{1.25}$$

$$dS_2 = \frac{1}{T_2}dU_2 + \frac{P_2}{T_2}dV_2 - \frac{\mu_{a2}^c}{T_2}d\mathcal{N}_{a2} - \frac{\mu_{b2}^c}{T_2}d\mathcal{N}_{b2} - \frac{\mu_{c2}^c}{T_2}d\mathcal{N}_{c2} + \frac{1}{T_2}dW_{rev2} \tag{1.26}$$

The constraints on the overall system are the same as in Equations 1.6 through 1.10. The total system must have a constant volume, mass, and internal energy. The equilibrium requirement remains unchanged. We must have the entropy maximized. Substituting into Equation 1.3, we obtain

$$dS = \left(\frac{1}{T_1} - \frac{1}{T_2}\right)dU_1 + \left(\frac{P_1}{T_1} - \frac{P_2}{T_2}\right)dV_1 - \left(\frac{\mu_{a1}^c}{T_1} - \frac{\mu_{a2}^c}{T_2}\right)d\mathcal{N}_{a1} + \frac{1}{T_2}dW_{rev2} + \frac{1}{T_1}dW_{rev1} = 0 \tag{1.27}$$

The equilibrium constraints are satisfied when

$$T_1 = T_2 \qquad P_1 = P_2 \qquad \mu_{a1}^c = \mu_{a2}^c \qquad dW_{rev1} = -dW_{rev2} \tag{1.28}$$

We see that in order for both subsystems to be in equilibrium, they must do work on one another of exactly the same magnitude, but not necessarily the same type. For instance, an electric field generated in subsystem 1 may perform work on subsystem 2 by polarizing it, while the polarization of subsystem 2 may place a stress on and change the shape of subsystem 1 (PV work).

1.6 FLUID STATICS

The thermodynamic analysis we considered previously was generic and holds for mass, energy, charge, and momentum transport. It is rather abstract though, and to obtain useful information for fluids, we must delve a little deeper into fluid equilibrium. Fluids at rest or in steady, uniform motion are in "mechanical" equilibrium with their surroundings. On Earth though, the fluid is exposed to a uniform gravitational field, and thus, there is a work interaction with the environment. This interaction exerts a force on the fluid and causes the pressure in the fluid to vary as a function of position. We can use the fluid's response to the gravitational field to build instruments that measure the atmospheric pressure, the altitude, the flow of fluids, or the speed of airplanes. Knowing how these fluids behave also aids in designing dams to retain water, designing submarines or underwater habitats, or understanding how sperm whales can survive dives of thousands of feet under the ocean. Our purpose here is simply to determine what the pressure field is inside a static fluid.

We consider a fluid system like that shown in Figure 1.13 and the associated control or sample volume within that fluid. The work interaction between the environment and the fluid can be determined by considering Newton's law of motion. Newton's law asserts that the sum of all the forces acting on the fluid must be balanced by mass times the acceleration of the fluid. The two forces acting on the fluid that we need to consider are surface forces and body forces. The acceleration of the fluid is zero at equilibrium, but we will also see how we can consider accelerating systems as well, if we require that the whole of the fluid moves as a rigid body, that is, constant acceleration.

The surface forces arise due to a difference in the pressure force between each pair of faces of the control volume. Thus, we have

$$[P(x)-P(x+\Delta x)]\Delta y\Delta z \qquad x\text{-face} \tag{1.29}$$

$$[P(y)-P(y+\Delta y)]\Delta x\Delta z \qquad y\text{-face} \tag{1.30}$$

$$[P(z)-P(z+\Delta z)]\Delta x\Delta y \qquad z\text{-face} \tag{1.31}$$

The body force is the force of gravity or the weight of the fluid, and that acts on the whole volume of fluid:

$$\rho\vec{\mathbf{g}}\Delta x\Delta y\Delta z \tag{1.32}$$

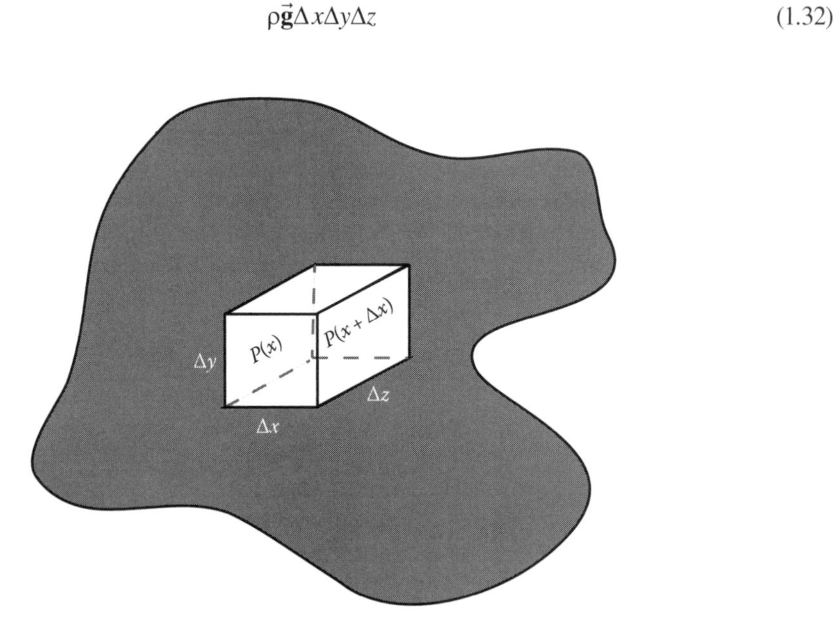

FIGURE 1.13 A control volume within a larger mass of fluid.

The sum of these forces must be zero since the fluid is stationary and there is no acceleration:

$$[P(x)-P(x+\Delta x)]\Delta y\Delta z\,\vec{\mathbf{i}}+[P(y)-P(y+\Delta y)]\Delta x\Delta z\,\vec{\mathbf{j}}$$
$$+[P(z)-P(z+\Delta z)]\Delta x\Delta y\,\vec{\mathbf{k}}+\rho\vec{\mathbf{g}}\Delta x\Delta y\Delta z=m\vec{\mathbf{a}}=0 \quad (1.33)$$

Now this equation operates over the small volume of fluid we defined. To have the equation defined over every point in the fluid, we must evaluate it in the limit as $\Delta x\Delta y\Delta z \to 0$. Dividing through by the volume of the control volume, $\Delta x\Delta y\Delta z$, and taking the limit gives

$$\lim_{\Delta x\Delta y\Delta z\to 0}\left[\frac{P(x)-P(x+\Delta x)}{\Delta x}\right]\vec{\mathbf{i}}+\lim_{\Delta x\Delta y\Delta z\to 0}\left[\frac{P(y)-P(y+\Delta y)}{\Delta y}\right]\vec{\mathbf{j}}$$
$$+\lim_{\Delta x\Delta y\Delta z\to 0}\left[\frac{P(z)-P(z+\Delta z)}{\Delta z}\right]\vec{\mathbf{k}}+\rho\vec{\mathbf{g}}=0 \quad (1.34)$$

Each of the terms in brackets is the definition of the derivative with respect to the denominator, and so Equation 1.34 becomes

$$-\frac{\partial P}{\partial x}\vec{\mathbf{i}}-\frac{\partial P}{\partial y}\vec{\mathbf{j}}-\frac{\partial P}{\partial z}\vec{\mathbf{k}}+\rho\vec{\mathbf{g}}=-\left(\frac{\partial P}{\partial x}\vec{\mathbf{i}}+\frac{\partial P}{\partial y}\vec{\mathbf{j}}+\frac{\partial P}{\partial z}\vec{\mathbf{k}}\right)+\rho\vec{\mathbf{g}}=0 \quad (1.35)$$

We generally call the term within parentheses the gradient (Table 1.2) and write it as

$$-\vec{\nabla}P+\rho\vec{\mathbf{g}}=0 \quad (1.36)$$

Equation 1.36 describes the pressure field in an arbitrary fluid under the influence of gravity. Normally, we make our lives simpler by imposing a coordinate system that lines up one axis with the direction of gravity and reduces the partial differential equation to a series of ordinary differential equations. In this way, we can reduce the complexity of Equation 1.36 to

$$\frac{dP}{dx}=0 \qquad \frac{dP}{dy}=0 \qquad \frac{dP}{dz}=-\rho g \quad (1.37)$$

Here we use $g_z=-g$ to insure that as z increases relative to the Earth's surface, the pressure decreases.

TABLE 1.2
The Gradient Operator in Cartesian, Cylindrical, and Spherical Coordinates

Coordinate System	Gradient Equation
Cartesian	$\vec{\nabla}\psi=\left(\frac{\partial}{\partial x}\vec{\mathbf{i}}+\frac{\partial}{\partial y}\vec{\mathbf{j}}+\frac{\partial}{\partial z}\vec{\mathbf{k}}\right)\psi$
Cylindrical	$\vec{\nabla}\psi=\left(\frac{\partial}{\partial r}\vec{\mathbf{e}}_r+\frac{1}{r}\frac{\partial}{\partial\theta}\vec{\mathbf{e}}_\theta+\frac{\partial}{\partial z}\vec{\mathbf{e}}_z\right)\psi$
Spherical	$\vec{\nabla}\psi=\left(\frac{\partial}{\partial r}\vec{\mathbf{e}}_r+\frac{1}{r}\frac{\partial}{\partial\theta}\vec{\mathbf{e}}_\theta+\frac{1}{r\sin\theta}\frac{\partial}{\partial\phi}\vec{\mathbf{e}}_\phi\right)\psi$

1.6.1 Pressure Variation in a Static Fluid: Manometers

We have just shown that we can simplify the pressure relation such that there is a variation over just one coordinate direction. In most instances close to the Earth, the acceleration due to gravity can be assumed constant, and the density of liquids such as water and oil is insensitive to the amount of pressure applied to them. We call these incompressible fluids, and these simplifications enable us to measure pressures and pressure differences using the relative heights of liquids. The devices that we use to make the measurement are called manometers.

Integrating $dP/dz = -\rho g$ gives

$$P - P_o = -\rho g(z - z_o) \tag{1.38}$$

where we have used the fact that at z_o, we know the pressure is P_o. This is the basic equation used to determine the pressure via a manometer.

A manometer is basically a U-shaped tube filled with a liquid of known density. The two ends of the tube are exposed to different pressures so that the fluid level is different at each end of the tube as shown in Figure 1.14. There are three basic types of manometers. We can classify them according to the following:

a. One end sealed, the other open to the environment—a barometer
b. One end sealed, the other exposed to a flowing fluid
c. Differential—both ends exposed to a flowing fluid

Referring to Figure 1.14, we draw a horizontal line that cuts through both arms of the manometer. According to Equation 1.37, the pressure variation exists only in the z-direction, so along line a–b, the pressure is constant. Thus, at any height, z, the pressure in the two arms of the manometer is equal. Using the densities of the fluids, ρ_1, ρ_2, and ρ_m, the heights of the fluids above line a–b (d_1, d_2, and h), and the pressures exerted on both tubes, P_1 and P_2, the statement of equality of the pressures in both arms becomes

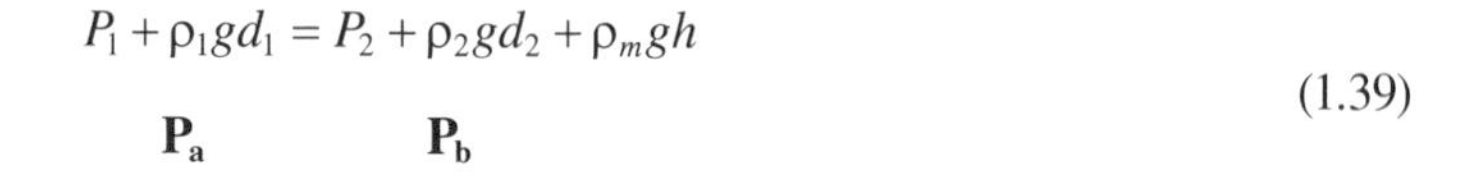

$$\underbrace{P_1 + \rho_1 g d_1}_{\mathbf{P_a}} = \underbrace{P_2 + \rho_2 g d_2 + \rho_m g h}_{\mathbf{P_b}} \tag{1.39}$$

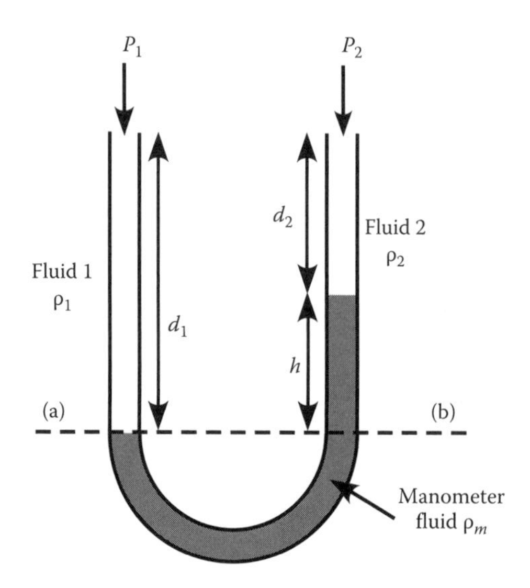

FIGURE 1.14 U-tube manometer with three fluids.

Example 1.1 U-Tube Manometer

A U-tube manometer uses three distinct liquids in its two arms. Fluid 1 is methanol, fluid 2 is water, and fluid m is a manometer fluid with a specific gravity of 1.4. The pressure P_2 = 120.0 kPa, d_2 = 30.0 cm, and h = 25.0 cm. The objective here is to calculate the pressure difference across the two arms of the manometer and so obtain P_1.

Using Equation 1.39, we first recognize that $d_1 = d_2 + h$. Substituting into Equation 1.39 and solving for P_1 gives

$$P_1 = P_2 + g[\rho_2 d_2 + \rho_m h - \rho_1(d_2 + h)] = 121{,}000\ \text{Pa} + 9.8\ \text{m/s}^2$$

$$\times\left[\left(1000\frac{\text{kg}}{\text{m}^3}\right)(0.3\ \text{m}) + \left(1400\frac{\text{kg}}{\text{m}^3}\right)(0.25\ \text{m}) - \left(792\frac{\text{kg}}{\text{m}^3}\right)(0.55\ \text{m})\right] = 123{,}101\ \text{Pa}$$

Here, we use the fact that water at a specific gravity of 1 has a density of 1000 kg/m^3.

Example 1.2 Water Flow Down an Inclined Tube

Water flows downward along a pipe that is inclined at 45° below the horizontal. The pressure difference, $\Delta P = P_2 - P_1$, is due in part to gravity and in part because a pressure difference is required to overcome friction with the walls of the tube. We can measure this pressure difference using a differential manometer. Determine an algebraic expression for ΔP and evaluate that difference if L = 0.5 m and h = 10 cm.

To analyze this situation, we first choose a reference height. Generally, we locate that reference line at an interface between manometer fluids. The pressure along the reference line is constant. Using Equation 1.39 as a model, we consider both legs of the manometer from the reference line to the centerline of the pipe where we want to know the pressure (Figure 1.15):

$$P_1 + \rho_w g L \sin\theta + \rho_w g d + \rho_w g h \qquad \text{Leg \#1}$$

$$P_2 + \rho_w g d + \rho_{hg} g h \qquad \text{Leg \#2}$$

Subtracting leg #2 from leg #1 and solving for ΔP gives

$$\Delta P = \rho_{hg} g h - \rho_w g L \sin\theta - \rho_w g h$$

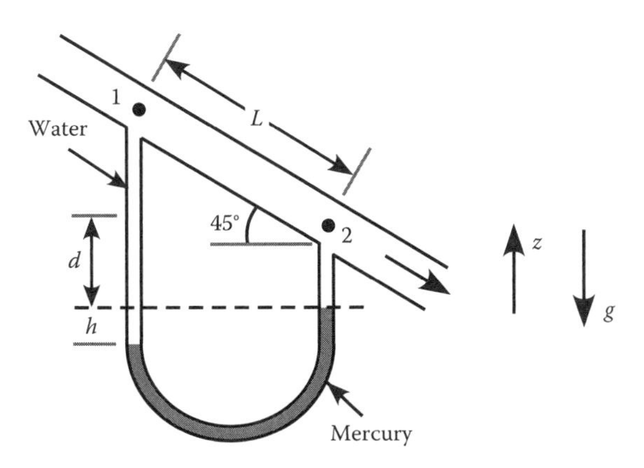

FIGURE 1.15 Water flowing down an inclined tube.

Using the numbers for the density of water and mercury from Appendix D, we find that

$$\Delta P = 8785\ \text{Pa}$$

Notice that unless the absolute pressure is defined at some point in the system, all we can measure via the manometer is a pressure difference between the two legs of the manometer. Later we will learn that we can determine the flow rate using this pressure difference.

The procedures we used to analyze a manometer can also be applied to gases. Unlike liquids, gases are compressible and their density is tied to the temperature, volume, and number of moles of gas. Suppose we were interested in the pressure variation in the atmosphere as a function of altitude so that we could perform calculations for jet engine combustion or determine minimum cooking and baking times. The basic pressure field equation remains the same, but now we must account for the variation in fluid density. As an example, let us take the standard atmosphere whose temperature is roughly a linear function of altitude for the first several miles above the Earth's surface. At the pressures we are likely to encounter, we can treat air as an ideal gas so that

$$PV = \frac{mRT}{M_w} \quad \text{or} \quad \rho = \frac{M_w P}{RT} \tag{1.40}$$

Substituting into Equation 1.37 gives

$$\frac{dP}{dz} = -\rho g = -\frac{M_w P}{RT} g \tag{1.41}$$

In the atmosphere, we are assuming

$$T = T_o - \left(\frac{\Delta T}{h}\right) z \tag{1.42}$$

Substituting and separating the P's from the z's, we have

$$\frac{dP}{P} = -\frac{M_w g}{R[T_o - (\Delta T/h)z]} dz \tag{1.43}$$

Integrating from $P = P_o$ at $z = 0$ to $P = P(z)$ gives

$$\ln\left(\frac{P}{P_o}\right) = \frac{M_w g}{(\Delta T/h)R} \ln\left[\frac{T_o - (\Delta T/h)z}{T_o}\right] \tag{1.44}$$

$$\frac{P}{P_o} = \left[\frac{T_o - (\Delta T/h)z}{T_o}\right]^{\frac{M_w g}{(\Delta T/h)R}} \tag{1.45}$$

and so as we go up in altitude, the pressure decreases.

Example 1.3 Atmospheric Pressure and Altitude during Air Travel

We can often use a pressure measurement to determine the altitude. Consider a jet airplane that is supposed to fly at 35,000 ft (10,670 m). The air temperature at sea level is 25°C and drops off at a rate of 6.5×10^{-3} K/m. If the pressure at sea level is 1.04×10^5 N/m² and the pressure measured by the airplane is 2.58×10^4 Pa, what is the airplane's altitude?

The molecular weight of air is 29 kg/kg mol. Thus, using Equation 1.44,

$$\ln\left(\frac{2.58\times10^4\ \text{Pa}}{1.04\times10^5\ \text{Pa}}\right)=\frac{(29\ \text{kg/kmol})(9.8\ \text{m/s}^2)}{(6.5\times10^{-3}\ \text{K/m})(8314\ \text{J/kmol K})}\ln\left(\frac{298-\left(6.5\times10^{-3}\right)z}{298}\right)$$

Solving for z shows that the altitude is 10,675 m.

1.7 BUOYANCY AND STABILITY

Objects immersed in a fluid feel not only their weight due to the gravitational force but also a force from the fluid, due to the weight of the fluid displaced. This force is termed buoyancy and is formally defined as the net vertical force acting on an object immersed in a liquid due to liquid pressure.

Consider the object totally immersed in a liquid as shown in Figure 1.16. At any point below the surface of the liquid, the pressure in the liquid is given by the hydrostatic equation. Keep in mind that h_1 and h_2 are negative since we are measuring distance below the surface:

$$\frac{dP}{dh}=-\rho g\rightarrow P=P_o-\rho gh \qquad h \text{ is negative} \tag{1.46}$$

The net vertical pressure force on the solid element depicted in Figure 1.16 is found from the difference in pressure force across the body. Thus,

$$dF=(P_o-\rho gh_2)\,dA-(P_o-\rho gh_1)\,dA=-\rho g(h_2-h_1)\,dA \tag{1.47}$$

Since $(h_2-h_1)\,dA$ is the volume of the element, dV, integrating over the entire body shows that the force, due to gravity, is

$$F=\int_V dF=\int \rho g dV=-\rho gV \tag{1.48}$$

and acts in the downward direction.

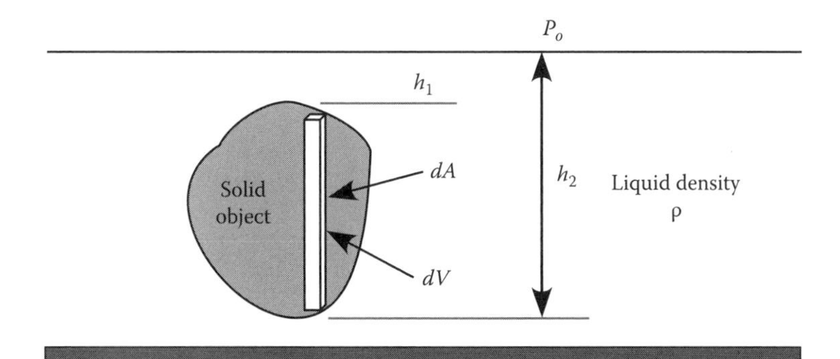

FIGURE 1.16 A solid object immersed in a liquid.

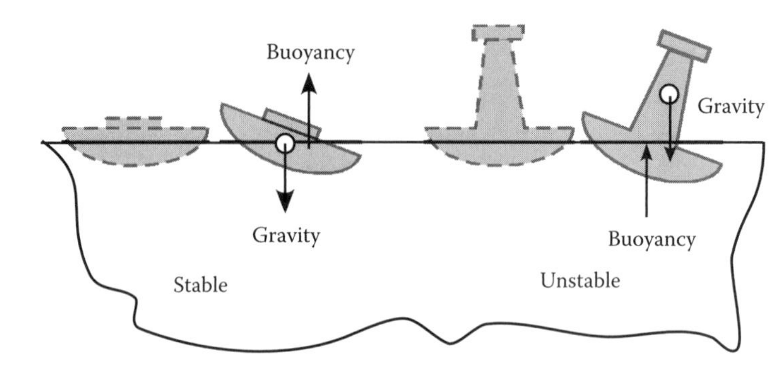

FIGURE 1.17 The stability of a floating craft. Effect of alignment of gravitational and buoyancy forces.

The net vertical force, or buoyancy, is equal to the force of gravity on the liquid displaced by the object or the *weight of the liquid displaced.* This is Archimedes' principle. Part of the force exerted on the solid object is used up by having to displace the liquid and change the elevation of the liquid. For a body that is not totally immersed, the buoyancy is equal to the weight of the object itself.

The gravitational body force on an object acts through the body's center of gravity. The buoyancy force acts through the centroid of the displaced volume. Depending on the shape of the object, the two forces may act through totally different locations. In certain circumstances, the forces form a couple that tends to keep a floating object upright, like a Bop Bag toy you cannot knock down. In other circumstances, the two forces can act to destabilize the craft and cause it to roll over. Figure 1.17 shows when the forces act to keep a craft stable and when they act to make the craft unstable. In the left-hand diagram, the couple formed between the buoyancy and gravitational forces acts to damp out any rocking motion. In the right-hand diagram, the couple formed between the forces actually amplifies any rocking motion. In layman's terms, the left-hand figure is bottom heavy and the right-hand figure is top heavy.

Example 1.4 Specific Gravity Required for a Spherical Plug

We are interested in plugging a hole in a leaking tank using a spherical plug that places itself automatically. A variation of this type of plug has been used for a long time to seal leaks in car radiators. A sphere of radius, r_o, and specific gravity, s.g., is submerged in a tank of water. The sphere is placed over a hole of radius, r_i, in the bottom of the tank. Find an expression for the range of sphere-specific gravities such that the sphere will remain on the bottom.

The sum of all the forces acting on the sphere must equal zero for the sphere to just sit on the bottom. According to Figure 1.18, this requires

$$\sum F = 0 = F_{air} + F_{buoy} - F_{press} - F_{grav}$$

Here, F_{air} is resisted by F_{press} and F_{buoy} is resisted by F_{grav}. F_{air} is the force of air on the area of the sphere of radius, r_i:

$$F_{air} = P_o \pi r_i^2$$

F_{buoy} is the net buoyant force on the sphere. It acts over all parts of the sphere in contact with the liquid. Thus, we exclude that part of the sphere that is over the hole of radius, r_i:

$$F_{buoy} = \rho_w g V_{net} = \rho_w g \left(\frac{4\pi r_o^3}{3} - 2\pi r_o r_i^2 \right)$$

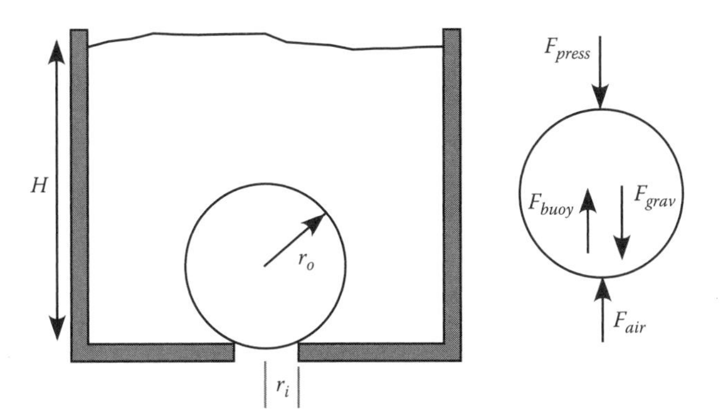

FIGURE 1.18 Spherical plug at the bottom of a tank.

Both F_{air} and F_{buoy} are approximate since they neglect the cap of the sphere over the hole, r_i. F_{press} is the total force on the area of the sphere of radius, r_i, at a depth of $h = H - 2r_o$:

$$F_{press} = [P_o + \rho_w g(H - 2r_o)]\pi r_i^2$$

F_{grav} is the force due to gravity on the entire sphere:

$$F_{grav} = \rho_w (\text{s.g.})\, g\left(\frac{4\pi r_o^3}{3}\right)$$

Combining all the terms gives

$$0 = P_o \pi r_i^2 + \rho_w g\left(\frac{4\pi r_o^3}{3} - 2\pi r_o r_i^2\right) - [P_o + \rho_w g(H - 2r_o)]\pi r_i^2 - \rho_w (\text{s.g.})\, g\left(\frac{4\pi r_o^3}{3}\right)$$

Solving for the specific gravity gives

$$\text{s.g.} = 1 - \frac{3}{4}\left(\frac{H}{r_o}\right)\left(\frac{r_i}{r_o}\right)^2$$

So for $H = 0.5$ m, $r_o = 25$ mm, $r_i = 5$ mm, we have s.g. = 0.4 less than that of water. Surprising?

Example 1.5 Partially Submerged Cylindrical Tank

A 1 m diameter hollow tank is made of steel (s.g. 7.85) with 5 mm wall thickness as shown in Figure 1.19. The tank is 1 m long. We want to use this as a makeshift buoy. How high will the tank float in water? How much weight must be added to the tank to make it neutrally buoyant?

The buoyant force felt by the tank is the weight of the displaced water. Here, that is the density of water, multiplied by the volume of the tank that is submerged. We must remember to include the tank's wall thickness:

$$F_{buoy} = \pi \rho_{water} g (0.505)^2 (1)\left(1 - \cos^{-1}\left(\frac{0.505 - h}{0.505}\right)\right) + \rho_{water} g (0.505 - h)(1)\sqrt{2(0.505)h - h^2}$$

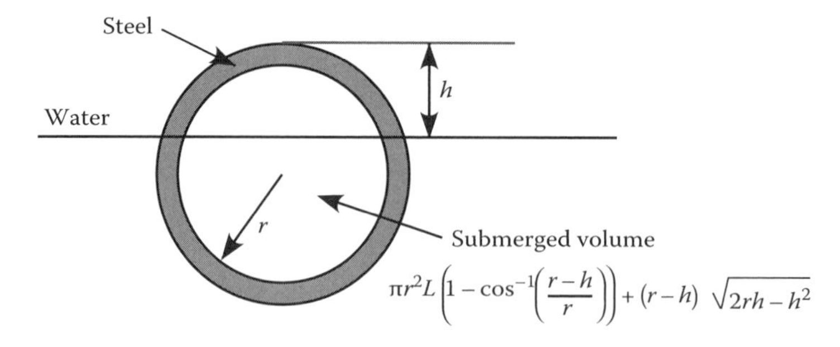

FIGURE 1.19 A cylindrical tank partially immersed in water.

The buoyant force is also equal to the weight of the steel tank if the tank is partially submerged:

$$F_{buoy} = W = \rho_{steel}V_{steel}g = 7850(0.024)(9.8) = 1846\text{ N}$$

$$V_{steel} = \pi(r+t)^2 L - \pi r^2 L + 2\pi(r+t)^2 t = 0.024\text{m}^3$$

Equating the two forces and solving via trial and error yields
$h = 0.199$ m

For neutral buoyancy, the weight of the tank plus the added weight must equal the weight of the water displaced by the entire tank. Thus, the whole tank is submerged:

$$W_{total} = 1831 + W_{added} = W_{water} = (1000)(9.8)\pi(0.505)^2(1)$$

$$W_{added} = 6021\text{ N}$$

The neutral buoyancy calculation is a critical one for scuba divers. If too little weight is added to the diving belt, one cannot remain submerged.

1.8 FLUIDS IN RIGID BODY MOTION

When we discussed how to determine the pressure field inside a fluid, we assumed the fluid was static and derived an expression for the force due to pressure and gravity:

$$d\vec{\mathbf{F}} = \left(-\vec{\nabla}P + \rho\vec{\mathbf{g}}\right)dV \qquad \frac{d\vec{\mathbf{F}}}{dV} = -\vec{\nabla}P + \rho\vec{\mathbf{g}} \tag{1.49}$$

If a fluid moves in rigid body motion, it behaves as if it were a solid and does not deform. Since there is no deformation, there are no shear forces and the only stress the fluid sees is that caused by the pressure. Once we accelerate the fluid though, dF/dV is no longer zero as it was in the static case. Newton's law defines how the fluid must respond and says that

$$\frac{d\vec{\mathbf{F}}}{dV} = \rho\vec{\mathbf{a}} = -\vec{\nabla}P + \rho\vec{\mathbf{g}} \tag{1.50}$$

This vector equation allows us to look at what the shape of the liquid would be when we accelerate it and it moves as a rigid body.

Example 1.6 Water Undergoing Uniform Acceleration Down an Incline

A rectangular container of water is pushed down an inclined plane so that it has a constant acceleration of 4 m/s² in the direction of the plane. Determine the slope of the free surface using the coordinate system shown in Figure 1.20 so we will know if it will spill.

The basic equation we need is (1.50) decomposed into its component form:

$$-\frac{\partial P}{\partial x}+\rho g_x=\rho a_x \qquad -\frac{\partial P}{\partial y}+\rho g_y=\rho a_y \qquad -\frac{\partial P}{\partial z}+\rho g_z=\rho a_z$$

Based on the coordinate system shown in Figure 1.20, we specify the acceleration due to gravity and due to the container motion g and a, respectively:

$$a_y=0 \qquad a_z=0 \qquad a_x=4$$

$$g_z=0 \qquad g_x=g\sin\theta \qquad g_y=-g\cos\theta$$

Using these components, we can substitute into the three components of the force balance and show that

$$\frac{\partial P}{\partial z}=0 \qquad \frac{\partial P}{\partial x}=\rho g\sin\theta-\rho a_x \qquad \frac{\partial P}{\partial y}=-\rho g\cos\theta$$

Since the pressure is uniform in the z-direction, $P = f(x,y)$, and from calculus we know that the total derivative of the pressure can be expressed as

$$dP=\left(\frac{\partial P}{\partial x}\right)_y dx+\left(\frac{\partial P}{\partial y}\right)_x dy$$

The key to determining the shape of the free surface of the liquid is to realize that along the free surface, the pressure is constant at atmospheric pressure. Thus, the total derivative of the pressure vanishes along the free surface, $dP = 0$, and so solving for dy/dx gives an equation for the slope of the surface:

$$\left.\frac{dy}{dx}\right|_{surface}=-\frac{(\partial P/\partial x)_y}{(\partial P/\partial y)_x}=\frac{\rho g\sin\theta-\rho a_x}{\rho g\cos\theta}=0.11$$

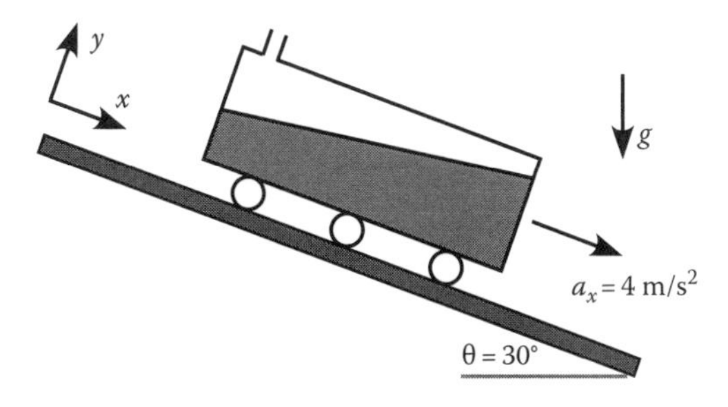

FIGURE 1.20 A fluid-containing cart rolling down an inclined plane.

PROBLEMS

MANOMETERS

1.1 Determine the pressure difference between water in the left-hand bulb and the oil in the right-hand bulb. The system is shown in Figure P1.1.

1.2 Determine the pressure difference between the two fluids in bulbs *a* and *b*. The system is shown in Figure P1.2.

1.3 A piston of area 0.1 m^2 sits atop a container filled with water. A U-tube manometer is connected to the container at one end. The other end is open to the atmosphere. If h_1 = 50 mm and h_2 = 120 mm, what is the force, *P*, acting on the piston? The weight of the piston is 1 kg (Figure P1.3).

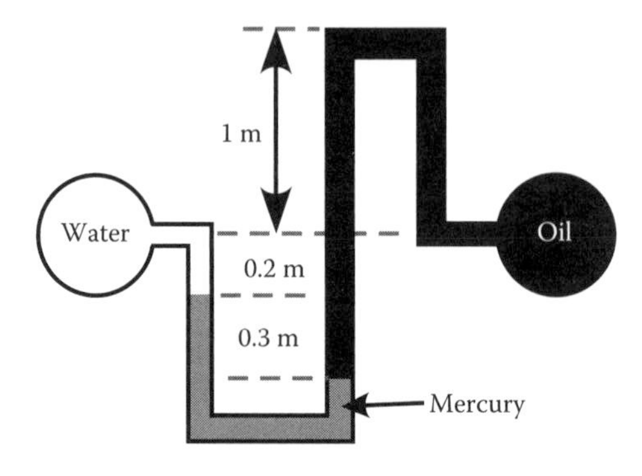

FIGURE P1.1

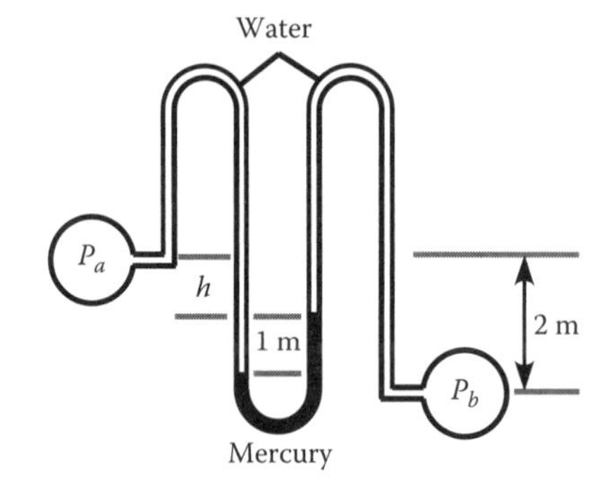

FIGURE P1.2

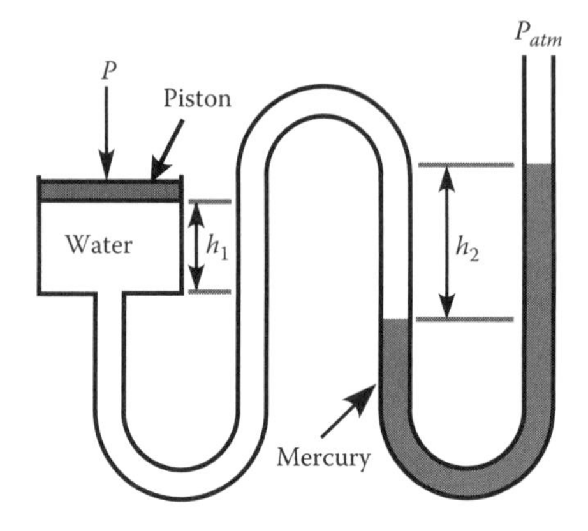

FIGURE P1.3

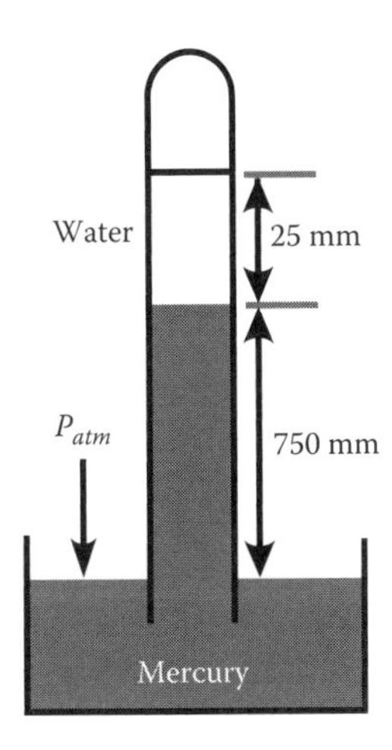

FIGURE P1.4

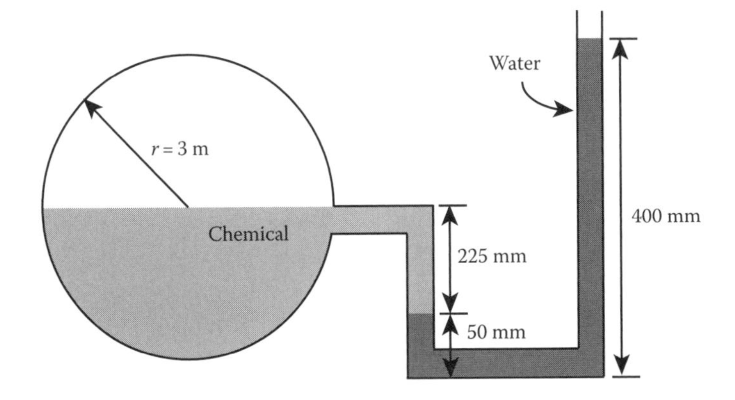

FIGURE P1.5

1.4 A barometer is contaminated with water. Atop the mercury of the barometer sits 25 mm of water. If the height of the mercury is 750 mm, what is the atmospheric pressure (Figure P1.4)?

1.5 An open water manometer is used to measure the pressure in a spherical tank. The tank is half-filled with 50,000 kg of a chemical that is immiscible with water. The manometer tube is partially filled with the chemical. What is the pressure in the tank if the situation is shown in Figure P1.5?

Atmospheric/Hydraulic Pressures

1.6 The U2 spy plane was designed to fly at altitudes of up to 21,000 m. Its wing span was so large that special wheels were incorporated within the wings to keep them from scraping the runway when taking off, landing, or being parked. If the air temperature at sea level is 25°C, drops off at a rate of 5×10^{-3} K/m, and the pressure at sea level is 1.03×10^5 N/m^2, what is the temperature and pressure at the airplane's height? The molecular weight of air is 29 kg/kg mol.

1.7 An atmospheric inversion has taken place such that the temperature at the surface is 5°C while the temperature at 4000 m is a balmy 20°C. Assuming a linear temperature drop over that distance and a barometric pressure of 750 mmHg at the surface, what is the pressure at 4000 m?

1.8 The atmosphere on Venus can be considered to behave as an ideal gas with a mean molecular weight of 44. The temperature varies little with depth but hovers at an incredible 350°C. The density of the atmosphere at the planet's surface is 2 kg/m^3 and the acceleration due to gravity is the same as the Earth's. Calculate how the density of the atmosphere varies as a function of height above the surface.

1.9 Water boils at 100°C at atmospheric pressure, that is, at sea level. The boiling point is defined as the temperature at which the vapor pressure of water is equal to the atmospheric pressure. This has consequences for cooking in places like Denver and Santa Fe where it takes quite a bit longer to make a hard-boiled egg than it would in New York. If the vapor pressure of water obeys the Antoine equation (shown in the succeeding text), and the albumen in an egg needs to reach 90°C for the protein to coagulate, at what height would it be impossible to hard-boil an egg?

$$\text{Antoine equation} \quad \ln P_{vap}(\text{kPa}) = 16.262 - \frac{3799.89}{T(°\text{C}) + 226.35}$$

1.10 The successor to Alvin, the submersible that was used to locate the Titanic, among its other research tools, is being built. The goal is to be able to reach the deepest depths in the Atlantic and Pacific oceans, about 11,000 m. Assuming water density is unchanged with depth

a. What is the pressure at the 11,000 m mark?
b. If the submersible is to be nominally 3 m in diameter, what force does the water exert on the craft?
c. How thick would the hull of the submersible have to be? The design equation for thick-walled spherical shells can be expressed as

$$t_h \cong r_i\left[\left(\frac{2S_d + 2P}{2S_d - P}\right)^{1/3} - 1\right]$$

where

t_h is the wall thickness
r_i is the inner radius of the shell (3 m)
S_d is the design stress for the wall material
P is the pressure the vessel is exposed to

For this example, the hull material is a titanium–molybdenum alloy whose design stress is approximately 275 MPa.

1.11 The Three Gorges Dam spanning the Yangtze River in China will be the largest dam in the world when completed. The dam is designed to be 101 m tall and, when filled, will contain water to a height of 91 m:

a. What would be the pressure at the bottom of the dam?
b. Dam design is a very complicated topic, but to simplify it a bit, we can just look at the pressure the water exerts at the dam bottom and the pressure the concrete exerts and compare the two numbers. Assuming the dam is 115 m thick at the bottom and 40 m thick at the top, what is the pressure per unit width at the bottom due to the weight of the concrete? Is that enough to hold back the water?

Specific Gravity and Buoyancy

1.12 One of the most accurate means of determining the fat content of a person is by measuring his specific gravity. The measurement is made by immersing the person in a tank of water and measuring his net weight. Derive an expression for a person's specific gravity in terms of their weight in air, net weight in water, and specific gravity of the water. Find values for the density of fat and muscle to develop your correlation. Is there anything you might have left out of the analysis that would affect your results?

1.13 Icebergs are dangerous because most of the ice lies below the surface of the water. The old rule of thumb is that 90% of the volume of the iceberg lies below the water. Assuming water at −4°C in equilibrium with the ice and a spherical iceberg, determine if this rule of thumb is accurate.

1.14 An air bubble rises slowly through maple syrup. Three forces act on the bubble, the weight of the bubble, the buoyant force, and a "drag" force due to fluid friction between the bubble and syrup as the bubble forces the syrup out of the way to rise. The drag force is given by $F_{drag} = 6\pi\mu v r_o$, where v is the velocity of the bubble and μ is the viscosity of the syrup. Assuming a 5 mm bubble, what is the velocity of the bubble? In which direction does the drag force act? Syrup properties are

$$\mu = 1\ \mathrm{N\,s/m^2} \qquad \rho = 1250\ \mathrm{kg/m^3}$$

1.15 A 100 kg man is preparing to spend the day on a dive. If we can approximate the man as a cylinder, 2 m long and 0.2 m in diameter, how much weight must he carry to insure he is neutrally buoyant?

Rigid Body Acceleration

1.16 An open rectangular tank, 0.5 m wide and 1 m long, contains water to a depth of 1 m. If the height of the tank is 1.25 m, what is the maximum horizontal acceleration (along a line parallel to the long axis of the tank) that can be sustained before the water spills?

1.17 An open, cylindrical tank of height 0.25 m is partially filled with water. The depth of the water is 0.1 m; the diameter of the tank is 0.5 m. If the tank is rotated about its vertical axis, what will be the shape of the surface of the liquid?

1.18 For the tank and water system of Problem 1.17, how fast can we spin the tank before the bottom surface of the tank is exposed? How fast can we spin the tank before water reaches the rim?

REFERENCES

1. Bird, R.B., Stewart, W.E., and Lightfoot, E.N., *Transport Phenomena*, John Wiley & Sons, New York (1960).
2. Hirschfelder, J.O., Curtiss, C.F., and Bird, R.B., *Molecular Theory of Gases and Liquids*, John Wiley & Sons, New York (1954).
3. Bennet, C.O. and Myers, J.E., *Momentum, Heat, and Mass Transfer*, McGraw-Hill, New York (1962).
4. Bejan, A., *Shape and Structure: From Engineering to Nature*, Cambridge University Press, Cambridge, U.K. (2000).
5. Garner, H.F., *The Origin of Landscapes*, Oxford University Press, New York (1974).
6. National Aeronautics and Space Administration, http://spaceplace.nasa.gov/hurricanes/hurricane_diagram_large.en.jpg (accessed on March 3, 2013).
7. Lehninger, A.L., *Biochemistry*, 2nd edn., Worth Publishers, New York (1975).
8. Bailey, J.E. and Ollis, D. F., *Biochemical Engineering Fundamentals*, 2nd edn., McGraw-Hill, New York (1986).
9. Cumming, S., Henderson, R., Horsfield, K., and Singhal, S.S., The functional morphology of the pulmonary circulation, in Fishman, A.P. and Hect, H.H., eds., *The Pulmonary Circulation and Interstitial Space*, The University of Chicago Press, Chicago, IL (1969).
10. Hauck, G., *The Aqueduct of Nemausus*, McFarland & Co., Jefferson, NC (1988).
11. Hawking, S.H., *A Brief History of Time*, Bantam Books, New York (1988).
12. Eisberg, R. and Resnick, R., *Quantum Physics of Atoms, Molecules, Solids,, Nuclei, and Particles*, 2nd edn., John Wiley & Sons, New York (1985).
13. Koretsky, M.D., *Engineering and Chemical Thermodynamics*, 2nd edn., John Wiley & Sons, New York (2013).
14. Holman, J.P., *Thermodynamics*, 4th edn., McGraw-Hill, New York (1988).
15. Modell, M. and Reid, R.C., *Thermodynamics and Its Applications*, Prentice Hall, Englewood Cliffs, NJ (1974).
16. Smith, J.M., Van Ness, H.C., and Abbott, M.M., *Introduction to Chemical Engineering Thermodynamics*, 6th edn., McGraw-Hill, New York (2001).

2 Flows, Gradients, and Transport Properties

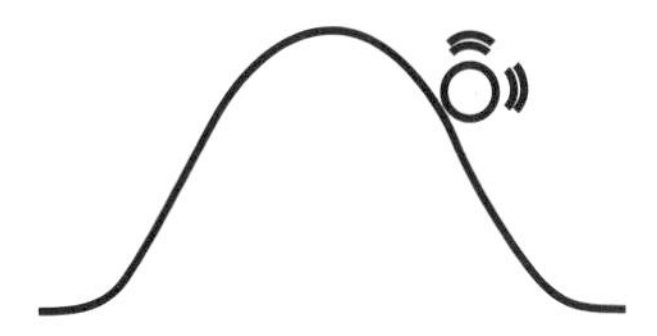

2.1 INTRODUCTION

When we disturb a system from its original equilibrium situation, we know that a transport process must occur to restore the original equilibrium state, to bring the system to some new equilibrium state, or to take the system to a new non-equilibrium state. The disturbance is always in the form of a change in one of the conserved quantities of the system (momentum, energy, mass, charge, etc.) and can be tracked by measuring a change in one of the state variables of the system such as the temperature, pressure, voltage, or mole fraction. These changes set up a spatial variation in the state variable also called a gradient. The particles, molecules, electrons, atoms, etc., move in response to this gradient in an attempt to diminish the gradient and restore equilibrium.

We can say that the flow of particles, the number of particles moving per unit time, is a function of this gradient. The flux of particles, the number of particles moving per unit time, per unit area, is proportional to the flow:

$$\text{Flow} = \text{Flux} \times \text{Area} = f(\text{Gradient}) \tag{2.1}$$

The functionality between flow, flux, and gradient may be quite complex, especially for systems composed of many different components or in systems where many state variables are changing simultaneously. Fortunately, for most engineering purposes, the flows and fluxes can be approximated quite accurately by making them linearly proportional to gradients in the state variables. The constants of proportionality in the gradient-state variable relationships we call the transport properties of the material.

In this chapter, we will consider the various flows and fluxes commonly encountered, their relationship to specific gradients, and the resulting constitutive relationships that define the transport properties of the material. The flow of particles may be related to any number of state variable gradients but one of these gradients is usually dominant. The dominant gradient gives rise to the *primary* transport property of the material and the other gradients give rise to *secondary* transport properties. We will consider primary transport properties first and draw analogies between the various transport processes. Toward the end of the chapter, we will discuss some secondary transport properties and will briefly discuss situations where the linear relationship between the flow or flux and the gradient fails.

2.2 MOMENTUM TRANSPORT: NEWTON'S LAW OF VISCOSITY

The flow quantity, potential gradient, and transport property relationship for momentum transport has its roots in experiments performed in the latter half of the seventeenth century by Robert Hooke (1635–1703). These experiments were concerned with the deformation of solid objects and

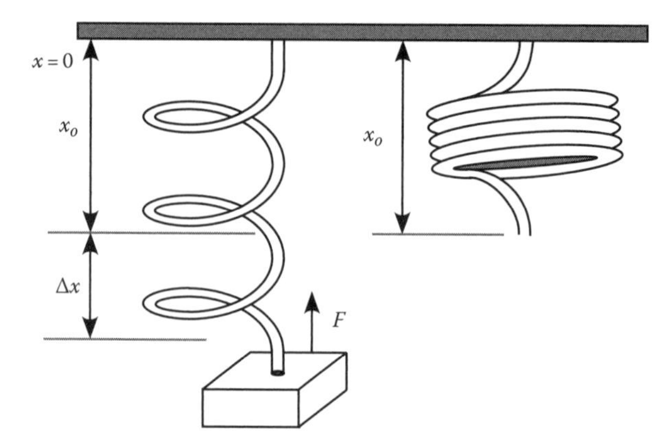

FIGURE 2.1 Extension of a spring under the application of a force. Hooke's law and the spring constant.

the behavior of springs. Hooke showed that if we apply a force, F, to the end of a spring by attaching a mass as shown in Figure 2.1, the elongation of that spring is directly proportional to the force:

$$F = -k_{sp}x \tag{2.2}$$

The constant of proportionality, k_{sp}, is termed the spring constant and Equation 2.2 is called Hooke's law. The minus sign reminds us that the force the spring exerts points in the opposite direction from the displacement of the spring.

Hooke extended this linear relationship to solid objects as well. A solid beam of length, L, cross-sectional area, A_c, and height, Δy, is shown in Figure 2.2. We bond the beam to a plate that holds the lower surface of the beam immobile. If we apply a shearing force to the upper surface and measure the displacement of that surface as a function of the beam's thickness, we find that the displacement is proportional to the force and inversely proportional to the thickness of the beam:

$$\frac{\text{Force}}{\text{Area}} = \text{Stress} = \tau \propto \frac{\Delta L}{\Delta y} \tag{2.3}$$

The constant of proportionality in Equation 2.3 is analogous to the spring constant of Equation 2.2. It is called the modulus of elasticity, Y, and has the units of N/m^2 or force per unit area:

$$Y = \frac{\text{Force}}{\text{Area}} = \frac{\text{N}}{\text{m}^2} \tag{2.4}$$

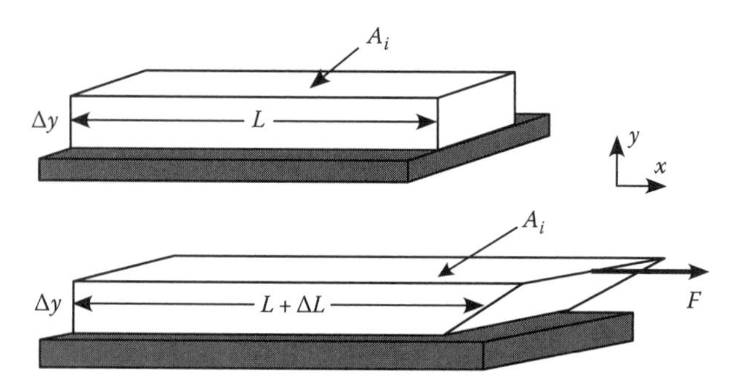

FIGURE 2.2 Deformation of a solid object. Hooke's law and the modulus of elasticity.

By allowing Δy to approach 0 in Equation 2.3, we can obtain the differential form of Hooke's law for elastic bodies, where τ_{yx} is the shear stress exerted on the beam, and ΔL is the displacement of the beam from its original, unstressed, state. Dividing ΔL by the unstressed length of the beam, L, defines a new quantity, γ, the strain:

$$\tau_{yx} = -Y\left(\frac{d\Delta L}{dy}\right) = Y\gamma \quad \text{Hooke's law} \tag{2.5}$$

Hooke's law provides an adequate description of the stress–strain behavior of most solid materials at room temperature. However, if we heat the solid to very high temperatures, or if we try to perform the experiment of Figure 2.2 on a liquid or gas, we will find that the application of the force results in a displacement that varies with the duration of the applied force. The material starts to flow, as if it were a fluid.

In the late seventeenth century, Isaac Newton (1642–1727) proposed that a fluid's resistance to motion was proportional to a velocity gradient and proved this by a series of experiments measuring the drag force on a sphere immersed in a flowing fluid. We name the analog of Hooke's law for fluids after Newton. A simpler experiment for determining the relationship between the stress on a fluid and the velocity gradient is shown in Figure 2.3. A fluid is placed between two very long, parallel plates. One plate is set in motion relative to the other. The force needed to pull one plate past the other depends upon the size of the plates, the gap between the plates, and the relative velocity between the plates:

$$\frac{\text{Force}}{\text{Area}} = \text{Stress}\left(\frac{\text{N}}{\text{m}^2}\right) = \tau_{yx} \propto \frac{\Delta V_x}{\Delta y} \tag{2.6}$$

The proportionality constant in Equation 2.6 is termed the viscosity of the fluid, μ, and has the units of N s/m^2 (mks) or poise (cgs):

$$\mu = \frac{\text{Force} \times \text{Time}}{(\text{Length})^2} = \frac{\text{Dyne} \times \text{s}}{\text{cm}^2} = \frac{\text{g}}{\text{cm} \times \text{s}} = \text{Poise} \tag{2.7}$$

For many fluids, the viscosity is a constant, independent of the shear stress applied and independent of the shear rate, $\Delta v_x/\Delta y$. Such fluids are termed *Newtonian.*

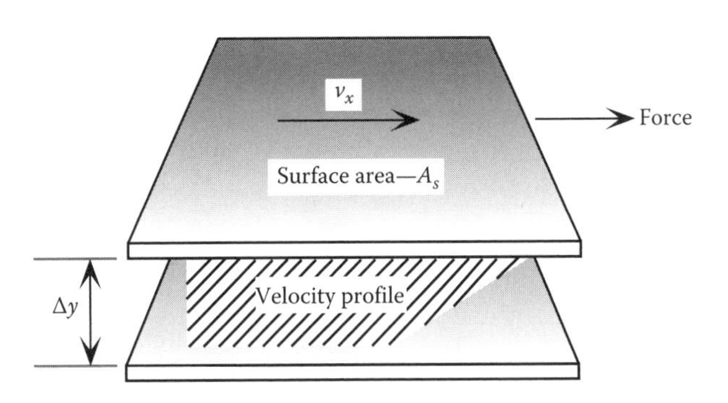

FIGURE 2.3 Fluid flow in response to an applied shear stress.

By letting Δy approach 0 in Equation 2.6, we obtain a differential relationship between the stress, τ_{yx}, the velocity, v_x, and the spatial coordinate, y, termed *Newton's law of viscosity*. This is a form of the flux state variable–gradient relationship we discussed in Equation 2.1:

$$\tau_{yx} = -\mu\left(\frac{dv_x}{dy}\right) \quad \text{Newton's law of viscosity} \tag{2.8}$$

The minus sign in Equation 2.8 indicates that the shear stress is in a direction opposing the change in the velocity of the fluid (analogous to Newton's third law of motion). Multiplying both sides of Equation 2.8 by the cross-sectional area gives the force exerted between the plate and the fluid. This force acts to oppose the change in velocity and is the analog of the flow–gradient relationship of Equation 2.1:

$$F_x = \tau_{yx}A = -\mu A\left(\frac{dv_x}{dy}\right) \tag{2.9}$$

If we divide the viscosity by the density of the fluid, we derive a new material property termed the *kinematic viscosity*, ν:

$$\nu = \frac{\mu}{\rho} = \frac{\text{m}^2}{\text{s}} \quad \text{Kinematic viscosity} \tag{2.10}$$

The kinematic viscosity has the units of (length^2/time) and, as such, is termed a diffusivity of momentum. Rearranging Newton's law to write it in terms of the kinematic viscosity shows that the force exerted on the fluid is proportional to a gradient in the momentum per unit volume of the fluid, hence the term *momentum transport*:

$$F_x = \tau_{yx}A_s = -\left(\frac{\mu A_s}{\rho}\right)\frac{d}{dy}(\rho v_x) = -\nu A_s\frac{d}{dy}(\rho v_x) \tag{2.11}$$

Newton's law is the simplest relationship between the flow or flux of momentum and the velocity gradient. Many others exist. The minus sign tells us that momentum flows from the higher velocity to the lower velocity.

Example 2.1 Calculating the Force and Stress between Flat Plates

Consider the Newtonian fluid of viscosity, μ, and density, ρ, flowing between two parallel plates of surface area, A_s. As shown in Figure 2.4, the fluid velocity profile is parabolic in shape and given by

$$v = ay(L - y)$$

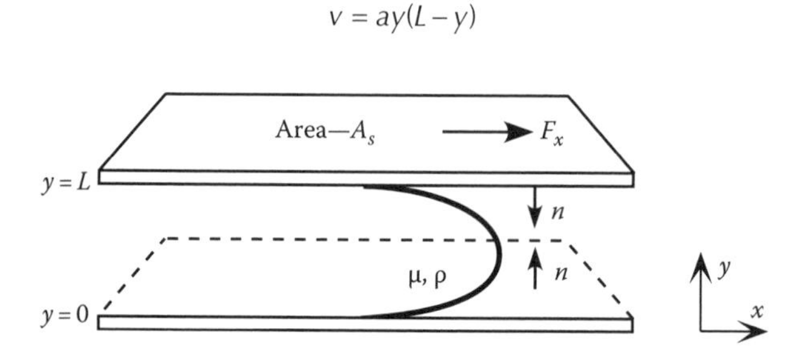

FIGURE 2.4 Fluid flowing in between two flat plates.

What is the shear stress and force exerted on the fluid at the two solid surfaces? What is the direction of the momentum flow at the two solid surfaces?

The shear stress and force are governed by Newton's law of viscosity:

$$\tau_{yx} = -\mu \frac{dv}{dy} = \mu a L \quad \text{Upper surface}$$

$$\tau_{yx} = -\mu \frac{dv}{dy} = -\mu a L \quad \text{Lower surface}$$

The sign on the shear stress gives the direction of the momentum flow. At the upper surface, the momentum flow is in the positive y direction and at the bottom surface it is in the negative y direction. In both cases, momentum is flowing from the higher velocity to the lower.

The force exerted on the fluid is just the stress multiplied by the surface area:

$$F_x = -\mu a L A_s \quad \text{Upper surface}$$

$$= -\mu a L A_s \quad \text{Lower surface}$$

To find the sign of the x-direction of the force, we compare the sign of the stress (momentum flow) with the sign of the inward facing surface normal (into the fluid). If both have the same sign, the force is in the positive direction. If they differ in sign, the force is in the negative direction. Thus, on both surfaces, the force points in the negative direction indicating the plates act to retard the flow of fluid.

2.3 ENERGY TRANSPORT: FOURIER'S LAW OF HEAT CONDUCTION

During the late eighteenth century, James Joule (1818–1889) first classified heat as a form of energy and proved that a given amount of mechanical work done on a system was equivalent to adding heat to the system. Jean-Baptiste Biot (1771–1862), and later Jean-Baptiste Joseph Fourier (1786–1830), performed and analyzed a series of experiments finally tying the flux quantity or heat flux, to a gradient in the state variable, the temperature. These experiments may have been akin to the one shown in Figure 2.5, where a material is covered in insulation and placed between two systems held at constant temperatures. By varying the reservoir temperatures and the length and cross-sectional area of the material between the reservoirs, one can show that the amount of heat flowing between the

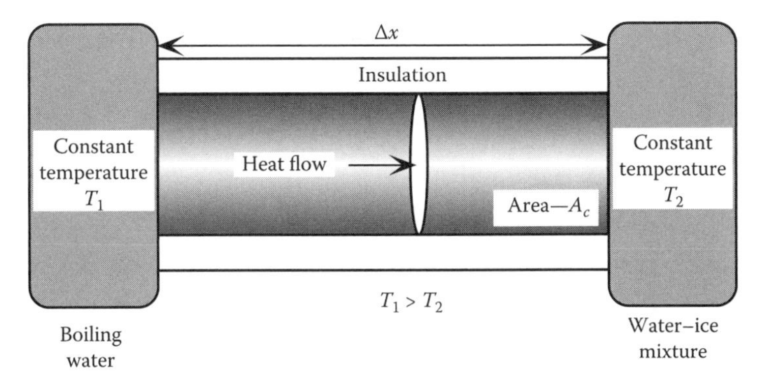

FIGURE 2.5 Heat flow caused by a temperature gradient.

reservoirs is proportional to the temperature difference and the cross-sectional area and inversely proportional to the length of the bar between the reservoirs:

$$q(\mathrm{W}) = q''\left(\frac{\mathrm{W}}{\mathrm{m}^2}\right) \times \text{Area} \propto \frac{\Delta T}{\Delta x} \times \text{Area} \tag{2.12}$$

The proportionality constant in Equation 2.12 is termed the *thermal conductivity*, k, and it has the units of W/m K:

$$k = \frac{\text{Energy}}{\text{Time} \times \text{Length} \times \text{Temperature}} = \frac{\mathrm{W}}{\mathrm{m\,K}} \tag{2.13}$$

Biot in 1804 first suggested the form for Equation 2.12.

By letting Δx approach 0 in Equation 2.12, we find a differential relationship between the heat flux, q_x'', and the temperature gradient, dT/dx. This relationship, developed by Fourier in 1822, is called Fourier's law of heat conduction:

$$q_x = q_x'' A_c = -kA_c\left(\frac{dT}{dx}\right) \quad \text{Fourier's law of heat conduction} \tag{2.14}$$

The minus sign reminds us that heat flows from the higher temperature to the lower temperature (down the gradient).

We can rearrange Fourier's law so that it is written in the form of a flux, a diffusivity, and a gradient. Dividing the thermal conductivity by the density, ρ, and heat capacity, C_p, of the conducting material gives us a new material property, α, called thermal diffusivity:

$$\alpha = \frac{k}{\rho C_p} - \frac{\text{Length}^2}{\text{Time}}\left(\frac{\mathrm{m}^2}{\mathrm{s}}\right) \tag{2.15}$$

The thermal diffusivity has the units of length squared divided by time and so is a true diffusivity. The flux can be expressed in terms of the gradient in the energy of the material, E:

$$q_x'' = -\alpha\left(\frac{d(\rho C_p T)}{dx}\right) = -\alpha\left(\frac{dE}{dx}\right) \quad \left(\frac{\text{Energy}}{\text{Flux}}\right) \tag{2.16}$$

Unlike Newton's law of viscosity, Fourier's law of heat conduction is the only relationship we need between the heat flux, q'', and the temperature or energy gradient.

Example 2.2 Fourier's Law and a Variable Thermal Conductivity

Assume steady-state 1-D conduction through the symmetric shape shown in Figure 2.6.

Derive an expression for the variation in thermal conductivity if the cross-sectional area, $A(x)$, the temperature profile, $T(x)$, and the heat flow, q, are given by

$$A_c(x) = (1 - x^2) \qquad T(x) = 450(1 - 2x - x^3) \qquad q = 7500\ \mathrm{W}$$

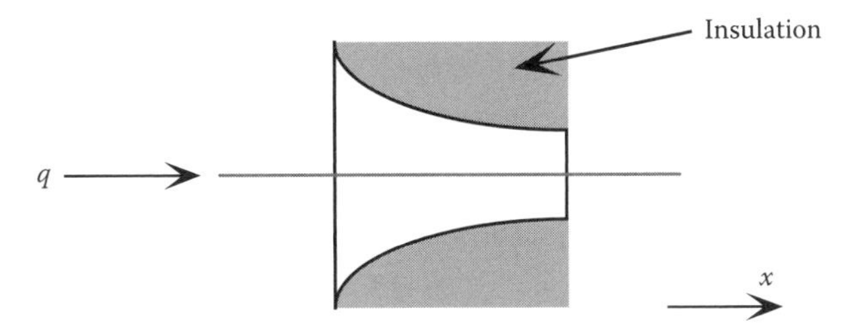

FIGURE 2.6 System cutaway view.

Most materials have a thermal conductivity that varies as a function of temperature. Here we will translate that into a thermal conductivity that varies as a function of position. Fourier's law states that $q = -kA_c(dT/dx)$. Inverting to obtain k gives

$$k = -\frac{q/A_c}{dT/dx} = -\frac{7500/(1-x^2)}{450(-2-3x^2)} = \frac{50}{3(2+x^2-3x^4)}$$

If we knew $T(x)$, we could get $k(T)$.

2.4 MASS TRANSPORT: FICK'S LAW OF DIFFUSION

Long after Newton's law, Fourier's law, and Ohm's law for electrical conductance were developed, people were still uncertain about the movement of chemical species from one location to another. Adolf Eugen Fick (1829–1901), a physiologist, performed a series of experiments tying the flux of chemical species to a gradient involving the mole/mass fraction or concentration/density of chemical species.

Figure 2.7 shows a simple apparatus designed to measure the motion, termed diffusion, of species a, through a medium. An "infinite" reservoir of liquid a is attached to a long tube of

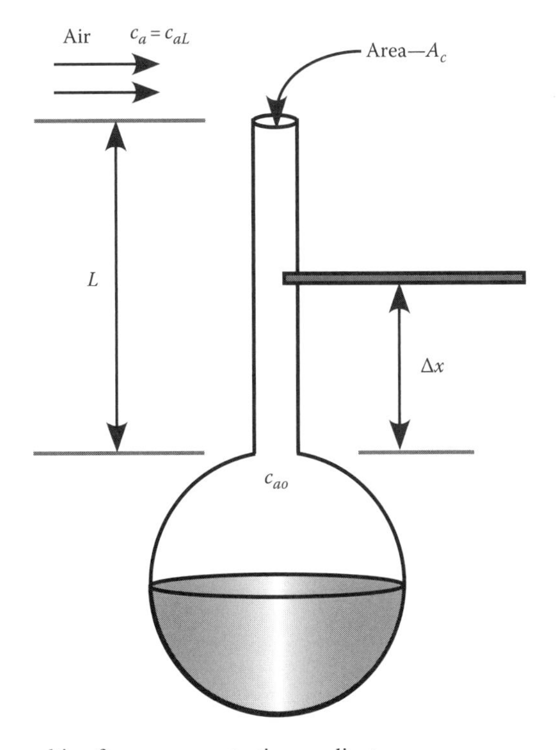

FIGURE 2.7 Diffusion resulting from a concentration gradient.

small cross-sectional area, A_c. The temperature and pressure of the system are held such that the concentration of species, a, in the vapor is very low. At the inlet to the small tube, near the interface between vapor and liquid, the concentration of a is held constant at c_{ao}. At the opposite end of the tube, pure air is blown by the opening to keep the concentration of a zero at the outlet. The small tube is fitted with ports so that a sensor can be inserted and used to measure the concentration of species a along the length of the tube. By varying the diameter of the small tube and the concentration of species a in the air stream, we find, as Fick did in his experiments in 1855, that the molar/mass flux of species a is directly proportional to the difference in concentration/density between the inlet and outlet of the tube, and inversely proportional to the height of the sample tube:

$$\dot{M}_a\left(\frac{\text{mol}"a"}{\text{s}}\right) = J_a\left(\frac{\text{mol}}{\text{m}^2\text{s}}\right)\times \text{Area} \propto \frac{\Delta c_a}{\Delta x}\times \text{Area} \tag{2.17}$$

The proportionality constant is termed the diffusivity of a through medium, b, or D_{ab}. The diffusivity has the units of length squared divided by time:

$$D_{ab} = \frac{(\text{Length})^2}{\text{Time}} = \frac{\text{m}^2}{\text{s}} \tag{2.18}$$

By letting Δx approach 0, we obtain a differential relationship between M_a, J_a, c_a, and x that is called *Fick's first law of diffusion*:

$$\dot{M}_{ax} = J_{ax}A_c = -D_{ab}A_c\left(\frac{dc_a}{dx}\right) \quad \text{Fick's law of diffusion} \tag{2.19}$$

The minus sign reminds us that mass flows from the higher concentration to the lower concentration. In terms of Equation 2.19, the state variable can be the concentration/density or mole/mass fraction and flux is the mass or molar flux.

Like Fourier's law and Newton's law, the relationship between the flux quantity and the gradient is a linear one when the amount of a is small. Fick's law has an additional complication because the relationship between the flux quantity and the state variable gradient may not be linear. In general, the state variable of interest for diffusion is the activity of a and not the concentration of a. Since the diffusion of mass always involves one species moving relative to all others present, the relationship between the mass flux and the gradient will depend upon the stoichiometry of the diffusion process and on the observer's frame of reference.

2.5 CHARGE TRANSPORT: OHM'S LAW OF CONDUCTION

The field of electric circuits was virtually unexplored in the early part of the nineteenth century. The relationship between current, voltage, and the resistance of a conductor was unknown before Georg Simon Ohm (1789–1854) began a series of experiments to determine if such a relationship existed. The types of experiments that Ohm performed were related to the system shown in Figure 2.8. Here we apply an electrical potential difference, a voltage, across a conductor of known size and shape (Ohm used the recently developed Voltaic cell or battery to supply the potential). We measure the current flowing through the conductor as a function of its length, Δx, its cross-sectional area, A_c, and as a function of the magnitude of the potential difference applied to the two ends of the test specimen.

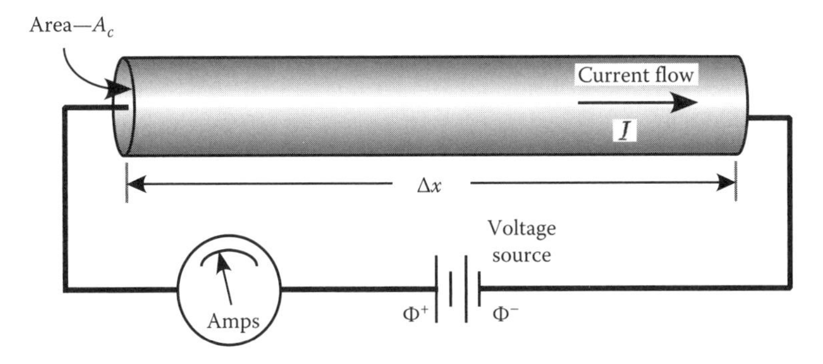

FIGURE 2.8 Current flow under an applied electric potential.

If we plot the current, I, as a function of the applied potential for each specimen, we will find, as Ohm discovered in 1826, that the current density, $J = I/A_c$, is directly proportional to the applied potential (electric field) and inversely proportional to the length of the conductor:

$$I(\mathrm{A}) = \text{Current} = \text{Rate of change flow} = J\left(\frac{\mathrm{A}}{\mathrm{m}^2}\right) \times \text{Area} \propto \frac{\Delta\Phi}{\Delta x} \times \text{Area} \tag{2.20}$$

The proportionality constant in Equation 2.20 is termed the *conductivity* of the medium, σ, and it has the units of charge and time per mass and length. The units of conductivity are amps per volt-meter or Siemens per meter:

$$\sigma = \frac{\text{Charge}^2 \cdot \text{Time}}{\text{Mass} \times \text{Length}^3} = \frac{\mathrm{coul}^2 \cdot \mathrm{s}}{\mathrm{kg} \cdot \mathrm{m}^3} = \frac{\mathrm{A}}{\mathrm{V} \cdot \mathrm{m}} = \frac{\mathrm{S}}{\mathrm{m}} \tag{2.21}$$

By letting Δx approach 0, we obtain a differential relationship between I_x, J_x, Φ, and x that we now refer to as Ohm's law:

$$I_x = J_x A_c = \sigma \mathcal{E}_x A_c = -\sigma A_c \left(\frac{d\Phi}{dx}\right) \quad \text{Ohm's law} \tag{2.22}$$

The electric field strength, $\mathcal{E}_x$, in Equation 2.22 is equivalent to the state variable gradient. The minus sign reminds us that charges flow down this gradient. Negative charges flow from the negative (cathode) to positive (anode) potential (voltage) and positive charges flow from the positive to the negative potential. Nearly everyone is more familiar with the integrated form of Ohm's law, written in terms of the resistance of a material. For our simple 1-D case, the resistance of the conductor would be

$$R_e = \frac{l}{\sigma A_c} \tag{2.23}$$

where l is the length of the conducting medium. In terms of the resistance, the integrated form of Ohm's law for a conductor of uniform cross-sectional area is

$$\Delta\Phi = IR_e \tag{2.24}$$

We can rewrite Ohm's law in terms of a flux quantity, the current density, J, the gradient of our state variable, the electric field strength, and a diffusivity. First, however, we must transform the conductivity into a diffusivity by multiplying by $RT/(z_e\mathcal{F}_a)^2$ where $\mathcal{F}_a$ is Faraday's constant, 96,500 C/mol, and z_e is the valence of the charge carrier. The new material property is called the charge diffusivity, $D_\pm$, and has the units of a diffusivity, m^2/s:

$$D_\pm = \left[\frac{RT}{(\mathcal{F}_a z_e)^2}\right]\sigma = \frac{\text{Length}^2}{\text{Time}} = \frac{\text{m}^2}{\text{s}} \tag{2.25}$$

Ohm's law now becomes

$$J_x = -D_\pm\left[\frac{d}{dx}\left(\frac{(\mathcal{F}_a z_e)^2\Phi}{RT}\right)\right] \tag{2.26}$$

and the state variable is the charge of the system. Notice that like Fourier's, Newton's, and Fick's laws, Ohm's law states that the flux is linearly related to the gradient. Ohm's law, written in the form of Equation 2.26, also looks suspiciously like Fick's law. The two laws are intimately related for Ohm's law is a *diffusion* of charged particles under the application of an electric field. The law is often rewritten in terms of a material property called the mobility, the electric potential gradient, and the concentration of charge carriers:

$$J_x = -\mu_e c_e \frac{d\Phi}{dx} \tag{2.27}$$

If the concentration of charge carriers is small, we can relate the mobility and mass diffusivity via the Einstein relationship (2.28). Like Newton's law, Ohm's law holds for special cases only. Non-Ohmic conductors, like the transistor and diode, are the rule rather than the exception:

$$D_\pm = \frac{\mu_e k_b T}{e z_e} \quad \text{Einstein relation} \tag{2.28}$$

Example 2.3 Point Source of Heat and the Temperature

Point sources represent a useful mathematical abstraction in the analysis of many engineering situations. How energy, momentum, or field strength behaves in the presence of a point source is a fundamental building block. Here we will determine the temperature field associated with a point heat source of strength, Q_o (W).

The simplest way of calculating the potential is to surround the point source with a hypothetical sphere of radius, r, as shown in Figure 2.9, calculate the heat flow, and then use Fourier's law of conduction to determine the potential.

In this spherically symmetric system, Fourier's law, considering the change of temperature with radius, is written as in Equation 2.14 but with x replaced by r. Separating and integrating gives the temperature profile:

$$Q_o = -kA\frac{dT}{dr} = -k\left(4\pi r^2\right)\frac{dT}{dr}$$

$$T = -\frac{Q_o}{4\pi kr} + C_1$$

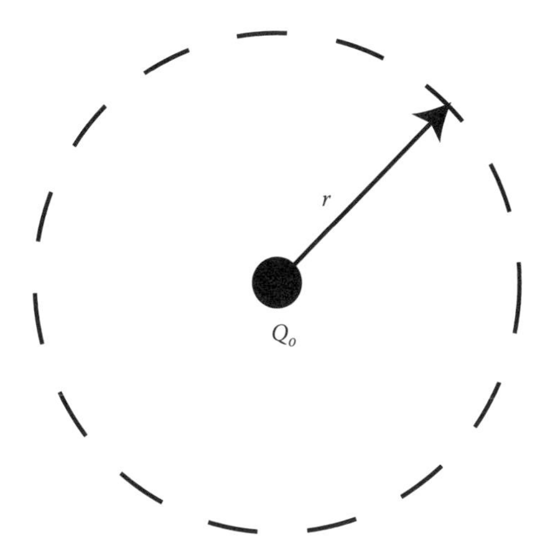

FIGURE 2.9 Point source of heat surrounded by a hypothetical sphere of radius, r.

TABLE 2.1
Potential Distributions Arising from Point Sources

Flux	Point Source	Potential
Gravitational flux density	Mass (m)	$\Phi_g = -\frac{m}{4\pi\varepsilon_g r}$
Electric flux density	Charge (Q)	$\Phi = -\frac{Q}{4\pi\varepsilon r}$
Heat flux	Heat flow (q)	$T = -\frac{q}{4\pi k r}$
Momentum flux	Force (F)	$v = -\frac{F}{4\pi\mu r}$
Mass flux	Mass (m_a)	$w_a = -\frac{m_a}{4\pi D_{ab} r}$

Infinitely far from the point source, the temperature is zero and so $C_1 = 0$:

$$T = -\frac{Q_o}{4\pi k r}$$

This form for the temperature field also holds for the electric field distribution due to a point source of charge, the velocity field due to a point source of momentum, and the concentration field due to a point source of mass. Table 2.1 shows the analogy.

2.6 DRIVING FORCE: RESISTANCE CONCEPTS

To this point, we have focused our attention on the laws, typified by Equation 2.1 describing the flows or fluxes as functions of a gradient in some state variable or potential. Though Newton's law of viscosity and Fourier's law of heat conduction both predate Ohm's law, the field of electrical circuits developed faster than the other transport processes and so formed a basis for comparison. This historical trend has led to an electrical analogy for these laws.

TABLE 2.2
Transport Analogies: Driving Forces and Resistances

Transport Quantity	DC Electric Current	Momentum	Heat	Mass
Transport coefficient	Conductivity σ (Siemens/m)	Viscosity μ (Poise)	Thermal conductivity k (W/m K)	Diffusivity D_{ab} (m^2/s)
Potential difference	Electromotive force (EMF) Φ (V)	Velocity v (m/s)	Temperature T(°C K)	Concentration mole fraction $c_a, x_a\left(\frac{\text{mol}}{\text{m}^3}\right)$
Driving force	Voltage gradient $\mathcal{E}\left(\frac{V}{m}\right) d\Phi/dx$	Velocity gradient $\frac{dv}{dx}\left(\frac{\text{m/s}}{\text{m}}\right)$	Temperature gradient $\frac{dT}{dx}\left(\frac{°\text{C}}{\text{m}}\right)$	Concentration gradient $\frac{dc_a}{dx}\left(\frac{\text{mol/s}}{\text{m}}\right)$
Flux quantity	Current density $\mathcal{J}$(A/m^2)	Stress τ(N/m^2)	Heat flux $\mathbf{q}''$(W/m^2)	Mass flux $N_a\left(\frac{\text{mol/s}}{\text{m}}\right)$
Resistance (planar)	$\frac{\Delta x}{\sigma A_c}$	$\frac{\Delta x}{\mu A_c}$	$\frac{\Delta x}{k A_c}$	$\frac{\Delta x}{D_{ab} A_c}$

Each of the flux laws can be represented as a driving force divided by a resistance:

$$\text{Flow} = \text{Flux} \times \text{Area} = \frac{\text{Driving force}}{\text{Resistance}} \tag{2.29}$$

The prototypical form for Equation 2.29 is, of course, Ohm's law:

$$\mathcal{I} = \frac{\Delta\Phi}{\mathcal{R}_e} = \frac{\Delta\Phi}{(\Delta x/\sigma A_c)} - \frac{\text{Driving force}}{\text{Resistance}} \tag{2.30}$$

If we consider momentum transport, for example, Newton's law of viscosity would become

$$\tau_{xy} A_c = F = \frac{\Delta v_x}{(\Delta y/\mu A_c)} - \frac{\text{Driving force}}{\text{Resistance}} \tag{2.31}$$

Table 2.2 lists the various fluxes, driving forces, and their resistances for all the transport laws discussed so far. Keep in mind that these all presume a *planar geometry and Cartesian coordinates*. We will derive relationships for other geometries later.

2.7 FLUX LAWS IN TWO AND THREE DIMENSIONS

We can extend the flux–gradient relationships to two and three dimensions in a fairly straightforward manner. All fluxes and quantities derived from them have the properties of magnitude and direction. Therefore, the fluxes are at least vector quantities but may be tensor quantities as well. With the exception of momentum transport, all of the state variables such as temperature, concentration or mole fraction, and electric potential (voltage) are scalar quantities. They have no directional component yet may vary as a function of position throughout the system of interest. Their gradients have a direction associated with them and so are vectors. The state variable for momentum transport is the velocity, a vector quantity. Applying the gradient operator to a vector adds yet another directional component and so the gradient of the velocity vector is a second rank tensor. It consists of nine separate components in three dimensions. If the gradient is a tensor quantity, then the flux law, Newton's law of viscosity, requires that the flux also be a tensor. Therefore, in three dimensions, we should expect the stress to have nine components. Table 2.3 shows the 3-D forms for the flux laws in Cartesian coordinates.

Up to this point, we have assumed that all the flux–gradient relationships were based on Cartesian coordinates. Often we will have to analyze a system whose boundaries are such that the application of the conservation laws and flux–gradient relationships in Cartesian coordinates leads to a very complicated description of the problem. Fortunately, if the system is one of eleven regular geometric

TABLE 2.3
Flux–Gradient Relationships in Cartesian Coordinates

Flux Law	Vector Representation	Expanded Representation (Isotropic Material, Constant Properties)
Newton's law	$\vec{\tau} = -\mu\left(\vec{\nabla}\mathbf{v} + \left(\vec{\nabla}\mathbf{v}\right)^t\right) + \left(\frac{2}{3}\mu - K_\mu\right)\left(\vec{\nabla}\cdot\vec{\mathbf{v}}\right)\vec{\mathbf{I}}$	$\tau_{xx} = -\left[2\mu\frac{\partial v_x}{\partial x} - \left(\frac{2}{3}\mu - \kappa_\mu\right)\right]\left(\vec{\nabla}\cdot\vec{v}\right)$
		$\tau_{yy} = -\left[2\mu\frac{\partial v_y}{\partial y} - \left(\frac{2}{3}\mu - \kappa_\mu\right)\right]\left(\vec{\nabla}\cdot\vec{v}\right)$
		$\tau_{zz} = -\left[2\mu\frac{\partial v_z}{\partial z} - \left(\frac{2}{3}\mu - \kappa_\mu\right)\right]\left(\vec{\nabla}\cdot\vec{v}\right)$
		$\tau_{xy} = \tau_{yx} = -\mu\left(\frac{\partial v_x}{\partial y} + \frac{\partial v_y}{\partial x}\right)$
		$\tau_{xz} = \tau_{zx} = -\mu\left(\frac{\partial v_x}{\partial z} + \frac{\partial v_z}{\partial x}\right)$
		$\tau_{yz} = \tau_{zy} = -\mu\left(\frac{\partial v_y}{\partial z} + \frac{\partial v_z}{\partial y}\right)$
Flux Law	**Vector Representation**	**Expanded Representation (Isotropic Material)**
Fourier's law	$\vec{\mathbf{q}}'' = -\vec{\nabla}(kT)$	$q''_x = -k\left(\frac{\partial T}{\partial x}\right)$
		$q''_y = -k\left(\frac{\partial T}{\partial y}\right)$
		$q''_z = -k\left(\frac{\partial T}{\partial z}\right)$
Fick's law	$\vec{\mathbf{J}}_a = -\vec{\nabla}(c_t D_{ab} x_a)$	$J_{ax} = -c_t D_{ab}\left(\frac{\partial x_a}{\partial x}\right)$
		$J_{ay} = -c_t D_{ab}\left(\frac{\partial x_a}{\partial y}\right)$
		$J_{az} = -c_t D_{ab}\left(\frac{\partial x_a}{\partial z}\right)$
Ohm's law	$\vec{J} = \sigma\vec{\mathcal{E}} = -\vec{\nabla}(\sigma\Phi)$	$J_{ax} = -\sigma\left(\frac{\partial \Phi}{\partial x}\right)$
		$J_{ay} = -\sigma\left(\frac{\partial \Phi}{\partial y}\right)$
		$J_{az} = -\sigma\left(\frac{\partial \Phi}{\partial z}\right)$

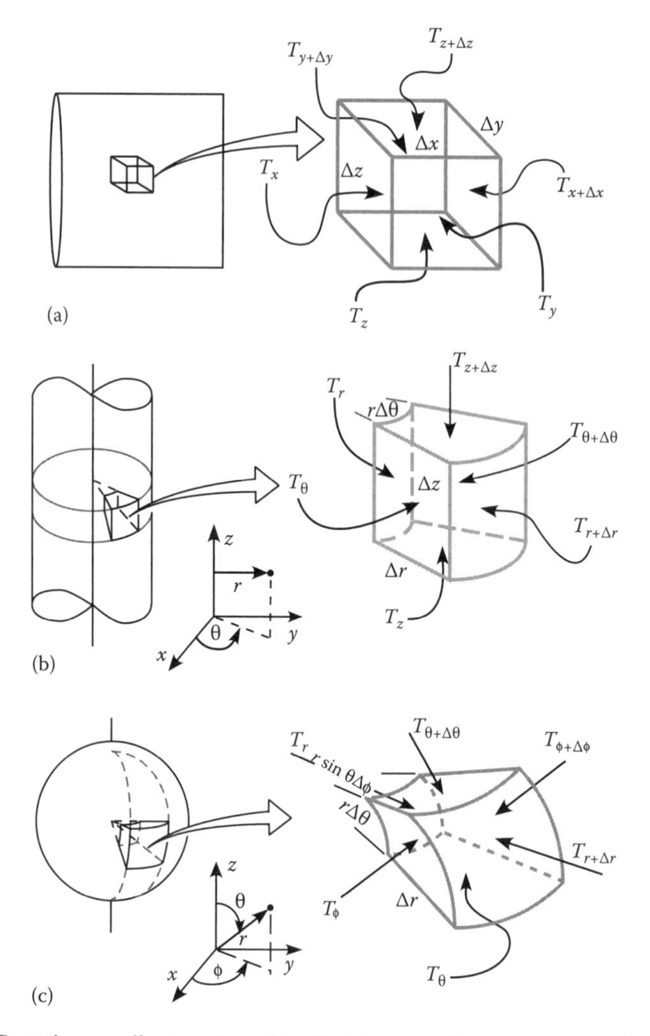

FIGURE 2.10 (a) Cartesian coordinate system, (b) cylindrical coordinate system, and (c) spherical coordinate system.

shapes, we can change coordinate systems and, in so doing, simplify the description of the problem and its solution considerably. For our purposes, only the three most commonly encountered coordinate systems will be discussed: Cartesian coordinates, cylindrical coordinates, and spherical coordinates. We think in terms of Cartesian coordinates most often and so our usual task will be to perform a conversion from the Cartesian frame, Figure 2.10a, to one of the other systems shown in Figure 2.10b and c. The relations between the Cartesian coordinates, x, y, z, and the polar, r, θ, z, or spherical coordinates, ρ, θ, and ϕ, are given in Equations 2.32 and 2.33.

Cylindrical coordinates

$$x = r\cos(\theta) \qquad y = r\sin(\theta) \qquad z = z \tag{2.32}$$

Spherical coordinates

$$x = r\cos(\theta)\cos(\phi) \qquad y = r\cos(\theta)\sin(\phi) \qquad z = r\cos(\theta) \tag{2.33}$$

We can transpose all components of the flux laws given in Table 2.3 using these relationships and obtain the entries in Table 2.4. Transformation laws for the other eight coordinate systems can be found in general physics or math texts [1,2].

TABLE 2.4
Flux–Gradient Relationships in Curvilinear Coordinates

Flux Law	Expanded Representation [Cylindrical Coordinates] (Isotropic Material)
Newton's law	$\tau_{rr} = -\left[2\mu\frac{\partial v_r}{\partial r} - \left(\frac{2}{3}\mu - \kappa_\mu\right)\left(\vec{\nabla}\cdot\vec{\mathbf{v}}\right)\right]$
	$\tau_{\theta\theta} = -\left[2\mu\left(\frac{1}{r}\frac{\partial v_\theta}{\partial \theta} + \frac{V_r}{r}\right) - \left(\frac{2}{3}\mu - \kappa_\mu\right)\left(\vec{\nabla}\cdot\vec{\mathbf{v}}\right)\right]$
	$\tau_{zz} = -\left[2\mu\frac{\partial v_z}{\partial z} - \left(\frac{2}{3}\mu - \kappa_\mu\right)\left(\vec{\nabla}\cdot\vec{\mathbf{v}}\right)\right]$
	$\tau_{r\theta} = \tau_{\theta r} = -\mu\left(r\frac{\partial}{\partial r}\left(\frac{v_\theta}{r}\right) + \frac{1}{r}\frac{\partial v_z}{\partial \theta}\right)$
	$\tau_{\theta z} = \tau_{z\theta} = -\mu\left(\frac{\partial v_\theta}{\partial z} + \frac{1}{r}\frac{\partial v_z}{\partial \theta}\right)$
	$\tau_{rz} = \tau_{zr} = -\mu\left(\frac{\partial v_z}{\partial r} + \frac{\partial v_r}{\partial z}\right)$
	$\vec{\nabla}\cdot\vec{\mathbf{v}} = \frac{1}{r}\frac{\partial}{\partial r}(r v_r) + \frac{1}{r}\frac{\partial v_\theta}{\partial \theta} + \frac{\partial v_z}{\partial z}$
	Expanded Representation [Spherical Coordinates] (Isotropic Material)
Newton's law	$\tau_{rr} = -\left[2\mu\frac{\partial v_r}{\partial r} - \left(\frac{2}{3}\mu - \kappa_\mu\right)\left(\vec{\nabla}\cdot\vec{\mathbf{v}}\right)\right]$
	$\tau_{\theta\theta} = -\left[2\mu\left(\frac{1}{r}\frac{\partial v_\theta}{\partial \theta} + \frac{v_r}{r}\right) - \left(\frac{2}{3}\mu - \kappa_\mu\right)\left(\vec{\nabla}\cdot\vec{\mathbf{v}}\right)\right]$
	$\tau_{\phi\phi} = -\left[\begin{array}{c} 2\mu\left(\frac{1}{r\sin(\theta)}\frac{\partial v_\phi}{\partial \phi} + \frac{v_r}{r} + \frac{v_\theta\cot(\theta)}{r}\right) \\ -\left(\frac{2}{3}\mu - \kappa_\mu\right)\left(\vec{\nabla}\cdot\vec{\mathbf{v}}\right) \end{array}\right]$
	$\tau_{r\theta} = \tau_{\theta r} = -\mu\left(r\frac{\partial}{\partial r}\left(\frac{v_\theta}{r}\right) + \frac{1}{r}\frac{\partial v_r}{\partial \theta}\right)$
	$\tau_{\theta\phi} = \tau_{\phi\theta} = -\mu\left(\frac{\sin(\theta)}{r}\frac{\partial}{\partial \theta}\left(\frac{v_\phi}{\sin(\theta)}\right) + \frac{1}{r\sin(\theta)}\frac{\partial v_\theta}{\partial \phi}\right)$
	$\tau_{r\phi} = \tau_{\phi r} = -\mu\left(\frac{1}{r\sin(\theta)}\frac{\partial v_r}{\partial \phi} + r\frac{\partial}{\partial r}\left(\frac{v_\phi}{r}\right)\right)$
	$\vec{\nabla}\cdot\vec{\mathbf{v}} = \frac{1}{r^2}\frac{\partial}{\partial r}(r^2 v_r) + \frac{1}{r\sin(\theta)}\frac{\partial}{\partial \theta}(v_\theta\sin(\theta)) + \frac{1}{r\sin(\theta)}\frac{\partial v_\phi}{\partial \phi}$

(continued)

TABLE 2.4 (continued)
Flux–Gradient Relationships in Curvilinear Coordinates

Flux Law	Expanded Representation [Cylindrical Coordinates] (Isotropic Material)	Expanded Representation [Spherical Coordinates] (Isotropic Material)
Fourier's law	$q''_r = -k\left(\frac{\partial T}{\partial r}\right)$	$q''_r = -k\left(\frac{\partial T}{\partial r}\right)$
	$q''_\theta = -\frac{k}{r}\left(\frac{\partial T}{\partial \theta}\right)$	$q''_\theta = -\frac{k}{r}\left(\frac{\partial T}{\partial \theta}\right)$
	$q''_z = -k\left(\frac{\partial T}{\partial z}\right)$	$q''_\phi = -\frac{k}{r\sin\theta}\left(\frac{\partial T}{\partial \phi}\right)$
Fick's law	$J_{ar} = -c_t D_{ab}\left(\frac{\partial x_a}{\partial r}\right)$	$J_{ar} = -c_t D_{ab}\left(\frac{\partial x_a}{\partial r}\right)$
	$J_{a\theta} = -\frac{c_t D_{ab}}{r}\left(\frac{\partial x_a}{\partial \theta}\right)$	$J_{a\theta} = -\frac{c_t D_{ab}}{r}\left(\frac{\partial x_a}{\partial \theta}\right)$
	$J_{az} = -c_t D_{ab}\left(\frac{\partial x_a}{\partial z}\right)$	$J_{a\phi} = -\frac{c_t D_{ab}}{r\sin\theta}\left(\frac{\partial x_a}{\partial \phi}\right)$
Ohm's law	$J_r = -\sigma\left(\frac{\partial \Phi}{\partial r}\right)$	$J_r = -\sigma\left(\frac{\partial \Phi}{\partial r}\right)$
	$J_\theta = -\frac{\sigma}{r}\left(\frac{\partial \Phi}{\partial \theta}\right)$	$J_\theta = -\frac{\sigma}{r}\left(\frac{\partial \Phi}{\partial \theta}\right)$
	$J_z = -\sigma\left(\frac{\partial \Phi}{\partial z}\right)$	$J_\phi = -\frac{\sigma}{r\sin\theta}\left(\frac{\partial \Phi}{\partial \phi}\right)$

The flux laws allow the transport coefficients or transport properties to be either scalars or tensors. For the most part, we will concern ourselves with the case where the transport properties are scalars. One may ask what difference it makes if the transport properties are scalar or tensor quantities? If we look at Fourier's law in three dimensions with a scalar thermal conductivity, we have

$$\vec{\mathbf{q}}'' = -k \cdot \vec{\nabla} T \tag{2.34}$$

or in expanded form

$$q''_x\vec{\mathbf{i}} + q''_y\vec{\mathbf{j}} + q''_z\vec{\mathbf{k}} = -k\frac{\partial T}{\partial x}\vec{\mathbf{i}} - k\frac{\partial T}{\partial y}\vec{\mathbf{j}} - k\frac{\partial T}{\partial z}\vec{\mathbf{k}} \tag{2.35}$$

Equation 2.35 states that the heat flux, $\vec{\mathbf{q}}''$ in any direction, is parallel to the temperature gradient in that direction. If we were to plot lines of constant temperature and lines of constant heat flux on one graph, we would find the two lines intersect at right angles. In the isotropic material defined by a scalar thermal conductivity, lines of constant heat flux are perpendicular to lines of constant temperature.

If the thermal conductivity were a tensor quantity, the material becomes anisotropic. Fourier's law is expressed as

$$\begin{bmatrix} q''_x \\ q''_y \\ q''_z \end{bmatrix} = -\begin{bmatrix} k_{xx} & k_{xy} & k_{xz} \\ k_{yx} & k_{yy} & k_{yz} \\ k_{zx} & k_{zy} & k_{zz} \end{bmatrix} \cdot \begin{bmatrix} \dfrac{\partial T}{\partial x} \\ \dfrac{\partial T}{\partial y} \\ \dfrac{\partial T}{\partial z} \end{bmatrix} \tag{2.36}$$

Notice that each component of the heat flux now depends upon all components of the temperature gradient. Even if there is no temperature gradient in the y or z direction, there may still be a heat flux in that direction. The heat flux in a given direction is also no longer strictly parallel to the temperature gradient in that direction for the anisotropic material. Plotting lines of constant heat flux and lines of constant temperature will not result in a grid where the lines intersect at right angles. An example of this is shown in Figure 2.11. Here, the finite element software, Comsol®, was used to calculate and generate the profiles though many other forms of software can handle the same task.

There are many materials where the variation of a transport property with direction is an important aspect of the material's behavior. Examples of these common anisotropic materials include minerals, wood, bone, sea shells, and hair. Other materials such as single crystal silicon, polymeric membranes, or fiber-reinforced composites are engineered with a particular asymmetric structure in mind. Fortunately, we do not have to deal with Equation 2.36 in all its detail. It is possible to define

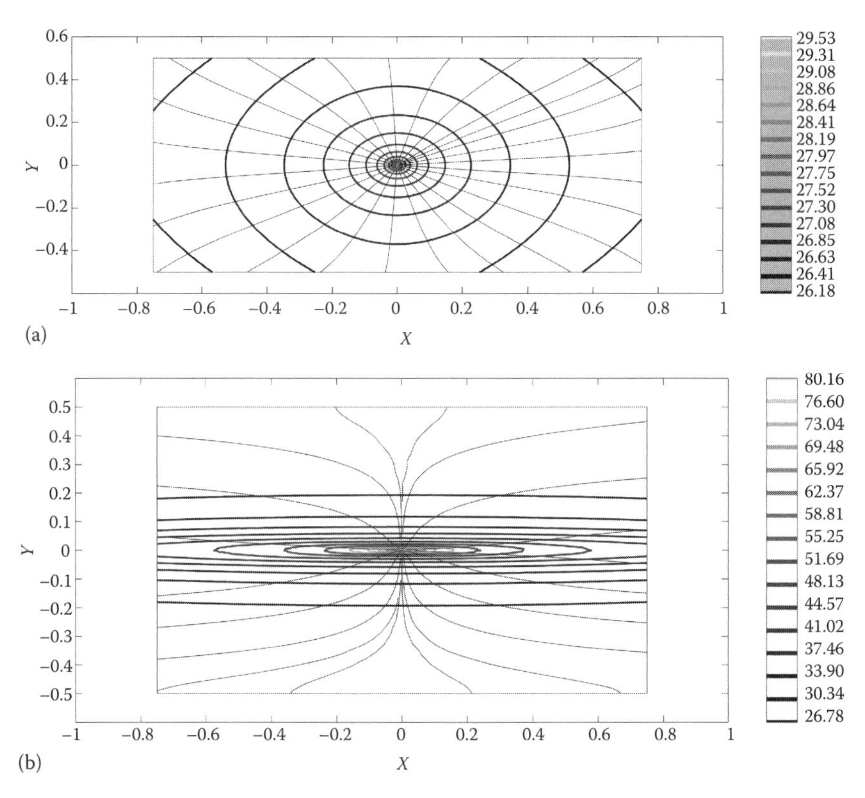

FIGURE 2.11 Isotherm flux plots for an (a) isotropic material (amorphous carbon) and an (b) anisotropic material (graphite). Notice the angle between flux and isotherm. Scale is temperature in °C.

a new set of coordinates, ξ, χ, ω, by rotation of the x, y, and z axes that will reduce Equation 2.36 to the form of Equation 2.35.

The three new transport property coefficients are termed the principal transport properties of the material and the new coordinates, ξ, χ, and ω define the principal axes (directions) of the material:

$$\begin{vmatrix} k_{xx}-\lambda & k_{xy} & k_{xz} \\ k_{yx} & k_{yy}-\lambda & k_{yz} \\ k_{zx} & k_{zy} & k_{zz}-\lambda \end{vmatrix} = 0 \tag{2.37}$$

Evaluating the determinant in Equation 2.37 yields a cubic equation for λ, the principal transport property coefficients. Several simplifications to Equation 2.36 are still possible. It is known from the Onsager symmetry rules of non-equilibrium thermodynamics [3,4] that all nine transport property coefficients in Equation 2.36 are not independent. In fact, the maximum number of independent coefficients is 6 and the off-diagonal elements exhibit the following reciprocal relationships:

$$k_{xy} = k_{yx} \qquad k_{xz} = k_{zx} \qquad k_{yz} = k_{zy} \tag{2.38}$$

Most natural and engineered materials possess one or more centers of symmetry. The symmetry properties provide additional relationships between the various components of the transport properties and so further reduce the complexity of finding the principal axes and transport properties of the materials.

Example 2.4 Principal Axes and Thermal Conductivities for a Monoclinic Crystal

As an example of the simplification process, let us determine the principal axes and thermal conductivities for a hypothetical monoclinic crystal having the following thermal conductivities in Cartesian coordinates:

$$k = \begin{bmatrix} k_{xx} & 0 & k_{xz} \\ 0 & k_{yy} & 0 \\ k_{xz} & 0 & k_{zz} \end{bmatrix} \qquad \begin{cases} k_{xx} = 2.0 & k_{xz} = 1.0 \\ k_{yy} = 5.0 & k_{zz} = 1.0 \end{cases}$$

The monoclinic crystal class [5] is a "triaxial" crystal so that $k_{xx} \neq k_{yy} \neq k_{zz}$. Gypsum is a common material in the monoclinic class. It does have a center of inversion symmetry so that the only off-diagonal elements are $k_{xz} = k_{zx}$. Looking at the thermal conductivity tensor, we see that the Cartesian y-axis is already one of the principal axes (only k_{yy} exists) so we need only find two others. The eigenvalue equation for the principal conductivities is

$$\begin{vmatrix} 2-\lambda & 0 & 1 \\ 0 & 5-\lambda & 0 \\ 1 & 0 & 1-\lambda \end{vmatrix} = 0$$

Solution of the cubic equation gives

$$\lambda_1 = 2.618 \qquad \lambda_2 = 5 \qquad \lambda_3 = 0.382$$

These are the principal conductivities with λ_2 being k_{yy}. To determine the directions of the principal axes, we return to our transport property tensor. We let ξ_1, ξ_2, and ξ_3 be the coordinates of the first principal axis direction vector, ω_1, ω_2, and ω_3 be the coordinates of the second principal axis vector, and χ_1, χ_2, and χ_3 be the coordinates for the third principal axis direction vector. We set up three equations to determine the three vectors in the following way:

$$\begin{bmatrix} 2 & 0 & 1 \\ 0 & 5 & 0 \\ 1 & 0 & 1 \end{bmatrix}\begin{bmatrix} \xi_1 \\ \xi_2 \\ \xi_3 \end{bmatrix} = \begin{bmatrix} \lambda_1\xi_1 \\ \lambda_1\xi_2 \\ \lambda_1\xi_3 \end{bmatrix} \qquad \begin{bmatrix} 2 & 0 & 1 \\ 0 & 5 & 0 \\ 1 & 0 & 1 \end{bmatrix}\begin{bmatrix} \chi_1 \\ \chi_2 \\ \chi_3 \end{bmatrix} = \begin{bmatrix} \lambda_2\chi_1 \\ \lambda_2\chi_2 \\ \lambda_2\chi_3 \end{bmatrix}$$

$$\begin{bmatrix} 2 & 0 & 1 \\ 0 & 5 & 0 \\ 1 & 0 & 1 \end{bmatrix}\begin{bmatrix} \omega_1 \\ \omega_2 \\ \omega_3 \end{bmatrix} = \begin{bmatrix} \lambda_3\omega_1 \\ \lambda_3\omega_2 \\ \lambda_3\omega_3 \end{bmatrix}$$

These equations reduce to

$$\xi_1 = 1.618\,\xi_3 \qquad \xi_2 = 0$$

$$\chi_1 = 0 \qquad \chi_2 = 1 \qquad \chi_3 = 0$$

$$1.618\,\omega_1 = -\omega_3 \qquad \omega_2 = 0$$

We have one degree of freedom for each direction vector so we can choose ξ_1 and ω_1 to be 1. We then calculate the remaining components and normalize the vectors to have a length of 1. The eigenvectors and hence the directions of the principal axes are

$$\xi = \begin{bmatrix} 0.851 \\ 0 \\ 0.526 \end{bmatrix} \qquad \chi = \begin{bmatrix} 0 \\ 1 \\ 0 \end{bmatrix} \qquad \omega = \begin{bmatrix} 0.526 \\ 0 \\ -0.851 \end{bmatrix}$$

2.8 MECHANISTIC DIFFERENCES BETWEEN THE TRANSPORT PHENOMENA

There are many similarities between the flux–gradient relationships among the transport phenomena, and we have highlighted them so far. There are fundamental differences too that lend a particular flavor to each transport process. The first fundamental difference involves the mechanism of transport. Mass, heat, momentum, and charge transport are predominantly governed by diffusive mechanisms. An initial disturbance tends to spread and decay toward zero. Momentum transport on the other hand can be conservative too. A conservative transport can be viewed to proceed in a wavelike fashion without diffusion.

A second point of departure between the transport processes involves what occurs at a boundary between dissimilar media. Mass, heat, and charge transport allow for jump discontinuities in their state variables at the boundary. In heat transfer, we refer to this as a contact resistance. It arises because no two surfaces can be in perfect contact and hence a finite temperature difference exists between them. Even if two surfaces are in perfect contact, the interface between different materials exhibits an interfacial resistance we will talk about more in the next chapter. In mass transfer, we refer to the discontinuity in terms of an "equilibrium" constant or partition function tying together the allowed concentration of a solute in the two media. A similar quantity exists in charge transfer. The complication is that the partition coefficient (charge transfer) is also affected by the presence of an electric field, set up due to the charge imbalance formed by the partition coefficient.

Momentum transport does not normally allow for a discontinuous state variable. We generally assume that there is no "slip" at the interface between two fluids or between a fluid and solid surface. This insures that the state variable, the velocity, is continuous across the interface. A "slip" coefficient however is often used to account for phenomena such as the movement of a contact line separating liquid, vapor, and solid when a liquid evaporates or rewets a solid surface, or the flow of a polymer near the wall of a pipe or mold. While a controversial topic, the existence of slip is becoming more accepted and documented [6]. In most instances, the no-slip condition will yield adequate results indistinguishable from experimental measurements.

Discontinuities in flux are generally related to some material property difference, like contact between two media with different thermal conductivities. The flux relationships become most complicated for conservative transport. In this wavelike transport, we can have transmission, reflection, or absorption at the boundary. Thus, there are discontinuities in the magnitude of the potential and the flux and also in the direction of transport.

The final difference between the transport phenomena concerns the "medium" through which transport is occurring and the effect of the observer's frame of reference on the specification of the flux. The flux laws apply only when we have a physical medium through which the transport occurs. Fick's law has an added complication due to the requirement that at least two separate components move relative to one another. This means that the specification of Fick's law will change depending on one's frame of reference and how one counts the moving particles. If more than two components are being transferred at the same time, it is even possible that one of those components can move against its concentration gradient (in apparent violation of Fick's law) without any external work or forces applied.

Fundamentally, the flux is defined as the product of a species' "concentration" or "density" and its "velocity":

$$\text{Flux} = \begin{pmatrix} \text{Concentration} \\ \text{or} \\ \text{Density} \end{pmatrix} \cdot \text{Velocity} \tag{2.39}$$

There are six common ways to specify the mass flux in a system because we can base mass transfer on a purely mass basis or on a molar basis. On a mass basis, the mass flux of species, a, relative to a set of stationary coordinates (defined by the user) is defined in terms of the density, ρ_a:

$$n_a = \rho_a v_a \left(\frac{\text{kg}}{\text{m}^2\text{s}}\right) \tag{2.40}$$

The velocity, v_a, is not the velocity of an individual molecule of species, a, but rather the mean velocity of all a molecules within the volume of interest. The corresponding molar flux of species, a, relative to the same set of stationary coordinates is defined in terms of the concentration of species, a:

$$N_a = c_a v_a \left(\frac{\text{mol}}{\text{m}^2\text{s}}\right) \tag{2.41}$$

If one wants to measure diffusion coefficients directly, the set of stationary coordinates is not a convenient one to work with. It is easier to use a coordinate system that moves with the bulk flow. We can define two bulk velocities—a molar average velocity and a mass average velocity:

$$v_{Mole} = \frac{\sum_{i=1}^{n} c_i v_i}{\sum_{i=1}^{n} c_i} \quad \text{Molar average velocity} \tag{2.42}$$

$$v_{mass} = \frac{\sum_{i=1}^{n} \rho_i v_i}{\sum_{i=1}^{n} \rho_i} \quad \text{Mass average velocity} \tag{2.43}$$

Using these velocities, we can define two "diffusion" velocities for species, a:

$v_a - v_{mass}$ Diffusion velocity relative to mass average velocity (2.44)

$v_a - v_{Mole}$ Diffusion velocity relative to molar average velocity (2.45)

Two new fluxes are readily apparent, each based on a diffusion velocity:

$J_i = c_i(v_i - v_{Mole})$ Molar flux relative to molar average velocity (2.46)

$j_i = \rho_i(v_i - v_{mass})$ Mass flux relative to mass average velocity (2.47)

These diffusion velocities are related to the variance about the mean flow of particles. Both the flux definitions in Equations 2.46 and 2.47 are used frequently; however, we can define two other fluxes that are used sparingly:

$J_i^\rho = c_i(v_i - v_{mass})$ Molar flux relative to mass average velocity (2.48)

$j_i^c = \rho_i(v_i - v_{Mole})$ Mass flux relative to molar average velocity (2.49)

The mass and molar average velocities, and the fluxes based on them, are in common usage. Still other average velocity definitions, like a volume average velocity, are possible and fluxes may be defined based on them as well. With the six definitions for the mass or molar flux we have developed, we can create the six different versions of Fick's first law shown in Table 2.5.

TABLE 2.5
Equivalent Expressions of Fick's Law of Binary Diffusion

Flux	Gradient	Diffusion Law
j_a	$\vec{\nabla} w_a$	$\vec{\mathbf{j}}_a = -\rho D_{ab} \vec{\nabla} w_a$
j_a^c	$\vec{\nabla} x_a$	$\vec{\mathbf{j}}_a^c = -\left(\frac{c_t^2}{\rho}\right) M_{wa} M_{wb} D_{ab} \vec{\nabla} x_a$
J_a	$\vec{\nabla} x_a$	$\vec{\mathbf{J}}_a = -c_t D_{ab} \vec{\nabla} x_a$
J_a^ρ	$\vec{\nabla} w_a$	$\vec{\mathbf{J}}_a^\rho = -\left(\frac{\rho^2}{c_t M_{wa} M_{wb}}\right) z D_{ab} \vec{\nabla} w_a$
n_a	$\vec{\nabla} w_a$	$\vec{\mathbf{n}}_a = -\rho D_{ab} \vec{\nabla} w_a + w_a(\vec{\mathbf{n}}_a + \vec{\mathbf{n}}_b)$
N_a	$\vec{\nabla} x_a$	$\vec{\mathbf{N}}_a = -c_t D_{ab} \vec{\nabla} x_a + x_a(\vec{\mathbf{N}}_a + \vec{\mathbf{N}}_b)$

Example 2.5 Relationship between J_a and N_a

We often have to interconvert between mass fluxes to design a process or to analyze experimental data. Determine the relationship between the molar flux of species, *a*, based relative to the molar average velocity, J_a, and the molar flux, based relative to a set of stationary coordinates, N_a.

Our definition for J_a is

$$J_a = c_a(v_a - v_{Mole})$$

Expanding this expression and substituting our definition for v_{Mole} gives J_a in terms of terms we can relate to N:

$$J_a = c_a v_a - \frac{c_a}{c_t}\sum_{i=1}^{n} c_i v_i$$

Now $c_a v_a$ is just N_a and likewise, the $c_i v_i = N_i$. Substituting these expressions into the definition for J_a gives the final relationship between J_a and N_a:

$$J_a = N_a - x_a \sum_{i=1}^{n} N_i$$

Notice that while J_a is related linearly to the concentration or mole fraction gradient, N_a may not be because it depends on the fluxes of the other components in the solution.

2.9 PRIMARY AND SECONDARY FLUXES [7–9]

In the previous sections, we considered the flux–gradient relationships to apply to pure components or, in the case of mass transfer, to binary mixtures. We considered only the primary flux–gradient relationship, the situation most commonly encountered in engineering problems. Nature is more complicated than that. In a multicomponent system, or a system where more than one gradient in a state variable may exist, a flux may respond to each of the gradients. In response to a temperature gradient for instance, we may have a mass flux. This phenomenon, called thermodiffusion, or the Soret effect, is an example of a secondary flux and gives rise to a secondary transport property, the thermal diffusion coefficient. Table 2.6 summarizes the fluxes we have encountered along with their primary and some of their more common secondary flux–gradient relationships.

The primary fluxes are unique in that the single transport coefficient associated with them only applies to that particular flux–gradient relationship. The secondary fluxes comprise the off-diagonal elements in the general flux–gradient matrix and a reciprocal relationship exists between those elements and their associated transport coefficients. In this vein, the ionic mobility and ionic diffusivity are mathematically related, so are the Dufour and Soret effect coefficients. Similar relationships exist between the Peltier, Thomson, and Seebeck effect coefficients and between the galvanomagnetic and magnetogalvanic effect coefficients. For more detailed information, the reader should turn to the wide variety of texts on the thermodynamics of irreversible processes [3,4]. Herein, we will survey only a few of the most common secondary fluxes and coefficients.

2.9.1 MOMENTUM

We can separate momentum transport from heat and mass transport because the flux of momentum is a tensor and not a vector quantity. Thus, the flux of momentum does not depend linearly upon gradients in temperature, or concentration (there is no transport property associated with them), and the fluxes of mass and energy do not depend linearly on gradients in velocity (again there is no

TABLE 2.6
Primary and Secondary Fluxes, Gradients, and Transport Properties

Driving Force	Velocity Gradient	Temperature Gradient	Concentration Gradient	Voltage Gradient	Pressure Gradient
Momentum (second-order tensor)	**Newton's law** (μ, κ_μ)				Darcy's law (k_{perm}/μ)
Energy (vector)		**Fourier's law** (**k**)	Dufour effect ($\mathbf{D}^T$)	Peltier effect ($\pi_T = \alpha_T T$) Thomson effect (τ_e)	Pressure thermal effect (k_p)
Mass (vector)		Soret effect ($\mathbf{D}^T$)	**Fick's law ($\mathbf{D}_{ab}$)**	Electro-diffusion $\left(\mu_e = \frac{D_{ab} z_e}{RT}\right)$	Pressure diffusion effect
Charge (vector)		Seebeck effect $\left(\alpha_T = \frac{\pi_T}{T}\right)$	Electro-diffusion $\left(D_{ab} = \frac{\mu_e RT}{z_e}\right)$	**Ohm's law** (σ)	

Note: Bold represents primary laws.

specific transport property associated with such a gradient). The momentum flux, $\vec{\tau}$, for a Newtonian fluid consists of two parts:

$$\vec{\tau} = -\mu\left[\vec{\nabla}\vec{\mathbf{v}} + \left(\vec{\nabla}\vec{\mathbf{v}}\right)^t\right] + \left(\frac{2}{3}\mu - \kappa_\mu\right)\left(\vec{\nabla}\cdot\vec{\mathbf{v}}\right)\vec{\mathbf{I}} \tag{2.50}$$

Here, μ and κ_μ, are two phenomenological coefficients that are called the coefficient of "shear" and "bulk" viscosity, respectively. The coefficient of shear viscosity is important in flows where successive layers of fluid molecules move with respect to one another. This is the common situation encountered in most fluid flow problems. The coefficient of bulk viscosity is important in the pure expansion or contraction of a fluid and is important in explosions and supersonic/hypersonic flows. It can be derived from the kinetic theory of gases when we consider the vibrational and rotational contributions to the molecule's overall energy and momentum. Since these contributions are generally much smaller than translational contributions at room temperature and pressure, the coefficient of bulk viscosity is small and can generally be neglected. If the fluid is incompressible, then only the translation component of momentum is important, the divergence of the velocity, $\vec{\nabla}\cdot\vec{\mathbf{v}}$, is identically zero, and the bulk viscosity does not enter into the model formulation.

2.9.2 Energy [7,8]

The flux of energy (with respect to a mass average velocity) can be decomposed into many separate components, six of which are shown in the following:

$$\vec{\mathbf{q}}'' = \vec{\mathbf{q}}''^T + \vec{\mathbf{q}}''^H + \vec{\mathbf{q}}''^D + \vec{\mathbf{q}}''^R + \vec{\mathbf{q}}''^E + \vec{\mathbf{q}}''^P \tag{2.51}$$

The first term, $\vec{\mathbf{q}}''^T$, is the familiar conductive energy flux, the primary flux–gradient relationship, which is related to the temperature gradient:

$$\vec{\mathbf{q}}''^T = -k\cdot\vec{\nabla}T \tag{2.52}$$

where k is the instantaneous value of the thermal conductivity of the mixture.

The energy flux caused by interdiffusion, $\vec{\mathbf{q}}''^H$, is defined for a fluid containing n species by

$$\vec{\mathbf{q}}''^H = \sum_{i=1}^{n} \frac{\overline{H}_i}{M_{wi}} \vec{\mathbf{j}}_i = \sum_{i=1}^{n} \overline{H}_i \vec{\mathbf{J}}_i \tag{2.53}$$

where
$\overline{H}_i$ is the partial molar enthalpy of species i
$\vec{\mathbf{J}}_i$ and $\vec{\mathbf{j}}_i$ are the molar and mass fluxes defined by Equations 2.46 and 2.47
M_{wi} is its molecular weight

You are familiar with this form of heat flux normally called the heat of mixing. $\vec{\mathbf{q}}''^H$ may be the dominant flux component in some systems, such as those that are chemically reacting or when mixing acids and bases.

Example 2.6 Simultaneous Heat and Mass Transfer

Consider the case shown in Figure 2.12. A well-mixed gaseous system containing HCl vapor in contact with water (stagnant) forms the basis of a small scrubbing system for vapor phase production of optical waveguides. The HCl is absorbed into the liquid and an equilibrium partition coefficient, m_k, exists between HCl in the liquid and vapor. The dissolution of HCl in water is accompanied by a tremendous amount of heat. We want to know the steady-state temperature profile in the liquid if the overall system is adiabatic. The concentration profile of HCl in the liquid has been measured and found to be

$$\frac{1-x_a}{1-y_{ao}/m_k} = \left(\frac{1-x_{aL}}{1-y_{ao}/m_k}\right)^{Z/L}$$

Using Fick's first law to determine the steady-state flux of a, we find

$$N_{az} = -c_t D_{ab} \frac{dx_a}{dz} = \frac{c_t D_{ab}}{L} \ln\left(\frac{1-x_{aL}}{1-y_{ao}/m_k}\right)$$

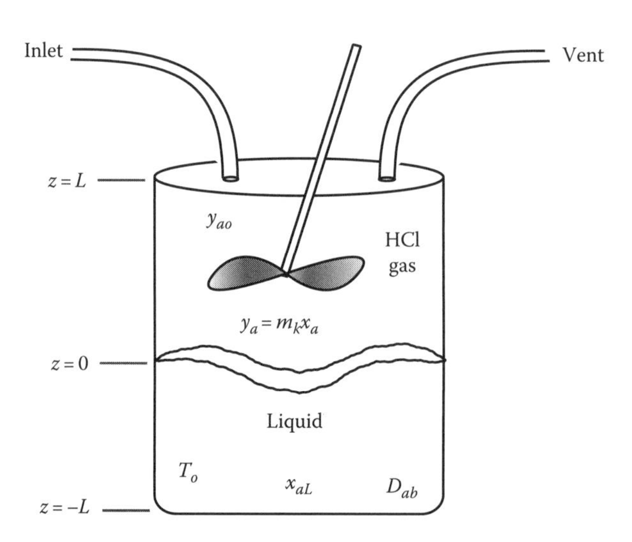

FIGURE 2.12 Heat and mass transfer resulting from a heat of solution process.

The heat flux is driven by the temperature gradient and the concentration gradient (through interdiffusion). Thus, we need to know N_{az} and N_{bz}:

$$q_z'' = -k\frac{dT}{dz} + N_{az}\bar{H}_a + N_{bz}\bar{H}_b$$

Since the water is stagnant, $N_{bz} = 0$, and we can express $\overline{H_a}$ in terms of a temperature difference and the partial molar heat capacity at constant pressure:

$$\bar{H}_a = \bar{C}_{pa}(T - T_{ref})$$

Letting $\theta = T - T_{ref}$ and substituting into the heat flux equation gives

$$k\frac{d\theta}{dz} - N_{az}\bar{C}_{pa}\theta = 0 \qquad \frac{dT}{dz} = \frac{d\theta}{dz}\frac{dT}{d\theta}$$

when the heat flux is zero. Since the flux, N_{az}, is constant, we can separate and integrate the differential equation to give an exponential temperature profile:

$$\frac{\theta}{\theta_o} = \exp\left(\frac{N_{az}\bar{C}_{pa}}{k}z\right)$$

Here we have used a boundary condition that states that we know the temperature at the fluid surface $(z = z_o, \theta = \theta_o)$.

$\vec{\mathbf{q}}''^R$ is the radiant heat flux. It is given by the Stefan-Boltzmann equation:

$$\vec{\mathbf{q}}''^R = \sigma^r \varepsilon_1^r F_{12}(T_1^4 - T_2^4) \tag{2.54}$$

with σ^r being the Stefan–Boltzmann constant, ε_1^r is the emissivity of surface "1," and F_{12} is the view factor between surfaces "1" and "2."

$\vec{\mathbf{q}}''^E$ is the heat flux that results from imposing an electric potential gradient across a substrate (charge carriers are required to carry the heat) as shown in Figure 2.13. This component of the heat flux can take a number of forms depending on whether the substrate is a homogeneous material (Thompson effect) or whether the substrate is heterogeneous like a thermocouple (Peltier and Seebeck effects) and the heat is localized at the junction [7]. The Seebeck effect is the most convenient one to treat first and is the one most often encountered (thermocouples). If the total resistance of the wires, a and b, is known, a measurement of the current, I, flowing through the circuit will allow an evaluation of the voltage driving force. Experimentally, we would find that there is an additional voltage, Φ_{ab}, other than the applied voltage, Φ_o, that appears in the circuit. This additional voltage is only a function of materials, a and b, and the temperature difference, $T_h - T_c$. The voltage, Φ_{ab}, is termed the Seebeck voltage after its discoverer and is related to the temperature difference by

$$\Phi_{ab} = \alpha_T(T_h - T_c) \quad \alpha_T, \quad \text{Seebeck coefficient} \tag{2.55}$$

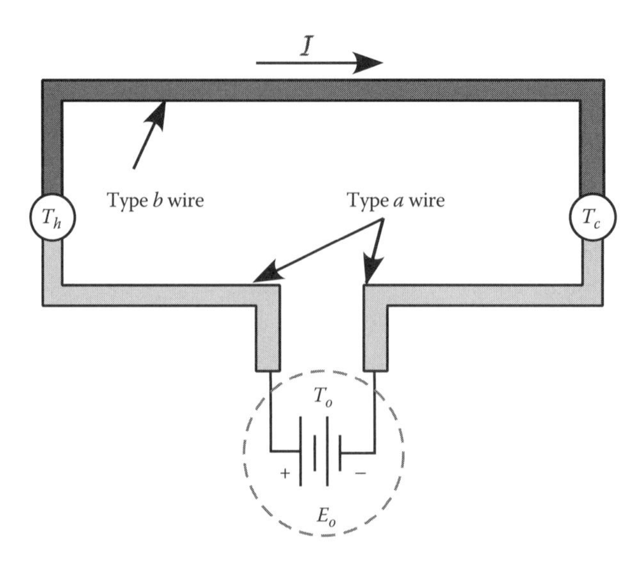

FIGURE 2.13 A simple thermoelectric circuit consisting of two materials, *a* and *b*. The junctions at T_h, T_c, and T_o are isothermal.

If the total resistance can be considered constant, we can relate the current density, *J*, to the temperature difference via

$$\vec{J} = -\left(\frac{\alpha_T}{R_e}\right)\vec{\nabla}\mathbf{T} \tag{2.56}$$

As the current flows through the junction, heat must be exchanged with the surroundings if the junction is to be at a constant temperature. The amount of heat exchanged depends upon the direction of current flow. This phenomenon is called the Peltier effect and the heat exchanged is given by

$$\vec{\mathbf{q}}''^{E} = -\Pi_T\vec{J} = -\left(\frac{\Pi_T}{R_e}\right)\vec{\nabla}\Phi \qquad \Pi_T,\text{Peltier coefficient} \tag{2.57}$$

The Seebeck and Peltier coefficients are related through the Onsager reciprocity relations [3,4,7]:

$$\Pi_T = T\alpha_T \tag{2.58}$$

The final important heat flow phenomenon is termed the Thompson effect after its discoverer William Thompson (Lord Kelvin). If we pass a current through wire *b* only and simultaneously impose a temperature gradient along the wire, we will find that heat must be exchanged with the surroundings to maintain the original temperature gradient. The magnitude of this heat flow is proportional to the current flow and temperature gradient:

$$\vec{\mathbf{q}}''^{E} = \tau_e\vec{J}\cdot\vec{\nabla}T \qquad \tau_e,\text{Thompson coefficient} \tag{2.59}$$

The Thompson coefficient and the Seebeck coefficient are related by [3,4,7]

$$\tau_{ea} - \tau_{eb} = T\frac{d\alpha_T}{dT} \tag{2.60}$$

The thermoelectric heat fluxes are negligible in heat transfer problems involving dielectric materials and are also small in heat transfer problems involving conductors. Their chief usefulness is in temperature measurement by thermocouples and in fabricating thermoelectric heaters or refrigerators for controlling the temperature of electronic components such as integrated circuits (ICs) or charge coupled devices (CCDs).

Example 2.7 Thermoelectric Effects

Figure 2.14 shows a very crude apparatus for measuring the Thompson coefficient in the conducting bar. The process involves imposing a temperature gradient on the bar, flowing a current through it, and then adjusting the current to achieve zero heat flux from the bar to the surroundings. We want to know how the voltage varies as a function of position in the bar.

In this situation, we will have two heat fluxes: the first set up by the temperature gradient in the bar and the second due to the thermoelectric effect. At our balance point, the sum will be zero. Using Equations 2.52 and 2.59 gives

$$\mathbf{q}''_x = 0 = -k\frac{dT}{dx} + \frac{\tau_e}{\mathcal{R}_e}\left(\frac{dT}{dx}\right)\left(\frac{d\Phi}{dx}\right)$$

Since the temperature gradient is not zero, we can factor it out of the equation and see that the voltage gradient must be a linear function of position:

$$\frac{d\Phi}{dx} = \frac{k\mathcal{R}_e}{\tau_e} \qquad \Phi = \Phi_{ref} + \left(\frac{k\mathcal{R}_e}{\tau_e}\right)x$$

Since the Thompson coefficient is generally small, large voltages would be necessary to balance any kind of thermal conduction.

$\vec{\mathbf{q}}''^D$ represents the heat flux arising from a concentration gradient. This is the Dufour effect and is generally so small as to be immeasurable. The explicit form for the Dufour energy flux is pretty complicated, but we can cast it in the form of Fourier's law and state that

$$\vec{\mathbf{q}}''^D = -D_i^T\vec{\nabla}c_i \tag{2.61}$$

Finally, we need to consider all manner of external, mechanical forces acting on the system. These generally translate into some form of pressure gradient, and the pressure gradient can cause

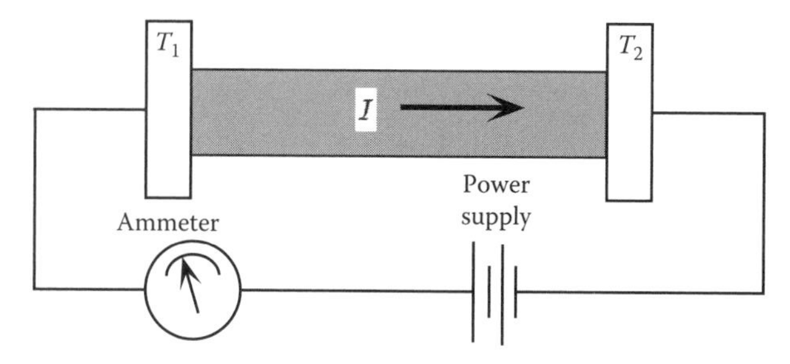

FIGURE 2.14 Thermoelectric effects occurring in a nonisothermal conductor.

a heat flux. Again, we can write the flux potential relationship in the form of Fourier's law where k_p can be considered as the "pressure conductivity" of the material:

$$\vec{\mathbf{q}}''^P = -k_p \vec{\nabla} P \tag{2.62}$$

The components of the heat flux can be based on a set of stationary coordinates too. In such cases, we must consider the convective components of the energy flux that arise from fluid passing through the system. This topic is considered further in Chapter 10.

2.9.3 Mass [8,9]

The flux of mass (with respect to the mass average velocity), $\vec{j}_i$, in a multicomponent system can be expressed in terms of two contributions associated with mechanical driving forces, a contribution associated with the thermal driving force, and another contribution associated with an electrical driving force (charged species required). The fluxes as written in the following are strictly applicable to gases but a simplified form is also used for liquids and solids:

$$\vec{\mathbf{j}}_i = \vec{\mathbf{j}}_i^D + \vec{\mathbf{j}}_i^P + \vec{\mathbf{j}}_i^T + \vec{\mathbf{j}}_i^E \tag{2.63}$$

$\vec{\mathbf{j}}_i^D$ is the mass flux called ordinary or normal diffusion and driven by a concentration gradient. It is expressed in terms of the mole fraction gradient, ∇x_k, as

$$\vec{\mathbf{j}}_i^D = -\frac{c_t^2}{\rho RT} \sum_{j=1}^{n} M_{wi} M_{wj} \mathcal{D}_{ij} \left[x_j \sum_{\substack{k=1 \\ k \neq j}}^{n} \left(\frac{\partial \overline{G}_j}{\partial x_k} \right)_{T,P,x} \vec{\nabla} x_k \right] \quad \text{Gas} \tag{2.64}$$

Notice that the change in partial molar Gibbs energy, $\overline{G}_j$, with mole fraction is included in the flux. This indicates that the mass flux is proportional to the concentration gradient only in ideal solutions. The fugacity coefficient plays a role in gases and the activity coefficient in liquids and solids for non-ideal systems.

$\vec{\mathbf{j}}_i^P$ is the pressure driven diffusion term and is important only in systems where high accelerations are encountered such as in centrifuges. It was the fundamental mechanism exploited in the centrifugal method for separating U^{235} from U^{238}. The pressure diffusion flux for gases is given by

$$\vec{\mathbf{j}}_i^P = -\frac{c_t^2}{\rho RT} \sum_{j=1}^{n} M_{wi} M_{wj} \mathcal{D}_{ij} \left[x_j M_{wj} \left(\frac{\overline{V}_j}{M_{wj}} - \frac{1}{\rho} \right) \vec{\nabla} P \right] \tag{2.65}$$

Pressure diffusion is assisted when the species have large changes in partial molar volume, $\overline{V}_j$, upon mixing. Pressure diffusion is negligible in liquids because by and large, they are incompressible.

$\vec{\mathbf{j}}_i^T$ is the mass flux that results from the imposition of a temperature gradient. It is referred to as the Soret effect and is the analog of the Dufour effect. The Soret coefficient is identical to the Dufour coefficient. The mass flux resulting from thermal diffusion is given by

$$\vec{\mathbf{j}}_i^T = -\mathcal{D}_i^T \vec{\nabla} \ln T \tag{2.66}$$

The thermal diffusion flux depends on the gradient of the log of temperature and hence is very small. It has also been used for isotope separation.

$\vec{\mathbf{j}}_i^E$ is the mass flux of charged species responding to a potential gradient. It is very important in many systems where there are charge carriers such as electrons and holes in semiconductors, proteins in electrophoresis separation units, neurotransmitters in axons and along nerve pathways, and ions in batteries. The expression for this electro-diffusive flux is

$$\vec{j}_i^E = -\frac{c_t^2 e}{\rho}\sum_{j=1}^{n} M_{wi}M_{wj}\left[x_j\mu_{ej}z_{ej} - \sum_{\substack{k=1\\k\neq j}}^{n}\frac{\rho_k}{\rho}\mu_{ek}z_{ek}\right]\vec{\nabla}\Phi \tag{2.67}$$

where μ_e is the mobility of the charge carrying species and in dilute solutions can be related to the diffusion coefficient, $\mathcal{D}_{ij}$, via the Einstein relation (2.28).

The $\mathcal{D}_{ij}$ appearing in the previous equations are the multicomponent diffusion coefficients and the $\mathcal{D}_i^T$ are the multicomponent thermal diffusion coefficients. The $\mathcal{D}_{ij}$ and $\mathcal{D}_i^T$ have the following properties:

$$\mathcal{D}_{ii} = 0 \qquad \sum_{i=1}^{n}\mathcal{D}_i^T = 0 \qquad \sum_{i=1}^{n}\{M_{wi}M_{wl}\mathcal{D}_{il} - M_{wi}M_{wk}\mathcal{D}_{ik}\} = 0 \tag{2.68}$$

Notice that when the number of components exceeds 2, the D_{ij} and $\mathcal{D}_{ij}$ are not necessarily equal. In such instances, there is also no simple relationship between the binary diffusivity we are used to using, D_{ij}, and the binary diffusivity, $\mathcal{D}_{ij}$, found in a multicomponent mixture.

If we have a simple binary system, $\mathcal{D}_{ab}$ and $\mathcal{D}_{ba}$ are equivalent to D_{ab}. In a binary system, we can substitute for the partial molar Gibbs energy and make use of the activity of the species instead:

$$(\partial\overline{G}_a)_{T,P} = RTd\ln(a_a) \tag{2.69}$$

It also simplifies matters to define a thermal diffusion ratio

$$k^T = \left(\frac{\rho}{c_t^2 M_{wa}M_{wb}}\right)\frac{D_a^T}{D_{ab}} \tag{2.70}$$

so that the flux equations become

$$\begin{aligned}\vec{\mathbf{j}}_a = -\vec{\mathbf{j}}_b = &-\left(\frac{c_t^2}{\rho}\right)M_{wa}M_{wb}D_{ab}\left[\left(\frac{\partial\ln a_a}{\partial\ln x_a}\right)_{T,p}\vec{\nabla}_{x_a} + k^T\vec{\nabla}\ln T\right]\\ &-\left(\frac{c_t^2}{\rho}\right)M_{wa}M_{wb}D_{ab}\left[\frac{M_{wa}x_a}{RT}\left(\frac{\overline{V}_a}{M_{wa}} - \frac{1}{\rho}\right)\vec{\nabla}P\right]\\ &-\frac{M_{wa}\rho_b x_a e}{\rho}(\mu_{ea}z_{ea} - \mu_{eb}z_{eb})\vec{\nabla}\Phi\end{aligned} \tag{2.71}$$

We can simplify Equation 2.71 further if we recast the driving force for diffusion as a gradient in chemical potential. Then, if we know how the chemical potential varies as a function of temperature, pressure, mole fraction, and any other thermodynamic state variables we require, we can recast the fluxes as

$$\vec{\mathbf{j}}_a = -D_{ab}x_a\vec{\nabla}\mu_a^c \qquad \vec{\mathbf{j}}_b = -D_{ab}x_b\vec{\nabla}\mu_b^c \tag{2.72}$$

Life is much more complicated for multicomponent mixtures. Let us look at flux component $\vec{\mathbf{j}}_i^D$ for an ideal gas (solution) mixture. In such cases, the activity of component i in the gas (solution) is equal to its mole fraction so Equation 2.64 becomes

$$\vec{\mathbf{j}}_i^D = \frac{c_t^2}{\rho}\sum_{j=1}^{n} M_{wi}M_{wi}\mathcal{D}_{ij}\vec{\nabla}x_i \tag{2.73}$$

The relationship between the $\mathcal{D}_{ij}$ of the ideal gas mixture is known. All the $\mathcal{D}_{ij}$ are different and all depend upon the concentrations of all other species present in the solution. This makes Equation 2.73 very difficult to use. Fortunately, we can turn these equations for $\vec{\mathbf{j}}_i^D$ inside out and express them in terms of $\vec{\nabla}x_i$:

$$\vec{\nabla}x_i = \sum_{j=1}^{n}\frac{c_i c_j}{c_t^2 D_{ij}}\left(\vec{\mathbf{v}}_j - \vec{\mathbf{v}}_i\right) = \sum_{j=1}^{n}\frac{1}{c_t D_{ij}}\left(x_i\vec{\mathbf{N}}_j - x_j\vec{\mathbf{N}}_i\right) \tag{2.74}$$

The D_{ij} are the regular diffusion coefficients you are familiar with and $D_{ij} = D_{ji}$. They are essentially independent of composition. The set of Equations 2.74 are the normal starting point for dealing with multicomponent diffusion and are referred to as the *Stefan–Maxwell* equations. In this chapter, we will deal with binary systems, but the Stefan–Maxwell equations are essential for dealing with multicomponent systems [10].

Example 2.8 Partitioning in a Deep Well

A binary solution of a in b fills a deep well, like that shown in Figure 2.15. We are interested in how the composition of the solution might change as a function of depth. This can be important in the deepest oil wells, for example. The solution density is assumed to be only a function of composition and we neglect changes in the partial molar volumes of the constituents as a function of composition. We will also assume that the activity coefficients of the two components are independent of composition.

Here we will have a mass flux caused by the pressure gradient in the well that is resisted by a mass flux due to the induced concentration gradient. At equilibrium, these two fluxes will be equivalent and so the overall mass flux will be zero. The pressure gradient in the y direction is due to gravity acting in that direction:

$$\frac{dp}{dy} = -\rho g$$

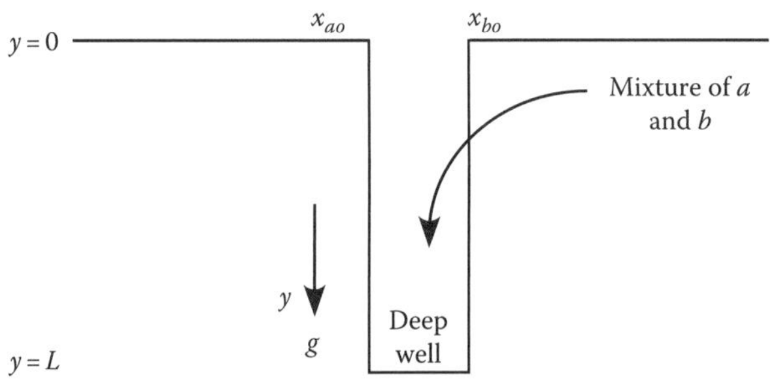

FIGURE 2.15 Pressure diffusion of a through b in a deep well.

Referring to Equation 2.71, the fluxes for a and b are as follows:

$$\underline{\mathbf{j}}_a = 0 = -\left(\frac{c_t^2}{\rho}\right) M_{wa} M_{wb} D_{ab} \left[\left(\frac{\partial \ln a_a}{\partial \ln x_a}\right)_{T,p} \vec{\nabla} x_a + x_a \left(\frac{M_{wa}}{RT}\right)\left(\frac{\bar{V}_a}{M_{wa}} - \frac{1}{\rho}\right)\vec{\nabla} P\right]$$

$$\underline{\mathbf{j}}_b = 0 = -\left(\frac{c_t^2}{\rho}\right) M_{wa} M_{wb} D_{ab} \left[\left(\frac{\partial \ln a_b}{\partial \ln x_b}\right)_{T,p} \vec{\nabla} x_b + x_b \left(\frac{M_{wb}}{RT}\right)\left(\frac{\bar{V}_b}{M_{wb}} - \frac{1}{\rho}\right)\vec{\nabla} P\right]$$

Using our assumption for the activities (ideal solution so $a_a = x_a$) and getting rid of as many constants as possible, we find

$$\frac{d \ln x_a}{dy} = -\frac{g}{RT}(M_{wa} - \rho \bar{V}_a) \qquad \frac{d \ln x_b}{dy} = -\frac{g}{RT}(M_{wb} - \rho \bar{V}_b)$$

Now we multiply the equation for species a by $\bar{V}_b$ and the equation for species b by $\bar{V}_a$ and subtract a from b to find

$$\frac{g}{RT}(M_{wb}\bar{V}_a - M_{wa}\bar{V}_b)dy = \bar{V}_b \frac{dx_a}{x_a} - \bar{V}_a \frac{dx_b}{x_b}$$

Assuming the only variables are x_a and x_b, we can integrate the equation to give

$$\left(\frac{x_a}{x_{ao}}\right)^{\bar{V}_b} \left(\frac{x_{bo}}{x_b}\right)^{\bar{V}_a} = \exp\left[(\bar{V}_a M_{wb} - \bar{V}_b M_{wa})\frac{gy}{RT}\right]$$

Notice that the effect of pressure will be small unless g is very large or the well is very deep, or the partial molar volume difference is appreciable.

Example 2.9 Thermal Diffusion

Two bulbs are joined together by a thin, insulated tube as shown in Figure 2.16. The bulbs are initially filled with the same mixture of ideal gases, a and b. The tube is small enough that convection currents do not form. Each of the bulbs is held at a different temperature and we want to develop an expression relating the differences in composition between the two bulbs to the temperature difference between them.

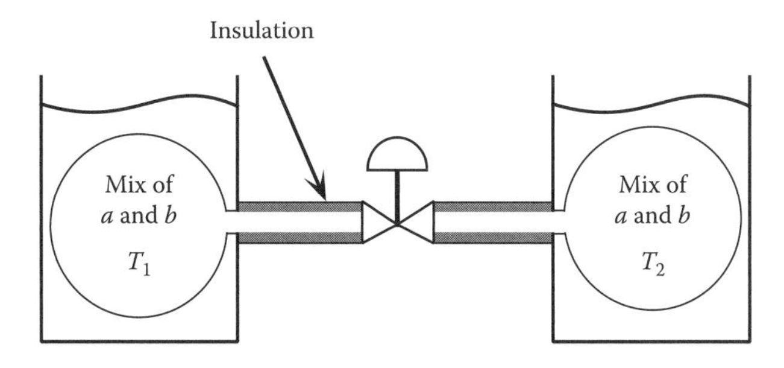

FIGURE 2.16 Thermal diffusion of a mixture of a and b from one container to another.

At time zero, we open the valve separating the two bulbs and allow the gases to mix. The temperature gradient will induce a flux of matter and that flux will be resisted by the developing concentration gradient. When we reach equilibrium, the sum of the fluxes will be zero since no net mass transfer will occur. Simplifying Equation 2.71 to apply to this situation leaves us with

$$j_a^D + j_a^T = -\frac{c_t^2}{\rho} M_{wa} M_{wb} D_{ab}\left[\frac{dx_a}{dy} + \frac{k^T}{T}\frac{dT}{dy}\right] = 0$$

If k^T is positive, then a moves from the hot region toward the cooler region, and if k^T is negative, then a moves in the opposite direction. Since composition, density, molecular weight, and diffusivity are all nonzero, we can rewrite this equation as

$$\frac{dx_a}{dy} = -\frac{k^T}{T}\frac{dT}{dy}$$

Removing dy lets us separate the equation and integrate to give the equilibrium composition:

$$x_{a2} - x_{a1} = -k^T \ln\frac{T_2}{T_1}$$

In systems with high temperature gradients, we can achieve substantial separation of material. This is the basis of a number of gas separation systems. In general, k_T is a fairly strong function of temperature and a usual practice is to use a k_T evaluated at the mean temperature. The recommended mean temperature for this purpose is

$$T_m = \frac{\ln T_2 - \ln T_1}{1/T_1 - 1/T_2}$$

Example 2.10 Electrophoresis

Suppose we have the electrophoretic cell shown in Figure 2.17 with a dissolved protein sample in it. Assume the protein has a positive charge of valence z_{ea}. The solution is isothermal and a constant electric field is applied across the solution. What is the concentration profile of solute a after a very long time?

At long times we may safely assume the system is at equilibrium and the flux of protein due to the applied electric field will be just balanced out by the flux due to the concentration gradient.

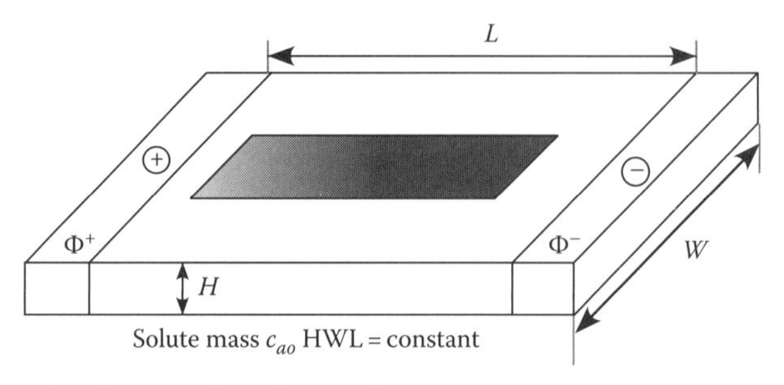

FIGURE 2.17 Electrophoretic separation of a protein.

Simplifying Equation 2.71 again and assuming dilute, ideal, solutions gives the following equation for the overall mass flux of protein:

$$j_{ax} = -D_{aw}\left(\frac{dc_a}{dx} + \frac{z_{ea}\mathcal{F}_a}{RT}c_a\frac{d\Phi}{dx}\right) = 0$$

Notice that the diffusivity is unimportant since at equilibrium, no net diffusion is really occurring. Multiplying through by dx and dividing through by c_a gives

$$\frac{dc_a}{c_a} = -\frac{z_{ea}\mathcal{F}_a}{RT}d\Phi = -\beta_e d\Phi$$

This is easily solved to give

$$c_a = A\exp[-\beta_e\Phi]$$

The boundary condition for this problem states that the total amount of protein must be conserved so

$$\frac{\int_{\Phi^-}^{\Phi^+} c_a d\Phi}{\int_{\Phi^-}^{\Phi^+} d\Phi} = c_{ao}$$

The final solution is

$$c_a = \left(\frac{c_{ao}\beta_e(\Phi^+ - \Phi^-)}{\exp[-\beta_e\Phi^-] - \exp[-\beta_e\Phi^+]}\right)\exp[-\beta_e\Phi]$$

The concentration profile decays exponentially with a characteristic length of $1/\beta_e$. This decay length does not depend on any transport property, such as the diffusivity or mobility, nor does it depend on the strength of the electric field. Those factors only influence the speed at which equilibrium is attained. The exponential concentration profile explains why one observes such tight, dark bands in electrophoretic separation.

We can also look at this separation in terms of isoelectric focusing. Here we develop a pH gradient along with using the electric field. Since proteins will change their charge state at different pH values, we can use the combination to provide better separation than using electrophoresis alone. As the protein moves through the pH gradient driven by the electric field, it reaches a point where it becomes neutrally charged. There it is fixed by the action of the electric field and becomes concentrated. Since the charge on the protein changes with pH, the exponential tail of concentration profile we derived changes slope along the medium thereby providing a more efficient separation.

2.9.4 Charge

Charge transport and mass transport are intimately related because charges are carried by physical particles such as electrons, holes, or ions, and these can be modeled to move about a medium as molecules do. The flux of charge can be related to the same mechanisms we had for the flux of mass:

$$\vec{j}_i = \vec{j}_i^D + \vec{j}_i^E + \vec{j}_i^T \tag{2.75}$$

$\vec{j}_i^E$ is the flux of ions due to the field force and gives rise to Ohm's law. There are many ways to write this flux and here we choose to write it in terms of the mobility of the charge carrier so that the flux will apply to solids, liquids, gases, and even plasmas:

$$\vec{j}_i^E = -\frac{c_t^2 e}{\rho}\sum_{j=1}^{n} M_{wi}M_{wj}\left[x_j\mu_{ej}z_{ej} - \sum_{\substack{k=1\\k\neq j}}^{n}\frac{\rho_k}{\rho}\mu_{ek}z_{ek}\right]\vec{\nabla}\Phi \tag{2.76}$$

Notice that this expression is identical to the expression we used for mass transfer. This is not a coincidence but arises from the fact that charge transport is intimately related with mass transfer. Equation 2.76 applies strictly only for gases, but can be simplified to apply to liquids and solids.

$\vec{j}_i^D$ is the flux of ions due to their concentration gradient and this is identical to the flux of molecules we considered previously:

$$\vec{j}_i^D = -\frac{c_t^2}{\rho RT}\sum_{j=1}^{n} M_{wi}M_{wj}\mathcal{D}_{ij}\left[x_j\sum_{\substack{k=1\\k\neq j}}^{n}\left(\frac{\partial \overline{G}_j}{\partial x_k}\right)_{T,P,x}\vec{\nabla}x_k\right] \tag{2.77}$$

Here the partial molar Gibbs energy of the ions plays a much more important role in the diffusion process than it does when considering uncharged species. The reason is due to the interaction of the charged species with the solvent, usually water. The high fields surrounding the charge carriers affect the structure of the solvent and change the effective velocity of the particles through the solvent.

We can also have pressure and thermal diffusion of ions and the expressions for these fluxes follow the same forms as Equations 2.65 and 2.66. The effects are much less pronounced in charge transport because the electrical forces are so large. Any imbalance in the ionic flux due to a concentration, pressure, or thermal gradient has to overcome the electrical forces acting on the ions as they are displaced from one another.

Finally, we also need to consider thermoelectric effects when dealing with conducting systems. In general, Equation 2.56, describing the Seebeck effect, may often be of interest in these systems, especially when all other driving forces are small:

$$\vec{J} = -\left(\frac{\alpha_T}{R_e}\right)\vec{\nabla}T \tag{2.56}$$

2.10 FAILURE OF THE LINEAR FLUX: GRADIENT LAWS

In nearly every instance discussed so far, we have considered the flux potential–gradient relationship to be a linear one. This represents a good approximation to the behavior of physical systems under most conditions. Unfortunately, nature is not a linear system and these laws break down often. As a flavor of how and when these laws break down, we will now briefly consider two common instances: non-Newtonian fluids and non-Ohmic devices.

2.10.1 NON-NEWTONIAN FLUIDS [11]

Our flux law for fluids can be expressed in the following form:

$$\tau \propto f(\dot{\gamma}) \qquad \text{with} \qquad \dot{\gamma} = \frac{dv_x}{dy} \tag{2.78}$$

where τ, the shear stress, is some function of $\dot{\gamma}$, the shear rate. If the functional relationship is a linear one, then the fluid is called Newtonian and the proportionality constant is the viscosity. Gases, water, and most organic liquids are Newtonian fluids. Everyday experience shows us that there are many fluids whose behavior is nowhere near Newtonian. Mixing cornstarch with water creates a fluid whose apparent viscosity increases dramatically with the rate at which it moves. This type of fluid, whose viscosity increases with increasing shear rate, is called a dilatant fluid since in most instances these fluids also show volumetric increases when they move as well. If the viscosity decreases with increasing shear rate, it is a pseudoplastic fluid. Pseudoplastic fluids act like Newtonian fluids at low and high shear rates and have a variable viscosity in between. Paint is a common example.

A Bingham plastic or Bingham fluid is a Newtonian fluid that has a yield stress. Below the yield stress level, the fluid will not flow, yet above it, it flows with constant viscosity:

$$\tau = -\mu_o \dot{\gamma} \pm \tau_o \tag{2.79}$$

where τ_o is positive if $\dot{\gamma}$ is positive and vice versa. This type of fluid describes everyday pastes, ketchup, very fine suspensions such as rouge in water, and concrete before it has cured.

An Ostwald–de Waele model (Figure 2.18) can display both pseudoplastic and dilatant behavior. Its constitutive relation is

$$\tau = -m\left|\dot{\gamma}\right|^{n-1}\dot{\gamma} \qquad \begin{matrix} n>1 & \text{dilatant} \\ n<1 & \text{pseudoplastic} \end{matrix} \tag{2.80}$$

In a flow where the shear rate changes with time, we can have different phenomena. A thixotropic fluid has a viscosity that decreases with time under a sudden applied stress. One can imagine a process of bond breakage occurring where as time increases, more bonds are broken and so the viscosity decreases. A rheopectic fluid has a viscosity that increases with time under the applied stress. Here we can envision generating entanglements within the fluid making it difficult to move.

Finally, a viscoelastic fluid has memory, like rubber. Its behavior in flowing from a nozzle looks like Figure 2.19. A prime example of a viscoelastic fluid is egg white. All these common fluid types point out that Newtonian flow is the special case whereas non-Newtonian flow is the rule (Table 2.7).

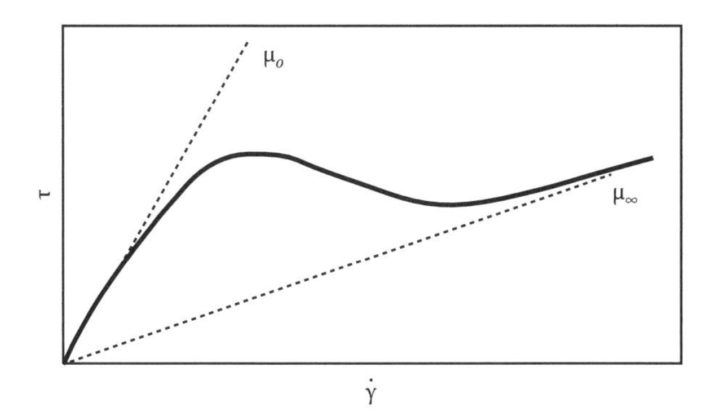

FIGURE 2.18 Pseudoplastic viscous behavior.

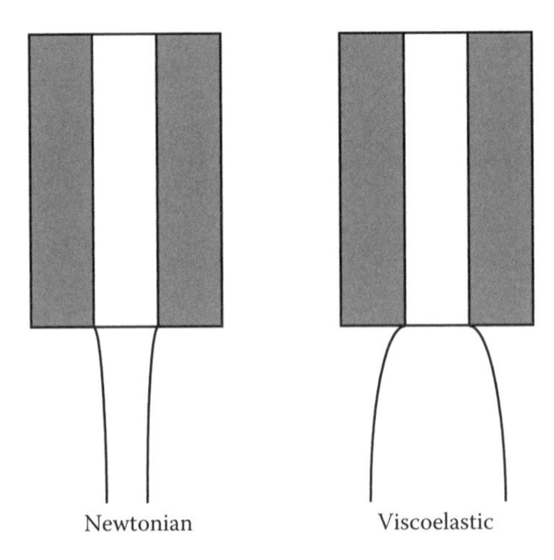

FIGURE 2.19 Newtonian versus viscoelastic flow behavior.

TABLE 2.7
General Flux-Gradient Relationships for Non-Newtonian Fluids

Name	Law	Example Fluids
Bingham	$\tau = -\mu_\infty \dot{\gamma} \quad \tau > \tau_o$ $\dot{\gamma} = 0 \quad \tau < \tau_o$	Ketchup, toothpaste, concrete
Dilatant or shear thickening	$\tau = -K\left(\lvert\dot{\gamma}\rvert\right)^{n-1}\left(\dot{\gamma}\right) \; n > 1$	Cornstarch, quicksand
Pseudoplastic or shear thinning	$\tau = -K\left(\lvert\dot{\gamma}\rvert\right)^{n-1}\left(\dot{\gamma}\right) \; n < 1$	Polymers, paints, paper pulp, honey
Carreau fluid	$\tau = -\mu_o\left[1+\left(\lambda\dot{\gamma}\right)^2\right]^{\frac{n-1}{2}}\dot{\gamma}$	Jack of all trades
Cross fluid	$\tau = -\left[\mu_\infty + \dfrac{(\mu_o - \mu_\infty)}{1+\lambda\dot{\gamma}^m}\right]\dot{\gamma}$	Jack of all trades
Viscoelastic (Maxwell model)	$\dot{\gamma} = \dfrac{\tau}{\mu} + \dfrac{1}{Y}\left(\dfrac{d\tau}{dt}\right)$	Y—Young's modulus Polymers, egg white, and fluids that undergo creep

2.10.2 Non-Ohmic Electronic Components [12]

Ohm's law is the general relationship describing the voltage–current relationship that exists in conductors and resistors. It is not the only relationship and more commonly, we encounter situations where the voltage–current behavior of a conducting or semiconducting device does not obey Ohm's law. Common examples are diodes, transistors, and vacuum tubes. Figure 2.20 shows the I–Φ characteristics of a typical Zener diode and a tunnel diode. Notice that the resistance is a function of the applied voltage. We will consider this failure of Ohm's law and derive the I–Φ characteristics for a typical diode when we discuss charge transport in semiconductor devices.

Though non-Newtonian fluids and non-Ohmic devices have very little in common physically, they show that nonlinearity is the rule. Fortunately, the nonlinearity can often be represented in a

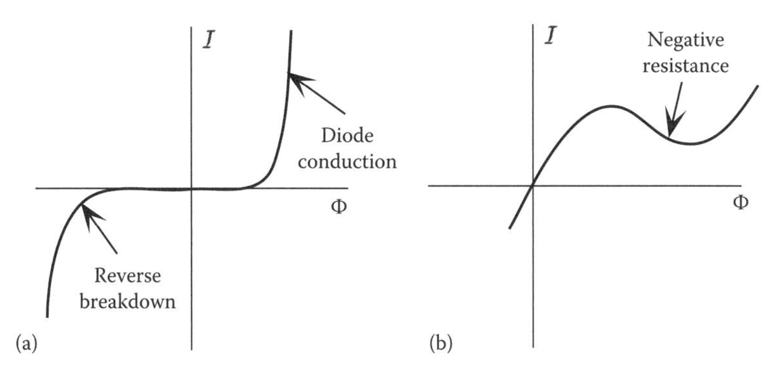

FIGURE 2.20 Circuit elements that do not obey Ohm's law: (a) Zener diode and (b) tunnel diode.

common mathematical formulation that allows us to draw parallels between the phenomena. The analogy between the types of transport phenomena will be used repeatedly in this text and provides a very powerful tool for the analysis of physical processes.

2.11 SUMMARY

In this chapter, we introduced several flux potential–gradient laws and showed the relationships and analogies that exist between the various transport phenomena. We separated the fluxes into primary and secondary sources. The primary fluxes are most commonly encountered and their relationship between flux and potential gradient is the strongest. For example, Fourier's law relates the flux of heat to a temperature gradient. This is the primary flux quantity yet heat can be transferred via molecular diffusion as in the Dufour effect. The Dufour effect, a secondary transport mechanism, is characteristic of all secondary mechanisms. It is smaller than the primary mechanisms and is only appreciable in certain circumstances (when the secondary gradient is extraordinarily high). Moreover, the secondary fluxes have a reciprocal relationship with another transport system. The Dufour effect, causing a heat flux due to a concentration gradient, has a sister mechanism in mass transport called the Soret effect. This effect forces a mass flux in response to a temperature gradient. Moreover, the transport properties associated with each effect are the same and the laws of irreversible thermodynamics assert that this must be the case. Such symmetry is fundamental to nature and natural transport processes.

PROBLEMS

Momentum Transport

2.1 A fluid has the following velocity profile in three dimensions:

$$v_r = v_o\left(r^2\theta + \theta^2 z - r^2 z\right) \qquad v_\theta = v_o\left(\theta^2 z - \theta^2 r + \theta^3\right)$$

$$v_z = v_o\left(zr - z^2\theta\right)$$

a. What are the nine stresses for this fluid?
b. Is the fluid incompressible?

2.2 In 1856, H. Darcy published a paper where he described experiments showing that the flow of fluid through a porous medium was linearly related to the pressure drop across the medium:

$$\vec{v} = -\frac{k}{\mu}\vec{\nabla}P \quad \text{Darcy's law}$$

where k is the permeability of the porous medium. In three dimensions, Darcy's law can be written as

$$v_x = -\frac{k}{\mu}\left(\frac{\partial P}{\partial x}\right) \qquad v_y = -\frac{k}{\mu}\left(\frac{\partial P}{\partial y}\right) \qquad v_z = -\frac{k}{\mu}\left(\frac{\partial P}{\partial z}\right)$$

Show that if the fluid is incompressible, the pressure must obey Laplace's equation ($\nabla^2 P = 0$).

2.3 Many fluids are non-Newtonian. A classic non-Newtonian material is low-fat mayonnaise. The reason behind this is the modified starches and xanthan gum used to stabilize the water and oil emulsion. The following data set is representative of one brand of mayonnaise:

a. Plot the data and discuss whether mayonnaise is a shear-thinning or shear-thickening fluid.

b. If you fit the data to a power law expression, what is the exponent you determine?

Apparent Viscosity (Pa·s)	Shear Rate (1/s)	Apparent Viscosity (Pa·s)	Shear Rate (1/s)
0.396	249.096	2.207	51.413
0.598	167.090	2.413	47.971
0.790	133.307	2.580	44.767
0.988	108.264	2.790	41.064
1.194	91.034	3.017	38.983
1.412	77.892	3.227	37.010
1.596	67.828	3.450	35.138
1.805	61.132	3.649	33.363
1.996	55.107	3.859	31.678
2.207	51.413	4.080	30.078

Source: Donatella, P. et al., *J. Food Eng.*, 35, 409, 1998.

2.4 The Carreau–Yasuda model is a popular formulation for representing non-Newtonian fluid behavior. The model is

$$\mu = \left[\mu_\infty + (\mu_0 - \mu_\infty)\left(1 + \left(K|\dot{\gamma}|\right)^a\right)^{\frac{n-1}{a}}\right]$$

where

$\dot{\gamma}$ is the shear rate

μ_0 is the viscosity at zero shear rate

μ_∞ is the viscosity at infinite shear rate

One of the composite materials that the Carreau–Yasuda model has been used for is to determine the rheology of blood. Given the following data, fit the Carreau–Yasuda model and determine the values of the parameters.

Viscosity (N s/m²)	Shear Rate (1/s)	Viscosity (N s/m²)	Shear Rate (1/s)
0.2	0	0.0254	50
0.048	3.33333	0.0244	60
0.04	6.66667	0.0236	70
0.037	10	0.0229	80
0.034	13.3333	0.0223	90
0.033	16.6667	0.0218	100
0.031	20	0.0208	120
0.030	23.3333	0.0202	140
0.029	26.6667	0.0196	160
0.028	30	0.0190	180
0.028	33.3333	0.0186	200
0.027	36.6667	0.0181	230
0.0268	40	0.0176	260
0.0263	43.3333	0.0172	290
0.0259	46.6667	0.0167	320

Mass Transport

2.5 Prove that all forms of Fick's law agree with thermodynamics, that is, "at equilibrium, the concentration of all species throughout the system should be uniform."

2.6 Prove that in a binary mixture whose total concentration of species, c_t, remains constant, there is only one diffusion coefficient: $D_{ab} = D_{ba}$.

2.7 A solid containing species a has been analyzed and the mole fraction profile has been found to obey the following function of y alone:

$$x_a = a_o \sqrt[3]{y - K y_o} \qquad a_o,\ y_o,\ K,\ \text{constants}$$

a. Assuming a constant value for the diffusivity, $D_{ab} = D_{abo}$, and a dilute solution of a in b has the system reached a steady state, does it obey the continuity equation in one dimension?

b. Assuming a diffused through stagnant b ($N_b \approx 0$) and has reached steady state, what can you say about how the diffusivity varies as a function of composition?

2.8 A new lithium sulfate (Li_2SO_4) electrolyte has been developed for battery applications. Assuming ideal solutions and a stationary coordinate system, what does Fick's law for the flux (N_{Li}, N_{SO_4}) for both lithium and sulfate species look like? (Hint: The fluxes for the ions must obey local electroneutrality so that we do not encounter regions where we have large excesses of one charge over another.)

2.9 We defined a molar flux relative to the molar average velocity as

$$\vec{\mathbf{J}}_i = -c_t D_{ij} \vec{\nabla} x_i$$

We could just as easily have defined the flux relative to the volume average velocity, v^v:

$$\vec{\mathbf{J}}_i^v = -c_i(\vec{\mathbf{v}}_i - \vec{\mathbf{v}}^v) = -D_{ij}^v \vec{\nabla} c_i$$

Show that the two diffusivities are equal even if the molar concentration, c_t, is not constant.

2.10 The mass flux of a species can be written using the chemical potential as a driving force. Consider the simple case of binary diffusion in an ideal mixture of liquids. If the chemical potential is given by

$$\mu_i^c = \mu_{io}^c + RT\ln x_i$$

Prove that the total flux $\mathbf{j}_a + \mathbf{j}_b = 0$. What must hold true if the chemical potentials are given by the following equation and the sum of the fluxes is to be zero?

$$\mu_i^c = \mu_{io}^c + RT\ln(\gamma_i x_i) \quad \gamma_i \text{ is the activity coefficient}$$

Heat Transport

2.11 Thermocouples attached to a truncated, conical roller bearing show that the temperature profile and heat flow rate are (Figure P2.11)

$$T(x) = 450(2 - 3x + x^2 - x^3) \qquad q = 7500 \text{ W}$$

If the cross-sectional area of the bearing is $A(x) = 0.04\pi(1 - x)\text{m}^2$

a. What is the thermal conductivity as a function of x?
b. What is the heat flux at $x = 0$? $x = 0.2$?
c. Where is the heat flux, highest $(0 < x < 0.2)$?

2.12 The 2-D object shown in Figure P2.12 is insulated with the exception of two flat portions that are exposed to two different temperatures. The temperature gradient at surface I is measured and found to be $\partial T/\partial x = 45$ K/m. What are $\partial T/\partial x$ and $\partial T/\partial y$ at surface II?

2.13 A spherical shell of inner radius r_i, outer radius r_o, and thermal conductivity k is being used to dissipate heat. At a particular time, the temperature profile within the shell is measured and found to be

$$T(r) = \frac{C_1}{r} + C_2$$

a. Is the heat transfer at steady-state, that is, is the rate constant?
b. How does the heat flow rate vary with position?

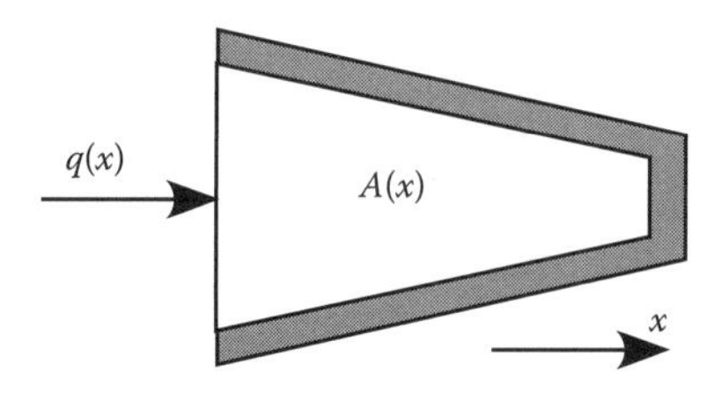

FIGURE P2.11

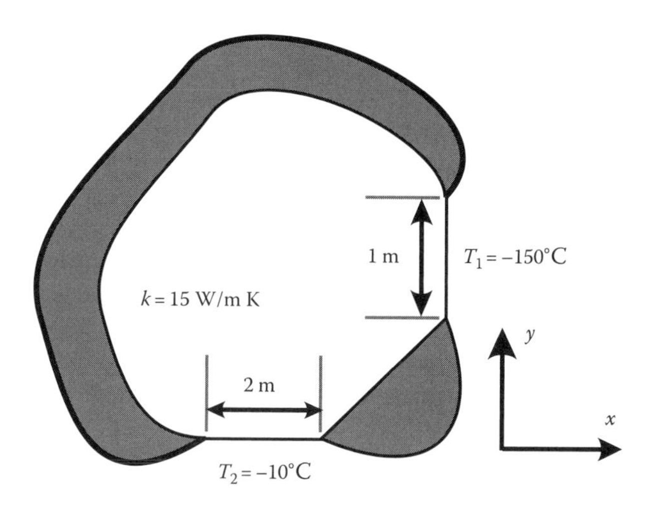

FIGURE P2.12

2.14 The heat flux, $\vec{q}''$, due to a volume source distribution (expressed in spherical coordinates) is given by

$$\vec{q}_r'' = C r^2 \sin(\alpha r)\vec{\mathbf{e}}_r$$

a. What is the temperature gradient for this system?
b. If the temperature at $r = 0$ is $T = T_o$, what is the temperature profile?
c. At what value of r does the solution become aphysical?

Charge Transport

2.15 The voltage (V)–current (I) behavior for a new material was measured. The material was fashioned into a wire 1 mm in diameter and 1 m long.

a. What is the conductivity of the material?
b. What is the conductivity of the material at $V = 10$ V?
c. What is the diffusivity of the charge carriers at $V = 10$ V and $T = 298$ K?
d. If the valence of the charge carriers is 1, what is their mobility?

I (A)	*V* (V)
0	0
0.001	2
0.003	5
0.005	10
0.012	20
0.050	50
0.200	100

Primary and Secondary Fluxes

2.16 We showed that one of the most often used forms of Fick's law for multicomponent systems could be written as

$$\vec{\mathbf{J}}_i = \underbrace{-D_{ijo}\left(1+\frac{\partial \ln \gamma_i}{\partial \ln x_i}\right)}_{D_{ij}}\vec{\nabla} c_i = c_i(\vec{\mathbf{v}}_i - \vec{\mathbf{v}}_{Mole})$$

Kinetic theory derivations of the flux equation lead to an expression for the gradient in chemical potential, $\nabla\mu_i$, of the form

$$\vec{\nabla}\mu_i^c = \frac{RTx_j}{D_{ijo}}(\vec{\mathbf{v}}_j - \vec{\mathbf{v}}_i)$$

Show that these two forms are equivalent representations for the binary case with species i and j. Remember that the chemical potential for species i is given by $\mu_i^c = \mu_{io}^c + \mathrm{RT}\ln(\gamma_i x_i)$. Hint: Use expressions for J_i to solve for v_i and v_j.

2.17 Activities in solution can often be correlated by the Margules equations:

$$\ln\gamma_1 = x_2^2[A_{12} + 2(A_{21} - A_{12})x_1] \qquad \ln\gamma_2 = x_1^2[A_{21} + 2(A_{12} - A_{21})x_2]$$

For 2,4-dimethylpentane (1) and benzene (2), the coefficients A_{12} and A_{21} are 1.96 and 1.48, respectively. Using the relations developed from Problem 2.16, plot how D_{ij} will depend upon composition for this binary system.

2.18 A student tries to dissolve a congealed mass of NaOH by adding water to the beaker. The concentration profile of NaOH in the water above it was measured and found to be

$$1 - x_a = (1.143)^{z/L-1}$$

The initial and reference temperature of the water is 0°C. c_t = 63,055 mol/m^3; $D_{ab} = 1.0\times10-9\text{m}^2/\text{s}$; L = 5 cm. Assume the water properties are constant at the reference temperature values. Data on the partial molar enthalpies of solution as a function of temperature and hence the partial molar heat capacities can be obtained crudely from the enthalpy concentration diagram [14]. What is the temperature profile?

2.19 A young engineer has the bright idea of trying to separate methanol and water by centrifugation. The target system is an antifreeze consisting of 30 mol% methanol. The partial molar volumes and pure component molar volumes at 25°C are

$$\overline{V_m} = 38.632\ \text{cm}^3/\text{mol} \qquad \overline{V_w} = 17.765\ \text{cm}^3/\text{mol}$$

$$V_m = 40.727\ \text{cm}^3/\text{mol} \qquad V_w = 18.068\ \text{cm}^3/\text{mol}$$

Assuming a centrifuge like that in Figure P2.19 operates at 20,000 rpm and a temperature of 25°C, what would be the concentration of m at r = 0.2 m? What is the maximum separation

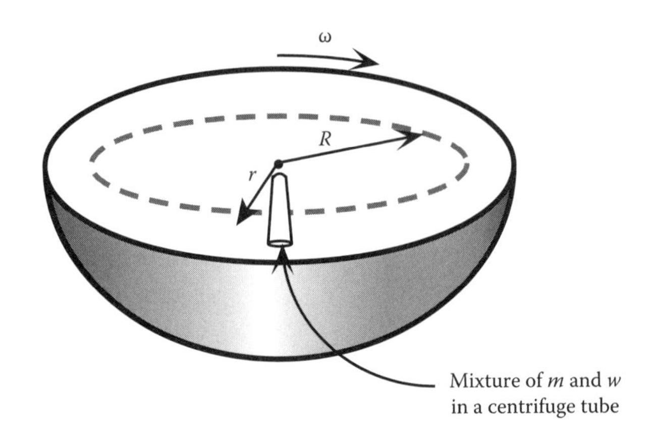

FIGURE P2.19

ratio defined by x_{mL}/x_{wL}? Assume the pressure gradient, $\partial P/\partial r = 4\pi^2\omega^2\rho r$, where ω is the revolution frequency and ρ is the fluid density.

2.20 Often we have a multicomponent mixture of gases and do not want to deal with the diffusion coefficients for every pair of gases. We would like to define a pseudo-binary diffusion coefficient for each species relative to the mixture.

a. Show how using Fick's law in the form

$$\vec{\mathbf{N}}_i = -c_t D_{im} \vec{\nabla} x_i + x_i \sum_{j=1}^{n} \vec{\mathbf{N}}_j$$

and the Stefan–Maxwell relations, Equation 2.74, we can define such a coefficient based on individual binary diffusion coefficients for each pair of gases. D_{im} is the pseudo-binary diffusion coefficient.

b. Show in the limit where $x_1 \approx x_2 \approx 0$ that

$$D_{3m} = \frac{D_{31}D_{32}}{D_{31} + D_{32}} \qquad D_{2m} = D_{23} \qquad D_{1m} = D_{13}$$

c. Use your result to calculate the pseudo-binary diffusion coefficient for each species in the system H_2O, He, N_2:

$$H_2O - He \quad 0.908 \times 10^{-9}\ m^2/s \qquad H_2O - N_2 \quad 0.256 \times 10^{-9}\ m^2/s$$

$$He - N_2 \quad 0.687 \times 10^{-9}\ m^2/s$$

d. Can your results be extended to higher order mixtures?

2.21 A mixture of gases, *a* in *b*, is to be separated from one another by thermal diffusion. A set of experiments is conducted at various temperatures to measure the thermal diffusion coefficients. Defining the separation ratio as k_a^T/k_b^T, what are the best temperature conditions with which to operate the device?

Experimental Data

T_1	T_2	$x_{a2} - x_{a1}$	$x_{b2} - x_{b1}$
25	50	0.05	0.03
	75	0.08	0.05
	100	0.10	0.06
50	75	0.07	0.01
	100	0.120	0.017
	150	0.190	0.027
100	150	0.06	−0.02
	200	0.103	−0.034
	300	0.163	−0.054

2.22 In sintering of materials, we can have mass flow in the absence of a concentration gradient via *surface diffusion*. In this process, surface molecules redistribute themselves driven by a gradient in surface curvature. This phenomenon causes metals to bead up on surfaces when heated and other materials to redistribute themselves over time. It is of fundamental importance to the semiconductor industry. If we express the mass flux as

$$\vec{\mathbf{N}}_a = -\vec{\nabla}(\gamma\kappa) \qquad \kappa = \frac{\pm\dfrac{d^2y}{dx^2}}{[1+(dy/dx)^2]^{3/2}}$$

where
γ is a surface energy parameter
κ is the curvature, what are the two possible equilibrium surface shapes?

2.23 Let us put some numbers to our well example from the text. Fracking for natural gas promises to be a clean technology because we are probing so deeply for the gas. Such deep wells mean there might be a considerable difference in gas composition from the bottom to the top. The well contains methane and propane and is 2000 m deep. You may assume the mixture behaves ideally, but as the well gets deeper, the temperature rises, 20°C for every kilometer in depth. At the surface, the composition is 80% methane, 20% propane, and the surface temperature is 25°C. What is the composition at the bottom of the well?

2.24 Our electrophoresis sample needs some numbers to make sense of the concentration that can be achieved. Let us assume that we are operating at room temperature, 298 K. We have adjusted the pH of the solution so that the protein we are trying to separate has a valence charge now of –5.

a. If the initial protein concentration is 10 mmol and we apply a voltage of +100 V at the anode and ground (0 V) the cathode, plot the concentration profile.
b. What is the maximum concentration that can be achieved?

REFERENCES

1. Wylie, C.R., *Advanced Engineering Mathematics,* 4th edn., McGraw-Hill, New York (1975).
2. Morse, P.M. and Feschbach, H., *Methods of Theoretical Physics Parts I and II*, McGraw-Hill, New York (1953).
3. Haase, R., *Thermodynamics of Irreversible Processes,* Addison-Wesley, Reading, MA (1969).
4. De Groot, S.R. and Mazur, P., *Non-Equilibrium Thermodynamics*, North-Holland Publishing Co., Amsterdam, the Netherlands (1962).
5. Nye, J.F., *Physical Properties of Crystals*, Oxford University Press, London, U.K. (1957).
6. Fetzer, R., Munch, A., Wagner, B., Rauscher, M., and Jacobs, K., Quantifying hydrodynamic slip: A comprehensive analysis of deweting profiles, *Langmuir*, **23**, 10559 (2007).
7. Heikes, R.R. and Ure, Jr. R.W., *Thermoelectricity: Science and Engineering*, Interscience, New York (1961).
8. Bird, R.B., Stewart, W.E., and Lightfoot, E.N., *Transport Phenomena*, John Wiley & Sons, New York (1960).
9. Hirschfelder, J.O., Curtiss, C.F., and Bird, R.B., *Molecular Theory of Gases and Liquids,* John Wiley & Sons, New York (1954).
10. Taylor, R. and Krishna, R., *Multicomponent Mass Transfer*, John Wiley & Sons, New York (1993).
11. Brodkey, R.S. and Hershey, H.C., *Transport Phenomena: A Unified Approach*, McGraw-Hill, New York (1988).
12. Streetman, B.G., *Solid State Electronic Devices,* 3rd edn., Prentice Hall, Englewood Cliffs, NJ (1990).
13. Donatella, P., Alessandro, S., and Bruno de, C., Rheological characterization of traditional and light mayonnaises, *J. Food Eng*., **35**, 409–417 (1998).
14. McCabe, W.L., *Trans.* The enthalpy-concentration chart: A useful device for chemical engineering calculations, *AIChE,* **31**, 129 (1935).

3 Transport Properties of Materials

3.1 INTRODUCTION

In the previous chapters, we saw why we call the processes involving the exchange of mass, momentum, energy, and charge, transport phenomena. Though transport processes occur via the exchange of particles, the fundamental quantity exchanged can be viewed as information. A particle's state contains quite a bit of information. Photons and other massless particles possess information in the form of their frequency, velocity, and phase. One can derive all the important physical properties of electromagnetic, matter, or gravitational waves through that information. Particles having mass such as atoms and molecules possess useful information from a chemical engineering perspective in four basic areas outlined in Table 3.1.

Kinetic energy is exchanged by a collision between particles. Potential energy can be converted to kinetic energy and exchanged by a collision, or it can be produced or converted by interaction with a field. State information can be exchanged during solvation, phase change, quantum interaction, or chemical reaction. All these processes involve some release or absorption of energy, and so state and energy are intimately related. Molecular composition information is usually exchanged through chemical reaction, though it can also be exchanged by an interaction with an external field (spectroscopy).

Since information is the fundamental quantity involved in transport phenomena, one can ask, how fast can this information be exchanged? In free space or a vacuum, the information can be exchanged no faster than the speed of light in a vacuum. In an arbitrary medium, electromagnetic information generally travels at the speed of light in the medium, whereas molecular information can be thought to travel as fast as the speed of sound in the medium.

In this chapter, we will investigate the use of historical models for gases, liquids, and solids and derive theoretical expressions for the viscosity, thermal conductivity, diffusivity, and conductivity of a material. These will be ideal approximations to the true transport properties of the material. For the most accurate predictive correlations, the reader is referred to texts on the transport properties of liquids and gases [1] and to texts on condensed matter physics [2,3].

3.2 VISCOSITY OF GASES

A theoretical expression for the viscosity of gases can be obtained by considering the transport of momentum via molecular collisions between gas particles. The classical approach is based on one of the most successful and long-lived models in physics; the kinetic theory of gases. Before deriving an expression for the viscosity, we must review several of the key results from kinetic theory. The discussion that follows is based on treatments of the subject by Sears [4], Curtis et al. [5], and Bird et al. [6].

3.2.1 Kinetic Theory Development [4–6]

The kinetic theory of gases is based on the assumption that the transport and thermodynamic properties of a collection of gas particles can be obtained from a knowledge of their masses, number density, and velocity distribution. Each gas particle is assumed to be a rigid, spherical,

TABLE 3.1
Particle Information

	Energy			
Momentum	**Kinetic**	**Potential**	**State**	**Composition**
Linear	Rotational	Interaction with field:	Liquid	Atomic arrangement
Angular	Vibrational	Gravitational	Solid	
	Translational	Electromagnetic	Gas	
			Charge	
			Quantum	

non-attracting, billiard ball of diameter, d_p, and mass, m. The spacing between particles must also be large enough so that we can approximate the particles as points. Collisions between particles are assumed to be perfectly elastic, so total momentum and energy is conserved upon collision. These assumptions presume ideal gases, and the original purpose of kinetic theory was to derive the ideal gas law from first principles. The two basic equations of kinetic theory describe the pressure exerted by the gas on the walls of a container and the velocity distribution of the gas particles at a given temperature, T.

The velocity distribution for the ideal gas was first developed by James Clerk Maxwell (1831–1879) in 1859 and later proven theoretically by Ludwig Eugen Boltzmann (1844–1906) in 1872. It bears both their names and defines the probability, P_v, that a gas particle has a velocity between v and $v + dv$, where $v = \sqrt{v_x^2 + v_y^2 + v_z^2}$:

$$\mathcal{P}_v(v) = \left(\frac{m}{2\pi k_b T}\right)^{3/2} \exp\left[-\frac{m\left(v_x^2 + v_y^2 + v_z^2\right)}{2k_b T}\right] \quad \text{Maxwell–Boltzmann velocity distribution} \tag{3.1}$$

Boltzmann's constant, $k_b = 1.38 \times 10^{-23}$ J/K, arises in the formulation and is the gas constant, R, divided by Avogadro's number, $\mathcal{N}_{av}$. The pressure of the gas is determined by the total change in momentum that occurs as the gas particles bounce off the walls of the container. In a 3-D box, assuming no preferential velocity direction, $v_x \approx v_y \approx v_z \approx v$, and the pressure is

$$P = \frac{1}{3}\left(\frac{\mathcal{N}}{V}\right) m\overline{v^2} \tag{3.2}$$

where
$\mathcal{N}$ is the number of gas particles
V is the volume

Evaluating the pressure requires that we find the mean square velocity, $\overline{v^2}$, and for other calculations, we will need the mean velocity, $\overline{v}$, and the root-mean-square velocity, $\sqrt{\overline{v^2}}$. All three velocity measures are determined using the velocity distribution and the concept of moments about a distribution. The mean square velocity is defined by

$$\overline{v^2} = \int_0^\infty v^2 \mathcal{P}_v(v)\, dv = \frac{3k_b T}{m} \quad \text{Mean-square velocity} \tag{3.3}$$

The mean velocity is defined by

$$\bar{v} = \int_0^\infty v\mathcal{P}(v)\,dv = \left(\frac{8k_bT}{\pi m}\right)^{1/2} \quad \text{Mean velocity} \tag{3.4}$$

and the root–mean–square velocity is

$$\sqrt{\overline{v^2}} = \left(\frac{3k_bT}{m}\right)^{1/2} \quad \text{Root-mean-square velocity} \tag{3.5}$$

As a particle moves, it will encounter other particles and collide with them. We consider two particles to have collided if their centers pass within one particle diameter of one another. To determine the frequency of such collisions, we fix all particles in space and time with the exception of one. As this free particle travels, it sweeps out a collision cylinder of volume, $\sigma_c\bar{v}\Delta t$, shown in Figure 3.1. The number of particles with their centers inside this cylinder is $\sigma_c\bar{v}\Delta t\mathcal{N}/V$. The number of collisions per unit time, or the collision frequency of the particle, is $\sigma_c\bar{v}\mathcal{N}/V$. The true velocity involved in the collision is not the velocity of the free particle, but the relative velocity between that particle and its colliding neighbor, or $\sqrt{2}\,\bar{v}$. With this in mind, we define the collision frequency, ω_c, as

$$\omega_c = \sqrt{2}\,\bar{v}\sigma_c\left(\frac{\mathcal{N}}{V}\right) \quad \text{Collision frequency} \tag{3.6}$$

The collision frequency we just calculated is the number of collisions a single particle makes. If we want to know the total number of collisions all particles make, we must multiply the collision frequency by one-half the total number of particles per unit volume. For a collection of identical particles of mass, m, this overall collision frequency is

$$\Omega_c(a,a) = \frac{1}{2}\omega_c\left(\frac{\mathcal{N}}{V}\right) = \frac{1}{\sqrt{2}}\bar{v}\sigma_c\left(\frac{\mathcal{N}}{V}\right)^2 \tag{3.7}$$

Using the value of $\bar{v}$ and σ_c already calculated, the overall collision frequency becomes a function of the absolute temperature, the particle mass, and the particle diameter:

$$\Omega_c(a,a) = \frac{\pi}{\sqrt{2}}d_p^2\left(\frac{8k_bT}{\pi m}\right)^{1/2}\left(\frac{\mathcal{N}}{V}\right)^2 \quad \text{Overall collision frequency} \tag{3.8}$$

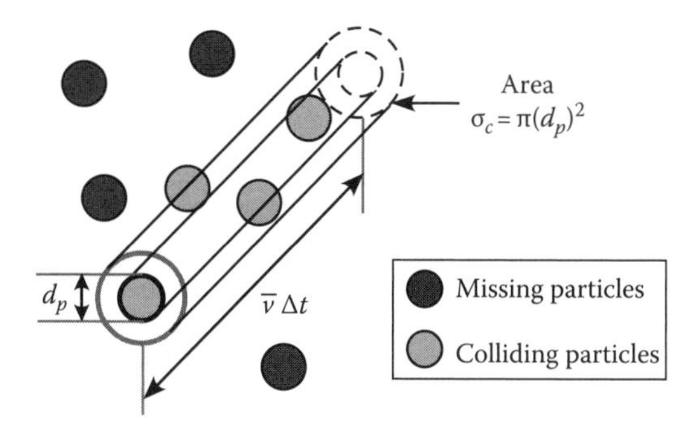

FIGURE 3.1 The collision cross section and collision tube for a particle of diameter, d_p.

If we need to know the overall collision frequency between dissimilar particles of masses, m_a and m_b, and diameters, d_a and d_b, the relative velocity is determined based on the reduced mass of the system, m_{ab}, and the mean diameter, d_{ab}:

$$m_{ab} = \frac{m_a m_b}{m_a + m_b} \quad \text{Reduced mass} \tag{3.9}$$

$$d_{ab} = \frac{d_a + d_b}{2} \quad \text{Mean diameter} \tag{3.10}$$

$$\Omega_c(a,b) = \frac{\pi}{4}(d_a + d_b)^2 \left(\frac{8k_bT}{\pi m_{ab}}\right)^{1/2} \left(\frac{\mathcal{N}}{V}\right)^2 \tag{3.11}$$

Knowing the collision frequency allows us to determine one of the most important concepts in kinetic theory, the mean free path between collisions, λ_f. If a particle is colliding with a frequency ω_c, and traveling at an average velocity $\bar{v}$, then it spends a period in time equal to $1/\omega_c$ between collisions and travels a distance of $\bar{v}/\omega_c$ during that period. The mean free path is just this traveling distance or

$$\lambda_f(a,a) = \frac{\bar{v}}{\omega_c} = \frac{1}{\sqrt{2}\,\pi d_p^2 \left(\frac{\mathcal{N}}{V}\right)} \quad \text{Mean free path} \tag{3.12}$$

$$\lambda_f(a,b) = \frac{\bar{v}}{\omega_c} = \frac{\pi}{4\sqrt{2}\,(d_a + d_b)^2 \left(\frac{\mathcal{N}}{V}\right)} \tag{3.13}$$

The collision frequency and velocity in Equation 3.13 have been modified to reflect the presence of two dissimilar particles. The mean free path does not tell the entire story of particle travel between collisions. We will also need to know the distribution of mean free paths for all the particles, so we can determine how many particles have a certain mean free path between collisions. We let $\mathcal{N}_c$ be the number of particles in our group of $\mathcal{N}$ that have not made a collision within recent memory. These $\mathcal{N}_c$ particles will have traveled some distance, z, along their free path. In the next instant, they travel a distance, dz, and within that distance, a fraction of the particles will collide with neighboring particles. The number of particles colliding in that instant should be proportional to the number of particles that have not yet made a collision, $\mathcal{N}_c$. Since each collision removes a particle from the group of uncollided particles, the change in $\mathcal{N}_c$, $d\mathcal{N}_c$, as a function of distance traveled, dz, can be written as

$$d\mathcal{N}_c = -\mathcal{P}_c(T,P)\mathcal{N}_c dz \tag{3.14}$$

where $\mathcal{P}_c(T, P)$ is a proportionality constant defining the probability that a collision occurs. $\mathcal{P}_c(T, P)$ depends upon the state of the gas, but not on $\mathcal{N}_c$ or z. Integrating Equation 3.14 using the boundary condition that when $z = 0$, $\mathcal{N}_c = \mathcal{N}$, gives the fraction of particles at any position, z, that have yet to collide:

$$\frac{\mathcal{N}_c}{\mathcal{N}} = \exp(-\mathcal{P}_c(T,P)z) \quad \text{Survival equation} \tag{3.15}$$

This equation is often referred to as the survival equation [4] since it describes the number of particles that have survived traveling a distance, z, without colliding. Substituting Equation 3.15 back into Equation 3.14 gives us the fraction of particles whose mean free path lies between z and $z + dz$:

$$\frac{d\mathcal{N}_c}{\mathcal{N}} = -\mathcal{P}_c(T,P)\exp(-\mathcal{P}_c(T,P)z)\,dz \tag{3.16}$$

We have yet to define the collision probability, $\mathcal{P}_c(T, P)$, but it seems that it should be related to the mean free path; collision probabilities should be smaller for longer free paths. We can use the definition of a weighted mean to derive the mean free path using Equation 3.16. The global average mean free path is just the sum of all free paths for each particle divided by the total number of particles:

$$\lambda_f = \frac{1}{\mathcal{N}}\int_0^\infty \mathcal{P}_c(T,P)\,\mathcal{N}\exp(-\mathcal{P}_c(T,P)z)\,z\,dz = \frac{1}{\mathcal{P}_c(T,P)} \tag{3.17}$$

Not surprisingly, the collision probability is the inverse of the mean free path. Now we can write the survival equation in terms of the mean free path:

$$d\mathcal{N}_c = -\frac{\mathcal{N}}{\lambda_f}\exp\left(-\frac{z}{\lambda_f}\right)dz \tag{3.18}$$

Finally, we are often interested in the collision frequency of particles with a wall, surface, or hypothetical plane rather than their collision frequency with each other. The collision frequency with respect to a hypothetical plane will drive our development of expressions for the transport coefficients. Collision frequencies with walls and surfaces are important in many low-pressure processes (semiconductor industry) and in high-pressure processes where the size of the confining space is smaller than the mean free path of the particle (diffusion and reaction in heterogeneous catalysts). Consider a wall or plane of area A_s, perpendicular to the particles' motion in the z direction. In the container, there are $\mathcal{N}/V$ particles per unit volume. If a particle has a velocity, v_z, it will strike the wall in a time, Δt, if it lies within a distance of $v_z\Delta t$ from the wall. The total number of collisions, on average, in the time interval, Δt, is given by

$$\begin{matrix}\text{Number of}\\ \text{collisions}\end{matrix} = A_s\Delta t\left(\frac{\mathcal{N}}{V}\right)\bar{v}_z = A_s\Delta t\left(\frac{\mathcal{N}}{V}\right)\int_0^\infty v_z\mathcal{P}_v(v_z)\,dv_z \tag{3.19}$$

Using the Maxwell–Boltzmann distribution for the velocities, we can evaluate Equation 3.19:

$$\begin{matrix}\text{Number of}\\ \text{collisions}\end{matrix} = A_s\Delta t\left(\frac{\mathcal{N}}{V}\right)\int_0^\infty v_z\exp\left(-\frac{mv_z^2}{2k_bT}\right)dv_z = A_s\Delta t\left(\frac{\mathcal{N}}{V}\right)\left(\frac{k_bT}{2\pi m}\right)^{1/2} \tag{3.20}$$

The collision frequency per unit area, Ω_w, is

$$\Omega_w = \left(\frac{k_bT}{2\pi m}\right)^{1/2}\left(\frac{\mathcal{N}}{V}\right) = \frac{1}{4}\left(\frac{\mathcal{N}}{V}\right)\bar{v} \tag{3.21}$$

Using the information on average velocities and collision frequencies, we can now develop a theoretical expression for the viscosity of a gas based on an analysis of momentum transport.

3.2.2 Kinetic Theory [5,6]

We consider a set of rigid particles, identical to the particles we discussed previously, confined between two long, parallel plates. The situation is shown in Figure 3.2. One plate is held stationary, and the other is moving at a constant velocity, v_o. The particles have a diameter d_p, a mass m, and are separated by a sufficient distance so that we can view them as points in space. We concentrate on the small highlighted region surrounding one of the particles and see how that particle exchanges momentum with the other particles in the system.

The moving plates create a macroscopic velocity gradient, dv_z/dy, in the gas phase. At the molecular level, the gas particles are free to move in whatever direction they desire. We have particles crossing the plane at y from both sides at an arbitrary angle, θ, with respect to the velocity gradient. To determine the net particle flux across the plane at y and hence the net momentum flux, we subtract the z-momentum of those particles traveling at lower velocities from the z-momentum of those traveling at higher velocities. Thus, momentum flows from the higher velocity to the lower velocity particles. The change in momentum of the particles is related to the shear stress exerted between the particles and the plates:

$$\frac{F}{A_s} = \tau_{yz} = \Omega_w m\, v_z\Big|_{y-\delta} - \Omega_w m\, v_z\Big|_{y+\delta} \tag{3.22}$$

On average, the particles travel a distance equivalent to one mean free path between collisions. As Figure 3.2 shows, the distance the particle travels between collisions, parallel to the velocity gradient, is dependent upon the particle's direction of travel and need not be one mean free path. On the average, particles reaching a certain velocity plane have had their last collision a distance, δ, from the plane where δ is some fraction of the mean free path. We can compute this distance using the collision frequencies, Ω_w and Ω_c, and our knowledge of the survival rate of particles traveling a distance, r, between collisions. Following the procedure outline by Sears [4], Figure 3.3 shows a small patch of volume, dV, a distance, r, away from an element of area, dA_s.

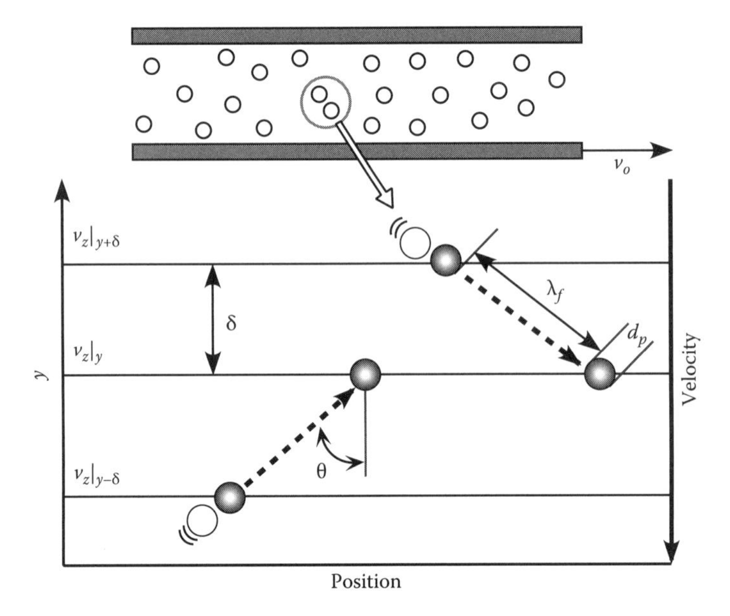

FIGURE 3.2 Gas particles flowing between parallel plates and exchanging momentum by collision. (Adapted from Bird, R.B. et al., *Transport Phenomena*, John Wiley & Sons, New York, 1960. With permission.)

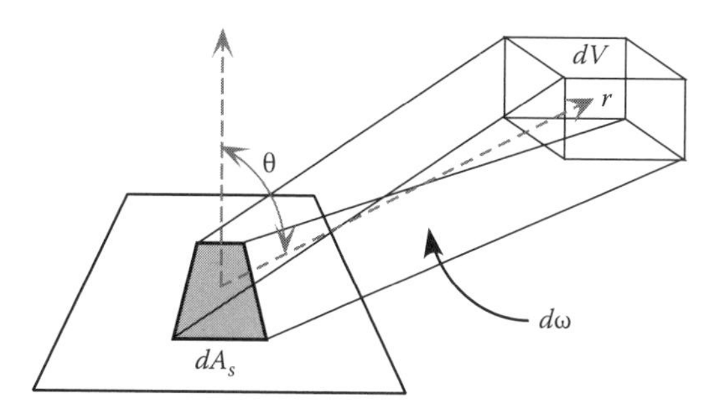

FIGURE 3.3 Schematic for calculating the number of particles colliding at dV that reach dA_s before colliding again.

The volume element makes an angle of θ with respect to the normal to dA_s. We assume that particles are colliding in dV with a frequency, Ω_c. The number of collisions is large enough that we can assume the particles rebound in all possible directions with equal probability. At any instant in time, the number of collisions within dV is $\Omega_c dV$. Since there are two particles per collision, the number of particles heading toward dA_s is just twice the collision rate multiplied by the fraction of space dA_s occupied when viewed from the point of collision in dV. This fraction of space is called the *view factor*, and we will see it again in Chapter 15:

$$\begin{matrix}\text{Number of particles headed} \\ \text{toward } dA_s \text{ with starting point at } r,\,\theta\end{matrix} = (2\Omega_c dV)\left(\frac{\cos\theta}{4\pi r^2}dA_s\right) \tag{3.23}$$

Equation 3.23 just counts the number of particles headed toward dA_s. The number that actually makes it to dA_s is a function of how far away from dA_s they are and is given by the survival equation (3.18). Incorporation of the information from (3.18) just requires multiplying Equation 3.23 by $\exp(-r/\lambda_f)$:

$$\begin{matrix}\text{Number of particles reaching } dA_s \\ \text{with starting point at } r,\,\theta\end{matrix} = (2\Omega_c dV)\left(\frac{\cos\theta}{4\pi r^2}\right)\exp\left(-\frac{r}{\lambda_f}\right)dA_s \tag{3.24}$$

To determine the average height, δ, of these particles above dA_s, we multiply the height above dA_s, $r\cos\theta$, by the number of particles surviving without collision from that height, and then divide by the total number of particles reaching dA_s. The total number of particles reaching dA_s is the wall collision frequency, Ω_w. Expressing dV in terms of spherical coordinates, $r^2 \sin\theta\, d\theta\, d\phi\, dr$, and integrating over the hemisphere above dA_s, gives

$$\delta = \frac{\dfrac{2}{4\pi}\Omega_c dA_s \displaystyle\int_0^{\pi/2}\sin\theta\cos^2\theta\, d\theta \int_0^{2\pi} d\phi \int_0^{\infty} r\exp\left(-\frac{r}{\lambda_f}\right)dr}{\Omega_w} = \frac{2}{3}\lambda_f \tag{3.25}$$

Therefore, the average height of a particle crossing an arbitrary plane in the y-direction of Figure 3.2 is just $2\lambda_f/3$.

To complete the description of momentum transport in a gas, we assume that the particles do not lose their identity (momentum) instantaneously and that all the particles crossing an arbitrary plane due to a collision have velocities representative of where they came from. Though the velocity gradient need not be linear, we will assume that it is so over a distance of at least 3δ. This restriction enables us to represent the velocities at $y + \delta$ and $y - \delta$ in terms of the velocity at y using a Taylor series expansion truncated after the second term:

$$v_z\Big|_{y-\delta} = v_z\Big|_y - \delta\frac{dv_z}{dy} = v_z\Big|_y - \frac{2}{3}\lambda_f\frac{dv_z}{dy}$$
$$v_z\Big|_{y+\delta} = v_z\Big|_y + \delta\frac{dv_z}{dy} = v_z\Big|_y + \frac{2}{3}\lambda_f\frac{dv_z}{dy} \tag{3.26}$$

Substituting these velocity functions into Equation 3.22, and using the definition for the wall collision frequency, Ω_w, lets us eliminate the velocity, $v_z\big|_y$, and determine the shear stress, τ_{yz}, to be

$$\tau_{yz} = -\frac{1}{3}\left(\frac{\mathcal{N}}{V}\right)m\bar{v}\lambda_f\frac{dv_z}{dy} = -\mu\frac{dv_z}{dy} \tag{3.27}$$

This result is analogous to Newton's law of viscosity if we define the viscosity to be

$$\mu = \frac{1}{3}\left(\frac{\mathcal{N}}{V}\right)m\bar{v}\lambda_f = \frac{1}{3}\rho\bar{v}\lambda_f \qquad \text{Viscosity}\left(\frac{\text{N}\cdot\text{s}}{\text{m}^2}\right) \tag{3.28}$$

By substituting for the average velocity, the density, and the mean free path, we find the viscosity depends on the particle mass, its size, and the absolute temperature:

$$\mu = \frac{2}{3\pi^{3/2}}\frac{\sqrt{mk_bT}}{d_p^2} \tag{3.29}$$

Note that this representation requires an experimental point to fix the particle diameter, d_p. Equation 3.29 predicts that the viscosity is independent of pressure, a result that is reasonable for pressures up to about 10 atmospheres. Experimental evidence shows that the temperature dependence of $T^{1/2}$ is not strong enough. To predict the temperature dependence of the viscosity more accurately, one must take into account the force field between particles and their finite size.

3.2.3 Chapman–Enskog Theory [5–9]

The Chapman–Enskog theory was developed independently by Sydney Chapman in England [8] and David Enskog in Sweden [9]. It relates the transport properties of gases in terms of the potential energy, Φ_u, of interaction between a pair of particles in the gas. The theory considers both long-range and short-range forces acting between the particles and abandons the concept that the particles are just rigid, point-like, spheres. The force between the particles is given by

$$F = -\frac{d\Phi_u}{dr} \tag{3.30}$$

where r is the distance between the particles. The exact functional form for $\Phi_u(r)$ is not always known, though in principle, it can be calculated using quantum mechanics. A wealth of

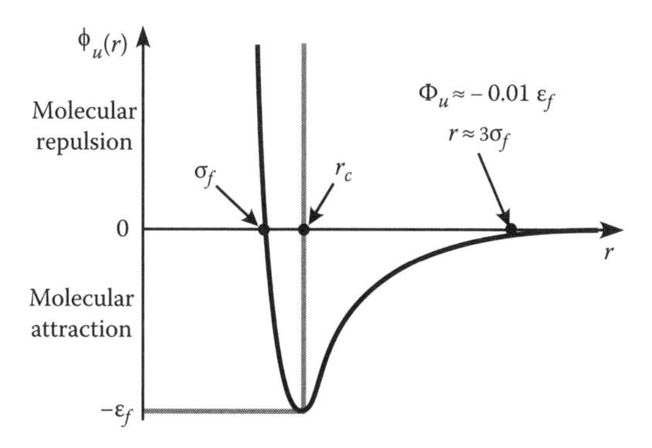

FIGURE 3.4 Lennard–Jones (6-12) potential function.

experimental evidence has shown that the Lennard–Jones (6-12) potential is a good engineering approximation to the true nature of the interaction:

$$\Phi_u(r) = 4\varepsilon_f \left[\left(\frac{\sigma_f}{r} \right)^{12} - \left(\frac{\sigma_f}{r} \right)^{6} \right] \quad \text{Lennard-Jones potential} \tag{3.31}$$

where
σ_f is the characteristic diameter of the particle (its collision cross section)
ε_f is a characteristic energy of interaction between the particles

The Lennard–Jones potential function is shown in Figure 3.4. It displays the characteristic features of molecular interactions including a weak attraction at large separations and a strong repulsion at small separations. In general, this function represents *nonpolar* particles well. Polar particles have other long-range interactions that render a simple model like (3.31) inadequate. Values for σ_f and ε_f have been tabulated for many molecules, and their values can be found in any number of handbooks [1,4]. If a particular compound is not listed in one of these references, the Lennard–Jones parameters can be obtained based on knowledge of the critical properties from the gaseous state, the normal boiling point from the liquid state, or the freezing point from the solid state:

$$\frac{\varepsilon_f}{k_b}(K) = 0.77 T_{critical} \qquad \sigma_f(m) = 0.841 \times 10^{-10} \sqrt[3]{V_{critical}} \qquad \text{Gas} \tag{3.32}$$

$$\frac{\varepsilon_f}{k_b} = 1.15 T_{boiling} \qquad \sigma_f = 1.166 \times 10^{-10} \sqrt[3]{V_{boiling}} \qquad \text{Liquid} \tag{3.33}$$

$$\frac{\varepsilon_f}{k_b} = 1.92 T_{fusion} \qquad \sigma_f = 1.222 \times 10^{-10} \sqrt[3]{V_{fusion}} \qquad \text{Solid} \tag{3.34}$$

In Chapman–Enskog theory, the viscosity for a pure monatomic gas of molecular weight M_w can be written in terms of σ_f, ε_f, k_b, and the absolute temperature, T, as

$$\mu = 2.669 \times 10^{-26} \frac{\sqrt{M_w T}}{\sigma_f^2 \Omega_\mu} \quad (\text{N s/m}^2) \tag{3.35}$$

Ω_μ is a slowly varying function of the dimensionless temperature, $k_b T/\varepsilon_f$, and its functional form varies between $T^{0.6} \to T^{1.0}$. Equation 3.35 holds well for monatomic gases and surprisingly well for

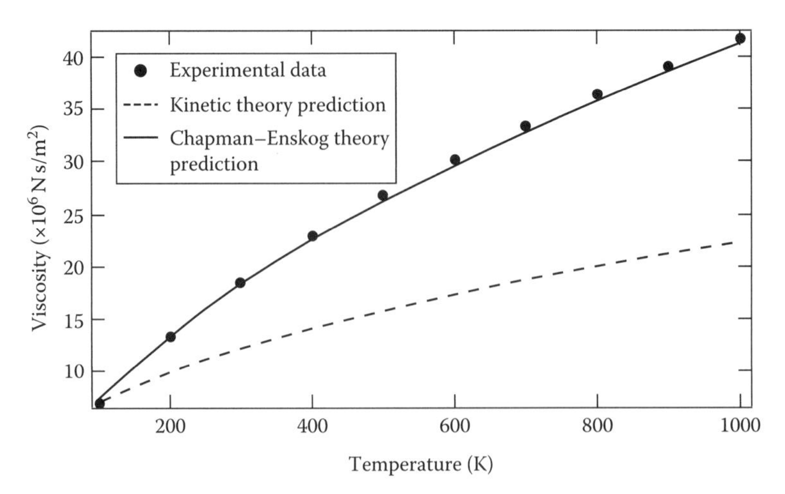

FIGURE 3.5 Viscosity of air as a function of temperature at 1 atm. Success of molecular theories in predicting the viscosity. (Data from Weast, R.C., ed., *Handbook of Chemistry and Physics*, 55th edn., CRC Press, Cleveland, OH, 1974.)

polyatomic gases, once the parameters, σ_f and ε_f, are known. If the gas were made of rigid spheres, the temperature function, Ω_μ, would be identically 1. Thus, Ω_μ can be thought of as a deviation from rigid sphere behavior.

At higher pressures, the Chapman–Enskog model also fails because it predicts no pressure dependence. As pressures become higher, the distances between particles are small enough that two-body interactions (the only type of collisions considered by the kinetic and Chapman–Enskog theories) are not the only ones responsible for the behavior of the gas. Many-body and cluster interactions must be included. Since these interactions depend upon the pressure, neither the kinetic nor Chapman–Enskog theory can be used at high pressures. Figure 3.5 shows how well the Chapman–Enskog theory predicts the low-pressure viscosity of air at virtually any temperature. Figure 3.6 shows how both the kinetic theory and Chapman–Enskog theory fail for gases at higher pressures. A more detailed discussion of how the viscosity of gases can be estimated can be found in Reid et al. [1].

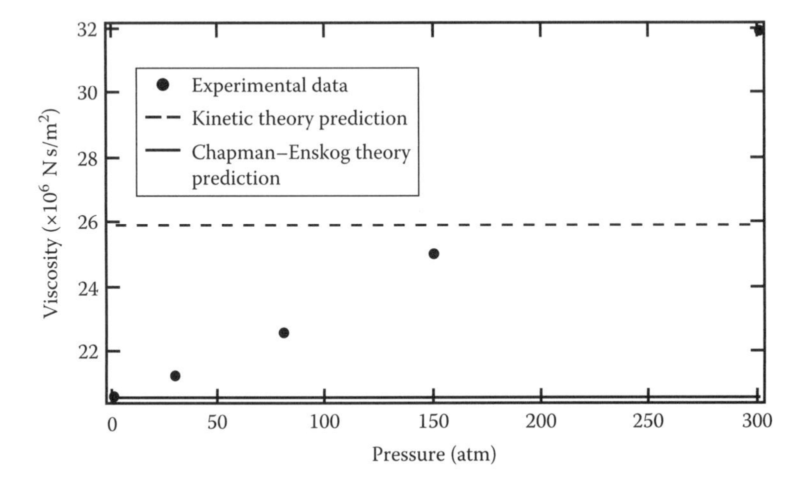

FIGURE 3.6 Viscosity of oxygen as a function of pressure. Kinetic and Chapman–Enskog theories fail. (Data from Weast, R.C., ed., *Handbook of Chemistry and Physics*, 55th edn., CRC Press, Cleveland, OH, 1974.)

TABLE 3.2
Physical Properties of Gas Particles

Gas	*Mw* (g/mol)	d_p (m × 10^9)	σ_f (m × 10^{10})	ε_f/k_b
H_2	2	0.4386	2.82	59.7
Ne	20.18	0.3784	2.82	32.8
C_6H_6	78.11	0.7152	5.34	412.3

TABLE 3.3
Calculated versus Experimental Gas Viscosities

Gas	Kinetic Theory Viscosity (N s/m²)	Chapman–Enskog Viscosity (N s/m²)	% Difference (Exp versus CE)	Experimental Viscosity (N s/m²)
Ne	9.850×10^{-6}	3.11×10^{-5}	34	4.708×10^{-5}
H_2	7.300×10^{-5}	8.86×10^{-6}	36	1.386×10^{-5}
C_6H_6	5.425×10^{-6}	7.56×10^{-6}	49	1.484×10^{-5}

Example 3.1 Comparison of Kinetic and Chapman–Enskog Predictions

Using the data provided in Appendix F, compare predictions for the viscosity of H_2, Ne, and C_6H_6 at 300 K and 1 atm. using the kinetic theory of gases and the Chapman–Enskog theory.

The kinetic theory of gases states that the viscosity can be given by

$$\mu = \frac{2}{3\pi^{3/2}} \frac{\sqrt{mk_bT}}{d_p^2}$$

The Chapman–Enskog theory gives the viscosity as

$$\mu = 2.669 \times 10^{-26} \frac{\sqrt{M_wT}}{\sigma_f^2 \Omega_\mu}$$

The basic physical information we need for all the gases is the molecular weight, the molecular diameter, and the parameters for the Chapman–Enskog theory, σ_f and ε_f. The Chapman–Enskog parameters for these gases are given in Appendix F and are listed in Table 3.2. Table 3.3 shows the results of the calculations.

The molecular diameter is harder to find. One general way of obtaining it is to use the Van der Waals parameter, b, for the gas. This gives the volume for each mole of gas particles [10].

3.3 VISCOSITY OF LIQUIDS: FREE VOLUME THEORY

The example calculation we performed in the previous section showed that we have a good theoretical understanding of the molecular basis for gas viscosity. Our understanding of the viscosity of liquids is largely empirically based. Molecular dynamics calculations have greatly improved the situation, but a simple kinetic theory of liquids is not available since we cannot calculate, in any analytical way, the extent of the intermolecular interactions in dense systems

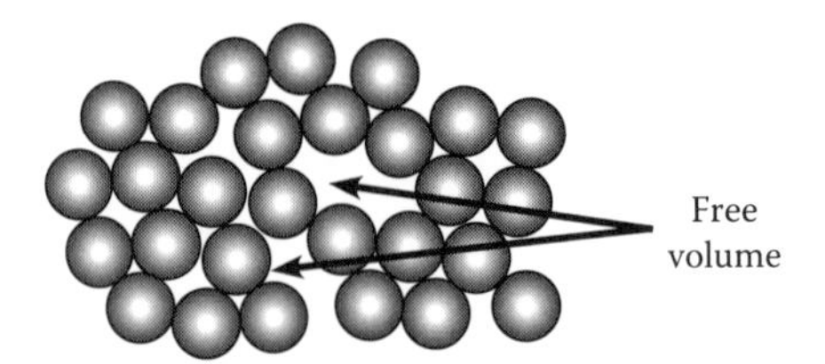

FIGURE 3.7 Structure of a static liquid—free volume.

such as liquids. Still a number of theories of liquid viscosity have been developed, and it is instructive to consider one of the more successful; a phenomenological, rate-based, model that provides a feel for the mechanism involved and can be used to give a rough estimate for the viscosity. This model was originally developed by Henry Eyring and coworkers during the 1940s and 1950s, and it has been extended to a generalized model called the free volume or rate model [11,12].

The particles of a pure liquid *at rest* are in a constant state of motion similar to a gas. Since the particles are so closely packed, the motion is largely vibrational in scope and limited to a cage-like area formed by the nearest neighbor particles. The situation is shown in Figure 3.7 where the vacant sites, or holes, are referred to as free volume. This *free* volume may redistribute itself throughout the liquid, requiring no energy to do so. Eventually, free volume collects to form a space the size of a liquid particle, and so a particle may decide to hop from its current position to the neighboring free volume region.

The cage surrounding the hopping particle represents a barrier of height $\Delta G^{\dagger}/\mathcal{N}_{av}$ where $\Delta G^{\dagger}$ is an activation energy per mole of liquid and $\mathcal{N}_{av}$ is Avogadro's number. On average, particles breaking out of their cages do so by leaping free in hops of length, ℓ, and at a frequency, ω_o. The frequency of such hops is given by the rate equation:

$$\omega_o = \frac{k_b T}{\hbar}\exp\left[-\frac{\Delta G^{\dagger}}{RT}\right] \tag{3.36}$$

where
- k_b is Boltzmann's constant
- $\hbar$ is Planck's constant
- R is the gas constant
- T is the absolute temperature

When the fluid is set into motion under an applied stress, say τ_{yz}, the energy barrier is distorted and is lower for particles moving in the direction of the applied stress and higher for particles moving against the applied stress. The activation energy for molecular motion, as shown in Figure 3.8, can adjust to account for the applied stress:

$$-\Delta G = -\Delta G^{\dagger} \pm \left(\frac{\ell}{\delta}\right)\left(\frac{\tau_{yz}V_m}{2}\right) \tag{3.37}$$

where
- V_m represents the volume of a mole of liquid
- $(\ell/\delta)(\tau_{yz}V_m/2)$ represents the work done on the particles as they climb to the top of the barrier

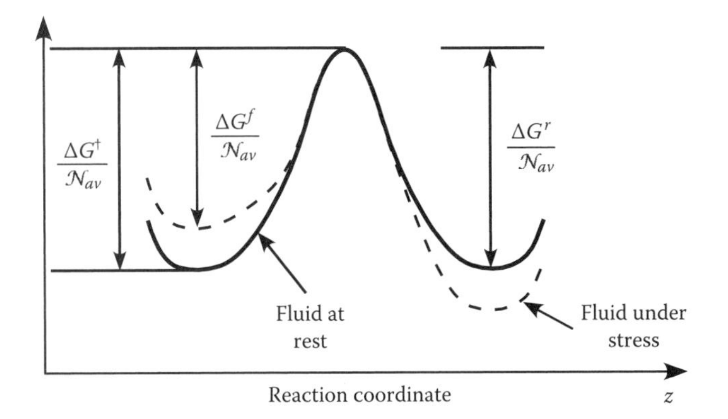

FIGURE 3.8 Energy barrier to fluid motion. The effect of an applied stress is to alter the barrier height and make its magnitude directional.

The plus sign is chosen for particles moving with the applied stress and the minus sign for particles moving against the stress. δ is an arbitrary length designed to represent the length scale over which the stress changes appreciably.

We can define ω_f as the frequency of forward hops (following the stress) and ω_r as the frequency of backward hops (opposing the stress):

$$\omega_f = \left(\frac{k_b T}{\hbar}\right)\exp\left[-\Delta G^\dagger + \left(\frac{\ell}{\delta}\right)\left(\frac{\tau_{yz} V_m}{2}\right)\right] \tag{3.38}$$

$$\omega_r = \left(\frac{k_b T}{\hbar}\right)\exp\left[-\Delta G^\dagger - \left(\frac{\ell}{\delta}\right)\left(\frac{\tau_{yz} V_m}{2}\right)\right] \tag{3.39}$$

The net velocity of the particles slipping past one another is just the average hop length multiplied by the net frequency of forward hops:

$$v_{zf} - v_{zr} = \ell(\omega_f - \omega_r) \tag{3.40}$$

If we only consider a very small distance between fluid layers undergoing the induced stress, the velocity profile will be linear, and we can express the velocity gradient in the y-direction in terms of the velocity difference and fluid layer separation, δ:

$$-\frac{dv_z}{dy} = \frac{v_{zf} - v_{zr}}{\delta} = \frac{\ell}{\delta}(\omega_f - \omega_r) \tag{3.41}$$

Substituting in for the hopping frequencies, ω_f and ω_r, gives

$$\begin{aligned} -\frac{dv_z}{dy} &= \left(\frac{\ell}{\delta}\right)\left(\frac{k_b T}{\hbar} e^{-\Delta G^\dagger/RT}\right)\left[\exp\left(\frac{\ell \tau_{yz} V_m}{2\delta RT}\right) - \exp\left(-\frac{\ell \tau_{yz} V_m}{2\delta RT}\right)\right] \\ &= \left(\frac{\ell}{\delta}\right)\left(\frac{k_b T}{\hbar} e^{-\Delta G^\dagger/RT}\right)\left[2\sinh\left(\frac{\ell \tau_{yz} V_m}{2\delta RT}\right)\right] \end{aligned} \tag{3.42}$$

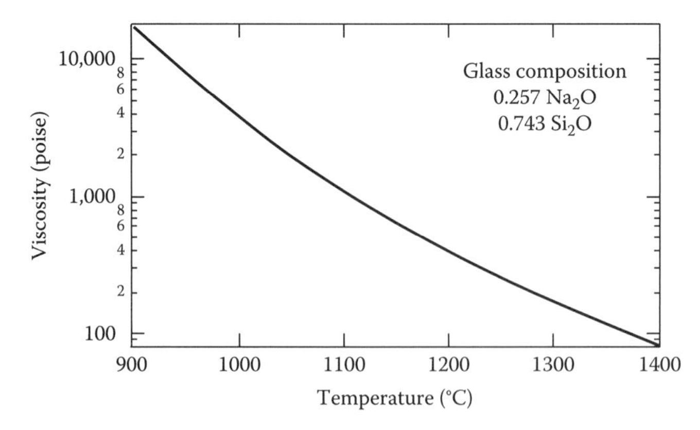

FIGURE 3.9 Viscosity–temperature profile for a sodium silicate glass. (Data from Lillie, H.R., *J. Am. Ceram. Soc.*, 22, 367, 1939.)

Notice that Equation 3.42 predicts non-Newtonian flow for all fluids. If the argument for the sinh function is small ($\tau_{yz} << RT$), then we can use the approximation sinh $(x) = x$, and the velocity gradient expression is consistent with Newton's law of viscosity. The predicted viscosity is

$$\mu = \left(\frac{\ell}{\delta}\right)^2 \left(\frac{\mathcal{N}_{av}\hbar}{V_m} e^{\Delta G^\dagger / RT}\right) \tag{3.43}$$

We lose no loss of generality if we assume $\delta = \ell$ because $\Delta G^\dagger$ must be determined empirically to make the equation fit the experimental data. Equation 3.43 predicts that the viscosity should decrease exponentially with increasing temperature. This behavior is observed for most liquids with the striking example being glasses and ceramics (Figure 3.9).

3.3.1 Viscosity of Suspensions

Suspensions of particles in a liquid are found in many common products including paints, cosmetics, concrete, foodstuffs, and pharmaceuticals. The behavior of these suspensions is critical to their performance. For example, ketchup and chocolate both get their "mouth feel" from careful management of the type of particulate suspension used. Paint sticks on the wall without dripping based on careful design of its flow properties or rheology. Concrete can be pumped and made to flow around rebar by careful choice of the particulate sizes and liquid-suspending matrix. Foundation cosmetics are designed in much the same way as paint. All these design properties fall under the heading of rheology.

The rheology of a dilute suspension of rigid, noninteracting spherical particles was first calculated by Einstein [14]. By approximating fluid flow around a set of spherical particles and assuming the fluid suspension can be described by a single, homogeneous value of viscosity, Einstein determined that the suspension viscosity is given by

$$\mu_r = \frac{\mu}{\mu_0} = 1 + 2.5\phi \tag{3.44}$$

where

μ_0 is the velocity of the fluid in the absence of particles
ϕ is the particle volume fraction

This is a simplification for small ϕ, of the equation [15]

$$\mu_r = \frac{1+0.5\phi}{(1-\phi)^2} \tag{3.45}$$

Neither of these equations holds as ϕ gets large and so, since the original result, many other expressions have been developed to try to correlate and even predict what the viscosity of a suspension of particles of arbitrary shape would be as a function of particle volume fraction. Some of these models use a straightforward polynomial extension of Einstein's original equation:

$$\mu_r = 1+2.5\phi + A\phi^2 + \ldots \qquad 4.375 < A < 14.1 \tag{3.46}$$

Here, Batchelor derived a theoretical estimate for $A = 6.2$ [16]. Others alter the original form to fit a concentrated suspension (3.45) [17]:

$$\mu_r = \frac{1-0.5\phi}{(1-\phi)^3} \tag{3.47}$$

Still others, like Robinson [18], attempt to realistically extend the original formulation to high particle volume fractions by incorporating a critical packing fraction. The critical packing fraction occurs when particles are so closely packed that the fluid cannot move. For example, if we have all spherical particles of the same size, the critical packing fraction is hexagonally closed packed where $\phi_c = 0.74$.

$$\mu_r = 1+B\left(\frac{\phi}{1-\phi/\phi_c}\right) \qquad \phi_c\text{—Critical packing fraction} \tag{3.48}$$

3.4 THERMAL CONDUCTIVITY OF GASES

We begin our attempt to predict the thermal conductivity of gases in the same manner we treated the viscosity of gases; we assume that thermal energy is exchanged in the form of collisions between gas particles and molecules. Each collision transfers kinetic energy between the particles, changing the particles' velocity and hence its temperature.

3.4.1 Kinetic Theory [4–6]

The kinetic theory of gases was very successful at predicting the form of Newton's law of viscosity and the viscosity of simple gases. It exhibits similar success in reproducing Fourier's law of conduction and estimating the thermal conductivity of simple gases. We assume we have a pure, monatomic gas consisting of rigid, spherical, non-attracting, molecules of diameter, d_p, and mass, m, present in a concentration of N molecules per volume, V. Here, we consider the gas as a whole to have no net velocity although the individual particles will be moving quite rapidly in a random fashion. All the internal energy, U, of the gas is assumed to be in the form of translational energy.

The system we will be considering is shown in Figure 3.10. The mean translational kinetic energy per molecule is

$$U = \frac{1}{2}m\overline{v^2} = \frac{3}{2}k_bT \tag{3.49}$$

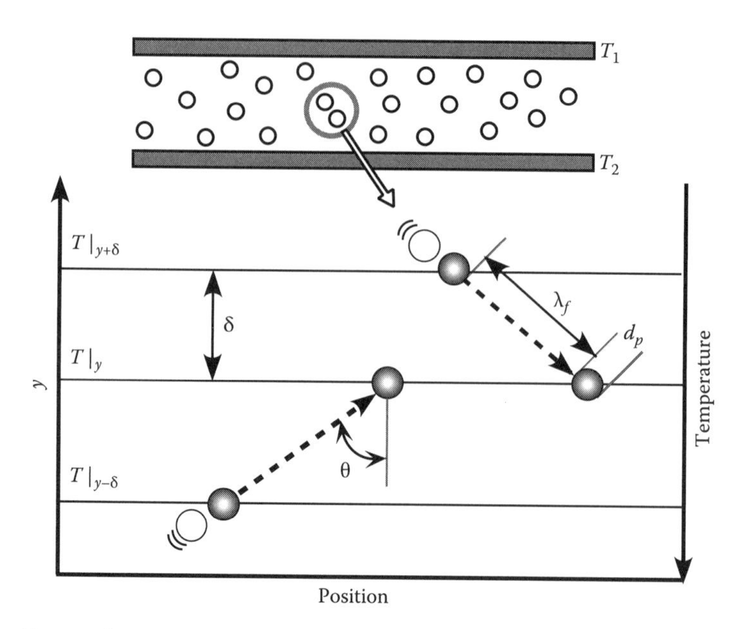

FIGURE 3.10 Gas particles confined between heated, parallel plates. (Adapted from Bird, R.B. et al., *Transport Phenomena*, John Wiley & Sons, New York, 1960. With permission).

The heat capacity per mole of an ideal gas at constant volume is defined by the change in internal energy as a function of temperature. For an ideal gas, all its energy is kinetic and so

$$C_v = \left(\frac{\partial U}{\partial T}\right)_V = \mathcal{N}_{av}\frac{d}{dT}\left(\frac{1}{2}m\overline{v^2}\right)_V = \frac{3}{2}R \tag{3.50}$$

The heat capacity at constant pressure is found analogously:

$$C_p = \left(\frac{\partial H}{\partial T}\right)_P = \left(\frac{\partial}{\partial T}(U+PV)\right)_P \qquad C_v + R = \frac{5}{2}R \tag{3.51}$$

To determine the thermal conductivity, we investigate the behavior of the gas particles as they move with respect to a temperature gradient, dT/dy. Again, we have molecules crossing a hypothetical temperature plane from both sides. To determine the net particle flow across the plane at y and hence the net energy flux, we subtract the kinetic energy of those particles crossing in the negative y-direction from the kinetic energy of those crossing in the positive y-direction:

$$q''_y = \Omega_w\left(\frac{1}{2}m\overline{v^2}\right)_{y-\delta} - \Omega_w\left(\frac{1}{2}m\overline{v^2}\right)_{y+\delta} \tag{3.52}$$

$$q''_y = \frac{3}{2}k_b\Omega_w\left(T\Big|_{y-\delta} - T\Big|_{y+\delta}\right) \tag{3.53}$$

Here, we assume that the particles do not lose their identity immediately, so all the particles crossing the plane have velocities/energies representative of where they came from. We will also assume the temperature profile is linear over a distance ≥3δ – 4δ, so a two-term Taylor series approximation is valid:

$$T\big|_{y-\delta} = T\big|_y - \delta\frac{dT}{dy} = T\big|_y - \frac{2}{3}\lambda_f\frac{dT}{dy} \tag{3.54}$$

$$T\big|_{y+\delta} = T\big|_y + \delta\frac{dT}{dy} = T\big|_y + \frac{2}{3}\lambda_f\frac{dT}{dy} \tag{3.55}$$

Plugging in for Ω_w from Equation 3.21, for the temperatures from Equations 3.54 and 3.55, and eliminating $T\big|_y$, Equation 3.53 becomes

$$q''_y = -\frac{1}{2}\left(\frac{\mathcal{N}}{V}\right)k_b\bar{v}\lambda_f\frac{dT}{dy} = -k\frac{dT}{dy} \tag{3.56}$$

This result is analogous to Fourier's law of heat conduction if the thermal conductivity is defined as

$$k = \frac{1}{2}\left(\frac{\mathcal{N}}{V}\right)k_b\bar{v}\lambda_f = \frac{1}{3}\rho C_v\bar{v}\lambda_f \qquad \text{Thermal conductivity}\left(\frac{\text{W}}{\text{m K}}\right) \tag{3.57}$$

where $\rho = \mathcal{N}$ m/V, the mass density of the gas, and we have used Equation 3.45 to remove Boltzmann's constant. By substituting for the average velocity and the mean free path using (3.4) and (3.12), we obtain an expression for the thermal conductivity in terms of the absolute temperature, the mass of the molecule, and the molecular diameter:

$$k = \frac{1}{d_p^2}\sqrt{\frac{k_b^3T}{\pi^3m}} \quad \text{(W/m K)} \tag{3.58}$$

Note that this representation also requires one experimental point to fix d_p. Equation 3.58 predicts that the thermal conductivity is independent of pressure. This is acceptable for pressures up to about 10–20 atm. At higher pressures, the possibility of multi-body collisions becomes important, and this alters the thermal conductivity of the gas. The square-root temperature dependence is approximate, and experimental evidence shows that this dependence is too weak. To predict the temperature dependence of the thermal conductivity more accurately, one must take into account the force field between molecules and their finite diameter.

3.4.2 Chapman–Enskog Theory

We can return to Chapman–Enskog theory to get a better approximation for the thermal conductivity. In this case, we have the following expression where the collision function, Ω_k, is identical to that for the viscosity, Ω_μ. σ_f represents the collision cross section for the particle and M_w is the molecular weight:

$$k = 8.332\times10^{-22}\frac{\sqrt{T/M_w}}{\sigma_f^2\Omega_k} \quad \text{(W/m K)} \tag{3.59}$$

Equation 3.59 has been found to be very accurate for monatomic gases and reasonably accurate for polyatomic gases as well. Again, it does not apply at pressures higher than 10–20 atm.

TABLE 3.4
Prandtl Number Predictions (288 K)

Gas	Kinetic Theory	Chapman–Enskog Theory	Experimental
Air	1.67	0.67	0.715
H_2	1.67	0.67	0.90
H_2O	1.67	0.67	0.842
CO_2	1.67	0.67	0.781
C_2H_6	1.67	0.67	0.815

Source: Data from Perry, R.H. and Chilton, C.H., eds., *Chemical Engineer's Handbook*, 5th edn., McGraw-Hill, New York, 1973.

The Chapman–Enskog and kinetic theory expressions for the viscosity and thermal conductivity are similar. The respective relationships between k and μ can be written in terms of the heat capacities:

$$k = \frac{15}{4}\frac{R}{M_w}\mu = \frac{5}{2}C_v\mu \quad \text{Chapman–Enskog} \tag{3.60}$$

$$k = C_v\mu \quad \text{Rigid sphere} \tag{3.61}$$

The Prandtl number, which we will later see is a ratio of momentum to thermal diffusivity, can also be estimated from this relationship. Notice that for ideal gases, $C_p = C_v + R$ and $C_p/C_v = 5/3$ so that the Prandtl number can be written as

$$\Pr = \frac{C_p\mu}{k} \approx \frac{2}{5}\frac{C_p}{C_v} = 0.67 \quad \text{Chapman–Enskog} \tag{3.62}$$

$$\Pr = \frac{C_p\mu}{k} = \frac{C_p}{C_v} = 1.67 \quad \text{Rigid sphere} \tag{3.63}$$

If we apply these expressions to predict the Prandtl number for real gases [19], we find that the kinetic theory grossly overpredicts the Prandtl number, while the Chapman–Enskog theory underpredicts the Prandtl number. Keep in mind that we did not correctly account for the intermolecular force field in deriving the relationship between C_p and C_v for use in Equation 3.62. Incorporating the intermolecular forces into Equations 3.50 and 3.51 would have brought us much closer to the correct value. Equations 3.62 and 3.63 apply strictly to monatomic gases. Thus the discrepancy between predicted and experimental values in Table 3.4 is not surprising.

3.5 THERMAL CONDUCTIVITY OF LIQUIDS

The nature of the thermal conductivity of liquids is more obscure than that of gases. To date, there is no fundamental theory that gives accurate predictions for most fluids. We can view the transmission of energy in a fluid, in the same manner as we viewed its transmission in a gas. Therefore, the kinetic theory approximation we derived for the thermal conductivity of a gas can be modified to apply to liquids. The average velocity of gas particles can be written in terms of the speed of sound in the gas rather than in terms of the absolute temperature and mass of the particles:

$$\bar{v} = \left(\frac{8k_bT}{\pi m}\right)^{1/2} = \left(\frac{8C_v}{\pi C_p}\right)^{1/2} c_{\lambda g} \tag{3.64}$$

Here, $c_{\lambda g}$ is the speed of sound of the gas. By analogy, the speed of sound in the liquid, $c_{\lambda l}$, and hence the average velocity of the liquid particles, is

$$\overline{v_{liq}} = \left(\frac{8C_v}{\pi C_p}\right)^{1/2} c_{\lambda l} \tag{3.65}$$

In a liquid, the heat capacity at constant volume, C_v, is very nearly $3R/M_w$ rather than the $3R/2M_w$ for a monatomic gas since the liquid molecules can be thought of as moving in a simple, 3-D harmonic potential (a cage). We approximate the mean free path in the liquid as the cube root of the volume per molecule of liquid:

$$\lambda_f \approx \sqrt[3]{\frac{M_w}{\rho \mathcal{N}_{av}}} \tag{3.66}$$

Substituting these values for the mean free path, mean velocity, and heat capacity at constant volume into Equation 3.65 gives us the required approximation for the thermal conductivity:

$$k_{liq} = \left(\frac{\rho^2 k_b^3 \mathcal{N}_{av}^2}{M_w^2}\right)^{1/3} \left(\frac{8C_v}{\pi C_p}\right)^{1/2} c_{\lambda l} \tag{3.67}$$

Equation 3.67 is a reasonable approximation for monatomic liquids at atmospheric pressure but is not as accurate as Chapman–Enskog theory is for gases.

Example 3.2 Thermal Conductivity Estimates for Liquids

We use Equation 3.67 to estimate the thermal conductivities of various liquids. We need data on the speed of sound in the material, its density, and molecular weight. For estimation purposes, the heat capacity ratio, C_v/C_p, is approximately 1. Table 3.5 shows the comparison for several fluids: polar, hydrogen bonded, and metallic. The predictions are fairly good considering Equation 3.67 is not accurate for polyatomic molecules. Notice the striking failure for the liquid metal, mercury. There is obviously another mechanism for heat conduction in the liquid metal and that mechanism is the conduction electrons. We will discuss this contribution in the next section when we consider the thermal conductivity of solids and will show that the electronic contribution may be upward of 20–30 times that of the dielectric contribution we have just calculated.

TABLE 3.5
k_{exp} versus k_{calc} for Several Liquids

Liquid	Density (ρ), (kg/m³)	Molecular Weight, (M_w) (kg/kmol)	Sonic Velocity, c_λ/(m/s)	k_{exp} (W/m K)	k_{calc} (W/m K)
H_2O	1,000	18	1497.6	0.610	0.342
CH_3COCH_3	790	58	1174	0.190	0.105
CH_3CH_2OH	790	46	1207	0.167	0.126
$CHCl_3$	1,490	119.35	987	0.109	0.083
Hg	13,500	200.6	1450	8.69	0.377

Source: Data on density and sonic velocity obtained from Weast, R.C., ed., *Handbook of Chemistry and Physics*, 55th edn., CRC Press, Cleveland, OH, 1974.

3.6 THERMAL CONDUCTIVITY OF SOLIDS

Modern theories of the thermal conductivity of solids are about as successful as the preceding theory developed for liquids. The nature of the thermal conductivity in solids is very complicated owing to the wide variation in the molecular arrangement of solid materials, the enormous number and type of defects present in their structure, and the different mechanisms by which energy can be transported. So, while the basics of the theory are sound, matching it up with experiment is exceedingly difficult. Most modern theories make a clear distinction between energy transport in metals and dielectrics.

3.6.1 Dielectric Materials

The thermal conductivity of dielectric materials is due entirely to thermal vibrations of the crystalline lattice. In amorphous materials, like glasses, an effective lattice size, on the order of a single silicon tetrahedron in a glass, or the chain spacing in a polymer, would be used. The lattice vibrations can be regarded as a superposition of acoustic waves. If the crystal has two faces at different temperatures, heat is transferred from the hotter face to the colder face by acoustic radiation, a form analogous to transmission of energy by electromagnetic radiation.

The theory of thermal conductivity in dielectric materials is attributed to Debye [20]. His theory considered sound waves (phonons) arising from vibrations in the solid lattice to be the fundamental carriers of thermal energy. The frequency (ω) of lattice waves with velocity (v) covers a wide range, and the thermal conductivity (k), in general form, can be written in terms of a superposition of these waves:

$$k = \frac{1}{3}\int \rho C_p(\omega) v \lambda_f(\omega) d\omega \tag{3.68}$$

where

$C_p(\omega)$ is the contribution to the specific heat per frequency interval for lattice waves of that frequency

$\lambda_f(\omega)$ is the attenuation length or the mean free path for the lattice waves

The mean free path can be very long for perfect crystals like diamond or relatively short in glassy materials, for example. At temperatures in excess of 50 K, heat transfer through disordered, dielectric materials can be described by a diffusion process. The phonon mean free paths are very short (~few Å), and the temperature (T)-dependent thermal diffusivity (α) can be written as

$$\alpha(T) = \frac{k}{\rho C_p} = \frac{1}{3} v \lambda_f(T) \tag{3.69}$$

where the macroscopic density ρ, the specific heat, C_p, of the material, the transport velocity, v, of the lattice waves (or phonons), and the phonon mean free path, λ_f, are the factors that determine the thermal conductivity. The temperature dependence is very weak for ρ, C_p, and v in dielectric materials, and so the mean free path is the major factor affecting the conductivity. If λ_f is equal to the separation distance between constituent atoms, the resulting conductivity is referred to as the *minimum thermal conductivity* [21]. Conductivities in excess of the minimum arise from additional mechanisms for heat transport through the solid. For amorphous materials like glasses, the mean free path is almost constant at room temperature [22] and is limited to several interatomic spacings. The mean free path of fused silica at room temperature is 5.6 Å [23], which is about the size of an elementary silicate ring in glass [24]. The maximum wavelength of a phonon that could exist in a dielectric is always less than or equal to the characteristic dimension of the solid, generally defined as the volume divided by the surface area.

There are two regimes of heat transport governed by lattice vibrations. At low vibrational energies, the lattice vibrations are wavelike, and phonons exist with well-defined wavelengths (λ), wave vectors (κ_λ) = $2\pi/\lambda$, and velocities (v). In this regime, we can use kinetic theory-like approximations from previous sections as the basis for developing an expression for the thermal conductivity of solids. We first replace the mean velocity of the gas in Equation 3.57 by the speed of sound in the solid. The mean free path in the solid is the smallest linear dimension of the crystal. Thus, the thermal conductivity for the dielectric solid in terms of its heat capacity, density, speed of sound, and the phonon mean free path is

$$k_s = \frac{1}{3}\rho C_v c_{\lambda s} \lambda_f \tag{3.70}$$

The phonon mean free path is affected by a number of factors. The dominant factor in most solids is inelastic phonon–phonon scattering (Umklapp processes). Such scattering is absent in amorphous materials with no long-range order such as silica glass. However, various other factors cause scattering. These factors include interactions between phonons and any defects or imperfections in the materials such as interfaces, microvoids, microcracks, particles, and pores [25–27]. The presence of different size atoms and impurities also leads to increased phonon scattering. The increased scattering is due to differences in the mass of the elements, differences in the binding force of the substituted atom, and the elastic strain field around the substituted atom.

The effective mean free path is found by adding the contribution of each of the aforementioned effects as a sum of resistances in parallel (i.e., adding $1/\lambda_f$ for each mechanism). Thus, the more factors that cause anharmonicities, or the more disorder, the shorter the effective mean free path and the lower the conductivity of the material. At sufficiently high temperatures, generally above room temperature, scattering due to imperfections is independent of the temperature and frequency. If imperfections are reduced or eliminated, the thermal conductivity will increase. The disorder in a film and hence its thermal conductivity can be radically altered by changes in process conditions.

3.6.2 Metallic Solids

The thermal conductivity of metals and semiconductors depends upon two mechanisms, energy transfer by lattice vibrations or phonon conduction, and energy transfer by the motion of free electrons in the conduction bands of the metal. Phonon conduction in metals or semiconductors is essentially the same mechanism we previously discussed for dielectric materials. Since it is independent of energy transport by the motion of free electrons, the overall thermal conductivity of a metallic solid can be written as the sum of contributions from phonon and electron transport:

$$k = k_{ph} + k_e \tag{3.71}$$

At room temperature, the electric contribution to the thermal conductivity in pure metals is nearly 30 times larger than the phonon contribution [28]. The same can be said of liquid metals, hence the large conductivity of mercury. Metallic defects and impurities lead to scattering of the electrons in a manner similar to phonon scattering in dielectric solids, and the phonon and electron contributions to the thermal conductivity may become of the same order of magnitude. The problem is particularly acute in microprocessors where surface scattering in thin metal lines leads to increased resistivity and poor thermal conduction.

TABLE 3.6
Wiedemann–Franz Law Predictions (300 K)

Metal	Thermal Conductivity, k (W/m K)	Electronic Conductivity, (σ/Ω m)	k_e (W/m K)	k_{ph} (W/m K)
Copper	401	5.80×10^7	426.3	—
Gold	317	4.10×10^7	301.4	15.6
Nickel	90.7	1.28×10^7	94.1	—
Mercury	8.69	1.044×10^6	7.67	1.02
Silicon	148	1000	0.007	148

In 1853, Wiedemann and Franz used an electron gas theory of metals to develop a relationship between the electric conductivity of metals and their thermal conductivity. The Wiedemann–Franz law is still used and states that when phonon transport is negligible (at room temperature and above), the thermal conductivity can be related to the electric conductivity via

$$k_e = \mathcal{L}o\, \sigma T \quad \text{Wiedemann–Franz law} \tag{3.72}$$

where
σ is the electric conductivity
$\mathcal{L}o = 2.45 \times 10^{-8}$ W Ω/K^2 is the Lorenz constant
T is the absolute temperature

We may assume that the Wiedemann–Franz law describes the electronic contribution to the thermal conductivity, and this gives us a means for calculating the phonon contribution. Combining Equations 3.71 and 3.72, we have

$$k_{ph} = k - \mathcal{L}o\, \sigma T \tag{3.73}$$

Example 3.3 Thermal Conductivity of Common Conductors

Table 3.6 shows predictions for the thermal conductivity of conductors using the Wiedemann–Franz law compared with actual measurements. The key experimental quantity required is the electrical conductivity of the material. Using Equation 3.72, we calculate the electronic contribution to the thermal conductivity. Equation 3.73 can then be used to extract the phonon contribution from a knowledge of the electronic contribution and the actual value of the material's thermal conductivity.

For copper and gold, both excellent conductors at room temperature, the Wiedemann–Franz law is within 10% of the actual value. The prediction for Ni is also very good even though Ni is a much poorer conductor than copper or gold. The electronic contribution to the thermal conductivity for Si is only a fraction of its total conductivity. Here, we can assume that unless high doped, electron conduction is very small, and so the phonon contribution to the thermal conductivity is appreciable.

3.7 DIFFUSIVITY OF GASES [4–6]

Fick's law of diffusion is perhaps the simplest flux law, conceptually, to view in terms of the kinetic theory of gases and particle–particle collisions. The kinetic theory result is often accurate (within 5%) for monatomic gases, but as the gas structure gets more complicated, the intermolecular force fields become dominant, and the hard sphere approach of kinetic theory becomes unsuitable.

3.7.1 Kinetic Theory

The kinetic theory approximation for the diffusivity of a gas assumes we have a monatomic gas consisting of a mixture (a and a^*) of rigid, spherical, non-attracting, molecules of diameter, d_a, and mass, m_a. The molecules are present in a concentration of N molecules per volume, V. Both a and a^* molecules are identical, so we will be calculating the *self-diffusivity* of a molecules. The gas moves with a velocity v_y^* throughout the volume of interest. To determine the diffusivity, D_{aa^*}, we first calculate the flux, N_{aa^*}. The situation is shown in Figure 3.11.

We consider the motion of species a in the y-direction under the influence of a concentration gradient dx_a/dy. We assume the temperature, T, pressure, P, and the total molar concentration, c_t, of the gas are constant. The molar flux of species a, N_{ay}, across any plane of constant, y, is found by counting the number of molecules that cross a unit area of the plane in the positive y-direction and subtracting from that the number that cross the same area of the plane in the negative y-direction. We have contributions to the flux from the overall motion of the fluid at a mean molar average velocity, v_{My}, and the diffusion of species a:

$$N_{ay} = \left(\frac{\mathcal{N}}{V\mathcal{N}_{av}}\right) x_a\, v_{My}\Big|_y + \frac{1}{\mathcal{N}_{av}}\left\{\Omega_w\, x_a\Big|_{y-\delta} - \Omega_w\, x_a\Big|_{y+\delta}\right\} \tag{3.74}$$

Notice that the bulk velocity of the gas forces a molecules to cross the plane at y in unidirectional motion, whereas the *mixing* or back and forth crossing of the plane arises solely from diffusion. We now make the assumptions that all particles crossing the plane have velocities representative of where they came from and that the concentration profile is linear over the extent of Figure 3.11. A Taylor series approximation defines the mole fractions at $y + \delta$ and $y - \delta$ in terms of the mole fraction at y:

$$x_a\Big|_{y-\delta} = x_a\Big|_y - \delta\frac{dx_a}{dy} = x_a\Big|_y - \frac{2}{3}\lambda_f\frac{dx_a}{dy} \tag{3.75}$$

$$x_a\Big|_{y+\delta} = x_a\Big|_y + \delta\frac{dx_a}{dy} = x_a\Big|_y + \frac{2}{3}\lambda_f\frac{dx_a}{dy} \tag{3.76}$$

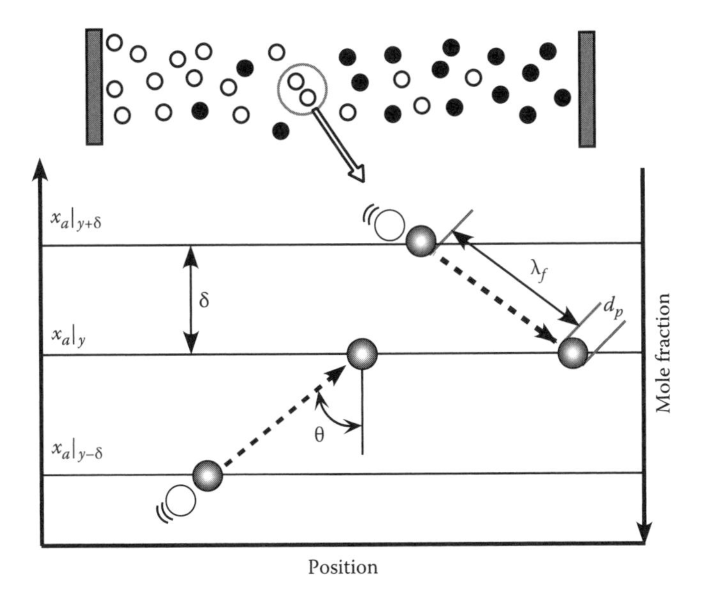

FIGURE 3.11 Collision of molecules resulting in a net diffusion of species along a concentration gradient. (Adapted from Bird, R.B. et al., *Transport Phenomena*, John Wiley & Sons, New York, 1960. With permission.)

Plugging in for Ω_w from Equation 3.21 and for the mole fractions from Equations 3.75 and 3.76, the molar flux of Equation 3.74 becomes

$$N_{ay} = x_a(N_{ay} + N_{a^*y}) - \frac{1}{3}c_t\bar{v}\lambda_f \frac{dx_a}{dy} = x_a(N_{ay} + N_{a^*y}) - c_t D_{aa^*} \frac{dx_a}{dy} \tag{3.77}$$

Since the total concentration, c_t, is constant, and only two species are moving, a and a^*, we can relate the bulk velocity to the individual fluxes by

$$c_t v_y = \mathcal{N}_{ay} + N_{a^*y} \tag{3.78}$$

Notice that Equation 3.77 is an analog of Fick's law of diffusion if the diffusion coefficient, D_{aa^*}, is written as

$$D_{aa^*} = \frac{1}{3}\bar{v}\lambda_f \qquad \text{Tracer or self diffusivity} \quad \left(\frac{\text{m}^2}{\text{s}}\right) \tag{3.79}$$

Our assumption of rigid spheres lets us use the ideal gas law $P = c_t RT = (\mathcal{N}/V)k_b T$ to rewrite the diffusion coefficient in terms of the absolute temperature, T, the mass of species, a, m_a, the molecular diameter, d_a, and the pressure, P:

$$D_{aa^*} = \frac{2}{3}\left(\frac{k_b^3}{\pi^3 m_a}\right)^{1/2}\left(\frac{T^{3/2}}{Pd_a^2}\right) \tag{3.80}$$

Like the kinetic theory representations for viscosity and thermal conductivity, this representation for diffusivity requires one experimental point to fix d_a. The result in Equation 3.80 is valid for two species of identical mass and size and so is referred to as a tracer or self-diffusivity. For species of differing mass and size, the mean free paths, velocities, and collision frequencies will be different. The derivation of the diffusivity is more complicated but gives

$$D_{ab} = \frac{2}{3}\left(\frac{k_b}{\pi}\right)^{3/2}\left(\frac{1}{2m_a} + \frac{1}{2m_b}\right)^{1/2}\left(\frac{T^{3/2}}{P\left(\frac{d_a + d_b}{2}\right)^2}\right) \qquad \text{Binary diffusivity} \tag{3.81}$$

Equations 3.80 and 3.81 predict that diffusivity varies with the inverse of the pressure and with temperature to the 3/2 power. The pressure dependence is reasonable for pressures up to about 10–20 atm. At higher pressures, multi-body collisions become important and the pressure dependence is greater. Experimental evidence shows that a temperature dependence of $T^{1.5}$ is too weak, leading us to the Chapman–Enskog approach.

Example 3.4 Binary Diffusivities and Variation with Molecular Size

Calculate the diffusivity of H_2 in He, ethane, butane, and cyclohexane at 298 K and 1 atm pressure. Compare the kinetic theory predictions with actual experimental data.

Data for the kinetic theory calculation and the comparison with experiment are shown in Table 3.7. Equation 3.81 predicts the diffusivity. As shown in the table, the calculated diffusivity values are quite a bit higher than the experimental values. The kinetic theory does seem to account fairly well for the variation of diffusivity with size and mass of the molecules.

TABLE 3.7
Comparison of Experimental and Kinetic Theory-Derived Diffusivities for H_2–Gas Pairs

Gas	Collision Diameter (m × 10^9)	Molecular Mass (kg)	Actual Binary Diffusivity (m^2/s)	Calculated Binary Diffusivity (m^2/s)
H_2	0.141	3.32×10^{-27}	1.02×10^{-4}	2.73×10^{-4}
He	0.113	6.64×10^{-27}	1.13×10^{-4}	2.91×10^{-4}
C_2H_6	0.222	4.98×10^{-26}	5.37×10^{-5}	1.20×10^{-4}
C_4H_{10}	0.234	9.63×10^{-26}	3.61×10^{-5}	1.11×10^{-4}
C_6H_{14}	0.309	1.395×10^{-25}	3.19×10^{-5}	7.67×10^{-5}

3.7.2 Chapman–Enskog Theory

To predict the diffusivity more accurately, one must take into account the force field between molecules and their finite diameter. Chapman–Enskog theory provides for a more accurate temperature dependence than kinetic theory. In this case, we find the following expression for self- and binary diffusivities, respectively:

$$c_t D_{aa^*} = 3.203 \times 10^{-31} \frac{\sqrt{T/M_{wa}}}{\sigma_f^2 \Omega_D} \left(\frac{\text{kg-mol}}{\text{ms}} \right) \tag{3.82}$$

$$c_t D_{ab} = 2.265 \times 10^{-31} \frac{\sqrt{T\left(\dfrac{1}{M_{wa}} + \dfrac{1}{M_{wb}}\right)}}{\sigma_f^2 \Omega_D} \left(\frac{\text{kg-mol}}{\text{ms}} \right) \tag{3.83}$$

The collision function, Ω_D, while not completely analogous to that for the viscosity and thermal conductivity, is very close in magnitude to those quantities. If we approximate the total concentration, c_t, using the ideal gas law, we can rewrite the diffusivity in terms of the temperature and pressure (in atmospheres):

$$D_{ab} = 1.88 \times 10^{-22} \frac{\sqrt{T^3((1/M_{wa}) + (1/M_{wb}))}}{P \sigma_f^2 \Omega_D} \left(\frac{\text{m}^2}{\text{s}} \right) \tag{3.84}$$

$$\sigma_{fa} = \frac{\sigma_{f1} + \sigma_{f2}}{2} \qquad \Omega_D = f\left(\frac{\sqrt{\varepsilon_1 \varepsilon_2}}{k_b} \right) \tag{3.85}$$

Chapman–Enskog theory also gives an inverse relationship between diffusivity and pressure, but temperature dependence is more accurate.

In the previous section dealing with thermal conductivity, we established a relationship between μ and k. An analogous relationship between μ and D_{aa^*} exists and is given by

$$\frac{\mu}{\rho D_{aa^*}} = \frac{\nu}{D_{aa^*}} = \frac{5}{6} \frac{\Omega_D}{\Omega_\mu} \tag{3.86}$$

The quantity, $\mu/\rho D_{aa^*}$, is better known as the Schmidt number, Sc, and is the analog of the Prandtl number for heat transfer. The relationship in (3.86) shows that self-diffusivity and kinematic viscosity are of the same order of magnitude for gases at low pressures. The relationship between ν and D_{ab} is not so simple because kinematic viscosity may vary with composition, whereas the Schmidt number remains in a relatively restricted range for most gas pairs.

3.8 DIFFUSION IN LIQUIDS

The diffusivity of liquid particles is generally much smaller than that of gaseous particles. The difference may be 4 or 5 orders of magnitude or more. Table 3.8 shows the diffusivity of several gases in liquid water and water vapor. A general rule of thumb is that gaseous diffusivities are in a range between 10^{-3} and 10^{-5} m²/s, while liquid phase diffusivities lie in a range of 10^{-9}–10^{-10} m²/s. Table 3.9 shows how the interaction between solvent and solute affects the diffusivity in liquids. These interactions are quite complicated and involve the viscosity of the solvent and the intermolecular interactions between the solvent and solute molecules. The problem is particularly acute with water molecules whose hydrogen bonding interactions can greatly affect its diffusivity in different

TABLE 3.8
Diffusivity of Gases in Water and Water Vapor

Solute	Gaseous Diffusivity ($\times 10^4$ m²/s)	Liquid Diffusivity ($\times 10^9$ m²/s)
CH_4	0.292	1.49
CO_2	0.202	1.92
H_2	0.915	4.50
C_2H_4	0.204	1.87

Sources: Data from Cussler, E.L., *Multicomponent Diffusion*, Elsevier, Amsterdam, the Netherlands, 1976; Cussler, E.L., *Diffusion: Mass Transfer in Fluid Systems*, University Press, Cambridge, U.K., 1984; Sherwood, T.K. et al., *Mass Transfer*, McGraw-Hill, New York, 1975.

TABLE 3.9
Diffusivity of Water in Various Solvents

Solvent	Liquid Diffusivity ($\times 10^{-9}$ m²/s)
Acetone	4.56
n-Butyl alcohol	0.56
Ethanol	1.24
Ethyl acetate	3.2

Source: Selected data from Reid, R.C. et al., *The Properties of Gases and Liquids*, 4th edn., McGraw-Hill, New York, 1987.

solvents or diffusivities of substances in water. Though the diffusivities of liquids and gases are much different, the rates of mass transfer in both media may be comparable. Since the density of a liquid is about one thousand times that of a gas, the flux of particles in the liquid may be high even though the diffusivity of an individual particle is low. Unlike the kinetic theory of gases or the Chapman–Enskog theory, there is no rigorous theory of diffusivity in liquids. However, there are two commonly employed approximate theories that are accurate to about 50% or better.

The hydrodynamic theory of diffusivity, attributed to Stokes (1850) and Einstein (1905), considers the particles to be spheres moving through a continuous fluid. This model presumes that the diffusing particle is much larger than the particles making up the medium through which it is moving. As such, the model works best when analyzing the diffusion of large objects like micelles, colloids, globular proteins, or polymer particles. Still, it remains the basis for estimating molecular diffusion coefficients and is the standard upon which all other theoretical models are compared.

The physical situation is shown in Figure 3.12. The actual structure of the liquid is a close-packed, yet random, aggregation of molecules in continuous motion. A tagged molecule diffuses through this fluid by a random walk. We expect the velocity of the particle to be proportional to the force that is exerted on it so that the particle moves at a constant velocity called the *terminal velocity* (force ∝ particle velocity). The constant of proportionality we define as a coefficient of friction, f:

$$F = f \cdot v \tag{3.87}$$

If we assume that the net velocity of the particle is very slow, we can use the results of Stokes' solution for creeping flow about a sphere to define the coefficient of friction. There are two cases to consider. The first case assumes that the fluid molecules stick to the surface of the diffusing particle (Stokes' no-slip formulation), and the second considers the particle to *slip* past the fluid molecules (an extension attributed to Lamb [32] and perhaps more appropriate when all particles are of comparable size):

$$f = 6\pi r_p \mu \quad \text{Sticking molecules} \tag{3.88}$$

$$f = 4\pi r_p \mu \quad \text{Slipping molecules} \tag{3.89}$$

In ordinary diffusion, we know from Fick's law that the flux of matter is in response to a gradient in the concentration of the species. We can think of this in terms of a hypothetical force giving the particles a push. Einstein defined this force as the gradient in the chemical potential of the diffusing species:

$$\vec{\mathbf{F}} = -\vec{\nabla}\mu^c \tag{3.90}$$

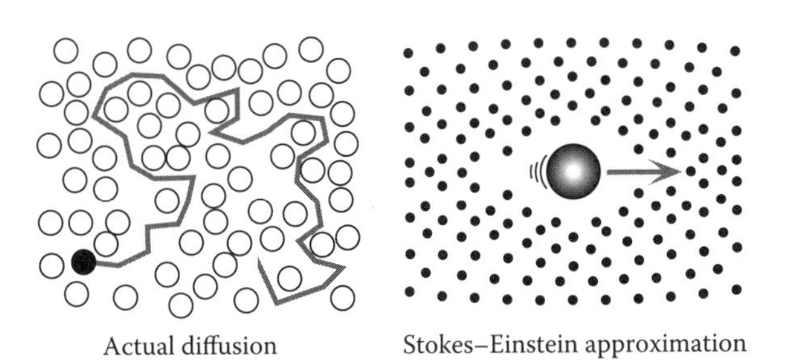

FIGURE 3.12 Diffusion in a liquid: actual process and Stokes–Einstein hydrodynamic model.

If the solution is dilute, it can be considered ideal, and the chemical potential is related to the concentration or mole fraction of the diffusing species:

$$\mu^c = \mu_o^c + k_b T \ln(x_p) = \mu_o^c + k_b T \ln\left(\frac{c_p}{c_t}\right) \tag{3.91}$$

where

c_p is the particle concentration
c_t is the total concentration of all species

In the dilute solution, $c_t >> c_p$, so we can assume c_t is constant. The gradient in chemical potential, and hence the force acting on the diffusing particle, is

$$F = -\frac{d\mu^c}{dx} = -\left(\frac{k_b T}{c_p}\right)\frac{dc_p}{dx} \tag{3.92}$$

We can combine this expression for the applied force, with Equation 3.87 defining the velocity, v, and our definition for the molar flux of particles, $J_p = c_p v$, to obtain the flux in terms of the concentration gradient, the viscosity of the fluid, the particle size, and the absolute temperature:

$$J_p = -D_p \frac{dc_p}{dx} = -\left(\frac{k_b T}{6\pi r_p \mu}\right)\frac{dc_p}{dx} \quad \text{No-slip formulation} \tag{3.93}$$

$$J_p = -D_p \frac{dc_p}{dx} = -\left(\frac{k_b T}{4\pi r_p \mu}\right)\frac{dc_p}{dx} \quad \text{Slip allowed} \tag{3.94}$$

It is clear from these equations that the diffusion coefficient is defined by

$$D_p = \frac{k_b T}{6\pi r_p \mu} \quad \text{or} \quad \frac{k_b T}{4\pi r_p \mu} \quad \text{Stokes Einstein} \tag{3.95}$$

Equation 3.95 is the Stokes–Einstein equation for diffusivity and is rarely accurate to better than 20%. Figure 3.13 shows how diffusivity varies with solvent viscosity in an actual system.

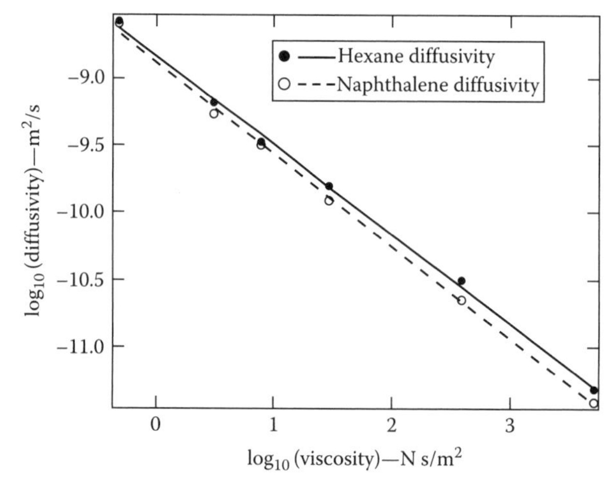

FIGURE 3.13 Viscosities of small solutes in high viscosity solvents. (From Hiss, T.G. and Cussler, E.L., *AIChE J.*, 19, 698, 1973.)

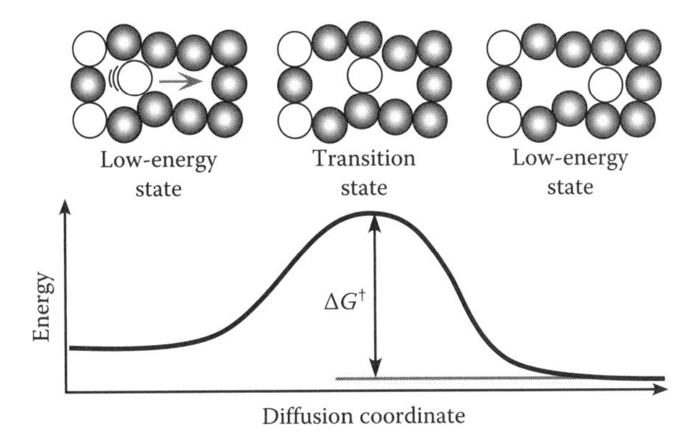

FIGURE 3.14 Liquid diffusion—Eyring rate theory.

The diffusivities of *n*-hexane and naphthalene were measured in various hydrocarbon solvents of known viscosity. The diffusivity is not inversely proportional to the viscosity as predicted by the Stokes–Einstein equation but is proportional to $\mu^{-2/3}$. The reason behind this is that instead of having a large solute molecule diffusing in a sea of small solvent molecules, we have small solute molecules diffusing in a sea of large solvent molecules.

The second major, historical model for liquid diffusivity can simulate this observed behavior a bit better because its dependence upon the viscosity can be adjusted depending upon the exact model one assumes for the liquid state. The model is based on the rate theory of Eyring and coworkers. The concepts behind the rate theory for diffusivity are essentially the same as those we used to develop the rate theory for the viscosity of liquids. The free volume in the liquid redistributes itself until there exists an opening or hole large enough to accommodate the diffusing particle. At this point, the particle must jump over an energy barrier separating it from the free space. The situation is illustrated in Figure 3.14.

Applying the rate theory concept, we view the motion of the particles as a chemical reaction. The rate of this reaction is a function of the activation energy for the jumps. The presence of the concentration gradient, like the presence of the velocity gradient we discussed when we considered liquid viscosity, distorts the activation energy. The activation energy is lowered in the direction of decreasing concentration, and so the frequency of jumps in that direction is increased. To be rigorous, we express the distortion of the activation energy in terms of the gradient in chemical potential or Gibbs energy. We can express the frequency of forward and reverse jumps in the following form where ℓ is the jump distance (a mean free path analog), and the factor of 1/2 results from half the particles traveling to the right and half to the left:

$$\omega_f = \omega_o \exp\left[-\frac{\Delta G^\dagger - \frac{1}{2}\ell\nabla G^\dagger}{k_b T}\right] \tag{3.96}$$

$$\omega_r = \omega_o \exp\left[-\frac{\Delta G^\dagger + \frac{1}{2}\ell\nabla G^\dagger}{k_b T}\right] \tag{3.97}$$

The net frequency of jumps, $\omega_{net} = \omega_f - \omega_r$, is

$$\omega_{net} = \omega_o \exp(-\Delta G^\dagger)\left[\exp\left(\frac{\ell\nabla G^\dagger}{2k_b T}\right) - \exp\left(-\frac{\ell\nabla G^\dagger}{2k_b T}\right)\right] = 2\omega_o \exp(-\Delta G^\dagger)\sinh\left[\frac{\ell\nabla G^\dagger}{2k_b T}\right] \tag{3.98}$$

In general, the gradient in chemical potential or Gibbs energy is very small compared to the thermal energy of the particles. Therefore, sinh $(x) \approx x$, and the net jump frequency is

$$\omega_{net} = 2\omega_o \exp(-\Delta G^\dagger)\left[\frac{\ell \nabla G^\dagger}{2k_b T}\right] \tag{3.99}$$

In a dilute, ideal solution, we relate the gradient in Gibbs energy to the concentration gradient:

$$\nabla G^\dagger = -\frac{k_b T}{c_p}\frac{dc_p}{dx} \tag{3.100}$$

Substituting into Equation 3.99, we obtain

$$\omega_{net} = -\frac{\ell\omega_o}{c_p}\exp(-\Delta G^\dagger)\frac{dc_p}{dx} \tag{3.101}$$

The net velocity of the diffusing particles is just the net jump frequency multiplied by the jump distance:

$$v = \ell\omega_{net} \tag{3.102}$$

This translates into a net particle flux of

$$J_p = \ell\omega_{net}c_p = -\ell^2\omega_o \exp(-\Delta G^\dagger)\frac{dc_p}{dx} \tag{3.103}$$

If we compare this to Fick's law, we immediately see that the particle diffusivity is given by

$$D_p = \ell^2\omega_o \exp(-\Delta G^\dagger) \tag{3.104}$$

Finally, if we assume that the jump distance, ℓ, intrinsic frequency, ω_o, and activation energy, $\Delta G^\dagger$, for diffusion and viscosity are essentially the same, we can express the diffusivity in terms of the viscosity:

$$D_p = \frac{k_b T}{2\mu r_p} \tag{3.105}$$

Though similar to the Stokes–Einstein equation, the rate theory expression is generally less accurate. The rate theory implicitly assumes there is some structure, a lattice that can be used to describe the liquid state. Such a structure does not truly exist, but the ability to postulate and *tweak* the structure makes the rate model a more flexible tool than the Stokes–Einstein model. As a result, even though generally less accurate, the rate model can be made to fit the data of Figure 3.13 [33]; Stokes–Einstein cannot.

The similarity between the two expressions reinforces the idea that the fundamental relationship between diffusivity and viscosity expressed by the Stokes–Einstein equation is correct and that all empirical relationships should start with the Stokes–Einstein relation as a basis [34].

3.9 DIFFUSION IN SOLIDS

Diffusion in solids is complicated due to the many diffusion mechanisms present and the wide range of solid forms that exist. Ideally, diffusion takes place as a random walk within a crystal lattice. Since there are 32 different crystal classes, there would be 32 different random walks and 32 different diffusivities. Any defects that may exist in the crystal, any impurities, dopants, or alloys will change the nature of the random walk and hence the expression for the diffusivity. Add to this the proliferation of semicrystalline solids and amorphous solids, and we have still more representations for the diffusivity. As examples, diffusion coefficients in a variety of solid systems are presented in Table 3.10. The wide variability in diffusion coefficient shows that a general theory for solid-state diffusion would be extremely hard to develop.

We can separate the diffusion mechanisms into two categories, bulk diffusion and surface diffusion. Two predominant mechanisms exist in bulk diffusion, and these are shown in Figure 3.15. In the interstitialcy mechanism, the diffusing species occupies a space between atoms making up the lattice. Diffusion occurs when the diffusing species hops between these spaces. This mechanism allows for diffusion within a perfect crystal. The vacancy mechanism requires a defect in the solid. Diffusion occurs as the diffusing species hops from one defect site to another.

Surface diffusion and grain boundary diffusion in solids are the other predominant mechanisms of diffusive transport. They are important because they may be the dominant transport

TABLE 3.10
Diffusivities in the Solid State

System	Temperature (°C)	Diffusivity (m^2/s)
Hydrogen in iron	50	1.14×10^{-12}
Hydrogen in SiO_2	200	6.5×10^{-14}
Cadmium in copper	20	2.7×10^{-19}
Aluminum in copper	20	1.3×10^{-34}
Aluminum in copper	850	2.2×10^{-13}
Silver in aluminum	50	1.2×10^{-13}
Gold in lead	285	4.6×10^{-10}
Uranium in tungsten	1727	1.3×10^{-15}

Sources: Data extracted from Barrer, R.M., *Diffusion in and through Solids*, MacMillan, New York, 1941; Shewmon, P., *Diffusion in Solids*, 2nd edn., Minerals, Metals and Materials Society, Warrendale, PA, 1989; Kingery, W.D. et al., *Introduction to Ceramics*, 2nd edn., John Wiley & Sons, New York, 1976.

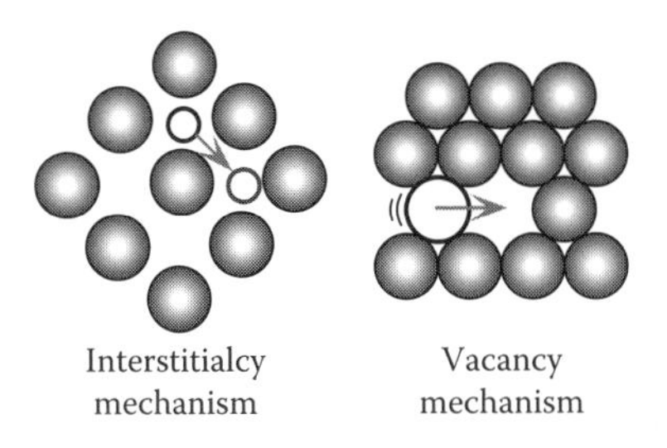

FIGURE 3.15 Bulk diffusion mechanisms in solids. Example shows face-centered cubic crystal structure.

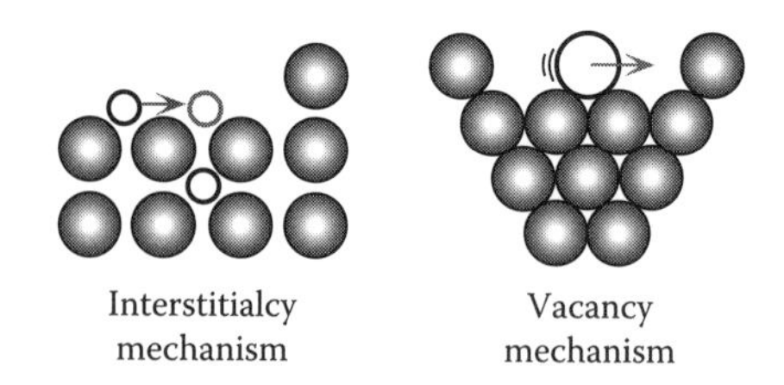

FIGURE 3.16 Mechanisms of surface diffusion. Lattice distortion is minimal when compared to bulk diffusion.

mechanisms in materials processing operations such as sintering. Surface diffusion, shown in Figure 3.16, may occur at a solid–liquid or solid–gas interface, and grain boundary diffusion may occur at the interface between different grain or crystallite boundaries in a solid. Since the surface area involved in these transport mechanisms may be extremely large, mass transfer rates may be high even though the diffusivities are low. Both surface and grain boundary diffusions are referred to as low energy processes because less displacement of the crystal lattice(s) must occur for particles to diffuse.

The Eyring rate theory of diffusion in liquids can be applied to bulk and surface diffusion in solids. The rate theory works much better when applied to solids than it did for liquids. The lattice structure required in rate theory actually exists in solids leading to the improved results. The pre-exponential factor and enthalpy of activation are usually fit to experimental data:

$$D_s = \lambda_f^2 \omega_o \exp\left[-\frac{\Delta H}{RT}\right] \tag{3.106}$$

3.10 CONDUCTIVITY, MOBILITY, AND RESISTIVITY

The conductivity of a material depends on the movement of charge carriers. These charge carriers may be ions, electrons, or holes, depending upon the nature of the conducting medium. Their electric interactions complicate the transport process and the nature of their transport coefficients. Charged particles respond to two forces: the electric force resulting from an applied or induced electric field, and the virtual, chemical potential force that we discussed in the last section on diffusion. These two responses force us to deal with two transport properties simultaneously: the diffusivity in a concentration gradient and the mobility in an electric field. From a phenomenological point of view, we can treat the motion of charged species in the solid, liquid, or gaseous state using a common model. We will concentrate on ions in liquid solution and electrons or holes in solids, but keep in mind that the concepts we develop can be applied to charged species in gases (plasmas), liquids, or solids.

3.10.1 Ionic Mobility and Conductivity of Solutions

An ion in solution is subject to a number of forces. The first force results from the random bombardment the ion gets from solvent molecules. Since this bombardment is coming from all directions, the ion remains essentially motionless in a uniform solution. The second force is attributed to the chemical potential gradient. If there are concentration gradients, the ions will spread from regions of high concentration to regions of lower concentration much as a neutral molecule will. Finally, since ions are charged species, they also respond to electric forces.

Figure 3.17 shows a cell consisting of two electrodes in an ionic solution. There is a potential gradient, $\Delta\Phi/L$, set up between the two electrodes. Positive ions slide down the potential gradient and negative ions climb up. We view the movement of these ions as being generated by an applied

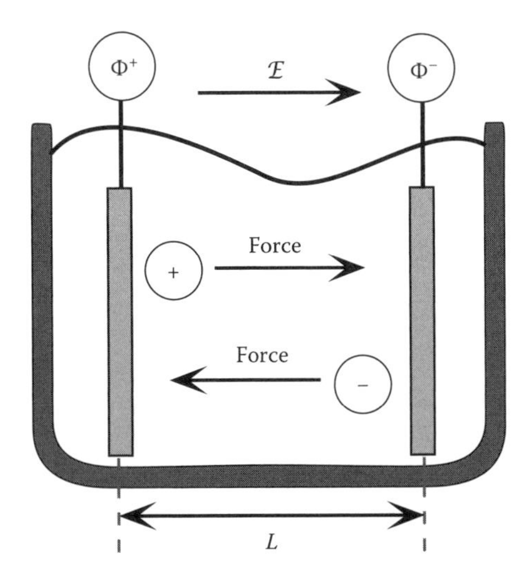

FIGURE 3.17 A positively charged (cation) and negatively charged (anion) ion in the presence of an electric field.

force, F. This force is proportional to the potential gradient, or the electric field strength, E, and is also proportional to the charge on the ion:

$$F = z_e e \mathcal{E} \tag{3.107}$$

In Equation 3.107, z_e is the valence of the ion, e is the fundamental charge on an electron, and E is the electric field strength. Since the electric field strength is the negative of the potential gradient

$$\mathcal{E}_x = -\frac{d\Phi}{dx} \tag{3.108}$$

$$F = -z_e \mathcal{E}\frac{d\Phi}{dx} = -\frac{z_e \mathcal{E}\Delta\Phi}{L} \tag{3.109}$$

Notice that if $\Delta\Phi$ is negative, the force on the positive ions (cations) pushes them from left to right, whereas the negative ions (anions) get moved from right to left. In solid-state conductors or semiconductors, the cations represent holes, and the anions represent electrons.

The force acting on the ions gives them an acceleration. As the ions move through the liquid, a frictional force, drag, retards them. In a solid, scattering off of atoms in the lattice provides the same drag force on electrons or holes. Eventually, the ions will reach a terminal velocity such that the frictional force and the electric force are equal. This terminal velocity is termed the drift velocity, v_d. If we assume, by analogy with diffusion that the frictional force can be given by Stokes' law (only valid for large ions or particles), $F_f = 6\pi\mu v r_\pm$, where $r_\pm$ is the radius of an ion, v is its velocity, and μ is the viscosity of the liquid, then we may combine Equation 3.107 with Stokes' law and define the drift velocity as

$$z_e e \mathcal{E} = 6\pi\mu r_\pm v_d \tag{3.110}$$

$$v_d = \frac{z_e e \mathcal{E}}{6\pi\mu r_\pm} \quad \text{Drift velocity} \tag{3.111}$$

TABLE 3.11
Infinite Dilution Ionic Conductivities (25°C)

Cation	Conductivity ($\times 10^2$ m^2/Ω·mol)	Anion	Conductivity ($\times 10^2$ m^2/Ω·mol)
H^+	3.50	OH^-	1.98
Li^+	0.387	Cl^-	0.763
Na^+	0.501	Br^-	0.784
K^+	0.735	I^-	0.768
Mg^{2+}	1.06	NO_3^-	0.714
Ca^{2+}	1.19	CO_2^{-3}	0.409
Ba^{2+}	1.27	SO_4^{2-}	1.60

Source: Data on mobility obtained from Harned, H.S. and Owen, B.B., *The Physical Chemistry of Electrolytic Solutions*, ACS Monograph, Washington, BC, Vol. 95, 1950.

The drift velocity governs the rate at which current may be conducted in a solution or in a solid. Based on our analysis, we expect that the conductivity of a solution will decrease with increasing viscosity and increasing ionic size.

The experimental evidence supports the first assertion but not the second. The conductivity of a gelatin solution, for instance, decreases as the solution sets. The experimental data in Table 3.11 show that in some instances, replacing Li^+ for K^+, for example, the conductivity actually increases though the ionic size is much larger. The discrepancy between ionic size and solution conductivity arises because smaller ions tend to have lower drift velocities. Every ion in solution is surrounded by a set of associated solvent molecules. The solvent molecules are attracted to the ions by virtue of the ions' electric field, and so when the ions move, the associated solvent molecules move. We can view the ion and solvent molecules as a composite particle and define a hydrodynamic radius, r_h, for purposes of calculating the conductivity. The magnitude of the hydrodynamic radius is governed by the strength of the electric field surrounding the ion. This field is stronger for smaller ions, so the smaller ions have larger hydrodynamic radii and slower drift velocities. The hydrodynamic radius can be estimated if we know how many solvent (water) molecules are associated with each ion:

$$r_h = \left[r_\pm^3 + \frac{3n_h}{4\pi} \left(\frac{\overline{V_{H_2O}}}{\mathcal{N}_{av}} \right) \right]^{1/3} \tag{3.112}$$

where

$r_\pm$ is the ionic radius
n_h is the hydration number
$\overline{V_{H_2O}}$ is the molar volume of water

This analysis applies to all ions with the exceptions of protons, hydronium ions, or hydroxyl ions. With these ions, transport is very rapid because the ions react with water molecules and transfer their charge from one water molecule to the next. Since water is dense, and the mean free path between molecules is very small, the drift velocities for these *reactive* charges are very high.

The drift velocity of an ion is proportional to the magnitude and direction of the applied electric field. We generally prefer to deal with a transport property that is independent of

TABLE 3.12
Ionic Mobilities and Diffusivities at Infinite Dilution in Aqueous Solution (25°C)

	Mobility ($m^2/s \cdot V$) × 10^8	Diffusivity (m^2/s) × 10^9
Cation		
H^+	36.30	9.31
Li^+	4.01	1.03
Na^+	5.19	1.33
K^+	7.62	1.96
Rb^+	7.92	2.07
NH_4^+	7.60	1.96
Ca^{2+}	6.16	0.79
La^{3+}	7.21	0.62
Anion		
OH^-	20.50	5.28
F^-	5.70	1.47
Cl^-	7.91	2.03
Br^-	8.13	2.08
I^-	7.95	2.05
CO_3^{2-}	7.46	0.92
SO_4^{2-}	8.25	1.06
$CH_3CO_2^-$	4.23	1.09

Source: Data obtained from Harned, H.S. and Owen, B.B., *The Physical Chemistry of Electrolytic Solutions*, ACS Monograph, Washington, BC, Vol. 95, 1950.
Note: Diffusivities calculated using the Einstein relation.

such external forces, so we relate the drift velocity to the electric field by defining a new transport property, the mobility, μ_e:

$$v_d = \mu_e \mathcal{E} = -\mu_e \frac{d\Phi}{dx} \tag{3.113}$$

The mobility is defined as the speed of the ion when a field of unit strength is applied. Mobilities in liquids are generally in the range of 10^{-8} m²/sV. Their usefulness lies in the fact that they can be related to measurable quantities such as the conductivity of the solution. Table 3.12 shows ionic mobilities and diffusivities in aqueous solution.

Consider how we would express the flux of ions across any given plane. In Figure 3.18, we assume we have a uniform solution of an ionic solvent in water. A drift current in the solution is established as ions are attracted to the oppositely charged electrodes. The flux of positive ions across the membrane (1/2 the current density) can be written as

$$J^+ = \upsilon^+ z_{e+} e \mathcal{N}_{av} v_{d+} c_p \tag{3.114}$$

where

c_p is the concentration of solute in the tank
υ^+ is the stoichiometric coefficient for the cations in the solute
z_{e+} is the valence of the cation

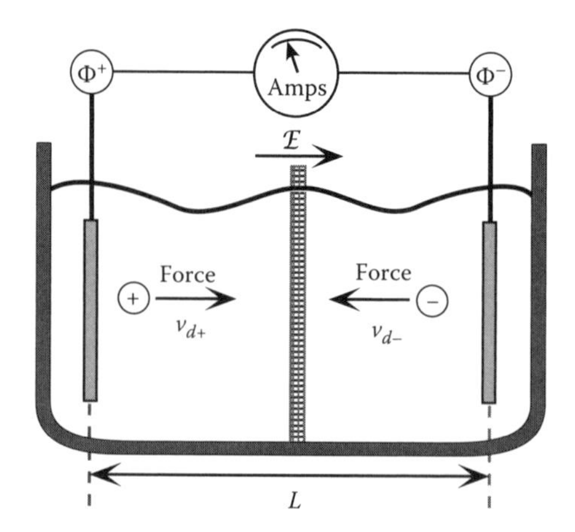

FIGURE 3.18 Ionic current flowing in an aqueous solution. A permeable membrane separates two halves of the solution.

Negative ions flow in the opposite direction and form the second half of the total current density. Their flux is given by

$$J^- = \upsilon^- z_{e-} e \mathcal{N}_{av} v_{d-} c_p \tag{3.115}$$

The total flux is given by the sum of cation and anion currents:

$$\begin{aligned} J = J^+ + J^- &= (\upsilon^+ z_{e+} v_{d+} + \upsilon^- z_{e-} v_{d-}) e \mathcal{N}_{av} c_p \\ &= (\upsilon^+ z_{e+} \mu_{e+} + \upsilon^- z_{e-} \mu_{e-}) \mathcal{F}_a \mathcal{E} c_p \\ &= -(\upsilon^+ z_{e+} \mu_{e+} + \upsilon^- z_{e-} \mu_{e-}) \mathcal{F}_a c_p \frac{d\Phi}{dx} \end{aligned} \tag{3.116}$$

where $\mathcal{F}_a$ is Faraday's constant. Notice that J represents a charge flux, not a mass flux. We determine the current flowing through this solution by multiplying the current density, J, by the area of the electrodes, $I = JA_c$:

$$I = -(\upsilon^+ z_{e+} \mu_{e+} + \upsilon^- z_{e-} \mu_{e-}) \mathcal{F}_a c_p A_c \frac{\Delta\Phi}{L} \tag{3.117}$$

L represents the distance between the electrodes shown in Figure 3.18. We can also express the current in terms of Ohm's law:

$$I = \frac{\Delta\Phi}{\mathcal{R}_e} = \sigma \Delta\Phi \frac{A_c}{L} \tag{3.118}$$

We can derive an identity using Equations 3.117 and 3.118 that relates the conductivity of the solution to the mobilities of the cations and anions:

$$\sigma = (\upsilon^+ z_{e+} \mu_{e+} - \upsilon^- z_{e-} \mu_{e-}) \mathcal{F}_a c_p \tag{3.119}$$

The mobility and diffusivity should be related in some fashion. We can use Einstein's view that the gradient in the chemical potential of the species can be thought of as a force pushing the ions down that gradient. We can then relate this force to the flux of matter and derive a relationship between the diffusivity and mobility. Assuming an ideal solution, the chemical potential is defined by

$$\mu^c = \mu_o^c + RT \ln c_p \tag{3.120}$$

and the virtual force pushing the molecule is

$$F = -\frac{d\mu^c}{dx} = -\left(\frac{RT}{c_p}\right)\frac{dc_p}{dx} \tag{3.121}$$

The flux of molecules, J, is composed of the product of the number of molecules moving across a plane and their average drift velocity. This is how we defined the flux in Chapter 2:

$$J = v_d c_p = -D_\pm \frac{dc_p}{dx} \tag{3.122}$$

Solving for the drift velocity from Equation 3.122 and substituting in for dc_p/dx from Equation 3.121 let us write the drift velocity in terms of the hypothetical driving force, F:

$$v_d = -\left(\frac{D_\pm}{c_p}\right)\frac{dc_p}{dx} = \left(\frac{D_\pm}{k_b T}\right)F \tag{3.123}$$

Thus, in response to a unit force, the drift velocity is equal to $D_\pm/k_b T$.

We know that the mobility is also related to a force and that force is related to the electric field strength:

$$v_d = \mu_e \mathcal{E} = \left(\frac{\mu_e}{z_e \mathcal{E}}\right) z_e e \mathcal{E} = \left(\frac{\mu_e}{z_e \mathcal{E}}\right)F \tag{3.124}$$

Therefore, the drift speed under a unit electric force is given by $\mu_e/z_e e$. In reality, the nature of the driving force is immaterial. In a situation where an electric field induces a force that drives a concentration gradient, we can identify the two drift speeds and equate them to give

$$\frac{\mu_e}{z_e \mathcal{E}} = \frac{D_\pm}{k_b T} \tag{3.125}$$

Rearranging Equation 3.125 achieves our goal of relating the ionic mobility and diffusivity. This relationship is termed the Einstein relation:

$$D_\pm = \frac{\mu_e k_b T}{z_e \mathcal{E}} \quad \text{or} \quad \mu_e = \frac{D_\pm z_e \mathcal{F}_a}{RT} \quad \text{Einstein relation} \tag{3.126}$$

Finally, we can use the Einstein relation to relate the conductivity of the material to its ionic diffusivities. Substituting for the mobility in Equation 3.119 gives

$$\sigma = \left[\frac{\upsilon^+ z_{e+}^2 e D_+}{k_b T} - \frac{\upsilon^- z_{e-}^2 e D_-}{k_b T}\right]\mathcal{F}_a c_p \tag{3.127}$$

Multiplying top and bottom of (3.127) by Avogadro's number and dividing through by the concentration of charge carriers, c_p, give the molar conductivity of the solution. This relationship is often termed the Nernst–Einstein relation:

$$\sigma_m = \sum_{i=1}^{n} \left(\upsilon_i \left| z_{ei} \right|^2 D_i \right) \frac{\mathcal{F}_a^2}{RT} \quad \text{Nernst–Einstein relation} \tag{3.128}$$

Both the Nernst–Einstein relation and the Einstein relation are based on the assumption of ideal solutions. Therefore, they are applicable only in dilute solutions. As the concentration of charged species increases, the electrostatic interactions between the ions and solvent and between individual ions become more involved (a many-body problem), and the solutions become less ideal. Any relationship between the mobility and diffusivity soon disappears, and Equations 3.126 and 3.128 no longer apply.

3.10.2 Charge Mobility, Conductivity, and Resistivity in Solids

The analysis we completed for the mobility of ions in liquids can also be used to treat the motion of charge carriers in solids. We need only replace cations by holes and anions by electrons to deal with metallic conductors or semiconductors. For conduction in salts, like NaCl, or conduction and diffusion in dielectrics, like glasses, the ionic analysis is directly applicable.

In metallic conductors, like copper, the charge carriers are electrons, and they are not bound to individual copper atoms, but are free to move about the lattice. These electrons are envisioned to occupy a conduction band in the solid. Each metal atom contributes a certain number of its valence electrons to this conduction band. Table 3.13 lists the number of such *free* electrons per atom in several metallic conductors. Negative values indicate that the charge carriers *appear* to be holes (vacancies in the valence band) rather than electrons. We can express the mobility and conductivity of these metals using a model similar to the kinetic theory of gases. We assume that the electrons, or holes, travel through the metal lattice with an average velocity, $\bar{v}$, colliding with the core of the metal atoms and executing a random walk throughout the metal. The average velocity for electrons in copper, for example, is on the order of 10^6 m/s. In keeping with the kinetic theory analogy, we

TABLE 3.13
Number of Free Electrons per Metal Atom

Metal	Free Electrons per Atom
Na	0.99
K	1.1
Cu	1.3
Ag	1.3
Al	3.5
Be	−2.2
Zn	−2.9
Cd	−2.5

Source: Data from Eisborg, R. and Resnick, R., *Quantum Physics*, 2nd edn., John Wiley & Sons, New York, 1985.

must also define a mean free path, λ_f, between collisions. In copper, the mean free path is on the order of 4×10^{-8} m or 200 atomic diameters.

When we apply an electric field to the metal, the electrons alter their random motion and begin to drift along the potential gradient with a drift velocity, v_d, that is much less than the mean velocity, $\bar{v}$. The drift velocity can be determined by the magnitude of the applied electric field, the mean velocity, and the mean free path between collisions. The applied electric field exerts a force on each electron and gives the electron an acceleration, a. Using Newton's second law, that acceleration can be written as

$$a = \frac{e\mathcal{E}}{m_e} \tag{3.129}$$

where m_e is the mass of an electron. At each collision with the stationary lattice atoms, the electron changes direction and begins to move again under the force exerted by the field. We define the drift velocity in terms of the field-induced acceleration and mean time between collisions:

$$v_d = a\left(\frac{\lambda_f}{\bar{v}}\right) = \frac{e\mathcal{E}\lambda_f}{m_e\bar{v}} \tag{3.130}$$

The drift velocity determines the flux of charge carriers across any plane in the metal and hence determines the current density flowing through the metal. The current is defined by the same relationship we used in ionic transport:

$$J = \left(\frac{\mathcal{N}}{V}\right)ev_d \tag{3.131}$$

where

$\mathcal{N}$ is the number of charge carriers
V is the volume of metal

Combining Equation 3.130 with Equation 3.131 and realizing that the conductivity of the metal is just the current density divided by the electric field (Ohm's law)

$$\sigma = \frac{J}{\mathcal{E}} \tag{3.132}$$

gives a relationship between the conductivity, average electron velocity, and mean free path:

$$\sigma = \frac{m_e\bar{v}}{e^2\lambda_f}\left(\frac{V}{\mathcal{N}}\right) \tag{3.133}$$

The mobility is defined as the drift velocity per unit electric field, $v_d/\mathcal{E}$, and so we define a relationship between the mobility and electron velocity as we did for the conductivity and drift velocity:

$$\mu_e = \frac{e\lambda_f}{m_e\bar{v}} \tag{3.134}$$

TABLE 3.14
Effective Masses of Charge Carriers in Semiconducting Materials

Material	Effective Mass of an Electron, m_{eff} (m_e)	Effective Mass of a Hole, m_{heff} (m_e)
Si	0.26	0.38
Ge	0.35	0.56
GaAs	0.068	0.50

Source: Data obtained from Sze S.M., *Physics of Semiconductor Devices*, Wiley-Interscience, New York, 1969.

So far, we have assumed that one species carries all the current. In many cases such as semiconductors, both positive and negative charges are involved in transporting the current.

The conductivity can then be expressed in terms of the mobilities of both charge carriers, similar to Equation 3.119:

$$\sigma = \mathcal{N}_{+}\mu_{e+} + \mathcal{N}_{-}\mu_{e-} \tag{3.135}$$

The relationships we have just derived suffer from the same limitations as the kinetic theory expressions we derived for the diffusivity, thermal conductivity, and viscosity of gases. They do not consider any interaction between the electrons and the atoms with which they collide. As such, they work reasonably well for true metallic conductors such as silver or copper but break down seriously for semiconductor materials and metals with high resistivities. In these instances, we speak of an effective electron mass, m_{eff}, or effective hole mass, m_{heff}, that is different for each material and serves to correct Equation 3.135. Table 3.14 lists some conductivity-based, effective electron and hole masses in the three most common semiconducting materials.

Finally, we can express the diffusivity of the charge carriers in terms of the mobility using the Einstein relation. This measure of diffusivity has the same limitations as it did in liquid solutions. It is strictly applicable in infinite dilution but is often used at all charge carrier concentrations and corrected by the effective mass. Several values for diffusivity and mobility of charge carriers in semiconductors are shown in Table 3.15. Notice that the mobility and diffusivity all obey Einstein's relation and that the diffusivities are all eight to nine orders of magnitude higher than similar ionic diffusivities in solids.

TABLE 3.15
Diffusivities and Mobilities of Charge Carriers in Semiconductors

Material	D_- (m²/s)	μ_{e-} (m²/V · s)	D_+ (m²/s)	μ_{e+} (m²/V · s)
Si	0.0035	0.135	0.00125	0.048
Ge	0.010	0.390	0.0050	0.190
GaAs	0.022	0.850	0.001	0.04

Source: Data obtained from Sze, S.M., *Physics of Semiconductor Devices*, Wiley-Interscience, New York, 1969.

3.11 SUMMARY

We have seen how the flux–gradient relationships we established in Chapter 2 can be derived from a number of simple models of material structures and how the transport properties of these materials arise from the properties of these materials and the nature of their interaction with one another. The models presented here are only the simplest representations for these properties and are often not accurate. They serve as the starting point for most of the more sophisticated empirical and theoretical correlations and, as such, are important in any discussion of the properties of materials.

The next chapters will be concerned with taking the background knowledge we developed in Chapters 1 through 3 and applying that knowledge to the solution of transport problems. Similarities between the phenomena will be highlighted, and any differences will be emphasized. We will begin with simple, 1-D, steady-state systems and proceed to move to transient systems and finally to multidimensional systems. Chapter 4 in particular will consider 1-D, steady-state diffusive interactions.

PROBLEMS

VISCOSITY

3.1 Compare predictions for the viscosity of He, CCl_4, and C_6H_{14} at 300 K and 1 atm. using the kinetic theory of gases and the Chapman–Enskog theory. How do the predictions stack up to actual experimental data (you will have to find data for this part)? What reasons related to the chemical or structural nature of the molecules or atoms in the list you do believe may be responsible for the discrepancies?

3.2 Using the results of free volume theory for calculating the viscosity of liquids, determine $\Delta G^{\dagger}$ for the following liquids (20°C < T < 50°C): water, ethylene glycol, refrigerant 134a, and engine oil. Is there any relationship between their molecular structure and interactions and $\Delta G^{\dagger}$?

3.3 The speed of sound in dry air at 20°C is about 343 m/s:

a. Calculate the average speed of individual air molecules using the Maxwell–Boltzmann velocity distribution.

b. How do the two velocities compare? What is the ratio of molecular to sonic velocity?

c. How do you think you might account for the discrepancy? What might prevent an aggregate of molecules from moving as fast as an individual molecule? (Hint: the ratio of velocities $v_{sound}/v_{molecule}$ is proportional to $\sqrt{C_v/8C_p}$.)

3.4 The viscosity of suspensions has been the subject of much study. Knowing its value is critical to the processing of items such as paints, ketchup, cosmetics, concrete, and pharmaceuticals. One of the most famous relationships between the viscosity of a suspension and the volume fraction of particles in the suspension was developed by Einstein. His original equation is given by

$$\mu_r = \frac{\mu}{\mu_o} = \frac{1+0.5\phi}{(1-\phi)^2} \qquad \mu_o = \text{viscosity of pure fluid}$$

Use a one-term, Taylor series expansion for small ϕ to show that

$$\mu_r \approx 1 + \left(\frac{\partial \mu_r}{\partial \phi}\right)_{\phi=0} \qquad \phi = 1 + 2.5\phi$$

3.5 The Einstein formula in Problem 3.4 only applies when the volume fraction of particles is very small. Many times, we deal with very concentrated suspensions such as in fermentation of yeast cells. An alternative expression for the viscosity of such a suspension is [41]

$$\mu_r = \frac{1-0.5\kappa\phi}{(1-\kappa\phi)^2(1-\phi)} \qquad \kappa \sim 1+0.6\phi$$

a. Plot both the original Einstein relation and the expression given here, and discuss the difference as ϕ gets large, say beyond 10%.
b. The following data show the reduced viscosity for 5 μm particles in water. How does the viscosity stack up to the model? If you fit for κ, what value do you get?

Volume Fraction Particles, (ϕ)	Reduced Viscosity, (μ_r)
0	1
0.05	1.07
0.1	1.2
0.15	1.3
0.2	1.6
0.3	1.8
0.4	2.9
0.5	4.6

Thermal Conductivity

3.6 Compare predictions for the thermal conductivity of He, C_5H_{12}, and C_6H_6 at 300 K and 1 atm. using the kinetic theory of gases and the Chapman–Enskog theory. How do the predictions stack up to actual experimental data (you will have to find data for this part)? What reasons related to the chemical or structural nature of the molecules or atoms in the list you do believe may be responsible for the discrepancies?

3.7 Liquid thermal conductivities may be calculated using the following equation from Weber:

$$k = \frac{3.59\times10^{-8} C_p \rho^{4/3}}{M_w^{1/3}} \qquad \text{Weber equation}$$

where k has the unit of W/m K. Determine the thermal conductivity of the liquids in Table 3.5. How does Weber's formula compare with the procedure outlined in that section [42]?

3.8 A key aspect of conductors for integrated circuits is their ability to remove heat. Several materials are commonly used including aluminum, copper, and tungsten, and lately, silver is being considered. Calculate the electronic thermal conductivity for each at 298 K using literature values for the electric conductivity. Define a figure of merit, σ/k, for each conductor, and compare your figure of merit with that calculated from experimental data. Does silver turn out to be the clear winner?

3.9 Diamond has an extraordinarily high thermal conductivity given the fact that it is an excellent insulator. With reference to the derivation in Section 3.6, can you say something about diamond that leads it to have such a high value for thermal conductivity?

3.10 The steady-state heat flux through the 3/4 inch argon gap in an insulated window is to be measured. Assuming pure conduction through the gas and windowpane temperatures of 20°C and −10°C, what is the heat flux if

a. The thermal conductivity is given by kinetic theory?
b. The thermal conductivity is given by Chapman–Enskog theory?

3.11 The thermal conductivity of gases in confined spaces or at very low pressures is quite different from diffusion at atmospheric pressure. The motion of the gas is governed by collisions of the gas particles with the container walls, not with other molecules. For a container of radius, r_o, derive an expression for the thermal conductivity of a gas using kinetic theory and assuming that wall collisions are all important. What implications does your result have for designing insulation materials?

3.12 Composite materials offer the possibility of engineering materials with properties that cannot be achieved using a single material. One such class of materials are metal-filled epoxies that can be used as replacements for pure metal solders or as thermal interface materials to enhance the thermal conductivity in the joint between two elements such as between a transistor and heat sink [43]. Assume we have a metal-filled epoxy material where contact between the metal particles yields an electric resistivity of 5×10^{-5} Ω cm at 298 K. Assuming the heat will all be transferred by conduction electrons in the metal fraction, provide an estimate for the thermal conductivity of the composite.

3.13 Composite materials are a very common and important class of substances. Many theories have been developed to help explain how their properties depend upon their composition. One class of models are called effective medium theories since they provide an average estimate of a material's property based on its overall composition and a geometrical picture of its structure. Assuming a random, heterogeneous composite composed of freely overlapping spherical inclusions, a representation for the thermal conductivity of such a composite can be obtained using the Hashin–Shtrikman formalism [44]:

$$k_m\left(1+\frac{3\phi(k_f-k_m)}{3k_m+(1-\phi)(k_f-k_m)}\right) < k < \left(1-\frac{3(1-\phi)(k_f-k_m)}{3k_f-\phi(k_f-k_m)}\right)k_f$$

where

k_m is the thermal conductivity of the matrix material
k_f is the thermal conductivity of the filler material

Assuming the matrix is an epoxy ($k = 0.2$ W/m K) and the filler is silver ($k = 428$ W/m K), plot both bounds for filler volume fractions between 10% and 90%. If a real material is made that has a thermal conductivity of 40 W/m K at a filler volume fraction of 48%, does the material follow the upper or lower bound you plotted?

Diffusivity

3.14 Compare predictions for the diffusivity of He, C_6H_{14}, and H_2O in air at 298 K and 1 atm using the kinetic theory of gases and the Chapman–Enskog theory. How do the predictions stack up to actual experimental data (you will have to find data for this part)? What reasons related to the chemical or structural nature of the molecules or atoms in the list you do believe may be responsible for the discrepancies?

3.15 Compare experimental and calculated values of the diffusivity (at 298 K) for the following compounds in water at infinite dilution: methanol, propane, oxygen, helium, and hemoglobin. Which version of the Stokes–Einstein relation comes closest, no slip (sticking) or no stress (slipping)? How do these values compare with the free volume result?

3.16 An empirical correlation for diffusion coefficients was developed by Wilke and Chang. For water as the solvent, this formula is

$$D = \frac{5.06\times10^{-16}T}{\mu_w \bar{V}_s^{0.6}} \quad \frac{\text{m}^2}{\text{s}} \quad \text{Wilke–Chang equation}$$

where
T is the absolute temperature
μ_w is the viscosity of water
V_s is the molar volume of the solute at its boiling point

Compare values calculated from Problem 3.15 with those calculated here with the exception of hemoglobin [45].

3.17 Another empirical correlation for predicting the diffusion coefficient of large molecules and polymers was given by Polson and later used to look at biopolymers by Young et al.:

$$D = \frac{9.4 \times 10^{-15} T}{\mu M_w^{1/3}} \quad \text{Polson equation}$$

where D has the unit of m^2/s. Use this expression to calculate the diffusivity of fluorescein (M_w, 332), IgG (M_w, 150,000), and phycoerythrin (M_w, 240,000) in water at 25°C. Compare with data taken for diffusion in gels (mostly water) by de Beer et al. How close are the predicted and measured values [46–48]?

3.18 Mass diffusion of gases in confined spaces or at very low pressures, called Knudsen diffusion, is quite different from diffusion at atmospheric pressure. The motion of the gas is governed by collisions of the gas particles with the container walls, not with other molecules. For a container of radius, r_o, derive an expression for the self-diffusivity of a gas using kinetic theory and assuming that wall collisions are all important. Show that this expression is independent of the gas pressure.

3.19 If we look at the kinetic theory expressions for the viscosity, thermal conductivity, and diffusivity, we can express them in the following form:

$$3\frac{\mu}{\rho} = 3\frac{k}{\rho C_v} = 3D_{aa^*} = \overline{v}\lambda_f$$

The product of velocity and mean free path has the dimensions of length^2/time or diffusivity. In the preceding equation, all quantities on the left-hand side represent a diffusivity, specifically the same diffusivity. Comment on what this means for the mechanism whereby mass, energy, and momentum are transported in gas. Would you expect the same type of mechanism to operate in a liquid? In a solid? What would be required for the same mechanism to operate?

Conductivity and Mobility

3.20 What is the ionic conductivity of a 0.01 M sulfuric acid solution at 25°C? What is the mobility of the H^+ ion? The SO_4^{2-} ion? If one were to apply an electric field of 10 V/cm across the solution, what would be the drift velocity of the ions?

3.21 The solubility product constant for magnesium hydroxide, $Mg(OH)_2$, in water is $K_{sp} = 1.2 \times 10^{-11}$ (kg mol/m)^3. If 5 g of magnesium hydroxide were put into 50 mL of water and allowed to dissolve to equilibrium, estimate the molar conductivity of the solution at 298 K. (The mobility and diffusivity of Mg^{2+} is not given in the tables. However, data for Ca^{2+} exist in Table 3.12, and data for the infinite dilution conductivity of a Mg-based and Ca-based solution exist in Table 3.11. Based on that information, what is your best estimate for the mobility and diffusivity of Mg^{2+}?)

3.22 In a solid, like Si, the mobile charge carriers are in constant motion due to their thermal energy. This energy, a classic kinetic energy, is $3k_bT/2$:

a. What is the thermal velocity, v_{th}, of the charge carriers at room temperature, 300 K? Use the effective mass given in Table 3.14.

b. What is the drift velocity of those carriers, v_d, if their mobility, μ_e, is 1000 cm^2/V s, and the applied electric field, E, is 1000 V/cm?
c. Compare the results of (a) and (b). Can we cause gas molecules to move faster than their thermal velocity using a pressure gradient? Can we get the charges to move faster than their thermal velocity?

REFERENCES

1. Reid, R.C., Prausnitz, J.M., and Poling, B.E., *The Properties of Gases and Liquids*, 4th edn., McGraw-Hill, New York (1987).
2. Kittel, C., *Introduction to Solid-State Physics*, 6th edn., John Wiley & Sons, New York (1986).
3. Ashcroft, N.W. and Merman, N.D., *Solid-State Physics*, Holt, Rinehart & Winston, New York (1976).
4. Sears, F.W., *An Introduction to Thermodynamics, the Kinetic Theory of Gases, and Statistical Mechanics*, 2nd edn., Addison-Wesley, Reading, MA (1953).
5. Hirschfelder, J.O., Curtiss, C.F., and Bird, R.B., *The Molecular Theory of Gases and Liquids*, John Wiley & Sons, New York (1954).
6. Bird, R.B., Stewart, W.E., and Lightfoot, E.N., *Transport Phenomena*, John Wiley & Sons, New York (1960).
7. Chapman, S. and Cowling, T.G., *The Mathematical Theory of Non-Uniform Gases,* 3rd edn, University Press, Cambridge, U.K. (1970).
8. Chapman, S., On the law of distribution of velocities and on the theory of viscosity and thermal conduction in a non-uniform, simple, monatomic gas, *Phil. Trans. Royal Soc. A*, **216**, 279 (1916).
9. Enskog, D., The kinetic theory of phenomena in fairly rare gases, PhD dissertation, Upsala, Sweden (1917).
10. Weast, R.C., ed., *Handbook of Chemistry and Physics*, 55th edn., CRC Press, Cleveland, OH (1974).
11. Glasstone, S., Laidler, K.J., and Eyring, H., *Theory of Rate Processes*, McGraw-Hill, New York (1941).
12. Eyring, H., Viscosity, plasticity, and diffusion as examples of absolute reaction rates, *J. Chem. Phys.*, **4**, 283 (1936).
13. Lillie, H.R., High temperature viscosities of soda-silica glasses, *J. Am. Ceram. Soc.*, **22**, 367 (1939).
14. Einstein, A., Eine Neue Bestimmung der Molekuldimensionen, *Ann. Der Physik*, 324, 289–306 (1906).
15. Einstein, A., Elementary consideration of the thermal conductivity of dielectric solids, *Ann. Phys.* **34**, 591–592 (1911).
16. Batchelor, G.K., The effect of Brownian motion on the bulk stress in a suspension of spherical particles, *J. Fluid Mech.,* **83**, 97–117 (1977).
17. Toda, K. and Furuse, H., Extension of Einstein's viscosity equation to that for concentrated dispersions of solutes and particles, *J. Biosci. Bioeng.*, **102**, 524–528 (2006).
18. Robinson, J.V., The viscosity of suspensions of spheres. *J. Phys. Chem.,* **53**, 1042–1056 (1949).
19. Perry, R.H. and Chilton, C.H., eds., *Chemical Engineer's Handbook*, 5th edn., McGraw-Hill, New York (1973).
20. Debye, P., *Vortrage Über der Kinetische Theorie der Materie und der Elektrizität,* Tuebner, Berlin, Germany (1941).
21. Cahill, D.G., in *Microscale Energy Transport*, eds. C.-L. Tien, A. Majumdar, and F.M. Gerner, Chapter 2, Heat transport in dielectric thin films and at solid-solid interfaces, Taylor & Francis, Washington, DC, p. 95 (1998).
22. Ziman, M., *Electrons and Phonons*, Oxford University Press, London, U.K. (1960).
23. Kingery, W.D., Bowen, H.K., and Uhlmann, D.R., *Introduction to Ceramics*, 2nd edn., Wiley, New York, p. 617 (1976).
24. Kittel, C., *Introduction to Solid State Physics*, 7th edn., Wiley, New York, p. 534 (2000).
25. Goodson, K.E., Ju, Y.S., and Asheghi, M., in *Microscale Energy Transport*, eds. C.-L. Tien, A. Majumdar, and F.M. Gerner, Chapter 7, Thermal phenomena in semiconductor devices and interconnects, Taylor & Francis, Washington, DC, p. 229 (1998).
26. Hrubesh, L.W. and Pekala, R.W., Thermal properties of organic and inorganic aerogels, *J. Mater. Res.*, **9**, 731 (1994).
27. Srivastava, G.P., Theory of thermal conduction in nonmetals, *MRS Bull.*, **26**, 445, (2001).
28. Eckert, E.R.G. and Drake, R.M., *Analysis of Heat and Mass Transfer*, McGraw-Hill, New York (1972).
29. Cussler, E.L., *Multicomponent Diffusion*, Elsevier, Amsterdam, the Netherlands (1976).

30. Cussler, E.L., *Diffusion: Mass Transfer in Fluid Systems*, University Press, Cambridge, U.K. (1984).
31. Sherwood, T.K., Pigford, R.L., and Wilke, C.R., *Mass Transfer*, McGraw-Hill, New York (1975).
32. Lamb, H., *Hydrodynamics,* 6th edn., Dover, New York (1945).
33. Hiss, T.G. and Cussler, E.L., Diffusion in high viscosity liquids, *AIChE J.*, **19**, 698 (1973).
34. Ghai, R.K., Ertl, H., and Dullien, F.A.L., Liquid diffusion in non-electrolytes, Part 1, *AIChE J.*, **19**, 881 (1973).
35. Barrer, R.M., *Diffusion in and through Solids*, MacMillan, New York (1941).
36. Shewmon, P., *Diffusion in Solids*, 2nd edn., Minerals, Metals and Materials Society, Warrendale, PA (1989).
37. Kingery, W.D., Bowen, H.K., and Uhlmann, D.R., *Introduction to Ceramics*, 2nd edn., John Wiley & Sons, New York (1976).
38. Harned, H.S. and Owen, B.B., *The Physical Chemistry of Electrolytic Solutions*, ACS Monograph, Washington, BC, vol. 95 (1950).
39. Eisborg, R. and Resnick, R., *Quantum Physics*, 2nd edn., John Wiley & Sons, New York (1985).
40. Sze, S.M., *Physics of Semiconductor Devices*, Wiley-Interscience, New York (1969).
41. Toda, K. and Furuse, H., Extension of Einstein's viscosity equation to that for concentrated dispersions of solutes and particles, *J. Biosci. Bioeng.*, **102**, 524–528 (2006).
42. Weber, H.F., Investigations into the Thermal Conductivity of Liquids, *Wied. Ann. Phys. Chem.*, **10**, 103 (1880).
43. Zhang, R., Moon, K.-S., Lin, W., and Wong, C.P., Preparation of highly conductive polymer nanocomposites by low temperature sintering of silver nanoparticles, *J. Mater. Chem.*, **20**, 2018–2023 (2010).
44. DeVera, A. Jr. and Strieder, W., Upper and lower bounds on the thermal conductivity of a random, two-phase material, *J. Phys. Chem.*, **81**, 1783–1790 (1977).
45. Wilke, C.R. and Chang, P.C., Correlation of diffusion coefficients in dilute solutions, *AIChE J.*, **1**, 264 (1955).
46. Polson, A.J., The some aspects of diffusion in solution and a definition of a colloidal particle, *J. Phys. Chem.*, **54**, 649 (1950).
47. Young, M.E., Carroad, P.A., and Bell, R.L., Estimation of diffusion coefficients of proteins, *Biotechnol. Bioeng.*, **22**, 947 (1980).
48. de Beer, D., Stoodley, P., and Lewandewski, Z., Measurement of local diffusion coefficients in biofilms by microinjection and confocal microscopy, *Biotechnol. Bioeng.*, **53**, 151 (1997).

4 1-D, Steady-State, Diffusive Transport

In − Out + Gen = Acc

4.1 INTRODUCTION

In previous chapters we dealt with background information. Beginning here we enter into the main subject of study, the rate of transport processes. We will be considering the simplest form of transport first—the 1-D, steady-state, diffusive process. Fortunately, most aspects of transport processes can be understood using 1-D systems. Our starting point is the general balance equation or conservation law:

$$\text{In} - \text{Out} + \text{Gen} = \text{Acc} \tag{4.1}$$

This equation applies to a predetermined system of interest and accounts for all mechanisms that might produce a change in the state of the system. The balance states that the accumulation of our quantity of interest within the system is increased by the amount of that quantity entering the system, decreased by the amount that leaves the system, and increased (decreased) by the generation (consumption) of our quantity via sources (sinks) within the system. This balance equation will be applied to all our transport processes and can be applied to all processes where we can identify a specific component that is conserved. A personal example might include the accumulation of knowledge in our brains. The *In* terms represent all information coming into our brains from our senses; the *Out* terms represent everything we forget; the *Gen* terms account for all our original thoughts and ideas; and the *Acc* term is the net knowledge we retain, or wisdom. A more general example includes the national budget where *In* terms represent taxes and tariffs, *Out* terms represent government spending, *Gen* terms represent the printing of currency, and the *Acc* term, usually negative, implies deficit spending. We can dream up many more examples from energy to entropy to body weight and even to the accumulation of neuroses or other forms of psychological baggage. Equation 4.1 is the most important equation in this text.

As we discussed in Chapter 1, transport phenomena relies on the general balance equation to express four primary conservation laws: conservation of momentum (linear and angular), conservation of energy, conservation of mass, and conservation of charge. Our focus in this chapter will be the first two terms in the balance, the *In* and the *Out* terms.

4.2 BOUNDARY CONDITIONS

To uniquely define a transport process, we must understand what happens on the boundaries of the system. This means specifying a set of boundary conditions, often the most difficult part of the analysis. There are two basic forms of boundary conditions, *specification of the dependent variable* or *specification of the flux of the transported quantity*. Both types of boundary conditions can exist in several forms depending on the nature of the boundary and the mechanism of transport. In this chapter and the next, we will construct a catalog of common boundary conditions. Our first task will be to consider all boundary conditions that only involve the *In* and *Out* terms, that is, no boundary conditions that arise from a process involving generation. To accomplish this we must first preview two of the more important transport mechanisms, convection and radiation, and present them in a simplified, approximate form useful as a boundary condition.

4.2.1 Convection Preview

Convective transport occurs in a fluid. The quantity being transported is carried along by the fluid as it moves. Energy, mass, momentum, and charge are all quantities that may be convectively transported from one location to another. Convective boundary conditions always occur at the interface between two phases and one phase must be a fluid, in motion, relative to the other phase. To develop a boundary condition, our system of interest focuses on a small region of both phases that includes the interface. In all situations encountered in this text, we will assume the interface has no or negligible thickness and so has no capacity to store heat, mass, charge, or momentum. (If interfacial dynamics are of interest, the reader is referred to texts on interfacial transport [1].) Therefore, the boundary conditions will not have an accumulation term. In a purely convective boundary condition, there is no mechanism for generating information at the interface.

Consider the most common form of convective situation—a solid object immersed in a fluid that flows past it at a steady velocity, v_∞. The situation is shown later in Figure 4.1. The interface between the fluid and solid defines a plane, and there exists some flux, f'', across this plane into the fluid. ϑ may be any one of the state variables: temperature, velocity, mole fraction, charge, etc.

At the interface between the wall and the fluid, we know we must balance the flux through the wall and to the interface by diffusion, f''_w, with the flux from the interface and into the fluid by convection, f''_f. The diffusive flux through the wall is given by one of the flux–gradient relationships from Chapter 2:

$$f''_w = -\mathcal{D}_w \left.\frac{d\vartheta}{dx}\right|_w \quad \text{Diffusive flux through the wall} \tag{4.2}$$

We can make a crude estimate of the flux in the fluid using an expression similar to Equation 4.2. We assume that the state variable difference within the fluid, $\vartheta_o - \vartheta_\infty$, occurs within a small layer or film thickness of width, Δ_f. The gradient within the fluid can be approximated by the change in state variable divided by the film thickness:

$$\left.\frac{d\vartheta}{dx}\right|_f = \frac{\vartheta_o - \vartheta_\infty}{\Delta_f} \quad \text{Gradient approximation} \tag{4.3}$$

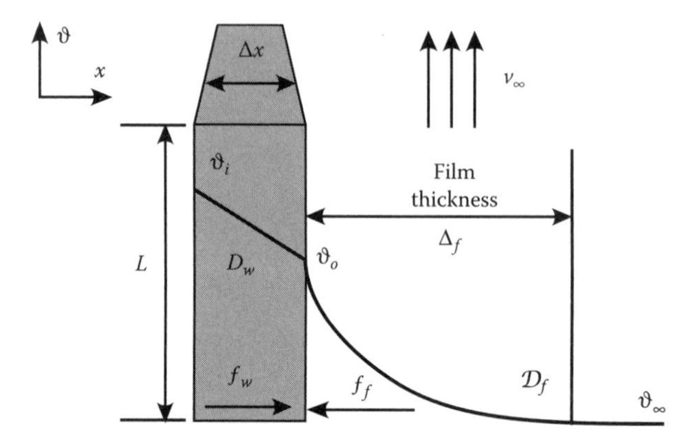

FIGURE 4.1 Interface between a solid wall and a fluid. Convective transport occurs across the interface and extends into the fluid at a distance, Δ_f.

We now write Equation 4.2 with reference to the liquid (at the liquid–wall interface) and substitute Equation 4.3 for the gradient to define the convective flux within the fluid:

$$f_f'' = -\mathcal{D}_f \left.\frac{d\vartheta}{dx}\right|_f = \frac{\mathcal{D}_f}{\Delta_f}(\vartheta_o - \vartheta_\infty) \quad \text{Convective flux within the fluid} \tag{4.4}$$

We refer to $\mathcal{D}_f/\Delta_f$ as a convection coefficient, C. It generally takes the form of a heat transfer coefficient, h (W/m^2 K), or a mass or charge transfer coefficient, k_c (m/s). Momentum transport uses a slightly different formulation for the flux that we will introduce later to arrive at a dimensionless friction factor, C_f. The convection coefficient incorporates everything we do not know about the flow field and conditions at the interface and gives us an empirical way of measuring and referring to a transport *resistance* within the fluid. The convective boundary condition is formulated by applying the general balance about the interface. The *In* term is represented by Equation 4.2 and the *Out* term by Equation 4.4:

$$\begin{aligned} &\text{In} && = && \text{Out} \\ &-\mathcal{D}_w \left.\frac{d\vartheta}{dx}\right|_w && = && C(\vartheta - \vartheta_\infty) \quad \text{Convective boundary condition} \end{aligned} \tag{4.5}$$

In writing this boundary condition, we have made the implicit assumption that we do not know ϑ at the interface but must use the boundary condition to find it.

Δ_f, the film thickness, depends on the fluid properties, the velocity distribution in the fluid, the wetting characteristics of the surface, etc. It cannot be calculated directly unless we know the gradient within the fluid. Therefore, in convective problems, we must couple momentum and one or more of our other transport processes to arrive at a full solution. This is often a task too complicated to do, especially in real process equipment. Convective coefficients, C, are often determined by experiment.

In engineering analysis, we routinely refer to a set of dimensionless groups that contain the convection coefficients. We define a Nusselt number, Nu, in heat transfer and a Sherwood number, Sh, in mass transfer. These dimensionless numbers are actually the dimensionless gradients at the interface, within the fluid, and are used to express correlations for transport coefficients. To derive the groups, we define a dimensionless quantity, S:

$$S = \frac{\vartheta - \vartheta_\infty}{\vartheta_o - \vartheta_\infty} \tag{4.6}$$

and a dimensionless length scale, $\mathcal{L}$:

$$\mathcal{L} = \frac{x}{\Delta_f} \tag{4.7}$$

We then rewrite the approximation for flux within the fluid, Equation 4.4, in terms of the dimensionless numbers:

$$-\left.\frac{dS}{d\mathcal{L}}\right|_{\mathcal{L}=0} = \frac{C\Delta_f}{\mathcal{D}_f} = \text{Nu or Sh} \tag{4.8}$$

We will consider these groupings and how they behave in more detail when we delve into convection in later chapters.

4.2.2 Equilibrium Relations and Mass Transfer Coefficients

Convective heat transfer coefficients or friction factors for momentum transport are rather straightforward concepts in their implementation. The temperature difference that defines the heat transfer coefficient or the velocity difference that defines the friction factor is unambiguous. Mass transfer coefficients, by contrast, are more complicated beasts. The complication arises because mass transfer may occur between phases, and there is often a jump discontinuity in the concentration of a particular species across a phase boundary (Figure 4.2). For example, if we try to transfer toluene from an aqueous to an organic phase, the toluene will have different solubilities in the two phases leading to a jump in concentration at the interface between the two. This discontinuity leads to a whole host of mass transfer coefficients, depending upon the choice of concentration driving force used. In this section we will attempt to describe the various forms that these mass transfer coefficients take and explain the conventions upon which they are based. First, we must review some simple equilibrium relationships that are normally applied to gas–liquid and liquid–liquid systems. These relationships allow us to calculate equilibrium compositions in one phase knowing the composition in the other.

The partial pressure of a component in a gas phase is defined by

$$p_a = y_a P \tag{4.9}$$

where
- p_a is the partial pressure
- P is the total pressure on the system
- y_a is the mole fraction in the gas phase

We can determine the equilibrium partial pressure of a constituent in the vapor phase as a function of its mole fraction in the liquid phase and the vapor pressure of that component at the system temperature. The simplest relationship between vapor pressure, liquid phase mole fraction, and partial pressure is called Raoult's law:

$$p_a = x_a p_a^v \quad \text{Raoult's law} \tag{4.10}$$

where
- x_a is the mole fraction in the liquid phase
- p_a^v is the vapor pressure of component a at the temperature of the system

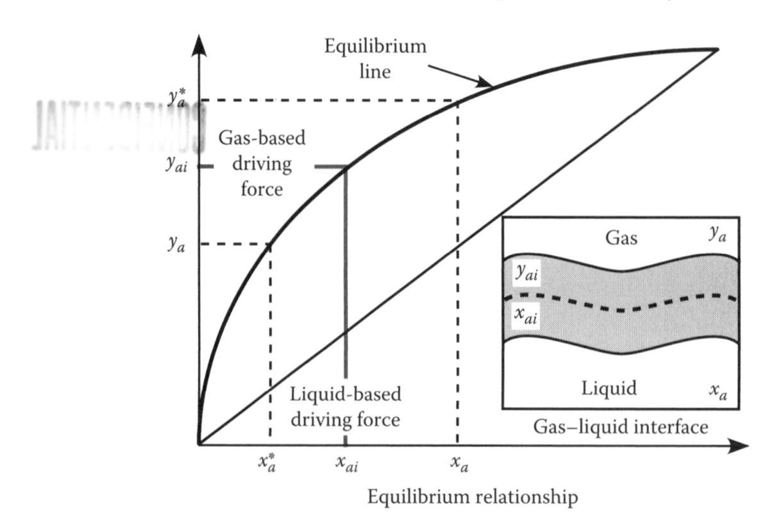

FIGURE 4.2 Interface between a gas and a liquid. Partitioning of solute, a, between the phases.

Raoult's law is most useful when the component is present in high concentration in the liquid because it assumes an ideal solution. When a is present in dilute solution, we use Henry's law:

$$p_a = x_a H_a \quad \text{Henry's law} \tag{4.11}$$

where H_a is the Henry's law constant and is found experimentally. Henry's law also assumes an ideal solution. Finally, the generic form of Henry's law or Raoult's law for any system is usually written as

$$y_a = m_k x_a \tag{4.12}$$

where m_k is a partition coefficient sometimes written as an equilibrium constant, K_{eq}. This coefficient may be a function of temperature, pressure, and composition and so can handle any system regardless of whether it is ideal.

We encounter a number of conventions when defining the mass transfer coefficient. If we have a system consisting of a solute dissolving at a steady state from a solid surface, for example and into a liquid, the mass transfer coefficient is defined by the following equation:

$$N_a = c_t k_{cx}(x_{as} - x_a) \quad \text{Dissolving solid} \tag{4.13}$$

where
- N_a is the flux of solute from the surface
- x_{as} is the mole fraction of solute in the liquid, at the interface, *in equilibrium with the solid* (not the mole fraction of a in the solid) at the temperature and pressure of the system
- x_a is the mole fraction of solute at some reference point in the fluid phase
- k_{cx} is the mass transfer coefficient with the units of m/s

Generally x_a is taken as the free stream value, $x_{a\infty}$, but in a closed system or a system without a uniform free stream (flow in a tube for instance), x_a would change with position and this must be taken into account. For flow in a tube, with solute dissolving from the tube wall, x_a is often taken as the average mole fraction at the axial point of interest in the tube. This average mole fraction is referred to as the *mixing-cup* value.

Mass transfer from a solid phase to a gas stream, for example, is described by the following convective equation:

$$N_a = c_t k_{cy}(y_{as} - y_a) \quad \text{Solid to gas} \tag{4.14}$$

where
- y is used to denote a gas phase mole fraction
- k_{cy} is the gas phase mass transfer coefficient in units of m/s
- y_{as} is the mole fraction of a in the gas, at the interface, *in equilibrium with the solid* at the temperature and pressure of the system
- y_a is a gas phase mole fraction of a at some reference point

Normally this reference point is taken as the free stream where $y_a = y_{a\infty}$, but the same considerations we used for the liquid phase mole fraction, x_a, in closed systems or systems without a free stream apply here as well.

Mass transfer between a liquid and a gas phase can be represented in one of two ways depending upon whether we want to view the system from the liquid phase or from the gas phase point of view:

$$N_a = c_t k_{cx}(x_{ai} - x_a) \quad \text{Liquid phase based} \tag{4.15}$$

$$N_a = c_t k_{cy}(y_a - y_{ai}) \quad \text{Gas phase based} \tag{4.16}$$

For simplicity, we assume the total concentrations are equal in both phases. The fluxes are equal in magnitude but opposite in direction. x_{ai} represents the liquid mole fraction of species a on the liquid side of the interface. y_{ai} represents the gas phase mole fraction of species a on the vapor side of the interface (Figure 4.2).

Though y and x are used to denote mole fractions in the gas and liquid phases, respectively, they are also used to denote transport between two liquid phases. In such a case, y usually represents the mole fraction in the dispersed phase and x, the mole fraction in the continuous phase.

4.2.3 Overall Mass Transfer Coefficients

So far, we have defined mass transfer coefficients that are based on a compositional driving force evaluated from the point of view of a single phase. When dealing with mass transfer equipment and interphase mass transfer, it is often more convenient to use *overall mass transfer coefficients*; coefficients that include interfacial resistances to mass transfer as well as resistances within a given phase. These mass transfer coefficients afford us the luxury of writing the mass flux in terms of the known composition of both phases, x_a and y_a, and equilibrium information without having to know what goes on at the real interface.

The overall mass transfer coefficient may be defined by either of the following two equations.

$$N_a = c_t k_{cy}(y_a - y_a^*) \quad \text{Gas phase based} \tag{4.17}$$

$$N_a = c_t k_{cx}(x_a^* - x_a) \quad \text{Liquid phase based} \tag{4.18}$$

where y_a^* and x_a^* represent equilibrium concentrations. Specifically, y_a^* represents the mole fraction of solute in the gas *in equilibrium with a liquid of composition* x_a. Similarly, x_a^* is the mole fraction of solute in the liquid *in equilibrium with a gas of composition* y_a. Therefore, $(y_a - y_a^*)$ and $(x_a^* - x_a)$ represent the overall driving force for mass transfer between the two phases.

We assume the simplest equilibrium relationship between x_{ai} and y_{ai} in terms of a *partition* coefficient or proportionality constant, m_k:

$$y_{ai} = m_k x_{ai} \tag{4.19}$$

We now substitute for x_{ai} in Equation 4.15 using the equilibrium relationship of Equation 4.19:

$$\frac{y_{ai}}{m_k} - x_a = \frac{N_a}{c_t k_{cx}} \tag{4.20}$$

Multiplying through by m_k gives:

$$y_{ai} - m_k x_a = \frac{N_a m_k}{c_t k_{cx}} \tag{4.21}$$

We now add this expression to Equation 4.16, removing y_{ai} to obtain

$$y_a - m_k x_a = \frac{N_a}{c_t}\left(\frac{1}{k_{cy}} + \frac{m_k}{k_{cx}}\right) \tag{4.22}$$

Now $m_k x_a$ represents the mole fraction of solute in the gas that would be in equilibrium with a liquid of composition x_a. It is identical to y_a^*. Thus, we can compare Equations 4.17 and 4.22 to see that the overall mass transfer coefficient is actually

$$\underbrace{\frac{1}{K_{cy}}}_{\substack{\text{Overall}\\\text{resistance}}} = \underbrace{\frac{1}{k_{cy}}}_{\substack{\text{Gas-phase}\\\text{resistance}}} + \underbrace{\frac{m_k}{k_{cx}}}_{\substack{\text{Interfacial and}\\\text{liquid-phase}\\\text{resistance}}} \quad \text{Gas phase based} \tag{4.23}$$

In our terminology, we relate the flux of a, N_a, to a driving force divided by a resistance. The resistance is the reciprocal of the mass transfer coefficient. Looking at Equation 4.23, we see that the overall mass transfer coefficient is a sum of resistances and considers the resistance to mass transfer in the gas phase as well as the resistance to transport in the liquid phase and across the interface.

We can perform the same mathematical manipulations starting with Equation 4.16 to obtain the overall liquid phase-based coefficient:

$$\frac{1}{K_{cx}} = \frac{1}{k_{cx}} + \frac{1}{m_k k_{cy}} \quad \text{Liquid phase based} \tag{4.24}$$

Even though the individual mass transfer coefficients may be considered as constants, the overall coefficients may vary since the equilibrium relationship, denoted by m_k, may be a function of temperature, pressure, and composition. In textbooks we try to deal with dilute solutions, so that Equations 4.23 and 4.24 are valid. If m_k is a function of the state variables, the analysis presented here would result in large errors and we would have to redefine our overall mass transfer coefficient by making further assumptions. As a note of terminology, if $1/k_{cx} \gg 1/m_k k_{cy}$ in Equation 4.24, we refer to the operation as *liquid phase controlled* because the resistance to mass transfer is highest in that phase. If $1/k_{cx} \ll 1/m_k k_{cy}$ in Equation 4.24, we refer to the operation as *gas phase controlled* for an analogous reason. We will illustrate the use of these various mass transfer coefficients in later sections of this chapter.

4.2.4 Radiation Preview

In heat transfer primarily, we often have surface conditions that promote the transfer of energy by radiation through free space. This leads to another variant of the flux specification condition. Consider two simple surfaces with areas A_{s1} and A_{s2} respectively as shown in Figure 4.3. These surfaces have emissivities $\varepsilon_1^r = \varepsilon_2^r = 1$ and are at temperatures T_1 and T_2.

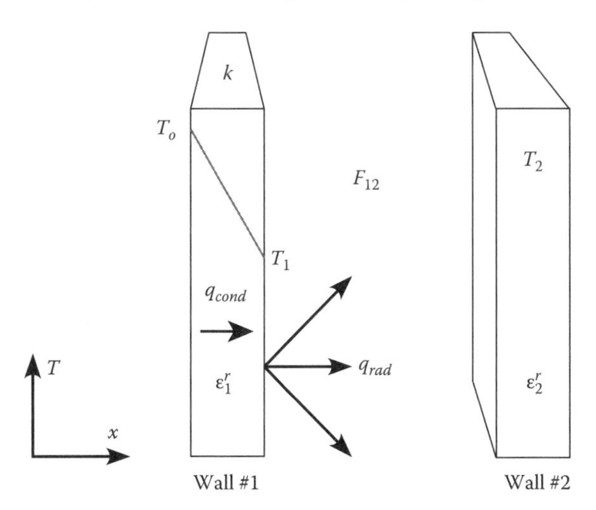

FIGURE 4.3 Two solid objects separated by a vacuum. Heat transfer occurs by radiation from wall 1 to wall 2. Wall 1 has a surface area A_{s1} and wall 2 has a surface area A_{s2}.

Our system of interest includes the interfacial plane defined by the surface of wall 1. As before, an overall energy balance about the system contains only *In* or *Out* terms. The *In* term is represented by heat conduction through the wall and to the interface. This flux is given by Equation 4.2 written in terms of the temperature and the thermal conductivity:

$$q''=-k\left.\frac{dT}{dx}\right|_w \quad \text{Diffusive flux through the wall} \tag{4.25}$$

The *Out* term results from radiative heat transfer to wall 2. The radiative flux is given by the Stefan–Boltzmann law:

$$q''=\sigma^r\varepsilon_1^r F_{12}\left(T_1^4-T_2^4\right) \quad \text{Stefan–Boltzmann law} \tag{4.26}$$

The radiation boundary condition arises by substituting Equations 4.25 and 4.26 into the balance about the interface. We determine the surface temperature of wall 1 from the boundary condition:

$$\text{In} = \text{Out}$$

$$-k\left.\frac{dT}{dx}\right|_w=\sigma^r\varepsilon_1^r F_{12}\left(T^4-T_2^4\right)=\sigma^r\varepsilon_1^r F_{12}\left(T_1^4-T_2^4\right) \quad \text{Radiation boundary condition} \tag{4.27}$$

4.3 BOUNDARY CONDITION CATALOG

Having briefly discussed convection and radiation, we are now in a position to catalog the major types of boundary conditions encountered so far. Specification of the dependent variable at the boundary is perhaps the most common form of boundary condition. It may be referred to as a *hard* or a *Dirichlet* boundary condition after the French mathematician who made a study of these types of boundary conditions in solving differential equations. Regardless of the name, specification of the dependent variable can be written in the following generic form:

$$x=x_o \qquad \vartheta=K_{eq}\vartheta_o \tag{4.28}$$

Here we are stating that at position x_o, the value of ϑ is $K_{eq}\vartheta_o$, where K_{eq} is an equilibrium or partition coefficient and ϑ_o is the value of ϑ on the other side of the boundary. Equilibrium or partition coefficients are in common use in all transport problems. The partition coefficient in momentum transport is often referred to as a slip or slip friction coefficient. Generally, we assume an infinite coefficient of sliding friction between a moving fluid and another phase so the slip coefficient is one. This means the fluid takes on the velocity of the material to which it is attached. Often, experimental data cannot be reconciled with a boundary condition of this type (e.g., rewetting of a surface), and better agreement between theory and experiment is obtained when we assume that the two faces of the interface can slip past one another [2].

A more common example is found in heat transfer. The partition coefficient in that case accounts for the fact that no two surfaces can ever be in perfect contact. Some microscopic amount of roughness exists and this inhibits heat transfer between the two surfaces. Therefore, the two faces of the interface may not be at the same temperature.

The most common examples of partition coefficients exist in mass or charge transfer. When we consider mass transfer, it is common practice to use Henry's law constants for calculating the solubility of gases, and partition coefficients in liquid–liquid and solid–solid systems are the rule rather than the exception. Finally, when we consider electromagnetic radiation, (thermal radiation) for example, the partition coefficient or coefficients take into account the fact that the radiation can be reflected, transmitted, or absorbed in the medium.

Examples of the property specification boundary condition for various transport processes are as follows:

$$x = x_o \qquad v = v_o \quad \text{Momentum} \tag{4.29}$$

$$x = x_o \qquad c_a = K_{eq} c_{ao} \quad \text{Mass transfer} \tag{4.30}$$

$$x = x_o \qquad T = T_o \quad \text{Heat transfer} \tag{4.31}$$

Flux specification at the boundary may be referred to as a *soft*, a *natural*, or a *Neumann* boundary condition. This type of boundary condition is usually written in the following generic form:

$$x = x_o \qquad -\mathcal{D}\frac{d\vartheta}{dx} = f'' \tag{4.32}$$

Here we are stating that at position x_o, we know the flux to be f''. This flux condition may take on many forms. Some simple examples include specifying the heat flux, q'', at the interface; specifying the stress, τ, at the interface; or specifying the mass flux, N_a, at the interface:

$$x = x_o \qquad -k\frac{dT}{dx} = q'' \quad \text{Heat transfer} \tag{4.33}$$

$$x = x_o \qquad -\mu\frac{dv}{dx} = \tau \quad \text{Momentum transfer} \tag{4.34}$$

$$x = x_o \qquad -D_{ab}\frac{dc_a}{dx} = N_a \quad \text{Mass transfer} \tag{4.35}$$

A special variant of the conditions in Equations 4.33 through 4.35 occurs when q'', τ, or N_a is zero. We refer to these conditions as *adiabatic, inviscid, or impermeable* in heat, momentum, or mass transfer, respectively. If we considered electromagnetic radiation, we would refer to a similar boundary condition as perfectly absorbing. Two other important variants of Equations 4.33 through 4.35 occur when the flux is given in terms of convection or radiation. Examples of these types of boundary conditions include

$$x = x_o \qquad -k\frac{dT}{dx} = h(T - T_\infty) \quad \text{Heat transfer} \tag{4.36}$$

$$x = x_o \qquad -\mu\frac{dv}{dx} = C_f \rho v_\infty^2 \quad \text{Momentum transfer} \tag{4.37}$$

$$x = x_o \qquad -D_{ab}\frac{dc_a}{dx} = k_c(c_a - c_{a\infty}) \quad \text{Mass transfer} \tag{4.38}$$

$$x = x_o \qquad -k\frac{dT}{dx} = \sigma^r \varepsilon_1^r F_{12}\left(T^4 - T_2^4\right) \quad \text{Heat transfer radiation} \tag{4.39}$$

4.4 1-D, STEADY-STATE DIFFUSIVE TRANSPORT

We begin the study of transport phenomena by considering 1-D steady-state systems. We envision the transport to occur by a diffusive mechanism only, and that means transport cannot occur by bulk flow of material but must occur by molecular motions. We will consider momentum, heat, mass, and charge transport in turn and look at several examples of the boundary conditions we discussed earlier. Our purposes in this section are fourfold:

1. To apply conservation laws to a small control volume of interest and derive the differential equations governing transport within that system
2. To apply the flux–gradient relationships within our differential balance equation to determine the profiles of our system or state variables
3. To see how the previously mentioned profiles vary as we apply different boundary conditions to our systems
4. To see how the profiles vary for systems with differing geometry

4.4.1 Momentum Transport: Fluid Flow

Consider the case of flow between two very long parallel plates as shown in Figure 4.4. The plates each have a surface area, A_s, in contact with the fluid. The lower plate moves with a velocity v_o and the upper moves with a velocity v_∞. The fluid is Newtonian with a constant viscosity, μ. Momentum is transported between the plates solely by the shearing action of individual fluid layers, a situation called laminar flow. To determine the shear stress and the velocity profiles inside the fluid, we need to perform a momentum balance on a small volume of fluid of height, Δz, and area, A_s, located between the plates. Since the plates are effectively infinite in extent (much wider and longer than the thickness of the fluid), the velocity profile will only change in the z-direction and the shear stress will have one component, τ_{zx}.

The momentum balance is written in terms of the forces acting on the control volume. We have no generation of momentum nor any accumulation, and so the balance equation considers momentum that is transferred due to the forces acting on the two faces of our control volume. At the lower face, the force, $F_x(z)$, is given in terms of the shear stress there, $(\tau_{zx})_z$, multiplied by the area through which the momentum is being transferred, A_s. On the upper face, the force, $F_x(z + \Delta z)$, is given by a similar term involving the shear stress on the upper face, $(\tau_{zx})_{z+\Delta z}$, and the area, A_s. Substituting into our balance equation gives

$$\begin{aligned} &\text{In} \quad - \quad \text{Out} \quad + \text{Gen} = \text{Acc} \\ &F_x(z) - F_x(z+\Delta z) \ + \ 0 \ = \ 0 \end{aligned} \tag{4.40}$$

Formally, the momentum balance equation states that *the rate of change of momentum within the control volume is equal to the net force applied across the upper and lower faces.*

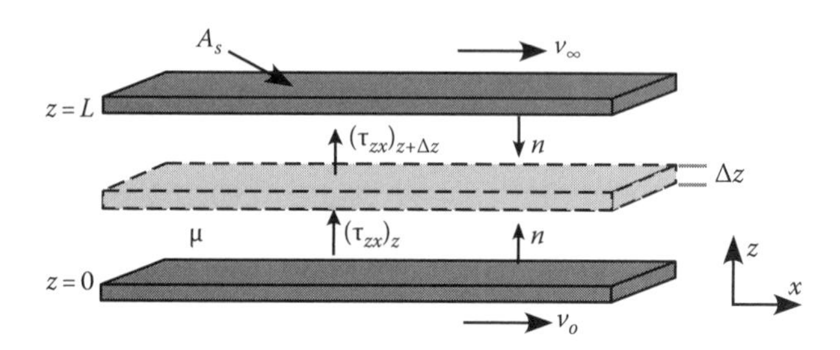

FIGURE 4.4 Flow of a Newtonian fluid between parallel, infinite plates. Arrows indicate the direction of the flow of momentum through the control volume.

At present, Equation 4.40 is a difference equation that applies over a macroscopic control volume. To derive the differential equation that applies at every point in the fluid, we divide by Δz and take the limit as $\Delta z \to 0$:

$$\lim_{\Delta z \to 0}\left(\frac{F_x(z) - F_x(z+\Delta z)}{\Delta z}\right) = \lim_{\Delta z \to 0}\left(\frac{0}{\Delta z}\right) \tag{4.41}$$

Equation 4.41 is the definition of the derivative of F_x with respect to z and so becomes a differential equation:

$$-\frac{dF_x}{dz} = 0 \tag{4.42}$$

We can replace the force by the stress multiplied by the surface area over which that stress acts:

$$-\frac{d}{dz}(A_s \tau_{zx}) = 0 \tag{4.43}$$

The differential equations (4.42) and (4.43) state that the force applied across the fluid is constant. Since the surface area of the system is also constant, the stress field within the fluid is constant. We should have expected this since the momentum balance indicated that we could have no accumulation of momentum within the control volume.

To determine the velocity profile, we must use the flux–gradient relationship, Newton's law of viscosity, to relate the shear stress, τ_{zx}, to the velocity gradient:

$$\tau_{zx} = -\mu \frac{dv_x}{dz} \quad \text{Newton's law of viscosity} \tag{4.44}$$

Substituting into Equation 4.43 and eliminating the viscosity and the constant area, A_s, gives the required differential equation in terms of the velocity:

$$\mu A_s \frac{d}{dz}\left(\frac{dv_x}{dz}\right) = \mu A_s \frac{d^2 v_x}{dz^2} = 0 \tag{4.45}$$

Equation 4.45 can be separated and integrated to show that if the stress on the system is constant, and the area and viscosity are constant, then the velocity profile in the fluid must be linear:

$$v_x = C_1 z + C_2 \tag{4.46}$$

The boundary conditions for the problem are taken from the fact that we know the fluid must take on the velocity of the bounding plates (no-slip condition):

$$z = 0 \qquad v_x = v_o \tag{4.47}$$

$$z = L \qquad v_x = v_\infty \tag{4.48}$$

Application of these boundary conditions to Equation 4.46 gives us the final velocity profile:

$$v_x = \left(\frac{v_\infty - v_o}{L}\right) z + v_o \quad \text{Velocity profile} \tag{4.49}$$

The stress is found by substituting the velocity profile back into Newton's law of viscosity. *The stress shows that momentum flows from the higher velocity to the lower velocity*:

$$\tau_{zx} = -\mu \frac{dv_x}{dz} = -\mu\left(\frac{v_\infty - v_o}{L}\right) \quad \text{Stress} \tag{4.50}$$

We will find it convenient later to rewrite the stress or force in the form of a driving force divided by a resistance:

$$\tau_{zx} = \frac{\text{Driving force}}{\text{Resistance/Area}} = -\frac{v_\infty - v_o}{(L/\mu)} \tag{4.51}$$

$$F_x = \frac{\text{Driving force}}{\text{Resistance}} = -\frac{v_\infty - v_o}{(L/\mu A_s)} \tag{4.52}$$

The driving force is the velocity difference between the plates and the resistance, for the planar system of parallel plates, is the separation of the plates divided by the viscosity of the fluid and area of the plane. We will derive parallels to Equations 4.51 and 4.52 for the other transport processes and geometries that we will discuss. *Keep in mind that the area we use is the area through which the momentum flows.* In Figure 4.4, this is the surface area of the plate.

4.4.2 Energy Transport: Heat Transfer

The energy balance equation is written in terms of the heat flow, q, not the heat flux q''. We will see why this distinction is important when we consider heat transfer through the wall of the long, cylindrical tube shown in Figure 4.5. If the tube length, L, is very long compared to its diameter and if we hold the inner and outer walls of the tube at uniform temperatures, T_i and T_o, respectively, throughout their length, the heat transfer problem will be reduced to 1-D conduction in the radial direction. The tube is made of a material with a thermal conductivity, k, and has inner and outer radii of r_i and r_o respectively.

To write the balance equation, we define a thin, cylindrical subsystem of width Δr somewhere within the wall of the tube. We have an amount of heat, q_r, flowing through the inner face of the

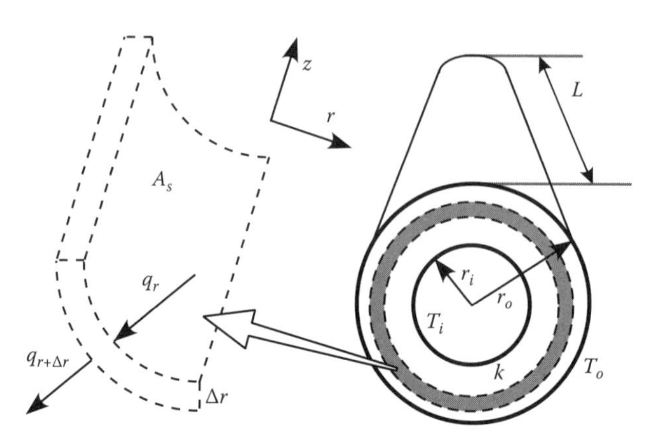

FIGURE 4.5 Heat conduction through the wall of a long, cylindrical tube.

system and $q_{r+\Delta r}$ flowing out of the outer face as shown in Figure 4.5. The balance equation, specifying no heat generation or energy accumulation, is

$$\begin{aligned} &\text{In} \quad - \quad \text{Out} \quad + \text{Gen} = \text{Acc} \\ &q(r) - q(r+\Delta r) \; + \; 0 \; = \; 0 \end{aligned} \tag{4.53}$$

The energy balance equation states that *the rate of change in the energy of the control volume is equal to the net amount of heat flowing through the two faces.* The energy balance reflects the fact that any accumulation of energy in the system must occur due to a differential flow of heat through the two faces of the system. The difference equation we derived in (4.53) can be converted to a differential equation describing the heat flow at any radial position by dividing the equation through by Δr and taking the limit as $\Delta r \to 0$:

$$\lim_{\Delta r \to 0} \left(\frac{q(r) - q(r+\Delta r)}{\Delta r} \right) = 0 \tag{4.54}$$

The left-hand side of Equation 4.54 is the definition of the derivative of q with respect to r and so the balance equation becomes

$$-\frac{dq}{dr} = 0 \tag{4.55}$$

Equation 4.55 states that the heat flow throughout the tube wall is a constant. This has to be the case because we have no generation or accumulation to affect the flow of energy through the wall.

To determine the temperature profile, we must use Fourier's law, to relate the heat flow to the temperature gradient:

$$q = -kA_s \frac{dT}{dr} \quad \text{Fourier's law} \tag{4.56}$$

Now we can see why it is important to write the energy balance in terms of the heat flow and not the heat flux. The area through which the heat is flowing is a function of radial position within the tube wall: $A_s = 2\pi rL$. Though the heat flow through the wall is a constant, the heat flux is not. In fact, the heat flux decreases as we move from the inner to the outer wall of the tube because the surface area of the tube available for heat transfer is increasing.

Substituting Fourier's law into Equation 4.55 and eliminating the constant terms, $2\pi kL$, gives the required differential equation describing the temperature profile:

$$\frac{d}{dr}\left(r \frac{dT}{dr} \right) = 0 \tag{4.57}$$

This differential equation can be separated and integrated twice to obtain the following logarithmic profile:

$$T = C_1 \ln(r) + C_2 \tag{4.58}$$

The boundary conditions for the problem are specifications of the temperature at the inner and outer tube radii:

$$r = r_i \qquad T = T_i \tag{4.59}$$

$$r = r_o \qquad T = T_o \tag{4.60}$$

Using these conditions, the final temperature profile can be written as

$$T = T_i + \frac{T_o - T_i}{\ln(r_o/r_i)} \ln\frac{r}{r_i} \quad \text{Temperature profile} \tag{4.61}$$

Notice that the temperature profile is nonlinear. This is a requirement to keep the heat flow constant throughout the wall. We can use Fourier's law again to calculate the heat flow from Equation 4.61 and see that it is indeed constant. The heat flow also shows that heat flows from the higher temperature to the lower temperature:

$$q = -kA_s\frac{dT}{dr} = -k(2\pi rL)\frac{dT}{dr} = \frac{2\pi kL(T_i - T_o)}{\ln(r_o/r_i)} \quad \text{Heat flow} \tag{4.62}$$

Later, we will find it convenient to express the heat flow in terms of a driving force divided by a resistance. In this cylindrical system, the driving force is the temperature difference between the bounding surfaces and the resistance involves the thermal conductivity of the tube wall, the ratio of inner and outer wall radii, and the overall length of the tube:

$$q = \frac{\text{Driving force}}{\text{Resistance}} = \frac{T_i - T_o}{\dfrac{\ln(r_o/r_i)}{2\pi kL}} \tag{4.63}$$

Example 4.1 Convective Boundary Condition

Suppose we consider steady-state heat conduction in the cylindrical tube of Figure 4.5, but instead of having a constant temperature on the outer surface, we expose the tube to a fluid such that we have convective heat transfer from the outer surface. The situation is shown in Figure 4.6.

Since the geometry of the situation remains the same as in Figure 4.5, we still have 1-D conduction through the wall of the tube. The heat flow and temperature distribution within the wall are still described by the differential equations (4.55) and (4.57). The only aspect of the problem that is different is the boundary condition at the outer surface. Since we have convection on the

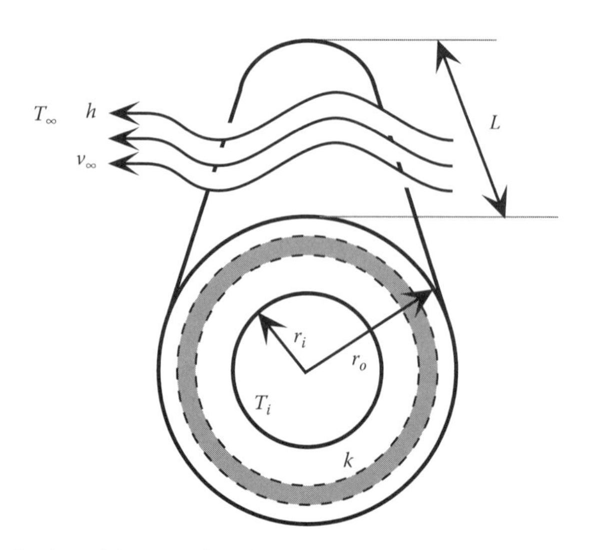

FIGURE 4.6 Heat conduction with convection through the wall of a long cylinder.

outer surface of the tube, we do not know the temperature there and must determine it from the convective boundary condition:

$$r = r_i \qquad T = T_i$$

$$r = r_o \qquad -kA_s\frac{dT}{dr} = hA_s(T - T_\infty)$$

Remember the convective boundary condition for heat transfer involves equating Fourier's law of conduction at the solid surface with Newton's law of cooling in the fluid. The temperature profile within the tube wall is still governed by Equation 4.58:

$$T = C_1\ln(r) + C_2$$

Applying the boundary conditions, we find that the constants C_1 and C_2 are given by

$$C_1 = -\frac{T_i - T_\infty}{\ln\left(\frac{r_o}{r_i}\right) + \frac{k}{hr_o}} \qquad C_2 = T_i + \frac{T_i - T_\infty}{\ln\left(\frac{r_o}{r_i}\right) + \frac{k}{hr_o}}\ln(r_i)$$

and the temperature profile is

$$T = T_i - \left[\frac{T_i - T_\infty}{\ln(r_o/r_i) + (k/hr_o)}\right]\ln\left(\frac{r}{r_i}\right)$$

The temperature profile has essentially the same form as that given by Equation 4.61 but contains an extra term arising from the convective boundary condition. If we calculate the heat flow through the tube wall using Fourier's law, we will see that this extra term is a second resistance to heat transfer arising at the tube–fluid interface. This resistance represents the difficulty in transferring heat from the tube wall and into the fluid:

$$q = \frac{\text{Driving force}}{\text{Resistance}} = -k(2\pi rL)\frac{dT}{dr} = \frac{T_i - T_\infty}{\frac{\ln(r_o/r_i)}{2\pi kL} + \frac{1}{2\pi r_o Lh}}$$

The resistance to heat transfer is the sum of two resistances in series. The conduction resistance through the wall is the same as we found before in Equation 4.63. The convection resistance is given by $1/hA_s$ where A_s is the area exposed to the fluid, that is, the area through which the heat flows. The sum-of-resistance concept is a very important one in transport phenomena and we will encounter it again many times.

4.4.3 Mass Transport

The case for mass transfer can be visualized by looking at a spherical geometry. This system is shown in Figure 4.7 and represents a room air freshener. Perfume, component a, is fed into the spherical wicking structure and travels through that medium with a diffusivity, D_{ab}. At the feed point, the mole fraction of perfume a is x_{ao} and remains constant for all time. At the wick–air interface, air motion in the room insures that the mole fraction of solute a is held at a constant value of x_{aR}. We will assume that the total concentration of all species in the wick remains constant at c_t.

The analysis of mass transfer is more complicated than the analyses we considered for heat transfer or fluid flow. The complexity arises because in mass transfer we generally have at least two components moving with respect to one another. They may move counter to one another in a one-to-one fashion (toluene diffusing through benzene) or in some other fixed ratio (Na^+ moving

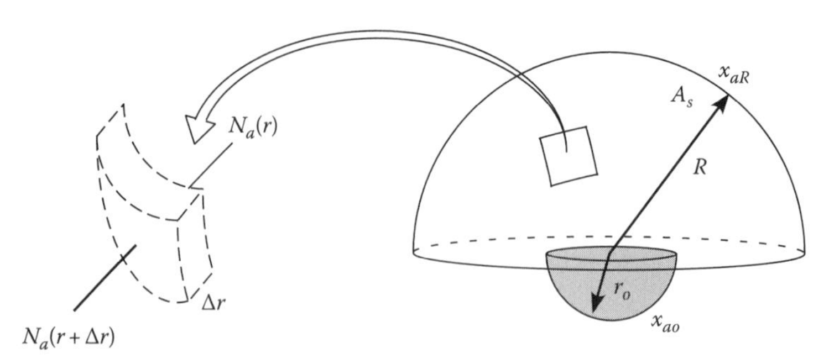

FIGURE 4.7 Diffusion of solute *a* through solvent *b*.

though SO_4^{2-}). One component may appear stationary because it is actually stagnant or because it is present in large excess. We begin with a balance on the mass flow ($\dot{m}_a$) or molar flow ($\dot{M}_a$) (not flux) of *a* through the system. For the 1-D situation with no generation or consumption of *a* and no way of accumulating *a* in the system, the balance equation (molar basis) about the differential element in Figure 4.7 is

$$\begin{array}{cccc} \text{In} & - \quad \text{Out} & + \text{Gen} = & \text{Acc} \\ \dot{M}_a(r) & - \dot{M}_a(r+\Delta r) & + \ 0 \ = & 0 \end{array} \tag{4.64}$$

The mass balance equation, written here on a molar basis, states that *the rate of change of mass (moles) in our control volume is equal to the net amount of material diffusing through the two faces.* Remembering that the area through which the material is diffusing depends upon the radial position and applying the familiar limiting process gives us a differential equation describing the molar flux of species *a* through the system:

$$\frac{d\dot{M}_a}{dr} = \frac{d}{dr}(A_c N_a) = 0 \qquad \text{or} \qquad \frac{d}{dr}\left[\left(4\pi r^2\right)N_a\right] = 0 \tag{4.65}$$

The differential balance states that the molar flow rate of *a* must be a constant throughout the wick and is subject to boundary conditions specifying a known mole fraction at the bottom and top of the wick:

$$r = r_i \qquad x_a = x_{ao} \tag{4.66}$$

$$r = r_o \qquad x_a = x_{aR} \tag{4.67}$$

Equation 4.64 was written using the wick structure as the reference for a stationary coordinate system. Species *a* diffuses through this unmoving coordinate system defined by the wick material, *b*. Under these conditions, N_a is given by Fick's law for stationary coordinates:

$$N_a - x_a(N_a + N_b) = -c_t D_{ab}\frac{dx_a}{dr} \quad \text{Fick's law} \tag{4.68}$$

and the molar flux of *a* depends not only on the concentration gradient but also on the molar flux of species *b*.

If x_a is small we can neglect the term involving the total molar flux of species a and b, $x_a(N_a + N_b)$, because the convective (bulk flow) component of species a, represented by $x_a(N_a + N_b)$, is so small. This is termed the *dilute solution approximation*. Fick's law and the differential balance become

$$N_a = -c_t D_{ab} \frac{dx_a}{dr} \quad \text{Fick's law for dilute solutions} \tag{4.69}$$

$$\frac{d}{dr}\left[c_t D_{ab} r^2 \frac{dx_a}{dr}\right] = \frac{d}{dr}\left[r^2 \frac{dx_a}{dr}\right] = 0 \tag{4.70}$$

The balance equation can be separated and integrated directly to determine the final solution for the mole fraction profile. The molar flux is found by substituting the profile back into Fick's law:

$$x_a = \left[\frac{x_{ao} - x_{aR}}{\left(\frac{1}{r_i} - \frac{1}{r_o}\right)}\right]\frac{1}{r} + \left[\frac{\frac{x_{ao}}{r_o} - \frac{x_{aR}}{r_i}}{\frac{1}{r_i} - \frac{1}{r_o}}\right] \tag{4.71}$$

$$N_a = -c_t D_{ab} \frac{dx_a}{dr} = c_t D_{ab}\left[\frac{x_{ao} - x_{aR}}{\left(\frac{1}{r_i} - \frac{1}{r_o}\right)}\right]\frac{1}{r^2} \quad \text{Molar flux of } a \tag{4.72}$$

We can write the molar flow rate in terms of a driving force for mass transfer divided by a resistance. The driving force is the overall mole fraction or concentration difference and the resistance incorporates the diffusivity and cross-sectional area available for diffusion of a:

$$\dot{M}_a = A_c N_a = \frac{\text{Driving force}}{\text{Resistance}} = \left[\frac{c_t\left(x_{ao} - x_{aR}\right)}{\frac{1}{4\pi D_{ab}}\left(\frac{1}{r_i} - \frac{1}{r_o}\right)}\right] \tag{4.73}$$

Notice that the molar flow rate is constant throughout the geometry exactly as Equation 4.65 stated.

Fick's law for dilute solutions, the resulting differential balance, and the mole fraction profile obtained earlier also apply for another important mass transfer situation—*equimolar counterdiffusion*. In this situation the flux of species b is equal in magnitude and opposite in direction to the flux of species a; $N_b = -N_a$. For this situation, the term involving the total molar flux also vanishes and the flux of a is again directly related to the concentration gradient.

The general solution to Equation 4.65 is more complicated and requires that we know the exact flux relationship between N_a and N_b, that is, $N_b = f(N_a)$. Though there are many cases where we know the relationship between N_a and N_b from stoichiometric or other requirements (electroneutrality), one of the most common situations arises when fluid or substance b can be assumed stagnant; $N_b = 0$. In *diffusion through a stagnant medium*, we solve for N_a so that Fick's law for the flux of solute, a becomes:

$$N_a = -\frac{c_t D_{ab}}{1 - x_a}\frac{dx_a}{dr} \quad \text{Stagnant medium} \tag{4.74}$$

Substituting into the differential balance, (4.65) gives

$$\frac{d}{dr}\left[\frac{c_t A_c D_{ab}}{1-x_a}\frac{dx_a}{dr}\right] = \frac{d}{dr}\left[\frac{4\pi c_t D_{ab}}{1-x_a} r^2 \frac{dx_a}{dr}\right] = 0 \tag{4.75}$$

Integrating once shows

$$\frac{r^2}{1-x_a}\frac{dx_a}{dr} = C_1 \quad \text{Constant flow} \tag{4.76}$$

Separating and integrating again shows that the mole fraction profile obeys an exponential form:

$$-\ln(1-x_a) = -\frac{C_1}{r} + C_2 \tag{4.77}$$

Applying the boundary conditions of Equations 4.66 and 4.67 yields the final solution:

$$-\ln(1-x_a) = \frac{\ln\left[\dfrac{1-x_{aR}}{1-x_{ao}}\right]}{\dfrac{1}{r_i}-\dfrac{1}{r_o}}\left(\frac{1}{r}\right) + \left[\frac{\dfrac{\ln(1-x_{ao})}{r_o} - \dfrac{\ln(1-x_{aR})}{r_i}}{\dfrac{1}{r_i}-\dfrac{1}{r_o}}\right] \tag{4.78}$$

Using Fick's law *in the form of Equation 4.74* allows us to calculate the molar flux of a.

$$N_a = \frac{c_t D_{ab}\ln\left[\dfrac{1-x_{aR}}{1-x_{ao}}\right]}{\dfrac{1}{r_i}-\dfrac{1}{r_o}}\left(\frac{1}{r^2}\right) \quad \text{Molar flux} \tag{4.79}$$

How does the magnitude of this flux compare with that for equimolar counterdiffusion assuming the same total concentration and bounding mole fractions?

If we write the molar flow in terms of a driving force and resistance, we see that the resistance is the same as the case for equimolar counterdiffusion or dilute solutions. The driving force has changed and is no longer a simple difference in mole fractions from inlet to outlet but is a logarithmic difference between these mole fractions:

$$\dot{M}_a = A_c N_a = \frac{\text{Driving force}}{\text{Resistance}} = \frac{c_t \ln\left[\dfrac{1-x_{aL}}{1-x_{ao}}\right]}{\left[\left(\dfrac{1}{4\pi D_{ab}}\right)\left(\dfrac{1}{r_i}-\dfrac{1}{r_o}\right)\right]} \tag{4.80}$$

Notice that mass transfer is fundamentally different from fluid flow or heat transfer. Complicated profiles can occur even in situations where we have uniform 1-D transport with no generation. The differential mass or mole balance requires a constant flow rate that is maintained even though the general solution may be logarithmic.

We have written the solutions for the flow of our transported substance in terms of a driving force divided by a resistance. In at least one instance, we have shown how multiple resistances to transport can arise and can be handled in terms of this driving force–resistance concept. We will now continue in this vein and explore one of the most important concepts in transport phenomena, composite systems and the sum of resistances.

4.5 COMPOSITE MEDIA

The ability to analyze steady-state transport in composite systems, systems where different materials coexist or different transport mechanisms are occurring sequentially or simultaneously, is essential if we are to deal with most realistic situations. Examples of these situations include calculating the heat flow through a multilayered, insulated wall, determining the rate of water vapor transport through the layers of a Gore-Tex laminate or through a layered piece of plastic food packaging, evaluating transdermal drug delivery systems, assessing the performance of evaporation and condensation systems, or calculating the current flow through an electric circuit.

Though we instinctively view these systems as being physically multilayered, in reality, they represent a situation where we have individual resistances to transport, arranged in a series or parallel configuration. We already encountered this type of series resistance arrangement when we considered the convective boundary condition around the circular cylinder in Section 4.4.2. The concept of series and parallel resistances is one of the cornerstones of transport phenomena theory.

4.5.1 Planar Systems

The simplest composite system to analyze is a planar one, like that shown in Figure 4.8. In this example we have a system where three immiscible fluids are flowing between two moving plates. All the fluids are Newtonian and have constant properties, and conditions are at steady state with no generation. The

$$\rho_1 > \rho_2 > \rho_3 \qquad v_\infty > v_o$$

two bounding plates are long enough that we can consider the velocity profile to be 1-D; $v = f(y)$.

To analyze a system of this type, we select a differential element within each fluid layer and write the conservation of momentum equation there:

$$\begin{array}{lcl} \text{In} \quad - \quad \text{Out} & + \text{Gen} = \text{Acc} & \\ F_{x1}(y) - F_{x1}(y+\Delta y) & + \quad 0 \quad = \quad 0 & \text{Layer } \#1 \end{array} \tag{4.81}$$

$$F_{x2}(y) - F_{x2}(y+\Delta y) + 0 = 0 \quad \text{Layer } \#2 \tag{4.82}$$

$$F_{x3}(y) - F_{x3}(y+\Delta y) + 0 = 0 \quad \text{Layer } \#3 \tag{4.83}$$

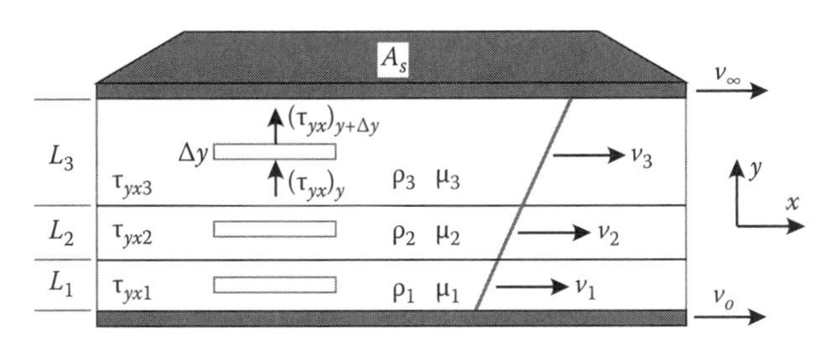

FIGURE 4.8 Flow of a composite fluid between two parallel plates. The arrows represent the flow of momentum through the control volume.

Allowing the thickness of the differential elements to approach zero, writing the forces in terms of the stresses and areas through which the momentum flows, and using Newton's law gives three differential equations, each representing the velocity profile in a fluid layer:

$$\underbrace{\frac{d^2v_1}{dy^2}=0}_{\text{Layer 1}} \quad \underbrace{\frac{d^2v_2}{dy^2}=0}_{\text{Layer 2}} \quad \underbrace{\frac{d^2v_3}{dy^2}=0}_{\text{Layer 3}} \tag{4.84}$$

These differential equations indicate that the shear stress in each layer is a constant and that the velocity profile in each layer must be linear. From an overall momentum balance, we know that since both plates are moving with a constant velocity, the stress in each fluid layer should not only be constant but also be the same. Whatever momentum flows through one layer must also flow through the others to maintain the steady state. This means that the differential equations, (4.84), are dependent and must be coupled to one another via boundary conditions that insure continuity of momentum and velocity (no slip) at the interfaces between the fluids.

The velocity profiles are linear in each layer (refer back to Equation 4.46) and so we need six boundary conditions to determine the constants of integration:

$$v_1 = C_1y + C_2 \tag{4.85}$$

$$v_2 = C_3y + C_4 \tag{4.86}$$

$$v_3 = C_5y + C_6 \tag{4.87}$$

At the surfaces of the two moving plates, the fluid must obey the no-slip condition and travel at the velocity of the plates. The first two boundary conditions reflect this:

$$y = 0 \qquad v_1 = v_o \tag{4.88}$$

$$y = L_1 + L_2 + L_3 \qquad v_3 = v_\infty \tag{4.89}$$

The other four boundary conditions are evaluated at the interfaces between the fluids. At each interface we invoke the no-slip condition and so the velocities at those interfaces must be equal:

$$y = L_1 \qquad v_1 = v_2 \tag{4.90}$$

$$y = L_1 + L_2 \qquad v_2 = v_3 \tag{4.91}$$

Newton's third law of motion provides us with the last two boundary conditions: the force exerted by fluid 1 on fluid 2 must be equal to the force exerted by fluid 2 on fluid 1. Hence, the stresses must be equal at the interface. Using Newton's law for each fluid, these boundary conditions are

$$y = L_1 \qquad -\mu_1 A_s\frac{dv_1}{dy} = -\mu_2 A_s\frac{dv_2}{dy} \tag{4.92}$$

$$y = L_1 + L_2 \qquad -\mu_2 A_s\frac{dv_2}{dy} = -\mu_3 A_s\frac{dv_3}{dy} \tag{4.93}$$

Solving Equations 4.85 through 4.87 subject to our boundary conditions (4.88) through (4.93) allows us to determine the three velocity profiles:

$$v_1 = \frac{1}{\mu_1}\left[\frac{v_\infty - v_o}{\frac{L_1}{\mu_1}+\frac{L_2}{\mu_2}+\frac{L_3}{\mu_3}}\right]y + v_o \tag{4.94}$$

$$v_2 = \frac{1}{\mu_2}\left[\frac{v_\infty - v_o}{\frac{L_1}{\mu_1}+\frac{L_2}{\mu_2}+\frac{L_3}{\mu_3}}\right]\left[y - \frac{L_1(\mu_1-\mu_2)}{\mu_1}\right] + v_o \tag{4.95}$$

$$v_3 = \frac{1}{\mu_3}\left[\frac{v_\infty - v_o}{\frac{L_1}{\mu_1}+\frac{L_2}{\mu_2}+\frac{L_3}{\mu_3}}\right][y-(L_1+L_2+L_3)] + v_\infty \tag{4.96}$$

If we calculate the stress in each layer and hence the force exerted by the fluids, we find that the force is determined by the overall velocity driving force and the sum of the resistances to momentum transport within each fluid layer:

$$F_x = -\frac{v_\infty - v_o}{\left[\frac{L_1}{\mu_1 A_s}+\frac{L_2}{\mu_2 A_s}+\frac{L_3}{\mu_3 A_s}\right]} \qquad \frac{\text{Overall driving force}}{\text{Sum of resistances}} \tag{4.97}$$

We can cast Equation 4.97 in the form of a typical electric circuit composed of resistances in series since momentum must flow through each fluid sequentially. The circuit is shown in Figure 4.9. The nodes correspond to velocities, the currents flowing through the resistors correspond to the shear stresses acting on the fluid layers, and the resistors are the individual transport resistances associated with each fluid layer:

$$\tau_{yx_1} = \tau_{yx2} = \tau_{yx3} = \tau_{yx}$$

If we consider the analogous heat transfer case, we have the situation illustrated in Figure 4.10. From the earlier analysis for momentum transport, we know that at steady state with no generation and a constant thermal conductivity, the temperature profile within each layer should be linear. Since *In* – *Out* = 0 in each layer, we also know that whatever heat flows in one layer must flow out that layer and into the next. Therefore, $q_3 = q_2 = q_1 = q$. The boundary conditions for the problem state that we know the temperatures, T_1 and T_2, at the top and bottom of the composite,

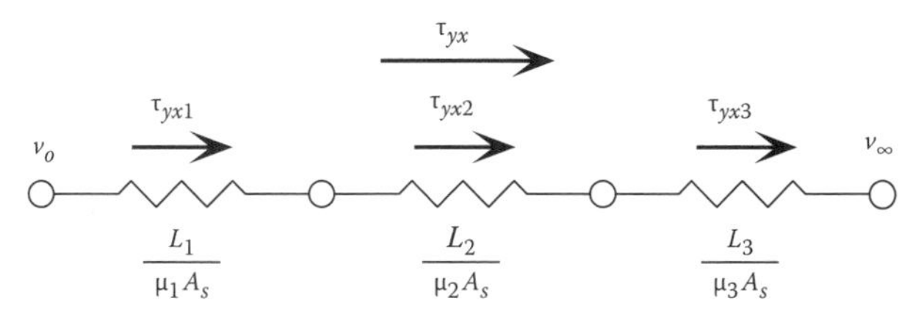

FIGURE 4.9 Electrical analogy with series transport resistances.

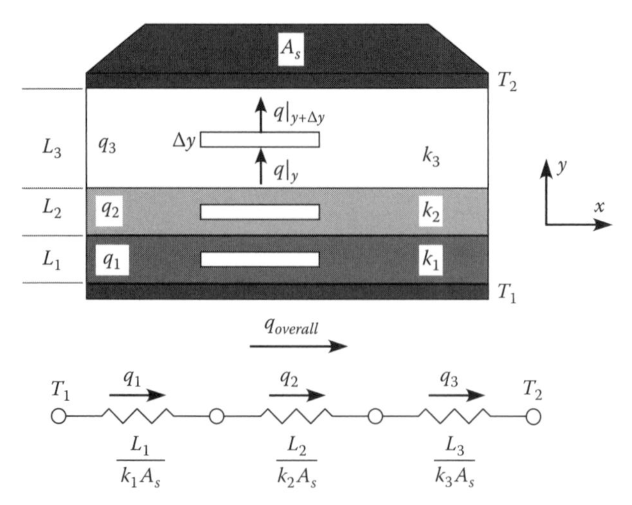

FIGURE 4.10 Steady–state heat transfer through a planar composite.

and at the interface between each layer, we must have continuity of temperature and continuity of heat flow. These are the same boundary conditions we used for the momentum transport case. The solution to the heat transfer problem leads to the same formulation in terms of the overall temperature driving force and sum of thermal resistances. For the planar case, the thermal resistance is the thickness of the layer divided by the thermal conductivity of the layer and area through which the heat flows:

$$q = -\frac{T_2 - T_1}{\left[\dfrac{L_1}{k_1 A_s} + \dfrac{L_2}{k_2 A_s} + \dfrac{L_3}{k_3 A_s}\right]} \qquad \frac{\text{Overall driving force}}{\text{Sum of resistances}} \tag{4.98}$$

The planar mass transfer case that is analogous to the planar momentum or heat transfer case is shown in Figure 4.11. The analysis leading to the mass flow equations is exactly the same as we used for the heat transfer or momentum transfer case. One only need to remember that to arrive at Equation 4.99 we assume equimolar counterdiffusion or dilute solutions within the composite, while to arrive at Equation 4.100 we assume diffusion through stagnant media.

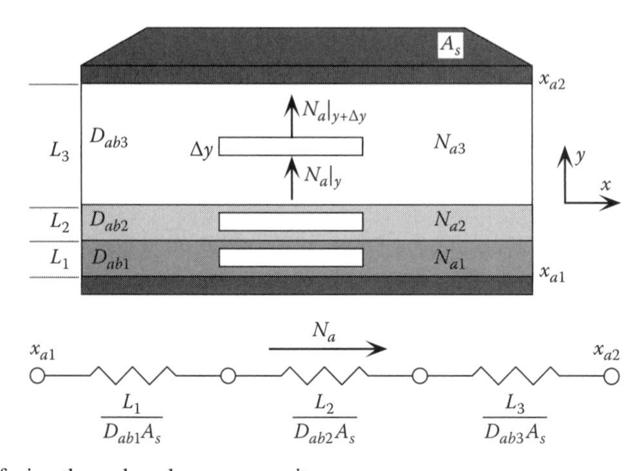

FIGURE 4.11 Diffusion through a planar composite.

Equimolar counterdiffusion or dilute solutions

$$\dot{M}_a = N_a A_s = -\frac{c_t(x_{a2} - x_{a1})}{\left[\dfrac{L_1}{D_{ab1}A_s} + \dfrac{L_2}{D_{ab2}A_s} + \dfrac{L_3}{D_{ab3}A_s}\right]} \quad \frac{\text{Overall driving force}}{\text{Sum of resistances}} \tag{4.99}$$

Diffusion through stagnant media

$$\dot{M}_a = N_a A_s = -\frac{c_t \ln\left(\dfrac{1 - x_{a1}}{1 - x_{a2}}\right)}{\left[\dfrac{L_1}{D_{ab1}A_s} + \dfrac{L_2}{D_{ab2}A_s} + \dfrac{L_3}{D_{ab3}A_s}\right]} \tag{4.100}$$

We can treat the electric fields in much the same way. If we consider a multilayer capacitor, for instance, we may have two charged plates and three or more dielectric layers sandwiched between as shown in Figure 4.12a. In such a case, the electric displacement can be written in terms of an overall potential driving force and a sum of resistances involving the thickness of the layers and the permittivity of each layer:

$$D_e A_s = -\frac{\Phi_2 - \Phi_1}{\left[\dfrac{L_1}{\varepsilon_1 A_s} + \dfrac{L_2}{\varepsilon_2 A_s} + \dfrac{L_3}{\varepsilon_3 A_s}\right]} \quad \frac{\text{Overall driving force}}{\text{Sum of resistances}} \tag{4.101}$$

For parallel plate capacitors of the kind we are talking about, with generic interfacial area, A_i, we speak of capacitances instead of transport resistances. We define the capacitance, C, of each layer in terms of the resistance to transporting the field force:

$$C_i = \frac{\varepsilon_i A_i}{L_i} \tag{4.102}$$

The total capacitance of the composite is the sum of resistances or

$$\frac{1}{C} = \sum_i \frac{1}{C_i} = \sum_i \frac{L_i}{\varepsilon_i A_i} \tag{4.103}$$

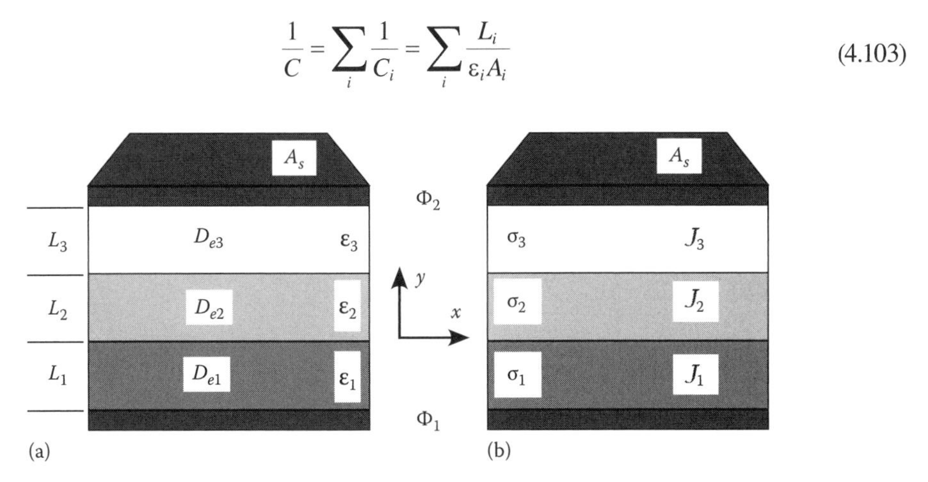

FIGURE 4.12 Multilayer capacitance (a) and conductance (b) networks.

Finally, we can look at conservation of charge and current flowing through a composite conductor. Here we have multiple layers of conducting media sandwiched between two plates as shown in Figure 4.12b. In this case, the current density can be written in terms of an overall potential driving force and a sum of resistances involving the thickness of the layers and the conductivity of each layer:

$$I = -\frac{\Phi_2 - \Phi_1}{\left[\dfrac{L_1}{\sigma_1 A_s} + \dfrac{L_2}{\sigma_2 A_s} + \dfrac{L_3}{\sigma_3 A_s}\right]} \qquad \frac{\text{Overall driving force}}{\text{Sum of resistances}} \tag{4.104}$$

For multiple conductors with areas, A_i, we define the electrical resistance, $\mathcal{R}_e$, of each layer as

$$\mathcal{R}_{ei} = \frac{L_i}{\sigma_i A_i} \tag{4.105}$$

The total electrical resistance of the composite is

$$\mathcal{R}_e = \sum_i \mathcal{R}_{ei} = \sum_i \frac{L_i}{\sigma_i A_i} \tag{4.106}$$

The relationship between the resistance, driving force, and current density is Ohm's law.

Example 4.2 Heat Transfer with Radiation and Convection

Consider a planar situation similar to Figure 4.10. At one, exposed surface we will have radiation and convection occurring. We can presume this represents the outer wall of a furnace. The three layers of the wall represent firebrick, fiberglass insulation, and a painted metal shell. The physical properties of the materials and the environmental conditions are

Physical properties

$k_1 - 1.1\,\text{W/m K}$	$k_2 - 0.038\,\text{W/m K}$	$k_3 - 60.5\,\text{W/m K}$
$L_1 - 0.05\,\text{m}$	$L_2 - 0.075\,\text{m}$	$L_3 - 0.003\,\text{m}$
$T_o - 1000°\text{C}$	$T_\infty - 25°\text{C}$	$h - 10\,\text{W/m}^2\,\text{K}$
$\varepsilon^r - 0.85$	$A_s - 1\,\text{m}^2$	

The physical situation is shown in Figure 4.13:

$$\begin{array}{cccc} \text{In} \;-\; \text{Out} & +\text{Gen} & = & \text{Acc} \\ q_i(y) - q_i(y+\Delta y) + & 0 & = & 0 \end{array} \qquad \text{Layer \#}i$$

Dividing by Δy and taking the limit as $\Delta y \to 0$ gives a differential equation for the heat flow, q_i. The differential equation says the heat flow is a constant and that all q_i's are equal:

$$-\frac{dq_i}{dy} = 0$$

Substituting in Fourier's law allows us to express the differential equations in terms of the temperatures. Assuming constant thermal properties and constant cross-sectional area,

$$\frac{d^2 T_i}{dy^2} = 0$$

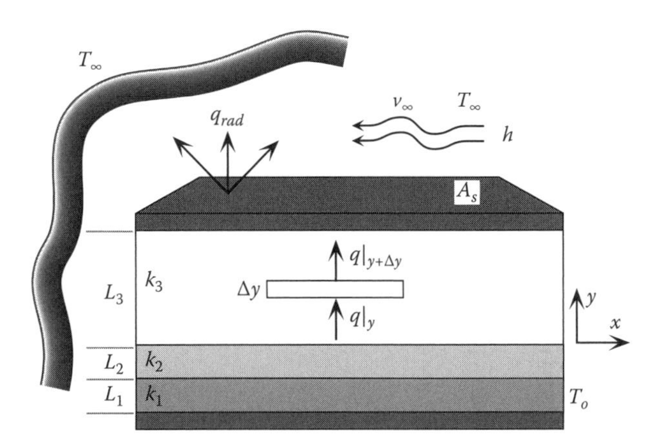

FIGURE 4.13 Heat conduction through a composite wall with convection and radiation.

The boundary conditions for this problem specify a constant temperature at $y = 0$, continuity of temperature and heat flow at the interfaces between individual material layers, and finally a heat flow balance between conduction, radiation, and convection at the upper surface. The last boundary condition is a combination of our convective and radiative boundary conditions:

$$y = 0 \quad T_1 = T_o$$

$$y = L_1 \quad T_1 = T_2 \quad -k_1 A_s \frac{dT_1}{dy} = -k_2 A_s \frac{dT_2}{dy}$$

$$y = L_1 + L_2 \quad T_2 = T_3 \quad -k_2 A_s \frac{dT_2}{dy} = -k_3 A_s \frac{dT_3}{dy}$$

$$y = L_1 + L_2 + L_3 \quad -k_3 A_s \frac{dT_3}{dy} = hA_s(T_3 - T_\infty) + \sigma' \varepsilon' A_s \left(T_3^4 - T_\infty^4\right)$$

The final boundary condition is nonlinear because it involves radiation and the Stefan–Boltzmann law. This makes the problem much more difficult, and it would be impossible to express the result in terms of an overall driving force and sum of resistances. We reduce the complexity by defining a radiation heat transfer coefficient, h_r, through factoring the Stefan–Boltzmann law:

$$\sigma' \varepsilon' A_s \left(T_3^4 - T_\infty^4\right) = \sigma' \varepsilon' A_s (T_3^2 + T_\infty^2)(T_3 + T_\infty)(T_3 - T_\infty)$$
$$= h_r A_s (T_3 - T_\infty)$$

Now the final boundary condition can be written as

$$y = L_1 + L_2 + L_3 \quad -k_3 A_s \frac{dT_3}{dy} = (h + h_r) A_s (T_3 - T_\infty)$$

Notice that the radiation heat transfer coefficient depends upon the surface temperature so rewriting the problem makes finding the solution an iterative process.

After solving our equations subject to the boundary conditions, we can express the result for the heat flow in terms of an overall driving force and sum of resistances:

$$q = \frac{T_0 - T_\infty}{\left[\dfrac{L_1}{k_1 A_s} + \dfrac{L_2}{k_2 A_s} + \dfrac{L_3}{k_3 A_s} + \left(\dfrac{1}{h + h_r}\right)\dfrac{1}{A_s}\right]}$$

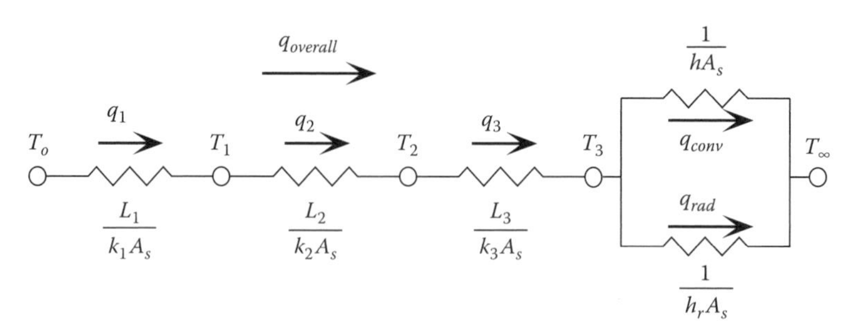

FIGURE 4.14 Circuit diagram corresponding to Figure 4.13.

In this situation, we have five resistances. The three conduction resistances are in series with a parallel combination of resistances arising from convection and radiation at the upper surface. The radiation and convection resistances are in parallel because both heat transfer mechanisms are occurring simultaneously and independently at the upper surface. Thus, the electrical analogy leads to the circuit diagram shown in Figure 4.14.

We can use an iterative technique to solve this problem for the overall heat flow. The heat flow, q, between nodes T_o and T_3 is

$$q = \frac{T_0 - T_3}{\left[\dfrac{L_1}{k_1 A_s} + \dfrac{L_2}{k_2 A_s} + \dfrac{L_3}{k_3 A_s}\right]}$$

We first guess a value for T_3, calculate h_r, and use the solution for the entire system to evaluate the heat flow, q. With this value for q, we can use the earlier solution (between T_o and T_3) to determine a new value for T_3 and repeat the entire procedure until we converge on a solution. Assuming an initial T_3 of 30°C, we find a final value for T_3 of 54.4°C and q = 468.3 W.

The major heat transfer resistance lies in the fiberglass insulation and this controls the overall heat flow. Looking at each resistance in the diagram lets us determine the temperatures at the interfaces between the insulation layers:

$$q = \frac{T_o - T_1}{\left[\dfrac{L_1}{k_1 A_s}\right]} = \frac{T_o - T_2}{\left[\dfrac{L_1}{k_1 A_s} + \dfrac{L_2}{k_2 A_s}\right]}$$

Solving gives T_1 = 978.7°C and T_2 = 54.5°C.

Example 4.3 Mass Transfer Coefficients

Composite media present an excellent opportunity to look at overall mass transfer coefficients. Consider the composite system shown in Figure 4.15. We have a liquid–liquid contacting system with a well-defined interface between the liquids. At the bottom of the contactor, a solute dissolves in liquid #1. The dissolution maintains a constant mole fraction of solute at the solid surface. The solute diffuses through liquid #1 until it reaches the interface. At the interface, the solute partitions between the liquids. Then the solute diffuses through the second liquid phase. A steady stream of air flows over the top of the device maintaining the solute concentration at the top constant. We would like to determine the mole fraction profiles in the liquids and the mass transfer coefficients across the interface. We will assume a constant concentration in both liquids, the same overall total concentration in both liquids, and dilute solutions.

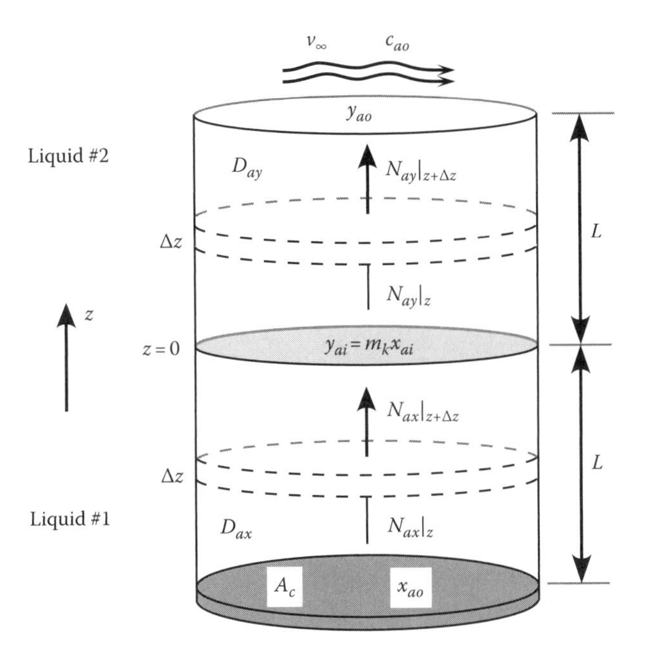

FIGURE 4.15 Gas-liquid contactor.

To determine the mole fraction profiles, we must first create a control volume in each phase. We then write a conservation of mass law about each control volume, substitute in Fick's law for dilute solutions, and solve for the profile. The differential equations describing the profiles are

$$\frac{d^2 x_a}{dz^2} = 0 \qquad \frac{d^2 y_a}{dz^2} = 0$$

The mole fraction profiles in each fluid are linear:

$$x_a = a_1 z + a_2 \qquad y_a = b_1 z + b_2$$

The constants of integration must be determined from the boundary conditions. Here we know the solute mole fraction at the lower boundary and the solute mole fraction at the upper boundary:

$$z = -L \qquad x_a = x_{ao} \qquad z = L \qquad y_a = y_{ao}$$

At the interface, we know we have a partitioning of the solute according to the equilibrium law:

$$z = 0 \qquad y_a = m_k x_a$$

Here we assume that at the mathematical interface we have equilibrium between the liquids ($x_a = x_{ai}$ and $y_a = y_{ai}$). Since the system is at steady state, whatever solute flows to the interface from liquid #1 must flow from the interface and into liquid #2, that is, we must equate the fluxes there. For simplicity, we assume the total concentrations in the two phases are equal:

$$z = 0 \qquad -c_t D_{ax} \frac{dx_a}{dz} = -c_t D_{ay} \frac{dy_a}{dz}$$

The four boundary conditions enable us to obtain the mole fraction profiles and the fluxes:

$$x_a = \left[\frac{D_{ay}(y_{ao} - m_k x_{ao})}{L(D_{ax} + m_k D_{ay})}\right] z + \frac{y_{ao} D_{ay} + x_{ao} D_{ax}}{D_{ax} + m_k x_{ay}}$$

$$y_a = \left[\frac{D_{ax}(y_{ao} - m_k x_{ao})}{L(D_{ax} + m_k D_{ay})}\right] z + \frac{m_k(y_{ao} D_{ay} + x_{ao} D_{ax})}{D_{ax} + m_k x_{ay}}$$

$$N_{ax} = N_{ay} = \frac{c_t D_{ax} D_{ay}}{L(D_{ax} + m_k D_{ay})}(y_{ao} - m_k x_{ao})$$

Representative concentration profiles are shown in Figure 4.16 for a system where the ratio of diffusivities is 1:5 and the ratio of layer thicknesses is 2:3. These results were obtained using Comsol® module TwoRegionDiffusion.

Now we can begin to look at different mass transfer coefficient definitions. When multiplied by the appropriate driving force, each coefficient should yield the mass flux through the system. The first two coefficients we considered were local coefficients based on a viewpoint located within one of the two phases. We define these coefficients as

$$N_{ax} = c_t k_{cx}(x_{ao} - x_{ai}) \qquad x_{ai} = x_a(z = 0)$$

$$N_{ay} = c_t k_{cy}(y_{ai} - y_{ao}) \qquad y_{ai} = y_a(z = 0)$$

Using the mole fraction profiles and the flux we calculated from them, we can solve for k_{cx} and k_{cy} to show

$$k_{cx} = \frac{D_{ax}}{L} \qquad k_{cy} = \frac{D_{ay}}{L}$$

The local mass transfer coefficients in this example are just the diffusivities in the individual phases divided by the thickness of each layer (one over the transport resistance in the layer multiplied by the cross sectional area, A_c). Only one resistance is involved in these mass transfer coefficients because we know the composition at the interface exactly.

If we do not know the composition at the interface, we must use an overall mass transfer coefficient. We defined our overall mass transfer coefficients to be

$$N_{ax} = c_t K_{cx}\left(x_{ao} - \frac{y_{ao}}{m_k}\right)$$

$$N_{ay} = c_t K_{cy}(m_k x_{ao} - y_{ao})$$

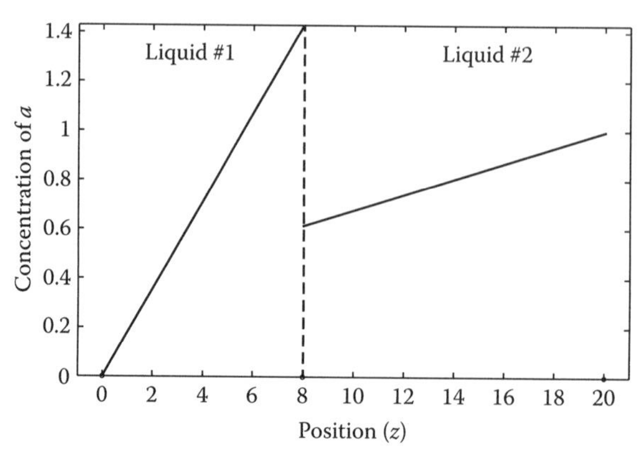

FIGURE 4.16 Concentration profiles in a liquid-liquid system. Comsol® simulation using: $K_{eq} = 2.5$, $D_{ax}/D_{ay} = 1/5$, $L_x/L_y = 2/3$.

Using the flux, these overall coefficients are

$$K_{cx} = \left(\frac{1}{\dfrac{L}{m_k D_{ay}} + \dfrac{L}{D_{ax}}} \right) \quad K_{cy} = \left(\frac{1}{\dfrac{L}{D_{ay}} + \dfrac{m_k L}{D_{ax}}} \right)$$

Notice that both coefficients are different representations of the sum of resistances (multiplied by area) in the system. Each contains diffusion resistances for both phases and a resistance to getting material across the interface.

Example 4.4 Interfacial Thermal Resistance

No two materials can be in perfect contact. The atomic force microscope image in Figure 4.17a illustrates how jagged a real surface can be. Bringing two such surfaces into contact leaves pockets of air trapped in the interstices that manifest themselves as an interfacial resistance. We can use a simple model in one dimension that is remarkably similar to the situation we just studied in the previous example to represent this phenomenon. Figure 4.17b is a schematic of the situation.

At steady state, in each material we know that the temperature profile will be governed by

$$\frac{d^2T_1}{dy^2} = 0 \quad \text{and} \quad \frac{d^2T_2}{dy^2} = 0$$

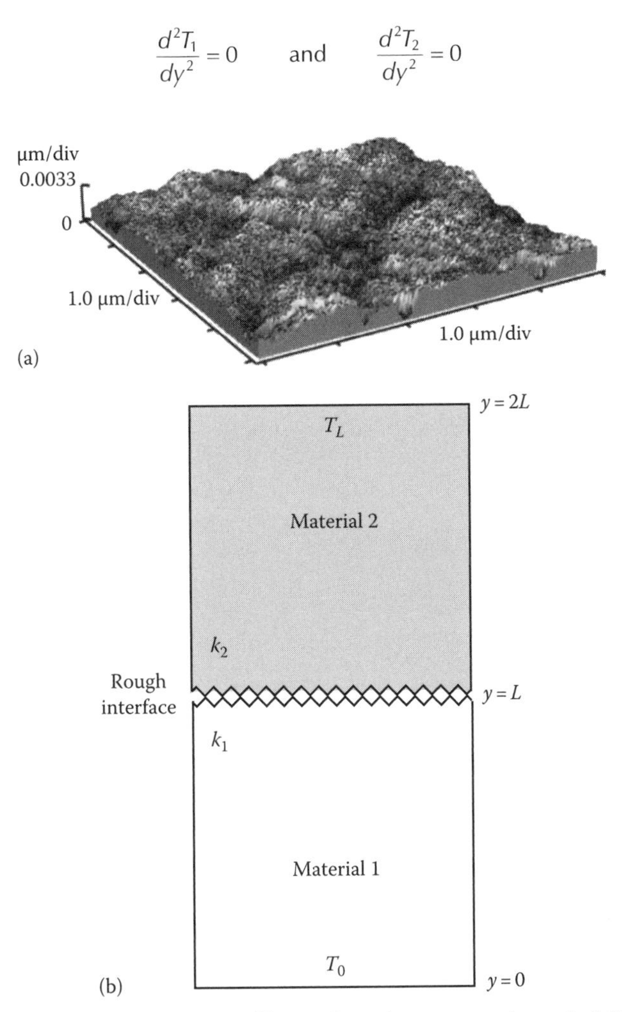

FIGURE 4.17 (a) AFM image of a polished silica surface. Average roughness is 1.5 nm. (b) Schematic of contact between two materials with an interfacial resistance.

At the lower surface of material 1, we have the condition that

$$y = 0 \qquad T_1 = T_0$$

At the upper surface of material 2, we have

$$y = 2L \qquad T = T_L$$

The heart of the problem lies at the interface between the two materials. Here we have the interfacial resistance. Though we have a resistance, the heat that flows from material 1 must flow into material 2. Thus, one condition at the interface must be

$$y = L \qquad -k_1 \frac{dT_1}{dy} = -k_2 \frac{dT_2}{dy}$$

The other condition actually embodies the interface resistance. This condition specifies that there is a temperature jump across the interface. We can specify any functional form between T_1 and T_2, within the bounds of what is physically possible, but the following form is likely complicated enough:

$$y = L \qquad T_1 = \alpha T_2 + \beta$$

The solution for the temperature profiles is

$$T_1 = \frac{k_2(\alpha T_L + \beta - T_0)}{(k_2 + \alpha k_1)L} y + T_0$$

$$T_2 = \frac{k_1(\alpha T_L + \beta - T_0)}{(k_2 + \alpha k_1)L}(y - 2L) + T_L$$

The heat flow shows that the effect of the interfacial resistance is to alter the driving force and the composite resistance. In the situation here where $T_0 > T_L$, α is a number greater than one that decreases the driving force and increases the resistance:

$$q = \frac{T_0 - (\alpha T_L + \beta)}{\left[\dfrac{L}{k_1 A_s} + \dfrac{\alpha L}{k_2 A_s}\right]} \qquad \frac{\text{Driving force}}{\text{Resistance}}$$

4.5.2 Cylindrical Systems

Many important applications of composite media and sum-of-resistance concepts occur in a cylindrical geometry. Examples include multiply insulated pipes for carrying heated or refrigerated fluids, multilayer resistors and capacitors, and composite hollow fiber membranes. The application we will consider, as an illustration of applying the sum-of-resistance concept to cylindrical systems, is the critical insulation thickness required around a nerve bundle.

Nerve bundles travel throughout the body, relaying signals from the brain to individual muscle groups, to individual internal organs, and to sensory organs as well. Since the body is filled with an electrolyte solution having a fairly large electrical conductivity, it is important for the nerve fibers to be well insulated. The insulation is a protein or biopolymer called myelin. If inadequate insulation is provided, signal integrity will be lost, and the human or animal will be paralyzed. Such diseases of the insulation are uncommon but manifest themselves as multiple sclerosis and, other more rare, degenerative myelopathies.

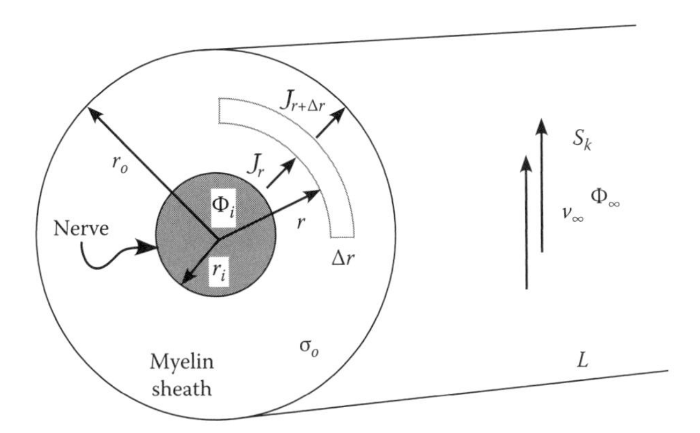

FIGURE 4.18 Schematic of a nerve fiber immersed in body fluid.

We can model the nerve fiber as a cylindrical conductor of radius, r_i, covered with an insulating sheath of radius, r_o, and conductivity, σ_o. The nerve bundle will be assumed to have an infinite conductivity so that it will be charged to a uniform potential, Φ_i, when the brain sends a command. The nerve bundle is immersed in body fluid that has an indefinite extent and may be flowing past the nerve bundle. We define a convection coefficient, S_k, that describes the leakage of current through the insulation and out into the body fluid.

Figure 4.18 shows the physical situation. Our task is to determine the amount of insulation needed to insure we minimize current flow from the nerve to the body fluid and hence proper signal conduction. We write the charge balance equation over a small control volume located within the insulation. We assume that the system is at steady state and we have no generation terms to consider. It is important to remember that we must write the equation in terms of the current flow, I, and not the current flux or current density, J. The current density will not be a constant in this case because the surface area available for charge transfer changes with position:

$$\begin{aligned} &\text{In} \quad - \quad \text{Out} \quad + \text{Gen} = \text{Acc} \\ &I(r) - I(r+\Delta r) + \quad 0 \quad = 0 \end{aligned} \tag{4.107}$$

We divide Equation 4.107 by Δr and take the limit as $\Delta r \to 0$ to obtain the differential equation:

$$-\frac{d}{dr}(I) = 0 \tag{4.108}$$

Next, we substitute for the current density using Ohm's law and rewrite the differential equation in terms of the electric potential difference between the nerve and surrounding fluid:

$$I = JA_s = -\sigma_o A_s \frac{d\Phi}{dr} \quad \text{Ohm's law} \tag{4.109}$$

Substituting Equation 4.109 into the differential equation along with the surface area of a cylinder, $A_s = 2\pi rL$ (the area through which the charge will flow), and eliminating the constant term, $2\pi L\sigma_o$, gives the required second-order differential equation:

$$\frac{d}{dr}\left(r\frac{d\Phi}{dr}\right) = 0 \tag{4.110}$$

The previously presented differential equation states that the current flow throughout the composite system is a constant but the current density is not because the area available for transport increases with r. The general solution to Equation 4.110 is

$$\Phi = a_1 \ln(r) + a_2 \tag{4.111}$$

The boundary conditions reflect the facts that the nerve is charged to a uniform potential, Φ_i, and that the current flow due to conduction through the nerve–insulation composite must be equal to the current flow at the surface of the insulation and into the fluid:

$$r = r_i \qquad \Phi = \Phi_i \tag{4.112}$$

$$r = r_o \qquad -2\pi L r_o \sigma_o \frac{d\Phi}{dr} = 2\pi L r_o S_k (\Phi - \Phi_\infty) \tag{4.113}$$

Evaluating the constants gives the required voltage profile:

$$a_1 = -\frac{(\Phi_i - \Phi_\infty)}{\dfrac{1}{\sigma_o}\ln\left(\dfrac{r_o}{r_i}\right) + \dfrac{1}{r_o S_k}} \tag{4.114}$$

$$a_2 = \Phi_i + \left[\frac{(\Phi_i - \Phi_\infty)}{\dfrac{1}{\sigma_o}\ln\left(\dfrac{r_o}{r_i}\right) + \dfrac{1}{r_o S_k}}\right]\ln(r_i) \tag{4.115}$$

$$\Phi = \Phi_i + \left[\frac{(\Phi_i - \Phi_\infty)}{\dfrac{1}{\sigma_o}\ln\left(\dfrac{r_o}{r_i}\right) + \dfrac{1}{r_o S_k}}\right]\ln\left(\frac{r_i}{r}\right) \tag{4.116}$$

Using Ohm's law we determine the current flowing from the nerve through the insulation:

$$I = -\sigma_o(2\pi r L)\frac{d\Phi}{dr} = \frac{(\Phi_i - \Phi_\infty)}{\underbrace{\dfrac{1}{2\pi L\sigma}\ln\left(\dfrac{r_o}{r_i}\right)}_{\text{Conduction}} + \underbrace{\dfrac{1}{2\pi L r_o S_k}}_{\text{Convection}}} \tag{4.117}$$

Notice that we have the sum of two resistances: one due to conduction through the insulating material and a second resistance due to convection at the edge of the insulation. As we begin to add insulation to the nerve, we increase the surface area available for charge transfer. While we may be increasing the conduction resistance by adding insulation, we are decreasing the resistance from the insulation into the fluid at the same time. At some point, the insulating power of the insulation will just balance the increase in current flow due to the increase in surface area.

This point of maximum current flow is called the *critical insulation thickness*. Mathematically we find this point by taking the derivative of the current flow, I, with respect to the outer insulation radius, r_o, and setting the derivative to zero:

$$\frac{dI}{dr_o} = \frac{d}{dr_o}\left(\frac{\Phi_i - \Phi_\infty}{\frac{1}{2\pi L\sigma_o}\ln\left(\frac{r_o}{r_i}\right) + \frac{1}{2\pi L r_o S_k}}\right) = 0 \tag{4.118}$$

Solving Equation 4.118 for r_o shows that the critical insulation thickness is reached when

$$r_o = \frac{\sigma_o}{S_k} \tag{4.119}$$

If $r_o < \sigma_o/S_k$, then the current flow increases with increasing amounts of insulation, and if $r_o > \sigma_o/S_k$, then the current flow decreases with increasing amounts of insulation. What would r_o need to be to insure less leakage current than would exist from a bare nerve?

The critical insulation thickness also translates to optical, electrical, and thermal systems. Optical fibers have a thick cladding layer that prevents light from leaking out from the core. Without sufficient cladding, we would need to boost the signal in optical communications system much more frequently than we do now. Without sufficiently thick electrical insulation, current would leak from a conductor. This could lead to electrocution, or in microprocessors, this leakage leads to cross-talk between conductors and misfiring of transistors. Most pipes carrying hot fluids have a thick layer of insulation. If the insulation is not thicker than the critical thickness, more heat is lost than would occur without insulation at all. In our bodies, all nerve bundles must have a protein insulation in an amount much greater than the critical thickness for proper muscle and sensory action. In autoimmune diseases where the insulating sheath is attacked, portions of the bundle may have an insulating thickness below the critical value and any information flowing through that bundle will be distorted (tremors) or leaked away (paralysis).

4.5.3 Spherical Systems

As a final example, we will now consider the process in spherical coordinates and calculate the evaporation rate from a liquefied natural gas tank. Figure 4.19 shows a spherical tank, 3 m in diameter, filled with methane at a temperature of –164°C. The tank is lagged (covered) with half a meter

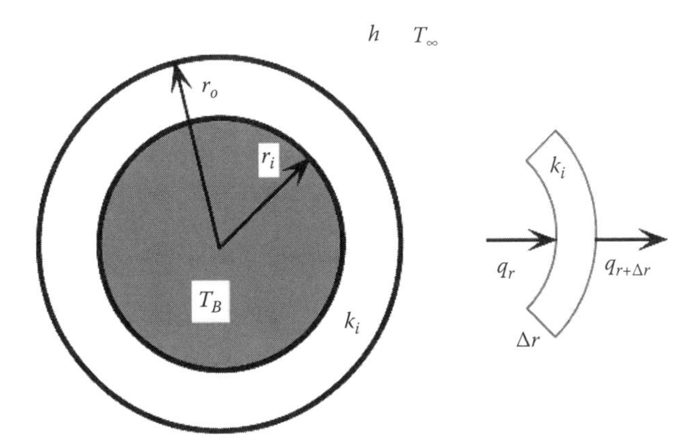

FIGURE 4.19 Spherical container filled with liquid natural gas.

of urethane insulation (k_i = 0.023 W/m K). Air flows over the tank and the tank is designed for a convective heat transfer coefficient of 25 W/m^2 K at a free stream temperature; T_∞ = 35°C. The heat of vaporization of methane is 510 kJ/kg.

We write the balance equation in terms of the heat flow, q, about a small control volume located within the insulation:

$$\begin{aligned} &\text{In} - \text{Out} + \text{Gen} = 0 \\ &q(r) - q(r+\Delta r) + 0 = 0 \end{aligned} \tag{4.120}$$

Dividing by Δr and taking the limit as $\Delta r \to 0$ gives

$$-\frac{d}{dr}(q) = 0 \tag{4.121}$$

Using Fourier's law and accounting for the change in surface area of the control volume with r, $A_s = 4\pi r^2$, we have

$$\frac{d}{dr}\left[4\pi k_i r^2 \frac{dT}{dr}\right] = \frac{d}{dr}\left[r^2 \frac{dT}{dr}\right] = 0 \tag{4.122}$$

The solution to this equation is

$$T = -\frac{a_1}{r} + a_2 \tag{4.123}$$

The boundary conditions for this situation state that we know the temperature of the liquid methane and that we have convective heat transfer from the surface:

$$r = r_i \qquad T = T_B \tag{4.124}$$

$$r = r_o \qquad -k_i \frac{dT}{dr} = h(T - T_\infty) \tag{4.125}$$

Evaluating the constants a_1 and a_2 yields the final solution for the temperature profile:

$$T = \left(\frac{1}{k_i}\right)\left(\frac{T_B - T_\infty}{\dfrac{1}{hr_o^2} + \dfrac{1}{k_i}\left(\dfrac{1}{r_i} - \dfrac{1}{r_o}\right)}\right)\left(\frac{1}{r_i} - \frac{1}{r}\right) + T_B \tag{4.126}$$

The heat flow is given either by Fourier's law or Newton's law of cooling evaluated at the tank/air interface. Using Fourier's law we find

$$q = -k_i\left(4\pi r_o^2\right)\frac{dT}{dr}\bigg|_{r=r_o} = 4\pi\left(\frac{T_\infty - T_B}{\dfrac{1}{hr_o^2} + \dfrac{1}{k_i}\left(\dfrac{1}{r_i} - \dfrac{1}{r_o}\right)}\right) \quad \frac{\text{Driving force}}{\text{Resistance}} \tag{4.127}$$

All this heat input goes into evaporating the liquefied methane. Thus, the rate of evaporation per unit area is

$$\dot{m}\left(\frac{\text{kg}}{\text{s}}\right) = \frac{q}{h_{fg}} = \frac{4\pi}{h_{fg}}\left(\frac{T_\infty - T_B}{\dfrac{1}{hr_o^2} + \dfrac{1}{k_i}\left(\dfrac{1}{r_i} - \dfrac{1}{r_o}\right)}\right) \tag{4.128}$$

4.6 VARIABLE TRANSPORT PROPERTIES, COUPLED TRANSPORT, AND MULTIPLE FLUXES

In the previous sections of this chapter, we have considered simplified problems where the transport properties of the material were constants. Nature is never that cooperative and all transport properties vary with composition, temperature, applied field, and even with the shear stress applied to the system. In this section we will consider some examples of variable transport coefficients, how these variations yield coupled transport equations, and how multiple fluxes also couple the transport equations. We will discuss the implications of variable transport coefficients on the balance (conservation) equations and what these variable coefficients and multiple fluxes imply for the distribution of the potentials (state variables) in the system.

4.6.1 Composition-Dependent Diffusivity

In this first example, we consider a common case where the diffusivity of a medium depends upon the composition of the medium and where the diffusion process itself is the agent that changes this composition. In Figure 4.20, we show a situation where we allow the diffusivity of species *a* to vary linearly with the amount of species *a* present in the solution. The situation is one where we will have equimolar counterdiffusion of species *a* and *b*.

We analyze this situation in exactly the same manner we used when considering mass transfer with constant diffusivity. The starting point involves writing the balance equation (conservation of mass) over a small control volume of area, A_c, and width, Δy:

$$\begin{array}{ccccc} \text{In} & - & \text{Out} & +\text{Gen} = & \text{Acc} \\ \dot{M}_a(y) & - & \dot{M}_a(y+\Delta y) & +\ 0 \ = & 0 \end{array} \tag{4.129}$$

Dividing by $\Delta y A_c$, and letting $\Delta y \to 0$ yields the differential equation stating that the molar flux is a constant throughout the system:

$$\frac{dN_a}{dy} = 0 \tag{4.130}$$

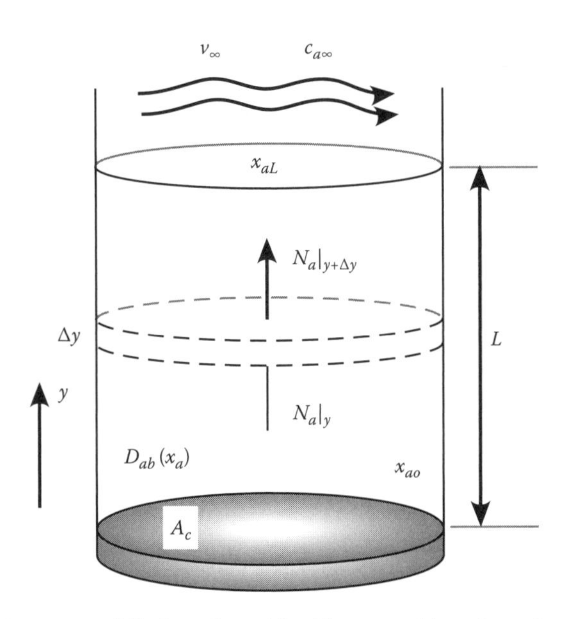

FIGURE 4.20 Equimolar counterdiffusion of *a* and *b* with composition–dependent binary diffusivity.

Since we are interested in the mole fraction profile throughout the system, we transform the differential equation by substituting for N_a using Fick's law. Assuming equimolar counterdiffusion and an incompressible solution,

$$N_a = -c_t D_{ab} \frac{dx_a}{dy} \tag{4.131}$$

The diffusivity is a linear function of composition:

$$D_{ab} = D_{abo}(1 + \gamma_x x_a) \tag{4.132}$$

Substituting Equations 4.131 and 4.132 into the differential equation and removing all constant terms (c_t and D_{abo}) results in a differential equation for the mole fraction profile:

$$\frac{d}{dy}\left[\left(1 + \gamma_x x_a\right)\frac{dx_a}{dy}\right] = 0 \tag{4.133}$$

Equation 4.133 is easily solved by separating variables and integrating twice:

$$(1 + \gamma_x x_a)\frac{dx_a}{dy} = a_o \tag{4.134}$$

$$x_a + \frac{\gamma_x}{2} x_a^2 = a_o y + a_1 \tag{4.135}$$

The boundary conditions for this problem specified the mole fractions of species a at both the inlet and outlet (see Figure 4.20) of the device:

$$y = 0 \qquad x_a = x_{ao} \tag{4.136}$$

$$y = L \qquad x_a = x_{aL} \tag{4.137}$$

Substituting these conditions into the solution, Equation 4.135, and solving for a_o and a_1 gives:

$$a_o = \frac{x_{aL} - x_{ao}}{L} + \frac{\gamma_x}{2L}\left(x_{aL}^2 - x_{ao}^2\right) = \frac{\Delta x}{L} + \gamma_x\left(\frac{\Delta x}{L}\right)x_{avg} \tag{4.138}$$

$$a_1 = x_{ao} + \frac{\gamma_x}{2} x_{ao}^2 \tag{4.139}$$

Where Δx is the overall mole fraction difference between the inlet and outlet of our system and x_{avg} is the arithmetic average of the inlet and outlet mole fractions:

$$\Delta x = x_{aL} - x_{ao} \qquad x_{avg} = \frac{x_{aL} + x_{ao}}{2} \tag{4.140}$$

Using the definitions for Δx and x_{avg}, the final mole fraction profile and molar flux can be written as

$$(x_a - x_{ao}) + \frac{\gamma_x}{2}\left(x_a^2 - x_{ao}^2\right) = \frac{\Delta x}{L}(1 + \gamma_x x_{avg})y \tag{4.141}$$

$$\dot{M}_a = A_s N_a = \frac{\Delta x}{L/D_{abo}(1 + \gamma_x x_{avg})A_s} \qquad \frac{\text{Driving force}}{\text{Resistance}} \tag{4.142}$$

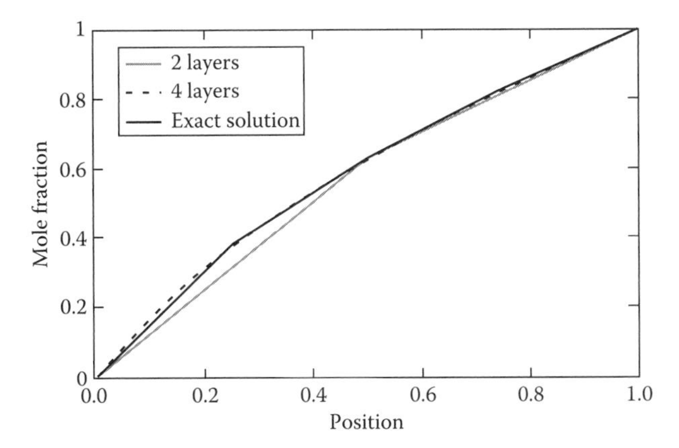

FIGURE 4.21 Comparison of exact solution for diffusion with variable diffusivity with solutions generated by assuming sum-of-resistances in discrete media.

The only difference between this flux solution for the variable diffusivity and that for the constant diffusivity case lies in the definition for the diffusivity and the resistance to mass transfer. In the present case, we can define a *constant* diffusion coefficient if we evaluate the variable diffusion coefficient at x_{avg}.

The mole fraction profile is shown in Figure 4.21. Also shown in Figure 4.21 is the mole fraction profile obtained by approximating the variable diffusivity material as a series of constant diffusivity materials linked using a sum-of-resistance approach. One can see that the sum-of-resistance concept can be used to solve variable transport coefficient problems if we know the functionality of the transport coefficient. Very few *slices* or regions of constant diffusivity are required to obtain an accurate approximation to the full solution.

The concentration-dependent diffusivity concept we have just solved arises in a number of other scenarios too. One such occurs during diffusion in solid materials. In this case we have a solid that is essentially incompressible. Diffusing material in leads to a force that resists the further introduction of solute. This so-called "elastic stress" concept manifests itself as a concentration-dependent diffusivity. We will explore this concept in the problems section using Comsol®.

4.6.2 Forced Ionic Diffusion

Forced, ionic diffusion in a battery is a prime example of steady-state, 1-D transport under the control of two flux–gradient laws simultaneously. Consider a battery consisting of a quiescent aqueous solution of a salt $PbSO_4$ between two flat, parallel plates of lead, Pb. We pass a constant direct current between the plates such that the metal is dissolved (oxidized) at the anode, flows through the solution, and is deposited (reduced) on the cathode. We want to know the concentration profile of $PbSO_4$ in the solution and the maximum possible current that we can draw. The solution of $PbSO_4$ is very dilute and all material properties, temperature, and pressure are constant in the system. A diagram of the battery is shown in Figure 4.22.

$PbSO_4$ is a 1:1 salt so we can consider the solution to be a ternary mixture of Pb^{2+}, water, and SO_4^{2-}. Since we have dissolution and deposition of Pb^{2+} on the plates, we can consider Pb^{2+} to be the only current-carrying species and the overall fluxes of SO_4^{2-} and water are then zero. This presumes we have global electroneutrality and that ion information travels instantaneously throughout the solution. Thus, the concentrations of Pb^{2+} and SO_4^{2-} are equal everywhere. In reality, the concentrations near the plates (in a boundary layer) would be different than that in the bulk. We may satisfy global electroneutrality but not local electroneutrality. Since we assume that the solution is dilute, the total concentration is constant and we can deal with either concentrations or mole fractions.

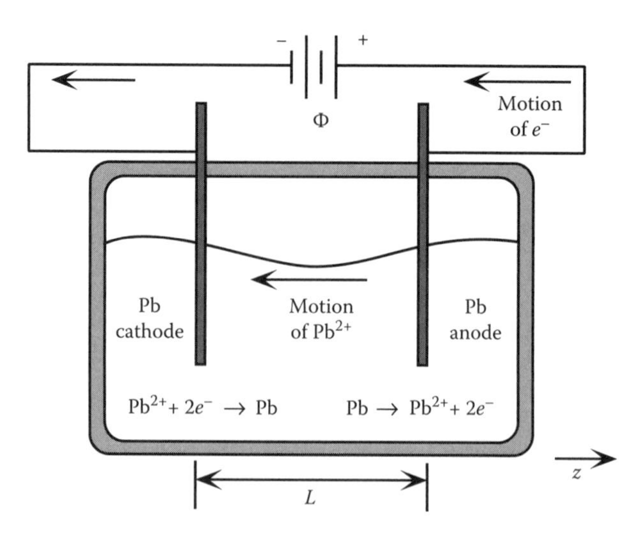

FIGURE 4.22 Schematic diagram of a battery consisting of an aqueous salt solution, $PbSO_4$, confined between parallel plates of lead, Pb.

The starting point for this problem is the small control volume we isolate from a region between the anode and cathode. We consider conservation of mass within this control volume for both cations (Pb^{2+}) and anions $\left(SO_4^{2-}\right)$. There are two flux components for each ion—one arising from the concentration gradient and the second arising from the applied electric field. The control volume and fluxes are shown in Figure 4.23. The flux of cations leads to a current, I, while the solvent (water) and anions are assumed to be stagnant.

With this in mind, we can write down the fluxes for the ions and water and hence their respective currents: $PbSO_4$

$$I = \left[J_x(\text{Pb}^{2+}) + J_e(\text{Pb}^{2+})\right]A_c \tag{4.143}$$

$$0 = \left[J_x\left(\text{SO}_4^{2-}\right) + J_e\left(\text{SO}_4^{2-}\right)\right]A_c \tag{4.144}$$

$$0 = J_x(\text{H}_2\text{O})A_c \tag{4.145}$$

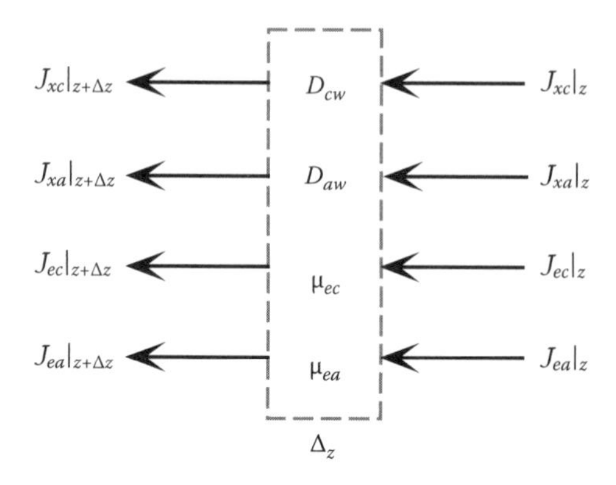

FIGURE 4.23 Control volume within the battery detailing the fluxes arising from the concentration gradients and the applied electric field. Cation (Pb = c); anion ($SO_4 = a$).

If we assume that the ions are dilute enough so that they do not exert any forces on each other, then we can assume that we have two pseudo-binary systems: cation in water and anion in water. We need two diffusivities, D_{cw} and D_{aw}, and two mobilities, μ_{ec} and μ_{ea}, respectively. Using results from Chapter 2, Equation 2.67, to describe the fluxes of the ions in terms of the mole fraction and voltage gradients yields

$$I = -A_c z_e e\left[c_t D_{cw}\frac{dx_c}{dz} + c_t \mu_{ec} x_c \frac{d\Phi}{dz}\right] \tag{4.146}$$

$$0 = -A_c z_e e\left[c_t D_{aw}\frac{dx_a}{dz} - c_t \mu_{ea} x_a \frac{d\Phi}{dz}\right] \tag{4.147}$$

where we have included the equal but opposite charges on the ions in writing the equations. The mole fractions of the individual ions are equal ($x_c = x_a$) since we have a 1:1 salt and we can use that fact to remove the field gradient, $d\Phi/dz$, from the two equations and obtain an expression for the current in terms of dx_c/dz:

$$I = -c_t A_c z_e e\left[D_{cw} + \frac{\mu_{ec}}{\mu_{ea}} D_{aw}\right]\frac{dx_c}{dz} \tag{4.148}$$

We define a pseudo-binary diffusivity, D_{ca}, which accounts for the different diffusivities and mobilities of the ions:

$$D_{ca} = \left[D_{cw} + \frac{\mu_{ec}}{\mu_{ea}} D_{aw}\right] \tag{4.149}$$

In a dilute, ideal solution, both the total concentration, c_t, and the diffusivity, D_{ca}, will be constant. Thus, we can integrate Equation 4.148 to obtain the mole fraction profile of cations in terms of the current flowing through the battery:

$$\frac{I}{c_t A_c z_e e D_{ca}} = \frac{x_c - x_o}{z} \tag{4.150}$$

Here, we have incorporated a boundary condition that specifies the mole fraction of cations at the cathode surface:

$$z = 0 \qquad x_c = x_o \tag{4.151}$$

Equation 4.150 shows that the concentration gradient in the cell is linear. The maximum current that can be drawn, I_{max}, is reached when we have an infinite deposition rate. An infinite deposition rate can only occur when the concentration of Pb^{2+} at the cathode is zero ($x_o = 0$). Using this constraint and evaluating the equation at $z = L$ to account for the complete resistance, I_{max} becomes

$$I_{max} = \frac{c_t A_c z_e e D_{ca} x_{avg}}{L} \tag{4.152}$$

Example 4.5 Maximum Current from an Ag–$AgNO_3$ Battery

Consider a battery composed of a silver metal anode and cathode immersed in a water solution containing ($AgNO_3$) silver nitrate. The electrodes are spaced 1 mm apart and have a total surface area of 10 cm². When initially assembled, the battery was filled with its electrolyte solution such that the specific gravity of the solution, at 20°C, was 1.45. We want to determine the maximum current that can flow in a battery of this type.

To obtain a solution for this problem, we need to determine the average concentration of silver ions in the solution and the diffusivities and mobilities for the silver and nitrate ions in solution. The first quantity we can find from one of many handbooks since data on the specific gravities of electrolyte solutions are common. From the *Handbook of Chemistry and Physics*, we find that the concentration of silver nitrate corresponding to the specific gravity of 1.45 is 3.3 mol/L. Since this is a 1:1 electrolyte, the concentration of silver ions and the concentration of nitrate ions are both 3.3 mol/L, so $c_{ag,avg} = 3300$ mol/m³. In terms of the number of charges per unit volume, $c_t x_{avg} = 3300(ez^+)\mathcal{N}_{av} = 3.179 \times 10^8$ C/m³. The diffusivities and mobilities for silver and nitrate ions can be found from data in Appendix D:

$$D_{Ag} = 1.65 \times 10^{-9}\,\text{m}^2/\text{s} \qquad \mu_{Ag} = 6.41 \times 10^{-8}\,\text{m}^2/\text{s}\cdot\text{V}$$

$$D_{NO_3} = 1.90 \times 10^{-9}\,\text{m}^2/\text{s} \qquad \mu_{NO_3} = 7.40 \times 10^{-8}\,\text{m}^2/\text{s}\cdot\text{V}$$

Using Equations 4.149 and 4.152, the overall diffusivity and maximum current are

$$D_{ca} = \left[1.65 \times 10^{-9} + \left(\frac{6.41 \times 10^{-8}}{7.40 \times 10^{-8}}\right)1.90 \times 10^{-9}\right]$$

$$= 3.30 \times 10^{-9}\,\text{m}^2/\text{s}$$

$$\mathcal{I}_{max} = \frac{(0.001)(3.179 \times 10^8)(3.3 \times 10^{-9})}{0.001} = 1.05\,\text{A}$$

In actual batteries, the total surface area may be on the order of 0.1–1 m². Large currents can be drawn under ideal situations. Notice how the diffusivity varies as a function of temperature in Appendix D. How would this behavior affect the performance of electric cars, for example, in northern latitudes?

4.7 SUMMARY

We have discussed steady-state, 1-D transport in planar, cylindrical, and spherical geometries and showed by application of the conservation laws that the flow of our transported quantity must be a constant throughout the system. We also showed how we can express the flow of heat, mass, momentum, charge, etc., in terms of a driving force divided by a resistance to transport. The resistances to flow of heat, mass, momentum, and charge are shown in Table 4.1 for several common geometries. This concept was applied to composite media and we showed how we can express the flow of our transported quantity in terms of an overall driving force across the system, divided by the sum of all resistances to transport throughout the system. This sum-of-resistance concept could be expressed in terms of an electrical network containing resistances in series and in parallel. Finally, we discussed some more realistic instances of transport processes where the transport coefficients were no longer constant but depended on one or more of the state variables in the system.

TABLE 4.1
Resistances to Flow of Momentum, Heat, Mass, and Charge

	Resistances			
	Planar		Cylindrical	
Transport Process	**Diffusive**	**Convective**	**Diffusive**	**Convective**
Momentum	$\frac{L}{\mu A}$	See Chapter 10	$\frac{\ln(r_o/r_i)}{2\pi\mu L}$	See Chapter 10
Heat	$\frac{L}{kA}$	$\frac{1}{hA}$	$\frac{\ln(r_o/r_i)}{2\pi kL}$	$\frac{1}{2\pi r_o Lh}$
Mass	$\frac{L}{D_{ab}A}$	$\frac{L}{k_c A}$	$\frac{\ln(r_o/r_i)}{2\pi D_{ab}L}$	$\frac{1}{2\pi r_o Lk_c}$
Charge	$\frac{L}{\sigma A}$	$\frac{1}{k_\pm A}$	$\frac{\ln(r_o/r_i)}{2\pi\sigma L}$	$\frac{1}{2\pi r_o Lk_\pm}$
	Spherical			
	Diffusive	**Convective**		
Momentum	$\frac{(1/r_i - 1/r_o)}{4\pi\mu}$	See Chapter 10		
Heat	$\frac{(1/r_i - 1/r_o)}{4\pi k}$	$\frac{1}{4\pi r_o^2 h}$		
Mass	$\frac{(1/r_i - 1/r_o)}{4\pi D_{ab}}$	$\frac{1}{4\pi r_o^2 k_c}$		
Charge	$\frac{(1/r_i - 1/r_o)}{4\pi\sigma}$	$\frac{1}{4\pi r_o^2 k_\pm}$		

PROBLEMS

Momentum Transport

4.1 The space between two very long concentric cylinders is filled with water. The inner cylinder is stationary while the outer cylinder is pulled along axially with a velocity v_o (Figure P4.1).

a. Derive the differential equation to give the velocity profile of the water.
b. Solve the differential equation for the velocity profile and the stress at the surface of the inner cylinder. What are the boundary conditions?
c. What is the volumetric flow rate of water between the cylinders at any axial location?

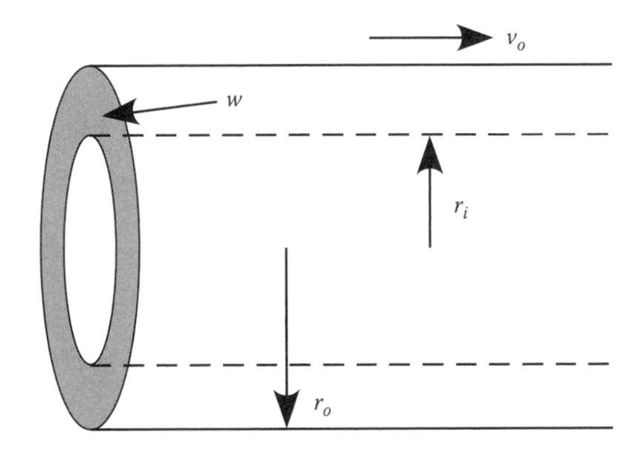

FIGURE P4.1

4.2 A Newtonian fluid of viscosity, μ, and density, ρ, fills the gap between two very long concentric cylinders. The inner cylinder ($r = r_i$) is stationary while the outer cylinder ($r = r_o$) rotates at a constant angular velocity, ω_o. Determine the velocity profile of the fluid.

4.3 The Ostwald–de Waele equation is often used to correlate the shear stress and shear rate relationship of shear-thickening fluids such as a cornstarch suspension. Here,

$$\tau_{xy} = K\left(-\frac{dv_x}{dy}\right)^n \qquad n > 1$$

Consider a cornstarch solution in the gap between two long parallel plates. One of the plates is stationary and the other is dragged along with a constant velocity, v_o. Solve for the steady-state velocity profile assuming a gap thickness of t_h. What force per unit area is required to yank the top plate at the prescribed velocity? What happens as the velocity is increased?

4.4 Plate glass is made by the float process. A molten glass slab rides on a pool of liquid metal. The other side of the glass is exposed to the air. To produce a striated, textured plate glass, the molten glass is sheared between two plates as shown in Figure P4.4. Derive an expression for the temperature and velocity profiles within the glass. Determine the force necessary to shear the glass as a function of the temperature difference between the upper and lower plates. Assume the plates are infinitely long and that the thermal conductivity of the glass is constant.

4.5 An engineer is attempting to measure the force needed to shear two plates past one another when the fluid inside is a slurry. The slurry is relatively dilute and obeys the following equation:

$$\mu_r = \frac{\mu}{\mu_o} = \frac{1+0.5\phi}{(\phi_c-\phi)^2} \qquad \begin{array}{l}\mu_o = \text{viscosity of pure fluid}\\ \phi_c = \text{critical volume fraction}\end{array}$$

a. Derive an expression for the velocity profile between the plates assuming the lower plate is stationary; the upper plate moves with a velocity, v_0; and the gap between the plates is d.
b. Determine an expression for the shear stress between the plates.
c. What is the stress if $d = 1$ mm, $v_0 = 1$ cm/s, $\mu_o = 1$ Pa·s, $\phi = 0.05$, and $\phi_c = 0.35$?
d. When $\phi \to \phi_c$, particles can span the space between the plates leading to jamming. Plot the shear stress between the plates as a function of ϕ for the conditions in (c). When does the shear stress become a problem, that is, an order of magnitude larger than that for the pure fluid?

4.6 A slurry is contained between the walls of the concentric cylinder in problem 1. The slurry is dilute and so obeys the linear form of the Einstein relation ($\mu = \mu_0(1 + 2.5\phi)$). Determine the shear stress between the cylinder walls and the fluid's velocity profile.

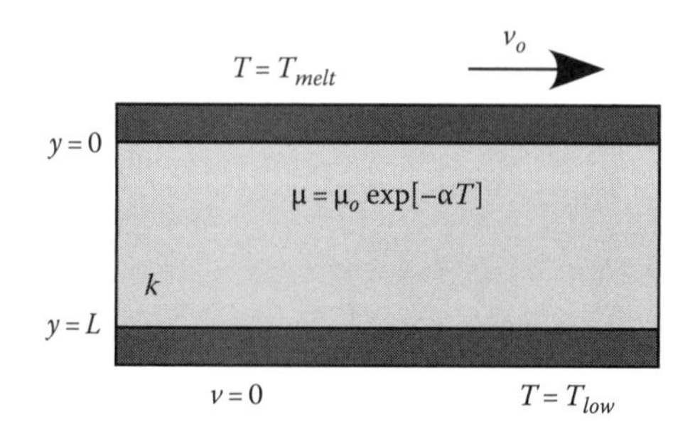

FIGURE P4.4

Heat Transport

4.7 Calculate the steady-state heat loss from the cup of cappuccino shown in Figure P4.7. You may assume that the cappuccino (not the foam) is held at a constant temperature of 91.67°C, the perfect brewing temperature. At the top of the cup, you may neglect the rough texture of the *head* and the small amount of area occupied by the rim of the cup.

4.8 Wind chill, the phenomenon experienced on a cold, windy day, is related to the increased heat transfer from exposed human skin to the surrounding atmosphere. Consider a layer of skin, 3 mm thick (k = 0.37) whose interior surface is maintained at a temperature of 36°C. On a calm day the heat transfer coefficient at the outer surface is 10 W/m^2 K, but with a 30 km/h wind, it reaches 25 W/m^2 K. In both cases the air temperature is −15°C.

a. What is the ratio of the heat loss per unit area from the skin for the calm day to that for the windy day?
b. What temperature would the air have to assume on the calm day to produce the same heat loss occurring with the air temperature at −15°C on the windy day?
c. What thickness of fat (k_{fat} = 0.2 W/m K) underneath the skin layer would be necessary to counteract the increased heat transfer on the windy day? Consider the same overall temperature difference.

4.9 As a building contractor, you are designing the walls of a house. Each wall is a three-component (minimum) composite: plasterboard, insulation, and plywood shown in Figure P4.9. You are to design the wall based on the following information and constraints:

1. Minimum outside air temperature, −25°C
2. Outside heat transfer coefficient, 100 W/m^2 K
3. Inside heat transfer coefficient, 50 W/m^2 K
4. Inside air temperature, 20°C
5. Maximum heat flux through the wall, 12 W/m^2
6. Maximum wall thickness, 0.175 m
7. Maximum wall cost, $25/m^2
   a. Neglecting the vinyl siding, the vapor barrier, and the top and sole plates (boards), design a wall to meet these specifications. You can obtain pricing information for the various components from local home improvement stores.
   b. Suppose you wanted to use metal studs instead of wood. Would you have to add extra insulation to keep your heat flux within specifications?

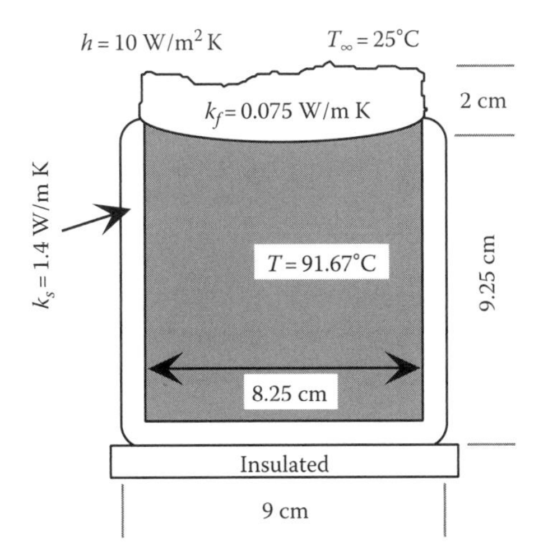

FIGURE P4.7

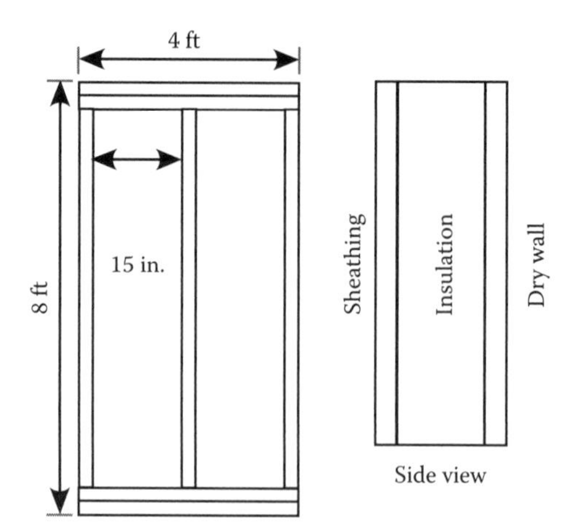

FIGURE P4.9

4.10 Please use the resistance diagram in Figure P4.10 to construct the heat transfer situation. The diagram represents a composite plane wall. Assume steady-state operation and no generation. Draw the wall and label all the heat flows and mechanisms (conduction, convection, or radiation).

4.11 Tempered glass substrates for high-speed compact disks and disk drives are made by healing surface cracks that form during manufacture. The disks are 10 cm in diameter, 3 mm thick, and are illuminated from below using a CO_2 laser that locally heats surface. The laser can produce a heat flux on the order of 10 kW/m^2. If the surface of glass must be heated to 750°C, what temperature must we heat the bottom of the substrate to?

4.12 Water boils in a paper cup over an open flame.

a. Ignoring radiative heat transfer, determine a numerical value for the maximum thickness of the bottom of the cup.

Data: Paper burns at 250°C, h_o(underside of cup) = 50 W/m^2 K
k(paper) = 0.5 W/m K h_i(inside of cup) = 4000 W/m^2 K
T_f(flame temp) = 750°C T_b(boiling temp) = 100°C

b. Explain carefully why a *large* (rather than *small*) thickness will promote burning of the cup.

c. By introducing radiative heat transfer from the flame, would you expect a decrease or an increase in the numerical value for the maximum cup bottom thickness? Why?

4.13 Consider a hollow sphere designed to hold a cryogenic fluid. The fluid inside the sphere is at a uniform temperature T_i. Insulation with thermal conductivity, k, is added to the outer surface. Convective heat loss to the environment is controlled by a heat transfer coefficient, h. What is the critical insulation thickness for this geometry? With reference to the cylindrical geometry, is there a pattern that can be extended to dimensions higher than 3?

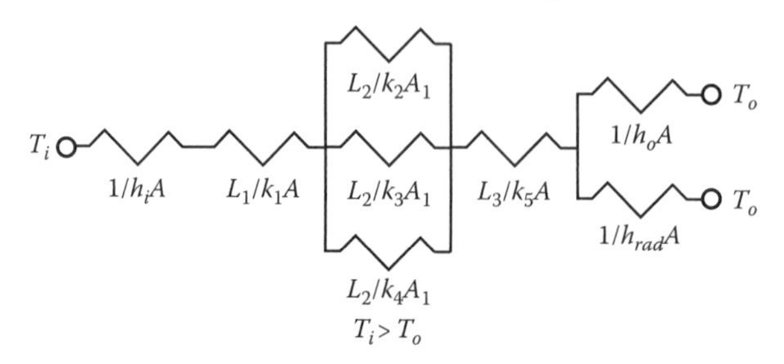

FIGURE P4.10

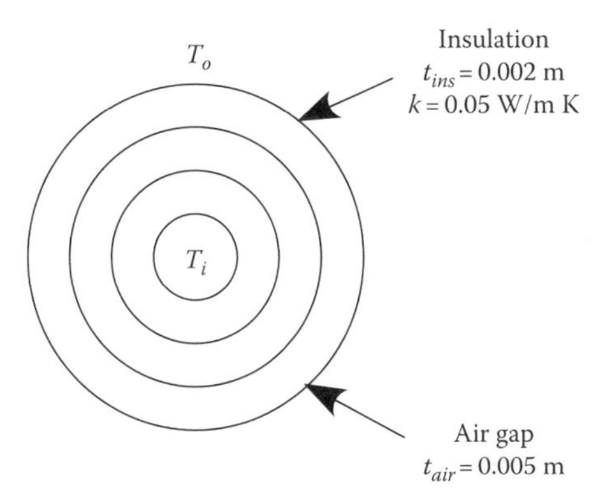

FIGURE P4.14

4.14 Many forms of insulation, especially for pipes are formed as periodic structures consisting of layers of insulation separated by air gaps. For the structure in Figure P4.14, calculate the thermal resistance of the insulation as a function of the number of layers of material.

4.15 A fired heater is designed to produce superheated steam at a temperature of 300°C and a pressure of 15 atm. The steam is carried in tubes (4 cm inside diameter) composed of a high-temperature alloy (k = 40 W/m K) 3 mm thick. The tube walls are heated by radiative exchange with the walls of the furnace (T_{wall} = 1100°C). What is heat flux into the fluid assuming the inside wall temperature is 300°C? You will probably have to solve this problem numerically. Assume that the heat transfer via radiation obeys

$$q_{rad} = \sigma^r A_s \left(T_{wall}^4 - T_{pipe}^4\right)$$

4.16 In the test example we considered a thermal interface resistance in rectangular coordinates. Consider the cylindrical coordinate version shown in Figure P4.16.

a. Derive an expression for the temperature profile and heat flow through the two materials.
b. If one increases the interfacial area by increasing the length of the cylinder, does the heat flow increase?
c. If one increases the interfacial area by increasing the outer diameter, does the heat increase or decrease? How does the increase or decrease compare with a similar system having no interfacial resistance?

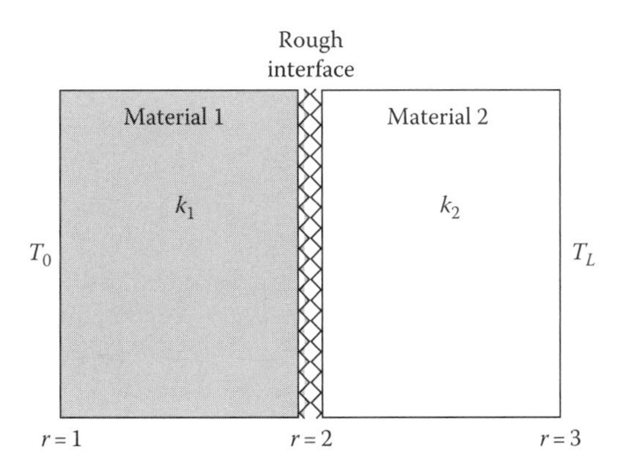

FIGURE P4.16

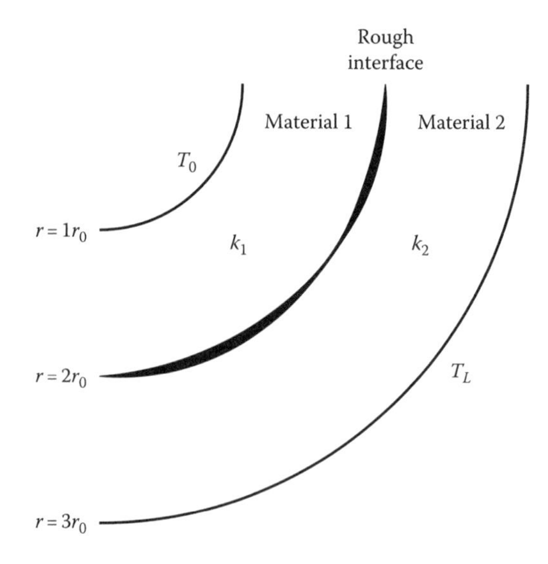

FIGURE P4.17

4.17 In the test example we considered a thermal interface resistance in rectangular coordinates. Consider the spherical coordinate version shown in Figure P4.17.

a. Derive an expression for the temperature profile and heat flow through the two materials.
b. If one increases the interfacial area by increasing the outer diameter, does the heat increase or decrease? How does the increase or decrease compare with a similar system having no interfacial resistance?

Mass Transport

4.18 Compucon Inc. has developed a new bottle that it claims will store volatile materials for a long time without a cover (Figure P4.18). The key is that the cross-sectional area of the device is designed to obey the function $A = A_o e^{-\alpha z}$, where z is the distance from the bottom. To test

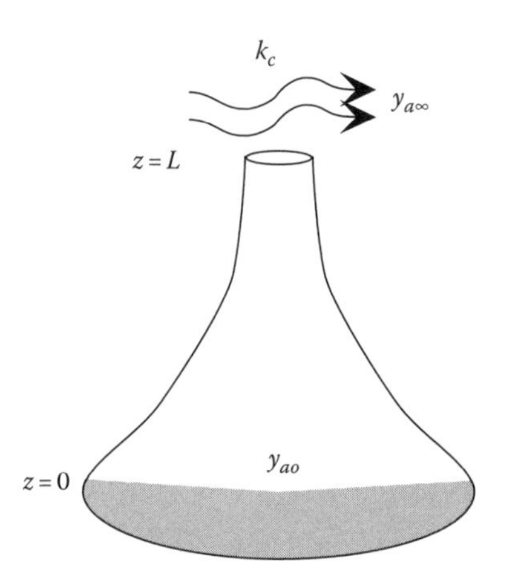

FIGURE P4.18

the concept you must perform the experiment shown in Figure P4.18. Assuming dilute solutions, constant total concentration, constant diffusivity, and steady-state operation, provides answers to the following

a. Derive the differential equation for the mole fraction profile of a.
b. What are the boundary conditions?
c. Solve the differential equation for the mole fraction profile and flux.
d. Can the mole fraction of a exceed y_{ao} at any point within the flask? What does this have to say about Compucon's claims (what is the effect of the bizarre shape)?

4.19 Oxygen transfer from the atmosphere to the interior regions of the eye depends enormously on whether one wears contact lenses. Treating the eye as a composite spherical system, determine the mass transfer rate from the atmosphere through the cornea with and without contact lenses. You may assume dilute solutions. The oxygen concentration in the interior is maintained at a uniform level by circulation of the fluid inside the eye as shown in Figure P4.19.

Data: $x_{a\infty} = 0.2$ $\quad x_{ao} = 0.05$ $\quad r_1 = 0.012$ m

$r_2 = 0.0127$ m $\quad r_3 = 0.014$ m $\quad D_{a1} = 1\times10^{-6}\,\text{m}^2/\text{s}$

$D_{a2} = 5\times10^{-6}\,\text{m}^2/\text{s}$ $\quad k_c = 0.001$ m/s $\quad c_t = 0.008\ \text{mol/m}^3$

4.20 A timed-release drug is dissolving in the intestine of a modern humanoid. As a steady-state approximation, we may assume that the drug is a rod of overall radius r_o (m) and length L (m). The timed-release action is accomplished by putting an inert coating on the drug through which the drug diffuses with a diffusivity D_{ab}. At the inner edge of the coating, (r_i), the composition (mole fraction) of the drug is x_{ao}. On the outer surface, the digestive juices provide for mass transfer with a mass transfer coefficient of k_{ac} (m/s). The amount of drug within the intestine can be approximated as $x_{a\infty} \approx 0$. Derive an expression for the steady-state, mole fraction profile of the drug through the coating. You may assume that the total concentration of all species within the coating is c_t and that the partition coefficient between the drug in the coating and the fluid is 1.

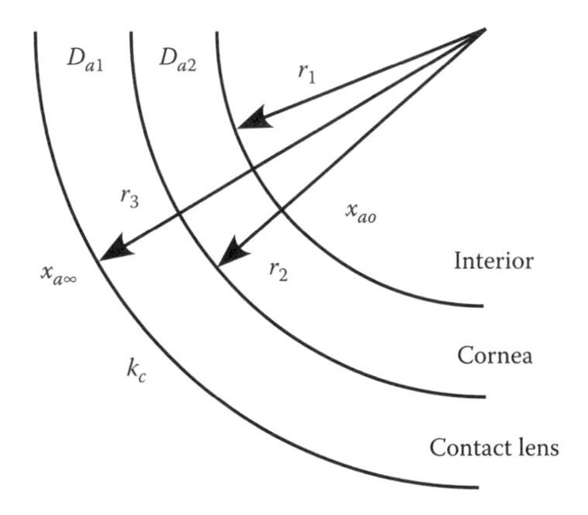

FIGURE P4.19

4.21 Oxygen diffuses through the wall of drug containers and oxidizes many drugs rendering them inactive. It is the oxygen diffusion/reaction scenario that limits the shelf life of many pharmaceutical products. To limit the oxidation of drugs, oxygen scavengers, like sodium bisulfite ($NaHSO_3$), are often added. The reaction to remove oxygen is

$$2HSO_3^- + O_2 \rightarrow SO_4^{2-} + H_2O$$

Consider a liquid drug stored in a cylindrical polyethylene container. The container is 15 cm high and has an inner diameter of 6 cm. The initial concentration of $NaHSO_3$ in the drug formulation is 1 g/L. How thick must the walls of the container be to ensure that 90% of the $NaHSO_3$ remains after 1 year? Neglect diffusion through the top and bottom of the container and assume that the $NaHSO_3$ reacts instantly with the O_2 so that the O_2 concentration in the drug is always 0. The effective diffusivity of O_2 through polyethylene is 9×10^{-13} m²/s and you may assume the partition coefficient is 1, $P = 1$ atm, $T = 20°C$, and the process operates at steady state (Enever, Lewandowski, and Taylor).

4.22 You detect the smell of gas in your house. Following the odor you discover a small gas leak in your furnace. Assuming a steady state concentration profile of gas, diffusion of the gas through air, and radial diffusion, how much gas has leaked? (Scheid, Fleming, Preap)
Data: You detect gas from your furnace.
Mole fraction detection limit for the nose is $y_{gas} = 1 \times 10^{-6}$
Mole fraction of odor component in gas

$$y_{gas,o} = 1 \times 10^{-3} \text{ at } r = 0.1\,\text{m}$$

$$T = 25°\text{C} \qquad P = 1\text{ atm} \qquad D_{gas} = 6.55 \times 10^{-6}\,\text{m}^2/\text{s}$$

Size of house ~20 m (equivalent diameter).

4.23 Ion exchange in glasses is one method for producing optical waveguides and gradient index lenses. Relatively high concentrations of dopant ions are required to produce a gradient index lens and the dopant presence alters the diffusivity in the substrate. Derive an expression for the steady-state concentration profile within a planar substrate assuming the diffusivity varies linearly with dopant concentration and the glass substrate acts as a stagnant medium. Refer to Figure P4.23.

4.24 A nicotine patch is designed to deliver 24 mg of nicotine over a 24 h period. The patch is 5 cm in diameter as shown in Figure P4.24. The steady-state nicotine concentration in the blood is designed to be 1 μg/L. Assuming the nicotine must pass through the skin, a layer of fat just beneath the skin, and finally into the blood, what is the concentration at the patch–skin interface? The partition coefficient between the patch and skin is $K_{eq} = c_{patch}/c_{skin} = 10$ and between the fat and blood is 1000. All other partition coefficients are 1. The diffusivities of nicotine in the skin and fat layer are
Data: $D_{ns} = 1.0 \times 10^{-10}$ m²/s $D_{nf} = 1.0 \times 10^{-9}$ m²/s

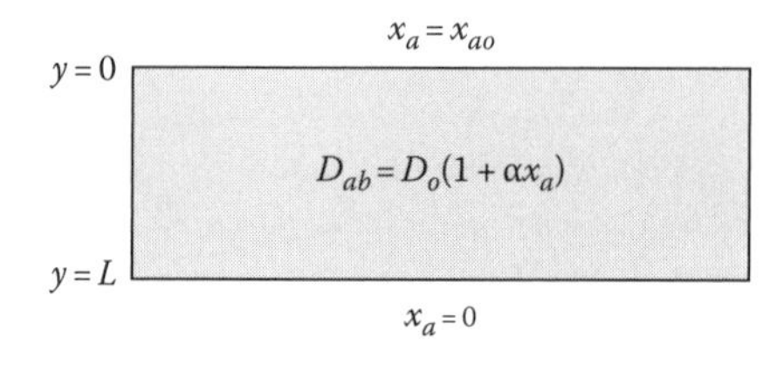

FIGURE P4.23

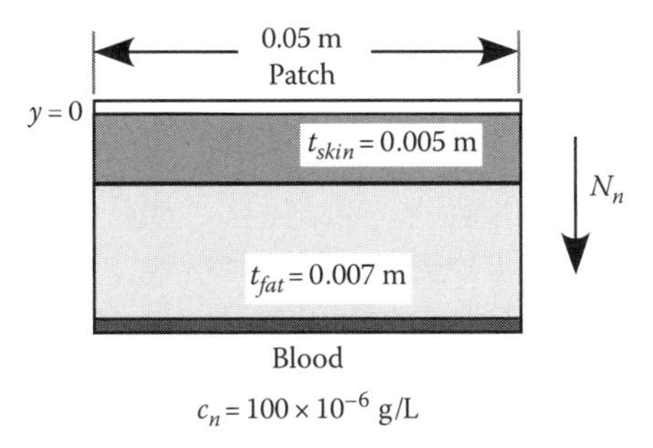

FIGURE P4.24

4.25 Consider the continuous adsorption situation shown in the Figure P4.25. The adsorbate is present in the gas phase at a concentration, c_{ao}. It adsorbs uniformly on the surface of the adsorbent and then diffuses through the liquid phase. Different adsorption isotherms are in common use including the

$$\text{Langmuir} \quad q = q_o\left(\frac{c_{ag}}{K_a + c_{ag}}\right) \quad \text{and}$$

$$\text{Freundlich} \quad q = K_a(c_{ag})^n$$

isotherms. Here q represents the moles of adsorbate adsorbed per gram of adsorbent. q_o, n, and K_a are constants that must be measured experimentally. Using each of these isotherms and assuming $K_{eq} = 1$, provide answers to the following:

a. Solve for the concentration profile of adsorbate in the liquid phase.
b. Derive an expression for the mass transfer coefficient based on the concentration driving forces: $(c_e - c_{aL})$; $(c_{ao} - c_{aL})$.

Here c_e is the concentration of adsorbate in the adsorbent. You may assume the adsorbent has a mass, m_o; a density, ρ_o; a molecular weight, M_{wo}; and a volume, v_o.

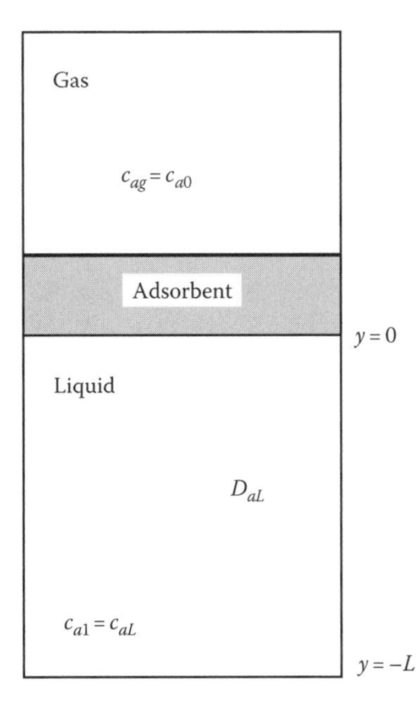

FIGURE P4.25

4.26 Diffusion of dopants into solids under the application of an electric field often requires the introduction of a composition-dependent diffusivity to account for the phenomenon of "elastic drift" where the solid resists the introduction of additional atoms into its lattice by a kind of mechanical, or spring like, force [3]. Following the discussion in Section 4.6.1, assume that we can express elastic drift using a linear expression for diffusivity as a function of concentration. Initial impression might suggest that the diffusivity should decrease with increasing dopant concentration. Show that this idea cannot work, and that to actually increase the resistance, the diffusivity must increase with dopant concentration (i.e., we need a less steep profile at equilibrium). The total flux consists of two components, one due to diffusion and another due to drift within the electric field (flux = $v_{electric}c$), that can be written simply as a velocity times a concentration.

Charge Transport

4.27 A new type of battery uses a gel electrolyte. Ions move through the gel via diffusion and drift through the electric field but the gel remains stagnant. Determine the maximum current through such a battery based on the data for the Ag–$AgNO_3$ battery used in the example in Section 4.6.2. Assume the silver nitrate electrolyte is 10 mol % of the gel. How does it compare with the equimolar counter diffusion result?

REFERENCES

1. Slattery, J.C., Sagis, L., and Oh, E.-S., *Interfacial Transport*, 2nd edn., Springer, New York (2007).
2. Fetzer, R., Munch, A., Wagner, B., Rauscher, M., and Jacobs, K., Quantifying hydrodynamic slip: A comprehensive analysis of dewetting profiles, *Langmuir*, **23**, 10559 (2007).
3. Achanta, R.S., Plawsky, J.L., and Gill, W.N., Copper ion transport induced dielectric failure: Inclusion of elastic drift and consequences for reliability, *J. Vac. Sci. Technol. A: Vac. Surf.*, **26**, 1497–1500 (2008).

5 Generation

In − Out + Gen = Acc

5.1 INTRODUCTION

In Chapter 5, we continue our discussion of 1-D, steady-state, diffusive transport by considering systems where generation of our transported quantity within the system or on the boundaries of the system is allowed. Generation pervades the natural world, and virtually, every transport process we encounter will contain some sort of generation. The presence of generation in a system precludes our analysis of transport in that system in terms of a random walk or correlated random walk. Generation provides a *force* that perturbs the system and alters its course from its original random walk. This fact provides a roundabout means of defining what generation actually is. We define generation a la Sherlock Holmes, by first eliminating everything we know not to be generation. Whatever remains must be the generation terms. To start, anything that flows through the system boundaries is related to the *In – Out* terms not to the *Gen* terms. Any term involving a change in a state variable with time is related to the *Acc* term and also cannot be a *Gen* term. Finally, generation occurring within the boundaries of the system must be represented by terms that are in units of the transported quantity per unit volume (N/m^3, W/m^3, $(mol/s)/m^3$, $(kg/s)/m^3$, C/m^3, etc.). Generation occurring on the boundaries of the system must be represented by terms with the units of the transported quantity per unit area (N/m^2, W/m^2, etc.). Generation can be positive or negative, and so we make no distinction between *production* and *consumption*. In the next few sections, we will catalog some of the more common forms generation terms take for each of our transport processes.

5.1.1 Generation in Transport Processes

Momentum transport: In momentum transport, the generation terms result from forces acting on the fluid. Most commonly, these are pressure forces, body forces such as gravity, or surface forces. These forces accelerate or decelerate the fluid and so generate momentum in the system. Though we often think of body forces as just the force of gravity on the fluid (fluid statics), there are other examples. In natural convection, heat and momentum transport interact through the density of the fluid causing fluid motion. Hot fluid, which is less dense, sees a buoyant force that accelerates it and causes the fluid to rise. Cold, denser fluid feels an opposite force and sinks. These buoyant forces are formally generation and involve not only gravity but also the temperature gradient. In magnetohydrodynamics, small metal particles are mixed with the fluid, and the system is placed in an electromagnetic field. The interaction of the field with the particles generates a force on the fluid that accelerates the fluid.

Surface forces are important at the interface between a fluid phase and another phase. We define the strength of the surface force in terms of an interfacial tension that measures how strongly the molecules of one phase prefer to interact with the molecules of the other phase. High values of interfacial tension signify adverse interaction between the phases, whereas low values indicate no preference for interaction. Surface forces are responsible for fluid climbing up a drinking straw when the straw is inserted in the fluid, for fluid wetting or beading up on a solid surface, and for the turbulent motion induced in a fluid by a dissolving solute such as salt. These surface forces accelerate the fluid and so are also classified as generation terms.

Energy transport: Generation in energy transport most often means the production or consumption of heat. The generation term in the energy equation is given its own special symbol, $\dot{q}$, since it may take many forms, and it may not be a function of outside sources. Heat generation may come from radioactive decay, chemical reaction, friction, electric heating, absorption of radiation, phase changes, or from viscous dissipation in a fluid, where *friction* between the fluid molecules generates the heat. Significant heat generated by fluid *friction* occurs in cases such as turbulent flow and flow in closely moving parts such as oil lubricating a journal bearing.

Mass transport: Generation in mass transfer is easy to visualize since it involves the production or consumption of chemical species in the system. The generation term in mass transfer is different from the other cases we will consider because it involves the addition or depletion of a specific component of the total system. The generation term is normally restricted to chemical reactions or phase changes. For chemical reactions, when the continuity equations for all components are added together, the generation terms must cancel to satisfy conservation of mass. In nuclear reactions, we must also take into account matter–energy conversions by coupling the mass and energy transport equations. Volumetric additions of one component over time are not examples of generation in mass transport; they are accumulation terms.

Charge transport: The form of the generation term in the charge transport equation is essentially equivalent to the generation term in the species continuity equations for mass transfer. Here, we are talking about the continuity of charge rather than mass, but in most situations, it is some material component that is transferring the charge (ion, electron, etc.). The generation term can be a recombination of positive and negative charges (a chemical reaction) or can be a true generation term like current generated in a vacuum tube by thermionic emission or current generated in a photodetector as light ejects electrons from a photosensitive cathode.

5.2 GENERATION ON THE BOUNDARY: BOUNDARY CONDITIONS

Before we begin to consider transport processes involving generation terms, it is important to discuss the various boundary conditions that we can pigeonhole into our generation concept. There are a great many of these types of boundary conditions, and so we can only list a few. A common thread among all these conditions is that they involve specifying the flux at a boundary of our system. The key to the generation boundary condition is that the flux does not pass through the interface, but originates there.

5.2.1 Momentum Transport

Boundary conditions involving generation in momentum transport all involve specifying a source of momentum at the boundary. The source can be a classical mathematical entity such as a point, line, or plane source, but the most common boundary condition of the generation type involves the application of an external force, namely, the force due to interfacial tension between two phases. As the interfacial tension changes from point to point along a surface either due to changes in temperature or composition, it generates a force that tends to pull at the surface generating momentum. The mere fact of a surface having a recognizable curvature is also enough to generate a pressure difference within the fluid that can alter fluid flow.

In thermodynamics, we learned that if we look at an interface between two phases where we have a known interfacial tension, γ_{ij}, the pressures and normal stresses to that interface will be different depending on which side of the interface we are on (Figure 5.1). If we know the radii of curvature of

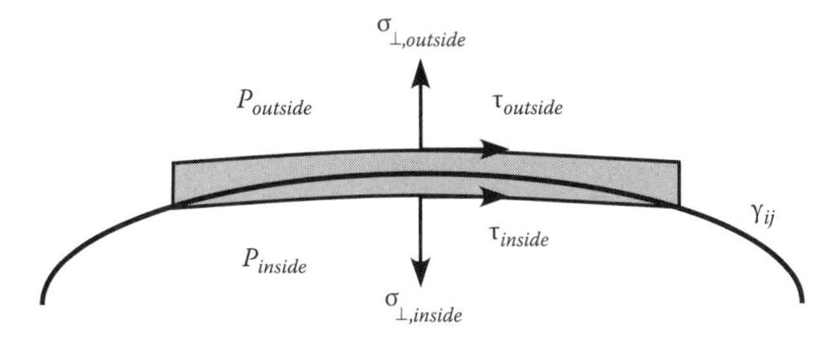

FIGURE 5.1 Interface between two phases with interfacial tension, σ_{ij}.

the interface, we can determine the pressure and normal stress difference. Formally, they are related to the force generated by the interfacial tension via

$$\sigma_{\perp,inside} = \sigma_{\perp,outside} + \gamma_{ij}\left(\frac{1}{R_1} + \frac{1}{R_2}\right) \tag{5.1}$$

where R_1 and R_2 are the radii of curvature in any two planes perpendicular to the surface. Any change in surface curvature changes the internal pressure and hence can drive fluid motion. In most instances, the normal stress in a fluid is the pressure, $\sigma_\perp = P$.

Tangential stress differences can also occur if there is a change in the surface tension along the interface. This phenomenon is called the Marangoni effect and mathematically can be stated via Equation 5.2:

$$\tau_{outside} - \tau_{inside} = \vec{\mathbf{t}} \cdot \vec{\nabla}\gamma_{ij} \tag{5.2}$$

Again, the change in surface tension exerts a force along the interface that can drive fluid motion.

5.2.2 Energy Transport

Boundary conditions arising from generation are very common in heat transfer. They are all variants of the flux boundary condition and refer to the manner in which the heat is generated. Two of the most common boundary conditions involving heat generation involve a tie to mass transfer.

A heat flux can be generated on the boundary due to a chemical reaction there or a heat of mixing effect (Figure 5.2a). In either case, we have some mass transfer process governing the rate at which heat evolves on the boundary. In Figure 5.2b, we see the second form of generation conditions where we have a phase change occurring. In this instance, we are generating a heat flux on the boundary by virtue of a latent heat release or absorption. In all cases illustrated by Figure 5.2, the heat flux resulting from the generation process may be positive or negative.

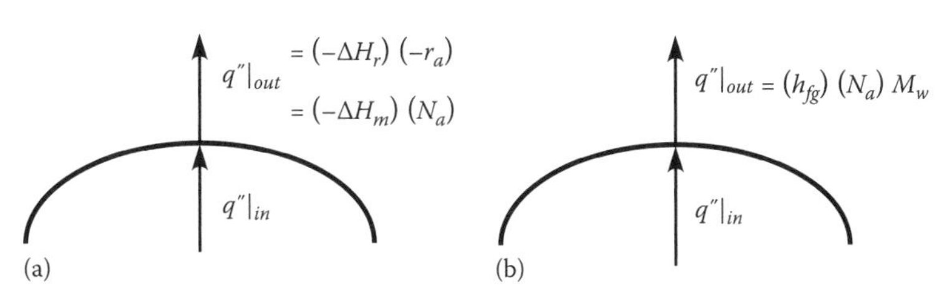

FIGURE 5.2 (a) Heat generation on the boundary due to a chemical reaction or heat of mixing effect. (b) Heat generation on the boundary due to a phase change.

The actual boundary conditions equate Fourier's law on one face of the boundary, with the appropriately generated heat flux on the other face:

$$-k\frac{dT}{dx} = (-\Delta H_r)(-r_{as}) \quad \text{Chemical reaction} \tag{5.3}$$

$$-k\frac{dT}{dx} = (-\Delta H_m)M_w N_a \quad \text{Heat of mixing} \tag{5.4}$$

$$-k\frac{dT}{dx} = \pm(h_{fg})M_w N_a \quad \text{Phase change} \tag{5.5}$$

In these equations, $-\Delta H_r$ represents the heat of reaction per mole of species a consumed, $-r_{as}$ is the rate of reaction of species a per unit of boundary surface area, $-\Delta H_m$ is the heat of mixing of species a with species b, N_a is the molar flux of species a at the interface, and h_{fg} is the heat of vaporization, fusion, or sublimation. The plus and minus sign in Equation 5.5 corresponds to the direction of the phase change and is positive for phase changes from gas to liquid, liquid to solid, or gas to solid. It is negative in the reverse direction. This reflects the fact that to change from a solid to a gas requires heat input from the denser material while the opposite phase change liberates heat.

5.2.3 Mass Transport

The boundary conditions involving generation for the mass transfer case are similar to those we just considered for heat transfer. We most commonly run into situations where we have a phase change or chemical reaction on the boundary that we must consider. The boundary conditions formally involve equating a form of Fick's law at one face of the boundary to an appropriately generated mass flux or mass flow on the opposite side:

$$-c_t D_{ab}\frac{dx_a}{dy} = (-r_{as}) \quad \text{Chemical reaction} \tag{5.6}$$

$$-c_t D_{ab}\frac{dx_a}{dy} = \pm\frac{q''}{h_{fg}M_w} \quad \text{Phase change} \tag{5.7}$$

In Equation 5.6, $-r_{as}$ is the rate of surface reaction of species a and can represent formation or destruction of a. Notice that when we refer to Equation 5.7, we couple heat and mass transfer since we need to know the total heat flux to determine how much material may evaporate, condense, or sublimate from the surface.

5.2.4 Charge Transport

The boundary conditions involving generation in charge transport reflect the similarity between it and mass transfer. The two basic *generation* mechanisms are reaction (neutralization) between charges of opposite sign and charge formation due to some external action (photo-generated, thermionic emission, etc.).

The reaction-type mechanism uses a boundary condition identical to Equation 5.6 where now we refer to a reaction rate of charged species. Such a situation may occur at the junction between p- and n-type semiconductor materials, for example:

$$-\sigma\frac{d\Phi}{dy} = (-r_e) \tag{5.8}$$

The second boundary condition specifies a flux at the interface that may depend upon the temperature (thermionic emission) or the light intensity (photo injection). We equate Ohm's law to the generated flux. Examples of these include

$$-\sigma\frac{d\Phi}{dy}=J_o\exp\left[-\frac{E_a}{RT}\right]\quad \text{Thermionic emission} \tag{5.9}$$

$$-\sigma\frac{d\Phi}{dy}=(-r_{optical})G^r\quad \text{Photo-generated}\quad r_{optical}(\text{Coul/W}) \tag{5.10}$$

5.3 ONE-DIMENSIONAL TRANSPORT WITH GENERATION AT THE BOUNDARY

Here, we look at two examples of how the techniques for solving problems covered in Chapter 4 can be applied when we have boundary conditions involving generation. We begin with chemical reaction at the surface.

5.3.1 Diffusion with Heterogeneous Reaction

Consider a situation, like that shown in Figure 5.3, where we have diffusion of species a from a source located at the top of a large tank. The lower surface of the tank is made of a catalytic material such that a is converted to b at that surface at a rate proportional to the concentration of a right above the surface, $-r_{as}$ *(moles/m²s)* $= c_t k_s'' x_a$. One mole of a reacts to form one mol of b. *b diffuses back through the fluid*, and so we have equimolar counterdiffusion throughout the system. We will also assume that the total concentration of species does not change.

Since the chemical reaction occurs on the boundary, if we consider a mass balance around a small control volume, we find a situation identical to that we discussed in Chapter 4:

$$\begin{aligned} &In \quad - \quad Out \quad + Gen = Acc \\ &\dot{M}_a(z) - \dot{M}_a(z+\Delta z) + \ 0 \ = \ 0 \end{aligned} \tag{5.11}$$

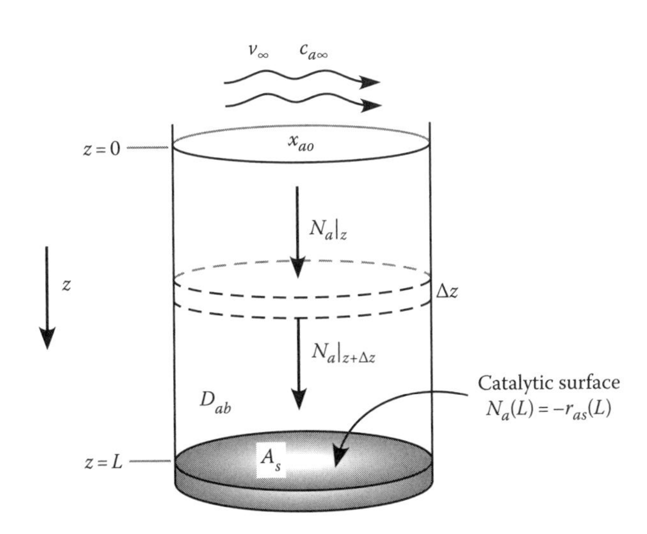

FIGURE 5.3 Diffusion with heterogeneous reaction on the boundary.

Using our limiting process and Fick's law for equimolar counterdiffusion gives the differential equation in terms of the mole fraction of species a, x_a:

$$\frac{d^2 x_a}{dz^2} = 0 \tag{5.12}$$

Remember this differential equation states that the flux of a throughout the system will be constant and that the mole fraction profile will be linear:

$$x_a = a_1 z + a_2 \tag{5.13}$$

The boundary conditions for the problem assert that we know the mole fraction of species a at the top of the tank, and that at the bottom, whatever a we flow to that surface, gets consumed by the reaction. This means we must equate Fick's law for the flux of a to the reaction rate at the lower surface:

$$z = 0 \qquad x_a = x_{ao} \tag{5.14}$$

$$z = L \quad N_a = -r_{as} \qquad \text{or} \qquad -c_t D_{ab}\frac{dx_a}{dz} = c_t k_s'' x_a \tag{5.15}$$

where k_s'' is the reaction rate constant at the surface and has the unit of meters per second (m/s). Applying the first boundary conditions specifies $a_2 = x_{ao}$. The second boundary conditions gives

$$a_1 = -\frac{k_s'' x_{ao}}{k_s'' L + D_{ab}} \tag{5.16}$$

The mole fraction profile and molar flux of species a are now

$$\frac{x_a}{x_{ao}} = 1 - \left(\frac{k_s''}{k_s'' L + D_{ab}}\right) z \tag{5.17}$$

$$\dot{M}_a = A_s N_a = \frac{c_t\left(x_{ao} - 0\right)}{\left(L/A_s D_{ab}\right) + \left(1/k_s'' A_s\right)} \qquad \frac{\text{Driving force}}{\text{Resistance}} \tag{5.18}$$

Equation 5.18 is written in terms of a familiar driving force–resistance concept. Notice that the effect of the chemical reaction is to *add* to the effective resistance to transport. However, at the same time, it increases the driving force. Mass transfer is enhanced because the reaction, though increasing the resistance, also increases the driving force by forcing the mole fraction at $z = L$ to be zero. In most instances, the increase in driving force more than offsets the increase in resistance since it takes a very slow reaction to be slower than diffusion. Reaction is one mechanism used to model what is often referred to as *facilitated* transport, a process that biological organisms (cells, bacteria, etc.) use to enhance transport across their cell membranes and the membranes within their organelles.

If $L/D_{ab} \gg 1/k_s''$, we refer to the system as *mass transfer controlled*. If $1/k_s'' \gg L/D_{ab}$, we refer to the system as *reaction rate controlled*. What happens if $k_s'' \to \infty$, if $k_s'' \to 0$? What happens at equilibrium if $x_a(z = L) = x_{eq}$?

5.3.2 Heat Transfer with Evaporation

Our next example of generation boundary conditions involves heat transfer through a fluid with evaporation from the surface. We would like to determine the temperature profile within the fluid, the temperature of the upper surface, and the rate at which the free surface recedes. The physical situation is shown in Figure 5.4. The rate of heat removal by evaporation depends on the temperature of the free surface and the rate of mass transfer there. As we will see, the rate of mass transfer and the temperature of the surface are coupled.

We must first determine the differential equation governing the temperature profile within the liquid. We perform an energy balance about the small element we isolated in Figure 5.4:

$$\begin{aligned} &In - Out + Gen = Acc \\ &q(z) - q(z+\Delta z) + 0 = 0 \end{aligned} \tag{5.19}$$

Using the limiting process to shrink the control volume and substituting for q using Fourier's law gives the differential equation in terms of the temperature:

$$\frac{d^2T}{dz^2} = 0 \tag{5.20}$$

Equation 5.20 again states that the heat flow is a constant and that the temperature profile is linear:

$$T = a_1 z + a_2 \tag{5.21}$$

The boundary conditions for this problem involve a specified temperature at the bottom of the fluid and loss of heat at the upper surface via evaporation. To specify the second boundary condition, we use Equation 5.7 and equate Fourier's law describing the flow of heat into the upper surface

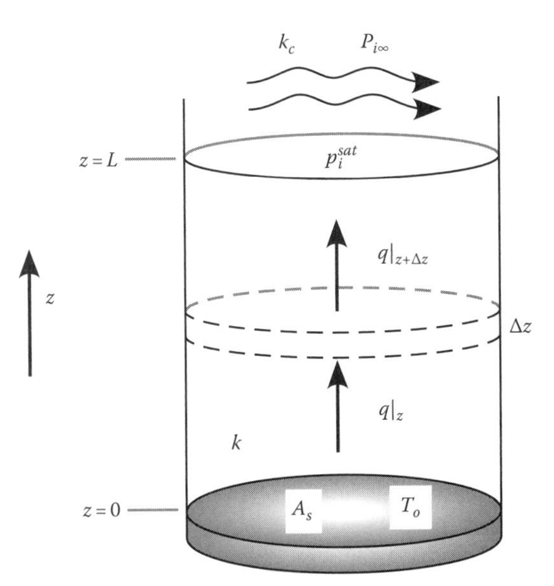

FIGURE 5.4 Heat conduction with evaporation at the boundary.

by conduction within the fluid to the heat removed via evaporation. The mass flow rate of evaporating fluid is $\dot{m}$, and each kilogram of evaporating fluid removes h_{fg} joules of heat:

$$z = 0 \qquad T = T_o \tag{5.22}$$

$$z = L \qquad -kA_s\frac{dT}{dz} = \dot{m}h_{fg} \tag{5.23}$$

The first boundary condition states that a_2 must be T_o. To apply the second boundary condition, we must calculate the mass flow rate. We were given a convection coefficient in the problem that lets us write the mass flow rate in terms of the convection coefficient and the driving force for mass transfer. This driving force is the difference between the partial pressure of water in the free stream, $p_{i,\infty}$, and the partial pressure of water in equilibrium with the fluid, p_i^{sat} . P is the total pressure:

$$\dot{m} = \frac{c_t k_c A_s M_w}{P}\left(p_i^{sat} - p_{i,\infty}\right) \tag{5.24}$$

Knowing p_i^{sat}, we can determine a_1 and the temperature profile:

$$a_1 = -\frac{c_t k_c h_{fg} M_w}{k}\frac{\left(p_i^{sat} - p_{i,\infty}\right)}{P} \tag{5.25}$$

$$T = T_o - \left[\frac{c_t k_c h_{fg} M_w}{k}\frac{\left(p_i^{sat} - p_{i,\infty}\right)}{P}\right]z \tag{5.26}$$

The temperature at the interface is found by setting $z = L$. We use Fourier's law to determine the heat flow and then the driving force and resistance to heat transfer. Here, the heat transfer is controlled by how much evaporation we have from the interface. The resistance to heat transfer is $(1/k_c h_{fg} A_s)$. The driving force for heat transfer is that for mass transfer, the pressure difference, $c_t(p_i^{sat} - p_{i,\infty})/P$:

$$T(L) = T_o - \left[\frac{c_t k_c h_{fg} M_w}{k}\frac{\left(p_i^{sat} - p_{i,\infty}\right)}{P}\right]L \tag{5.27}$$

$$q = -kA_s\frac{dT}{dz} = \left[\frac{c_t k_c h_{fg} A_s M_w\left(p_i^{sat} - p_{i,\infty}\right)}{P}\right] \tag{5.28}$$

The rate at which the interface recedes is given by the evaporation rate divided by the density of the liquid and area of the interface:

$$v_{sur} = \frac{\dot{m}}{\rho A_s} = \frac{c_t k_c M_w}{\rho P}\left(p_i^{sat} - p_{i,\infty}\right) \tag{5.29}$$

At the beginning of the problem, we stated that the temperature of the free surface and the mass transfer rate were coupled. This is not immediately apparent from our equations, but the coupling occurs through the saturation pressure, as the example in the following will illustrate.

Example 5.1 Evaporation of Water into Dry Air at 25°C

We consider the same problem we have just solved, but now, look directly at the evaporation of water into dry air. We assume that the column is so constructed that we have no gross movement of the fluid, and so heat transfer through the fluid takes place by conduction only. In general, all physical parameters for the water are a function of temperature. The two most important parameters are the heat of vaporization, h_{fg}, and the saturation vapor pressure, p_i^{sat}. Since these two parameters depend upon the temperature of the interface, we have a problem that can be solved via iteration or by reducing the problem to a single nonlinear equation whose root is the temperature we seek.

The parameters for the problem are

$$L = 0.1\,\text{m} \qquad A_s = 0.001\,\text{m}^2 \qquad T_o = 60°\text{C}$$
$$k_c = 1\times10^{-2}\,\text{m/s} \qquad k = 10\,\text{W/mK} \qquad p_{i,\infty} = 0$$

Data for the heat of vaporization and vapor pressure between 300 and 325 K are given by the following correlations:

$$h_{fg}\,(\text{kJ/kg}) = 3158 - 2.4T$$

$$p_i^{sat}\,(\text{Pa}) = (1\times10^5)\left\{0.77 - 0.74\exp\left[-\left(\frac{T-292.7}{82.89}\right)^2\right]\right\}$$

Combining all our data leaves us with the following equation to solve:

$$T - 333 + \frac{(1\times10^{-2})(3158\times10^3 - 2.4\times10^3 T)(18)}{(10)(8314)T}$$
$$\times\left\{(1\times10^5)\left\{0.77 - 0.74\exp\left[-\left(\frac{T-292.7}{82.89}\right)^2\right]\right\}\right\}(0.1) = 0$$

The final surface temperature appears to be 44.8°C. With this, we can determine the mass flow rate of evaporating fluid:

$$\dot{m} = \frac{c_t k_c A_s M_w}{P}\left(p_i^{sat} - p_{i,\infty}\right) = \frac{k_c A_s M_w}{RT}\left(p_i^{sat} - p_{i,\infty}\right)$$
$$= \frac{(0.01)(0.001)(18)}{(8314)(317.8)}(9319 - 0) = 6.4\times10^{-7}\,\text{kg/s}$$

and the velocity of the receding evaporating surface:

$$v_{sur} = \frac{\dot{m}}{\rho A_s} = \frac{6.4\times10^{-7}}{(991.08)(0.001)} = 6.4\times10^{-7}\,\text{m/s}$$

5.4 CONSTANT GENERATION TERMS

So far, we have discussed generation solely in terms of how it applies to boundary conditions. Now, we incorporate generation into the conservation laws and see how generation affects the transport process and the properties of the solutions we derive for the state variable. We will see two primary effects. The driving force/resistance concept will no longer apply since the flow or flux of the transport quantity is no longer a constant throughout the system. The maximum or minimum value of the state variable need no longer occur on a system boundary as it did in Chapter 4.

5.4.1 Momentum Transport: Flow between Inclined Plates

Constant generation terms in momentum transport take two common forms: a body force due to gravity or an overall, differential pressure force applied across the system. In this case, we will first consider generation due to the acceleration of gravity. We assume we have a viscous fluid trapped between parallel plates. To drain the fluid, we incline the plates at an angle of θ with respect to the ground. If the plates are long enough, and wide enough, and we wait for a sufficient length of time, then fluid flow in the middle of the plates will approximate steady-state conditions and we will have a 1-D problem. We would like to calculate the velocity profile in the fluid as a function of the angle of inclination and also determine the mass flow rate of fluid so we can estimate how long it will take to drain out.

Our first task is to define a small control volume over which we will apply the balance (conservation of momentum). The control volume is labeled in Figure 5.5, and notice that there are three processes occurring. Momentum flows into the face, $\tau_{xz}(x)$, and out of the face, $\tau_{xz}(x + \Delta x)$. This flow of momentum arises due to the stresses on both faces. Since we have placed the system in a gravitational field, we have an extra force on the control volume due to the weight of the fluid. This force is the generation term and has the units of momentum flow per unit volume, or force per unit volume (ρg). Notice that the gravitational acceleration points straight down, but the plane is inclined with respect to the acceleration by an angle θ. The portion of the total applied force that acts to generate momentum in the system is that component parallel to the flow, (+)ρg cos θ:

$$\begin{aligned} &In \quad - \quad Out \quad + \quad Gen \quad = Acc \\ &F_z(x) - F_z(x+\Delta x) + (\rho g \cos\theta) A_s \Delta x = \ 0 \end{aligned} \tag{5.30}$$

Since the generation term has the units of force per unit volume, *we must multiply the generation term by the volume of the control element,* $\Delta V = A_s \Delta x$.

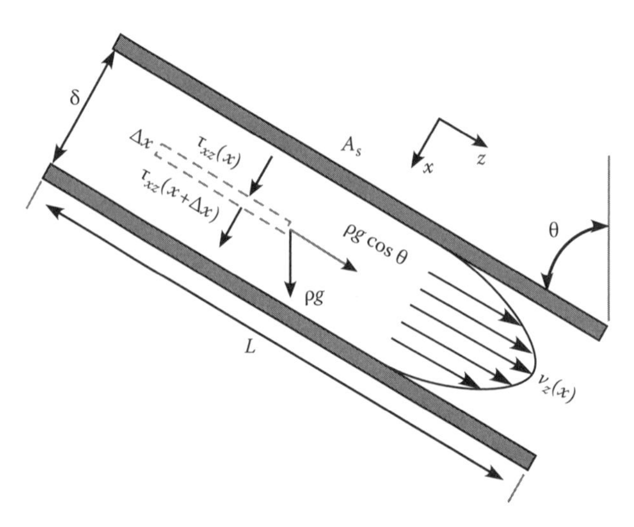

FIGURE 5.5 Steady flow of a viscous fluid between *infinite*, inclined, flat plates. *Arrows show momentum flow through the control volume.*

Equation 5.30 applies over our finite control volume. To have it applied at every point within the fluid, we convert it to a differential equation by dividing through by the volume of the control element, $A_s\Delta x$, and taking the limit as $\Delta x \to 0$:

$$\lim_{\Delta x\to 0}\left\{\frac{F_z(x)-F_z(x+\Delta x)}{A_s\Delta x}\right\}+\lim_{\Delta x\to 0}\{\rho g\cos\theta\}=0 \tag{5.31}$$

The first term in Equation 5.31 is the definition of the derivative of the force per unit area with respect to x, and the second term is independent of x. We can rewrite the equation in terms of the derivative of the stress, τ_{xz}:

$$-\frac{d}{dx}(\tau_{xz})+\rho g\cos\theta=0 \tag{5.32}$$

Equation 5.32 states that the stress, τ_{xz}, is no longer constant throughout the system, but changes with position due to the external force applied to the fluid. τ_{xz} will thus be an increasing function of x.

To determine the velocity profile and hence the mass flow rate of fluid through the faces of the control volume, we must substitute for the stress using Newton's law of viscosity:

$$\tau_{xz}=-\mu\frac{dv_z}{dx} \tag{5.33}$$

$$\frac{d}{dx}\left(\mu\frac{dv_z}{dx}\right)+\rho g\cos\theta=0 \tag{5.34}$$

We can separate this equation and integrate twice (assuming a constant viscosity) to obtain an expression for the velocity profile:

$$v_z=-\left(\frac{\rho g\cos\theta}{\mu}\right)\frac{x^2}{2}+a_1x+a_2 \tag{5.35}$$

The boundary conditions for this problem involve specifying the velocity at the two plates. We know that since the plates are stationary, the velocity must be zero there due to the no-slip condition:

$$x=0 \qquad v_z=0 \tag{5.36}$$

$$x=\delta \qquad v_z=0 \tag{5.37}$$

Evaluating the constants of integration gives

$$a_1=\left(\frac{\rho g\cos\theta}{2\mu}\right)\delta \qquad a_2=0 \tag{5.38}$$

where a_1 is a function of the mean gravitational force acting on the fluid divided by the fluid's viscosity. The overall velocity profile is

$$v_z=\frac{\rho g\cos\theta}{2\mu}(\delta-x)x \tag{5.39}$$

The volumetric flow rate is defined by integrating the velocity profile between the plates and multiplying by the width of the plates, W:

$$\dot{V} = W\int_0^{\delta} v_z dx = \frac{W\rho g\cos\theta}{2\mu}\int_0^{\delta}(\delta - x)x\,dx = \left(\frac{\rho g W\cos\theta}{12\mu}\right)\delta^3 \tag{5.40}$$

Notice that once we add generation into the balance equation, the minimum and maximum values of the state variable (the velocity) no longer need to fall on the system boundaries. In the present case, though the minimum velocities exist on the system boundaries, the maximum velocity lies at the center when $x = \delta/2$.

Let us now look at flow where Newton's law of viscosity does not apply everywhere, and see how the velocity and stress profiles are affected.

Example 5.2 Flow of a Bingham Fluid through a Tube

Consider the flow of a Bingham fluid through a very long tube. A Bingham fluid is one that has a yield stress, τ_o, below which the fluid will not flow (shear). Applying a stress in excess of τ_o allows the fluid layers to slide past one another, and so the fluid flows in a Newtonian manner with viscosity, μ. Examples of Bingham fluids include slurries, mayonnaise, ketchup, and paint. In this example, we will consider an oil-based paint having a yield stress, $\tau_o = 9.2$ N/m², and a viscosity of 0.389 N s/m². Other physical properties for the system include

$$\Delta p = 2.5\times 10^5\ \text{N/m}^2 \qquad r_o = 0.01\,\text{m} \qquad L = 25\,\text{m}$$

Since we must exert a yield stress before the fluid will flow, and since we will have a stress distribution across the tube, we expect a fundamentally different velocity profile than we would have obtained had the fluid been Newtonian. The profile of the Newtonian fluid will be parabolic like the example considering flow between two plates. The highest stress levels will occur at the pipe surface since the velocity must be zero there. In the case of the Bingham fluid, the flow will look parabolic in the high stress region, but toward the center, where the stress may be below the yield stress, the fluid will resist shearing and will appear to flow as a plug with zero velocity gradient. The expected shape is shown in Figure 5.6.

Our analysis begins with the control volume shown in Figure 5.7, and we apply conservation of momentum within that volume. We have a flow of momentum through the two faces of the control volume and a momentum generator within the control volume due to the pressure drop along the tube. The balance equation is

$$\mathit{In} \quad - \quad \mathit{Out} \quad + \quad \mathit{Gen} \quad = \mathit{Acc}$$

$$F_z(r) - F_z(r+\Delta r) + \left(\frac{\Delta P}{L}\right)(2\pi rL)\Delta r = 0$$

where

$2\pi rL$ is the surface area of the control volume through which the momentum flows
$2\pi rL\Delta r$ is the volume of the control element

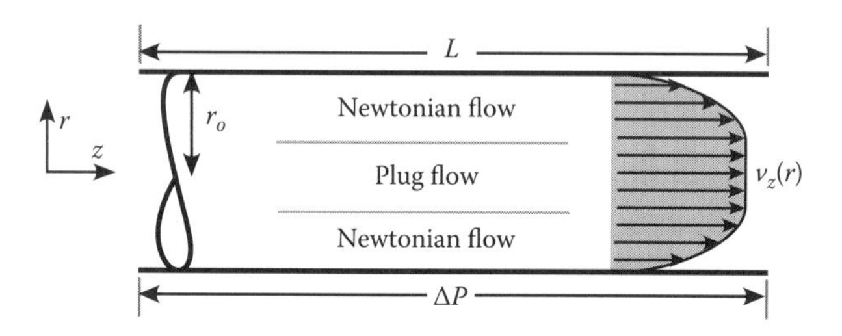

FIGURE 5.6 Flow of a Bingham fluid in a tube. Velocity profile showing regions of Newtonian flow and plug flow.

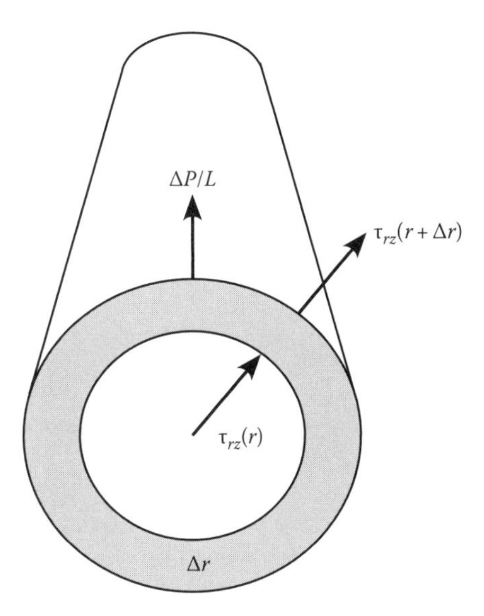

FIGURE 5.7 Control volume for flow of a Bingham fluid in a tube.

We have assumed that the pressure drop $\Delta P/L$ is a constant. The balance equation applies to the finite control volume, so to convert it to apply at each point in the fluid, we rewrite the force as $F(r) = 2\pi rL\tau_{rz}$, divide by Δr, and take the limit as $\Delta r \to 0$:

$$\lim_{\Delta r \to 0}\left\{\frac{r\tau_{rz}(r) - r\tau_{rz}(r+\Delta r)}{\Delta r}\right\} + \lim_{\Delta r \to 0}\left\{\left(\frac{\Delta P}{L}\right)r\right\} = 0$$

The first term is just the derivative of $r\tau_{rz}$ with respect to r. The second term is independent of Δr, and so the differential equation becomes

$$-\frac{d}{dr}(r\tau_{rz}) + \left(\frac{\Delta P}{L}\right)r = 0$$

Remember we cannot remove r from the definition of the derivative because the surface area through which the momentum flows changes with radial position. To express the differential equation in terms of the velocity, we substitute in the augmented form of Newton's law of viscosity for the Bingham fluid:

$$\tau_{rz} = \tau_o - \mu\frac{dv_z}{dr}$$

$$-\frac{d}{dr}\left(r\tau_o - \mu r\frac{dv_z}{dr}\right) + \left(\frac{\Delta P}{L}\right)r = 0$$

This differential equation is subject to boundary conditions that state that at the walls of the tube, $r = r_o$, we must have our no-slip condition:

$$r = r_o \qquad v_z = 0$$

At the centerline of the tube, we have a point of symmetry. If we stand on that point, all momentum appears to flow out from that point to the walls. Therefore, the center point is impermeable to the flow of momentum. The boundary condition there can be stated in one of two ways:

$$r = 0 \qquad v_z = \text{finite} \qquad \text{or} \qquad \frac{dv_z}{dr} = 0$$

both resulting in the same final solution. The second form, involving the derivative of v_z with respect to r, asserts that the centerline is impermeable to momentum flow.

Separating and integrating the differential equation gives

$$v_z = -\left(\frac{\Delta P}{4\mu L}\right)r^2 - \left(\frac{a_1}{\mu}\right)\ln(r) + \left(\frac{\tau_o}{\mu}\right)r + a_2$$

The boundary condition at $r = 0$ forces $a_1 = 0$ because we cannot have an infinite velocity at the center. The boundary condition at $r = r_o$ shows that a_2 must be

$$a_2 = \left(\frac{\Delta P}{4\mu L}\right)r_o^2 - \left(\frac{\tau_o}{\mu}\right)r_o$$

and so the final velocity profile is

$$v_z = \left(\frac{\Delta P}{4\mu L}\right)\left(r_o^2 - r^2\right) - \left(\frac{\tau_o}{\mu}\right)(r_o - r)$$

We plot both Bingham fluid and Newtonian behavior ($\mu = 0.389$ N s/m^2) in Figure 5.8. Under these conditions, the stress at the center of the tube lies below the yield stress of the fluid, and we have plug flow behavior there. The critical radius where plug flow occurs is determined by solving

$$\frac{dv_z}{dr} = 0 = -\left(\frac{\Delta P}{2\mu L}\right)r_c + \left(\frac{\tau_o}{\mu}\right) \qquad r_c = \frac{2L\tau_o}{\Delta P}$$

for r_c. Here $r_c = 0.00184$ m. The plug flow region extends from $-r_c < r < r_c$.

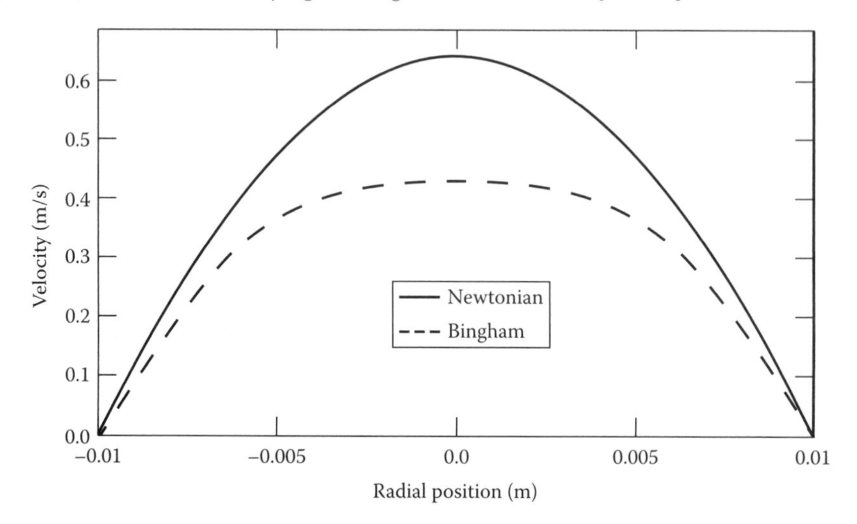

FIGURE 5.8 Velocity profiles for a Newtonian fluid and a Bingham fluid having the same shear viscosity.

5.4.2 Energy Transport: Joule Heating in an Insulated Wire

Constant sources of heat generation are very common occurrences. They may result from radioactive decay or more commonly from Joule heating. Joule heating occurs when we pass a current through a material having a measurable electric resistance. The nature of the problem illustrating this form of generation is shown in Figure 5.9. Here, we have a wire of resistivity, $\rho_r = 1.0 \times 10^{-7}$ $\Omega \cdot$m, and diameter, $d = 0.0033$ m, carrying a current I. The metal wire has a thermal conductivity of 20 W/m K. The wire is covered with a layer of electric insulation that has an extremely low electric conductivity but whose thermal conductivity, k, is on the order of 0.05 W/m K. We would like to calculate the critical insulation thickness to insure the maximum heat transfer rate from the wire. We must assume that the wire will be carrying the maximum current allowed by the National Electric Code, $I = 35$ A. The wire is also exposed to an environment where the average heat transfer coefficient to the atmosphere is on the order of 10 W/m² K.

We have to consider two regions separately, the conductor with heat generation and the insulation. As shown in Figure 5.9, we set up two small control volumes, one within each region, and write our balance equations:

$$\begin{aligned} In \quad - \quad Out \quad + \quad Gen \quad &= Acc \\ q_c(r) - q_c(r+\Delta r) + \dot{q}(2\pi r L)\Delta r &= 0 \end{aligned} \tag{5.41}$$

$$q_i(r) - q_i(r+\Delta r) + 0 = 0 \tag{5.42}$$

The heat generation rate, $\dot{q}$, is a function of the resistance of the wire ($\rho_r L/A_c$) and the current flowing through it:

$$\dot{q} = \frac{I^2 \rho_r L}{A_c} = \frac{I^2 \rho_r L}{\pi r_i^2} \tag{5.43}$$

Dividing Equations 5.41 and 5.42 through by Δr and following the limiting procedure as $\Delta r \to 0$ yield two differential equations for the heat flows in the conductor and insulation:

$$-\frac{dq_c}{dr} + \left(\frac{2L^2 I^2 \rho_r}{r_i^2}\right) r = 0 \tag{5.44}$$

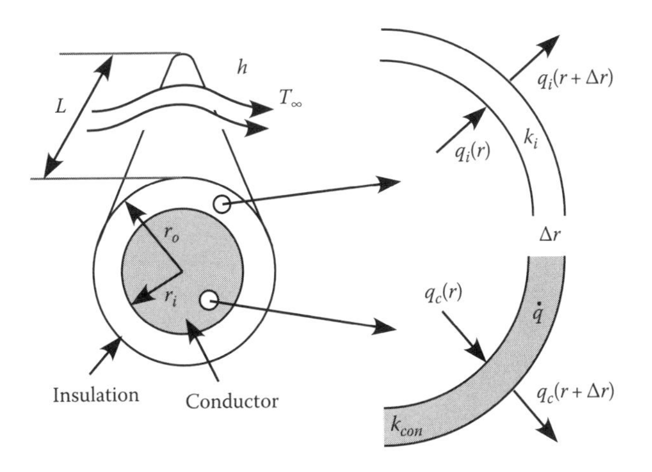

FIGURE 5.9 Heat flow with generation inside an insulated wire.

$$-\frac{dq_i}{dr} = 0 \tag{5.45}$$

Using Fourier's law for q_c and q_i transforms the equations to yield the temperature profile.

$$\frac{d}{dr}\left(r\frac{dT_c}{dr}\right) + \left(\frac{\rho_r I^2 L}{\pi r_i^2 k_{con}}\right) r = 0 \tag{5.46}$$

$$\frac{d}{dr}\left(r\frac{dT_i}{dr}\right) = 0 \tag{5.47}$$

These equations are subject to boundary conditions involving a point of symmetry at the center ($r = 0$), *continuity of temperature and heat flow at the junction* of the conductor and insulation ($r = r_i$), and a balance between conduction and convection at the outer surface of the insulation ($r = r_o$):

$$r = 0 \qquad \frac{dT_c}{dr} = 0 \tag{5.48}$$

$$r = r_i \qquad T_c = T_i \qquad -k_{con}\frac{dT_c}{dr} = -k_i\frac{dT_i}{dr} \tag{5.49}$$

$$r = r_o \qquad -k_i\frac{dT_i}{dr} = h(T_i - T_\infty) \tag{5.50}$$

Integrating Equations 5.46 and 5.47 gives

$$T_c = -\left(\frac{\rho_r I^2 L}{\pi r_i^2 k_{con}}\right)\frac{r^2}{4} + a_1 \ln(r) + a_2 \tag{5.51}$$

$$T_i = a_3 \ln(r) + a_4 \tag{5.52}$$

Evaluating the four constants using the boundary conditions gives us the final temperature profiles:

$$T_c = -\left(\frac{\rho_r I^2 L}{\pi r_i^2 k_{con}}\right)\left(\frac{r^2}{4} - \frac{r_i^2}{4}\right) + \left(\frac{\rho_r I^2 L}{2\pi k_i}\right)\ln\left(\frac{r_o}{r_i}\right) + \frac{\rho_r I^2 L}{2\pi r_o h} + T_\infty \tag{5.53}$$

$$T_i = \left(\frac{\rho_r I^2 L}{2\pi k_i}\right)\ln\left(\frac{r_o}{r}\right) + \frac{\rho_r I^2 L}{2\pi r_o h} + T_\infty \tag{5.54}$$

We are searching for a solution that will give us the minimum centerline temperature. To determine this critical insulation thickness, we must evaluate T_c at $r = 0$ and then determine what value of r_o makes this T_c a minimum:

$$\frac{d}{dr_o}(T_c(r = 0)) = \left(\frac{\rho_r I^2 L}{2\pi k_i}\right)\left(\frac{1}{r_o}\right) - \frac{\rho_r I^2 L}{2\pi r_o^2 h} = 0 \tag{5.55}$$

Solving for r_o gives us the critical insulation thickness. This occurs when $r_o = k_i/h$. At values of r_o smaller than this, our surface area available for heat transfer is too small and the centerline temperature is too high. At values of r_o higher than this value, the insulation we installed provides too great a resistance to heat transfer and so increases the centerline temperature. This problem should look familiar. It is the heat transfer analog of the critical insulation thickness problem we discussed last chapter. Notice that the critical insulation thickness does not depend upon the generation rate.

There is a fundamental difference between the solution for the temperature profile with generation and the solution we presented earlier for steady-state operation with no sources. In this system, we can have a solid cylindrical conductor operating at steady state. In the other circumstance, the cylinder had to be hollow because we needed heat flow into the inner wall to sustain a steady state. When we have generation, the material may provide its own heat flow at the centerline, and so a solution exists for $r = 0$.

We have seen that a constant generation term can be a common occurrence in transport systems. The important points to remember are the following:

1. The addition of a generation term alters the dissipative nature of the transport process, and in steady-state transport, the flow of the transported quantity is no longer independent of position.
2. Generation terms are always referenced per unit volume of the material. This is the primary method for recognizing generation terms in a process.
3. The addition of generation means that the maximum or minimum value of the state variable need not occur on the boundary, but may often occur somewhere inside the system.

In the next few sections, we will look at some more complicated examples of generation and allow the generation term to be a function of the dependent variable. This will let us look at some more realistic problems including diffusion with homogeneous chemical reaction, natural convection, and the operation of a diode.

5.5 VARIABLE GENERATION AND COUPLED TRANSPORT

We have considered cases where the generation term was a constant quantity. Often, nature is not that simplistic, and the generation term varies as a function of position, time, or one of the state variables of our system. In this section, we will consider a few of the more common examples of variable generation and show how the generation term can couple transport processes much as the variable transport coefficients did in Chapter 4.

5.5.1 Simultaneous Diffusion and Reaction

Our first example of variable generation and coupled transport involves diffusion, an exothermic chemical reaction, and heat generation in a fluid system. The physical situation is shown in Figure 5.10. Two fluids are separated by a membrane permeable only to species a. Fluid I is well mixed. The concentration of species a is c_{ao} everywhere within this fluid, and the temperature in fluid I is held constant at T_o. Species a diffuses through stagnant fluid II and undergoes a first-order reaction ($a \rightarrow b$) to form species b within the fluid. Fluid I is replenished with a, while fluid II has an outlet (not shown) to remove excess a and b.

We assume constant properties in both phases. The reaction occurs with a rate constant, k'', and is exothermic, yielding $-\Delta H_r$ joules per mole of a reacted. The whole system is insulated from the surroundings, and all sides of the vessel are impermeable to both fluids. We want to predict both the concentration distribution and the temperature distribution in the system. To simplify the problem to something manageable, we will assume that solute a is only sparingly soluble in fluid II so that we can assume dilute solutions and keep the overall volume of fluid II constant. The partition

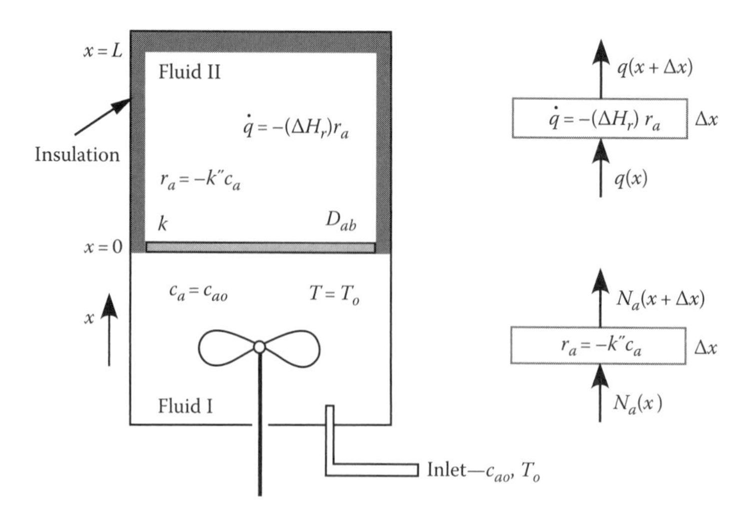

FIGURE 5.10 Diffusion and reaction in an insulated fluid system.

coefficient across the membrane is K_{eq}. Since we have a dilute solution, temperature differences are not great enough to set up any thermally induced diffusive effects or any density-driven convective effects. To determine the temperature and concentration profiles, we need to construct two balances: one for the continuity of species a and one for the energy transport through the system. Using the control volumes set up in Figure 5.10, the mass balance on species a in fluid II is

$$
\begin{aligned}
&In \quad - \quad Out \quad + \quad Gen \quad = Acc \\
&\dot{M}_a(x) - \dot{M}_a(x+\Delta x) - k''c_a A_s \Delta x = 0
\end{aligned}
\tag{5.56}
$$

Notice that the generation term involves the reaction rate multiplied by the volume of the control volume. Factoring out the constant area and performing the limiting process with respect to x gives

$$
-\frac{dN_a}{dx} - k''c_a = 0 \tag{5.57}
$$

Since a forms a dilute solution in fluid II, we can approximate the flux as

$$
N_a = -D_{ab}\frac{dc_a}{dx} \tag{5.58}
$$

The final form for the continuity equation is

$$
\frac{d^2c_a}{dx^2} - \frac{k''}{D_{ab}}c_a = 0 \tag{5.59}
$$

with boundary conditions

$$
x = 0 \qquad c_a = K_{eq}c_{ao} \tag{5.60}
$$

$$
x = L \qquad \frac{dc_a}{dx} = 0 \tag{5.61}
$$

These boundary conditions reflect the well-mixed system in fluid I, the partition across the membrane, and that the concentration of species a no longer changes beyond the vessel wall at $x = L$.

The energy balance is

$$\begin{aligned} &In \quad - \quad Out \quad + \quad Gen \quad = Acc \\ &q(x) - q(x+\Delta x) + k''c_a(-\Delta H_r)A_s\Delta x = 0 \end{aligned} \tag{5.62}$$

Following the limiting procedure yields the differential equation for the heat flow:

$$-\frac{dq}{dx} + k''c_a(-\Delta H_r)A_s = 0 \tag{5.63}$$

The generation term in the energy equation depends upon the reaction rate and hence, upon the solution to the mass transfer problem. We use Fourier's law to transform the energy equation to an equation in terms of the temperature:

$$\frac{d^2T}{dx^2} + \frac{k''(-\Delta H_r)}{k} c_a = 0 \tag{5.64}$$

The boundary conditions for the heat transform problem are direct analogs of the boundary conditions we used in the mass balance on a:

$$x = 0 \qquad T = T_o \tag{5.65}$$

$$x = L \qquad \frac{dT}{dx} = 0 \tag{5.66}$$

and reflect the uniformity of fluid I and that the reaction stops at $x = L$, and so the temperature should not change beyond that point.

It is often very useful to put the conservation equations in dimensionless form. This procedure simplifies the boundary conditions, forces the problem to extend over a range between 0 and 1 (scales the equations), and leads to groupings of system parameters that have important physical significance. We can put the balance equations for this problem in dimensionless form by defining three new variables:

$$\chi = \frac{c_a}{K_{eq}c_{ao}} \qquad \theta = \frac{T - T_o}{T_o} \qquad \xi = \frac{x}{L} \tag{5.67}$$

Here, it is important to think about what scaling parameter one wants to use for each variable. Choosing the right scaling variables often leads to a problem that is easier to solve analytically or numerically. We use the chain rule for differentiation to translate from our regular variables, c_a, T, and x, to the new dimensionless variables. Examples of this procedure for dc_a/dx and d^2T/dx^2 are

$$\frac{dc_a}{dx} = \frac{d\chi}{d\xi} \cdot \frac{dc_a}{d\chi} \cdot \frac{d\xi}{dx} = \frac{d\chi}{d\xi}\left(\frac{K_{eq}c_{ao}}{L}\right) \tag{5.68}$$

$$\frac{d^2T}{dx^2} = \frac{d}{d\xi}\left(\frac{d\theta}{d\xi} \cdot \frac{dT}{d\theta} \cdot \frac{d\xi}{dx}\right) \cdot \frac{d\xi}{dx} = \frac{d^2\theta}{d\xi^2}\left(\frac{T_o}{L^2}\right) \tag{5.69}$$

Substituting back into Equations 5.63 and 5.64, the transformed equations now become

$$\frac{d^2\chi}{d\xi^2} - \left[\frac{k''L^2}{D_{ab}}\right]\chi = 0 \quad \text{Mass balance} \tag{5.70}$$

$$\frac{d^2\theta}{d\xi^2} + \left[\frac{k''c_{ao}K_{eq}L^2(-\Delta H_r)}{kT_o}\right]\chi = 0 \quad \text{Energy balance} \tag{5.71}$$

The boundary conditions are similarly transformed to read

$$\xi = 0 \qquad \chi = 1 \qquad \theta = 0 \tag{5.72}$$

$$\xi = 1 \qquad \frac{d\chi}{d\xi} = 0 \qquad \frac{d\theta}{d\xi} = 0 \tag{5.73}$$

As promised, the procedure yields two dimensionless quantities:

$$\frac{k''L^2}{D_{ab}} \quad - \quad \begin{array}{c}\text{Damkohler}\\ \text{Number}\\ \text{(Da)}\end{array} \quad \frac{\textit{Diffusion time}}{\textit{Reaction time}} \quad \left(\frac{L^2/D_{ab}}{1/k''}\right) \tag{5.74}$$

$$\frac{k''K_{eq}c_{ao}(-\Delta H_r)}{kT_o/L^2} \quad - \quad \begin{array}{c}\text{Generation}\\ \text{Number}\\ \text{(Gn)}\end{array} \quad \frac{\textit{Rate of heat generation}}{\textit{Rate of heat removal}} \tag{5.75}$$

The Damkohler number reflects the competition between the rate at which species a is supplied to fluid II by diffusion and the rate at which that species is consumed in fluid II by the reaction. High Damkohler numbers imply that the reaction rate is so fast we cannot supply enough of species a to satisfy it. We refer to this situation as *mass transfer controlled.* Low Damkohler numbers imply that the system is controlled by how fast we can consume species a. We refer to this regime as *reaction rate controlled.*

The generation number, Gn, reflects the competition between the amount of heat generated per unit volume by the reaction and the amount of heat per unit volume that can be removed from the system via conduction. High generation numbers imply the heat flow in the system is controlled by the amount of heat generated. Exothermic reaction systems with high generation numbers and high reaction rates tend to be unstable. Small generation numbers imply that the system can remove more heat than can be generated.

The two balance equations are coupled. To solve for the concentration profile and the temperature profile, we must first solve the mass balance for the concentration field. The general solution to Equation 5.70 cannot be obtained by separating and integrating. It is found by substituting a trial solution, $Ae^{r\xi}$, into the equation. This procedure yields a quadratic equation for r:

$$r^2 - \text{Da} = 0 \qquad r_{1,2} = \pm\sqrt{\text{Da}} \tag{5.76}$$

The solution can be written as exponentials, or the exponentials can be arranged in the form of hyperbolic sine and hyperbolic cosine functions:

$$\chi = b_1 e^{\eta\xi} + b_2 e^{-\eta\xi} = a_1 \cosh(\sqrt{\text{Da}}\,\xi) + a_2 \sinh(\sqrt{\text{Da}}\,\xi) \tag{5.77}$$

According to our boundary conditions,

$$\xi = 0 \qquad \chi = 1 \;\rightarrow\; a_1 = 1 \tag{5.78}$$

$$\xi = 1 \quad \sqrt{\text{Da}}\sinh(\sqrt{\text{Da}}) + a_2\sqrt{\text{Da}}\cosh(\sqrt{\text{Da}}) = 0$$

$$a_2 = -\tanh(\sqrt{\text{Da}}) \tag{5.79}$$

Using these constants, the concentration profile is

$$\chi = \cosh(\sqrt{\text{Da}}\,\xi) - \tanh(\sqrt{\text{Da}})\sinh(\sqrt{\text{Da}}\,\xi) \tag{5.80}$$

The temperature profile can now be obtained by substituting for χ in Equation 5.71. The resulting differential equation can be separated and integrated twice to yield

$$\theta = -\frac{\text{Gn}}{\text{Da}}[\cosh(\sqrt{\text{Da}}\,\xi) - \tanh(\sqrt{\text{Da}})\sinh(\sqrt{\text{Da}}\,\xi)] + a_3\xi + a_4 \tag{5.81}$$

Using the boundary conditions Equations 5.72 and 5.73, we find that

$$a_3 = 0 \qquad a_4 = \frac{\text{Gn}}{\text{Da}} \tag{5.82}$$

and the resulting temperature profile is

$$\theta = \frac{\text{Gn}}{\text{Da}}[1 - \cosh(\sqrt{\text{Da}}\,\xi) + \tanh(\sqrt{\text{Da}})\sinh(\sqrt{\text{Da}}\,\xi)] \tag{5.83}$$

The full concentration and temperature profiles are shown in Figure 5.11. For convenience, the generation number Gn = 2, and the Damkohler number Da = 1. Since the ratio of Gn to Da is greater than one, the temperature profile is steeper than the concentration profile.

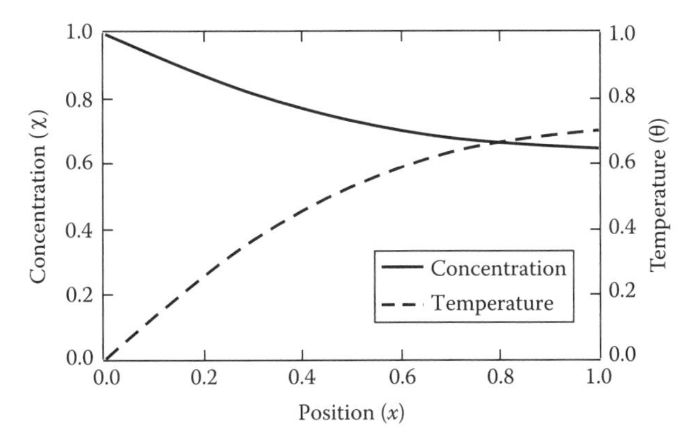

FIGURE 5.11 Concentration and temperature profiles in fluid II of Figure 5.10. Since Gn/Da > 1, the temperature increases faster than the concentration decreases.

Example 5.3 Simultaneous Diffusion and Enzymatic Reaction

We can use a system, like that shown in Figure 5.10, to run an enzymatic reaction for producing a drug, foodstuff, or ethanol for car fuel. The only difference between an enzymatic reaction and the first-order reaction we just analyzed is the nature of the kinetic expression. Most enzymatic reactions obey Michaelis–Menten kinetics and are inhibited by high concentrations of the reactant. The rate of the enzymatic reaction can be modeled as

$$r_a = -\frac{k''c_a}{K_m + c_a}$$

where

k'' is the rate constant (general proportional to enzyme concentration)
K_m is the Michaelis constant

The physical parameters for this problem are

$$c_{ao} = 100\,\text{mol/m}^3 \qquad k'' = 1.0\times 10^{-6}/\text{s} \qquad K_m = 667\,\text{mol/m}^3$$

$$D_{ab} = 1\times 10^{-9}\,\text{m}^2/\text{s} \qquad L = 1\,\text{m}$$

Referring to Figure 5.12, the mass balance on species a is

$$\begin{array}{ccccccc} In & - & Out & + & Gen & = & Acc \\ \dot{M}_a(x) & - & \dot{M}_a(x+\Delta x) & - & \left(\dfrac{k''c_a}{K_m + c_a}\right)A_s\Delta x & = & 0 \end{array}$$

Removing the area and performing the limiting procedure give

$$-\frac{dN_a}{dx} - \frac{k''c_a}{K_m + c_a} = 0$$

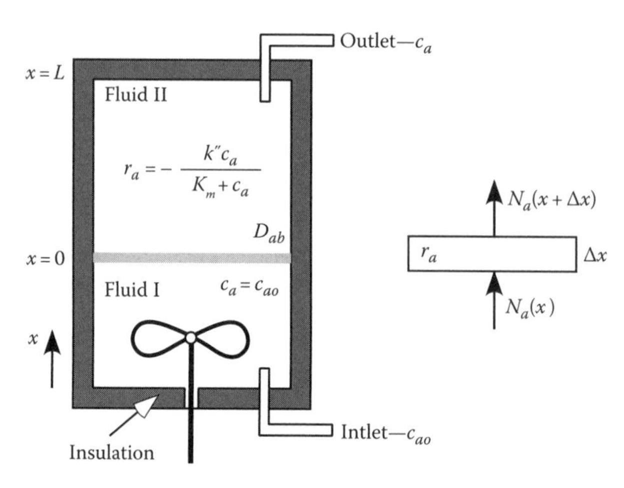

FIGURE 5.12 Simultaneous diffusion and enzymatic chemical reaction. The reaction follows Michaelis–Menten kinetics.

We will assume dilute solutions again, and so Fick's law is written solely in terms of the concentration gradient:

$$N_a = -D_{ab}\frac{dc_a}{dx}$$

The final form for the continuity equation is

$$\frac{d^2c_a}{dx^2} - \left(\frac{1}{D_{ab}}\right)\frac{k''c_a}{K_m + c_a} = 0$$

In its present form, the continuity equation is too difficult to solve analytically. However, we can get an idea of the complete solution by looking at two limiting cases; $c_a \ll K_m$; $K_m \gg c_a$. The mass balances for these two limits are

$$\frac{d^2c_{a1}}{dx^2} - \frac{k''}{K_m D_{ab}}c_{a1} = 0 \qquad c_a \ll K_m$$

$$\frac{d^2c_{a2}}{dx^2} - \frac{k''}{D_{ab}} = 0 \qquad c_a \gg K_m$$

subject to boundary conditions that state

$$x = 0 \qquad c_{a2} = c_{ao} \qquad \text{assume } c_a \gg K_m \text{ at } x = 0$$

$$x = L \qquad \frac{dc_{a1}}{dx} = 0 \qquad \text{assume } c_a \ll K_m \text{ at } x = L$$

and that splice the two solutions together at $c_a = K_m$. The splicing insures continuity of concentration and mass flux at the junction. This may not be a realistic splice point, and we will compare our spliced solution to the actual solution to see how close we are:

$$x = x_i \qquad c_a = K_m \qquad c_{a1} = c_{a2} \qquad \frac{dc_{a1}}{dx} = \frac{dc_{a2}}{dx}$$

Solutions to the differential equations for c_{a1} and c_{a2} are

$$c_{a1} = a_1\cosh\left(\sqrt{\frac{k''}{K_m D_{ab}}}\,x\right) + a_2\sinh\left(\sqrt{\frac{k''}{K_m D_{ab}}}\,x\right)$$

$$c_{a2} = \frac{k''}{2D_{ab}}x^2 + b_1 x + b_2$$

Evaluating the constants of integration using the first two boundary conditions gives

$$c_{a1} = a_1\left[\cosh\left(\sqrt{\frac{k''}{K_m D_{ab}}}\,x\right) - \tanh\left(\sqrt{\frac{k''}{K_m D_{ab}}}\,L\right)\sinh\left(\sqrt{\frac{k''}{K_m D_{ab}}}\,x\right)\right]$$

$$c_{a2} = \frac{k''}{2D_{ab}}x^2 + b_1 x + c_{ao}$$

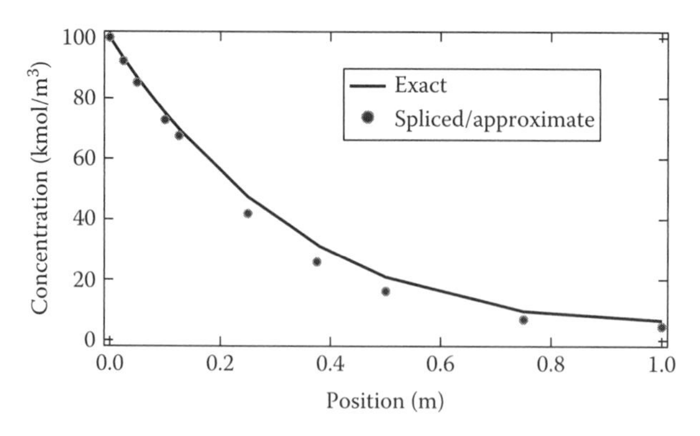

FIGURE 5.13 Concentration profiles for the enzymatic reaction.

To evaluate a_1 and b_1, we need to use the splicing conditions. This actually involves solving for a_1, b_1, and the interfacial position where the splice occurs, x_i. We end up with three nonlinear algebraic equations to solve:

$$K_m = \frac{k''}{2D_{ab}} x_i^2 + b_1 x_i + c_{ao}$$

$$K_m = a_1\left[\cosh\left(\sqrt{\frac{k''}{K_m D_{ab}}}\, x_i\right) - \tanh\left(\sqrt{\frac{k''}{K_m D_{ab}}}\, L\right)\sinh\left(\sqrt{\frac{k''}{K_m D_{ab}}}\, x_i\right)\right]$$

$$\frac{k''}{D_{ab}} x_i + b_1 = a_1\sqrt{\frac{k''}{K_m D_{ab}}}\left[\sinh\left(\sqrt{\frac{k''}{K_m D_{ab}}}\, x_i\right) - \tanh\left(\sqrt{\frac{k''}{K_m D_{ab}}}\, L\right)\cosh\left(\sqrt{\frac{k''}{K_m D_{ab}}}\, x_i\right)\right]$$

Using the values for our physical constants, a_2, b_1, and x_i are

$$x_i = 0.129\text{m} \qquad b_1 = -322.29 \qquad a_1 = 109.96$$

The corresponding profile is shown in Figure 5.13. Our approximate solution turns out to be a reasonable approximation to the exact solution.

5.5.2 Simultaneous Heat and Momentum Transport: Natural Convection

The previous section showed how mass transfer and heat transfer could be coupled via a generation term. The current problem shows a common example of the same phenomenon serving to couple momentum and heat transfer. We refer to this coupling as natural or free convection since a temperature difference drives the inherent motion of a fluid. Natural convection is responsible for wind, the cellular surface patterns observed on the sun, the great red spot of Jupiter, and what is called radiant heat in most homes and apartments.

The problem we consider is an idealization of flow between two vertical plates, one heated and the other cooled. A fluid with density, ρ, and viscosity, μ, is placed between two walls that are separated by a distance, $2L$. The wall at $y = -L$ is held at T_1, and the wall at $y = L$ is held at a lower temperature, T_2. Due to the temperature gradient, the fluid in contact with the hot wall will rise and that near the cold wall will sink. It is assumed that the system is so constructed (infinitely long plates) that the volume rate of flow in the upward stream is the same as that in the downward moving stream. This means temperature and velocity vary only in the y-direction.

Since momentum and heat transfer are coupled via the momentum generation term, our first task is to determine the temperature distribution within the fluid. If we assume the fluid moves very slowly and only in the z-direction, we can assert that heat is transported through the system by conduction only. We perform an energy balance on a small section of the fluid shown in Figures 5.14 and 5.15:

$$\begin{aligned} &In \;-\; Out \;+ Gen = Acc \\ &q(y) - q(y+\Delta y) + 0 = 0 \end{aligned} \tag{5.84}$$

Dividing by Δy, taking the limit as $\Delta y \to 0$, and substituting in Fourier's law give

$$kA_s \frac{d^2T}{dy^2} = 0 \tag{5.85}$$

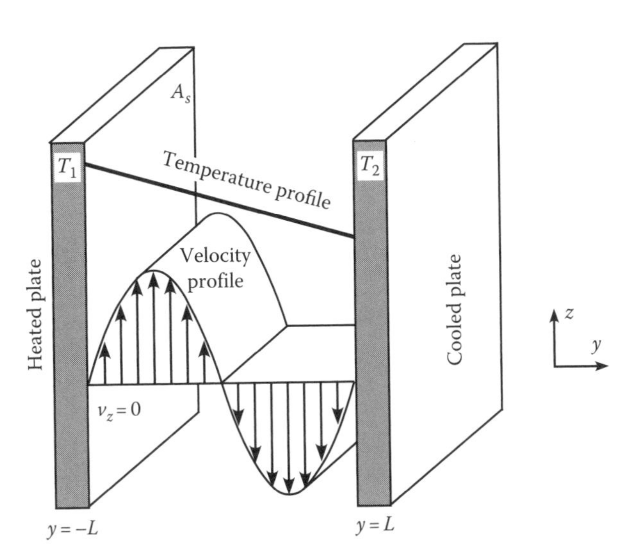

FIGURE 5.14 Simultaneous momentum and heat transport. Momentum is generated by a buoyant force arising from the different densities of the hot and cold fluids.

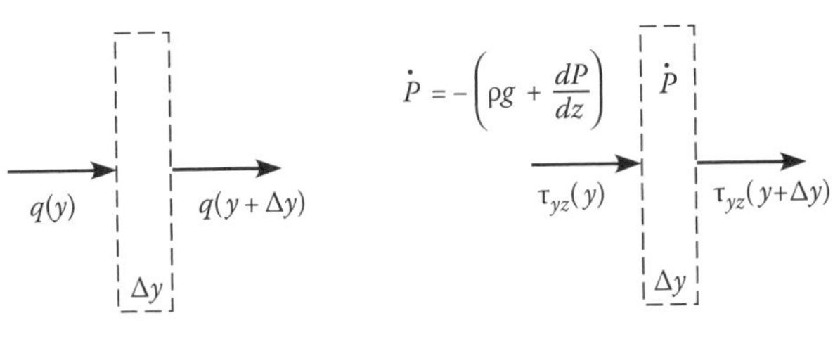

FIGURE 5.15 Control volumes for writing conservation of energy and momentum within the fluid. Arrows represent the flow of heat and momentum through the control volume.

The boundary conditions specified a constant temperature at both walls:

$$y = -L \qquad T = T_1 \tag{5.86}$$

$$y = L \qquad T = T_2 \tag{5.87}$$

We have seen in Chapter 4 that the solution to Equation 5.85 is a straight line. We can write this solution as follows:

$$T = T_m - \frac{1}{2}\Delta T\left(\frac{y}{L}\right) \tag{5.88}$$

where

ΔT is the temperature difference between the two walls
T_m is the arithmetic mean temperature of the fluid

$$\Delta T = T_1 - T_2 \qquad T_m = \frac{T_1 + T_2}{2} \tag{5.89}$$

Now that we have determined the temperature profile, we can write down the momentum balance referring to Figure 5.15. The momentum balance over the same small region must consider the pressure force due to the weight of the fluid and the body force due to gravity. These two forces represent generation terms and are related to the density gradient within the fluid:

$$\begin{gathered} In \quad - \quad Out \quad + \quad Gen \quad = Acc \\ F_z(y) - F_z(y+\Delta y) - \left(\frac{dP}{dz} + \rho g\right)A_S\Delta y = 0 \end{gathered} \tag{5.90}$$

Again, expressing the force as $F_z(y) = A_S\tau_{yz}$, dividing by Δy, taking the limit as $\Delta y \to 0$, and substituting in Newton's law of viscosity gives

$$\mu\frac{d^2v_z}{dy^2} - \frac{dP}{dz} - \rho g = 0 \tag{5.91}$$

We know that the density of the fluid changes with temperature, causing the motion of the fluid. We assume that over a suitably small range of temperatures, we can represent the temperature dependence of the density as a linear function where

$$\rho = \bar{\rho} - \bar{\rho}\beta\left(T - T_{ref}\right) \tag{5.92}$$

T_{ref} is as yet unspecified. β is the coefficient of volume expansion, a thermodynamic quantity defined by

$$\beta = -\frac{1}{\bar{\rho}}\left(\frac{\partial \rho}{\partial T}\right)_P \tag{5.93}$$

and $\bar{\rho}$ is the density of the fluid at T_{ref}. Substituting the expression for ρ into the differential equation, we obtain the classic *Boussinesq approximation* that has been used for over a 100 years to represent natural convection problems:

$$\mu \frac{d^2 v_z}{dy^2} - \frac{dP}{dz} - \bar{\rho} g + \bar{\rho} g \beta \left(T - T_{ref}\right) = 0 \tag{5.94}$$

If the pressure gradient in the system is due solely to the weight of the fluid (a reasonable approximation as we will see for steady flow with no acceleration of the fluid), then we can replace dP/dz by the average *weight* of the fluid per unit volume:

$$\frac{dP}{dz} = -\bar{\rho} g \tag{5.95}$$

In reality, the pressure gradient is due to the weight of the fluid at each y location. Unfortunately, this means that for the less dense fluid, the pressure drop is smaller than for the heavier fluid, and so there would be a flow in the y-direction to compensate. Using our simplified pressure gradient approximation, the momentum equation remains 1-D:

$$\mu \frac{d^2 v_z}{dy^2} + \bar{\rho} g \beta (T - T_{ref}) = 0 \tag{5.96}$$

The physical meaning of this equation is that the viscous forces (left-hand side) are just balanced by the buoyant forces (right-hand side). We now include the expression for the temperature distribution into Equation 5.96:

$$\mu \frac{d^2 v_z}{dy^2} + \bar{\rho} g \beta \left[T_m - T_{ref} - \frac{1}{2} \Delta T \left(\frac{y}{L} \right) \right] = 0 \tag{5.97}$$

The boundary conditions for this problem assert the no-slip condition at both walls:

$$y = -L \qquad v_z = 0 \tag{5.98}$$

$$y = L \qquad v_z = 0 \tag{5.99}$$

The solution to Equation 5.97 is found by separating and integrating twice. After applying the boundary conditions, we are left with

$$v_z = \frac{\bar{\rho} g \beta \Delta T L^2}{12\mu} \left[\left(\frac{y}{L} \right)^3 - C \left(\frac{y}{L} \right)^2 - \left(\frac{y}{L} \right) + C \right] \tag{5.100}$$

$$C = \frac{6(T_m - T_{ref})}{\Delta T} \tag{5.101}$$

Having applied the boundary conditions, we are still left with one unknown, C, or more precisely, T_{ref}. We determine this constant by requiring that the net volume flow of fluid be zero (what goes up must come down):

$$\int_{-L}^{L} v_z A_s dy = -\frac{2}{3}C + 4C = 0 \tag{5.102}$$

Solving for C_1 shows

$$C = 0 \quad \rightarrow \quad T_{ref} = T_m \tag{5.103}$$

The final expression for v_z is

$$v_z = \frac{\bar{\rho} g \beta \Delta T L^2}{12\mu}\left[\left(\frac{y}{L}\right)^2 - 1\right]\left(\frac{y}{L}\right) = v_{zo}\left[\left(\frac{y}{L}\right)^2 - 1\right]\left(\frac{y}{L}\right) \tag{5.104}$$

The velocity profile is cubic in y and does show warm fluid rising and cool fluid descending. By defining a dimensionless position, $\xi = y/L$, and rearranging the constants, Equation 5.104 can also be written in the following form:

$$\text{Re} = \frac{1}{12}\text{Gr}\left(\xi^2 - 1\right)\xi \tag{5.105}$$

The velocity profile is now expressed in the form of two dimensionless groups, the Reynolds number, Re, and the Grashof number, Gr. These two groupings have specific physical meaning for the problem at hand. The Reynolds number is a ratio of the strength of inertial to viscous forces involved in the flow of the fluid. The Grashof number represents a ratio of the strength of the buoyant forces driving the flow to the viscous forces trying to retard the flow:

$$\text{Gr} = \frac{\bar{\rho}^2 g \beta \Delta T L^3}{\mu^2} \qquad \frac{\text{Buoyant forces}}{\text{Viscous forces}} \tag{5.106}$$

$$\text{Re} = \frac{\bar{\rho} v_z L}{\mu} \qquad \frac{\text{Inertial forces}}{\text{Viscous forces}} \tag{5.107}$$

There is an analog of natural convection in mass transfer. If we replace the walls of the system in Figure 5.14 by semipermeable membranes separating the system from reservoirs of species a having mole fractions x_{a1} and x_{a2}, we arrive at the system shown in Figure 5.16. The fluid within the system will have a compositionally dependent density similar to the temperature-dependent density we just considered. If we assume dilute solutions, we can express this density in terms of a coefficient of compositional expansion, γ_x:

$$\rho = \bar{\rho} - \bar{\rho}\gamma_x\left(x_a - x_{a,ref}\right) \tag{5.108}$$

Our problem for natural convection in the mass transfer situation is formally identical to the problem we just solved for the heat transfer case. Assuming fluid with a high a content is lighter than fluid with a low a content, we will have a cubic velocity profile identical in shape

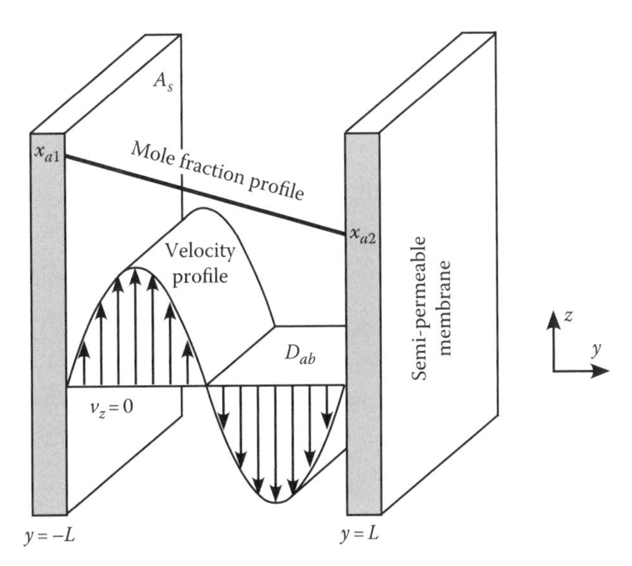

FIGURE 5.16 Natural convection in a mass transfer system. Momentum is generated by a buoyant force that arises due to the dependence of fluid density on composition.

to Equation 5.104. We can express this velocity profile in terms of the Reynolds number and a Grashof number for mass transfer:

$$\mathrm{Re} = \frac{1}{12}\mathrm{Gr}_{ab}(\xi^2 - 1)\xi \tag{5.109}$$

where

$$\mathrm{Gr}_{ab} = \frac{\bar{\rho} g \gamma_x \Delta x_a L^3}{\mu^2} \quad \frac{\text{Buoyant forces}}{\text{Viscous forces}} \tag{5.110}$$

and the Reynolds number is defined as before.

5.5.3 Diffusion, Generation, and Recombination in a Semiconductor

One of the most important transport processes in modern technology concerns the diffusion of charge carriers (electrons or holes) in semiconductor devices. The behavior of a transistor or simple p–n junction is, in part, governed by the same mass transport processes as the diffusion of perfume in the air. If we inject a pulse of electrons into a semiconductor, for example, the electrons will diffuse through the material due to inherent thermal motion and scattering off lattice planes or impurities in the material. If positively charged carriers (holes) are present, a fraction of these electrons will encounter holes, recombine with them, and both species will disappear. There is no difference between this process and the diffusion and reaction process we discussed in Section 5.5.1. We can treat the entire system as a pseudo-binary one, since the semiconductor material acts as a *solvent* for the charge carriers.

Consider the diffusion of holes in the p–n junction shown in Figure 5.17. The diode consists of two regions. The left-hand region is p-type material and is doped to produce an excess of positive charge carriers or holes, signified by c_p. This is accomplished by replacing silicon atoms, Si^{+4}, with gallium, Ga^{+3}, or more commonly by boron atoms, B^{+3}. The excess positive charge is *free* to move about the semiconductor and is referred to as a *hole*. In the p-type regions, the holes are called the

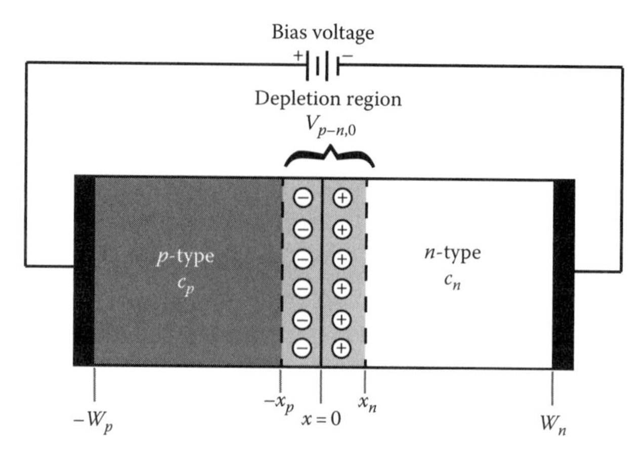

FIGURE 5.17 Schematic diagram of a p–n diode.

majority charge carrier. The right-hand region is n-type material that has been doped to provide an excess of electrons, signified by c_n. To generate n-type material, silicon atoms are replaced by arsenic atoms, As^{+5}, or phosphorous atoms, P^{+5}. The excess negative charge, the electron, is also free to move about the material. The electrons in the n-type region are the majority charge carriers.

A diode is operated by applying an external bias voltage across the junction. Under what is termed forward bias, the diode current is due to carrier recombination. Under the opposite conditions, reverse bias, the diode current is due to carrier generation. The latter is facilitated by light leading to the creation of a photodiode. We will calculate the ideal diode current for the case where generation or recombination occurs in the quasi-neutral region of the diode. This is often called the long diode case because the quasi-neutral region is much longer than the diffusion length or mean free path of the charge carriers before they recombine. Recombination can also occur at the ohmic contacts to the diode when the quasi-neutral region is much shorter than the diffusion length. This is called the short diode case.

The large concentration difference between electrons and holes at the p–n junction cannot be sustained in the absence of an applied potential. Thus, to achieve thermal equilibrium, mobile electrons and holes close to the p–n junction diffuse across the junction into the p-type/n-type region and recombine, neutralizing one another. This process leaves fixed, ionized donors (acceptors) behind; positive, nonmobile charge in the n-type region and negative, nonmobile charge in the p-type region. The region around the junction, with no mobile charge carriers, is called the depletion region. The depletion region cannot grow indefinitely and the nonmobile charge left behind due to the ionized donors and acceptors induces an electric field that opposes the motion of mobile charge carriers and limits the depletion layer width. Equilibrium is established when the drift current due to the induced field balances the diffusion current due to the concentration gradients leading to net zero current. The internal voltage that exists due to the induced field is called the built-in potential, $V_{p-n,0}$:

$$V_{p-n,0} = V_{thermal} \ln\left[\frac{c_p c_n}{2}\right] = \left(\frac{e}{k_B T}\right) \ln\left[\frac{c_p c_n}{2}\right] \quad (5.111)$$

When we apply a positive potential, $V_{applied}$, at the anode (p-type material) and negative potential at the cathode (n-type material) as shown in Figure 5.17, we refer to this as forward biased. The opposite case, applying a positive voltage at the cathode and negative voltage at the anode, is called reverse biased, and in that case, $V_{applied}$ will be opposite in sign to that shown in Figure 5.17. In either case, the net potential, V_{net}, applied across the diode can be represented as

$$V_{net} = V_{p-n,0} - V_{applied} \quad (5.112)$$

If $V_{net} < 0$, the depletion layer thickness decreases because the p-type material repels holes and drives them toward the interface, while the n-type material repels the electrons and drives them toward the interface. When $V_{net} > 0$, the process is reversed, and the depletion layer thickness increases.

Though complicated, the operation of the diode can be described as a classic diffusion with chemical reaction problem similar to the one we analyzed in Section 5.5.1. For example, if we apply a forward bias, the depletion layer becomes thin, and electrons can be injected across the depletion layer and into the p-type region. They can only diffuse a short distance beyond the depletion layer before recombining with holes. Though the electrons penetrate only a short distance into the p-type region before recombining, at steady state, a current continues to flow because the majority carriers in that region begin to flow in the opposite direction. The total current due to the motion of electrons and holes in both regions remains balanced to avoid charge imbalance.

With this in mind, we can look at a very simplified case and apply conservation of charge to small regions outside the depletion zone to determine the mole fraction profile of the minority carriers, electrons in the region $-W_p < x < -x_p$, and holes in the region $x_n < x < W_n$. If the bias voltage is not far removed from the built-in voltage, the electrons and holes will be dilute species.

The control volumes within the two regions are shown in Figure 5.18. D_+ represents the diffusion coefficient for the holes in the n-type region, D_- is the diffusion coefficient for the electrons in the p-type region, and k'' is the recombination rate constant. The electron-hole recombination process is inherently a bimolecular one. However, outside of the depletion region, the majority charge carriers (electrons in n-type region) outnumber the minority charge carriers (holes in the n-type region) by a large margin. Therefore, the concentration of majority carriers remains unchanged, and we can represent the recombination rate as pseudo first-order in either the electron or hole mole fraction. The balance (conservation of charge) about the control volumes is

$$\begin{aligned} &In \quad - \quad Out \quad + \quad Gen \quad = Acc \\ &z_e e\dot{M}_p(x) - z_e e\dot{M}_p(x+\Delta x) - z_e e k'' c_p A_c \Delta x = 0 \end{aligned} \tag{5.113}$$

Holes

$$-\frac{d}{dx}(z_e e A_c N_p) - k'' z_e e c_p A_c = 0 \tag{5.114}$$

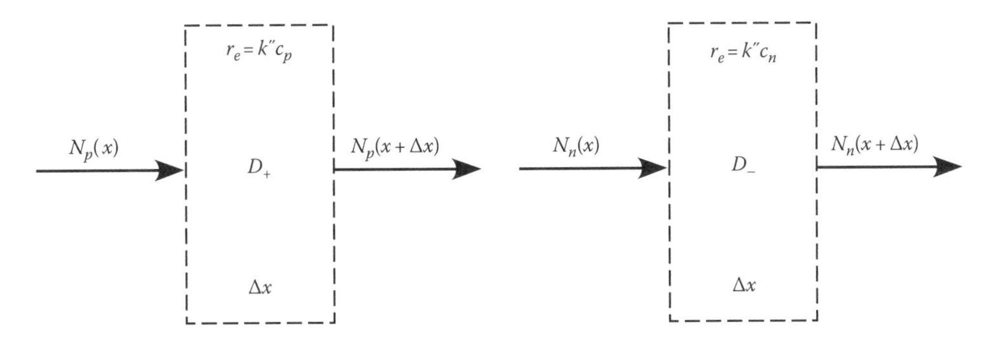

FIGURE 5.18 Control volume for diffusion in p and n regions of a diode.

$$
\begin{array}{ccccccc}
In & - & Out & + & Gen & = & Acc \\
z_e e\dot{M}_n(x) & - & z_e e\dot{M}_n(x+\Delta x) & - & z_e e k'' c_n A_c \Delta x & = & 0
\end{array}
\tag{5.115}
$$

Electrons

$$
-\frac{d}{dx}\left(z_e e A_c N_n\right) - k'' z_e e c_n A_c = 0 \tag{5.116}
$$

where

z_e is the valence of the charge carrier (+/–1 here)
e is the charge on an electron

We use Fick's law for dilute solutions to express c_p or c_n in terms of the mole fraction gradient. The differential equations for the mole fraction profiles assuming constant diffusivities are

$$
\frac{d^2 c_p}{dx^2} - \left(\frac{k''}{D_+}\right) c_p = 0 \tag{5.117}
$$

$$
\frac{d^2 c_n}{dx^2} - \left(\frac{k''}{D_-}\right) c_n = 0 \tag{5.118}
$$

This equation is subject to boundary conditions stating that we know the excess carrier concentration at the edge of the depletion region and at the metal contact. At the edges of the depletion region, the minority carrier concentrations are

$$
x = x_n \qquad c_p = c_{p0} \exp\left[\frac{V_{applied}}{V_{thermal}}\right] \tag{5.119}
$$

$$
x = -x_p \qquad c_n = c_{n0} \exp\left[\frac{V_{applied}}{V_{thermal}}\right] \tag{5.120}
$$

At the contacts, we assume thermal equilibrium applies and so

$$
x = W_n \qquad c_p = c_{p0} \tag{5.121}
$$

$$
x = -W_p \qquad c_n = c_{n0} \tag{5.122}
$$

It is again instructive to put Equations 5.117 and 5.118 in dimensionless form. We define a dimensionless hole mole fraction, χ, a dimensionless electron fraction, ψ, and dimensionless position coordinates, ξ and ζ, as

$$
\chi = \frac{c_p - c_{p0}}{c_{p0}} \qquad \xi = \frac{x - x_n}{W_n - x_n} \tag{5.123}
$$

$$
\psi = \frac{c_n - c_{n0}}{c_{n0}} \qquad \zeta = \frac{x + x_p}{x_p - W_p} \tag{5.124}
$$

Equations 5.117 and 5.118 now read

$$\frac{d^2\chi}{d\xi^2} - \underbrace{\left(\frac{k''(W_n - x_n)^2}{D_+}\right)}_{\mathrm{Da}_p}\chi = 0 \tag{5.125}$$

$$\frac{d^2\psi}{d\zeta^2} - \underbrace{\left(\frac{k''(x_p + W_p)^2}{D_-}\right)}_{\mathrm{Da}_n}\psi = 0 \tag{5.126}$$

The dimensionless groups, $k''(W_n - x_n)^2/D_+$ and $k''(x_p - W_p)^2/D_-$, we have encountered before. They are the Damkohler numbers, for the holes (Da_p) and electrons (Da_n). The solution to Equations 5.125 and 5.126 is given in terms of hyperbolic sines and cosines as it was in Section 5.5.1:

$$\chi = a_1 \cosh(\sqrt{\mathrm{Da}_p}\,\xi) + a_2 \sinh(\sqrt{\mathrm{Da}_p}\,\xi) \tag{5.127}$$

$$\psi = b_1 \cosh(\sqrt{\mathrm{Da}_n}\,\zeta) + b_2 \sinh(\sqrt{\mathrm{Da}_n}\,\zeta) \tag{5.128}$$

Evaluating the constants in light of our boundary conditions, (5.119) through (5.122), gives the concentration profile in the base region. Remember we must recast the boundary conditions in dimensionless form:

$$\begin{aligned} \xi = 0 \qquad & \chi = \exp\left[\frac{V_{applied}}{V_{thermal}}\right] - 1 \\ \zeta = 0 \qquad & \psi = \exp\left[\frac{V_{applied}}{V_{thermal}}\right] - 1 \end{aligned} \tag{5.129}$$

$$\begin{aligned} \xi = 1 \qquad & \chi = 0 \\ \zeta = 1 \qquad & \psi = 0 \end{aligned} \tag{5.130}$$

$$\chi = \left[\exp\left(\frac{V_{applied}}{V_{thermal}}\right) - 1\right]\left\{\cosh(\sqrt{\mathrm{Da}_p}\,\xi) - \frac{\sinh(\sqrt{\mathrm{Da}_p}\,\xi)}{\tanh(\sqrt{\mathrm{Da}_p})}\right\} \tag{5.131}$$

$$\psi = \left[\exp\left(\frac{V_{applied}}{V_{thermal}}\right) - 1\right]\left\{\cosh(\sqrt{\mathrm{Da}_n}\,\zeta) - \frac{\sinh(\sqrt{\mathrm{Da}_n}\,\zeta)}{\tanh(\sqrt{\mathrm{Da}_n})}\right\} \tag{5.132}$$

Now that we know the concentrations of the minority charge carriers in each region, we can calculate the current flowing. The total current is the sum of the maximum electron current in the p-type region, the maximum hole current in the n-type region, and the current due to recombination in the depletion region. The maximum currents outside of the depletion region occur at the

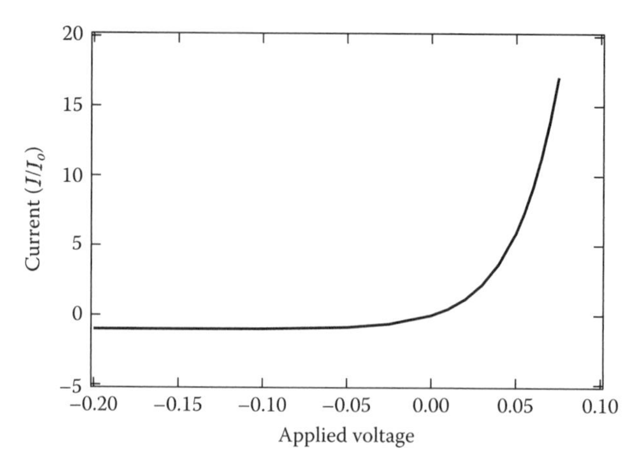

FIGURE 5.19 Current response of a p–n junction diode.

boundaries of the depletion region. The currents inside the depletion region we do not know, but for now, we can assume they are negligible. Using Equations 5.131 and 5.132, we find

$$I = I_p + I_n = -z_e e A_c D_- \left.\frac{dc_n}{dx}\right|_{x=-x_p} - z_e e A_c D_+ \left.\frac{dc_p}{dx}\right|_{x=x_n}$$

$$= -\left(\frac{z_e e A_c D_- c_{n0}}{x_p - W_p}\right)\left.\frac{d\psi}{d\zeta}\right|_{\zeta=0} - \left(\frac{z_e e A_c D_+ c_{p0}}{W_n - x_n}\right)\left.\frac{d\chi}{d\xi}\right|_{\xi=0} \tag{5.133}$$

Evaluating the derivatives using Equations 5.131 and 5.132 gives

$$I = \left[\exp\left(\frac{V_{applied}}{V_{thermal}}\right) - 1\right]\left\{-\left(\frac{z_e e A_c D_- c_{n0}}{x_p - W_p}\right)\left(-\frac{\sqrt{\mathrm{Da}_n}}{\tanh(\sqrt{\mathrm{Da}_n})}\right) - \left(\frac{z_e e A_c D_+ c_{p0}}{W_n - x_n}\right)\left(-\frac{\sqrt{\mathrm{Da}_p}}{\tanh(\sqrt{\mathrm{Da}_p})}\right)\right\}$$

$$= \left[\exp\left(\frac{V_{applied}}{V_{thermal}}\right) - 1\right]\left\{\left(\frac{e A_c D_- c_{n0}}{W_p - x_p}\right)\left(\frac{\sqrt{\mathrm{Da}_n}}{\tanh(\sqrt{\mathrm{Da}_n})}\right) + \left(\frac{e A_c D_+ c_{p0}}{W_n - x_n}\right)\left(\frac{\sqrt{\mathrm{Da}_p}}{\tanh(\sqrt{\mathrm{Da}_p})}\right)\right\} \tag{5.134}$$

Notice that the current is completely controlled by the Damkohler numbers and the applied voltage. If the applied voltage is zero, no current flows. If there is no recombination, the current is zero. If the applied voltage is negative, we have a small current, and if it is positive, the current can be very large. Thus, we have reproduced the diode response curve shown in Figure 5.19.

5.5.4 Diffusion, Drift, and Recombination

The model in the previous section neglected several complicating mechanisms that we can explore now via analytical and numerical computation. Since there is an electric field applied across the diode, the charge carriers move in response to that field, a process called *drift*. We encountered this before in Chapter 3 when we discussed electrophoresis. If we consider just the holes moving through the n-type region this time around, then the change in net charge concentration through the n-type region sets up a local electric field that opposes the diffusion and drift of the holes. This local field is governed by Poisson's equation, the mathematical prototype for a balance equation with generation. Our mass balance remains essentially the same except that the flux of holes now contains two terms, the diffusion term described by Fick's law and the drift term where μ_+ is the mobility of the holes in the field:

$$N_p = -D_+ \frac{dc_p}{dx} + \mu_+ E c_p \qquad E = -\frac{dV}{dz} \tag{5.135}$$

This new flux transforms our balance equation, (5.114), to

$$D_+ \frac{d^2 c_p}{dx^2} + \frac{d}{dx}\left(\mu_+ c_p \frac{dV}{dx}\right) - k'' c_p = 0 \tag{5.136}$$

The electric field, $E(x)$, and voltage profile, $V(x)$, are governed by Poisson's equation, where F_a is Faraday's constant, ε is the dielectric constant of the material, and ε_o is the permittivity of a vacuum. As we pump positive charges into the n-type region, we establish an opposing electric field that inhibits the injection of more charge:

$$\frac{d^2V}{dz^2} = \frac{z_e F_a e}{\varepsilon \varepsilon_o} c_p \quad \text{Poisson's equation} \tag{5.137}$$

Assuming we apply a voltage of V_o at the emitter–base interface and ground the base–collector interface, the boundary conditions for this problem become

$$x = x_n \qquad c_p = c_{p0} \exp\left[\frac{V_{applied}}{V_{thermal}}\right] \qquad V = V_{applied} \tag{5.138}$$

$$x = W_n \qquad c_p = c_{p0} \qquad V = 0 \tag{5.139}$$

Putting the equations in dimensionless form helps understand the process a bit better. We define the following new variables:

$$\xi = \frac{x - x_n}{W_n - x_n} \qquad \chi = \frac{c_p - c_{p0}}{c_{p0}} \qquad \phi = \frac{V}{V_{applied}} \tag{5.140}$$

and assume constant material properties so that Equations 5.136 through 5.139 become

$$\frac{d^2\chi}{d\xi^2} + \left(\frac{\mu_+ V_{applied}}{D_+}\right) \frac{d}{d\xi}\left(\chi \frac{d\phi}{d\xi}\right) - \left(\frac{k''(W_n - x_n)^2}{D_+}\right)\chi = 0 \tag{5.141}$$

$$\frac{d^2\phi}{d\xi^2} = \left[\frac{z_e e F_a (W_n - x_n)^2 c_{p0}}{\varepsilon \varepsilon_o V_{applied}}\right]\chi \tag{5.142}$$

$$\xi = 0 \qquad \chi = \exp\left(\frac{V_{applied}}{V_{thermal}}\right) - 1 \qquad \phi = 1 \tag{5.143}$$

$$\xi = 1 \qquad \chi = 0 \qquad \phi = 0 \tag{5.144}$$

The dimensionless groups appearing in these equations are the Damkohler number

$$\mathrm{Da} = \frac{k''(W_n - x_n)^2}{D_+} \quad \frac{\text{Diffusion time}}{\text{Reaction time}} \tag{5.145}$$

the Peclet number

$$\mathrm{Pe} = \frac{\mu_+ V_{applied}}{D_+} \quad \frac{\text{Drift velocity}}{\text{Diffusion velocity}} \tag{5.146}$$

and for lack of a better name, the field number.

$$F_E = \frac{z_e e F_a (W_n - x_n)^2 c_{p0}}{\varepsilon\varepsilon_o V_{applied}} = \frac{z_e e F_a (W_n - x_n) c_{p0}/\varepsilon\varepsilon_o}{V_{applied}/(W_n - x_n)} \quad \frac{\text{Induced field}}{\text{Applied field}} \tag{5.147}$$

The Peclet number is a measure of the strength of the ionic drift, a form of convection to the strength of ionic diffusion. In later chapters, we will see the analogy between convection and drift in the presence of a field. Large Peclet numbers mean the process is dominated by drift. The field number represents a ratio of the induced electric field due to the charge separation to the applied field due to the applied voltages. Small field numbers mean the electric field is relatively constant throughout the medium.

These equations are a nonlinear set due to the drift term and generally would have to be solved numerically. We can solve the equations analytically if we assume that the field is a constant, E_o. The analytical solution is

$$\frac{d^2\chi}{d\xi^2} + \mathrm{Pe}\frac{d\chi}{d\xi} - \mathrm{Da}\chi = 0 \tag{5.148}$$

$$\chi = \left[\exp\left(\frac{V_{applied}}{V_{thermal}}\right) - 1\right]\left\{e^{r_-\xi} + \left(\frac{e^{r_-}}{e^{r_+} - e^{r_-}}\right)\left(e^{r_-\xi} - e^{r_+\xi}\right)\right\}$$

$$r_\pm = \frac{-\mathrm{Pe} \pm \sqrt{\mathrm{Pe}^2 - 4\,\mathrm{Da}}}{2} \tag{5.149}$$

The solution in Equation 5.149 shows that if the Peclet number is large, we have nearly constant hole concentration across the n-type region. If $\mathrm{Pe} < 2\sqrt{\mathrm{Da}}$, all the holes are consumed essentially at $x = x_n$. The hole current is determined from the flux:

$$N_p = -D_+ \left.\frac{dc_p}{dx}\right|_{x=x_n} + \mu_+ E c_p = -\left(\frac{D_+ c_{p0}}{W_n - x_n}\right)\left.\frac{d\chi}{d\xi}\right|_{\xi=0} + \mu_+ c_{p0} E_0 \chi \qquad E_0 < 0 \tag{5.150}$$

$$N_p = c_{p0}\left[\exp\left(\frac{V_{applied}}{V_{thermal}}\right) - 1\right]\left\{\mu_+ E_0 - \left(\frac{D_+}{W_n - x_n}\right)\left(r_- + \frac{e^{r_-}(r_- - r_+)}{e^{r_+} - e^{r_-}}\right)\right\} \tag{5.151}$$

The full solution to Equations 5.141 through 5.144 is plotted in Figures 5.20 and 5.21 along with the solution given by Equation 5.149.

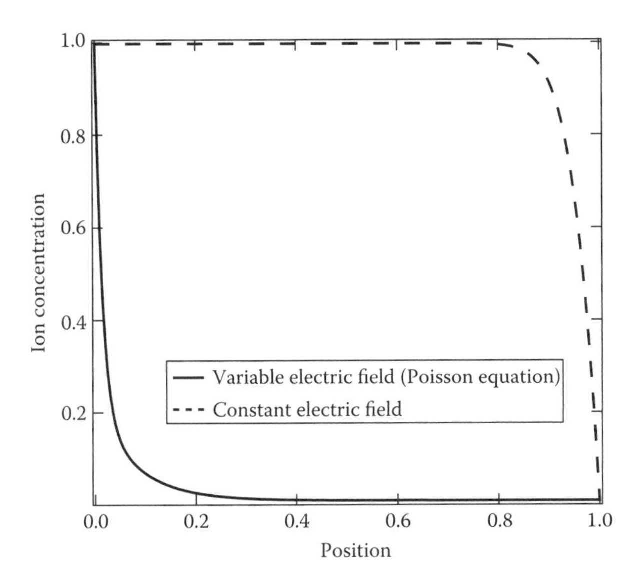

FIGURE 5.20 Dimensionless ion concentration. Allowing the electric field to vary sets up an opposing force that completely changes the concentration profile.

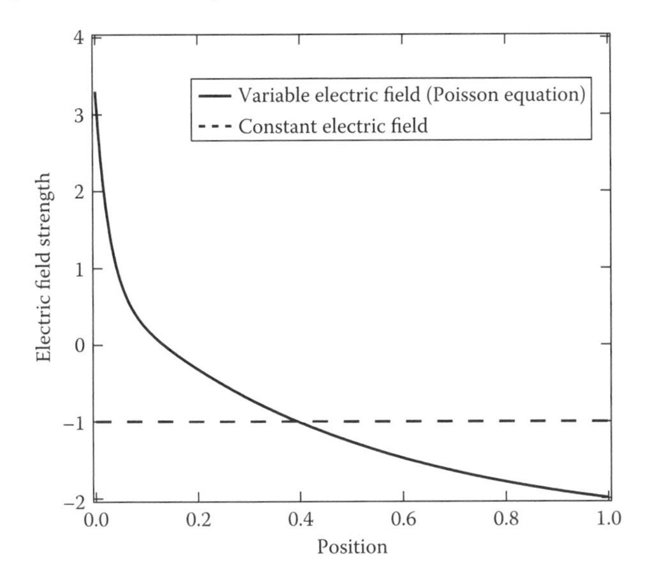

FIGURE 5.21 Electric field strength through the dielectric. The effect of incorporating Poisson's equation is evident in the figure. Notice the change in the sign of the electric field where the concentration is highest.

5.6 SUMMARY

We have investigated the role of generation in the balance equations and have shown how generation affects the nature and rate of a transport process. The generation process can occur on the boundary of a system in which case it does not affect the structure of the conservation equation but does affect the type of boundary condition used in the problem. Generation as part of a boundary condition augments the rate of the transport process and, in some cases, can reduce the resistance to transport considerably. Chemical reactions at the interface of cell walls can increase the transport of species across those walls in a process called facilitated transport. Evaporation or condensation on a boundary can dramatically reduce the resistance to heat transfer from a surface and hence increase the heat transfer rate. Such processes are used extensively to cool critical surfaces such as rocket nozzles, semiconductor chips, or the human body.

The most complicated instances of generation involve homogeneous generation terms. These terms appear in the balance equation itself. We have seen how the presence of a generation term affects the profile of the state variable in the system and how it can shift the location of the maximum or minimum in this state variable away from the boundaries of the system. We have also seen how generation can couple transport processes.

In the next chapter, we will explore the last term in the general balance: accumulation. Since many transport processes do not occur in the steady state, the addition of an accumulation term is most often the rule, not the exception. We deal with accumulation last because the addition of the accumulation term brings us into the realm of partial differential equations and thereby complicates the analysis of our transport process.

PROBLEMS

Momentum Transport

5.1 A fluid flows between two plates separated by a distance d. The velocity profile is measured, and the data can be fit to the following function. The fluid viscosity is a constant, μ, and the system length is L:

$$v_z(x) = C_1(x-d) + C_2(x-d)^2 + C_3(x-d)^3$$

a. Determine an expression for the pressure drop (dP/dz) driving the flow.
b. What is the force exerted on the fluid at $x = 0$?

5.2 Two immiscible fluids are flowing between parallel plates separated by a distance, d. Each fluid occupies 50% by volume of the pipe. For the pressure drop given in Figure P5.2, determine the velocity profiles within the two liquids and the stress at the interface (Pieracci, Savoca, Ibrahim).

5.3 An incompressible, Newtonian, fluid is in steady, laminar flow within the annular region between two concentric cylinders of radii r_i and r_o. The cylinders are inclined at an angle of $\theta°$ from the vertical as shown in Figure P5.3. Determine the velocity distribution and the volumetric flow rate from the cylinders.

5.4 A fluid of density, ρ, and viscosity, μ, drains from between very long parallel flat plates oriented vertically with respect to gravity. One plate is held stationary, and the other travels vertically with a velocity, v_o.

a. What is the differential equation governing the flow of fluid between the plates?
b. What are the boundary conditions for the problem?
c. Solve the equation for the velocity profile $v_z(x)$.
d. If we are to have no net volumetric flow (m³/s) between the plates, what value should v_o have and what is its direction? Assume the plates extend 1 m deep in the y-direction.

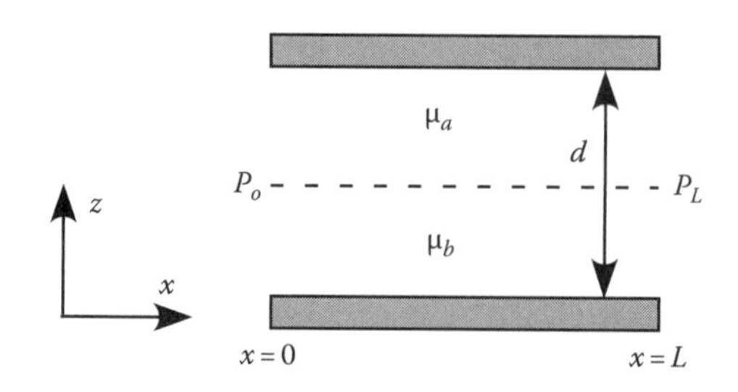

FIGURE P5.2

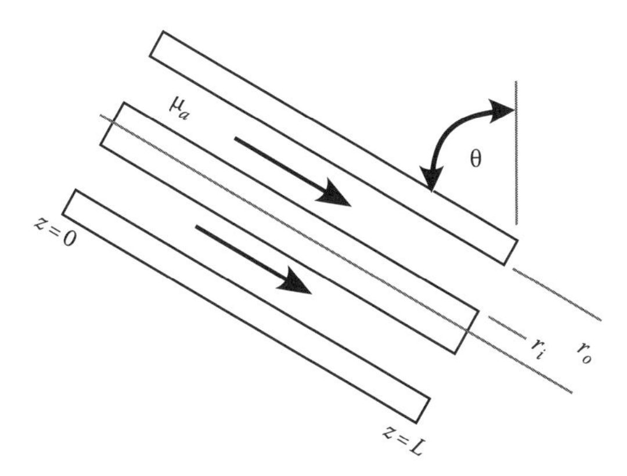

FIGURE P5.3

5.5 A Bingham fluid flows of viscosity, μ, and yield stress, τ_0, through a horizontal pipe of radius, R, and length, L. The flow is driven by a pressure drop, $\Delta P/L$, along the length of the pipe.

a. Derive an expression for the velocity profile in the pipe. There is a critical radius, r_c, within which the fluid moves as a slug, so the velocity profile will be composed of two parts. Within r_c, the velocity will be a constant, and outside of r_c, the profile will depend upon r.

b. Determine the volumetric flow rate for flow through the pipe using the velocity profile from part (a).

c. If $\tau_0 = 1000$ N/m^2, $R = 0.002$ m, $L = 1$ m, and $\mu = 0.1$ N s/m^2, use the expression from part (b) to determine the pressure drop required to deliver a volumetric flow rate of 2×10^{-7} m^3/s.

5.6 A fluid is sheared between two parallel plates as we saw in Chapter 4. If the gap between the plates is small and the shearing rate is high, friction between the fluid molecules will generate heat. Energy will flow by conduction in the y-direction from the hot plate toward the cold plate. As we will see later in Chapter 10, the heat generation rate will be given by

$$\dot{q}\left(\frac{W}{m^3}\right) = \mu\left(\frac{dv_x}{dy}\right)^2$$

a. For a fluid of viscosity, μ, a gap width between the plates of δ, and an upper plate velocity of v_0, what is the heat generation rate within the fluid?

b. If the upper surface of the plate is held at a temperature of T_0 and the lower surface at a temperature of T_1, what is the temperature profile between the plates?

c. For the parameters

$$\mu = 0.25\,\text{N s/m}^2 \qquad \delta = 0.001\,\text{m}$$
$$v_0 = 1\,\text{m/s} \qquad T_0 = 325\,\text{K} \qquad T_1 = 300\,\text{K}$$

evaluate the heat generation rate and the temperature profile.

Heat Transport

5.7 Consider the case of steady-state, simultaneous heat and mass transfer with chemical reaction as discussed in the notes. The reaction is autocatalytic (self-catalyzed) and *takes off* at a temperature of 75°C. For the reactor and reaction data given in the following, pick an organic solvent that will insure safe, controllable, steady-state operation.

Reactor length	$L = 0.05$ m
Reactor temperature	$T_o = 50°C$
Diffusivity of A	$D_a = 3 \times 10^{-5}$ m²/s
Rate constant	$k'' = 0.15$ s⁻¹
Heat of reaction	$\Delta H = 52{,}500$ J/mol
Initial concentration	$c_{ao} = 10$ mol/m³

5.8 A truncated solid cone is of circular cross section with a diameter that varies along its length according to $d(x) = ae^x$ where $a = 0.8$ m and x is in meters. The cone has a length, $L = 1.8$ m, and a thermal conductivity, $k = 8$ W/m K, and is shown in Figure P5.8. It also exhibits a uniform volumetric rate of heat generation, $\dot{q} = 1993$ W/m³. The sides of the cone are insulated, and one end surface ($x = 0$) is held at $T_o = 300°C$ and has a heat flow rate, $q = 500$ W. Determine the temperature at the other end surface ($x = L$) and the heat transfer rate there.

5.9 Consider the composite wall shown in Figure P5.9 exhibiting an interfacial resistance due to the roughness of the interface. Derive the steady-state temperature distributions in both walls, and based on the flux of heat obtained, define local heat transfer coefficients based on the driving force within a given medium and an overall heat transfer coefficient based on the overall driving force. Comment on the relationship between local coefficients for the heat and mass transfer cases and overall coefficients for the heat and mass transfer cases. Comment on what would happen if medium y included a constant volumetric rate of heat generation, $\dot{q}$.

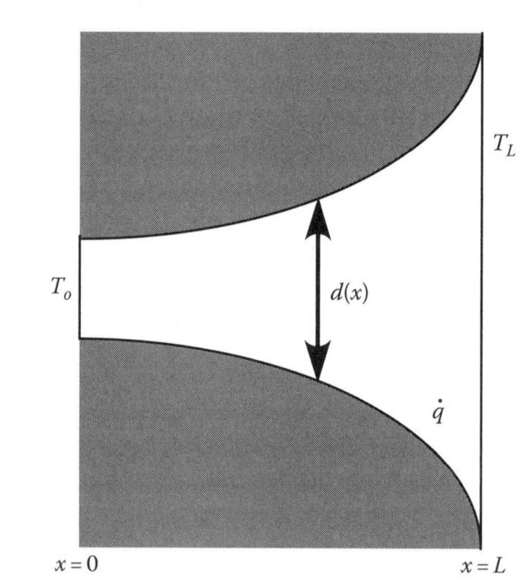

FIGURE P5.8

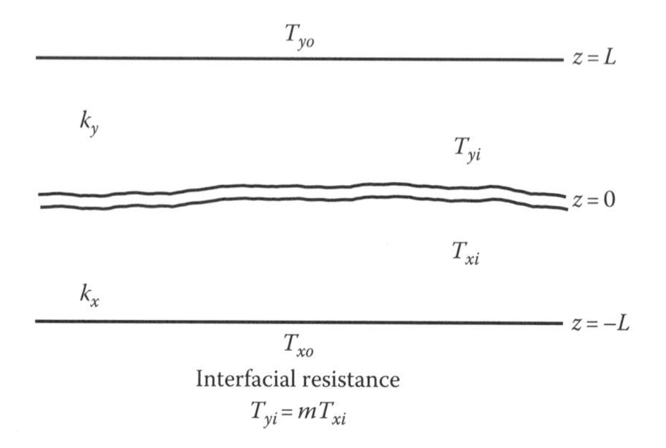

FIGURE P5.9

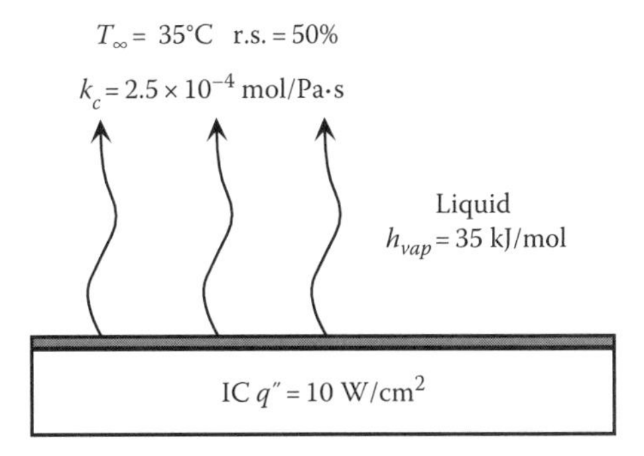

FIGURE P5.10

5.10 One way to cool off integrated circuits efficiently is by having a film of liquid evaporating from the surface of the chip. Figure P5.10 shows the general situation. Evaporation takes place at the surface of the liquid into an atmosphere with a relative humidity of 50%. The temperature of the environment is held at 35°C, and the dynamics of the environment provide for a mass transfer coefficient $k_c = 2.5 \times 10^{-4}$ mol/Pa·s. The computer chip dissipates 10 W/cm^2.

a. If the heat of vaporization of the liquid is 35 kJ/mol, what is the flux of vapor from the surface of the liquid at steady state?

b. The vapor pressure of the liquid can be represented by the following:

Antoine equation

$$\log_{10} P^{sat}\,(\text{mmHg}) = 7.96681 - \frac{1668.21}{T + 228.0}$$

What is the temperature at the surface of the liquid?

c. What is the temperature at the surface of the chip assuming film thickness of 100 μm and a thermal conductivity for the liquid of 0.5 W/m K?

d. If we modify our machine by upping the clock speed by 10%, upping the heat dissipation by 10% as well, can we still operate the system without exceeding the chip's maximum temperature of 125°C?

5.11 A solid rod is manufactured from a new superconducting material, Coldmet, and is shown in Figure P5.11. The material has the property that when a current is passed through it, it actually gets colder. The heat generation rate in the rod is $\dot{q} = -J^2\rho_r$ where $\rho_r = 2 \times 10^{-3}$ Ω – m; J (A/m^2). Consider a rod, 3 cm in diameter and 1 m long. One end ($z = 0$) is insulated, and the other end ($z = L$) is held at 77 K.

a. Derive the differential equation describing the axial temperature profile in the rod. Assume steady state.

b. Solve the differential equation to determine the temperature profile (no numbers needed).

c. For a current of 1000 A and a thermal conductivity of 20 W/m K, what is the temperature at the insulated end? What is the heat flow at the other end?

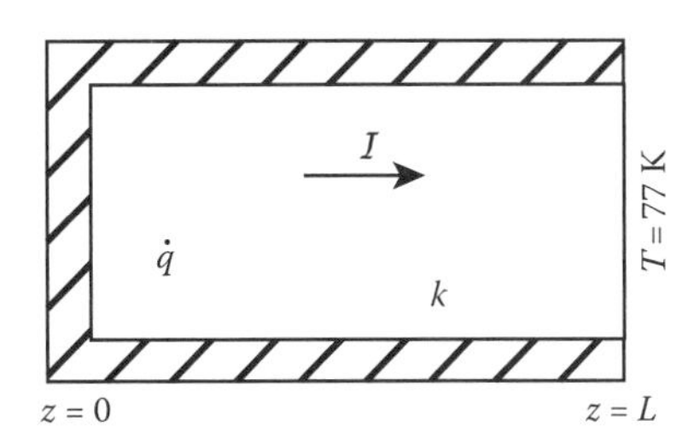

FIGURE P5.11

5.12 The exposed surface ($x = 0$) of a plane wall of thermal conductivity, k, is subjected to microwave radiation that causes the heat generation rate within the wall to vary as

$$\dot{q}(x) = \dot{q}_o\left(1 - \frac{x}{L}\right)$$

The boundary at $x = L$ is insulated, while the wall at $x = 0$ is maintained at $T = T_o$.

a. Derive the differential equation for the temperature profile.
b. What are the boundary conditions for this problem?
c. Solve the differential equation to obtain the temperature profile.

5.13 A cylindrical nuclear fuel rod contains a core of active, fissionable material and a cladding formed from a high temperature metal (Figure P5.13). The heat generation rate within this core is a function of position that we will approximate via a parabolic function:

$$\dot{q} = q_o\left[1 + \alpha\left(\frac{r}{r_i}\right)^2\right]$$

The fuel rod is exposed to a coolant that circulates at a temperature of T_∞ and provides a heat transfer coefficient of h_o (W/m^2 K). Determine

a. The temperature profiles in the fuel and cladding
b. The heat flow through the fuel rod
c. The critical cladding thickness yielding maximum heat transfer

5.14 Consider natural convection between two infinite parallel plates like the system discussed in Section 5.5.2. Instead of using a Newtonian fluid between the heated walls, what would happen if a Bingham fluid were used?

a. Determine the velocity profile assuming a yield stress, τ_o, and viscosity, μ_o, for the fluid.
b. Where does the fluid stop flowing?

5.15 New generations of safer nuclear reactors including the pebble bed reactor use spherical nuclear fuel pellets that consist of a fissionable core of radius r_{core} and thermal conductivity k_{core} that generates an energy per volume $\dot{q}$. To handle the material and prevent problems if coolant is lost, the spheres are encased in an inert graphite coating of thermal conductivity

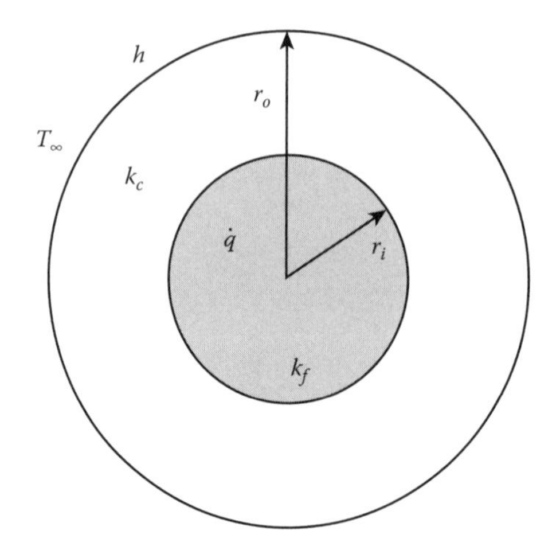

FIGURE P5.13

k_{coat} and outer radius r_{coat}. The fuel pellets are immersed in a fluid of temperature T_∞ and heat transfer coefficient, h.

a. Derive the energy balances for the core and coating of the fuel pellet.
b. What are the boundary conditions at the core of the pellet? The surface of the pellet in contact with the fluid? The interface between the core and cladding of the pellet?
c. Solve for the temperature distribution for the core and the shell. What is the heat flow rate from a pellet into the fluid?
d. If we are interested in maximizing the heat transfer to the fluid and in turn maximizing reactor safety, how thick should the cladding layer be?

Mass Transport

5.16 In the production of superconducting wire tube, the alloy is reacted with oxygen to produce the appropriate ceramic phase. Oxygen at a concentration c_o flows through the inner tube and diffuses into the metal, undergoing a chemical reaction to produce the ceramic as shown in Figure P5.16. The reaction rate has been measured as a function of position and follows

$$-r_a(\text{mol/s}) = k''c_o\left(\frac{r_o - r}{r_o - r_i}\right)^2$$

Pure nitrogen flows on the outside of the tube to scavenge away any oxygen that does not react by the time it reaches the surface.

a. Draw the control volume for deriving the mass balance for the system.
b. Using the general balance equation, derive the differential mass balance giving the concentration profile of oxygen in the superconductor.
c. What are the boundary conditions for this problem?
d. Solve the differential equation for the concentration profile.
e. What is the flux of a at $r = r_i$?

5.17 Mitochondria are the centers for energy conversion in the body. Enzymes attached to the internal membranes oxidize sugars to produce energy. To get an idea about mitochondria performance, the system shown in Figure P5.17 was developed. The substrate diffuses through the liquid phase until it reaches the membrane surface where it reacts. The rate of reaction is governed by Michaelis–Menten kinetics:

$$-r_s(\text{mol/s}) = \frac{k_s''c_s}{1 + K_m c_s}$$

a. Develop an expression for the concentration profile, mass flux through the liquid, and mass transfer coefficient to the solid phase.
b. What happens when $K_m c_s \ll 1$? When $K_m c_s$ is $\gg 1$?

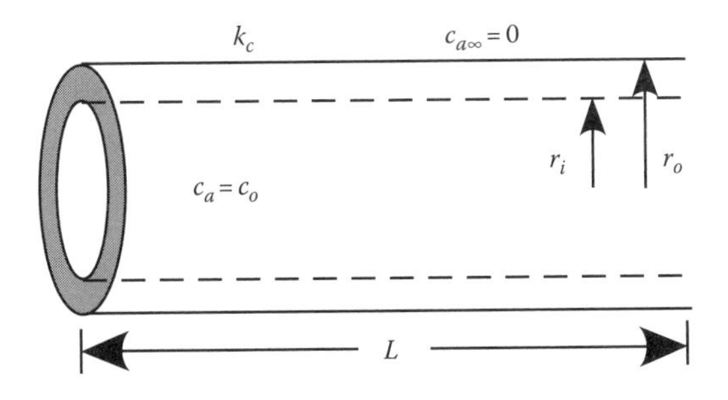

FIGURE P5.16

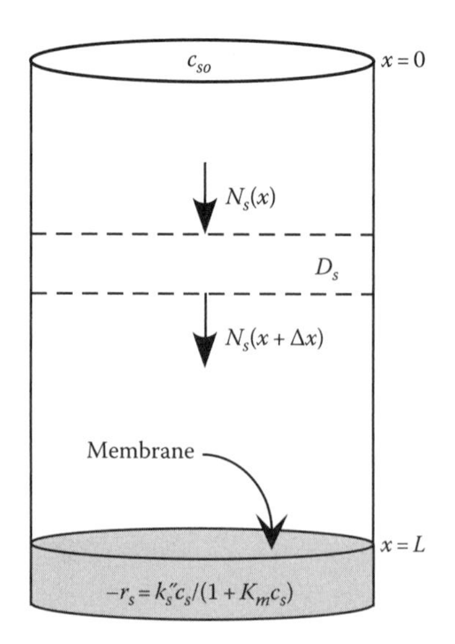

FIGURE P5.17

5.18 A semiconductor photodiode consists of a light-sensitive layer of thickness, d, coupled to electronic circuitry that measures the current produced in that layer (Figure P5.18). Electrons are generated at a rate, $\dot{M}$, by light striking the sensitive layer and are directly proportional to the light intensity. The light gets absorbed by the layer, so the intensity varies exponentially with depth:

$$I = I_o \exp(-\alpha x) \qquad \dot{M}(\text{mol/m}^3\text{s}) = m_o I$$

Once generated, the electrons diffuse toward the circuitry, and a signal is issued proportional to the electron flux at $x = d$.

a. Determine the concentration profile of electrons in the light-sensitive layer and how the flux of electrons depends upon the current. (Hint: Use boundary conditions $x = 0$ $c_e = 0$; $x = d$ $\dot{M}_e = A_c \int_0^d \dot{M}\, dx$).

b. All photodiodes have a dark current, I_d, electrons formed by random thermal means. This dark current is proportional to the volume of light-sensitive material, $I_d = I_0 V$, and represents noise in the system. Develop an expression for the signal to noise ratio. Comment on how to operate the device.

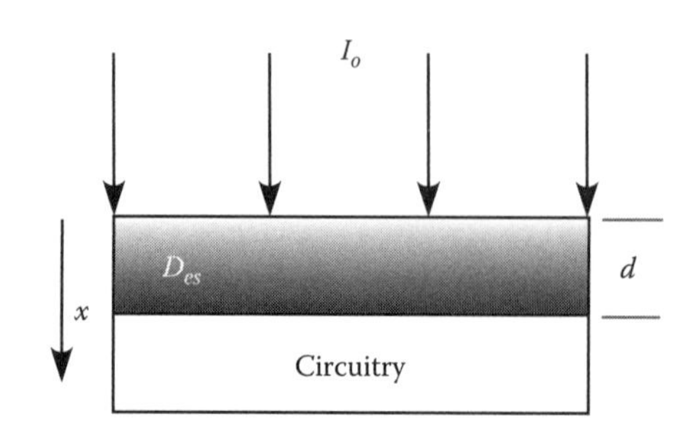

FIGURE P5.18

5.19 The speed of semiconductor photodetectors is increased by applying a bias voltage. We can investigate this effect by assuming a constant electric field, E_o, is applied across the light-sensitive layer and that this field forces the electrons to flow via a drift mechanism toward the detector circuitry. The mobility of the electrons can be estimated using the Nernst–Einstein relationship. Using the physical data and configuration in the problem previously presented, determine the concentration profile of electrons in the light-sensitive layer and the flux of electrons leaving that layer. Use boundary conditions from Problem 5.18.

5.20 Avalanche photodiodes work on a bit different mechanism from that discussed in Problem 5.18. Here, a thin photosensitive layer is used as the source of electrons (Figure P5.20). Those electrons enter a second layer or layers where they generate more electrons by collision with the material. It is akin to having an autocatalytic reaction proportional to the electron concentration:

$$-r_e = -k''c_e$$

a. Assuming that electrons diffuse through the second layer as they collide with that layer to generate new electrons, determine the concentration profile and flux of electrons into the circuitry. You can use the same boundary conditions as in Problems 5.18 and 5.19.
b. Assuming we also apply a constant electric field across the device, determine the concentration profile and flux into the circuitry.

5.21 A membrane separates two fluids. One of the fluids is well mixed, and the other is unmixed as shown in Figure P5.21. On the side of the membrane adjacent to the unmixed fluid, an enzyme is attached that can react with substrate present in the fluids. The rate of reaction per unit surface area is given by

$$-r_s = \frac{k_s''K_sc_s}{K_iK_m + K_ic_s}$$

where
K_m is the Michaelis constant
K_i is an inhibition constant

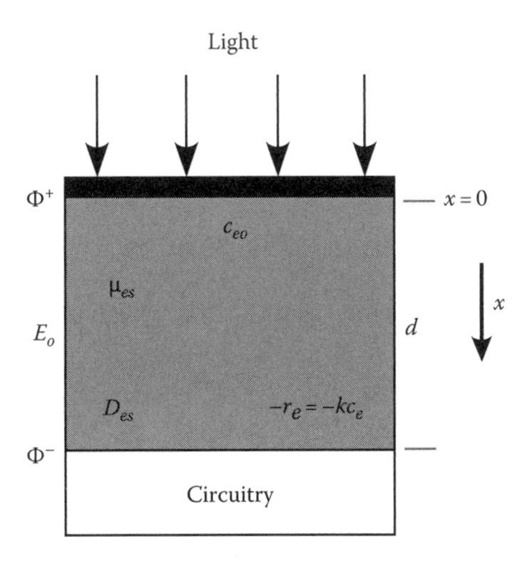

FIGURE P5.20

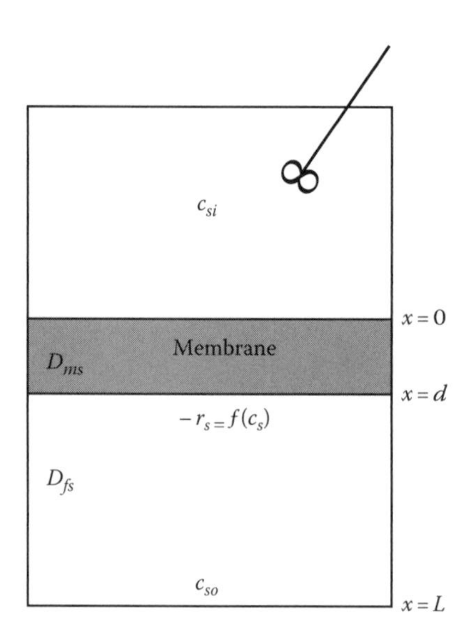

FIGURE P5.21

Excess substrate at the membrane surface can diffuse through the membrane to the well-mixed side.

a. Derive the differential equation governing the concentration profiles in the unmixed fluid and in the membrane. Assume all interfacial resistances are 1 and that we have dilute solutions.
b. What are the boundary conditions for this problem?
c. Is it possible to have $c_{so} = c_{si}$ yet still have a flux through the membrane from the unmixed to the well-mixed fluid? From the well-mixed to the unmixed fluid? Why?

5.22 You are working for a contract research firm, DATATEC, and your first job is to determine the diffusion coefficient and reaction rate constant for a decomposition reaction (A → C). You propose to do this in a diffusion cell where you introduce a small amount of A into a cell filled with solvent, B. The A diffuses through the cell and reacts along the way. At the exit of the cell, the reaction is complete. You have a spectrophotometer that lets you isolate a slice of the cell and measure how much A is remaining there. Assuming you know the concentration of A at the inlet to the cell:

a. What type of model will you use to analyze the data? Formulate and solve it.
b. What assumptions must you make to use your model?
c. Can you determine the diffusion coefficient and reaction rate constant independently? Why?
d. Given the experimental data shown in the succeeding text, what can you tell about the diffusion coefficient and the reaction rate constant?

$$c_{ao} = 1.0 \times 10^{-3}\,\text{mol/m}^3 \qquad L = 0.06\,\text{m} \qquad A_o = 1.0 \times 10^{-4}\,\text{m}^2$$

Position (m)	Mole Fraction of *a*
0	0.02
0.01	0.015
0.02	0.012
0.03	0.008
0.04	0.005
0.05	0.002

5.23 Reconsider the first example in the chapter regarding heterogeneous reaction on the boundary. In that example, we used the simplest boundary condition for the reaction. In reality, the reaction occurs on the surface, and we need species a to adsorb on that surface prior to reaction. In most instances of heterogeneous catalytic reaction, we relate the surface concentration to the bulk concentration using a Langmuir isotherm:

$$x_{a,surface} = \frac{Kx_a}{1+Kx_a}$$

a. Reevaluate the example assuming that the reaction at the surface can be given by

$$-r_a = k_s'' x_{a,surface}$$

b. What is the mass flow rate to the surface if

$L = 0.05\,\text{m}$	$D_{ab} = 1\times10^{-9}\,\text{m}^2/\text{s}$	$A_s = 7.85\times10^{-7}\,\text{m}^2$	
$K = 10$	$k_s'' = 1\times10^{-6}\,\text{m/s}$	$x_{ao} = 0.5$	$c_t = 100\,\text{kg-mol/m}^3$

(Hint: To choose the value of the integration constant found in (a), compare choices with the value obtained from Equation 5.16)

5.24 Reconsider the first example in the text a final time. However, in this instance, let us assume we have a first-order chemical reaction $a \xrightarrow{k_a''} b$ occurring in the liquid. Species a diffuses through the liquid, and at the catalytic surface, species a undergoes an instantaneous reaction that destroys it.

a. Develop the differential equation that describe the concentration profile of a.
b. What are the boundary conditions for this problem?
c. Solve the problem to determine the concentration profiles of a.

5.25 The simple analysis of the junction diode shows that if the Damkohler number, Da = 0, no current will flow, and so recombination is the main source of current flow (Equation 5.134). However, if we add a drift component as in Equation 5.151, do we arrive at the same conclusion? What happens in Equation 5.151 if Da = 0, and does it agree with the solution to Equation 5.148 when Da = 0?

6 Accumulation

In − Out + Gen = Acc

6.1 INTRODUCTION

Our previous discussions in Chapters 4 and 5 have dealt almost exclusively with systems that were operating at a steady state. While this is a convenient situation to deal with and is a most simplifying assumption, no system on Earth is at a steady state. In fact, if the Earth operated at a steady state, we would probably not exist and would certainly have no conception of time. At steady state, we would never have time for an original thought, for changing clothes, for holding a conversation, or for reading a book. Instead, we would be eating, eliminating waste, and performing countless other tasks simultaneously. Life would be at once boring and bizarre, if not impossible. Either time exists because the universe is not at steady state or the universe is not at steady state because time exists.

As engineers or scientists, we design processes to operate at a steady state because such systems are efficient, continuous, easily controlled, easily operated, easily analyzed and satisfy an innate desire in our species to resist change in our environment. More often than not, it will be your job, as an engineer or scientist, to decide when a system can be analyzed as a steady-state operation and when fluctuations are fast enough or strong enough to require a transient analysis. As an example, suppose we are able to measure the momentum of our system at any time. If the system were not operating at a steady state, we would notice that our momentum measurements would change over time. We refer to these changes as accumulation. Though we tend to view accumulation as an increasing process, in the jargon of transport phenomena, accumulation can represent either an increase or a decrease in our transport quantity.

Most transient processes are dissipative so that the accumulation term starts out negative and steadily approaches zero. Negative accumulation is the nature of a diffusive system that smooths out gradients over time. If we start out with a system removed from equilibrium and allow it to go its own way, a transport process will occur to return the system to an equilibrium state (uniformity). Since we measure the accumulation of our transport quantity in terms of a state variable, this means the state variable will approach a constant value throughout time and space.

The mere addition of time as another variable in our analysis complicates the problems physically and mathematically. The physical complexity involves charting the course of the system over space and time. The mathematical complexity takes the form of partial differential equations. In this chapter, we will introduce partial differential equations into our analysis and these will become the mathematical staple for the rest of the text. We will necessarily review several methods for solving partial differential equations, but of course, the reader is referred to the many excellent texts on partial differential equations for a complete discussion of their solution [1–3].

Partial differential equations are an order of magnitude more difficult to solve than ordinary differential equations. Numerical methods, such as Comsol®, Fluent®, Ansys®, MATLAB®, Maple®, and other such programs, provide invaluable tools for solving transient problems in one or multiple space dimensions. In this chapter, we will use Comsol® for solving transient problems once we have discussed and understood the analytical methods.

6.2 LUMPED CAPACITANCE

We will begin a discussion of transient transport processes using a lumped analysis approach. The lumped analysis method assumes that the interior of the system is uniform and can be characterized by a single value of the state variable (temperature, concentration, etc.). The state variable itself is a function of time only. This approach allows us to visualize the system as a single entity and to characterize its behavior over time using ordinary, first-order, differential equations rather than partial differential equations. For this reason, all analyses of transient transport problems should begin with a consideration of whether the lumped analysis approach is valid.

Historically the lumped analysis is called a *lumped capacitance* approach because we have storage (accumulation of heat, mass, etc.) and dissipation (exchange of heat, mass, etc.) occurring at the same time. Early on, investigators found that many transient transport processes could be simulated by an analogy with the transport of charge in an electric circuit. All they needed was a mechanism for storage (a capacitor) and a mechanism for exchange or dissipation (a resistor) to develop a kind of analog computer. The state variable in the system was the voltage, and the transport quantity was the charge, measured as the current flowing through the system. Using an appropriately chosen capacitor and resistor, one could simulate any of the transient transport processes (satisfying the lumped formulation) by charging the capacitor, attaching it to the resistor, and monitoring the voltage across the capacitor as a function of time. An example of the *V*–*t* output of one of these simulations is shown in Figure 6.1.

6.2.1 CHARGE TRANSPORT: CHARGE DECAY IN A CONDUCTOR

Consider the case of a conductor that has excess free charge placed in it by hooking it up to a battery for a short period of time (Figure 6.2). The medium external to the conductor we assume to be a perfect dielectric (insulator) so it has zero conductivity. This means that no charge can leave the conductor through its surface to be exchanged with the environment. The instant that the conductor is attached to the battery by closing switch 1, the excess free charge in the conductor sets up coulomb forces that will induce the charges to move and a current will flow from the battery through the conductor. This current is a transient phenomenon. Once we remove the conductor from the source of excess charge (by opening switch 1), we must have electrostatic conditions, and at steady state, the electric field within the conductor must be zero. Since no charge can leave the conductor, during the transient period, excess free charge travels to the surface and accumulates there as a surface charge density. If you rub your feet on a carpet, for example, you generate excess charge that resides on the surface of your skin. If you touch a conductor, the conductor and you form an insulated system and charge moves from the surface of your skin to the surface of the conductor–skin complex. The shock you get is the transient current flow. The removal of the charge from the interior of our conductor (Figure 6.2) continues until all the charge is located on the surface. The time required for this

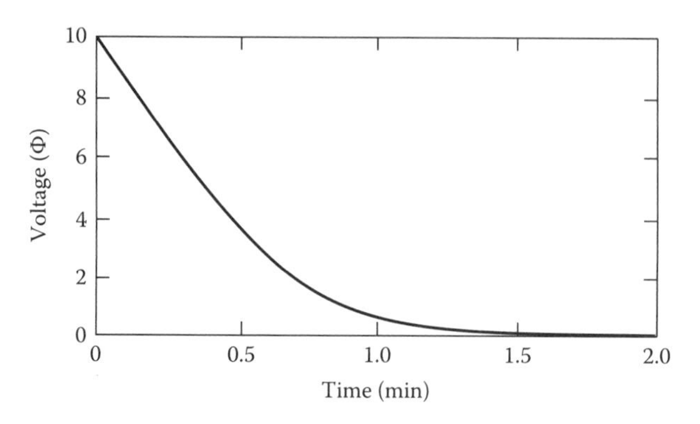

FIGURE 6.1 Discharge of a capacitor through a resistor.

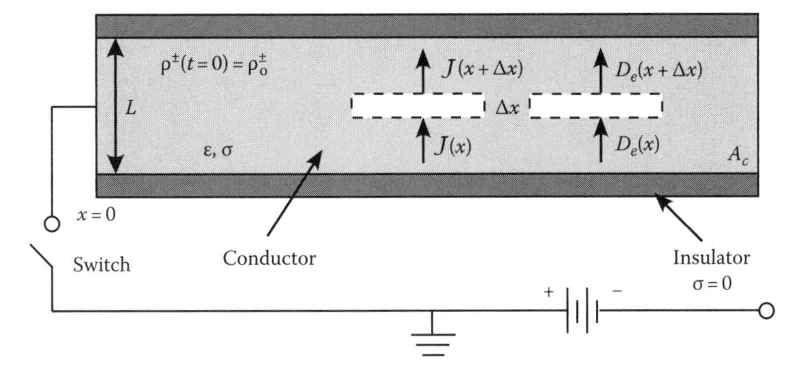

FIGURE 6.2 Charging of a conductor: a case of transient charge transport.

to occur is called the relaxation time, and for good conductors, it is very, very short, perhaps 10^{-15} s or so. We can determine the relaxation time and transient current flow using our charge transport equation and Gauss' law.

To determine the current flow, we perform a charge balance on the conductor considering the current flow and the electric field inside. We set up two differential elements within the conductor and write the balance equations around them. Considering charge conservation first, we have charge flowing into, and out of, the control volume on its way to the surface of the conductor. We have no mechanism for generating charge within the conductor so we have no generation term. Since we are storing excess charge on the conductor, we have an accumulation term. The accumulation term is written in terms of the time rate of change of the total charge present in the control volume.

Conservation of charge (current flow)

$$\begin{aligned} &In \quad - \quad Out \quad + \; Gen \; = \; Acc \\ &A_cJ(x) - A_cJ(x+\Delta x) + \quad 0 \quad = \frac{\partial}{\partial t}\left(\rho^{\pm}A_c\Delta x\right) \end{aligned} \tag{6.1}$$

The electric field will change as the charge density, $\rho^{\pm}$, in the conductor changes, but there is no accumulation of an electric field per se.

Conservation of charge $Q^{\pm}$ (electric field): Poisson's equation

$$\begin{aligned} &In \quad - \quad Out \quad + \; Gen \; = Acc \\ &Q^{\pm}(x) - Q^{\pm}(x+\Delta x) + \rho^{\pm}A_c\Delta x = \; 0 \end{aligned} \tag{6.2}$$

In Equation 6.2, the generation term is positive implying that the charges present in the conductor contribute to the electric field there. After dividing through both Equations 6.1 and 6.2 by Δx, taking the limit as $\Delta x \to 0$, and substituting for $Q^{\pm}$ and J using the constitutive relations,

$$Q^{\pm} = A_cD_e = -\varepsilon A_c\frac{d\Phi}{dx} \qquad J = -\sigma\frac{d\Phi}{dx} \tag{6.3}$$

we find that

$$\frac{\partial^2\Phi}{\partial x^2} = \frac{1}{\sigma}\frac{\partial\rho^{\pm}}{\partial t} \quad \text{Charge conservation} \tag{6.4}$$

$$\frac{d^2\Phi}{dx^2} = -\frac{\rho^{\pm}}{\varepsilon} \quad \text{Poisson's equation field} \tag{6.5}$$

Subtracting Equation 6.5 from Equation 6.4 gives a single, first-order, differential equation describing the variation of the charge density, $\rho^{\pm}$, inside the conductor, as a function of time. This equation states that we can assume the charge density in the interior of the conductor is a function of time only:

$$\frac{\partial \rho^{\pm}}{\partial t} + \frac{\sigma}{\varepsilon}\rho^{\pm} = 0 \tag{6.6}$$

Equation 6.6 is subject to the initial condition where the charge density in the interior is $\rho_o^{\pm}$, everywhere. The solution to Equation 6.6 is an exponential decay with $\rho^{\pm} = \rho_o^{\pm}$ at $t = 0$:

$$\rho^{\pm} = \rho_o^{\pm} \exp\left[-\left(\frac{\sigma}{\varepsilon}\right)t\right] \tag{6.7}$$

The solution for the field, $\mathcal{E}$, is found by substituting Equation 6.7 into (6.5) and includes the boundary condition that the field must be zero at the center of the conductor, $x = 0$:

$$\mathcal{E} = -\frac{d\Phi}{dx} = \frac{\rho_o^{\pm}}{\varepsilon} \exp\left[-\left(\frac{\sigma}{\varepsilon}\right)t\right]x \tag{6.8}$$

Notice that the charge density inside the conductor decays exponentially with a time constant, τ_t, given by

$$\tau_t = \frac{\varepsilon}{\sigma} \quad \text{RC time constant} \tag{6.9}$$

We define the resistance of the conductor as $\mathcal{R}_e = L/\sigma A$ and the capacitance of the conductor is given by $C = \varepsilon A/L$. τ_t is then just the $\mathcal{R}_e C$ time constant for the conductor. What we have done is to *lump* all the spatial variation in charge density into the capacitance of the system. We could have analyzed the system by treating the entire conductor as the control volume and representing the system as a resistor and capacitor hooked together in parallel (we will take this approach in Chapter 11). In such an instance, we equate the current flowing through the resistor to that flowing through the capacitor. The situation is shown in Figure 6.3.

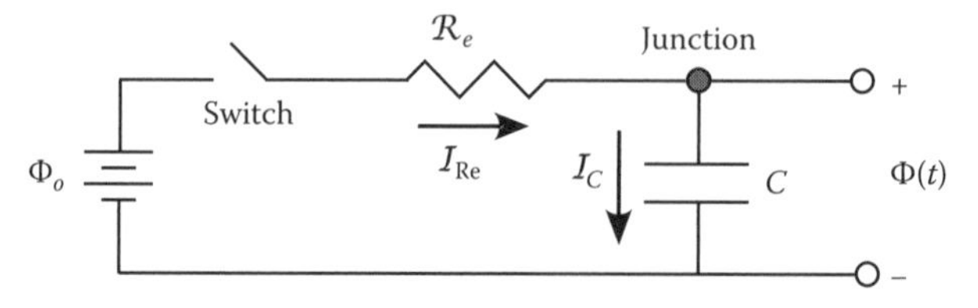

FIGURE 6.3 Discharge of an $\mathcal{R}_e C$ system. Initially, the switch is closed and the capacitor is charged to Φ_o. When the switch is opened, the system discharges, and the voltage, Φ, is recorded as a function of time.

At the junction, $I_C = I_{Re}$, so using Ohm's law for the resistor and the $\Phi - I$ relationship for a capacitor, we have

$$I = C\frac{d\Phi}{dt} = \frac{\Phi}{\mathcal{R}_e} \tag{6.10}$$

This differential equation is solved with reference to the initial condition where $\Phi(t = 0) = \Phi_o$:

$$\Phi = \Phi_o \exp\left[-\frac{t}{\mathcal{R}_e C}\right] \tag{6.11}$$

Again, we see the exponential decay whose rate is governed by the time constant, $\mathcal{R}_e C$. Notice that we can obtain Equation 6.11 using the result of Equation 6.7 and the definition of capacitance, $C = Q^{\pm}/\Phi$:

$$\Phi = \frac{Q^{\pm}}{C} = \frac{\rho^{\pm} A_c L}{C} = \left(\frac{\rho_o^{\pm} A_c L}{C}\right)\exp\left[-\left(\frac{\sigma}{\varepsilon}\right)t\right] \tag{6.12}$$

$\rho_o^{\pm} A_c L/C$ is just Φ_o the initial voltage the system was charged to.

In the next few sections, we will show how the $\mathcal{R}_e C$ time constant is related to the time constants for the decay of our other state variables involved in transport processes.

6.2.2 Energy Transport: Cooling of a Ball Bearing

The classic *lumped capacitance* heat transfer problem involves cooling a hot ball bearing in air or water. The situation is shown in Figure 6.4. We assume that the ball bearing is at a uniform temperature, T_i, initially, and that the bearing is small enough or of a good conducting material so that it remains at a uniform temperature throughout as it cools. This latter assumption is what makes the lumped capacitance approach applicable.

We start the analysis with an energy balance. Since we have assumed that there are no internal temperature gradients, the entire bearing is the control volume. Notice that we cannot have both an *In* term and an *Out* term. By choosing the entire bearing as the control volume, we force the heat

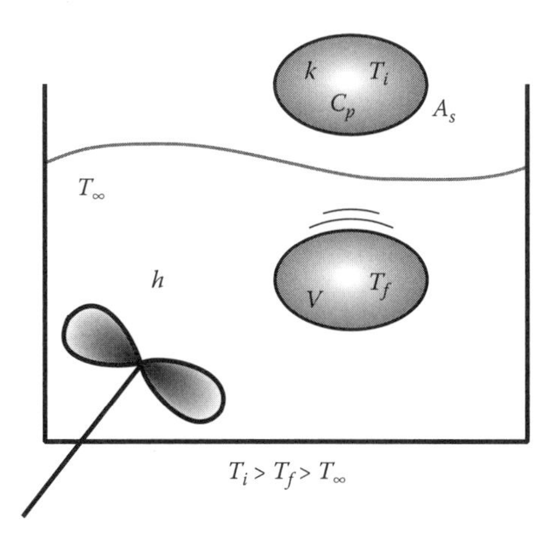

FIGURE 6.4 Cooling of a hot sphere in water. The sphere is assumed to have a uniform temperature so $T = f(t)$ only.

to flow into or out of the bearing only. In the present case, the bearing is at a higher temperature so that we only have an *Out* term. For now, we will also assert that we have no generation within the bearing. Finally, we must consider the accumulation term. It is written in terms of the change in total energy of the ball bearing, E_{tot}, with time:

$$\begin{array}{ccccccccc} In & - & Out & + & Gen & = & Acc \\ 0 & - & hA_s\left(T - T_\infty\right) & + & 0 & = & \dfrac{dE_{tot}}{dt} \end{array} \quad (6.13)$$

The total energy of the bearing is composed of the particle's internal energy, U, its kinetic energy, $K.E.$, and its potential energy, $P.E.$ In our case, both the potential and kinetic energies are negligible (the bearing has near-zero velocity and does not change height in the gravitational field). Moreover, we treat the bearing as a rigid object so that any PV work that could be done on the bearing as it exchanges energy with the surroundings will also be negligible. Using thermodynamics, we can relate the energy of the bearing to something we can measure, the temperature:

$$\frac{dE_{tot}}{dt} \approx \frac{dU}{dt} = \rho V C_v \frac{dT}{dt} \approx \rho V C_p \frac{dT}{dt} \quad (6.14)$$

We could substitute Equation 6.14 directly into (6.13) and solve the resulting equation. However, to avoid having to solve an inhomogeneous differential equation, we define a new temperature variable, $\theta = T - T_\infty$, and transform Equations 6.13 and 6.14 to

$$\frac{d\theta}{dt} = -\frac{hA_s}{\rho V C_p}\theta \quad (6.15)$$

Notice that this equation looks exactly like Equation 6.6 describing the charge density in the conductor. The solution to this problem is again an exponential decay subject to the initial condition, $\theta(t = 0) = \theta_i$:

$$\frac{\theta}{\theta_i} = \exp\left[-\left(\frac{hA_s}{\rho V C_p}\right)t\right] = \exp[-\mathrm{Bi} * \mathrm{Fo}] \quad (6.16)$$

We often write the solution in terms of two new dimensionless numbers: the Biot number, Bi, and the Fourier number, Fo. Both these numbers have particular physical significance to the lumped capacitance analysis:

$$\mathrm{Bi} = \frac{h\lambda_v}{k} = \frac{\lambda_v / kA_s}{1/hA_s} \quad \frac{\text{Conduction resistance}}{\text{Convection resistance}} \quad (6.17)$$

$$\mathrm{Fo} = \frac{kt}{\rho C_p \lambda_v^2} = \frac{\alpha t}{\lambda_v^2} \quad \text{Dimensionlesss time} \quad (6.18)$$

$$\lambda_v = \frac{V}{A_s} \quad \begin{array}{c}\text{Characteristic length} \\ \text{for heat diffusion}\end{array} \quad (6.19)$$

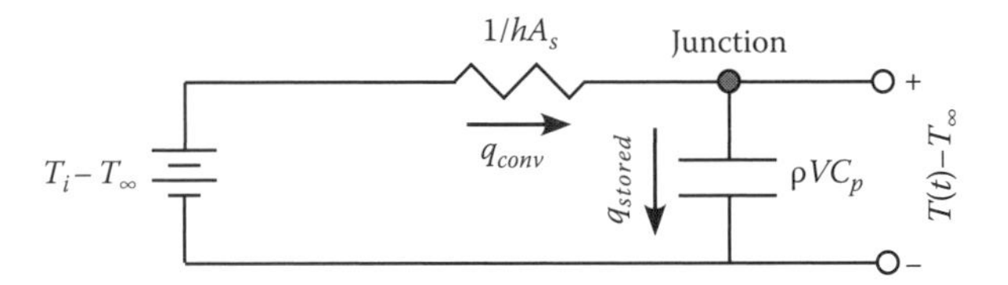

FIGURE 6.5 Electrical analogy for lumped capacitance heat transfer.

$$\alpha = \frac{k}{\rho C_p} \quad \text{Thermal diffusivity} \tag{6.20}$$

$$\tau_t = \frac{\rho V C_p}{h A_s} \quad \text{Time constant – Relaxation time} \tag{6.21}$$

We can consider ρVC_p as the capacitance of the system (the amount of heat that can be stored) and $1/hA_s$ as the resistance of the system to heat transfer from the pellet. The electric circuit corresponding to this heat transfer problem is shown in Figure 6.5.

The Fourier number represents a dimensionless time coordinate for the problem. The Biot number represents a ratio of the resistance to heat flow within the pellet to the resistance to heat flow from the pellet and into the fluid. If the internal resistance to heat flow is very small or the external resistance is very large, then the temperature inside the pellet will be uniform and the lumped capacitance formulation will be a valid approximation. Both situations correspond to a small Biot number. If the internal resistance is high or the external resistance is very small, then the Biot number will be large and the temperature inside the pellet will no longer be uniform. *By comparison with exact solutions, it has been found that the lumped capacitance approximation is valid whenever* $\text{Bi} \leq 0.10$.

Example 6.1 Cooling of a Composite Sphere

As an example of applying the lumped capacitance method and an extension of the Biot number concept, we consider the cooling of a hot, stainless steel, sphere that is covered by a thin, insulating, dielectric layer. The situation is shown in Figure 6.6. The sphere comes from the coating process at a uniform temperature of 450°C and it is dumped into a recirculating oil bath held at a temperature of 100°C. We would like to find the time required for the sphere's center to reach a temperature of 150°C.

To analyze this problem, we must make a few assumptions. First, we naively assume that the lumped capacitance approach will be valid so the sphere has no temperature gradients within it.

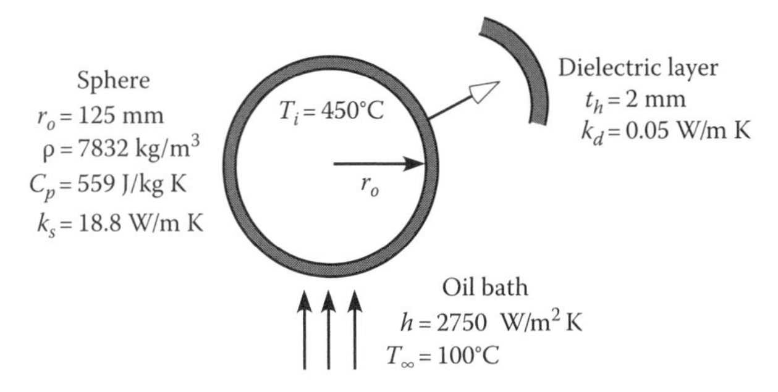

FIGURE 6.6 Cooling of a composite sphere. Stainless steel sphere covered in an insulating, dielectric coating is exposed to an oil bath at a temperature of 100°C.

The dielectric layer is thin compared to the sphere so we may assume that it stores little, if any, energy. Though the sphere starts out at a very high temperature, we will neglect radiative heat transfer and will assume that the materials' properties remain constant.

Our first calculation is a determination of the Biot number. We define the Biot number in terms of a resistance to heat flow within the system of interest versus a resistance to heat flow from the system:

$$\text{Bi} = \frac{\text{Resistance to heat flow within the system}}{\text{Resistance to heat flow from the system}}$$

To calculate the Biot number, we have to choose a system. In this instance, it makes a great deal of difference how we pick the system. Initially, we can choose the metal sphere as the system and lump the dielectric coating and external oil bath together to form the surroundings. The external resistance to heat transfer is composed of the conductive resistance through the dielectric coating, $\mathcal{R}_d$, and the convective resistance from the surface of the dielectric into the oil bath, $\mathcal{R}_c$. This overall external resistance is

$$\sum \mathcal{R}_t = \mathcal{R}_d + \mathcal{R}_c = \frac{t_h}{k_d A_s} + \frac{1}{hA_s}$$

$$= \frac{0.002\,\text{m}}{(0.05\,\text{W/mK})(0.196\,\text{m}^2)} + \frac{1}{(2750\,\text{W/m}^2\,\text{K})(0.196\,\text{m}^2)} = 0.206\,\text{K/W}$$

Notice that we have assumed no curvature in calculating this resistance. The dielectric is so thin that the difference between inner and outer radius of curvature is negligible, and we can relate the conduction resistance solely to the planar thickness of the dielectric. The external resistance in this case seems small.

The internal resistance to heat transfer is given in terms of the characteristic length of the system (V/A_s) and the thermal conductivity of the system:

$$\mathcal{R}_i = \frac{\lambda_v}{k_s} = \frac{V}{k_s A_s} = \frac{r_o}{3k_s} = \frac{0.125\,\text{m}}{3(18.8\,\text{W/mK})} = 2.216 \times 10^{-3}\,\text{K/W}$$

The Biot number is

$$\text{Bi} = \frac{\mathcal{R}_i}{\sum \mathcal{R}_t} = \frac{2.216 \times 10^{-3}}{0.206} = 1.08 \times 10^{-2} < 0.1 \quad \text{Lumps OK}$$

and since it is less than 0.1, we know the lumped capacitance approach is valid for this system. To determine the cooling time, we invert the solution presented in Equation 6.16 to obtain the Fourier number and, hence, the time:

$$\text{Fo} = \frac{\alpha t}{\lambda_v^2} \qquad \lambda_v = \frac{V}{A_s} \qquad \alpha = \frac{k_s}{\rho C_p}$$

$$\frac{\theta}{\theta_i} = \exp(-\text{Bi} \cdot \text{Fo}) \qquad \text{so} \qquad t = \left[-\frac{\ln[T_f - T_\infty / T_i - T_\infty]}{\text{Bi}} \right] \cdot \frac{\lambda_v^2}{\alpha}$$

$$t = \frac{(7832)(559)}{(1.08 \times 10^{-2})(18.8)} \cdot \left(\frac{0.125}{3}\right)^2 \cdot \ln\left[\frac{450 - 100}{150 - 100}\right]$$

$$= 7.285 \times 10^4\,\text{s} \qquad \text{or} \qquad 20.2\,\text{h}$$

Our analysis showed that most of the resistance to heat transfer occurred in the dielectric, even though the dielectric stored no heat and was very thin. Let us reconsider the problem assuming the dielectric wasn't applied to the metal. In this case, the internal and external resistances are

$$\mathcal{R}_i = \frac{\lambda_V}{k_s} = \frac{V}{k_s A_s} = \frac{r_o}{3k_s} = \frac{0.125\text{m}}{3(18.8\,\text{W/mK})} = 2.216\times10^{-3}\,\text{K/W}$$

$$\sum R_t = \frac{1}{hA_s} = 0.00018\text{ K/W} \quad \text{much smaller than before.}$$

The Biot number now becomes

$$\text{Bi} = \frac{0.00222}{0.00018} = 12.3 > 0.1$$

and since it is so large, the lumped capacitance approach can no longer be used. Without the dielectric, the resistance to heat transfer from the surface of the sphere is low enough to afford strong temperature gradients within the metal. This clearly illustrates how the proper choice of system is crucial to the lumped capacitance formulation of the problem.

A paradox of the lumped capacitance approach is evident from an examination of the solution presented in Equation 6.16. The final temperature of the pellet will exponentially approach the bath temperature, but as we see from the solution, the final pellet temperature can never equal the bath temperature. Intuitively, we know that this is untrue and that the pellet will indeed reach the bath temperature and in a finite time as well. The key to solving the paradox lies in Heisenberg's uncertainty principle. According to Heisenberg, if we try to measure the energy, E, of a particle at a given time, t, we will find that we cannot measure both quantities with absolute precision. Attempting to pin down the energy will lead to an uncertainty in the time associated with the measurement and attempting to fix the time will lead to an uncertainty in the energy. The relationship between these uncertainties is given by

$$\Delta E \Delta t = \frac{\hbar}{2\pi} \qquad \hbar = \text{Planck's constant} \tag{6.22}$$

With a bit of scientific license, we can express the energy of our body in terms of its volume, density, heat capacity, and temperature, $\Delta E \approx \rho V C_p \Delta T$. Substituting into Equation 6.22 gives the relationship between our temperature and time measurements:

$$\Delta T \Delta t \approx \frac{\hbar}{2\pi\rho V C_p} \tag{6.23}$$

Notice that we are unable to determine the precise temperature and time. For all practical purposes, the uncertainties for any macroscopic pellet are extraordinarily low and so our pellet does reach the bath temperature exactly.

6.2.3 Mass Transport: Adsorption by a Pellet

The lumped capacitance approach is not limited to charge transfer or heat transfer. We can formulate a mass transfer analog too. Consider a porous adsorbent particle, like activated carbon, placed into a contaminated medium. The particle contains some of the contaminant at an initial concentration of c_{ai} that is very close to zero. The particle falls into a recirculating bath where the concentration of contaminant is very high and uniform at a value of $c_{a\infty}$. Solute diffuses into the pellet with a diffusivity D_{ap} and is transferred from the fluid at a rate governed by the concentration

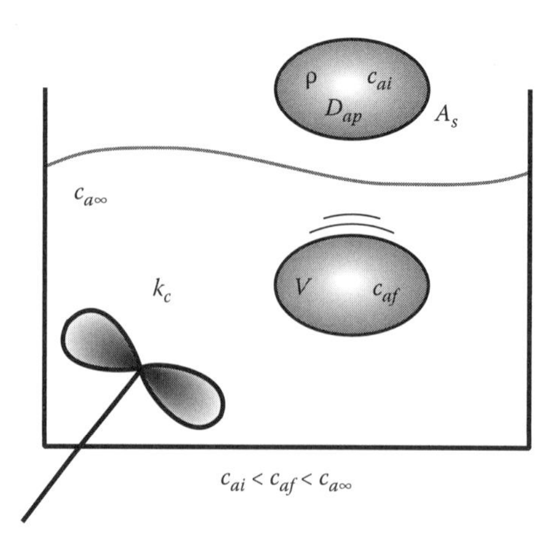

FIGURE 6.7 Mass transfer from a pellet. An application of the lumped capacitance approach for mass transfer.

difference between the pellet and bath and the convective mass transfer coefficient, k_c. The situation is shown in Figure 6.7. Our goal is to determine the time required for the solute concentration within the pellet to reach c_{af}.

The heart of the lumped capacitance approach for mass transfer is to assume no concentration gradients within the pellet. For this to occur, the resistance to mass transfer within the pellet must be much lower than the resistance from the pellet and into the fluid. Picking the pellet as the system and applying the mass balance gives

$$\begin{array}{ccccccc} In & - & Out & + & Gen & = & Acc \\ k_c A_s \left(c_{a\infty} - c_a\right) & - & 0 & + & 0 & = & \dfrac{dM_a}{dt} \end{array} \tag{6.24}$$

In this case, we have no *Out* term since the pellet is accumulating contaminant from the solution. Mass flows from the fluid to the pellet, and so all mass transfer occurs via the *In* term. Likewise, we have assumed no generation. The accumulation term is written as an accumulation of system mass, M_a (moles). We can relate the moles of a in the system, M_a, to the concentration via the volume:

$$\frac{dM_a}{dt} = \frac{d}{dt}(Vc_a) = V\frac{dc_a}{dt} \tag{6.25}$$

At this point, we define a new concentration variable, $\chi_a = c_{a\infty} - c_a$, to avoid solving an inhomogeneous equation. The transformed equation is

$$\frac{d\chi_a}{dt} = -\frac{k_c A_s}{V}\chi_a \tag{6.26}$$

The solution is again an exponential decay given in terms of a Biot number for mass transfer, Bi_m, and a corresponding Fourier number, Fo_m:

$$\frac{\chi_{af}}{\chi_{ai}} = \exp[-\mathrm{Bi}_m \cdot \mathrm{Fo}_m] \tag{6.27}$$

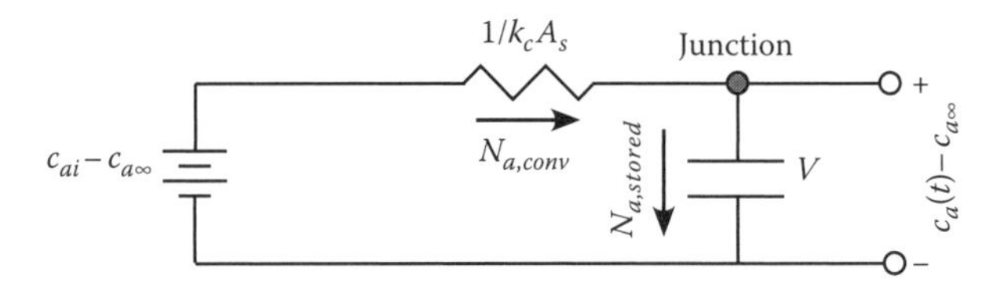

FIGURE 6.8 Electrical analogy for lumped capacitance mass transfer situations. In this example, we are looking at charging up the circuit, not discharging it.

The analogous terms to Equations 6.17 through 6.21 for the heat transfer case are

$$\text{Bi}_m = \frac{k_c \lambda_v}{D_{ap}} \quad \frac{\text{Diffusion resistance}}{\text{Convection resistance}} \tag{6.28}$$

$$\text{Fo}_m = \frac{D_{ap} t}{\lambda_v^2} \quad \text{Dimensionless time} \tag{6.29}$$

$$\lambda_v = \frac{V}{A_s} \quad \text{Characteristic length for diffusion} \tag{6.30}$$

$$\tau_m = \frac{V}{k_c A_s} \quad \text{Relaxation time} \tag{6.31}$$

Solutions are formulated and analyzed in exactly the same manner we applied for heat transfer. The capacitance of the system is its volume, V, and the resistance is the reciprocal of the mass transfer coefficient times the surface area of the particle, $1/k_cA_s$. The equivalent circuit is shown in Figure 6.8. Here, the lumped capacitance approach is valid when $\text{Bi}_m < 0.10$.

Example 6.2 Dissolution into a Tank of Finite Volume

We would like to consider mass transfer to a porous cylinder from a solution similar to the process we analyzed previously. The only difference is that the tank, into which the particle is placed, has a finite volume. This means that the concentration of solute in the tank, while spatially uniform, will change with time as mass transfer takes place. Our goal is to determine the concentration of solute a in the cylinder and the tank after 3 min. The situation is shown in Figure 6.9.

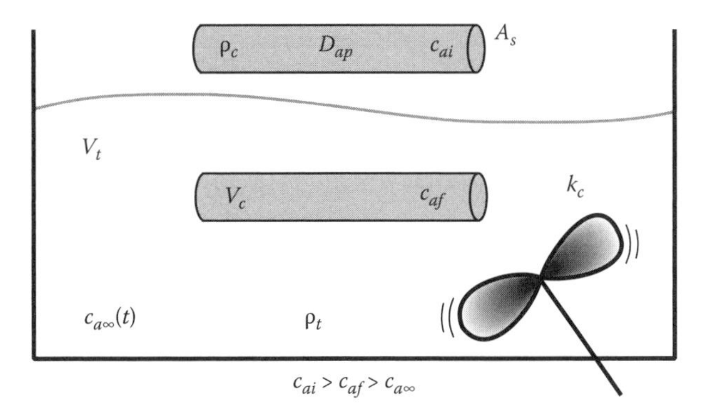

FIGURE 6.9 Mass transfer from a small cylinder into a tank of finite volume.

We first assume that there are no concentration gradients within the particle. Since the tank is well mixed, there are no concentration gradients there either. We also assume that solute a is present in low enough concentrations that it does not alter the densities of the solutions in the pellet or in the tank. The physical properties of the particle and the fluid in the tank are

$$c_{a\infty i} = 10 \text{ mol/m}^3 \qquad c_{ai} = 0 \text{ mol/m}^3 \qquad D_{ap} = 2\times 10^{-9} \text{ m}^2\text{/s}$$

$$r = 0.001 \text{ m} \qquad L = 0.05 \text{ m} \qquad V_t = 1\times 10^{-6} \text{ m}^3$$

$$k_c = 1\times 10^{-7} \text{ m/s} \qquad t_f = 500 \text{ s}$$

We must first calculate the Biot number to determine if we can proceed with the lumped capacitance analysis:

$$\text{Bi}_m = \frac{k_c\lambda_V}{D_{ap}} = \frac{k_c(r/2)}{D_{ap}} = \frac{(1\times 10^{-7})(5\times 10^{-4})}{2\times 10^{-9}} = 0.025 \quad \text{Lumps} \quad \text{OK}$$

Since the concentration of solute in the tank will be changing with time, we just cannot proceed and use Equation 6.27 to determine the final concentration. We must start again from the basic mass balance. We perform a mass balance for solute a about the two systems, the cylinder and the tank:

In	–	*Out*	+	*Gen*	=	*Acc*	
0	–	$k_cA_s(c_a - c_{a\infty})$	+	0	=	$\frac{dM_{ac}}{dt}$	Cylinder
$k_cA_s(c_a - c_{a\infty})$	–	0	+	0	=	$\frac{dM_{at}}{dt}$	Tank

Notice that the *Out* term in the cylinder mass balance is equal to the *In* term in the tank mass balance and that the two balances, when added together, show that the overall mass of solute a is conserved. Notice also that the *In* and *Out* terms are both negative since the concentration in the tank is larger than that in the pellet. As long as we are consistent with our sign convention, the solution will be correct. We can relate the mass of solute in the tank and cylinder to their respective volumes, molecular weights, and concentrations:

$$\frac{dc_a}{dt} = -\frac{k_cA_s}{V_c}(c_a - c_{a\infty}) \qquad \frac{dc_{a\infty}}{dt} = \frac{k_cA_s}{V_{tank}}(c_a - c_{a\infty})$$

The two mass balances are coupled since both concentrations change with time. To solve the equations, we first find a relationship between c_a and $c_{a\infty}$ by dividing the mass balances and integrating

$$\frac{dc_{a\infty}/dt}{dc_a/dt} = \frac{dc_{a\infty}}{dc_a} = -\frac{V_c}{V_{tank}} = -V_r$$

$$c_{a\infty} = -V_rc_a + K$$

Since we know the two initial concentrations, $t = 0$, $c_a = c_{ai}$ and $c_{a\infty} = c_{a\infty i}$,

$$c_{a\infty i} - c_{a\infty} = V_r(c_a - c_{ai})$$

Using this relationship to substitute for c_a in the tank mass balance, we obtain a single equation for $c_{a\infty}$:

$$\frac{dc_{a\infty}}{dt} + \left(\frac{k_c A_s}{V_{tank}}\right)\left(\frac{1+V_r}{V_r}\right)c_{a\infty} - \left(\frac{k_c A_s}{V_{tank}}\right)\left(\frac{V_r c_{ai} + c_{a\infty i}}{V_r}\right) = 0$$

This equation is inhomogeneous. We solve it by using an integrating factor method. The integrating factor is defined as

$$\text{Integrating factor} = \exp\left[\int\left(\frac{k_c A_s}{V_{tank}}\right)\left(\frac{1+V_r}{V_r}\right)dt\right] = \exp\left[\left(\frac{k_c A_s}{V_{tank}}\right)\left(\frac{1+V_r}{V_r}\right)t\right]$$

We now multiply our differential equation through by the integrating factor and rearrange it using the chain rule for differentiation to yield

$$\frac{d}{dt}\left\{\exp\left[\left(\frac{k_c A_s}{V_{tank}}\right)\left(\frac{1+V_r}{V_r}\right)t\right]c_{a\infty}\right\} - \left(\frac{k_c A_s}{V_{tank}}\right)\left(\frac{V_r c_{ai} + c_{a\infty i}}{V_r}\right)\cdot \exp\left[\left(\frac{k_c A_s}{V_{tank}}\right)\left(\frac{1+V_r}{V_r}\right)t\right] = 0$$

Separating, integrating both sides of the equation, and dividing through by the integrating factor gives

$$c_{a\infty} = \left(\frac{c_{a\infty i} + V_r c_{ai}}{1+V_r}\right) + a_1 \exp\left[-\left(\frac{k_c A_s}{V_{tank}}\right)\left(\frac{1+V_r}{V_r}\right)t\right]$$

The integration constant, a_1, is determined from the initial condition on $c_{a\infty}$:

$$t = 0 \qquad c_{a\infty} = c_{a\infty i}$$

Using this condition, we find that a_1 and the full solutions for c_a and $c_{a\infty}$ are

$$a_1 = \left(\frac{V_r}{1+V_r}\right)(c_{a\infty i} - c_{ai})$$

$$c_a = \frac{c_{a\infty i} + V_r c_{ai}}{1+V_r} - \left(\frac{c_{a\infty i} - c_{ai}}{1+V_r}\right)\exp\left[-\left(\frac{k_c A_s}{V_{tank}}\right)\left(\frac{1+V_r}{V_r}\right)t\right]$$

$$c_{a\infty} = \frac{c_{a\infty i} + V_r c_{ai}}{1+V_r} + \left(\frac{V_r}{1+V_r}\right)(c_{a\infty i} - c_{ai})\exp\left[-\left(\frac{k_c A_s}{V_{tank}}\right)\left(\frac{1+V_r}{V_r}\right)t\right]$$

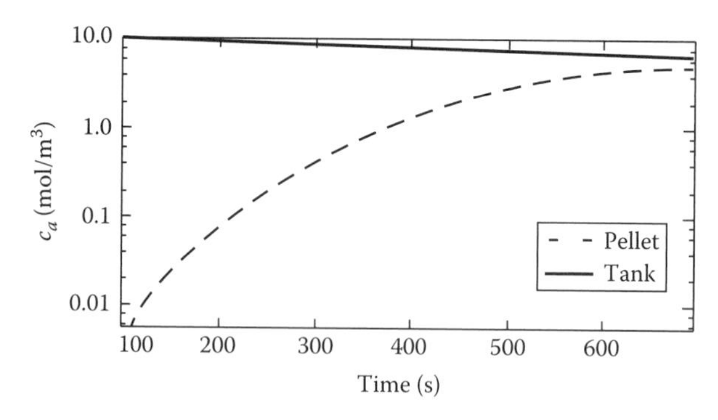

FIGURE 6.10 Concentrations of solute a in the tank and cylinder as a function of contact time.

TABLE 6.1
Lumped Capacitance Parameters

	Resistance	Capacitance	Characteristic Length	
Charge	$\frac{L}{\sigma A}$	$\frac{\varepsilon A}{L}$	Plane wall	$\frac{L}{2}$
Heat	$\frac{1}{hA}$	$\rho V C_p$	Cylinder	$\frac{r}{2}$
Mass	$\frac{1}{k_c A}$	V	Sphere	$\frac{r}{3}$
			Triangular bar	$\frac{HL}{2(3L+H)}$
			Cube	$\frac{L}{6}$

The concentrations in the pellet and tank are shown in Figure 6.10. Notice that as time increases, the two concentrations approach their equilibrium value:

$$c_{eq} = \left(\frac{c_{a\infty i} + V_r c_{ai}}{1+V_r}\right) = \left(\frac{10+(0.157)0}{1+0.157}\right) = 8.64\,\text{mol/m}^3$$

This concludes our discussion of lumped capacitance solutions. Table 6.1 contains a listing of the resistances, capacitances, and characteristic lengths for a variety of transport situations and geometries.

6.3 INTERNAL GRADIENTS AND GENERALIZED SOLUTIONS

The key assumption in the lumped capacitance analysis was that the internal transport resistance was so low that any internal gradients in the state variable were negligible. We quantified the ratio of internal to external resistance in the form of the Biot number. The key question we want to answer in this section is what do we do when the Biot number is larger than 0.1, internal gradients are important, and we cannot use the lumped capacitance approximation.

6.3.1 Systems Dominated by External Convection on the Boundary

If internal gradients exist, we must resort to a full solution of the transient problem. Fortunately, many of these solutions have been found for simple geometries if the differential equations are linear and are presented in graphical [4] or analytical form. Several excellent references including

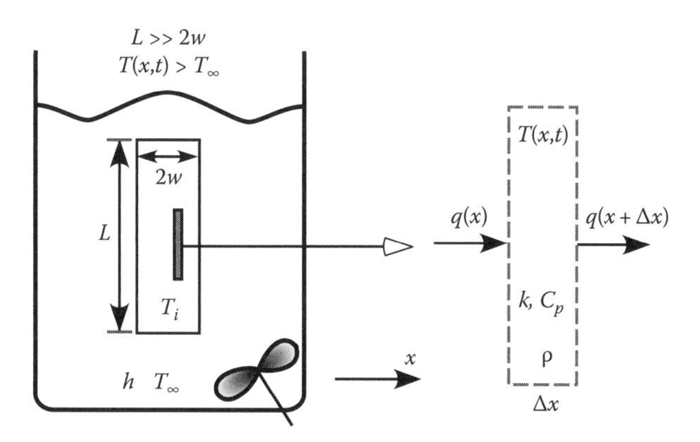

FIGURE 6.11 Cooling of a thin wall. The wall is thin enough so that heat is transferred only in the x-direction, but thick enough so Bi > 0.1.

those by Crank [5] and Carslaw and Jaeger [6] present these solutions. As an example of how the solutions are found and presented, we will consider the problem of a plane wall immersed in a recirculating fluid bath. The situation is shown in Figure 6.11. The wall is long enough that we may assume 1-D conduction in the x-direction. It is thick enough for the Biot number to be greater than 0.1. The energy balance over the control volume states

$$\begin{array}{ccccccccc} In & - & Out & + & Gen & = & Acc \\ q(x) & - & q(x+\Delta x) & + & 0 & = & \dfrac{\partial E}{\partial t} = \rho C_p A_c \Delta x \dfrac{\partial T}{\partial t} \end{array} \tag{6.32}$$

We have implicitly assumed that there is no generation in the system and that all physical properties are constant. Dividing through by Δx and taking the limit as $\Delta x \to 0$ gives us the differential equation for the heat flow, q:

$$-\frac{\partial q}{\partial x} = \rho C_p A_c \frac{\partial T}{\partial t} \tag{6.33}$$

Lastly, we substitute for the heat flow, q, using Fourier's law to obtain a second-order, partial differential equation in terms of the temperature:

$$\alpha \frac{\partial^2 T}{\partial x^2} = \frac{\partial T}{\partial t} \tag{6.34}$$

The initial condition for this equation states that we have a uniform temperature throughout the slab. The boundary conditions state that we have symmetry at the slab's center ($x = 0$) and that we have convective heat transfer from the surface ($x = w$) into the fluid:

$$t = 0 \qquad T(x,t) = T_i \tag{6.35}$$

$$x = 0 \qquad \frac{\partial T}{\partial x} = 0 \tag{6.36}$$

$$x = \pm w \qquad -k\frac{\partial T}{\partial x} = h(T(x,t) - T_\infty) \tag{6.37}$$

The solution to this equation may be found using separation of variables or Laplace transforms. Here, we will illustrate the use of separation of variables. It is well worth our effort to put Equation 6.34 into dimensionless form by defining the following new variables:

$$\xi = \frac{x}{w} \qquad \theta = \frac{T - T_\infty}{T_i - T_\infty} \qquad \tau = \frac{\alpha t}{w^2} \tag{6.38}$$

We use the chain rule of differentiation to transform from $x - T - t$ coordinates to $\xi - \theta - \tau$. The resulting differential equation and initial and boundary conditions are

$$\frac{\partial^2 \theta}{\partial \xi^2} = \frac{\partial \theta}{\partial \tau} \tag{6.39}$$

$$\tau = 0 \qquad \theta = 1 \tag{6.40}$$

$$\xi = 0 \qquad \frac{\partial \theta}{\partial \xi} = 0 \tag{6.41}$$

$$\xi = \pm 1 \qquad -\frac{\partial \theta}{\partial \xi} = \left(\frac{hw}{k}\right)\theta = \mathrm{Bi}\,\theta \tag{6.42}$$

Notice that τ, the dimensionless time variable, is the Fourier number we encountered previously and that hw/k is a form of Biot number for this problem.

The separation of variables technique begins by postulating that we can separate θ into two components, one a function of time only and the second a function of position only:

$$\theta = E(\xi)F(\tau) \tag{6.43}$$

We substitute Equation 6.43 into (6.39)

$$F\frac{\partial^2 E}{\partial \xi^2} = E\frac{\partial F}{\partial \tau} \tag{6.44}$$

and divide through by EF

$$\left(\frac{1}{E}\right)\frac{\partial^2 E}{\partial \xi^2} = \left(\frac{1}{F}\right)\frac{\partial F}{\partial \tau} = \pm\lambda^2 \tag{6.45}$$

Since the right-hand side of Equation 6.45 is only a function of τ, and the left-hand side is only a function of ξ, the only way the two sides can be equivalent is if both are equal to the same *constant*, $\pm\lambda^2$. The separation of variables procedure leads to two ordinary differential equations and we hope these equations are easier to solve than the original partial differential equation.

Our first obstacle involves picking the correct sign for λ^2. If we choose $\lambda = 0$, then we no longer have a time-dependent problem. If we choose the positive sign, separate and integrate the equation for F, the model predicts the temperature within the wall increases exponentially. This is the exact opposite behavior from what we intuitively know is going to happen. Therefore, we have to reject the positive sign for λ^2 and must use the negative sign. The resulting ordinary differential equations for E and F are

$$\frac{d^2E}{d\xi^2} + \lambda^2 E = 0 \tag{6.46}$$

$$\frac{dF}{F} = -\lambda^2 d\tau \tag{6.47}$$

Equation 6.46 is solved by using the substitution $E \propto \exp(r\xi)$ and solving for r. Equation 6.47 can be integrated directly. Alternatively, we can also look back into any differential equations text to see that the solution to Equation 6.46 involves sines and cosines:

$$E = a_1 \sin(\lambda\xi) + a_2 \cos(\lambda\xi) \tag{6.48}$$

$$F = b_1 \exp(-\lambda^2\tau) \tag{6.49}$$

As written, we have four unknowns, three integration constants and the separation constant, λ. There are only three boundary and initial conditions. If we reconstitute the solution for θ by multiplying F and E together, we can lump b_1 with a_1 and a_2 so that we have only three constants to solve for:

$$\theta = F \cdot E = [c_1 \sin(\lambda\xi) + c_2 \cos(\lambda\xi)]\exp(-\lambda^2\tau) \tag{6.50}$$

From the symmetry condition involving the flux at $\xi = 0$, we find

$$c_1\lambda = 0 \tag{6.51}$$

and so the constant, $c_1 = 0$. The boundary condition at $\xi = 1$ states that

$$c_2[\lambda \sin(\lambda) - \mathrm{Bi}\cos(\lambda)] = 0 \tag{6.52}$$

We cannot satisfy this condition with $c_2 = 0$ since c_1 was already set to zero. Therefore, we require that the term in parentheses vanish:

$$\lambda \tan(\lambda) = \mathrm{Bi} \tag{6.53}$$

For any given Bi number, there will be an infinite number of λs that will solve this equation. A valid solution to the problem exists at each λ_n, and since the differential equation is linear (all θs appear to the first power), the sum of all λ_n solutions is also a solution. In fact, we will need to use this sum of solutions because the basic solution, $c_2 \cos(\lambda\xi)\exp(-\lambda^2\tau)$, cannot satisfy the initial condition by itself:

$$\theta = \sum_{n=0}^{\infty} c_{2n} \cos(\lambda_n\xi)\exp(-\lambda_n^2\tau) \tag{6.54}$$

The constants, c_{2n}, are determined using the initial condition:

$$\sum_{n=0}^{\infty} c_{2n}\cos(\lambda_n\xi) = 1 \tag{6.55}$$

To extract c_{2n}, we multiply both sides of Equation 6.55 by $\cos(\lambda_m\xi)$ and integrate over all ξ from 0 to 1:

$$\int_0^1 \cos(\lambda_m\xi)d\xi = \sum_{n=1}^{\infty}\int_0^1 c_{2n}\cos(\lambda_n\xi)\cos(\lambda_m\xi)d\xi \tag{6.56}$$

The right-hand side of Equation 6.56 only exists when $\lambda_n = \lambda_m$ because the cosine functions are orthogonal to one another. The integral is zero for any other combination. The integrations for c_{2n} yield the final solution:

$$c_{2n} = \frac{4\sin(\lambda_n)}{2\lambda_n + \sin(2\lambda_n)} \tag{6.57}$$

$$\theta = \sum_{n=1}^{\infty}\left[\frac{4\sin(\lambda_n)}{2\lambda_n + \sin(2\lambda_n)}\right]\cos(\lambda_n\xi)\exp\left(-\lambda_n^2\tau\right) \tag{6.58}$$

where λ_n are given by the roots to Equation 6.53.

Example 6.3 Temperature Profiles in a Plane Wall

As an example of the use of the solution given in Equation 6.58, we will calculate and plot the temperature profiles in a concrete wall having the following thermophysical properties:

$w = 0.05$ m	$k = 1$ W/m K	$\rho = 2000$ kg/m^3	
$C_p = 0.88$ J/kg K	$T_i = 125°C$	$T_\infty = 25°C$	$h = 20$ W/m^2 K

A quick calculation of the Biot number (hw/k) shows that Bi = 1 and so lumped capacitance is invalid here. The solution requires we first determine the λ_n from Equation 6.53 that, in this case, is

$$\lambda_n \tan(\lambda_n) = 1$$

The simplest way to solve this equation is using a computer package such as Mathematica®, Maple®, and MathCad®. The first five roots are given in Table 6.2. With λ_n determined, we can calculate the constants, c_{2n}. The first five are shown in Table 6.2 as well. Notice that the values for the c_{2n} do not decrease very rapidly. In general, it takes several terms in the series to converge to the proper solution, but as λ_n gets larger, the exponential term gets smaller very quickly and so convergence can be fast. Figure 6.12 shows the full solution for various values of the Fourier number, τ. Even though the Biot number is a full ten times that required for the lumped capacitance approach to apply, the temperature gradients are still fairly small and concentrated toward the center of the wall.

TABLE 6.2
Roots of $\lambda_n \tan(\lambda_n) = 1$

n	λ_n	c_{2n}
1	0.8603	1.119
2	3.4256	−0.1517
3	6.4373	0.0466
4	9.5293	−0.0217
5	12.6453	0.0124

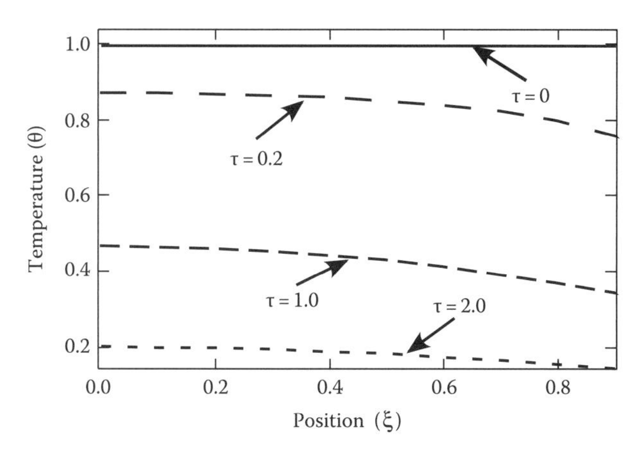

FIGURE 6.12 Temperature profiles for the infinite plane wall immersed in a fluid bath (Bi = 1).

TABLE 6.3
Translation Table between Heat and Mass Transfer

Symbol	Description	Heat Transfer	Mass Transfer
θ	Ratio at any location, x	$\dfrac{T-T_\infty}{T_i-T_\infty}$	$\dfrac{c_a-c_{a\infty}}{c_{ai}-c_{a\infty}}$
Bi	Biot number	$\left(\dfrac{hL}{k}\right)$ or $\left(\dfrac{hr_o}{k}\right)$	$\left(\dfrac{k_cL}{D_{ab}}\right)$ or $\left(\dfrac{k_cr_o}{D_{ab}}\right)$
Fo	Fourier number	$\left(\dfrac{\alpha t}{L^2}\right)$ or $\left(\dfrac{\alpha t}{r_o^2}\right)$	$\left(\dfrac{D_{ab}t}{L^2}\right)$ or $\left(\dfrac{D_{ab}t}{r_o^2}\right)$

Often one can get by using only one or two terms in the series solution. Though most of the original solutions were developed for heat transfer problems, due to the mathematical similarity between the differential equations governing heat and mass transfer, these solutions can be applied to mass transfer as well. The translation is given in Table 6.3. Keep in mind that the solutions for mass transfer only apply to cases where we can reasonably assume dilute solutions or equimolar counterdiffusion.

Example 6.4 Comparison of Convection and Conventional Ovens

Convection ovens assist in the cooking of food by having a fan force hot air inside the oven to flow across the food at a relatively high velocity. As we will see in later chapters, this forced convection results in efficient heat transfer. A conventional oven, on the other hand, is assisted in heating the food solely by natural convection. The hot air inside the oven escapes and is replaced by cooler air entering from the room. The manufacturer asserts that the convection oven provides for a threefold increase in the heat transfer coefficient inside the oven and will cut cooking time in half. Is this reasonable for an 8 kg turkey? Details of the turkey's and oven's thermal properties are given as follows:

$\rho = 900$ kg/m^3 $\quad C_p = 3800$ J/kg K $\quad k = 0.5$ W/m K

$h_{nc} = 10$ W/m^2 K $\quad h_{fc} = 50$ W/m^2 K $\quad r_o = 0.1285$ m

$T_f = 140°C$ $\quad T_i = 15°C$ $\quad T_\infty = 165°C$

Turkey = sphere 0.257 m diameter

We calculate the Biot number for the turkey first considering natural convection, but assume that lumped capacitance will not be valid:

$$\text{Bi} = \frac{hr_o}{3k} = \frac{10(0.1285)}{3(0.5)} = 0.86 \quad \text{Lumps are invalid}$$

In this problem, we know both the Biot number and the final centerline temperature. What we need to compare for the two cases are the Fourier numbers. We can use the solution in Table 6.4 to estimate the result even though that solution assumes a cylinder of infinite length:

$$\frac{T_f - T_\infty}{T_i - T_\infty} = \frac{140 - 165}{15 - 165} = 0.167$$

$$\text{Bi} = \frac{hr_o}{k} = 2.57 \quad \text{Conventional oven}$$

$$= 12.85 \quad \text{Convection oven}$$

The first three eigenvalues for each Bi are as follows:

Bi	1	2	3
2.57	2.192	5.016	8.047
12.85	2.901	5.825	8.786

We can use all three to find the time, solving the equation numerically, or just take the first term for a quick-and-dirty answer:

$$\theta = \left[\frac{4(\sin(2.192) - 2.192\cos(2.192))}{2(2.192) - \sin(2(2.192))}\right]\left(\frac{\sin(2.192 \times 0)}{2.192 \times 0}\right)\exp\left(-(2.192)^2\tau\right)$$

$$0.167 = 1.568\exp(-4.805\tau) \quad \tau = 0.466$$

$$\theta = \left[\frac{4(\sin(2.901) - 2.901\cos(2.901))}{2(2.901) - \sin(2(2.901))}\right]\left(\frac{\sin(2.901 \times 0)}{2.901 \times 0}\right)\exp\left(-(2.901)^2\tau\right)$$

$$0.167 = 1.951\exp(-8.416\tau) \quad \tau = 0.292$$

In this case, the ratio of dimensionless times is very close to 2:1, and given the nature of our approximation, we can give the manufacturer the benefit of the doubt.

TABLE 6.4
Analytical Solutions for Simple Geometries

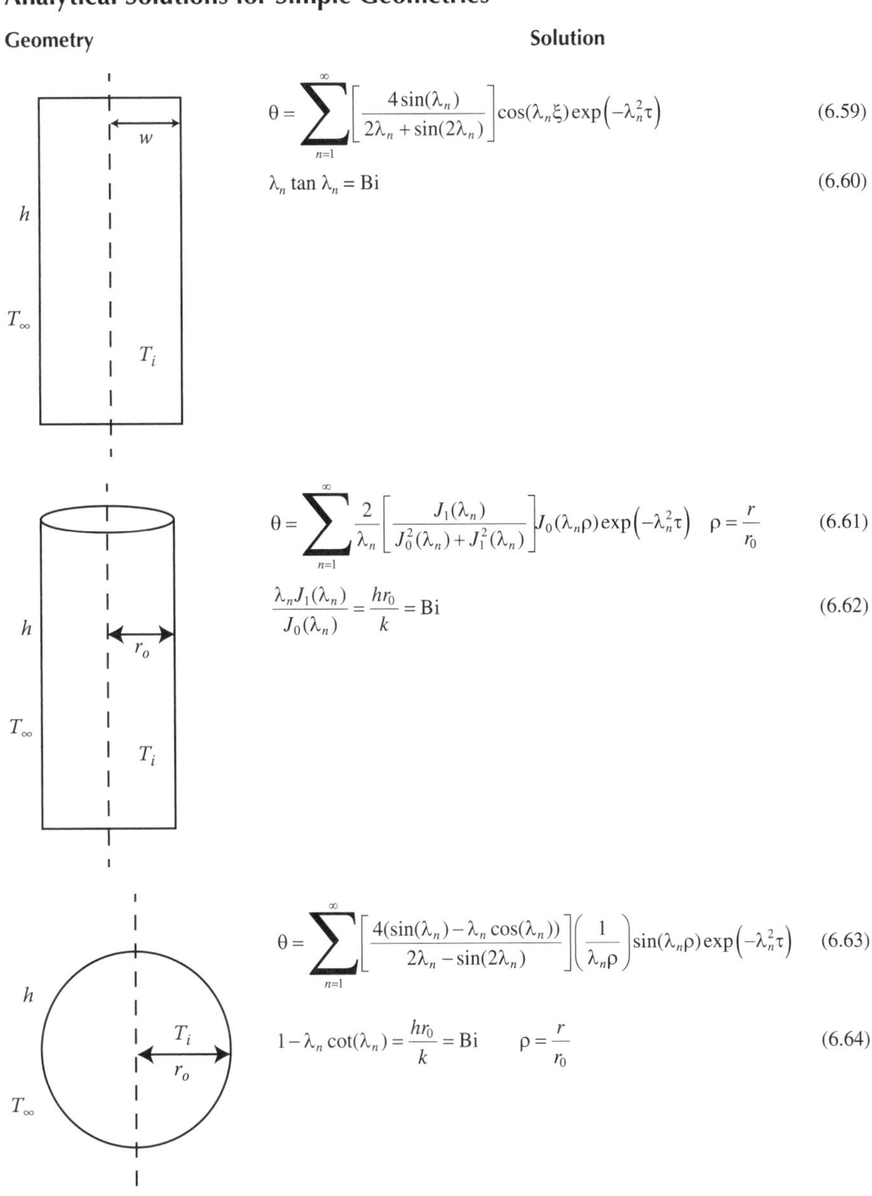

$$\theta = \sum_{n=1}^{\infty} \left[\frac{4\sin(\lambda_n)}{2\lambda_n + \sin(2\lambda_n)} \right] \cos(\lambda_n \xi) \exp\left(-\lambda_n^2 \tau\right) \tag{6.59}$$

$$\lambda_n \tan \lambda_n = \mathrm{Bi} \tag{6.60}$$

$$\theta = \sum_{n=1}^{\infty} \frac{2}{\lambda_n} \left[\frac{J_1(\lambda_n)}{J_0^2(\lambda_n) + J_1^2(\lambda_n)} \right] J_0(\lambda_n \rho) \exp\left(-\lambda_n^2 \tau\right) \quad \rho = \frac{r}{r_0} \tag{6.61}$$

$$\frac{\lambda_n J_1(\lambda_n)}{J_0(\lambda_n)} = \frac{h r_0}{k} = \mathrm{Bi} \tag{6.62}$$

$$\theta = \sum_{n=1}^{\infty} \left[\frac{4(\sin(\lambda_n) - \lambda_n \cos(\lambda_n))}{2\lambda_n - \sin(2\lambda_n)} \right] \left(\frac{1}{\lambda_n \rho} \right) \sin(\lambda_n \rho) \exp\left(-\lambda_n^2 \tau\right) \tag{6.63}$$

$$1 - \lambda_n \cot(\lambda_n) = \frac{h r_0}{k} = \mathrm{Bi} \qquad \rho = \frac{r}{r_0} \tag{6.64}$$

6.3.2 Analytical Solutions for Transient Problems Involving Generation

The tabulated series solutions in Table 6.4 are very useful when we have an object being cooled or heated by convection. Several terms of the series solution can be used if more accurate answers are required. These solutions are not very useful for other types of boundary conditions such as constant surface flux or radiative boundary conditions. Nor can they be readily used for fluid mechanics problems or problems involving any sort of generation term. In such instances, we have no recourse but to derive the balance equations and solve the partial differential equations ourselves by analytical or numerical techniques. In the next few pages, we will consider two problems: diffusion of a solute through a cylindrical object with a homogeneous first-order chemical reaction and designing a heat shield for aerocapture.

6.3.2.1 Diffusion and Reaction in a Cylinder

The mass transfer problem we wish to consider is one where we have a long cylindrical tube. The cylinder is heated and impregnated with a substance, a, at a concentration of c_{ao}. At the outer surface of the tube, we flow air across the surface effectively keeping the concentration of a at the outer surface zero. The material, a, decomposes according to a first-order chemical reaction. If the cylinder is very long compared to its diameter, we can assume that diffusion takes place only in the radial direction. To make life even simpler, we will also assume that the properties of the rod are constant and that a can only diffuse through the cylinder with a diffusivity, D_{ab}. The physical situation is shown in Figure 6.13.

We begin our analysis by formulating a mass balance on species a about the control volume considering diffusion, generation, and accumulation:

$$\begin{array}{ccccccccc} In & - & Out & + & Gen & = & Acc \\ \dot{M}_a(r) & - & \dot{M}_a(r+\Delta r\Delta) & - & k''c_a\Delta V & = & \dfrac{\partial M_a}{\partial t} = \Delta V\dfrac{\partial c_a}{\partial t} \end{array} \tag{6.65}$$

The volume of the control volume can be written in terms of its surface area and its width; $\Delta V = 2\pi rL\Delta r$. Substituting into Equation 6.65, expressing the mass flow rate in terms of a flux and area, dividing through by Δr, and taking the limit as $\Delta r \to 0$ gives

$$-\frac{\partial}{\partial r}(2\pi rLN_a) - k''c_a(2\pi rL) = (2\pi rL)\frac{\partial c_a}{\partial t} \tag{6.66}$$

Using Fick's law to express the mass flux of a in terms of its concentration gradient gives

$$\frac{\partial}{\partial r}\left(2\pi rLD_{ab}\frac{\partial c_a}{\partial r}\right) - k''c_a(2\pi rL) = (2\pi rL)\frac{\partial c_a}{\partial t} \tag{6.67}$$

Upon dividing through by the area and factoring out all constant terms, we obtain

$$\frac{D_{ab}}{r}\frac{\partial}{\partial r}\left(r\frac{\partial c_a}{\partial r}\right) - k''c_a = \frac{\partial c_a}{\partial t} \tag{6.68}$$

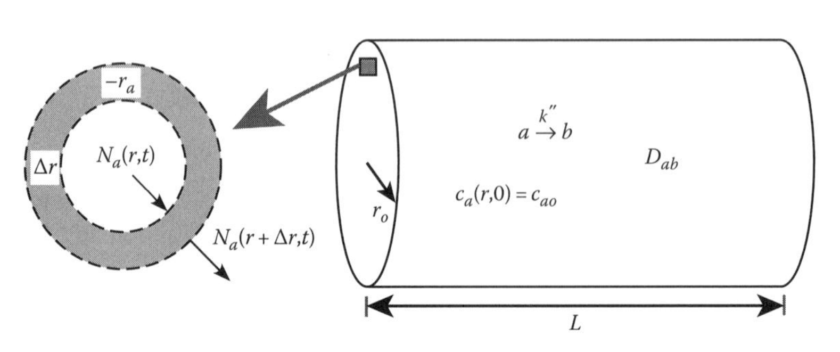

FIGURE 6.13 Diffusion and reaction inside a cylinder.

Our differential mass balance, Equation 6.68, is subject to boundary conditions that require the concentration of a to be c_{ao} at the outer surface and to have symmetry at the center:

$$r = 0 \qquad \frac{dc_a}{dr} = 0 \tag{6.69}$$

$$r = r_o \qquad c_a = 0 \tag{6.70}$$

The initial condition states that the rod contains a uniform amount of species a:

$$t = 0 \qquad c_a = c_{ao} \tag{6.71}$$

To solve Equation 6.68, we again use separation of variables. It helps to define three new dimensionless variables:

$$\xi = \frac{r_i}{r_{oi}} \qquad \chi = \frac{c_a}{c_{ao}} \qquad \tau = \frac{D_{ab}t}{r_o^2} \tag{6.72}$$

Substituting these variables into the differential equation, initial, and boundary conditions gives

$$\frac{1}{\xi}\frac{\partial}{\partial \xi}\left(\xi\frac{\partial \chi}{\partial \xi}\right) - \left(\frac{k'' r_o^2}{D_{ab}}\right)\chi = \frac{\partial \chi}{\partial \tau} \tag{6.73}$$

$$\xi = 0 \qquad \frac{d\chi}{d\xi} = 0 \tag{6.74}$$

$$\xi = 1 \qquad \chi = 0 \tag{6.75}$$

$$\tau = 0 \qquad \chi = 1 \tag{6.76}$$

where the dimensionless group appearing in Equation 6.73, $k'' r_o^2 / D_{ab}$, is the Damkohler number for a cylindrical system and first-order chemical reaction.

To solve Equation 6.73, we again use the technique of separation of variables:

$$\chi = E(\xi)F(\tau) \tag{6.77}$$

Substituting into Equation 6.73 gives

$$F\frac{1}{\xi}\frac{\partial}{\partial \xi}\left(\xi\frac{\partial E}{\partial \xi}\right) - \text{Da}EF = E\frac{\partial F}{\partial \tau} \tag{6.78}$$

Dividing through by EF shows that we can separate the equations and set them equal to a constant, $-\lambda^2$, to insure we get a physically meaningful solution that approaches equilibrium as time progresses:

$$\frac{1}{E}\frac{1}{\xi}\frac{\partial}{\partial \xi}\left(\xi\frac{\partial E}{\partial \xi}\right) - \text{Da} = \frac{1}{F}\frac{\partial F}{\partial \tau} = -\lambda^2 \tag{6.79}$$

We now split the partial differential equation into two ordinary differential equations:

$$\xi^2 \frac{d^2E}{d\xi^2} + \xi \frac{dE}{d\xi} + (\lambda^2 - \text{Da})\xi^2 E = 0 \tag{6.80}$$

$$\frac{dF}{F} = -\lambda^2 d\tau \tag{6.81}$$

The solution to Equation 6.81 is an exponential decay:

$$F = a_1 \exp(-\lambda^2 \tau) \tag{6.82}$$

The solution to Equation 6.80 is more difficult to obtain and is related to the differential equation

$$x^2 \frac{d^2y}{dx^2} + x\frac{dy}{dx} + (a^2x^2 - k^2)y = 0 \tag{6.83}$$

first solved by Bessel using a series expansion technique. The series he obtained are referred to now as Bessel functions of order k. The solution to Equation 6.80 is determined by substituting the series

$$E = a_o(\upsilon\xi)^r + \sum_{n=1}^{\infty} a_n(\upsilon\xi)^{r+n} \qquad \upsilon = \sqrt{\lambda^2 - \text{Da}} \tag{6.84}$$

into the differential equation to give

$$\sum_{n=0}^{\infty} a_n[(n+r)(n+r-1)+(n+r)]\big(\upsilon\xi\big)^{r+n} + \sum_{n=0}^{\infty} a_n(\upsilon\xi)^{r+n+2} = 0 \tag{6.85}$$

Isolating the a_o and a_1 terms yields two algebraic equations to solve:

$$\begin{aligned} a_o\big[r(r-1)+r\big](\upsilon\xi)^r + a_1\big[r(r+1)+(r+1)\big](\upsilon\xi)^{r+1} \\ + \sum_{n=2}^{\infty}\big\{a_n[(n+r)(n+r-1)+(n+r)] + a_{n-2}\big\}(\upsilon\xi)^{r+n} = 0 \end{aligned} \tag{6.86}$$

The solution to equation set (6.86) is found by setting each coefficient of ξ^r to zero. Valid values of r are determined from the roots of the *indicial* equation, the a_o term:

$$r(r-1)+r = r^2 = 0 \tag{6.87}$$

Here, there are two solutions with $r_1 = r_2 = 0$. Using these values, we can determine that $a_1 = 0$ and that for all other coefficients to be zero, we need the following recurrence relation:

$$a_n(r) = -\frac{a_{n-2}(r)}{(n+r)^2} \qquad n \geq 2 \tag{6.88}$$

To determine the first solution $E_1(\upsilon\xi)$, we set $r = 0$. From Equation 6.88 and $a_1 = 0$, we see that all the odd coefficients, $a_3 = a_5 = a_7 \ldots a_{2n+1} = 0$ and

$$a_n(0) = -\frac{a_{n-2}(0)}{n^2} \qquad n = 2,4,6\ldots \tag{6.89}$$

Letting $2n = m$, we can represent all the a_n in terms of a_o:

$$a_{2m}(0) = -\frac{a_{2m-2}(0)}{(2m)^2} = \frac{a_{2m-4}(0)}{(2m)^2(2m-2)^2} = \frac{(-1)^m a_o}{2^{2m}(m!)^2} \qquad m = 1,2,3\ldots \tag{6.90}$$

The solution for $E_1(\upsilon\xi)$ is now

$$E_1(\upsilon\xi) = a_o\left[1 + \sum_{m=1}^{\infty}\frac{(-1)^m(\upsilon\xi)^{2m}}{2^{2m}(m!)^2}\right] \qquad \upsilon\xi > 0 \tag{6.91}$$

E_1 is known as the *Bessel function of the first kind of order zero* and is given the symbol, $J_o(\upsilon\xi)$. Several orders of Bessel function of the first kind are shown in Figure 6.14.

The second solution, $E_2(\upsilon\xi)$, is found using a special theorem since the roots of the indicial equation were equal. Thus,

$$E_2(\upsilon\xi) = E_1(\upsilon\xi)\ln(\upsilon\xi) + \sum_{n=1}^{\infty} b_n(0)(\upsilon\xi)^n \tag{6.92}$$

Substituting into Equation 6.80 and remembering that J_o (E_1) is a solution to the differential equation, we can show that

$$\sum_{n=2}^{\infty} n(n-1)b_n(\upsilon\xi)^n + \sum_{n=1}^{\infty} nb_n(\upsilon\xi)^n + \sum_{n=1}^{\infty} b_n(\upsilon\xi)^{n+2} + 2(\upsilon\xi)\frac{dJ_o(\upsilon\xi)}{d\xi} = 0 \tag{6.93}$$

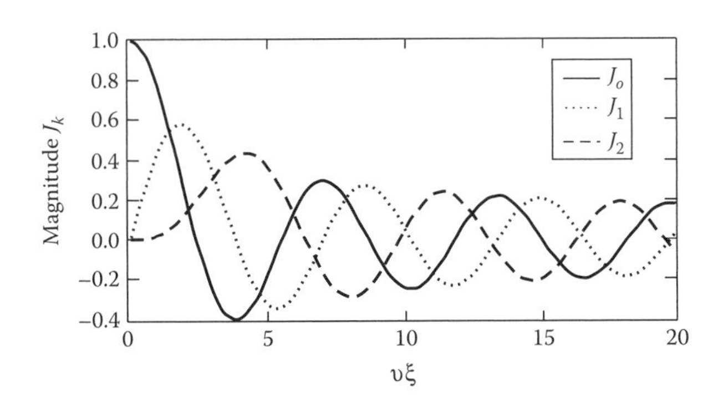

FIGURE 6.14 The Bessel functions, J_k.

Taking the derivative of the series representation for J_o, we can rearrange Equation 6.93 to

$$b_1(\upsilon\xi)+2^2b_2(\upsilon\xi)^2+\sum_{n=3}^{\infty}\left(n^2b_n+b_{n-2}\right)(\upsilon\xi)^n=-2\sum_{n=1}^{\infty}\frac{(-1)^n2n(\upsilon\xi)^{2n}}{2^{2n}(n!)^2} \tag{6.94}$$

Notice that only even powers of $(\upsilon\xi)$ appear on the right-hand side of Equation 6.94, which means all the odd b_n, b_1, b_3, b_5... = 0. The even b_n can be shown to follow the relation

$$b_{2n}=\frac{(-1)^{n+1}R_n}{2^{2n}(n!)^2}\qquad R_n=1+\frac{1}{2}+\frac{1}{3}+\cdots+\frac{1}{n} \tag{6.95}$$

Generally, we do not use E_2 directly but combine it with J_o to form the Bessel function of the second kind of order zero, Y_o:

$$Y_o(\upsilon\xi)=\frac{2}{\pi}[E_2(\upsilon\xi)+(0.5772-\ln 2)J_o(\upsilon\xi)] \tag{6.96}$$

The Y functions are shown in Figure 6.15.

The solution to Equation 6.80 can now be written as

$$E=c_1J_o(\upsilon\xi)+c_2Y_o(\upsilon\xi) \tag{6.97}$$

At the center of the rod, we have a point of symmetry so the concentration must be finite there. Since Y_o approaches infinity as we reach the center of the rod, c_2 must be zero. Reconstituting χ leaves us with one constant of integration to determine, λ, the *eigenvalues* for the system:

$$\chi=c_3J_o(\upsilon\xi)\exp(-\lambda^2\tau)\qquad \upsilon=\sqrt{\lambda^2-\mathrm{Da}} \tag{6.98}$$

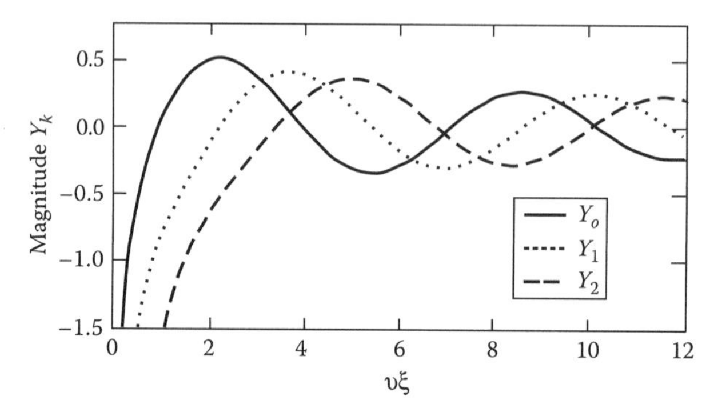

FIGURE 6.15 The Bessel functions, Y_k.

The boundary condition at $\xi = 1$ states that the concentration must be zero there:

$$c_3 J_o(\upsilon)\exp(-\lambda^2\tau) = 0 \tag{6.99}$$

Obviously, c_3 cannot be zero or the concentration would be zero everywhere. Similarly, we cannot rely on this boundary condition to apply only when $\tau \to \infty$. Therefore, we need to have $J_o(\upsilon) = 0$ and this can occur at any number of values of υ as shown in Figure 6.14.

To determine c_3, we use the initial condition. Using just one value of λ_n will not reproduce the initial condition, so we superimpose an infinite number of solutions together in hopes that the series will represent χ at $t = 0$:

$$\sum_{i=1}^{\infty} c_{3n} J_o(\upsilon_n\xi)\exp\left(-\lambda_n^2\tau\right) = 1 \tag{6.100}$$

The Bessel functions are an orthogonal set of polynomials, and so like sines and cosines, they obey an integration relationship:

$$\int_0^x xJ_k(\alpha x)J_k(\beta x)dx = 0 \qquad \alpha \neq \beta$$

$$= \int_0^x xJ_k^2(\alpha x)dx \qquad \alpha = \beta \tag{6.101}$$

To determine the c_{3n}, we multiply both sides of Equation 6.100 by $\xi J_o(\upsilon\xi)$ and integrate from 0 to 1:

$$\sum_{n=1}^{\infty} \int_0^1 c_{3n}\xi J_o(\upsilon_n\xi)J_o(\upsilon_m\xi)d\xi = \int_0^1 c_{3n}\xi J_o(\upsilon_m\xi)d\xi \tag{6.102}$$

Since the integral on the left-hand side of Equation 6.102 exists only when $n = m$, the sum consists of only a single term and each c_{3n} can be evaluated from

$$c_{3n} = \frac{\int_0^1 \xi J_o(\upsilon_n\xi)d\xi}{\int_0^1 \xi J_o^2(\upsilon_n\xi)d\xi} = \left(\frac{2}{\upsilon_n}\right)\frac{J_1(\upsilon_n)}{J_o^2(\upsilon_n) + J_1^2(\upsilon_n)} \tag{6.103}$$

To perform the integrations, we have used two other identities for the Bessel functions:

$$\int_0^1 xJ_o(\alpha x)dx = \frac{J_1(\alpha)}{\alpha} \tag{6.104}$$

$$\int_0^1 xJ_o^2(\alpha x)dx = \frac{1}{2\alpha}\left[J_o^2(\alpha) + J_1^2(\alpha)\right] \tag{6.105}$$

Since the $\upsilon_n = \sqrt{\lambda_n^2 - \mathrm{Da}}$ are the roots of J_o, the final solution for χ is

$$\chi = \sum_{n=1}^{\infty} \frac{2}{\sqrt{\lambda_n^2 - \mathrm{Da}}} \left[\frac{J_o\left(\sqrt{\lambda_n^2 - \mathrm{Da}}\,\xi\right)}{J_1^2\left(\sqrt{\lambda_n^2 - \mathrm{Da}}\right)} \right] \exp\left(-\lambda_n^2 \tau\right) \tag{6.106}$$

This solution describes how the presence of the chemical reaction consuming species a affects the concentration profile. Had the reaction not been present, $Da = 0$, and we would have had a profile similar to that seen in the convection problem with $Bi = \infty$. A nonzero value of Da forces the eigenvalues toward higher, positive values. This greatly affects the time dependence of the solution and heightens the decay of the profile toward uniformity. These effects are shown in Figure 6.16, which plots the concentration profile at $\tau = 0.25$ for various values of the Damkohler number. Table 6.5 shows the roots of J_o along with the values of λ_n for each choice of Damkohler number.

Often, we would also like to determine the flux of material passing through a surface within the cylinder. As in any other circumstance where we have calculated the state variable profile, we use a flux law to determine the flux or flow of our transport quantity. Here, Fick's law in dimensionless form is

$$N_a = -\frac{D_{ab} c_{ao}}{r_o} \frac{d\chi}{d\xi} \tag{6.107}$$

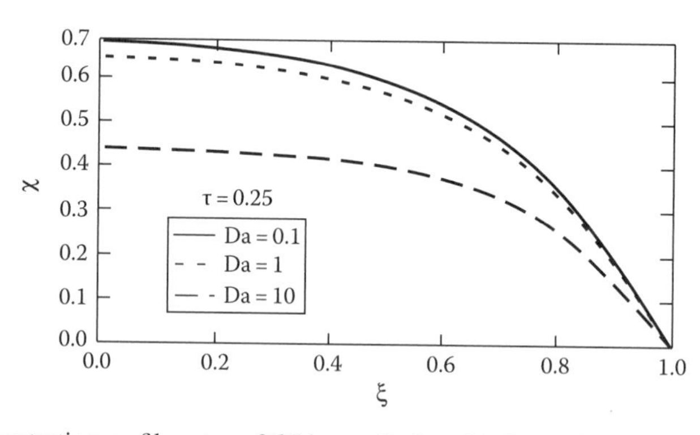

FIGURE 6.16 Concentration profiles at $\tau = 0.25$ in a cylinder with first-order chemical reaction occurring.

TABLE 6.5
λ_n and the First Five Roots of J_o

Root of J_o	λ_n (Da = 0.1)	λ_n (Da = 1)	λ_n (Da = 10)
2.405	2.426	2.605	3.973
5.520	5.529	5.610	6.362
8.654	8.660	8.712	9.214
11.79	11.794	11.832	12.207
14.93	14.933	14.963	15.261

The derivative of the Bessel function, J_o, can be found from the identity ($k = 0$):

$$\frac{d}{dx}\left(x^{-k}J_k(x)\right) = -x^{-k}J_{k+1}(x) \tag{6.108}$$

At the surface of the cylinder, $\xi = 1$, and at the halfway point, the fluxes become

$$N_a = \frac{D_{ab}C_{ao}}{r_o}\sum_{n=1}^{\infty} 2\exp\left(-\lambda_n^2\tau\right) \tag{6.109}$$

$$N_a = \frac{D_{ab}C_{ao}}{r_o}\sum_{n=1}^{\infty} 2\left[\frac{J_1\left(\frac{1}{2}\sqrt{\lambda_n^2 - \text{Da}}\right)}{J_1^2\left(\sqrt{\lambda_n^2 - \text{Da}}\right)}\right]\exp\left(-\lambda_n^2\tau\right) \tag{6.110}$$

respectively. As expected, the flux is highest at the surface.

6.3.2.2 Aerocapture

The solutions we have presented offer valuable guidance for how transient systems behave. They are often an engineer's first attempt at estimating how a system is to behave. To look at more realistic systems, we must resort to numerical simulation.

Aerocapture is an experimental maneuver designed to insert a spacecraft into planetary orbit in an economical way. Instead of using rocket motors for active braking or skimming through the upper regions of a planet's atmosphere over a period of several months, the spacecraft uses friction with a planet's atmosphere to decelerate the craft in one pass. Since the spacecraft may be traveling upward of 25 km/s, the craft must penetrate deep into the atmosphere to generate enough friction. A great deal of heat is generated, and the delicate instruments within the craft need to be protected by a sophisticated insulation system.

The Comsol® "aerocapture" module is designed to simulate entry into the atmosphere of Titan, Saturn's largest moon. The outer surface temperature of the craft during entry, the temperature of the moon's atmosphere, and the properties of the heat shield (insulation materials) as a function of temperature are all known and included in the module. The goal of the module is to design an appropriate composite insulation system that protects the internal contents and provides minimum weight and maximum internal volume for the craft.

The heat shield contains two basic materials: *fiberform*, a porous amorphous carbon, and *aerogel*, a highly porous silica. The maximum temperatures the fiberform and aerogel can withstand are 3000 and 1200 K, respectively, while to protect the internal contents of the craft, the bond line temperature at the interface between the aerogel insulation and the aluminum skin of the craft cannot exceed 250°C.

Figure 6.17 shows a NASA-Ames simulation of the expected heat flux at the surface of the spacecraft. Clearly, this type of input into a model cannot be handled via an analytical solution. Figure 6.18 shows the basic structure of the insulation, an example of how we can tie transient and composite systems together. Since the surface coating is so thin, it is neglected in the thermal simulation.

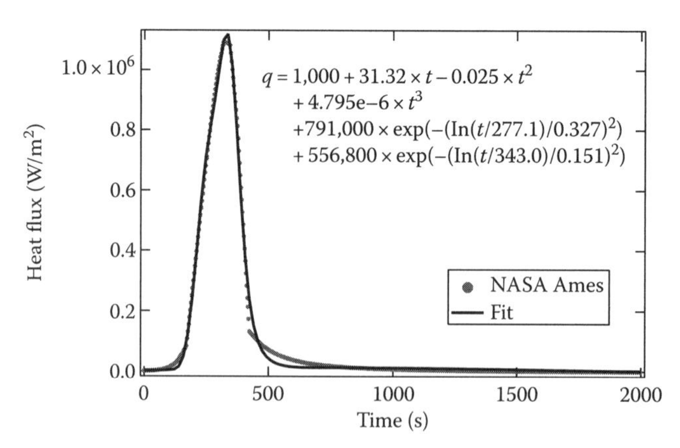

FIGURE 6.17 Heat flux profile a spacecraft is expected to encounter upon aerocapture at Titan.

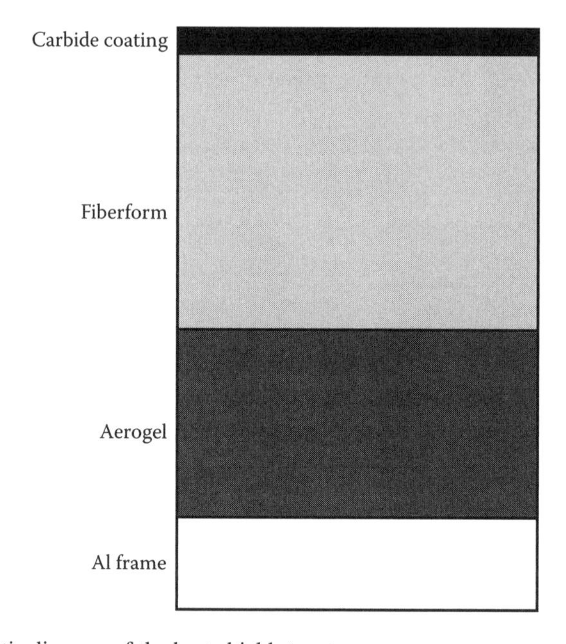

FIGURE 6.18 Schematic diagram of the heat shield structure.

The thermal properties of the actual materials are complicated functions of temperature documented in the numerical module. We can simulate the system using average properties or input the full temperature variations into the model. The following figures show the results for both scenarios. The constant property case could, in principle, be solved analytically. The variable property case must be solved numerically. The maximum temperature reached at the surface of the heat shield occurs approximately 5 min following entry into the planetary atmosphere. Figure 6.19 shows the temperature profile throughout the heat shield assuming the actual physical properties of the material and what we would have predicted had we assumed some average property for each material. Figure 6.20 shows the maximum bond line temperatures reached during the aerocapture maneuver. Had we only considered constant, average properties, we would have overestimated the bond line temperature and hence either made the craft unnecessarily heavy by using too much heat shield material or restricted the volume of the payload to keep the weight constant. So, the constant property case, while a useful physical tool for assessing the feasibility of the idea, would clearly be inadequate for a final system design.

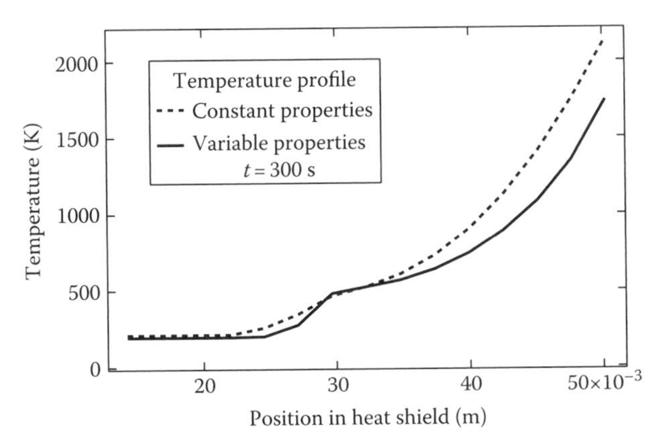

FIGURE 6.19 Temperature profile throughout the heat shield of a spacecraft at the point of maximum heat shield temperature.

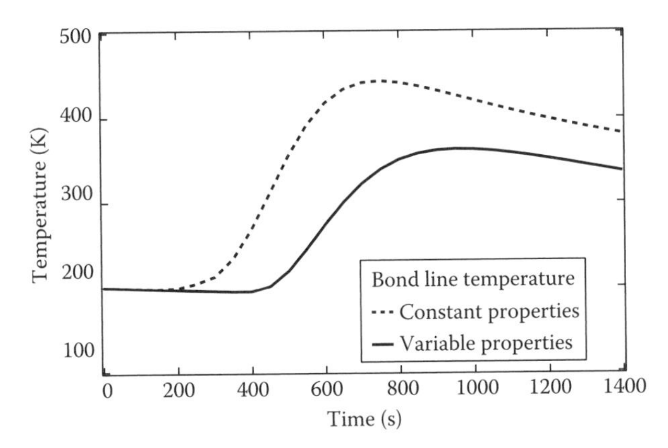

FIGURE 6.20 Temperature profile at the bond line between the heat shield and spacecraft structure as a function of time during aerocapture.

6.4 SEMI-INFINITE SYSTEMS

So far, we have discussed time-dependent transport situations concentrating only on systems that have well-defined boundaries. Many important transport processes occur in situations where we are interested in what happens at relatively short times or the system is very large and one boundary either does not exist or is indistinct. Examples include calculating the fluid velocity profile on the surface of a lake between a moving speedboat and the shoreline, determining the temperature profile in the Earth as the seasons change, and predicting the concentration profile of ions diffused into a crystal substrate. In such cases, we can assume that the system is so large it extends, essentially, to infinity. These situations are referred to as semi-infinite problems because we know precisely where one boundary is but have no second boundary to consider. Semi-infinite systems are characterized by our complete knowledge of what occurs at one boundary of the system and the certainty that *nothing* occurs at the system boundary located infinitely far away. To see how semi-infinite systems operate, we will consider several diffusive transport processes ranging from motion in a fluid to freezing at the surface of a lake.

6.4.1 Fluid Flow Next to a Plate Suddenly Set into Motion

One of the most commonly encountered semi-infinite situations involves the motion of a fluid near a plate or wall suddenly set into motion. The physical system is shown in Figure 6.21. A flat plate

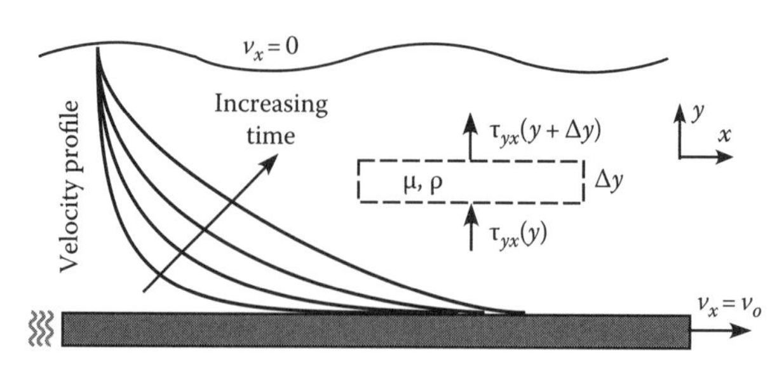

FIGURE 6.21 Flow near a plate suddenly set into motion. *Arrows show the flow of momentum through the control volume.*

is immersed in an infinite pool of fluid. Due to the symmetry of the problem, what occurs in the fluid above the plate also occurs to the fluid below. We need only consider one half. The fluid and plate are initially at rest, and at time, $t = 0$, the plate begins to move with a steady velocity, v_o. We would like to determine the velocity of the fluid for a short time after the initial starting motion of the plate but before bulk convection of the fluid becomes important (once bulk convection starts, the viscous forces driving fluid motion become negligible with respect to inertial forces). The profile we calculate will be indicative of the velocity profile of air above an airplane as it starts to taxi down a runway, the velocity profile of fluid above a photograph as the photograph is pulled from the developer solution, or the velocity profile of liquid against a ship's hull as the ship starts to move through the water.

We need some simplifying assumptions to be able to solve the problem. We assert that the properties of the fluid remain constant, that the plate is essentially infinitely long so the velocity varies only in the y direction, that there are no pressure or body forces (neglect the weight of the fluid), and that there is no bulk fluid motion (convection). We set up a small, differential control volume like the one shown in Figure 6.24 and construct a momentum balance. The balance equation states that the rate of change of momentum in the control volume is equal to the difference in forces exerted on the two faces of the control volume. Representing the force in terms of the stress and area, dividing by $A_s\Delta y$, taking the limit as $\Delta y \to 0$, and using Newton's law to substitute for τ_{yx} gives us a differential equation describing the velocity profile in the fluid:

$$\begin{array}{ccccccc} In & - & Out & + & Gen & = & Acc \\ F(y) & - & F(y+\Delta y) & + & 0 & = & A_s\Delta y \dfrac{\partial}{\partial y}(\rho v_x) \end{array} \tag{6.111}$$

$$\nu \frac{\partial^2 v_x}{\partial y^2} = \frac{\partial v_x}{\partial t} \qquad \nu = \frac{\mu}{\rho} \quad \text{Kinematic viscosity} \tag{6.112}$$

The second derivative with respect to y requires we have two boundary conditions. The time derivative requires that we have one initial condition. The initial condition for this problem states that the fluid and plate have zero velocity. The boundary conditions specify no slip at the plate surface ($v_x = v_o$) and that very far from the plate, the fluid is unaffected by the plate's motion ($v_x = 0$):

$$v_x(y,0) = 0 \quad \text{Initial condition} \tag{6.113}$$

$$v_x(0,t) = v_o \tag{6.114}$$

Boundary conditions

$$v_x(\infty,t) = 0 \tag{6.115}$$

We will find it helpful during the course of this analysis to define a dimensionless velocity that varies between 0 at the plate surface and 1 at the fluid surface:

$$u = \frac{v_o - v_x}{v_o} \tag{6.116}$$

Our goal is always to try to avoid having to solve a partial differential equation. Ideally, we would like to transform the partial differential equation into an ordinary differential equation. We can accomplish this in semi-infinite systems using a quantity called a similarity variable. The similarity variable is a particular, dimensionless, combination of the independent variables and parameters of the problem. The reason the similarity variable works is because the fluid extends toward infinity. There is no hard upper boundary communicating with the plate. Thus, the velocity profile will always resemble the profiles shown in Figure 6.21. It will be self-similar in time and space, so no matter where or when we decide to look at the profile, the velocity will always appear to be the same function of y and t. In this case, the similarity variable, ε, will be a function of y, t, and ν:

$$u = u(\varepsilon) \qquad \varepsilon = C(\nu)^a (y)^m (t)^n \quad \text{Similarity variable} \tag{6.117}$$

To determine the exact functional form for the similarity variable, we group the parameters so that they become dimensionless and we pit the independent variables, time and space, against one another. The dimensions of ν, y, and t are

$$\nu[=]\text{m}^2/\text{s} \qquad y[=]\text{m} \qquad t[=]\text{s}$$

and we place either y or t in the numerator with the other in the denominator. The grouping leads to the following form for the similarity variable:

$$a = -\frac{1}{2}m \qquad n = a \tag{6.118}$$

$$\varepsilon = C\left(\frac{y}{\sqrt{\nu t}}\right)^m \rightarrow \varepsilon = \frac{y}{\sqrt{4\nu t}} \tag{6.119}$$

We are free to choose both C and m, so we picked $m = 1$ and $C = 1/2$ (for convenience as will be seen later) and obtained the similarity variable shown in Equation 6.119. Having determined the form of the similarity variable, we now use the chain rule for differentiation to transform the coordinates of the problem from y and t to ε:

$$\frac{\partial}{\partial t} = \frac{d}{d\varepsilon}\left(\frac{\partial \varepsilon}{\partial t}\right) = -\frac{1}{2}\left(\frac{y}{\sqrt{4\nu t}}\right)\left(\frac{1}{t}\right)\frac{d}{d\varepsilon} = -\frac{\varepsilon}{2t}\frac{d}{d\varepsilon} \tag{6.120}$$

$$\frac{\partial^2}{\partial y^2} = \frac{d}{d\varepsilon}\left(\frac{\partial \varepsilon}{\partial y}\right)\left[\frac{d}{d\varepsilon}\left(\frac{\partial \varepsilon}{\partial y}\right)\right] = \frac{1}{4\nu t}\frac{d^2}{d\varepsilon^2} \tag{6.121}$$

Substituting these definitions into the differential equation leads to a second-order ordinary differential equation of the form

$$\frac{d}{d\varepsilon}\left(\frac{du}{d\varepsilon}\right)+2\varepsilon\frac{du}{d\varepsilon}=0 \tag{6.122}$$

Luckily, the similarity transformation left us with an equation we can solve. This is not always the case. The ordinary differential equation is often just as hard to solve as the partial differential equation. Having transformed the partial differential equation, we must also transform the boundary conditions:

$$\varepsilon=\infty \qquad u=1 \quad \text{Initial condition} \tag{6.123}$$

$$\varepsilon=0 \qquad u=0 \tag{6.124}$$

Boundary conditions

$$\varepsilon=\infty \qquad u=1 \tag{6.125}$$

Notice that the initial condition, Equation 6.123, and the boundary condition at $\varepsilon = \infty$, Equation 6.125, are the same. In effect, we have incorporated the initial condition in the structure of the similarity variable and so only need the two boundary conditions. To solve the differential equation we let

$$u'=\frac{du}{d\varepsilon} \tag{6.126}$$

This transforms the second-order differential equation into a first-order differential equation that we can integrate

$$\frac{du'}{d\varepsilon}+2\varepsilon u'=0 \tag{6.127}$$

$$u'=a_1\exp(-\varepsilon^2) \tag{6.128}$$

Integrating Equation 6.128 gives

$$u=a_1\int_0^{\varepsilon}\exp(-\eta^2)d\eta+a_2 \tag{6.129}$$

From the initial condition, $\varepsilon = \infty \rightarrow u = 1$, we have

$$1=a_1\int_0^{\infty}\exp(-\eta^2)\,d\eta+a_2 \quad\rightarrow\quad \frac{\sqrt{\pi}}{2}a_1+a_2 \tag{6.130}$$

where the integral was evaluated to be $\sqrt{\pi}/2$. The second boundary condition, $\varepsilon = 0 \rightarrow u = 0$, gives

$$0=a_1\int_0^{0}\exp(-\eta^2)d\eta+a_2 \tag{6.131}$$

and so the integration constants are

$$a_1 = \frac{2}{\sqrt{\pi}} \qquad a_2 = 0 \tag{6.132}$$

Since the integral occurring in Equation 6.129 occurs so commonly in transport and other areas, we have given it a name: the error function, *erf*. The solution can now be written in terms of this new function:

$$u = \frac{2}{\sqrt{\pi}} \int_0^{\varepsilon} \exp(-\eta^2)\, d\eta = erf(\varepsilon) = erf\left(\frac{y}{\sqrt{4\nu t}}\right) \tag{6.133}$$

A plot of the error function is shown in Figure 6.22. At any time, t, we call the value of x when $erf(\varepsilon) = 0.99$ the penetration depth, δ_μ. This is the depth to which an imposed disturbance is felt.

In our current example, to find the depth to which the imposed velocity is felt, we set $u = 0.99$ and solve for δ_μ:

$$u = 0.99 = erf\left(\frac{\delta_\mu}{\sqrt{4\nu t}}\right) \quad \rightarrow \quad \delta_\mu = 3.643\sqrt{\nu t} \tag{6.134}$$

Example 6.5 Penetration Depths in Water

As an example illustrating the magnitude of penetration phenomena, let's consider the motion of a submarine in water. If the diameter of the submarine is large, compared to the penetration depth, we can view the submarine as a flat plate and consider how far the velocity disturbance extends into the water surrounding the sub. We may assume that the vessel is in the Caribbean and the water temperature is 25°C. The sub will be moving at an average speed of 19.5 knots (10 m/s) and we wish to determine the penetration depth at $t = 1$ and 100 s.

Since we are approximating the sub to be a flat plate, we can use Equation 6.134 to determine the penetration depth. All we need to know is the kinematic viscosity of the water at 25°C (8.947×10^{-7} m²/s):

$$\delta_\mu(t = 1) = 3.643\sqrt{\nu t} = 3.643\sqrt{(8.947 \times 10^{-7})(1)} = 3.446 \times 10^{-3}\ \text{m}$$

$$\delta_\mu(t = 100) = 3.643\sqrt{(8.947 \times 10^{-7})(100)} = 3.446 \times 10^{-2}\ \text{m}$$

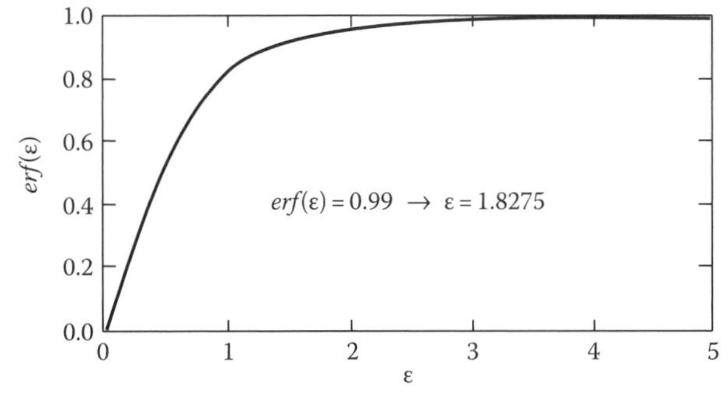

FIGURE 6.22 The error function, *erf*. Also shown in the value of ε when *erf* is 0.99. This value of ε defines the penetration depth, δ_μ.

Penetration depths in water are exceedingly small so our assumption of flat plate performance for the submarine is justified. We will see in later chapters that the convective boundary layer thickness surrounding an object like a submarine is much more important to the transport of momentum than the diffusive penetration depth we just calculated.

6.4.2 Heat Transfer Analogs

The same penetration-type phenomena can be found in heat transfer too. The heat transfer analog is one very familiar to anyone who has noticed a water main break after a long cold spell. In such instances, the engineers in charge of determining how deep to bury the water main made a miscalculation. In the heat transfer system, we impose a temperature at a surface and we want to know how long before that temperature is felt, at a distance of z meters from the surface.

The governing equation and boundary conditions are found by substituting temperature, T, for the velocity and the thermal diffusivity, α, for the kinematic viscosity, ν, in Equation 6.112:

$$\alpha\frac{\partial^2 T}{\partial z^2}=\frac{\partial T}{\partial t} \qquad \alpha=\frac{k}{\rho C_p} \tag{6.135}$$

The boundary and initial conditions are also quite similar to those for the developing velocity profile. Initially, we specify the temperature to be uniform throughout the solid. At time, $t = 0$, we impose the temperature T_o on the surface (Figure 6.23). Far away from the surface, the imposed temperature is never felt and remains at T_i:

$$t=0 \qquad T(z,0)=T_i \tag{6.136}$$

$$z=0 \qquad T(0,t)=T_o \tag{6.137}$$

$$z=\infty \qquad T(\infty,t)=T_i \tag{6.138}$$

We define a dimensionless temperature in terms of the initial and imposed temperatures, so the differential equation is identical in form to Equation 6.112:

$$\alpha\frac{\partial^2 \theta}{\partial z^2}=\frac{\partial \theta}{\partial t} \qquad \theta=\frac{T-T_i}{T_o-T_i} \tag{6.139}$$

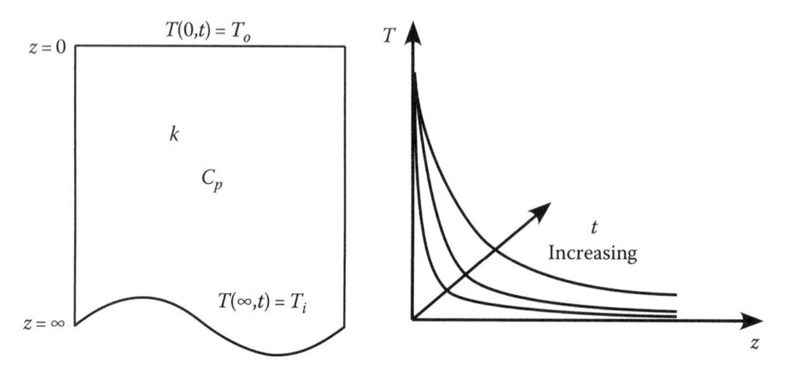

FIGURE 6.23 Constant surface temperature imposed upon a semi-infinite solid. The graph illustrates how the imposed temperature penetrates into the solid as time progresses.

This means we can develop a similarity variable similar to the variable we used in the moving plate problem. In fact, all we need to do is replace the kinematic viscosity by the thermal diffusivity to generate the new similarity variable:

$$\varepsilon = \frac{z}{\sqrt{4\alpha t}} \tag{6.140}$$

Substituting the similarity variable into Equation 6.139 leads to an ordinary differential equation of the same form as Equation 6.122 and an error function solution:

$$\theta = a_1 \int_0^{\varepsilon} \exp(-\eta^2)d\eta + a_2 = a_1 \frac{\sqrt{\pi}}{2} erf(\varepsilon) + a_2 \tag{6.141}$$

The boundary conditions are a bit different. Applying the condition at $\varepsilon = 0$ shows that

$$1 = a_1 \frac{\sqrt{\pi}}{2} erf(0) + a_2 \tag{6.142}$$

and so $a_2 = 1$ since $erf(0) = 0$. The boundary condition at $\varepsilon = \infty$ shows that

$$0 = a_1 \frac{\sqrt{\pi}}{2} erf(\infty) + 1 \tag{6.143}$$

so $a_1 = -2/\sqrt{\pi}$. The temperature profile is now

$$\theta = 1 - \frac{2}{\sqrt{\pi}} \int_0^{\varepsilon} \exp(-\eta^2)d\eta = 1 - erf(\varepsilon) = erfc\left(\frac{z}{\sqrt{4\alpha t}}\right) \tag{6.144}$$

We call the term $1 - erf(\varepsilon)$ the complementary error function (*erfc*) because it appears so frequently in many transport problems. As in the fluid flow example, the penetration depth, δ_T, varies as the square root of time and is given by

$$\delta_T = 3.643\sqrt{\alpha t} \tag{6.145}$$

Example 6.6 Wine Cellar Depth

If we were to live in the Finger Lakes region of New York State, we would be required to have a wine cellar in our home. To avoid heating costs in the winter and cooling costs in the summer, we decide to build the cellar underground, where the temperature of the Earth is nearly constant at the ideal wine temperature of 10°C. The temperature at the surface is affected by the change of seasons, so we need to determine how deep we must dig to insure a constant temperature for the wine. Assuming we could have 6 months where temperatures are above 10°C or 6 months below 10°C, the depth of the cellar will be determined from the penetration depth for that time period:

$$6 \text{ months} \approx 1.577 \times 10^7 \text{ s} \qquad \alpha_{earth} = 1.39 \times 10^{-7} \text{ m}^2/\text{s}$$

The wine cellar depth should be

$$\delta_T = 3.643\sqrt{(1.39\times10^{-7})(1.577\times10^{7})} = 5.39\text{m}$$

Often it is important to consider semi-infinite conditions where we have convection at the surface. This is particularly true for problems involving the climate such as determining the depth to dig before laying a water pipe or pouring the foundation for a home or building. The governing equation is (6.139) with temperature θ, defined using the free stream temperature of the convecting fluid, T_∞:

$$\alpha\frac{\partial^2\theta}{\partial z^2} = \frac{\partial\theta}{\partial t} \qquad \theta = \frac{T - T_\infty}{T_i - T_\infty} \tag{6.146}$$

The boundary conditions are changed to reflect convection at the surface:

$$t = 0 \qquad \theta = 1 \tag{6.147}$$

$$z = \infty \qquad \theta = 1 \tag{6.148}$$

$$z = 0 \qquad \frac{d\theta}{dz} = -\frac{h}{k}\theta \tag{6.149}$$

The solution to this problem is quite involved algebraically but can be found in Carslaw and Jaeger [6]. The final temperature profile within the solid is

$$\theta = 1 - erfc\left(\frac{z}{\sqrt{4\alpha t}}\right) + \exp\left(\frac{hz}{k} + \frac{h^2\alpha t}{k^2}\right)\left[erfc\left(\frac{z}{\sqrt{4\alpha t}} + \frac{h\sqrt{\alpha t}}{k}\right)\right] \tag{6.150}$$

As the heat transfer coefficient approaches ∞, the solution approaches the case for constant surface temperature where now $T_o = T_\infty$.

6.4.3 Mass Transfer Analogs

As you probably have realized, we can develop an analog of the semi-infinite solid situation for mass transfer too. The correspondence with the previous heat transfer problems will be exact if we assume dilute solutions or equimolar counterdiffusion. In such instances, the differential equation is

$$D_{ab}\frac{\partial^2 c_a}{\partial z^2} = \frac{\partial c_a}{\partial t} \tag{6.151}$$

The similarity variable appropriate for this situation is

$$\varepsilon = \frac{z}{\sqrt{4D_{ab}t}} \tag{6.152}$$

and the solution to Equation 6.151 is again given in terms of the error function.

Constant surface concentration

$$\chi = \frac{c_a - c_{ao}}{c_{ai} - c_{ao}} = 1 - \frac{2}{\sqrt{\pi}}\int_0^\varepsilon \exp(-\eta^2)d\eta = erfc\left(\frac{z}{\sqrt{4D_{ab}t}}\right) \tag{6.153}$$

Convection at the surface

$$\chi = \frac{c_a - c_{a\infty}}{c_{ai} - c_{a\infty}} = 1 - erfc\left(\frac{z}{\sqrt{4D_{ab}t}}\right) + \exp\left(\frac{k_c z}{D_{ab}} + \frac{k_c^2 t}{D_{ab}}\right)\left[erfc\left(\frac{z}{\sqrt{4D_{ab}t}} + \frac{k_c\sqrt{t}}{\sqrt{D_{ab}}}\right)\right] \quad (6.154)$$

The penetration depth is also analogous to the heat transfer or momentum transfer case with ν or α replaced by the diffusivity, D_{ab}:

$$\delta_x = 3.643\sqrt{D_{ab}t} \quad (6.155)$$

An interesting case arises when we consider mass transfer from a sphere of radius, r_o, into an infinite medium. This process is important in many situations and can be used to describe dissolution of a pill in an empty stomach, the dissolution of a sugar cube in a cup of tea, or the evaporation of a drop of water in the air. We will make our problem simple by first assuming dilute solutions, the absence of generation, and constant properties. Prior to $t = 0$, the sphere has a uniform concentration of solute c_{ao}. At $t = 0$ it is suspended in a large tank and is slowly leached so that at r_o, at least for relatively short times, the concentration is always c_{ao}. We would like to know the concentration of solute within the tank as a function of radial position. Figure 6.24 shows the overall situation and the control volume for developing the differential mass balance.

We write a transient mass balance about the control volume in Figure 6.24:

$$\begin{array}{ccccccccc} In & - & Out & + & Gen & = & Acc \\ \dot{M}_a(r) & - & \dot{M}_a(r+\Delta r) & + & 0 & = & (4\pi r^2 \Delta r)\dfrac{\partial c_a}{\partial t} \end{array} \quad (6.156)$$

Expressing the mass flow in terms of the flux and area, dividing through by Δr, and taking the limit as $\Delta r \rightarrow 0$ leads to a differential equation for the molar flux of solute a:

$$-\frac{\partial}{\partial r}\left(4\pi r^2 N_a\right) = (4\pi r^2)\frac{\partial c_a}{\partial t} \quad (6.157)$$

Substituting Fick's law for dilute solutions into Equation 6.157 leads to a final partial differential equation in terms of c_a:

$$\frac{D_{ab}}{r^2}\frac{\partial}{\partial r}\left(r^2\frac{\partial c_a}{\partial r}\right) = \frac{\partial c_a}{\partial t} \quad (6.158)$$

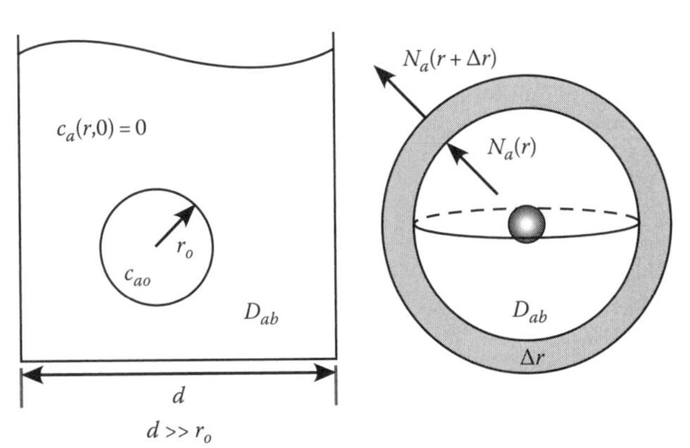

FIGURE 6.24 Transient mass transfer in a semi-infinite, spherical system.

The initial and boundary conditions for the problem stipulate a constant concentration at the pill's surface and effectively zero concentration of solute outside the pill at $t = 0$ and at $r = \infty$:

$$t = 0 \qquad c_a = 0 \tag{6.159}$$

$$r = r_o \qquad c_a = c_{ao} \tag{6.160}$$

$$r = \infty \qquad c_a = 0 \tag{6.161}$$

Since this is a semi-infinite problem, a similarity solution should exist. Unfortunately, finding a similarity variable here appears difficult. The key to solving this problem is a change of variable. If we let $\chi = c_a r$, we find that Equation 6.158 gets transformed to the familiar equation for diffusion in a rectangular semi-infinite solid (6.151):

$$D_{ab}\frac{\partial^2 \chi}{\partial r^2} = \frac{\partial \chi}{\partial t} \tag{6.162}$$

$$r = r_o \qquad \chi = c_{ao} r_o = \chi_o \tag{6.163}$$

$$r = \infty \qquad \chi = 0 \tag{6.164}$$

$$t = 0 \qquad \chi = 0 \tag{6.165}$$

We already know the solution to Equation 6.162 is given in terms of the error function and similarity variable, ε:

$$\chi = a_1 \frac{\sqrt{\pi}}{2} erf(\varepsilon) + a_2 \qquad \varepsilon = \frac{r}{\sqrt{4D_{ab}t}} \tag{6.166}$$

Applying the boundary conditions leads to two algebraic equations for a_1 and a_2:

$$\varepsilon = \varepsilon_o \qquad \chi = \chi_o \qquad \chi_o = a_1 \frac{\sqrt{\pi}}{2} erf(\varepsilon_o) + a_2 \tag{6.167}$$

$$\varepsilon = \infty \qquad \chi = 0 \qquad 0 = a_1 \frac{\sqrt{\pi}}{2} + a_2 \tag{6.168}$$

Solving for a_1 and a_2 leads to the final solution

$$\chi = \frac{\chi_o}{erfc(\varepsilon_o)} erfc(\varepsilon) \tag{6.169}$$

or in dimensional units

$$c_a = \frac{c_{ao}}{erfc\left(r_o/\sqrt{4D_{ab}t}\right)}\left[\left(\frac{r_o}{r}\right) erfc\left(\frac{r}{\sqrt{4D_{ab}t}}\right)\right] \tag{6.170}$$

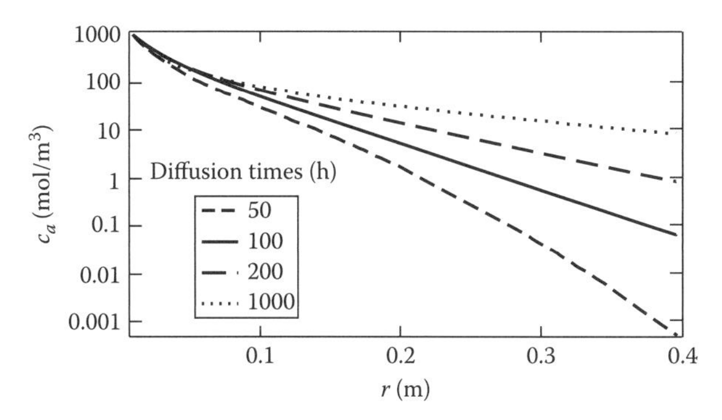

FIGURE 6.25 Concentration profiles from a pill dissolving into a semi-infinite medium.

Figure 6.25 shows the concentration profiles for various times assuming the following physical constants:

$$r_o = 0.01\,\text{m} \qquad c_{ao} = 1000\,\text{mol/m}^3 \qquad D_{ab} = 1\times 10^{-7}\,\text{m}^2/\text{s}$$

The change of variable we used in this problem is applicable to all problems in spherical systems and is a general method for reducing problems in spherical coordinates to the corresponding problem in rectangular coordinates.

6.5 MOVING BOUNDARY PROBLEMS

In this last section of the chapter, we would like to present and analyze several more complicated transient, diffusive transport problems. These examples will illustrate some more basic transport problems and allow us to remove our assumption of equimolar counterdiffusion or dilute solutions for mass transfer. We will look into mass transfer in semi-infinite media with chemical reaction occurring, we will consider a number of problems involving moving boundaries between phases, and finally, we will discuss a transient problem in fluid flow with a periodic boundary condition that leads to oscillatory motion.

6.5.1 Transient Evaporation through a Stagnant Vapor [7]

One of the earlier restrictions on transient mass transfer was that we had equimolar counterdiffusion or that we had dilute solutions. These assumptions enabled us to write Fick's law in the same form as Newton's law or Fourier's law, and so we use our knowledge about transient heat or momentum transport to say something about the corresponding mass transfer case. Here, we would like to relax that restriction and consider an evaporation process where we are evaporating a liquid of species a into a stagnant vapor of species b.

The system consists of a liquid, initially pure a, in contact with a vapor, initially pure b. Both liquid and vapor are at a constant temperature as shown in Figure 6.26. We will assume that the two species form an ideal vapor mixture so that we can keep the total molar concentration in the vapor, c_t, constant. We will also specify that species b is essentially insoluble in liquid a so that at the surface, $N_{b0} = 0$. We begin the analysis by writing transient mass balances about the control volume in the vapor space. We need mass balances for both species, a and b:

$$\begin{array}{ccccc} In & - & Out & + Gen = & Acc \\ \dot{M}_a(y) & - & \dot{M}_a(y+\Delta y) & + \;\; 0 & = A_c \Delta y c_t \dfrac{\partial x_a}{\partial t} \end{array} \tag{6.171}$$

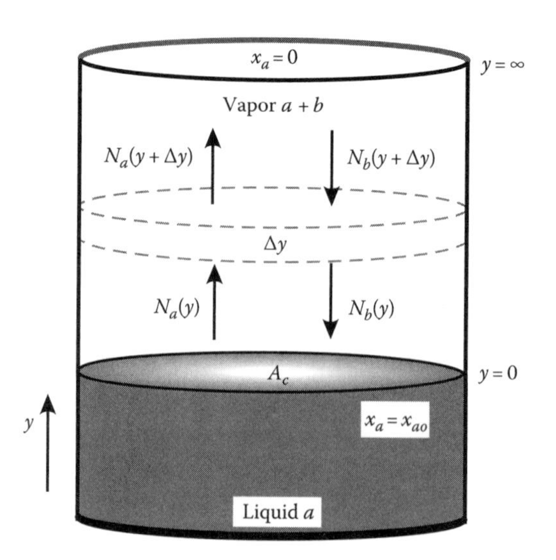

FIGURE 6.26 Diffusion of species a into a semi-infinite vapor space of species b.

$$\begin{array}{ccccccccc} In & - & Out & + & Gen & = & Acc \\ \dot{M}_b(y+\Delta y) & - & \dot{M}_b(y) & + & 0 & = & -A_c\Delta y c_t \dfrac{\partial x_b}{\partial t} \end{array} \tag{6.172}$$

We switch signs on the balance for component b since it is traveling in the opposite direction to that of a. Expressing the mass flow in terms of the flux, dividing both Equations 6.171 and 6.172 by $A_c\Delta y$, and taking the limit as $\Delta y \to 0$ yields two differential equations in terms of the mole fractions and fluxes of a and b:

$$-\frac{\partial N_a}{\partial y} = c_t \frac{\partial x_a}{\partial t} \tag{6.173}$$

$$-\frac{\partial N_b}{\partial y} = c_t \frac{\partial x_b}{\partial t} \tag{6.174}$$

Upon adding Equations 6.173 and 6.174 while noting that $x_a + x_b = 1$, we find that the total flux of species a and b must be a constant throughout the system (does not vary with y) and so can only be a function of time:

$$-\frac{\partial}{\partial y}(N_a + N_b) = c_t \frac{\partial}{\partial t}(x_a + x_b) = 0 \tag{6.175}$$

At $y = 0$ the flux of b, N_{bo}, across the interface is zero since b is insoluble in a. The flux of a at that interface is found using Fick's law. In this case, where $N_{bo} = 0$, Fick's law at the interface becomes

$$N_{ao} = -c_t D_{ab} \left.\frac{\partial x_a}{\partial y}\right|_{y=0} + x_{ao}\left(N_{ao} + \cancel{N_{bo}}^{0}\right) = -\left[\frac{c_t D_{ab}}{1 - x_{ao}}\right]\left.\frac{\partial x_a}{\partial y}\right|_{y=0} \tag{6.176}$$

If $N_a + N_b$ is only a function of time as Equation 6.175 shows, then $N_a + N_b$ at the interface should be the same as $N_a + N_b$ everywhere else:

$$N_a + N_b = N_{ao} = -\left[\frac{c_t D_{ab}}{1 - x_{ao}}\right] \frac{\partial x_a}{\partial y}\bigg|_{y=0} \tag{6.177}$$

The flux of species a anywhere within the vapor is defined by Fick's law

$$N_a = -c_t D_{ab} \frac{\partial x_a}{\partial y} + x_a (N_a + N_b) \tag{6.178}$$

and using Equation 6.177, we have

$$N_a = -c_t D_{ab} \frac{\partial x_a}{\partial y} - x_a \left[\frac{c_t D_{ab}}{1 - x_{ao}} \frac{\partial x_a}{\partial y}\bigg|_{y=0}\right] \tag{6.179}$$

Substituting into Equation 6.173 puts the equation in terms of x_a only:

$$\frac{\partial x_a}{\partial t} = D_{ab} \frac{\partial^2 x_a}{\partial y^2} + \left[\frac{D_{ab}}{1 - x_{ao}} \frac{\partial x_a}{\partial y}\bigg|_{y=0}\right] \frac{\partial x_a}{\partial y} \tag{6.180}$$

This equation is subject to semi-infinite boundary conditions that specify the composition at $y = 0$ and state that x_a must remain at 0 far away from the interface:

$$y = 0 \qquad x_a(0, t) = x_{ao} \tag{6.181}$$

$$y = \infty \qquad x_a(\infty, t) = 0 \tag{6.182}$$

The initial condition states that $x_a = 0$ throughout the vapor space:

$$t = 0 \qquad x_a(y, 0) = 0 \tag{6.183}$$

Even though Equation 6.180 is not exactly in the form of Equation 6.151, we should still be able to use the similarity variable approach since we have a semi-infinite domain and a linear differential equation. The similarity variable given in Equation 6.152 will make a good first guess for the solution. We first define a dimensionless mole fraction $\chi(y, t)$ and similarity variable as

$$\chi = \frac{x_a}{x_{ao}} \qquad \varepsilon = \frac{y}{\sqrt{4 D_{ab} t}} \tag{6.184}$$

Substituting the variables of Equation 6.184 into Equation 6.180 transforms the partial differential equation into an ordinary differential equation

$$\frac{d^2\chi}{d\varepsilon^2} + 2(\varepsilon - \varepsilon_c)\frac{d\chi}{d\varepsilon} = 0 \qquad \varepsilon_c = -\frac{1}{2}\left[\frac{x_{ao}}{1 - x_{ao}}\right] \frac{d\chi}{d\varepsilon}\bigg|_{\varepsilon=0} \tag{6.185}$$

where ε_c is a constant. The initial and boundary conditions for the problem are transformed to

$$\chi = 1 \qquad \varepsilon = 0 \tag{6.186}$$

$$\chi = 0 \qquad \varepsilon = \infty \tag{6.187}$$

Since ε_c is a constant, we can integrate Equation 6.185 easily (using $\chi' = d\chi/d\varepsilon$) to obtain

$$\chi = a_1 erf(\varepsilon - \varepsilon_c) + a_2 \tag{6.188}$$

Evaluating the integration constants using the boundary conditions gives

$$\chi = \frac{1 - erf(\varepsilon - \varepsilon_c)}{1 + erf(\varepsilon_c)} \tag{6.189}$$

and x_{ao} is given by

$$x_{ao} = \frac{1}{1 + \dfrac{1}{\sqrt{\pi}(1 + erf(\varepsilon_c))\varepsilon_c \exp\left(\varepsilon_c^2\right)}} \tag{6.190}$$

We can gain some more insight into the effect of including bulk fluid motion by calculating the flux of a at the interface. N_{ao} is given by

$$\begin{aligned} N_{ao} &= -\left[\frac{c_t D_{ab}}{1 - x_{ao}}\right] \frac{\partial x_a}{\partial y}\bigg|_{y=0} = -\frac{c_t}{2}\sqrt{\frac{D_{ab}}{t}}\left[\frac{x_{ao}}{1 - x_{ao}}\right] \frac{d\chi}{d\varepsilon}\bigg|_{\varepsilon=0} \\ &= \sqrt{\frac{D_{ab}}{4\pi t}}\left(\frac{c_t x_{ao}}{1 - x_{ao}}\right)\left[\frac{\exp\left(-\varepsilon_c^2\right)}{1 + erf\left(\varepsilon_c\right)}\right] \end{aligned} \tag{6.191}$$

Had we solved the problem with the restriction that we had dilute solutions, the flux would have been

$$\begin{aligned} N_{ao} &= -c_t D_{ab} \frac{\partial x_a}{\partial y}\bigg|_{y=0} = -\frac{c_t x_{ao}}{2}\sqrt{\frac{D_{ab}}{t}} \frac{\partial \chi}{\partial \varepsilon}\bigg|_{\varepsilon=0} \\ &= c_t x_{ao}\sqrt{\frac{D_{ab}}{4\pi t}} \end{aligned} \tag{6.192}$$

Table 6.6 shows how ε_c varies as a function of x_{ao} and Figure 6.27 shows the fluxes assuming dilute solutions and stagnant b as a function of x_{ao}. Even though the vapor space is infinite in extent, the dilute solution approximation is valid in only a narrow range and underestimates the actual flux. This must be considered during any evaporation measurements.

TABLE 6.6
Values of ε_c for Various x_{ao}

ε_c	x_{ao}	ε_c	x_{ao}
0.0	0.0	0.8	0.824
0.1	0.166	1.0	0.899
0.2	0.311	2.0	0.997
0.4	0.543	∞	1.0

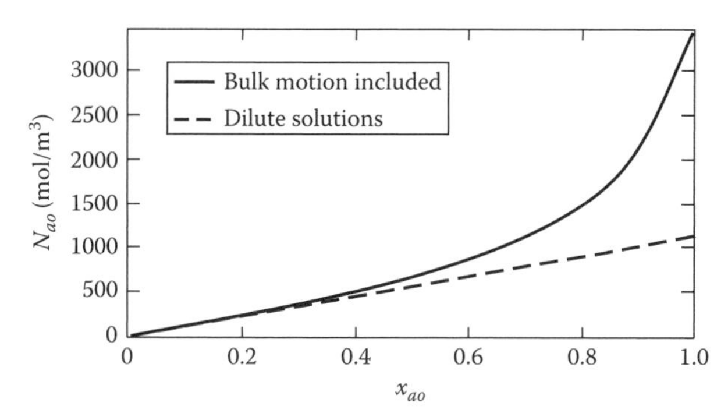

FIGURE 6.27 Flux of a at the interface as a function of the mole fraction of a in the liquid. $D_{ab} = 1 \times 10^{-9}$ m^2/s; $c_t = 1000$ mol/m^3; $t = 1 \times 10^9$ s.

6.5.2 Transient Absorption with Instantaneous Reaction [8]

In Chapter 5, we saw that having a chemical reaction on the boundary could increase the flux of material through that surface. We would now like to consider that process in a transient setting. Suppose we have a gaseous component, a, that is absorbed by a solvent, s. The solvent contains a solute b that reacts instantaneously with a via the reaction $a + b \rightarrow ab$. This type of reaction is common in systems that involve scrubbing acids or bases from gaseous streams such as HCl vapor in optical fiber manufacturing or SO_2 from flue gas in a power plant. We assume the concentrations of a, b, and ab are very small and all solutions are ideal. Therefore, the total concentration of all species remains constant and Fick's law can be written solely in terms of the concentration or mole fraction gradients.

Since we have an instantaneous reaction, there will be an advancing front (located at y') that separates a region devoid of a from a region devoid of b. This means that the boundary condition at the interface will be time dependent and we will have to calculate the velocity of the interface to determine the interface's location.

We begin by writing conservation of mass balances for a and b about the control volume shown in Figure 6.28. The solvent is assumed inert and stagnant. We do not need to consider the reaction in the continuity equations because the reaction is confined only to the boundary separating regions of a and b. The continuity equation for a is written for that region of the system between the original interface, $y = 0$, and the new interface, $y = y'(t)$. Similarly, the continuity equation for b is written between the new interface, $y = y'(t)$ and $y = \infty$:

$$\frac{\partial c_a}{\partial t} = D_{as}\frac{\partial^2 c_a}{\partial y^2} \qquad 0 < y < y'(t) \tag{6.193}$$

$$\frac{\partial c_b}{\partial t} = D_{bs}\frac{\partial^2 c_b}{\partial y^2} \qquad y'(t) < y < \infty \tag{6.194}$$

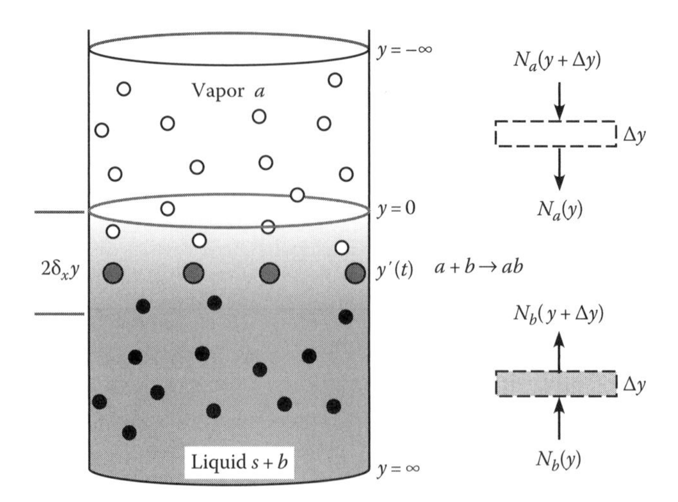

FIGURE 6.28 Absorption of vapor a into a solvent containing a reactant b. The reaction is instantaneous resulting in a moving reaction front.

D_{as} and D_{bs} are pseudo-binary diffusion coefficients and implicitly assume that the presence of a and b does not affect their numerical value. We have already solved equations of this type, (6.151), and so can try similar solutions:

$$\frac{c_a}{c_{ao}} = a_1 + a_2 erf\left(\frac{y}{\sqrt{4D_{as}t}}\right) \tag{6.195}$$

$$\frac{c_b}{c_{bo}} = b_1 + b_2 erf\left(\frac{y}{\sqrt{4D_{bs}t}}\right) \tag{6.196}$$

where

c_{ao} is the interfacial liquid-phase concentration of a
c_{bo} is the original concentration of b in the solvent

The boundary conditions depend upon the position of the interface so, to solve the problem, we must locate the boundary. At the boundary, we know that no a will be present. All of it will have reacted with b:

$$c_a(y', t) = 0 \tag{6.197}$$

The total change in c_a as a function of y' and t is given by the total differential of c_a, dc_a. At the interface, this is zero because no a remains ($c_a = 0$ there forever though there are changes in c_a with both y' and t):

$$dc_a = \left(\frac{\partial c_a}{\partial y'}\right)_t dy' + \left(\frac{\partial c_a}{\partial t}\right)_{y'} dt = 0 \tag{6.198}$$

Equation 6.198 allows us to solve for the front velocity. We evaluate the partial derivatives using Equation 6.195:

$$\frac{dy'}{dt} = \frac{-\left(\frac{\partial c_a}{\partial t}\right)_{y'}}{\left(\frac{\partial c_a}{\partial y'}\right)_t} = \frac{y' \exp\left(-\frac{y'^2}{4D_{as}t}\right) \Big/ 2t^{3/2}\sqrt{\pi D_{as}}}{\exp\left(-\frac{y'^2}{4D_{as}t}\right) \Big/ t^{1/2}\sqrt{\pi D_{as}}} = \frac{y'}{2t} \quad (6.199)$$

We are left with a differential equation for the interface position, y', subject to the requirement that at $t = 0$, our interface is located at $y = 0$. Integrating gives

$$y' = \sqrt{4\gamma t} \quad (6.200)$$

where γ is a constant of integration that we will have to evaluate. Overall, we now have five integration constants to determine, and we can use the following boundary and initial conditions to determine them:

$$t = 0 \qquad c_b = c_{bo} \qquad \text{No reaction yet} \quad (6.201)$$

$$y = 0 \qquad c_a = c_{ao} \qquad \text{Reaction front has moved into solvent} \quad (6.202)$$

$$y = y'(t) \qquad c_a = c_b = 0 \qquad \text{Instantaneous reaction at the interface} \quad (6.203)$$

$$y = y'(t) \qquad -D_{as}\frac{\partial c_a}{\partial y} = D_{bs}\frac{\partial c_b}{\partial y} \qquad \text{Fluxes of reactants must be equal to feed the reaction at the interface} \quad (6.204)$$

The four constants, a_1, a_2, b_1, b_2 are given explicitly as

$$a_1 = 1 \qquad a_2 = -\frac{1}{erf\left(\sqrt{\gamma/D_{as}}\right)} \quad (6.205)$$

$$b_1 = 1 - \frac{1}{1 - erf\left(\sqrt{\gamma/D_{bs}}\right)} \qquad b_2 = \frac{1}{1 - erf\left(\sqrt{\gamma/D_{bs}}\right)} \quad (6.206)$$

The constant, γ, cannot be obtained explicitly but must be found by solving the following implicit equation that comes from Equation 6.204:

$$1 - erf\left(\sqrt{\frac{\gamma}{D_{bs}}}\right) = \frac{c_{bo}}{c_{ao}}\sqrt{\frac{D_{bs}}{D_{as}}}\, erf\left(\sqrt{\frac{\gamma}{D_{as}}}\right) \exp\left[\frac{\gamma}{D_{as}} - \frac{\gamma}{D_{bs}}\right] \quad (6.207)$$

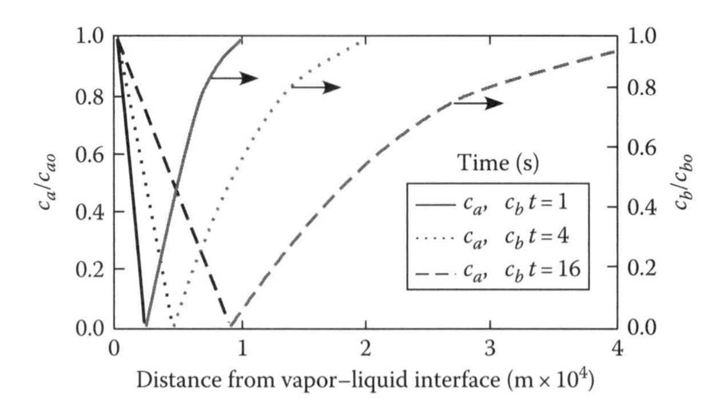

FIGURE 6.29 Concentration profiles of a and b as a function of time. Interface is noted by $c_a = c_b = 0$. ($\gamma = 1.58 \times 10^{-10}$ m²/s.)

The instantaneous flux of a through the original interface at $y = 0$ is given by evaluating Fick's law there:

$$N_a\Big|_{y=0} = -D_{as}\frac{\partial c_a}{\partial y}\Big|_{y=0} = \frac{c_{ao}}{erf(\sqrt{\gamma/D_{as}})}\sqrt{\frac{4D_{as}}{\pi t}} \tag{6.208}$$

Integration of Equation 6.208 over time yields the average flux over the period. Since the flux varies as $t^{-1/2}$, the average flux works out to be just twice the instantaneous flux at time, t. The concentration profiles for a and b are shown in Figure 6.29 with key parameters being

$$c_{ao} = 1000\,\text{mol/m}^3 \qquad c_{bo} = 1250\,\text{mol/m}^3$$

$$D_{as} = 1\times10^{-9}\,\text{m}^2/\text{s} \qquad D_{bs} = 2\times10^{-9}\,\text{m}^2/\text{s} \qquad \gamma = 1.58\times10^{-10}\,\text{m}^2/\text{s}$$

The abscissa represents the distance from the original vapor–liquid interface at $y = 0$. Notice how the interface penetrates into the liquid.

6.5.3 Melting and Solidification [6,9]

Our discussion of diffusion with instantaneous reaction on the boundary provides the perfect introduction for a discussion of the next, important, transport process involving a moving boundary, melting or solidification. In this specific example, we will be discussing the solidification of a liquid, but the corresponding melting problem can be treated in the same way. Consider the surface of a lake or pond shown in Figure 6.30. We suppose that the liquid originally fills the entire space, $y > 0$,

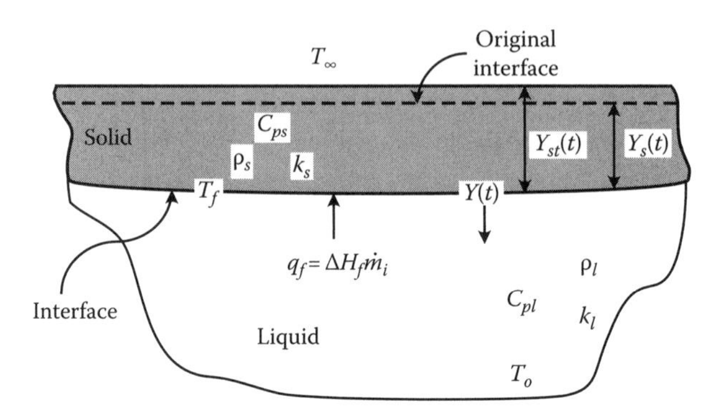

FIGURE 6.30 Solidification at a planar interface. Solid and liquid have different thermal properties and different densities.

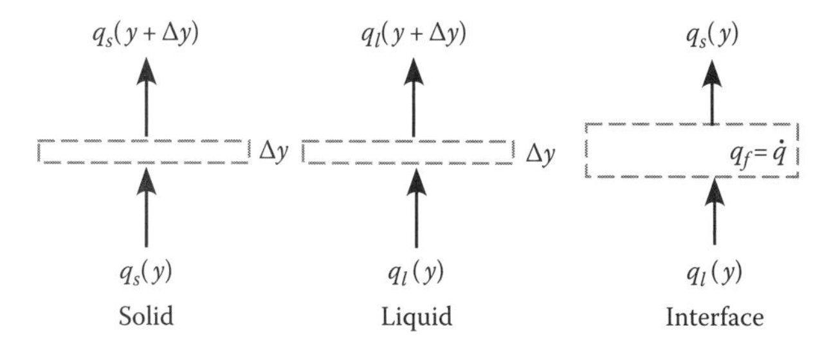

FIGURE 6.31 Control volumes in the liquid and solid at the interface. $\Delta y = 0$ at the interface.

and is being solidified from the exposed surface by removal of heat. The exposed surface must be at a temperature below the freezing point, T_f, and we will assume that the upper surface is held at a constant temperature equal to the ambient air temperature, T_∞. The liquid is initially at a uniform temperature above its freezing point, T_o, and remains at that temperature far away from the surface.

We begin our analysis by considering control volumes located within the solid and liquid. These are shown in Figure 6.31. In both the solid and liquid, we have heat transfer by conduction only, with no generation. If we write an energy balance about the control volumes, assuming 1-D conduction, we arrive at two partial differential equations identical in form to Equation 6.135:

$$\alpha_l \frac{\partial^2 T_l}{\partial y^2} = \frac{\partial T_l}{\partial t} \qquad \alpha_l = \frac{k_l}{\rho_l C_{pl}} \tag{6.209}$$

$$\alpha_s \frac{\partial^2 T_s}{\partial y^2} = \frac{\partial T_s}{\partial t} \qquad \alpha_s = \frac{k_s}{\rho_s C_{ps}} \tag{6.210}$$

We use two different coordinate systems in the liquid and solid, y_l and y_s, because the solid, having a different density than the liquid, may expand or contract as it grows over time. In Figure 6.30, we denote the thickness of the solid as Y_{st} and the depth to which the solid has penetrated the liquid as Y_s. We can express the relationship between these two measures in terms of the density of each phase. We define a volume expansion ratio, ϖ, as ρ_l/ρ_s so

$$\frac{Y_{st}}{Y_s} = \frac{\rho_l}{\rho_s} = \varpi \tag{6.211}$$

Our initial conditions for this problem state that the liquid is at T_o and solid is at T_f:

$$t = 0 \qquad T_l = T_o \tag{6.212}$$

$$t = 0 \qquad T_s = T_f \tag{6.213}$$

The boundary conditions for the problem involve constant temperature at the solid surface,

$$y_s = 0 \qquad T_s = T_\infty \tag{6.214}$$

a semi-infinite condition at the bottom of the lake or pond,

$$y_l = \infty \qquad T_l = T_o \tag{6.215}$$

conditions at the interface where both T_l and T_s equal the freezing temperature, T_f,

$$y_l = Y_s(t) \qquad T_l = T_f \tag{6.216}$$

$$y_s = Y_{st}(t) \qquad T_s = T_f \tag{6.217}$$

and an energy balance. The energy balance considers heat into and out of the interface via conduction and heat generation due to the latent heat of fusion. If ΔH_f is the latent heat of fusion, then when the interface grows, an amount of heat, q_f, must be liberated:

$$q_f = \rho_l \Delta H_f A_c \frac{dY_s}{dt} = \rho_s \Delta H_f A_c \frac{dY_{st}}{dt} \tag{6.218}$$

The energy balance at the interface equates conduction into and out of the interface and generation at the interface:

$$\begin{aligned} &In \quad - \quad Out \quad + Gen = Acc \\ &-k_l \frac{\partial T_l}{\partial y_l} - (-k_s)\frac{\partial T_s}{\partial y_s} + \quad q_f \; = \; 0 \end{aligned} \tag{6.219}$$

A little physical intuition will help solve this problem in a rather convenient way. We know that the liquid is a semi-infinite medium, and so our solution for the temperature profile there should be in the form of the error function. The solid *penetrates* into the liquid and should tend to grow according to the penetration depth of the interface (i.e., penetration of the line $T = T_f$). Therefore, the temperature distribution in the solid can be represented as an error function solution as well. With this in mind, we write

$$\frac{T_l - T_o}{T_f - T_o} = a_1 erf\left(\frac{y_l}{\sqrt{4\alpha_l t}}\right) + a_2 \tag{6.220}$$

$$\frac{T_s - T_\infty}{T_f - T_\infty} = b_1 erf\left(\frac{y_s}{\sqrt{4\alpha_s t}}\right) + b_2 \tag{6.221}$$

Evaluating the boundary conditions leads to four algebraic equations for the integration constants:

$$a_1 + a_2 = 0 \qquad \text{from (6.215)} \tag{6.222}$$

$$b_2 = 0 \qquad \text{from (6.214)} \tag{6.223}$$

$$a_1 erf\left(\frac{Y_s}{\sqrt{4\alpha_l t}}\right) + a_2 = 1 \qquad \text{from (6.216)} \tag{6.224}$$

$$b_1 erf\left(\frac{Y_{st}}{\sqrt{4\alpha_s t}}\right) + b_2 = 1 \qquad \text{from (6.217)} \qquad (6.225)$$

$$-\frac{k_l}{\sqrt{\pi\alpha_l t}}\exp\left(-\frac{Y_s^2}{4\alpha_l t}\right)a_1\left(T_f - T_o\right) + \frac{k_s}{\sqrt{\pi\alpha_s t}}\exp\left(-\frac{Y_{st}^2}{4\alpha_s t}\right)b_1\left(T_f - T_\infty\right)$$

$$+\rho_s \Delta H_f \frac{dY_{st}}{dt} = 0 \qquad \text{from (6.219)} \qquad (6.226)$$

To finally determine the integration constants, we need to know, approximately, how Y_s and Y_{st} grow with time. Since the solid penetrates into the liquid, the solid should move into the liquid as our reaction front moved into the liquid in the last problem. By analogy we presume

$$Y_s = \sqrt{4Kt} \qquad (6.227)$$

and using Equation 6.211, we must have

$$Y_{st} = \varpi\sqrt{4Kt} \qquad (6.228)$$

for the solid. Equations 6.227 and 6.228 are substituted into the boundary condition equations. We can now solve for a_1, a_2, and b_1 in terms of K. This leaves us with one implicit equation for K, similar to the implicit equation we derived in the last example for γ:

$$a_1 = \frac{1}{erf\left(\sqrt{K/\alpha_l}\right) - 1} \qquad a_2 = -\frac{1}{erf\left(\sqrt{K/\alpha_l}\right) - 1} \qquad (6.229)$$

$$b_1 = \frac{1}{erf\left(\varpi\sqrt{K/\alpha_s}\right)} \qquad (6.230)$$

$$-\frac{\left(k_l/\sqrt{\pi\alpha_l}\right)\exp\left(-K/\alpha_l\right)}{erf\left(\sqrt{K/\alpha_l}\right) - 1}\left(T_f - T_o\right) + \frac{\left(k_s/\sqrt{\pi\alpha_s}\right)\exp\left(-\varpi\dfrac{K}{\alpha_s}\right)}{erf\left(\varpi\sqrt{K/\alpha_s}\right) - 1}\left(T_f - T_\infty\right) + \rho_s \varpi \Delta H_f \sqrt{K} = 0$$

$$(6.231)$$

The final solution is given implicitly in terms of K:

$$\frac{T_l - T_o}{T_f - T_o} = \frac{erfc\left(y_l/\sqrt{4\alpha_l t}\right)}{erfc\left(\sqrt{K/\alpha_l}\right)} \qquad Y_s \le y_l \le \infty \qquad (6.232)$$

$$\frac{T_s - T_\infty}{T_f - T_\infty} = \frac{erf\left(y_s/\sqrt{4\alpha_s t}\right)}{erf\left(\varpi\sqrt{K/\alpha_s}\right)} \qquad 0 \le y_s \le Y_{st} \qquad (6.233)$$

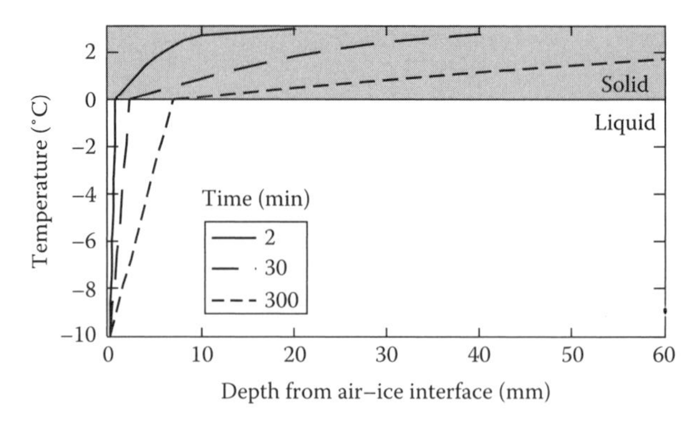

FIGURE 6.32 Freezing of water on the surface of a lake or pond. Ice penetrates into the water, simultaneously expanding by 9% as it transforms from liquid to solid.

Using water as an example, we have the following physical properties. Notice that the solid phase is less dense than the liquid phase, a peculiar characteristic of water that allows life to exist when water is below its freezing point:

$\alpha_l = 1.349 \times 10^{-7}$ m²/s	$\alpha_s = 1.002 \times 10^{-6}$ m²/s
$\varpi = 1.087$	$\Delta H_f = 333.8$ kJ/kg
$k_l = 0.569$ W/m K	$k_s = 1.88$ W/m K
$\rho_l = 1000$ kg/m³	$\rho_s = 920$ kg/m³

The other parameters of our problem necessary for a solution are the various temperatures. We will assume 1 atmosphere pressure so that $T_f = 0°C$. Furthermore, we will assume the air temperature, $T_\infty = -5°C$ and the initial water temperature to be just above freezing at $T_o = 2°C$. With these values, we can solve for $K \sim 4.8 \times 10^{-10}$ m²/s. The temperature profiles are plotted in Figure 6.32.

6.5.4 Traveling Waves and Front Propagation

We first encountered convective mass transfer when we considered drift of charge carriers in Chapter 5. The marriage of generation and accumulation allows for a kind of convection that we will explore more fully in Chapter 7 but introduce, in a way, here. If we look at the growth of a population, like a bacterial colony in a petri dish and plot the size of the colony as a function of time, we eventually see that after a time, the colony grows with a characteristic velocity. This occurs even though there is no true convection; the cells are not flowing. The behavior of the colony boundary has historically been referred to as a traveling wave, but it is really a front that moves with some characteristic velocity rather than a true wave with a characteristic wavelength.

Let's look at the particular situation shown in Figure 6.33. Assume we have a tube filled with growth medium of cross-sectional area, A_c, and radius, r_0, that is much longer than it is wide. At the center, we inoculate it with a strain of non-motile bacteria. Within a short period of time, the bacteria grow in all directions until they span the radial dimension of the tube. Thereafter, they can only spread down the axis of the tube and appear to do so at a constant velocity. We assume the bacteria

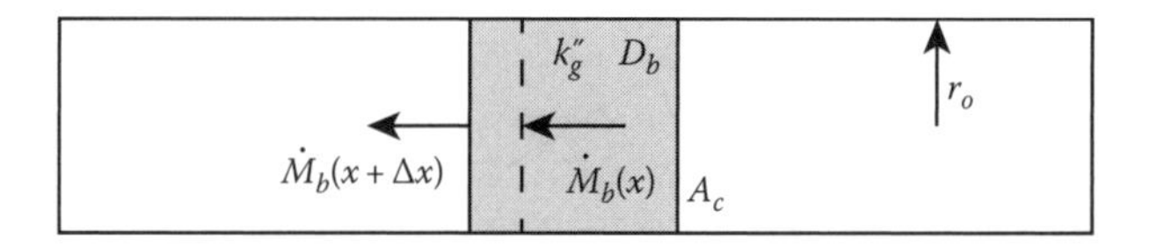

FIGURE 6.33 Bacteria growing and diffusing in a tube.

are moving via diffusion only, since they are non-motile, and if there is enough food and space, they grow exponentially. Exponential growth is equivalent to a first-order reaction. If we take a small slice in the axial direction, we can write a balance equation for the bacterial population:

$$\begin{array}{ccccccc} In & - & Out & + & Gen & = & Acc \end{array}$$

$$\dot{M}_b(x) - \dot{M}_b(x+\Delta x) + k_g'' A_c \Delta x c_b = A_c \Delta x \frac{dc_b}{dt} \tag{6.234}$$

where

- $\dot{M}_b$ is the flow of bacteria in #cells/s
- c_b is the bacterial cell population density in #cells/m^3
- k_g'' is the growth rate constant

Since the cells spread via diffusion, we can express the flux of cells, N_b, in terms of Fick's law. Generally, the cell densities are not large relative to molecular concentrations and so we can use a dilute solution approximation. Dividing Equation 6.234 by Δx and taking the limit as $\Delta x \to \infty$ yields

$$-\frac{\partial \dot{M}_b}{\partial x} + k_g'' A_c c_b = A_c \frac{\partial c_b}{\partial t} \tag{6.235}$$

If we insert the version of Fick's law in for $\dot{M}_b$, we find

$$-\frac{\partial}{\partial x}(N_b A_c) + k_g'' A_c c_b = \frac{\partial}{\partial x}\left(D_b A_c \frac{\partial c_b}{\partial x} \right) + k_g'' A_c c_b = A_c \frac{\partial c_b}{\partial t} \tag{6.236}$$

$$D_b \frac{\partial^2 c_b}{\partial x^2} + k_g'' c_b = \frac{\partial c_b}{\partial t} \tag{6.237}$$

To understand how convection might evolve, we first put the equation in dimensionless form. Defining the variables

$$\xi = \frac{x}{r_0} \qquad \chi = \frac{c_b}{c_{b0}} \qquad \tau = \frac{D_b t}{r_0^2} \tag{6.238}$$

leads to the differential equation

$$\frac{\partial^2 \chi}{\partial \xi^2} + \underbrace{\left(\frac{k_g'' r_0^2}{D_b} \right)}_{\text{Da}} \chi = \frac{\partial \chi}{\partial \tau} \tag{6.239}$$

This equation may allow for the propagation of traveling waves moving with a velocity, v. If so, we can define a new variable, $\varepsilon = \xi - v\tau$, that the equation will adhere to. Substituting into our differential equation transforms it to

$$\frac{d^2 \chi}{d\varepsilon^2} + \underbrace{v \frac{d\chi}{d\varepsilon}}_{Convection} + \text{Da}\chi = 0 \tag{6.240}$$

Now we can clearly see how this has transformed our equation from a time-dependent partial differential equation to a steady, second-order, ordinary differential equation that has a convective component. What we don't know is the velocity, but we can solve the equation and get a bound on what that velocity might be. For a linear, second-order equation like (6.240), the solution is of the form

$$\chi = Ae^{r_+\varepsilon} + Be^{r_-\varepsilon} \tag{6.241}$$

where r_+ and r_- are the roots of

$$r^2 + vr + \mathrm{Da} \qquad r_\pm = \frac{-v \pm \sqrt{v^2 - 4\mathrm{Da}}}{2} \tag{6.242}$$

While the magnitude of v may be very large, indicating motion to the right or left, there is a minimum velocity that v can assume and still yield a physically realistic answer. If $|v| < \sqrt{4\mathrm{Da}}$, then $r_\pm$ is imaginary and the solution for the bacterial concentration loses physical significance. This represents the minimum velocity and a good ballpark estimate of the true velocity.

Our original differential equation, (6.237), is also a bit unrealistic since it assumes that populations grow exponentially while they spread. If we were to look at the point $x = 0$, the cell density would increase over time without bound. Real populations do not behave that way. They are limited by outside forces such as food or space availability, and thus, any given environment has a carrying capacity or maximum number of individuals it can support. We can modify our differential equation to account for this, but in the process, it becomes nonlinear and so we must solve it numerically. To include the carrying capacity, we need to replace our exponential growth model with what is called a *logistic* growth model. The logistic growth model assumes that individuals compete to naturally limit the size of their populations. The new differential equation becomes

$$D_b \frac{\partial^2 c_b}{\partial x^2} + k_g''\left(c_b - \frac{c_b^2}{K}\right) = \frac{\partial c_b}{\partial t} \tag{6.243}$$

If we solve this model, we see that the population grows exponentially from a small number as it spreads. Once it reaches the carrying capacity, the population moves as a front, continuously invading new space. This is shown for a test case in Figure 6.34. If we plot the speed at which the midpoint of the population density moves (mean cell density), we find that it eventually approaches a constant velocity that we define as v (Figure 6.35).

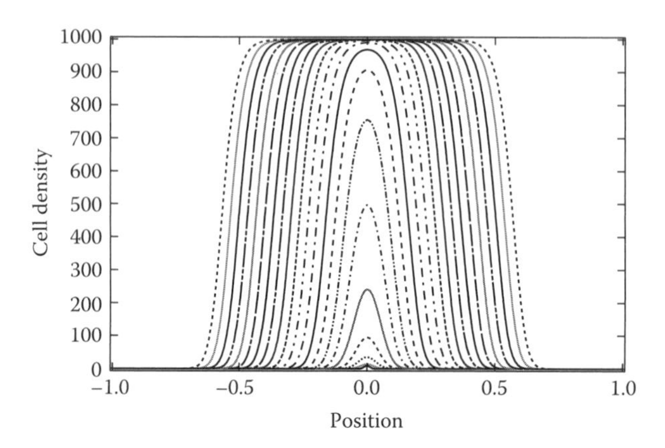

FIGURE 6.34 Spread of a population seeded at $x = t = 0$. The population grows exponentially until it reaches its carrying capacity. Once that capacity is reached, the population spreads at a constant rate.

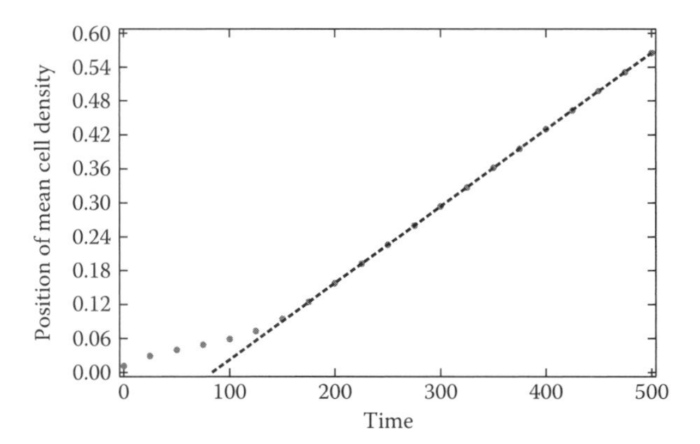

FIGURE 6.35 Position of the mean cell density as a function of time. The slope of the plot gives the spreading velocity of the population front.

6.5.5 Kirkendall Effect

In the past few sections, we have considered the appearance of a front that forms from reaction or solidification and how that front moves through a material. We can also look at a similar kind of process in the solid state that is important in materials processing. For example, if we join a bar of copper to a bar of brass and then heat the system, the copper and zinc will interdiffuse to form an interfacial region. One might think that the interfacial region would grow in both directions equally so that it would be symmetric about the original joint. However, that is not what is observed experimentally. In a classic experiment, Smigelskas and Kirkendall [10] joined two systems, one a region of pure copper and the other a region of brass that was 70% copper and 30% zinc. They inserted a molybdenum tracer at the joint. As the materials interdiffused, if the fronts penetrated equally into both materials, the molybdenum tracer would have remained fixed at the original joint location. However, it became clear that the front moved further into the brass region as shown in Figure 6.36. The diffusion process and movement of the species occurs via an interstitial process whereby the diffusing species moves through the space between atoms in the host lattice (see Chapter 3). If two species are moving and one happens to diffuse faster than the other, vacant space can be left behind. This process can be exploited to form novel nanostructured materials [11].

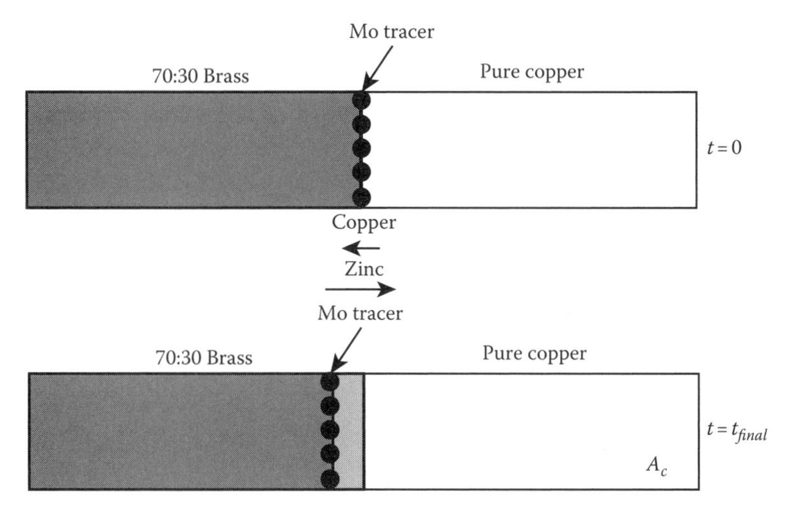

FIGURE 6.36 Schematic of Kirkendall's original experiment. Difference in rates of diffusion causes the molybdenum tracer to move further into the brass region.

The original copper/brass experiment can be approached in the same way we handled the absorption/reaction or the freezing problem. If we take a small slice from within the brass region, we can perform a mass balance there on copper and zinc. We know that there are only two species and that the total concentration will have to remain a constant, c_t. Since the system is a solid, we also have a fixed geometry:

$$\begin{aligned} &In \quad - \quad Out \quad + Gen = \quad Acc \\ &\dot{M}_{Cu}(y) - \dot{M}_{Cu}(y+\Delta y) + \quad 0 \quad = A_c \Delta y c_t \frac{\partial x_{Cu}}{\partial t} \end{aligned} \tag{6.244}$$

$$\begin{aligned} &In \quad - \quad Out \quad + Gen = \quad Acc \\ &\dot{M}_{Zn}(y) - \dot{M}_{Zn}(y+\Delta y) + \quad 0 \quad = A_c \Delta y c_t \frac{\partial x_{Zn}}{\partial t} \end{aligned} \tag{6.245}$$

Dividing through by $A_c\Delta y$ and taking the limit as $\Delta y \to 0$ gives

$$-\frac{\partial N_{Cu}}{\partial y} = c_t \frac{\partial x_{Cu}}{\partial t} \tag{6.246}$$

$$-\frac{\partial N_{Zn}}{\partial y} = c_t \frac{\partial x_{Zn}}{\partial t} \tag{6.247}$$

The fluxes of copper and zinc are different in both magnitude and direction and so are given by

$$N_{Cu} = -c_t D_{Cu} \frac{\partial x_{Cu}}{\partial y} + x_{Cu}(N_{Cu} + N_{Zn}) \tag{6.248}$$

$$N_{Zn} = -c_t D_{Zn} \frac{\partial x_{Zn}}{\partial y} + x_{Zn}(N_{Cu} + N_{Zn}) \tag{6.249}$$

Using the definitions from Chapter 3, we can show that $N_{Cu} + N_{Zn} = c_t v_M$. Substituting into Equations 6.248 and 6.249 and then into the differential equations (6.246) and (6.247) we find

$$\frac{\partial}{\partial y}\left(D_{Cu} \frac{\partial x_{Cu}}{\partial y} \right) - \frac{\partial}{\partial y}(x_{Cu} v_M) = \frac{\partial x_{Cu}}{\partial t} \tag{6.250}$$

$$\frac{\partial}{\partial y}\left(D_{Zn} \frac{\partial x_{Zn}}{\partial y} \right) - \frac{\partial}{\partial y}(x_{Zn} v_M) = \frac{\partial x_{Zn}}{\partial t} \tag{6.251}$$

Adding Equations 6.250 and 6.251 and realizing that $x_{Cu} + x_{Zn} = 1$ allows us to determine the molar average velocity:

$$v_M = D_{Cu} \frac{\partial x_{Cu}}{\partial y} + D_{Zn} \frac{\partial x_{Zn}}{\partial y} = (D_{Cu} - D_{Zn}) \frac{\partial x_{Cu}}{\partial y} \tag{6.252}$$

Now we can substitute Equation 6.252 into Equation 6.250 and combine terms to give

$$\frac{\partial}{\partial y}\left(D_{Cu}\frac{\partial x_{Cu}}{\partial y}\right)-\frac{\partial}{\partial y}\left(x_{Cu}\left(D_{Cu}-D_{Zn}\right)\frac{\partial x_{Cu}}{\partial y}\right)=\frac{\partial}{\partial y}\left[\left(x_{Zn}D_{Cu}+x_{Cu}D_{Zn}\right)\frac{\partial x_{Cu}}{\partial y}\right]=\frac{\partial x_{Cu}}{\partial t} \quad (6.253)$$

The result of these operations is to show that we have an effective diffusivity defined by

$$D_{eff}=x_{Cu}D_{Zn}+x_{Zn}D_{Cu} \quad (6.254)$$

Equations 6.252 and 6.254 are referred to as Darken's equations and can be solved simultaneously to yield the individual diffusion coefficients from experimental data. Equation 6.254 assumes we have an ideal solution. As we showed in Chapter 3, if the solution is non-ideal, Equation 6.254 becomes a function of the activity of the solute or

$$D_{eff}=[x_{Cu}D_{Zn}+x_{Zn}D_{Cu}]\left[1+\ln\left(\frac{\gamma_{Cu}}{x_{Cu}}\right)\right] \quad (6.255)$$

The differential equation for the copper or zinc concentration profile is too difficult to solve analytically, but if we assume that D_{eff} is a constant, we can obtain a reasonable approximation:

$$D_{eff}\frac{\partial^2 x_{Cu}}{\partial y^2}=\frac{\partial x_{Cu}}{\partial t} \quad (6.256)$$

The boundary conditions for this problem assume that far from the interface in the copper region, the concentration is constant for all time. At $t = 0$, we also know the composition. Finally, we know that since we are assuming that D_{eff} and c_t are constants, the copper profile will have to be symmetric about an interfacial composition defined by

$$x_{Cui}=\frac{x_{Cu1}+x_{Cu2}}{2} \quad (6.257)$$

The boundary conditions are

$$x_{Cu}(y,0)=x_{Cu2}$$

$$x_{Cu}(\infty,t)=x_{Cu2}$$

$$x_{Cu}(0,t)=x_{Cui} \quad (6.258)$$

We can recast our equation in terms of a similarity variable and scaled mole fraction profile using,

$$\chi=\frac{x_{Cu2}-x_{Cu}}{x_{Cu2}-x_{Cui}} \qquad \varepsilon=\frac{y}{\sqrt{4D_{eff}t}} \quad (6.259)$$

show that the solution is,

$$\chi=a_0+a_1 erf(\varepsilon) \quad =1-erf(\varepsilon) \quad (6.260)$$

and, due to the way it is defined, applies across the entire domain. The molar velocity can now be calculated to be

$$v_M = (D_{Cu} - D_{Zn})\frac{\partial x_{Cu}}{\partial y} = -(D_{Cu} - D_{Zn})\left[\frac{x_{Cu2} - x_{Cui}}{\sqrt{4D_{eff}t}}\right]\frac{d\chi}{d\varepsilon}$$

$$= -(D_{Cu} - D_{Zn})\left[\frac{x_{Cu2} - x_{Cui}}{\sqrt{4D_{eff}t}}\right]\exp\left[-\frac{y^2}{4D_{eff}t}\right] \quad (6.261)$$

6.6 PERIODIC FLOW IN A ROTATING CYLINDRICAL SYSTEM

Prior to this example, we considered transient systems where the temperature, velocity, concentration, etc., profile decayed over time to reach some steady-state or equilibrium value. Though most transient situations conform to this view, there are numerous instances where we have periodic variations in time. Examples include diurnal fluctuations in atmospheric temperature, tidal flows, blood flow, mass transfer in the lungs, and charge transport in a neuron or transistor. We would like to consider one such periodic example: periodic flow of a rotating cylinder. In particular, we are interested in how the velocity profile varies as a function of the rotation frequency. This problem has application, for instance, in calculating the velocity profile in a washing machine.

The physical situation is shown in Figure 6.37. Here, we have two smooth, concentric cylinders. The inner cylinder rotates with an angular velocity, Ω_v (rev/s), and this velocity changes with time from clockwise to counterclockwise according to $\Omega_v \cos \omega t$. The outer cylinder is stationary. Both cylinders are quite large compared to the gap between them. The fluid between the cylinders is Newtonian, and we will stipulate that the flows are small enough so that bulk convection or turbulence is negligible.

We begin our analysis by performing a momentum balance on the control volume shown in Figure 6.37. Remember we write the balance in terms of the rate of change of momentum or forces acting on the control volume. Here, we will say we have no generation and that all material properties will be constant:

$$\begin{array}{ccccccccc} In & - & Out & + & Gen & = & Acc \\ F(r) & - & F(r+\Delta r) & + & 0 & = & (2\pi r\Delta r L\rho)\dfrac{\partial v_\theta}{\partial t} \end{array} \quad (6.262)$$

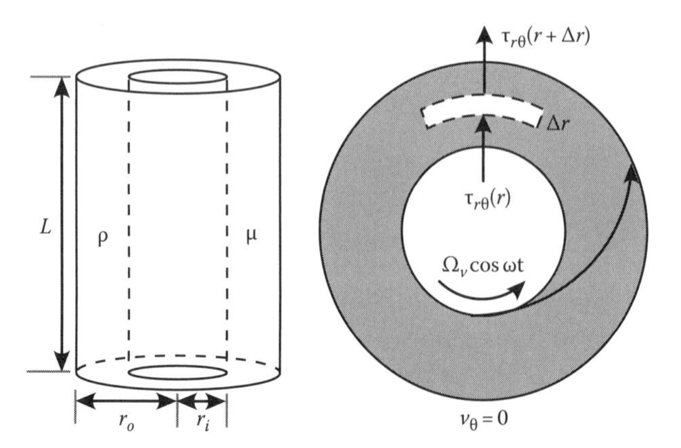

FIGURE 6.37 Flow occurring between cylinders, one of which undergoes periodic reversals in its direction of rotation.

Expressing the force as $F(r) = 2\pi rL\tau_{r\theta}$, dividing through by Δr, and taking the limit as $\Delta r \to 0$ gives

$$-(2\pi L)\frac{\partial}{\partial r}(r\tau_{r\theta}) = (2\pi rL\rho)\frac{\partial v_\theta}{\partial t} \tag{6.263}$$

Substituting for $\tau_{r\theta}$ using Newton's law yields the differential equation for the velocity profile:

$$\frac{\nu}{r}\frac{\partial}{\partial r}\left(r\frac{\partial v_\theta}{\partial r}\right) = \frac{\partial v_\theta}{\partial t} \tag{6.264}$$

We know that the inner cylinder will have a periodic velocity and we are interested in periodic solutions for v_θ. Since we know the time dependence of v_θ, we can substitute that into the differential equation. Instead of using the cosine function for v_θ, we substitute a complex representation into the differential equation:

$$v_\theta = v(r)\exp(i\omega t) = v(r)[\cos(\omega t) + i\sin(\omega t)] \tag{6.265}$$

This makes the formulation of the problem much simpler and provides for a more general solution to the problem:

$$\frac{\nu}{r}\frac{\partial}{\partial r}\left(r\frac{\partial v}{\partial r}\right) = i\omega t \tag{6.266}$$

Equation 6.266 is now the differential equation governing $v(r)$. Before going any further with this problem, it is helpful to place everything in dimensionless form. We define two dimensionless variables using the radius of the inner cylinder and the linear velocity of the inner cylinder as the scales:

$$\xi = \frac{r}{r_i} \qquad u = \frac{v}{2\pi r_i \Omega_v} \tag{6.267}$$

Transforming Equation 6.266 gives

$$\frac{1}{\xi}\frac{d}{d\xi}\left(\xi\frac{du}{d\xi}\right) - i\left(\frac{\omega}{\nu/r_i^2}\right)u = 0 \tag{6.268}$$

We refer to ν/r_i^2 as the *characteristic or natural frequency of the system*, ω_o. Below this frequency, the fluid can respond to changes in the inner cylinder's rotation. Above this frequency, the fluid can no longer respond and changes in the inner cylinder's velocity are damped out by the fluid.

The solution to Equation 6.268 is given in terms of the Bessel functions, J and K. The arguments of these Bessel functions are complex numbers:

$$u = a_1 J_o\left(\sqrt{-\frac{i\omega}{\omega_o}}\,\xi\right) + a_2 K_o\left(\sqrt{-\frac{i\omega}{\omega_o}}\,\xi\right) \tag{6.269}$$

We can perform a number of algebraic manipulations of the J and K Bessel function series and split them into their corresponding real and imaginary parts. The resulting real and imaginary series for

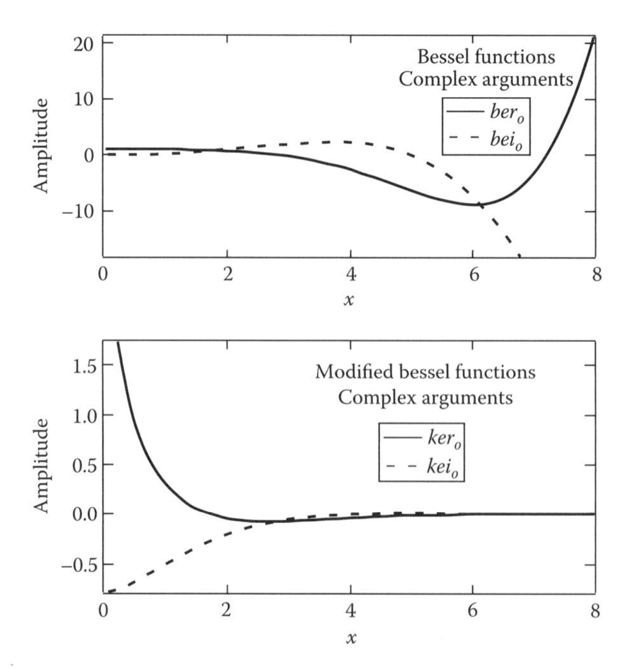

FIGURE 6.38 Bessel functions with complex arguments: ber_o, bei_o, ker_o, and kei_o.

the *J* Bessel function are called *ber* and *bei*, respectively, and for the *K* Bessel function, we have *ker* and *kei*. These functions are shown in Figure 6.38.

We can now write our solution in terms of these new Bessel functions:

$$u = b_1\left[ber_o\left(\frac{\omega}{\omega_o}\xi\right) + i\,bei_o\left(\frac{\omega}{\omega_o}\xi\right)\right] + b_2\left[ker_o\left(\frac{\omega}{\omega_o}\xi\right) + i\,kei_o\left(\frac{\omega}{\omega_o}\xi\right)\right] \tag{6.270}$$

We evaluate the integration constants using the boundary conditions:

$$\xi = 1 \qquad u = 1$$

$$1 = b_1\left[ber_o\left(\frac{\omega}{\omega_o}\right) + i\,bei_o\left(\frac{\omega}{\omega_o}\right)\right] + b_2\left[ker_o\left(\frac{\omega}{\omega_o}\right) + i\,kei_o\left(\frac{\omega}{\omega_o}\right)\right] \tag{6.271}$$

$$\xi = \frac{r_o}{r_i} \qquad u = 0$$

$$0 = b_1\left[ber_o\left(\frac{\omega}{\omega_o}\frac{r_o}{r_i}\right) + i\,bei_o\left(\frac{\omega}{\omega_o}\frac{r_o}{r_i}\right)\right] + b_2\left[ker_o\left(\frac{\omega}{\omega_o}\frac{r_o}{r_i}\right) + i\,kei_o\left(\frac{\omega}{\omega_o}\frac{r_o}{r_i}\right)\right] \tag{6.272}$$

Since the symbolic solution for b_1 and b_2 is very complicated, we will illustrate the solution by looking at a particular case using water as the working fluid. The properties of the cylindrical system and water are

$$r_i = 0.001\,\text{m} \qquad r_o = 0.002\,\text{m} \qquad \nu = 8.53\times10^{-7}\,\text{m}^2/\text{s}$$

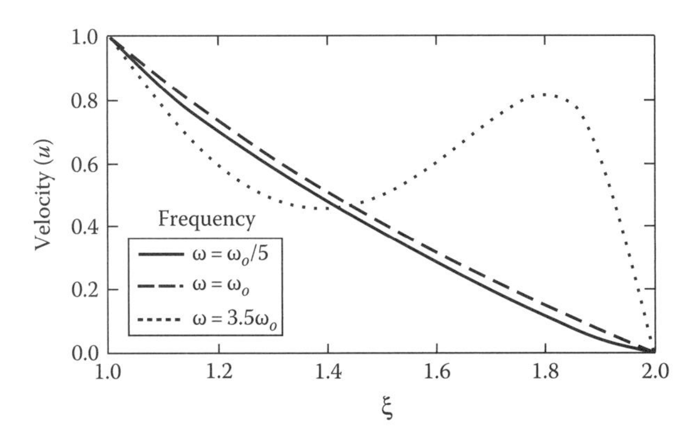

FIGURE 6.39 Velocity profile between two cylinders as a function of inner cylinder frequency.

Thus, the characteristic frequency for the system is $\omega_o = 8.53 \times 10^{-1}$/s. The fluid cannot respond to changes faster than this, and any attempt to force it to do so will result in dampening of the signal and eventually periodic and finally turbulent flow.

Figure 6.39 shows the velocity profile between the cylinders as a function of time for three frequencies, $\omega = \omega_o/5$, $\omega = \omega_o$, $\omega = 3.5\,\omega_o$. Notice that for $\omega \leq \omega_o$, the fluid velocity profile responds well and in-phase with the motion of the inner cylinder. This is evidenced by the smooth decrease in velocity from the inner to the outer cylinder. By contrast, when $\omega > \omega_o$, the fluid cannot keep up with the changing motion of the inner cylinder, and therefore, parts of the fluid toward the outer cylinder are still responding to past changes. In effect, the fluid is responding as a low-pass filter. Only signals with low frequency are transmitted through the fluid toward the outer cylinder. Higher frequencies are attenuated. You can perform an experiment similar to this type by stirring a glass of water clockwise then counterclockwise. The faster you switch directions, the more violent the behavior of the fluid. That is because you have tried to get the fluid to respond faster than its characteristic frequency will allow.

6.6.1 Pulsatile Flow

The periodic solution technique used in the previous discussion can also be used to look at pulsatile flow, such as occurs in an artery. The system is shown in Figure 6.40. We assume a tube of length, L, and radius, r_o. Blood is flowing through the tube under the action of an oscillating pressure. The pressure function can be assumed to vary according to

$$P = P_o[1 + \cos(\omega t)] \tag{6.273}$$

The Comsol® module "PulsatileBlood" simulates this situation. Two versions are considered. The first assumes a Newtonian fluid. The solution to this problem can actually be obtained analytically. The second version of the problem looks at a system that cannot be solved analytically, that of a non-Newtonian fluid whose viscosity varies as

$$\mu = \mu_\infty + (\mu_o - \mu_\infty)\left(1 + l_{da}\left|\frac{dv_x}{dr}\right|^a\right)^n \tag{6.274}$$

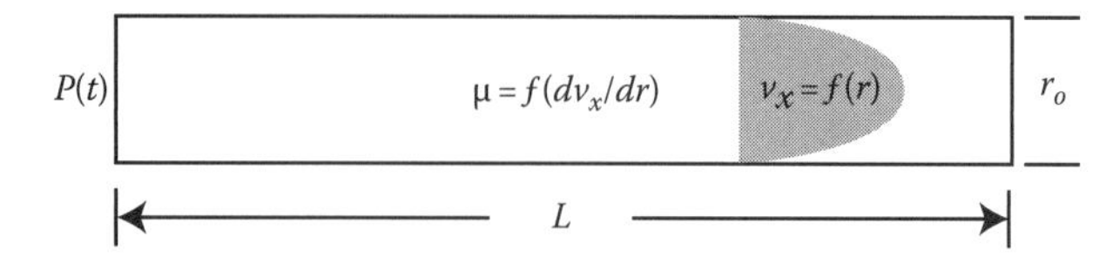

FIGURE 6.40 Schematic of pulsatile flow in an artery.

If the flow is slow enough and the frequency of oscillation is also slow, then the velocity profile at any time will be parabolic. The flow exiting the tube will then be the primary variable of interest. We expect that if the oscillation frequency is very slow, there should be little difference between the time-dependent solution and the steady-state solution. As the oscillation frequency increases, the fluid should not be able to follow the pressure variations and the result will be a lower flow rate from the tube. Figures 6.41 through 6.43 show results of the simulations. Figure 6.41 shows results for the non-Newtonian fluid at two frequencies. Indeed as the frequency increases, the maximum flow rate decreases, though at these frequencies the decay is slow. Figure 6.42 shows the flow rate and pressure fluctuations for the non-Newtonian fluid. Notice that there is a lag between the pressure and flow. As the frequency gets higher, this lag will grow since the fluid will not be able to respond to the pressure fluctuation. How fast can the pressure fluctuation travel through the fluid? What would happen if the tube was very long and how would this model have to be modified? Finally,

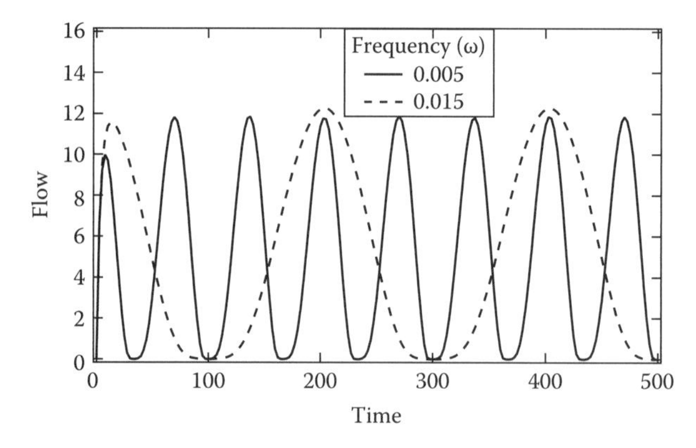

FIGURE 6.41 Flow of the non-Newtonian fluid (blood) at two different frequencies. Higher frequencies lead to a decrease in the maximum flow rate.

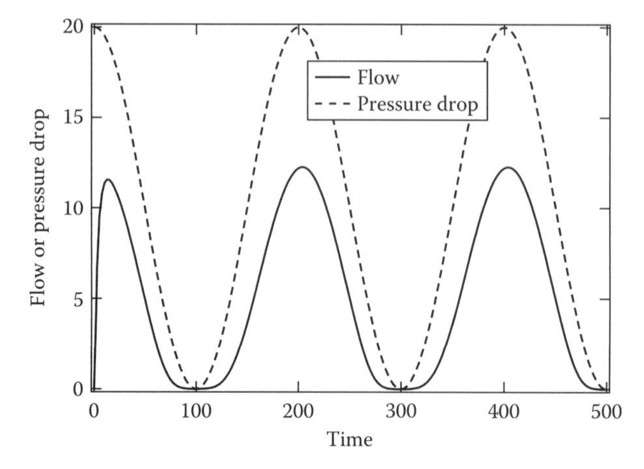

FIGURE 6.42 Flow and pressure drop for the blood flow. The flow lags behind the change in the pressure as the fluid cannot keep pace with the fluctuations.

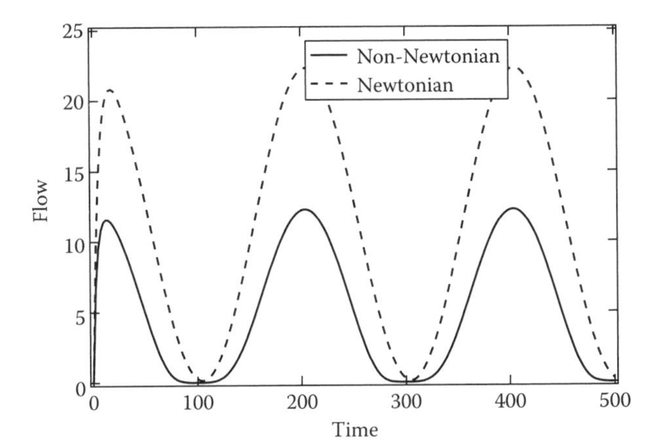

FIGURE 6.43 Flow of a Newtonian and non-Newtonian version of blood. The Newtonian flow rate is higher than the non-Newtonian due to the changes in viscosity of the latter fluid. Notice that the profile at low pressures is also quite different due to the magnitude of the viscosity at low shear rates.

Figure 6.43 plots the flow for the non-Newtonian and Newtonian fluids. Notice the difference in magnitude of the flows. Even more noticeable is the difference in shape of the flow versus time curve. Since the viscosity of the non-Newtonian fluid changes at low values of the shear stress, when the pressure is a minimum, we have a pronounced flattening of the profile.

6.7 SUMMARY

We have discussed transient, diffusive transport and have illustrated the resulting complexities involved in increasing the sophistication of our analysis. Most transient systems you will come in contact with will involve some form of convection on the boundary. As such, the first question to consider in any transient system is whether the lumped capacitance approximation is a valid mechanism for attacking the problem. If the lumped capacitance method fails, our next recourse is the general solutions handled in Table 6.4 and Appendix C. Only when these two options fail do we truly need to solve the problem in all its complexity. Keep in mind the similarities between the transport phenomena and how those similarities are exploited to generate solutions to a wide variety of problems.

The next chapter will focus on the last relationship between elements of our balance equation. We will be considering balances where the primary relationship is one where *Gen* = *Acc*. This relationship is the hallmark of conservative systems, and so we will move from dissipative systems where fluctuations in our state variable are damped out to conservative systems where disturbances propagate as waves throughout the medium.

PROBLEMS

LUMPED CAPACITANCE

6.1 Consider the operation of the *lava* lamp. Polymer is heated at the base and due to buoyant forces, begins to rise through a colored liquid in the form of *small* spheres (r_o = 1.5 cm). Assume that the colored liquid has a density of $\rho_o = 910$ kg/m³ and a uniform temperature of $T_\infty = 325$ K. The polymer sphere has a temperature-dependent density given by

$$\rho = \rho_o - \beta(T - T_\infty) \qquad \rho_o = 910\,\text{kg/m}^3 \qquad \beta = 2\,\text{kg/m}^3\,\text{K}$$

If the polymer liquid begins its ascent at a temperature of 350 K, how long before it begins to fall back to be reheated? You may assume

$$h = 20\,\text{W/m}^2\,\text{K} \qquad C_p(\text{sphere}) = 400\,\text{J/kg K} \qquad k(\text{sphere}) = 10\,\text{W/m K}$$

6.2 A polymer manufacturer is looking to cut cost. An engineer has the idea of changing the geometry of their "pellets" from spherical to cubes while keeping the "pellet" volume the same. This would allow them to cool the pellet off faster and increase throughput. If

$$\rho = 800\frac{\text{kg}}{\text{m}^3} \qquad C_p = 530\frac{\text{J}}{\text{kg K}} \qquad h = 10\frac{\text{W}}{\text{m}^2\,\text{K}}$$

$$k = 5\frac{\text{W}}{\text{m K}} \qquad V = 1\times 10^{-5}\,\text{m}^3$$

is the engineer's idea a good one?

6.3 A somewhat sadistic yet normal child, James decides to pull out his magnifying glass and torture an ant. The ignition temperature of an ant is roughly 300°C. James' low-power magnifying glass can deliver a heat flux of 50 W/cm^2. The surrounding breeze provides for a heat transfer coefficient of 50 W/m^2 K and the ambient air temperature is 20°C. If we can describe an ant as a cylinder, 5 mm long and 1 mm in diameter with the properties of hair (α = 0.14 mm^2/s, k = 0.15 W/m K), how long must the ant remain in the crosshairs for James to be successful?

6.4 The lumped capacitance formulation can be used when a system is being cooled by thermal radiation too. Consider a spherical object of radius 1 m in low Earth orbit at an initial temperature of T_i = 600 K being exposed to the vacuum of space where the temperature is effectively T_∞ = 100 K. The emissivity of the object is ε = 0.95, its thermal conductivity is k = 350 W/m K, its density is 2000 kg/m^3, and its heat capacity is 500 J/kg K:

a. Derive the differential equation governing the temperature of the object assuming that the lumped capacitance formulation is valid.

b. Unlike convection, there is no clean Biot number that falls out of the equation. However, we can define one using the radiation heat transfer coefficient

$$h_{rad} = \sigma^r \varepsilon^r \left(\overline{T}^2 + T_\infty^2\right)\left(\overline{T} + T_\infty\right)$$

At the initial phase of operation, is the lumped capacitance formulation valid? At what initial object temperature would the lumped capacitance formulation become feasible?

c. Assuming the lumped capacitance formulation is valid over the entire range of cooling, how long would it take to cool the object from 500 to 300 K?

6.5 The makers of rapid beverage coolers, like the Cooper Cooler™, claim that their device can cool a bottle of wine from 25°C to 4°C in 6 min and a soda or beer can in 1 min. The cooler uses an ice–water mixture at 0°C. Assuming that the lumped capacitance formulation is valid, evaluate the manufacturer's claim by providing an estimate for what the heat transfer coefficient would have to be.

6.6 A sugar cube (L_o = 1 cm) is dissolving in a cup of tea. The mass transfer coefficient, k_c, to the tea is 5.0 × 10^{-5} m/s. You may assume lumped capacitance is valid. The molecular weight of sugar is 342.3 and its density is 1580.6 kg/m^3. The concentration of sugar in the tea, c_∞, is effectively zero. How long will it take the cube to dissolve away?

6.7 A steel cube, 2.0 cm on a side, is initially at a temperature of 300°C. The cube is dropped into a cooling bath (T_∞ = 25°C, h = 100 W/m^2°C) to reduce its temperature so that it can be handled:

$$k = 16\,\text{W/m K} \qquad \rho C_p = 3.636 \times 10^6\ \text{J/m}^3\,\text{K}$$

a. What is the temperature of the cube after it has been in the bath for 5 min?
b. How much heat has the cube lost by that time?
c. What is the heat flux from the surface of the cube at that time?
d. For the conditions specified in this problem, what is the maximum stainless steel cube size that can be analyzed by lumped capacitance?

6.8 The development of *safe* nuclear fuel pellets is critical to future reactor designs. One project the engineers have is to look into what happens when the reactor cooling system fails and the heat transfer mechanism changes from forced convective cooling of the pellets to free convective cooling (coefficient—h_{fc}). The reactor cooling fluid is essentially infinite in extent and remains at a uniform temperature of T_o. The pellet generates heat at a rate that varies with the pellet temperature:

$$q^* = q_o(T - T_o)\,\text{W/m}^3$$

$$\rho = 1500\ \text{kg/m}^3 \qquad k = 50\ \text{W/m K} \qquad C_p = 450\ \text{J/kg K}$$

$$h_{fc} = 55\ \text{W/m}^2\ \text{K} \qquad T_1 = 125°\text{C} \qquad T_o = 20°\text{C}$$

$$q_o = 55{,}000\ \text{W/m}^3\ \text{K} \qquad r_{pellet} = 4\ \text{mm}$$

a. Assuming the lumped capacitance approach is appropriate, derive the transient differential heat balance and solve the resulting differential equation (pellet temperature at $t = 0$ is T_1).
b. How would you define the Biot number for this situation?
c. From the following information, what will be the temperature of the pellet at t = 10 s following cooling system failure?
d. Using the information previously presented, how large must a pellet be before the lumped capacitance approach fails?

6.9 You are attempting to model the growth of bacteria and decrease in nutrient over time in a well-stirred flask of volume, V. You want to chart the total number of cells per unit volume (cell concentration), N, and the concentration of nutrient, c_n. An initial inoculate of cells, N_o, is charged to the flask. The initial nutrient concentration in the flask is c_{no}. You may assume the reproductive frequency of the bacteria, $k'' = k_o'' c_n$ is directly proportional to the concentration of nutrient in the flask. The reproductive rate, $r_p = k''N$, is a product of the reproductive frequency and the number of cells per unit volume. For every new cell that is formed, α units of nutrient are consumed $-r_n = \alpha r_p$ where r_n is the rate of nutrient consumption.

a. Using this information, derive the two differential equations describing the cell concentration and the concentration of nutrient.
b. Show that these two equations can be collapsed into a single equation for the cell concentration:

$$\frac{dN}{dt} = k_o'' N[c_{no} - \alpha(N - N_o)]$$

c. Solve the set of equations to determine the cell concentration and nutrient concentration as a function of time.
d. What about cell death?

6.10 Carbon steel can be converted into martensite by heating the steel above its eutectic temperature and then rapidly quenching it. In a particular experiment, a steel pellet, 1 cm in diameter at room temperature, was heated in a furnace at 800°C to its eutectoid temperature and then quenched in water at 25°C to a temperature of 200°C in 5.4 s to produce martensite. Calculate the eutectoid temperature and the time of furnace heating assuming the sphere started out at 25°C (Pisupatti and Kumar):

$$\rho_{steel} = 7500\ \text{kg/m}^3 \qquad C_{p,steel} = 500\ \text{J/kg K} \qquad k_{steel} = 46\ \text{W/m K}$$

$$h_{furnace} = 180\ \text{W/m}^2\ \text{s} \qquad h_{water} = 1000\ \text{W/m}^2\ \text{s}$$

6.11 A cylindrical roll of cloth was soaked in a dye mixture to color it. After the dye binds to the cloth, the cloth is soaked in a well-mixed tank of solvent to remove the unbound dye. The concentration of dye in the cloth after soaking is 1000 mol/m³. This concentration is to be reduced to 10% of its starting value. If the solvent initially contains no dye and the soaking process can take no more than 20 min, what volume of solvent is required? The cloth roll is 0.1 m long and 0.04 m in diameter. The diffusivity of dye in the cloth is $D_{dc} = 2 \times 10^{-7}$ m²/s and the mass transfer coefficient in the tank is $k_c = 2 \times 10^{-5}$ m/s. Assume the solvent is well mixed at all times (Enever, Lewandowski, and Taylor).

Internal Gradients

6.12 In preparation to attend a concert on a hot summer day, you straighten your hair to give yourself a new *do*. Unfortunately, it is very humid outside. Water vapor will penetrate your hair eventually causing it to *spring* back into its original shape. Springback occurs when the concentration of water vapor at the cuticle/cortex boundary exceeds 0.85 mol/m³. Using the following data and Figure P6.12, how long can you attend the concert before you *frizz out*?

$$c_{w\infty} = 1\ \text{mol/m}^3 \qquad c_{wo} = 0.01\ \text{mol/m}^3$$
$$k_c = 5.0 \times 10^{-9}\ \text{m/s} \qquad D_{wh} = 1.0 \times 10^{-14}\ \text{m}^2/\text{s}$$

6.13 As a master chef, chemical engineer, and employee of Pillsbury, you are devising a new devil's food cake mix. You would like to have some idea of how long to bake the thing, and your only guideline is that you know the center of the cake must reach a temperature of at least 125°C for 15 min. Of course you realize that the physical properties of the cake will change with time as it goes from batter to fully baked. Assuming you have the following physical

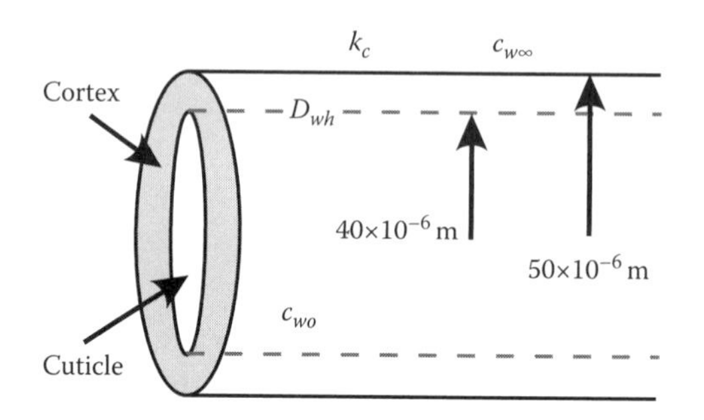

FIGURE P6.12

property data, the oven temperature is 175°C and the cake can be treated as a *horizontal slab*, how long (approximately) must you bake it?

Cake Temperature (K)	Density (kg/m³)	Thermal Conductivity (W/m K)	Heat Capacity (J/kg K)
300	720	0.223	3300
340	575	0.175	2750
370	400	0.150	2250
410	280	0.121	1750

You may assume a sheet cake 0.33 m (l) × 0.23 m (w) × 0.025 m (d) initially. The heat transfer coefficient in the oven is low, 5 W/m² K, and convective heat transfer occurs from both sides of the cake. How small a heat transfer coefficient would we need to solve the problem using lumped capacitance? Comment on how long it would take to bake the cake.

6.14 The thermal oxidation of a silicon wafer to form a thin film of protective oxide on the surface can be modeled as follows:

a. Oxygen from the atmosphere is transported to the surface of the wafer and diffuses through the thin oxide layer to reach the silicon substrate. The partition coefficient between the film and air is $K_{eq} = c_{film}/c_{air} = 10$.

b. Oxygen reacts with the silicon at the interface. Since the silicon is in great excess, the reaction can be modeled as first order in O_2 concentration.

Show that the thickness of the oxide at any time, t, can be represented by

$$h_{ox}^2 + a_1 h_{ox} = a_2 t \qquad a_1 = 2D_{og}\left(\frac{1}{k_s''} + \frac{1}{k_c}\right) \qquad a_2 = \frac{2D_{og} c_{og} M_{wg}}{\rho_g}$$

where

D_{og} is the oxygen diffusivity
ρ_g is the density of SiO_2
M_{wo} is the molecular weight of O_2
k_s'' is the rate constant (m/s)
k_c is the mass transfer coefficient (m/s)
c_{og} is the O_2 concentration
M_{wg} is the molecular weight of SiO_2

You may assume diffusion occurs at steady state and couple that with a transient mass balance about the oxide.

6.15 Consider the case of diffusion and reaction inside a slab. Initially, the concentration of solute in the slab is c_{ao}. The slab is made of a material that when heated catalyzes the reaction of a so that a decomposes with first-order kinetics and a reaction rate constant, k''. A heating fluid bathes the slab so that at the outer surfaces, the concentration of a is always 0. Determine the concentration profile within the slab as a function of time so that we will know when the reaction is essentially complete and we can remove the slab from the heating fluid (Figure P6.15).

6.16 Surgical implants are cylindrical pellets of powdered drug that are inserted into the body for controlled release of medication. These implants must be sterilized, and since they are water soluble, the drugs are sterilized using dry heat. To insure complete sterilization, the pellets must be exposed to a temperature of at least 140°C for 2 h. Consider a testosterone implant 1.0 in. long and 0.3 in. in diameter. The drug has a melting point of 152°C and so the oven

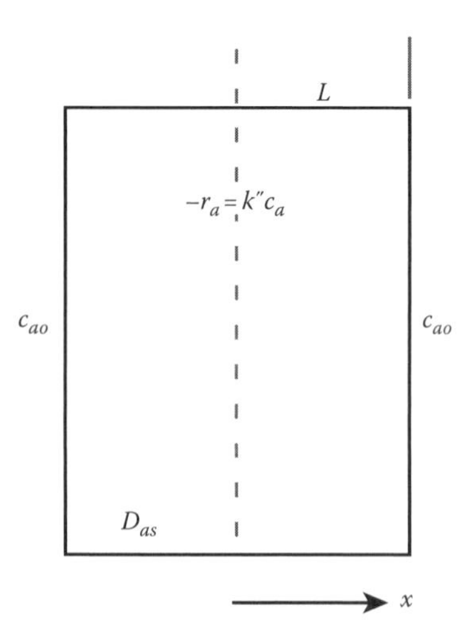

FIGURE P6.15

temperature is set at 145°C. How long will sterilization take? The pellet is initially at 25°C and its properties are

$$\rho = 0.9\,\mathrm{g/cm^3} \qquad k = 0.02\,\mathrm{W/m\,K} \qquad C_p = 2000\,\mathrm{J/kg\,K}$$

The heat transfer coefficient in the sterilizer is h = 10 W/m^2 K (Enever, Lewandowski, and Taylor).

6.17 Toothbrush manufacturers have been impregnating their bristles with dye to use as wear indicators. Each bristle is 0.1 mm in diameter and 12 mm long. Assuming the dye is lost by diffusion from the tip of the bristle into the mouth, that is, the side wall is impermeable and the bristle is *clear* when $c(t)/c_o < 0.01\ c_o$, perform the following:

a. Determine the concentration profile within the bristle.
b. Assuming continuous brushing, and a diffusion coefficient $D = 7 \times 10^{-9}$ m^2/s, how long before the toothbrush must be replaced (i.e., half the bristle is clear)?
c. Manufacturer's claims are that the toothbrush will last 3 months. Assuming the average human brushes his or her teeth twice a day for 3 min each time, will the toothbrush last that long (Dvorozniak, Medeiros, Miller, and Spilker)?

6.18 A porous cylinder absorbent was used to soak up a toxic material. This material is now to be destroyed by immersing the cylinder in a neutralizing bath (Figure P6.18). The toxic liquid diffuses from the cylinder and reacts with the neutralizer at the outer surface:

$$N_{toxin} + N_{eut} \rightarrow N_{nontoxic}$$

The neutralizer is present in great excess so the reaction is essentially first order in toxin concentration. We can assume dilute solutions within the absorbent as a first approximation. Determine the concentration profile of nerve gas in the membrane as a function of time.

6.19 Polishing operations involve rubbing two plates past one another with an intervening fluid. The fluid holds abrasive for the polishing and keeps the surfaces cool. One theory asserts that the degree of polishing is proportional to the shear stress at the surface of the object to be polished. Consider two such oscillating plates moving past one another. The bottom plate is stationary, while the top plate moves with a velocity given by

$$v = v_o \sin(\omega t + \phi)$$

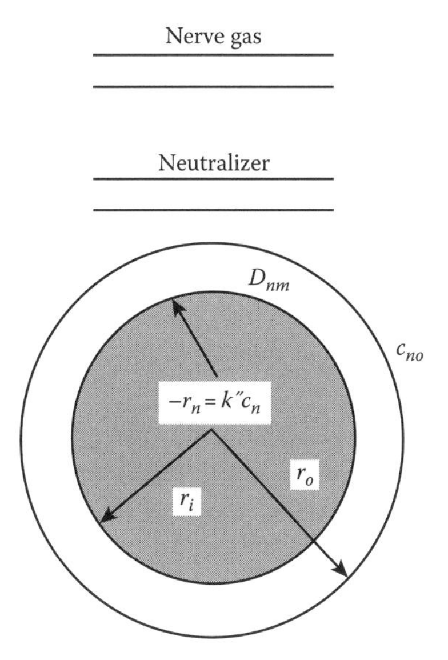

FIGURE P6.18

For the system shown in Figure P6.19, determine the velocity profile in the fluid as a function of time and the shear stress exerted on the lower plate.

6.20 A new method for polishing the inside of ferrules used in optical fiber connectors involves a wire that reciprocates through the ferrule opening. We would like to determine the velocity profile of the abrading fluid surrounding the wire. The geometry is shown in Figure P6.20. Assume that the wire/ferrule combination is so long that end effects on the flow are negligible. You may also neglect the fact that the wire spins about its axis.

Semi-Infinite Systems

6.21 Pyrethrin is an insecticide derived from the pyrethrum plant. Polymeric supports are impregnated with pyrethrin and the insecticide slowly diffuses through the polymer and into a room, for example (Shell's No-Pest Strip™). The lethal concentration for most insects is about 5×10^{-4} mg/m^3. A spherical pyrethrin delivery device is suspended from the ceiling of a room. The concentration of pyrethrin in the polymer is 1 wt% and may be assumed constant over time. The density of the delivery device is 0.96 g/cm^3 and its radius is 6 mm. The diffusivity of pyrethrin in air is 0.03 cm^2/s and its molecular weight is 328 g/mol. How long will it take for the lethal concentration to be reached at a point 3 m from the delivery device? Assume no interfacial resistance in getting the pyrethrin from the polymer and into the air (Enever, Lewandowsk, and Taylor).

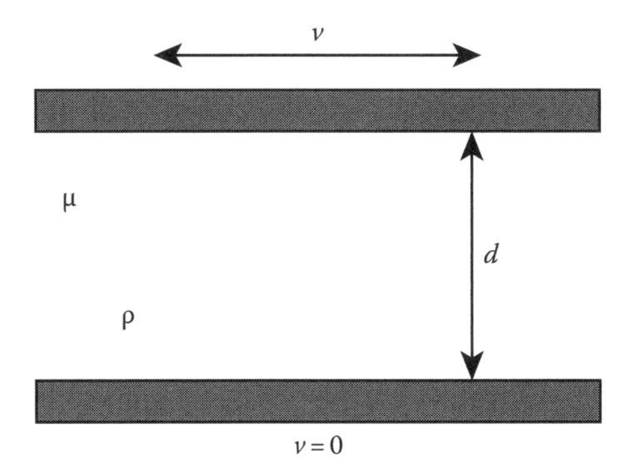

FIGURE P6.19

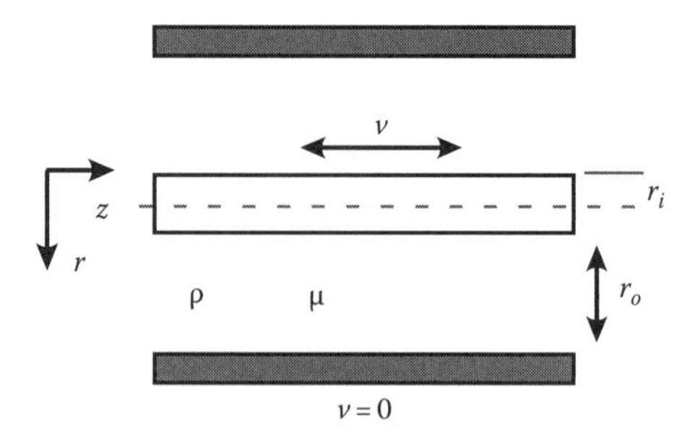

FIGURE P6.20

6.22 In Fahrenheit 451, Ray Bradbury writes about firemen whose job is to burn books. The 451 refers to the temperature (232.8°C) at which paper starts to burn. Assume we have a 4 cm thick book thrown atop a fire burning from below. The fire provides a temperature of 650°C and the pile limits the heat transfer coefficient to the book to be 450 W/m^2 K.
The properties of paper are

$$C_p = 850 \frac{\text{J}}{\text{kg K}} \qquad k = 0.7 \frac{\text{W}}{\text{m K}} \qquad \rho = 950 \frac{\text{kg}}{\text{m}^3}$$

a. Assuming the book starts out at 25°C, how long will it take the surface of the book in contact with the pile to ignite?
b. Based on your answer to part (a), what is the temperature on the side of the book exposed to the atmosphere? Is the semi-infinite approximation valid?

6.23 Revisit the example of diffusion and instantaneous reaction of Section 6.5.2. There we determined the flux of species a through the original vapor–liquid interface at $y = 0$. However, we would really like to know the flux at the interface defined by the reaction front y'. Using the solution and the values for the parameters given in the section, determine the flux of a at y'. Does the flux of b equal that of a at y'?

6.24 As the infrastructure engineer for the City of Troy, the coming cold wave has made you nervous about the water main on 4th street. The situation is shown in Figure P6.24. Current temperature at the center of the main, $T_{g\infty}$, is 10°C (50°F). The projected average temperature for the cold wave, T_∞, is −23°C. A 1 in. (2.54 cm) layer of packed snow covers the street:
a. What is the governing equation describing the temperature profile in the ground?
b. What are the boundary and initial conditions?
c. After looking in texts, you discover the following solution for conduction in a semi-infinite system with convection at the surface:

$$\frac{T(x,t)-T_i}{T_\infty - T_i} = \left[1 - erf\left(\frac{x}{\sqrt{4\alpha t}}\right)\right] - \left[\exp\left(\frac{hx}{k} + \frac{h^2\alpha t}{k^2}\right)\right]$$
$$\times \left[1 - erf\left(\frac{x}{\sqrt{4\alpha t}} + \frac{h\sqrt{\alpha t}}{k}\right)\right]$$

What will it take to adapt this solution to your problem, that is, show the new formula?
d. Can the main survive a 3-day cold spell?

6.25 Consider the transfer of substance a between water (1) and isooctane (2). The two solvents are immiscible and both can be assumed to extend out toward infinity. The concentrations of a

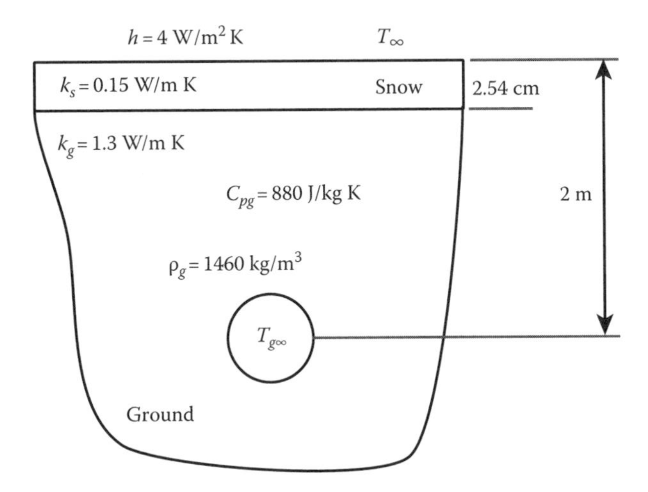

FIGURE P6.24

in both solvents are very small so that we can write the transient diffusion equations in both phases in terms of the concentration of a:

$$\frac{\partial c_{a1}}{\partial t} = D_1 \frac{\partial^2 c_{a1}}{\partial z^2} \qquad -\infty < z < 0$$

$$\frac{\partial c_{a2}}{\partial t} = D_2 \frac{\partial^2 c_{a2}}{\partial z^2} \qquad 0 < z < +\infty$$

There is also an equilibrium partition coefficient between the two phases so that $c_{a2} = mc_{a1}$. Initially, the concentrations in the two phases are c_{a1o} and c_{a2o}:

a. What are the boundary conditions for this problem?
b. Solve the equations and determine the concentration gradients.
c. We can specify the transfer of a (flux) from water to isooctane in terms of a mass transfer coefficient, k_c, and the concentration difference in the isooctane phase, $(c_{a2}(0) - c_{a2}(\infty))$. Determine the instantaneous mass transfer coefficient.

6.26 Though it's only October, you have purchased your copy of the Farmer's Almanac and are excited to learn that the average air temperature predicted for the first 2 weeks of January is −15°C. You are dreaming of taking your snowmobile out on Lake Ontario and crossing over from Rochester to Toronto. To be safe, the ice must be at least 0.25 m thick. Do you think you will be able to make the crossing in time to see the Maple Leafs play on January 14?

6.27 Anyone suffering from a sports injury knows that one must put ice on the area immediately. Moreover, the recommended icing time is 20-min intervals. Assuming the structure for a muscular region shown in Figure P6.27, is there any reason behind this 20-min rule? How deep does the ice temperature penetrate into the body (please see the solution found in Carslaw and Jaeger for inspiration [6, p. 321])?

Properties

$k_{skin} = 0.37$ W/m K $C_{pskin} = 3600$ J/kg K $\rho_{skin} = 998$ kg/m^3

$k_{fat} = 0.37$ W/m K $C_{pfat} = 3600$ J/kg K $\rho_{fat} = 998$ kg/m^3

$k_{muscle} = 0.41$ W/m K $C_{pmuscle} = 3750$ J/kg K $\rho_{muscle} = 1100$ kg/m^3

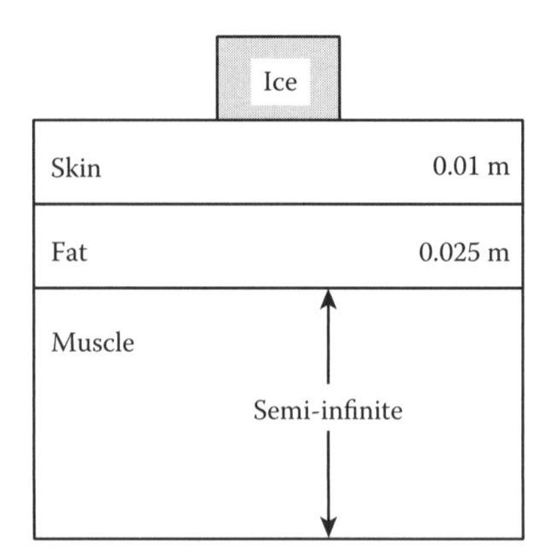

FIGURE P6.27

6.28 Semiconductor devices are prepared by doping silicon to create the source, gate, and drain regions used in a typical transistor. This doping is done primarily by diffusion processes and occurs in two steps, predeposition and drive-in. In the predeposition stage, we expose the pure silicon substrate to the dopant assuming that the final doped layer will be much thinner than the substrate. We also assume a constant value of the diffusivity, D_{ds}, and that we expose the substrate to a constant surface concentration of dopant, c_{do}:

a. Determine the dopant concentration profile in the silicon substrate as a function of time.
b. How much dopant was deposited per unit area of substrate?
(At 900°C, the diffusivity of boron and phosphorous in silicon is about 1.5×10^{-19} m²/s and the solubility of boron and phosphorous in silicon is about 3.7×10^{26} atoms/m³ and 6×10^{26} atoms/m³, respectively.)

6.29 The second diffusion step in semiconductor processing is drive-in where we no longer expose the wafer to dopant but allow the dopant to redistribute itself within the substrate. To model this region, we assume that the drive-in depth is much smaller than the substrate thickness and that we do not lose any dopant in the process:

a. What is the differential equation describing the process?
b. What are the initial condition and the boundary conditions?
c. Verify the following solution holds, where M represents the moles of material originally present in the substrate before drive-in commenced:

$$c_d = \frac{M}{A_c\sqrt{\pi D_{ds} t}}\exp\left(-\frac{x^2}{4D_{ds}t}\right) \qquad M = \int_0^{\infty} c_d(t=0)A_c dx$$

6.30 Optical waveguides can be fabricated in glass or crystal substrates by dopant diffusion as shown in Figure P6.30. The process is similar to fabricating a transistor but diffusion depths are deeper, and since the diffusivities are small, an electric field is applied. Assuming a constant diffusivity, a constant applied electric field, the applicability of the Nernst–Einstein relationship, and diffusion depths small compared to substrate thickness:

a. Derive the following differential equation describing the dopant concentration profile:

$$\frac{\partial c_d}{\partial t} = D\frac{\partial^2 c_d}{\partial x^2} - \mu E\frac{\partial c_d}{\partial x}$$

b. What are the boundary conditions for this problem?

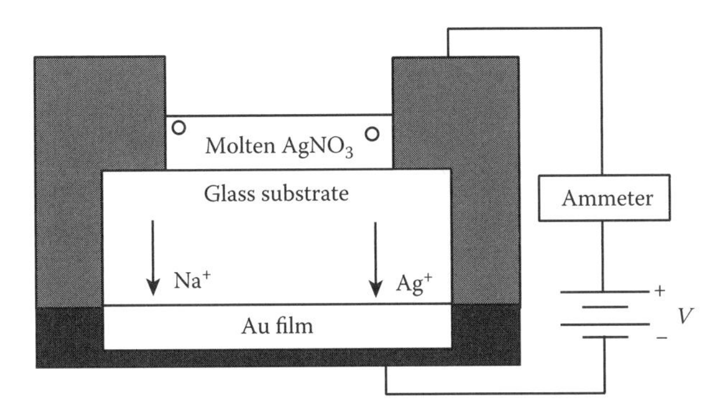

FIGURE P6.30

c. Show that the solution

$$c_d(x,t) = \frac{c_{do}}{2}\left[\exp\left(\frac{\mu Ex}{D}\right) erfc\left(\frac{x+\mu Et}{\sqrt{4Dt}}\right) + erfc\left(\frac{x-\mu Et}{\sqrt{4Dt}}\right)\right]$$

satisfies the differential equation and the boundary conditions.

d. In this operation, either diffusion can control the spread of dopant into the substrate or drift via the electric field can control it. What does the profile look like if drift within the field dominates?

6.31 In 1951, Skellam [12] suggested that the spread of a population was similar to the spread of a chemical component via diffusion. Let's consider the spread of Native Americans across North America. For simplicity, we assume 1-D diffusion of the population across the plane of North America from west to east. The population spreads via diffusion with a constant value of diffusivity, D. We account for births (b) and deaths (d) by assuming that they are akin to chemical reactions that are first order in the population density, P, with some *reaction constants*, k''_b and k''_d:

a. What is the differential equation describing the population density, P?

b. Assuming an initial, exploratory, population, P_o, at $x = 0$, $t = 0$, and a slow spread of that population across the continent, what are the boundary conditions for this problem?

c. Show that the following solution satisfies the differential equation:

$$P(x,t) = \frac{P_o}{\sqrt{4\pi Dt}}\exp\left(k''t - \frac{x^2}{4Dt}\right) \qquad k'' = k''_b - k''_d$$

d. If we consider regions of equal population density, we can determine the speed at which the population spreads. Show that if P_i is a particular value of the population, such a contour is represented by

$$\frac{x}{t} = \pm\left[4k''D - \frac{2D}{t}\ln(4\pi Dt) - \frac{4D}{t}\ln\left(\frac{P_i}{P_o}\right)\right]^{1/2}$$

e. Demonstrate that for long times, the solution in part (d) simplifies to a constant speed given by

$$\frac{x}{t} = \pm\sqrt{4k''D}$$

f. If we assume it took 1000 years for humans to spread from Alaska to Tierra del Fuego, what was their speed and hence $k''D$?

6.32 We are often interested in modeling the growth and spread of a population of bacteria. While we often have a good idea of bacterial reproduction rates, getting a handle on how fast they spread is equally important. Segel et al. [13] developed a simple way to measure the motility/diffusion coefficient for a bacterial population using a capillary tube experiment. Imagine a capillary tube of area, A, with one end sealed, initially filled with pure growth media. At time, $t = 0$, the open end of the tube is placed in a bacterial suspension containing N_0 cells/m^3. The tube is removed from the suspension at a time $t = T$, and the number of bacteria in the tube is counted and determined to be N cells. The mobility can be calculated from

$$\mu_{cells} = \frac{\pi N_{tot}^2}{4N_0^2 A^2 T}$$

based on the governing differential equation that assumed only diffusion occurring in the capillary and that the cells climb only a short distance up the capillary tube in the time, T:

$$\frac{\partial N}{\partial t} = \mu_{cell} \frac{\partial^2 N}{\partial x^2}$$

a. Given that the mobility has units of diffusivity, is the equation dimensionally correct?
b. At the open end of the tube, what is the flux of bacteria?
c. Based on your answer to part (b), derive Segel et al.'s result for the mobility.
d. Suppose the radius of the capillary tube is 100 μm, the bacterial suspension at the tube opening is 7×10^{13} cells/m^3, and the time history of the cell population in the tube is as follows:

Time (s)	N_{tot}
120	1800
300	3700
600	4800
750	5500
900	6700
1200	8000

what is the value of μ_{cell} for each time?

e. Assuming we only measure cell numbers at $T = 600$ s, what is the value for μ_{cell} given the following data?

N_0 (cells/m^3) × 10^{-7}	N_{tot}
2.5	1350
4.6	2300
5.0	3400
12.0	6200

6.33 Experimentally, it has been observed that rust advances through a highly susceptible metal at a rate proportional to $t^{1/2}$. Explain why this should be so assuming:

a. Oxygen diffuses through the rust layer to get to the active metal surface.
b. The oxygen reacts essentially instantaneously with the active metal.
c. The thickness of the metal is much larger than that of the rust.

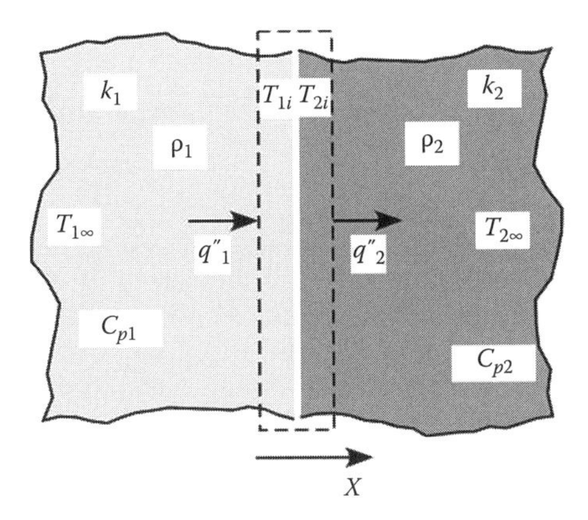

FIGURE P6.34

6.34 Two semi-infinite solids, initially at temperatures T_1 and T_2, are suddenly placed in contact. Find the resulting temperature profiles and the temperature at the interface. See Figure P6.34.

6.35 Experiments like Kirkendall's are performed at elevated temperature. One such experiment using a 70:30 mol% copper mixture on one side and pure zinc on the other was conducted at 785°C and yielded individual diffusion coefficients for copper and zinc in the alloy of

$$D_{Cu} = 2.2\times10^{-13}\,\frac{\text{m}^2}{\text{s}} \qquad D_{Zn} = 5.1\times10^{-13}\,\frac{\text{m}^2}{\text{s}}$$

a. What is the nominal effective diffusivity in the alloy, D_{eff}?
b. Where does the maximum molar velocity occur?
c. Given your answers to (a) and (b), plot the molar velocity as a function of time.
d. If we define the front location, that is, where the tracer particles lie as the point where $0.5 = \exp[-(y^2/4D_{eff}t)]$, how does that front progress as a function of time? Plot it.

6.36 Meat left exposed to the air slowly turns brown due to reaction with atmospheric oxygen. You have purchased a steak and inadvertently left it out on the counter for 3 h. Upon cutting the steak in half, you discover the brown color extends to a depth of 3 mm from the surface (the penetration depth). If the diffusivity of oxygen in the meat is 10^{-12} m²/s, what is the reaction rate constant for the meat/oxygen reaction? The solution for the oxygen concentration profile is

$$\frac{c}{c_o} = \frac{1}{2}\exp\left(-x\sqrt{\frac{k''}{D}}\right)\left[erfc\left\{\frac{x}{\sqrt{4Dt}} - \sqrt{k''t}\right\} + erfc\left\{\frac{x}{\sqrt{4Dt}} + \sqrt{k''t}\right\}\right]$$

REFERENCES

1. Boyce, W.E. and DiPrima, R.C., *Elementary Differential Equations and Boundary Value Problems*, 2nd edn., John Wiley & Sons, New York (1969).
2. Duff, G.F.D. and Naylor, D., *Differential Equations of Applied Mathematics*, John Wiley & Sons, New York (1966).
3. Zauderer, E., *Partial Differential Equations of Applied Mathematics*, John Wiley & Sons, New York (1983).
4. Heisler, M.P., Temperature charts for induction and constant temperature heating, *Trans. ASME*, **69**, 227 (1947).
5. Crank, J., *Mathematics of Diffusion*, Clarendon Press, Oxford, U.K. (1956).

6. Carslaw, H.S. and Jaeger, J.C., *Conduction of Heat in Solids*, 2nd edn., Oxford University Press, London, U.K. (1959).
7. Arnold, J.H., Studies in diffusion: III. Unsteady-state vaporization and absorption, *Trans. AIChE*, **40**, 361–378 (1944).
8. Sherwood, T.K. and Pigford, R.L., *Absorption and Extraction*, McGraw-Hill, New York, pp. 332–337 (1952).
9. Eckert, E.R.G. and Drake, Jr., R.M., *Analysis of Heat and Mass Transfer*, McGraw-Hill, New York (1972).
10. Smigelskas, A.D. and Kirkendall, E.O., Zinc diffusion in alpha brass, *Trans. AIME*, **171**, 130–142 (1947).
11. Gonzalez, E., Arbiol, J., and Puntes, V.F., Carving at the nanoscale: Sequential galvanic exchange and kirkendall growth at room temperature, *Science*, **334**, 1377–1380 (2011).
12. Skellam, J.G., Random dispersal in theoretical populations, *Biometrika*, **38**, 196–218 (1951).
13. Segel, L.A., Chet, I., and Henis, Y., A simple quantitative assay for bacterial motility, *Microbiology*, 98, 329–337 (1977).

7 Conservative Transport and Waves

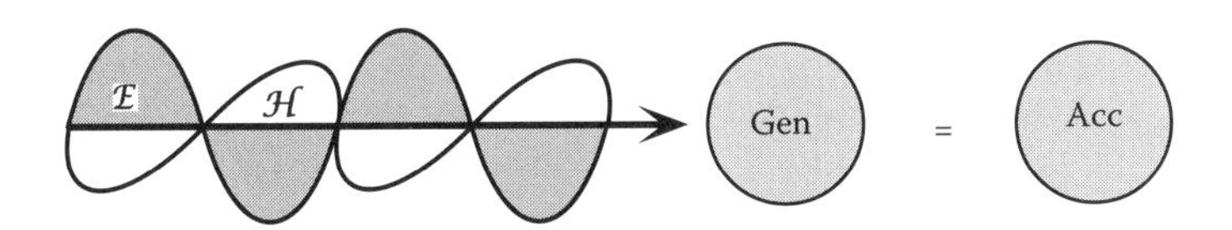

7.1 INTRODUCTION

Diffusion, as a process, can be modeled by considering random walks of particles along a line. We can show that if we had a totally uncorrelated random walk (equal probability of a particle taking a step forward or backward), we could predict the location of the particle at any time by solving the diffusion equation, (7.1). This is true even if we superimpose some background velocity on the particles:

$$\frac{\partial \Psi}{\partial t} = -v_x \frac{\partial \Psi}{\partial x} + \frac{1}{2} D_{ab} \frac{\partial^2 \Psi}{\partial x^2} \quad \text{1-D Diffusion equation} \tag{7.1}$$

where

ψ represents the probability that the particle is located at $x \pm \Delta x$ at a time $t \pm \Delta t$
v_x is some average background velocity of particles upon which the random walk is superimposed
D_{ab} is the diffusivity

The diffusion equation involves a single time derivative. As such, it predicts that an initial disturbance in an otherwise homogeneous system would be felt instantaneously throughout the system no matter how far away from the source of the disturbance the point of interest in the system lies. The diffusion equation also predicts that the disturbance dissipates over time in proportion to $\sqrt{t}$.

The diffusion equation is remarkably useful for describing a whole host of phenomena. Examples range from the dispersion of perfume molecules in air, to the fully developed, laminar flow of a fluid down a flat plate or in a tube, to the temperature profile in an object whose surfaces are exposed to different conditions. We have solved each of these problems and many more in the preceding chapters. Unfortunately, the diffusion equation cannot be used to represent all observed transport phenomena. In fact, it fails in some of the most important transport processes, the conservative transport of momentum and energy. The diffusion equation cannot predict how electromagnetic waves travel through media. It also cannot describe the action of water waves or sound waves. The transport of energy, mass, etc., in these situations does not involve particles executing a pure random walk. Instead, the particles exhibit correlated motions, and that means if a particle takes a step in one direction, it prefers to or may only take, subsequent steps in that direction.

We can predict the motion of a particle undergoing a correlated random walk by using the telegrapher's equation, a form of the wave equation involving a second-order time derivative and originally developed to describe the propagation of telegraph signals. The velocity of the wave is c_λ, and the wave dissipates with a *diffusivity*, c_λ^2/μ, where μ is a *damping* coefficient in this context:

$$\frac{\partial^2 \Psi}{\partial t^2} - c_\lambda^2 \frac{\partial^2 \Psi}{\partial x^2} + 2\mu \frac{\partial \Psi}{\partial t} = 0 \quad \text{Telegrapher's equation} \tag{7.2}$$

This motion described by the telegrapher's equation is oscillatory in nature and simulates the action of all waves (electromagnetic, water, sound, etc.) well. The properties of an initial disturbance described by the telegrapher's equation (when $\mu = 0$) are conserved, and the disturbance propagates as a wave through the medium at a finite speed, c_λ, and with constant amplitude.

7.2 MOMENTUM TRANSPORT

Our first look at wavelike motion will begin with some of the more common waves we encounter, water/tidal waves and sound or acoustic waves. Both these forms of waves have one thing in common; they involve the transport of momentum in a conservative fashion. We will show how wavelike motion arises from the momentum balance equations and relate that motion to the balance between generation and accumulation terms.

7.2.1 Water/Tidal Waves

Consider water waves found in an open channel of depth, z_o. These are the simplest type of momentum waves to consider because the fluid is incompressible and the flow is assumed to be irrotational. The last assumption is extremely important since the flow involving bulk fluid rotations tends to oppose wavelike motion. The moment a water wave begins to crest, for example, the motion is no longer irrotational and the rotation within the flow causes the wave to break and collapse.

There are two types of water waves to consider. The first case, called tidal waves, results when the wavelength of the motion is much larger than the depth of the channel. The amplitude of the wave is also much less than the channel depth. In these cases, the vertical acceleration of the fluid is negligible when compared to the horizontal acceleration so all the fluid in a vertical plane moves synchronously. The second type of wave is called a surface wave and its wavelength is much less than the depth of the fluid. Here, the disturbance leading to wave motion is confined to the surface, and the vertical acceleration of the fluid can no longer be neglected. This situation corresponds to wind-generated waves or surface-tension waves and is inherently a 2-D problem. We can treat tidal waves as 1-D.

We impose some kind of initial disturbance on the fluid that creates a tidal wave moving to the right with velocity, c_λ. We assume that the fluid velocities in the x-, or horizontal, and z-, or vertical, directions are small. We also assume that the height of the wave is small enough that we can neglect velocity variations in the z-direction altogether (notice that in a breaking wave on the shoreline, the velocity in the z-direction exceeds that in the horizontal direction and the wave grows and collapses). To describe the motion of the fluid, we need two balance equations, one to describe conservation of mass, the continuity equation, and another to describe conservation of momentum, the x-component of the momentum equation. It is helpful to have a picture of the situation and to consider a small section of the wave as shown in Figure 7.1.

To write the balance equations, we pick a small control volume of width Δx. We assume that the channel extends a distance, b, into the page. Notice that the cross-sectional area through which

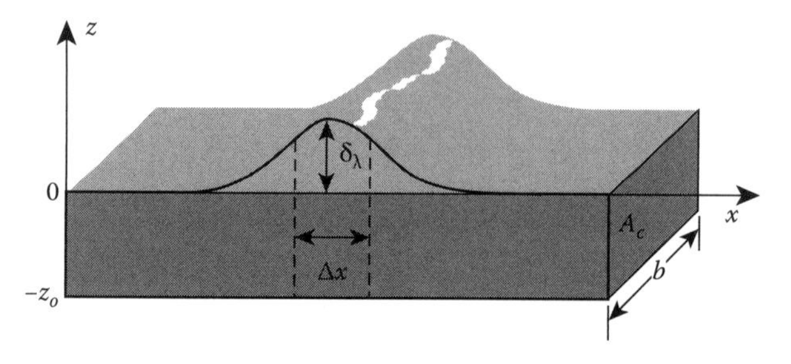

FIGURE 7.1 Water wave traveling in a channel of depth z_o.

the fluid is flowing is not constant because the height of the wave above the waterline ($z = 0$) is not constant. We let A_c be the cross-sectional area of the channel and b be its width. The continuity equation is

Continuity

$$\begin{array}{ccccccc} In & - & Out & + & Gen & = & Acc \\ [\rho(A_c + b\delta_\lambda)v_x]_x & - & [\rho(A_c + b\delta_\lambda)v_x]_{x+\Delta x} & + & 0 & = & \frac{\partial}{\partial t}[\rho(A_c + b\delta_\lambda)\Delta x] \end{array} \tag{7.3}$$

Each term in the continuity equation has the units of mass flow, kg/s. We considered the changing cross-sectional area within the wave and the fact that for the wave to form, fluid must accumulate above the waterline. Dividing through by Δx, taking the limit as $\Delta x \to 0$, and dividing by ρ gives the required continuity equation:

$$-\frac{\partial}{\partial x}[A_c v_x] - \underbrace{\frac{\partial}{\partial x}[b\delta_\lambda v_x]}_{\text{small}} = \underbrace{\frac{\partial}{\partial t}[A_c]}_{0} + \frac{\partial}{\partial t}[b\delta_\lambda] \tag{7.4}$$

Our approximation states that the wave velocity, v_x, and wave height, δ_λ, are both small. Therefore, we can neglect the second term on the left-hand side of Equation 7.4 with respect to the first.

The first term on the right-hand side of Equation 7.4 is identically zero for the channel and can be eliminated. Since the cross-sectional area of the channel does not depend upon x and the width of the channel does not depend upon time, we can remove both those quantities from the derivatives and obtain a simplified form of the continuity equation:

$$-A_c \frac{\partial v_x}{\partial x} = b\frac{\partial \delta_\lambda}{\partial t} \tag{7.5}$$

The momentum equation uses the same control volume. Since we neglected the velocity in the z-direction in the continuity equation, we will neglect momentum in the z-direction as well. In the x-direction, body forces due to gravity are also negligible. The pressure on the system will be due solely to the weight of the fluid underneath the wave and so will depend upon x. We neglect the viscous terms in this conservative system because there is no surface upon which frictional forces can act. Since the flow is termed *irrotational*, $v \neq f(z, y)$. The overall momentum balance thus contains only two terms, one considering the pressure forces exerted on the fluid (generation) and the other accounting for the accumulation of momentum in the control volume:

Momentum

$$\begin{array}{ccc} Gen & = & Acc \\ [(A_c + b\delta_\lambda)P]_x - [(A_c + b\delta_\lambda)P]_{x+\Delta x} & = & \Delta x\frac{\partial}{\partial t}[(A_c + b\delta_\lambda)\rho v_x] \end{array} \tag{7.6}$$

Dividing through by $\rho\Delta x$ and taking the limit as $\Delta x \to 0$ gives

$$-\frac{1}{\rho}\frac{\partial}{\partial x}(A_c P + b\delta_\lambda P) = \frac{\partial}{\partial t}[A_c v_x + b\delta_\lambda v_x] \tag{7.7}$$

The pressure within the wave is just the atmospheric pressure at the wave surface plus the pressure due to the weight of the wave above the channel:

$$P = P_{atm} + \rho g\delta_\lambda \tag{7.8}$$

Plugging the representation for the pressure into the momentum balance, Equation 7.7, and factoring out the density, ρ, gives

$$-\frac{\partial}{\partial x}\left[\frac{A_c P_{atm}}{\rho}+\frac{b\delta_\lambda P_{atm}}{\rho}+A_c\delta_\lambda g+b\delta_\lambda^2 g\right]=\frac{\partial}{\partial t}[A_c v_x+b\delta_\lambda v_x] \tag{7.9}$$

I II III IV V VI

Since the wave height is very small, we can neglect term IV with respect to terms I–III. Following the same reasoning, $b\delta_\lambda \ll A_c$, so we can also neglect term VI with respect to term V. The consequence of neglecting these terms is that the equation we derive for wave motion will show that all waves in the channel will travel at the same velocity regardless of their frequency. This result conflicts with reality since changes in velocity with frequency, called dispersion, are the norm (dispersion is the process that enables us to separate sunlight into a rainbow using a prism). Fortunately, dispersion is often quite small, and the results we derive here will provide a good back-of-the-envelope approximation to what actually occurs.

We can simplify Equation 7.9 further. Term I is a constant and so does not change with position. Term II can be neglected since $b\delta_\lambda P_{atm}/\rho \ll A_c\delta_\lambda g$ and the wave amplitude, δ_λ, is small. This leaves a momentum equation with only two terms:

$$-\frac{\partial}{\partial x}[A_c g\delta_\lambda]=\frac{\partial}{\partial t}[A_c v_x] \tag{7.10}$$

Our last simplification results from dividing Equation 7.10 through by the channel area, A_c. The final forms of the momentum and continuity equations are

$$-g\frac{\partial \delta_\lambda}{\partial x}=\frac{\partial v_x}{\partial t} \quad \text{Momentum} \tag{7.11}$$

$$-A_c\frac{\partial v_x}{\partial x}=b\frac{\partial \delta_\lambda}{\partial t} \quad \text{Continuity} \tag{7.12}$$

Notice that Equations 7.11 and 7.12 are coupled. Equation 7.11 states that a time rate of change in wave velocity gives rise to a change or generation of wave height, δ_λ. This is intuitively obvious since if we retard the motion of the fluid in the horizontal direction, the fluid must flow in the vertical direction. Remember how a wave coming into shore begins to grow and then breaks. It is the sea floor retarding the horizontal wave motion that causes the wave to rise up. Conversely, we can also say that any change in fluid level will result in an accumulation of horizontal velocity. If we look at the continuity equation, we see that a time rate of change of wave height translates into a change in horizontal velocity, whereas any variation in horizontal velocity causes the wave height to accumulate. Again, think of this in terms of a wave coming to shore. Another way to think of this is balancing the kinetic and potential energy of the fluid in the wave. We can convert between the two, but the total energy is conserved.

The generation and accumulation terms of the momentum equation are bootstrapped to one another through the continuity equation. There is no way we can adjust the height or velocity independently. This is the hallmark of wavelike transport. To generate the familiar wave equation, we need to eliminate v_x or δ_λ from Equations 7.11 and 7.12. Differentiating the momentum equation by x and interchanging the order of differentiation gives

$$-g\frac{\partial^2 \delta_\lambda}{\partial x^2}=\frac{\partial}{\partial t}\left[\frac{\partial v_x}{\partial x}\right] \tag{7.13}$$

We can now substitute for $\partial v_x/\partial x$ using the continuity equation. This leads to the wave equation, Equation 7.14, with second derivatives in both space and time. Compare Equation 7.14 with the "traveling wave" equation from Chapter 6 that while having a restoring force to give it a velocity, had only a single time derivative:

$$\left[\frac{gA_c}{b}\right]\frac{\partial^2\delta_\lambda}{\partial x^2} = \frac{\partial^2\delta_\lambda}{\partial t^2} \quad \text{Wave equation} \tag{7.14}$$

The dimensions of the individual terms in Equation 7.14 show that the velocity of the wave is a function of the depth of the channel ($A_c = z_o \times b$) but not the width into the page, b:

$$c_\lambda = \sqrt{\frac{gA_c}{b}} = f(g, z_o) \tag{7.15}$$

Deeper channels can support faster waves.

Example 7.1 Design of a Wave Pool

You are the designer of a new wave pool at Disney's latest theme park. The pool is to be 40 m wide and 1 m deep over most of its length. To determine the appropriate frequency of wave formation, you need to know the velocity of the waves. For a pool of this size, what would the wave velocity be?

Here, the wave velocity is easy to calculate from the depth of the pool and its width:

$$c_\lambda = \sqrt{\frac{gA_c}{b}} = \sqrt{\frac{(9.8)(40)(1)}{40}} = 3.1\,\text{m/s}$$

The wave equation is second order in time and in space. To solve it, we must specify two initial conditions and two boundary conditions. We will assume that the wave has zero height at $x = 0$ and that at some periodic distance, L, its height has returned to its value at $x = 0$. Thus, Figure 7.1 represents only one cycle of the wave. The full wave is an infinite train of cycles that repeats every L meters:

$$x = 0 \quad \delta_\lambda = 0 \tag{7.16}$$

$$x = L \quad \delta_\lambda = 0 \tag{7.17}$$

The initial conditions require us to specify some nonzero function for the wave profile at $t = 0$ and to specify how fast the wave height changes at $t = 0$. We will provide some nonspecific function for the wave shape, $f(x)$, and specify that the height of the wave does not change until $t = 0$:

$$t = 0 \quad \delta_\lambda = f(x) \tag{7.18}$$

$$t = 0 \quad \frac{\partial\delta_\lambda}{\partial t} = 0 \tag{7.19}$$

We can solve the wave equation by separation of variables or any other method used for partial differential equations. Letting

$$\delta_\lambda = F(x)G(t) \tag{7.20}$$

substituting into the differential equation, and dividing through by $F(x)G(t)$ leads to

$$\frac{F''}{F} = \frac{G''}{c_\lambda^2 G} = -\lambda^2 \tag{7.21}$$

Since the left-hand side of the equation depends only on position and the right-hand side depends only on time, the only way the two can be equal is if they both are equal to a constant. We set this constant to $-\lambda^2$. You should convince yourself that other possibilities lead to aphysical solution or trivial solution with the wave amplitude equal to zero (see Chapter 6). The separation of variables procedure yields two ordinary differential equations:

$$\frac{d^2F}{dx^2} + \lambda^2 F = 0 \tag{7.22}$$

$$\frac{d^2G}{dt^2} + c_\lambda^2 \lambda^2 G = 0 \tag{7.23}$$

Each of these equations has solutions given in terms of sines and cosines:

$$F = a_1 \cos(\lambda x) + a_2 \sin(\lambda x) \tag{7.24}$$

$$G = b_1 \cos(c_\lambda \lambda t) + b_2 \sin(c_\lambda \lambda t) \tag{7.25}$$

The separation procedure has left us with five constants and only four boundary conditions to evaluate them. Reconstructing the overall solution lets us combine constants and evaluate the four remaining constants using the boundary conditions:

$$\delta_\lambda = [a_1 \cos(\lambda x) + a_2 \sin(\lambda x)][b_1 \cos(c_\lambda \lambda t) + b_2 \sin(c_\lambda \lambda t)] \tag{7.26}$$

The application of the first boundary condition at $x = 0$ shows that $a_1 = 0$. The application of the second boundary condition at $x = L$ leaves us with

$$a_2 \sin(\lambda L)[b_1 \cos(c_\lambda \lambda t) + b_2 \sin(c_\lambda \lambda t)] = 0 \tag{7.27}$$

Now we can have neither $a_2 = 0$ nor b_1 and b_2 both be zero because this leads to a trivial solution. The only recourse is to have $\sin(\lambda L) = 0$ and this requires that

$$\lambda = \frac{n\pi}{L} \tag{7.28}$$

The initial condition for the slope of Equation 7.14 leads to

$$a_2 \sin\left(\frac{n\pi x}{L}\right) b_2 \left(\frac{n\pi c_\lambda}{L}\right) = 0 \tag{7.29}$$

Since $a_2 \neq 0$, we must have $b_2 = 0$. We are now left with one term. We can collapse a_2 and b_1 into a single constant, a_3, that we evaluate using the last initial condition:

$$\delta_\lambda = a_3 \sin\left(\frac{n\pi x}{L}\right) \cos\left(\frac{n\pi c_\lambda}{L} t\right) \tag{7.30}$$

Unfortunately, a single term of the form of Equation 7.30 will not mimic the arbitrary function, $f(x)$, we specified at $t = 0$. Since every solution of the form (7.30) is a solution to the differential

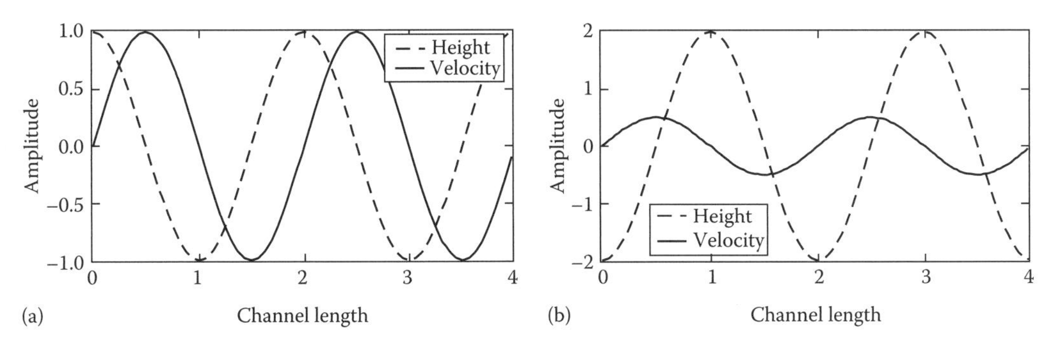

FIGURE 7.2 Wave height and velocity ($L = 1, f(x) = \sin(\pi x)$). (a) $t = 0$ and (b) $t = ½$.

equation, a sum of those solutions is also a solution and we can use that sum to approximate $f(x)$. We take this sum over all n:

$$\delta_\lambda = \sum_{n=-\infty}^{\infty} a_{3n} \sin\left(\frac{n\pi x}{L}\right)\cos\left(\frac{n\pi c_\lambda t}{L}\right) \tag{7.31}$$

Applying the final initial condition lets us solve for the a_{3n}:

$$f(x) = \sum_{n=-\infty}^{\infty} a_{3n} \sin\left(\frac{n\pi x}{L}\right) \tag{7.32}$$

We have solved such a set before in Chapter 6 by multiplying both sides of Equation 7.32 by $\sin(m\pi x/L)$ and integrating with respect to x. Since the sine function is orthogonal to itself, the only term left on the right-hand side after the multiplication and integration procedure will be the term where $n = m$. The a_{3n} are

$$a_{3n} = \frac{2}{L}\int_0^L f(x)\sin\left(\frac{n\pi x}{L}\right)dx \tag{7.33}$$

and this completes the solution to our problem. The solution for $L = 1$ and $f(x) = \sin(\pi x)$ is shown in Figure 7.2. Notice that the sine function is an odd function so that $\sin(-x) = -\sin(x)$. What does this imply about our solution for $n < 0$? (Hint: If $n < 0$, in what direction does the wave travel?)

Example 7.2 Tidal Waves in a Sloped Channel

In this example, we look at tidal waves as they travel toward the beach. We assume that we have a channel with a sloping base, similar to the one shown in Figure 7.3. The shoreline has a linear slope and the depth of the channel obeys the following function: $z_o = h_o x$. The slope is quite shallow so that $h_o \ll 1$. We would like to determine what the waves traveling toward the shore would look like. Since the slope of the beach is assumed to be very shallow, any change in cross-sectional area, A_c, occurs over length scales much longer than the wavelength of the tidal wave. This means that we can write Equations 7.5 and 7.10 as

$$-gA_c\frac{\partial \delta_\lambda}{\partial x} = \frac{\partial}{\partial t}(A_c v_x) \quad \text{Momentum}$$

$$-\frac{\partial}{\partial x}(A_c v_x) = b\frac{\partial \delta_\lambda}{\partial t} \quad \text{Continuity}$$

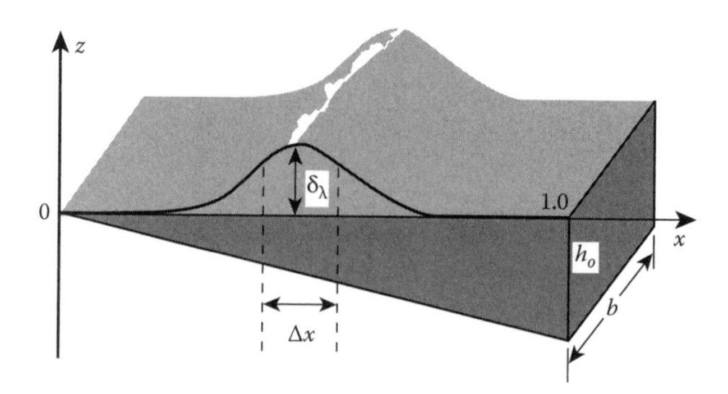

FIGURE 7.3 Wave traveling on a sloping shore.

To obtain the wave equation for this problem, we differentiate the momentum equation by x and continuity equation by t and subtract the two. We also substitute in for A_c using the nominal depth and breadth of the channel:

$$h_o g \frac{\partial}{\partial x}\left[x\frac{\partial \delta_\lambda}{\partial x}\right] = \frac{\partial^2 \delta_\lambda}{\partial t^2}$$

Removing and collecting the constants and expanding the derivatives, we have

$$x\frac{\partial^2 \delta_\lambda}{\partial x^2} + \frac{\partial \delta_\lambda}{\partial x} = \frac{1}{h_o g}\frac{\partial^2 \delta_\lambda}{\partial t^2}$$

We solve this equation by separation of variables. Letting $\delta\lambda = F(x)G(t)$, we substitute into the differential equation and show that both sides must be equal to a constant, $-\lambda^2$:

$$x\frac{F''}{F} + \frac{F'}{F} = \frac{1}{h_o g}\frac{G''}{G} = -\lambda^2$$

We can separate this equation into two ordinary differential equations:

$$x\frac{d^2F}{dx^2} + \frac{dF}{dx} + \lambda^2 F = 0$$

$$\frac{d^2G}{dt^2} + (h_o g)\lambda^2 G = 0$$

The second of these two equations has a solution represented by Equation 7.25:

$$G = b_1\cos\left(\sqrt{h_o g}\,\lambda t\right) + b_2\sin\left(\sqrt{h_o g}\,\lambda t\right)$$

The first equation is a bit more complicated but we can show that it is a form of the Bessel equation, an equation we first encountered in Chapter 6. The solution is given in terms of the Bessel functions of the first kind, of order 0 [1]:

$$F = a_1 J_o\left(\lambda\sqrt{x}\right) + a_2 Y_o\left(\lambda\sqrt{x}\right)$$

We can use the same boundary conditions that we used in the previous solution, namely,

$$x = 0, L \qquad \delta_\lambda = 0$$

$$t = 0 \qquad \delta_\lambda = f(x) \qquad \frac{\partial \delta_\lambda}{\partial t} = 0$$

Since the height of the wave must tend to zero as $x \to 0$, the constant a_2 must vanish ($Y_o \to -\infty$). Similarly, with $\partial\delta_\lambda/\partial t = 0$ at $t = 0$, we must have $b_2 = 0$. We are left with two constants to solve for, the product of a_1 and b_1 that we call a_3, and λ_n:

$$\delta_{\lambda n} = a_{3n} J_o\left(\lambda_n \sqrt{x}\right) \cos\left(\sqrt{h_o g}\, \lambda_n t\right)$$

We require $\delta_\lambda = 0$ when $x = L$, and since we cannot have $a_{3n} = 0$, we must have $J_o = 0$. This can only occur if the argument of J_o corresponds to a root of J_o. The λ_n must be

$$\lambda_1 = 2.4048 \frac{1}{2\sqrt{L}} \qquad \lambda_2 = 5.5201 \frac{1}{2\sqrt{L}} \qquad \lambda_3 = 8.6537 \frac{1}{2\sqrt{L}} \ldots$$

Notice that the wavelength of these waves increases as x increases. Like waves at the beach, these waves appear to become compressed as they approach shore.

Finally, we require $\delta_\lambda = f(x)$ at $t = 0$. For this condition to be met, we must sum over all solutions. Considering only left-to-right moving waves,

$$f(x) = \sum_{n=1}^{\infty} a_{3n} J_o\left(\lambda_n \sqrt{x}\right) \cos\left(\sqrt{h_o g}\, \lambda_n t\right)$$

To extract the a_{3n}, we multiply both sides of the equation by $J_o(\lambda_m \sqrt{x})$ and integrate from $x = 0$ to $x = L$. Since the Bessel functions are orthogonal to one another, the only term in the integral that survives is the term when $n = m$. Thus,

$$a_{3n} = \frac{2}{L}\left[\frac{\int_o^L f(x) J_o(\lambda_n \sqrt{x})\, dx}{\left[J_1(\lambda_n \sqrt{L})\right]^2}\right]$$

and the solution is complete.

Example 7.3 Reflection and Refraction of a Tidal Wave

We return to the original flat-bottomed system and look at what happens if we have a discontinuity in the depth of the channel in the direction of wave propagation. The situation is illustrated in Figure 7.4. As the tidal wave travels through the interface between the depths,

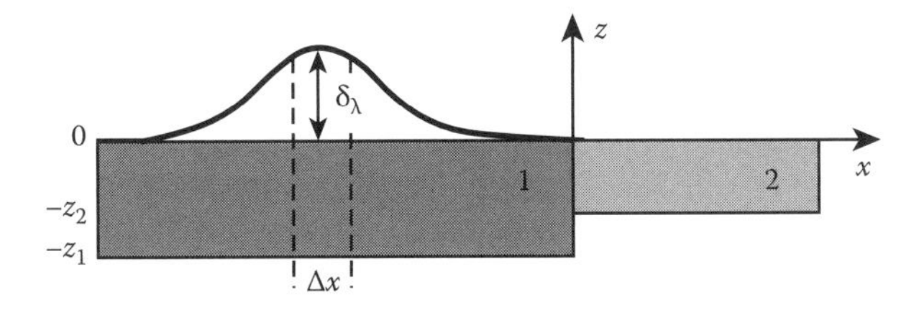

FIGURE 7.4 Tidal wave traveling from one depth into another.

we expect several changes. A portion of the wave's energy will be reflected from the interface and a corresponding portion will pass through the interface to form a wave in the next channel. We have three waves to consider and must use a control volume in each region to analyze them. Fortunately, we have derived the differential equations for the waves before and need only write them down for the two regions:

$$\left[\frac{gA_{c1}}{b}\right]\frac{\partial^2\delta_{\lambda i}}{\partial x^2} = \frac{\partial^2\delta_{\lambda i}}{\partial t^2} \quad \text{Region 1 incident wave}$$

$$\left[\frac{gA_{c1}}{b}\right]\frac{\partial^2\delta_{\lambda r}}{\partial x^2} = \frac{\partial^2\delta_{\lambda r}}{\partial t^2} \quad \text{Region 1 reflected wave}$$

$$\left[\frac{gA_{c2}}{b}\right]\frac{\partial^2\delta_{\lambda t}}{\partial x^2} = \frac{\partial^2\delta_{\lambda t}}{\partial t^2} \quad \text{Region 2 transmitted wave}$$

Notice that the reflected and incident waves have the same velocity. This is true because they travel in the same medium and in a portion of the system with the same geometry as the incident wave. To obtain more information about the waves, we must solve the wave equations. We can go about this via separation of variables, or we can use the fact that having solved the wave equation in the past, we know its generic solution. In fact, the following functions satisfy the differential equations ($\exp(\pm ix) = \cos(x) \pm i\sin(x)$):

$$\delta_i = a_1 \exp[i(2\pi\omega_i t - k_i x)] + a_2 \exp[i(2\pi\omega_i t + k_i x)]$$

$$\delta_r = b_1 \exp[i(2\pi\omega_r t - k_r x)] + b_2 \exp[i(2\pi\omega_r t + k_r x)]$$

$$\delta_t = c_1 \exp[i(2\pi\omega_t t - k_t x)] + c_2 \exp[i(2\pi\omega_t t + k_t x)]$$

where
- ω is the frequency of the wave
- k is the propagation constant, $2\pi/\lambda$
- λ is the wavelength

The speed of the wave $c_\lambda = \omega/k$.

The equations we have just proposed represent waves traveling in the $+x$ and $-x$ directions. Clearly, the incident and transmitted waves travel in one direction and the reflected wave travels in the other. Thus, if we say the incident wave travels in the $+x$ direction, then we can immediately set a_2, b_1, and c_2 to zero. Since the reflected and incident waves travel in the same region, their velocities are the same and so $k_r = k_i$ and $\omega_r = \omega_i$:

$$\delta_i = a_1 \exp[i(2\pi\omega_i t - k_i x)]$$

$$\delta_r = b_2 \exp[i(2\pi\omega_i t + k_i x)]$$

$$\delta_t = c_1 \exp[i(2\pi\omega_t t - k_t x)]$$

What remains is to find a relationship between the amplitudes of the waves and between their propagation constants, k. For this, we need the boundary conditions at the interface between the two regions. These conditions must preserve continuity of the wave height and slope at the interface and continuity of the frequency of the wave since the medium is linear and cannot generate another wave at a different frequency. So, at the interface,

$$\delta_i + \delta_r = \delta_t \quad \text{Amplitude continuity}$$

$$\frac{\partial \delta_i}{\partial x} + \frac{\partial \delta_r}{\partial x} = \frac{\partial \delta_t}{\partial x} \quad \text{Slope continuity}$$

$$\omega_i = \omega_r = \omega_t \quad \text{Frequency preservation}$$

Substituting the expressions for the waves into the first two boundary conditions and solving simultaneously lets us specify the amplitudes of the reflected and transmitted waves in terms of the amplitude of the incident wave and the propagation constants of the waves:

$$\frac{b_2}{a_1} = \frac{k_i - k_t}{k_r + k_t} \qquad \frac{c_1}{a_1} = \frac{k_i + k_r}{k_r + k_t} \qquad k_r = k_i$$

The amplitudes of the reflected and transmitted waves depend heavily upon the velocities of the waves. If the transmitted velocity is the same as the incident velocity, we have no impediment to wave motion and so no reflected wave. If we place an obstruction at the interface such that $z_2 = 0$, then the transmitted velocity is zero and we have a purely reflected wave. This behavior is seen in all waves, most commonly in electromagnetic waves.

7.2.2 Acoustic Waves

Previously, we showed how velocity disturbances in an incompressible fluid lead to wave propagation. This is the mechanism for generating water waves. Another mechanism in momentum transport also leads to wave propagation. These waves are called acoustic or sound waves and come from changes in pressure causing changes in the density of the material. Momentum propagates in the form of a pressure or density wave. We will consider the simple case of a long horizontal tube of constant cross-sectional area, similar to the pipes in a pipe organ.

We set up a control volume of width Δx and cross-sectional area A_c (Figure 7.5). A pressure disturbance is initiated at $x = 0$ and propagates down the length of the tube until it reaches $x = L$. The continuity equation reflects conservation of mass and so must account for the density differences within the wave:

Continuity

$$\begin{array}{ccccccc} In & - & Out & + & Gen & = & Acc \\ (\rho A_c v_x)_x & - & (\rho A_c v_x)_{x+\Delta x} & + & 0 & = & \dfrac{\partial}{\partial t}(\rho A_c \Delta x) \end{array} \quad (7.34)$$

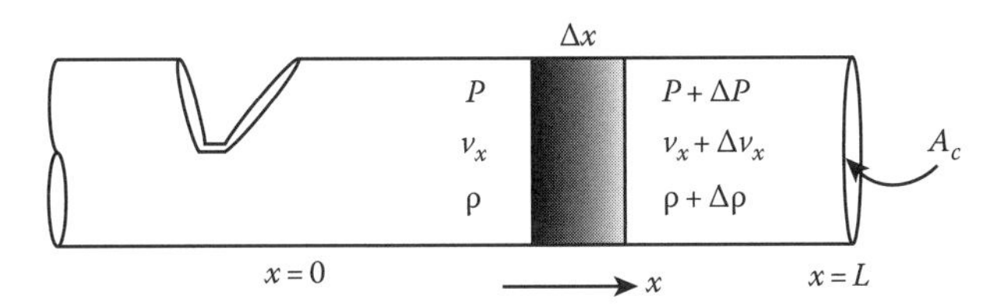

FIGURE 7.5 Acoustic waves in an organ pipe.

Choosing A_c to be constant and Δx independent of time, dividing through by $A_c\Delta x$, and taking the limit as $\Delta x \to 0$ gives the following differential equation:

$$-\frac{\partial}{\partial x}(\rho v_x) = \frac{\partial \rho}{\partial t} \tag{7.35}$$

Notice that both velocity and density may vary with position in the tube. The density may also be a function of time.

The momentum balance must consider pressure forces along the axis of the tube and the accumulation of momentum within the wave. The first term can be thought of as a generation term (see Chapter 5):

Momentum

$$\begin{aligned} &Gen &&= &&Acc \\ &[A_cP(x) - A_cP(x+\Delta x)] &&= &&\Delta x \frac{\partial}{\partial t}(A_c\rho v_x) \end{aligned} \tag{7.36}$$

Again, dividing by $A_c\Delta x$ and taking the limit $\Delta x \to 0$ gives

$$-\frac{\partial P}{\partial x} = \frac{\partial}{\partial t}(\rho v_x) \tag{7.37}$$

If we are dealing with atmospheric pressure systems, we can assume an ideal gas. A change in pressure is related to a change in density by

$$dP = \left(\frac{\partial P}{\partial \rho}\right) d\rho \tag{7.38}$$

At constant temperature, there is a direct proportionality between pressure and density arising from the ideal gas law, ($P = \rho RT/M_w$). Unfortunately, acoustic waves rarely travel in an isothermal environment. There just isn't a source or sink of heat to hold the temperature constant. Normally, these waves travel isentropically (exchanging no heat with the surroundings, and so $\Delta S = 0$). The pressure and density for an isentropic process are related by a power law relationship:

$$P = K\rho^{\gamma} \tag{7.39}$$

where $\gamma = C_p/C_v$ for an ideal gas and

$$dP = \left(\frac{\partial P}{\partial \rho}\right)_{\Delta S=0} d\rho \tag{7.40}$$

Substituting for $\partial P/\partial x$ in the momentum equation using Equation 7.40 gives our new momentum equation in terms of the density:

$$-\left[\left(\frac{\partial P}{\partial \rho}\right)_{\Delta S=0}\right]\frac{\partial \rho}{\partial x} = \frac{\partial}{\partial t}(\rho v_x) \quad \text{Momentum} \tag{7.41}$$

$$-\frac{\partial}{\partial x}(\rho v_x) = \frac{\partial \rho}{\partial t} \quad \text{Continuity} \tag{7.42}$$

To formulate the wave equation, we differentiate Equation 7.41 by x and substitute in for $\partial(\rho v_x)/\partial x$ from the continuity equation. We implicitly assume that $(\partial P/\partial \rho)|_{\Delta S=0}$ is a constant. The result of this exercise is a single, second-order partial differential equation describing the changes in density as a function of position and time:

$$\left[\left(\frac{\partial P}{\partial \rho}\right)_{\Delta S=0}\right]\frac{\partial^2 \rho}{\partial x^2} = \frac{\partial^2 \rho}{\partial t^2} \quad \text{Wave equation} \tag{7.43}$$

This is the familiar form of the wave equation where the wave velocity (speed of sound) is given by

$$c_\lambda^2 = \left(\frac{\partial P}{\partial \rho}\right)_{\Delta S=0} \tag{7.44}$$

By assuming the change in pressure with density to be a constant, we imply that the speed of sound is independent of the frequency. Like tidal waves or electromagnetic waves, acoustic waves also exhibit dispersion and so our simplification will leave us unable to simulate this effect. The dispersion of acoustic waves is a very important consideration in the design of loudspeakers, recital halls, musical instruments, and electronic equipment to synthesize sound. It is also the key component of the Doppler effect.

The momentum and continuity equations have the identical relationship to one another that we discussed in the section on tidal waves. The momentum equation exhibits the feedback relationship between generation and accumulation that is the hallmark of a conservative process. As before, we cannot separate the two effects. There is no arrow that unambiguously points to a given procedural direction.

We can derive the acoustic wave equation in terms of the width of the pressure pulse, that is, the wavelength of the pressure disturbance, instead of the density. This formulation is a bit easier to physically visualize. The pulse width, $\delta_{\lambda p}$, the region of varying pressure, is related to the density via

$$d\rho = -\rho_o d\delta_{\lambda p} \tag{7.45}$$

where ρ_o is a reference density or equilibrium density of the gas at the temperature and pressure of the system. Substituting into Equation 7.43 leads to the alternate form of the wave equation:

$$\left[\left(\frac{\partial P}{\partial \rho}\right)_{\Delta S=0}\right]\frac{\partial^2 \delta_{\lambda p}}{\partial x^2} = \frac{\partial^2 \delta_{\lambda p}}{\partial t^2} \quad \text{Wave equation} \tag{7.46}$$

Example 7.4 Acoustic Waves and the Horn Tweeter

We are going to consider what occurs in sound production in a horn tweeter, the kind of loudspeaker you often see in public address systems. The key difficulty we encounter is that the cross-sectional area of the tweeter varies and so the acoustic wave will propagate in an enclosure whose cross-sectional area varies slowly in the direction of propagation. A typical horn tweeter is shown in Figure 7.6.

Since the cross-sectional area through which the wave will propagate is a function of position, we cannot use the solutions we developed previously but must resort to another derivation. We cut the horn across its x-axis and define a differential element of width Δx. The horn tweeter is

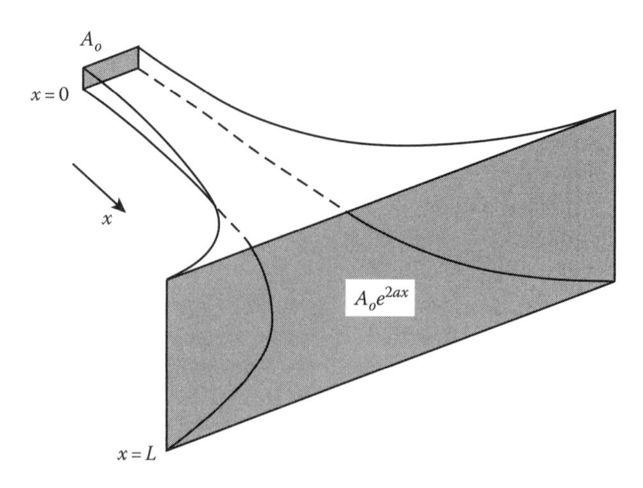

FIGURE 7.6 Schematic diagram of a horn tweeter.

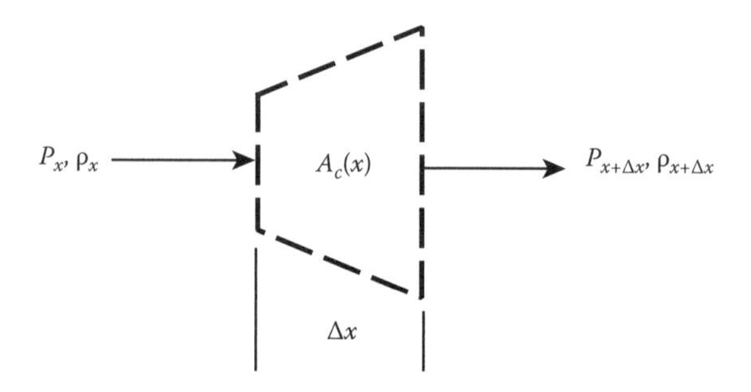

FIGURE 7.7 Differential control volume for the horn tweeter.

designed to be symmetrical about the x-axis. The continuity equation for this element reflects the fact that both density and area change:

$$\begin{array}{ccccc} In & - & Out & = & Acc \\ (\rho A_c v_x)_x & - & (\rho A_c v_x)_{x+\Delta x} & = & \dfrac{\partial}{\partial t}(\rho A_c \Delta x) \end{array}$$

Dividing through by Δx, taking the limit as $\Delta x \to 0$, and recognizing that both the density and area change with position gives

$$-\frac{\partial}{\partial x}(\rho A_c v_x) = \frac{\partial}{\partial t}(\rho A_c)$$

The momentum equation is similarly derived. The accumulation of momentum is related to the change in pressure force felt by the gas as it expands down the length of the horn (Figure 7.7):

$$-\frac{\partial}{\partial x}(PA_c) = \frac{\partial}{\partial t}(\rho A_c v_x)$$

For an isentropic expansion (no exchange of heat) of the gas, we relate the pressure and density by

$$P = K\rho^{\gamma} \qquad \gamma = \frac{C_p}{C_v}$$

and so define the velocity as

$$c_\lambda^2 = \left.\frac{\partial P}{\partial \rho}\right|_{\Delta S=0} = \gamma K \rho^{\gamma-1}$$

We know that the cross-sectional area, A_c, of the horn does not change with time. Differentiating the momentum equation by x gives

$$-\frac{\partial}{\partial x}\left[\frac{\partial}{\partial x}(PA_c)\right] = \frac{\partial}{\partial t}\left[\frac{\partial}{\partial x}(\rho A_c v_x)\right]$$

We remove v_x from the right-hand side by substituting in from the continuity equation while remembering that the cross-sectional area of the horn does not change with time:

$$-\frac{\partial}{\partial x}\left[\frac{\partial}{\partial x}(PA_c)\right] = \frac{\partial}{\partial t}\left[\frac{\partial}{\partial t}(\rho A_c)\right] = A_c \frac{\partial^2 \rho}{\partial t^2}$$

Expanding the x-derivative on the left-hand side and substituting for P using its relationship to the density gives

$$\frac{\partial}{\partial x}\left[A_c \frac{\partial P}{\partial x} + P\frac{\partial A_c}{\partial x}\right] = -\frac{\partial}{\partial x}\left[c_\lambda^2 A_c \frac{\partial \rho}{\partial x} + \frac{\rho c_\lambda^2}{\gamma}\frac{\partial A_c}{\partial x}\right]$$

Now we specified that the area would be a slowly varying function of x. Therefore, $\partial A_c/\partial x \ll A_c$, and so is a small quantity. Since the density is also small compared to the change in density with x, $\rho \ll \partial\rho/\partial x$, the $\partial A_c/\partial x$ term can be neglected when compared to $\partial\rho/\partial x$ term and the wave equation becomes

$$c_{\lambda o}^2 \left[A_c \frac{\partial^2 \rho}{\partial x^2} + \frac{\partial A_c}{\partial x}\frac{\partial \rho}{\partial x}\right] = A_c \frac{\partial^2 \rho}{\partial t^2} \qquad c_{\lambda o}^2 = \gamma K \rho_o^{\gamma-1}$$

where we have expanded the spatial derivative, neglected dispersion (c_λ is constant), and used the fact that the area, A_c, does not change with time. Neglecting dispersion prevents us from simulating the full behavior of the loudspeaker but makes the equation solvable analytically and gives us the gross features of the loudspeaker.

The horn tweeter has a special design. The area obeys the following function:

$$A_c = A_o \exp(2ax)$$

Plugging the area function into the wave equation yields the following differential equation for ρ:

$$c_{\lambda o}^2 \left[\frac{\partial^2 \rho}{\partial x^2} + 2a\frac{\partial \rho}{\partial x}\right] = \frac{\partial^2 \rho}{\partial t^2}$$

Notice that this equation is more complicated than the wave equations we previously derived. We will see that the additional term affects the types of waves that can propagate in the horn. Waves of a certain frequency will be attenuated as they pass through, a process termed cutoff.

The wave equation we just derived is still a linear equation and so we can solve it by separation of variables. Substituting $\rho = T(t)\chi(x)$ into the differential equation and separating like terms leads to two ordinary differential equations:

$$\frac{d^2T}{dt^2} + \omega^2 T = 0$$

$$\frac{d^2\chi}{dx^2} + 2a\frac{d\chi}{dx} + \frac{\omega^2}{c_{\lambda o}^2}\chi = 0 \qquad \omega\text{— separation constant}$$

The solution to the first differential equation can be written in terms of sines and cosines:

$$T = a_1 \cos(\omega t) + a_2 \sin(\omega t)$$

The solution to the second equation can be found by substituting $\chi = Be^{mx}$ and solving for m:

$$\chi = b_1 \exp(m_1 x) + b_2 \exp(m_2 x)$$

$$m_1 = -a + \frac{\omega}{c_{\lambda o}}\sqrt{\frac{ac_{\lambda o}^2}{\omega^2} - 1} \qquad m_2 = -a - \frac{\omega}{c_{\lambda o}}\sqrt{\frac{ac_{\lambda o}^2}{\omega^2} - 1}$$

We can combine the two solutions using exponential notation. Remembering that $\exp(i\theta) = \cos\theta + i\sin\theta$, we reconstruct the solution for ρ and combine into

$$\rho = \exp(-ax)\left\{c_1 \exp\left[i\left(\omega t - \frac{\omega}{c_{\lambda o}} x\sqrt{1 - \frac{a^2 c_{\lambda o}^2}{\omega^2}}\right)\right] + c_2 \exp\left[i\left(\omega t + \frac{\omega}{c_{\lambda o}} x\sqrt{1 - \frac{a^2 c_{\lambda o}^2}{\omega^2}}\right)\right]\right\}$$

Here, the first term represents a wave moving from left to right in Figure 7.6 and the second term represents a wave moving from right to left. The waves move along the horn at a velocity, c_h, given by

$$c_h = \frac{c_{\lambda o}}{\sqrt{1 - \frac{a^2 c_{\lambda o}^2}{\omega^2}}}$$

Notice that the speed becomes imaginary when $a^2 c_{\lambda o}^2/\omega^2 > 1$. We define the point where $\omega = ac_{\lambda}o$ as the cutoff point or cutoff frequency for the horn. Frequencies lower than this will not be transmitted. Acoustic waves can be sent out along the horn with a speed that is essentially independent of frequency, which is great news for audiophiles. The horn tweeter has very low dispersion. The ability to have a cutoff frequency, and hence limit the response of the speaker, does have its costs. There will be an attenuation in the wave amplitude across all wavelengths represented by the term $\exp(-ax)$. If the horn is properly designed, this attenuation can be minimized. (What does this have to say about minimum power requirements to drive speakers?)

Though we have looked at a horn in a Cartesian framework, the properties of the horn are universal and have been exploited for centuries. Look at ram's horn instruments and modern instruments such as trumpets, clarinets, and tubas. How do their shapes affect the wavelength range of operation [1]?

We discussed dispersion and how neglecting the smaller terms in our momentum and continuity equations prevented us from observing it. The Comsol® module "DispersiveWave" allows us to investigate some aspects of dispersion. The generic equation that is solved in the module is

$$\frac{\partial^2 u}{\partial t^2} = \frac{\partial}{\partial x}\left(c\frac{\partial u}{\partial x}\right) \qquad c = f(u) \tag{7.47}$$

We can look at a variety of forms for c, but in this case, we look at two specific cases related to sound waves in air. In such a case, c is a function of u raised to the power $C_p/C_v - 1$ giving us

$$\frac{\partial^2 u}{\partial t^2} = \frac{\partial}{\partial x}\left(\alpha u^{0.4}\frac{\partial u}{\partial x}\right) \tag{7.48}$$

with boundary conditions specifying a periodic function of density at $x = 0$ and no flux at $x = L$. Of course, we are free to choose other boundary conditions. Figure 7.8 shows a plot of the amplitude of u as a function of position at time equals 3 s. The effect of dispersion is clearly evident since it affects both the amplitude and velocity of the wave. Notice that as the frequency increases, the amplitude decays more sharply and the wave travels more slowly. Figure 7.9 shows how the speed varies with position. Notice that speed variations are not too great even though we have a large change in amplitude.

The effects of dispersion are even more pronounced if we allow ourselves the freedom to choose an alternative dispersion function. If, for example, we borrow a constitutive relation from non-Newtonian fluids and allow the dispersion relation to be a function of the derivative of u, rather than u, we have

$$\frac{\partial^2 u}{\partial t^2} = \frac{\partial}{\partial x}\left[\alpha\left|\frac{\partial u}{\partial x}\right|^{0.4}\frac{\partial u}{\partial x}\right] \tag{7.49}$$

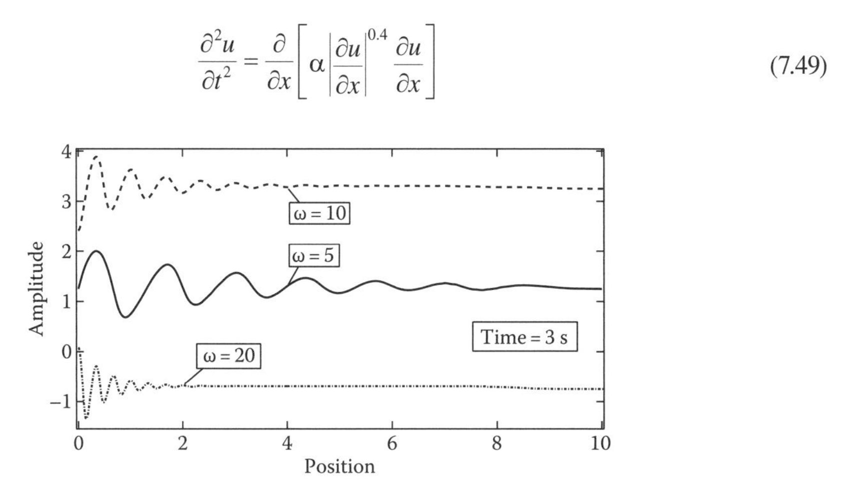

FIGURE 7.8 Snapshot of the waveform at $t = 3$ s for three different frequencies. The amplitude and velocity of the wave change as the frequency increases.

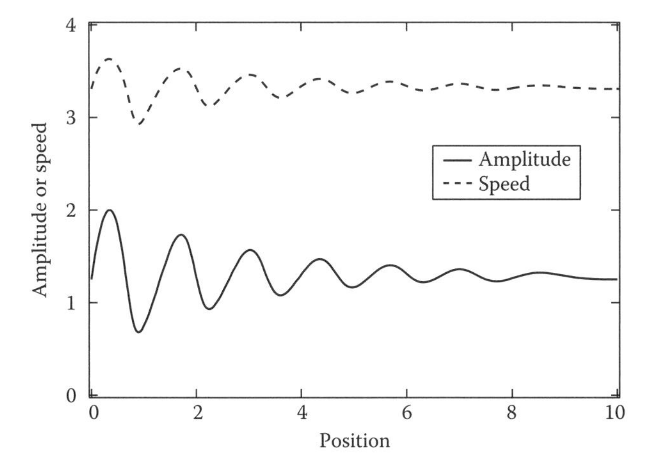

FIGURE 7.9 Speed and amplitude variations for a wave at $t = 3$ s. Changes in speed are pretty much in phase with amplitude changes since the dispersion is a function of u.

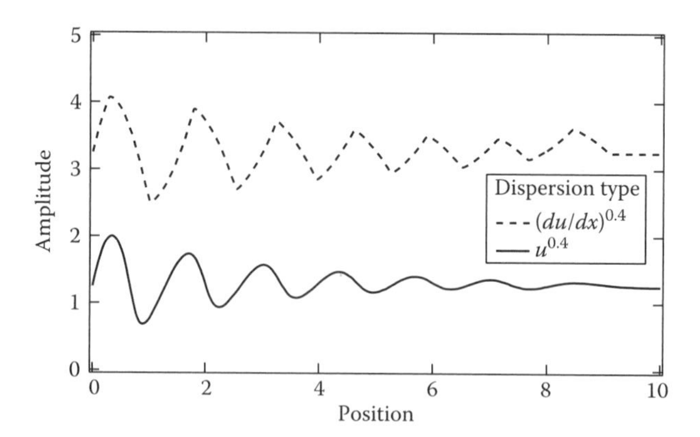

FIGURE 7.10 Snapshot of the waveform at $t = 3$ s. The dispersion function is based on either u or its derivative. Basic dispersion on the derivative affects the shape of the wave as well as the amplitude response.

Figure 7.10 shows the results of these calculations. The boundary conditions remain the same as those used for solving Equation 7.48, and the graph looks at a frequency $\omega = 5\ \mathrm{s}^{-1}$ at a snapshot in time of 3 s. Clearly, dispersion based on the derivative of u sharpens the waveform making it more "sawtooth" like and also decreases the rate of amplitude decay.

7.3 SUMMARY

In this chapter, we have seen how wavelike transport originates from a feedback mechanism involving *Gen* and *Acc* terms in the balance equations. This feedback results in the conservation of the transport quantity (momentum, energy) without dissipation and hence wave motion. We have demonstrated this mechanism for acoustic waves and for tidal waves. We have also shown how dissipation affects the wave and causes its amplitude to decrease over time and space.

There are many other wave systems that can be found [2] including gravitational waves, waves resulting from oscillating chemical reactions [3], waves propagating through chromatography columns, seismic waves, and optical solitons (waves that propagate indefinitely through optical fibers with no loss in amplitude or change in shape) [4]. We could not go into much more detail than what was presented, but the study of waves is a fascinating area of interest in many fields.

At present, we have gone through the general balance equation term by term. We have said nothing about designing systems for efficient transport operations. As the flux gradient relationships showed, the flow of a transport quantity is a function of the properties of the medium where transport is occurring, the gradient in state variable driving the transport, and the area through which the transport occurs. Often, we cannot change the properties of the medium nor can we make the gradients larger. Thus, to achieve more efficient transport, we must find a way to increase the area. This is the subject of the next chapter on extended surfaces.

PROBLEMS

Tidal Waves

7.1 Often a linearly sloped beach is not a good model. Waves coming in from deep water up a gradually shelving beach are better approximated by assuming a parabolic shape like that shown in Figure P7.1. If $h \propto x^2$,

a. Derive the differential equation describing the height of the wave as a function of space and time.

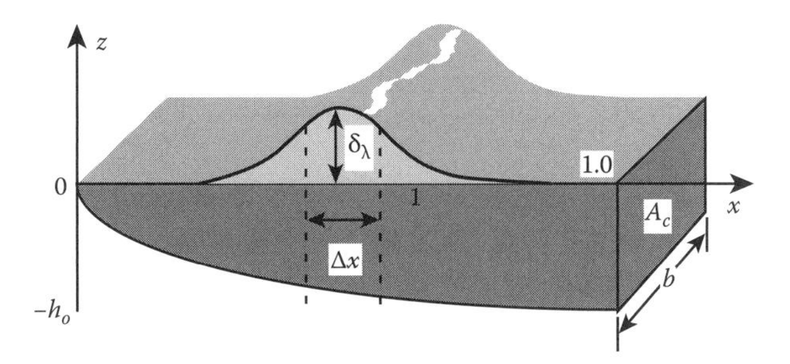

FIGURE P7.1

b. Solve the equation for an arbitrary initial surface, $f(x)$. Assume the wave height is zero at $x = 0$, 1 and periodic with period 1 and that the wave height is constant for all time $t < 0$. Seek solutions of the form

$$\delta_\lambda = x^n\left[a_n \cos\left(n\sqrt{h_o g}\ t\right) + b_n \sin\left(n\sqrt{h_o g}\ t\right)\right]$$

7.2 We can look at a slight variant of the problem earlier (Problem 7.1). Given the differential equation

$$\frac{\partial^2 \delta}{\partial t^2} = \frac{\partial}{\partial x}\left(gh\frac{\partial \delta}{\partial x}\right) \qquad h = \frac{x^2}{2b}$$

show that the following wave solution satisfies the following differential equation:

$$\delta = \frac{A}{\sqrt{x}}\cos\left\{p\left(\sqrt{\frac{2b}{g} - \frac{1}{4p^2}}\ \ln x + t\right) + \alpha\right\}$$

7.3 It is possible to consider the effect of small disturbing forces acting on water waves in a canal of depth h. Since the height of a wave is assumed to be very small, the only component of force that we need to consider is that component acting along the canal in the x-direction. In essence, we add the force into the momentum equation before combining it with the continuity equation. Show that for an arbitrary force per unit mass, $f(x)$, and a canal like that shown in Figure 7.1 in the text, the differential equation describing the amplitude of the wave is given by

$$\left[\frac{gA_c}{b}\right]\frac{\partial^2 \delta_\lambda}{\partial x^2} - \left[\frac{A_c}{b}\right]\frac{\partial f}{\partial x} = \frac{\partial^2 \delta_\lambda}{\partial t^2}$$

7.4 Consider a shallow lake on the equator and assume that the only tide-raising force is due to the Moon. In Figure P7.4, O represents the center of the Earth. M is the point on the equator directly under the Moon. G represents 0° longitude at the equator, and P is the point we are interested in at θ° longitude. The Moon moves with a constant angular velocity, v_θ, so that at any time, t, the angle between the Moon and G is $v_\theta t + \phi$. The tide-raising force, per unit mass of water, at P is $f = f_o \sin(v_\theta t + \phi + \theta)$.

a. Using the results of the previous problem (Problem 7.3), show that the differential equation describing the wave height can be written as

$$\frac{\partial^2 \delta_\lambda}{\partial t^2} = \left[\frac{c_\lambda^2}{r_o^2}\right]\frac{\partial^2 \delta_\lambda}{\partial \theta^2} + \left[\frac{2h_o f_o}{r_o}\right]\cos(v_\theta t + \phi + \theta)$$

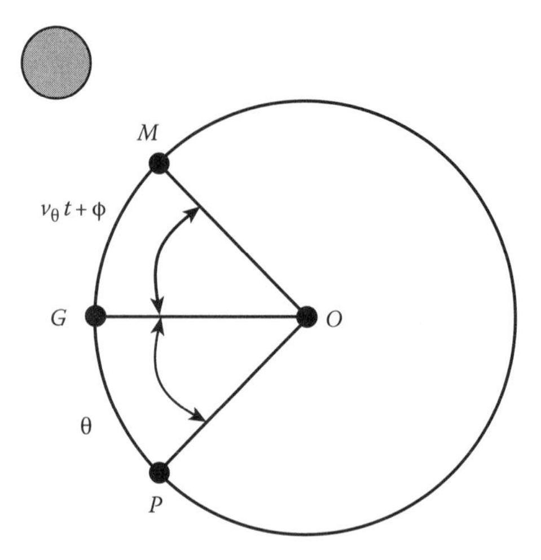

FIGURE P7.4

where
r_o is the radius of the Earth
h_o is the depth of the lake

b. The solution can be obtained in two parts: a solution to the unforced equation and a particular solution that gives the oscillations due to the tides. Determine this particular solution $\delta_\lambda = A\cos(v_\theta t + \phi + \theta)$.

7.5 Consider a spherical wave formed from a disturbance at the center of a shallow pool. At time $t = 0$, a small, steady stream of droplets begins hitting the pool. They cause a disturbance of amplitude, δ_o, with slope 0 at the center that spreads out in all directions. This disturbance results in conditions at the center so that the slope, height, and speed of the wave remain constant there.

a. Derive the differential equation describing the wave propagation for an infinite pool of depth, h_o.

b. Convert the differential equation to pseudo–Cartesian coordinates by forming a new height variable, $\eta = r\delta_\lambda$. Solve the differential equation for the wave height, δ_λ.

7.6 Consider an isolated wave traveling in a shallow canal. The wave strikes the end of the canal. We have shown that the wave is reflected with only a change in phase. Show that during the impact of the wave on the wall, the water rises to twice the normal height of the isolated wave.

7.7 We can transform the wave equation from a partial differential equation to an ordinary differential equation by considering sustained harmonic motion of the type cos $(nt + \phi)$. Using the equation for the wave height in a channel of varying cross section,

$$\left[\frac{g}{b}\right]\frac{\partial}{\partial x}\left(hb\frac{\partial \delta_\lambda}{\partial x}\right) = \frac{\partial^2 \delta_\lambda}{\partial t^2}$$

consider an estuary, like the Hudson River, for which $b = \beta x/\alpha$, $h = \gamma x/a$, and $0 < x < a$. The estuary communicates with the open sea at $x = a$ where a sustained tidal oscillation of

$$\delta_\lambda = K\cos(nt + \phi)$$

occurs. Show that the tidal waves of the estuary are given by

$$\delta_\lambda = K\frac{J_1\left(\sqrt{4\kappa x}\right)}{J_1\left(\sqrt{4\kappa a}\right)}\sqrt{\frac{a}{x}}\cos(nt + \varepsilon)$$

Sound Waves

7.8 Many musical instruments operate by the formation of plane waves inside a cylindrical tube. To describe them, we seek periodic solutions to the wave equation of the form

$$\delta_\lambda = f(x)\cos(nt)$$

These solutions represent stationary waves in the tube.

a. Show that the stationary wave solutions to the wave equation must obey the following ordinary differential equation:

$$\frac{d^2 f}{dx^2} + \frac{n^2}{c_\lambda^2} f = 0$$

where c_λ is the wave velocity.

b. The ends of our tube may be open or closed. If closed, the velocity vanishes there ($\partial\delta_\lambda/\partial x = 0$). Consider a tube of length, L, closed at both ends. Determine the generic solution for such a case. What is the lowest frequency note (fundamental) that can be sounded in such a configuration?

c. If a tube end is open, the pressure must equal atmospheric pressure at the end and so $\partial\delta_\lambda/\partial t = 0$. Consider a tube of length L, whose end at $x = 0$ is closed and whose end at $x = L$ is open. Determine the generic solution for this case. What is the fundamental frequency?

d. Is there any difference in fundamental frequency for a tube with both ends closed or both ends open?

7.9 Sound waves can deeply influence our emotional state and certain combinations of waves sound particularly appealing. Chords in general are composed of three separate frequencies. A major chord is composed of a fundamental tone, a tone a major third above the fundamental and a tone a major fifth above. If the fundamental tone is of frequency 1, the major chord consists of frequencies 1:1.25:1.5. A minor chord is likewise 1:1.2:1.5:

$$\delta = A\exp[i(2\pi\omega_i t - k_i x)] \qquad k_i = \frac{2\pi}{\lambda_i}$$

Sum up the amplitudes for a major chord assuming A is the same for each, the fundamental frequency is 220 Hz, and the speed of sound is 343.2 m/s and plot the waveforms you observe at $x = 0$; 1.

7.10 Some feel that chords exist because they sound more pleasing than single tones.

a. Compare the waveforms for the major and minor harmonic chords with something that does not generally sound so pleasing like the dissonant chord, 1:1.12:1.25. What do you observe? (Use the same information as in Problem 7.9.)

b. We assume that the amplitude of all the individual tones is the same. Can we make a dissonant chord more pleasing by altering the amplitudes?

7.11 A point source of sound gives rise to a wave whose amplitude can be described by

$$\delta = \left(\frac{B}{r}\right)\cos(\omega t - kr)$$

If the energy density per unit radial position associated with the wave is given by

$$E = \frac{1}{2}\rho A\left(\frac{\partial\delta}{\partial t}\right)^2$$

where A is the area through which the wave flows, show that the mean rate of transmission of energy across the surface of a sphere is given by

$$2\pi\rho c k^2 A^2$$

General 1-D Waves

7.12 Consider an effectively infinite transmission line (telephone line). Initially the line is dead, with both potential, Φ, and current, $d\Phi/dt$, being equal to zero. At $t = 0$, we impose a signal at the origin of the line:

$$\Phi(0,t) = \Phi_o \cos(\omega t)$$

Solve the telegrapher's equation:

$$\frac{\partial^2 \Phi}{\partial x^2} = LC\frac{\partial^2 \Phi}{\partial t^2} + (RC + GL)\frac{\partial \Phi}{\partial t} + RG\Phi$$

for the transient and steady-state responses. Here, L is the inductance, C is the capacitance, R is the resistance, and G is the impedance of the line.

7.13 Alternating current flowing through a conductor prefers to flow close to the surface, so the current density in the conductor increases with radial position. If we assume the current is of the form $I = q_e''(r)\exp(i\omega t)$, the differential equation describing the current density, q_e'', becomes

$$\frac{d^2 q_e''}{dr^2} + \frac{1}{r}\frac{dq_e''}{dr} - i\left(\frac{4\pi\omega^2}{\rho}\right)q_e'' = 0$$

where ρ is the resistivity of the conductor. Determine the current density as a function of radial position for a cylindrical conductor of radius, r_o, when the current density at the surface is q_e''.

References

1. Fletcher, N.H. and Rossing, T.D., *The Physics of Musical Instruments*, Springer-Verlag, New York (1991).
2. Bertram, B. and Sandiford, D.J., Second sound in solid helium, *Sci. Am.*, **222**, 92 (1970).
3. Simoyi, R.H., Wolf, A., and Swinney, H.L., One dimensional dynamics in a multicomponent chemical reaction, *Phys. Rev. Lett.*, **49**, 245 (1982).
4. Guenther, R.D., *Modern Optics*, John Wiley & Sons, New York (1990).

8 Transport Enhancement Using Extended Surfaces

8.1 INTRODUCTION

Our discussions of transport phenomena have shown that the quantity of information we can transport depends, in part, upon the amount of surface through which this information can flow. Thus, the heat flow, q, is directly related to the surface area through which the heat flux, q'', passes. All our examples have focused on surfaces or interfaces having very simple geometrical configurations: spheres, cylinders, or plane walls. In most real situations, these simple surfaces do not provide enough surface area for efficient transport to occur. To enhance the transport, we alter the amount of surface by attaching protuberances such as fins or by corrugating the surface, or by making a porous material. Such systems can look very complex but have the high surface area required for the job. Look at the back of an air conditioner or dehumidifier, for example.

Of course, we were not the first to encounter this problem of limited transport area. Nature ran into the same problem a billion or so years before us and solved it successfully by using extended surfaces. Examples of extended surfaces commonly found in organisms include gills in fish to enhance the surface area available for oxygen transport and bronchioli to enhance the exchange of oxygen and carbon monoxide in the lungs of land animals. Leaves and/or needles on trees provide sufficient surface area for energy absorption, and capillaries and blood vessels in animals provide the necessary nutrients to cells, assist in removing wastes, and aid in regulating body temperature. Even the plates of a stegosaurus or the ears of an elephant were primarily used to enhance the surface area available for heat transfer and help the huge beast regulate its body temperature. Virtually every system in nature, even inanimate ones such as a river delta or lightning bolt, exploits an extended surface to enhance or dissipate heat, mass, momentum, or charge.

Humans also discovered that extended surfaces could be used to enhance transport. Our first achievements in this area may have been sails to capture the wind, fins on arrows to stabilize their flight, windmills to provide power, and kindling to make starting fires easier. In this chapter, we will discuss three such examples in detail. We will first consider the use of fins to aid in heat transfer and then the way nature has incorporated fin-like projections in the intestine to enhance nutrient absorption (mass transfer). We will also consider how man has used extended surfaces in the development of catalyst materials to increase mass transfer and hence the rate of reaction in petroleum refining, polymer manufacture, etc.

Though we do not treat momentum transfer in detail, please keep in mind that extended surfaces are used extensively. Examples include water wheels, windmills, fins in fishes and aquatic mammals, screw-type pumps, bird/airplane wings, feathers, and dandelion or maple seeds. In fact, flight would not be possible without extended surfaces. For example, the lifting force, L_f, experienced by a wing is defined by

$$L_f = C_L A \left[\frac{\rho v_s^2}{2} \right] \tag{8.1}$$

and is directly proportional to A, the chord area of the lifting vane. Anyone who has held a hand out from the window of a car has experienced the use of a moveable extended surface for manipulating momentum transport in this fashion.

8.2 HEAT TRANSFER: FINNED SURFACES

One of the most useful ways of enhancing heat transfer from a surface is to increase its surface area by the use of projections or fins. Finned surfaces are used in automobile radiators, home heating radiators, semiconductor chip packages, and motor housings to name but a few applications. They are generally long, thin extensions to a surface and come in a variety of profiles ranging from cylindrical pins, to rectangular bars, to triangular or pyramidal shapes. Several are shown in Figure 8.1.

8.2.1 Rectangular Fins: Constant Cross-Sectional Area

We can begin an analysis of heat transfer in a fin by considering one of the most common geometries, the rectangular fin. Since fins are almost always placed within a fluid (here on Earth) and their extra surface area is always aimed at increasing convective heat transfer from the surface, we have two modes of heat transfer occurring simultaneously: conduction through the fin and convection from the fin's surface. Assuming the fin's thickness, t_h, is very small compared to its surface area, we can treat the problem as a case of 1-D conduction. The cross-sectional area of the fin is labeled A_c and the perimeter, P_d.

The use of A_c and P_d lets us extend our results to fins of other geometries. The particular fin geometry we have chosen in Figure 8.2 shows the fin immersed on five sides by a fluid. Four other cases are normally encountered:

1. The tip of the fin may be insulated rather than exposed to the fluid.
2. The fin may act as a bridge between two surfaces.
3. The fin tip may be kept at a constant temperature.
4. The fin may be so long that its tip temperature reaches that of the surrounding fluid (an infinite fin).

Fins are generally passive elements so we have no generation to consider. We will also assume that the fin properties and fluid properties do not depend upon temperature and that the system

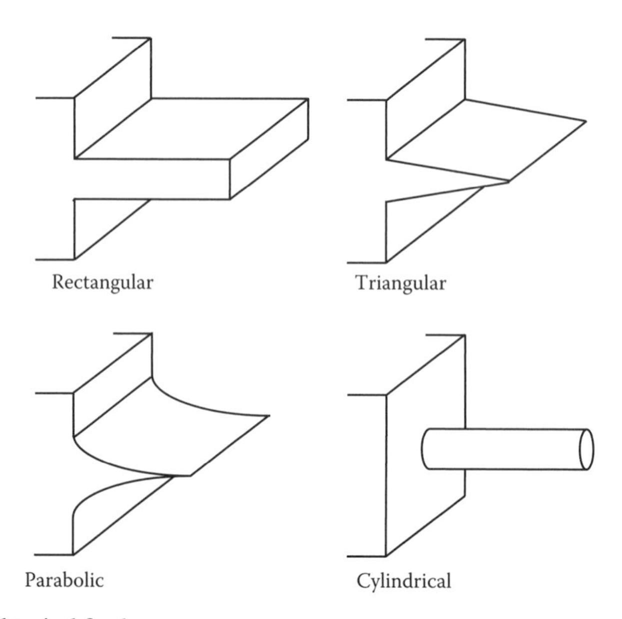

FIGURE 8.1 Several typical fin shapes.

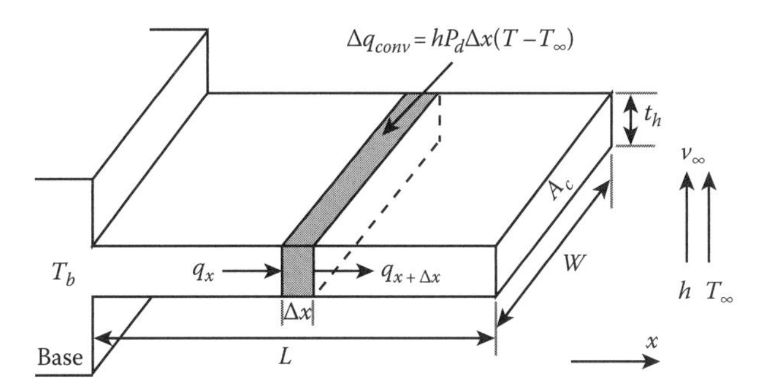

FIGURE 8.2 A rectangular fin immersed on all sides. Convection occurs from all fin surfaces including the tip.

operates at a steady state. We write an energy balance about a differential control volume located somewhere within the fin:

$$\begin{aligned} In \quad - \quad & Out \quad = Acc \\ q(x) - q(x+\Delta x) - & q_{conv} = 0 \end{aligned} \tag{8.2}$$

$$q(x) - q(x+\Delta x) - hP_d\Delta x(T - T_\infty) = 0 \tag{8.3}$$

Notice there is one *In* term due to conduction within the fin but there are two *Out* terms, one representing conduction through the control volume and the other representing convection from the surface of the control volume. Dividing through by Δx and taking the limit as $\Delta x \to 0$ gives us a differential equation that shows the heat flow decreases along the length of the fin:

$$-\frac{dq}{dx} - hP_d(T - T_\infty) = 0 \tag{8.4}$$

Using Fourier's law, $q = -kA_c dT/dx$, allows Equation 8.4 to be written in terms of the temperature along the length of the fin:

$$kA_c\frac{d^2T}{dx^2} - hP_d(T - T_\infty) = 0 \tag{8.5}$$

The boundary conditions for this problem consider what occurs at the base of the fin, $x = 0$, and at the tip of the fin, $x = L$. We usually specify a specific, constant temperature at the base, and in this example, we have a convective boundary condition at the tip:

$$x = 0 \qquad T = T_b \tag{8.6}$$

$$x = L \qquad -k\frac{dT}{dx} = h(T - T_\infty) \tag{8.7}$$

At this point, we could solve the equation directly and obtain the temperature profile. However, we can gain some more insight into the important variables governing the performance of a fin by putting Equation 8.5 into dimensionless form. We define two new variables:

$$\theta = \frac{T - T_\infty}{T_b - T_\infty} \qquad \xi = \frac{x}{L} \tag{8.8}$$

Substituting into Equation 8.5 gives

$$\frac{d^2\theta}{d\xi^2} - f^2\theta = 0 \tag{8.9}$$

The dimensionless parameter, f, can be called the *fin number* and may be thought of as a modified Biot number for the fin:

$$f = \sqrt{\frac{hP_dL^2}{kA_c}} = \sqrt{\frac{L}{kA_c}\Big/\frac{1}{hP_dL}} = \sqrt{\frac{\mathcal{R}_{conduction}}{\mathcal{R}_{convection}}} \tag{8.10}$$

The fin number represents the ratio of conduction resistance within the fin to convection resistance from the fin and identifies which heat transfer mechanism is controlling. Small fin numbers indicate heat transfer along the fin axis is easier than heat transfer to the surroundings. Fins made of material with high thermal conductivity are in this category. High fin numbers indicate significant internal resistance to heat conduction. Fins made of dielectric materials fall in this class.

To find a solution to Equation 8.9, we also transform the boundary conditions to read:

$$\xi = 0 \qquad \theta = 1 \tag{8.11}$$

$$\xi = 1 \qquad -\frac{d\theta}{d\xi} = \frac{hL}{k}\theta = \lambda_a f^2\theta \tag{8.12}$$

where $\lambda_a = A_c/LP_d$ is the ratio of cross-sectional to surface area. In most fins, $\lambda_a \ll 1$. The solution to Equation 8.9 is found by substituting $\theta = Ae^{rx}$ into the differential equation and is given in terms of hyperbolic sine and cosine or, alternatively, exponential functions:

$$\theta = a_1\left[\frac{\exp(f\xi)+\exp(-f\xi)}{2}\right] + a_2\left[\frac{\exp(f\xi)-\exp(-f\xi)}{2}\right]$$

$$= a_1\cosh(f\xi) + a_2\sinh(f\xi) \tag{8.13}$$

Evaluating the integration constants gives

$$\theta = \cosh(f\xi) - \left[\frac{\lambda_a f\coth(f)+1}{\coth(f)+\lambda_a f}\right]\sinh(f\xi) \tag{8.14}$$

At steady state, we can treat the fin as a single object and perform an overall energy balance on the entire structure. This balance states the following:

Heat passing through the base = Heat removed by convection

The heat flowing through the base, q_b, is given by Fourier's law evaluated at $x = 0$:

$$q_b = -kA_c\frac{dT}{dx}\bigg|_{x=0} = hL(T_b - T_\infty)\int_0^1 \theta\, d\xi$$

$$= -\frac{kA_c}{L}(T_b - T_\infty)\frac{d\theta}{d\xi}\bigg|_{\xi=0} = (T_b - T_\infty)\frac{\sqrt{hP_dkA_c}}{f}\frac{d\theta}{d\xi}\bigg|_{\xi=0} \tag{8.15}$$

Evaluating q_b using the temperature profile, Equation 8.14, gives

$$q_b = \sqrt{hP_dkA_c}\left[\frac{\lambda_a f \coth(f)+1}{\coth(f)+\lambda_a f}\right](T_b - T_\infty) \tag{8.16}$$

Knowing q_b, we ask how much extra heat is removed by the fin. This extra heat is defined as the ratio between q_b and q if the fin were not present. In the absence of the fin, only that patch of surface covered by the base of the fin would be exposed to the fluid and available for heat transfer. We define the ratio of heat removed with the fin to heat removed in the absence of the fin as the *fin effectiveness*:

$$\frac{q_b}{q_{without\ fin}} = \text{Fin effectiveness} = \frac{\sqrt{hP_dkA_c}\left[\frac{\lambda_a f \coth(f)+1}{\coth(f)+\lambda_a f}\right](T_b - T_\infty)}{hA_c(T_b - T_\infty)}$$

$$= \frac{1}{\lambda_a f}\left[\frac{\lambda_a f \coth(f)+1}{\coth(f)+\lambda_a f}\right] \tag{8.17}$$

We have now calculated the benefits accrued by having a finned surface. Unfortunately, all fin materials have a finite thermal conductivity, and so we do not have the maximum driving force $(T_b - T_\infty)$ for heat removal existing over the entire length of the fin. A *perfect fin* would have this maximum driving force over its entire length:

$$q_{perfect} = \underset{\text{Fin sides}}{hP_dL(T_b - T_\infty)} + \underset{\text{Fin tip}}{hA_c(T_b - T_\infty)}$$

$$= h(P_dL + A_c)(T_b - T_\infty) \tag{8.18}$$

In reference to the perfect fin, the real fin operates with a certain *efficiency* defined as

$$\eta_f = \frac{q_f(T)}{q_{perfect}} = \frac{\sqrt{hP_dkA_c}}{h(A_c + P_dL)}\left[\frac{\lambda_a f \coth(f)+1}{\coth(f)+\lambda_a f}\right] \tag{8.19}$$

We can simplify this expression a bit by realizing that

$$\frac{\sqrt{hP_dkA_c}}{h(A_c + P_dL)} = \frac{1}{\sqrt{\frac{hP_dL^2}{kA_c}} + \sqrt{\frac{hA_c}{kP_d}}} = \frac{1}{(1+\lambda_a)f} \tag{8.20}$$

$$\eta_f = \frac{1}{(1+\lambda_a)f}\left[\frac{\lambda_a f \coth(f)+1}{\coth(f)+\lambda_a f}\right] \quad \text{Fin efficiency} \tag{8.21}$$

The fin number, f, is the controlling parameter in the differential equation governing fin performance. It is reasonable to ask what happens to the fin efficiency and effectiveness as we vary the fin number to signify no conduction resistance in the fin or no convective resistance to heat flow from the fin:

No conduction resistance $f \to 0$:

$$\eta_f \to 1 \quad \text{small } f \text{ desirable} \tag{8.22}$$

No convective resistance $f \rightarrow \infty$:

$$\eta_f \rightarrow 0 \quad \text{large } f \text{ undesirable} \tag{8.23}$$

Clearly, decreasing the resistance to conduction through the fin results in much higher fin efficiencies. If we turn our attention to the heat removed, q_b, we find the following:

No conduction resistance $f \rightarrow 0$

$$\text{Effectiveness} \rightarrow \infty \quad \text{small } f \text{ desirable} \tag{8.24}$$

No convection resistance $f \rightarrow \infty$

$$\text{Effectiveness} \rightarrow 0 \quad \text{large } f \text{ undesirable} \tag{8.25}$$

Notice that fins operate best in situations where the external resistance to heat transfer is higher than the internal resistance. It does not pay to use fins in areas where the intrinsic heat transfer coefficients are so high that any fin number would be too high to be practical.

Example 8.1 Composite Fins

Let us consider the example of a composite fin shown in Figure 8.3. A carbon steel fin of circular cross section is coated with a thin dielectric layer for corrosion protection. The dielectric layer serves as a protective shield. The composite fin is immersed in a brine solution whose velocity is such that the heat transfer coefficient, $h = 1000$ W/m^2 K. The base temperature of the fin is 150°C and the temperature of the brine is 20°C. Based on the physical dimensions of the fin and the thermal properties of the steel and dielectric, we would like to calculate the heat removed per fin, the fin effectiveness, and the fin efficiency:

$$L = 0.05 \text{ m} \qquad d_s = 0.01 \text{ m} \qquad t_h = 0.001 \text{ m}$$

$$k_s = 45 \text{ W/m K} \qquad k_d = 0.5 \text{ W/m K}$$

We treat this fin as we did the composite sphere in Chapter 4. Our first task is to decide what we wish to view as the fin. Remember to use the fin analysis we need to have a 1-D problem. This means that we want the fin number to be small. If we take both the steel core and dielectric cladding as the fin, the internal resistance to heat transfer will be too high and the fin number will be high. Internal gradients will be present and the analysis will fail. On the other hand, if we restrict our analysis to the steel core, the fin number will be low and we can obtain an accurate solution. The energy balance about the control volume of Figure 8.3 is

$$\begin{array}{ccccccccc} In & - & Out & & + & Gen & = & Acc \\ q(x) & - & q(x+\Delta x) & - \; q_{conv} & + & 0 & = & 0 \end{array}$$

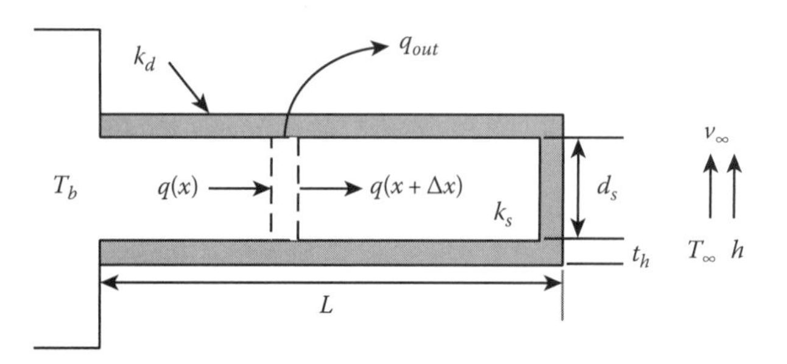

FIGURE 8.3 Composite fin having a steel core and dielectric cladding.

The second *Out* term, q_{conv}, considers all the heat being removed from the fin's surface. Clearly, we have two modes of heat transfer occurring at the surface, conduction through the dielectric layer and convection from the surface. Since we are operating at a steady state, we can use the sum of resistance concept to calculate the overall, areal-based resistance (m^2 K/W) to heat transfer from the steel core and hence this *Out* term. If the cladding is thin, we can neglect its curvature and write the overall resistance as a combination of a convective resistance and conduction through a plane wall:

$$\mathcal{R}_t = \frac{1}{h} + \frac{t_h}{k_d}$$

We define an overall heat transfer coefficient as

$$\mathcal{U} = \frac{1}{\mathcal{R}_t} = \frac{1}{(1/1000 + 0.001/0.5)} = 333\ \text{W/m}^2\,\text{K}$$

so q_{conv} and the balance equation become

$$q_{conv} = \mathcal{U} P_d \Delta x (T - T_\infty)$$

$$q(x) - q(x + \Delta x) - \mathcal{U} P_d \Delta x (T - T_\infty) = 0$$

Dividing through by Δx, taking the limit as $\Delta x \to 0$, and substituting Fourier's law to express the heat flow in terms of the temperature gradient gives

$$\frac{d^2T}{dx^2} - \frac{\mathcal{U} P_d}{k_s A_c}(T - T_\infty) = 0$$

subject to the boundary conditions that

$$x = 0 \qquad T = T_b$$

$$x = L \qquad \frac{dT}{dx} = 0$$

This equation is a familiar problem. We need only define a dimensionless temperature and spatial coordinate to express the solution in terms of hyperbolic sines and cosines. Defining temperature and position as

$$\theta = \frac{T - T_\infty}{T_b - T_\infty} \qquad \xi = \frac{x}{L}$$

allows us to write the differential equation and solution as

$$\frac{d^2\theta}{d\xi^2} - f^2\theta = 0 \qquad f^2 = \frac{\mathcal{U} P_d L^2}{k_s A_c}$$

$$\theta = a_1 \sinh(f\xi) + a_2 \cosh(f\xi)$$

The boundary conditions become transformed to

$$\xi = 0 \qquad \theta = 1$$

$$\xi = 1 \qquad \frac{d\theta}{d\xi} = 0$$

Using these boundary conditions, the final solution is

$$\theta = \cosh(f\xi) - \tanh(f)\sinh(f\xi)$$

The heat flowing through the base is determined from Fourier's law:

$$q(T_b) = -k_s A_c \left.\frac{dT}{dx}\right|_{x=0} = -\frac{k_s A_c (T_b - T_\infty)}{L} \left.\frac{d\theta}{d\xi}\right|_{\xi=0}$$

$$= \frac{k_s A_c (T_b - T_\infty)}{L} f \tanh(f) = 64.73 \text{ W}$$

Using the definitions for the fin effectiveness and fin efficiency, we find

$$\text{Fin effectiveness} = \frac{\dfrac{k_s A_c f}{L}\tanh(f)}{hA_c} = 1.585$$

$$\eta_f = \frac{\dfrac{k_s A_c f}{L}\tanh(f)}{hP_d L} = 0.317$$

Notice that we define the effectiveness and efficiency based on the heat transfer coefficient, h, not the overall coefficient, $\mathcal{U}$. In the case of effectiveness, the base of the fin has no cladding on it. When we consider the efficiency, the perfect fin would have the temperature at the dielectric equal to the base temperature, and so we would have no resistance due to the dielectric.

8.2.2 Cylindrical Fins: Constant Cross-Sectional Area

One of the most common fin geometries is the circumferential fin shown in Figure 8.4. This type of fin is used in many heating and cooling applications. If one has forced hot water, baseboard heat in a home, the chances are very good that the tubing carrying the water has been fitted with circumferential fins. We are interested in the amount of heat removed per fin and the fin efficiency and fin effectiveness. We can use the same form of analysis we applied to describe the rectangular fin to obtain the information we want. Here we take a thin, annular slice of width Δr through

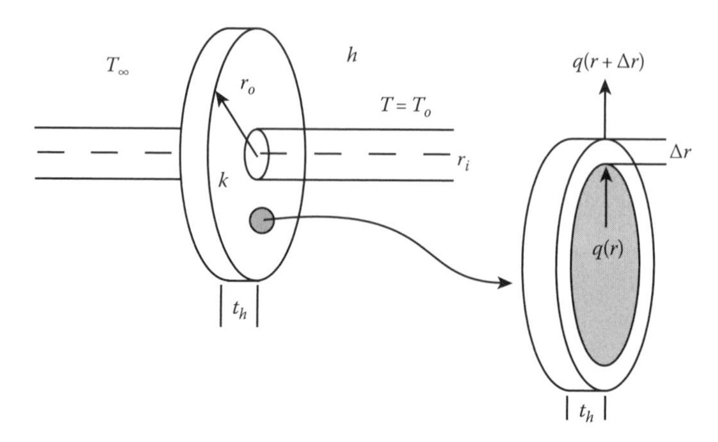

FIGURE 8.4 Circumferential fin surrounding a tube.

the fin. There are two heat transfer mechanisms working: conduction radially through the fin and convection from the exposed faces. The balance equation takes the following form:

$$\begin{array}{ccccccccc} In & - & Out & & & + & Gen & = & Acc \\ q(r) & - & q(r+\Delta r) & - & h(2\pi r\Delta r)(T-T_\infty) & + & 0 & = & 0 \end{array} \tag{8.26}$$

Dividing by Δr and taking the limit as $\Delta r \to 0$ gives

$$-\frac{dq}{dr} - 2\pi r h(T - T_\infty) = 0 \tag{8.27}$$

Substituting Fourier's law for the heat flow to write the differential equation in terms of the temperature,

$$q = -k(2\pi r t_h)\frac{dT}{dr} \tag{8.28}$$

gives the final form of the following differential equation:

$$\frac{d}{dr}\left(r\frac{dT}{dr}\right) - \left(\frac{h}{t_h k}\right) r(T - T_\infty) = 0 \tag{8.29}$$

If the fin is very thin and extends for at least one tube diameter beyond the tube surface, we can approximate the fin as being infinitely long. The boundary conditions for this situation specify the base temperature as the tube wall temperature, while the tip temperature is equal to the ambient fluid temperature:

$$r = r_i \qquad T = T_o \tag{8.30}$$

$$r = r_o \qquad T = T_\infty \tag{8.31}$$

Letting

$$\theta = \frac{T - T_\infty}{T_o - T_\infty} \qquad \xi = \frac{r}{r_i} \tag{8.32}$$

we can solve the differential equation in terms of modified Bessel functions introduced in Chapter 6:

$$\theta = a_1 I_o(f\xi) + a_2 K_o(f\xi) \tag{8.33}$$

The boundary conditions become

$$\xi = 1 \qquad \theta = 1 \tag{8.34}$$

$$\xi = \xi_o = \frac{r_o}{r_i} \qquad \theta = 0 \tag{8.35}$$

and $f = \sqrt{hr_i^2/t_h k}$ is the fin number for the circumferential fin. Applying the boundary conditions yields the final temperature profile:

$$\theta = \frac{I_o(f\xi)K_o(f\xi_o) - K_o(f\xi)I_o(f\xi_o)}{I_o(f)K_o(f\xi_o) - K_o(f)I_o(f\xi_o)} \tag{8.36}$$

The heat removed by the fin is defined in terms of the heat passing through the base and is given by Fourier's law. The fin efficiency and fin effectiveness are defined exactly as we defined them for the rectangular fin:

$$q = -k(2\pi r_i t_h)\left.\frac{dT}{dr}\right|_{r_i} = -k(T_o - T_\infty)(2\pi t_h)\left.\frac{d\theta}{d\xi}\right|_1$$

$$= -kf(T_o - T_\infty)(2\pi t_h)\left[\frac{I_1(f)K_o(f\xi_o) + K_1(f)I_o(f\xi_o)}{K_o(f)I_o(f\xi_o) - I_o(f)K_o(f\xi_o)}\right] \tag{8.37}$$

$$eff = \frac{-k\left.\frac{d\theta}{d\xi}\right|_1}{hr_i} = -\frac{kf}{hr_i}\left[\frac{I_1(f)K_o(f\xi_o) + K_1(f)I_o(f\xi_o)}{K_o(f)I_o(f\xi_o) - I_o(f)K_o(f\xi_o)}\right] \tag{8.38}$$

$$\eta_f = \frac{-(kt_h)\left.\frac{d\theta}{d\xi}\right|_1}{h\left(r_o^2 - r_i^2\right)} = -\frac{kft_h}{h\left(r_o^2 - r_i^2\right)}\left[\frac{I_1(f)K_o(f\xi_o) + K_1(f)I_o(f\xi_o)}{K_o(f)I_o(f\xi_o) - I_o(f)K_o(f\xi_o)}\right] \tag{8.39}$$

The fin efficiency for the circumferential fin and fins of other geometry are shown in Figure 8.5. These graphs assume an insulated fin tip boundary condition but are corrected for use with a convective tip condition. The correction is in the form of an alternative length, L_c, that is the cross-sectional area, A_c, divided by the perimeter, P_d.

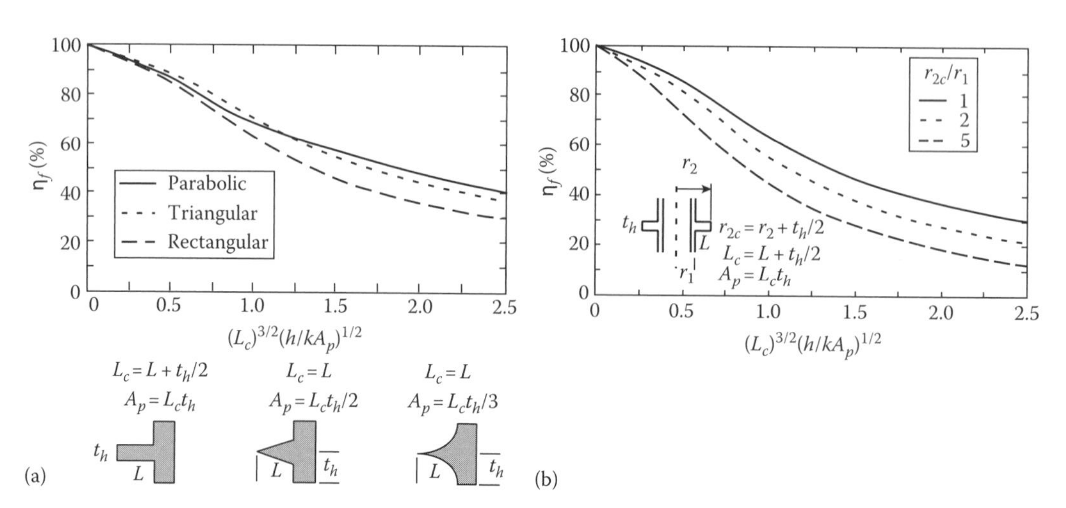

FIGURE 8.5 (a) Fin efficiency for rectangular, triangular, and parabolic profiles. (b) Efficiency for a circumferential fin of rectangular profile. (Adapted from Incropera, F.P. and DeWitt, D.P., *An Introduction to Heat Transfer*, 3rd edn., John Wiley & Sons, New York, 1996. With permission.)

8.3 MASS TRANSFER: GILLS, LUNGS, ETC.

The use of extended surfaces to enhance the rate of mass transfer is a very common occurrence especially in biological organisms. Consider the absorption of nutrients in your small intestine. The small intestine is a long, convoluted tube whose inside surface is covered by hairlike protrusions called villi. These villi, shown schematically in Figure 8.6, are actually mechanisms for increasing the surface area available for the absorption of nutrients and the elimination of waste.

The same type of analysis that we employed to describe finned surface heat transfer we can use for extended surface mass transfer. As an example, we can model the small intestine as a long tube with cylindrical fins protruding from it. The nutrient we want to absorb, b, is carried by convection in the space outside the villi, and this nutrient will be exchanged for a waste item, a. Figure 8.7 shows the model system in more detail, considering a single villus. Species b moves through the villus with diffusivity, D_{ab}. There is a resistance to convection characterized by a mass transfer coefficient, k_c, and a partition coefficient between the villus and free stream characterized by an equilibrium constant, K_{eq}. The partition coefficient will take any transport resistance across the villus membrane into account. We can write the balance equation exactly as we did for the fin problem considering the system at a steady state and having no generation:

$$\begin{array}{ccccccccc} In & - & Out & & + & Gen & = & Acc \\ \dot{M}_b(x) & - & \dot{M}_b(x+\Delta x) & - \; \dot{M}_{b,conv} & + & 0 & = & 0 \end{array} \tag{8.40}$$

Substituting for $\dot{M}_b$ and $\dot{M}_{b,conv}$, dividing through by Δx, and taking the limit as $\Delta x \to 0$ yields a differential equation for N_b:

$$-\frac{d}{dx}(A_c N_b) - k_c P_d(c_b - K_{eq} c_{b\infty}) = 0 \tag{8.41}$$

To simplify the problem and to show how the fin and villus problem are related, we must assume that the nutrient, b, is either very dilute or that a and b migrate through the villus via equimolar

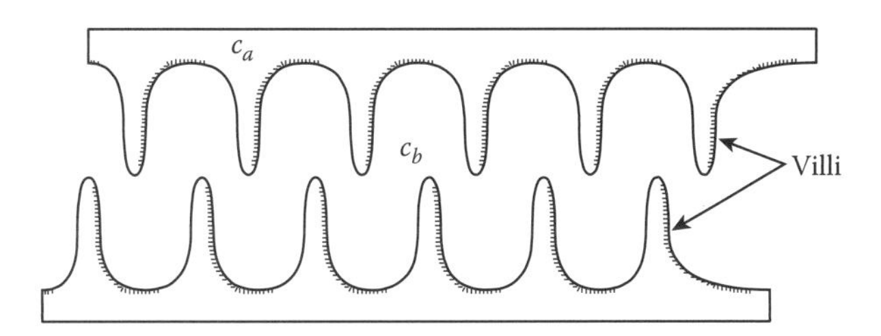

FIGURE 8.6 Schematic diagram of villi inside the small intestine (not to scale). Villi are the mass transfer analog of heat transfer fins.

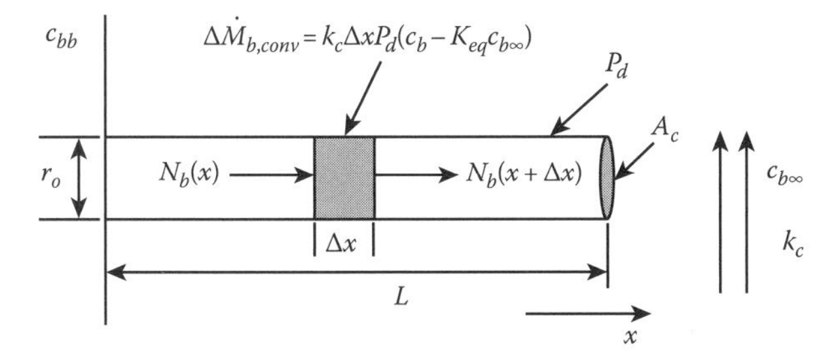

FIGURE 8.7 Diagram of a single villus modeled as a cylindrical fin.

counterdiffusion. This allows us to use the simple form for Fick's law and to convert the balance equation into a differential equation in terms of the concentration of species b:

$$\frac{d^2 c_b}{dx^2} - \frac{k_c P_d}{A_c D_{ab}}(c_b - K_{eq} c_{b\infty}) = 0 \tag{8.42}$$

We put this equation in dimensionless form by defining a new concentration and new position variable:

$$\chi = \frac{c_b - K_{eq} c_{b\infty}}{c_{bb} - K_{eq} c_{b\infty}} \qquad \xi = \frac{x}{L} \tag{8.43}$$

Substituting into Equation 8.42 gives

$$\frac{d^2 \chi}{d\xi^2} - f_m^2 \chi = 0 \tag{8.44}$$

This equation has exactly the same form as the fin Equation 8.9. The villi number, f_m, represents the ratio of the diffusion resistance through the villi to the convection resistance outside the villi. It is nearly identical in form and meaning to the fin number, f, and is given by

$$f_m = \sqrt{\frac{k_c P_d L^2}{A_c D_{ab}}} = \sqrt{\frac{L}{A_c D_{ab}} \bigg/ \frac{1}{k_c P_d L}} = \sqrt{\frac{\mathcal{R}_{diffusion}}{\mathcal{R}_{convection}}} \tag{8.45}$$

Small villi numbers lead to the best mass transfer performance.

We will use the following boundary conditions assuming that the tip of the villus is somehow cut off from allowing mass transfer to occur (impermeable tip). Physiologists often refer to this condition as a sealed-end condition:

$$\xi = 0 \qquad \chi = 1 \tag{8.46}$$

$$\xi = 1 \qquad \frac{d\chi}{d\xi} = 0 \tag{8.47}$$

The solution to Equation 8.44 using these boundary conditions is

$$\chi = \cosh(f_m \xi) - \tanh(f_m)\sinh(f_m \xi) \tag{8.48}$$

It is important to realize that we can now define a villus effectiveness and a villus efficiency in the same manner in which we defined the fin effectiveness and fin efficiency:

$$eff = \frac{\dot{M}_{bb}}{\dot{M}_{b,without\ fin}} = \frac{\tanh(f_m)}{\lambda_a f_m} \quad \text{Villus effectiveness} \tag{8.49}$$

$$\eta_f = \frac{\dot{M}_{bb}}{\dot{M}_{b,perfect\ fin}} = \frac{\tanh(f_m)}{f_m} \quad \text{Villus efficiency} \tag{8.50}$$

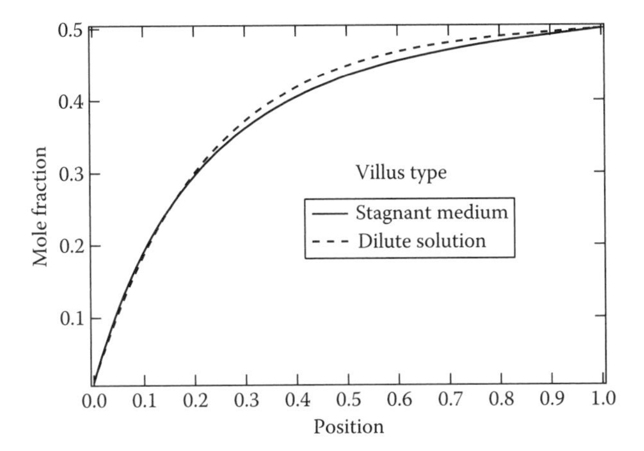

FIGURE 8.8 Comparison of villus equations assuming the dilute solution or stagnant medium approximations. Here, $K_{eq} = 0.5$.

The villus equation we have just solved was based on the assumption of dilute solutions or equimolar counterdiffusion. If we relax these restrictions and assume that the material comprising the villus behaves as a stagnant medium, the differential equation becomes nonlinear and we can no longer solve it analytically. The new villus equation is

$$\frac{d}{d\xi}\left(\frac{1}{1-x_b}\frac{dx_b}{d\xi}\right) - f_m^2(x_b - K_{eq}x_{b\infty}) = 0 \tag{8.51}$$

where

f_m has the same definition as in Equation 8.45
x_b is the mole fraction of the diffusing species

Applying the boundary conditions of Equations 8.46 and 8.47, we can solve Equation 8.51 using Comsol® or similar program and compare that solution with Equation 8.48. Figure 8.8 shows this comparison. Notice that the overall shape of both curves is the same. The absolute magnitude is also not very different between the two solutions mainly because the differential term in Equation 8.51 is only significantly different as $x_b \to 1$. As K_{eq} gets larger, this term then becomes significant. For Figure 8.8, $K_{eq} = 0.5$.

Example 8.2 Facilitated Transport

So far, we have been dealing with extended surfaces that are passive in operation. In many instances, just increasing the available surface area is not enough to enhance transport to the required level. This is particularly true for mass transport where diffusivities are much smaller than thermal conductivities. Biological organisms often couple extended surface transport with a facilitated mechanism to enhance mass transport. We have talked about facilitated transport in the past and have shown how we can model it using a chemical reaction at the surface. We can combine extended surface and facilitated transport by looking at a villus that employs a first-order chemical reaction to enhance the rate of nutrient absorption. Figure 8.9 shows a diagram of the villus and the control volume we will be using. Again, we assume a cylindrical, pin fin geometry. The physical parameters for the system are given in the following:

$L = 0.0001$ m $r = 0.0005$ m $D_{ab} = 1\times10^{-7}$ m²/s

$c_{b\infty} = 100$ mol/m³ $c_{bb} = 10$ mol/m³ $k_c = 1\times10^{-4}$ m/s

$k_s'' = 1\times10^{-5} - 1\times10^{-4}$ m/s $K_{eq} = 1$

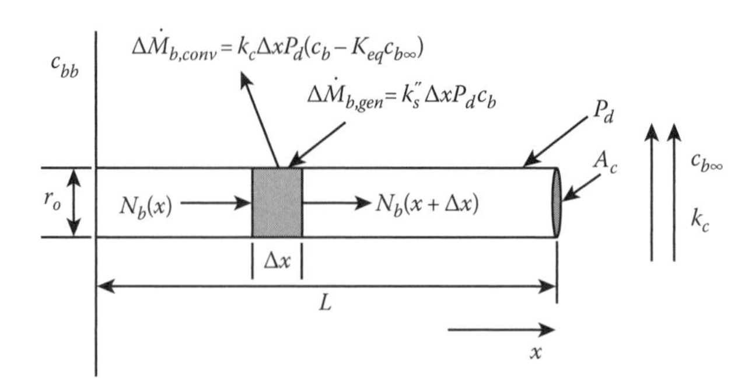

FIGURE 8.9 Single villus with facilitated transport at the surface. k_s'' is the surface reaction rate constant.

The balance equation must consider the generation term on the surface as well as diffusion through the villus and convective mass transfer from the surface. Here, the generation term is positive because the reaction enhances the absorption of b into the villus:

$$\begin{array}{ccccccccc} In & - & Out & & & + & Gen & = & Acc \\ \dot{M}_b(x) & - & \dot{M}_b(x+\Delta x) & - & k_c P_d \Delta x(c_b - K_{eq}c_{b\infty}) & + & k_s'' P_d \Delta x c_b & = & 0 \end{array}$$

Dividing through by Δx, substituting for $\dot{M}_b$, and using the limiting procedure yields a differential equation for the mass flux, N_b:

$$-\frac{d}{dx}(A_c N_b) - k_c P_d(c_b - K_{eq}c_{b\infty}) + k_s'' P_d c_b = 0$$

Assuming we transport one mole of a out of the villus for each mole of b absorbed, we can express Fick's law in terms of the concentration gradient alone. Our differential equation is expressed in terms of the concentration of b:

$$\frac{d^2 c_b}{dx^2} - \frac{k_c P_d}{A_c D_{ab}}(c_b - K_{eq}c_{b\infty}) + \frac{k_s'' P_d}{A_c D_{ab}} c_b = 0$$

We can solve this equation by brute force by first solving the homogeneous equation and then determining a particular solution, or we can play a game and place this equation in dimensionless form by defining new concentration and new position variables. If we let

$$\chi = \frac{1}{c_{bb}}\left[c_b - \left(\frac{k_c}{k_c - k_s''}\right)c_{b\infty}\right] \qquad \xi = \frac{x}{L}$$

then

$$\frac{d^2\chi}{d\xi^2} - \left(\underbrace{\frac{k_c P_d L^2}{A_c D_{ab}}}_{f_m^2} - \underbrace{\frac{k_s'' P_d L^2}{A_c D_{ab}}}_{\mathrm{Da}^2}\right)\chi = 0$$

The first dimensionless group is the villus number, f_m, we defined in Equation 8.45. The second is the Damkohler number, Da, comparing how fast we deliver b to the surface via convection versus how fast we absorb b via the surface reaction. Notice that the effectiveness and efficiency of the villus will depend upon the relative magnitude of the mass transfer coefficient and the reaction

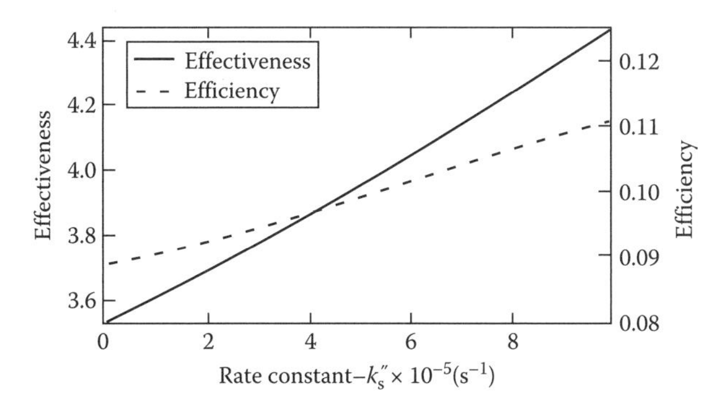

FIGURE 8.10 Villus effectiveness and efficiency for facilitated transport.

rate constant. Too fast a reaction and the villus will be starved for *b*. Too slow a reaction and we achieve no enhancement from the facilitated system.

If we assume the sealed-end condition again, the solution to the balance equation is given in a form similar to Equation 8.48:

$$\chi = \left[1 - \left(\frac{k_c}{k_c - k_s''}\right)\frac{c_{b\infty}}{c_{bb}}\right]\left\{\cosh(f_m'\xi) - \tanh(f_m')\sinh(f_m'\xi)\right\}$$

$$f_m' = \sqrt{\frac{P_d L^2}{A_c D_{ab}}(k_c - k_s'')}$$

and the effectiveness and efficiency by

$$eff = \frac{D_{ab} f_m' \tanh(f_m')}{k_c L(c_{bb} - c_{b\infty})}\left[c_{bb} - \left(\frac{k_c}{k_c - k_s''}\right)c_{b\infty}\right]$$

$$\eta_v = \frac{D_{ab} A_c f_m' \tanh(f_m')}{k_c P_d L^2 (c_{bb} - c_{b\infty})}\left[c_{bb} - \left(\frac{k_c}{k_c - k_s''}\right)c_{b\infty}\right]$$

Figure 8.10 shows a plot of the effectiveness and efficiency as a function of the reaction rate constant, k_s''. Notice that both quantities increase as the rate constant increases. What happens if the reaction rate term in the villus number is larger than the mass transfer term? Is it physically realistic to have an imaginary fin or villus number and what does that mean for the concentration inside the villus? Is the model we derived still valid in this case?

8.3.1 Transient Response

The villus problem we just considered assumed steady-state operation. As anyone who has ever eaten a meal knows, the absorption of nutrients does not proceed in a steady-state manner but is cyclic, depending upon the number of meals one eats in a given day. Therefore, we must concern ourselves not only with the steady-state response of extended surface systems but also with their transient response. In effect, we need to know the effectiveness and efficiency of such systems when they are forced to respond to a periodic stimulus. In the present case, we presume that the free stream concentration will vary as $c_{b\infty} = c_{b\infty o}e^{i\omega t}$ and that the concentration within the villus will be forced to follow the same form, $c_b(x)e^{i\omega t}$. Neglecting any type of facilitated transport so that the

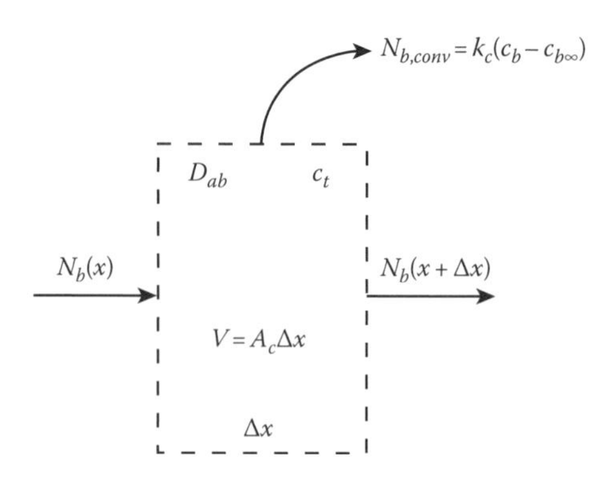

FIGURE 8.11 Control volume for the transient operation of an extended surface mass transfer system.

geometry is that of Figure 8.7, we can derive the mass balance about the differential control volume shown in Figure 8.11. We assume $K_{eq} = 1$:

$$\begin{array}{ccccc} In & - & Out & = & Acc \\ \dot{M}_b(x) & - & \dot{M}_b(x+\Delta x) - k_c P_d \Delta x (c_b - c_{b\infty}) & = & \Delta V \dfrac{\partial c_b}{\partial t} \end{array} \tag{8.52}$$

The balance states that the accumulation of b within the control volume is related to the amount of b flowing into and out of the control volume via diffusion and convection. Dividing through by Δx, taking the limit as $\Delta x \to 0$, and substituting in Fick's law for the flux, N_b, gives

$$\frac{\partial^2 c_b}{\partial x^2} - \frac{k_c P_d}{A_c D_{ab}}(c_b - c_{b\infty}) = \frac{1}{D_{ab}}\frac{\partial c_b}{\partial t} \tag{8.53}$$

Here, we used the fact that $\Delta V = A_c \Delta x$. To make the equation easier to solve, we put it into dimensionless form by defining the variables:

$$\chi = \frac{c_b - c_{b\infty}}{c_{bb} - c_{b\infty}} \qquad \xi = \frac{x}{L} \tag{8.54}$$

The differential equation becomes

$$\frac{\partial^2 \chi}{\partial \xi^2} - \frac{k_c P_d L^2}{A_c D_{ab}}\chi = \frac{L^2}{D_{ab}}\frac{\partial \chi}{\partial t} \tag{8.55}$$

The concentration variable, χ, is still a time-varying quantity proportional to $\exp(i\omega t)$. All we have done here is to insure that its magnitude varies from 0 to 1. Knowing that $\chi = \chi_b(\xi)\exp(i\omega t)$, we can substitute this into our partial differential equation to obtain an ordinary differential equation for the concentration response in the villus:

$$\frac{\partial^2 \chi_b}{\partial \xi^2} - \left(f_m^2 + i\frac{\omega L^2}{D_{ab}} \right)\chi_b = 0 \tag{8.56}$$

f_m is the villus number we defined previously. The second dimensionless term, $\omega L^2/D_{ab}$, defines the characteristic frequency of the system. In this case, the characteristic frequency $\omega_o = D_{ab}/L^2$.

Changes in the free stream concentration that occur at frequencies below the characteristic frequency can be followed by the villus and appear as changes in the base concentration. Changes that occur above the characteristic frequency cannot be followed and their variations will be attenuated throughout the system.

We will use the same boundary conditions we used in the previous example,

$$\xi = 0 \qquad \chi_b = 1 \tag{8.57}$$

$$\xi = 1 \qquad \frac{d\chi_b}{d\xi} = 0 \tag{8.58}$$

so the solution to the differential equation is in the form of hyperbolic sines and cosines:

$$\chi = \left\{ \cosh\left(\sqrt{f_m^2 + i\frac{\omega L^2}{D_{ab}}}\xi\right) - \tanh\left(\sqrt{f_m^2 + i\frac{\omega L^2}{D_{ab}}}\right)\sinh\left(\sqrt{f_m^2 + i\frac{\omega L^2}{D_{ab}}}\xi\right)\right\}\exp(i\omega t) \tag{8.59}$$

Notice that the magnitude of the concentration variable depends not only on the position within the villus but also on the frequency of the changes in the external medium. Since $\exp(i\omega t)$ is a complex number, the only portion of the solution we are interested in is the real part. Let's see how this villus performs as a function of frequency of the input by considering a numerical example.

Example 8.3 Effectiveness and Efficiency as a Function of Frequency

We would like to consider the effectiveness and efficiency of the villus discussed earlier as changes in the external medium occur with increasing frequency. The physical parameters for the fin are

$$L = 0.0001\ \text{m} \qquad r_o = 5\times10^{-5}\ \text{m} \qquad D_{ab} = 1\times10^{-7}\ \text{m}^2/\text{s}$$

$$c_{b\infty o} = 1000\ \text{mol/m}^3 \qquad k_c = 1\times10^{-5}\ \text{m/s}$$

and so the villus number, f_m, and characteristic frequency are

$$f_m = \sqrt{\frac{k_c P_d L^2}{A_c D_{ab}}} = 0.2 \qquad \omega_o = \frac{D_{ab}}{L^2} = 10\ \text{s}^{-1}$$

When we try to evaluate the effectiveness and efficiency of this structure, we must realize that under transient operation, not all the material we transfer to or from the villus from the external fluid makes it to the base and into the host body. Some of that material is merely stored in the villus body itself. To determine the effectiveness and efficiency, we must isolate only that fraction of material transported through the villus' surface. We do this by integrating the concentration driving force over the length of the villus and multiplying it by the surface area and mass transfer coefficient to obtain the actual amount of material transferred in. Since the efficiency and effectiveness are spatial quantities, we can work with χ_b. With that in mind, the definitions for effectiveness and efficiency are

$$eff = \frac{k_c P_d L(c_{bo} - c_{b\infty})\int_0^1 \chi_b d\xi}{k_c A_c (c_{bo} - c_{b\infty})} = \frac{2L}{r_o}\frac{\tanh\left(\sqrt{f_m^2 + i\frac{\omega L^2}{D_{ab}}}\right)}{\sqrt{f_m^2 + i\frac{\omega L^2}{D_{ab}}}}$$

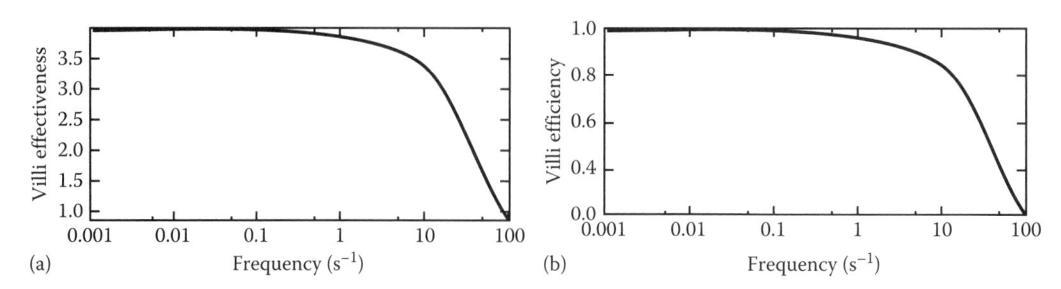

FIGURE 8.12 (a) Villi effectiveness and (b) efficiency as a function of external concentration changes. Response is akin to that of a low-pass electronic filter.

$$\eta_v = \frac{k_c P_d L(c_{bo} - c_{b\infty}) \int_0^1 \chi_b d\xi}{k_c P_d L(c_{bo} - c_{b\infty})} = \frac{\tanh\left(\sqrt{f_m^2 + i\dfrac{\omega L^2}{D_{ab}}}\right)}{\sqrt{f_m^2 + i\dfrac{\omega L^2}{D_{ab}}}}$$

A plot of how these two functions depend upon the frequency is shown in Figure 8.12a and b. We plot only the real parts of these functions. Notice that below the characteristic frequency, the effectiveness and efficiency are at their maximum values. Above this frequency, they drop off sharply indicating that the villus is ineffectual in that range. The response is identical to that of a low-pass electronic filter. Such response is favored in living systems to avoid noise upsetting the rhythms of the organism.

8.4 DIFFUSION AND REACTION IN A CATALYST PELLET

One of the most important uses of extended surfaces is in the area of catalyzed chemical reactions. Many important chemical processes, such as the conversion of crude oil to gasoline, the synthesis of hydrocarbons from CO and H_2O, or the hydrogenation of vegetable oils, are carried out using catalysts, and the reaction takes place on the catalyst surface. Since the reaction occurs on a surface, the rate is proportional to the surface area available for reaction. Therefore, most catalysts are made to be porous and consist of interpenetrating networks of micro-, meso-, and macropores (Figure 8.13).

Macropores are large-sized voids sometimes nearly large enough to be visible to the naked eye. Mesopores are very small and often cannot be seen even with conventional optical microscopes. Micropores are even smaller, and the lower limit of micropores, such as those found in zeolites, may be the size of the molecules themselves (on the order of 10 Å in size or smaller).

Most catalytic reactions, using catalysts of the type shown in Figure 8.13, are carried out in the gas phase to take advantage of the high diffusivity of gases. The diffusion of gases through the bulk gas phase, the catalyst macropores, and the micropores represent totally different phenomena. We have concentrated so far on regular or bulk diffusion through an essentially infinite gas or liquid phase. Bulk diffusion is controlled by molecular collisions and so the mean free path of the molecules is the important physical parameter. Using kinetic theory, we determined that the diffusivity should vary with the square root of temperature, $T^{0.5}$, and that it should be inversely proportional to the total pressure, P^{-1}. As the pore size in the catalyst pellet gets smaller, we eventually reach a situation where the pore diameter is smaller than the mean free path of the molecules. Here molecule–molecule collisions are less important and of more significance is the rate of molecule–wall collisions. The transition to this regime, called the Knudsen diffusion regime, can also occur for gases at less than atmospheric pressure. Such situations arise under vacuum conditions most notably in semiconductor processing equipment. In the Knudsen regime, gas diffusivity is roughly the same order of magnitude as liquid diffusivities. Knudsen diffusion is not important for liquids since the mean free path between liquid molecules is on the order of molecular size, much smaller than anything normally encountered in a porous solid.

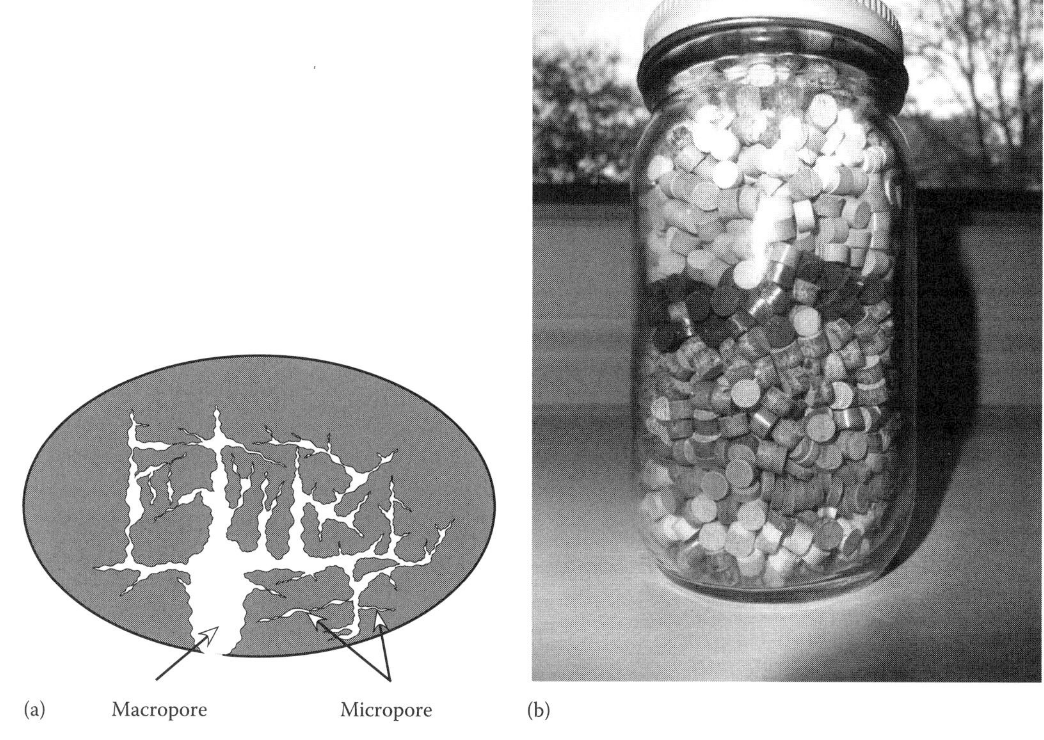

FIGURE 8.13 (a) Schematic diagram of a porous catalyst pellet showing one macropore and its associated micropore network. (b) Actual catalyst pellets used to speed the same reaction. The shading difference represents alterations in the catalyst composition.

We defined the self-diffusivity of gases in the bulk phase using the kinetic theory in Chapter 3:

$$D_{aa*} = \frac{1}{3}\bar{v}\lambda_f \tag{8.60}$$

where
 $\bar{v}$ is the mean velocity of the gas molecules
 λ_f is the mean free path between collisions

In Knudsen diffusion, the mean free path of gas molecules in the catalyst pore is d, the diameter of the pore. Defining $\bar{v}$ by

$$\bar{v} = \sqrt{\frac{8RT}{\pi M_{wa}}} \tag{8.61}$$

as we did before and substituting into Equation 8.60 gives the following expression for the Knudsen diffusivity:

$$D_k = \frac{4}{3} r_o \sqrt{\frac{2RT}{\pi M_{wa}}} \quad \text{Knudsen diffusivity} \tag{8.62}$$

where
 M_{wa} is the molecular weight of the diffusing species
 r_o is the pore radius

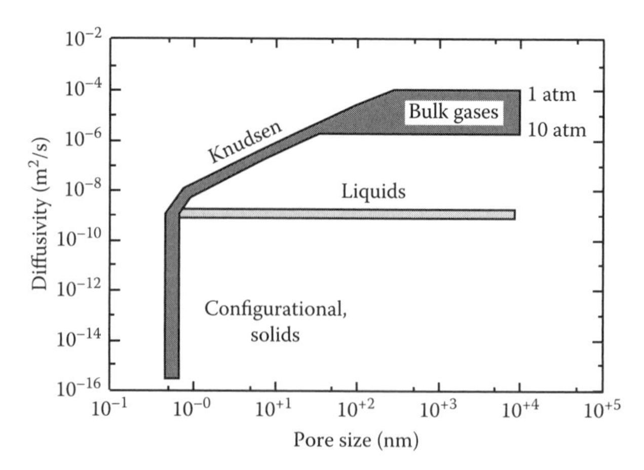

FIGURE 8.14 Diffusivity as a function of pore size showing bulk, Knudsen, and configurational regimes.

The diffusivity is again proportional to $T^{0.5}$, but it no longer depends upon the pressure in the system. This is a key feature of Knudsen diffusion.

If we reduce the size of the pores still further, we eventually reach a regime where the pore size is on the order of the molecular size. These types of pores are quite common in zeolites and some membranes designed for separating gases. In such systems, molecules will be excluded from pores according to their size and diffusivities are extremely small. We call this the configurational regime because molecules must rearrange or reconfigure themselves in order to pass through the pores. Figure 8.14 shows the relationship between the various regimes and the relative magnitudes of the expected diffusivities.

Keeping in mind these vast differences in diffusion mechanisms and diffusivities, we can now turn our attention to diffusion and reaction in catalyst pores or pellets. There are two major differences between the fin/villus system we discussed before and the catalyst system. In every catalyst operation, we have a generation term in the equation. This means that we will have competition between the generation/reaction rate, the rate of mass transport via diffusion, and the rate of mass transport via convection (external to the particle). Any one of these three mechanisms may control the overall performance of the system. We also have a different geometry to consider. Most fins or villi are attached to the external surface of an object. They protrude into the external medium and are totally exposed to the flowing fluid. In heterogeneous catalysis, the active surface area (surface participating in the reaction) is located almost entirely on the inside of the particle, and this extended area is not exposed to the flowing fluid at all. It is as if we turned the fin/villus inside out. All convective resistances lie in a boundary layer about the catalyst particle. Diffusion and reaction alone occur in the interior.

To handle the multitude of diffusional processes that may occur within the pores of a given catalyst pellet, we define an effective diffusivity for the catalyst, D_{eff}. This effective diffusivity not only considers the relative importance of bulk, Knudsen, and configurational diffusion but also accounts for how tortuous the individual pores may be. As such, it is a complicated function that varies with each type of catalyst and each catalyst configuration (pellets, spheres, loose cracked). The effective diffusivity can rarely be calculated. It is generally found by experiment.

Consider reaction in a single, cylindrical catalyst pore shown in Figure 8.15. We have a first-order decomposition reaction occurring

$$a \rightarrow \text{Products} \tag{8.63}$$

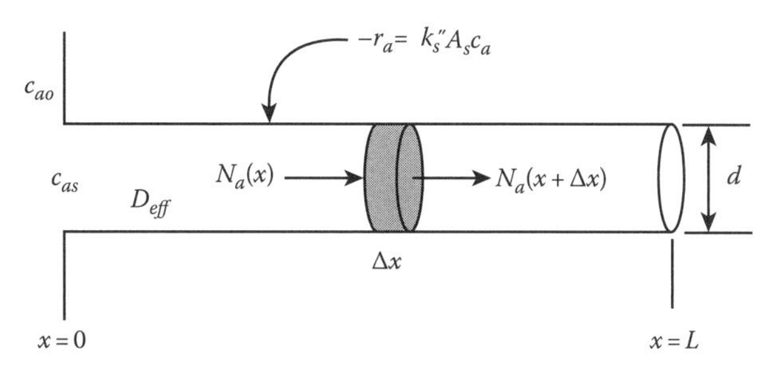

FIGURE 8.15 Cylindrical catalyst pore with reaction on the pore surface.

on the surface of the catalyst whose rate can be expressed in terms of

$$\dot{M}_a = A_s N_a = -k_s'' A_s c_a = -k'' V_p c_a \tag{8.64}$$

$$k_s'' = k'' \left(\frac{V_p}{A_s} \right) = k'' \lambda_v \tag{8.65}$$

where
- A_s is the surface area of the pore
- V_p is the volume of the pore
- λ_v is the characteristic dimension of the pore

The reaction rate constant, k_s'', is a rate per unit surface area of the catalyst and is related to the volumetric reaction rate constant, k'', which is based on a unit volume of catalyst. It is important to remember that we can define these rate constants based on a per pore basis or an overall particle basis. Due to the drastically simplified geometry in this example, we will be using the per pore basis for the remainder of the discussion.

At steady state, we start out with the general mass balance assuming steady-state operation:

$$\begin{array}{ccccccc} In & - & Out & + & Gen & = & Acc \\ \dot{M}_a(x) & - & \dot{M}_a(x+\Delta x) & - & (2\pi r \Delta x) k_s'' c_a & = & 0 \end{array} \tag{8.66}$$

Dividing through by Δx, letting $\Delta x \to 0$, substituting for $\dot{M}_a$ and N_a using Fick's law, and assuming a dilute solution of a ($N_a = -D_{eff} dc_a/dx$) gives us the differential equation for the concentration profile in the catalyst pore:

$$\frac{d^2 c_a}{dx^2} - \left(\frac{k_s'' P_d}{A_c D_{eff}} \right) c_a = 0 \tag{8.67}$$

We prefer to use k'', the volume-based rate constant, over k_s'' since in a laboratory setting, we tend to measure k'', the intrinsic rate of reaction. The ratio of cross-sectional to surface area, A_c/P_d, is $r_o/2$, the characteristic length, λ_v, for diffusion. Using Equation 8.65, we rewrite Equation 8.67 as

$$\frac{d^2 c_a}{dx^2} - \left(\frac{k''}{D_{eff}} \right) c_a = 0 \tag{8.68}$$

Equation 8.68 is subject to a symmetry boundary condition that states that the end of the pore is impermeable since the pore extends to the center of the pellet. Another boundary condition assumes that the resistance to mass transfer at the pore entrance is negligible (concentration of a at the surface is the same as in bulk):

$$x = 0 \qquad c_a = c_{as} = c_{ao} \tag{8.69}$$

$$x = L \qquad \frac{dc_a}{dx} = 0 \tag{8.70}$$

As we did in the fin/villus analysis, it is advantageous to put Equation 8.68 in dimensionless form by defining the following variables:

$$\xi = \frac{x}{L} \qquad \chi = \frac{c_a}{c_{ao}} \tag{8.71}$$

so

$$\frac{d^2\chi}{d\xi^2} - \left(\frac{k''L^2}{D_{eff}}\right)\chi = 0 \tag{8.72}$$

Equation 8.72 has the general solution

$$\chi = a_1 \cosh(\phi_p \xi) + a_2 \sinh(\phi_p \xi) \tag{8.73}$$

where the dimensionless parameter, ϕ_p, is the ratio of a characteristic diffusion time to a characteristic reaction time and can be thought of as a diffusion time divided by a reaction time. It is essentially identical to the Damkohler number we discussed in previous lectures but is given the name Thiele modulus. It also serves the same function as the fin number:

$$\phi_p = \sqrt{\frac{k''L^2}{D_{eff}}} = \sqrt{\frac{L^2/D_{eff}}{1/k''}} \sqrt{\frac{\text{Diffusion time}}{\text{Reaction time}}} \tag{8.74}$$

Using the boundary conditions to evaluate a_1 and a_2 yields the following concentration profile. Notice that this profile is essentially identical to the villus profile we calculated earlier:

$$\chi = \cosh(\phi_p \xi) - \tanh(\phi_p)\sinh(\phi_p \xi) \tag{8.75}$$

Equation 8.75 is plotted in Figure 8.16 for a variety of values of the Thiele modulus.

The results show that the diffusional resistance inside the pellet causes a concentration gradient to develop since reactants cannot diffuse in fast enough to saturate all the active reaction sites. As the diffusional resistance decreases (as D_{eff} gets larger), the curves flatten out and reaction is rapid and efficient everywhere (reaction rate controlled). Since the rate of reaction is proportional to $k''c_a$, the concentration profile causes a decreased average rate of reaction relative to that if the entire catalyst volume were exposed to the concentration of a in the bulk solution. The decrease in the rate of reaction is usually more than offset by the whopping increase in surface area afforded by using a

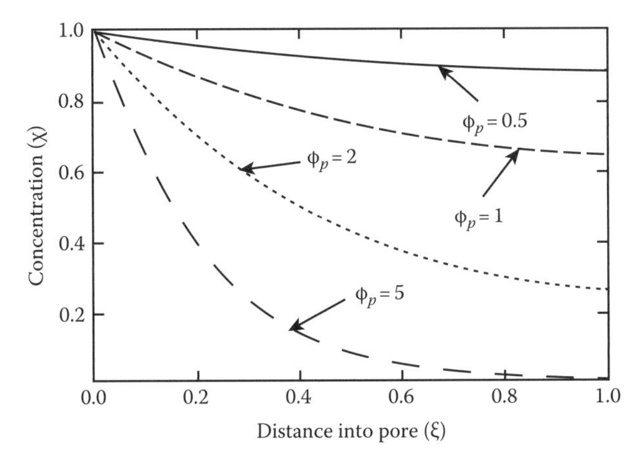

FIGURE 8.16 Concentration profile inside a cylindrical catalyst pore as a function of the Thiele modulus. (Adapted from Levenspiel, O., *Chemical Reaction Engineering*, John Wiley & Sons, New York, 1962. With permission.)

catalyst pellet rather than a flat surface. To quantify this concept, we can define a catalyst efficiency, more commonly referred to as an *effectiveness factor*, η_p, as

$$\eta_p = \frac{\text{Reaction rate with pore diffusion resistance}}{\text{Reaction rate at surface conditions}}$$

$$= \frac{(r_a)_{observed}}{(r_a)_{\max}} = \frac{\int r_a(c_a)dV_p}{V_p r_a(c_{as})} \tag{8.76}$$

The effectiveness factor is analogous to the fin or villus efficiency. It is the ratio of the actual overall rate of reaction to the maximum possible overall rate of reaction. This maximum rate occurs when there is no diffusional resistance and the pore sees a concentration of a, c_{as} everywhere. For the cylindrical catalyst pellet with constant circular cross section, the effectiveness factor becomes

$$\eta_p = \frac{\tanh(\phi_p)}{\phi_p} \quad \text{Effectiveness factor} \tag{8.77}$$

The effectiveness factors for systems of other geometries (cylindrical pellets, spherical pellets, etc.) are shown in Table 8.1 and plotted in Figure 8.17. In these geometries, the characteristic dimension is defined as the volume of the catalyst particle divided by the area available for the diffusion of the reactant into the particle. The surface area needed here is the overall surface area (macroscopic surface area) of the pellet.

8.4.1 External Mass Transfer Resistance

The catalyst effectiveness factors listed in Table 8.1 all assume that there is no resistance to mass transfer from the bulk solution to the catalyst pellet. Unless the pellet is somehow being vigorously agitated, there will be some resistance to mass transfer, and the concentration of reactant at the pellet surface will not be the same as the concentration in the bulk solution. If we have significant external mass transfer resistance to the pellet, we also have a concentration profile in a thin boundary layer close to the pellet's surface. We can account for this external resistance in the model of

TABLE 8.1
Effectiveness Factors for Catalyst Pellets (First-Order Chemical Reaction)

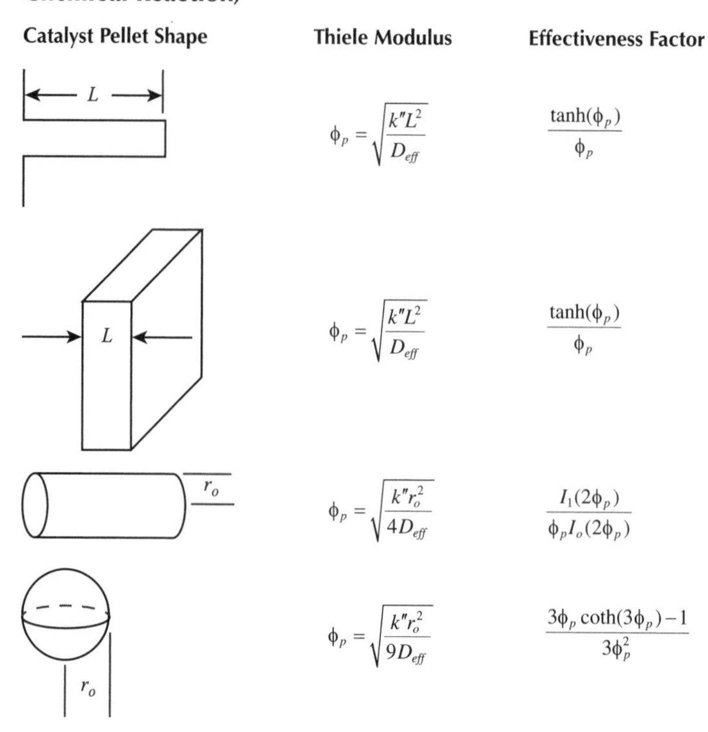

Catalyst Pellet Shape	Thiele Modulus	Effectiveness Factor
L	$\phi_p = \sqrt{\frac{k''L^2}{D_{eff}}}$	$\frac{\tanh(\phi_p)}{\phi_p}$
L	$\phi_p = \sqrt{\frac{k''L^2}{D_{eff}}}$	$\frac{\tanh(\phi_p)}{\phi_p}$
r_o	$\phi_p = \sqrt{\frac{k''r_o^2}{4D_{eff}}}$	$\frac{I_1(2\phi_p)}{\phi_p I_o(2\phi_p)}$
r_o	$\phi_p = \sqrt{\frac{k''r_o^2}{9D_{eff}}}$	$\frac{3\phi_p \coth(3\phi_p)-1}{3\phi_p^2}$

Source: Adapted from Levenspiel, O., *Chemical Reaction Engineering*, John Wiley & Sons, New York, 1962. With permission.

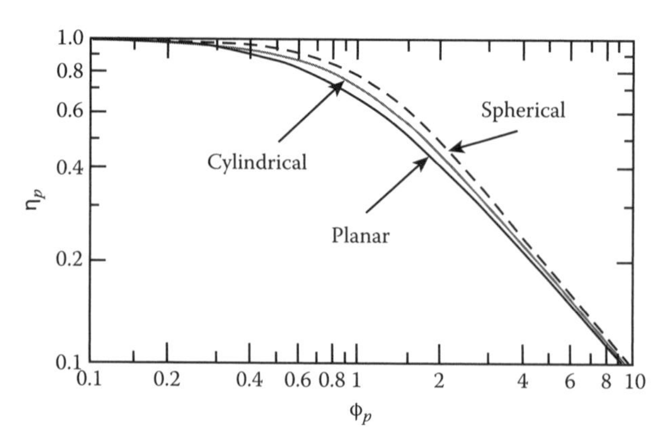

FIGURE 8.17 Effectiveness as a function of the Thiele modulus for catalysts of varying shape and first-order chemical reaction. (Adapted from Levenspiel, O., *Chemical Reaction Engineering*, John Wiley & Sons, New York, 1962. With permission.)

the cylindrical catalyst pore by changing the boundary condition at the pore mouth ($\xi = 0$) to a convective boundary condition with mass transfer coefficient, k_c:

$$\xi = 0 \qquad -\frac{d\chi}{d\xi} = \frac{k_c L}{D_{eff}}(1-\chi) \tag{8.78}$$

$$\xi = 1 \qquad \frac{d\chi}{d\xi} = 0 \tag{8.79}$$

The solution to Equation 8.73 with these boundary conditions is

$$\chi = \left[\frac{\dfrac{k_c L}{\phi_p D_{eff}}}{\tanh(\phi_p) + \dfrac{k_c L}{\phi_p D_{eff}}} \right] \left\{ \cosh(\phi_p \xi) - \tanh(\phi_p)\sinh(\phi_p \xi) \right\} \tag{8.80}$$

In this case, the effectiveness factor is defined in terms of the bulk fluid concentration c_{ao}, not the concentration at the particle surface, c_{as}. The effectiveness factor becomes a global effectiveness factor, η_g, since it contains details concerning the external mass transfer resistance:

$$\eta_g = \frac{\int r_a(c_a)\, dV_p}{V_p r_a(c_{ao})} = \frac{\tanh(\phi_p)/\phi_p}{1 + \left(\dfrac{\phi_p D_{eff}}{k_c L} \right) \tanh(\phi_p)} \tag{8.81}$$

We can write Equation 8.81 more effectively if we rearrange it into the following form:

$$\frac{1}{\eta_g} = \frac{1}{\eta_p} + \frac{\phi_p^2}{Bi_m} = \frac{\phi_p}{\tanh(\phi_p)} + \lambda_v \left(\frac{k''}{k_c} \right) \tag{8.82}$$

where Bi_m is the Biot number for mass transfer we introduced in Chapter 6:

$$\mathrm{Bi}_m = \frac{k_c \lambda_v}{D_{eff}} \tag{8.83}$$

η_p is termed the local effectiveness factor. *For first-order reactions*, we can use a sum of resistance concept to account for both internal and external mass transfer resistances. Notice that the resistances are in parallel and that the effect of an additional mass transfer resistance is to make the global effectiveness factor, $\eta_g < \eta_p$.

8.4.2 Thermal Effects

One final complication can arise during reactions in a catalyst pellet. The catalyst pellet material has a finite thermal conductivity, and so the heat generated during the course of the reaction will not be dissipated immediately. Thermal gradients develop, and these must be known so that accurate reaction rates can be defined and reactors designed. Considering the simplest geometry possible, a slab-type pore (1D), the mass and energy balance equations can be written as

$$\frac{d}{dx}\left(D_{eff}\frac{dc_a}{dx}\right)=r_a(c_a,T) \tag{8.84}$$

$$\frac{d}{dx}\left(k_{eff}\frac{dT}{dx}\right)=-(-\Delta H_r)r_a(c_a,T) \tag{8.85}$$

where k_{eff} is the effective thermal conductivity of the particle and where we have made no special restrictions on the form of the reaction rate. These two equations must be solved simultaneously due to their coupling through the reaction rate term. If we divide Equation 8.85 through by $(-\Delta H_r)$ and add it to Equation 8.84, we obtain

$$\frac{d}{dx}\left(D_{eff}\frac{dc_a}{dx}+\frac{k_{eff}}{(-\Delta H_r)}\frac{dT}{dx}\right)=0 \tag{8.86}$$

Integrating from the center ($x = 0$) to the outer edge of the pellet gives

$$D_{eff}\frac{dc_a}{dx}+\frac{k_{eff}}{(-\Delta H_r)}\frac{dT}{dx}=\text{constant}=0 \tag{8.87}$$

where we have used the symmetry boundary conditions stating that the concentration and thermal gradients must be zero at the centerline of the pellet:

$$x=0 \qquad \frac{dc_a}{dx}=\frac{dT}{dx}=0 \tag{8.88}$$

Remember, at the center of the symmetrical pellet, we see unfavorable gradients for transport in every direction. It is analogous to placing an object at the center of the Earth and expecting it to fall to the surface. Another integration of Equation 8.87 gives

$$D_{eff}c_a+\frac{k_{eff}}{(-\Delta H_r)}T=\text{constant} \tag{8.89}$$

We can use a second boundary condition on the surface stating that we know the temperature and concentration there:

$$x=L \qquad c_a=c_{as} \qquad T=T_s \tag{8.90}$$

to give

$$T-T_s=\frac{D_{eff}(-\Delta H_r)}{k_{eff}}(c_{as}-c_a) \tag{8.91}$$

This equation can be used to eliminate either T or c_a from either the mass or energy balance giving only one nonlinear equation to solve. Equation 8.91 is useful for determining the maximum temperature rise that can occur in the particle. Assuming complete reaction so that $c_a = 0$ shows that the maximum temperature rise is

$$\frac{\Delta T_{max}}{T_s} = \frac{(-\Delta H_r) D_{eff} c_{as}}{k_{eff} T_s} \tag{8.92}$$

ΔT_{max} is referred to as the adiabatic temperature rise and Equation 8.92 is valid for any particle of any geometry.

If the full, transient mass and energy balance equations are written, we can cast them in dimensionless form and derive several important dimensionless groups governing the performance of the catalyst system. Assuming a first-order chemical reaction,

$$\frac{\partial c_a}{\partial t} = D_{eff} \frac{\partial^2 c_a}{\partial z^2} - A_o \exp\left(-\frac{E_a}{RT}\right) c_a \tag{8.93}$$

$$\rho_s C_{ps} \frac{\partial T}{\partial t} = k_{eff} \frac{\partial^2 T}{\partial z^2} + (-\Delta H_r) A_o \exp\left(-\frac{E_a}{RT}\right) c_a \tag{8.94}$$

Here, we have incorporated the temperature dependence of the reaction rate constant. E_a is the activation energy for reaction and A_o is the pre-exponential factor. Using the following dimensionless variables,

$$\chi = \frac{c_a}{c_{as}} \qquad \theta = \frac{T}{T_s} \qquad \zeta = \frac{z}{L} \qquad \tau_p = \frac{D_{eff} t}{L^2} \tag{8.95}$$

transforms Equations 8.93 and 8.94 to

$$\frac{\partial \chi}{\partial \tau_p} = \frac{\partial^2 \chi}{\partial \zeta^2} - \phi_p^2 \chi \exp\left[\gamma\left(1 - \frac{1}{\theta}\right)\right] \tag{8.96}$$

$$\frac{1}{Lw'} \frac{\partial \theta}{\partial \tau_p} = \frac{\partial^2 \theta}{\partial \zeta^2} - \kappa \phi_p^2 \chi \exp\left[\gamma\left(1 - \frac{1}{\theta}\right)\right] \tag{8.97}$$

where

$$\phi_p^2 = \frac{L^2 A_o}{D_{eff}} \exp\left(-\frac{E_a}{RT_s}\right) \tag{8.98}$$

$$\gamma = \frac{E_a}{RT_s} \tag{8.99}$$

$$Lw' = \frac{Sc'}{Pr'} = \frac{k_{eff}}{\rho_s C_{ps} D_{eff}} \tag{8.100}$$

$$\kappa = \frac{D_{eff} c_{as} (-\Delta H_r)}{T_s k_{eff}} \tag{8.101}$$

The steady-state solutions to these equations will only be a function of the Thiele modulus, ϕ_p; the dimensionless activation energy, γ; the dimensionless temperature rise, κ; and the Lewis number, Lw'. One of the most interesting results is that for an exothermic reaction where $\kappa > 0$, there are regions where the effectiveness factor $\eta_p > 1$. This occurs because for sufficient temperature rise in the particles due to heat transfer limitations, the increase in the reaction rate constant more than offsets any decrease in concentration due to mass transfer limitations. The other interesting feature is that for large κ and a rather narrow range of ϕ_p values, there are three possible steady states. This results from the nonlinear nature of the equations introduced via the reaction rate constant. Only the highest or lowest steady states are inherently stable and can be observed. Figure 8.18 shows the behavior.

One can also consider the maximum temperature rise in the pellet when there are external resistances to mass and heat transfer. At steady state, the governing equations are

$$D_{eff}\frac{\partial^2 c_a}{\partial z^2} - A_o\exp\left(-\frac{E_a}{RT}\right)c_a = 0 \tag{8.102}$$

$$k_{eff}\frac{\partial^2 T}{\partial z^2} + (-\Delta H_r)A_o\exp\left(-\frac{E_a}{RT}\right)c_a = 0 \tag{8.103}$$

with boundary conditions

$$z = L \qquad D_{eff}\frac{\partial c_a}{\partial z} = k_c(c_{a\infty} - c_{as}) \tag{8.104}$$

$$z = L \qquad k_{eff}\frac{\partial T}{\partial z} = h(T_\infty - T_s) \tag{8.105}$$

We can combine Equations 8.102 and 8.103 in the same manner we combined (8.84) and (8.85) to yield

$$\frac{d^2}{dz^2}\left[D_{eff}c_a + \frac{k_{eff}}{(-\Delta H_r)}T\right] = 0 \tag{8.106}$$

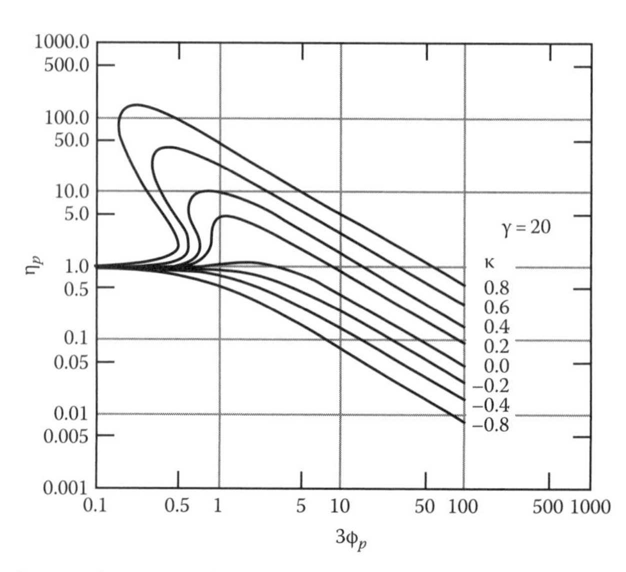

FIGURE 8.18 Effectiveness factor with first-order reaction in a spherical, nonisothermal catalyst pellet. (Adapted from Weisz, P.B. and Hicks, J.S., *Chem. Eng. Sci.*, 17, 265, 1962. With permission.)

Integration using the boundary conditions gives

$$T - T_\infty = (T_s - T_\infty) - \frac{D_{eff}(-\Delta H_r)}{k_{eff}}(c_a - c_{as})$$

$$= (-\Delta H_r)\frac{k_c}{h}(c_{a\infty} - c_{as}) + (-\Delta H_r)\frac{D_{eff}}{k_{eff}}(c_{as} - c_a) \quad (8.107)$$

The right-hand side of Equation 8.107 is the sum of the external and internal temperature differences. The maximum temperature difference occurs at complete conversion when $c_a = 0$:

$$\frac{T_{max} - T_\infty}{T_\infty} = \kappa_g \frac{\text{Bi}_m}{\text{Bi}}\left(1 - \frac{c_{as}}{c_{a\infty}}\right) + \kappa_g \frac{c_{as}}{c_{a\infty}} \quad (8.108)$$

where

$$\kappa_g = \frac{(-\Delta H_r) D_{eff} c_{a\infty}}{k_{eff} T_\infty} \quad (8.109)$$

and Bi_m and Bi are the Biot numbers for mass and heat transfer. One can see quite clearly that there is a competition between the rate of reaction, the rate of mass transfer to the reaction sites, and the rate of heat removal from the particle in terms of the temperature rise and catalyst effectiveness. In this sense, the situation is quite similar to multiple steady states in continuous stirred tank reactors.

Example 8.4 Determination of Effectiveness and Adiabatic Temperature Rise in a Spherical Catalyst Pellet

A set of experiments using a series of sizes of crushed catalyst spheres was performed to determine the influence of pore diffusion on the performance of the catalyst. The reaction chosen for the study was first order and irreversible. The concentration of reactant at the surface of the pellet was $c_s = 500$ mol/m³. Reaction rate data as a function of particle size are shown later. We would like to determine the intrinsic reaction rate, k''; the effective diffusivity, D_{eff}; and the adiabatic temperature rise if the heat of reaction $\Delta H_r = -175$ kJ/mol and the effective thermal conductivity of the particle is 0.05 W/m K. We would also like to know what the effectiveness would be for a spherical catalyst pellet of diameter, 0.003 m.

Diameter of particle (m)	2.5×10^{-3}	7.5×10^{-4}	2.5×10^{-4}	7.5×10^{-5}
r_{obs} (mol/m³ s)	61.1	194.4	444.4	666.7

It is clear from the data that the observed rate of reaction depends upon particle size and that pore diffusion would be a problem. Though we have data at small particle sizes (7.5×10^{-5}), there is no guarantee that pore diffusion at that size is not a problem still. Plotting reaction rate as a function of particle size shows no clear asymptote and so that route is also no help. The only way to get a handle on the problem is to use what we know about reaction and diffusion in catalysts.

There are two primary quantities governing the reaction and diffusion, the Thiele modulus and the effectiveness factor. For the spherical or nearly spherical particles used in the experiments, these quantities are

$$\phi_p = \frac{d}{6}\sqrt{\frac{k''}{D_{eff}}} \qquad \eta_p = \frac{r_{obs}}{r_{int}} = \frac{3\phi_p \coth(3\phi_p) - 1}{3\phi_p^2}$$

The intrinsic reaction rate, r_{int}; the reaction rate constant, k''; and the effective diffusivity, D_{eff}, should be the same no matter what the size of the particle. Therefore, to determine k'' and D_{eff}, we can use a ratio of effectiveness factors at two particle sizes to solve for $\sqrt{k''/D_{eff}}$:

$$\frac{\eta_{p1}}{\eta_{p2}} = \left(\frac{\phi_{p2}}{\phi_{p1}}\right)^2 \left[\frac{3\phi_{p1}\coth(3\phi_{p1}) - 1}{3\phi_{p2}\coth(3\phi_{p2}) - 1}\right]$$

$$\frac{\eta_{p1}}{\eta_{p2}} = \left(\frac{d_2}{d_1}\right)^2 \left[\frac{\left(\frac{d_1}{2}\sqrt{\frac{k''}{D_{eff}}}\right)\coth\left(\frac{d_1}{2}\sqrt{\frac{k''}{D_{eff}}}\right) - 1}{\left(\frac{d_2}{2}\sqrt{\frac{k''}{D_{eff}}}\right)\coth\left(\frac{d_2}{2}\sqrt{\frac{k''}{D_{eff}}}\right) - 1}\right]$$

Using all six combinations of particle sizes and then averaging the results, we find that

$$\sqrt{\frac{k''}{D_{eff}}} = 26174.6\frac{1}{\text{m}}$$

The actual reaction rate can be determined by calculating an effectiveness factor:

$$\phi_p(7.5\times10^{-5}) = \frac{7.5\times10^{-5}}{6}(26174.6) = 0.327$$

$$\eta_p(7.5\times10^{-5}) = \frac{3(0.327)\coth[3(0.327)] - 1}{3(0.327)^2} = 0.939$$

This smallest of catalyst sizes sees very little diffusional resistance. The intrinsic (surface conditions) reaction rate and rate constant can now be found from

$$r_{int} = k''c_s = \frac{r_{obs}}{\eta_p} = 710\frac{\text{mol}}{\text{m}^3\,\text{s}} \qquad \text{so} \qquad k'' = 1.42\ \text{s}^{-1}$$

Now we can calculate the effective diffusivity:

$$D_{eff} = \frac{k''}{(26174.6)^2} = 2.07\times10^{-9}\ \frac{\text{m}^2}{\text{s}}$$

Notice that we are in the Knudsen diffusion regime.

The adiabatic temperature rise is found from Equation 8.92:

$$\Delta T = \frac{(-\Delta H)D_{eff}c_s}{k_{eff}} = \frac{(175{,}000)(2.07\times10^{-9})(500)}{0.05} = 3.6\ \text{K}$$

In semiconductor processing, the etching of silica is accomplished using fluorocarbon containing species such as CHF_3 in a plasma reactor. Under certain reaction conditions, such as at higher gas pressures, instead of etching the silica, the deposition of fluorocarbon polymer can occur. This is a useful situation when porous dielectric materials use the interconnect system linking transistors as the pores in the dielectric are troublesome when depositing metal interconnect lines. The Comsol® module *PoreDiffMM* uses a deforming mesh to show how the polymer develops and how the deposition can be used to seal off the pore. Figure 8.19 is an image at t = 100 s

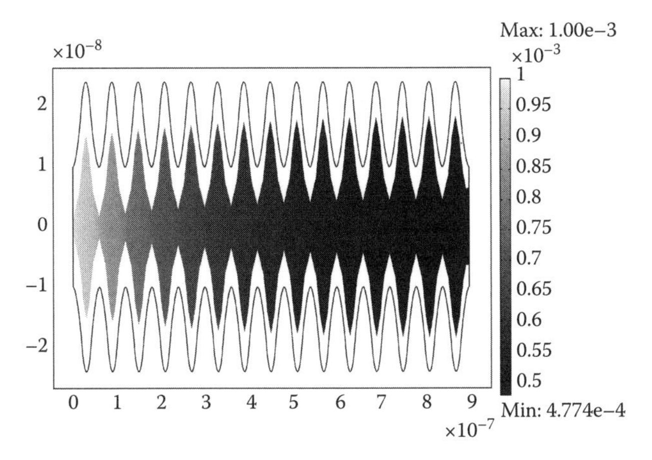

FIGURE 8.19 Image of a catalyst pore with polymer deposition on the surface.

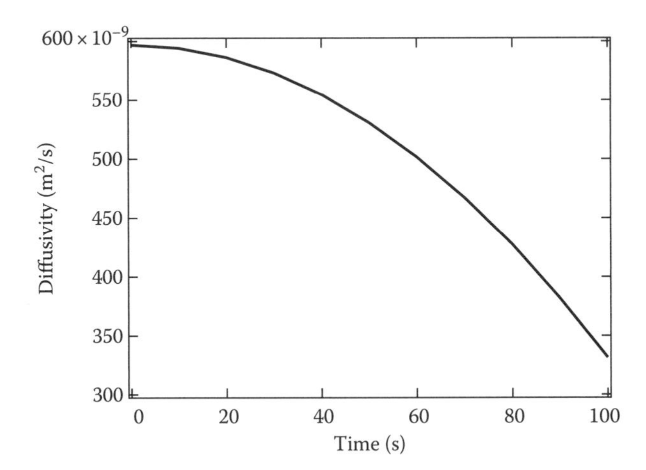

FIGURE 8.20 Knudsen diffusivity at the pore mouth as a function of time.

showing the deposited polymer on the surface of the pore (pores appear as multi-bottleneck structures) as well as the concentration gradient in the pore. The depositing species undergoes Knudsen diffusion within the pore of the dielectric, and Figure 8.20 shows how the Knudsen diffusion coefficient at the pore mouth behaves as the pore mouth begins closing under the action of polymer deposition. This kind of simulation is useful for understanding how to set reactor conditions and is an example of a complicated situation that cannot be handled analytically. Keep in mind that like any computer simulation, an analytical approximation can be obtained that allows one to determine whether the computer simulation is valid [4]. Here, the analytical approximation states that the pore mouth should close in about 1970s, close enough to give confidence in the Comsol® result.

8.5 SUMMARY

We have seen how extended surfaces are used to enhance the rates of transport. Many solutions to engineering could not be accomplished without these surfaces. Life itself would be impossible without the extended surfaces engineered into all scales of systems from mitochondria to trees.

We have shown when extended surfaces are useful, how to measure their usefulness in terms of the effectiveness and efficiency, and how the ratio of internal to external transport resistance governs their performance. An important point to remember is that whether the extended surface area is external or internal, the same governing mechanisms apply.

In the next chapter, we will consider what changes we need to make in our analyses when we must consider transport in multiple dimensions.

PROBLEMS

Heat Transfer Fins

8.1 A copper rod, 15 mm in diameter, is attached to a wall that is at a temperature of 200°C. The opposite end of the rod is insulated. The rod is 100 mm long and is immersed in air at a temperature of 25°C. The heat transfer coefficient for the air is $h = 10$ W/m^2 K and the thermal conductivity of the rod is 393 W/m K.

a. How much heat is dissipated by the fin?
b. What is the fin efficiency?
c. How does the heat dissipated by this fin compare to that which would be dissipated by an infinite fin of the same diameter?

8.2 Consider the composite pin fin shown in Figure P8.2. Half of the fin is formed of a material with thermal conductivity, k_1, and the other half is formed of material with thermal conductivity, k_2. The base is held at a constant temperature, T_b, and the tip is insulated. The heat transfer coefficient to the environment is h.

a. Determine the temperature profile in the fin.
b. What is the fin effectiveness? How does it depend upon the fin numbers for the two regions?

8.3 A flat fin of length, L; width, w; and thickness, d, is exposed to a uniform heat flux of q''_o along its length (see Figure P8.3). The base of the fin is attached to a fluid stream whose free stream

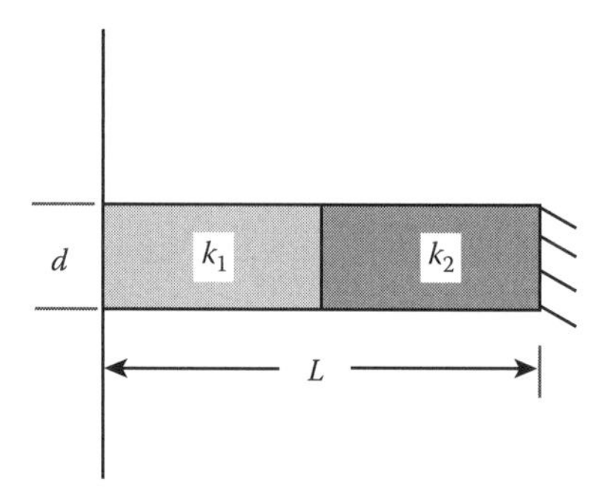

FIGURE P8.2

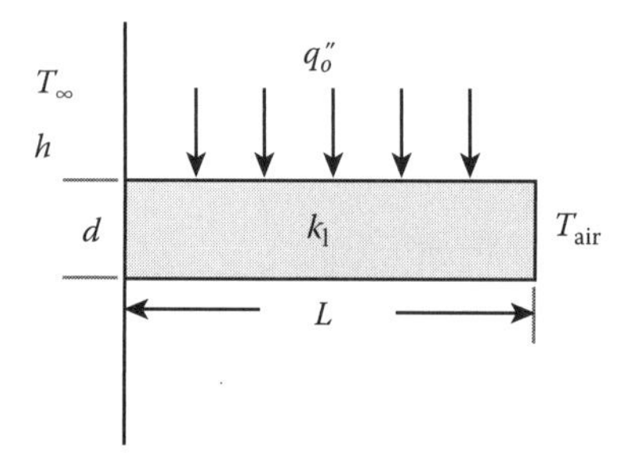

FIGURE P8.3

temperature is T_∞ and whose heat transfer coefficient is h. The opposite end of the fin is held at the air temperature of T_{air}.

a. Determine the temperature profile within the fin.
b. What are the heat fluxes at the ends?
c. What is the fin effectiveness?

8.4 Springs are often annealed by passing a current through them. Care must be taken to insure that the temperature along the spring does not exceed the desired annealing range. Consider a thin wire of thermal conductivity, k; diameter, d; and length $2L$. The current passed through the wire results in a uniform volumetric rate of heat generation, $\dot{q}$ (W/m^3). The wire is exposed to the atmosphere at a temperature of T_∞, and convection about the wire is sufficient to provide for a heat transfer coefficient, h. The wire ends are also maintained at T_∞. Determine the temperature profile, $T(x)$, and the magnitude and location of the maximum temperature.

8.5 A rod of diameter 25 mm and thermal conductivity $k = 40$ W/m K is part of a handle that protrudes from a furnace wall. The wall temperature is $T_w = 200°C$ and the first portion of the rod is covered by a thick blanket of insulation. To be useful as a handle, the end of the rod must be no more than 50°C. The ambient air temperature is $T_\infty = 25°C$ and the heat transfer coefficient to the environment is $h = 40$ W/m^2 K.

a. Derive an expression for the temperature of the rod. Assume the exposed portion of the rod acts as an infinite fin.
b. How long must the insulated portion of the rod be to meet specifications?

8.6 A microprocessor has an aluminum heat sink attached that consists of a series of parallel, rectangular fins. Each fin is 2 mm thick, 25 mm long, 25 mm high, and spaced on 6 mm centers. When the microprocessor is turned on, it begins producing heat at a rate of 1 mW/cm^2 of surface area. A fan within the computer circulates air at $T_\infty = 25°C$ through the machine providing a heat transfer coefficient of 150 W/m^2 K. The fin tip can be assumed to be exposed to the convecting fluid.

a. Determine the steady-state temperature profile in the fin.
b. What is the temperature of the base?

8.7 A heat pipe is basically a hollow fin that is partially filled with fluid. The fluid evaporates at the hot end, travels the length of the fin, and condenses at the cold end. Capillary forces return the liquid to the hot end. We can model the heat pipe by modifying the fin equation using an internal heat transfer coefficient that describes whether heat is gained or lost by the fin via evaporation whose driving force is given by the temperature difference between the fin wall temperature, T, and the fluid's boiling temperature, T_v.

a. Derive the differential equation describing the operation of the heat pipe including an out term defined by heat transfer to the surroundings at T_∞ and an internal convection term describing the evaporation and condensation process.
b. Solve the differential equation assuming that the fin base temperature is T_b and the fin tip is held at the temperature of the surroundings. (It may help to separate the constant terms and use the trial solution, $T = A\exp(rx)$.)

8.8 Since heat pipes are sealed objects, whatever liquid gets evaporated must be condensed. Thus, if one were to integrate the internal heat transfer rate, $h(T - T_v)$, over the entire length of the heat pipe, the sum should be zero.

a. Solve the heat pipe equation using the integral of internal heat transfer rate as the second boundary condition.
b. What is the fin tip temperature in this case?

8.9 The other way to operate a heat pipe is to fix the tip temperature as we did before and to use the internal integral to define the "operating temperature" of the pipe (T_v).

a. Solve the heat pipe equation assuming that the tip is held at the temperature of the surroundings and use the internal integral to obtain the operating temperature.

b. For an Al, cylindrical heat pipe that is 3 mm in diameter and 40 mm long, using pentane as an operating fluid, the external and internal heat transfer coefficients are 15 and 300 W/m^2 K. Assuming a base temperature of 100°C and a free stream air temperature of 20°C, what is the operating temperature of the heat pipe?

c. Since the heat pipe is filled with a pure fluid, use the Antoine equation and coefficients for pentane to determine the internal operating pressure of the heat pipe based on your results from part (b).

8.10 Consider the transient performance of the fin in Problem 8.6 when the microprocessor is cycled with a frequency ω_o.

a. What is the differential equation describing the transient situation?

b. The differential equation can be solved by assuming the transient response solution can be written as

$$\theta = T - T_\infty = \theta_o(x)e^{i\omega_o t}$$

The transient solution must obey the following conditions:

$$x = 0 \qquad -\frac{k}{L}\frac{d\theta}{dx} = q_o'' e^{i\omega_o t}$$

$$x = L \qquad \theta = 0 \qquad t = 0 \qquad \theta = 0$$

where we assume that the fin behaves as if it is infinitely long.

c. Using the information in part (b), solve the differential equation.

8.11 The stegosaurus had two long rows of armored plates running along its spine. Some paleontologists have suggested that this dinosaur used these plates to regulate body temperature (like present-day elephants do with their ears). We can model each plate as a single fin of parabolic profile and determine the heat lost or gained (depending upon the time of day). The relevant parameters are (Dvorozniak, Medeiros, Miller and Spilker)

T_∞ = 30°C (it was hotter back then) T_{dino} = 38°C

k_{eff} = 50 W/m K (accounts for blood flow inside the fin)

h = 10 W/m^2 K

Calculate the heat loss per stegosaurus plate given the data earlier and Figure P8.11.

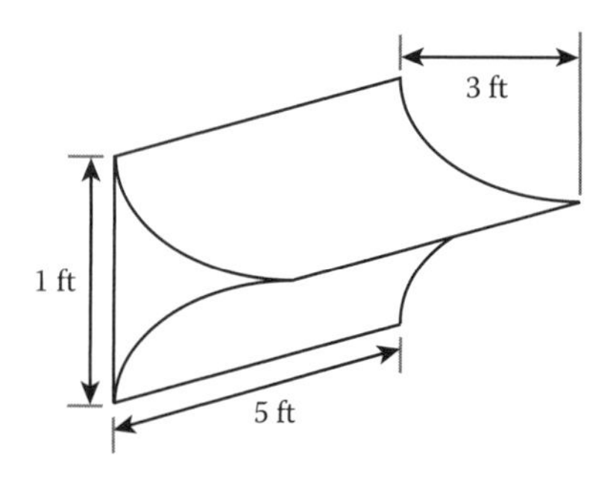

FIGURE P8.11

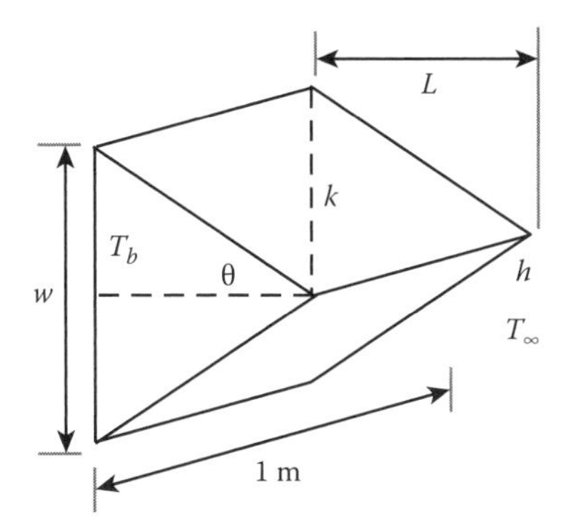

FIGURE P8.12

8.12 Triangular-shaped fins offer high heat transfer capability and low weight when compared to other geometries. Consider the triangular fin whose base is held a constant temperature of T_b. Since the cross-sectional area of the fin is zero at its tip, the location poses a problem. Resolve it by changing the coordinate system to make the fin tip at $x = 0$ and its base at $x = L$. The free stream temperature is T_∞. For the fin dimensions shown in Figure P8.12, show that the temperature profile is given by

$$\theta = \frac{T - T_\infty}{T_b - T_\infty} = \frac{I_o\left(2\alpha\sqrt{x}\right)}{I_o(2\alpha)} \qquad \alpha^2 = \frac{hL}{k\sin\theta}$$

Mass Transfer Fins

8.13 Consider a fin incorporating facilitated transport only at its base. The reaction occurring there is given by $-r_a = k''c_a$ (mol/m^{3}s). The fin is essentially infinitely long and has a diameter, r_o, and species a diffuses through it with a diffusivity, D_a. It is immersed in a fluid containing species a at a concentration of $c_{a\infty}$ ($c_{a\infty} > c_a$ throughout the fin). Assuming constant properties, dilute solutions, and steady-state operation and letting $\chi = c_a/c_{a\infty}$,

a. Determine the concentration profile within the fin.
b. What is the flux and concentration at the base of the fin?

8.14 The alveoli in your lungs can be modeled as conventional pin fins. Each fin is 10 μm in diameter and 100 μm long. The fin is designed to pick up oxygen from the air and transport it to the base where it reacts instantaneously with hemoglobin. The fin tip is impermeable to oxygen. Assuming air at 25°C and 1 atm; a mass transfer coefficient, k_c, of 1×10^{-5} m/s; and a diffusion coefficient for O_2 in the fin of 3×10^{-9} m^2/s,

a. What is the governing equation for this problem?
b. What are the boundary conditions?
c. Solve the equation to obtain the concentration profile?
d. What is the effectiveness for this fin?

8.15 One way to increase the flux of ionic materials through a mass transfer fin is to apply an electric field along its length as shown in Figure P8.15. The fin has a diameter, d, and length L. The base of the fin is exposed to a constant concentration of a, c_{ao}. The tip of the fin is sealed so that the flux of a is zero there. The free stream concentration of a is $c_{a\infty}$ and the mass transfer coefficient is k_c. Both the diffusivity of a and its mobility are constant throughout the fin.

a. Assuming a constant applied electric field, E_o, derive the differential equation for the concentration profile. Assume dilute solutions and that the Nernst–Einstein equation holds.
b. Determine the concentration profile within the fin.
c. How does the fin effectiveness depend upon the applied electric field?

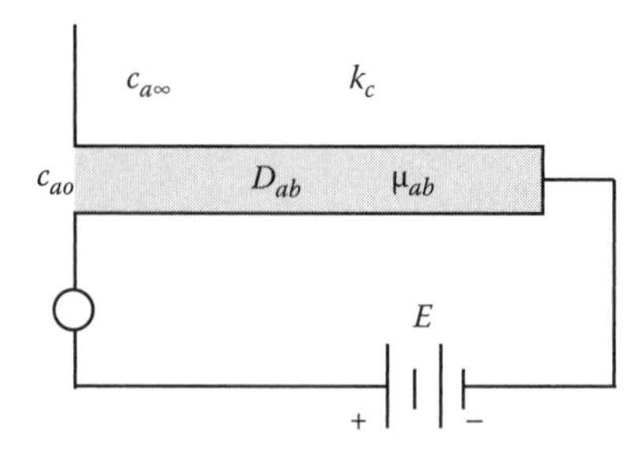

FIGURE P8.15

8.16 A villus of diameter, d, and length, L, is attached to a membrane. At the base of the villus, the membrane has enzymes immobilized on it that catalyze the transformation of the substrate according to the rate law:

$$-r_s = \frac{k''K_s c_s}{K_i K_m + K_i c_s}$$

The tip of the villus is held at the free stream concentration of substrate, $c_{s\infty}$. The motion of the fluid about the villus provides for a mass transfer coefficient of k_c. The diffusivity of substrate through the villus material is D_{vs}.

a. Determine the substrate concentration profile within the villus.
b. What is its efficiency?

8.17 Consider a highly idealized model of the root of a leguminous plant shown in Figure P8.17. A root of diameter, d, and length, L, is composed of two sections. The section near the base is catalytic and consumes substrate along its length to produce a more useful fuel. The catalytic section absorbs no nutrients through its surface. This part corresponds to the root nodules where bacteria convert nitrogen into nitrate for absorption by the plant. The partition coefficient between the root and the soil is assumed to be one for convenience. The substrate undergoes a first-order conversion:

$$-r_s = k''c_s$$

The outer region of the root collects the substrate for the catalytic part. The base of the root is free of substrate, while the tip of the root is impermeable.

a. Determine the substrate concentration profile in both root sections. You may assume the diffusivity of substrate is the same in both sections.
b. How does the rate of reaction in the nodule or catalytic part control the mass transfer in the passive section, that is, how does the root's effectiveness depend upon k''?

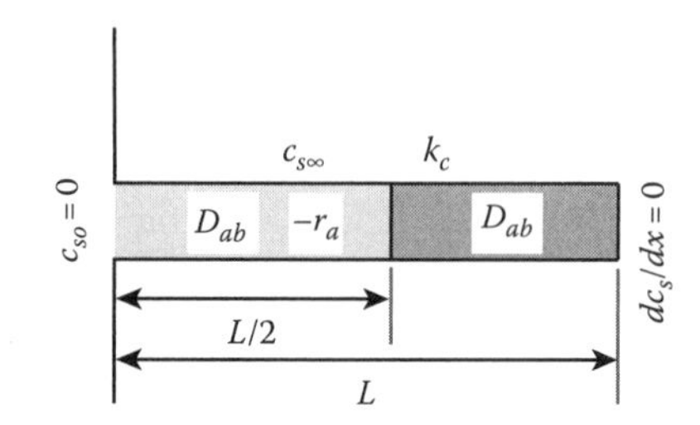

FIGURE P8.17

8.18 The alveoli of the lungs can be modeled as pin fins with a diameter of 20 μm and a length of 250 μm. The asthma drug isoproterenol is administered by inhalation. The patient inhales fully and holds his breath for ~30 s. The normal dose of isoproterenol is 120 μg ($M_w = 247.7$), and the volume of the average male's lungs aside from the alveoli is 2.25 L. The tissue of the alveoli is approximately 0.1 μm thick and has an isoproterenol diffusivity of 1×10^{-9} m²/s. The diffusion coefficient of isoproterenol inside an alveolus is 3×10^{-9} m²/s. The drug is picked up by the blood and rapidly removed, so the concentration of isoproterenol at the base of the alveolus is zero. The mass transfer coefficient to the alveolus is 1×10^{-5} m/s. The alveolus tip is impermeable.

a. Find the initial flux of isoproterenol into the blood assuming pseudo-steady-state operation.
b. Find the effectiveness and efficiency of an alveolus.
c. If the human lungs were simply two spheres with no extended surfaces, what radius would they need to have to achieve the same mass transport rate? There are some 700,000,000 alveoli in the lungs.

8.19 Consider the absorption of water through a plant root like that in Figure P8.19. The plant is a desert plant that has just experienced its first rainstorm in over 4 months. The surface of the ground is saturated, but because of the dry conditions, the water does not seep into the ground very well. In fact, the concentration of water in the ground varies linearly with depth. Assuming the geometry shown later and physical properties given in the problem,

a. Derive an expression for the water concentration profile in the root. You may assume a constant concentration at the root base, an impermeable root tip, and steady-state operation.
b. Derive an expression for the effectiveness and efficiency of this root. For the physical properties listed, what is the effectiveness and efficiency?

$$L = 5 \text{ cm} \qquad d = 0.005 \text{ m} \qquad c_{ao} = 50 \text{ mol/m}^3$$
$$c_{\infty L} = 10 \text{ mol/m}^3 \qquad c_{\infty o} = 1000 \text{ mol/m}^3$$
$$D_{ab} = 1 \times 10^{-7} \text{m}^2/\text{s} \qquad k_c = 1 \times 10^{-6} \text{m/s}$$

8.20 A neuron is designed to provide a stimulus to a muscle at a frequency of 2 s⁻¹. The neuron's axon can be modeled as a cylinder with a permeable tip. For the stimulus to be effective, the concentration of signal ions at the tip can be no less than 50% of that at the soma (base). Using the following parameters, what is the maximum axon length and required axon diameter?

$$D_{ab} = 1 \times 10^{-8} \text{ m}^2/\text{s} \qquad k_c = 1 \times 10^{-6} \text{ m/s}$$
$$c_b = 1000 \text{ mol/m}^3 \qquad c_\infty = 10 \text{ mol/m}^3$$

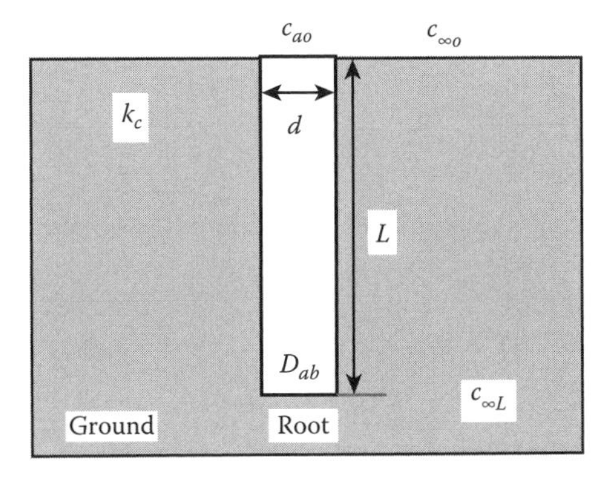

FIGURE P8.19

Catalysts and Reaction/Diffusion

8.21 A cylindrical catalyst pellet of radius r_o and length L has a chemical reaction occurring in it. In the free stream of the fluid, the concentration of reactant is $c_{a\infty}$ and the flux of reactant to the pellet's surface is characterized by a mass transfer coefficient, k_c. The partition coefficient is defined as $K_{eq} = c_{a\infty}/c_{a,pellet} = 10$. The length of the pellet is much larger than its diameter, so the mass transfer is essentially 1-D in the radial direction.

a. Derive the differential equation for the concentration profile inside the catalyst assuming a zero-order reaction with rate constant k_o and diffusivity D_o.
b. Derive the differential equation for the concentration profile inside the catalyst assuming a first-order reaction with rate constant k''.
c. What are the boundary conditions?
d. Solve the equations for the concentration profile.
e. Calculate the flux at the pellet's surface.
f. How does the external mass transfer affect the solution?
g. How do I know if I am
   1. Reaction rate controlled?
   2. Internal mass transfer controlled?
   3. External mass transfer controlled?

8.22 The following rates were observed for a first-order, irreversible reaction, carried out on a spherical catalyst:

$$d_{p1} = 0.625 \text{ cm} \qquad r_{obs,1} = 0.09 \text{ mol/g cat h}$$

$$d_{p2} = 0.1 \text{ cm} \qquad r_{obs,2} = 0.275 \text{ mol/g cat h}$$

Strong diffusional limitations were observed in both cases. Determine the true rate of reaction. Is the diffusional resistance still important when $d_p = 0.05$ cm?

8.23 Consider a catalyst pellet in the shape of a Raschig ring (a hollow cylinder of inner radius, r_i, and outer radius, r_o). A first-order reaction occurs inside the catalyst:

$$A \xrightarrow{k''} B \qquad k''_{eff} = 1.5 \times 10^{-5} \text{ s}^{-1}$$

a. If the effective diffusivity of A inside the catalyst is $D_{eff} = 2 \times 10^{-9}$ m²/s, we have dilute solutions and 1-D mass transfer in the radial direction, and there is no external mass transfer resistance, find the concentration profile and effectiveness of the catalyst.
b. If the outer radius is 1.5 cm and the inner radius is 1.0 cm, find the effectiveness. Compare with the effectiveness for a solid cylindrical catalyst of the same outer dimensions (Jain, Rogojevic and Shukla).

8.24 A first-order heterogeneous irreversible reaction is taking place within

$$A \xrightarrow{k''} B$$

a spherical catalyst pellet 200 μm in diameter. The reactant concentration halfway between the surface and the pellet's center is 1/5 the concentration at the surface. There is no external mass transfer resistance, so the surface concentration of A is the same as the bulk, $c_{as} = 1$ mol/m³. The effective diffusivity of A within the pellet is 0.1 cm²/s. To solve the differential equation for the pellet, use the transformation $\chi = rc_a$.

a. What is the concentration of the reactant at a distance of 20 μm from the pellet's surface?
b. If we need an effectiveness factor of 0.75, what diameter should the pellet be? (Pisupatti and Kumar)

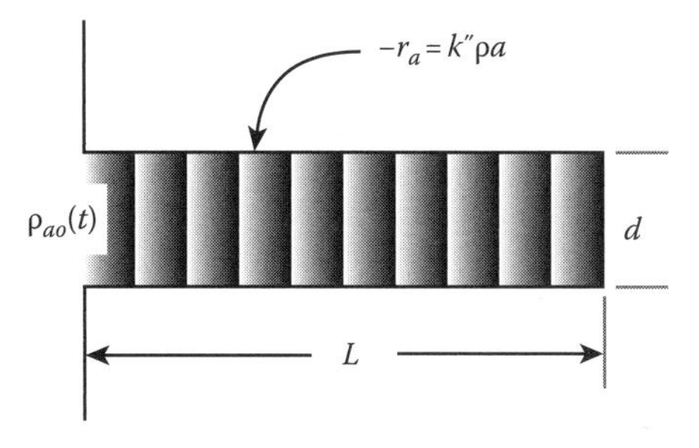

FIGURE P8.25

8.25 An inventor has the bright idea of running a gas–solid catalytic reaction with an ultrasonic assist. The concept is to induce acoustic waves inside the pores of the solid catalyst and therefore increase the reaction rate and effectiveness of the catalyst. As a patent examiner, you must evaluate the idea shown in Figure P8.25. Consider a cylindrical catalyst pore of diameter, d, and length, L.

a. Assuming that diffusion is negligible compared to the speed of the acoustic wave, show that the differential equation describing the concentration/density profile within the pore for an nth order reaction is given by

$$\frac{\partial^2 \rho_a}{\partial t^2} = c^2 \frac{\partial^2 \rho_a}{\partial x^2} - k''\rho_a^n$$

b. Considering only a first-order reaction and periodic solutions of the form $\rho = A(x)\exp(i\omega t)$, solve the differential equation for the concentration/density profile. Assume the pore has an impermeable tip.

c. What is the pore effectiveness for this scheme for a first-order reaction? For a zero-order reaction?

8.26 Consider a chemical reaction, $A \rightarrow B$, taking place in a bed of spherical catalyst pellets. The true reaction rate is nth order in the concentration of A and it has an activation energy of E_a. You decide to measure this in your reactor and find that the reaction order is n' with activation energy E_a'. You know that the Thiele modulus for your pellets is large so that you have severe diffusion limitations and wonder if that has any effect on the kinetics.

a. What is the Thiele modulus for a spherical catalyst pellet assuming an nth order reaction?

b. For large values of the Thiele modulus, the effectiveness factor for a spherical catalyst pellet is

$$\eta = \frac{3}{\phi_n}\sqrt{\frac{2}{n+1}}$$

What is the relationship between the observed reaction rate, $-r_{a,obs}$, and the actual reaction rate, $-r_a$, using this approximation? Does it affect n?

8.27 An elementary reaction is taking place inside a cylindrical catalyst pore. The process contains a small amount of a catalyst poison, P, and would like to know how this affects the effectiveness of your catalyst. The poison has severe diffusional limitations inside the pore and so can be modeled as deactivating the pore up to a distance, z_d, from the pore mouth. The pore has a radius, r_o, and a length, L. The diffusivity of reactant within the pore is D_{ab} and a first-order reaction occurs on the pore walls with a rate constant k''.

a. Derive expressions for the reactant concentration in the poisoned and active regions of the catalyst. Assume only diffusion takes place in the poisoned portion of the pore.

b. What is the effectiveness factor for the poisoned pore?

8.28 A first-order isomerization reaction is taking place in a nonisothermal, spherical catalyst pellet. Determine the effectiveness factor and the maximum temperature rise in the pellet for the following conditions:

$\Delta H = -750$ kJ/mol	$D_{eff} = 1 \times 10^{-7}$ m^2/s	$c_{as} = 15$ mol/m^3
$T_s = 425$ K	$E_a = 125$ kJ/mol	$k_{eff} = 0.025$ W/m K
$d = 0.01$ m	$k_s'' = 0.01$ m/s@425 K	

REFERENCES

1. Incropera, F.P. and DeWitt, D.P., *An Introduction to Heat Transfer*, 3rd edn., John Wiley & Sons, New York (1996).
2. Levenspiel, O., *Chemical Reaction Engineering*, John Wiley & Sons, New York (1962).
3. Weisz, P.B. and Hicks, J.S., The behavior of porous catalyst particles in view of internal mass and heat diffusion effects, *Chem. Eng. Sci.*, **17**, 265 (1962).
4. Cho, W., Saxena, R., Rodriguez, O., Achanta, R., Ojha, M., Plawsky, J.L., and Gill, W.N., Polymer penetration and pore sealing in nanoporous silica by CHF_3 plasma exposure, *J. Electrochem. Soc.*, **152**, F61–F65 (2005).

Part II

Multidimensional, Convective, and Radiative Transport

9 Multidimensional Effects, Potential Functions, and Fields

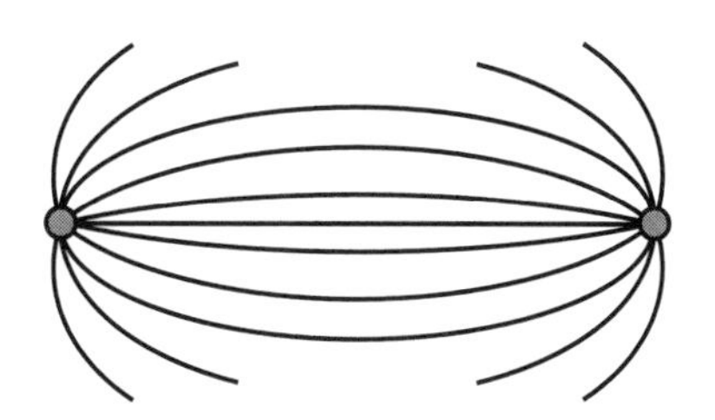

9.1 INTRODUCTION

In the previous chapters, we wandered through the general balance equation

$$In - Out + Gen = Acc$$

and highlighted the effects of adding each subsequent term. We focused on 1-D problems to illustrate the properties of the equations and solutions. There are many instances though when a 1-D formulation cannot represent the physical situation accurately.

Consider the classic fin problem introduced in Chapter 8 and shown in Figure 9.1. To perform the fin analysis, we assumed the fin was thin so that at any point along its length, the fin had a constant temperature throughout its cross section. We derived the fin number, a dimensionless grouping that assessed the relative resistance of conduction along the length of the fin to convection from its surface:

$$f_x = \sqrt{\left(\frac{L}{kA_c}\right)\Big/\left(\frac{1}{hA_s}\right)} = \sqrt{\frac{hP_dL^2}{kA_c}} - \frac{\mathcal{R}_{conduction}}{\mathcal{R}_{convection}} \tag{9.1}$$

Since the fin number is a kind of Biot number, we can also define the fin number so that it measures the conduction to convection resistance in a direction perpendicular to the fin axis. Using the fin of Figure 9.1, we would have

$$f_y = \sqrt{\left(\frac{y_o}{kA_s}\right)\Big/\left(\frac{1}{hA_s}\right)} = \sqrt{\frac{hy_o}{k}} \tag{9.2}$$

If f_y is very small, as is the case in a properly designed fin, then the temperature throughout the cross section is uniform. When f_y is large, then the resistance to conduction is high and there is a temperature gradient in the y-direction as well as in the x-direction. Under those circumstances, we cannot use a 1-D equation to describe the heat transfer but must resort to a 2-D formulation at least.

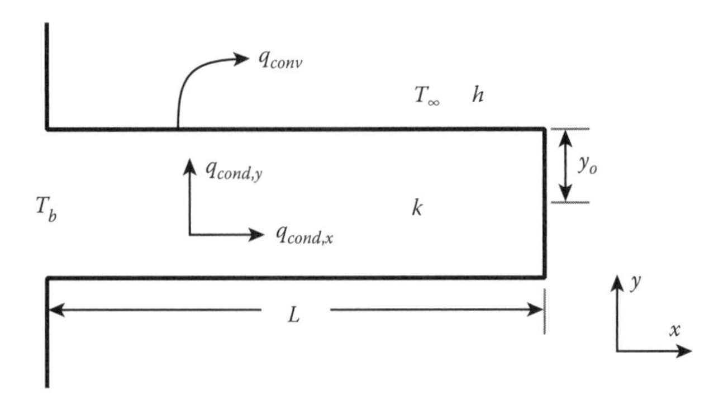

FIGURE 9.1 Straight fin showing conduction in two dimensions and convection at the surface.

Any one of the 1-D problems introduced in Chapters 4 through 8 can be recast into a form where gradients in more than one direction are important. In most engineering situations, multidimensional transport will be the norm, not the exception, and it is up to the engineer to determine when such effects are important and how to include them in the analysis.

The next few sections will discuss multidimensional effects in general and how to incorporate them into the general balance equation. We will first consider just *In* and *Out* terms and will then add on generation, *Gen* terms. For the most part, we will only consider 2-D problems. 3-D situations are inherently the same but more mathematically complicated. The same consideration extends to adding the accumulation, *Acc* term. The transient form of multidimensional problems is much more algebraically intensive to solve but is inherently no more difficult to solve than the steady-state problem. Several excellent texts including Carslaw and Jaeger [1] and Crank [2] discuss solutions for 3-D, transient problems in detail.

9.2 LAPLACE'S EQUATION AND FIELDS

The need to analyze multidimensional conduction was evident in the fin problem discussed in the previous section. This type of situation arises in many contexts. The conduction of heat in a fin is just one example. The diffusion of a drug from a small patch into the skin is an example of a multidimensional mass transfer analog. Creeping flow of a viscous liquid about a sphere, the flow of air about an aircraft wing, recirculating flow in a channel, and turbulent flow of all kinds are momentum transport analogs. The distribution of an electric field, the distribution of a magnetic field, and current flow through a semiconductor are all examples of electromagnetic analogs. In this section, we will consider only those problems exhibiting pure diffusive transport. The general balance equation will contain only *In* and *Out* terms in up to three dimensions. The equations we will derive were all first considered by the great, French mathematician, P.S. Laplace and so are called Laplace's equation. Their solutions lead to temperature, concentration, velocity, and voltage or current distributions that we refer to as spatial fields.

9.2.1 Revisiting the Fin Problem

We presume that the fin shown in Figure 9.2 is not constructed well so that there are temperature gradients in the x- and y-directions. Our first task is to look for symmetry points, lines, or planes that will enable us to cut down on the problem size. Here, we need only consider half the fin length since the fin connects two walls at the same temperature. We also only need consider half the thickness since there is another line of symmetry running down the axis of the fin.

We choose a small section of the fin and place a control volume within it as shown in Figure 9.3. All properties including the heat transfer coefficient, h, are assumed constant. The energy balance must consider flow in two dimensions this time. We treat each dimension separately and add the

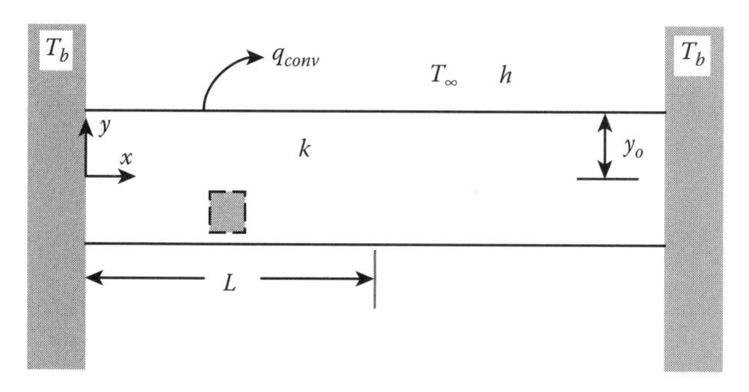

FIGURE 9.2 Fin with 2-D control volume.

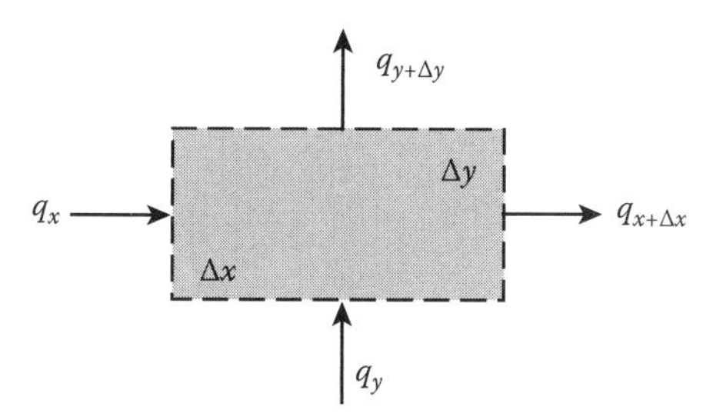

FIGURE 9.3 Control volume for the fin of Figure 9.2.

heat flows since the total heat flowing into or out of the control volume is the sum of what flows in the x- and y-directions. Assuming a square fin of thickness, t_h, into the page we have

$$\begin{array}{ccccc} (In-out)_x & + (In-out)_y & + Gen & = & Acc \\ [q(x)-q(x+\Delta x)] & +[q(y)-q(y+\Delta y)] & + \; 0 & = & 0 \end{array} \tag{9.3}$$

$$[q''(x)-q''(x+\Delta x)]\Delta y t_h + [q''(y)-q''(y+\Delta y)]\Delta x t_h = 0 \tag{9.4}$$

Dividing through by the volume of the control element, $\Delta x \Delta y t_h$, and taking the limit as Δx, $\Delta y \to 0$ leads to a first-order differential equation for the heat fluxes:

$$-\frac{\partial q_x''}{\partial x} - \frac{\partial q_y''}{\partial y} = 0 \tag{9.5}$$

To turn this into a differential equation describing the temperature distribution, we use Fourier's law in the x- and y-directions:

$$q_x'' = -k\frac{\partial T}{\partial x} \qquad q_y'' = -k\frac{\partial T}{\partial y} \tag{9.6}$$

Substituting (9.6) into Equation 9.5 and dividing through by the constant, k, gives

$$\frac{\partial^2 T}{\partial x^2} + \frac{\partial^2 T}{\partial y^2} = 0 \quad \text{Laplace's equation} \tag{9.7}$$

Since Equation 9.7 is a second-order, partial differential equation in both x and y, there must be four boundary conditions to complete the solution. The base of the fin is held at a constant temperature, T_b, and the whole fin is symmetric about $x = L$. The x boundary conditions are

$$x = 0 \qquad T = T_b \tag{9.8}$$

$$x = L \qquad \frac{\partial T}{\partial x} = 0 \tag{9.9}$$

The fin is also symmetric about the plane $y = 0$. Convective heat transfer occurs at $y = y_o$:

$$y = 0 \qquad \frac{\partial T}{\partial y} = 0 \tag{9.10}$$

$$y = y_o \qquad -k\frac{\partial T}{\partial y} = h(T - T_\infty) \tag{9.11}$$

We can solve Laplace's equation by many methods, one being the familiar separation of variables technique introduced in Chapter 6. We let

$$\theta(x, y) = \frac{T - T_b}{T_\infty - T_b} = \chi(x)\,\psi(y) \tag{9.12}$$

and substitute it into Equation 9.7:

$$\psi\frac{\partial^2\chi}{\partial x^2} + \chi\frac{\partial^2\psi}{\partial y^2} = 0 \tag{9.13}$$

Dividing through by $\chi\psi$ leaves a single equation in which the right-hand side depends only on y, while the left-hand side depends only on x:

$$\frac{1}{\chi}\frac{\partial^2\chi}{\partial x^2} = -\frac{1}{\psi}\frac{\partial^2\psi}{\partial y^2} = -\lambda^2 \tag{9.14}$$

Both sides must be equal to a constant, $-\lambda^2$. You can verify that only a constant temperature (equilibrium) solution exists if $\lambda = 0$ and that the temperature blows up if we choose $+\lambda^2$ as the constant. The process separates the original partial differential equation into two, second-order, ordinary differential equations:

$$\frac{d^2\chi}{dx^2} + \lambda^2\chi = 0 \tag{9.15}$$

$$\frac{d^2\psi}{dy^2} - \lambda^2\psi = 0 \tag{9.16}$$

The solutions to Equations 9.15 and 9.16 are given in terms of sine and cosine functions and hyperbolic sine and cosine functions, respectively.

$$\chi = a_1\sin(\lambda x) + a_2\cos(\lambda x) \tag{9.17}$$

$$\psi = b_1 \sinh(\lambda y) + b_2 \cosh(\lambda y) \tag{9.18}$$

θ is found by reconstituting the temperature from the two solutions:

$$\theta = [a_1 \sin(\lambda x) + a_2 \cos(\lambda x)][b_1 \sinh(\lambda y) + b_2 \cosh(\lambda y)] \tag{9.19}$$

The boundary condition at $x = 0$ requires $a_2 = 0$. The boundary condition at $x = L$ requires either $a_1 = 0$ or $\lambda = (n - 1/2)\pi/L$. Clearly, only the latter is allowed if we want a nontrivial solution. The condition at $y = 0$ requires $b_1 = 0$. The remaining partial solution for θ is

$$\theta = a_3 \sin(\lambda_n x) \cosh(\lambda_n y) \qquad \lambda_n = \frac{(n - 1/2)\pi}{L} \tag{9.20}$$

Since a single value of a_3 cannot reproduce the boundary condition at $y = y_o$ for all x, the general solution requires a sum over all the individual solutions. Remember this summation is allowed because the differential equation is linear, and if one solution satisfies the equation, a sum of similar solutions will also satisfy the differential equation:

$$\theta = \sum_{n=1}^{\infty} a_{3n} \sin(\lambda_n x) \cosh(\lambda_n y) \tag{9.21}$$

The last boundary condition is used to solve for a_{3n} and states the following:

$$-k \sum_{n=1}^{\infty} a_{3n} \lambda_n \sin(\lambda_n x) \sinh(\lambda_n y_o) - h \sum_{n=1}^{\infty} a_{3n} \sin(\lambda_n x) \cosh(\lambda_n y_o) = -h \tag{9.22}$$

To extract a_{3n}, we use the orthogonality properties of the sine function, multiply both sides by $\sin(\lambda_m x)$, and integrate from $x = 0$ to $x = L$:

$$\begin{aligned} &-k \int_0^L \sum_{n=1}^{\infty} a_{3n} \lambda_n \sin(\lambda_n x) \sin(\lambda_m x) \sinh(\lambda_n y_o)\, dx \\ &-h \int_0^L \sum_{n=1}^{\infty} a_{3n} \sin(\lambda_n x) \sin(\lambda_m x) \cosh(\lambda_n y_o)\, dx = -h \int_0^L \sin(\lambda_m x)\, dx \end{aligned} \tag{9.23}$$

The only terms involved in the integrals on the left-hand side are those occurring when $\lambda_m = \lambda_n$. Thus,

$$-k \int_0^L a_{3n} \lambda_n \sin^2(\lambda_n x) \sinh(\lambda_n y_o)\, dx - h \int_0^L a_{3n} \sin^2(\lambda_n x) \cosh(\lambda_n y_o)\, dx = -h \int_0^L \sin(\lambda_n x)\, dx \tag{9.24}$$

Solving for a_{3n} we find

$$a_{3n} = \frac{4h}{L\lambda_n [k\lambda_n \sinh(\lambda_n y_o) + h \cosh(\lambda_n y_o)]} \tag{9.25}$$

and the solution is complete.

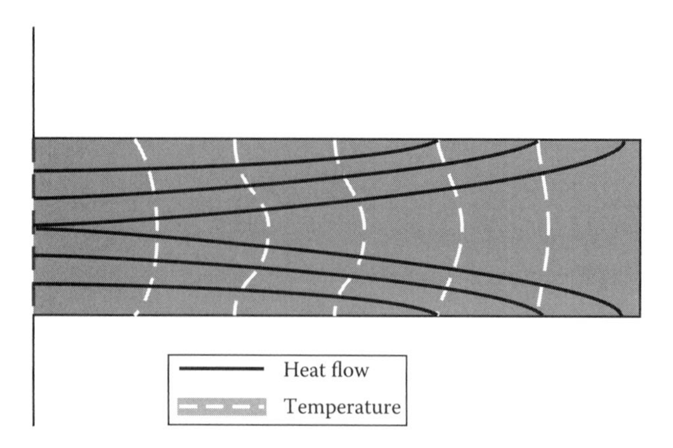

FIGURE 9.4 Fin showing lines of constant temperature and lines of constant heat flow.

The lines of constant temperature and lines of constant heat flow (from Fourier's law) are given by

$$\theta_i = \sum_{n=1}^{\infty} a_{3n}\sin(\lambda_n x)\cosh(\lambda_n y) \tag{9.26}$$

$$\vec{\mathbf{q}}_i = q_{xi}\vec{\mathbf{x}} + q_{yi}\vec{\mathbf{y}} = y_o t_h \sum_{n=1}^{\infty} a_{3n}\lambda_n \cos(\lambda_n x)\cosh(\lambda_n y)\vec{\mathbf{x}} + L t_h \sum_{n=1}^{\infty} a_{3n}\lambda_n \sin(\lambda_n x)\sinh(\lambda_n y)\vec{\mathbf{y}} \tag{9.27}$$

Both lines are orthogonal to one another. We discussed this first in Chapter 2, but it is easier to see now that they can be plotted. A sketch of these fields for the fin is shown in Figure 9.4. The heat flow lines start out perpendicular to the plane $x = 0$ since the base is held at a constant temperature. They intersect the plane $y = \pm\, y_o$ with a prescribed slope since there is a prescribed heat flow from that surface due to convection. With no heat flow through the plane at $x = L$, the heat flow lines must intersect there with zero slope. Throughout the body, the lines of heat flow and lines of constant temperature are perpendicular to one another. This situation holds because the thermal conductivity is isotropic. If the fin were made of a crystalline material, lines of constant heat flow and lines of constant temperature might not be perpendicular to one another.

The 2-D temperature field can also be interpreted in terms of functions of a complex variable [3]. Here we specify

$$\phi(z) + i\psi(z) = f(x + iy) = f(z) \tag{9.28}$$

where ϕ and ψ are conjugate functions of the complex variable, z. As conjugate functions, they obey the Cauchy–Riemann relations,

$$\frac{\partial \phi}{\partial x} = \frac{\partial \psi}{\partial y} \tag{9.29}$$

$$\frac{\partial \phi}{\partial y} = -\frac{\partial \psi}{\partial x} \tag{9.30}$$

and so they are orthogonal to one another. This makes the entire function $f(z)$, *analytic* [3]. If we differentiate Equation 9.29 by x and Equation 9.30 by y and then add the two equations, we find

$$\frac{\partial^2 \phi}{\partial x^2} + \frac{\partial^2 \phi}{\partial y^2} = 0 \tag{9.31}$$

a statement of Laplace's equation. Similarly, if we differentiate Equation 9.29 by y and Equation 9.30 by x and subtract the two, we obtain another version of Laplace's equation, this time in terms of ψ:

$$\frac{\partial^2 \psi}{\partial x^2} + \frac{\partial^2 \psi}{\partial y^2} = 0 \tag{9.32}$$

We refer to f as the *complex potential function* and ϕ as the *potential function* [4]. Thus, we can assign the real part of f, the function ϕ, to be the temperature, T. The curves $\phi = const.$ represent isotherms. Since ψ is orthogonal to ϕ, it is related to the heat flow, $\vec{\mathbf{q}}$. ψ is referred to as the *stream function*, and the curves where $\psi = const.$ are streamlines that show the path of the heat flow. We can see this clearly if we try to calculate the heat flow in the x-direction. Taking a thin slice through the fin in the y-direction, we have

$$dq_x = -k(t_h dy)\frac{\partial T}{\partial x} = -k(t_h dy)\frac{\partial \phi}{\partial x} = -k(t_h dy)\frac{\partial \psi}{\partial y} \tag{9.33}$$

Integrating Equation 9.33 over y from 0 to y_o

$$q_x = -kt_h \int_0^{y_o} \left(\frac{\partial \psi}{\partial y}\right) dy = -kt_h[\psi(y_o) - \psi(0)] \tag{9.34}$$

we see that ψ is in fact related to the heat flow per unit depth per unit thermal conductivity. It is important to realize that the complex variable approach is inherently a 2-D one since the complex variable representation maps the solution from the x–y plane to the potential-stream function plane.

9.2.2 Scalar and Vector Fields

Laplace's equation arises in many different areas: heat transfer, mass transfer, fluid flow, electromagnetic fields, and mechanics. For each case, we can define a potential function and a stream function. As we saw in the fin example, the temperature represented the potential function and the stream function was related to the heat flow. A similar case would exist in mass transfer where the potential function would be the concentration or mole fraction and the stream function would be related to the mass flow. Table 9.1 shows the potential and stream functions for a variety of transport operations involving what we refer to as *scalar* fields. There are two keys to defining a scalar field. The first is that the potential function is the variable that we would measure experimentally. The second is that the flow of the transported quantity is solely determined by the gradient of the scalar potential.

We can contrast this with momentum transport and the magnetic field. The velocity/stream function relationships in Cartesian and other coordinate systems can be found in Chapter 10. In both these instances, the measured quantity defines a vector, not a scalar field. If ψ represents the stream function, then the velocity components in Cartesian coordinates are defined by

$$v_x = \frac{\partial \psi}{\partial y} \qquad v_y = -\frac{\partial \psi}{\partial x} \tag{9.35}$$

TABLE 9.1
Transport Operations, Potential Function, and Stream Function Relationships

Transport Operation	Potential Function	Stream Function
Heat	Temperature	Heat flow
Mass	Concentration	Mass flow
Charge	Concentration	Current
Electric field	Voltage	Electric field force
Gravitational field	Mass	Gravity

The stream function is tangent to the velocity at any point. We speak of the velocity field in an analogous way with the temperature or concentration field. However, since velocity is a vector with three components, the velocity field is a *vector* field, and to truly describe it, we need a vector potential function in addition to the scalar potential function.

When we work with a scalar field, the problem is completely specified once we know the potential function. In a vector field such as the velocity field, we have two or more measured quantities, v_x and v_y, for example, that may interact. This interaction means that the stream function will obey Laplace's equation only under special circumstances. Normally, the differential equation describing the behavior of the stream function will have a generation term and will obey a modified form of Laplace's equation termed Poisson's equation.

In momentum transport, this generation term is called the *vorticity*, and when we consider the magnetic field, it is called the circulation. In either case, it describes the circulation of fluid or current around closed paths that exist within the field and is the vector potential function needed to complete the description of the vector field. The vorticity in Cartesian coordinates is defined by

$$\vec{\omega}_\mu = \vec{\nabla}\times\vec{\mathbf{v}} = \left(\frac{\partial v_y}{\partial z}-\frac{\partial v_z}{\partial y}\right)\vec{\mathbf{i}}+\left(\frac{\partial v_z}{\partial x}-\frac{\partial v_x}{\partial z}\right)\vec{\mathbf{j}}+\left(\frac{\partial v_x}{\partial y}-\frac{\partial v_y}{\partial x}\right)\vec{\mathbf{k}} \tag{9.36}$$

in three dimensions. Definitions in other coordinate systems are given in Chapter 10. Substituting the definitions of the velocity from Equation 9.35 into Equation 9.36, we find that in two dimensions, Poisson's equation for the stream function is

$$\frac{\partial^2\psi}{\partial x^2}+\frac{\partial^2\psi}{\partial y^2}=\omega_{\mu z} \tag{9.37}$$

If $\omega_{\mu z} = 0$, a condition referred to as *irrotational* flow in fluid mechanics, the stream function and velocity potential obey Laplace's equation. Irrotational flow occurs only in the absence of a shear stress, that is, when there is no viscosity. We can summarize how to express the generalized vector field using a relationship found by Helmholtz. *Helmholtz's relationship* states that the vector field is represented by the gradient of a scalar potential plus the curl of a vector potential. Considering momentum transport, we have

$$\vec{\mathbf{v}} = \vec{\nabla}\phi + curl(\vec{\omega}_\mu) = \vec{\nabla}\phi + \vec{\nabla}\times(\vec{\nabla}\times\vec{\mathbf{v}}) \tag{9.38}$$

In momentum transport, we refer to the scalar potential as the velocity potential. The vector potential is the vorticity. If the flow is *irrotational*, the vorticity vanishes and the velocity can be determined solely from the scalar velocity potential. The following sections will consider such irrotational situations and illustrate solutions to representative problems by the method of separation of variables and using complex functions. Later in the text, we will reintroduce viscosity and see how the solutions change when we introduce vorticity.

9.3 SOLUTIONS OF LAPLACE'S EQUATION

Solutions to Laplace's equation can be obtained by many methods. We have already seen an example of separation of variables. Other common methods include Green's functions [5–7], the use of complex variables [3,8,9], and numerical methods [10,11]. Numerical techniques have supplanted most of the analytical methods; however, it is instructive to solve several linear problems using the analytical techniques to get a feel for the solutions. In this section, we will look at several examples of solutions to Laplace's equation. Table 9.2 shows the form of Laplace's equation in three dimensions for Cartesian, cylindrical, and spherical coordinates.

9.3.1 Cylindrical Coordinates: Diffusion in a Villus

We have already seen an example of solving Laplace's equation in Cartesian coordinates when we discussed the fin problem. Let us look at what happens in cylindrical coordinates. Consider a cylindrical mass transfer fin (a villus) of radius, r_o, and height, L, immersed in the intestine. The fin is essentially infinitely long. The surface at $z = 0$ is held at a constant concentration of nutrient, c_{ao}, that is much less than the concentration in the free stream, $c_{a\infty}$. The mass transfer coefficient to the fin is k_c and is constant. The diffusivity of nutrient inside the fin is also a constant, D_{ab}. We would like to determine the steady-state concentration profile inside the cylinder. Figure 9.5 shows the cylinder and specifications.

TABLE 9.2
Laplace's Equation

Coordinate System	Laplace's Equation
Cartesian	$\frac{\partial^2 f}{\partial x^2} + \frac{\partial^2 f}{\partial y^2} + \frac{\partial^2 f}{\partial z^2} = 0$
Cylindrical	$\frac{\partial^2 f}{\partial r^2} + \frac{1}{r}\frac{\partial f}{\partial r} + \frac{1}{r^2}\frac{\partial^2 f}{\partial \theta^2} + \frac{\partial^2 f}{\partial z^2} = 0$
Spherical	$\frac{\partial^2 f}{\partial r^2} + \frac{2}{r}\frac{\partial f}{\partial r} + \frac{1}{r^2}\frac{\partial^2 f}{\partial \theta^2} + \frac{\cos\theta}{r^2 \sin\theta}\frac{\partial f}{\partial \theta} + \frac{1}{r^2 \sin^2\theta}\frac{\partial^2 f}{\partial \phi^2} = 0$

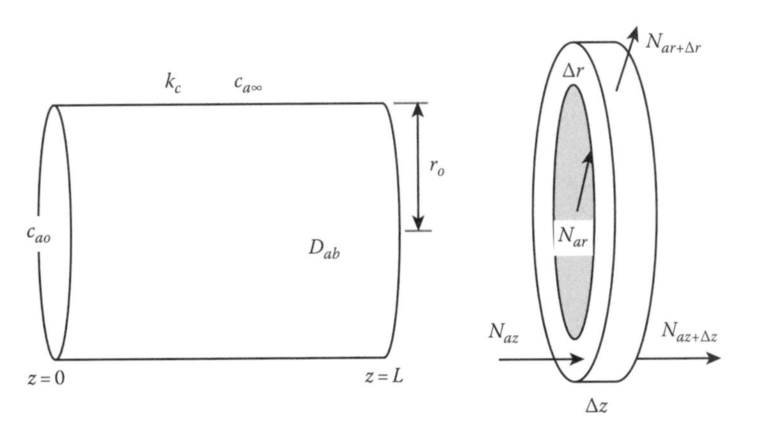

FIGURE 9.5 Cylindrical villus with mass transfer from the outer surface.

Since the entire cylinder is exposed to a uniform fluid, there will be no angular concentration dependence for this situation. Substituting into the general balance from Figure 9.5,

$$\begin{array}{ccccc} In & - & Out & = & Acc \\ \dot{M}_a(r) & - & \dot{M}_a(r+\Delta r)+\dot{M}_a(z)-\dot{M}_a(z+\Delta z) & = & 0 \end{array} \tag{9.39}$$

$$A_s N_a(r) - A_s N_a(r+\Delta r) + A_c N_a(z) - A_c N_a(z+\Delta z) = 0 \tag{9.40}$$

Substituting in for the surface area, $A_s \approx 2\pi r \Delta z$, and the cross-sectional area, $A_c \approx 2\pi r \Delta r$; dividing through by $\Delta r \Delta z$; and taking the limit as Δr, $\Delta z \rightarrow 0$ leads to a differential equation for the flux, N_a:

$$-\frac{1}{r}\frac{\partial}{\partial r}(rN_{ar}) - \frac{\partial}{\partial z}(N_{az}) = 0 \tag{9.41}$$

Using Fick's law to replace the flux with the concentration gradient (dilute solutions or equimolar counterdiffusion assumed) leads to Laplace's equation in cylindrical coordinates:

$$N_{ar} = -D_{ab}\frac{\partial c_a}{\partial r} \qquad N_{az} = -D_{ab}\frac{\partial c_a}{\partial z} \tag{9.42}$$

$$\frac{\partial^2 c_a}{\partial r^2} + \frac{1}{r}\frac{\partial c_a}{\partial r} + \frac{\partial^2 c_a}{\partial z^2} = 0 \tag{9.43}$$

The boundary conditions for this problem state the following:

$$z = 0 \qquad c_a = c_{ao} \tag{9.44}$$

$$z = L \qquad c_a = c_{a\infty} \quad \text{Infinite approximation} \tag{9.45}$$

$$r = 0 \qquad \frac{\partial c_a}{\partial r} = 0 \quad \text{Symmetry} \tag{9.46}$$

$$r = r_o \qquad -D_{ab}\frac{\partial c_a}{\partial r} = k_c(c_a - c_{a\infty}) \tag{9.47}$$

Before solving the equation, it is helpful to replace the concentration variable c_a with $\chi = (c_a - c_{a\infty})\,(c_{ao} - c_{a\infty})$. The differential equation becomes

$$\frac{\partial^2 \chi}{\partial r^2} + \frac{1}{r}\frac{\partial \chi}{\partial r} + \frac{\partial^2 \chi}{\partial z^2} = 0 \tag{9.48}$$

We can use separation of variables to solve this equation. Letting $\chi(r, z) = R(r)Z(z)$, substituting into the equation, and dividing through by RZ leads to

$$\frac{1}{R}\frac{\partial^2 R}{\partial r^2} + \frac{1}{R}\left(\frac{1}{r}\frac{\partial R}{\partial r}\right) = -\frac{1}{Z}\frac{\partial^2 Z}{\partial z^2} \tag{9.49}$$

Setting both sides equal to $-\lambda^2$ results in two ordinary differential equations:

$$\frac{d^2R}{dr^2}+\frac{1}{r}\frac{dR}{dr}+\lambda^2R=0 \tag{9.50}$$

$$\frac{d^2Z}{dz^2}-\lambda^2Z=0 \tag{9.51}$$

Equation 9.51 is easily solved in terms of hyperbolic sines and cosines:

$$Z=a_1\sinh(\lambda z)+a_2\cosh(\lambda z) \tag{9.52}$$

We encountered a form of Equation 9.50 in Chapter 6. Its solution is given in terms of the Bessel functions:

$$R=a_3J_o(\lambda r)+a_4Y_o(\lambda r) \tag{9.53}$$

The transformed boundary conditions for this problem state that

$$r=0 \qquad \frac{\partial\chi}{\partial r}=0 \tag{9.54}$$

$$r=r_o \qquad -D_{ab}\frac{\partial\chi}{\partial r}=k_c\chi \tag{9.55}$$

$$z=0 \qquad \chi=1 \tag{9.56}$$

$$z=L \qquad \chi=0 \tag{9.57}$$

Applying the boundary condition at $r=0$, we find that $a_4=0$ since Y_o approaches $-\infty$ there. At $r=r_o$, we find

$$-D_{ab}\lambda J_1(\lambda r_o)=k_cJ_o(\lambda r_o) \tag{9.58}$$

where we have used the derivative property of the Bessel function discussed in Chapter 6. For this equation to hold, λ must be a root of Equation 9.58. At $z=L$, we must have $a_2=-a_1\tanh(\lambda L)$. Thus, combining the remaining integration constants into a_5, the partial solution is

$$\chi=\sum_{n=1}^{\infty}a_{5n}J_o(\lambda_n r)[\sinh(\lambda_n z)-\tanh(\lambda_n L)\cosh(\lambda_n z)] \tag{9.59}$$

A single value of a_{5n} or λ_n cannot reproduce the boundary condition at $z=0$. We must use a sum of solutions and hope the sum can fit the boundary condition. To determine the values of a_{5n}, we apply the boundary condition at $z=0$ and use the orthogonal properties of the Bessel functions. Multiplying both sides of Equation 9.59 evaluated at $z=0$ by $rJ_o(\lambda_m r)$ and integrating from $r=0$ to $r=r_o$, we find that the only integral that survives is the one when $n=m$:

$$\int_0^{r_o}rJ_o(\lambda_n r)dr=-\tanh(\lambda_n L)a_{5n}\int_0^{r_o}rJ_o^2(\lambda_n r)dr \tag{9.60}$$

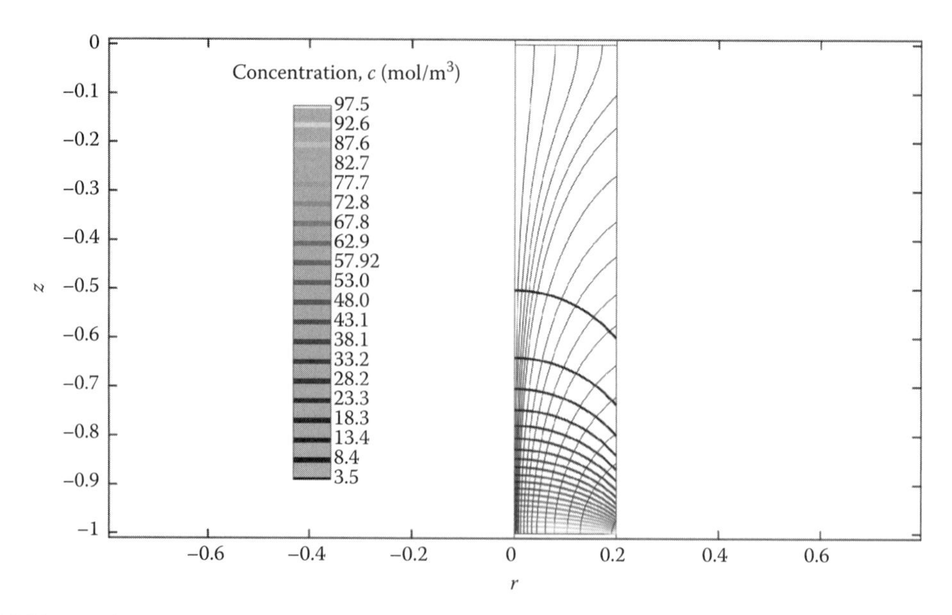

FIGURE 9.6 Concentration and flux line plot for the villus produced using Comsol® module Villus. Notice how lines of constant concentration and lines of constant mass flux intersect at a right angle.

The a_{5n} are

$$a_{5n} = -\frac{2J_1(\lambda_n r_o)}{\lambda_n r_o \tanh(\lambda_n L)[J_o^2(\lambda_n r_o) + J_1^2(\lambda_n r_o)]} \tag{9.61}$$

and full solution is

$$\chi = -\sum_{n=1}^{\infty} \frac{2J_1(\lambda_n r_o)J_o(\lambda_n r)[\sinh(\lambda_n z) - \tanh(\lambda_n L)\cosh(\lambda_n z)]}{\lambda_n r_o \tanh(\lambda_n L)\left[J_o^2(\lambda_n r_o) + J_1^2(\lambda_n r_o)\right]} \tag{9.62}$$

Again, because J_o and J_1 are orthogonal to one another, the lines of constant concentration and those for constant mass flow are perpendicular (Figure 9.6). Figure 9.4 applies to this situation as well since in the infinite fin, the temperature does not change at the fin tip either.

9.3.2 Spherical Coordinates: Heat Transfer in a Hemisphere

Consider a hemispherical surface of radius $r = r_o$, exposed to infrared radiation of uniform intensity. The flat portion of the hemisphere is maintained at a temperature, $T = 0$, while the curved surface is exposed to a heat flux of q''_o. We would like to determine the temperature distribution in the interior. The system is shown in Figure 9.7.

The system is symmetric in the ϕ direction and so the problem is 2-D. Laplace's equation in spherical coordinates is

$$\frac{\partial^2 T}{\partial r^2} + \frac{2}{r}\frac{\partial T}{\partial r} + \frac{1}{r^2}\frac{\partial^2 T}{\partial \theta^2} + \frac{\cos\theta}{r^2 \sin\theta}\frac{\partial T}{\partial \theta} = 0 \tag{9.63}$$

The boundary conditions for this problem are

$$r = 0 \qquad \frac{\partial T}{\partial r} = 0 \quad \text{Symmetry} \tag{9.64}$$

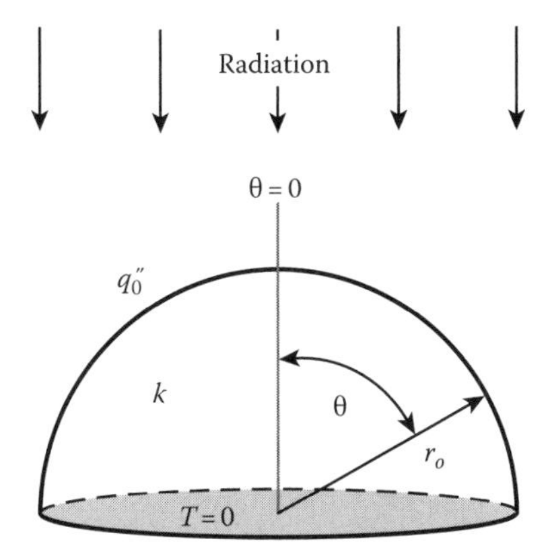

FIGURE 9.7 Hemispherical surface with upper surface exposed to constant heat flux and lower surface exposed to constant temperature.

$$r = r_o \qquad -k\frac{\partial T}{\partial r} = q''_o \tag{9.65}$$

$$\theta = 0 \qquad \frac{\partial T}{\partial \theta} = 0 \tag{9.66}$$

$$\theta = \frac{\pi}{2} \qquad T = 0 \tag{9.67}$$

The technique of separation of variables can also be used to solve this problem since the equation, despite its complexity, is linear. Letting $T = R(r)\Theta(\theta)$, substituting into the differential equation, multiplying the equation through by r^2, and then dividing through by $R\Theta$ leads to

$$\left(\frac{1}{\Theta}\right)\frac{\partial^2\Theta}{\partial\theta^2} + \left(\frac{\cos\theta}{\Theta\sin\theta}\right)\frac{\partial\Theta}{\partial\theta} = -\left(\frac{r^2}{R}\right)\frac{\partial^2 R}{\partial r^2} - \left(\frac{2r}{R}\right)\frac{\partial R}{\partial r} = -\lambda^2 \tag{9.68}$$

We separate the two ordinary differential equations:

$$r^2\frac{d^2R}{dr^2} + 2r\frac{dR}{dr} - \lambda^2 R = 0 \tag{9.69}$$

$$\sin\theta\frac{d^2\Theta}{d\theta^2} + \cos\theta\frac{d\Theta}{d\theta} + (\lambda^2\sin\theta)\Theta = 0 \tag{9.70}$$

The first differential equation is an example of Euler's equation [12]. Setting $R \propto r^m$, substituting into the differential equation, and solving the resulting quadratic form leads to the solution. In this case,

$$m(m-1) + 2m - \lambda^2 = 0 \tag{9.71}$$

The exponent, m, takes on particularly simple values if we let $\lambda^2 = n\,(n+1)$. Since we never specified the form for λ^2 and a solution for $n^2 + n - \lambda^2 = 0$ always exists, making the change from λ to n is merely a convenience. The solution to Equation 9.69 can now be written as

$$R = a_1 r^n + \frac{a_2}{r^{n+1}} \tag{9.72}$$

The boundary condition at $r = 0$ requires $a_2 = 0$ for the solution to exist.

Equation 9.70 appears complicated. It was first studied by French mathematician, A.M. Legendre, in its algebraic form (set $x = \cos\theta$ and change variables) and is known as the associated Legendre equation. Like the Bessel equation, it was solved by a series solution technique, and the set of orthogonal polynomials that arise from the solution are called Legendre polynomials, P_n. The first few polynomials in the series are given in Table 9.3 and their trigonometric form is shown graphically in Figure 9.8. Notice that each function oscillates between ±1:

$$P_n(\cos\theta) = \frac{1\cdot3\cdots(2n-1)}{2\cdot4\cdots(2n)} 2\cos(n\theta) + 2\frac{1}{2}\frac{1\cdot3\cdots(2n-3)}{2\cdot4\cdots(2n-2)}\cos((n-2)\theta)$$
$$+ 2\frac{1\cdot3}{2\cdot4}\frac{1\cdot3\cdots(2n-5)}{2\cdot4\cdots(2n-4)}\cos((2n-4)\theta) + \cdots \tag{9.73}$$

TABLE 9.3
Legendre Polynomials

Polynomial	Algebraic	Trigonometric
P_o	1	1
P_1	x	$\cos\theta$
P_2	$\frac{3x^2-1}{2}$	$\frac{3\cos(2\theta)+1}{4}$
P_3	$\frac{5x^3-3x}{2}$	$\frac{5\cos(3\theta)+3\cos\theta}{8}$
P_4	$\frac{35x^4-30x^2+3}{8}$	$\frac{35\cos(4\theta)+20\cos(2\theta)+9}{64}$
P_5	$\frac{63x^5-70x^3+15x}{8}$	$\frac{63\cos(5\theta)+35\cos(3\theta)+30\cos\theta}{128}$

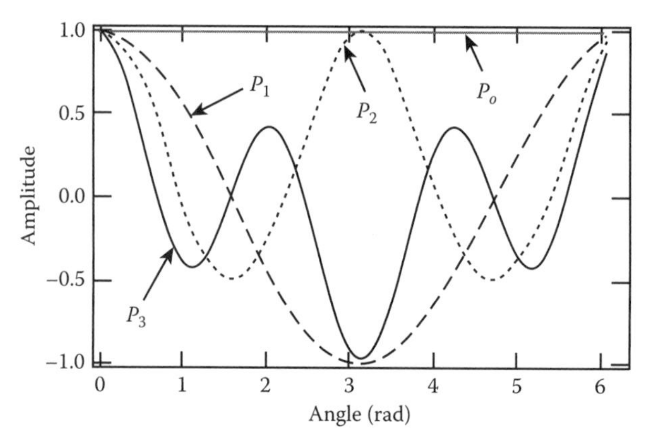

FIGURE 9.8 The first four Legendre functions.

The solution to Equation 9.70 can be written in terms of these functions. We generally separate the functions into their odd and even components:

$$\Theta = a_3 P_{2n-1}(\cos\theta) + a_4 P_{2n-2}(\cos\theta) \quad n = 1\ldots \tag{9.74}$$

The boundary condition at $\theta = \pi/2$ sets $T = 0$. Looking at the Legendre functions previously, it is clear that only the odd values will fit this condition so $a_4 = 0$. The boundary condition at $\theta = 0$ is satisfied automatically. Reconstituting the temperature function in terms of a series since, in general, one term will not satisfy the boundary conditions, yields

$$T = \sum_{n=1}^{\infty} a_{5(2n-1)} r^{2n-1} P_{2n-1}(\cos\theta) \tag{9.75}$$

Applying the boundary condition at $r = r_o$ gives

$$q_0'' = -k \sum_{n=1}^{\infty} a_{5(2n-1)} (2n-1) r_o^{2n-2} P_{2n-1}(\cos\theta) \tag{9.76}$$

To extract the $a_{5(2n-1)}$, we use the fact that the P_n are orthogonal with respect to the weighting function $P_m \sin\theta$. Thus, for any value of m and n,

$$\int_0^{\pi} P_m(\cos\theta) P_n(\cos\theta) \sin\theta \, d\theta = \begin{cases} 0 & m \neq n \\ 2/(2n+1) & m = n \end{cases} \tag{9.77}$$

Multiplying both sides of Equation 9.76 by $\sin\theta \, P_m(\cos\theta)$, where m is odd because the P_{2n-1} are odd functions, and integrating from $\theta = 0$ to $\theta = \pi/2$, we extract the $a_{5(2n-1)}$:

$$a_{5(2n-1)} = -\left(\frac{2n+1}{2n-1}\right) \frac{kq_0''}{r_o^{2n-2}} \int_0^{\pi/2} \sin\theta P_{2n-1}(\cos\theta) \, d\theta \tag{9.78}$$

The complete solution is

$$T = -kq_0'' \sum_{n=1}^{\infty} \frac{(2n+1)}{(2n-1)} \left(\frac{r}{r_o}\right)^{2n-2} r P_{2n-1}(\cos\theta) \int_0^{\pi/2} \sin\theta P_{2n-1}(\cos\theta) d\theta \tag{9.79}$$

and is shown in Figure 9.9. The streamlines and equipotential lines are orthogonal to one another and look somewhat like latitude and longitude lines.

9.3.3 Laplace Equation Solutions Using Complex Variables

The series solutions for 2-D problems are exact but often very difficult to evaluate. Many times the series are slow to converge and so it is time consuming to obtain a solution. The use of complex variables to solve 2-D Laplace's equations avoids these problems since series solutions are not normally involved. Not all problems that can be solved with the series solution technique can be solved using complex variables, but for those that can be solved, the solution is usually simpler to obtain and express.

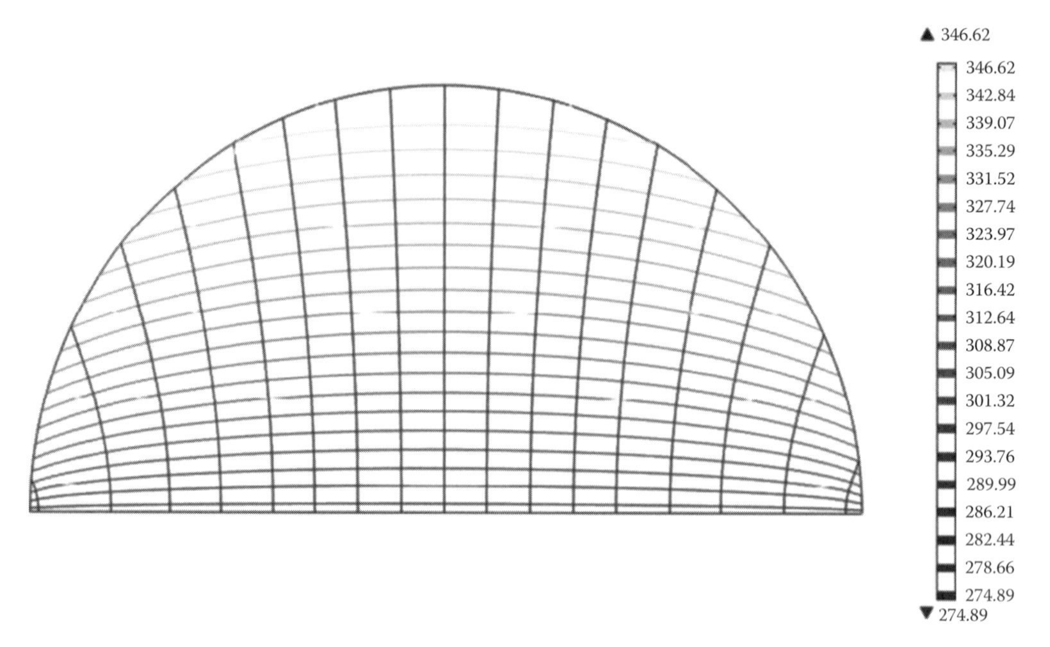

FIGURE 9.9 Equipotential lines (shaded) and streamlines (uniform) for the heated sphere problem. Solution developed using Comsol® module Hemisphere.

The use of complex variables to solve Laplace's equation is an example of an inverse problem. The procedure involves transforming a solution that we know from a simple geometry, usually a plane, to a more complicated geometry by the use of *conformal* mapping. This technique requires finding a suitable *analytic* function that performs the mapping transformation [3]. It is much easier to derive analytic functions and then see what field geometry it represents than it is to take a given geometry and try to develop an analytic function to take us there. Thus, solving Laplace's equation for a particular geometry is a trial and error procedure much like trying to solve a partial differential equation using a similarity variable.

Our first example considers flow in a corner as shown in Figure 9.10. Far away from the wall in the y-direction, we have unidirectional flow with $v_y = v_o$ and $v_x = 0$. Far away from the wall in the x-direction, we have the opposite condition, $v_x = v_o$, $v_y = 0$. The flow is irrotational and we will stipulate that it is inviscid. Inviscid means the fluid can slip along the walls as though it was frictionless. The friction between the fluid and the wall would produce vorticity and lead to circulation, and Laplace's equation would no longer apply. Here, we see another difference between the vector and scalar field. If we were to solve Laplace's equation for the analogous mass transfer situation,

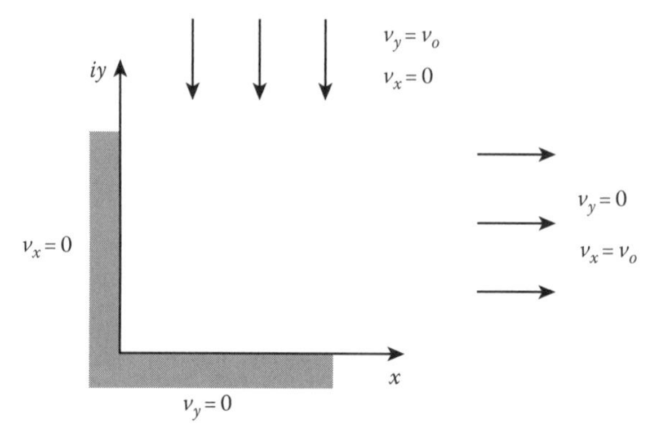

FIGURE 9.10 Inviscid and irrotational flow around a corner.

we could take the solution for the potential function, substitute into Fick's law, and obtain the mass flux anywhere. Forcing the fluid to be inviscid means we can no longer use the potential function to calculate the shear stresses. These stresses are identically zero for an inviscid fluid. In a bizarre way by making the fluid inviscid and irrotational, we allow the velocity to be determined using a scalar potential, yet we can no longer calculate a stress. Later, we will see that we can use the potential function at points far removed from solid objects to obtain the flow field but we must supplement that solution, using boundary layer theory, to obtain the stress and transport of momentum near the surface of the object.

Along the boundary defined by $x = 0$, we know that the velocity $v_x = 0$. Along the boundary at $y = 0$, $v_y = 0$. To solve this problem, we imagine the corner to be in the complex plane where $z = x + iy$ as shown in Figure 9.10. If we flatten out the corner so that it represents the half plane above $y = 0$, we can define a complex potential as

$$F(z) = Az = Ax + iAy \tag{9.80}$$

where the velocity potential and stream function are

$$\phi(x, y) = Ax \qquad \psi(x, y) = Ay \tag{9.81}$$

The velocity at any point is defined by the complex conjugate of the derivative of the potential function or:

$$v = \overline{\frac{dF}{dz}} = A \tag{9.82}$$

This represents uniform motion in the half plane from left to right with a velocity, $v_x = A$ ($v_y = 0$ since $\phi = f(x)$). To solve the problem of the corner, we need to find an analytic function that will take the half plane and bend it 90° at $y = 0$ to form the corner of Figure 9.10. Fortunately, such a transformation has been found in the form of a power law function, $w = z^n$. Specifically, replacing w by z^2 where

$$w = \upsilon + i\eta = z^2 = (x + iy)^2 = x^2 - y^2 + i2xy \tag{9.83}$$

completes the transformation and folds the plane. The complex potential function becomes

$$F(z) = Az^2 = \underbrace{A(x^2 - y^2)}_{\text{Potential function}} + i\underbrace{2Axy}_{\text{Stream function}} \tag{9.84}$$

and the potential and stream function are defined by the real and imaginary parts, respectively, of the complex potential.

$$\phi(x, y) = A(x^2 - y^2) \tag{9.85}$$

$$\psi(x, y) = 2Axy \tag{9.86}$$

Lines where $\psi(x, y) = C$ define a series of rectangular hyperbolas that show what the flow looks like within the corner (Figure 9.11).

The velocity is the complex conjugate of the derivative of F:

$$v = \overline{\frac{dF}{dz}} = 2A\bar{z} = 2A(x - iy) \tag{9.87}$$

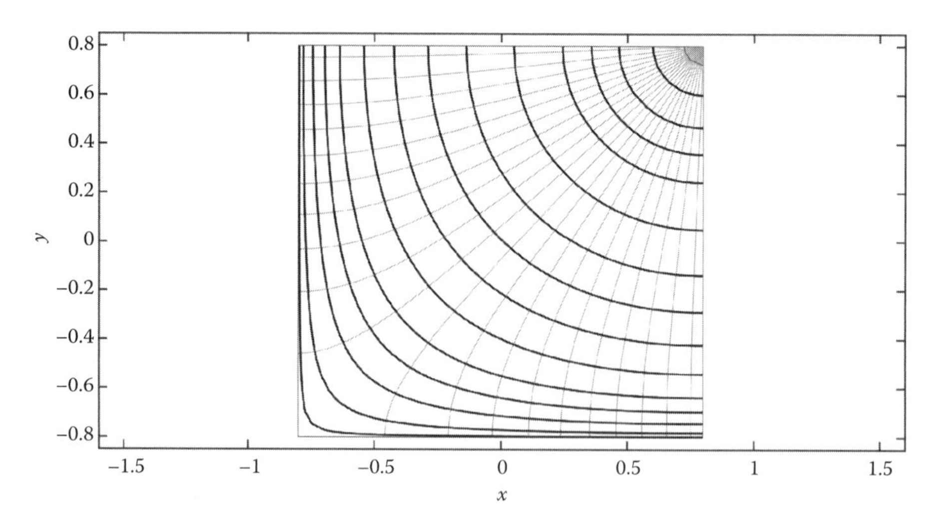

FIGURE 9.11 Streamlines (black) for flow of an inviscid fluid around a corner formed using Comsol® module Corner.

The speed or magnitude of the velocity is defined as the dot product of v with its complex conjugate, $v \cdot \bar{v}$:

$$|v| = \sqrt{4A^2(x+iy)(x-iy)} = 2A\sqrt{x^2+y^2} \tag{9.88}$$

Thus, we see that the speed of the flow is directly proportional to the distance from the origin of the corner.

The streamlines we just calculated for this irrotational flow also apply to a wide variety of other transport conditions involving scalar potentials. If we consider the heat transfer case, Figure 9.10 becomes a large block whose upper surface is held at a high temperature, T_{hi}, whose right-hand side is held at a low temperature, T_{lo}, and whose corner is insulated as shown in Figure 9.12. Figure 9.12 also shows mass transfer and electric field analogs of the flow in a corner. Notice how the boundary conditions requiring zero stress (zero momentum flux) in the fluid flow problem become zero flux (insulated or impermeable) boundary conditions in the scalar field cases. The power law transformation used to describe the 90° corner can be used to describe flow inside a corner or over a wedge of arbitrary angle. As the corner angle becomes more acute, the power in the power law increases. Flow inside corners requires the exponent, $n > 1$. Flow over wedges requires $n < 1$.

Another classic potential flow problem where the velocity can be described by a scalar velocity potential is flow around a cylinder. We will treat this problem as a mass transfer example. The situation is shown in Figure 9.13. The problem concerns a very long slab of material whose leftmost end is held at a concentration c_{ao} and whose rightmost end is held at a concentration of c_{aL}. In the middle of the slab, we cut an impermeable circular hole of radius r_o. Far away from the hole, diffusion occurs uniformly as if the system were 1-D and the flux of mass is

$$N_{a\infty} = \frac{D_{ab}(c_{ao} - c_{aL})}{L} \tag{9.89}$$

Near the hole, the 2-D effects become apparent and the flow of mass distorts to move material around the hole.

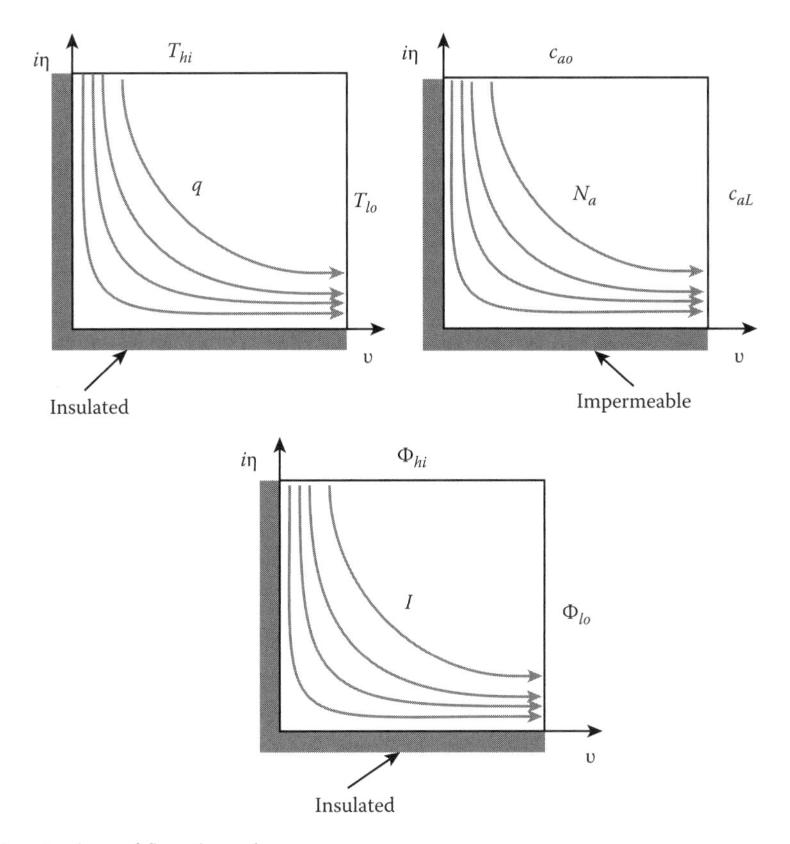

FIGURE 9.12 Analogs of flow through a corner.

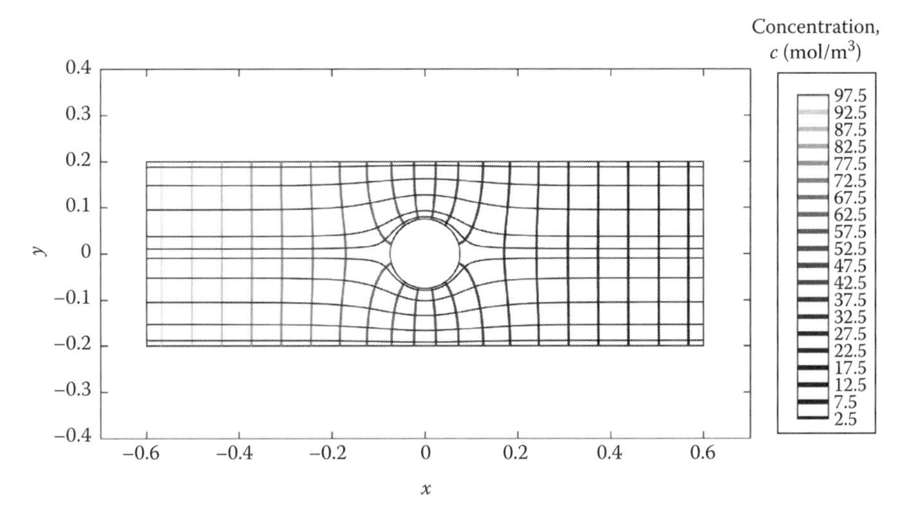

FIGURE 9.13 Equipotentials and streamlines for flow about a cylinder. Comsol® module Hole.

Complex analysis is well suited to dealing with problems like this where a hole or holes are distributed throughout a region because it allows you to bypass the hole by mapping the hole to be outside the region of interest. For this problem, the transformation that performs that mapping is

$$F(w) = N_{a\infty} w \rightarrow F\left(z + \frac{r_o}{z}\right) = N_{a\infty}\left(z + \frac{r_o}{z}\right) \tag{9.90}$$

and takes uniform mass flow in the complex plane, $F = N_{a\infty}w$, and transforms it to mass flow about the region with the hole of radius r_o. To look at the potential and stream functions, we split $F(z + r_o/z)$ into its real and imaginary parts:

$$F\left(z+\frac{r_o}{z}\right) \rightarrow F(x+iy) = \underbrace{N_{a\infty}x\left(1+\frac{r_o^2}{x^2+y^2}\right)}_{\text{Potential function}} + i\underbrace{N_{a\infty}y\left(1-\frac{r_o^2}{x^2+y^2}\right)}_{\text{Stream function}} \tag{9.91}$$

The stream function is the imaginary part

$$\psi = N_{a\infty}y\left(1-\frac{r_o^2}{x^2+y^2}\right) \tag{9.92}$$

and the potential function is the real part

$$\phi = N_{a\infty}x\left(1+\frac{r_o^2}{x^2+y^2}\right) \tag{9.93}$$

Plotting ϕ = *constant* and ψ = *constant* gives the lines of constant concentration and constant mass flow shown in Figure 9.13. The equipotential lines and streamlines of Figure 9.13 describe fluid flow around a cylinder when the flow field is irrotational and inviscid. The same diagram also describes what happens when a conducting cylinder is placed in a uniform electric field or when we replace the concentrations by temperatures simulating a whole cut in a metal plate or other object conducting heat.

The potential function, ϕ, for the cylinder obeys Laplace's equation, and so we can obtain a solution for it using separation of variables. Since we are interested in the field, say the velocity field, near the surface of the cylinder, it makes sense to use cylindrical coordinates. Laplace's equation for the velocity potential in this system is

$$\frac{\partial^2\phi}{\partial r^2} + \frac{1}{r}\frac{\partial\phi}{\partial r} + \frac{1}{r^2}\frac{\partial^2\phi}{\partial\theta^2} = 0 \tag{9.94}$$

We have a uniform velocity in the x-direction, v_o, far away from the cylinder. In Chapter 2, we discussed fluxes and potential functions and stated that a positive flux was always in the direction of a decreasing potential. Thus, heat flowed from a higher to a lower temperature, mass flowed from a higher to a lower concentration, and velocity must be in the direction of a higher to a lower velocity potential. Thus, we have

$$v_x = -\frac{\partial\phi}{\partial x} \qquad v_y = -\frac{\partial\phi}{\partial y} \tag{9.95}$$

For this problem, we know that far away from the cylinder, we must have

$$-\frac{\partial\phi}{\partial x} = v_x = v_o \tag{9.96}$$

Integrating Equation 9.96 gives the potential function in rectangular and cylindrical coordinates:

$$\phi = -v_o x = -v_o r\cos\theta \tag{9.97}$$

The boundary conditions for this problem state that we must have uniform flow in the x-direction far away from the cylinder. This requires the velocity components, v_r and v_θ, to be

$$r \to \infty \qquad v_r = -\frac{\partial \phi}{\partial r} = v_o \cos\theta \tag{9.98}$$

$$r \to \infty \qquad v_\theta = -\frac{1}{r}\frac{\partial \phi}{\partial \theta} = -v_o \sin\theta \tag{9.99}$$

at $r = \infty$. No fluid can pass through the surface of the cylinder so

$$r \to r_o \qquad v_r = -\frac{\partial \phi}{\partial r} = 0 \tag{9.100}$$

Finally, since we have angular symmetry, the velocity at every point, (r, θ), is the same as that at $(r, \theta+2n\pi)$. We can enforce this via the following conditions:

$$\frac{\partial \phi}{\partial r}(r,\theta+2n\pi) = \frac{\partial \phi}{\partial r}(r,\theta) \tag{9.101}$$

$$\frac{\partial \phi}{\partial \theta}(r,\theta+2n\pi) = \frac{\partial \phi}{\partial \theta}(r,\theta) \tag{9.102}$$

Letting $\phi = R(r)T(\theta)$, we separate the partial differential equation into

$$r^2\frac{d^2R}{dr^2} + r\frac{dR}{dr} - \lambda^2 R = 0 \tag{9.103}$$

$$\frac{d^2T}{d\theta^2} + \lambda^2 T = 0 \tag{9.104}$$

Equation 9.103 is Euler's equation again, but unlike some of our other examples, $\lambda^2 = 0$ leads to a viable result for this problem. Thus, we must solve the equation for two separate situations. When $\lambda^2 \neq 0$, the solution to Equation 9.103 is

$$R_n = a_1 r^{\lambda} + a_2 r^{-\lambda} \tag{9.105}$$

When $\lambda^2 = 0$, we have

$$R_o = a_3 + a_4 \ln r \tag{9.106}$$

We perform the same two-step solution procedure with Equation 9.104 solving it for $\lambda^2 \neq 0$ and $\lambda^2 = 0$. The solutions are

$$T_o = b_1\theta + b_o \tag{9.107}$$

$$T_n = b_2\cos(\lambda\theta) + b_3\sin(\lambda\theta) \tag{9.108}$$

and the reconstituted solution for ϕ is

$$\phi = R_oT_o + R_nT_n = (a_3 + a_4\ln r)(b_1\theta + b_o) + (a_1r^{\lambda} + a_2r^{-\lambda})[b_2\cos(\lambda\theta) + b_3\sin(\lambda\theta)] \tag{9.109}$$

Here, $b_o = 0$ since ϕ has no constant component. The finite velocity boundary conditions at $r = \infty$ force $a_4 = 0$. Likewise, Equation 9.97 requires $b_o = b_3 = 0$. The harmonic boundary conditions of Equations 9.101 and 9.102 require $\lambda = \pm 1, \pm 2, \pm 3\ldots$, but the boundary conditions describing the flow at $r = \infty$, Equations 9.98 and 9.99, force $\lambda = 1$. Thus, we are left with

$$\phi = c_1\theta + \left(c_2 r + \frac{c_3}{r}\right)\cos\theta \tag{9.110}$$

Now applying conditions (9.97) and (9.100) in a quantitative manner, we can solve for the constants c_2 and c_3. c_1, it appears, can take on any value and not affect whether the boundary conditions are satisfied:

$$\phi = c_1\theta - v_o\left(r + \frac{r_o^2}{r}\right)\cos\theta \tag{9.111}$$

This solution is quite similar to the solution we derived for the mass transfer problem. That solution, when applied to a flow problem and written in cylindrical coordinates so that $\vec{v} = -\vec{\nabla}\phi$, is

$$\phi = -v_o r\cos\theta\left(1 + \frac{r_o^2}{r^2}\right) = -v_o\left(r + \frac{r_o^2}{r}\right)\cos\theta \tag{9.112}$$

The only difference between the solutions lies in the first term in Equation 9.111 involving $c_1\theta$. The velocity corresponding to $c_1\theta$ is

$$v_\theta = -\frac{1}{r}\frac{\partial\phi}{\partial\theta} = -\frac{c_1}{r} \tag{9.113}$$

This represents pure rotational motion about the center of the cylinder. The velocity dies out at $r \to \infty$ and so the uniform flow far away from the cylinder is unaffected. This solution arises in separation of variables because we were not careful enough to require that the *circulation* of fluid about the cylinder to be zero. Circulation, in its mathematical form, was defined by Lord Kelvin in 1869 as the line integral about a simple closed curve. In this case, the simple closed curve runs about the cylinder and must include the cylinder in it. Thus, the circulation is

$$K = \oint (v_r\vec{\mathbf{e}}_r + v_\theta\vec{\mathbf{e}}_\theta)\cdot(dr\,\vec{\mathbf{e}}_r + rd\theta\,\vec{\mathbf{e}}_\theta) = \oint v_r dr + \oint v_\theta r d\theta = 2\pi c_1 \tag{9.114}$$

To force the separation of variables solution to be exactly like that developed using the complex variables approach, we must specify that the circulation be zero or that $c_1 = 0$. c_1 represents the *vorticity* of the flow or the rate of angular deformation of a fluid element. *Thus, the circulation about a closed curve is equal to the vorticity enclosed by it.*

The motion of the liquid about the cylinder, the circulation, provides a lift force that tends to push the cylinder in one direction or another. If the circulation is clockwise, we generate a force that pushes the cylinder in the positive y-direction. This is shown in Figure 9.14. Counterclockwise circulation has the opposite effect. There is a famous theorem in aerodynamics, the Kutta–Joukowski theorem that asserts that any body that causes circulation in an otherwise uniform flow will experience lift. According to the theorem, the lift on our cylinder with circulation, $2\pi c_1$, would be

$$F_L = -F_y = \rho v_o(2\pi c_1)L \qquad c_1 = \omega r_o^2 \tag{9.115}$$

where L is the length of the cylinder.

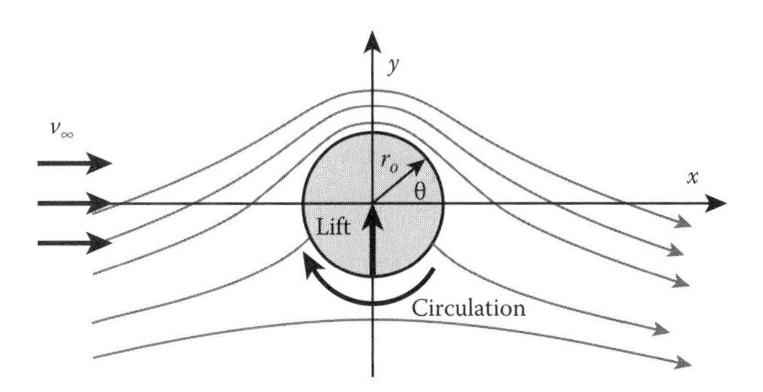

FIGURE 9.14 Streamlines about a cylinder with circulation. Counterclockwise circulation produces lift in the positive y-direction.

Notice that the circulation we discussed only makes sense in terms of a fluid flowing or a current circulating and generating a magnetic field [5]. This is because in those two cases, we are dealing with a vector field. In the case of mass or heat transfer, this type of circulation is not possible because it would force heat or mass to flow against the temperature or concentration gradient. The usual situations in which we have circulation occurring in a heat or mass transfer context are those where we also have fluid circulation. It is the heat or mass carried along by the fluid (convection) that results in net circulation.

9.4 GENERATION, SOURCES, SINKS, AND POISSON'S EQUATION

Laplace's equation is satisfied for simple, steady-state situations involving no generation. It is the 2-D analog of the material discussed in Chapter 4. As soon as generation enters into the problem, Laplace's equation is no longer satisfied throughout the domain. We must then deal with material like that presented in Chapter 5 on generation. In this section, we will start with generation on the boundary as we did in Chapter 5 and then move on to volumetric generation.

9.4.1 Sources, Sinks, and Generation on the Boundary

The 2-D fields calculated in the last few examples can also be analyzed in the context of generation on the boundary using three basic building blocks: a line source, a line sink, and a point vortex. This arises because the flow of our transport quantities, be they heat, mass, momentum, charge, etc., in the absence of internal, volumetric generation, must appear to start from a source and terminate at a sink. Though these sources formally represent generation terms, they exist outside the domain of our analysis. Within the region of interest, potential flow of heat or fluid, for example, still exists and Laplace's equation is still valid. Thus, the source/sink formulation is akin to having generation on the boundary.

The line source is a line between two parallel planes spaced a unit length apart. Fluid or heat or charge flows radially out from the line source. The strength of the source, say q, is related to the flow from it. A line heat source, for example, is given by

$$\frac{q}{2\pi r} = -\frac{\partial \phi}{\partial r} = -\frac{1}{r}\frac{\partial \psi}{\partial \theta} \quad \text{Line heat source} \tag{9.116}$$

where

q represents the heat flow

ϕ and ψ represent the potential and stream functions

The functional form for the line source is derived by realizing that all transport must occur through a cylindrical surface of radius, r, that surrounds the line source. The surface area through which the flow, q, moves is $2\pi rL$ ($L = 1$). The definition in Equation 9.116 also applies to fluid flow where q would represent the volumetric flow rate of fluid $\dot{v}$. In the case of mass flow, q would represent the mass flow rate of a given species, $\dot{m}$. Integrating Equation 9.116 gives the stream function and potential function for such a source. Thus,

$$\phi = -\frac{q}{2\pi}\ln r = -\frac{q}{2\pi}\ln\left(\sqrt{x^2+y^2}\right) \quad \text{Source} \tag{9.117}$$

$$\psi = -\frac{q}{2\pi}\theta = -\frac{q}{2\pi}\tan^{-1}\left(\frac{y}{x}\right) \quad \text{Source} \tag{9.118}$$

The equipotential lines and streamlines are orthogonal to one another because the line source exists outside the domain where the potential and stream functions are defined.

A sink is just the negative of the source and so we change the signs to give

$$\phi = \frac{q}{2\pi}\ln r = \frac{q}{2\pi}\ln\left(\sqrt{x^2+y^2}\right) \quad \text{Sink} \tag{9.119}$$

$$\psi = \frac{q}{2\pi}\theta = \frac{q}{2\pi}\tan^{-1}\left(\frac{y}{x}\right) \quad \text{Sink} \tag{9.120}$$

The vortex, or more precisely, the irrotational vortex, since the flow outside the center of the vortex is irrotational, is described as the mirror image of the line source. Thus, the potential function for the line source is the stream function for the vortex and the stream function for the line source is the potential function for the vortex:

$$\phi = -C\theta = -C\tan^{-1}\left(\frac{y}{x}\right) \quad \text{Vortex} \tag{9.121}$$

$$\psi = -C\ln r = -C\ln\left(\sqrt{x^2+y^2}\right) \quad \text{Vortex} \tag{9.122}$$

Changing the signs of the stream and potential functions changes the direction of rotation of the vortex. The equipotential lines and streamlines for the line source and vortex are shown in Figure 9.15. Notice that the potential function for the vortex, Equation 9.121, is exactly the missing piece we needed to match up the separation of variables solution for flow about the cylinder with the complex variable solution. The presence of circulation about the cylinder as specified by Equation 9.111 is exactly like having an irrotational vortex source of strength $2\pi c_1$ at $r = 0$. Had we taken our complex variable representation and included the vortex, we would have arrived at the same streamline configuration shown in Figure 9.14.

Example 9.1 Electrical Doublet and the Field about a Conducting Cylinder

The electrical doublet consists of two line charge sources of opposite polarity separated by a distance, L. The doublet is so arranged that the product of its strength, $\rho^{\pm}$, and the distance, r_o, remains a constant. The configuration is shown in Figure 9.16.

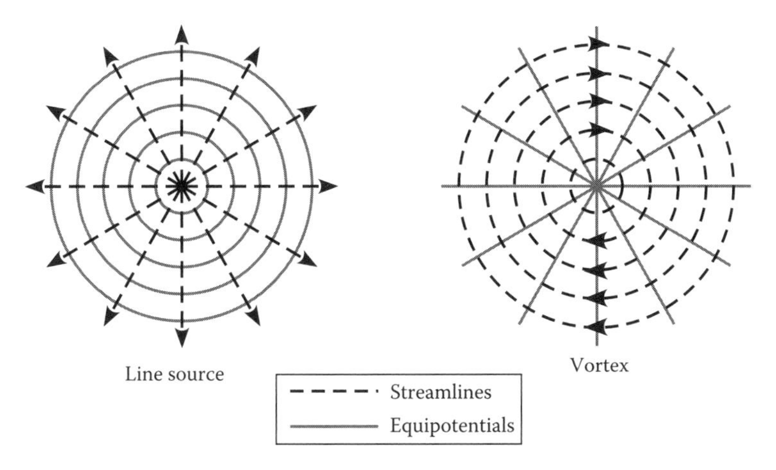

FIGURE 9.15 Equipotential and streamlines for the line source and vortex.

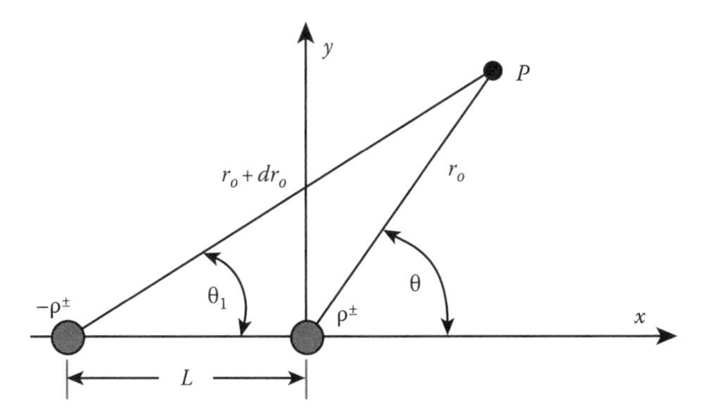

FIGURE 9.16 The electric doublet formed from two line charge sources.

The electric potential at point, P, due to the two line sources can be found by evaluating Equation 9.117 for each source separately and summing:

$$\phi = -\frac{\rho^{\pm}}{2\pi\varepsilon}\ln r_o + \frac{\rho^{\pm}}{2\pi\varepsilon}\ln(r_o + dr_o) = \frac{\rho^{\pm}}{2\pi\varepsilon}\ln\left(1 + \frac{dr_o}{r_o}\right)$$

For separations between the source and sink that are small, $L \ll r_o$, we can expand the log term in a power series:

$$\phi = \frac{\rho^{\pm}}{2\pi\varepsilon}\left[\frac{dr_o}{r_o} - \frac{1}{2}\left(\frac{dr_o}{r_o}\right)^2 + \cdots\right]$$

Keeping only the linear term and realizing that $\theta_1 \approx \theta$ for small $L \ll r_o$, we can express dr_o in terms of L and θ to obtain a simpler form for the potential function for the doublet:

$$\phi = \frac{\rho^{\pm} L\cos\theta}{2\pi\varepsilon r_o}$$

To fix the field about the conducting cylinder, we combine a uniform electric field in the x-direction with the doublet. The uniform field is described by a potential function akin to the uniform flow:

$$\phi_u = -\phi_o x$$

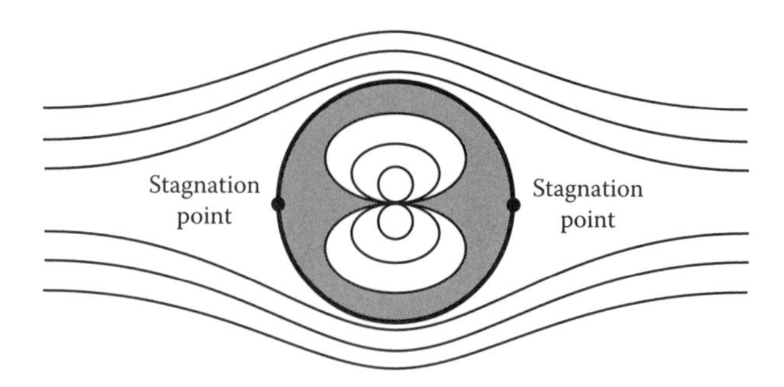

FIGURE 9.17 Field lines about a conducting cylinder immersed in a uniform field.

Adding to the doublet of strength $\phi_o L^2$ gives

$$\phi = -\phi_o x - \phi_o L^2 \frac{\cos\theta}{r_o} = -\phi_o r_o \cos\theta - \phi_o L^2 \frac{\cos\theta}{r_o}$$

$$\psi = -\phi_o r_o \sin\theta - \phi_o L^2 \frac{\sin\theta}{r_o}$$

Field lines for this combination formally look like those shown in Figure 9.17. Notice that the field lines inside the cylinder are also resolved so that two problems are solved. The second problem would be to find the field inside a conducting cylinder due to the presence of a doublet inside.

This combination of uniform field and doublet can also be used to describe the flow about a cylinder. The points along the flow axis are the stagnation points that in fluid flow are the points of highest pressure. Such a representation can also be used to describe a lifting body if a vortex is added to the uniform field and doublet:

$$\psi = v_\infty r_o \sin\theta - R^2 v_\infty \left(\frac{\sin\theta}{r_o}\right) + C\ln r_o \quad R - \text{cylinder radius}$$

Example 9.2 Tornado

A tornado may be approximated using a sink and a vortex. This solution is a good approximation everywhere except at the origin. A sum of Equations 9.119 with 9.121 and 9.120 with 9.122 defines the potential and stream function for the tornado:

$$\phi = \frac{q}{2\pi}\ln r - C\theta = \frac{q}{2\pi}\ln\left(\sqrt{x^2+y^2}\right) - C\tan^{-1}\left(\frac{y}{x}\right)$$

$$\psi = \frac{q}{2\pi}\theta - C\ln r = C\ln\left(\sqrt{x^2+y^2}\right) + \frac{q}{2\pi}\tan^{-1}\left(\frac{y}{x}\right)$$

Figure 9.18 displays the streamlines and equipotentials for the tornado.

9.4.2 Poisson's Equation and Generation

The situations analyzed in the previous section were fairly simplified because the generation terms only occurred on the boundary. Thus, Laplace's equation is valid throughout the interior of the domain of interest. When true volumetric generation terms exist, they must be included in

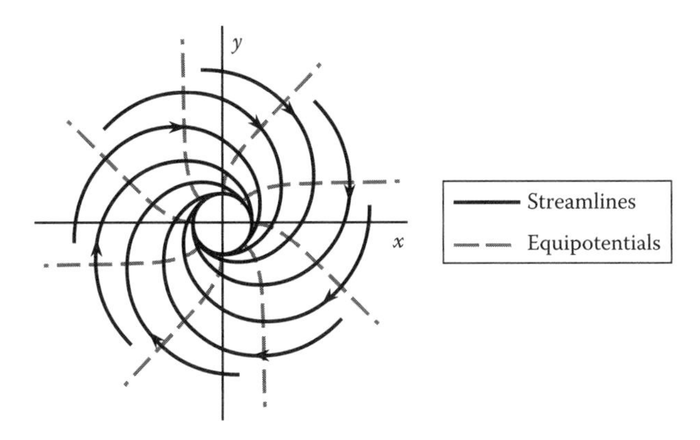

FIGURE 9.18 Tornado composed of a vortex and a sink. All flow is directed toward the center of the vortex in a counterclockwise rotation.

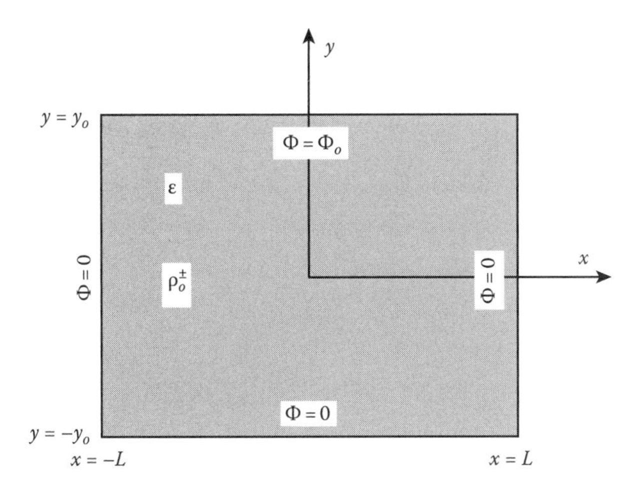

FIGURE 9.19 Solid object between two electrodes having a uniform charge distribution inside.

the differential equation. The general set of partial differential equations including generation is an example of Poisson's equation. To derive the classic Poisson's equation, we consider a material with an internal charge distribution, $\rho^{\pm}(x, y)$, as shown in Figure 9.19. The presence of internal charge produces an electric field distribution inside the material. The situation is analogous to the heat transfer case where a material has a distribution of radioactive particles generating heat inside.

The material of interest is square; of thickness, t_h; dielectric constant, ε; and sandwiched between two electrodes. One electrode is grounded. The sides of the material are also grounded. The change in electric field within the sample is due to the charge present inside. Writing a charge balance in terms of the electric field and dielectric constant about the control volume, we have

$$\begin{array}{ccccc} In \quad - & Out & + & Gen & = Acc \\ \varepsilon[\mathcal{E}(x) - \mathcal{E}(x+\Delta x)]\Delta y t_h & + \varepsilon[\mathcal{E}(y) - \mathcal{E}(y+\Delta y)]\,\Delta x t_h & + & \rho^{\pm}(x,y)\Delta x \Delta y t_h & = \;\; 0 \end{array} \tag{9.123}$$

Dividing through by $\Delta x \Delta y t_h$ and letting the control volume shrink to zero gives

$$-\frac{\partial \mathcal{E}}{\partial x} - \frac{\partial \mathcal{E}}{\partial y} + \frac{\rho^{\pm}(x,y)}{\varepsilon} = 0 \tag{9.124}$$

The electric field is in response to a change in voltage. Substituting for the electric field in terms of the voltage gradient gives

$$\mathcal{E}_x = -\frac{\partial \Phi}{\partial x} \qquad \mathcal{E}_y = -\frac{\partial \Phi}{\partial y} \tag{9.125}$$

$$\frac{\partial^2 \Phi}{\partial x^2} + \frac{\partial^2 \Phi}{\partial y^2} + \frac{\rho^{\pm}(x,y)}{\varepsilon} = 0 \quad \text{Poisson's equation} \tag{9.126}$$

Solving Poisson's equation is quite a bit more difficult than solving Laplace's equation. The analytical method of choice involves the use of Green's functions and a wide variety of Poisson-type problems can be solved this way. Interested readers can search out texts by Carslaw and Jaeger [1] and Crank [2] for more information. A more limited range of problems can be solved using a technique we considered for ordinary differential equations, namely, the method of undetermined coefficients. This requires that we formulate a good guess for the particular solution of the problem based on the functional form for $\rho^{\pm}(x, y)$. For this particular example, consider a uniform charge density so that $\rho^{\pm}(x,y) = \rho_0^{\pm}$. Poisson's equation becomes

$$\frac{\partial^2 \Phi}{\partial x^2} + \frac{\partial^2 \Phi}{\partial y^2} + \frac{\rho_0^{\pm}}{\varepsilon} = 0 \tag{9.127}$$

The boundary conditions based on Figure 9.19 are

$$x = -L \qquad \Phi = 0 \tag{9.128}$$

$$x = L \qquad \Phi = 0 \tag{9.129}$$

$$y = -y_o \qquad \Phi = 0 \tag{9.130}$$

$$y = y_o \qquad \Phi = \Phi_o \tag{9.131}$$

To solve this equation, we first find a general solution to the homogeneous part, Laplace's equation, and then find a particular solution that obeys the differential equation and boundary conditions too. The general solution to Laplace's equation for this problem that satisfies the boundary conditions in the x-direction and at $y = -y_o$ is

$$\Phi_h = a_o \cos\left(\frac{(2n+1)\pi x}{2L}\right)\sinh\left(\frac{(2n+1)\pi(y+y_o)}{2L}\right) \tag{9.132}$$

Before the last boundary condition can be used, we must find a particular solution to the problem. Determining the particular solution requires a bit of intuition and experience. Since the generation term is a constant, we need to find a function that when differentiated twice leaves a constant behind and that will satisfy the boundary conditions in the x-direction. The homogenous solution will satisfy the conditions on y, so the particular solution need only be a function of x. Several quadratic functions of the form

$$\Phi_p = b_o(L^2 - x^2) \tag{9.133}$$

will work and fit the boundary conditions on x. The streamlines befitting this Laplace's equation problem would look something like those shown in Figure 9.20. The field radiates out from the

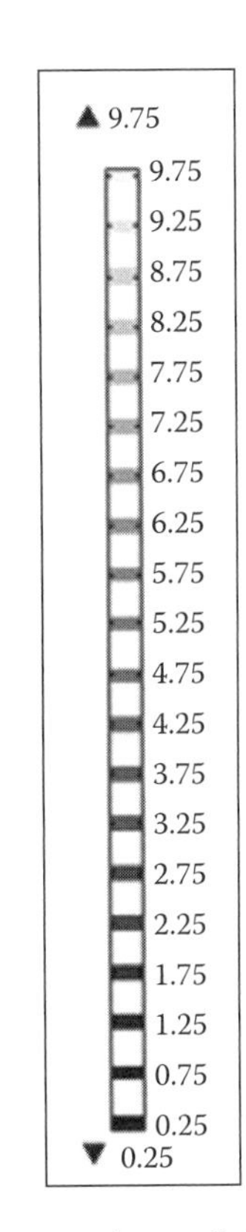

FIGURE 9.20 Streamlines and equipotential lines for the field problem without space charge distribution. Solution from Comsol® module Field.

surface at $y = y_o$, terminating at one of the other three walls. Changes in the x- and y-directions are important since all the streamlines are curved. The presence of the charge distribution inside the material will change the slope of the streamlines shown in Figure 9.20, but their basic character will remain the same.

We substitute the particular solution into Equation 9.127 to solve for b_o:

$$b_o = \frac{\rho_0^{\pm}}{2\varepsilon} \tag{9.134}$$

The generic overall solution is

$$\Phi = \Phi_h + \Phi_p$$

$$= a_o \cos\left(\frac{(2n+1)\pi x}{2L}\right) \sinh\left(\frac{(2n+1)\pi(y+y_o)}{2L}\right) + \frac{\rho_0^{\pm}}{2\varepsilon}(L^2 - x^2) \tag{9.135}$$

The last task is to apply the boundary condition at $y = y_o$. Clearly, the solution for Φ cannot match the y boundary condition over all x, so we need to use a sum of solutions for the homogeneous part in addition to the particular solution. Evaluating the final boundary condition gives

$$\Phi_o = \sum_{n=0}^{\infty}\left[a_n \cos\left(\frac{(2n+1)\pi x}{2L}\right) \sinh\left(\frac{(2n+1)\pi y_o}{L}\right)\right] + \frac{\rho_0^{\pm}}{2\varepsilon}(L^2 - x^2) \tag{9.136}$$

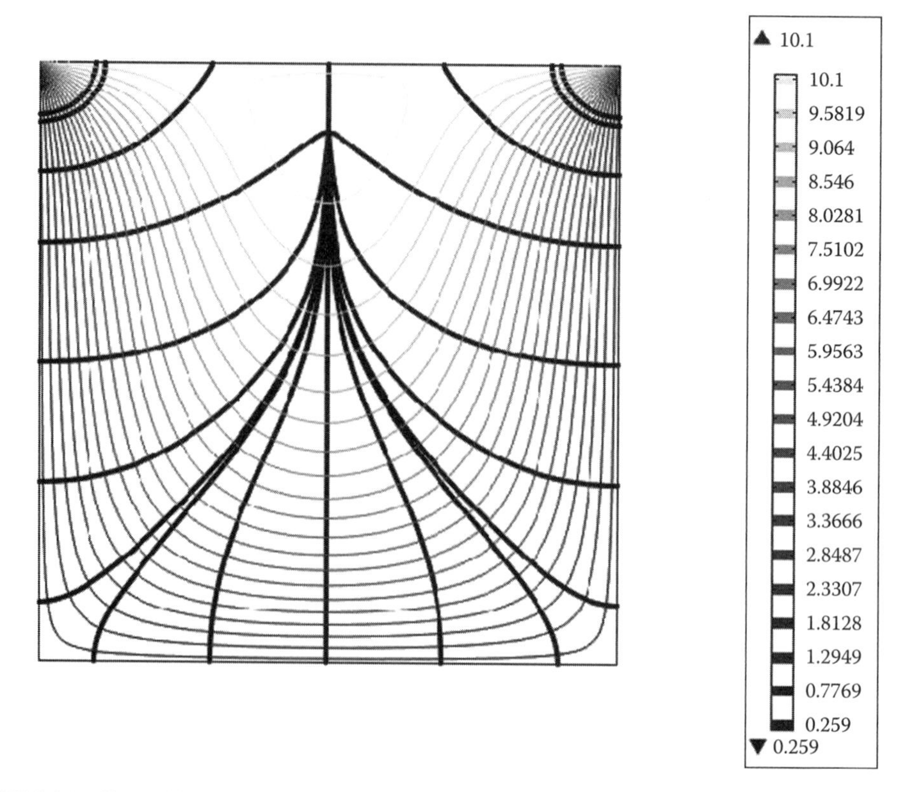

FIGURE 9.21 Streamlines and equipotential lines for the complete field problem. Solution from Comsol® module Field.

We extract the a_n using the orthogonality property of the cosine function. Thus,

$$\int_0^L \left(\Phi_o - \frac{\rho_0^{\pm}}{2\varepsilon}(L^2 - x^2)\right)\cos\left(\frac{(2n+1)\pi x}{2L}\right)dx = \int_0^L \left[a_n \sinh\left(\frac{(2n+1)\pi y_o}{L}\right)\cos^2\left(\frac{(2n+1)\pi x}{2L}\right)\right]dx \tag{9.137}$$

$$a_n = -(1)^n \frac{4}{\sinh\left(\frac{n\pi y_o}{L}\right)}\left[\frac{\Phi_o\pi^2(2n+1)^2 - 4\rho_0^{\pm}\varepsilon L^2}{(2n+1)^3\pi^3}\right] \tag{9.138}$$

extracts the a_n. The presence of the charge distribution causes the potential to decay much more quickly than would be the case if the solid were charge-free. The equipotential lines are more highly concentrated toward the center of the upper surface than what appears in Figure 9.20 and the streamlines also seem to collapse more quickly toward the center as shown in Figure 9.21.

Example 9.3 Heat Flow in a Wound Magnetic Coil

A slightly more complicated problem can be posed in cylindrical coordinates. A magnetic coil of radius, r_o, and height, L, is being used as a solenoid controller. The surfaces of the coil are all exposed to a coolant that keeps the temperature at T_∞. Current flowing through the coil provides for a constant rate of heat generation, q_o W/m³. We would like to know what the temperature is within the rod to help determine the solenoid's useful lifetime. The physical situation is shown in Figure 9.22.

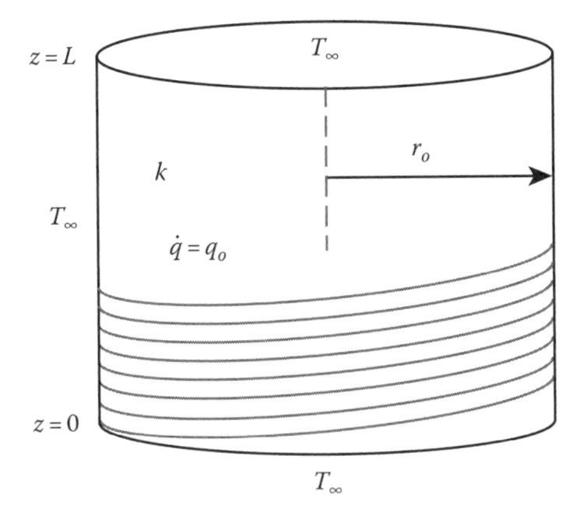

FIGURE 9.22 Solenoid with electrical heat generation.

This geometry has symmetry in the θ-direction, and so the governing equation is Laplace's equation in cylindrical coordinates with the generation term added:

$$\frac{\partial^2 T}{\partial r^2} + \frac{1}{r}\frac{\partial T}{\partial r} + \frac{\partial^2 T}{\partial z^2} + \frac{q_o}{k} = 0$$

The boundary conditions specify the temperature at $z = L$ and at $z = 0$:

$$z = 0, L \qquad T = T_\infty$$

The temperature is also set at $r = r_o$:

$$r = 0 \qquad \frac{\partial T}{\partial r} = 0$$

$$r = r_o \qquad T = T_\infty$$

First we solve the homogeneous equation, Laplace's equation. Letting $\theta = T - T_\infty$, the governing equation becomes

$$\frac{\partial^2 \theta}{\partial r^2} + \frac{1}{r}\frac{\partial \theta}{\partial r} + \frac{\partial^2 \theta}{\partial z^2} = 0$$

The general solution to Laplace's equation for this system is given by Equations 9.52 and 9.53. However, in this situation, it is easier to set the separation constant equal to $+\lambda^2$ and use the alternate solution:

$$Z = a_1 \sin(\lambda z) + a_2 \cos(\lambda z)$$

$$R = a_3 I_o(\lambda r) + a_4 K_o(\lambda r)$$

This solution allows us to fit the z boundary conditions readily. The boundary condition at $z = 0$ forces $a_2 = 0$, while the boundary condition at $r = 0$ forces $a_4 = 0$ ($K_o \to \infty$). The homogeneous solution boils down to

$$\theta_h = a_5 I_o(\lambda r)\sin(\lambda z)$$

Applying the boundary condition at $z = L$ requires λ to satisfy the equation

$$\lambda = \frac{n\pi}{L}$$

Since a single value for θ_h in conjunction with the particular solution will not be able to satisfy the boundary condition at $r = r_o$, we must use a sum:

$$\theta_h = \sum_{n=0}^{\infty} a_{5n} I_o(\lambda_n r)\sin(\lambda_n z)$$

The particular solution for this problem involves some guesswork. As before, we will let the homogeneous solution handle the r-dependence and find a particular solution that is a function of z and will satisfy the z boundary conditions. The following function satisfies those requirements:

$$\theta_p = \alpha_o z(L - z)$$

Substituting this into the equation and solving for α_o shows that the particular solution is

$$\theta_p = \frac{q_o}{2k} z(L - z)$$

The overall solution for θ becomes

$$\theta = \theta_h + \theta_p = \sum_{n=0}^{\infty} a_{5n} I_o(\lambda_n r)\sin(\lambda_n z) + \frac{q_o}{2k} z(L - z)$$

The boundary condition at $r = r_o$ allows the extraction of a_{5n}:

$$0 = \sum_{n=0}^{\infty} a_{5n} I_o(\lambda_n r_o)\sin(\lambda_n z) + \frac{q_o}{2k} z(L - z)$$

We use the orthogonality relation for the sine function to extract the a_{5n}. Multiplying both sides by $\sin(\lambda_m z)$ and integrating from $z = 0$ to $z = L$,

$$-\frac{q_o}{2k}\int_0^L z(L - z)\sin(\lambda_n z)dz = \int_0^L a_{5n} I_o(\lambda_n r_o)\sin^2(\lambda_n z)dz$$

Using integration by parts, we can evaluate both sides of the equation. Notice that the generation term again leads to steeper gradients in temperature:

$$a_{5n} = \frac{4L^2 q_o}{n^3\pi^3 k\, I_o(n\pi r_o/L)}$$

The analytical solutions derived so far have pointed out one important fact about the transport processes. As long as the differential equations describing the transport process are not nonlinear, adding generation to the system does not greatly change the fundamental character of the solution. Generation may augment the gradients and alter how fast the measured variable changes, but the homogeneous solution still provides enough of a description of the physics of the process to develop an *engineering feel* for the process with generation.

9.4.3 Generation as a Function of the Dependent Variable

One of the more common and important examples of Poisson's equation occurs when the generation term is a function of the dependent variable. An important example of this occurs when we look at diffusion and chemical reaction in two dimensions. Consider what happens when a hormone-based drug diffuses into the skin from a patch as shown in Figure 9.23.

In this situation, the patch provides a form of the hormone that decomposes once it diffuses into the skin to yield its active form. The active form can then proceed into the bloodstream and be distributed throughout the body. We will simplify the situation a bit by assuming a rectangular geometry and specifying that the hormone, as delivered from the patch (before decomposition), cannot penetrate past the skin layer. The decomposition reaction is first order in hormone concentration.

If we put a small control volume (Figure 9.24) within the skin, we can derive the form of Poisson's equation relevant to this situation:

$$\begin{array}{ccccccc} In & - & Out & + & Gen & = & Acc \\ {}[\dot{M}_h(x) - \dot{M}_h(x+\Delta x)] & + & [\dot{M}_h(y) - \dot{M}_h(y+\Delta y)] & - & k''c_h \Delta x \Delta y t_h & = & 0 \end{array} \tag{9.139}$$

$$[N_h(x) - N_h(x+\Delta x)]\Delta y t_h + [N_h(y) - N_h(y+\Delta y)]\Delta x t_h - k''c_h \Delta x \Delta y t_h = 0 \tag{9.140}$$

Dividing through by the volume of the control element and letting that volume shrink to zero gives

$$-\frac{\partial N_{hx}}{\partial x} - \frac{\partial N_{hy}}{\partial y} - k''c_h = 0 \tag{9.141}$$

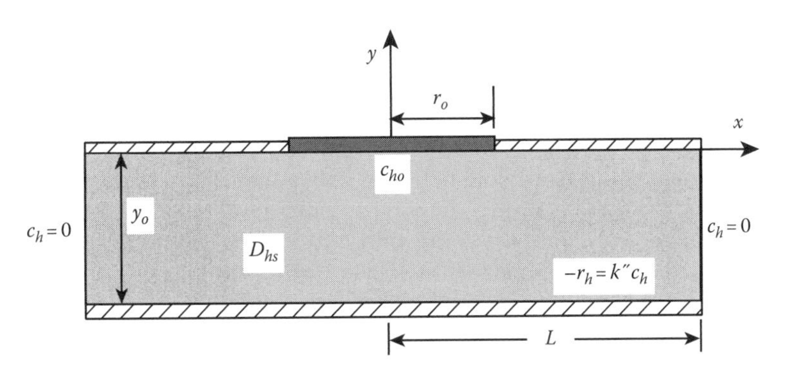

FIGURE 9.23 Hormone diffusing into the skin and decomposing into its active form.

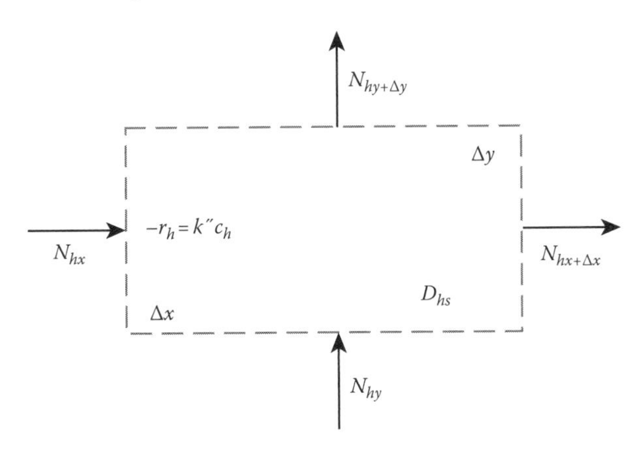

FIGURE 9.24 Control volume for the patch problem.

We will assume dilute solutions for this problem since the patch is not designed to deliver much hormone. Substituting in Fick's law gives the required partial differential equation:

$$N_{hx} = -D_{hs}\frac{\partial c_h}{\partial x} \tag{9.142}$$

$$N_{hy} = -D_{hs}\frac{\partial c_h}{\partial y} \tag{9.143}$$

$$\frac{\partial^2 c_h}{\partial x^2} + \frac{\partial^2 c_h}{\partial y^2} - \frac{k''}{D_{hs}}c_h = 0 \quad \text{Helmholtz's equation} \tag{9.144}$$

Equation 9.144 is often referred to as Helmholtz's equation for the physicist who first analyzed the equation in conjunction with transmission lines. Equation 9.144 is subject to boundary conditions that state the following:

$$x = 0, L \qquad c_h = 0 \tag{9.145}$$

$$y = 0 \qquad c_h = c_{ho}(-r_o < x < r_o) \qquad \frac{\partial c_h}{\partial y} = 0\left(|x| > r_o\right) \tag{9.146}$$

$$y = y_o \qquad \frac{\partial c_h}{\partial y} = 0 \tag{9.147}$$

At this point, we could put Equation 9.144 in dimensionless form. This requires that we decide on a length scale for the problem. We could choose the thickness of the skin, y_o, or the width of the patch, r_o. The choice depends upon what aspect of the problem we want to highlight. Here, it seems clear that the goal is to transport hormone through the skin and into the blood, so we would like to know how much of that actually occurs. Perhaps a length scale based on the thickness of the skin would then be appropriate. Setting

$$\phi = \frac{c_h}{c_{ho}} \qquad \xi = \frac{x}{y_o} \qquad \eta = \frac{y}{y_o} \tag{9.148}$$

transforms Equation 9.144 to

$$\frac{\partial^2 \phi}{\partial \xi^2} + \frac{\partial^2 \phi}{\partial \eta^2} - \underbrace{\left(\frac{k'' y_o^2}{D_{hs}}\right)}_{Da}\phi = 0 \tag{9.149}$$

and we recover the Damkohler number (Da) as the single dimensionless group. Since the differential equation is linear and three of four boundary conditions are homogeneous (=0), the separation of variables can be used to solve the equation. Letting

$$\phi = \chi(\xi)\psi(\eta) \tag{9.150}$$

and substituting into the equation gives

$$\psi \frac{\partial^2 \chi}{\partial \xi^2} + \chi \frac{\partial^2 \psi}{\partial \eta^2} - \text{Da}\chi\psi = 0 \tag{9.151}$$

Dividing through by $\chi\psi$ shows

$$\frac{1}{\chi}\frac{\partial^2 \chi}{\partial \xi^2} = -\frac{1}{\psi}\frac{\partial^2 \psi}{\partial \eta^2} + \text{Da} \tag{9.152}$$

Since the left-hand side only depends upon ξ and the right-hand side only depends upon η, both sides must be equal to a constant, $-\lambda^2$:

$$\frac{1}{\chi}\frac{\partial^2 \chi}{\partial \xi^2} = -\frac{1}{\psi}\frac{\partial^2 \psi}{\partial \eta^2} + \text{Da} = -\lambda^2 \tag{9.153}$$

We can now split this partial differential equation into two ordinary differential equations:

$$\frac{\partial^2 \chi}{\partial \xi^2} + \lambda^2 \chi = 0 \tag{9.154}$$

$$\frac{\partial^2 \psi}{\partial \eta^2} - [\text{Da} + \lambda^2]\psi = 0 \tag{9.155}$$

The solutions to these equations are

$$\chi = a_1 \sin(\lambda\xi) + a_2 \cos(\lambda\xi) \tag{9.156}$$

$$\psi = a_3 \sinh[(\text{Da} + \lambda^2)^{1/2}\eta] + a_4 \cosh[(\text{Da} + \lambda^2)^{1/2}\eta] \tag{9.157}$$

The boundary conditions on ξ indicate that

$$a_2 = 0 \qquad \text{and} \qquad \lambda_n = \frac{n\pi y_o}{L} \tag{9.158}$$

The boundary condition at $\eta = 1$ indicates that

$$a_3 = -a_4 \tanh[(\text{Da} + \lambda^2)^{1/2}] \tag{9.159}$$

The full solution is

$$\phi = a_5 \sin\left(\frac{n\pi y_o}{L}\xi\right)[\cosh(\beta\eta) - \tanh(\beta)\sinh(\beta\eta)] \tag{9.160}$$

where

$$\beta = \left[\text{Da} + \left(\frac{n\pi y_o}{L}\right)^2\right]^{1/2} \tag{9.161}$$

Since a single value of the solution for ϕ cannot handle the boundary condition at $\eta = 0$, we must use a sum of solutions. The boundary condition states that $\phi = 1$ up to $\xi = r_o/y_o$ and from there on, $\partial\phi/\partial\eta = 0$:

$$1 = \sum_{n=1}^{\infty} a_{5n} \sin\left(\frac{n\pi y_o}{L}\xi\right) \qquad 0 < \xi < r_o/y_o \tag{9.162}$$

$$0 = \sum_{n=1}^{\infty} a_{5n} \sin\left(\frac{n\pi y_o}{L}\xi\right)[\beta_n \tanh(\beta_n)] \qquad r_o/y_o < \xi < L/y_o \tag{9.163}$$

Here we have used the fact that the problem is symmetric about $x = 0$. To determine the a_{5n}, we use our orthogonality trick and integrate over ξ:

$$\int_0^{r_o/y_o} \sin\left(\frac{n\pi y_o}{L}\xi\right) d\xi = \int_0^{r_o/y_o} a_{5n} \sin^2\left(\frac{n\pi y_o}{L}\xi\right) d\xi \tag{9.164}$$

$$0 = \int_{r_o/y_o}^{L/y_o} a_{5n} \sin^2\left(\frac{n\pi y_o}{L}\xi\right)[\beta_n \tanh(\beta_n)] d\xi \qquad r_o/y_o < \xi < L/y_o \tag{9.165}$$

Equation 9.165 requires that all the values of $a_{5n} = 0$. Thus, the part of the skin without the patch covering it is a totally passive component in the solution. Solving Equation 9.164 gives

$$a_{5n} = \frac{2L\left[1 - \cos\left(\frac{n\pi r_o}{L}\right)\right]}{n\pi r_o - \frac{L}{2}\sin\left(\frac{2n\pi r_o}{L}\right)} \tag{9.166}$$

and the full solution is

$$\phi = \sum_{n=1}^{\infty}\left[\frac{2L[1 - \cos(n\pi r_o/L)]}{n\pi r_o - \frac{L}{2}\sin(2n\pi r_o/L)}\right] \sin\left(\frac{n\pi y_o}{L}\xi\right)[\cosh(\beta_n\eta) - \tanh(\beta_n)\sinh(\beta_n\eta)] \tag{9.167}$$

This solution is very reminiscent of the fin problem discussed earlier. Of course, the presence of generation again causes the gradients to be much steeper. The streamlines are shown in Figure 9.25.

9.5 TRANSIENT SYSTEMS

The addition of time dependence to any particular situation is conceptually no more difficult than adding another spatial dimension. Practically, the solutions become more difficult to obtain analytically and numerically. Two flavors of time dependence can be added. If we add a first-order time derivative to Laplace's equation, we introduce an exponentially decaying function of time into the solution that leads to a dissipative process. Adding a second-order time derivative introduces a periodic time function into the solution and leads to a wavelike oscillatory process. Adding both leads to a damped traveling wave solution.

The preferred method of obtaining analytic solutions to time-dependent transport problems involves using Green's functions [6]. The economy of using Green's functions stems from the fact that multidimensional solutions can be built up from a product of 1-D, time-dependent Green's

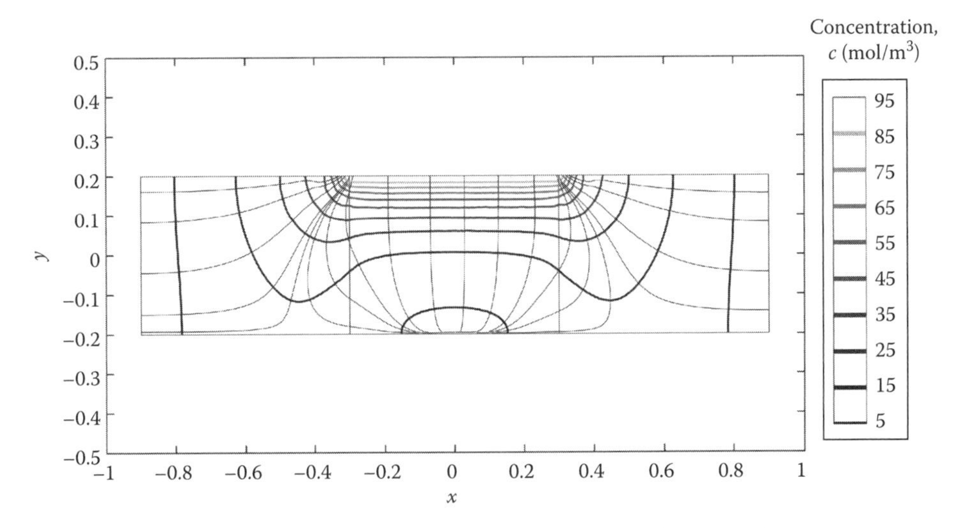

FIGURE 9.25 Streamlines and concentration profile for mass transfer from the patch. Solution from Comsol® module Hormone.

functions. To date, there is no preferred method of solving these equations numerically. Several techniques can be found in Finlayson [11], but each equation has its own peculiarities, and the numerical method used depends upon which method can yield a solution, the accuracy desired, and the patience of the person waiting for the computer to run the program. This is particularly true for nonlinear equations.

9.5.1 Pattern Formation [13]

One of the most interesting problems we can consider combines diffusion, generation, and a transient solution. Though we view diffusion as a stabilizing mechanism designed to smooth or eliminate gradients, it can also trigger an instability when combined with a generation term. The original idea behind this is attributed to Alan Turing [14], an English mathematician. He studied reaction–diffusion equations that contained nonlinear reaction terms and discovered that these equations could be unstable to small perturbations. The result of the instability was the formation of patterns that Turing called the chemical basis for morphogenesis. Morphogenesis is what gives animals their color pattern, describes the development of embryos, and can account for a number of features such as the pattern of teeth, pattern of limbs, and the development of branches in plants.

The main idea of the work was that a system of chemical substances, called morphogens, reacting together and diffusing through tissue was sufficient to describe the phenomena of morphogenesis. The morphogens are proposed to work by activating or suppressing certain genes and one for butterflies has recently been found [15]. Thus, one of the key features of morphogenesis is the ability of the system to work according to what is termed an activator/inhibitor concept.

We can look at a simplified, yet descriptive system that considers a substrate-inhibition mechanism where an enzyme is immobilized on an artificial membrane. A substrate (s) and co-substrate (a) react and diffuse on this membrane. Systems of this type were studied experimentally by Thomas and theoretically by Murray [13]. The arrangement leads to a reaction–diffusion model that can be expressed as

$$\frac{\partial a}{\partial \tau} = \left\{ \underbrace{\left(\frac{\alpha L^2}{D_{sb}}\right)}_{Pe_1}(1-a) - \underbrace{\left(\frac{k'' L^2 c_{a0}}{D_{sb}}\right)}_{\text{Da}} \frac{sa}{1 + c_{a0}s + K c_{a0}^2 s^2} \right\} + \underbrace{\left(\frac{D_{ab}}{D_{sb}}\right)}_{\beta} \nabla^2 a \tag{9.168}$$

$$\frac{\partial s}{\partial \tau} = \left\{ \underbrace{\left(\frac{\gamma L^2}{D_{sb}} \right)}_{Pe_2} (1-s) - \left(\frac{k'' L^2 c_{a0}}{D_{sb}} \right) \frac{sa}{1 + c_{a0} s + K c_{a0}^2 s^2} \right\} + \nabla^2 s \tag{9.169}$$

The $1 - s$ and $1 - a$ terms represent fluxes to the reaction surface membrane. The remainder of the generation term is the empirical uptake term, which for a given a concentration, for example, is like a Michaelis uptake for small s but which exhibits inhibition for large s with K as the inhibition parameter. We can define the dimensionless parameters as

$$a = \frac{c_a}{c_{a0}} \qquad s = \frac{c_s}{c_{a0}} \qquad \tau = \frac{D_{sb} t}{L^2} \qquad \beta = \frac{D_{ab}}{D_{sb}} \tag{9.170}$$

That leaves three dimensionless groups that are akin to Damkohler numbers:

$$\mathrm{Pe}_1 = \frac{\alpha L^2}{D_{sb}} \qquad \mathrm{Da} = \frac{k'' L^2 c_{a0}}{D_{sb}} \qquad \mathrm{Pe}_2 = \frac{\gamma L^2}{D_{sb}} \tag{9.171}$$

To see what is required for the equation system to be unstable, we first need to find the steady states where the composition is uniform. These are found from the solutions of

$$\begin{aligned} \mathrm{Pe}_1(1-a) - \mathrm{Da}\left(\frac{sa}{1 + c_{a0} s + K c_{a0}^2 s^2} \right) &= 0 \\ \mathrm{Pe}_2(1-s) - \mathrm{Da}\left(\frac{sa}{1 + c_{a0} s + K c_{a0}^2 s^2} \right) &= 0 \end{aligned} \tag{9.172}$$

Without specific numbers, all we can say about these equations is that

$$\mathrm{Pe}_1(1-a) = \mathrm{Pe}_2(1-s) \tag{9.173}$$

and, in general, we have cubic equations to determine a and s. Cubic equations have either three real solutions or one real and two complex solutions. We are interested in the case where we have one real solution that we will denote by $\bar{a}, \bar{s}$. Let us see what happens when we look at the stability of this particular solution. We assume a small perturbation about $\bar{a}, \bar{s}$ and substitute that into Equations 9.168 and 9.169. For simplicity, we will consider just the 1-D case. Let us write those equations in the following form:

$$\begin{aligned} \frac{\partial a}{\partial \tau} &= R_a + \beta \frac{\partial^2 a}{\partial \xi^2} \\ \xi &= \frac{x}{L} \\ \frac{\partial s}{\partial \tau} &= R_s + \frac{\partial^2 s}{\partial \xi^2} \end{aligned} \tag{9.174}$$

These are nonlinear equations. To look at their stability, we put them in a *linearized* form. The idea is to express the concentrations as $a = \bar{a} + a'$ and $s = \bar{s} + s'$. We then expand the reaction terms in

Taylor series about the steady-state solutions keeping only up to the linear terms in the series. Since the steady-state values for a and s satisfy Equations 9.172 and the time derivatives of $\bar{a}$ and $\bar{s}$ are identically zero, the newly "linearized" version of the equations is

$$\frac{\partial a'}{\partial \tau} = d_{11}a' + d_{12}s' + \beta\frac{\partial^2 a'}{\partial \xi^2}$$
$$\frac{\partial s'}{\partial \tau} = d_{21}a' + d_{22}s' + \frac{\partial^2 s'}{\partial \xi^2} \tag{9.175}$$

where

$$d_{11} = \left.\frac{\partial R_a}{\partial a}\right|_{\bar{a},\bar{s}} = -\mathrm{Pe}_1 - \frac{\mathrm{Da}\,\bar{s}}{1 + c_{a0}\bar{s} + Kc_{a0}^2\bar{s}^2}$$
$$d_{12} = \left.\frac{\partial R_a}{\partial s}\right|_{\bar{a},\bar{s}} = \frac{\mathrm{Da}(Kc_{a0}^2\bar{s}^2 - 1)}{(1 + c_{a0}\bar{s} + Kc_{a0}^2\bar{s}^2)^2} \tag{9.176}$$

$$d_{21} = \left.\frac{\partial R_s}{\partial a}\right|_{\bar{a},\bar{s}} = -\frac{\mathrm{Da}\,\bar{s}}{1 + c_{a0}\bar{s} + Kc_{a0}^2\bar{s}^2}$$
$$d_{22} = \left.\frac{\partial R_s}{\partial s}\right|_{\bar{a},\bar{s}} = -\mathrm{Pe}_2 + \frac{\mathrm{Da}(Kc_{a0}^2\bar{s}^2 - 1)}{(1 + c_{a0}\bar{s} + Kc_{a0}^2\bar{s}^2)^2} \tag{9.177}$$

Equations 9.175 are linear and can be solved via any number of techniques. One set of possible solutions, good enough for our purposes, is

$$a' = A\cos(q\xi)e^{\sigma\tau}$$
$$s' = S\cos(q\xi)e^{\sigma\tau} \tag{9.178}$$

Substituting Equation 9.178 into Equation 9.175, we have

$$A\sigma = Ad_{11} + Sd_{12} - A\beta q^2$$
$$S\sigma = Ad_{21} + Sd_{22} - Sq^2 \tag{9.179}$$

We are only interested in cases where the small perturbations grow with time. Rewriting Equation 9.179 yields

$$A(\sigma - d_{11} + \beta q^2) + S(-d_{12}) = 0$$
$$A(-d_{21}) + S(\sigma - d_{22} + q^2) = 0 \tag{9.180}$$

One solution to this set of equations is when $A = S = 0$, but this is trivial. The nontrivial solution only exists if the determinant of the coefficients of Equation 9.180 is zero. This requires

$$\sigma^2 + \sigma[-d_{22} + q^2 - d_{11} + \beta q^2] + [(d_{11} - \beta q^2)(d_{22} - q^2) - d_{12}d_{21}] = 0 \tag{9.181}$$

For the system to be unstable, we need for the solutions for σ to be positive or have a positive real part. To concentrate solely on diffusion as the destabilizing influence, we will assume that in the absence of diffusion, the mixture and process are stable. This means that if we set $D_a = D_s = 0$, we would obtain only negative values for σ or for the real part of σ. If we eliminate the diffusivities, then we can show that all the terms involving q disappear and we have

$$\sigma^2 + \sigma[-d_{22} - d_{11}] + [d_{11}d_{22} - d_{12}d_{21}] = 0 \tag{9.182}$$

and so the conditions for stability without diffusion are

$$d_{11} + d_{22} < 0 \qquad d_{11}d_{22} - d_{12}d_{21} > 0 \tag{9.183}$$

Now, if we consider the full equation, the conditions for stability would be

$$\begin{gathered} d_{11} + d_{22} - (1+\beta)q^2 < 0 \\ (d_{11} - \beta q^2)(d_{22} - q^2) - d_{12}d_{21} > 0 \end{gathered} \tag{9.184}$$

If the first condition of (9.183) holds, then the first condition of (9.184) must also hold since 1 + β, and q^2 are positive quantities. The only way diffusion can destabilize the system is if the second condition in (9.184) is violated. If we expand this condition, we see

$$\beta(q^2)^2 - (\beta d_{22} + d_{11})q^2 + (d_{11}d_{22} - d_{12}d_{21}) < 0 \tag{9.185}$$

Equation 9.185 is an equation for an upward facing parabola. The minimum point on this parabola occurs when

$$q_{min}^2 = \frac{1}{2}\left(d_{22} + \frac{d_{11}}{\beta}\right) \tag{9.186}$$

Substituting back into (9.185), we reach the condition that

$$(d_{11}d_{12} - d_{12}d_{21}) - \frac{1}{4}\left(\frac{\beta d_{22} + d_{11}}{\beta}\right) < 0 \tag{9.187}$$

or

$$(\beta d_{22} + d_{11}) > 2(\beta)^{1/2}(d_{11}d_{12} - d_{12}d_{21})^{1/2} > 0 \tag{9.188}$$

The physical significance of what we have shown can be described as follows:

1. From the first of condition (9.183), at least one of the two coefficients, d_{11} or d_{22}, must be negative. If d_{22} is negative, then

$$\frac{\partial R_s}{\partial s} < 0 \quad \text{Inhibitor} \tag{9.189}$$

and s inhibits its own rate of formation. Thus, we call s an *inhibitor*.

FIGURE 9.26 Simulated patterns on a starfish. Equations 9.168 and 9.169 were solved using Comsol module Starfish. Parameter values were:

$$\alpha = 1.5 \quad \gamma = 4 \quad k'' = 15 \quad K = 0.1$$

$$c_{ao} = 75 \quad c_{so} = 102 \quad D_{ab} = 7 \quad D_{sb} = 1$$

2. From (9.188),

$$\beta d_{22} + d_{11} > 0 \tag{9.190}$$

so that both d_{22} and d_{11} cannot both be negative. This means

$$\frac{\partial R_a}{\partial a} > 0 \quad \text{Activator} \tag{9.191}$$

and a promotes its own formation. a is called an *activator*. If we consider (9.190), it becomes clear that for diffusive instability to hold, the diffusivities of the activator and the inhibitor must be different and often quite different in magnitude.

These observations lend to a rather simple picture for how the diffusive instability arises. Due to thermal or some other form of fluctuation, we presume a small peak in the activator concentration arises. This leads to an enhanced production of inhibitor in that location. Diffusion should return things to normal, but since the inhibitor diffuses away more rapidly than the activator, activation outstrips inhibition and the peak grows. Eventually the imbalance grows large enough that diffusion takes over and maintains a steady imbalance in activator and inhibitor concentrations.

Example 9.4 Patterns on a Starfish

Let's return to Equations 9.168 and 9.169. We cannot solve these equations analytically but can solve them numerically. The Comsol module Starfish solves them on a geometry that resembles a 6-pointed starfish. Depending on the values we choose for the Damkohler numbers, we can generate uniform, banded, or dotted pattern formations as shown in Figure 9.26. The software integrates the equations to steady state starting from a small perturbation from a uniform solution where $a = 1$ and $s = c_{s0}/c_{a0}$. As expected, the solutions are very sensitive to the initial conditions, the parameters, and the boundary conditions. Here, the boundaries are all assumed to be impermeable.

9.6 SUMMARY

Multidimensional problems arise whenever we cannot define a particular length scale or transport direction that characterizes the situation of interest. These problems are often encountered in practice, but the first attempt at solving them generally includes trying to reduce them to a single dimension. When this is not possible, we can try several solution methods including separation of variables, Green's functions, complex analysis, Fourier's transforms, and numerical methods.

We have shown that multidimensional, steady-state problems involving no generation are described by Laplace's equation. The solution to Laplace's equation is a potential field that describes how the concentration, temperature, voltage, or charge distribution behaves throughout space. We separated the fields into two classes. The first are scalar fields where a single potential function describes the behavior of the measured variable throughout space. The second class of field was the vector field. Here, we need a potential function for each component of the vector and so generally require one Laplace equation for each component. The potential is not what is measured; the vector itself is the measured variable in this case.

When generation is included, the differential equation describing the system is a form of Poisson's equation. Solutions to Poisson's equation can be obtained in the same manner as Laplace's equation, albeit with a bit more work. If one understands the solution to Laplace's equation for a particular situation, one can generally develop a *feel* for how Poisson's equation will behave.

In the remaining chapters of this text, we will expand our horizons and discuss what happens when we allow fluid motion to occur simultaneously with other forms of transport. The interaction of the vector field describing fluid motion with the scalar field describing the temperature, concentration, etc., provides some fundamentally new phenomena that we must be able to analyze and harness.

PROBLEMS

LAPLACE'S EQUATION

9.1 Consider the fin of Section 9.2.1 and its solution given by Equation 9.21. Develop expressions for the effectiveness and efficiency of such a fin.

a. Estimate the effectiveness for the following data:

$k = 10$ W/m K $\quad h = 500$ W/m^2 K $\quad y_o = 0.005$ m

$L = 40$ mm $\quad T_b = 100°$C $\quad T_\infty = 25°$C

b. How does your calculated value compare to the 1-D approximation?

9.2 A miniature heat pipe is a heat transfer device that operates via an evaporation/condensation mechanism. The heat source evaporates fluid at one end of the device. The fluid travels the length of the device condensing on the walls and at the opposite end to reject heat. The situation is shown in Figure P9.2. Consider such a device with a square cross section. The temperature profile within the wall of the device near the evaporating end is of interest. We can treat the evaporating section as a slab of length, L, and width, w, and thickness t_h. At the start of the device, a known amount of heat, q_o, is injected. At the end of the evaporator, a constant temperature, T_a, is reached so that heat flow in the x-direction is zero. Convective boundary conditions exist on the inside and outside of the device. Determine an expression for the temperature distribution within the wall in the x- and y-directions.

9.3 One of the body's ways of increasing heat transfer with the environment is the use of goose bumps. The small raised regions on the skin offer increased surface area for transport. We would like to calculate the effectiveness of a goose bump. Referring to the Figure P9.3,

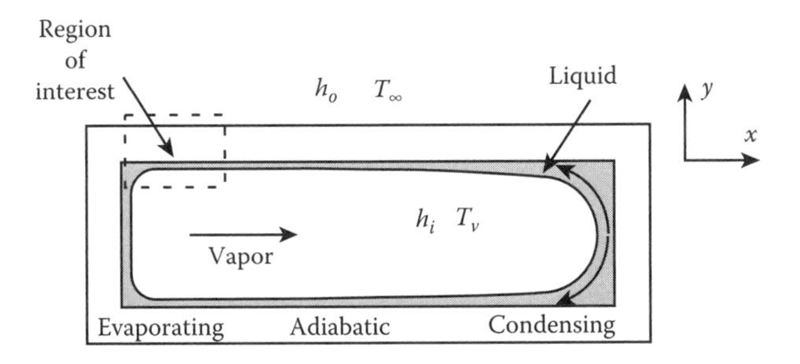

FIGURE P9.2

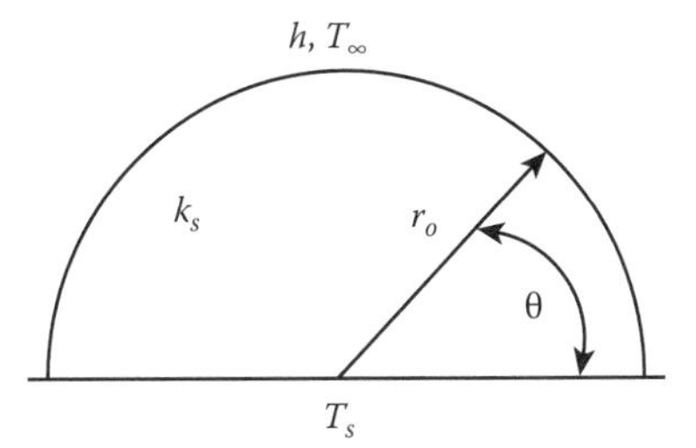

FIGURE P9.3

a. Determine the temperature profile within the goose bump.
b. Estimate the effectiveness of such a goose bump and comment on its usefulness as a heat transfer device:

$$h = 100 \text{ W/m}^2\text{ K} \qquad T_s = 37°\text{C} \qquad T_\infty = 30°\text{C}$$

$$k_s = 0.37 \text{ W/m K} \qquad r_o = 0.001 \text{ m}$$

9.4 A thin slab of metal is formed in the shape of a square. The x–y vertices of the square are (0,0), (1,0), (0,1), and (1,1). The faces of the slab in the z-direction are insulated so that all heat transfer occurs in the x–y plane. The initial temperature distribution in the plane is $T(x, y, 0) = T_0(x, y)$. Assuming that there are no sources of heat in the slab, that the x-faces are insulated, and that the y-faces are held at a temperature $T = 0$,

a. Derive the generic differential equation describing the situation.
b. Show that the equation

$$T_{mn}(x, y, t) = G_{mn} \sin(m\pi x)\cos(n\pi y)\exp[-\alpha(m^2 + n^2)\pi^2 t]$$

satisfies the differential equation.
c. A single value of G_{mn} cannot represent an arbitrary initial condition so we resort to using a sum. In this case, the G_{mn} are given by

$$G_{mn} = 4\int_0^1\int_0^1 T_0(x, y)\cos(n\pi y)\sin(m\pi x)\,dy\,dx$$

If $T_0(x, y) = xy$, determine the full solution for T.

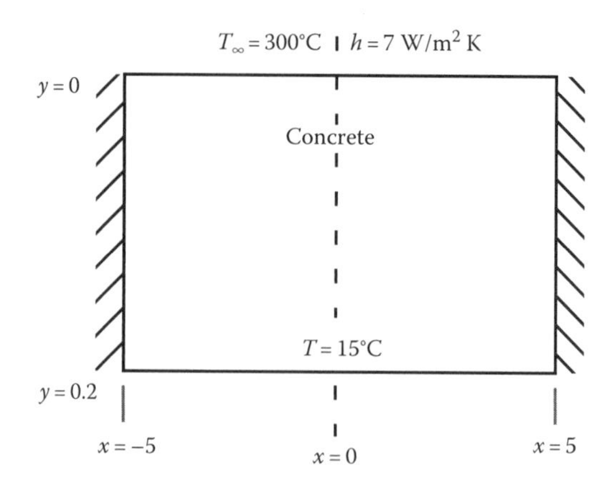

FIGURE P9.5

9.5 We are interested in determining the temperature distribution inside a concrete slab at a stoplight. The slab is exposed to cars' catalytic converters whose surface temperatures are 300°C. Figure P9.5 shows all the other important details.

a. Derive the partial differential equation governing the temperature profile inside the slab.

b. Solve the differential equation for the temperature field. How hot does the surface at $y = 0$ get (use only 1 or 2 terms in the series)?

9.6 Consider one 3-D, time-dependent problem to illustrate the power of a Green's function solution. We are interested in doping a semi-infinite boule of initially pure, single crystal silicon. By exposing the surface at $z = 0$ to the dopant, we manage to introduce an amount, m_o, of material into the silicon.

a. Demonstrate that the governing equation for this process is

$$\frac{\partial c_a}{\partial t} = D_{ab}\left(\frac{\partial^2 c_a}{\partial x^2} + \frac{\partial^2 c_a}{\partial y^2} + \frac{\partial^2 c_a}{\partial z^2}\right)$$

b. What are the boundary conditions, the initial condition, and any other necessary constraints for this problem?

c. Show that the following Green's function satisfies the differential equation and the boundary conditions:

$$c_a = \frac{2m_o}{(4\pi D_{ab}t)^{3/2}}\exp\left[-\frac{(x-x')^2+(y-y')^2+(z-z')^2}{4D_{ab}t}\right]$$

Complex Variable Transformations, Sources, and Sinks

9.7 Find the temperature profile in the semi-infinite solid defined by Figure P9.7. The side at $x = 0$ is held at a temperature $T = 0$. The side defined by $y = 0$ is insulated for $0 < x < 1$ and held at a temperature $T = 1$ for $x > 1$. Use the transformation $z = \sin(w)$ where $z = x + iy$ and $w = \phi + i\psi$. This transformation takes the region $x = 0$ and maps it to $\phi = 0$. It takes the region $0 < x < 1$ and maps it to $\psi = 0$, $0 \le \phi \le \pi/2$ and takes the region $x > 1$ and maps it to $\phi = \pi/2$.

9.8 Show that the stream function for flow in the corner of Figure P9.8 is given by

$$\psi(r,\theta) = Ar^4\sin(4\theta)$$

Use the transformation $z = w^4$ where $z = x + iy$ and $w = \phi + i\psi$.

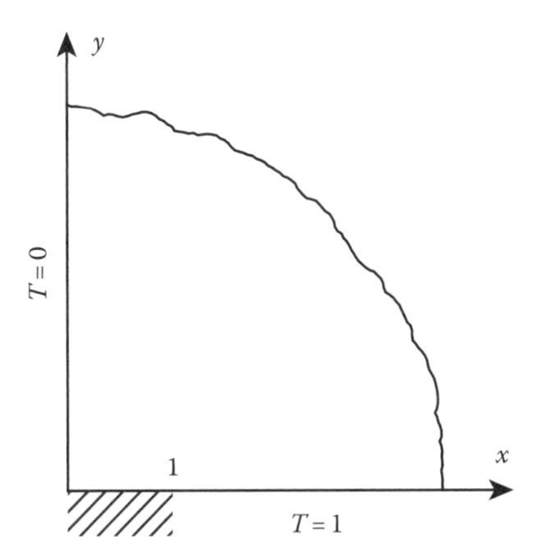

FIGURE P9.7

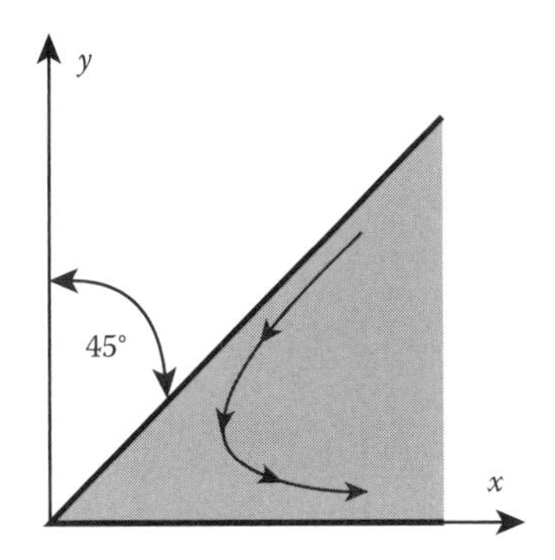

FIGURE P9.8

9.9 Using the transformation $w = \sin(z)$, determine the equation for the stream function for flow inside the semi-infinite region $y \geq 0$, $-\pi/2 \leq x \leq \pi/2$ shown in Figure P9.9.

9.10 Consider the flow formed by placing a source of strength, q_o, at a distance, d, from an infinitely long wall as shown in Figure P9.10. The velocity potential for this incompressible and irrotational flow field is

$$\phi = \frac{q_o}{4\pi}\left\{\ln[(x-d)^2 + y^2] + \ln[(x+d)^2 + y^2]\right\}$$

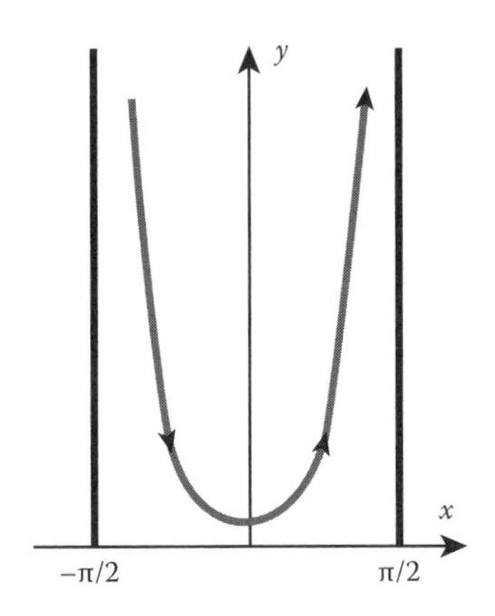

FIGURE P9.9

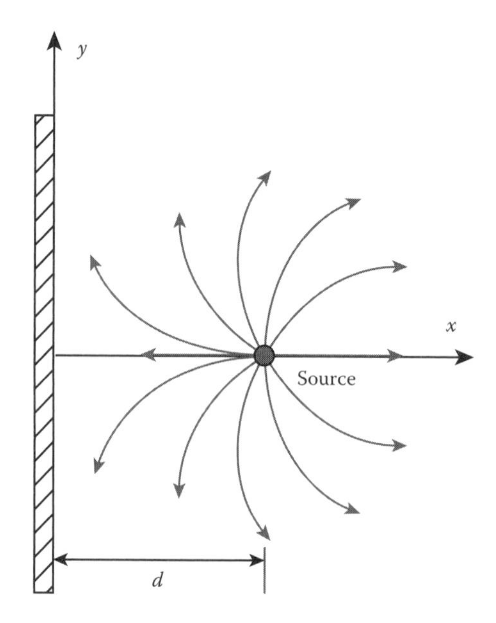

FIGURE P9.10

a. Demonstrate that there is no flow through the wall.
b. How does the velocity vary along the length of the wall?
c. In 1738, Daniel Bernoulli showed that for an incompressible fluid exhibiting an irrotational flow field,

$$P + \frac{1}{2}\rho v^2 + \rho g z = \text{constant} \quad \text{Bernoulli's equation}$$

Here, z is the height of the fluid above some reference and $v^2 = v_x^2 + v_y^2$. Using this result (we derive it in Chapter 11) and the fact that $P = P_o$ far away from the source, determine the pressure distribution along the wall.
d. What does the stream function for this flow look like?

9.11 A stream function is given by

$$\psi = \sin\left(\frac{x}{L}\right)\sinh\left(\frac{y}{K}\right)$$

where L and K are constants, $0 \le x \le \pi L$ and $y \ge 0$.
a. Does ψ represent a potential flow? Is the flow incompressible too?
b. If $L = K = 1$, locate any possible stagnation points.

9.12 Previously, we solved Laplace's equation using separation of variables for the velocity potential about a cylinder with circulation. The solution was of the form

$$\phi = c_1\theta + v_o\left(r + \frac{r_o^2}{r}\right)\cos\theta$$

and c_1 was effectively the strength of the circulation.

a. Show that the transformation below yields the same result for the velocity potential. What does the equation for the streamlines look like?

$$w = v_o\left(z + \frac{r_o^2}{z}\right) + ic_1 \log\left(\frac{z}{r_o}\right) \qquad z = x + iy$$

b. The lift force on the cylinder is given by

$$F_x = -\int_0^{2\pi} (P_s \cos\theta) r\, d\theta \qquad F_y = -\int_0^{2\pi} (P_s \sin\theta) r\, d\theta$$

where P_s is the surface pressure. The surface pressure can be obtained from Bernoulli's equation (see Problem 9.7):

$$P_s + \frac{1}{2}\rho v_s^2 = P_o + \frac{1}{2}\rho v_o^2$$

where
P_o is a reference pressure
v_s is the fluid velocity on the surface of the cylinder

Show that the lift force is given by

$$F_{lift} = -2\pi\rho c_1 v_o^2$$

9.13 The *method of images* uses precisely oriented collections of sources, sinks, and vortices to establish *artificial* walls and so simulate more complicated flow fields. One such flow field is formed by positioning two counter rotating vortices at a distance, d, from the location of the hypothetical wall (Figure P9.13). The stream function for this flow is given by

$$\psi = \frac{q_o}{4\pi}\left\{\ln[(x-d)^2 + y^2] + \ln[(x+d)^2 + y^2]\right\}$$

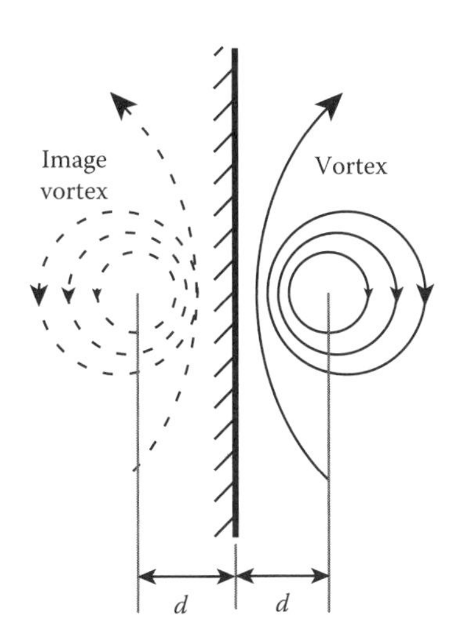

FIGURE P9.13

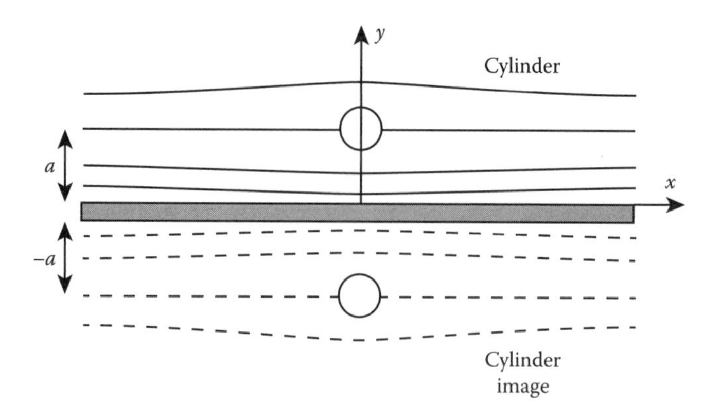

FIGURE P9.14

a. Demonstrate that there is no flow through the wall.
b. How does the velocity vary along the length of the wall?
c. What is the pressure distribution along the wall?

9.14 Another example of the method uses flow past a cylinder of radius, R, and its image to simulate the flow of fluid past a cylinder that lies close to a plane wall. The situation is shown in Figure P9.14.

The stream function for this flow is given by

$$\psi = -2v_\infty y + v_\infty R^2\left[\frac{y-a}{x^2+(y-a)^2} + \frac{y+a}{x^2+(y+a)^2}\right]$$

a. Determine the point where the maximum velocity is located.
b. The pressure within the irrotational flow field is

$$P + \frac{1}{2}\rho v^2 + \rho g z = \text{constant} \qquad \text{Bernoulli's equation}$$

Here, z is the height of the fluid above some reference (not important here, $z = 0$) and $v^2 = v_x^2 + v_y^2$. Using this result (we derive it in Chapter 11) and the fact that $P = P_o$ and $v = v_\infty$ far away from the source, determine the pressure distribution along the wall.

9.15 What would the stream function and velocity potential look like for the combination of plane flow and a tornado? Sketch the flow field. Is it realistic? Comment on the event horizon and escape velocity needed to resist being sucked in.

Poisson's Equation and Generation

9.16 A semitransparent slab of photoresist is being exposed to ultraviolet (UV) radiation. When exposed, the photoresist generates heat in direct proportion to the intensity of the radiation. The radiation intensity decreases exponentially as it penetrates into the slab. We would like to know the temperature distribution within the exposed portion of the slab. The situation is shown in Figure P9.16.

9.17 Reconsider the solenoid problem of Section 9.4.2 shown in Figure P9.17 but with convective heat transfer from the surface at $r = r_o$. The convection is characterized by a heat transfer coefficient, h, and the fluid is held at a temperature of T_∞.

9.18 The surface of a catalytic slab is exposed to a concentration distribution $c_{ao}(x)$. As material diffuses into the slab, it reacts according to a first-order chemical reaction ($A \rightarrow B$). The sides

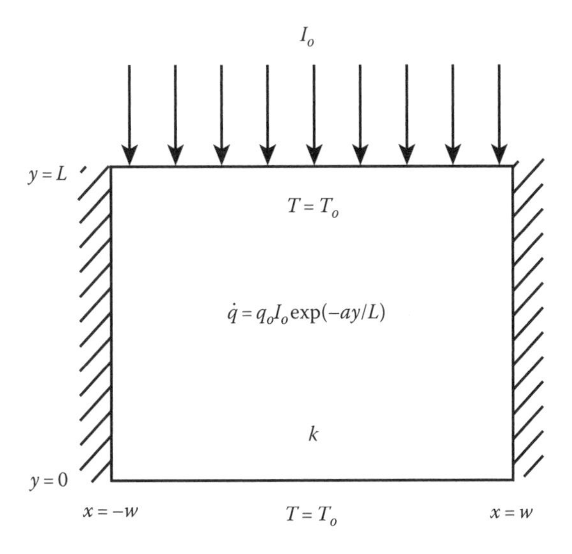

FIGURE P9.16

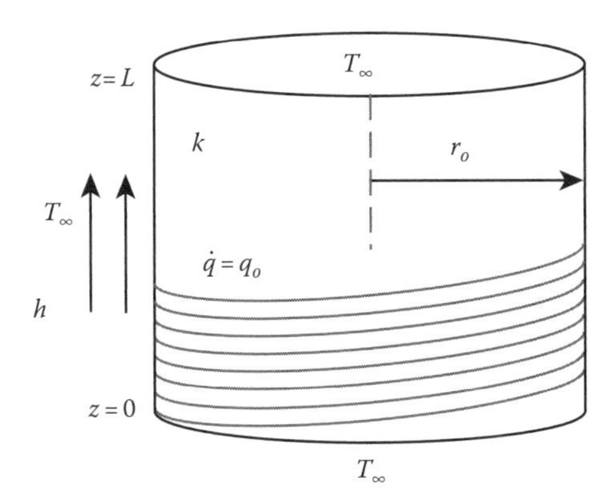

FIGURE P9.17

and bottom of the catalyst are impermeable and a exists in dilute solution in the slab as shown in Figure P9.18.

a. Derive the partial differential equation governing the concentration profile inside the slab.
b. Solve the differential equation for the concentration field.
c. What is the flux of material through the exposed surface?
d. How much unreacted material is actually inside the slab?

9.19 The cylindrical surface of a catalytic pellet is exposed to a concentration distribution $c_{ao}(z)$. The ends of the pellet are impermeable as shown in Figure P9.19. As material diffuses into the pellet, it reacts according to a first-order chemical reaction ($A \rightarrow B$). Equimolar counterdiffusion occurs inside the pellet.

a. Derive the partial differential equation governing the concentration profile.
b. Solve the differential equation for the concentration field.
c. What is the flux of material through the exposed surface?

9.20 Consider a thin circular sheet of water of radius, r_o, and depth, h, that is being excited by steady, periodic oscillations of its outer rim at a frequency, ω. We want to catalog the types of tidal waves that can form.

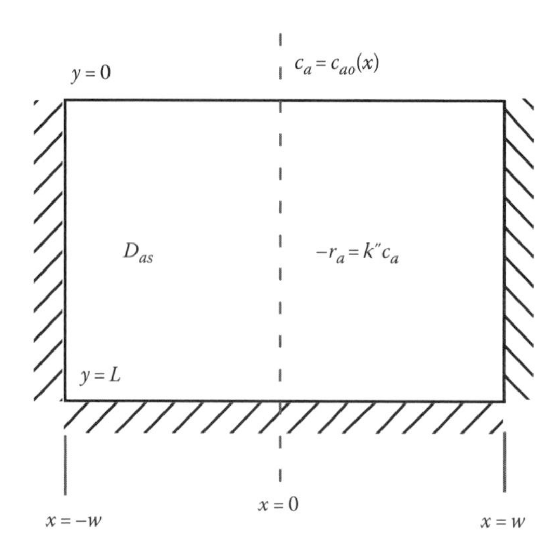

FIGURE P9.18

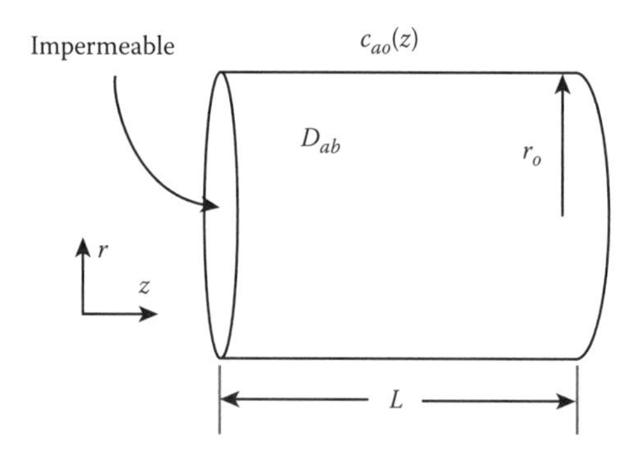

FIGURE P9.19

a. Show that the equation describing the wave height in two dimensions is

$$\frac{\partial^2 \delta_\lambda}{\partial r^2} + \frac{1}{r}\frac{\partial \delta_\lambda}{\partial r} + \frac{1}{r^2}\frac{\partial^2 \delta_\lambda}{\partial \theta^2} + k^2 \delta_\lambda = 0 \qquad k = \frac{\omega}{\sqrt{gh}}$$

b. Solve this equation in generic form using the following boundary conditions:

$$r = r_o \qquad \frac{\partial \delta_\lambda}{\partial r} = 0 \quad \text{wave rigidly follows the outer boundary}$$

$$r = 0 \qquad \delta_\lambda = \text{finite} \quad \text{wave height must be bounded at origin}$$

Show that the following solutions are valid:

$$\delta_\lambda = A_s J_s(kr)\cos(s\theta)\cos(\omega t) + B_s J_s(kr)\sin(s\theta)\cos(\omega t)$$

What are the allowable values of k?

c. Each value of s in the solutions previously yields a different *mode* of wave. Plot several streamlines for the first two $s = 0$ and $s = 1$ modes of the two families of solutions and note the location of the nodes.

9.21 Consider the diffusion situation described in Section 9.5. Instead of having pure diffusion, we put a catalyst into the fluid that forces a first-order chemical reaction $A \rightarrow B$ to occur. Resolve the transient problem for the concentration profile assuming dilute solutions, a constant surface concentration of a, and impermeable container walls.

REFERENCES

1. Wunsch, A.D., *Complex Variables with Applications*, Addison-Wesley Publishing, Reading, MA (1983).
2. Jeffrey, A., *Complex Analysis and Applications*, CRC Press Inc., Boca Raton, FL (1992).
3. Olson, R.M. and Wright, S.J., *Essentials of Engineering Fluid Mechanics*, 5th edn., Harper & Row Publishers, New York (1990).
4. Churchill, R.V., Brown, J.W., and Verhey, R.F., *Complex Variables and Applications,* 3rd edn., McGraw-Hill Inc, New York (1974).
5. Morse, P.M. and Feshbach, H., *Methods of Theoretical Physics*, Vol. 2, McGraw-Hill, New York (1953).
6. Greenberg, M.D., *Applications of Green's Functions in Science and Engineering*, Prentice Hall, Englewood Cliffs, NJ (1971).
7. Zauderer, E., *Partial Differential Equations of Applied Mathematics*, John Wiley & Sons, New York (1983).
8. Crank, J., *The Mathematics of Diffusion*, Oxford, New York (1956).
9. Carlslaw, H.S. and Jaeger, J.C., *Conduction of Heat in Solids*, 2nd edn., Oxford, New York (1959).
10. Setian, L., *Engineering Field Theory with Applications*, Cambridge University Press, Cambridge, U.K. (1992).
11. Finlayson, B.A., *Nonlinear Analysis in Chemical Engineering*, McGraw-Hill, New York (1980).
12. White, R.E., *An Introduction to the Finite Element Method with Applications to Nonlinear Problems*, John Wiley & Sons, New York (1985).
13. Murray, R.J., *Mathematical Biology II: Spatial Models and Biomedical Applications*, 3rd edn., Springer-Verlag, New York (2003).
14. Turing, A.M., The chemical basis for morphogenesis, *Philos. Trans. R. Soc. Lond. Ser. B Biol. Sci.*, **237**, 37–72 (1952).
15. Werner, T., Koshikawa, S., Williams, T.M., and Carroll, S.B., Generation of a novel colour pattern by the wingless morphogen, *Nature*, **464**, 1143–1148 (2010).

10 Convective Transport
Microscopic Balances

$$\vec{\nabla}\cdot(\text{In}-\text{Out}) + \text{Gen} = \text{Acc}$$

10.1 INTRODUCTION

We begin the second phase in our study of transport phenomena in this chapter by looking, in detail, at convective transport. All convective transport involves fluid motion and nearly every transport situation of practical importance involves convection. This is not unusual since 70% of the Earth's surface is covered by water, 65% of our bodies are water, and the atmosphere is a gas. All these fluids are in ceaseless motion, driven by energy released deep within the Earth and energy transported from the Sun.

Convection is a highly effective means of transport, and most processes could not occur efficiently without it. Try to imagine how different your life would be were it not for convective transport of nutrients within your body. A quick calculation of O_2 diffusion (penetration time) through a human assuming a maximum body thickness, 2Δ, of 18 cm and a fairly high diffusivity of 1×10^{-8} m^2/s

$$t = \frac{\Delta^2}{(3.643)^2 D_{ab}} = \frac{(0.09)^2}{(3.643)^2(1\times10^{-8})} = 61{,}000 \text{ s} \tag{10.1}$$

shows that sustaining a vigorous life in three dimensions would be impossible. That is why organisms exploiting O_2 diffusion alone to sustain respiration are often no larger than a small insect. The same considerations are true for engineering processes. We could not heat our homes without convection or operate a power plant, and just reading this page and moving your eyeballs back and forth create a convective pattern in the surrounding air.

So far we have been concerned with diffusive mechanisms like heat conduction, and when we needed to consider convection, we assumed we could characterize it in terms of a heat transfer coefficient, or a mass transfer coefficient, just some numbers given within a problem. We gave no thought to how we came upon these *magic* numbers or how they behave, but in this, and the next few chapters, we will look at this process in detail.

We begin our analysis with the driest subject in all transport: deriving the convective transport equations, in their simplest possible form, the form used in everyday practice. This is a tedious process but by starting out with the full conservation equations, we will better understand how the solutions and values for convective coefficients (such as from a handbook) have been derived and what simplifications were necessary to obtain them. As an engineer, understanding and evaluating how these simplifications affect your process designs and your products' performance will be the most important part of your job.

10.2 MOMENTUM TRANSPORT

We begin the derivations with momentum transport as the starting point. Our result will be the Navier–Stokes equations, a set of partial differential equations that are a statement of conservation of mass and Newton's law of motion, $F = ma$ for a fluid. We will derive these equations for a general fluid and then, for the most part, restrict ourselves to Newtonian fluids with constant properties.

10.2.1 Continuity

The continuity equation is an overall mass balance about a control volume. Consider the differential control volume shown in Figure 10.1. The mass balance equation incorporates what occurs in all three coordinate directions. It is written in terms of the mass flow rate (kg/s) moving through the control volume. Barring any nuclear reactions or other forms of matter/energy conversion, there are no *Gen* terms to consider:

In	–	*Out*	=	*Acc*
Rate of mass flow in by convection	–	Rate of mass flow out by convection	=	Rate of mass accumulated in the control volume

$$\underbrace{[\rho v_x(x)-\rho v_x(x+\Delta x)]\Delta y\Delta z}_{x\text{-}faces}+\underbrace{[\rho v_y(y)-\rho v_y(y+\Delta y)]\Delta x\Delta z}_{y\text{-}faces}+\underbrace{[\rho v_z(z)-\rho v_z(z+\Delta z)]\Delta x\Delta y}_{z\text{-}faces}$$
$$=\Delta x\Delta y\Delta z\frac{\partial\rho}{\partial t} \tag{10.2}$$

Dividing by $\Delta x\Delta y\Delta z$, the volume of the control volume, and taking the limit as $\Delta x\Delta y\Delta z \to 0$ gives us the following differential continuity equation:

$$-\left[\frac{\partial}{\partial x}(\rho v_x)+\frac{\partial}{\partial y}(\rho v_y)+\frac{\partial}{\partial z}(\rho v_z)\right]=\frac{\partial\rho}{\partial t} \tag{10.3}$$

If the fluid is incompressible, ρ is constant, the accumulation term vanishes, and we can factor out and divide through by the density to give

$$\left(\frac{\partial v_x}{\partial x}+\frac{\partial v_y}{\partial y}+\frac{\partial v_z}{\partial z}\right)=\underbrace{(\vec{\nabla}\cdot\vec{\mathbf{v}})}_{\text{Vector form}}=0 \tag{10.4}$$

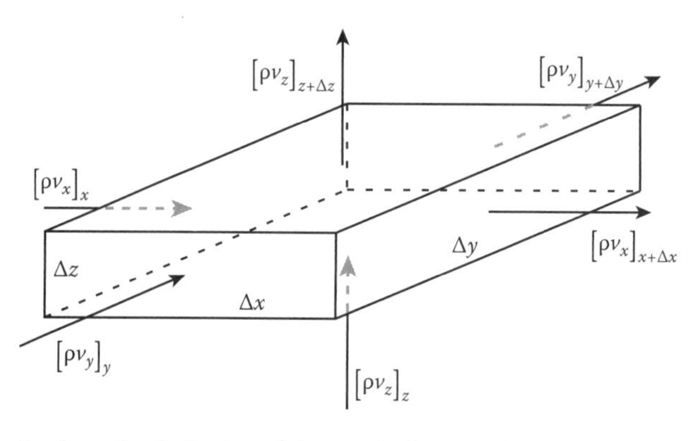

FIGURE 10.1 Control volume for derivation of the continuity equation.

Example 10.1 Cylindrical Coordinates

We can also derive the continuity equation in cylindrical coordinates without too much difficulty. The control volume in cylindrical coordinates is shown in Figure 10.2. We write the conservation equation in terms of the mass flow rate of fluid. The areas of each face through which the mass is flowing are given by

$$r\text{-}face - r\Delta\theta\Delta z \qquad \theta\text{-}face - \Delta r\Delta z \qquad z\text{-}face - r\Delta\theta\Delta r$$

The overall balance equation is

$$\begin{matrix} (In-Out)_r & + & (In-Out)_\theta & + & (In-Out)_z & = & Acc \end{matrix}$$

$$[\rho r v_r(r) - \rho[r+\Delta r]v_r(r+\Delta r)]\Delta\theta\Delta z + [\rho v_\theta(\theta) - \rho v_\theta(\theta+\Delta\theta)]\Delta r\Delta z + [\rho v_z(z) - \rho v_z(z+\Delta z)]r\Delta\theta\Delta r = \frac{\partial\rho}{\partial t} r\Delta r\Delta\theta\Delta z$$

where we have taken into account that the area through which fluid flows in the radial direction changes. Dividing by $r\,dr\,d\theta\,dz$, the volume of our element, and taking the limit as $dr, d\theta, dz \to 0$ gives the following continuity equation:

$$-\frac{1}{r}\left[\frac{\partial}{\partial r}(\rho r v_r) + \frac{\partial}{\partial\theta}(\rho v_\theta) + r\frac{\partial}{\partial z}(\rho v_z)\right] = \frac{\partial\rho}{\partial t}$$

For an incompressible fluid, density is again constant, accumulation is zero, and we have

$$\frac{1}{r}\frac{\partial}{\partial r}(r v_r) + \frac{1}{r}\frac{\partial}{\partial\theta}(v_\theta) + \frac{\partial}{\partial z}(v_z) = 0$$

The continuity equation in rectangular, cylindrical, and spherical coordinates is given in Table 10.1.

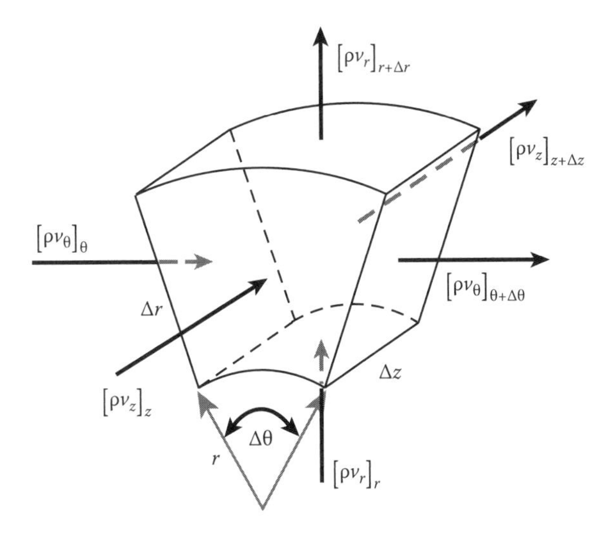

FIGURE 10.2 Control volume in cylindrical coordinates for deriving the continuity equation.

TABLE 10.1
Forms for the Continuity Equation

Coordinate System	Continuity Equation
Cartesian	$-\left[\frac{\partial}{\partial x}(\rho v_x)+\frac{\partial}{\partial y}(\rho v_y)+\frac{\partial}{\partial z}(\rho v_z)\right]=\frac{\partial \rho}{\partial t}$
Cylindrical	$-\frac{1}{r}\left[\frac{\partial}{\partial r}(\rho r v_r)+\frac{\partial}{\partial \theta}(\rho v_\theta)+r\frac{\partial}{\partial z}(\rho v_z)\right]=\frac{\partial \rho}{\partial t}$
Spherical	$-\frac{1}{r^2}\left[\frac{\partial}{\partial r}(\rho r^2 v_r)+\frac{r}{\sin\theta}\frac{\partial}{\partial \theta}(\rho v_\theta \sin\theta)+\frac{r}{\sin\theta}\frac{\partial}{\partial \phi}(\rho v_\phi)\right]=\frac{\partial \rho}{\partial t}$

Example 10.2 Use of the Continuity Equation

Given the following velocity profile for steady flow around a sphere, determine if the fluid is incompressible:

$$v_r = v_\infty \cos\theta\left[1-\frac{3}{2}\left(\frac{r_o}{r}\right)+\frac{1}{2}\left(\frac{r_o}{r}\right)^3\right] \qquad v_\theta = -v_\infty \sin\theta\left[1-\frac{3}{4}\left(\frac{r_o}{r}\right)-\frac{1}{4}\left(\frac{r_o}{r}\right)^3\right] \qquad v_\phi = 0$$

We use the continuity equation in spherical coordinates to determine if the fluid is incompressible. For an incompressible fluid at steady state, we should be able to remove the density from the continuity equation and have

$$\frac{1}{r^2}\frac{\partial}{\partial r}(r^2 v_r)+\frac{1}{r\sin\theta}\frac{\partial}{\partial \theta}(v_\theta \sin\theta)+\frac{1}{r\sin\theta}\frac{\partial}{\partial \phi}(v_\phi)=0$$

Since v_ϕ is zero, we have only two terms to evaluate:

$$\frac{1}{r^2}\frac{\partial}{\partial r}(r^2 v_r)=\frac{v_\infty\cos\theta}{r^2}\frac{\partial}{\partial r}\left[r^2-\frac{3}{2}(r_o r)+\frac{1}{2}\left(\frac{r_o^3}{r}\right)\right]$$

$$=v_\infty\cos\theta\left[\frac{2}{r}-\frac{3}{2}\frac{r_o}{r^2}-\frac{1}{2}\frac{r_o^3}{r^4}\right]$$

$$\frac{1}{r\sin\theta}\frac{\partial}{\partial \theta}(v_\theta \sin\theta)=-\frac{v_\infty}{r\sin\theta}\frac{\partial}{\partial \theta}\left[\sin^2\theta\left(1-\frac{3}{4}\left(\frac{r_o}{r}\right)-\frac{1}{4}\left(\frac{r_o}{r}\right)^3\right)\right]$$

$$=-v_\infty\cos\theta\left[\frac{2}{r}-\frac{3}{2}\frac{r_o}{r^2}-\frac{1}{2}\frac{r_o^3}{r^4}\right]$$

It is easy to see that the continuity equation for an incompressible fluid is satisfied.

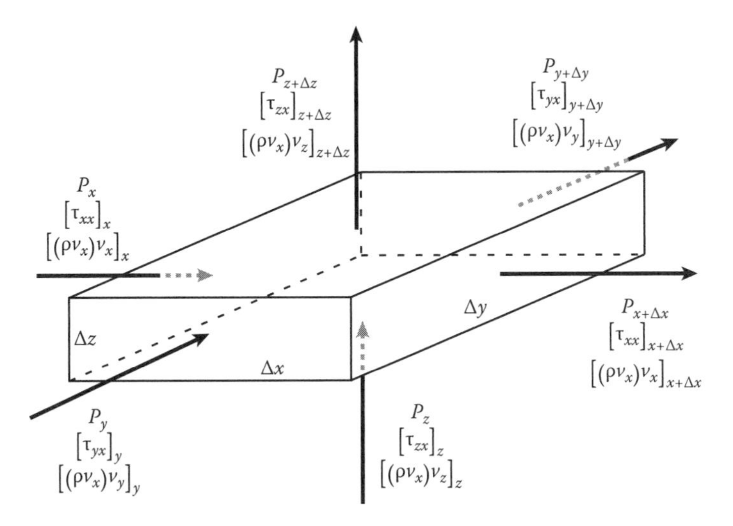

FIGURE 10.3 Control volume for deriving the x-component of the momentum equation.

10.2.2 Momentum Balance

The momentum balance is a statement of Newton's law of motion: $F = ma$. To derive it, we consider the same rectangular differential element we used in deriving the continuity equation (Figure 10.3). The balance equation for the momentum of the fluid element is

In	$-$	*Out*	$+$	*Gen*	$=$	*Acc*
Rate of momentum in by convection	$-$	Rate of momentum out by convection	$+$	Sum of forces acting on the body	$=$	Rate of momentum accumulated

Momentum may flow into and out of the control volume in a variety of ways: by convection, that is, by bulk fluid motion; by molecular transfer, that is, by velocity gradients; and by external forces acting on the body such as gravity or pressure. The first mechanism previously mentioned arises due to fluid acceleration through the control volume, while the second arises from shear stresses acting on the fluid. We consider only the x-component of momentum in detail and include gravitation as the only body force. Other body forces such as surface tension forces or electromagnetic forces may arise depending on the situation. Pressure is included as a separate surface force.

The components of our momentum equation include the following:

Bulk motion—momentum transport via convection

$$\begin{aligned}&(In-Out)_x+(In-Out)_y+(In-Out)_z\\&[(\rho v_x)v_x(x)-(\rho v_x)v_x(x+\Delta x)]\Delta y\Delta z+\\&[(\rho v_x)v_y(y)-(\rho v_x)v_y(y+\Delta y)]\Delta x\Delta z+\\&[(\rho v_x)v_z(z)-(\rho v_x)v_z(z+\Delta z)]\Delta x\Delta y\end{aligned} \tag{10.5}$$

Notice that each term has the units of force ($\text{kg}\cdot\text{m/s}^2 = N$) and that the x-component of momentum can be affected by fluid acceleration (changes in the fluid's velocity) in the y- or z-direction. Such effects are evident in recirculating flows such as flow about a sphere and flows within or around a corner.

Shear and normal stresses—internal fluid forces

Differences in shear and normal stresses across the control volume result in a flow of momentum from the higher to the lower stress. Shear and normal stresses generally arise from two sources: pressure differences in the fluid and velocity gradients. The pressure gives rise to a normal stress component, while velocity gradients give rise to shear stress components and can contribute to the total normal stress. We treat the pressure separately. Since these are forces, they are generation terms in the context of our momentum equation:

$$\begin{gathered} Gen \\ [\tau_{xx}(x)-\tau_{xx}(x+\Delta x)]\Delta y\Delta z+[\tau_{yx}(y)-\tau_{yx}(y+\Delta y)]\Delta x\Delta z+[\tau_{zx}(z)-\tau_{zx}(z+\Delta z)]\Delta x\Delta y \end{gathered} \tag{10.6}$$

External forces—the most important forces we encounter routinely are due to the pressure gradient and gravity. Both are formally generation terms:

$$\begin{gathered} Gen \\ [P_x(x)-P_x(x+\Delta x)]\Delta y\Delta z+\rho g_x\Delta x\Delta y\Delta z \end{gathered} \tag{10.7}$$

Accumulation—this term represents the accumulation of momentum within the control volume and, like the other accumulation terms we encountered, is a time-dependent quantity:

$$\begin{gathered} Acc \\ \Delta x\Delta y\Delta z\frac{\partial}{\partial t}(\rho v_x) \end{gathered} \tag{10.8}$$

If we combine Equations 10.5 through 10.8, divide by the volume element, $\Delta x\Delta y\Delta z$, and take the limit as $\Delta x\Delta y\Delta z \to 0$, we obtain the x-component of the momentum balance:

$$-\left\{\frac{\partial}{\partial x}[(\rho v_x)v_x]+\frac{\partial}{\partial y}[(\rho v_x)v_y]+\frac{\partial}{\partial z}[(\rho v_x)v_z]\right\}-\left[\frac{\partial\tau_{xx}}{\partial x}+\frac{\partial\tau_{yx}}{\partial y}+\frac{\partial\tau_{zx}}{\partial z}\right]-\frac{\partial P}{\partial x}+\rho g_x=\frac{\partial}{\partial t}(\rho v_x) \tag{10.9}$$

Similar equations can be written for the y- and z-components of the momentum. Equation 10.9 can be recast for all three components in vector form as

$$\begin{array}{ccc} In-Out \quad + & Gen & = \quad Acc \\ -[\vec{\nabla}\cdot(\rho\vec{\mathbf{v}})\vec{\mathbf{v}}] \;-\; & (\vec{\nabla}\cdot\vec{\tau})-\vec{\nabla}P+\rho\vec{\mathbf{g}} & = \;\frac{\partial}{\partial t}(\rho\vec{\mathbf{v}}) \end{array} \tag{10.10}$$

We can expand the first bracketed term on the left-hand side of Equation 10.9 via the chain rule and simplify it, using the continuity equation (10.3), to give a more compact form of the momentum equation:

$$-\left[\frac{\partial\tau_{xx}}{\partial x}+\frac{\partial\tau_{yx}}{\partial y}+\frac{\partial\tau_{zx}}{\partial z}\right]-\frac{\partial P}{\partial x}+\rho g_x=\rho\frac{Dv_x}{Dt} \tag{10.11}$$

Force per unit volume = Mass per unit volume × Acceleration

where we define the *substantial derivative*, *D/Dt*, as

$$\frac{Dv_x}{Dt}=\text{Substantial derivative}=\left[\frac{\partial v_x}{\partial t}+v_x\frac{\partial v_x}{\partial x}+v_y\frac{\partial v_x}{\partial y}+v_z\frac{\partial v_x}{\partial z}\right] \tag{10.12}$$

The form of Equation 10.11 is an exact analog of Newton's law of motion. Each term on the left-hand side has the units of force/unit volume. The term on the right-hand side is the mass per unit volume, ρ, multiplied by the acceleration where the substantial derivative, Dv_x/Dt, represents the total acceleration of the fluid through the control volume. The substantial derivative contains terms in addition to the change in velocity with time, $\partial v_x/\partial t$, because a fluid element can accelerate by changing positions within a velocity field. As the fluid element moves to a new position within the field, it takes on a new velocity. Had we rode atop the fluid element as it moved through the control volume, the acceleration of that component (from our new vantage point) would have just been the change of velocity with time. We term the latter viewpoint *Lagrangian* and the viewpoint used to derive Equation 10.11, where we sit at a fixed location and watch the world go by, as *Eulerian*.

The complete set of momentum equations can be written, compactly, in vector form as

$$-\vec{\nabla}P - [\vec{\nabla}\cdot\vec{\tau}] + \rho\vec{\mathbf{g}} = \rho\frac{D\vec{\mathbf{v}}}{Dt} \tag{10.13}$$

where

$\rho\frac{D\vec{\mathbf{v}}}{Dt}$ is the product of mass per unit volume and acceleration
$\vec{\nabla}P$ is the pressure force on element per unit volume
$[\vec{\nabla}\cdot\vec{\tau}]$ is the viscous force on element per unit volume
$\rho\vec{\mathbf{g}}$ is the gravitational or body force on element per unit volume

Newtonian fluids—to use the equations of motion, we need some constitutive relation that ties the shear stress to the velocity. In Chapter 2, we discussed these relationships in detail for Newtonian fluids. In that chapter, we related the stress components to the viscosity of the fluid and the velocity gradient. For completeness, we repeat these relationships here for rectangular coordinates:

$$\tau_{xx} = -2\mu\frac{\partial v_x}{\partial x} + \frac{2}{3}\mu(\vec{\nabla}\cdot\vec{\mathbf{v}}) \tag{10.14}$$

$$\tau_{yy} = -2\mu\frac{\partial v_y}{\partial y} + \frac{2}{3}\mu(\vec{\nabla}\cdot\vec{\mathbf{v}}) \tag{10.15}$$

$$\tau_{zz} = -2\mu\frac{\partial v_z}{\partial z} + \frac{2}{3}\mu(\vec{\nabla}\cdot\vec{\mathbf{v}}) \tag{10.16}$$

$$\tau_{xy} = \tau_{yx} = -\mu\left(\frac{\partial v_x}{\partial y} + \frac{\partial v_y}{\partial x}\right) \tag{10.17}$$

$$\tau_{yz} = \tau_{zy} = -\mu\left(\frac{\partial v_z}{\partial y} + \frac{\partial v_y}{\partial z}\right) \tag{10.18}$$

$$\tau_{xz} = \tau_{zx} = -\mu\left(\frac{\partial v_z}{\partial x} + \frac{\partial v_x}{\partial z}\right) \tag{10.19}$$

In general, the action of a shear stress, τ_{xy}, τ_{xz}, or τ_{yz}, is to try and cause some rotation of the fluid. This is evidenced by the cross-derivatives in its definition [1].

The equations we have so far derived provide a complete description of fluid motion. In instances where we assume the fluid is an incompressible, Newtonian fluid with constant properties, the continuity and momentum equations become

$$\vec{\nabla} \cdot \vec{\mathbf{v}} = 0 \tag{10.20}$$

$$-\frac{1}{\rho}\vec{\nabla}P + \nu\nabla^2\vec{\mathbf{v}} + \vec{\mathbf{g}} = \frac{D\vec{\mathbf{v}}}{Dt} \tag{10.21}$$

where ν is the kinematic viscosity. Tables 10.2 and 10.3 present the momentum equation in rectangular, cylindrical, and spherical coordinates. Table 10.2 is a general statement, while Table 10.3 is restricted to Newtonian fluids with constant properties.

TABLE 10.2
Momentum Equations: Stress Formulation

Momentum Equation: Cartesian Coordinates

x $$-\left(\frac{\partial \tau_{xx}}{\partial x} + \frac{\partial \tau_{yx}}{\partial y} + \frac{\partial \tau_{zx}}{\partial z}\right) - \frac{\partial P}{\partial x} + \rho g_x = \rho\left(\frac{\partial v_x}{\partial t} + v_x\frac{\partial v_x}{\partial x} + v_y\frac{\partial v_x}{\partial y} + v_z\frac{\partial v_x}{\partial z}\right)$$

y $$-\left(\frac{\partial \tau_{xy}}{\partial x} + \frac{\partial \tau_{yy}}{\partial y} + \frac{\partial \tau_{zy}}{\partial z}\right) - \frac{\partial P}{\partial y} + \rho g_y = \rho\left(\frac{\partial v_y}{\partial t} + v_x\frac{\partial v_y}{\partial x} + v_y\frac{\partial v_y}{\partial y} + v_z\frac{\partial v_y}{\partial z}\right)$$

z $$-\left(\frac{\partial \tau_{xz}}{\partial x} + \frac{\partial \tau_{yz}}{\partial y} + \frac{\partial \tau_{zz}}{\partial z}\right) - \frac{\partial P}{\partial z} + \rho g_z = \rho\left(\frac{\partial v_z}{\partial t} + v_x\frac{\partial v_z}{\partial x} + v_y\frac{\partial v_z}{\partial y} + v_z\frac{\partial v_z}{\partial z}\right)$$

Momentum Equation: Cylindrical Coordinates

r $$-\left(\frac{1}{r}\frac{\partial}{\partial r}(r\tau_{rr}) + \frac{1}{r}\frac{\partial \tau_{\theta r}}{\partial \theta} - \frac{\tau_{\theta\theta}}{r} + \frac{\partial \tau_{zr}}{\partial z}\right) - \frac{\partial P}{\partial r} + \rho g_r = \rho\left(\frac{\partial v_r}{\partial t} + v_r\frac{\partial v_r}{\partial r} + \frac{v_\theta}{r}\frac{\partial v_r}{\partial \theta} - \frac{v_\theta^2}{r} + v_z\frac{\partial v_r}{\partial z}\right)$$

θ $$-\left(\frac{1}{r^2}\frac{\partial}{\partial r}(r^2\tau_{r\theta}) + \frac{1}{r}\frac{\partial \tau_{\theta\theta}}{\partial \theta} + \frac{\partial \tau_{z\theta}}{\partial z}\right) - \frac{1}{r}\frac{\partial P}{\partial \theta} + \rho g_\theta = \rho\left(\frac{\partial v_\theta}{\partial t} + v_r\frac{\partial v_\theta}{\partial r} + \frac{v_\theta}{r}\frac{\partial v_\theta}{\partial \theta} + \frac{v_r v_\theta}{r} + v_z\frac{\partial v_\theta}{\partial z}\right)$$

z $$-\left(\frac{1}{r}\frac{\partial}{\partial r}(r\tau_{rz}) + \frac{1}{r}\frac{\partial \tau_{\theta z}}{\partial \theta} + \frac{\partial \tau_{zz}}{\partial z}\right) - \frac{\partial P}{\partial z} + \rho g_z = \rho\left(\frac{\partial v_z}{\partial t} + v_r\frac{\partial v_z}{\partial r} + \frac{v_\theta}{r}\frac{\partial v_z}{\partial \theta} + v_z\frac{\partial v_z}{\partial z}\right)$$

Momentum Equation: Spherical Coordinates

r $$-\left(\frac{1}{r^2}\frac{\partial}{\partial r}(r^2\tau_{rr}) + \frac{1}{r\sin\theta}\frac{\partial}{\partial \theta}(\tau_{\theta r}\sin\theta) - \frac{\tau_{\theta\theta}+\tau_{\phi\phi}}{r} + \frac{1}{r\sin\theta}\frac{\partial \tau_{\phi r}}{\partial \phi}\right) - \frac{\partial P}{\partial r} + \rho g_r$$
$$= \rho\left(\frac{\partial v_r}{\partial t} + v_r\frac{\partial v_r}{\partial r} + \frac{v_\theta}{r}\frac{\partial v_r}{\partial \theta} - \frac{v_\theta^2 + v_\phi^2}{r} + \frac{v_\phi}{r\sin\theta}\frac{\partial v_r}{\partial \phi}\right)$$

θ $$-\left(\frac{1}{r^2}\frac{\partial}{\partial r}(r^2\tau_{r\theta}) + \frac{1}{r\sin\theta}\frac{\partial}{\partial \theta}(\tau_{\theta\theta}\sin\theta) + \frac{1}{r\sin\theta}\frac{\partial \tau_{\phi\theta}}{\partial \phi} + \frac{\tau_{r\theta}}{r} - \frac{\cot\theta}{r}\tau_{\phi\phi}\right) - \frac{1}{r}\frac{\partial P}{\partial \theta} + \rho g_\theta$$
$$= \rho\left(\frac{\partial v_\theta}{\partial t} + v_r\frac{\partial v_\theta}{\partial r} + \frac{v_\theta}{r}\frac{\partial v_\theta}{\partial \theta} + \frac{v_r v_\theta}{r} - \frac{v_\phi^2\cot\theta}{r} + \frac{v_\phi}{r\sin\theta}\frac{\partial v_\theta}{\partial \phi}\right)$$

ϕ $$-\left(\frac{1}{r^2}\frac{\partial}{\partial r}(r^2\tau_{r\phi}) + \frac{1}{r}\frac{\partial \tau_{\theta\phi}}{\partial \theta} + \frac{1}{r\sin\theta}\frac{\partial \tau_{\phi\phi}}{\partial \phi} + \frac{\tau_{r\phi}}{r} + \frac{2\cot\theta}{r}\tau_{\theta\phi}\right) - \frac{1}{r\sin\theta}\frac{\partial P}{\partial \phi} + \rho g_\phi$$
$$= \rho\left(\frac{\partial v_\phi}{\partial t} + v_r\frac{\partial v_\phi}{\partial r} + \frac{v_\theta}{r}\frac{\partial v_\phi}{\partial \theta} + \frac{v_r v_\phi}{r} + \frac{v_\theta v_\phi}{r}\cot\theta + \frac{v_\phi}{r\sin\theta}\frac{\partial v_\phi}{\partial \phi}\right)$$

TABLE 10.3
Momentum Equations: Newtonian Fluids, Constant Properties

	Momentum Equation: Cartesian Coordinates
x	$\mu\left[\frac{\partial^2 v_x}{\partial x^2}+\frac{\partial^2 v_x}{\partial y^2}+\frac{\partial^2 v_x}{\partial z^2}\right]-\frac{\partial P}{\partial x}+\rho g_x=\rho\left(\frac{\partial v_x}{\partial t}+v_x\frac{\partial v_x}{\partial x}+v_y\frac{\partial v_x}{\partial y}+v_z\frac{\partial v_x}{\partial z}\right)$
y	$\mu\left[\frac{\partial^2 v_y}{\partial x^2}+\frac{\partial^2 v_y}{\partial y^2}+\frac{\partial^2 v_y}{\partial z^2}\right]-\frac{\partial P}{\partial y}+\rho g_y=\rho\left(\frac{\partial v_y}{\partial t}+v_x\frac{\partial v_y}{\partial x}+v_y\frac{\partial v_y}{\partial y}+v_z\frac{\partial v_y}{\partial z}\right)$
z	$\mu\left[\frac{\partial^2 v_z}{\partial x^2}+\frac{\partial^2 v_z}{\partial y^2}+\frac{\partial^2 v_z}{\partial z^2}\right]-\frac{\partial P}{\partial z}+\rho g_z=\rho\left(\frac{\partial v_z}{\partial t}+v_x\frac{\partial v_z}{\partial x}+v_y\frac{\partial v_z}{\partial y}+v_z\frac{\partial v_z}{\partial z}\right)$
	Momentum Equation: Cylindrical Coordinates
r	$\mu\left[\frac{\partial}{\partial r}\left(\frac{1}{r}\frac{\partial}{\partial r}(rv_r)\right)+\frac{1}{r^2}\frac{\partial^2 v_r}{\partial\theta^2}-\frac{2}{r^2}\frac{\partial v_\theta}{\partial\theta}+\frac{\partial^2 v_r}{\partial z^2}\right]-\frac{\partial P}{\partial r}+\rho g_r=\rho\left(\frac{\partial v_r}{\partial t}+v_r\frac{\partial v_r}{\partial r}+\frac{v_\theta}{r}\frac{\partial v_r}{\partial\theta}-\frac{v_\theta^2}{r}+v_z\frac{\partial v_r}{\partial z}\right)$
θ	$\mu\left[\frac{\partial}{\partial r}\left(\frac{1}{r}\frac{\partial}{\partial r}(rv_\theta)\right)+\frac{1}{r^2}\frac{\partial^2 v_\theta}{\partial\theta^2}+\frac{2}{r^2}\frac{\partial v_r}{\partial\theta}+\frac{\partial^2 v_\theta}{\partial z^2}\right]-\frac{1}{r}\frac{\partial P}{\partial\theta}+\rho g_\theta=\rho\left(\frac{\partial v_\theta}{\partial t}+v_r\frac{\partial v_\theta}{\partial r}+\frac{v_\theta}{r}\frac{\partial v_\theta}{\partial\theta}+\frac{v_r v_\theta}{r}+v_z\frac{\partial v_\theta}{\partial z}\right)$
z	$\mu\left[\frac{1}{r}\frac{\partial}{\partial r}\left(r\frac{\partial v_z}{\partial r}\right)+\frac{1}{r^2}\frac{\partial^2 v_z}{\partial\theta^2}+\frac{\partial^2 v_z}{\partial z^2}\right]-\frac{\partial P}{\partial z}+\rho g_z=\rho\left(\frac{\partial v_z}{\partial t}+v_r\frac{\partial v_z}{\partial r}+\frac{v_\theta}{r}\frac{\partial v_z}{\partial\theta}+v_z\frac{\partial v_z}{\partial z}\right)$
	Momentum Equation: Spherical Coordinates
r	$\frac{\mu}{r^2}\left[\frac{\partial^2}{\partial r^2}(r^2 v_r)+\frac{1}{\sin\theta}\frac{\partial}{\partial\theta}\left(\sin\theta\frac{\partial v_r}{\partial\theta}\right)+\frac{1}{\sin^2\theta}\frac{\partial^2 v_r}{\partial\phi^2}\right]-\frac{\partial P}{\partial r}+\rho g_r$ $=\rho\left(\frac{\partial v_r}{\partial t}+v_r\frac{\partial v_r}{\partial r}+\frac{v_\theta}{r}\frac{\partial v_r}{\partial\theta}-\frac{v_\theta^2+v_\phi^2}{r}+\frac{v_\phi}{r\sin\theta}\frac{\partial v_r}{\partial\phi}\right)$
θ	$\frac{\mu}{r^2}\left[\frac{\partial}{\partial r}\left(r^2\frac{\partial v_\theta}{\partial r}\right)+\frac{\partial}{\partial\theta}\left(\frac{1}{\sin\theta}\frac{\partial}{\partial\theta}(v_\theta\sin\theta)\right)+\frac{1}{\sin^2\theta}\frac{\partial^2 v_\theta}{\partial\phi^2}+2\frac{\partial v_r}{\partial\theta}-\frac{2\cos\theta}{\sin^2\theta}\frac{\partial v_\phi}{\partial\phi}\right]-\frac{1}{r}\frac{\partial P}{\partial\theta}+\rho g_\theta$ $=\rho\left(\frac{\partial v_\theta}{\partial t}+v_r\frac{\partial v_\theta}{\partial r}+\frac{v_\theta}{r}\frac{\partial v_\theta}{\partial\theta}+\frac{v_r v_\theta}{r}-\frac{v_\phi^2}{r}\cot\theta+\frac{v_\phi}{r\sin\theta}\frac{\partial v_\theta}{\partial\phi}\right)$
ϕ	$\frac{\mu}{r^2}\left[\frac{\partial}{\partial r}\left(r^2\frac{\partial v_\phi}{\partial r}\right)+\frac{\partial}{\partial\theta}\left(\frac{1}{\sin\theta}\frac{\partial}{\partial\theta}(v_\phi\sin\theta)\right)+\frac{1}{\sin^2\theta}\frac{\partial^2 v_\phi}{\partial\phi^2}+\frac{2}{\sin\theta}\frac{\partial v_r}{\partial\phi}+\frac{2\cos\theta}{\sin^2\theta}\frac{\partial v_\theta}{\partial\phi}\right]-\frac{1}{r\sin\theta}\frac{\partial P}{\partial\phi}+\rho g_\phi$ $=\rho\left(\frac{\partial v_\phi}{\partial t}+v_r\frac{\partial v_\phi}{\partial r}+\frac{v_\theta}{r}\frac{\partial v_\phi}{\partial\theta}+\frac{v_r v_\phi}{r}+\frac{v_\theta v_\phi}{r}\cot\theta+\frac{v_\phi}{r\sin\theta}\frac{\partial v_\phi}{\partial\phi}\right)$

Example 10.3 Shape of the Surface of a Rotating Liquid

Here, we consider what happens to the shape of a liquid's surface if it is contained in a rotating vessel. Anyone who has tried this experiment knows that the fluid rises at the edges where the container is rotating and falls toward the center. We will now show that the shape of the surface is a paraboloid. The physical situation is shown in Figure 10.4.

We assume that the whole tank of fluid is rotating about its axis at a frequency of ω_o revolutions per second. Since the fluid has been rotating for some time, we will assume that it has reached its steady-state profile. We will also assume that the fluid is Newtonian and that all fluid properties

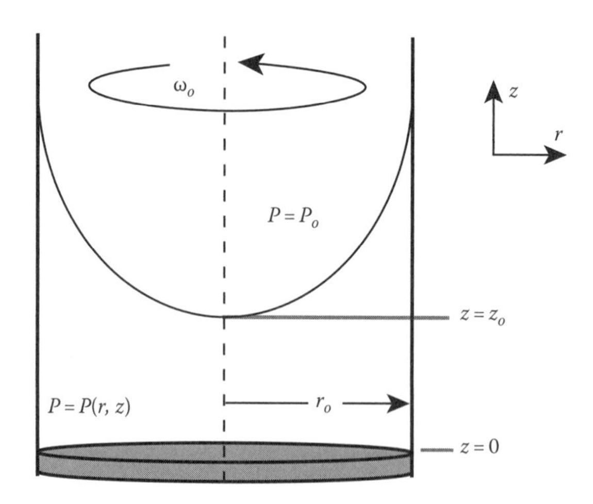

FIGURE 10.4 Shape of the surface of a rotating liquid.

are constant. Once the profile is steady, the fluid will exhibit only an angular velocity, v_θ, that depends upon how far away from the center we measure it. Any velocity in the radial or axial direction will long since have decayed to zero. The pressure in the fluid will be a function of axial position due to the weight of the fluid, and there will also be a radial component of the pressure since the act of rotating the fluid thrusts the fluid out toward the walls of the vessel.

To begin solving this problem, we first pick the most appropriate coordinate system, cylindrical in this case. We can refer to Table 10.3 and begin simplifying the momentum equations found there since we have derived them for an arbitrary problem. Taking the r-component first, we know that $v_r = 0$, $v_z = 0$, and $g_r = 0$. Furthermore, we also know that there is no variation of pressure or velocity in the angular direction; we are symmetric, so $(\partial/\partial\theta = 0)$. At steady state, the only terms remaining involve the variation of pressure with radial position and the *centrifugal force*, $\rho v_\theta^2/r$. The centrifugal force thrusts the fluid outward and results in a pressure gradient in the radial direction:

$$\rho\frac{v_\theta^2}{r} = \frac{\partial P}{\partial r}$$

Considering the θ-component, we have no component of gravity in that direction, so $g_\theta = 0$. We also have symmetry in the θ-direction, and under steady-state conditions, we know that $v_\theta = f(r)$ only. The remaining terms in the θ-momentum equation are

$$0 = \mu\frac{\partial}{\partial r}\left(\frac{1}{r}\frac{\partial}{\partial r}(rv_\theta)\right)$$

Finally, turning our attention to the z-component, the only terms remaining involve the change in pressure and the body force due to gravity. Pressure in the z-direction is due solely to the weight of the fluid:

$$0 = -\frac{\partial P}{\partial z} - \rho g_z$$

Our solution procedure will be to solve for v_θ first and then determine the pressure. Since the pressure depends upon r and z, we cannot integrate the z-component of the momentum equation without knowing v_θ and hence the centrifugal force first. Separating and integrating the θ-momentum equation for v_θ gives

$$v_\theta = \frac{a_o}{2}r + \frac{a_1}{r}$$

To fully evaluate v_θ, we need two boundary conditions. At the center of the tank, we know that v_θ cannot be infinite. In fact, it must be zero by symmetry, so $a_1 = 0$. At the edge of the tank, the linear velocity, v_θ, is given by the product of the distance traveled per revolution (perimeter) times the number of revolutions per second:

$$r = r_o \qquad v_\theta = 2\pi r_o \omega_o$$

and so the final solution for v_θ is

$$v_\theta = 2\pi\omega_o r$$

Now we can turn our attention to the pressure. The pressure is defined by the following two momentum equations:

$$\frac{\partial P}{\partial r} = 4\pi^2\omega_o^2\rho r \qquad \frac{\partial P}{\partial z} = -\rho g$$

Since pressure is a state variable of our system, it is an analytic function of position. The total change in pressure can be written in terms of a change in pressure in the radial direction multiplied by a small change in radial position plus a change in pressure in the vertical direction multiplied by a small change in height:

$$dP = \left(\frac{\partial P}{\partial r}\right)_z dr + \left(\frac{\partial P}{\partial z}\right)_r dz$$

Substituting our expressions for the derivatives of P and integrating gives

$$P = -\rho g z + 2\pi^2\omega_o^2\rho r^2 + a_2$$

The integration constant, a_2, is found by knowing the height of the fluid and pressure at any one radial position. We set $z = z_o$ and $P = P_o$ when $r = 0$, for example. This gives

$$a_2 = P_o + \rho g z_o$$

$$P - P_o = -\rho g(z - z_o) + 2\pi^2\omega_o^2\rho r^2$$

The surface of the fluid is always in contact with the surrounding atmosphere and so is always at a pressure of P_o. The locus of all points whose pressure is P_o gives the shape of the surface, in this case, a parabola.

$$z - z_o = \frac{2\pi^2\omega_o^2 r^2}{g}$$

Finally, the actual value of z_o can be determined from conservation of mass. Since we know the initial volume of fluid in the tank, $V_o = \pi r_o^2 H$, before we began rotating it, integrating the volume of fluid in the tank using the new shape and setting it equal to the initial volume yields z_o:

$$z_o = H - \frac{(\pi\omega_o r_o)^2}{g}$$

$$z = H + \frac{\pi^2\omega_o^2}{g}\left(2r^2 - r_o^2\right)$$

We can simulate the shape using finite element software and see how it compares, at least qualitatively, with the analytic solution. Figure 10.5 shows the results where the parabolic free surface shape is evident.

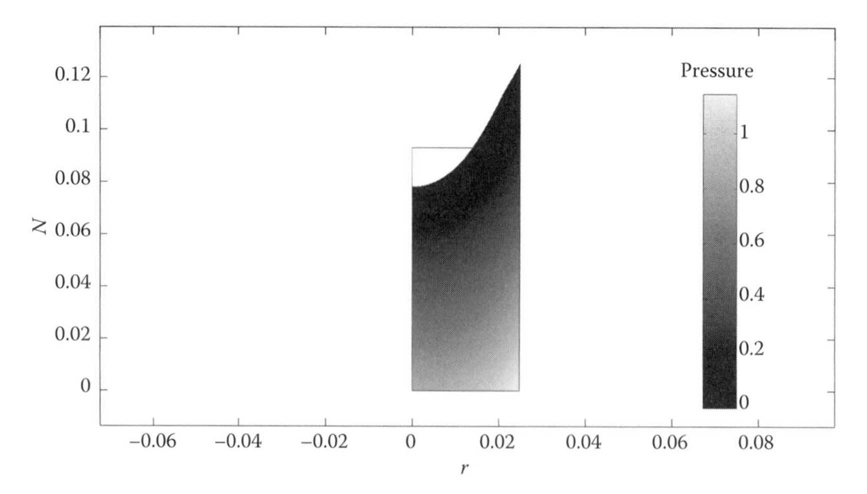

FIGURE 10.5 Shape of the surface of a rotating fluid. Comsol® solution showing the parabolic profile and the pressure distribution throughout the fluid volume.

Example 10.4 Creeping Flow about a Stationary Sphere

One of the classic examples of a solution to the momentum equations is creeping flow about a sphere. The problem was first solved by Stokes in 1853 [2] who was concerned about the time-keeping ability and run time of pendulums. Other more recent solutions can be found in Lamb [3], and the solution presented here follows that presented by Milne-Thompson [4]. Figure 10.6 shows the coordinate system used to describe the problem.

The sphere is suspended in the liquid and flow comes from below and is in the $+z$-direction. Since the sphere is motionless, the flow about the sphere is symmetric in the ϕ-direction, and the problem is 2-D. We assume a Newtonian, incompressible fluid for simplicity. In Chapter 9, we introduced the stream function, ψ, for 2-D problems that allowed us to satisfy the continuity equation. Use of the stream function also allows us to combine the two momentum equations into one, thereby eliminating the pressure. In this instance, we begin

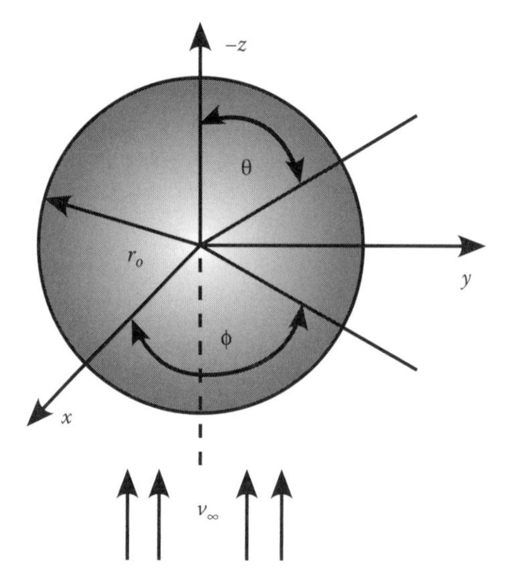

FIGURE 10.6 Creeping flow around a sphere.

with the momentum equations removing all reference to changes with respect to ϕ and noting that $v_\phi = 0$:

$$\frac{\mu}{r^2}\left[\frac{\partial^2}{\partial r^2}(r^2 v_r) + \frac{1}{\sin\theta}\frac{\partial}{\partial\theta}\left(\sin\theta\frac{\partial v_r}{\partial\theta}\right)\right] - \frac{\partial P}{\partial r} + \rho g_r = \rho\left(\frac{\partial v_r}{\partial t} + v_r\frac{\partial v_r}{\partial r} + \frac{v_\theta}{r}\frac{\partial v_r}{\partial\theta} - \frac{v_\theta^2}{r}\right) \quad r\text{-component}$$

$$\frac{\mu}{r^2}\left[\frac{\partial}{\partial r}\left(r^2\frac{\partial v_\theta}{\partial r}\right) + \frac{\partial}{\partial\theta}\left(\frac{1}{\sin\theta}\frac{\partial}{\partial\theta}(v_\theta\sin\theta)\right) + 2\frac{\partial v_r}{\partial\theta}\right] - \frac{1}{r}\frac{\partial P}{\partial\theta} + \rho g_\theta$$

$$= \rho\left(\frac{\partial v_\theta}{\partial t} + v_r\frac{\partial v_\theta}{\partial r} + \frac{v_\theta}{r}\frac{\partial v_\theta}{\partial\theta} + \frac{v_r v_\theta}{r}\right) \quad \theta\text{-component}$$

In steady creeping flow, we have a situation where there is negligible acceleration of the fluid. This means that the substantial derivatives of v_r and v_θ are zero, and so the right-hand sides of the momentum equations can be neglected. This leaves us with

$$\frac{\mu}{r^2}\left[\frac{\partial^2}{\partial r^2}(r^2 v_r) + \frac{1}{\sin\theta}\frac{\partial}{\partial\theta}\left(\sin\theta\frac{\partial v_r}{\partial\theta}\right)\right] - \frac{\partial P}{\partial r} + \rho g_r = 0$$

$$\frac{\mu}{r^2}\left[\frac{\partial}{\partial r}\left(r^2\frac{\partial v_\theta}{\partial r}\right) + \frac{\partial}{\partial\theta}\left(\frac{1}{\sin\theta}\frac{\partial}{\partial\theta}(v_\theta\sin\theta)\right) + 2\frac{\partial v_r}{\partial\theta}\right] - \frac{1}{r}\frac{\partial P}{\partial\theta} + \rho g_\theta = 0$$

In this situation, $v_\phi = \partial/\partial\phi = 0$ and the stream function is defined by (Table 10.4)

$$v_r = -\frac{1}{r^2\sin\theta}\frac{\partial\psi}{\partial\theta} \qquad v_\theta = \frac{1}{r\sin\theta}\frac{\partial\psi}{\partial r}$$

You can verify that the stream function satisfies the following continuity equation:

$$\frac{1}{r^2}\frac{\partial}{\partial r}(r^2 v_r) + \frac{1}{r\sin\theta}\frac{\partial}{\partial\theta}(v_\theta\sin\theta) = 0$$

Using Table 10.4, we can obtain the stream function representation of the momentum equations directly. Alternatively, you can derive it yourself by substituting in for the velocity components, differentiating the r-momentum equation by θ, differentiating the θ-momentum equation by r, and then subtracting one from the other to remove the pressure and body force terms. Since we have no inertial terms and are at steady state, the stream function equation simplifies to

$$\nu\left[\frac{\partial^2}{\partial r^2}(D^2\psi) + \frac{\sin\theta}{r^2}\frac{\partial}{\partial\theta}\left(\frac{1}{\sin\theta}\frac{\partial}{\partial\theta}(D^2\psi)\right)\right] = 0 \qquad D^2\psi = \frac{\partial^2\psi}{\partial r^2} + \frac{1}{r^2}\frac{\partial^2\psi}{\partial\theta^2} - \frac{\cot\theta}{r^2}\frac{\partial\psi}{\partial\theta}$$

Notice that by using the stream function representation, we have canceled out any dependence on the pressure and also the gravitational body force since both g_r and g_θ are constants.

The boundary conditions for this problem specify no slip at the sphere surface

$$v_r = -\frac{1}{r^2\sin\theta}\frac{\partial\psi}{\partial\theta} = 0 \qquad v_\theta = \frac{1}{r\sin\theta}\frac{\partial\psi}{\partial r} = 0$$

TABLE 10.4
Stream Function: Velocity Component Definitions

Type of Flow	Velocity Component Definitions	Stream Function
Planar Cartesian		
$v_z = 0$	$v_x = -\dfrac{\partial\psi}{\partial y}$	$\nu\left[\dfrac{\partial^2}{\partial x^2}(\nabla^2\psi) + \dfrac{\partial^2}{\partial y^2}(\nabla^2\psi)\right] - \dfrac{\partial\psi}{\partial x}\dfrac{\partial}{\partial y}(\nabla^2\psi) + \dfrac{\partial\psi}{\partial y}\dfrac{\partial}{\partial x}(\nabla^2\psi) = \dfrac{\partial}{\partial t}(\nabla^2\psi)$
$\dfrac{\partial}{\partial z} = 0$	$v_y = \dfrac{\partial\psi}{\partial x}$	$\nabla^2\psi = \dfrac{\partial^2\psi}{\partial x^2} + \dfrac{\partial^2\psi}{\partial y^2}$
Planar cylindrical		
$v_z = 0$	$v_r = -\dfrac{1}{r}\dfrac{\partial\psi}{\partial\theta}$	$\nu\left[\dfrac{\partial^2}{\partial r^2}(\nabla^2\psi) + \dfrac{1}{r}\dfrac{\partial}{\partial r}(\nabla^2\psi) + \dfrac{1}{r^2}\dfrac{\partial^2}{\partial\theta^2}(\nabla^2\psi)\right]$
$\dfrac{\partial}{\partial z} = 0$	$v_\theta = \dfrac{\partial\psi}{\partial r}$	$-\dfrac{1}{r}\left[\dfrac{\partial\psi}{\partial r}\dfrac{\partial}{\partial\theta}(\nabla^2\psi) - \dfrac{\partial\psi}{\partial\theta}\dfrac{\partial}{\partial r}(\nabla^2\psi)\right] = \dfrac{\partial}{\partial t}(\nabla^2\psi)$
		$\nabla^2\psi = \dfrac{\partial^2\psi}{\partial r^2} + \dfrac{1}{r}\dfrac{\partial\psi}{\partial r} + \dfrac{1}{r^2}\dfrac{\partial^2\psi}{\partial\theta^2}$
Axisymmetric cylindrical		
$v_\theta = 0$	$v_r = \dfrac{1}{r}\dfrac{\partial\psi}{\partial z}$	$\nu\left[\dfrac{\partial^2}{\partial r^2}(D^2\psi) - \dfrac{1}{r}\dfrac{\partial}{\partial r}(D^2\psi) + \dfrac{\partial^2}{\partial z^2}(D^2\psi)\right]$
$\dfrac{\partial}{\partial\theta} = 0$	$v_z = -\dfrac{1}{r}\dfrac{\partial\psi}{\partial r}$	$-\dfrac{1}{r}\left[\dfrac{\partial\psi}{\partial r}\dfrac{\partial}{\partial z}(D^2\psi) - \dfrac{\partial\psi}{\partial z}\dfrac{\partial}{\partial r}(D^2\psi)\right]$
		$-\dfrac{2}{r^2}\dfrac{\partial\psi}{\partial z}(D^2\psi) = \dfrac{\partial}{\partial t}(D^2\psi)$
		$D^2\psi = \dfrac{\partial^2\psi}{\partial r^2} - \dfrac{1}{r}\dfrac{\partial\psi}{\partial r} + \dfrac{\partial^2\psi}{\partial z^2}$
Axisymmetric spherical		
$v_\phi = 0$	$v_r = -\dfrac{1}{r^2\sin\theta}\dfrac{\partial\psi}{\partial\theta}$	$\nu\left[\dfrac{\partial^2}{\partial r^2}(D^2\psi) + \dfrac{\sin\theta}{r^2}\dfrac{\partial}{\partial\theta}\left(\dfrac{1}{\sin\theta}\dfrac{\partial}{\partial\theta}(D^2\psi)\right)\right]$
$\dfrac{\partial}{\partial\phi} = 0$	$v_\theta = \dfrac{1}{r\sin\theta}\dfrac{\partial\psi}{\partial r}$	$+\dfrac{2}{r^2\sin^2\theta}\left[\dfrac{\partial\psi}{\partial r}\cos\theta - \dfrac{\sin\theta}{r}\dfrac{\partial\psi}{\partial\theta}\right](D^2\psi)$
		$-\dfrac{2}{r^2\sin\theta}\left[\dfrac{\partial\psi}{\partial r}\dfrac{\partial}{\partial\theta}(D^2\psi) - \dfrac{\partial\psi}{\partial\theta}\dfrac{\partial}{\partial r}(D^2\psi)\right] = \dfrac{\partial}{\partial t}(D^2\psi)$
		$D^2\psi = \dfrac{\partial^2\psi}{\partial r^2} + \dfrac{1}{r^2}\dfrac{\partial^2\psi}{\partial\theta^2} - \dfrac{\cot\theta}{r^2}\dfrac{\partial\psi}{\partial\theta}$

Source: Bird et al., *Transport Phenomena*, John Wiley & Sons, New York, 1960.

and that $v_z = v_\infty$ far away from the sphere ($r = \infty$). To write this last boundary condition in terms of the stream function, we must relate v_r and v_θ to v_z. From trigonometry, we know that

$$v_r = v_x \sin\theta\cos\phi + v_y \sin\theta\sin\phi + v_z\cos\theta$$

$$v_\theta = v_x \cos\theta\cos\phi + v_y \cos\theta\sin\phi - v_z\sin\theta$$

Far from the sphere, both v_x and v_y vanish, $v_z = v_\infty$, and we have

$$v_r = -\frac{1}{r^2 \sin\theta}\frac{\partial\psi}{\partial\theta} = v_\infty \cos\theta$$

$$v_\theta = \frac{1}{r\sin\theta}\frac{\partial\psi}{\partial r} = -v_\infty \sin\theta$$

Integrating either equation for the stream function, ψ, shows the third boundary condition to be

$$r \to \infty \qquad \psi = -\frac{1}{2} v_\infty r^2 \sin^2\theta$$

This final boundary condition suggests that the stream function is a function of $\sin^2\theta$ and that it might be useful to substitute

$$\psi = f(r)\sin^2\theta$$

into the differential equation and solve for the function $f(r)$. The result of this procedure is a fourth-order, ordinary differential equation for f:

$$\frac{d^4 f}{dr^4} - \frac{4}{r^2}\frac{d^2 f}{dr^2} + \frac{8}{r^3}\frac{df}{dr} - \frac{8}{r^4} f = 0$$

This equation is an example of an *Euler* equation [6] and can be solved by substituting $f = Cr^n$ into the differential equation. The substitution leaves us with a quartic algebraic equation to solve for the permissible exponents of r:

$$n^4 - 6n^3 + 7n^2 + 6n - 8 = (n-1)(n+1)(n-2)(n-4) = 0 \qquad n = -1, 1, 2, 4$$

The solution for ψ must have the form

$$\psi = \left[\frac{a_1}{r} + a_2 r + a_3 r^2 + a_4 r^4\right]\sin^2\theta$$

The boundary condition at $r = \infty$ shows that the fourth-order term must vanish ($a_4 = 0$) and that a_3 must be

$$a_3 = -\frac{1}{2} v_\infty$$

Our other two boundary conditions lead to two algebraic equations:

$$\frac{a_1}{r_o^3} + \frac{a_2}{r_o} + a_3 = 0 \qquad -\frac{a_1}{r_o^3} + \frac{a_2}{r_o} + 2a_3 = 0$$

$$a_2 = \frac{3}{4} v_\infty r_o \qquad a_1 = -\frac{1}{4} v_\infty r_o^3$$

The full solution for the stream function is now

$$\psi = \left[-\frac{v_\infty r_o^2}{4}\frac{r_o}{r} + \frac{3 v_\infty r_o}{4} r - \frac{v_\infty}{2} r^2\right]\sin^2\theta$$

with the following velocity components:

$$v_r = v_\infty \left[1 - \frac{3}{2}\left(\frac{r_o}{r}\right) + \frac{1}{2}\left(\frac{r_o}{r}\right)^3\right]\cos\theta$$

$$v_\theta = -v_\infty \left[1 - \frac{3}{4}\left(\frac{r_o}{r}\right) - \frac{1}{4}\left(\frac{r_o}{r}\right)^3\right]\sin\theta$$

Figure 10.7 compares the creeping flow value for the stream function with the ideal flow pattern we solved for in Chapter 9. Notice that the effect of viscosity is to increase awareness of the particle further into the fluid, that is, the disturbance caused by the particle is felt further out in the fluid stream. The flow will no longer be irrotational once viscosity is introduced.

We can also find the pressure distribution about the sphere and calculate the forces acting on it. We use the momentum equations and the velocity components to calculate the radial and angular pressure variations. Noting that the pressure is an analytic function of position, and like the surface of the rotating liquid, we can write the pressure as

$$dP = \left(\frac{\partial P}{\partial r}\right)_\theta dr + \left(\frac{\partial P}{\partial \theta}\right)_r d\theta$$

Substituting in from the momentum equations, integrating, and specifying that far away from the sphere, at the plane where $z = 0$, we have $P = P_o$, we find

$$P = P_o - \rho g z - \frac{3}{2}\frac{\mu v_\infty}{r_o}\left(\frac{r_o}{r}\right)^2 \cos\theta$$

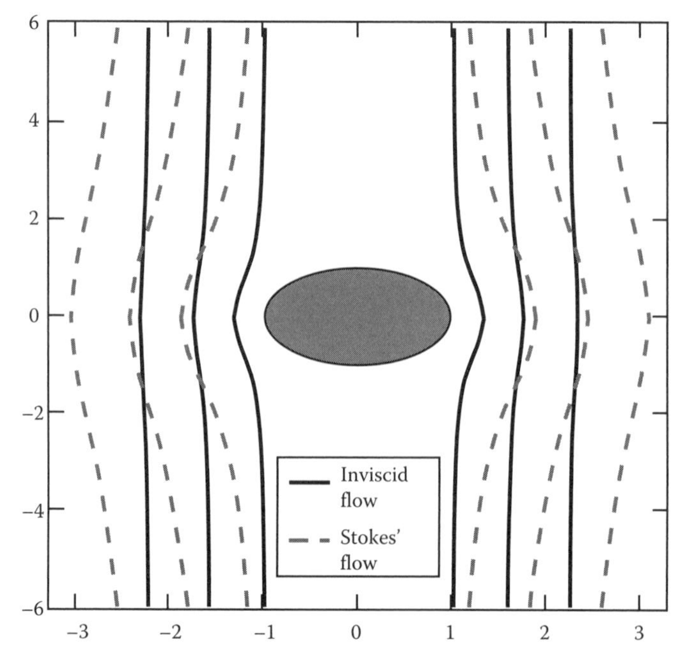

FIGURE 10.7 Streamlines for inviscid and Stokes' flow about a sphere.

If we look at the pressure distribution on the surface of the sphere ($z = r_o \cos\theta$) a bit more closely, we can determine where the maximum and minimum pressure points lie. Taking the derivative of pressure as a function of angle yields

$$\frac{dP}{d\theta} = \left[\rho g r_o + \frac{3}{2}\frac{\mu v_\infty}{r_o}\left(\frac{r_o}{r}\right)^2\right]\sin\theta = 0$$

and by setting this equation equal to 0, we can readily see that at $\theta = 0,\pi$ we have the minimum and maximum pressure, respectively. The point where the pressure is a maximum, $\theta = \pi$, is called the stagnation point and $v_\theta = 0$ there. The pressure is highest there because the radial velocity v_r opposes the bulk flow, v_∞. At the opposite point, $\theta = 0$, the pressure is lowest because the radial velocity, v_r, is in the same direction as v_∞. When $\theta = \pi/2$, the radial velocity vanishes and this is the critical point where we change from a favorable pressure gradient (v_r opposes v_∞) to an unfavorable one (v_r assists v_∞).

The pressure on the surface of the sphere exerts a force perpendicular to the surface. The z-component of that force (force in the direction of mean flow) is $-P\cos\theta$, and integration of that force over the surface area of the sphere will give us the total normal force experienced by the sphere:

$$F_\perp = -\int_0^{2\pi}\int_0^{\pi}\left[P_o\cos\theta - \rho g r_o\cos^2\theta - \frac{3}{2}\frac{\mu v_\infty}{r_o}\cos^2\theta\right] r_o^2\sin\theta\, d\theta\, d\phi$$

We can separate this expression into three integrals. The first integral involving P_o vanishes. The second integral involving the weight of the fluid gives us the buoyant force exerted on the sphere. Finally, the last integral involving the velocity of the fluid gives us what is often referred to as the *profile*, or *form* drag since it depends on the shape or form of the object. The total normal force is

$$F_\perp = \underbrace{\frac{4}{3}\pi r_o^3\rho g}_{\text{Buoyant force}} + \underbrace{2\pi\mu r_o v_\infty}_{\text{Form drag}}$$

Each point on the surface of the sphere also has a tangential force acting on it. This tangential force occurs due to the viscous stresses generated by having the fluid flow past the sphere and having the fluid stick to the surface of the sphere. The viscous stress acting here is $-\tau_{r\theta}$ and acts in the θ-direction. The z-component of this stress (i.e., in the direction of flow) is $(-\tau_{r\theta})(-\sin\theta)$. Integration of the stress over the surface of the sphere will yield the total tangential force. This force is often referred to as the *friction* or *skin-friction drag*:

$$F_\| = \int_0^{2\pi}\int_0^{\pi}[\tau_{r\theta}\sin\theta]\, r_o^2\sin\theta\, d\theta\, d\phi$$

$$= \int_0^{2\pi}\int_0^{\pi}\left[\frac{3}{2}\frac{\mu v_\infty}{r_o}\sin^2\theta\right] r_o^2\sin\theta\, d\theta\, d\phi = 4\pi\mu r_o v_\infty$$

The total force acting on the sphere is

$$F = \underbrace{\frac{4}{3}\pi r_o^3\rho g}_{\text{Buoyant force}} + \underbrace{2\pi\mu r_o v_\infty}_{\text{Form drag}} + \underbrace{4\pi\mu r_o v_\infty}_{\text{Friction drag}}$$

We can combine both drag forces into a single, total force often termed profile drag:

$$F = 6\pi\mu r_o v_\infty \qquad \text{Stokes' law}$$

The expression we just derived is called Stokes' law, originally derived to account for a pendulum's slowing and the decreasing amplitude of its swing. Stokes' law is valid only for very slow flows where the inertial forces in the fluid are negligible. Once the inertial terms (nonlinear terms) of the Navier–Stokes equations become important, the drag force is higher than that predicted by Stokes' law. Stokes' law is often written in terms of a drag coefficient, C_D, where the force is defined by

$$F = \left(\frac{1}{2}\rho v_\infty^2\right)\left(\pi r_o^2\right)C_D = 6\pi\mu r_o v_\infty$$

$$C_D = \frac{12\mu}{\rho v_\infty r_o} = \frac{24}{\text{Re}_D} \qquad \text{Re}_D = \frac{v_\infty D}{\nu} = \frac{\text{Inertial forces}}{\text{Viscous forces}}$$

Here, we use the projected area $\left(\pi r_o^2\right)$ in the direction of the flow, and Re_D is the Reynolds number, a quantity that provides a measure of the ratio of inertial to viscous forces in the fluid.

Example 10.5 Potential Flow and Lift on a Cylinder

The solution for flow about a sphere has a buoyant force but no lift. This to be expected since the Kutta–Joukowski theorem shows that circulation about the cylinder is needed to generate lift. We can use our potential flow solution that includes circulation along with the Navier–Stokes equations to calculate the lift. The stream function for flow about a rotating cylinder can be written as

$$\psi = v_\infty r\sin\theta - r_o^2 v_\infty\left(\frac{\sin\theta}{r}\right) + r_o v_\infty \ln\left(\frac{r}{r_o}\right)$$

According to the definitions of the stream function, the velocity components are

$$v_r = \frac{1}{r}\left(\frac{\partial\psi}{\partial\theta}\right) = v_\infty\left(1 - \frac{r_o^2}{r^2}\right)\cos\theta$$

$$v_\theta = -\left(\frac{\partial\psi}{\partial r}\right) = -v_\infty\left[\frac{r_o}{r} + \left(1 + \frac{r_o^2}{r^2}\right)\sin\theta\right]$$

We can obtain the pressure about the cylinder by revisiting the Navier–Stokes equation for the potential flow case. In this situation, all the terms containing viscosity in the momentum disappear, and we end up with a set of equations that, when coupled with the continuity equation, is normally referred to as the Euler equations:

$$\rho\left(\frac{\partial v_r}{\partial t} + v_r\frac{\partial v_r}{\partial r} + \frac{v_\theta}{r}\frac{\partial v_r}{\partial\theta} - \frac{v_\theta^2}{r} + v_z\frac{\partial v_r}{\partial z}\right) = -\frac{\partial P}{\partial r} + \rho g_r$$

$$\rho\left(\frac{\partial v_\theta}{\partial t} + v_r\frac{\partial v_\theta}{\partial r} + \frac{v_\theta}{r}\frac{\partial v_\theta}{\partial\theta} + \frac{v_r v_\theta}{r} + v_z\frac{\partial v_\theta}{\partial z}\right) = -\frac{1}{r}\frac{\partial P}{\partial\theta} + \rho g_\theta$$

$$\rho\left(\frac{\partial v_z}{\partial t} + v_r\frac{\partial v_z}{\partial r} + \frac{v_\theta}{r}\frac{\partial v_z}{\partial\theta} + v_z\frac{\partial v_z}{\partial z}\right) = -\frac{\partial P}{\partial z} + \rho g_z$$

If the flow depends only on r and θ, we only have two pressure components that contribute to the lift:

$$\rho\left(v_r\frac{\partial v_r}{\partial r}+\frac{v_\theta}{r}\frac{\partial v_r}{\partial \theta}-\frac{v_\theta^2}{r}\right)=-\frac{\partial P}{\partial r}$$

$$\rho\left(v_r\frac{\partial v_\theta}{\partial r}+\frac{v_\theta}{r}\frac{\partial v_\theta}{\partial \theta}+\frac{v_r v_\theta}{r}\right)=-\frac{1}{r}\frac{\partial P}{\partial \theta}$$

and we can determine the pressure components directly from the velocities

$$\frac{\partial P}{\partial r}=-\rho\left(v_r\frac{\partial v_r}{\partial r}+\frac{v_\theta}{r}\frac{\partial v_r}{\partial \theta}-\frac{v_\theta^2}{r}\right)$$

$$\frac{\partial P}{\partial \theta}=-\rho r\left(v_r\frac{\partial v_\theta}{\partial r}+\frac{v_\theta}{r}\frac{\partial v_\theta}{\partial \theta}+\frac{v_r v_\theta}{r}\right)$$

To calculate the lift, we need the pressure on the surface of the object. As in the surface shape problem, the pressure can be determined from

$$dP=\left(\frac{\partial P}{\partial r}\right)dr+\left(\frac{\partial P}{\partial \theta}\right)d\theta$$

Since we are interested in the surface pressure, $dr = 0$ and we can integrate to obtain

$$P(r_o)=P_\infty-2\rho v_\infty^2[\sin(\theta)+\sin^2(\theta)]$$

The lift force per unit length of cylinder is now the line integral of the y-component of the pressure over the surface of the cylinder:

$$F_L\left(\frac{\mathrm{N}}{\mathrm{m}}\right)=\int_0^{2\pi}-2r_o P\sin\theta d\theta=-\int_0^{2\pi}2r_o\left\{P_\infty-2\rho v_\infty^2[\sin(\theta)+\sin^2(\theta)]\right\}\sin\theta d\theta$$

$$=2\pi r_o\rho v_\infty^2$$

If we wish to keep an object from falling under its own weight, the lift force must be equal to the gravitational force acting on the object.

Example 10.6 Lubrication Flow

Very often we encounter flows where the fluid is either confined by a thin gap between surfaces or is flowing as a thin film over a surface. Examples include the layer of water between an ice skate and the ice, the synovial fluid in joints, the tear fluid in an eye, and the oil that lubricates the moving parts of an internal combustion engine. The latter example is why we refer to such situations as lubrication flows. Lubrication flows are normally characterized by incompressible fluids confined to thin gaps. The fluid flows at relatively small velocity and so the inertial terms in the Navier–Stokes equations are insignificant compared to the diffusive terms. Since the gaps or films are so thin, lubrication flows can be treated as 2-D. A classic example is shown in Figure 10.8.

To understand the flow, we must solve the continuity, x-momentum, and y-momentum equations:

$$\frac{\partial u}{\partial x}+\frac{\partial v}{\partial y}=0 \quad \text{Continuity}$$

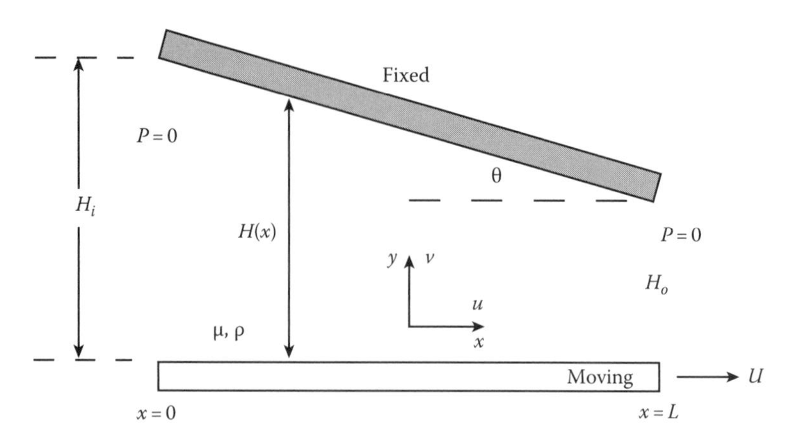

FIGURE 10.8 A lubrication flow occurring between the fixed and moving components of a journal bearing.

$$\rho\left(u\frac{\partial u}{\partial x}+v\frac{\partial u}{\partial y}\right)=-\frac{\partial P}{\partial x}+\mu\left(\frac{\partial^2 u}{\partial x^2}+\frac{\partial^2 u}{\partial y^2}\right)\quad x\text{-momentum}$$

$$\rho\left(u\frac{\partial v}{\partial x}+v\frac{\partial v}{\partial y}\right)=-\frac{\partial P}{\partial y}+\mu\left(\frac{\partial^2 v}{\partial x^2}+\frac{\partial^2 v}{\partial y^2}\right)\quad y\text{-momentum}$$

If the inertial terms are negligible, the momentum equations reduce to

$$\frac{\partial P}{\partial x}=\mu\left(\frac{\partial^2 u}{\partial x^2}+\frac{\partial^2 u}{\partial y^2}\right)$$

$$\frac{\partial P}{\partial y}=\mu\left(\frac{\partial^2 v}{\partial x^2}+\frac{\partial^2 v}{\partial y^2}\right)$$

Finally, if the gap or film is very thin, as it is in the orientation of Figure 10.8, there is little effect of gravity and so it has been neglected.

The velocity in the y-direction, v, is zero at the surfaces of both plates, and since the gap is so thin, $u \gg v$, rendering the y-momentum equation:

$$\frac{\partial P}{\partial y}=0$$

The thin gap approximation also means that the change in x-velocity in the y-direction is much larger than the corresponding change in the x-direction. This simplifies the x-momentum equation to:

$$\frac{\partial P}{\partial x}=\mu\frac{\partial^2 u}{\partial y^2}$$

and since the gap is so thin, we can assume that $\partial P/\partial x \neq f(y)$. We can integrate the x-momentum equation first and solve for u:

$$u=\left(\frac{1}{2\mu}\right)\frac{\partial P}{\partial x}y^2+C_o y+C_1$$

The boundary conditions specify that

$$y = 0 \qquad u = U_o$$
$$y = H(x) \qquad u = 0$$

and so the final form for u is

$$u = U_o + \left(\frac{1}{2\mu}\right)\frac{\partial P}{\partial x}y^2 - \left(\frac{U_o}{H(x)}\right)y - \left(\frac{H(x)}{2\mu}\right)\frac{\partial P}{\partial x}y$$

We know $H(x)$ from the geometry of the system, but must solve for $\partial P/\partial x$. We can determine $\partial P/\partial x$ from the constant volumetric flow rate of fluid passing through the bearing:

$$\dot{V} = \int_0^{H(x)} u\,dy = U_o y + \left(\frac{1}{6\mu}\frac{\partial P}{\partial x}\right)y^3 - \left(\frac{U_o}{2H(x)} + \frac{H(x)}{4\mu}\frac{\partial P}{\partial x}\right)y^2\Bigg|_0^{H(x)}$$

$$= \left(\frac{U_o}{2}\right)H(x) - \left(\frac{1}{12\mu}\frac{\partial P}{\partial x}\right)H^3(x)$$

Knowing the volumetric flow rate allows us to solve for the pressure drop:

$$\frac{\partial P}{\partial x} = \left[\left(\frac{U_o}{2}\right)H(x) - \dot{V}\right]\left(\frac{12\mu}{H^3(x)}\right)$$

Finally, we use the continuity equation to complete the problem and solve for the y-velocity, v. We need an expression for the shape of the bearing, $H(x)$, to get an explicit representation for v:

$$v = -\int_0^{H(x)} \frac{\partial u}{\partial x}dy$$

The pressure distribution exerts a vertical force against the bearing. The load per unit width that the bearing can support is given by the vertical component of the pressure force:

$$\text{Load} = \int_0^L P(x)\cos\theta\,dx$$

We can integrate the pressure drop equation to obtain the pressure and then solve for the load:

$$P(x) = \int_0^L \left[\left(\frac{U_o}{2}\right)H(x) - \dot{V}\right]\left(\frac{12\mu}{H^3(x)}\right)dx$$

At the edges of the bearings, the fluid is exposed to the atmosphere and so

$$x = 0 \qquad P = 0$$
$$x = L \qquad P = 0$$

All we need to know is how the height varies as a function of x. Letting

$$H(x) = \left(\frac{H_o - H_i}{L}\right)x + H_i \qquad \tan\theta = \frac{H_o - H_i}{L}$$

for example, provides us with the information we need to calculate the pressure distribution, the bearing load, and volumetric flow rate through the bearing:

$$P(x) = \frac{6x[H(x) - H_o]H_i^2}{H^2(x)[H_i + H_o]}$$

$$\dot{V} = U_o H_i \left[\frac{H_o}{H_i + H_o} \right]$$

$$\text{Load} = 6\mu U_o \left(\frac{\cos\theta}{\tan^2\theta} \right) \left[\ln\left(\frac{H_i}{H_o} \right) - 2\left(\frac{H_i - H_o}{H_i + H_o} \right) \right]$$

Notice that the load-bearing capacity increases with increased flow rate, increases as $\theta \to 0$, and increases with increasing fluid viscosity.

10.2.3 Mechanical Energy Balance

We can manipulate the momentum balance to obtain several other balances of importance in fluid mechanics. One of the most important is the mechanical energy balance describing how the kinetic energy of a fluid is redistributed through the action of inertial, viscous, and body forces. If we multiply the x-momentum Equation 10.11, by v_x, we obtain

$$-\rho v_x^2 \frac{\partial v_x}{\partial x} - \rho v_x v_y \frac{\partial v_x}{\partial y} - \rho v_x v_z \frac{\partial v_x}{\partial z} - v_x \left(\frac{\partial \tau_{xx}}{\partial x} + \frac{\partial \tau_{yx}}{\partial y} + \frac{\partial \tau_{zx}}{\partial z} \right) - v_x \frac{\partial P}{\partial x} + \rho v_x g_x = v_x \frac{\partial(\rho v_x)}{\partial t} \tag{10.22}$$

We can rearrange this equation using the chain rule to get an equation in v_x^2 that we can transform into an equation for the kinetic energy per unit volume of the fluid:

$$-\frac{1}{2}\rho \left[v_x \frac{\partial v_x^2}{\partial x} + v_y \frac{\partial v_x^2}{\partial y} + v_z \frac{\partial v_x^2}{\partial z} \right] - v_x \left(\frac{\partial \tau_{xx}}{\partial x} + \frac{\partial \tau_{yx}}{\partial y} + \frac{\partial \tau_{zx}}{\partial z} \right) - v_x \frac{\partial P}{\partial x} + \rho v_x g_x = \frac{\partial}{\partial t}\left(\frac{1}{2}\rho v_x^2 \right) \tag{10.23}$$

In its present form, Equation 10.23 does not have the physical meaning we require. We can fold v_x into the derivatives of pressure and stress and, using the chain rule, recast the equation so that each term has a defined thermodynamic meaning:

$$\underbrace{-\frac{1}{2}\rho \left[v_x \frac{\partial v_x^2}{\partial x} + v_y \frac{\partial v_x^2}{\partial y} + v_z \frac{\partial v_x^2}{\partial z} \right]}_{\text{I}} \underbrace{- \frac{\partial}{\partial x}(v_x \tau_{xx} + v_x \tau_{yx} + v_x \tau_{zx})}_{\text{II}}$$

$$+\underbrace{\left(\tau_{xx} \frac{\partial v_x}{\partial x} + \tau_{yx} \frac{\partial v_x}{\partial y} + \tau_{zx} \frac{\partial v_x}{\partial z} \right)}_{\text{III}} \underbrace{- \frac{\partial}{\partial x}(v_x P)}_{\text{IV}} + \underbrace{P\left(\frac{\partial v_x}{\partial x} \right)}_{\text{V}} + \underbrace{\rho v_x g_x}_{\text{VI}} = \underbrace{\frac{\partial}{\partial t}\left(\frac{1}{2}\rho v_x^2 \right)}_{\text{VII}} \tag{10.24}$$

The first term on the left-hand side of the equation, I, represents the net rate of input of kinetic energy (per unit volume) by bulk flow of the fluid through the control volume. The second term, II, describes the rate of work (per unit volume) done by viscous forces on the fluid element, and the third term, III, represents the rate of irreversible conversion of kinetic energy (per unit volume) to internal energy or to heat. This term is called *viscous dissipation*. If the system is isolated from its surroundings so that no heat interaction can take place, the viscous dissipation term is actually a conversion of work to internal energy since the fluid's temperature will rise. If the system is isothermal, then viscous dissipation results in a heat interaction with the surroundings. The fourth term, IV, represents the net rate of work (per unit volume) done by pressure on the fluid element. Term V represents the rate of reversible conversion of kinetic energy (per unit volume) to internal energy. It is formally one component of the $P\Delta V$ work generated by the fluid expanding or contracting. If the fluid is incompressible, this term vanishes. The sixth term, VI, represents the rate of work done by gravity on the fluid. In general, this term represents the work done by all body forces on the fluid. The final term, VII, on the right-hand side of Equation 10.24 represents the rate of accumulation of kinetic energy per unit volume of fluid. Equation 10.24 is a partial representation of the first law of thermodynamics for an open system where we have no explicit temperature gradients driving heat flow and where we are only interested in changes in the kinetic energy of the fluid.

To generate the full mechanical energy equation, we must multiply the y-momentum equation by v_y and the z-momentum equation by v_z, perform the same manipulations we did in Equations 10.22 through 10.24, and add all three components together. These three components are given in Table 10.5. We can express the equation much more succinctly in vector form:

$$\frac{\partial}{\partial t}\left(\frac{1}{2}\rho v^2\right) + \left(\vec{\nabla}\cdot\frac{1}{2}\rho v^2\vec{\mathbf{v}}\right) = (\vec{\tau}:\vec{\nabla}\vec{\mathbf{v}}) - \begin{bmatrix} P(\vec{\nabla}\cdot\vec{\mathbf{v}})+ \\ (\vec{\nabla}\cdot P\vec{\mathbf{v}})+ \\ (\vec{\nabla}\cdot[\vec{\tau}\cdot\vec{\mathbf{v}}])- \\ \rho(\vec{\mathbf{v}}\cdot\vec{\mathbf{g}}) \end{bmatrix} \tag{10.25}$$

$$\frac{\partial}{\partial t}\Big[\quad \Delta(\mathbf{K.E.}) = \mathbf{Q} - \mathbf{W}\Big]$$

TABLE 10.5
Mechanical Energy Equation: Rectangular Components

	Mechanical Energy Equation Components
x	$-\frac{1}{2}\left[v_x\frac{\partial v_x^2}{\partial x}+v_y\frac{\partial v_x^2}{\partial y}+v_z\frac{\partial v_x^2}{\partial z}\right]+\mu\frac{\partial}{\partial x}\left[v_x\frac{\partial v_x}{\partial x}+v_x\frac{\partial v_x}{\partial y}+v_x\frac{\partial v_x}{\partial z}\right]-\mu\mathcal{V}_\mu$ $-\frac{\partial}{\partial x}(v_xP)+P\left(\frac{\partial v_x}{\partial x}\right)+\rho v_xg_x=\frac{\partial}{\partial t}\left(\frac{1}{2}\rho v_x^2\right)$
y	$-\frac{1}{2}\left[v_x\frac{\partial v_y^2}{\partial x}+v_y\frac{\partial v_y^2}{\partial y}+v_z\frac{\partial v_y^2}{\partial z}\right]+\mu\frac{\partial}{\partial y}\left[v_y\frac{\partial v_y}{\partial x}+v_y\frac{\partial v_y}{\partial y}+v_y\frac{\partial v_y}{\partial z}\right]-\mu\mathcal{V}_\mu$ $-\frac{\partial}{\partial y}(v_yP)+P\left(\frac{\partial v_y}{\partial y}\right)+\rho v_yg_y=\frac{\partial}{\partial t}\left(\frac{1}{2}\rho v_y^2\right)$
z	$-\frac{1}{2}\left[v_x\frac{\partial v_z^2}{\partial x}+v_y\frac{\partial v_z^2}{\partial y}+v_z\frac{\partial v_z^2}{\partial z}\right]+\mu\frac{\partial}{\partial z}\left[v_z\frac{\partial v_z}{\partial x}+v_z\frac{\partial v_z}{\partial y}+v_z\frac{\partial v_z}{\partial z}\right]-\mu\mathcal{V}_\mu$ $-\frac{\partial}{\partial z}(v_zP)+P\left(\frac{\partial v_z}{\partial z}\right)+\rho v_zg_z=\frac{\partial}{\partial t}\left(\frac{1}{2}\rho v_z^2\right)$

TABLE 10.6
Viscous Dissipation Terms: Newtonian Fluid ($\mathcal{V}_\mu$)

Coordinate System	Viscous Dissipation Terms: $\mathcal{V}_\mu$ (Newtonian Fluid)
Cartesian	$2\left[\left(\frac{\partial v_x}{\partial x}\right)^2+\left(\frac{\partial v_y}{\partial y}\right)^2+\left(\frac{\partial v_z}{\partial z}\right)^2\right]+\left[\frac{\partial v_y}{\partial x}+\frac{\partial v_x}{\partial y}\right]^2+\left[\frac{\partial v_z}{\partial x}+\frac{\partial v_x}{\partial z}\right]^2+\left[\frac{\partial v_y}{\partial z}+\frac{\partial v_z}{\partial y}\right]^2-\frac{2}{3}\left[\frac{\partial v_x}{\partial x}+\frac{\partial v_y}{\partial y}+\frac{\partial v_z}{\partial z}\right]^2$
Cylindrical	$2\left[\left(\frac{\partial v_r}{\partial r}\right)^2+\left(\frac{1}{r}\frac{\partial v_\theta}{\partial \theta}+\frac{v_r}{r}\right)^2+\left(\frac{\partial v_z}{\partial z}\right)^2\right]+\left[r\frac{\partial}{\partial r}\left(\frac{v_\theta}{r}\right)+\frac{1}{r}\frac{\partial v_r}{\partial \theta}\right]^2+\left[\frac{1}{r}\frac{\partial v_z}{\partial \theta}+\frac{\partial v_\theta}{\partial z}\right]^2+\left[\frac{\partial v_r}{\partial z}+\frac{\partial v_z}{\partial r}\right]^2$ $-\frac{2}{3}\left[\frac{1}{r}\frac{\partial}{\partial r}(rv_r)+\frac{1}{r}\frac{\partial v_\theta}{\partial \theta}+\frac{\partial v_z}{\partial z}\right]^2$
Spherical	$2\left[\left(\frac{\partial v_r}{\partial r}\right)^2+\left(\frac{1}{r}\frac{\partial v_\theta}{\partial \theta}+\frac{v_r}{r}\right)^2+\left(\frac{1}{r\sin\theta}\frac{\partial v_\phi}{\partial \phi}+\frac{v_r}{r}+\frac{v_\theta\cot\theta}{r}\right)^2\right]+\left[r\frac{\partial}{\partial r}\left(\frac{v_\theta}{r}\right)+\frac{1}{r}\frac{\partial v_r}{\partial \theta}\right]^2$ $+\left[\frac{\sin\theta}{r}\frac{\partial}{\partial \theta}\left(\frac{v_\phi}{\sin\theta}\right)+\frac{1}{r\sin\theta}\frac{\partial v_\theta}{\partial \phi}\right]^2+\left[\frac{1}{r\sin\theta}\frac{\partial v_r}{\partial \phi}+r\frac{\partial}{\partial r}\left(\frac{v_\phi}{r}\right)\right]^2$ $-\frac{2}{3}\left[\frac{1}{r^2}\frac{\partial}{\partial r}(r^2 v_r)+\frac{1}{r\sin\theta}\frac{\partial}{\partial \theta}(v_\theta\sin\theta)+\frac{1}{r\sin\theta}\frac{\partial v_\phi}{\partial \phi}\right]^2$

where

$v^2 = v_x^2 + v_y^2 + v_z^2$ and all terms in the equation are referenced per unit volume of fluid

$\frac{\partial}{\partial t}\left(\frac{1}{2}\rho v^2\right)$ is the rate of accumulation of kinetic energy

$\left(\vec{\nabla}\cdot\frac{1}{2}\rho v^2\vec{\mathbf{v}}\right)$ is the net rate of input of kinetic energy by bulk flow

$P(\vec{\nabla}\cdot\vec{\mathbf{v}})$ is the rate of reversible conversion of pressure work to internal energy

$(\vec{\boldsymbol{\tau}}:\vec{\nabla}\vec{\mathbf{v}})$ is the rate of irreversible conversion of viscous work to internal energy (adiabatic system) or heat (isothermal system)

$(\vec{\nabla}\cdot P\vec{\mathbf{v}})$ is the rate of work done by external pressure on the fluid element

$(\vec{\nabla}\cdot[\vec{\tau}\cdot\vec{\mathbf{v}}])$ is the rate of work done by viscous forces on the fluid element

$\rho(\vec{\mathbf{v}}\cdot\vec{\mathbf{g}})$ is the rate of work done by body forces on the fluid element

Equation 10.25 is not often employed in its differential form. Though it is used in current models for describing the flow of turbulent fluids (K–ε and other models [7]), its most important applications arise in its integrated form (extended Bernoulli's equation). Table 10.6 shows expressions for the viscous dissipation term for Newtonian fluids in rectangular, cylindrical, and spherical systems. Notice that the viscous dissipation term is always positive representing an irreversible conversion of energy to heat.

10.2.4 Vorticity Equation

In Chapter 9, we introduced the velocity potential, ϕ; the stream function, ψ; and the vorticity, ω, for a 2-D flow. We define the vorticity as twice the average *local angular velocity* of the fluid element. Analysis of motion that has angular velocity components (weather forecasting, mixing, ocean currents, flow over an airfoil, etc.) is often enhanced by studying the transport of this angular velocity throughout the fluid. We can derive a transport equation for the vorticity by taking the curl of the Navier–Stokes equations:

$$\text{Vorticity transport equations} = Det\begin{vmatrix} \mathbf{i} & \mathbf{j} & \mathbf{k} \\ \partial/\partial x & \partial/\partial y & \partial/\partial z \\ \begin{pmatrix} x \\ \text{momentum} \\ \text{equation} \end{pmatrix} & \begin{pmatrix} y \\ \text{momentum} \\ \text{equation} \end{pmatrix} & \begin{pmatrix} z \\ \text{momentum} \\ \text{equation} \end{pmatrix} \end{vmatrix}$$

It is best to perform this operation using the vector form of the Navier–Stokes equations since this results in the simplest representation:

$$\nu\nabla^2\vec{\mathbf{v}} + \vec{\mathbf{v}}\cdot\vec{\nabla}\vec{\mathbf{v}} + \vec{\mathbf{g}} - \frac{1}{\rho}\vec{\nabla}P = \frac{\partial\vec{\mathbf{v}}}{\partial t} \tag{10.26}$$

The vorticity is the curl of the velocity, $\vec{\boldsymbol{\omega}}_\mu = \vec{\nabla}\times\vec{\mathbf{v}}$, and using this definition lets us rewrite the Navier–Stokes equations in the following vector form incorporating the vorticity directly:

$$\nu\nabla^2\vec{\mathbf{v}} - \frac{1}{2}\vec{\nabla}\vec{\mathbf{v}}\cdot\vec{\mathbf{v}} + \vec{\mathbf{v}}\times\vec{\omega}_\mu + \vec{\mathbf{g}} - \frac{1}{\rho}\vec{\nabla}P = \frac{\partial\vec{\mathbf{v}}}{\partial t} \tag{10.27}$$

Here, we have made use of the vector relation that

$$\vec{\mathbf{v}}\times(\vec{\nabla}\times\vec{\mathbf{v}}) = \vec{\mathbf{v}}\times\vec{\omega}_\mu = \frac{1}{2}\vec{\nabla}(\vec{\mathbf{v}}\cdot\vec{\mathbf{v}}) - \vec{\mathbf{v}}\cdot\vec{\nabla}\vec{\mathbf{v}} \tag{10.28}$$

If we take the curl ($\vec{\nabla}\times$) of both sides of Equation 10.27, we obtain

$$\nu\vec{\nabla}\times(\nabla^2\vec{\mathbf{v}}) - \frac{1}{2}\vec{\nabla}\times(\vec{\nabla}\vec{\mathbf{v}}\cdot\vec{\mathbf{v}}) + \vec{\nabla}\times(\vec{\mathbf{v}}\times\vec{\omega}_\mu) + \vec{\nabla}\times\vec{\mathbf{g}} - \frac{1}{\rho}\vec{\nabla}\times(\vec{\nabla}P) = \vec{\nabla}\times\frac{\partial\vec{\mathbf{v}}}{\partial t} \tag{10.29}$$

We can interchange the order of differentiation within the equation, and noting that

$$\vec{\nabla}\times\vec{\mathbf{g}} = 0 \qquad -\frac{1}{\rho}\vec{\nabla}\times(\vec{\nabla}P) = 0 \qquad \frac{1}{2}\vec{\nabla}\times(\vec{\nabla}\vec{\mathbf{v}}\cdot\vec{\mathbf{v}}) = 0$$

by vector identities, we have

$$\nu\nabla^2\vec{\omega}_\mu - \vec{\mathbf{v}}(\vec{\nabla}\cdot\vec{\omega}_\mu) + \vec{\omega}_\mu(\vec{\nabla}\cdot\vec{\mathbf{v}}) + \vec{\omega}_\mu\cdot\vec{\nabla}\vec{\mathbf{v}} - \vec{\mathbf{v}}\cdot\vec{\nabla}\vec{\omega}_\mu = \frac{\partial\vec{\omega}_\mu}{\partial t} \tag{10.30}$$

Since $\vec{\mathbf{v}}(\vec{\nabla}\cdot\vec{\boldsymbol{\omega}}_\mu) = \vec{\mathbf{v}}\vec{\nabla}\cdot(\vec{\nabla}\times\vec{\mathbf{v}}) = 0$ by the definition of the vorticity and a vector identity, Equation 10.30 reduces to

$$\nu\nabla^2\vec{\omega}_\mu + \vec{\omega}_\mu(\vec{\nabla}\cdot\vec{\mathbf{v}}) + \vec{\omega}_\mu\cdot\vec{\nabla}\vec{\mathbf{v}} - \vec{\mathbf{v}}\cdot\vec{\nabla}\vec{\omega}_\mu = \frac{\partial\vec{\omega}_\mu}{\partial t} \tag{10.31}$$

for a Newtonian fluid. If the fluid is also incompressible, then $\vec{\nabla} \cdot \vec{v} = 0$ and we can rewrite Equation 10.31 as

$$\underbrace{\nu \nabla^2 \vec{\omega}_\mu}_{\text{I}} + \underbrace{\vec{\omega}_\mu \cdot \vec{\nabla} \vec{v}}_{\text{II}} - \underbrace{\vec{v} \cdot \vec{\nabla} \vec{\omega}_\mu}_{\text{III}} = \underbrace{\frac{\partial \vec{\omega}_\mu}{\partial t}}_{\text{IV}} \tag{10.32}$$

$$\nu \nabla^2 \vec{\omega}_\mu + \vec{\omega}_\mu \cdot \vec{\nabla} \vec{v} = \frac{D \vec{\omega}_\mu}{Dt} \tag{10.33}$$

Notice that by describing how the flow changes in terms of its vorticity, we no longer have to consider the pressure or gravitational body forces. We can view the vorticity equation as describing how the angular momentum of a spherical fluid element changes at a rate determined by mechanical couples formed from the tangential viscous stresses. Term I describes the diffusion of vorticity throughout the fluid. Term II describes the rate of change of the moment of inertia, I, of the fluid element due to a change in the element's shape. It is zero in 2-D motion because the vorticity does not lie in the plane of fluid motion. The third term, III, describes the net rate of vorticity input via fluid convection through the control volume. Finally, the fourth term, IV, describes the net rate of accumulation of vorticity in the control volume. Table 10.7 shows the vorticity transport equation for an incompressible fluid in rectangular, cylindrical, and spherical coordinates.

The vorticity has an analog in the magnetic vector potential. In both cases, we can integrate along a line of force or a streamline to obtain the circulation of the magnetic field or fluid. To look at vorticity in a bit more detail, let's revisit our flow about a sphere problem and calculate the vorticity using the vorticity transport equation and the velocity profiles we obtained. We will then compare these values of the vorticity to that calculated based on our ideal flow condition.

Example 10.7 Vorticity in Stokes' Flow Past a Stationary Sphere

We can use the vorticity transport equation to investigate the circulation past a stationary sphere in Stokes' flow. We showed previously that the flow past the sphere was essentially 2-D, having no ϕ-dependence. A 2-D flow has only one vorticity component, perpendicular to the plane of flow. In this case, it must be the ϕ-component, ω_ϕ, that is defined by

$$\omega_{\mu\phi} = \frac{1}{2}\left[\frac{1}{r}\frac{\partial}{\partial r}(r v_\theta) - \frac{1}{r}\frac{\partial}{\partial \theta}(v_r)\right]$$

At steady state with no inertial effects present in the flow, the vorticity transport Equation 10.33 reduces to Laplace's equation for the vorticity:

$$\nabla^2 \omega_{\mu\phi} = 0$$

or using the expanded form from Table 10.7

$$\frac{1}{r^2}\frac{\partial}{\partial r}\left(r^2 \frac{\partial \omega_{\mu\phi}}{\partial r}\right) + \frac{1}{r^2 \sin\theta}\frac{\partial}{\partial \theta}\left(\sin\theta \frac{\partial \omega_{\mu\phi}}{\partial \theta}\right) - \frac{\omega_{\mu\phi}}{r^2 \sin^2\theta} = 0$$

Though this equation looks formidable, we can solve it by substituting in

$$\omega_{\mu\phi} = f(r)\sin\theta$$

TABLE 10.7
Vorticity Transport Equations: Incompressible, Newtonian Fluids

Vorticity Equation: Cartesian Coordinates

x

$$\mu\left[\frac{\partial^2\omega_{\mu x}}{\partial x^2}+\frac{\partial^2\omega_{\mu x}}{\partial y^2}+\frac{\partial^2\omega_{\mu x}}{\partial z^2}\right]$$
$$=\rho\left(\frac{\partial\omega_{\mu x}}{\partial t}+v_x\frac{\partial\omega_{\mu x}}{\partial x}+v_y\frac{\partial\omega_{\mu x}}{\partial y}+v_z\frac{\partial\omega_{\mu x}}{\partial z}\right)-\rho\left(\omega_{\mu x}\frac{\partial v_x}{\partial x}+\omega_{\mu y}\frac{\partial v_x}{\partial y}+\omega_{\mu z}\frac{\partial v_x}{\partial z}\right)$$

y

$$\mu\left[\frac{\partial^2\omega_{\mu y}}{\partial x^2}+\frac{\partial^2\omega_{\mu y}}{\partial y^2}+\frac{\partial^2\omega_{\mu y}}{\partial z^2}\right]$$
$$=\rho\left(\frac{\partial\omega_{\mu y}}{\partial t}+v_x\frac{\partial\omega_{\mu y}}{\partial x}+v_y\frac{\partial\omega_{\mu y}}{\partial y}+v_z\frac{\partial\omega_{\mu y}}{\partial z}\right)-\rho\left(\omega_{\mu x}\frac{\partial v_y}{\partial x}+\omega_{\mu y}\frac{\partial v_y}{\partial y}+\omega_{\mu z}\frac{\partial v_y}{\partial z}\right)$$

z

$$\mu\left[\frac{\partial^2\omega_{\mu z}}{\partial x^2}+\frac{\partial^2\omega_{\mu z}}{\partial y^2}+\frac{\partial^2\omega_{\mu z}}{\partial z^2}\right]$$
$$=\rho\left(\frac{\partial\omega_{\mu z}}{\partial t}+v_x\frac{\partial\omega_{\mu z}}{\partial x}+v_y\frac{\partial\omega_{\mu z}}{\partial y}+v_z\frac{\partial\omega_{\mu z}}{\partial z}\right)-\rho\left(\omega_{\mu x}\frac{\partial v_z}{\partial x}+\omega_{\mu y}\frac{\partial v_z}{\partial y}+\omega_{\mu z}\frac{\partial v_z}{\partial z}\right)$$

Vorticity Equation: Cylindrical Coordinates

r

$$\mu\left[\frac{\partial}{\partial r}\left(\frac{1}{r}\frac{\partial}{\partial r}(r\omega_{\mu r})\right)+\frac{1}{r^2}\frac{\partial^2\omega_{\mu r}}{\partial\theta^2}-\frac{2}{r^2}\frac{\partial\omega_{\mu\theta}}{\partial\theta}+\frac{\partial^2\omega_{\mu r}}{\partial z^2}\right]$$
$$=\rho\left(\frac{\partial\omega_{\mu r}}{\partial t}+v_r\frac{\partial\omega_{\mu r}}{\partial r}+\frac{v_\theta}{r}\frac{\partial\omega_{\mu r}}{\partial\theta}-\frac{v_\theta\omega_{\mu\theta}}{r}+v_z\frac{\partial\omega_{\mu r}}{\partial z}\right)-\rho\left(\omega_{\mu r}\frac{\partial v_r}{\partial r}+\frac{\omega_{\mu\theta}}{r}\frac{\partial v_r}{\partial\theta}-\frac{v_\theta\omega_{\mu\theta}}{r}+\omega_{\mu z}\frac{\partial v_r}{\partial z}\right)$$

θ

$$\mu\left[\frac{\partial}{\partial r}\left(\frac{1}{r}\frac{\partial}{\partial r}(r\omega_{\mu\theta})\right)+\frac{1}{r^2}\frac{\partial^2\omega_{\mu r}}{\partial\theta^2}+\frac{2}{r^2}\frac{\partial\omega_{\mu r}}{\partial\theta}+\frac{\partial^2\omega_{\mu\theta}}{\partial z^2}\right]$$
$$=\rho\left(\frac{\partial\omega_{\mu\theta}}{\partial t}+v_r\frac{\partial\omega_{\mu\theta}}{\partial r}+\frac{v_\theta}{r}\frac{\partial\omega_{\mu\theta}}{\partial\theta}+\frac{v_r\omega_{\mu\theta}}{r}+v_z\frac{\partial\omega_{\mu\theta}}{\partial z}\right)-\rho\left(\omega_{\mu r}\frac{\partial v_\theta}{\partial r}+\frac{\omega_{\mu\theta}}{r}\frac{\partial v_\theta}{\partial\theta}+\frac{v_r\omega_{\mu\theta}}{r}+\omega_{\mu z}\frac{\partial v_\theta}{\partial z}\right)$$

z

$$\mu\left[\frac{\partial}{\partial r}\left(r\frac{\partial\omega_{\mu z}}{\partial r}\right)+\frac{1}{r^2}\frac{\partial^2\omega_{\mu z}}{\partial\theta^2}+\frac{\partial^2\omega_{\mu z}}{\partial z^2}\right]$$
$$=\rho\left(\frac{\partial\omega_{\mu z}}{\partial t}+v_r\frac{\partial\omega_{\mu z}}{\partial r}+\frac{v_\theta}{r}\frac{\partial\omega_{\mu z}}{\partial\theta}+v_z\frac{\partial\omega_{\mu z}}{\partial z}\right)-\rho\left(\omega_{\mu r}\frac{\partial v_z}{\partial r}+\frac{\omega_{\mu\theta}}{r}\frac{\partial v_z}{\partial\theta}+\omega_{\mu z}\frac{\partial v_z}{\partial z}\right)$$

Vorticity Equation: Spherical Coordinates

r

$$\frac{\mu}{r^2}\left[\frac{\partial}{\partial r}\left(r^2\frac{\partial\omega_{\mu r}}{\partial r}\right)+\frac{1}{\sin\theta}\frac{\partial}{\partial\theta}\left(\sin\theta\frac{\partial\omega_{\mu r}}{\partial\theta}\right)+\frac{1}{\sin^2\theta}\frac{\partial^2\omega_{\mu r}}{\partial\phi^2}-2\left(\omega_{\mu r}+\omega_{\mu\theta}\cot\theta+\frac{\partial\omega_{\mu\theta}}{\partial\theta}+\frac{1}{\sin\theta}\frac{\partial\omega_{\mu\phi}}{\partial\phi}\right)\right]$$
$$=\rho\left(\frac{\partial\omega_{\mu r}}{\partial t}+v_r\frac{\partial\omega_{\mu r}}{\partial r}+\frac{v_\theta}{r}\frac{\partial\omega_{\mu r}}{\partial\theta}-\frac{v_\theta\omega_{\mu\theta}+v_\phi\omega_{\mu\phi}}{r}+\frac{v_\phi}{r\sin\theta}\frac{\partial\omega_{\mu r}}{\partial\phi}\right)$$
$$-\rho\left(\omega_{\mu r}\frac{\partial v_r}{\partial r}+\frac{\omega_{\mu\theta}}{r}\frac{\partial v_r}{\partial\theta}-\frac{v_\theta\omega_{\mu\theta}+v_\phi\omega_{\mu\phi}}{r}+\frac{\omega_{\mu\phi}}{r\sin\theta}\frac{\partial v_r}{\partial\phi}\right)$$

(continued)

TABLE 10.7 (continued)
Vorticity Transport Equations: Incompressible, Newtonian Fluids

Vorticity Equation: Spherical Coordinates

θ

$$\frac{\mu}{r^2}\left[\frac{\partial}{\partial r}\left(r^2\frac{\partial\omega_{\mu\theta}}{\partial r}\right)+\frac{1}{\sin\theta}\frac{\partial}{\partial\theta}\left(\sin\theta\frac{\partial\omega_{\mu\theta}}{\partial\theta}\right)+\frac{1}{\sin^2\theta}\frac{\partial^2\omega_{\mu\theta}}{\partial\phi^2}+2\left(\frac{\partial\omega_{\mu r}}{\partial\theta}-\frac{\omega_{\mu\theta}}{\sin^2\theta}-\frac{\cos\theta}{\sin^2\theta}\frac{\partial\omega_{\mu\phi}}{\partial\phi}\right)\right]$$

$$=\rho\left(\frac{\partial\omega_{\mu\theta}}{\partial t}+v_r\frac{\partial\omega_{\mu\theta}}{\partial r}+\frac{v_\theta}{r}\frac{\partial\omega_{\mu\theta}}{\partial\theta}-\frac{v_\phi\omega_{\mu\phi}\cot\theta-v_\theta\omega_{\mu r}}{r}+\frac{v_\phi}{r\sin\theta}\frac{\partial\omega_{\mu\theta}}{\partial\phi}\right)$$

$$-\rho\left(\omega_{\mu r}\frac{\partial v_\theta}{\partial r}+\frac{\omega_{\mu\theta}}{r}\frac{\partial v_\theta}{\partial\theta}-\frac{v_\phi\omega_{\mu\phi}\cot\theta-v_r\omega_{\mu\theta}}{r}+\frac{\omega_{\mu\phi}}{r\sin\theta}\frac{\partial v_\theta}{\partial\phi}\right)$$

ϕ

$$\frac{\mu}{r^2}\left[\frac{\partial}{\partial r}\left(r^2\frac{\partial\omega_{\mu\phi}}{\partial r}\right)+\frac{1}{\sin\theta}\frac{\partial}{\partial\theta}\left(\sin\theta\frac{\partial\omega_{\mu\phi}}{\partial\theta}\right)+\frac{1}{\sin^2\theta}\frac{\partial^2\omega_{\mu\phi}}{\partial\phi^2}-2\left(\frac{\omega_{\mu\phi}}{\sin^2\theta}-\frac{1}{\sin\theta}\frac{\partial\omega_{\mu r}}{\partial\phi}-\frac{\cos\theta}{\sin^2\theta}\frac{\partial\omega_{\mu\theta}}{\partial\phi}\right)\right]$$

$$=\rho\left(\frac{\partial\omega_{\mu\phi}}{\partial t}+v_r\frac{\partial\omega_{\mu\phi}}{\partial r}+\frac{v_\theta}{r}\frac{\partial\omega_{\mu\phi}}{\partial\theta}+\frac{v_\phi\omega_{\mu r}+v_\phi\omega_{\mu\theta}\cot\theta}{r}+\frac{v_\phi}{r\sin\theta}\frac{\partial\omega_{\mu\phi}}{\partial\phi}\right)$$

$$-\rho\left(\omega_{\mu r}\frac{\partial v_\phi}{\partial r}+\frac{\omega_{\mu\theta}}{r}\frac{\partial v_\phi}{\partial\theta}+\frac{v_r\omega_{\mu\phi}+v_\theta\omega_{\mu\phi}\cot\theta}{r}+\frac{\omega_{\mu\phi}}{r\sin\theta}\frac{\partial v_\phi}{\partial\phi}\right)$$

where $f(r)$ is a velocity function. Remember the magnitude of $\omega_{\mu\phi} \propto |v| \sin\theta$. Using the substitution for $\omega_{\mu\phi}$ results in an ordinary differential equation for f:

$$r^2\frac{d^2f}{dr^2}+2r\frac{df}{dr}-2f=0$$

This is another example of an Euler equation that we may solve by substituting $f = r^n$. Performing the algebra, we find that $n = \{1,-2\}$, so our solution for $\omega_{\mu\phi}$ is

$$\omega_{\mu\phi}=\left[\frac{a_1}{r^2}+a_2r\right]\sin\theta$$

We require two boundary conditions to solve for the two constants of integration. Far away from the sphere, at $r = \infty$, we know the flow is unidirectional and so there is no vorticity $\{r \to \infty, \omega_{\mu\phi} = 0\}$. This requires that $a_2 = 0$. Thus,

$$\omega_{\mu\phi}=\frac{a_1}{r^2}\sin\theta$$

Notice that the vorticity spreads out (diffuses) and decreases in amplitude with the square of the distance from the sphere. A quick inspection of the vorticity transport equation shows that the vorticity behaves in a similar fashion to the conduction of heat or the diffusion of mass. The magnitude of the vorticity is greatest at $\theta = \pi/2$, the point where we have a change in the sign of the pressure gradient along the surface of the sphere ($v_r = 0$, $|v_\theta| = \max$). Vorticity is smallest at the stagnation point and at the center of the wake behind the sphere, ($v_\theta = 0$, $|v_r| = \max$). To determine the magnitude of a_1, we have much more work to do. We do not know the vorticity right at the sphere's surface. All we know is that the velocity there is zero and this generates the vorticity. Thus, to obtain a_1, we must solve for the velocity components or the stream function.

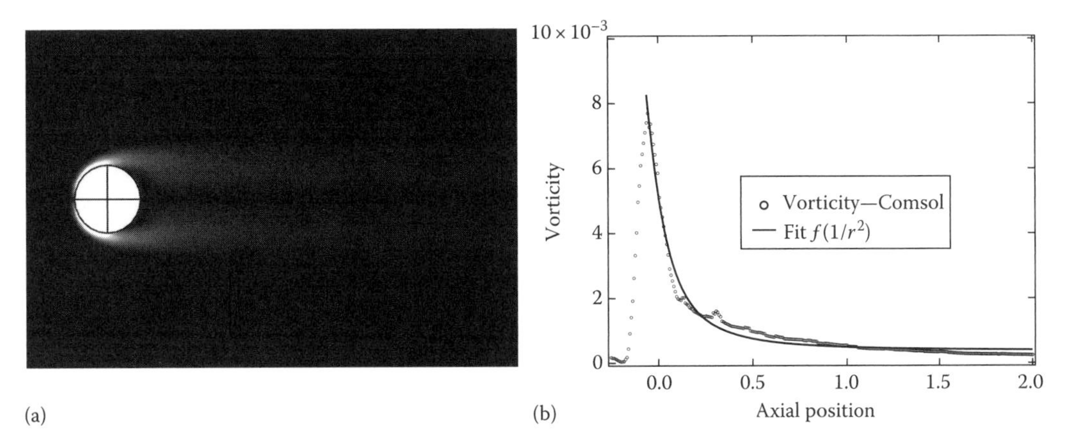

FIGURE 10.9 (a) Vorticity in flow about a sphere from Comsol® module sphere. Notice that the vorticity is generated at the solid's surface (white) and dissipates the farther one moves from the surface (dark). (b) Vorticity should dissipate as the square of distance from the sphere. A fit proportional to $1/r^2$ shows that the voriticity does decrease in a manner predicted by the analytical solution to the problem.

Since we have already obtained the velocity components, we can use them to obtain the magnitude of a_1 directly. If we substitute the velocity components directly into the definition of $\omega_{\mu\phi}$, we find

$$\omega_{\mu\phi} = \frac{1}{2}\left\{-\frac{1}{r}\frac{\partial}{\partial r}\left[v_\infty \sin\theta\left(r - \frac{3}{4}r_o - \frac{1}{4}\frac{r_o^3}{r^2}\right)\right] = -\frac{1}{r}\frac{\partial}{\partial\theta}\left[v_\infty \cos\theta\left(1 - \frac{3}{2}\left(\frac{r_o}{r}\right) + \frac{1}{2}\left(\frac{r_o}{r}\right)^3\right)\right]\right\}$$

$$= -\frac{3}{4}v_\infty r_o \frac{\sin\theta}{r^2}$$

and so $a_1 = -(3/4)v_\infty r_o$.

It is instructive to compare this result with the ideal flow case. Remember, in ideal or potential flow, we have no viscosity. The flow should be irrotational since there is no *no-slip* condition at the surface of the sphere to generate the vorticity. The potential flow velocity components are

$$v_r = -v_\infty \cos\theta\left[\left(\frac{r_o}{r}\right)^3 - 1\right] \qquad v_\theta = -v_\infty \sin\theta\left[1 + \frac{1}{2}\left(\frac{r_o}{r}\right)^3\right]$$

Substituting into our definition for $\omega_{\mu\phi}$ shows that $\omega_{\mu\phi} = 0$. This example illustrates why the streamlines for Stokes' flow were more curved and extended further out into the fluid. The vorticity generated at the surface of the sphere provided additional rotational motion to the fluid. This vorticity diffuses into the fluid and as it dissipates, the streamlines for Stokes' flow merge with that for potential flow. This action of the vorticity will be very important when we discuss boundary layer theory later. The development and extent of the boundary layer depends on the diffusion of vorticity from the solid surface into the free stream of the fluid. Integrating the vorticity over a streamline gives the circulation of a fluid element characterized by that streamline about the sphere (Figure 10.9).

One final note about the vorticity transport equations should be mentioned. In 2-D flows, there is only one component of vorticity directed outside the plane of flow and hence only one vorticity transport equation. This equation can be used to derive the stream function equation presented in Table 10.4. All one need do is to substitute $-\nabla^2\psi$ in for $\omega_{\mu z}$, or $\omega_{\mu\phi}$.

10.3 ENERGY TRANSPORT

In Section 10.2.3, we derived the mechanical energy balance considering the kinetic energy of the system. Though not appearing explicitly, the potential energy of the system was treated as a work term against the applied body force, gravity. To derive the full energy balance, we must include the internal energy of the fluid and must also account for heat flows that result due to temperature gradients within the system. Figure 10.10 shows the control volume for deriving the energy equation. We have energy flowing into and out of the control volume by convection and by diffusion due to a temperature gradient. We also have work being done by the fluid in the control volume against pressure forces, body forces, and viscous forces. In some instances, we may have energy flowing through the system as a result of mass diffusion though we will not consider this complication at the moment. The energy balance about the control volume, in its most common form, is

In	–	*Out*	+		*Gen*		=	*Acc*
Rate of internal and kinetic energy in by convection	–	Rate of internal and kinetic energy out by convection	+		–			
Rate of heat in by conduction	–	Rate of heat out by conduction	+	Net rate of heat addition by generation	–	Net rate of work done by system on surroundings against	[Pressure forces, Body forces, Viscous forces]	= Net rate of accumulation of internal and kinetic energy

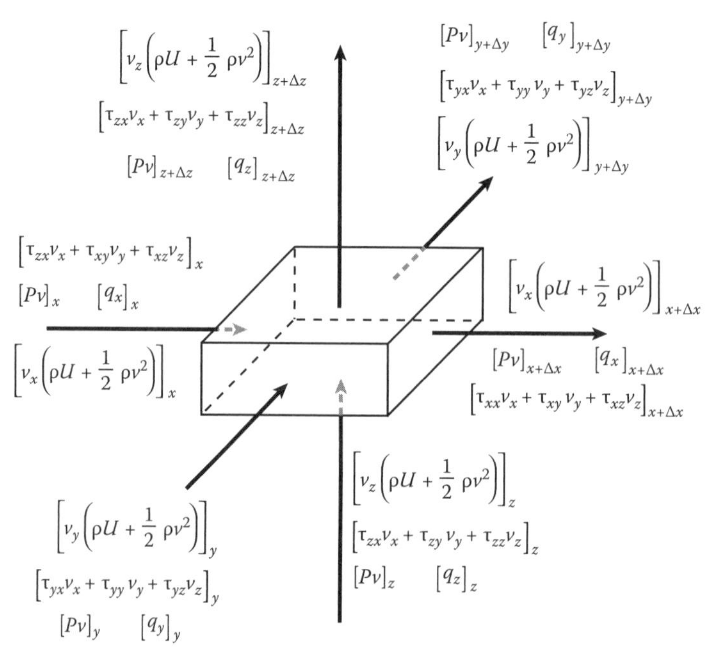

FIGURE 10.10 Control volume for derivation of the energy balance.

Rate of convection of internal and kinetic energy in and out

$$\Delta y \Delta z \left\{ \left[v_x \left(\rho U + \frac{1}{2}\rho v^2 \right) \right]_x - \left[v_x \left(\rho U + \frac{1}{2}\rho v^2 \right) \right]_{x+\Delta x} \right\} \tag{10.34}$$

$$\Delta x \Delta z \left\{ \left[v_y \left(\rho U + \frac{1}{2}\rho v^2 \right) \right]_y - \left[v_y \left(\rho U + \frac{1}{2}\rho v^2 \right) \right]_{y+\Delta y} \right\} \tag{10.35}$$

$$\Delta x \Delta y \left\{ \left[v_z \left(\rho U + \frac{1}{2}\rho v^2 \right) \right]_z - \left[v_z \left(\rho U + \frac{1}{2}\rho v^2 \right) \right]_{z+\Delta z} \right\} \tag{10.36}$$

Here we assume that $v^2 = v_x^2 + v_y^2 + v_z^2$.

Considering the energy that flows through the control volume as a result of imposed or induced temperature gradients, we find the following:

Rate of conduction of energy in and out

$$\Delta y \Delta z[q_x(x) - q_x(x+\Delta x)] + \Delta x \Delta z[q_y(y) - q_y(y+\Delta y)] + \Delta x \Delta y[q_z(z) - q_z(z+\Delta z)] \tag{10.37}$$

We must now consider the generation terms in the energy balance. These arise either as a volumetric rate of generation caused by chemical reaction, nuclear reaction, or the like, or in the form of work the system does against forces exerted on it.

Rate of internal energy generation

$$\Delta x \Delta y \Delta z[\dot{q}_x + \dot{q}_y + \dot{q}_z] \tag{10.38}$$

Rate of work against gravity

Work is defined as *force × distance.* It is done when the fluid velocity and gravitational forces act in opposite directions:

$$-\rho \Delta x \Delta y \Delta z[v_x g_x + v_y g_y + v_z g_z] \tag{10.39}$$

Rate of work against pressure

Here, work is again done when the fluid velocity and pressure forces act in opposite directions:

$$-\Delta y \Delta z[(v_x P)_x - (v_x P)_{x+\Delta x}] - \Delta x \Delta z[(v_y P)_y - (v_y P)_{y+\Delta y}] - \Delta x \Delta y[(v_z P)_z - (v_z P)_{z+\Delta z}] \tag{10.40}$$

Rate of work against viscous forces

Work is also performed against viscous forces where the convective motion of the fluid opposes the shear forces:

$$\begin{aligned}
&-\Delta y \Delta z[(\tau_{xx} v_x + \tau_{xy} v_y + \tau_{xz} v_z)_x - (\tau_{xx} v_x + \tau_{xy} v_y + \tau_{xz} v_z)_{x+\Delta x}] \\
&-\Delta x \Delta z[(\tau_{yx} v_x + \tau_{yy} v_y + \tau_{yz} v_z)_y - (\tau_{yx} v_x + \tau_{yy} v_y + \tau_{yz} v_z)_{y+\Delta y}] \\
&-\Delta x \Delta y[(\tau_{zx} v_x + \tau_{zy} v_y + \tau_{zz} v_z)_z - (\tau_{zx} v_x + \tau_{zy} v_y + \tau_{zz} v_z)_{z+\Delta z}]
\end{aligned} \tag{10.41}$$

Rate of accumulation

Finally, we come to the rate of accumulation of kinetic and internal energy within the control volume:

$$\Delta x\Delta y\Delta z\frac{\partial}{\partial t}\left(\rho U+\frac{1}{2}\rho v^2\right) \tag{10.42}$$

If we now divide all terms by the volume element, $\Delta x\Delta y\Delta z$, and take the limit of the entire equation as $\Delta x\Delta y\Delta z \to 0$, we have

$$-\frac{\partial}{\partial x}\left[v_x\left(\rho U+\frac{1}{2}\rho v^2\right)\right]-\frac{\partial}{\partial y}\left[v_y\left(\rho U+\frac{1}{2}\rho v^2\right)\right]-\frac{\partial}{\partial z}\left[v_z\left(\rho U+\frac{1}{2}\rho v^2\right)\right]$$

$$-\left[\frac{\partial q_x}{\partial x}+\frac{\partial q_y}{\partial y}+\frac{\partial q_z}{\partial z}\right]+\rho(v_xg_x+v_yg_y+v_zg_z)-\left[\frac{\partial}{\partial x}(v_xP)+\frac{\partial}{\partial y}(v_yP)+\frac{\partial}{\partial z}(v_zP)\right]$$

$$-\frac{\partial}{\partial x}(\tau_{xx}v_x+\tau_{xy}v_y+\tau_{xz}v_z)-\frac{\partial}{\partial y}(\tau_{yx}v_x+\tau_{yy}v_y+\tau_{yz}v_z)$$

$$-\frac{\partial}{\partial z}(\tau_{zx}v_x+\tau_{zy}v_y+\tau_{zz}v_z)+[\dot{q}_x+\dot{q}_y+\dot{q}_z]=\frac{\partial}{\partial t}\left(\rho U+\frac{1}{2}\rho v^2\right) \tag{10.43}$$

or in vector notation

$$-(\vec{\nabla}\cdot\vec{\mathbf{q}})+\rho(\vec{\mathbf{v}}\cdot\vec{\mathbf{g}})-(\vec{\nabla}\cdot P\vec{\mathbf{v}})-(\vec{\nabla}\cdot[\vec{\tau}\cdot\vec{\mathbf{v}}])+\dot{q}=\rho\frac{D}{Dt}\left(U+\frac{1}{2}v^2\right) \tag{10.44}$$

We can simplify Equation 10.44 using the continuity equation and the equation describing the rate of change of v^2 (mechanical energy) that we derived in Section 10.2.3 by taking the dot product of the momentum equation with the velocity vector. The equation for mechanical energy is

$$-(\vec{\mathbf{v}}\cdot\vec{\nabla}P)-(\vec{\mathbf{v}}\cdot[\vec{\nabla}\cdot\vec{\tau}])+\rho\left(\vec{\mathbf{v}}\cdot\vec{\mathbf{g}}\right)=\rho\frac{D}{Dt}\left(\frac{1}{2}v^2\right) \tag{10.45}$$

Subtracting Equation 10.45 from Equation 10.44 gives the most often used form of the *thermal* energy equation. Equation 10.46 represents a statement of the first law of thermodynamics for an open system:

$$-\vec{\nabla}\cdot\vec{\mathbf{q}}-(\vec{\tau}:\vec{\nabla}\vec{\mathbf{v}})+\dot{q}-P(\vec{\nabla}\cdot\vec{\mathbf{v}})=\rho\frac{DU}{Dt}$$

$$\frac{\partial}{\partial \mathbf{t}}\left[\mathbf{Q}\qquad-\qquad\mathbf{W}\quad=\quad\mathbf{U}\right] \tag{10.46}$$

Equation 10.46 is not in its most useful form since we directly measure temperature, not specific internal energy. From thermodynamics [8], we know that we can write the change in specific internal energy as

$$dU=\left(\frac{\partial U}{\partial V}\right)_T dV+\left(\frac{\partial U}{\partial T}\right)_V dT \tag{10.47}$$

$$dU=\left[-P+T\left(\frac{\partial P}{\partial T}\right)_V\right]dV+C_V dT \tag{10.48}$$

so the change in U can be rewritten as

$$\rho\frac{DU}{Dt}=\rho\left[-P+T\left(\frac{\partial P}{\partial T}\right)_V\right]\frac{DV}{Dt}+\rho C_V\frac{DT}{Dt} \quad (10.49)$$

Using the continuity equation allows us to substitute for the DV/Dt term in Equation 10.49:

$$\rho\frac{DV}{Dt}=-\frac{1}{\rho}\frac{D\rho}{Dt}=\vec{\nabla}\cdot\vec{\mathbf{v}} \quad (10.50)$$

Substituting (10.49) and (10.50) into the energy Equation 10.46, expressing the stress tensor $\vec{\tau}$ as a function of μ and velocity gradients (Newton's law of viscosity), and writing $\vec{\mathbf{q}}$ as a function of thermal conductivity, k, and temperature gradients (Fourier's law) give us the final form of the energy equation for a Newtonian fluid:

$$k\nabla^2T-T\left(\frac{\partial P}{\partial T}\right)_\rho(\vec{\nabla}\cdot\vec{\mathbf{v}})+\mu\mathcal{V}_\mu=\rho C_v\frac{DT}{Dt} \quad (10.51)$$

where $\mathcal{V}_\mu$ is the viscous dissipation term. This term is only important in turbulent flow or in flow where viscous forces (excessive shearing) dominate such as in a journal bearing or other lubrication problems.

Though we have expressed the energy equation in terms of the internal energy, we can write it in many other forms. Table 10.8 [5] lists various forms of the energy equation used in practice. Table 10.9 presents the energy equation in rectangular, cylindrical, and spherical coordinates, and Table 10.10 restricts the forms in Table 10.9 to Newtonian fluids having constant properties.

TABLE 10.8
Forms of the Energy Equation Used Frequently

Internal energy ($\hat{U}$)	$-\vec{\nabla}\cdot\vec{\mathbf{q}}-P(\vec{\nabla}\cdot\vec{\mathbf{v}})-(\vec{\tau}:\vec{\nabla}\vec{\mathbf{v}})+\dot{q}=\rho\frac{DU}{Dt}$
Kinetic energy $\left(K.E.=\frac{1}{2}\rho v^2\right)$	$-(\vec{\mathbf{v}}\cdot\vec{\nabla}P)-(\vec{\mathbf{v}}\cdot[\vec{\nabla}\cdot\vec{\tau}])+\rho(\vec{\mathbf{v}}\cdot\vec{\mathbf{g}})+\dot{q}=\rho\frac{D}{Dt}\left(\frac{1}{2}v^2\right)$
Internal and kinetic energy ($\hat{U}$ + $K.E.$)	$-(\vec{\nabla}\cdot\vec{\mathbf{q}})+\rho(\vec{\mathbf{v}}\cdot\vec{\mathbf{g}})-(\vec{\nabla}\cdot P\vec{\mathbf{v}})-\left(\vec{\nabla}\cdot[\vec{\tau}\cdot\vec{\mathbf{v}}]\right)+\dot{q}=\rho\frac{D}{Dt}\left(U+\frac{1}{2}v^2\right)$
Total energy $\hat{\mathbf{E}}$ ($\hat{U}$ + $K.E.$ + $P.E.$)	$-(\vec{\nabla}\cdot\vec{\mathbf{q}})-(\vec{\nabla}\cdot P\vec{\mathbf{v}})-(\vec{\nabla}\cdot[\vec{\tau}\cdot\vec{\mathbf{v}}])+\dot{q}=\rho\frac{D}{Dt}\left(U+\frac{1}{2}v^2+\Phi_g\right)$
Internal energy $d\hat{U}=C_vdT$	$-\vec{\nabla}\cdot\vec{\mathbf{q}}-T\left(\frac{\partial\rho}{\partial T}\right)_\rho(\vec{\nabla}\cdot\vec{\mathbf{v}})-(\vec{\tau}:\vec{\nabla}\vec{\mathbf{v}})+\dot{q}=\rho C_v\frac{DT}{Dt}$
Enthalpy $\hat{H}=\hat{U}+P$	$-\vec{\nabla}\cdot\vec{\mathbf{q}}-(\vec{\tau}:\vec{\nabla}\vec{\mathbf{v}})+\dot{q}=\rho\frac{D}{Dt}\left(H-\frac{P}{\rho}\right)$
Enthalpy $d\hat{H}=C_pdT$	$-\vec{\nabla}\cdot\vec{\mathbf{q}}+\left(\frac{\partial\ln V}{\partial\ln T}\right)_\rho\frac{DP}{Dt}-(\vec{\tau}:\vec{\nabla}\vec{\mathbf{v}})+\dot{q}=\rho C_p\frac{DT}{Dt}$

Source: Adapted from Bird et al., *Transport Phenomena*, John Wiley & Sons, New York, 1960. With permission.

TABLE 10.9
Energy Equation in Terms of the Heat Flows and Viscous Stresses

Coordinate System	Energy Equation
Cartesian	$-\left(\frac{\partial q_x}{\partial x}+\frac{\partial q_y}{\partial y}+\frac{\partial q_z}{\partial z}\right)-T\left(\frac{\partial P}{\partial T}\right)_\rho\left(\frac{\partial v_x}{\partial x}+\frac{\partial v_y}{\partial y}+\frac{\partial v_z}{\partial z}\right)-\left[\tau_{xy}\left(\frac{\partial v_x}{\partial y}+\frac{\partial v_y}{\partial x}\right)+\tau_{xz}\left(\frac{\partial v_x}{\partial z}+\frac{\partial v_z}{\partial x}\right)+\tau_{yz}\left(\frac{\partial v_z}{\partial y}+\frac{\partial v_y}{\partial z}\right)\right]$ $-\left(\tau_{xx}\frac{\partial v_x}{\partial x}+\tau_{yy}\frac{\partial v_y}{\partial y}+\tau_{zz}\frac{\partial v_z}{\partial z}\right)=\rho C_v\left(\frac{\partial T}{\partial t}+v_x\frac{\partial T}{\partial x}+v_y\frac{\partial T}{\partial y}+v_z\frac{\partial T}{\partial z}\right)$
Cylindrical	$-\left(\frac{1}{r}\frac{\partial}{\partial r}(rq_r)+\frac{1}{r}\frac{\partial q_\theta}{\partial \theta}+\frac{\partial q_z}{\partial z}\right)-T\left(\frac{\partial P}{\partial T}\right)_\rho\left(\frac{1}{r}\frac{\partial}{\partial r}(rv_r)+\frac{1}{r}\frac{\partial v_\theta}{\partial \theta}+\frac{\partial v_z}{\partial z}\right)$ $-\left[\tau_{r\theta}\left(r\frac{\partial}{\partial r}\left(\frac{v_\theta}{r}\right)+\frac{1}{r}\frac{\partial v_r}{\partial \theta}\right)+\tau_{rz}\left(\frac{\partial v_r}{\partial z}+\frac{\partial v_z}{\partial r}\right)+\tau_{\theta z}\left(\frac{1}{r}\frac{\partial v_z}{\partial \theta}+\frac{\partial v_\theta}{\partial z}\right)\right]$ $-\left[\tau_{rr}\frac{\partial v_r}{\partial r}+\tau_{\theta\theta}\frac{1}{r}\left(\frac{\partial v_\theta}{\partial \theta}+v_r\right)+\tau_{zz}\frac{\partial v_z}{\partial z}\right]=\rho C_v\left(\frac{\partial T}{\partial t}+v_r\frac{\partial T}{\partial r}+\frac{v_\theta}{r}\frac{\partial T}{\partial \theta}+v_z\frac{\partial T}{\partial z}\right)$
Spherical	$-\left(\frac{1}{r^2}\frac{\partial}{\partial r}(r^2q_r)+\frac{1}{r\sin\theta}\frac{\partial}{\partial \theta}(q_\theta\sin\theta)+\frac{1}{r\sin\theta}\frac{\partial q_\phi}{\partial \phi}\right)$ $-T\left(\frac{\partial P}{\partial T}\right)_\rho\left(\frac{1}{r^2}\frac{\partial}{\partial r}(r^2v_r)+\frac{1}{r\sin\theta}\frac{\partial}{\partial \theta}(v_\theta\sin\theta)+\frac{1}{r\sin\theta}\frac{\partial v_\phi}{\partial \phi}\right)$ $-\left[\tau_{r\theta}\left(r\frac{\partial}{\partial r}\left(\frac{v_\theta}{r}\right)+\frac{1}{r}\frac{\partial v_r}{\partial \theta}\right)+\tau_{r\phi}\left(r\frac{\partial}{\partial r}\left(\frac{v_\phi}{r}\right)+\frac{1}{r\sin\theta}\frac{\partial v_r}{\partial \phi}\right)+\tau_{\theta\phi}\left(\frac{1}{r}\frac{\partial v_\phi}{\partial \theta}+\frac{1}{r\sin\theta}\frac{\partial v_\theta}{\partial \phi}-\frac{\cot\theta}{r}v_\phi\right)\right]$ $-\left[\tau_{rr}\frac{\partial v_r}{\partial r}+\tau_{\theta\theta}\frac{1}{r}\left(\frac{\partial v_\theta}{\partial \theta}+v_r\right)+\tau_{\phi\phi}\left(\frac{1}{r\sin\theta}\frac{\partial v_\phi}{\partial \phi}+\frac{v_r}{r}+\frac{v_\theta\cot\theta}{r}\right)\right]$ $=\rho C_v\left(\frac{\partial T}{\partial t}+v_r\frac{\partial T}{\partial r}+\frac{v_\theta}{r}\frac{\partial T}{\partial \theta}+\frac{v_\phi}{r\sin\theta}\frac{\partial T}{\partial \phi}\right)$

Example 10.8 Falling Film Heat Transfer

We wish to determine the temperature profile and the rate of heat transfer from a heated wall into a falling film. We restrict the analysis to limited contact times, when the fluid has been in touch with the wall for a distance equal to a few film thicknesses. Figure 10.11 shows a schematic of the system. The fluid flows down the plate under the action of gravity. The surface of the plate is stationary and held at a constant temperature of T_s. The fluid enters the system at a temperature of T_o, and we may assume that the flow is fully developed that the fluid film is of constant thickness, Δ_f, and that all fluid properties are constant (ρ, μ, and k). The air in contact with the surface of the fluid will also be at a constant temperature of T_o.

To determine the temperature profile in the fluid as a function of position along the plate, we must first determine the steady-state velocity profile. If the profile is fully developed, then the velocity, v_x, will only be a function of distance from the plate (y). Using the x-momentum equation with $v_y = v_z = 0$ and variations in the y-direction only, we have

$$\mu\frac{d^2v_x}{dy^2}+\rho g_x=0$$

TABLE 10.10
Energy Equation for a Newtonian Fluid with Constant Properties

Coordinate System	Energy Equation
Cartesian	$k\left(\frac{\partial^2 T}{\partial x^2}+\frac{\partial^2 T}{\partial y^2}+\frac{\partial^2 T}{\partial z^2}\right)+2\mu\left[\left(\frac{\partial v_x}{\partial x}\right)^2+\left(\frac{\partial v_y}{\partial y}\right)^2+\left(\frac{\partial v_z}{\partial z}\right)^2\right]$ $+\mu\left[\left(\frac{\partial v_x}{\partial y}+\frac{\partial v_y}{\partial x}\right)^2+\left(\frac{\partial v_x}{\partial z}+\frac{\partial v_z}{\partial x}\right)^2+\left(\frac{\partial v_z}{\partial y}+\frac{\partial v_y}{\partial z}\right)^2\right]=\rho C_p\left(\frac{\partial T}{\partial t}+v_x\frac{\partial T}{\partial x}+v_y\frac{\partial T}{\partial y}+v_z\frac{\partial T}{\partial z}\right)$
Cylindrical	$k\left(\frac{1}{r}\frac{\partial}{\partial r}\left(r\frac{\partial T}{\partial r}\right)+\frac{1}{r^2}\frac{\partial^2 T}{\partial \theta^2}+\frac{\partial^2 T}{\partial z^2}\right)+2\mu\left[\left(\frac{\partial v_r}{\partial r}\right)^2+\left(\frac{1}{r}\frac{\partial v_\theta}{\partial \theta}+\frac{v_r}{r}\right)^2+\left(\frac{\partial v_z}{\partial z}\right)^2\right]$ $+\mu\left[\left(r\frac{\partial}{\partial r}\left(\frac{v_\theta}{r}\right)+\frac{1}{r}\frac{\partial v_r}{\partial \theta}\right)^2+\left(\frac{1}{r}\frac{\partial v_z}{\partial \theta}+\frac{\partial v_\theta}{\partial z}\right)^2+\left(\frac{\partial v_z}{\partial r}+\frac{\partial v_r}{\partial z}\right)^2\right]=\rho C_p\left(\frac{\partial T}{\partial t}+v_r\frac{\partial T}{\partial r}+\frac{v_\theta}{r}\frac{\partial T}{\partial \theta}+v_z\frac{\partial T}{\partial z}\right)$
Spherical	$k\left(\frac{1}{r^2}\frac{\partial}{\partial r}\left(r^2\frac{\partial T}{\partial r}\right)+\frac{1}{r^2\sin\theta}\frac{\partial}{\partial \theta}\left(\sin\theta\frac{\partial T}{\partial \theta}\right)+\frac{1}{r^2\sin^2\theta}\frac{\partial^2 T}{\partial \phi^2}\right)$ $+2\mu\left[\left(\frac{\partial v_r}{\partial r}\right)^2+\left(\frac{1}{r}\frac{\partial v_\theta}{\partial \theta}+\frac{v_r}{r}\right)^2+\left(\frac{1}{r\sin\theta}\frac{\partial v_\phi}{\partial \phi}+\frac{v_r}{r}+\frac{v_\theta\cot\theta}{r}\right)^2\right]$ $+\mu\left[\left(r\frac{\partial}{\partial r}\left(\frac{v_\theta}{r}\right)+\frac{1}{r}\frac{\partial v_r}{\partial \theta}\right)^2+\left(\frac{\sin\theta}{r}\frac{\partial}{\partial \theta}\left(\frac{v_\phi}{\sin\theta}\right)+\frac{1}{r\sin\theta}\frac{\partial v_\theta}{\partial \phi}\right)^2\right.$ $\left.+\left(\frac{1}{r\sin\theta}\frac{\partial v_r}{\partial \phi}+r\frac{\partial}{\partial r}\left(\frac{v_\phi}{r}\right)\right)^2\right]=\rho C_p\left(\frac{\partial T}{\partial t}+v_r\frac{\partial T}{\partial r}+\frac{v_\theta}{r}\frac{\partial T}{\partial \theta}+\frac{v_\phi}{r\sin\theta}\frac{\partial T}{\partial \phi}\right)$

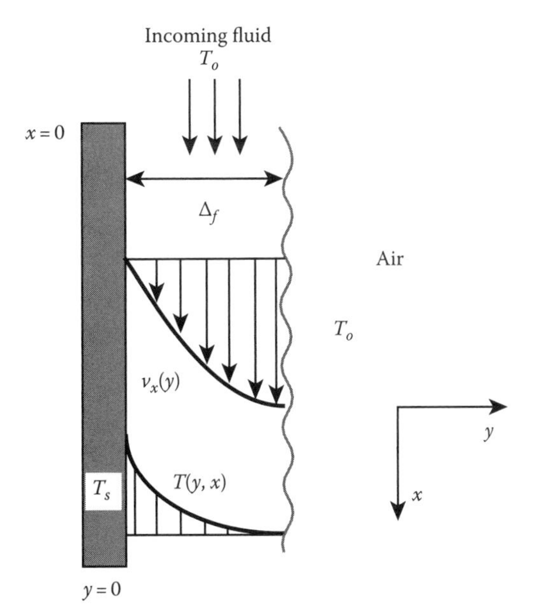

FIGURE 10.11 Heat flow into a falling liquid film. Surface temperature and film thickness are constant.

The momentum balance is subject to boundary conditions specifying no slip at the surface of the plate and no shear stress at the film–air interface (air viscosity is effectively 0):

$$y = 0 \qquad v_x = 0$$

$$y = \Delta_f \qquad \mu \frac{dv_x}{dy} = 0$$

The solution to this equation and associated boundary conditions is

$$v_x = \left(\frac{g_x}{2\nu}\right)[2\Delta_f y - y^2]$$

To determine the temperature profile, we begin with the energy equation for an incompressible fluid with constant properties. Since the fluid is incompressible, we may replace C_v with C_p:

$$k\left(\frac{\partial^2 T}{\partial x^2} + \frac{\partial^2 T}{\partial y^2} + \frac{\partial^2 T}{\partial z^2}\right) + 2\mu\left[\left(\frac{\partial v_x}{\partial x}\right)^2 + \left(\frac{\partial v_y}{\partial y}\right)^2 + \left(\frac{\partial v_z}{\partial z}\right)^2\right]$$
$$+\mu\left[\left(\frac{\partial v_x}{\partial y} + \frac{\partial v_y}{\partial x}\right)^2 + \left(\frac{\partial v_x}{\partial z} + \frac{\partial v_z}{\partial x}\right)^2 + \left(\frac{\partial v_z}{\partial y} + \frac{\partial v_y}{\partial z}\right)^2\right] = \rho C_p\left(\frac{\partial T}{\partial t} + v_x\frac{\partial T}{\partial x} + v_y\frac{\partial T}{\partial y} + v_z\frac{\partial T}{\partial z}\right)$$

We can make a number of simplifying assumptions to reduce this equation to a more tractable form. First, we do not have a high shear situation (shear stress at the film surface is zero), so all the viscous dissipation terms can be neglected. We also have no ***y***- or ***z***-components of velocity to consider and the system is at steady state. This leaves us with

$$k\left(\frac{\partial^2 T}{\partial x^2} + \frac{\partial^2 T}{\partial y^2} + \frac{\partial^2 T}{\partial z^2}\right) = \rho C_p v_x \frac{\partial T}{\partial x}$$

Our final assumptions consider the fact that heat transfer is occurring across the film, between the plate and the air. The highest temperature gradients will be in the direction of heat transfer, and so

$$\frac{\partial^2 T}{\partial y^2} \gg \frac{\partial^2 T}{\partial x^2}, \frac{\partial^2 T}{\partial z^2}$$

The final form for the energy equation becomes

$$\alpha\frac{\partial^2 T}{\partial y^2} = v_x\frac{\partial T}{\partial x}$$

Substituting in for the velocity profile gives

$$\alpha\frac{\partial^2 T}{\partial y^2} = \left(\frac{g}{2\nu}\right)[2\Delta_f y - y^2]\frac{\partial T}{\partial x}$$

This form of the energy equation is appropriate regardless of the length of the heated plate. However, it is very difficult to solve and we can make further simplifications since we are most concerned with what happens a short distance down the plate. At these short times, the heat will have penetrated only a small distance into the fluid. For our purposes, it will appear as if the air–fluid interface is infinitely far away. If the thermal penetration depth is much less than the film thickness, all heat transfer occurs across a distance, $y \ll \Delta_f$. Since $\Delta_f y \gg y^2$, we can neglect the curvature of the velocity profile:

$$\alpha \frac{\partial^2 T}{\partial y^2} = \left(\frac{g\Delta_f}{2\nu}\right) y \frac{\partial T}{\partial x}$$

The boundary conditions for this problem reflect that at the fluid–plate interface, the fluid temperature is the same as that of the plate, and at the fluid–air interface, the fluid temperature is that of the air. The starting condition states that the fluid contacts the plate at a uniform temperature of T_o:

$$y = 0 \qquad T = T_s$$

$$y = \infty \qquad T = T_o$$

$$x = 0 \qquad T = T_o$$

These boundary conditions should be reminiscent of the semi-infinite systems we discussed in Chapter 6. In fact, we can use a similarity solution to solve the energy balance. Since we have no length scale along the plate at short contact times, the temperature profile should be self-similar in x. The variables we must include in the similarity variable are

$$y - \text{length(m)} \qquad x - \text{length(m)} \qquad \frac{\alpha\nu}{g\Delta_f} = \ell - \text{length}^2\,(\text{m}^2)$$

Finding the similarity variable for this problem is the same as the process we used in Chapter 6. Knowing that we must form a single dimensionless variable pitting y against x, we obtain [9]

$$\varepsilon = \frac{y}{(9\ell x)^{1/3}}$$

Substituting into the energy balance leads to an ordinary differential equation of the form

$$\frac{d^2\theta}{d\varepsilon^2} + 3\varepsilon^2 \frac{d\theta}{d\varepsilon} = 0 \qquad \text{where } \theta = \frac{T - T_o}{T_s - T_o}$$

The boundary conditions become

$$\varepsilon = 0 \qquad \theta = 1$$

$$\varepsilon = \infty \qquad \theta = 0$$

We solve the ordinary differential equation with a familiar trick: letting $\theta' = d\theta/d\varepsilon$, separating, and integrating:

$$\theta' = a_1 \exp(-\varepsilon^3)$$

$$\theta = a_1 \int_0^{\varepsilon} \exp(-\eta^3)\,d\eta + a_2$$

Applying the boundary conditions shows that

$$1 = a_1 \int_0^0 \exp(-\eta^3)d\eta + a_2 \rightarrow a_2 = 1$$

$$0 = a_1 \int_0^\infty \exp(-\eta^3)d\eta + 1 \rightarrow a_1 = -\frac{9\Gamma(2/3)}{2\pi\sqrt{3}}$$

where Γ is the gamma function (something like the error function) and is tabulated in a number of mathematical handbooks [10]. The final temperature profile is

$$\theta = 1 - \frac{9\Gamma(2/3)}{2\pi\sqrt{3}} \int_0^\varepsilon \exp(-\eta^3)d\eta$$

Figure 10.12 shows the dimensionless temperature profile in the fluid. Notice that the profile is linear for small ε and then levels out as $\varepsilon > 1$. We can use this figure to define a penetration depth into the liquid. The penetration depth is the depth at which θ becomes 99% of the free stream value. This occurs when $\varepsilon \approx 1.5$, so the penetration depth is approximately

$$\delta_T = 3.12(\ell x)^{1/3}$$

Notice that the penetration depth here varies as $(x/v_{max})^{1/3}$, whereas in Chapter 6, the *analogous* penetration depth varied as $t^{1/2}$. Here, we can see that the role of convection is to increase the heat transfer by increasing the temperature gradient at the surface and thereby increasing the penetration depth for a given time or, in this case, distance down the plate.

We use Fourier's law to get the heat flow into the fluid. The heat flow is highest when the fluid first encounters the heated plate and thereafter decreases as the driving force for heat transfer decreases:

$$q = -kA_s \left.\frac{\partial T}{\partial y}\right|_{y=0} = -\frac{kA_s(T_s - T_o)}{(9\ell x)^{1/3}} \left.\frac{d\theta}{d\varepsilon}\right|_{\varepsilon=0}$$

$$= \frac{kA_s(T_s - T_o)}{(9\ell x)^{1/3}} \frac{9\Gamma(2/3)}{2\pi\sqrt{3}}$$

This is an important result. If we use the heat flow to define a heat transfer coefficient, h, we will see that the heat transfer coefficient is highest at the point of plate–fluid contact and decreases

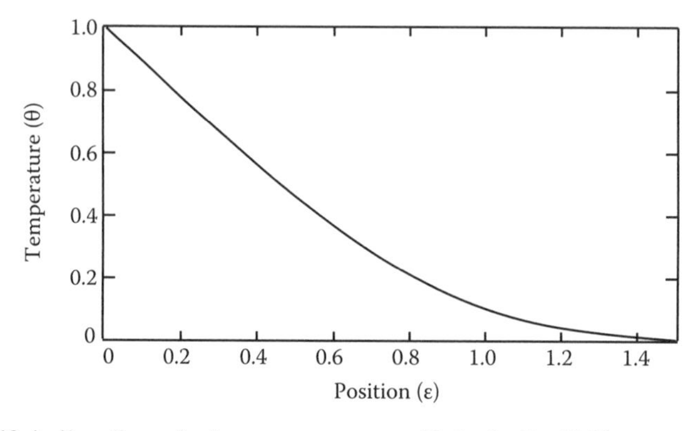

FIGURE 10.12 Self-similar, dimensionless temperature profile in the liquid film.

with position along the plate. We will see this behavior again when we consider thermal boundary layers in Chapter 11:

$$q = -\frac{kA_s(T_s - T_o)}{(9\ell x)^{1/3}} \left.\frac{d\theta}{d\varepsilon}\right|_{\varepsilon=0} = hA_s(T_s - T_o)$$

$$h = -\frac{k}{(9\ell x)^{1/3}} \left.\frac{d\theta}{d\varepsilon}\right|_{\varepsilon=0} = \frac{k}{(9\ell x)^{1/3}} \frac{9\Gamma(2/3)}{2\pi\sqrt{3}}$$

10.4 MASS TRANSPORT

Though the overall mass may be conserved, individual species may accumulate in the system through convection, diffusion, or chemical reaction. In this section, we will apply the conservation of mass to an individual species within a fixed volume element in space and thereby derive the species continuity equation. Figure 10.13 shows the control volume we will use. We implicitly assume the system is a binary mixture of *a* and *b*.

The balance equation considers convection in and out of the control volume, diffusion through the surfaces of the volume, generation of a species via reaction, and accumulation within the control volume:

In	–	*Out*	+	*Gen*	=	*Acc*
Rate of *a* in by convection	–	Rate of *a* out by convection	+			
Rate of *a* in by diffusion	–	Rate of *a* out by diffusion	+	Net rate of *a* produced	=	Rate of *a* accumulated

$$\begin{aligned} &\textit{In} \quad - \quad \textit{Out} \\ &[n_{ax}(x) - n_{ax}(x+\Delta x)]\Delta y\Delta z + [n_{ay}(y) - n_{ay}(y+\Delta y)]\Delta x\Delta z \\ &\qquad\qquad\qquad\qquad + \textit{Gen} \qquad = \textit{Acc} \\ &+[n_{az}(z) - n_{az}(z+\Delta z)]\Delta x\Delta y + r_a\Delta x\Delta y\Delta z = \Delta x\Delta y\Delta z\frac{\partial \rho_a}{\partial t} \end{aligned} \tag{10.52}$$

Dividing through by $\Delta x\Delta y\Delta z$ and letting $\Delta x\Delta y\Delta z \to 0$ gives the continuity equation in terms of the mass flux, n_a:

$$-\left(\frac{\partial n_{ax}}{\partial x} + \frac{\partial n_{ay}}{\partial y} + \frac{\partial n_{az}}{\partial z}\right) + r_a = \frac{\partial \rho_a}{\partial t} \tag{10.53}$$

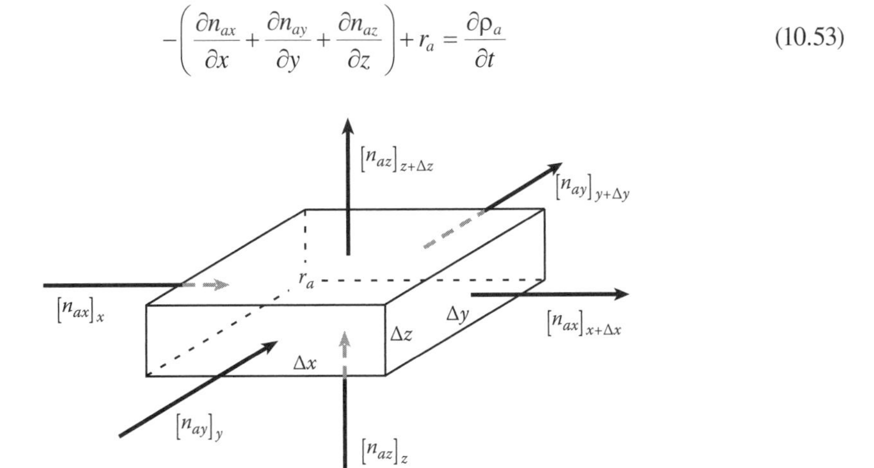

FIGURE 10.13 Control volume for derivation of the species continuity equation.

We apply the same technique to derive the species continuity equation for b:

$$-\left(\frac{\partial n_{bx}}{\partial x}+\frac{\partial n_{by}}{\partial y}+\frac{\partial n_{bz}}{\partial z}\right)+r_b=\frac{\partial \rho_b}{\partial t} \tag{10.54}$$

If we add Equations 10.53 and 10.54 together and assume that a and b interconvert, $a \rightleftharpoons b$, we obtain the following overall continuity equation:

$$-\left(\frac{\partial}{\partial x}(\rho v_x)+\frac{\partial}{\partial y}(\rho v_y)+\frac{\partial}{\partial z}(\rho v_z)\right)=\frac{\partial \rho}{\partial t}=-(\vec{\nabla}\cdot\rho\vec{\mathbf{v}}) \tag{10.55}$$

Here, we make use of the fact that $\vec{\mathbf{n}}_\mathbf{a}+\vec{\mathbf{n}}_\mathbf{b}=\rho\vec{\mathbf{v}}$ where $\vec{\mathbf{v}}$ is the mass average velocity, $r_a+r_b=0$, and $\rho_a+\rho_b=\rho$. In vector notation, the continuity equations for species a and b are

$$-(\vec{\nabla}\cdot\vec{\mathbf{n}}_\mathbf{a})+r_a=\frac{\partial \rho_a}{\partial t} \qquad (\vec{\nabla}\cdot\vec{\mathbf{n}}_b)+r_b=\frac{\partial \rho_b}{\partial t} \tag{10.56}$$

We can also write the continuity equations on a molar basis:

$$-\left(\frac{\partial N_{ax}}{\partial x}+\frac{\partial N_{ay}}{\partial y}+\frac{\partial N_{az}}{\partial z}\right)+R_a=\frac{\partial c_a}{\partial t} \tag{10.57}$$

$$-\left(\frac{\partial N_{bx}}{\partial x}+\frac{\partial N_{by}}{\partial y}+\frac{\partial N_{bz}}{\partial z}\right)+R_b=\frac{\partial c_b}{\partial t} \tag{10.58}$$

Adding these two equations yields an overall continuity equation, but unlike the overall continuity equation based on mass, the total number of moles may not be conserved and so $R_a+R_b\neq 0$:

$$-\left(\frac{\partial}{\partial x}(c_t v_{Mx})+\frac{\partial}{\partial y}(c_t v_{My})+\frac{\partial}{\partial z}(c_t v_{Mz})\right)+(R_a+R_b)=\frac{\partial c_t}{\partial t} \tag{10.59}$$

Here, $c_a+c_b=c_t$ and $\vec{\mathbf{N}}_\mathbf{a}+\vec{\mathbf{N}}_\mathbf{b}=c_t\vec{\mathbf{v}}_\mathbf{M}$ where $\vec{\mathbf{v}}_\mathbf{M}$ is the molar average velocity. In vector notation, we can rewrite Equation 10.59 to give

$$-(\vec{\nabla}\cdot c_t\vec{\mathbf{v}}_\mathbf{M})+(R_a+R_b)=\frac{\partial c_t}{\partial t} \tag{10.60}$$

and Equation 10.57

$$-(\vec{\nabla}\cdot\vec{\mathbf{N}}_\mathbf{a})+R_a=\frac{\partial c_a}{\partial t} \tag{10.61}$$

Equations 10.53 through 10.61 are complete but not the most useful forms for calculating concentration or density profiles. To obtain these descriptions, it is advantageous to decompose the fluxes, N_a and n_a, using Fick's laws (if appropriate):

$$\vec{\mathbf{N}}_\mathbf{a}=c_a\vec{\mathbf{v}}_\mathbf{M}-c_tD_{ab}\vec{\nabla}x_a \tag{10.62}$$

$$\vec{\mathbf{n}}_\mathbf{a}=\rho_a\vec{\mathbf{v}}-\rho D_{ab}\vec{\nabla}w_a \tag{10.63}$$

Substituting into the continuity equations, we find

$$-(\vec{\nabla}\cdot\rho_a\vec{\mathbf{v}})+(\vec{\nabla}\cdot\rho D_{ab}\vec{\nabla}w_a)+r_a=\frac{\partial\rho_a}{\partial t} \tag{10.64}$$

$$-(\vec{\nabla}\cdot c_a\vec{\mathbf{v}}_{\mathbf{M}})+(\vec{\nabla}\cdot c_t D_{ab}\vec{\nabla}x_a)+R_a=\frac{\partial c_a}{\partial t} \tag{10.65}$$

Finally, if we assume constant density, ρ, constant total concentration, c_t, and constant diffusivity, D_{ab}, we can simplify Equations 10.64 and 10.65. Treating Equation 10.64 first, we have by expansion

$$-\rho_a(\vec{\nabla}\cdot\vec{\mathbf{v}})-v\cdot\vec{\nabla}\rho_a+D_{ab}\nabla^2\rho_a+r_a=\frac{\partial\rho_a}{\partial t} \tag{10.66}$$

If the overall density, ρ, is constant, then $\vec{\nabla}\cdot\vec{\mathbf{v}}=0$ (from Equation 10.4) and the first term in Equation 10.66 vanishes:

$$-\vec{\mathbf{v}}\cdot\vec{\nabla}\rho_a+D_{ab}\nabla^2\rho_a+r_a=\frac{\partial\rho_a}{\partial t} \tag{10.67}$$

Dividing through by the molecular weight of *a*, M_{wa}, lets us rewrite Equation 10.67 in terms of the concentration of *a*:

$$-\vec{\mathbf{v}}\cdot\vec{\nabla}c_a+D_{ab}\nabla^2 c_a+R_a=\frac{\partial c_a}{\partial t} \tag{10.68}$$

This equation is very useful for describing the *concentration profile of dilute, liquid solutions containing a*. The solutions should be at constant temperature and pressure. The left-hand side of Equation 10.68 can be written in terms of the substantial derivative, Dc_a/Dt reflecting the overall accumulation of *a* as a result of convection taking place through our stationary control volume. Remember if we were to ride along with the mean flow of the fluid, our accumulation of *a* would be written solely in terms of $\partial c_a/\partial t$:

$$D_{ab}\nabla^2 c_a+R_a=\frac{Dc_a}{Dt} \tag{10.69}$$

Turning our attention to Equation 10.65, we find upon expansion

$$-c_a(\vec{\nabla}\cdot\vec{\mathbf{v}}_{\mathbf{M}})-\vec{\mathbf{v}}_{\mathbf{M}}\cdot\vec{\nabla}c_a+D_{ab}\nabla^2 c_a+R_a=\frac{\partial c_a}{\partial t} \tag{10.70}$$

If the total concentration is constant, then Equation 10.60 shows that

$$(\vec{\nabla}\cdot\vec{\mathbf{v}}_{\mathbf{M}})=\frac{1}{c_t}(R_a+R_b) \tag{10.71}$$

and we can rewrite Equation 10.70 as

$$-\vec{\mathbf{v}}_{\mathbf{M}}\cdot\vec{\nabla}c_a+D_{ab}\nabla^2 c_a+R_a-\frac{c_a}{c_t}(R_a+R_b)=\frac{\partial c_a}{\partial t} \tag{10.72}$$

TABLE 10.11
Species Continuity Equation: Flux Formulation

Coordinate System	Species Continuity Equation
Cartesian	$-\left(\frac{\partial N_{ax}}{\partial x}+\frac{\partial N_{ay}}{\partial y}+\frac{\partial N_{az}}{\partial z}\right)+R_a=\frac{\partial c_a}{\partial t}$
Cylindrical	$-\left(\frac{1}{r}\frac{\partial}{\partial r}(rN_{ar})+\frac{1}{r}\frac{\partial N_{a\theta}}{\partial \theta}+\frac{\partial N_{az}}{\partial z}\right)+R_a=\frac{\partial c_a}{\partial t}$
Spherical	$-\left(\frac{1}{r^2}\frac{\partial}{\partial r}\left(r^2 N_{ar}\right)+\frac{1}{r\sin\theta}\frac{\partial}{\partial \theta}(N_{a\theta}\sin\theta)+\frac{1}{r\sin\theta}\frac{\partial N_{a\phi}}{\partial \phi}\right)+R_a=\frac{\partial c_a}{\partial t}$

TABLE 10.12
Species Continuity Equation: Fick's Law and Constant Properties

Coordinate System	Species Continuity Equation
Cartesian	$D_{ab}\left(\frac{\partial^2 c_a}{\partial x^2}+\frac{\partial^2 c_a}{\partial y^2}+\frac{\partial^2 c_a}{\partial z^2}\right)+R_a=\left(\frac{\partial c_a}{\partial t}+v_x\frac{\partial c_a}{\partial x}+v_y\frac{\partial c_a}{\partial y}+v_z\frac{\partial c_a}{\partial z}\right)$
Cylindrical	$D_{ab}\left(\frac{1}{r}\frac{\partial}{\partial r}\left(r\frac{\partial c_a}{\partial r}\right)+\frac{1}{r^2}\frac{\partial^2 c_a}{\partial \theta^2}+\frac{\partial^2 c_a}{\partial z^2}\right)+R_a=\left(\frac{\partial c_a}{\partial t}+v_r\frac{\partial c_a}{\partial r}+\frac{v_\theta}{r}\frac{\partial c_a}{\partial \theta}+v_z\frac{\partial c_a}{\partial z}\right)$
Spherical	$D_{ab}\left(\frac{1}{r^2}\frac{\partial}{\partial r}\left(r^2\frac{\partial c_a}{\partial r}\right)+\frac{1}{r^2\sin\theta}\frac{\partial}{\partial \theta}\left(\sin\theta\frac{\partial c_a}{\partial \theta}\right)+\frac{1}{r^2\sin^2\theta}\frac{\partial^2 c_a}{\partial \phi^2}\right)+R_a$ $=\left(\frac{\partial c_a}{\partial t}+v_r\frac{\partial c_a}{\partial r}+\frac{v_\theta}{r}\frac{\partial c_a}{\partial \theta}+\frac{v_\phi}{r\sin\theta}\frac{\partial c_a}{\partial \phi}\right)$

This equation is useful for *low-density gases at a constant temperature and pressure.* We can rewrite the equation in terms of the substantial derivative:

$$D_{ab}\nabla^2 c_a+R_a-\frac{c_a}{c_t}(R_a+R_b)=\frac{Dc_a}{Dt} \tag{10.73}$$

Tables 10.11 and 10.12 show the binary species continuity equations in their most useful forms in rectangular, cylindrical, and spherical coordinates.

Example 10.9 Diffusion into a Falling Film

In this example, we wish to calculate the concentration profile and the rate of mass transfer from a gas containing species *a* into a falling film containing species *b*. We will concern ourselves with limited contact times, or when the fluid has been in touch with the gas for a distance equivalent in magnitude to a few film thicknesses. Within this time frame, species *a* will not be able to penetrate very far into the liquid film. Figure 10.14 shows a schematic of the system. The fluid flows down the plate under the action of gravity. The surface of the plate is stationary and the fluid enters the system containing no *a*. We may assume that the flow is fully developed, that the fluid film is of constant thickness, Δ_f, that all fluid properties are constant (ρ, μ, and D_{ab}), and that the partition coefficient between liquid and gas is $K_{eq} = 1$.

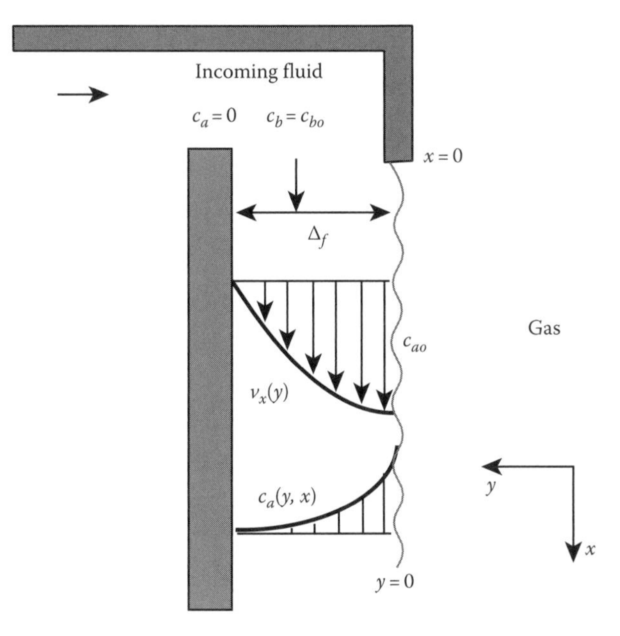

FIGURE 10.14 Diffusion into a falling film. Diffusing species, a.

As in the previous falling film example, our first task is to determine the velocity profile within the fluid. Since we have already solved for a profile of this type, we know it will be parabolic. Remember we assume that the velocity profile is fully developed so that it only varies in the y-direction. In this system, we reverse the coordinates so that the film–air interface is at $y = 0$ and the plate is at $y = \Delta_f$:

$$v_x = \left(\frac{g}{2\nu}\right)\left[\Delta_f^2 - y^2\right]$$

If a does not diffuse very far into the liquid ($y \ll \Delta_f$), it will be confined to a region where the velocity profile barely changes. In this region, we can approximate the velocity as

$$v_x = \frac{g\Delta_f^2}{\nu} = v_{max}$$

The species continuity Equation 10.69 with no chemical reaction is

$$D_{ab}\left(\frac{\partial^2 c_a}{\partial x^2} + \frac{\partial^2 c_a}{\partial y^2} + \frac{\partial^2 c_a}{\partial z^2}\right) = \left(\frac{\partial c_a}{\partial t} + v_x\frac{\partial c_a}{\partial x} + v_y\frac{\partial c_a}{\partial y} + v_z\frac{\partial c_a}{\partial z}\right)$$

We assume steady-state operation and with $v_y = v_z = 0$, we have

$$D_{ab}\left(\frac{\partial^2 c_a}{\partial x^2} + \frac{\partial^2 c_a}{\partial y^2} + \frac{\partial^2 c_a}{\partial z^2}\right) = v_x\frac{\partial c_a}{\partial x}$$

Since mass transfer is occurring from the gas into the liquid, $\partial c_a/\partial y \gg \partial c_a/\partial x$ or $\partial c_a/\partial z$ and we can simplify the species continuity equation to

$$D_{ab}\frac{\partial^2 c_a}{\partial y^2} = v_x\frac{\partial c_a}{\partial x} = v_{max}\frac{\partial c_a}{\partial x}$$

subject to the following boundary conditions:

$$x = 0 \qquad c_a = 0$$

$$y = 0 \qquad c_a = c_{ao}$$

$$y = \infty \qquad c_a = 0$$

This is a semi-infinite problem whose resemblance is clear, if we fold v_{max} into the x-coordinate to generate *a time* variable, x/v_{max}:

$$\frac{\partial c_a}{\partial (x/v_{max})} = D_{ab}\frac{\partial^2 c_a}{\partial y^2}$$

We can pirate the similarity variable from Chapter 6 to use in this problem. All we need is to include our new definition of *time*:

$$\varepsilon = \frac{y}{\sqrt{4D_{ab}(x/v_{max})}}$$

Changing variables in the mass balance leads to an ordinary differential equation of the following form:

$$\frac{d^2\chi}{d\varepsilon^2} + 2\varepsilon\frac{d\chi}{d\varepsilon} = 0 \qquad \chi = \frac{c_a}{c_{ao}}$$

with transformed boundary conditions of

$$\varepsilon = 0 \qquad \chi = 1$$

$$\varepsilon = \infty \qquad \chi = 0$$

The solution to the differential equation is given in terms of the error function, *erf*,

$$\chi = 1 - erf(\varepsilon)$$

and the penetration depth of *a* into the fluid is given by

$$\delta_x = 3.643\sqrt{D_{ab}\left(\frac{x}{v_{max}}\right)}$$

Unlike the heat transfer example, since we diffuse from the gas into the fluid in a region where the velocity gradient is effectively zero, we obtain no change in the form of the penetration depth by having convection present. The penetration depth still varies with $\sqrt{x/v_{max}}$ instead of $\sqrt{t}$, but we can now control t by changing the velocity of the liquid.

The rate of mass transfer is found by substituting the concentration profile into Fick's law evaluated at $y = \delta_x$:

$$N_a = -D_{ab}\left.\frac{\partial c_a}{\partial y}\right|_{y=\delta_x} = c_{ao}\sqrt{\frac{D_{ab}}{\pi(x/v_{max})}}$$

Notice that the rate of mass transfer is highest at the start of the plate and decreases with $\sqrt{x}$. We can define a mass transfer coefficient in the same way we previously defined a heat transfer coefficient. Evaluating the mass transfer coefficient shows that it too is highest at the beginning of the plate and steadily decreases as the fluid flows down the plate and the driving force for mass transfer decreases:

$$N_a = k_c(c_{ao} - c_a) = c_{ao}\sqrt{\frac{D_{ab}}{\pi(x/v_{max})}}$$

$$k_c = \sqrt{\frac{D_{ab}}{\pi(x/v_{max})}}$$

Example 10.10 Absorption from a Bubble Growing in a Liquid [11]

A gas bubble grows from the tip of a nozzle that is held in a stagnant liquid. The bubble is supplied with pure gas at a constant volumetric flow rate, $\dot{v}$, m³/s. Absorption of species a into the liquid, b, proceeds and so the bubble grows. Our purpose is to derive an approximate expression for the rate of diffusion into the liquid as a function of the time of bubble growth. The gas is slightly soluble in the liquid and the amount of gas dissolved is assumed to be small enough so that it does not affect the size of the bubble. Moreover, we assume that since the bubble is growing, convection within the bubble keeps the concentration of a there fixed at c_{ai}. The resistance to diffusion is confined to a thin film of liquid surrounding the growing bubble. The thickness of this film will then be much smaller than the size of the bubble, and far away from the bubble, the fluid will not have felt the effects of mass transfer so the concentration of a there will be c_{ao}. Figure 10.15 shows the situation. If the volume of the bubble grows at a constant rate, $\dot{v}$, then the bubble radius, r_o, can be determined from its volumetric growth rate by

$$r_o = bt^{1/3} \qquad b = \left(\frac{3\dot{v}}{4\pi}\right)^{1/3}$$

The velocity of the interface is the time rate of change of the bubble radius:

$$v_i = \frac{dr_o}{dt} = \left(\frac{b}{3}\right)t^{-2/3}$$

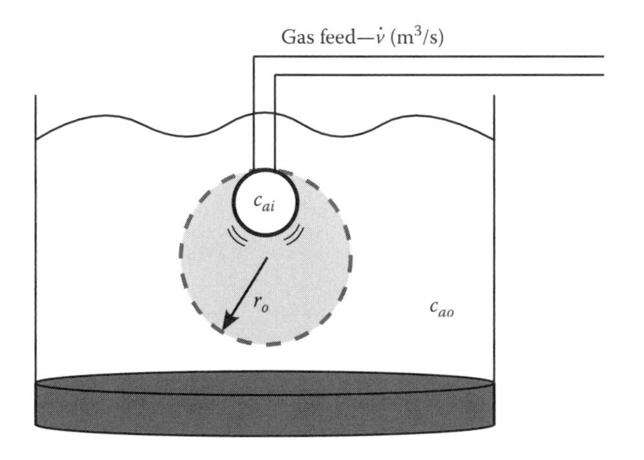

FIGURE 10.15 Gas absorption from a growing bubble.

and the radial velocity at any distance, r, from the center, v_r, is

$$v_r = v_i \left(\frac{r_o}{r} \right)^2$$

We have a transient process here where transport of the gas, a, takes place via two mechanisms: diffusion from the bubble into the liquid and convection due to the growing bubble carrying gas along with it into the fluid. Since the gas is hardly soluble in the liquid, we can write the transient diffusion equation solely in terms of the concentration of gas, c_a. With bubble growth in the radial direction only, we can reduce the overall mass balance to

$$\frac{\partial c_a}{\partial t} + v_r \frac{\partial c_a}{\partial r} = D_{ab} \left[\frac{\partial^2 c_a}{\partial r^2} + \left(\frac{2}{r} \right) \frac{\partial c_a}{\partial r} \right]$$

The boundary conditions in the liquid are

$t = 0 \qquad c_a(r,0) = c_{ao}$ Initial condition

$r = \infty \qquad c_a(\infty,t) = c_{ao}$ Semi-infinite liquid medium relative to the size of the bubble

$r = r_o \qquad c_a(r_o,t) = c_{ai}$ Constant composition inside the bubble. No intra-bubble gradients

Now, we assume the diffusional resistance always takes place inside a thin layer next to the bubble's surface, and since the growth of the bubble stretches that layer and makes it ever thinner, the influence of curvature on the solution represented by $(2/r)\partial c_a/\partial r$ is negligible. Furthermore, the velocity near the bubble's surface, $r = r_o$, can be represented accurately by using a linear approximation for the quadratic function:

$$v_r \approx v_i \left(1 - \frac{2y}{r} \right) = \frac{b}{3t^{2/3}} - \frac{2}{3}\frac{y}{t}$$

where $y = r - r_o$ and is the distance from the interface rather than the distance from the center of the bubble. Transforming the mass balance from r to y gives

$$\frac{\partial c_a}{\partial t} - \left(\frac{2y}{3t} \right) \frac{\partial c_a}{\partial y} = D_{ab} \frac{\partial^2 c_a}{\partial y^2}$$

To solve this equation, we use a trick that is not at all straightforward. We define two dimensionless variables composed of the independent variables of our problem: y, t, D_{ab}, and b. One variable, z, replaces y and the second, τ, replaces t:

$$z = \left(\frac{D_{ab}^3}{b^7} \right) t^{2/3} y \qquad \tau = \left(\frac{3}{7} \right) \left(\frac{D_{ab}^7}{b^{14}} \right) t^{7/3}$$

Substituting into the differential equation leaves us with a simple problem to solve:

$$\frac{\partial c_a}{\partial \tau} = \frac{\partial^2 c_a}{\partial z^2}$$

Our new boundary conditions read

$$\tau = 0 \qquad c_a(z,0) = c_{ao}$$

$$z = \infty \qquad c_a(0,\tau) = c_{ao}$$

$$z = 0 \qquad c_a(0,\tau) = c_{ai}$$

The solution is a familiar one, written in terms of the error function:

$$\frac{c_a - c_{ao}}{c_{ai} - c_{ao}} = 1 - erf\left(\frac{z}{2\sqrt{\tau}}\right)$$

The flux at the surface of the growing bubble is found from Fick's law evaluated at the bubble's surface:

$$N_a = -D_{ab}\left(\frac{\partial c_a}{\partial y}\right)_0 = -\left(\frac{D_{ab}^4}{b^7}\right)t^{2/3}\left(\frac{\partial c_a}{\partial z}\right)_0 = \sqrt{\frac{7D_{ab}}{3\pi\tau}}(c_{ai} - c_{ao})$$

Notice that this corresponds quite closely to penetration into a semi-infinite solid. The factor of $\sqrt{7/3}$ is due to fluid motion induced by the sphere's expansion. The increase in the flux of solute is due to a convective velocity near the interface greater than that farther out in the liquid (convection is zero there). This causes the diffusion boundary layer (film thickness) to be compressed as it is stretched around the spherical surface. Since the boundary layer is compressed, concentration gradients are sharper and mass transfer is increased. The total rate of mass transfer is obtained by multiplying the flux by the surface area:

$$\dot{M}_a = N_a A_s = 4\pi r_o^2 N_a = (4\pi)^{1/3}\sqrt{\frac{7D_{ab}}{3\pi}}(c_{ai} - c_{ao})(3\dot{v})^{2/3}t^{1/6}$$

Notice that the mass transfer rate increases as $t^{1/6}$. The appearance of $\sqrt{D_{ab}}$ should also be noted.

10.5 CHARGE TRANSPORT

Charge transport is intimately related to mass transport because all charges are transported via a carrier. This carrier may be an electron or hole in a semiconductor or it may be an ion in a battery or a protein in a biological system. Unlike the species continuity equation we just derived, the charged species continuity equation is a bit more complicated. This arises because there are two forms of convection for charged species: conventional convection driven by a velocity gradient and convection driven by charge motion under the influence of an electric field. To derive the charged species continuity equation, we refer to the control volume shown in Figure 10.16.

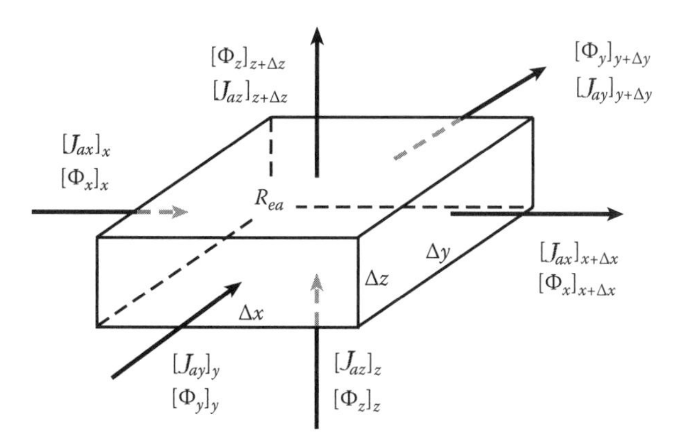

FIGURE 10.16 Control volume for the charged species continuity equation.

The balance equation on charged species *a* considers both forms of convection plus diffusion, generation due to a reaction or charge recombination, and accumulation of charged species. If charges accumulate, Poisson's equation must be used to determine the electric field distribution throughout the system:

In	–	*Out*	+	*Gen*	=	*Acc*
[Rate of *a* in by convection	–	Rate of *a* out by convection]	+			
[Rate of *a* in by diffusion	–	Rate of *a* out by diffusion]	+			
[Rate of *a* in by E-field drift	–	Rate of *a* out by E-field drift]	+	Net rate of *a* produced	=	Rate of accumulation of charged species *a*

Referring to our control volume, we have

$$[J_{ax}(x)-J_{ax}(x+\Delta x)]\Delta y\Delta z+[J_{ay}(y)-J_{ay}(y+\Delta y)]\Delta x\Delta z+[J_{az}(z)-J_{az}(z+\Delta z)]\Delta x\Delta y$$

$$+z_e\mathcal{F}_a R_{ea}\Delta x\Delta y\Delta z = z_e\mathcal{F}_a\Delta x\Delta y\Delta z\frac{\partial c_a}{\partial t} \tag{10.74}$$

Dividing by $\Delta x\Delta y\Delta z$ and letting the volume of the control volume approach zero, $\Delta x\Delta y\Delta z\to 0$, gives

$$-\left(\frac{\partial J_a}{\partial x}+\frac{\partial J_a}{\partial y}+\frac{\partial J_a}{\partial z}\right)+z_e\mathcal{F}_a R_{ea}=z_e\mathcal{F}_a\frac{\partial c_a}{\partial t} \tag{10.75}$$

The generation rate, R_{ea}, accounts for any recombination reactions that may occur, neutralizing the positive and negative charges.

In ordinary diffusion, the flux of matter is in response to a gradient in the chemical potential of the species. We showed in Chapter 2 that when we consider charged species, the motion of those charges in response to an electric field may be as large or larger than the motion that arises due to the concentration gradient. We must include this secondary transport mechanism in our balance equation. Using the form for Fick's law appropriate for charged species from Chapter 2, the flux of species *a* is

$$J_{ax}=-c_t z_e\mathcal{F}_a D_{ab}\left(\frac{dx_a}{dx}+\left(\frac{\mathcal{F}_a}{RT}\right)x_a\frac{d\Phi}{dx}\right)+v_{Mx}z_e\mathcal{F}_a c_a \tag{10.76}$$

where z_e is the valence of species *a* and $\mathcal{F}_{\hat{a}}$ is Faraday's constant. We have similar expressions for the *y*- and *z*-coordinate directions. Here, we have assumed dilute solutions so that the Einstein relationship between diffusivity, D_{ab}, and mobility, μ_{ea}, holds. Substituting for the flux in Equation 10.75 gives

$$\frac{\partial}{\partial x}\left[c_t z_e\mathcal{F}_a D_{ab}\left(\frac{dx_a}{dx}+\frac{\mathcal{F}_a}{RT}x_a\frac{d\Phi}{dx}\right)+v_{Mx}z_e\mathcal{F}_a c_a\right]+\frac{\partial}{\partial y}\left[c_t z_e\mathcal{F}_a D_{ab}\left(\frac{dx_a}{dy}+\frac{\mathcal{F}_a}{RT}x_a\frac{d\Phi}{dy}\right)+v_{My}z_e\mathcal{F}_a c_a\right]$$

$$+\frac{\partial}{\partial z}\left[c_t z_e\mathcal{F}_a D_{ab}\left(\frac{dx_a}{dz}+\frac{\mathcal{F}_a}{RT}x_a\frac{d\Phi}{dz}\right)+v_{Mz}z_e\mathcal{F}_a c_a\right]+z_e\mathcal{F}_a R_{ea}=z_e\mathcal{F}_a\frac{\partial c_a}{\partial t} \tag{10.77}$$

TABLE 10.13
Charged Species Continuity Equation: Flux Formulation

Coordinate System	Species Continuity Equation
Cartesian	$\left(\frac{\partial J_{ax}}{\partial x}+\frac{\partial J_{ay}}{\partial y}+\frac{\partial J_{az}}{\partial z}\right)+R_{ea}=\frac{\partial c_a}{\partial t}$
Cylindrical	$\left(\frac{1}{r}\frac{\partial}{\partial r}(rJ_{ar})+\frac{1}{r}\frac{\partial J_{a\theta}}{\partial \theta}+\frac{\partial J_{az}}{\partial z}\right)+R_{ea}=\frac{\partial c_a}{\partial t}$
Spherical	$\left(\frac{1}{r^2}\frac{\partial}{\partial r}(r^2 J_{ar})+\frac{1}{r\sin\theta}\frac{\partial}{\partial \theta}(J_{a\theta}\sin\theta)+\frac{1}{r\sin\theta}\frac{\partial J_{a\phi}}{\partial \phi}\right)+R_{ea}=\frac{\partial c_a}{\partial t}$

We can express this much more simply in the following vector notation:

$$-(\vec{\nabla}\cdot c_a\vec{\mathbf{v}}_{\mathbf{M}})+\left(\vec{\nabla}\cdot c_t D_{ab}\vec{\nabla}x_a+\frac{\mathcal{F}_a}{RT}(\vec{\nabla}\cdot c_t D_{ab}x_a\vec{\nabla}\Phi)\right)+R_{ea}=\frac{\partial c_a}{\partial t} \tag{10.78}$$

If the diffusivity and the total concentration of all species are constant, we can rewrite this equation as

$$-(\vec{\nabla}\cdot c_a\vec{\mathbf{v}}_{\mathbf{M}})+D_{ab}\left(\nabla^2 c_a+\frac{\mathcal{F}_a}{RT}(c_a\nabla^2\Phi+\vec{\nabla}\Phi\cdot\vec{\nabla}c_a)\right)+R_{ea}=\frac{\partial c_a}{\partial t} \tag{10.79}$$

Finally, if the fluid is incompressible so that $(\vec{\nabla}\cdot\vec{\mathbf{v}}_{\mathbf{M}})=1/c_t(R_a+R_b)$, then

$$D_{ab}\left[\nabla^2 c_a+\frac{\mathcal{F}_a}{RT}(c_a\nabla^2\Phi+\vec{\nabla}\Phi\cdot\vec{\nabla}c_a)\right]+\frac{c_a}{c_t}(R_{ea}+R_{eb})=\frac{Dc_a}{Dt} \tag{10.80}$$

where we have written the equation in terms of the substantial derivative, reflecting the accumulation of a due to convection within the control volume. Tables 10.13 and 10.14 show the charged species continuity equation written in terms of the fluxes and in terms of concentrations.

Equations 10.74 through 10.80 were written in terms of the concentration of charge-carrying species. If we multiply Equation 10.78 through by $z_e\,\mathcal{F}_a$, we can rewrite the equation in terms of the charge density from species a, $\rho_a^{\pm}$:

$$-(\vec{\nabla}\cdot\rho_a^{\pm}\vec{\mathbf{v}}_{\mathbf{M}})+[\vec{\nabla}\cdot(D_{ab}\vec{\nabla}\rho_a^{\pm})+\sigma_a(\vec{\nabla}\cdot(\rho_a^{\pm}\vec{\nabla}\Phi))]+r_{ca}=\frac{\partial\rho_a^{\pm}}{\partial t} \tag{10.81}$$

Here, σ_a is the conductivity of the medium due to the movement of a.

Example 10.11 Field-Assisted Coating of a Surface

Many metal parts are protected by coatings. Automobile bodies, for instance, are coated with rust-inhibiting primers and multilayer paint topcoats. To insure uniform coatings, without pinholes, the metal surface is charged to drive the deposition of the coating material over the entire surface. Here, we look at an electric field-assisted process for coating a metal plate. The plate is charged to a uniform potential. A fluid, containing the material to be deposited, is flowing over the surface of the plate in the form of a thin film. We will assume that the film is of uniform thickness, Δ_f,

TABLE 10.14
Charged Species Continuity Equation: Fick's Law and Constant Properties

Coordinate System	Species Continuity Equation
Cartesian	$D_{ab}\left(\frac{\partial^2 c_a}{\partial x^2}+\frac{\partial^2 c_a}{\partial y^2}+\frac{\partial^2 c_a}{\partial z^2}\right)+\frac{\mathcal{F}_a D_{ab}}{RT}\left[c_a\left(\frac{\partial^2 \Phi}{\partial x^2}+\frac{\partial^2 \Phi}{\partial y^2}+\frac{\partial^2 \Phi}{\partial z^2}\right)\right]$ $+\frac{\mathcal{F}_a D_{ab}}{RT}\left[\frac{\partial \Phi}{\partial x}\frac{\partial c_a}{\partial x}+\frac{\partial \Phi}{\partial y}\frac{\partial c_a}{\partial y}+\frac{\partial \Phi}{\partial z}\frac{\partial c_a}{\partial z}\right]+\frac{c_a}{c_t}(R_{ea}+R_{eb})=\left(\frac{\partial c_a}{\partial t}+v_x\frac{\partial c_a}{\partial x}+v_y\frac{\partial c_a}{\partial y}+v_z\frac{\partial c_a}{\partial z}\right)$
Cylindrical	$D_{ab}\left(\frac{1}{r}\frac{\partial}{\partial r}\left(r\frac{\partial c_a}{\partial r}\right)+\frac{1}{r^2}\frac{\partial^2 c_a}{\partial \theta^2}+\frac{\partial^2 c_a}{\partial z^2}\right)+\frac{\mathcal{F}_a D_{ab}}{RT}\left[c_a\left(\frac{1}{r}\frac{\partial}{\partial r}\left(r\frac{\partial \Phi}{\partial r}\right)+\frac{1}{r^2}\frac{\partial^2 \Phi}{\partial \theta^2}+\frac{\partial^2 \Phi}{\partial z^2}\right)\right]$ $+\frac{\mathcal{F}_a D_{ab}}{RT}\left[\frac{\partial \Phi}{\partial r}\frac{\partial c_a}{\partial r}+\frac{1}{r^2}\frac{\partial \Phi}{\partial \theta}\frac{\partial c_a}{\partial \theta}+\frac{\partial \Phi}{\partial z}\frac{\partial c_a}{\partial z}\right]+\frac{c_a}{c_t}(R_{ea}+R_{eb})=\left(\frac{\partial c_a}{\partial t}+v_r\frac{\partial c_a}{\partial r}+\frac{v_\theta}{r}\frac{\partial c_a}{\partial \theta}+v_z\frac{\partial c_a}{\partial z}\right)$
Spherical	$D_{ab}\left(\frac{1}{r^2}\frac{\partial}{\partial r}\left(r^2\frac{\partial c_a}{\partial r}\right)+\frac{1}{r^2\sin\theta}\frac{\partial}{\partial \theta}\left(\sin\theta\frac{\partial c_a}{\partial \theta}\right)+\frac{1}{r^2\sin^2\theta}\frac{\partial^2 c_a}{\partial \phi^2}\right)$ $+\frac{\mathcal{F}_a D_{ab}}{RT}\left[c_a\left(\frac{1}{r^2}\frac{\partial}{\partial r}\left(r^2\frac{\partial \Phi}{\partial r}\right)+\frac{1}{r^2\sin^2\theta}\frac{\partial}{\partial \theta}\left(\sin\theta\frac{\partial \Phi}{\partial \theta}\right)+\frac{1}{r^2\sin^2\theta}\frac{\partial^2 \Phi}{\partial \phi^2}\right)\right]$ $+\frac{\mathcal{F}_a D_{ab}}{RT}\left[\frac{\partial \Phi}{\partial r}\frac{\partial c_a}{\partial r}+\frac{1}{r^2}\frac{\partial \Phi}{\partial \theta}\frac{\partial c_a}{\partial \theta}+\frac{1}{r^2\sin^2\theta}\frac{\partial \Phi}{\partial \phi}\frac{\partial c_a}{\partial \phi}\right]+\frac{c_a}{c_t}(R_{ea}+R_{eb})$ $=\left(\frac{\partial c_a}{\partial t}+v_r\frac{\partial c_a}{\partial r}+\frac{v_\theta}{r}\frac{\partial c_a}{\partial \theta}+\frac{v_\phi}{r\sin\theta}\frac{\partial c_a}{\partial \phi}\right)$

the electric field across the film is uniform, the system is at steady state, and all properties are constant. The fluid will be Newtonian and incompressible. We would like to determine the concentration profile of dopant in the fluid and the average rate of deposition. The situation is shown schematically in Figure 10.17.

We begin first by determining the velocity profile within the fluid. This was solved in the previous falling film heat transfer example where we showed that v_x is

$$v_x=\rho g\left(\Delta_f y-\frac{y^2}{2}\right)$$

For short contact times, $y \ll \Delta_f$, the velocity profile is linear and we can write

$$v_x=v_{max}y \qquad \text{where} \qquad v_{max}=\frac{\rho g\Delta_f^2}{2}$$

If the flow is fully developed, convection in the x-direction outweighs diffusion in that direction, and we can write the species continuity equation for our deposited material, +, including the convective terms arising from the electric field as

$$(v_{max}y)\frac{\partial c_+}{\partial x}=D_+\frac{\partial^2 c_+}{\partial y^2}+D_+\beta_e\frac{\partial c_+}{\partial y}\frac{d\Phi}{dy} \qquad \beta_e=\frac{\mathcal{F}_a}{RT}=\text{constant}$$

Since $d\Phi/dy$ is a constant, we can replace it by $d\Phi/dy|_{y=0}$. The boundary conditions state that at $x=0$, we have uniform concentration of positive charge, +; that at $y=\Delta_f$, we are far enough away

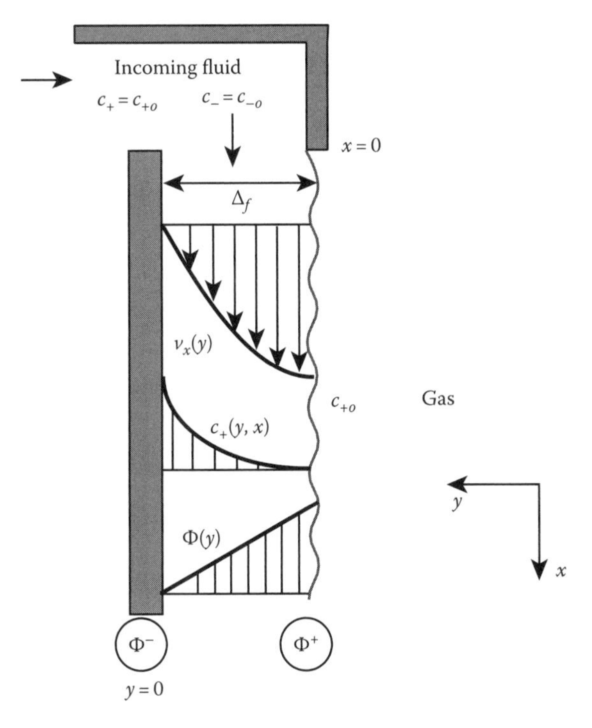

FIGURE 10.17 Field-assisted coating process. The electric field drives the positively charged coating material to the surface of the plate.

from the surface that the concentration of charged species remains $c_+ = c_{+o}$; and that the surface of the plate is impermeable $(J_+ = 0)$:

$$x = 0 \qquad c_+ = c_{+o}$$

$$y = 0 \qquad -D_+ \left.\frac{\partial c_+}{\partial y}\right|_{y=0} - D_+\beta_e c_+ \left.\frac{d\Phi}{dy}\right|_{y=0} = 0$$

$$y = \delta \qquad c_+ = c_{+o}$$

We can solve this equation using the similarity variable we tried for our previous heat transfer problem. Using the chain rule, we can substitute into the differential equation and reduce the partial differential equation to an ordinary differential equation:

$$\varepsilon = \frac{y}{(9D_+x/v_{max})^{1/3}}$$

$$\frac{d^2c_+}{d\varepsilon^2} + (3\varepsilon^2 + \varphi)\frac{dc_+}{d\varepsilon} = 0 \qquad \varphi = \beta_e \left[\frac{9D_+x}{v_{max}}\right]^{-1/3} \left.\frac{d\Phi}{d\varepsilon}\right|_{\varepsilon=0}$$

Repeated use of the chain rule allows us to write the differential equation as

$$\frac{d}{d\varepsilon}\left[\exp(\varepsilon^3 + \varphi\varepsilon)\frac{dc_+}{d\varepsilon}\right] = 0$$

Separating and integrating twice gives

$$c_+ = a_1 \int_0^{\varepsilon} \exp(-\eta^3 - \varphi\eta)d\eta + a_2$$

Our boundary conditions, in terms of ε, are

$$\varepsilon = 0 \quad -D_+\left[\frac{9D_+x}{v_{max}}\right]^{-1/3}\frac{dc_+}{d\varepsilon}\bigg|_{\varepsilon=0} - D_+\beta_e\left[\frac{9D_+x}{v_{max}}\right]^{-1/3} c_+\frac{d\Phi}{dy}\bigg|_{\varepsilon=0} = 0$$

$$-D_+\frac{dc_+}{d\varepsilon}\bigg|_{\varepsilon=0} - D_+\beta_e c_+\frac{d\Phi}{dy}\bigg|_{\varepsilon=0} = 0$$

$$\varepsilon = \infty \quad c_+ = c_{+o}$$

Applying these yields the following two algebraic equations for a_1 and a_2:

$$-D_+a_1 - a_2D_+\beta_e\frac{d\Phi}{d\varepsilon}\bigg|_{\varepsilon=0} = 0$$

$$a_1\int_0^{\infty}\exp(-\eta^3 - \varphi\eta)d\eta + a_2 = c_{+o}$$

Solving yields the following final solution:

$$\frac{c_+}{c_{+o}} = \frac{\displaystyle\int_0^{\varepsilon}\exp(-\eta^3 - \varphi\eta)d\eta - \frac{1}{\beta_e\frac{d\Phi}{d\varepsilon}\Big|_{\varepsilon=0}}}{\displaystyle\int_0^{\infty}\exp(-\eta^3 - \varphi\eta)d\eta - \frac{1}{\beta_e\frac{d\Phi}{d\varepsilon}\Big|_{\varepsilon=0}}}$$

To see this solution at work, we must look at a numerical example. Assuming the following physical parameters,

$$c_{+o} = 1000 \text{ mole/m}^3 \qquad g = 9.8 \text{ m/s}^2 \qquad \Delta_f = 0.002 \text{ m}$$
$$D_+ = 1\times10^{-9}\text{m}^2/\text{s} \qquad \Delta\Phi = 0.5 \text{ V}$$

we can use numerical methods to obtain

$$v_{max} = 0.02 \text{ m/s} \qquad D_+\beta_e\frac{d\Phi}{d\varepsilon}\bigg|_{\varepsilon=0} = D_+\frac{\mathcal{F}_a}{RT}\frac{d\Phi}{d\varepsilon}\bigg|_{\varepsilon=0} = 2.42\times10^{-5}$$

$$\varphi = \beta_e\frac{d\Phi}{d\varepsilon}\bigg|_{\varepsilon=0} = 7.41 \qquad \int_0^{\infty}\exp(-\eta^3 - \varphi\eta)d\eta = 0.133$$

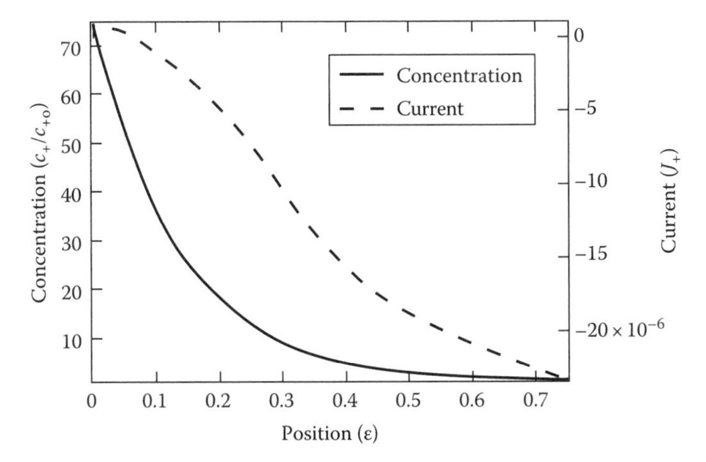

FIGURE 10.18 Effect of electric field on the concentration profile of charged species depositing on a flat plate.

Figure 10.18 shows how the dimensionless concentration, c_+/c_{+o}, and the current, J_+, vary as a function of ε. Clearly the presence of even a small electric field across the film greatly enhances the deposition rate and amount of solute at the interface. Finally, we can define a mass transfer coefficient for this case as well. Equating fluxes gives

$$-D_+\left[\frac{\partial c_+}{\partial y}-\beta_e c_+\frac{d\Phi}{dy}\right]_{y=0}=k_c(c_{+o}-0)$$

Substituting in for the concentration profile, we find that the mass transfer coefficient would be zero. This occurs because the plate is impermeable. To obtain a mass transfer coefficient, we approximate the flux

$$N_a=\frac{c_{+o}D_+\left(1-\dfrac{c_+(0)}{c_{+o}}\right)}{\Delta_f}$$

and see that the mass transfer coefficient depends upon the electric field strength

$$k_c=\frac{D_+\left(1+\dfrac{1}{\beta_e\left.\dfrac{d\Phi}{dy}\right|_{y=0}}\left(\dfrac{9D_+x}{v_{max}}\right)^{-1/3}\right)}{\displaystyle\int_0^\infty \exp(-\eta^3-\varphi\eta)d\eta-\frac{1}{\beta_e\left.\dfrac{d\Phi}{dy}\right|_{y=0}}\left(\frac{9D_+x}{v_{max}}\right)^{-1/3}}$$

10.6 SUMMARY

We have concentrated on deriving the general balance equations governing conservation of momentum, energy, mass, and charge in situations where we have convective transport occurring. We have considered how the inclusion of convection augments the transport and have looked at some simple cases where it was possible to solve the momentum transport equation exactly and obtain expressions for the convective transport coefficients.

All our derivations and solutions considered transport from a microscopic viewpoint. That is, we were describing profiles throughout the system, considering every coordinate direction and time. We often considered situations where symmetry considerations reduced the complexity of the problem, but we never resorted in a reduction in the complexity by performing some kind of average over one or more coordinate directions. In the next chapter, we will do just that. By looking at the system as a whole, that is, treating the system as a black box, we will artificially reduce the complexity of our problem and thereby develop *macroscopic* versions of the balance equations. These equations are of critical importance because they form the foundations of engineering analysis.

PROBLEMS

Momentum Transport

10.1 The surface of a rotating cylinder of fluid takes on a parabolic shape. Engineers have used that fact to produce large telescope mirrors using mercury as the liquid. You propose to make a 3 m telescope mirror from mercury using a similar approach.

a. If the focal length of the mirror is to be six times the mirror's diameter, how fast would one need to rotate the container of mercury?

b. What would be the minimum volume of mercury required to make the mirror? How much would it weigh?

10.2 A journal bearing of the form shown in Figure 10.8 is being designed to carry a load of 2 metric tons. The liquid being used is conventional lubricating oil with a viscosity of 1 Pa·s at 100°C. If the bearing angle is 0.05°, the length of the bearing is 10 cm, and the gap thickness at the thick end is 0.5 mm, what fluid velocity would be needed to support the load?

10.3 Consider the spin-coating process used to coat silicon wafers with photoresist, television picture tubes with phosphorescent layers, etc. In all cases, the process is designed to produce a very thin, uniform coating by spinning a viscous, Newtonian, liquid onto a substrate. The process has angular symmetry, the rotation rate is constant, and since the film is thin, there are no real pressure gradients or fluid accelerations to speak of. The thin film also moves with the substrate as if it were a rigid body, $v_\theta \neq f(z)$ (Figure P10.3).

a. Since we do not need to determine the pressure in the film, we only need to determine v_r and v_z. Show that the continuity and momentum equations reduce to

$$-\rho\frac{v_\theta^2}{r} = \mu\frac{d^2 v_r}{dz^2} \qquad \frac{1}{r}\frac{\partial}{\partial r}(r v_r) + \frac{\partial v_z}{\partial z} = 0$$

b. What are the boundary conditions for this problem?

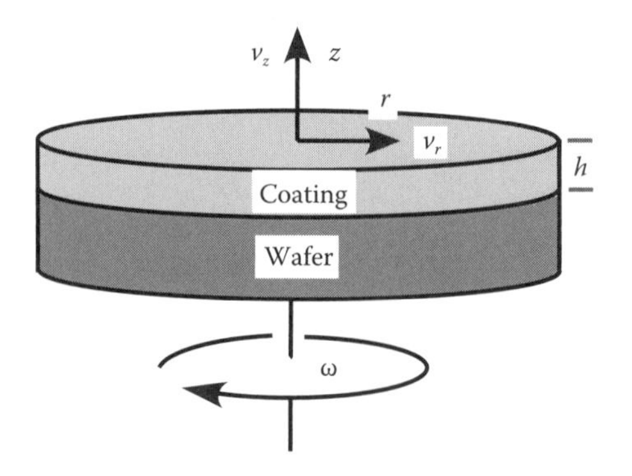

FIGURE P10.3

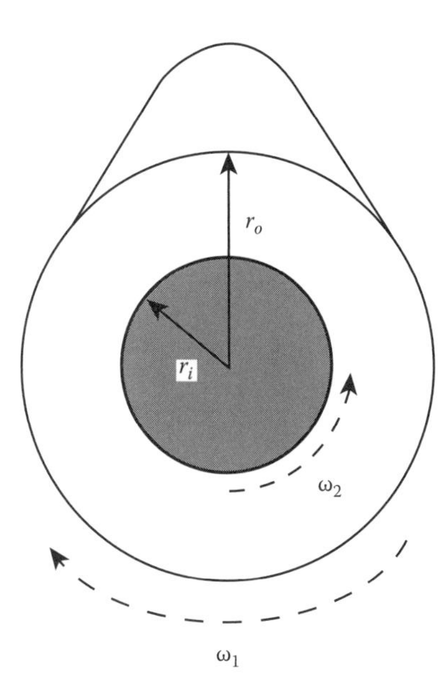

FIGURE P10.4

c. Solve the equations for v_r and v_z.

d. The velocity, v_z, at the film/air interface is just the change in film thickness with time. Use this to obtain a differential equation for h, and integrate this equation to obtain the following solution:

$$\frac{1}{2}\left(\frac{1}{h^2}+\frac{1}{h_o^2}\right)=\left(\frac{2\rho\omega^2}{3\mu}\right)t$$

10.4 Consider the system shown in Figure P10.4 of two concentric rotating cylinders. The two cylinders each rotate at constant but different angular velocities.

a. Determine the velocity profile $v_\theta(r)$ between the cylinders and the pressure distribution in the radial direction $P(r)$.

b. Determine a friction factor by calculating the force required to turn either of the two cylinders. Looking at the outer cylinder, we have

$$F=\tau_{r\theta}A=C_f\left(\frac{1}{2}\rho v_o^2\right)A$$

where v_o is the linear velocity of the outer cylinder in m/s.

10.5 An infinite rigid rod of radius, r_o, rotates in an infinite, incompressible, Newtonian fluid of viscosity, μ, and density, ρ, with a constant rotational velocity, ω. Calculate the resultant velocity distribution in the fluid and the radial pressure gradient.

10.6 In Section 10.2.2, we discussed how to find the shape of the surface of a rotating liquid. Consider the situation where the fluid is confined between two counterrotating cylinders as shown in Figure P10.6, and determine the shape of the free surface.

10.7 The stream function for flow in a 90° corner is

$$\psi=\frac{rv_o}{\pi^2-4}(2\pi\theta\sin\theta+4\theta\cos\theta-\pi^2\sin\theta)$$

a. What are the velocity components, v_r, v_θ?

b. Calculate the vorticity for this 2-D flow. Plot lines of constant vorticity.

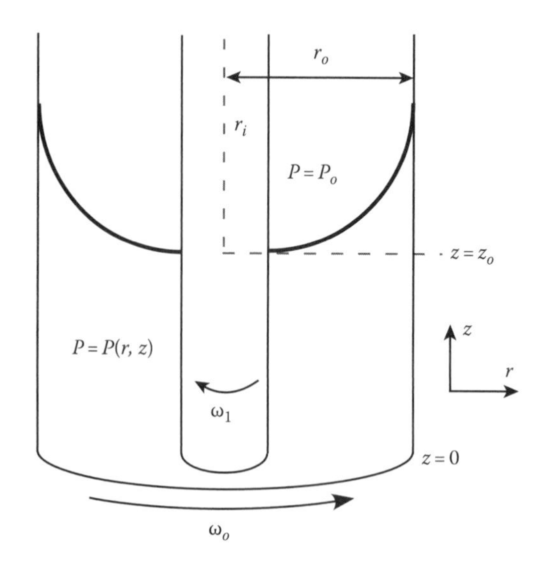

FIGURE P10.6

10.8 A 2-D flow exists between fixed boundaries at $\theta = \pi/4$ and $\theta = -\pi/4$. The flow is due to a source of strength, m, at $r = a$, $\theta = 0$, and a sink of equal strength at $r = b$, $\theta = 0$. The stream function for such a flow is given by

$$\psi = -m\tan^{-1}\left[\frac{r^4(a^4 - b^4)\sin 4\theta}{r^8 - r^4(a^4 - b^4)\cos 4\theta + a^4b^4}\right]$$

a. Determine the velocity components, v_r, v_θ.
b. Calculate the vorticity for this flow. Does it obey the vorticity transport theorem?

$$\frac{D\vec{\omega}}{Dt} = \vec{\omega}\cdot\vec{\nabla}\vec{\mathbf{u}}$$

10.9 If the ellipse

$$a(x^2 - y^2) + 2bxy - \frac{1}{2}\omega_o(x^2 + y^2) + c = 0$$

is full of liquid and is rotated about the origin with an angular velocity, ω_o, the stream function is

$$\psi = a(x^2 - y^2) + 2bxy$$

a. What are the velocity components?
b. Is there any vorticity associated with this flow?

10.10 A Newtonian fluid of viscosity, μ, and density, ρ, is contained in between two vertical pipes of diameters d_o and d_i. The situation is shown in Figure P10.10. If left to its own devices, the fluid would begin to drain from between the pipes under the influence of gravity. To stop the fluid from draining out, the inner pipe is lifted at some velocity v_i.

a. Derive the differential equation governing the fluid flow between the two pipes.
b. Specify the boundary conditions.
c. Solve the equation to get the velocity profile and the volumetric flow rate of fluid.

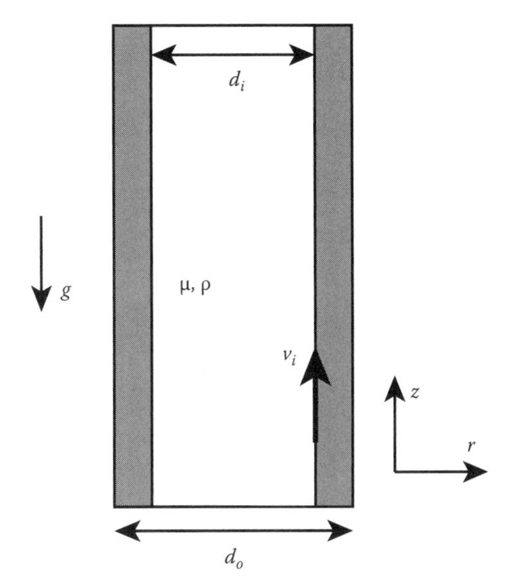

FIGURE P10.10

Hint: The following integral may help:

$$\int \left[a\left(r_o^2 - r^2\right) + b\ln\left(\frac{r}{r_o}\right) \right] r\,dr = \frac{1}{2}ar_o^2 r^2 - \frac{1}{4}r^4 + \frac{1}{2}br^2\ln\left(\frac{r}{r_o}\right) - \frac{1}{4}br^2$$

d. What inner pipe velocity is required to keep the fluid from draining out?

10.11 A fluid of viscosity, μ, flows down a vertical rod of radius, r_o. At some point down the rod, the flow reaches a steady condition where the film thickness, h, is constant and the velocity $v_z = f(r)$ only. Assuming no shear stress at the fluid–air interface,

a. Draw a picture of the situation and your control volume.
b. Derive the differential equation governing the velocity profile in the fluid film.
c. What are the relevant boundary conditions?
d. Solve the equation for the velocity profile.
e. Determine the volumetric flow rate of fluid leaving the rod.

10.12 Skimmers are used to remove viscous fluids, such as oil, from the surface of water. As shown on the diagram in Figure P10.12, a continuous belt moves upward at velocity V_o through the fluid, and the more viscous liquid (with density ρ and viscosity μ) adheres to the belt. A film with thickness h forms on the belt. Gravity tends to drain the liquid, but the upward belt velocity is such that net liquid is transported upward. Assume the flow is fully developed and laminar, with zero pressure gradient and zero shear stress at the outer film surface where air contacts it. Determine an expression for the velocity profile and flow rate. Use a differential analysis similar to that used for fully developed laminar flow through an inclined pipe. Clearly state the velocity boundary conditions at the belt surface and at the free surface.

10.13 In Section 10.2.2, we calculated the velocity field for flow past a stationary sphere. Such a sphere experienced a buoyant force but no lift. To generate lift, we need to rotate the sphere. We can calculate the velocity profile for the rotating sphere by changing the boundary condition on the velocity at the surface of the sphere. For a sphere rotating about the θ-direction at a rate of B revolutions per second (Hint: $v_\theta = 2\pi r_o B \sin\theta$):

a. What are the boundary conditions on the sphere surface?
b. Solve for the stream function.
c. What are the r- and θ-velocity components?

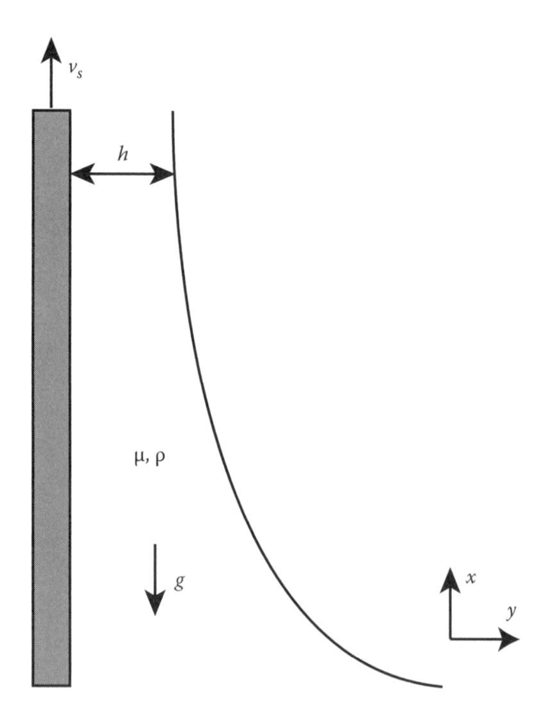

FIGURE P10.12

d. Substitute the velocity into the momentum equations and determine an expression for the pressure (many pages of algebra).
e. Using the results of part (d), integrate the normal component of the pressure, and by comparing with the stationary solution, isolate the lift component of the normal force (many more pages of algebra).

Energy Transport

10.14 A sliding bearing, modeled as flow between two plates separated by a distance, δ, is lubricated by a Newtonian fluid of viscosity, μ, and density, ρ. Both sides of the bearing are maintained at a constant temperature, T_o. The bearing lubricant heats up due to viscous dissipation. Find the temperature distribution and the maximum temperature assuming that the thermal conductivity of the fluid varies with temperature according to

$$k = \frac{1}{\mathrm{A} + \mathrm{B}T} \quad \mathrm{A,B} - \text{constants} \quad \text{(Jain, Rogojevic, and Shukla)}$$

10.15 We would like to look at a case of coupled transport, natural convection between horizontal walls. The situation is shown in Figure P10.15. We assume that the length of the device is much larger than its height, that is, $v_y \approx 0$ and $v_x = f(y)$.

a. Using the Boussinesq approximation for buoyant driven flow, show that the Navier–Stokes and energy equations can be reduced to the following set for this problem:

$$\frac{1}{\rho}\frac{\partial P}{\partial y} = \beta_T g(T - T_{ref}) \qquad \frac{1}{\rho}\frac{\partial P}{\partial x} = \nu\frac{\partial^2 v_x}{\partial y^2}$$

$$v_x \frac{\partial T}{\partial x} = \alpha\left(\frac{\partial^2 T}{\partial x^2} + \frac{\partial^2 T}{\partial y^2}\right)$$

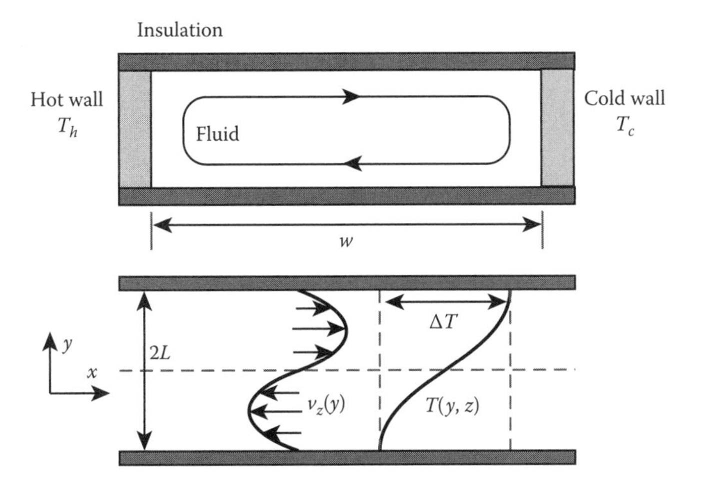

FIGURE P10.15

b. If the length of the device is much larger than its height, then

$$\frac{\partial T}{\partial x} \approx \text{constant (A)}, \quad \text{that is, } \Delta T \text{ is constant}$$

Eliminate pressure from the set of the previous equations and show that

$$\frac{\partial^3 v_x}{\partial y^3} = \frac{\beta_T g}{\nu} A$$

c. Use the results of part (b) to eliminate $\partial^2 T/\partial x^2$, and solve the energy and momentum equation to obtain the profiles.

10.16 Often, we have thermo-capillary convection when one surface is free. The temperature gradient induces a change in surface tension that can drive the motion of the liquid. Thus, at the free surface, there is a shear stress related to the temperature gradient:

$$y = L \qquad \mu \frac{\partial v_x}{\partial y} = \frac{\partial \gamma}{\partial T}\frac{\partial T}{\partial x}$$

Resolve the horizontal convection problem (P10.15) with this new condition assuming that the surface tension varies linearly with the temperature:

$$\frac{\partial \gamma}{\partial T} = -\beta_\gamma$$

10.17 Consider a gas metal arc-welding electrode as shown in Figure P10.17 where we are consuming the electrode as we weld (rod velocity = v_o). We are interested in the steady-state temperature distribution in the welding rod. Initially we assume that resistance heating of the rod is negligible and that heat loss via convection from the surface of the rod is also negligible. Since the rod is fairly thin, radial temperature gradients can also be neglected. The rod melts at its tip at a temperature, T_m, and is held at a constant temperature, T_b, at the base of the electrical contact.

a. What does the energy equation look like for this problem?
b. Solve the equation using the specified temperature boundary conditions to obtain the axial temperature profile within the rod.

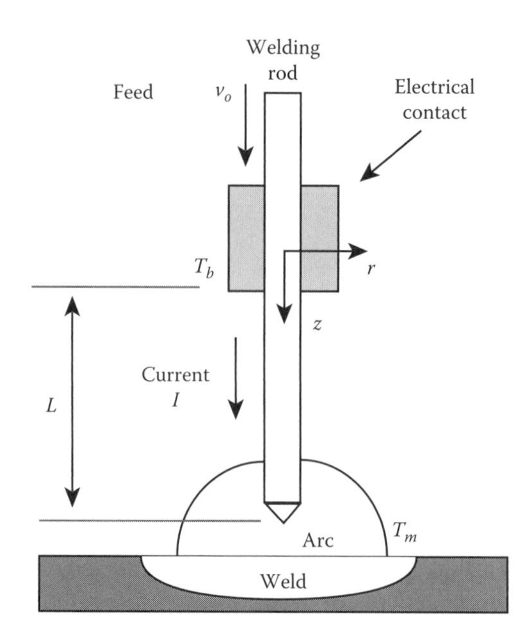

FIGURE P10.17

10.18 Consider the welding situation of the previous problem. This time let's consider the case where we have convective heat transfer from the rod's surface. For a rod of radius r_o and a heat transfer coefficient and free stream temperature of h and T_∞, respectively,

a. Derive the differential equation governing the temperature profile in the rod. Assume it acts as a fin with no radial temperature gradients
b. Solve the equation using the same boundary conditions as in the previous problem
c. Reconsider the problem if the rod also experiences resistance heating with a resistance, $\mathcal{R}_e$

10.19 Castrol's "Syntec" oil is touted by showing an ad in which a car engine runs with all its oil drained away. Aside from the combustion parts of the engine, critical failures could occur in the journal bearings of the crankshaft as heat is generated by viscous dissipation in the fluid. A typical journal bearing is shown in Figure P10.19. Assuming that the engine is idling at 1000 rpm, that the amount of heat removed by conduction in the metal parts is negligible, and that the engine starts out at temperature of 85°C,

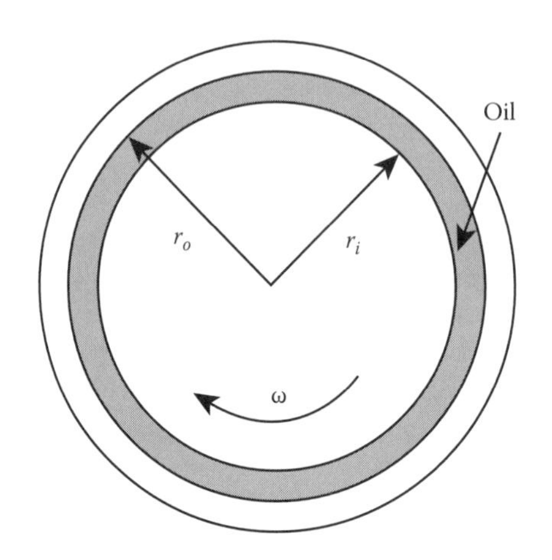

FIGURE P10.19

a. How fast does the temperature rise in the oil? Assume the oil is at a uniform temperature and use a parallel plate configuration to calculate the velocity profile with a thickness between the plates of $r_o - r_i$.
b. If oil breaks down at 250°C, how long before the engine dies?

$$\text{Oil properties: } \rho = 825 \text{ kg/m}^3 \qquad C_p = 2325 \text{ J/kg K}$$

$$k = 0.136 \text{ W/m K} \qquad \nu = 10.0 \times 10^{-6} \text{m}^2/\text{s}$$

$$\text{Bearing data}: r_o = 0.08 \text{ m} \qquad r_i = 0.07993 \text{ m}$$

10.20 Consider the heat transfer to a falling film problem of Section 10.3. The mass transfer analog is the dissolution of a solid wall into the falling film. Assuming the solid wall is composed of salt, NaCl, use heat transfer solution as a template and derive the corresponding mass transfer solution. How do the penetration depths for the two problems compare?

10.21 Vorticity is generated any time a fluid contacts a solid surface. The vorticity diffuses from the solid surface and is convected away by the fluid. Consider the heat transfer to a falling film problem of Section 10.3.
a. Write the vorticity transport equation analog of the falling film heat transfer problem.
b. Use heat transfer solution as a template and derive the corresponding solution for the vorticity problem. The vorticity is zero at the air/water interface and can be found at the solid interface using the velocity profile.
c. How do the penetration depths for the two problems compare?

Mass and Charge Transport

10.22 Smoluchowski's theory of coagulation—in this theory of coagulation, we focus on an individual sphere and assume that other like particles diffuse toward it. Once they reach the sphere, they collide and form a new spherical aggregate. Once stuck together, the diffusing particles are removed from the solution. Thus, there exists a concentration gradient of particles from the free stream to the sphere. Assume the sphere has a radius of R_o and is surrounded by fluid containing n_o particles/m^3, uniformly distributed. The particles have a diffusivity D and are in dilute solution.
a. By simplifying the continuity equations in spherical coordinates, show that the diffusion equation reduces to

$$\frac{\partial n}{\partial t} = D\frac{1}{r^2}\frac{\partial}{\partial r}\left(r^2\frac{\partial n}{\partial r}\right)$$

b. What are the boundary and initial conditions for this problem? The fluid is essentially infinite in extent. (Remember we are interested in what is beyond $r = R$.)
c. To solve the problem, we must use a trick or two. Setting

$$y = \left(\frac{(r-R)}{R}\right) \qquad \text{and} \qquad C = \frac{r(n_o - n)}{n_o} = \frac{(y+1)(n_o - n)}{n_o}$$

show that we can derive a similarity solution of the following form:

$$n = n_o\left[1 - \frac{R}{r} + \frac{R}{r}erf\left(\frac{r-R}{\sqrt{4Dt}}\right)\right]$$

d. Using Fick's law, show that the *flow* (# particles/time) of particles to the sphere's surface is

$$N = 4\pi D n_o R\left[1 + \frac{R}{\sqrt{\pi D t}}\right]$$

e. For $t \gg R^2/D$, we can express the flow of particles as $4\pi D n_o R$. To determine the coagulation rat we must account for the motion of our sphere. If that is allowed to diffuse with diffusivity D, we effectively double the diffusion coefficient of all particles to 2-D. Finally, the number of contacts per unit volume of dispersion is just the number of contacts per particle multiplied by the number of particles per unit volume or

$$N_t = 8\pi D R n_o^2$$

initially. If all the particles stick when they collide, the rate of change of the number of particles per unit volume, n, is

$$\frac{dn}{dt} = -8\pi D R n^2$$

Integrate this equation (for constant DR) and show that the characteristic coagulation time, T, is given by

$$T = \frac{1}{8\pi D R n_o}$$

f. The Earth is roughly 4.5 billion years old and was formed by accretion in roughly 500,000,000 years. Assuming we needed to reduce the original number of planetesimals to a fraction of 1×10^{-6} within that time to reach our current size, what was the time constant for bombardment?

10.23 As a biochemical engineer, you are evaluating a drug delivery system for an artificial protein to combat Alzheimer's. The protein is very large (200,000 molecular weight) and bulky and is sensitive to the stress level placed upon it (if it irreversibly elongates, it is of no therapeutic value). Laboratory measurements indicate that the yield stress for the protein is 100,000 N/m2. The company claims that they are delivering 0.016 mol/h of active protein with no more than 5% of the protein inactive. The concentration of protein in the solution is 1 mol/m^3 and the physical properties of the solution are essentially that of water with the viscosity being three times that of water. The pump is peristaltic and so squeezes the fluid to pump it through the catheter tubing. The pump supplies the equivalent of 3 atm pressure over a length of 0.1 mm and delivers the protein through a 1.6 mm diameter catheter tube. You may assume that the pump squeezes the fluid through an opening of 0.4 mm. The pressure drop through the rest of the 5 m long tubing is negligible. Are the claims of the company accurate? How much denatured protein do they actually deliver?

10.24 Consider the example in Section 10.4 about mass transfer to a growing bubble. Calculate the mass transfer coefficient for that system. Does it increase or decrease over time? Why should this be so?

10.25 Consider the mass transfer example concerning diffusion into a falling film shown in Figure P10.25. A new inventor claims that he can rig the device to operate such that the flux

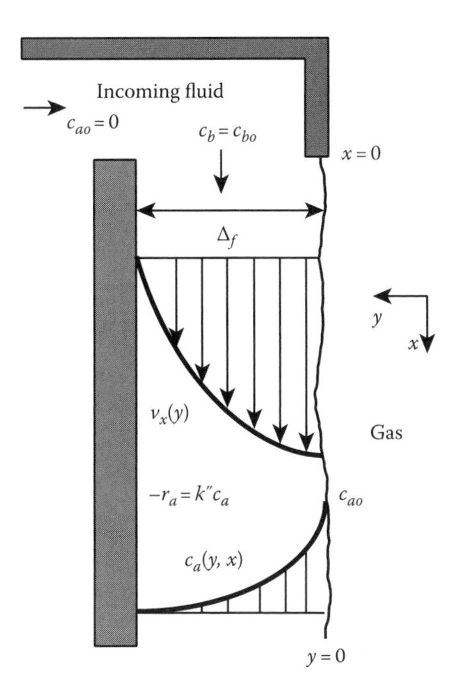

FIGURE P10.25

of a from the gas to the liquid is constant rather than the concentration at the interface being constant. He claims that this will provide a better mass transfer situation.

a. Rework the solution to this example assuming the mass flux at the interface is a constant, N_{ai}.
b. How does the mass transfer coefficient behave as a function of position along the length of the plate?
c. For typical values of material properties, calculate an average mass transfer coefficient and determine what the flux would have to be to break even on this new process.

$$D_{ab} = 1\times10^{-10}(\text{m}^2/\text{s}) \qquad L = 1\ (\text{m}) \qquad \delta = 100\ (\mu\text{m})$$

$$\nu = 1\times10^{-6}(\text{m}^2/\text{s}) \qquad c_{ao} = 10\ \text{mol/m}^3$$

10.26 Consider absorption into a falling film, where the film contains a reactive component in great excess. This results in reaction that is pseudo–first order in absorbate concentration. Dilute solutions prevail.

a. Assuming the same simplifications as in Section 10.4, pare down the species continuity equation and determine the partial differential equation governing steady-state operation of the device.
b. What are the boundary conditions?
c. Show that the solution for the concentration profile is

$$\frac{c_a}{c_{ao}} = \frac{1}{2}\exp\left(-y\sqrt{\frac{k''}{D_{ab}}}\right) erfc\left(\frac{y}{\sqrt{4D_{ab}x/v_{\max}}} - \sqrt{\frac{k''v_{\max}}{x}}\right)$$

$$+\frac{1}{2}\exp\left(-y\sqrt{\frac{k''}{D_{ab}}}\right) erfc\left(\frac{y}{\sqrt{4D_{ab}x/v_{\max}}} + \sqrt{\frac{k''v_{\max}}{x}}\right)$$

d. Show that the molar flux of a at the interface, $y = 0$, is

$$N_a = c_{ao}\sqrt{D_{ab}k''}\left[erf\left(\sqrt{\frac{k''x}{v_{max}}}\right) + \frac{\exp\left(-\frac{k''x}{v_{max}}\right)}{\sqrt{\frac{\pi k''x}{v_{max}}}}\right]$$

e. What is the mass transfer coefficient for this system?

10.27 Consider the motion of sodium and chloride ions in a membrane under the influence of an externally applied electric field. The situation is shown in the Figure P10.27. Charge transport in the presence of an electric field obeys the following equations:

$$\frac{\partial^2 V}{\partial y^2} = -\frac{q}{\varepsilon}(p - n)$$

$$\frac{\partial p}{\partial t} = D\frac{\partial^2 p}{\partial y^2} + \mu\frac{\partial}{\partial y}\left(p\frac{\partial V}{\partial y}\right)$$

V—voltage
q—charge on electron
ε—permittivity
p—Na
n—Cl
D—diffusvity
μ—mobility

$$\frac{\partial n}{\partial t} = D\frac{\partial^2 n}{\partial y^2} - \mu\frac{\partial}{\partial y}\left(n\frac{\partial V}{\partial y}\right)$$

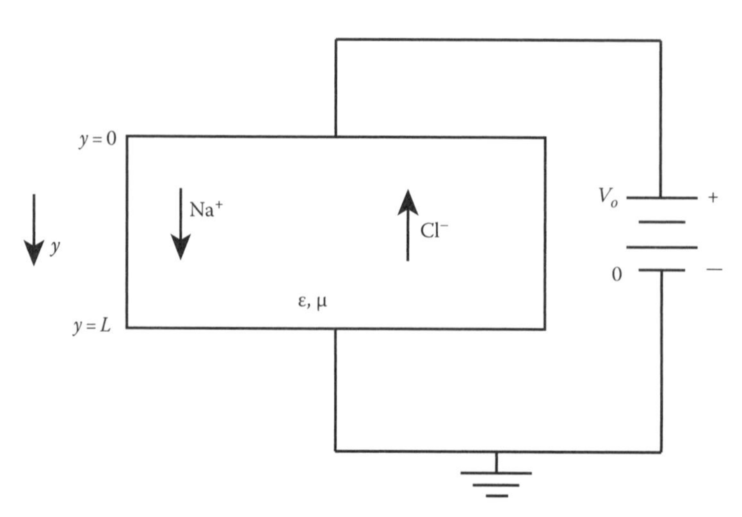

FIGURE P10.27

a. Assuming steady state and that diffusion is negligible, derive one conservation of mass equation for the excess charge in the device?
b. Solve the equations for the voltage distribution and excess charge inside the medium.
c. Assuming the charge imbalance is known and is Ψ_o, what boundary conditions would you impose on the system? Why?
d. What does your solution say about the relationship between the charge imbalance and the electric field ($E = -dV/dy$)?

REFERENCES

1. Batchelor, G.K., *Fluid Dynamics*, University Press, Cambridge, U.K. (1970).
2. Stokes, G.G., On the effect of internal friction of fluids on the motion of pendulums, *Trans. Camb. Phil. Soc.*, **9**, 8 (1855).
3. Lamb, H., *Hydrodynamics*, Dover, New York (1945).
4. Milne-Thompson, L.M., *Theoretical Hydrodynamics*, 5th edn., MacMillan, New York (1967).
5. Bird, R.B., Stewart, W.E., and Lightfoot, E.N., *Transport Phenomena*, John Wiley & Sons, New York (1960).
6. Boyce, W.E. and DiPrima, R.C., *Elementary Differential Equations and Boundary Value Problems*, 2nd edn., John Wiley & Sons, New York (1969).
7. Cho, J.R., and Chung, M.K., A k-ε-c Equation turbulence model, *J. Fluid Mech.*, **237**, 301 (1992).
8. Smith, J.M. and Van Ness, H.C., *Introduction to Chemical Engineering Thermodynamics*, 3rd edn., McGraw-Hill, New York (1975).
9. Pigford, R.L. Non-isothermal flow and heat transfer inside vertical tubes, *CEP Symp. Ser. 17*, **51**, 79 (1955).
10. Abramowitz, M. and Stegun, I.A., eds., *Handbook of Mathematical Functions*, Dover, New York (1972).
12. Sherwood, T.K., Pigford, R.L., and Wilke, G.R., *Mass Transfer*, McGraw-Hill, New York (1975).

11 Macroscopic or Engineering Balances

$$\iiint (In - Out + Gen = Acc)dV$$

11.1 INTRODUCTION

In the previous chapter, we derived the microscopic balance equations governing the conservation of mass, energy, momentum, and charge. These balances were very complicated, and actual engineering situations rarely lend themselves to analyses using the full set of balance equations. Flows are most likely turbulent and require a full 3-D momentum treatment. The physical systems are often highly irregular in shape, so mass or heat balances cannot be simplified. Finally, the media through which transport takes place are often multicomponent with properties varying as a function of position, temperature, and composition.

Fortunately, we may only need to know how the system of interest affects a transport operation and so can treat the system as a *black box*, operating on a fluid passing through. Such an approach is often used to analyze piping systems where a fluid is being pushed along by pumps or compressors and all this equipment is lumped together. In other circumstances, a transport process may occur along a predominant direction, and we may only need detailed information in that direction. Examples include knowing the temperature distribution along the length of a heat exchanger or the composition along the axis of a chemical reactor or piece of separations equipment. We can modify the microscopic balance equations to apply to these systems by integrating over nonessential coordinate directions. This *partial* integration leads to *macroscopic*, *engineering*, or *lumped analysis* balances. In this chapter, we will derive these balances and apply them to several physical systems involving momentum, energy, mass, and charge transport.

11.2 MACROSCOPIC CONTINUITY EQUATION

We begin a discussion of macroscopic balances by looking at the conservation of overall mass, the continuity equation. The system, shown in Figure 11.1, consists of a *black box* through which a fluid is flowing. All we know about the system are the inlet and outlet areas through which the fluid flows, the overall surface area and volume of the system, the physical properties of the fluid at the inlet and outlet, and the velocities of the fluid entering and leaving the system.

We know that the fluid will have some velocity profile in those portions of the system conveying the fluid to and from the *black box*. A detailed knowledge of that profile is generally unnecessary; we can do just as well knowing the average or *mixing-cup* velocity. The mixing-cup velocity is determined by placing a bucket at the inlet or outlet of the system and recording the volume of fluid that flows into the bucket over a designated period of time. Knowing the cross-sectional areas of the inlet and outlet orifices allows us to determine the average fluid velocity. The mathematical definition is

$$\overline{v} = \frac{\int_{A_c} v \, dA_c}{\int_{A_c} dA_c} \quad \text{Any cross-sectional shape} \qquad (11.1)$$

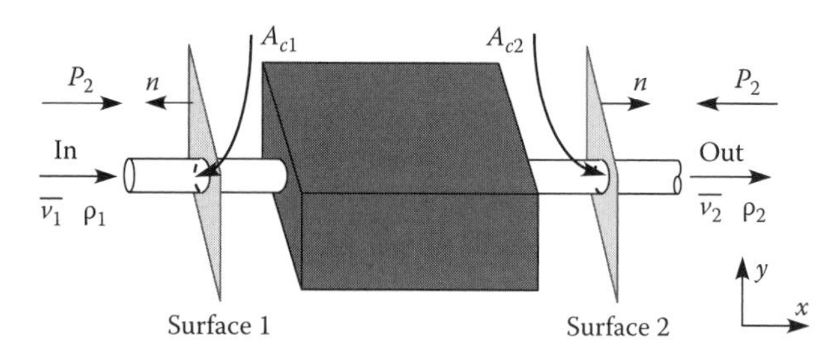

FIGURE 11.1 System for deriving the overall continuity equation.

$$\bar{v} = \frac{\int_0^{2\pi}\int_0^{r_o} vr\,dr\,d\theta}{\int_0^{2\pi}\int_0^{r_o} r\,dr\,d\theta} \quad \text{Pipes with circular cross-section} \tag{11.2}$$

To derive the macroscopic continuity equation, we integrate the microscopic continuity equation over the volume of the system. Notice that mass may enter or leave the system only by passing through the planes at either end of the system. Before we can integrate the microscopic continuity equation, we need to consider whether we want to use volume integrals, surface integrals, or some combination of both. The volume of the system inside the black box may be horribly complicated and the volume integral, very involved. Since mass must pass through surfaces at either end of the system, it seems that it would be much simpler to consider surface integrals over the inlet and outlet planes rather than a volume integral.

Gauss's law relates surface integrals to their corresponding volume integral counterparts [1]. We need three versions of Gauss's law, one for scalars like the pressure (P) or temperature (T), one for vectors like the heat flow ($\vec{\mathbf{q}}$) or velocity ($\vec{\mathbf{v}}$), and the other for tensors like the shear stress ($\vec{\tau}$). Equations 11.3 through 11.5 are versions of Gauss's law for these three cases:

$$\int_V (\vec{\nabla} s)\,dV = \int_A (\vec{\mathbf{n}} s)\,dA \quad \text{Scalars} \tag{11.3}$$

$$\int_V (\vec{\nabla}\cdot\vec{\mathbf{v}})\,dV = \int_A (\vec{\mathbf{n}}\cdot\vec{\mathbf{v}})\,dA \quad \text{Vectors} \tag{11.4}$$

$$\int_V (\vec{\nabla}\cdot\vec{\tau})\,dV = \int_A (\vec{\mathbf{n}}\cdot\vec{\tau})\,dA \quad \text{Tensors} \tag{11.5}$$

Here, $\vec{\mathbf{n}}$ is the unit vector in the direction of the outward normal to the surface.

In the course of our integrations, we will have to integrate the accumulation term. If we have the derivative of an integral or the integral of a derivative, we cannot simply reverse the order of integration and differentiation because the boundaries of the system may be changing over time. We must make use of Leibnitz's rule for differentiating an integral [1] to take this into account:

$$\frac{\partial}{\partial t}\int_V (\mathfrak{f})\,dV = \int_V \left(\frac{\partial \mathfrak{f}}{\partial t}\right) dV + \int_A (\vec{\mathbf{n}}\cdot\vec{\mathbf{v}}_{\mathbf{a}})\,dA \quad \text{Leibnitz's rule} \tag{11.6}$$

In Equation 11.6, the velocity of the surface element is $\vec{\mathbf{v}}_{\mathbf{a}}$ and f is a function of position and time. If the system is enclosed by rigid boundaries, the surface velocity $\vec{\mathbf{v}}_{\mathbf{a}} = 0$, and the surface integral vanishes.

Returning to the continuity equation, we assume that the physical properties of the material do not change over the cross section of the inlet and outlet bounding planes and that the average, or mixing-cup, velocity of the material is strictly parallel to the conduits through which the material is flowing. We use Gauss's law to transform the volume integral of the convective term to a surface integral over the inlet and outlet surfaces. The velocity we must use in this surface integral is the *velocity in the direction of the outward normal to the surface.* These normal vectors are shown in Figure 11.1. Integrating the continuity equation gives

$$\int_V (\vec{\nabla}\cdot\rho\vec{v})\,dV = \int_{A_c} (\rho v_n)\,dA_c = -\int_V \frac{\partial\rho}{\partial t}\,dV \tag{11.7}$$

where v_n is the velocity of the fluid *relative to the outward surface normal.* Since the inlet and outlet surface planes and the system volume are fixed in space and time, we can interchange the order of differentiation and integration in Equation 11.7 and integrate to get

$$-\rho_1\overline{v}_1 A_{c1} + \rho_2\overline{v}_2 A_{c2} = -\frac{\partial}{\partial t}\int_V \rho\,dV = -\frac{dm_{tot}}{dt} \tag{11.8}$$

The macroscopic mass balance states that the net rate of accumulation of mass in the system is equal to the difference in influx and efflux rates. There is a minus sign between the two mass flows because v_1 is in a direction opposite to the outward normal to the surface 1 while v_2 is in the same direction as the outward normal to surface 2. The process by which we arrived at Equation 11.8 is called applying the Reynolds transport theorem [1].

We derived Equation 11.8 assuming the system boundaries were impermeable. In many instances, we may have permeable system boundaries, and mass may pass through as a result of diffusion. In such instances, we must modify Equation 11.8 to take this into account:

$$-\dot{m}_d - \rho_1\overline{v}_1 A_{c1} + \rho_2\overline{v}_2 A_{c2} = -\frac{dm_{tot}}{dt} \tag{11.9}$$

$$\dot{m}_d = \int_{A_p} (\rho v_n)\,dA_p \tag{11.10}$$

where

A_p is the surface area of the control volume that is permeable
v_n is the velocity normal to that surface
$\dot{m}_d$ is defined as positive when material diffuses or permeates into the system

Example 11.1 Multiple Outlets and Inlets

An incompressible fluid flows at steady state through the rectangular duct shown in Figure 11.2. The depth of the duct into the page is 1 m. Determine the velocity of stream 3 given the data in Figure 11.2.

We apply the continuity equation first, keeping in mind the direction of the outward surface normal relative to the flow direction at each inlet or outlet. Flows in the same direction as the outward surface normal are positive and flows in the opposite direction of the outward surface normal are negative:

$$-\rho\overline{v}_1 A_{c1} - \rho\overline{v}_4 A_{c4} + \rho\overline{v}_2 A_{c2} + \rho\overline{v}_3 A_{c3} = 0$$

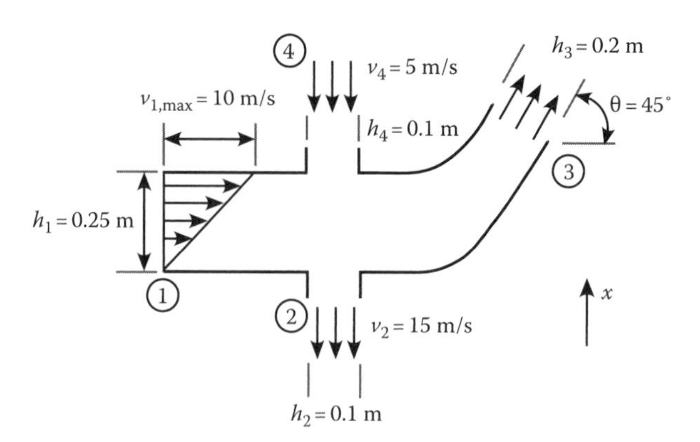

FIGURE 11.2 Flow through a duct with multiple outlets.

The average velocity over the inlet, surface 1, must be determined using the velocity profile:

$$v_1 = \left(\frac{v_{1,max}}{0.25}\right)x \qquad \text{so:} \qquad \bar{v}_1 = \frac{\int_0^{0.25}\left(\frac{v_{1,max}}{0.25}\right)x\,dx}{0.25} = \frac{v_{1,max}}{2}$$

Substituting into the continuity equation allows us to solve for $\bar{v}_3$:

$$\bar{v}_3 = \frac{(10/2)(0.25)(1) + 5(0.1)(1) - 15(0.1)(1)}{0.2(1)} = 1.25 \text{ m/s}$$

11.3 MACROSCOPIC MOMENTUM BALANCE

We can integrate the microscopic momentum balance over the control volume of Figure 11.1 to obtain a macroscopic version. Here, it is important to set up a coordinate system for the problem so we can reference pressures, gravity, etc., to that coordinate system. We keep the assumptions that the physical properties of the fluid do not change over A_1 and A_2 and that the velocity of the fluid is strictly parallel to the inlet and outlet conduits. Keeping Gauss's law and Leibnitz's rule in mind, we begin by integrating the entire momentum equation over the system volume:

$$\int_V (\vec{\nabla}\cdot(\rho\vec{\mathbf{v}})\vec{\mathbf{v}})dV + \int_V \vec{\nabla}P\,dV + \int_V (\vec{\nabla}\cdot\vec{\tau})dV - \int_V \rho\vec{\mathbf{g}}\,dV = -\int_V \frac{\partial}{\partial t}(\rho\vec{\mathbf{v}})dV \tag{11.11}$$

If we keep the system fixed in space and time, we can integrate term by term. First, we interchange the order of differentiation and integration ($v_a = 0$) in the accumulation term:

$$\int_V \frac{\partial}{\partial t}(\rho\vec{\mathbf{v}})\,dV = \frac{\partial}{\partial t}(L_{tot}) \qquad L_{tot} = \text{Total momentum} \tag{11.12}$$

Next, we work on the convection of momentum and transform the volume integral to a surface integral over the inlet and outlet planes:

$$\int_V (\vec{\nabla}\cdot(\rho\vec{\mathbf{v}})\vec{\mathbf{v}})\,dV = \int_{A_c} (\vec{\mathbf{n}}\cdot(\rho\vec{\mathbf{v}})\vec{\mathbf{v}})\,dA_c = \rho_2\overline{v_2^2}A_{c2} - \rho_1\overline{v_1^2}A_{c1} \tag{11.13}$$

Again, we have the difference between the two quantities because v_1, from the $\rho\vec{v}$ term is in a direction opposite to the outward surface normal. Considering the pressure term, we convert from a volume to a surface integral. The pressure is drawn so that its direction always points into the control volume. P_1 and P_2 are then referenced to the coordinate system imposed in the figure. Here, P_1 is positive and P_2 negative, and so the integrals become

$$\int_V \vec{\nabla} P\, dV = \int_{A_c} P_n\, dA_c = P_2 A_{c2} - P_1 A_{c1} \tag{11.14}$$

The integral over the stress represents the viscous forces acting on the system. These forces depend upon the flux–gradient relationship for the fluid:

$$\int_V (\vec{\nabla} \cdot \vec{\tau})\, dV = F_\mu \tag{11.15}$$

The integral over the gravitational body force is just the weight of the fluid:

$$\int_V (\rho \vec{\mathbf{g}})\, dV = m_{tot} g \tag{11.16}$$

Combining all the terms yields the desired macroscopic force balance:

$$\rho_2 \overline{v_2^2} A_{c2} - \rho_1 \overline{v_1^2} A_{c1} + P_2 A_{c2} - P_1 A_{c1} + F_\mu - m_{tot} g = -\frac{\partial L_{tot}}{\partial t} \tag{11.17}$$

If the system is operating at a steady state, the accumulation term vanishes, and we have the following force balance. Normally, we lump all forces, including external ones reacting to fluid-generated forces, into a single term and write the equation as

$$-\Delta\left(\rho \overline{v^2} A_c\right) - \Delta(P A_c) = F_\mu - m_{tot} g = \sum F \tag{11.18}$$

where $\Delta = Out - In$, a terminology you have probably encountered before. Often, the pressure forces are also included on the right-hand side of Equation 11.18. Finally, if the system boundaries are permeable and allow mass transfer across them, we can also have momentum transfer at the same time. The macroscopic momentum balance becomes

$$\Delta\left(\rho \overline{v^2} A_c\right) + \Delta(P A_c) + \sum F - \mathcal{F}_d = -\frac{\partial L_{tot}}{\partial t} \tag{11.19}$$

$$\mathcal{F}_d = \int_{A_p} (\vec{\mathbf{n}} \cdot (\rho \vec{\mathbf{v}}) \vec{\mathbf{v}})\, dA_p \tag{11.20}$$

where

$\mathcal{F}_d$ is positive for material entering the system

A_p is the permeable area

Example 11.2 Force on a Reducing Elbow

A 30° reducing elbow is shown in Figure 11.3. The fluid is water. We want to evaluate the components of force that must be provided by the adjacent pipes to keep the elbow in place.

In this problem, we must consider two components of the force, one in the x and the other in the y-direction. Thus, we must use two momentum equations plus the continuity equation. The system is at steady state, and so the continuity and generic momentum balances are

$$\rho_1 \bar{v}_1 A_{c1} = \rho_2 \bar{v}_2 A_{c2} = \rho \dot{V} \quad \text{continuity}$$

$$\sum F_x = \frac{\partial}{\partial t} \int_V \rho v_x \, dV + \int_{A_c} (\rho v_x) \vec{\mathbf{v}} \cdot d\vec{\mathbf{A}}_{\mathbf{c}} \quad x\text{-momentum}$$

$$\sum F_y = \frac{\partial}{\partial t} \int_V \rho v_y \, dV + \int_{A_c} (\rho v_y) \vec{\mathbf{v}} \cdot d\vec{\mathbf{A}}_{\mathbf{c}} \quad y\text{-momentum}$$

The continuity equation states that

$$\bar{v}_1 = \frac{\dot{V}}{A_{c1}} = 6.11\,\text{m/s} \qquad \bar{v}_2 = \frac{\dot{V}}{A_{c2}} = 13.75\,\text{m/s}$$

We now apply the momentum equations. There are some general rules to follow:

1. The pressures are always written so that they point into the control volume.
2. Both the pressure force and the gravity force are referenced relative to the coordinate directions set up in the problem.
3. The flow rates in and out of the control volume are referenced relative to the outward surface normals.
4. All other velocity components are referenced relative to the coordinate directions set up in the problem.
5. Finally, the sum of the forces includes the reaction forces necessary to hold the elbow in place, and those reaction forces (R) are referenced to the coordinate directions for the problem. We generally assume they are positive to start.

With that in mind, the momentum equations are

x-component

$$\underbrace{R_x + P_{1g} A_{c1} - P_{2g} A_{c2} \cos\theta}_{\text{Sum of all } x\text{-forces}} = -\rho \dot{V} \bar{v}_{x1} + \rho \dot{V} \bar{v}_{x2} = \rho \dot{V} (\bar{v}_{x2} - \bar{v}_{x1})$$

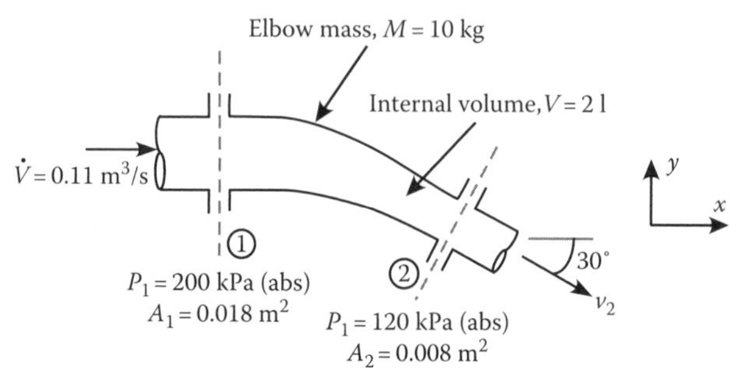

FIGURE 11.3 A 30° reducing elbow with water flowing at steady state.

$$R_x = (1000)(0.11)(13.75 - 6.11) - (200 - 101)\times 10^3(0.018) + (120 - 101)\times 10^3(0.008)\cos(30°)$$

$$= -810\,N$$

Since the atmospheric pressure force acts on all points of the system and the ends of the system are open to the environment, the pressures we use in the momentum balance are gauge pressures:

y-component

$$\underbrace{R_y + P_{2g}A_{c2}\sin\theta - Mg - \rho g V}_{\text{Sum of all } y\text{-forces}} = -\rho\dot{V}\,\bar{v}_{y1} + \rho\dot{V}\,\bar{v}_{y2} = \rho\dot{V}(0 - \bar{v}_2\sin\theta)$$

$$R_y = (10)(9.8) + (1000)(0.002)(9.8) - (120 - 101)\times 10^3(0.018)\sin(30°) - (1000)(0.11)(13.6)\sin(30°)$$

$$= -801.4\,N$$

where
- R_x is the horizontal reaction force and points to the left
- R_y is the vertical reaction force and points down

Example 11.3 Water Jet Striking a Moving Dish

The circular dish, whose cross section is shown in Figure 11.4, has an outside diameter of 0.15 m. A water jet strikes the dish concentrically and then flows outward along the surface of the dish. The jet speed is 45 m/s and the dish moves to the left at a constant velocity of 10 m/s. Find the thickness of the jet sheet at a radius of 75 mm from the jet axis. What horizontal force on the dish is required to maintain this motion?

This is a steady-state problem since the dish moves at a constant velocity. There are no pressure forces to consider since fluid enters and leaves the dish at atmospheric pressure. There are no body forces in the *x*-direction since the system is horizontal. Thus, the *x*-component of the momentum equation is

$$\sum F_x = R_x = \int_{cs} (\rho v_x)\vec{\mathbf{v}}\cdot d\vec{\mathbf{A}} = \rho\dot{V}(\bar{v}_{x2} - \bar{v}_{x1})$$

Now we must fill in for v_{x2} and v_{x1} keeping track of their orientation with respect to the coordinate system we set up. Since the dish moves away at a constant velocity, the velocities of the fluid in and out of the dish are relative to that velocity:

$$\bar{v}_{x2} = -(v - U)\cos\theta \qquad \bar{v}_{x1} = v - U$$

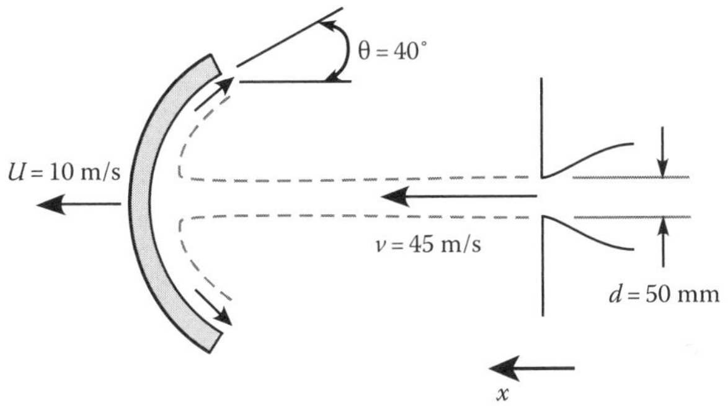

FIGURE 11.4 A water jet striking a circular dish. The dish moves at a constant velocity relative to the jet.

The volumetric flow rate of fluid into and out of the dish must also take into account that the vane moves away from the fluid source. Thus, what comes out of the nozzle is not what strikes the vane. Some fluid remains behind to lengthen the jet:

$$\dot{V} = \rho A_1(v - U)$$

Putting it all together gives

$$R_x = \rho\left(\frac{\pi d^2}{4}\right)(v-U)[-(v-U)\cos\theta - (v-U)]$$

$$= -\rho\left(\frac{\pi d^2}{4}\right)(v-U)^2[1+\cos\theta] = -4240\ N$$

To get the thickness of the sheet coming off the dish, we use the continuity equation. At steady state,

$$0 = \int_A \rho\vec{\mathbf{v}}\cdot\mathbf{d}\vec{\mathbf{A}} = \rho\bar{v}_{x2}A_2 - \rho\bar{v}_{x1}A_1$$

Now the velocities (relative to the outward surface normal) and densities in and out are equal, so we need only specify the areas to calculate the thickness:

$$A_1 = \frac{\pi d^2}{4} = A_2 = 2\pi Rt \qquad t = \frac{d^2}{8R} = 4.2\ \text{mm}$$

Example 11.4 Momentum Equation and Acceleration

The momentum equation can also be used in an accelerating system. The only caveat is that we need to be able to isolate a control volume with well-defined boundaries and with a corresponding velocity associated with it. Thus, a control volume such as a fluid, without defined boundaries, would be inappropriate (though a drop or bubble would be ok). Consider the system shown in Figure 11.5. Here, we have an assembly that is propelled by a liquid jet. The system slides along a track with a coefficient of friction, $\mu k = 0.3$. We would like to calculate the acceleration of the sliding object at the instant when the velocity of the object $U = 10$ m/s. We would also like to know the terminal speed of the slider.

First, we apply the continuity equation about a pseudo–control volume shown in the gray dotted lines:

$$-\rho_1\bar{v}_1A_1 + \rho_2\bar{v}_2A_2 + \rho_3\bar{v}_3A_3 = 0$$

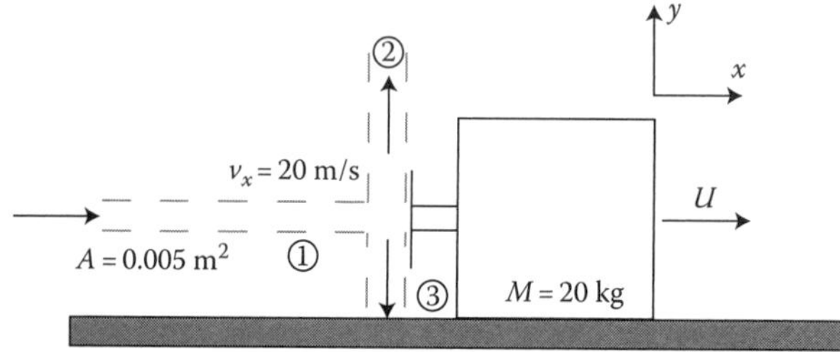

FIGURE 11.5 Slider system propelled by a fluid stream.

Since the fluid is incompressible and the object moves away from the water source,

$$\rho_1 = \rho_2 = \rho_3 = \rho \qquad \bar{v}_1 = v_x - U$$

Now we can apply the momentum equation. In an accelerating system, the x-momentum equation can be written as

$$\sum F_x = \sum F_{x,surface} + \sum F_{x,body} = \frac{\partial}{\partial t}\int_V \rho \bar{v}_x dV + \int_V \rho a_x dV + \int_A \bar{v}_x(\rho v \cdot dA)$$

$$\text{I} \qquad \text{II}$$

There are two acceleration terms. One considers momentum changes that occur due to a change in the shape of the control volume (I). The other considers the change in the velocity of the control volume itself (II). We choose the solid object as the system and it undergoes uniform acceleration. The time derivative term (I) is identically zero since the boundaries of the solid object are rigid. The second term is just the mass of the system times its acceleration.

The sum of the forces on the left-hand side of the equation considers surface forces and body forces, but since gravity acts in the y-direction, the body forces in the x-direction are zero. The surface forces in this case are friction given by

$$\sum F_{x,surface} = -fU = -\mu_k MgU$$

Substituting into the momentum equation, we have

$$-\mu_k MgU - a_x M = (-\rho\bar{v}_{1x}A_1)\bar{v}_{1x} + (\rho\bar{v}_{2x}A_2)\bar{v}_{2x} + (\rho\bar{v}_{3x}A_3)\bar{v}_{3x}$$

$$-\mu_k MgU - M\frac{dU}{dt} = \rho\bar{v}_{1x}^2 A = \rho(v_x - U)^2 A$$

If we substitute into the equation and solve for dU/dt when $U = 10$ m/s, we find

$$\frac{dU}{dt} = 9.4\frac{\text{m}}{\text{s}^2}$$

When the object reaches its terminal velocity, U_t, there is no longer any acceleration. Thus, we have

$$\frac{dU}{dt} = \frac{\rho(v_x - U_t)^2 A + \mu_k MgU_t}{M} = 0$$

$$U_t = \frac{2v_x + \dfrac{\mu_k Mg}{\rho A} \pm \sqrt{\left(2v_x + \dfrac{\mu_k Mg}{\rho A}\right)^2 - 4v_x^2}}{2}$$

Though we have two solutions, the only valid solution is the one where we choose the minus sign. (Why?) Here, $U_t = 9.45$ m/s.

11.4 MACROSCOPIC MECHANICAL ENERGY BALANCE: EXTENDED BERNOULLI'S EQUATION

We can derive the overall mechanical energy balance for the fluid system by integrating the microscopic equation over the volume of the system of interest. This integration, for a general, compressible fluid, is difficult to perform and tends to be confusing and cryptic. Therefore, we will derive the macroscopic mechanical energy equation for an incompressible fluid and then just present a form that applies to a generic fluid [2].

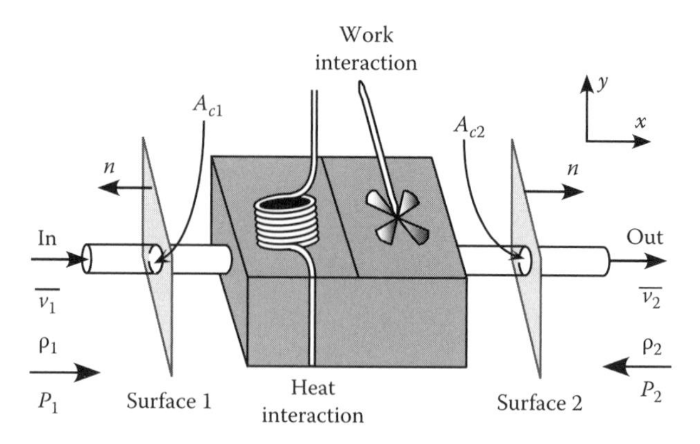

FIGURE 11.6 System for deriving the macroscopic mechanical energy balance.

When we consider the energy balance, we must consider heat and work interactions with the environment. The heat interaction may be necessary to keep the system isothermal. The work interaction is necessary because we may have turbines, compressors, pumps, etc., within our *black box*. Figure 11.6 shows a schematic diagram of the generic system.

Since we have moving surfaces in the system, we need to separate all the surfaces into four regions: the inlet area defined by the cut at surface 1, A_{c1}; the outlet area defined by the cut at surface 2, A_{c2}; the wetted surfaces that are stationary (system boundaries), A_s; and the wetted surfaces that are moving (pistons, propellers, etc.), A_m.

We consider the microscopic mechanical energy balance with the gravitational force written in terms of the gradient of a potential, $\vec{g} = -g\vec{\nabla}\Phi_g$. We will use the continuity equation to remove the $P(\vec{\nabla} \cdot \vec{\mathbf{v}})$ during the course of the calculations:

$$\left(\vec{\nabla}\cdot\left(\frac{1}{2}\rho v^2\right)\vec{\mathbf{v}}\right)+(\vec{\nabla}\cdot P\vec{\mathbf{v}})+(\vec{\nabla}\cdot[\vec{\tau}\cdot\vec{\mathbf{v}}])-(\vec{\tau}:\vec{\nabla}\vec{\mathbf{v}})+\rho g(\vec{\mathbf{v}}\cdot\vec{\nabla}\Phi_g)=-\frac{\partial}{\partial t}\left(\frac{1}{2}\rho v^2\right) \tag{11.21}$$

Integrating term by term over the volume of the system, we begin with the accumulation term. Assuming rigid boundaries,

$$\int_V \frac{\partial}{\partial t}\left(\frac{1}{2}\rho v^2\right)dV = \frac{d}{dt}\left(\int_V \left(\frac{1}{2}\rho v^2\right)dV\right) \tag{11.22}$$

Considering the term representing convection of energy next, we use Gauss's law to transform the volume integral to a surface integral:

$$\begin{aligned}\int_V \left(\vec{\nabla}\cdot\left(\frac{1}{2}\rho v^2\right)\vec{\mathbf{v}}\right)dV &= \int_{A_{c1}+A_{c2}} \left(\frac{1}{2}\rho v^2\right)v_n dA_c + \int_{A_s}\left(\frac{1}{2}\rho v^2\right)v_n dA_s + \int_{A_m}\left(\frac{1}{2}\rho v^2\right)v_n dA_m \\ &= -\frac{1}{2}\rho\overline{v_1^3}A_{c1} + \frac{1}{2}\rho\overline{v_2^3}A_{c2}\end{aligned} \tag{11.23}$$

The integral over all the wetted surfaces is zero due to no-slip conditions there, and no convection takes place on those surfaces. The integral over A_{c1} has a minus sign because the flow there is directed in the opposite direction of the outward surface normal.

Considering the pressure term, we have

$$\int_V (\vec{\nabla}\cdot P\vec{v})dV = \int_{A_{c1}+A_{c2}} Pv_n dA_c + \int_{A_s} Pv_n dA_s + \int_{A_m} Pv_n dA_m = -P\overline{v}_1 A_{c1} + P\overline{v}_2 A_{c2} + \dot{W}_p \tag{11.24}$$

W_p represents the rate of work done by the machinery on the fluid and is the integral over the moving surfaces. This work is accomplished by changing the pressure of the fluid. The integral over the stationary wetted surfaces is identically zero due to the no-slip condition on the fluid.

We also have a work term resulting from viscous forces pushing the fluid into and out of the system. In this integral, the only non-negligible component is that part on the moving surfaces that has to combat viscous drag forces to keep in motion. The small amount of work needed to overcome viscous forces and push the fluid into and out of the control volume can be neglected:

$$\int_V (\vec{\nabla}\cdot[\vec{\tau}\cdot\vec{v}])dV = \int_{A_{c1}+A_{c2}} [\vec{\tau}\cdot\vec{v}]_n dA_c + \int_{A_s} [\vec{\tau}\cdot\vec{v}]_n dA_s + \int_{A_m} [\vec{\tau}\cdot\vec{v}]_n dA_m = \dot{W}_\mu \tag{11.25}$$

We must also consider the conversion of mechanical energy to internal energy resulting from viscous dissipation. This is a generation term; it occurs over the entire system volume and so cannot be converted to a surface integral. We often refer to this term as the *lost work* term:

$$\int_V (\vec{\tau} : \vec{\nabla}\vec{v})dV = -\dot{E}_\mu \tag{11.26}$$

Finally, we must consider the work performed against gravity to move the fluid from one height to another:

$$\int_V \rho g(\vec{v}\cdot\vec{\nabla}\Phi_g)dV = \int_V \rho g(\vec{\nabla}\cdot\Phi_g\vec{v} - \Phi_g(\vec{\nabla}\cdot\vec{v}))dV = \int_V \rho g(\vec{\nabla}\cdot\Phi_g\vec{v})dV \tag{11.27}$$

Here, we have used the fact that the fluid is incompressible to remove the $\vec{\nabla}\cdot\vec{v}$ term. Transforming the volume integral into surface integrals gives

$$\int_V \rho g(\vec{\nabla}\cdot\Phi_g\vec{v})dV = \rho g \int_{A_{c1}+A_{c2}} \Phi_g v_n dA_c + \rho g \int_{A_s} \Phi_g v_n dA_s + \rho g \int_{A_m} \Phi_g v_n dA_m$$

$$= -\rho g \Phi_{g1}\overline{v}_1 A_{c1} + \rho g \Phi_{g2}\overline{v}_2 A_{c2} \tag{11.28}$$

The integral over all the wetted surfaces is zero since their position within the gravitational field is essentially constant.

Combining Equations 11.22 through 11.28 yields the macroscopic mechanical energy equation. This equation may be referred to as Bernoulli's equation though Bernoulli's equation is the macroscopic mechanical energy equation applied to a fluid with zero viscosity and no moving elements. Bernoulli's equation also applies only along a single streamline. With Δ = Out – In, we can write the equation as

$$\Delta\left(\frac{1}{2}\rho\overline{v^3}A_c\right) + \Delta(P\overline{v}A_c) + \dot{E}_\mu + \dot{W}_p + \dot{W}_\mu + \Delta(\overline{v}g\Phi_g A_c) = -\frac{d}{dt}\int_V \frac{1}{2}\rho\overline{v^2}dV = -\frac{d}{dt}(K.E.) \tag{11.29}$$

We can put Equation 11.29 in a more familiar form if we assume steady-state operation. Since the fluid is incompressible, $\rho_1\bar{v}_1A_{c1} = \rho_2\bar{v}_2A_{c2} = \rho\bar{v}A_c$, and we can divide through by the mass flow rate, $\rho\bar{v}A_c$, to obtain

$$\Delta\left(\frac{1}{2}\frac{\overline{v^3}}{\bar{v}}\right)+\frac{1}{\rho}\Delta P+\dot{E}_\mu+\dot{W}_p+\dot{W}_\mu+g\Delta\Phi_g=0 \tag{11.30}$$

Normally, we assume that the average of the cube of the velocity is just the average velocity cubed, $\overline{v^3}=\bar{v}^3$, and that $\Delta\Phi_g$ can be represented as the difference in height between A_{c1} and A_{c2}. The former is not a bad approximation for turbulent flow where the velocity profile is nearly uniform across A_{c1} or A_{c2} but may result in considerable error when the flow is laminar and there is a considerable velocity profile across A_{c1} or A_{c2}. Equation 11.31 is an additional term added into Equation 11.30 if mass is allowed to diffuse across the system boundaries.

The macroscopic energy balance for a general fluid is different than that for the incompressible fluid. We present the balance in Equation 11.32:

$$\dot{E}_d=\int_{A_p}\left(\vec{\mathbf{n}}\cdot\left[\frac{1}{2}\rho v^2+\rho g\Phi_g+P\right]\vec{\mathbf{v}}\right)dA_p \tag{11.31}$$

$$\Delta\left[\left(\frac{1}{2}\frac{\overline{v^3}}{\bar{v}}+g\Phi_g+\bar{G}\right)\rho\bar{v}A_c\right]+\dot{W}+\dot{E}_\mu+\dot{E}_d=-\frac{d}{dt}\left(\int_V\frac{1}{2}\rho\overline{v^2}dV+\int_V\rho g\Phi_g dV+\int_V\rho(\bar{U}-T\bar{S})dV\right) \tag{11.32}$$

where

$\bar{U}$ is the internal energy per unit mass
$\bar{S}$ is the entropy per unit mass
$\bar{G}$ is the Gibbs energy per unit mass
$\dot{W}$ is the rate of work done by the system on its surroundings ($\dot{W}_P+\dot{W}_\mu$)
$\dot{E}_\mu$ is the frictional loss (the rate at which mechanical energy is converted to thermal energy)
$\dot{E}_d$ is the kinetic energy exchanged through mass transfer if the system has permeable walls and permits this

Remember $\Delta U = T\Delta S - P\Delta V$ and $\Delta G = V\Delta P - S\Delta T$. With the three macroscopic equations we have just derived, we can analyze a wide variety of *isothermal* pumping and piping problems.

Example 11.5 Pressure Rise and Friction Loss in a Sudden Enlargement

Consider an incompressible fluid flowing through a circular tube of cross-sectional area A_{c1}, which empties into a larger tube of cross-sectional area A_{c2}. The fluid is assumed to be in turbulent or plug flow, and that means the velocity profile across the tube is essentially flat. We want to use the macroscopic balances to get an expression for the pressure change between planes A_{c1} and A_{c2} and the corresponding friction loss for the sudden enlargement. All velocities will be assumed to be average quantities over a cross section, and all fluid properties will be assumed constant over those cross-sectional areas. The steady-state situation is shown in Figure 11.7.

We begin the analysis with the macroscopic continuity equation. In this steady-state situation, we know that the mass flow rate through surface 1 must equal the mass flow rate through surface 2.

For an incompressible fluid, $\rho_1 = \rho_2$, and the continuity equation shows that the velocity in the larger tube must be smaller than the velocity in the smaller tube:

(a) Continuity

$$\rho_2\bar{v}_2A_{c2} - \rho_1\bar{v}_1A_{c1} = 0 \qquad \text{so:} \qquad \frac{\bar{v}_1}{\bar{v}_2} = \frac{A_{c2}}{A_{c1}}$$

(b) Momentum balance

The overall momentum balance in the direction of flow is

$$(\rho_2\bar{v}_2A_{c2})\bar{v}_2 - (\rho_1\bar{v}_1A_{c1})\bar{v}_1 = -R_x + P_1A_{c1} - P_2A_{c2} = \sum F_x$$

We must consider pressure differences between A_{c1} and A_{c2}, and we know there are velocity differences, but since the tubes are horizontal, we can safely neglect gravitational effects. The force, R_x, is the reaction force. It is made up of two parts: the viscous force on the cylindrical surfaces parallel to the flow direction and the pressure force on the washer-shaped surface just to the right of plane 1 and perpendicular to the flow axis. In this instance, we are looking at a small section of pipe about the sudden expansion. Since viscous forces in pipe flow are appreciable only in instances where the pipes are relatively long or flows are extremely rapid, we assume they are negligible with respect to what we will refer to as inertial forces for this problem. The force on the washer arises from recirculating flows that exist in the corners. These are shown in Figure 11.7. The force is determined by the pressure exerted on the washer multiplied by the washer area. Though the downstream pressure is P_2, we make a conservative estimate and assume the pressure exerted on the washer is the same pressure that exists at surface 1. Thus, the force can be approximated as

$$R_x = -P_1(A_{c2} - A_{c1})$$

This force will have to oppose the net pressure force between the inlet and outlet. Substituting into the momentum balance gives

$$(\rho_2\bar{v}_2A_{c2})\bar{v}_2 - (\rho_1\bar{v}_1A_{c1})\bar{v}_1 = P_1(A_{c2} - A_{c1}) + P_1A_{c1} - P_2A_{c2}$$

Solving for the pressure drop, $\Delta P = P_2 - P_1$, incorporating the continuity equation result, $\rho_1\bar{v}_1A_{c1} = \rho_2\bar{v}_2A_{c2}$, into the momentum equation, and realizing that the densities do not change from inlet to outlet gives

$$\Delta P = \rho_2\bar{v}_2(\bar{v}_1 - \bar{v}_2) = \rho\bar{v}_1^2\left(\frac{A_{c1}}{A_{c2}} - 1\right)$$

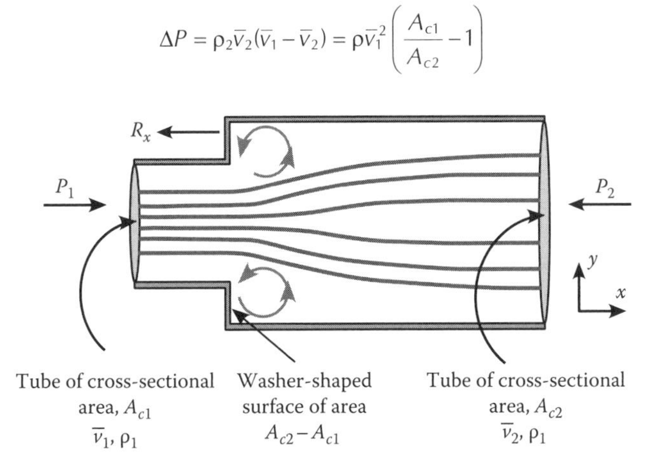

FIGURE 11.7 Sudden expansion of a fluid.

Notice that the momentum balance predicts a rise in pressure. This rise in pressure must occur because the fluid is exerting a force on the washer. The decrease in momentum upon passing through the sudden enlargement, $\bar{v}_2 < \bar{v}_1$, translates into the force on the washer. Since the pressure is higher downstream than it is upstream, the sudden enlargement works to impede the flow of fluid.

(c) Mechanical energy balance

The mechanical energy balance is used to calculate the lost work. A significant fraction of the potential work in this system is lost work since we have no mechanism for extracting it from the change in fluid momentum. The lost work gets dissipated as heat in the recirculating fluid behind the washer. Applying the mechanical energy balance yields

$$\frac{1}{2}\left[(\rho\bar{v}_2 A_{c2})\bar{v}_2^2 - (\rho\bar{v}_1 A_{c1})\bar{v}_1^2\right] + (\bar{v}_2 P_2 A_{c2} - \bar{v}_1 P_1 A_{c1}) + E_\mu = 0$$

Since the mass flow rate is constant throughout the system, we can divide the whole energy balance by $\rho_1\bar{v}_1 A_{c1}$ or $\rho_2\bar{v}_2 A_{c2}$ to obtain the mechanical energy balance incorporating the pressure drop we just calculated:

$$\frac{1}{2}\left(\bar{v}_2^2 - \bar{v}_1^2\right) + \frac{1}{\rho}(P_2 - P_1) + \overline{E_\mu} = 0$$

Solving for $\overline{E_\mu}$ using our results for the pressure drop and using the continuity equation for simplification gives

$$\overline{E_\mu} = \frac{1}{2}\bar{v}_2^2\left(\frac{A_{c2}}{A_{c1}} - 1\right)^2$$

Since $\overline{E_\mu}$ and ΔP are both positive, we can see that the kinetic energy of the fluid is being converted into reversible work (a rise in P) and heat (in the form of lost work, $\overline{E_\mu}$).

(d) Velocity averages in turbulent flow

In the mechanical energy balance we just used, we assumed that

$$\overline{v^3} = \bar{v}^3 \qquad \text{and} \qquad \overline{v^2} = \bar{v}^2$$

How much error did we introduce when we used these assumptions? For turbulent fluids in a circular pipe, we can approximate the flow field quite well using a 1/7th power law velocity profile [3]:

$$\frac{v}{v_{max}} = \left(1 - \frac{r}{r_o}\right)^{1/7}$$

We now need to calculate the velocity averages using the definition of our mixing-cup velocities:

$$\frac{\overline{v^2}}{\bar{v}^2} = \frac{\int_0^{2\pi}\int_0^{r_o}[1-(r/r_o)]^{2/7}\, r\,dr\,d\theta}{\int_0^{2\pi}\int_0^{r_o} r\,dr\,d\theta}\left[\frac{\int_0^{2\pi}\int_0^{r_o} r\,dr\,d\theta}{\int_0^{2\pi}\int_0^{r_o}[1-(r/r_o)]^{1/7}\, r\,dr\,d\theta}\right]^2$$

We can evaluate those integrals not containing the velocity profile easily:

$$\frac{\overline{v^2}}{\overline{v}^2} = \frac{\int_0^{r_o} [1-(r/r_o)]^{2/7} r\,dr}{\frac{r_o^2}{2}} \left[\frac{\frac{r_o^2}{2}}{\int_0^{r_o} [1-(r/r_o)]^{1/7} r\,dr} \right]^2$$

Since we are interested in a percentage difference, we can make our life easier and define a new dimensionless variable, $\eta = 1 - r/r_o$. The previous integrals now become

$$\frac{\overline{v^2}}{\overline{v}^2} = \frac{2\int_0^1 \eta^{2/7}(1-\eta)d\eta}{\left[2\int_0^1 \eta^{1/7}(1-\eta)d\eta\right]^2} = \frac{2\left(\frac{7}{9}-\frac{7}{16}\right)}{4\left(\frac{7}{8}-\frac{7}{15}\right)^2} = 1.02$$

By not using $\overline{v^2}$ in our momentum balance, we introduce a 2% error into the calculations. We can perform the same calculation for $\overline{v^3}$ and $\overline{v}^3$:

$$\frac{\overline{v^3}}{\overline{v}^3} = \frac{2\int_0^1 \eta^{3/7}(1-\eta)d\eta}{\left[2\int_0^1 \eta^{1/7}(1-\eta)d\eta\right]^3} = \frac{2\left(\frac{7}{10}-\frac{7}{17}\right)}{8\left(\frac{7}{8}-\frac{7}{15}\right)^3} = 1.06$$

By not using $\overline{v^3}$ in our energy calculations, we introduced an error of 6%. These errors are quite acceptable for turbulent flow. You should convince yourself that the errors get much larger when the velocity profile in the tubes is laminar.

Example 11.6 Hydraulic Jump

One important consequence of fluid flowing in an open channel is the phenomenon of the hydraulic jump. A hydraulic jump is an abrupt change in the depth of the flowing fluid that occurs as the mass flow rate is increased. The physical situation is shown in Figure 11.8. Our goal is to determine the depth of the fluid after the transition.

Our analysis of this situation will assume steady flow of an incompressible fluid. The flow is very fast, so we can use our approximations for $\overline{v^2}$ and $\overline{v^3}$. The channel is of square cross-sectional area, and the walls are high enough to contain the flow regardless of the depth of the water.

(a) Continuity—application of the continuity equation yields

$$\rho_2\overline{v}_2 A_{c2} - \rho_1\overline{v}_1 A_{c1} = 0$$

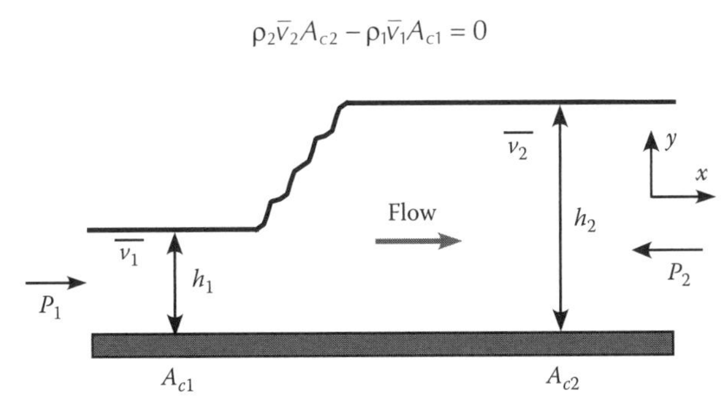

FIGURE 11.8 Steady flow in a square channel—the hydraulic jump.

Since the fluid is incompressible,

$$\frac{\bar{v}_1}{\bar{v}_2} = \frac{A_{c2}}{A_{c1}}$$

(b) Momentum—application of the momentum balance yields

$$\rho_2 \bar{v}_2^2 A_{c2} - \rho_1 \bar{v}_1^2 A_{c1} - P_1 A_{c1} + P_2 A_{c2} = 0$$

Unlike the previous example with the sudden enlargement, we have no analogous surface upon which a reaction force can be exerted ($R = 0$). The energy that went into exerting that force on the washer-shaped area of Figure 11.7 will be dissipated as lost work in the hydraulic jump. Using the continuity equation to simplify the momentum equation yields

$$\rho_1 \bar{v}_1^2 A_{c1} \left[\frac{A_{c1}}{A_{c2}} - 1 \right] - P_1 A_{c1} + P_2 A_{c2} = 0$$

To determine the height of the hydraulic jump, we make the approximation that the pressures, P_1 and P_2, are average hydrostatic pressures ($\rho g h/2$):

$$\rho_1 \bar{v}_1^2 A_{c1} \left[\frac{A_{c1}}{A_{c2}} - 1 \right] - \frac{\rho_1 g h_1}{2} A_{c1} + \frac{\rho_2 g h_2}{2} A_{c2} = 0$$

Dividing through by $\rho_1 A_{c1}$ because the fluid is incompressible and realizing that for a square channel $A_{c1}/A_{c2} = h_1/h_2$, we have a cubic equation for h_2:

$$\bar{v}_1^2 \left[\frac{h_1}{h_2} - 1 \right] - \frac{g h_1}{2} + \frac{g h_2}{2} \left(\frac{h_2}{h_1} \right) = 0$$

Let's suppose water at 25°C is flowing in a square channel at a velocity of 10 m/s. The density of water at this temperature is 1000 kg/m³ and the initial depth of the water is 15 cm. We would like to determine the depth and velocity following a hydraulic jump. Inserting the numerical values into our energy balance leads to the following cubic equation for h_2:

$$-32.667 h_2^3 + 100.735 h_2 - 1.5 = 0$$

The equation has three real solutions for h_2 but only one makes physical sense. Notice that the water level increased by a factor of over 10.

$$h_2 = 1.676 \text{ m} \qquad \text{and so} \qquad \bar{v}_2 = 0.895 \text{ m/s}.$$

Example 11.7 Pumping and Piping

A pump is transferring water from tank A to tank B at a rate of 0.02 m³/s. The tanks are both open to the atmosphere and tank A is 3 m above ground level. Tank B is 20 m above ground level. The pump is located at ground level, and the discharge pipe that enters tank B is 23 m above ground level at its highest point. All the piping is 5.08 cm in diameter, and the tanks are 6 m in diameter. The situation is shown in Figure 11.9. If we neglect friction losses in the piping, what size pump will we need? If the pump system is 70% efficient, what size motor would we need to drive the pump?

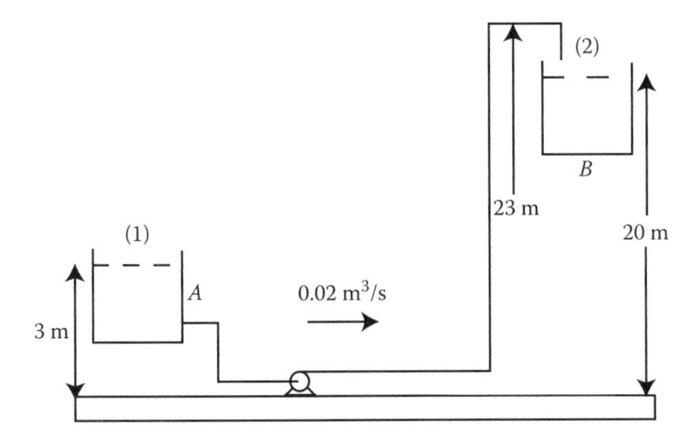

FIGURE 11.9 Pumping and piping between two tanks.

The first step is to apply the continuity equation, but in this case since what goes in goes out, the continuity equation is trivial. The work the pump must do on the fluid is determined by applying the mechanical energy balance between the two tanks. As a rule, the pump work goes on the left-hand side of the equation (supplier), while the lost work goes on the right-hand side (consumer):

$$\frac{P_1}{\rho g}+\frac{\bar{v}_1^2}{2g}+z_1+\frac{W_p}{\rho g}=\frac{P_2}{\rho g}+\frac{\bar{v}_2^2}{2g}+z_2+\frac{h_L}{\rho g}$$

Substituting what we know from the problem about the lost work ($h_L = 0$), the uniform piping diameter ($\bar{v}_1 = \bar{v}_2$), and the pressures ($P_1 = P_2 = 0$), we find

$$z_1+\frac{W_p}{\rho g}=z_2$$

$$W_p=\rho g(z_2-z_1)=1000(9.8)(23-3)=196{,}000\,\frac{\text{N}}{\text{m}^2}$$

The power required to push all that fluid is equal to the work required multiplied by the volumetric flow rate of the fluid:

$$\text{Power}=W_p*\dot{V}=196{,}000(0.02)=3920\,\text{W}\ (5.25\ \text{hp})$$

At 70% efficient, the motor would have to supply 5.25/0.7 = 7.5 hp.

Example 11.8 Liquid Drainage [4]

Our previous examples have focused on using the fully integrated versions of the macroscopic balances. We are not limited to this form of analysis but are free to mix and match microscopic and macroscopic balances. In this example, we look at one such case, considering how much liquid clings to the walls of a vessel once the bulk of the fluid has drained away. A schematic of the system is shown in Figure 11.10. We consider the fluid to be Newtonian with constant properties. We neglect inertial effects and consider long times after drainage has started. Therefore, we assume the flow field within the fluid to be laminar, even though the film thickness is constantly changing. We require a momentum balance for the drainage velocity, $v_z(x,\Delta_f)$, and a transient mass balance to determine the film thickness $\Delta_f(z,t)$.

The momentum balance considers viscous and gravitational forces. The microscopic momentum equation (see Chapter 5) describing the velocity profile is

$$\frac{d^2v_z}{dx^2}+\frac{\rho g}{\mu}=0 \qquad x=0 \qquad v_z=0 \qquad x=\Delta_f \qquad \frac{dv_z}{dx}=0$$

and is subject to boundary conditions specifying no slip at the fluid–plate interface and no stress at the fluid–air interface. The solution for the velocity profile is

$$v_z=\frac{\rho g}{\mu}x\left(\Delta_f-\frac{x}{2}\right)$$

We use a transient mass balance to determine the film thickness. Changes in the film thickness result from changes in the mass flow rate of fluid through the control volume shown in Figure 11.10. We are not concerned with the details of the velocity profile, but can use a macroscopic approach and use an average velocity, $\bar{v}_z$, through the surfaces of the control volume. This average velocity is a function of the film thickness and, hence, of vertical position at any time:

$$\bar{v}_z=\frac{1}{\Delta_f}\int_0^{\Delta_f}v_z(x)dx=\frac{\Delta_f^2}{3}\left(\frac{\rho g}{\mu}\right)$$

The mass balance is a hybrid, microscopic/macroscopic balance written in terms of the mass flow rate, $\dot{m}$, across the thin film:

$$\dot{m}(z)-\dot{m}(z+\Delta z)=\frac{\partial}{\partial t}(\Delta_f w\rho\Delta z)$$

Dividing through by Δz and taking the limit as $\Delta z \to 0$ give

$$-\frac{\partial}{\partial z}(\dot{m})=\rho\frac{\partial}{\partial t}(\Delta_f w)$$

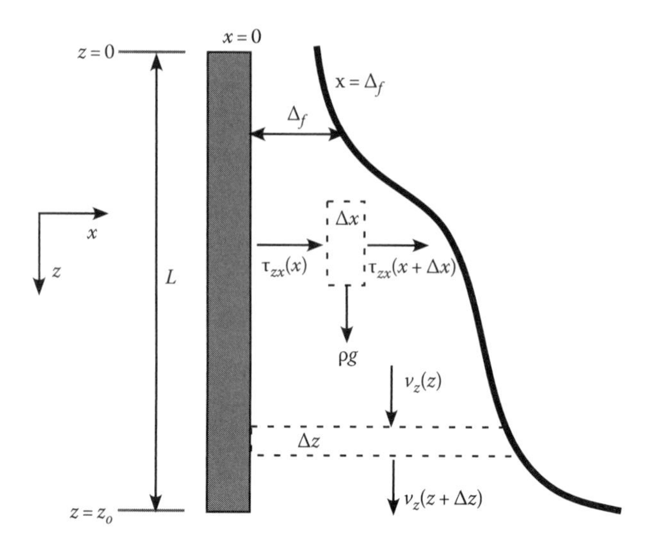

FIGURE 11.10 A Newtonian fluid draining from the walls of a large vessel.

Here, w is the width of the wall. We can substitute for the mass flow rate using the average velocity, the density, film thickness, and wall width:

$$-\rho \frac{\partial}{\partial z}(\Delta_f w \bar{v}_z) = \rho \frac{\partial}{\partial t}(\Delta_f w)$$

Substituting in for the velocity, $\bar{v}_z$, and removing the constant terms give

$$-\frac{\rho g}{\mu} \Delta_f^2 \frac{\partial}{\partial z}(\Delta_f) = \frac{\partial}{\partial t}(\Delta_f)$$

The solution is obtained using a similarity variable, $\varepsilon = z/t$. Notice that this implies an infinite film thickness at $t = 0$, a condition approximated when the tank is filled:

$$-\varepsilon \frac{\partial \Delta_f}{\partial \varepsilon} + \frac{\rho g}{\mu} \Delta_f^2 \frac{\partial \Delta_f}{\partial \varepsilon} = 0$$

Solving for Δ_f gives

$$\Delta_f = \sqrt{\frac{\mu}{\rho g}\varepsilon} = \sqrt{\left(\frac{\mu}{\rho g}\right)\frac{z}{t}}$$

and shows that the film thickness varies with the square root of distance down the plate and is inversely proportional to the square root of time. This behavior extends to film thicknesses as small as 200 Å [5].

11.5 MACROSCOPIC ENERGY BALANCE

Previously, we derived the macroscopic mechanical energy balance where we considered the system to be isothermal. We can relax that restriction by allowing temperature gradients within the system and integrate the microscopic energy balance to yield a macroscopic version. Referring back to Figure 11.7, we must realize that the heat interaction may include heat conduction, convection, or radiation. We assume that the physical properties of the fluid do not change over the areas A_{c1} or A_{c2}, that the flow is perpendicular to A_{c1} and A_{c2}, and that the overall system boundaries are stationary. We consider gravity as the only external body force, and since $\vec{g} = -g\vec{\nabla}\Phi_g$, we use the form of the energy equation that considers accumulation of total energy, E_{tot}. Here, U is the internal energy of the material:

$$\int_V \vec{\nabla}\cdot(\rho \bar{U}\vec{\mathbf{v}})\,dV + \int_V \vec{\nabla}\cdot\left(\frac{1}{2}\rho v^2\vec{\mathbf{v}}\right)dV + \int_V \vec{\nabla}\cdot(\rho g\Phi_g\vec{\mathbf{v}})\,dV + \int_V \vec{\nabla}\cdot\vec{\mathbf{q}}''\,dV$$

$$\int_V (\vec{\nabla}\cdot P\vec{\mathbf{v}})\,dV + \int_V [\vec{\nabla}\cdot(\vec{\tau}\cdot\vec{\mathbf{v}})]\,dV - \int_V \dot{q}\,dV$$

$$= -\int_V \frac{\partial}{\partial t}\left(\rho\bar{U} + \frac{1}{2}\rho v^2 + \rho g\Phi_g\right)dV = -\int_V \frac{\partial \overline{E_{tot}}}{\partial t}\,dV \tag{11.33}$$

Integrating the accumulation term gives us the time rate of change of the total energy (internal, kinetic, and potential) of the system:

$$\int_V \frac{\partial}{\partial t}(\overline{E_{tot}})\,dV = \frac{\partial}{\partial t}\int_V (\overline{E_{tot}})\,dV \tag{11.34}$$

Integrating the convective terms is best accomplished by using surface integrals. We then just have the difference in flow of internal, kinetic, and potential energies through our two surfaces, A_{c1} and A_{c2}. All the integrals over the wetted surfaces are zero due to the no-slip condition:

$$\int_V \vec{\nabla}\cdot(\rho\bar{U}\vec{\mathbf{v}})\,dV = \int_{A_c}\rho\bar{U}v_n dA_c + \int_{A_s}\rho\bar{U}v_n dA_s + \int_{A_m}\rho\bar{U}v_n dA_m$$

$$= \rho_2\bar{U}_2\bar{v}_2A_{c2} - \rho_1\bar{U}_1\bar{v}_1A_{c1} \tag{11.35}$$

$$\int_V \vec{\nabla}\cdot\left(\frac{1}{2}\rho v^2\vec{\mathbf{v}}\right)dV = \int_{A_c}\left(\frac{1}{2}\rho v^2 v_n\right)dA_c + \int_{A_s}\left(\frac{1}{2}\rho v^2 v_n\right)dA_s + \int_{A_m}\left(\frac{1}{2}\rho v^2 v_n\right)dA_m$$

$$= \frac{1}{2}\rho_2\overline{v_2^3}A_{c2} - \frac{1}{2}\rho_1\overline{v_1^3}A_{c1} \tag{11.36}$$

$$\int_V \vec{\nabla}\cdot(\rho g\Phi_g\vec{\mathbf{v}})\,dV = \int_{A_c}(\rho g\Phi_g v_n)dA_c + \int_{A_s}(\rho g\Phi_g v_n)dA_s + \int_{A_m}(\rho g\Phi_g v_n)\,dA_m$$

$$= \rho_2 g\Phi_{g2}\bar{v}_2A_{c2} - \rho_1 g\Phi_{g1}\bar{v}_1A_{c1} \tag{11.37}$$

The pressure term has components resulting from flow through surfaces A_{c1} and A_{c2} and from interaction with any *moving*, wetted surfaces that might be used to transfer work to or from the system. Separating the integral into integrals over all wetted surfaces and flow surfaces leads to

$$\int_V(\vec{\nabla}\cdot P\vec{\mathbf{v}})\,dV = \int_{A_c}(Pv_n)\,dA_c + \int_{A_s}(Pv_n)\,dA_s + \int_{A_m}(Pv_n)dA_m$$

$$= P_2\bar{v}_2A_{c2} + P_1\bar{v}_1A_{c1} + \dot{W}_p \tag{11.38}$$

The viscous term also leads to a work of pushing the fluid through the system against the viscous forces. This term is generally small, especially since we are using these equations primarily in high-speed or turbulent flow situations. If we separate the integrals over the wetted and flow surfaces, the only component that provides a non-negligible work term is the integral over the moving surface since these surfaces must overcome viscous drag to operate:

$$\int_V[\vec{\nabla}\cdot(\vec{\tau}\cdot\vec{\mathbf{v}})]\,dV = \int_{A_c}(\vec{\tau}\cdot\vec{\mathbf{v}})_n dA_c + \int_{A_s}(\vec{\tau}\cdot\vec{\mathbf{v}})_n dA_s + \int_{A_m}(\vec{\tau}\cdot\vec{\mathbf{v}})_n dA_m = \dot{W}_\mu \tag{11.39}$$

The heat flow term contains all components of heat flow arising from viscous dissipation, conduction, convection, and radiation. Its integral is over the entire surface area of the system, not just A_{c1} and A_{c2}, because heat may be exchanged across any surface that is not insulated:

$$\int_V \vec{\nabla}\cdot\vec{\mathbf{q}}''dV = \int_{A_s} q_n''dA_s = q_T \tag{11.40}$$

The heat generation term is a true volume integral and must be integrated as such. Later, we will lump this term in with our other heat terms:

$$\int_V \dot{q}\,dV = q_g \tag{11.41}$$

Combining Equations 11.34 through 11.41 yields the macroscopic energy balance:

$$\rho_2\bar{v}_2\bar{U}_2A_{c2} - \rho_1\bar{v}_1\bar{U}_1A_{c1} + \frac{1}{2}\rho_2\overline{v_2^3}A_{c2} - \frac{1}{2}\rho_1\overline{v_1^3}A_{c1} + \rho_2 g\bar{v}_2\Phi_g A_{c2} - \rho_1 g\bar{v}_1\Phi_g A_{c1} +$$

$$P_2\bar{v}_2A_{c2} - P_1\bar{v}_1A_{c1} + -q_g + q_T + \dot{W}_p + \dot{W}_\mu = -\frac{dE_{tot}}{dt} \tag{11.42}$$

We can write Equation 11.42 in more compact form by combining the heat terms, the work terms, and the convective terms:

$$\Delta\left[\left(\bar{U} + P\bar{V} + \frac{1}{2}\frac{\overline{v^3}}{\bar{v}} + g\Phi_g\right)\rho\bar{v}A_c\right] + q + \dot{W} = -\frac{dE_{tot}}{dt} \tag{11.43}$$

$\rho\bar{v}A_c$ is the mass flow rate of material. Notice that $\bar{H} = \bar{U} + P\bar{V}$ so that the macroscopic energy balance is a restatement of the *first law of thermodynamics* for an open system [6].

When we apply Equation 11.43, we often assume that the fluid is an ideal gas or that it is incompressible. In such instances, we can readily evaluate the enthalpy change, $\Delta\bar{H}$, in terms of the heat capacities of the mixture and the pressure change. If we have an ideal gas, $P\bar{V} = RT$ and $C_p - C_v = R$, so we evaluate the enthalpy change as

$$\Delta\bar{H} = \int_{T_1}^{T_2} C_p\,dT \quad \text{Ideal gas} \tag{11.44}$$

In terms of the heat capacity ratio, $\gamma = C_p/C_v$, we have

$$\Delta\bar{H} = \frac{R}{M_w}\int_{T_1}^{T_2} \frac{\gamma}{\gamma - 1}\,dT \quad \text{Ideal gas} \tag{11.45}$$

If we have an incompressible fluid, $C_p = C_v$, so the enthalpy change is

$$\Delta\bar{H} = \int_{T_1}^{T_2} C_p\,dT + \frac{1}{\rho}(P_2 - P_1) \quad \text{Liquid} \tag{11.46}$$

Finally, if we relax our restriction on the boundaries of the system and allow at least a portion of that boundary to be permeable to mass transfer, then the mixing of material diffusing into the control volume with material already present in the system may lead to a heat interaction. This heat interaction is a direct consequence of differences in the partial molar enthalpy of the individual species. The energy equation becomes

$$\Delta\left[\left(\bar{U} + P\bar{V} + \frac{1}{2}\frac{\overline{v^3}}{\bar{v}} + g\Phi_g\right)\rho\bar{v}A_c\right] + q - q_d + \dot{W} = -\frac{dE_{tot}}{dt} \tag{11.47}$$

Notice that we have a heat interaction, q_d, only when two different materials mix:

$$q_d = -\int_{A_p} \left(\mathbf{n} \cdot \sum_{i=1}^{n} c_i \mathbf{v}_i \bar{H}_i \right) dA_p \tag{11.48}$$

Example 11.9 Design of a Fireplace and Chimney

We would like to design a fireplace and chimney making sure the device operates reliably. The situation is shown in Figure 11.11. The air outside the chimney is stagnant and at 298 K and 1 atm. The chimney will be 10 m high, and we will have no viscous effects within the chimney. At the fire, the air is also at 1 atm, and we will assume we lose heat from the flue in an amount Q W/kg-m. The fire temperature is roughly 825 K and the areas A_{c1}, $A_{c2} \gg A_{c3}$ since the flow area within a chimney is small.

We assume a constant mass flow rate through the chimney and that the combustion gases are about the same molecular weight as air. The continuity equation is automatically satisfied and states $\bar{v}_1 \approx \bar{v}_2 \ll \bar{v}_3$. We use hydrostatics to relate the pressures at the fireplace and the chimney exit. For such a small height, we can assume constant density. Thus, we have

$$P_1 \approx P_2 = P_3 + \rho_s g h \quad \text{Hydrostatics}$$

Applying the mechanical and thermal energy equations between points 2 and 3 (beyond the abrupt change in the direction of gas), we have

$$\frac{1}{2}\rho_3 \bar{v}_3^2 - \frac{1}{2}\rho_2 \bar{v}_2^2 + P_3 - P_2 + g(y_3 - y_2) = 0 \quad \text{Mechanical}$$

$$\rho_2 \bar{v}_2 A_{c2}(\bar{H}_3 - \bar{H}_2) + Qh = \rho_2 \bar{v}_2 A_{c2} C_p (T_3 - T_2) + Qh = 0 \quad \text{Thermal}$$

We neglect any kinetic or potential energy effects in the thermal energy equation since heat loss dominates there. *If* $\bar{v}_2 \ll \bar{v}_3$, then

$$\bar{v}_3^2 = 2\left\{\frac{P_2 - P_3}{\rho_3}\right\} + 2\left\{\frac{\rho_2 g y_2 - \rho_3 g y_3}{\rho_3}\right\} = 0$$

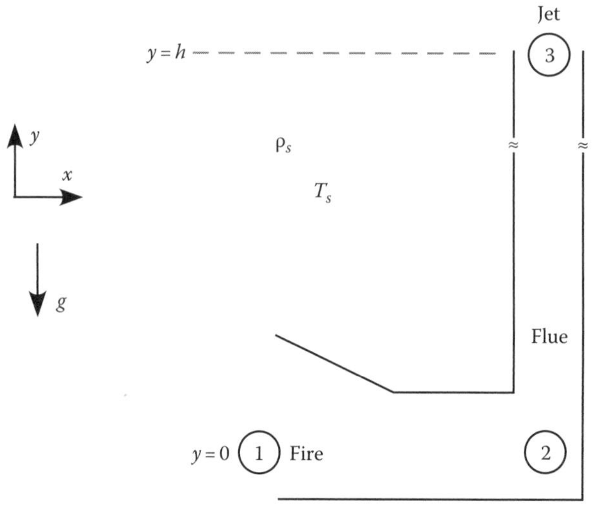

FIGURE 11.11 Fireplace and chimney.

Using the hydrostatics equation, we can solve for the velocity:

$$\bar{v}_3^2 = 2\left\{\frac{\rho_s}{\rho_3}gh\right\} + 2\left\{\frac{\rho_2 g(0) - \rho_3 gh}{\rho_3}\right\} = 2gh\left\{\frac{\rho_s}{\rho_3} - 1\right\}$$

If there is no heat loss, then $T_3 = T_2 = T_1$, and from the ideal gas law,

$$\frac{\rho_s}{\rho_3} = \frac{T_3}{T_s}$$

Solving for the velocity gives

$$\bar{v}_3 = \sqrt{2(2.768 - 1)(9.8)(10)} = 18.6 \text{ m/s}$$

The chimney exchanges heat with the surroundings leading to a heat loss, Q x h. The energy equation tells us that the temperature drops along the chimney and as that occurs, our exit velocity also drops. If we lose too much heat, we have insufficient *draft* for the fireplace and smoke backs up into the room. Too much heat loss also condenses water in the chimney ruining the liner. The key to a good fireplace is a tall, insulated chimney.

Example 11.10 Isentropic Flow of an Ideal Gas through a Nozzle

Consider the steady flow of an ideal gas from a large reservoir into a converging nozzle. Both the reservoir and nozzle are insulated so there is no heat interaction with the surroundings. There will also be no work interaction since we have no mechanisms for extracting work from the expanding gas. The system is shown in Figure 11.12.

We define T_1 and P_1 as the *absolute* temperature and pressure in the reservoir. These are called the stagnation conditions since $v_1 = 0$ in the reservoir. A_{c2} is the minimum cross-sectional area of the nozzle, called the *throat*. The mass flow rate of gas is $\dot{m}$. We wish to develop relationships between the temperature, pressure, and flow rate in the nozzle and the conditions existing in the reservoir. Particular interest will center on the variation in mass flow rate as the pressure, P_2, at the throat is decreased.

We first apply the macroscopic continuity equation to obtain a relationship between the areas, densities, and velocities of the gas in the reservoir and at the throat of the nozzle. It is important to remember that all three quantities, area, density, and velocity, change in going from surface 1 to surface 2. The mass flow rate of the gas is constant at steady state, and so the continuity equation becomes

$$\rho_1\bar{v}_1 A_{c1} = \rho_2\bar{v}_2 A_{c2} = \dot{m}$$

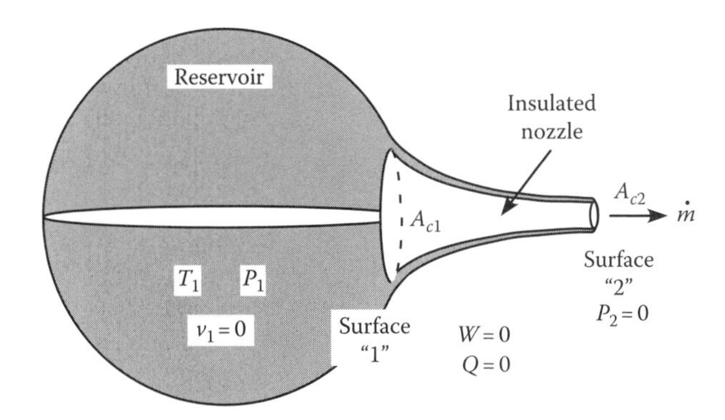

FIGURE 11.12 Isentropic flow of an ideal gas through a converging nozzle.

Next, we apply the energy balance to the gas. We have no shaft work being performed, no heat being transferred (insulated nozzle), and no gravity effects (horizontal system). With these simplifications in mind, the energy balance becomes

$$\Delta\left[\left(\bar{U} + P\bar{V} + \frac{1}{2}\bar{v}^2\right)\rho v A_c\right] = 0$$

Recognizing that $\bar{U} + P\bar{V} = \bar{H}$, we apply the energy balance between the two surfaces and multiply through by –1 to obtain

$$\bar{H}_2 - \bar{H}_1 + \frac{\bar{v}_2^2}{2} - \frac{\bar{v}_1^2}{2} = 0$$

If we replace $\bar{H}_2 - \bar{H}_1$ by $\bar{C}_p(T_2 - T_1)$, we have

$$\bar{C}_p T_1 = \bar{C}_p T_2 + \frac{\bar{v}_2^2}{2}$$

where we have included the fact that $\bar{v}_1 = 0$. The energy balance shows that any increase in velocity is accompanied by a decrease in temperature. In essence, we are transforming internal energy into kinetic energy.

The energy balance is not yet in a very useful form since we do not know the outlet temperature T_2. We would rather express the energy balance in terms of the outlet pressure, P_2, a quantity that we know. We can transform the balance by using the knowledge that the flow of the gas is *isentropic* and hence has no heat interaction with the outside. From thermodynamics, we know that a change in specific enthalpy is given by a corresponding change in pressure and specific entropy. The temperature and pressure must act in concert to keep the entropy constant in an isentropic process:

$$d\bar{H} = Td\bar{S} + \bar{V}dP$$

In this case, $d\bar{S} = 0$. Since the volume per unit mass, $\bar{V}$, is the reciprocal of the density, we can write the previous expression as

$$d\bar{H} = \bar{C}_p dT = \frac{1}{\rho}dP = \left(\frac{RT}{M_w P}\right)dP$$

where we have used the ideal gas law to express the density of the gas in terms of the temperature and pressure. For an ideal gas, we can also express the enthalpy as

$$\bar{H} = \bar{U} + P\bar{V} = \bar{U} + \frac{RT}{M_w}$$

and can express the heat capacity at constant pressure in terms of the heat capacity at constant volume using

$$\bar{C}_p - \bar{C}_v = \frac{R}{M_w} \quad \frac{\bar{C}_p}{\bar{C}_v} = \gamma \quad 1\,\text{kg mass basis}$$

$$\bar{C}_p = \frac{\gamma R}{M_w(\gamma - 1)}$$

Substituting in for $d\bar{H}$ provides a differential equation relating the pressure and temperature at any point within the system:

$$\left(\frac{\gamma}{\gamma-1}\right)\frac{dT}{T}=\frac{dP}{P}$$

Integrating and using the known values of temperature and pressure in the reservoir gives

$$\frac{T}{T_1}=\left(\frac{P}{P_1}\right)^{\frac{\gamma-1}{\gamma}}$$

We can use this relationship in the energy balance to express the velocity at the nozzle throat (smallest cross-sectional area) in terms of the pressure at the throat and the temperature and pressure in the reservoir. Plugging this relationship into the balance with $v_2 = v$ and $T_2 = T$, we have

$$\frac{\bar{v}^2}{2}=\frac{\gamma RT_1}{M_w(\gamma-1)}\left(1-\frac{T}{T_1}\right)=\frac{\gamma RT_1}{M_w(\gamma-1)}\left[1-\left(\frac{P}{P_1}\right)^{\frac{\gamma-1}{\gamma}}\right]$$

Notice that the velocity is a maximum at the throat where the pressure is low. The mass flow rate of the gas per unit area or mass flux at any point within the nozzle is

$$\frac{\dot{m}}{A_c}=\rho\bar{v}=\left(\frac{M_wP}{RT}\right)\bar{v}=\left(\frac{M_wP_1}{RT_1}\right)\left(\frac{P}{P_1}\right)\left(\frac{T_1}{T}\right)\bar{v}=\rho_1\left(\frac{P}{P_1}\right)\left(\frac{T_1}{T}\right)\bar{v}$$

Now, M_wP_1/RT_1 is the density, ρ_1, at the reservoir. We can express both $\bar{v}$ and the ratio T_1/T in terms of the pressure ratio P/P_1 using the energy balance and the expression for the relationship between T and P derived from the definition of the enthalpy. Substituting in we find

$$\frac{\dot{m}}{A_c}=\rho_1\sqrt{\frac{2\gamma RT_1}{M_w(\gamma-1)}}\left[\left(\frac{P}{P_1}\right)^{\frac{2}{\gamma}}-\left(\frac{P}{P_1}\right)^{\frac{\gamma+1}{\gamma}}\right]^{1/2}$$

$\dot{m}/A_c$ has its greatest value at the nozzle exit or throat where $P = P_2$ ($\gamma > 1$). As P decreases below P_1, $\dot{m}/A_c$ increases. However, since $\gamma > 1$, our expression predicts that a maximum $\dot{m}/A_c$ will eventually be attained. The critical pressure producing this maximum $\dot{m}/A_c$ is found by differentiation:

$$\frac{d(\dot{m}/A_c)}{d(P/P_1)}=0=\frac{\rho_1}{2}\sqrt{\frac{2\gamma RT_1}{M_w(\gamma-1)}}\left[\left(\frac{P}{P_1}\right)^{\frac{2}{\gamma}}-\left(\frac{P}{P_1}\right)^{\frac{\gamma+1}{\gamma}}\right]^{-1/2}\times\left[\frac{2}{\gamma}\left(\frac{P}{P_1}\right)^{\frac{2}{\gamma}-1}-\left(\frac{\gamma+1}{\gamma}\right)\left(\frac{P}{P_1}\right)^{\frac{1}{\gamma}}\right]$$

Solving for the critical pressure, P_c, gives

$$\frac{P_c}{P_1}=\left(\frac{2}{\gamma+1}\right)^{\frac{\gamma}{\gamma-1}}$$

which corresponds to a maximum mass flow per unit area of

$$\left(\frac{\dot{m}}{A_c}\right)_{\max}=\rho_1\sqrt{\frac{\gamma RT_1}{M_w}}\left(\frac{2}{\gamma+1}\right)^{\frac{\gamma+1}{2(\gamma-1)}}$$

If we substitute $P = P_c$ into the equation for the velocity, $\overline{v}_2$, we will find that

$$v_c = \sqrt{\frac{\gamma R T_1}{M_w}} \text{ – speed of sound at } T_1$$

and the critical velocity for maximum flow rate is the speed of sound of the gas. If we reduce the pressure at the throat, P_2, below P_c, we will find that the maximum flow rate will not increase. What will occur is an irreversible expansion of the gas from P_c to P_2 immediately beyond the nozzle since the gas cannot move faster than its own speed of sound. (What happens to the outlet temperature in such a case?) Another way of describing the speed of sound is that it is the velocity of the wave in the gas that represents the maximum interconversion of kinetic and potential energy.

One might ask how we would increase the velocity beyond the sonic limit. After all, jets and rockets routinely travel faster than sound. The key lies in the shape of the nozzle. If we look at the continuity equation first, we can put that in differential form as follows:

$$\ln(\rho A\overline{v}) = \ln(\rho) + \ln(A) + \ln(\overline{v}) = \text{constant}$$

$$\frac{d\rho}{\rho} + \frac{dA}{A} + \frac{d\overline{v}}{\overline{v}} = 0 \tag{11.49}$$

The energy equation can be written in a similar way:

$$d\overline{H} + \overline{v}d\overline{v} = Td\overline{S} + \frac{dP}{\rho} + \overline{v}d\overline{v} = 0 \tag{11.50}$$

If the process is isentropic, we can eliminate the entropy term and rewrite the pressure term as

$$\cancelto{0}{Td\overline{S}} + \frac{dP}{\rho} + \overline{v}d\overline{v} = \frac{1}{\rho}\left(\frac{\partial P}{\partial \rho}\right)_S d\rho + \overline{v}d\overline{v} = \overline{v}_c^2 \frac{d\rho}{\rho} + \overline{v}d\overline{v} = 0 \tag{11.51}$$

Plugging in for the density term using the continuity equation gives

$$-\overline{v}_c^2\left[\frac{dA_c}{A_c} + \frac{d\overline{v}}{\overline{v}}\right] + \overline{v}d\overline{v} = 0$$

$$\frac{dA_c}{A_c} = \left(\frac{\overline{v}^2}{\overline{v}_c^2} - 1\right)\frac{d\overline{v}}{\overline{v}} \tag{11.52}$$

We can identify the ratio of actual velocity to sonic velocity as the Mach number, $\mathcal{M}$, and look at the behavior of Equation 11.52. Configurations where $dA_c < 0$ are called nozzles, while when $dA_c > 0$, the devices are called diffusers. The behavior of each can be seen from Table 11.1.

Notice that nozzles and diffusers swap behavior as one moves from a Mach number less than one to one greater than one. That means in order to go faster than sound, one must combine the profiles so that at the throat of the nozzle, one switches to the profile of a diffuser. This is often referred to as a converging/diverging nozzle. The exiting velocity from the diffuser portion is

$$\frac{\overline{v}_e}{\overline{v}_c} = \mathcal{M} = \sqrt{\frac{2}{(\gamma - 1)}\left[1 - \left(\frac{P_e}{P_1}\right)^{\gamma-1/\gamma}\right]} \tag{11.53}$$

TABLE 11.1
Mach Number Behavior of Nozzles and Diffusers

		$\mathcal{M} < 1$	$\mathcal{M} > 1$
$dA_c<0$	$\bar{v}$	Increases	Decreases
	P	Decreases	Increases
	ρ	Decreases	Increases
	$\bar{H}(T)$	Decreases	Increases
$dA_c>0$	$\bar{v}$	Decreases	Increases
	P	Increases	Decreases
	P	Increases	Decreases
	$\bar{H}(T)$	Increases	Decreases

The pressure variation as a function of Mach number is

$$\frac{P_1}{P} = \left[1+\left(\frac{\gamma+1}{2}\right)\mathcal{M}^2\right]^{\gamma/\gamma-1} \tag{11.54}$$

Often, we design the engine based on a throat and an exit area. Using the continuity equation, we can show that the ratio of throat to exit area is given by

$$\frac{A_{throat}}{A_{exit}} = \left(\frac{\gamma+1}{\gamma}\right)^{1/\gamma-1}\left(\frac{P_e}{P_1}\right)^{1/\gamma}\sqrt{\left(\frac{\gamma+1}{\gamma-1}\right)\left[1-\left(\frac{P_e}{P_1}\right)^{\gamma-1/\gamma}\right]} \tag{11.55}$$

A few simple calculations using Equations 11.54 and 11.55 show that if we wanted to reach escape velocity, $M = 25$, the exit area would have to be nearly 46,000 times that of the throat area. This is something extraordinarily difficult to fabricate, thus the need for rockets to have multiple stages so we can exploit multiple P_1's.

Example 11.11 Continuous Heating of Maple Syrup in an Agitated Tank

Figure 11.13 shows maple syrup being heated by pumping it through a well-stirred heating tank. Heat is supplied by condensing steam in a heat exchanger. The inlet temperature of the syrup is T_i, and the temperature of the outer surface of the steam coil is T_s, the saturation temperature of the steam. The steam coil has a surface area, A_s, and heat transfer occurs via an overall heat transfer coefficient of $\mathcal{U}$. The volume of the tank is V_t and the mass flow rate of syrup is $\dot{m}$ (kg/s). The syrup has a density ρ and a heat capacity C_p. The syrup is agitated so that it has a uniform temperature $T(t)$ throughout its volume. This assumption also implies that the outlet temperature of the syrup is the same as the temperature in the tank.

Our objective is to determine the temperature of the syrup as a function of the amount of time it spends in the tank. This time, called the space time, is a function of the size of the tank and the flow rate of syrup. At a steady state, the continuity equation states that whatever we put into the steam and syrup tubes must come out:

$$\rho_1\bar{v}_1A_{c1} = \rho_2\bar{v}_2A_{c2} \quad \text{syrup}$$

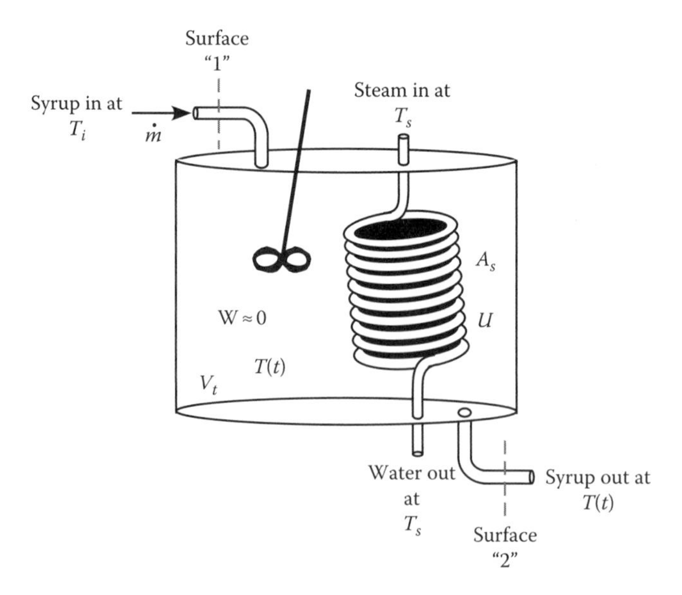

FIGURE 11.13 Continuous heating of maple syrup.

If we assume constant properties, $\rho_1 = \rho_2$, and specify the same inlet and outlet piping, then $\bar{v}_1 = \bar{v}_2$, $A_{c1} = A_{c2}$, and velocity differences are not important to our system. Finally, we will assume that the tank is not very tall so that height differences between the inlet and outlet are small.

To determine the temperature of the slurry as a function of the time it spends in the tank, the overall energy equation in its transient form is appropriate:

$$\frac{dE_{tot}}{dt} = -\Delta\left[\left(U + PV + \frac{1}{2}\frac{\overline{v^3}}{\bar{v}} + g\Phi_g\right)\rho\bar{v}A_c\right] - q - \dot{W}$$

Before we apply this equation, we must make a few additional assumptions. Gravity effects will be neglected. Since we are also neglecting velocity effects, this means that kinetic and potential energy changes between the inlet and outlet are negligible and so we are primarily concerned with variations in enthalpy. This generally is a good assumption whenever there are temperature differences involved. Small changes in enthalpy are equivalent to large changes in velocity or height. Furthermore, we will assume that the shaft work provided by the stirrer does not increase the fluid temperature and so the work term in the energy equation can be neglected.

We are then left with two major components: enthalpy added and removed from the syrup by convection of the syrup through the system and heat added to the system via the steam tube. With these assumptions in mind, the energy equation is simplified to

$$\rho C_p V_t \frac{dT}{dt} = \mathcal{U}A_s(T_s - T) + \dot{m}C_p(T_i - T)$$

where

$$q = \mathcal{U}A_s(T_s - T) \qquad -\Delta H_{conv} = \dot{m}C_p(T_i - T)$$

The energy balance equation should look familiar. It is a variant of the lumped capacitance equation we first looked at in Chapter 6. In this example, we have just applied the lumped capacitance approach to an open system where fluid can enter and leave the system via convection.

Integration of the energy equation is assisted by using an integrating factor:

$$\text{Integrating factor} = \exp[-\eta t] \qquad \eta = \frac{\mathcal{U}A_s + \dot{m}C_p}{\rho V_t C_p}$$

$$\frac{d}{dt}[T\exp(-\eta t)] = \left(\frac{\mathcal{U}A_s T_s + \dot{m}C_p T_i}{\mathcal{U}A_s + \dot{m}C_p}\right)\exp(-\eta t)$$

The solution to this equation is

$$T = \left(\frac{\mathcal{U}A_s T_s + \dot{m}C_p T_i}{\mathcal{U}A_s + \dot{m}C_p}\right) + a_o \exp(-\eta t)$$

and the constant, a_o is found from the initial condition:

$$t = 0 \quad T = T_i$$

$$T = \left(\frac{\mathcal{U}A_s T_s + \dot{m}C_p T_i}{\mathcal{U}A_s + \dot{m}C_p}\right) + \left\{\frac{\mathcal{U}A_s(T_i - T_s)}{\mathcal{U}A_s + \dot{m}C_p}\right\}\exp\left[-\left(\frac{\mathcal{U}A_s + \dot{m}C_p}{\rho V_t C_p}\right)t\right]$$

Notice that the initial condition is satisfied. We can define a thermal capacitance and resistance for our system as

$$\text{Resistance} = \frac{1}{\mathcal{U}A_s + \dot{m}C_p} \qquad \text{Capacitance} = \rho C_p V_t$$

If we look at the long time solution, we can see what the steady-state temperature will eventually be:

$$T_{eq} = \frac{(\mathcal{U}A_s/\dot{m}C_p)T_s + T_i}{1 + (\mathcal{U}A_s/\dot{m}C_p)}$$

As we increase the flow rate, the temperature approaches T_i. At very high flow rates, the residence time for the fluid in the tank is so short that negligible heat transfer occurs and the fluid exits at approximately the same temperature that it entered with.

Example 11.12 Double-Pipe Heat Exchanger and Differential Energy Balances

Heat exchange equipment is ubiquitous whether man-made, as in heat exchangers, or natural, as in the arrangement of arteries and veins in the human body or the evaporative structures used when dogs pant. We would like to analyze the steady-state behavior of one of the simplest pieces of equipment, a double-pipe heat exchanger, and illustrate how the macroscopic energy balance can be used in a differential form to predict the temperature profile along the axis of the exchanger. The double-pipe system is shown in Figure 11.14. Cold fluid enters from the left, on the tube side, at a constant mass flow rate of $\dot{m}_c$ and a heat capacity of C_{pc}. Hot fluid enters from the bottom, on the shell side, at a constant mass flow rate of $\dot{m}_h$ and a heat capacity of C_{ph}. The configuration of Figure 11.14 is called counterflow because the incoming cold fluid first contacts the outgoing hot fluid. The reverse situation is termed cocurrent flow. We assume constant properties for both fluids and incompressible flow.

Our analysis begins with the macroscopic versions of the continuity and energy balances applied to both fluids. The continuity equations for the cold and hot fluid yield

$$\rho_1 \bar{v}_1 A_{c1} = \rho_2 \bar{v}_2 A_{c2} = \dot{m}_c$$

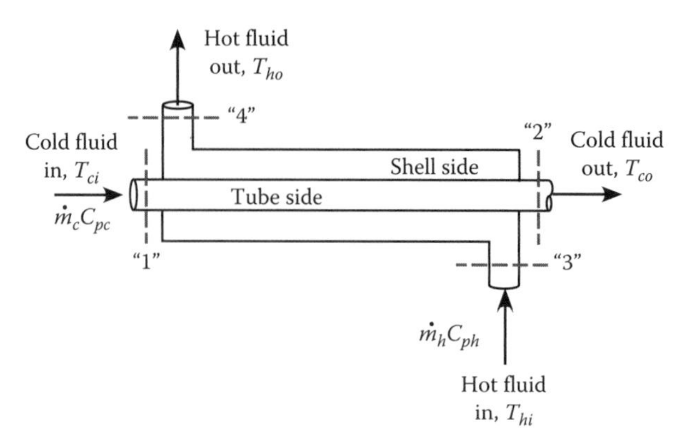

FIGURE 11.14 Double-pipe, counterflow heat exchanger.

$$\rho_3 \bar{v}_3 A_{c3} = \rho_4 \bar{v}_4 A_{c4} = \dot{m}_h$$

If densities are constant and inlet and outlet pipe areas are the same, then

$$\bar{v}_1 = \bar{v}_2 \qquad \text{and} \qquad \bar{v}_3 = \bar{v}_4$$

The energy balances are

$$\dot{m}_c\left[\bar{H}_1 - \bar{H}_2 + \bar{v}_1^2 - \bar{v}_2^2 + gz_1 - gz_2\right] - q_c = 0$$

$$\dot{m}_h\left[\bar{H}_3 - \bar{H}_4 + \bar{v}_3^2 - \bar{v}_4^2 + gz_3 - gz_4\right] - q_h = 0$$

where q_c and q_h represent heat removed from the cold fluid and gained by the hot fluid, respectively. With no change in inlet or outlet velocity (constant mass flow rates), we have no kinetic energy changes to consider. If we assume that the exchanger is horizontal and most are due to their size, we can also neglect all potential energy changes. The energy balances then simplify to

$$\dot{m}_c[\bar{H}_1 - \bar{H}_2] - q_c = 0 \qquad \dot{m}_h[\bar{H}_3 - \bar{H}_4] - q_h = 0$$

Expressing the changes in enthalpy as equivalent changes in heat capacity multiplied by temperature differences, we find the familiar balances

$$q = q_c = -q_h = \dot{m}_c C_{pc}(T_{co} - T_{ci}) = \dot{m}_h C_{ph}(T_{hi} - T_{ho}) \quad \text{Single phase fluid systems only}$$

stating that the heat transferred from the hot fluid is absorbed by the cold fluid.

The macroscopic form of the energy equation can provide us with a temperature profile, only if we know the heat transferred, q, as a function of position along the exchanger. This is generally not the case, and so we must resort to another description of energy flow between the phases to determine the temperature profiles. In general, we characterize the heat transferred in an exchanger in terms of an overall heat transfer coefficient between the phases, $\mathcal{U}$; the surface area of the exchanger, A_s; and a temperature driving force, $f(T_h - T_c)$. The overall heat transfer coefficient considers convective resistances in the two fluids, the conduction resistance through the walls of the tubing and any other conduction resistances that build up over time due to fouling of the tube surfaces (deposition of scale, etc.).

We can derive a differential form of the macroscopic energy balance that will lead us to a description of the temperature profiles by looking at a vertical slice cut through the exchanger of Figure 11.15 parallel to the exchanger's main axis. We are free to integrate over as few or as many

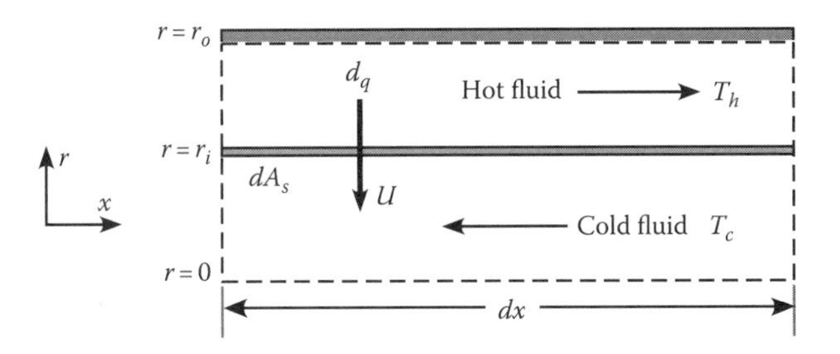

FIGURE 11.15 Differential slice through the main axis of the heat exchanger of Figure 11.14.

coordinate directions as we please. Since we are interested in axial temperature profiles, we integrate over the radial and angular coordinates. This means that the hot and cold temperatures at any axial position are really averages over the cross-sectional area of the tube or shell portion of the exchanger. Considering the differential control volume and the macroscopic energy balances we previously derived, the differential energy balances for the hot (shell-side) and cold (tube-side) fluids are

$$dq_c = \dot{m}_c C_{pc} dT_c = \mathcal{U} P_d (T_h - T_c) dx$$

$$dq_h = \dot{m}_h C_{ph} dT_h = -\mathcal{U} P_d (T_h - T_c) dx$$

where

P_d is the perimeter of the inner tube
q_c is the heat transferred to the cold fluid
q_h is the heat transferred from the hot fluid

Notice that the hot fluid balance has a minus sign included because in the direction of increasing x, dT_c is positive while dT_h is negative. If we isolate dT_c and dT_h then subtract the equation for dT_c from that for dT_h, we derive an equation for the temperature driving force (T_h – T_c) throughout the exchanger:

$$d(T_h - T_c) = -\left[\frac{1}{\dot{m}_c C_{pc}} + \frac{1}{\dot{m}_h C_{ph}}\right] \mathcal{U} P_d (T_h - T_c) dx$$

Separating and integrating from one end of the exchanger, surface 1, to any axial position, x, gives the temperature driving force at any point within the exchanger. Here, we have incorporated the boundary condition that $T_h = T_{ho}$ and $T_c = T_{ci}$ at $x = 0$:

$$\frac{T_h - T_c}{T_{ho} - T_{ci}} = \exp\left[-\left(\frac{1}{\dot{m}_c C_{pc}} + \frac{1}{\dot{m}_h C_{ph}}\right) \mathcal{U} P_d x\right]$$

We can use our result for the temperature difference and the differential energy balance to determine the total heat transferred by the exchanger:

$$q = \mathcal{U} P_d \int_0^L (T_h - T_c) dx$$

$$= \mathcal{U} P_d (T_{ho} - T_{ci}) \int_0^L \exp\left[-\left(\frac{1}{\dot{m}_c C_{pc}} + \frac{1}{\dot{m}_h C_{ph}}\right) \mathcal{U} P_d x\right] dx$$

Evaluating the integral, we have

$$q = \frac{(T_{ho} - T_{ci})}{\left(\dfrac{1}{\dot{m}_c C_{pc}} + \dfrac{1}{\dot{m}_h C_{ph}}\right)} \left\{ \exp\left[-\left(\frac{1}{\dot{m}_c C_{pc}} + \frac{1}{\dot{m}_h C_{ph}} \right) \mathcal{U} P_d L \right] - 1 \right\}$$

Using the result for the temperature difference at any point in the exchanger, we can replace the exponential term in the earlier equation. When $x = L$, $T_h = T_{ho}$ and $T_c = T_{ci}$:

$$q = \frac{(T_{ho} - T_{ci})}{\left(\dfrac{1}{\dot{m}_c C_{pc}} + \dfrac{1}{\dot{m}_h C_{ph}}\right)} \left\{ \frac{T_{hi} - T_{co}}{T_{ho} - T_{ci}} - 1 \right\} = \frac{(T_{hi} - T_{co}) - (T_{ho} - T_{ci})}{\left(\dfrac{1}{\dot{m}_c C_{pc}} + \dfrac{1}{\dot{m}_h C_{ph}}\right)}$$

Finally, we can invert the temperature profile equation and solve for the mass flow rate-heat capacity products in terms of the overall heat transfer coefficient and the total surface area of the exchanger. Substituting for the denominator in the heat flow equation earlier gives the heat exchanger design equation:

$$q = \mathcal{U} P_d L \left[\frac{(T_{hi} - T_{co}) - (T_{ho} - T_{ci})}{\ln\left(\dfrac{T_{hi} - T_{co}}{T_{ho} - T_{ci}} \right)} \right] = \mathcal{U} A_s \Delta T_{lm}$$

The heat exchanger design equation can be recast in terms of a driving force and resistance. The temperature function is termed the log-mean temperature difference and arises because a simple temperature difference between hot and cold fluids is not appropriate when the fluid temperatures change with position throughout the exchanger. The log-mean difference is an average over the entire exchanger:

$$q = \frac{\text{Driving force}}{\text{Resistance}} = \frac{\Delta T_{lm}}{1/\mathcal{U} A_s}$$

If we know the amount of heat transferred in the exchanger (a usual design parameter), we can solve for the required exchanger length:

$$L = \frac{q}{\mathcal{U} P_d \Delta T_{lm}} = \left(\frac{\dot{m}_c C_{pc}}{\mathcal{U} P_d} \right) \frac{(T_{co} - T_{ci})}{\Delta T_{lm}}$$

The grouping in brackets $\dot{m}_c C_{pc}/UP_d$ has the units of length and is referred to as the length of a transfer unit. The grouping $(T_{co} - T_{ci})/\Delta T_{lm}$ is referred to as the number of transfer units required to affect the exchange. Here, both are based on the cold fluid, but analogous equations can be derived based on the hot fluid:

$$\frac{\dot{m}_c C_{pc}}{\mathcal{U} P_d} \quad \text{— Length of a transfer unit}$$

$$\frac{(T_{co} - T_{ci})}{\Delta T_{lm}} \quad \text{— Number of transfer units}$$

11.6 MACROSCOPIC SPECIES CONTINUITY EQUATION

We derive the macroscopic species mass balance or species continuity equation for a fluid system by integrating the microscopic species continuity equation over the volume of the system of interest. Here, we must allow for the production or destruction of the species of interest within the black box by some overall reaction rate, $-r_a$. Figure 11.16 shows the situation.

We begin by integrating the microscopic species continuity equation over the system volume:

$$\int_V (\vec{\nabla}\cdot\rho_a\vec{\mathbf{v}})dV - \int_V (\vec{\nabla}\cdot\rho D_{ab}\vec{\nabla}w_a)dV - \int_V r_a dV = -\int_V \frac{\partial\rho_a}{\partial t}dV \quad (11.56)$$

where

ρ_a is the density of species a
w_a is the mass fraction of species a

If we specify that the boundaries of the system are rigid, we can begin to evaluate Equation 11.56 starting with the accumulation term and interchanging the order of differentiation and integration:

$$\int_V \frac{\partial\rho_a}{\partial t}dV = \frac{\partial}{\partial t}\int_V \rho_a dV = \frac{d}{dt}m_{a,tot} \quad (11.57)$$

Here, $m_{a,tot}$ represents the total amount of a present in the system. The convective term is easiest to integrate in the form of a surface integral over the inlet and outlet surface areas:

$$\int_V (\vec{\nabla}\cdot\rho_a\vec{\mathbf{v}})dV = \int_{A_c} \rho_a v_{an} dA_c = \rho_{a2}\bar{v}_2 A_{c2} - \rho_{a1}\bar{v}_1 A_{c1} \quad (11.58)$$

Mass transfer via diffusion occurs only through those surfaces that are permeable, A_p:

$$\int_V (\vec{\nabla}\cdot\rho D_{ab}\vec{\nabla}w_a)dV = \int_{A_p} (\vec{\mathbf{n}}\cdot\rho_a\vec{\mathbf{v}}_{\mathbf{a}})dA_p = \dot{m}_{da} \quad (11.59)$$

$\dot{m}_{da}$ is positive when mass is added to the system. Finally, we come to the generation term. Generation is on a per volume basis, and so we cannot convert the volume integral to a surface integral:

$$\int_V r_a dV = r_{a,tot} \quad (11.60)$$

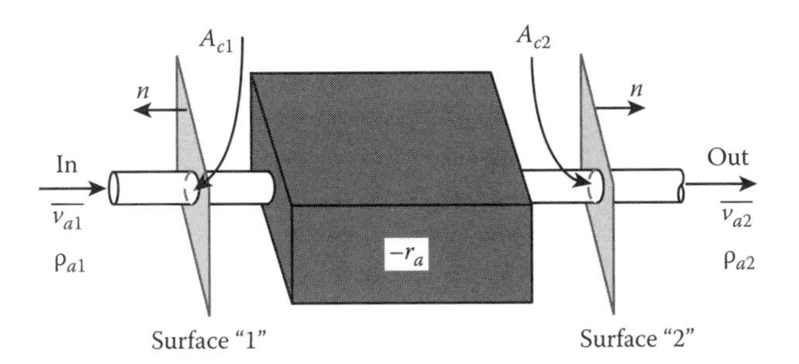

FIGURE 11.16 System for deriving the macroscopic species continuity equation.

Here, $r_{a,tot}$ represents the total production or destruction of species a by reaction.

Summing all the components in Equations 11.57 through 11.60 gives

$$\rho_{a2}\overline{v}_2 A_{c2} - \rho_{a1}\overline{v}_1 A_{c1} - \dot{m}_{da} - r_{a,tot} = -\frac{dm_{a,tot}}{dt} \quad (11.61)$$

We can also write Equation 11.61 in molar units:

$$c_{a2}\overline{v}_{M2} A_{c2} - c_{a1}\overline{v}_{M1} A_{c1} - \dot{M}_{da} - R_{a,tot} = -\frac{dM_{a,tot}}{dt} \quad (11.62)$$

where

$\overline{v}_M$ is the molar velocity of the fluid

the capital M, refers to the number of moles rather that the number of kilograms

Example 11.13 Gas–Liquid Extraction in a Packed Tower [7]

This example is the mass transfer analog of the heat exchanger. The system is the gas–liquid extractor shown in Figure 11.17. Systems like this one are used for scrubbing gases such as HCl or SO_2 from power plant or industrial exhaust. Gas containing solute, a, enters at the bottom of the tower and contacts a liquid flowing down from the top. The molar flow rates of gas and liquid are G and L (mol/s), respectively. If we stipulate dilute solutions for both the gas and liquid, then these molar flow rates will be essentially constant throughout the contactor. The contactor is designed to have a cross-sectional area, A_c, and is L meters high.

We are interested in the operational behavior of the column and, in particular, in the mole fraction of solute a in the gas and liquid phases. It is useful to define the flow of a in the gas and liquid on a *solvent-free* basis. To that end, we define new composite mole fractions, χ and ω, for the liquid and gas phases, respectively:

$$\text{Solute flow in the liquid} = \chi_{a1}\mathcal{L} = \left(\frac{x_{a1}}{1-x_{a1}}\right)\mathcal{L}$$

$$\text{Solute flow in the gas} = \omega_{a1}\mathcal{G} = \left(\frac{y_{a1}}{1-y_{a1}}\right)\mathcal{G}$$

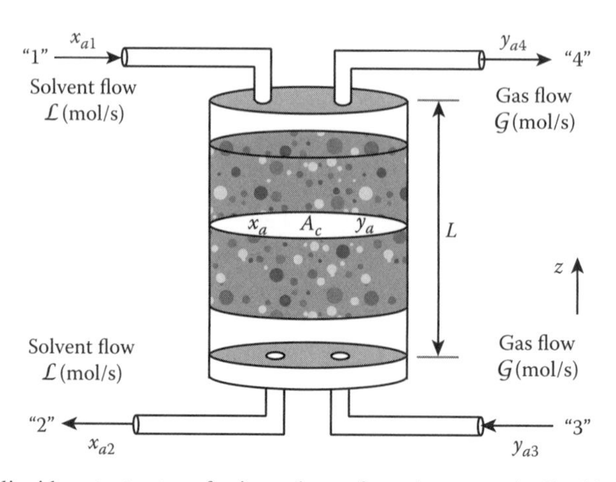

FIGURE 11.17 Gas–liquid contactor transferring solute a from the gas to the liquid phase.

With these definitions, we can apply the macroscopic species continuity equation to the gas and liquid phases. These equations state that the differences in molar flow rate of *a* between the inlet and outlet of either gas or liquid phase are equal to the amount of *a* transferred between phases:

$$\omega_{a3}\mathcal{G} - \omega_{a4}\mathcal{G} = N_{aG}A_i \qquad \chi_{a1}\mathcal{L} - \chi_{a2}\mathcal{L} = N_{aL}A_i$$

Here, A_i is the interfacial area between the phases. Much of the design of these systems focuses on maximizing A_i. Since we have no reaction occurring and the cross-sectional area of the column is constant, overall conservation of mass states that the fluxes of *a* in both phases are equal, $N_{aG} = -N_{aL}$. We can use this information to solve for χ_{a2}:

$$\chi_{a2} = \chi_{a1} + \frac{\mathcal{G}}{\mathcal{L}}(\omega_{a3} - \omega_{a4})$$

If we let χ_a and ω_a represent the mole fractions of *a* at any position in the column, then the previous mass balance is valid at any point in the column:

$$\chi_a = \chi_{a1} + \frac{\mathcal{G}}{\mathcal{L}}(\omega_a - \omega_{a4}) \quad \text{Operating line}$$

The relation earlier is referred to as the operating line for the contactor. We can perform equilibrium contacting experiments to determine the equilibrium distribution of *a* in the gas and liquid phases for any composition of either gas or liquid phases. This leads to a relationship of the form

$$\omega_{ae} = m_k\chi_{ae} \quad \text{Equilibrium relationship}$$

where m_k is the equilibrium partition coefficient for *a*. A plot of this equation defines the *equilibrium line* for the mass transfer process. The operating and equilibrium lines for the contactor are shown in Figure 11.18.

Though we did not discuss it, the heat exchanger of Figure 11.14 also has an operating and an equilibrium line. The operating and equilibrium lines are

$$T_c = T_{c1} + \left(\frac{\dot{m}_h C_{ph}}{\dot{m}_c C_{pc}}\right)(T_h - T_{h4}) \quad \text{Operating line}$$

$$T_c = T_h \quad \text{Equilibrium line}$$

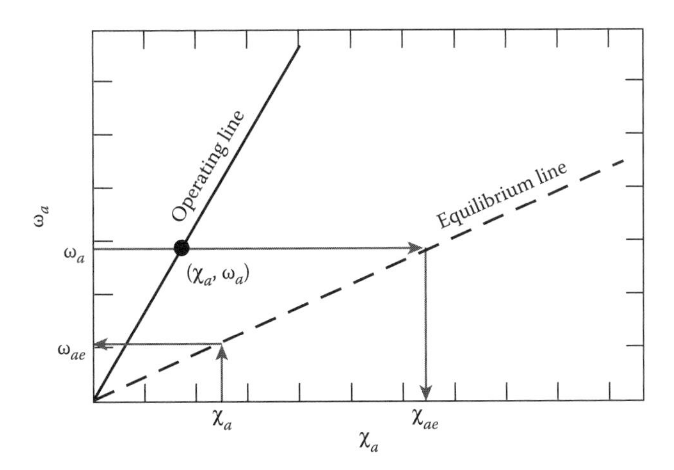

FIGURE 11.18 Operating and equilibrium lines for the gas–liquid contactor.

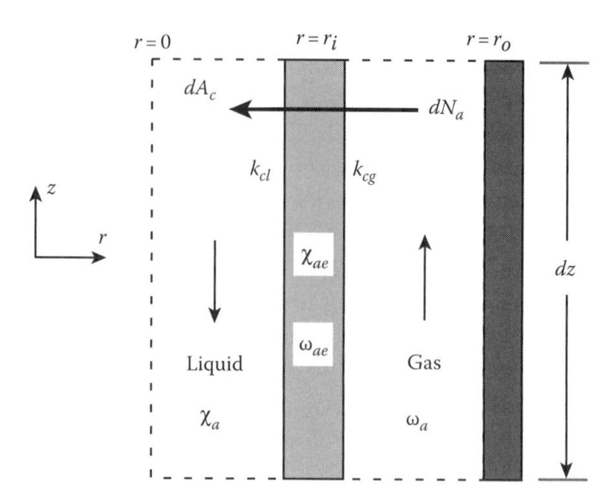

FIGURE 11.19 Differential slice along the main axis of the contactor of Figure 11.17. Equilibrium compositions exist at the interface between liquid and gas.

The equilibrium constant for heat exchange is 1, and the heat capacity rates, $\dot{m}C_p$, take the place of the molar flow rates in our mass balance.

We now apply a mass balance about a differential element for both the liquid and gas phases (Figure 11.19). This is a macroscopic balance because the gas and liquid mole fractions of *a* are averaged over the cross-sectional area of the column. Since we do not know the interfacial area, we translate it in terms of the cross-sectional area using

$$dA_i = a_v A_c dz \quad a_v \text{ is a measured quantity}$$

$$d(A_i N_{aL}) = k_{cL} a_v A_c (\chi_{ae} - \chi_a) dz$$

$$d(A_i N_{aG}) = k_{cG} a_v A_c (\omega_a - \omega_{ae}) dz$$

Here, we are using individual mass transfer coefficients for the liquid and gas phases. They are coupled to the function that translates from the cross-sectional area of the column to the interfacial area between the phases. As such, in real operating equipment, the mass transfer coefficient may vary with position in the column. The use of local coefficients means that the driving forces we must use are differences between equilibrium mole fractions that we assume exist at the interface between the two phases and the actual mole fractions in both phases. These mole fractions are shown in Figure 11.18.

We can use our equilibrium relationship between χ_a and ω_a and the fact that $d(A_i N_{aG}) - d(A_i N_{aL}) = 0$ to remove the equilibrium mole fractions. Solving for χ_{ae} lets us rewrite the differential mass balances using the mole fractions in each phase and an overall mass transfer coefficient based on the gas phase, K_{cG}. The overall mass transfer coefficient takes into account resistances in the gas and liquid phases as we discussed in Chapter 4:

$$d(A_i N_{aL}) = \frac{A_c}{\left(\frac{m_k}{k_{cL} a_v} + \frac{1}{k_{cG} a_v}\right)} (\omega_a - m_k \chi_a) dz$$

$$d(A_i N_{aG}) = \frac{A_c}{\left(\frac{m_k}{k_{cL} a_v} + \frac{1}{k_{cG} a_v}\right)} (\omega_a - m_k \chi_a) dz$$

$$\frac{1}{K_{cG}} = \frac{1}{\frac{m_k}{k_{cL}} + \frac{1}{k_{cG}}}$$

The change in mass flows between phases can be related to the molar flow rate of each phase (a constant) and a corresponding change in ω_a *or* χ_a:

$$d(A_i N_{aL}) = \mathcal{L} d\chi_a = -K_{cG} a_v A_c (\omega_a - m_k \chi_a) dz$$

$$d(A_i N_{aG}) = \mathcal{G} d\omega_a = K_{cG} a_v A_c (\omega_a - m_k \chi_a) dz$$

Multiplying the liquid phase balance by m_k, subtracting it from the gas phase balance, and isolating $(\omega_a - m_k\chi_a)$ lead to the following differential equation:

$$\frac{d(\omega_a - m_k \chi_a)}{(\omega_a - m_k \chi_a)} = -K_{cG} a_v A_c \left[\frac{1}{\mathcal{G}} - \frac{m_k}{\mathcal{L}}\right] dz$$

This equation is the analog of the balance we obtained for the heat exchanger describing the temperature difference at any axial position. We solve the mass balance equation in a way analogous to the way we handled the temperature equation. Separating and integrating yield

$$\frac{\omega_a - m_k \chi_a}{\omega_{a4} - m_k \chi_{a1}} = \exp\left\{-K_{cG} a_v A_c \left[\frac{1}{\mathcal{G}} - \frac{m_k}{\mathcal{L}}\right] z\right\}$$

We can use the result for the mole fraction driving force and the differential mass balance to determine the rate of moles of *a* transferred:

$$\dot{M}_a = K_{cG} a_v A_c \int_0^L (\omega_a - m_k \chi_a) dz$$

$$\dot{M}_a = K_{cG} a_v A_c (\omega_{a4} - m_k \chi_{a1}) \int_0^L \exp\left\{-K_{cG} a_v A_c \left[\frac{1}{\mathcal{G}} - \frac{m_k}{\mathcal{L}}\right] z\right\} dz$$

Using the same algebraic procedure we employed for the heat exchanger, we can integrate the previous equation and express the overall mass of *a* transferred in terms of the overall mass transfer coefficient, the cross-sectional area of the column, and the mole fractions of *a* in the gas and liquid phases at the inlet and outlet of the column:

$$\dot{M}_a = K_{cG} a_v A_c L \left[\frac{(\omega_{a3} - m_k \chi_{a2}) - (\omega_{a4} - m_k \chi_{a1})}{\ln\left(\frac{\omega_{a3} - m_k \chi_{a2}}{\omega_{a4} - m_k \chi_{a1}}\right)}\right] = K_{cG} a_v A_c L \Delta\chi_{lm}$$

The equation earlier can be recast in terms of a driving force and resistance. The composition function is termed the log-mean mole fraction difference and reflects the fact that a simple mole fraction driving force between gas and liquid phases is not appropriate because the compositions change with position throughout the contactor:

$$\dot{M}_a = \frac{\text{Driving force}}{\text{Resistance}} = \frac{\Delta\chi_{lm}}{1/K_{cG} a_v A_c L}$$

If we know the amount of a to be transferred in the contactor (a usual design parameter), we can solve for the required height of the contactor:

$$L = \frac{\dot{M}_a}{K_{cG} a_v A_c \Delta\chi_{lm}} = \left(\frac{\mathcal{G}}{K_{cG} a_v A_c}\right)\left(\frac{\omega_{a3} - \omega_{a4}}{\Delta\chi_{lm}}\right)$$

The grouping in brackets, $\mathcal{G}/K_{cG}a_vA_c$, has the units of length and is referred to as the height of a transfer unit. It is a direct analog of the length of a transfer unit we defined for the heat exchanger. The group $(\omega_{\alpha3} - \omega_{\alpha4})/\Delta\chi_{lm}$ is referred to as the number of transfer units required to affect the exchange. Analogous quantities can be derived based on the liquid phase. These are shown in the following set of equations:

$$\frac{\mathcal{G}}{K_{cG} a_v A_c}$$ – Height of a transfer unit (gas phase based)

$$\frac{\omega_{a3} - \omega_{a4}}{\Delta\chi_{lm}}$$ – Number of transfer units (gas phase based)

$$\frac{\mathcal{L}}{K_{cL} a_v A_c}$$ – Height of a transfer unit (liquid phase based)

$$\frac{\chi_{a2} - \chi_{a1}}{\Delta\chi_{lm}}$$ – Number of transfer units (liquid phase based)

$$\frac{1}{K_{cL} a_v} = \frac{1}{k_{cL} a_v} + \frac{1}{m_k k_{cG} a_v}$$ – Overall mass transfer coefficient (liquid phase based)

The mass transfer case is more complicated than the heat transfer case because one must preserve the distinction between the liquid and gas phases. For instance, if one defines the number of transfer units based on gas phase information, one must use the height of a transfer unit based on the gas phase. Any other combination would yield an incorrect value for the column height.

Example 11.14 Steady-State Operation of a Tubular Reactor with Continuous Sidestream Feed

Consider a tubular chemical reactor operating at a temperature, T_o. This reactor contains fluids a and b that react according to the following reactions:

$$a + b \xrightarrow{k_1''} 2r \qquad a + a \xrightarrow{k_2''} 2s$$

The desired product is r, but there is a side reaction producing s. This side reaction is appreciable if the concentration of a, at the reactor operating temperature, is allowed to be too high. Since we would like to maximize the production of r, we must keep b in excess all the way along the reactor. One way to accomplish this is to use a sidestream reactor like that shown in Figure 11.20. Here, we feed an excess amount of b at the entrance and supplement the reaction by feeding in a small sidestream containing a in high concentration. Our strategy is to keep c_a at a constant, but very low value, throughout the reactor ($c_a \sim 0$). Since the sidestream will feed in a at a very high concentration, c_{as}, the volumetric flow rate of the sidestream feed, $\dot{v}_s(x)$, will be very small, and the total volume of the reactor, V_t, and overall volume flow of the feed, $\dot{v}_o$, will be essentially constant.

Our goals are to determine the concentration profiles of a and b along the reactor and to determine the best way to control the sidestream feed. Since radial variations in concentration are not

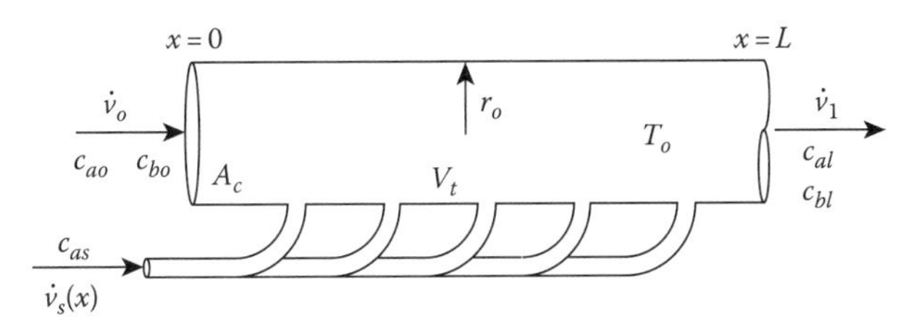

FIGURE 11.20 Tubular reactor with continuous sidestream feed.

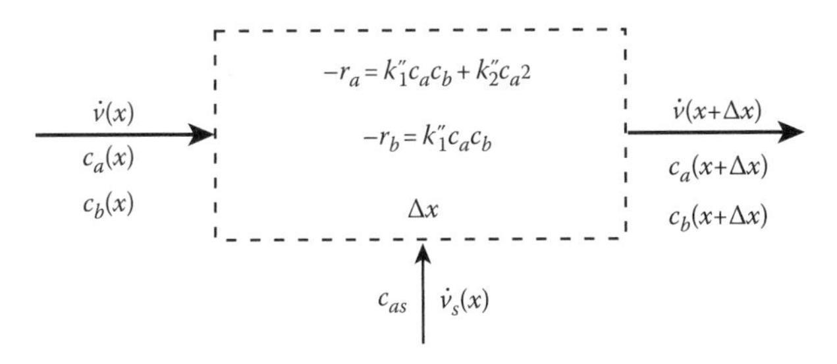

FIGURE 11.21 Differential slice of volume ΔV cut from the reactor.

a primary concern, we may as well define average concentrations at any axial position by using a mixing-cup formulation:

$$\bar{c}_a(x) = \frac{1}{\pi r_o^2}\int_0^{r_o} 2\pi c_a(r,x)\,r\,dr \qquad \bar{c}_b(x) = \frac{1}{\pi r_o^2}\int_0^{r_o} 2\pi c_b(r,x)\,r\,dr$$

We can now develop *differential-macroscopic* mass balances to determine the profiles. Figure 11.21 shows a small slice cut through the reactor. The overall mass balance or continuity equation states

$$\begin{array}{llll} In & -\,Out & +\,Gen & = Acc \\ \dot{v}(x) + \dot{v}_s(x)\Delta x & -\,\dot{v}(x+\Delta x) & +\quad 0 & = \quad 0 \end{array}$$

where
- $\dot{v}$ is a volumetric flow rate (m³/s)
- $\dot{v}_s$ is a volumetric flow rate per length of reactor (m³/s·m)

We implicitly assume constant density throughout the system. Dividing through by Δx and taking the limit as $\Delta x \to 0$ lead to

$$-\frac{d\dot{v}}{dx} + \dot{v}_s = 0$$

The species mass balance on *a* states

$$\begin{array}{llll} In & -\quad Out & +\quad Gen & = Acc \\ \dot{v}(x)\bar{c}_a(x) + \dot{v}_s(x)c_{as}\Delta x & -\,\dot{v}(x+\Delta x)\bar{c}_a(x+\Delta x) & -k_1''A_c\bar{c}_a(x)\bar{c}_b(x)\Delta x - k_2''A_c\bar{c}_a^{\,2}(x)\Delta x & = 0 \end{array}$$

Again, dividing through by Δx and allowing Δx to approach zero lead to

$$-\frac{d}{dx}(\dot{v}\bar{c}_a) + \dot{v}_s c_{as} - k_1 A_c\bar{c}_a\bar{c}_b - k_2'' A_c\bar{c}_a^{\,2} = 0$$

The species balance on b is

$$\begin{array}{llll} In & - \quad Out & + \quad Gen & = Acc \\ \dot{v}(x)\bar{c}_b(x) & -\dot{v}(x+\Delta x)\bar{c}_b(x+\Delta x) & -k_1'' A_c \bar{c}_a(x)\bar{c}_b(x)\Delta x & = 0 \end{array}$$

$$-\frac{d}{dx}(\dot{v}c_b) - k_1'' A_c c_a c_b = 0$$

Our assumptions about $\dot{v}$, $\dot{v}_s$, V_t, and $\bar{c}_a$ let us simplify the mass balances. The balance on a lets us determine the volumetric flow rate of the sidestream as a function of position along the reactor. Since we keep $\bar{c}_a$ constant,

$$\dot{v}_s = \frac{k_1'' A_c \bar{c}_a \bar{c}_b + k_2'' A_c \bar{c}_a^{\,2}}{c_{as}} \quad \text{Mass balance on } a$$

The balance on species b gives us the concentration of b at any point in the reactor:

$$\dot{v}\frac{d\bar{c}_b}{dx} + k_1'' A_c \bar{c}_a \bar{c}_b = 0 \quad \text{Mass balance on } b$$

$$\bar{c}_b = c_{bo} \exp\left[-\left(\frac{k_1'' A_c \bar{c}_a}{\dot{v}}\right)x\right] \quad \begin{array}{c}\text{Boundary condition} \\ x = 0 \quad \bar{c}_b = c_{bo}\end{array}$$

We can now substitute for $\bar{c}_b$ in the expression for $\dot{v}_s$ to determine the best feed strategy for the sidestream:

$$\dot{v}_s = \frac{k''_1 A_c \bar{c}_a c_{bo} \exp\left[-\left(\frac{k_1'' A_c \bar{c}_a}{\dot{v}}\right)x\right] + k_2'' A_c \bar{c}_a^{\,2}}{c_{as}}$$

Notice that $\dot{v}_s$ is an exponentially decreasing function of x. This is to be expected since the concentration of species b decreases throughout the reactor and therefore less a is needed. A more general solution for this problem, where $\dot{v}$ is not considered constant, has been solved and presented in [8].

This situation can be solved for laminar flow in Comsol®. The ability to solve the problem in multiple dimensions allows one to see how velocity and concentration gradients affect the effectiveness of the reactor. Figure 11.22 shows the concentration ratio of b to a as a function of radial and axial position. The ability to see both radial and axial changes allows one to optimize the reactor to minimize variations in both directions and would likely lead to a solution much different than that presented by the simplified model here.

Example 11.15 Startup of a Continuous, Stirred Tank Reactor

Consider a continuous stirred tank reactor where we are processing a compound, a, that undergoes a first-order decomposition reaction to produce species b. The reactor has a volume, V_t, and we feed material in at a concentration c_{ai} at a volumetric flow rate of $\dot{v}$ m^3/s. We assume the fluid is incompressible. The vessel will be well-stirred so that the concentration of a exiting the reactor is the same as that in the reactor. The whole system is shown later in Figure 11.23.

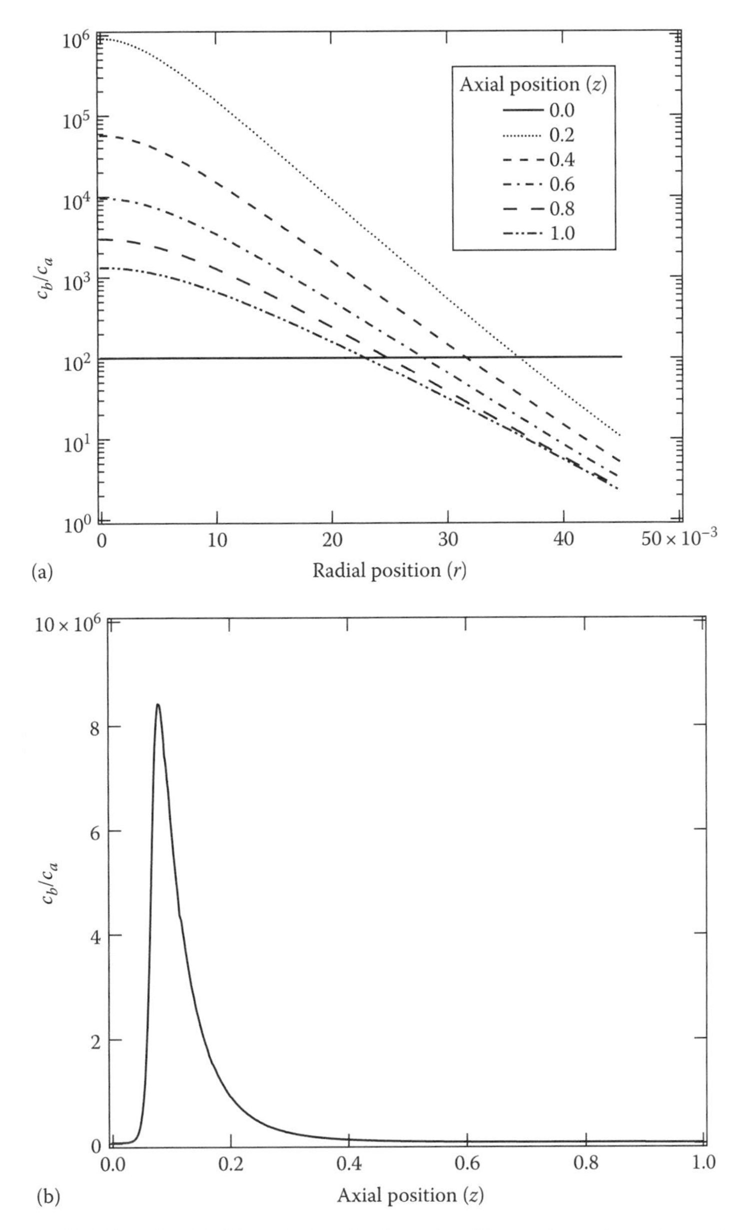

FIGURE 11.22 Radial (a) and axial (b) variations in the ratio of b to a throughout the reactor. The Comsol® module SideReactor allows one to investigate the effect of different reaction kinetics, different velocity profiles, and different feed scenarios for a and b on the formation of r and s.

Our goal is to calculate the concentration of species a exiting the reactor as a function of time and the feed flow rate. The transient mass balance is appropriate. We have no diffusion through the tank walls, and hence $\dot{M}_a = 0$:

$$C_{a2}\bar{V}_{M2}A_{c2} - C_{a1}\bar{V}_{M1}A_{c1} - \cancelto{0}{\dot{M}_a} - R_{a,tot} = -\frac{dM_{a,tot}}{dt}$$

The volumetric flow rate entering and leaving the tank is a constant and so

$$\bar{V}_{M1}A_{c1} = \bar{V}_{M2}A_{c2} = \dot{V}$$

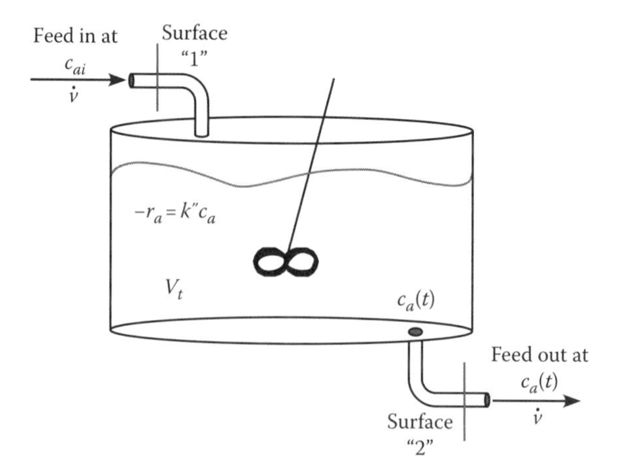

FIGURE 11.23 A continuous stirred tank reactor processing species a.

We can replace the mass of a in the tank, $M_{a,tot}$, by $V_t c_a$ and the generation rate, $R_{a,tot}$, by $k''V_t c_a$. Substituting into the mass balance yields

$$c_{ai}\dot{v} - c_a\dot{v} - k''V_t c_a = V_t \frac{dc_a}{dt}$$

where we have used the facts that the tank volume is constant and that the concentration of a in the tank is the same as the exit concentration.

We can define the space time for this system as

$$\tau = \left(\frac{\dot{v}}{V_t}\right)t$$

By solving the steady-state problem, we can define an *equilibrium* concentration, c_{eq}, and use it to define a dimensionless composition variable:

$$\chi = \frac{c_a}{c_{eq}} \qquad c_{eq} = \frac{\dot{v}c_{ai}}{\dot{v} + k''V_t}$$

Using these variables, the mass balance becomes

$$\frac{d\chi}{dt} = \chi_i - \left(1 + k''\frac{V_t}{\dot{v}}\right)\chi$$

subject to the initial condition

$$\tau = 0 \qquad \chi = 0$$

Separating, integrating, and applying the boundary condition lead to a solution for the composition, χ, as a function of time:

$$\chi = \frac{\chi_i\left\{1 - \exp\left[-\left(1 + \frac{k''V_t}{\dot{v}}\right)\tau\right]\right\}}{\left(1 + \frac{k''V_t}{\dot{v}}\right)}$$

Notice that like the heat transfer example, this is a lumped capacitance analysis applied to a flow system. The composition gradually settles down to its steady-state value.

11.7 MACROSCOPIC CHARGED SPECIES CONTINUITY EQUATION

It is possible to obtain a macroscopic version of the microscopic balance governing conservation of charge within a system. We begin with the version of the microscopic balance derived in Chapter 10 that assumes the total concentration and diffusivity are constants. $\rho_a^{\pm}$ is the charge density arising from species a:

$$\left(\vec{\nabla}\cdot\rho_a^{\pm}\vec{\mathbf{v}}_{\mathbf{M}}\right)-D_{ab}\vec{\nabla}\cdot\left[\vec{\nabla}\rho_a^{\pm}-\frac{z_{ea}\mathcal{F}\mathrm{a}}{RT}\left(\rho_a^{\pm}\vec{\mathcal{E}}\right)\right]-R_{ea}=-\frac{\partial\rho_a^{\pm}}{\partial t} \tag{11.63}$$

We can now integrate Equation 11.63 over the volume of the system (see Figure 11.16) to obtain the macroscopic charged species continuity equation:

$$\int_V \frac{\partial\rho_a^{\pm}}{\partial t}dV=\frac{\partial}{\partial t}\int_V \rho_a^{\pm}dV=\frac{dQ_{a,tot}}{dt} \tag{11.64}$$

Here, $Q_{a,tot}$ represents the total amount of charged species a present in the system:

$$\int_V\left(\vec{\nabla}\cdot\rho_a^{\pm}\vec{\mathbf{v}}_{\mathbf{M}}\right)dV=\int_{A_c}\rho_a^{\pm}v_{Mn}dA_c=\rho_{a2}^{\pm}\bar{v}_{M2}A_{c2}-\rho_{a1}^{\pm}\bar{v}_{M1}A_{c1} \tag{11.65}$$

$$D_{ab}\int_V\vec{\nabla}\cdot\left[\vec{\nabla}\rho_a^{\pm}\right]dV=-\int_{A_m}\left(\vec{\mathbf{n}}\cdot\rho_a^{\pm}\vec{\mathbf{v}}_{\mathbf{M}}\right)dA_m=q_a \tag{11.66}$$

$$D_{ab}\frac{z_{ea}\mathcal{F}\mathrm{a}}{RT}\int_V\nabla\cdot\left(\rho_a^{\pm}\mathcal{E}\right)dV=-D_{ab}\frac{z_{ea}\mathcal{F}\mathrm{a}}{RT}\int_{A_e}\mathbf{n}\cdot\left(\rho_a^{\pm}\mathcal{E}\right)dA_e=q_{ae} \tag{11.67}$$

In these integrals

- A_m represents that portion of the surface over which mass transfer occurs
- A_e represents that portion of the surface over which the voltage gradient (electric field) is applied

We assumed that both transfer surfaces are fixed in space and don't change. In essence, we are adding up the flux across the mass transfer surface. q_a and Q_a are positive when mass is added to the system:

$$\int_V R_{ea}dV=R_{ea,tot} \tag{11.68}$$

Here, $R_{ea,tot}$ represents the total production or destruction of species a by reaction, that is, neutralization. Summing all the components gives

$$-\frac{dQ_{a,tot}}{dt}=\rho_{a2}\bar{v}_{M2}A_{c2}-\rho_{a1}\bar{v}_{M1}A_{c1}-q_a-q_{ae}-R_{ea,tot} \tag{11.69}$$

One special case arises if we assume that diffusion is negligible and that we have no convection or generation. For such a situation, q_{ae} is equal to

$$q_{ae} = \sigma A_c \frac{\Delta\Phi}{L} = \frac{\Delta\Phi}{\mathcal{R}_e} \tag{11.70}$$

where $\mathcal{R}_e$ is the resistance of the medium between the surfaces of constant potential (voltage). The accumulation term involving the change of charge with time is just the current, $\mathcal{I}_a$, flowing through the system due to the motion of species a:

$$\frac{d\mathcal{Q}_{a,tot}}{dt} = \mathcal{I}_a \tag{11.71}$$

Combining Equations 11.69 through 11.71 gives Ohm's law for conductors:

$$\mathcal{I}_a = \frac{\Delta\Phi}{\mathcal{R}_e} \tag{11.72}$$

Example 11.16 Field-Assisted Absorption [9]

Much of the performance of the liquid–gas extractor we discussed previously was dependent upon the operating line and hence the flow rate of the two contacting streams. If we want to transfer a charged species from one phase to another, we can use an electric field to assist in controlling the operating line. Consider the field-assisted liquid–liquid extractor shown later in Figure 11.24. Here, we have two liquid streams in counterflow with a charged species being exchanged between them.

We will assume dilute solutions and incompressible fluids so that all the fluid properties are constant. The total concentration of species in either stream is also constant. We begin by defining the flow of solute in both phases using a solvent-free basis. The solute is carried along by bulk convection in each phase, but because of the presence of an electric field, there is an additional drift velocity, v_{el}, in the solvent 1 phase and v_{eg} in the solvent 2 phase. These velocities are proportional

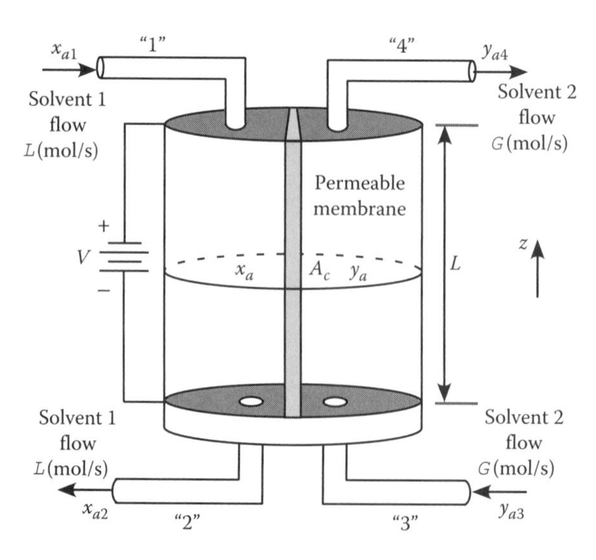

FIGURE 11.24 Absorption column with an applied electric field.

to the mobility of the charged species in the two solvents and the magnitude and direction of the electric field ($v_{el} = \mu_{el}\mathcal{E}$; $v_{eg} = \mu_{eg}\mathcal{E}$). Combining this knowledge with the total concentration of species in each phase lets us derive an expression for the solute flow in phase 1 and phase 2:

$$\text{Solute flow in liquid} = \chi_{a1}\mathcal{L}_e = \left(\frac{x_{a1}}{1-x_{a1}}\right)[\mathcal{L} - c_{o1}v_{e1}]$$

$$\text{Solute flow in gas} = \omega_{a1}\mathcal{G}_e = \left(\frac{y_{a1}}{1-y_{a1}}\right)[\mathcal{G} + c_{og}v_{eg}]$$

In these expressions, we have assumed that the field acts on the charged species in a direction opposite to the bulk flow of solvent 1. Notice that the electric field augments the total molar flow of both phases. We can now apply a mass balance on the solute for each phase:

$$\omega_{a3}\mathcal{G}_e - \omega_{a4}\mathcal{G}_e = N_{ag}A_i \qquad \chi_{a2}\mathcal{L}_e - \chi_{a1}\mathcal{L}_e = N_{al}A_i$$

where N_{ag} and N_{al} represent the flux of solute a between the phases. At steady state, both fluxes must be equal, and we can solve for the outlet mole fraction, χ_{a2}:

$$\chi_{a2} = \chi_{a1} + \left(\frac{\mathcal{G}_e}{\mathcal{L}_e}\right)(\omega_{a3} - \omega_{a4})$$

If we let χ_a and ω_a represent the mole fractions of a at any position in the column, then we can define an operating line for the contactor as

$$\chi_a = \chi_{a1} + \left(\frac{\mathcal{G} + c_{og}v_{eg}}{\mathcal{L} - c_{ol}v_{el}}\right)(\omega_a - \omega_{a4})$$

The electric field provides a way to *tweak* the performance of the column. For given flow rates of solvents, we can manipulate the electric field to correct the operation of the column if unforeseen changes occur in feed composition, solvent properties, etc. The electric field may but need not affect the equilibrium line, and so a diagram of operating and equilibrium lines for the contactor may look like Figure 11.25.

As in our previous contactor example, the overall mass balances tell us nothing about the concentration profiles in the column. Using an analogous differential balance on the contactor,

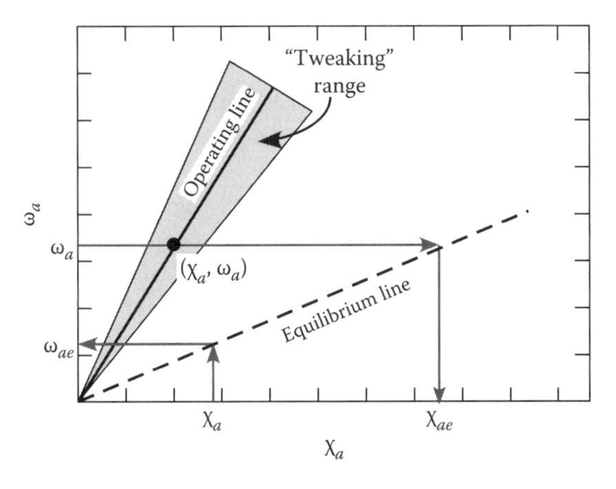

FIGURE 11.25 Operating diagram for the liquid–liquid contactor showing field-assisted adjustment range.

you should convince yourself that the mole fraction profiles, the heights of a transfer unit, and numbers of transfer units can be expressed as

$$\frac{\omega_a - m_k\chi_a}{\omega_{a4} - m_k\chi_{a1}} = \exp\left\{-\left[\frac{1}{\mathcal{G} + c_{og}v_{eg}} - \frac{m_k}{\mathcal{L} - c_{ol}v_{el}}\right]K_{cG}a_vA_cz\right\}$$

$$\frac{\mathcal{G} + c_{og}v_{eg}}{K_{cG}a_vA_c} \quad \text{— Height of a transfer unit (gas phase based)}$$

$$\frac{\omega_{a3} - \omega_{a4}}{\Delta\chi_{lm}} \quad \text{— Number of transfer units (gas phase based)}$$

where the overall mass transfer coefficient, $K_{cG}a_v$, and the log-mean concentration difference, $\Delta\chi_{lm}$, are defined as in the previous example.

11.8 SUMMARY

In this chapter, we have shown how we can use selective or total averaging procedures to derive macroscopic versions of the mass, momentum, energy, and charge balances. In complicated systems, these balances enable us to analyze the performance of the system and obtain some *engineering* insight into their behavior without having to resort to a full solution of the balance equations in three dimensions. In fact, the macroscopic balance equations presented here are generally the first line of attack for the practicing engineer. When phenomena appear that cannot be accounted for by macroscopic balances, we resort to a more complicated microscopic balance approach.

In the next chapter, we return to the microscopic balance and discuss the development of boundary layers on a flat plate. A knowledge of the behavior of these boundary layers will enable us to derive theoretical expressions for the friction factors, heat transfer coefficients, and mass transfer coefficients that form the basis for most analysis of transport processes.

PROBLEMS

Mass, Momentum, and Mechanical Energy Balances

11.1 A water jet pump has a jet area $A_j = 0.01$ m^2 and a jet velocity $v_j = 30$ m/s, which entrains a secondary stream of water having a velocity $v_s = 3$ m/s in a constant area pipe of total area $A = 0.75$ m^2. At section 2, the water is thoroughly mixed. Assume 1-D flow and neglect wall shear (Figure P11.1).

a. What is the average velocity of the mixed flow at section 2?
b. What is the pressure rise $(P_2 - P_1)$ assuming the pressure of the jet and secondary stream to be the same at section 1?

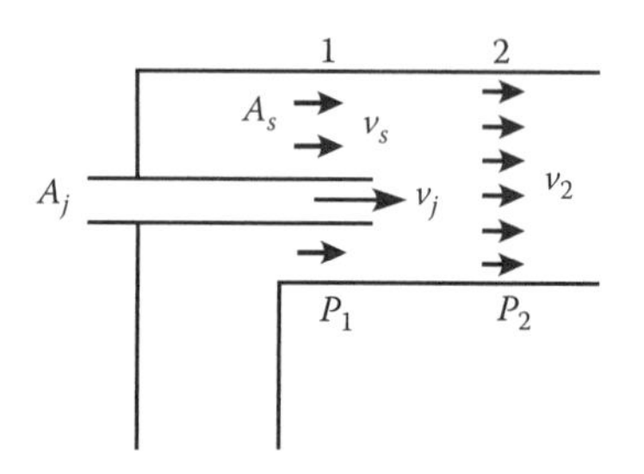

FIGURE P11.1

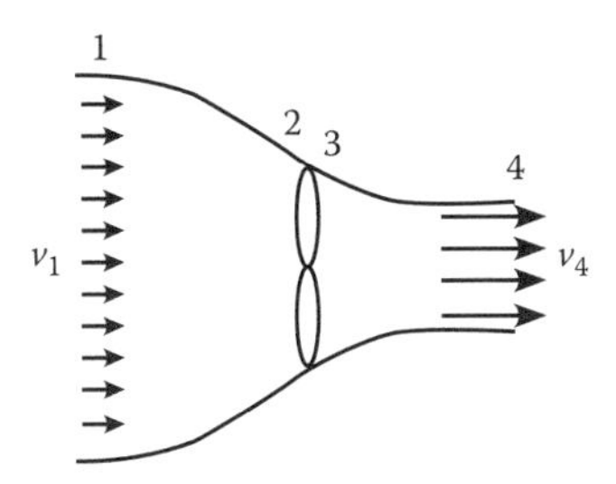

FIGURE P11.2

11.2 Fluid is accelerated as it passes through the slipstream of a propeller. Show that the average velocity of the fluid as it passes the plane of the propeller blades is the average of the upstream and downstream velocities, v_1 and v_4. Assume $P_1 = P_4$, $v_2 = v_3$ and neglect viscous effects (Figure P11.2).

11.3 Water flows steadily through a fire hose and nozzle. The hose diameter is 100 mm and the nozzle tip is 30 mm in diameter. The gauge pressure of the water in the hose is 525 kPa and the stream leaving the nozzle has a uniform velocity profile. The fluid exits the nozzle at a speed of 25 m/s.

a. Find the force transmitted by the coupling between the nozzle and the hose.

b. If the nozzle were pointed so that the water shot straight up into the air, how high would the water reach?

11.4 Water exits a pipe from a series of holes drilled into the side. The pressure at the inlet section is 35 kPa (gauge). Calculate the volumetric flow rate at the inlet section and the forces required at the coupling to hold the spray pipe in place. The pipe and water it contains weighs 5 kg (Figure P11.4).

11.5 A jet of water issuing from a nozzle at a volumetric flow rate, Q, with a velocity, v_j, strikes a series of vanes mounted on a wheel. This configuration is termed a Pelton wheel and it rotates with a peripheral velocity, $v_b = \omega R$. The jet is split in half by the vanes, each half of the jet being deflected through an angle, β. The entire system is at atmospheric pressure (Figure P11.5).

a. What is the power in the jet?

b. What power is given up by the jet to the blades?

c. What advantage results from splitting the jet in half?

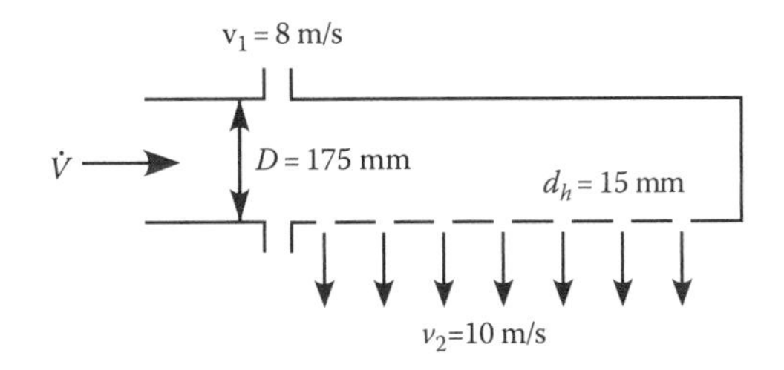

FIGURE P11.4

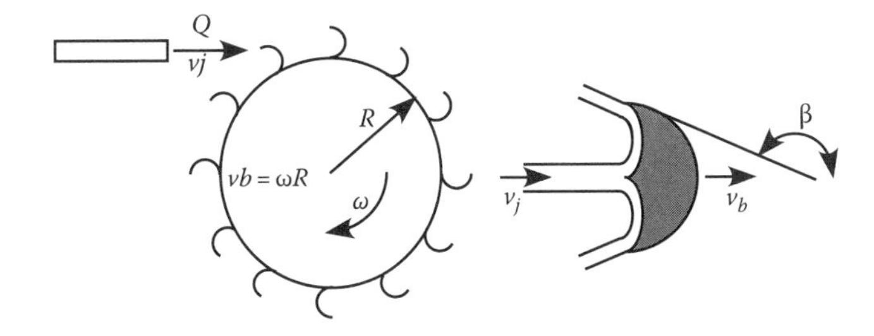

FIGURE P11.5

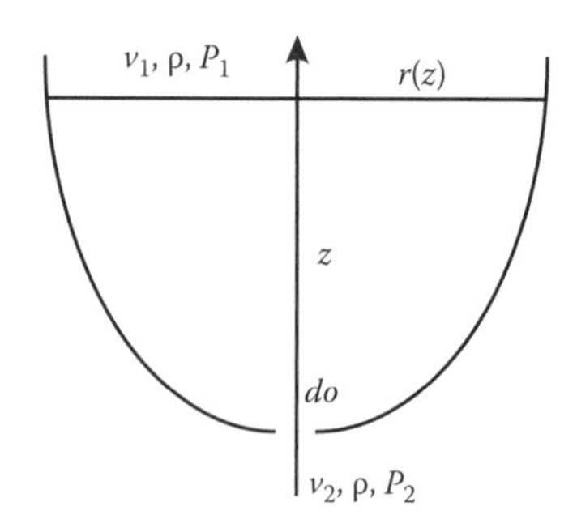

FIGURE P11.6

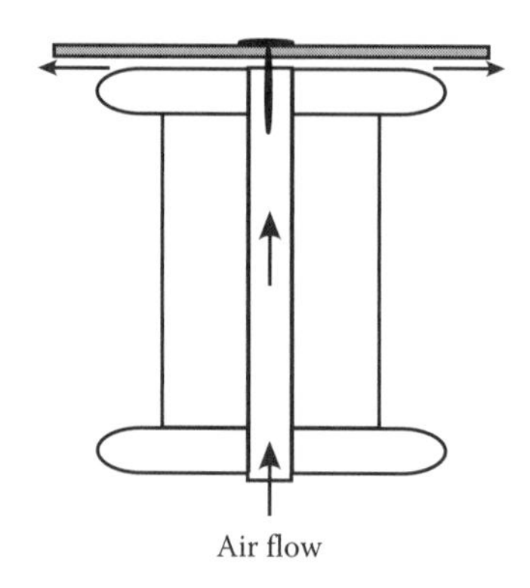

FIGURE P11.7

11.6 An ancient device for measuring the passage of time uses an axisymmetric vessel shaped so that the water level falls at a constant rate (see Figure P11.6). The top of the vessel is open to the atmosphere and water exists through a small hole (radius d_o) at the bottom. Determine the shape of the vessel, that is, radius, r, as a function of axial position, z, and its volume that allows us to measure time.

11.7 A small piece of 3 × 5 card can be held onto a spool of thread by blowing air through the hole in the center as shown in Figure P11.7. The harder one blows, the tighter the card hangs on. Explain why this is so. You need to put a nail or thumb tack through the card's center to make sure the card does not slide off the spool.

11.8 The depth of fluid following a hydraulic jump can be determined from an application of conservation of mass and conservation of momentum as discussed in Chapter 11. This change in height must also result in a loss of available energy (lost work) across the jump. Apply the macroscopic mechanical energy balance to the system shown in Figure P11.8 to derive the loss expression:

$$\text{Loss} = \frac{g(h_2 - h_1)^3}{4h_1h_2}$$

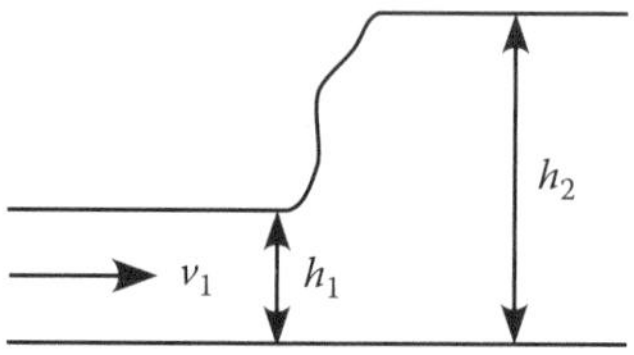

FIGURE P11.8

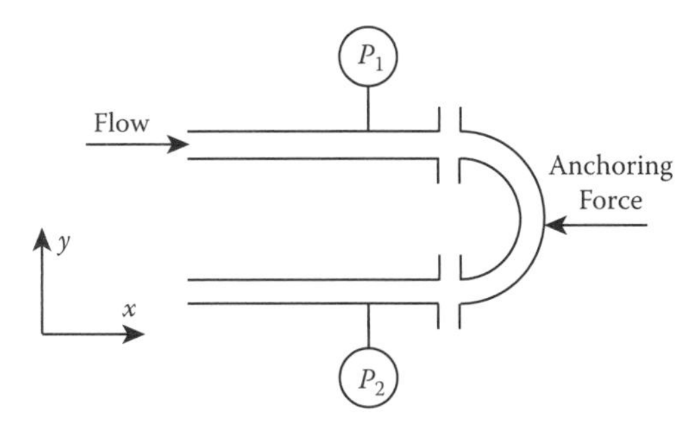

FIGURE P11.10

11.9 An advertisement from the manufacturer of a low-flow showerhead claims that their unique showerheads inject oxygen into the water to provide greater coverage with less water. They further claim to push water through an "accelerator fin" to increase the pressure. The result is a shower that uses 1.5 gal/min but feels as if it is delivering much more.

a. Based upon your knowledge of the continuity, momentum, and energy equations, what is wrong with the manufacturer's statement?

b. How would you word the ad to make it more representative of what is actually occurring in the device?

11.10 A new type of flow meter is shown in Figure P11.10. In it, we read the two pressure gauges and the force on the pipe bend (using strain gauges). From those three readings, we compute the fluid velocity in the pipe. The pipe diameter, the diameter of the couplings, and the diameter of the bend are all 2.54 cm. The fluid flowing through the system is water at room temperature. P_1 reads 150,000 Pa (gauge) and P_2 reads 130,000 Pa (gauge). The restraining force, measured by the strain gauge, is 200.2 N and acts in the $-x$ direction. Assume gravity has no affect on the system. What is the fluid velocity in the pipe?

11.11 Cooling water is pumped from a reservoir to drills at the site of a geothermal power plant. The drills are located 150 m above the level of the pump and the distance from the pump to the job site is 250 m. The flow rate of water, used to fracture the rock, is to be 2.5 m^3/min. At the job site, the water leaves the injection nozzle at a velocity of 35 m/s. The pipe diameter delivering the water is 10.16 cm.

a. Determine the minimum supply pressure required at the pump outlet.

b. Estimate the required power input to the pump if the pump efficiency is 65%.

11.12 Find the force components, F_x and F_y, required to hold the box below stationary. The fluid is oil with a specific gravity of 0.85. Neglect gravity and assume the pressure acting an all inlet and outlet faces is atmospheric (Figure P11.12).

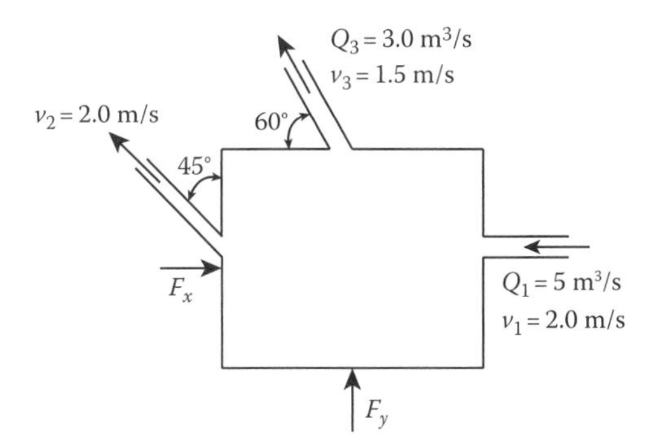

FIGURE P11.12

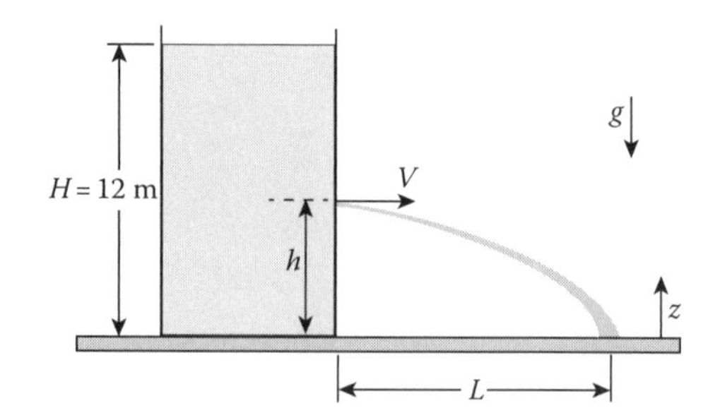

FIGURE P11.13

11.13 A hole is to be drilled into the side of the tank shown in Figure P11.13, so the liquid will travel the farthest horizontal distance, L. You may assume the liquid height in the tank remains constant, the flow is inviscid, and that gravity acts in the $-z$ direction.

a. What is the velocity of the water exiting the tank as a function of h?
b. What are the velocity components of the exiting water and how do they change over time as the stream falls to the ground? You may assume that the water stream is traveling purely horizontally as it exits the tank at $L = 0$.
c. Using your answers from (a) and (b), determine an expression for the distance L as a function of h, the velocity exiting the tank, and the height of liquid in the tank.
d. Where should the hole be drilled to insure L is a maximum? Why?

11.14 A plastic tube of 50 mm diameter is used to siphon water from the large tank as shown in Figure P11.14. If the pressure on the outside of the tube is 25 kPa greater than the pressure within the tube, the tube will collapse and the siphon will stop.

a. Where is the pressure in the tube lowest? Please prove it.
b. If the viscous effects are negligible, determine the minimum value of h allowed without the siphon stopping.

Overall Energy Balances

11.15 Consider the case of isentropic flow through a nozzle.

a. What is the critical temperature, T_c, for the flow?
b. Show that the average velocity at any point in the nozzle can be described by

$$\bar{v} = \sqrt{2C_pT_1\left[1-\left(\frac{P}{P_1}\right)^{\frac{\gamma-1}{\gamma}}\right]}$$

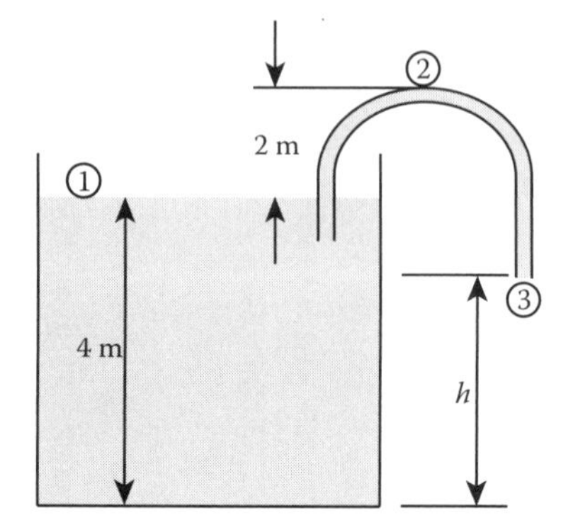

FIGURE P11.14

c. If the Mach number is defined as the ratio of the gas velocity to the velocity of sound at local temperature and pressure conditions, show that the temperature and pressure at any point in the nozzle is a function of the Mach number.

11.16 If we fail to insulate the nozzle in the gas flow system of Problem 11.15, we allow heat transfer to occur as the gas flows from the reservoir. Show under these conditions that the entropy of the gas at any location in the nozzle can be given by

$$S = S_1 + C_p \ln\left[\left(\frac{\rho_1}{\rho}\right)\left(\frac{P}{P_1}\right)^{1/\gamma}\right]$$

11.17 In developing our model for a heat exchanger, we assumed that the overall heat transfer coefficient, U, was a constant. In most instances, especially if the temperature difference between hot and cold fluids changes greatly through the exchanger, U will vary. Consider the case where U varies linearly with the temperature difference, $T_h - T_c$. Develop a new design equation for the heat exchanger with this information based on the analysis of Section 11.7.

11.18 Water is heated in a double-pipe heat exchanger from 15°C to 40°C. Oil (C_p = 2500 J/kg °C) with a mass flow rate of 0.03 kg/s and inlet and outlet temperatures of 80°C and 35°C, respectively, serves as the heating fluid.

a. Determine the required surface area for a counterflow exchanger if U = 300 W/m² °C.

b. What are the length of a transfer unit and the number of transfer units based on the water? On the oil?

11.19 A small steam condenser is designed to condense 1 kg/min of steam at 90 kPa with cooling water at 10°C. The exit water is not to exceed 60°C. The overall heat transfer coefficient for the exchanger is 3400 W/m². T_{sat} = 95.6°C. $h_{fg} = 2.27 \times 10^6$ J/kg.

a. Calculate the area required for a counterflow, double-pipe heat exchanger and the cooling water flow rate.

b. Calculate the area required for a parallel flow, double-pipe heat exchanger.

Species Mass Balances

11.20 A stirred tank has a capacity of V m³. Before time $t = 0$, the concentration of salt within the tank is c_i moles/m³. At time = 0, pure water is run in at a rate of Q m³/s, and brine is withdrawn from the bottom of the tank at the same rate. We assume negligible change in density of the fluid during dilution.

a. How long will it take for the concentration to be reduced to some final value, c_f?

b. Consider incomplete mixing so that the average salt concentration within the tank, c, is not the same as the outlet concentration c_o. Assume that c and c_o may be related by the following simple function containing one parameter, b, which is dependent on the stirrer speed and the geometry of the apparatus. How long will it take for c_i to be reduced to c_f in the tank?

$$\frac{c_o - c}{c_i - c} = e^{-bt}$$

11.21 The liquid phase reaction, 2A → B, is to be carried out in a 500 L well-mixed reactor. Pure A is to be fed to the reactor at a rate of 20 mol/h and at a temperature of 150°C. The endothermic reaction has a heat of reaction of 150,000 J/mol. Saturated steam flows through the reactor jacket that has an overall heat transfer coefficient of 5 W/m² K. The reactor is to be maintained at a temperature of 250°C at every point. A production rate of 7 mol/h of B is desired. The heat capacities for A and B are 10 J/mol K and 15 J/mol K, respectively. The steam jacket has a surface area of 2.5 m².

a. What is the temperature and pressure of the saturated steam required?
b. Assuming a perfectly insulated system, what is the minimum steam flow rate required? (Enever, Lewandowski, and Taylor)

11.22 A continuous, stirred tank reactor is initially filled with solvent. At $t = 0$, reactant, A, is fed into the tank at a rate of $\dot{v}_{ao}$ (m³/min). The concentration of A in the feed is c_{ao} (mol/m³). Compound A undergoes a first-order decomposition reaction with a characteristic rate constant of k'' (min^{-1}). For a tank of overall volume, V_o, what is the concentration of A in the tank as a function of time?

11.23 A tubular reactor is running a second-order addition reaction A + B → C in the liquid phase. The rate law is first order in the concentrations of A and B with rate constant k'' (mol/s). The reactor length is L, the cross-sectional area is A_c, and A and B are fed in stoichiometric ratio at a volumetric flow rate of $\dot{v}_o$ and concentration, $c_{ao} = c_{bo} = c_o$. Determine an expression for evaluating the concentration of A at any point within the reactor.

11.24 A radial flow reactor is often used for highly exothermic reactions. The high radial velocities at the reactor inlet compensate for any hot spots that might form in the reactor there. Consider the case where we are running a pseudo-first-order reaction in a radial reactor like that shown in Problem 11.15. Determine the concentration of A in the reactor as a function of radial position (Figure P11.24). The volumetric flow rate of reactant can be assumed constant, and the inlet concentration is c_{ao}. How would the results change if the reaction were second order in the concentration of a?

11.25 Often one absorbs a component from the gas phase into a liquid where a reaction occurs. The presence of the reaction increases the mass transfer and the loading the liquid can take. A prime example of this process is the absorption of CO_2 in basic solutions such as NaOH. Rework the gas–liquid absorption example considering a pseudo-first-order reaction of CO_2 in the liquid, $-r_a = k''c_{CO_2}$.
a. How do the balance equations change? Are the fluxes between phases still equal?
b. How does the overall mass transfer coefficient depend upon the reaction rate constant?
c. What are the new definitions for the height and number of transfer units?

11.26 Water is being used to absorb acetone in a packed tower whose cross-sectional area is 0.2 m². The inlet air contains 3 mole% acetone and the outlet contains 0.6%. The gas flow is 14.0 kg mol/h and the pure water inlet flow is 45.3 kg-mol/h. The local mass transfer coefficients have been measured to be

$$k_{cl} = 0.04 \text{ m/s} \quad k_{cg} = 0.0625 \text{ m/s}$$

The equilibrium relationship for acetone/water is $y_a = 1.186x_a$.

Determine the overall mass transfer coefficient and the tower height required.

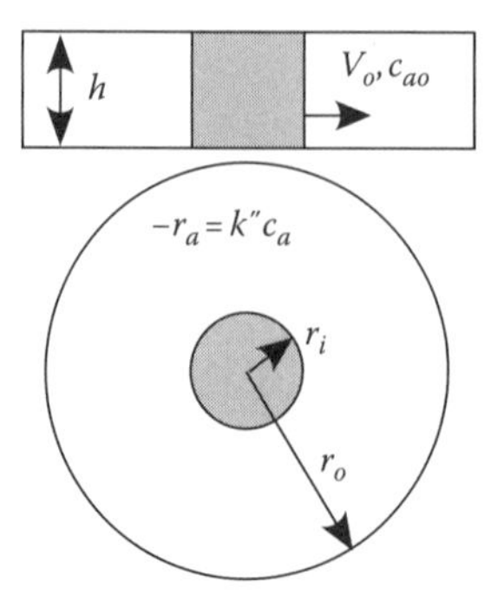

FIGURE P11.24

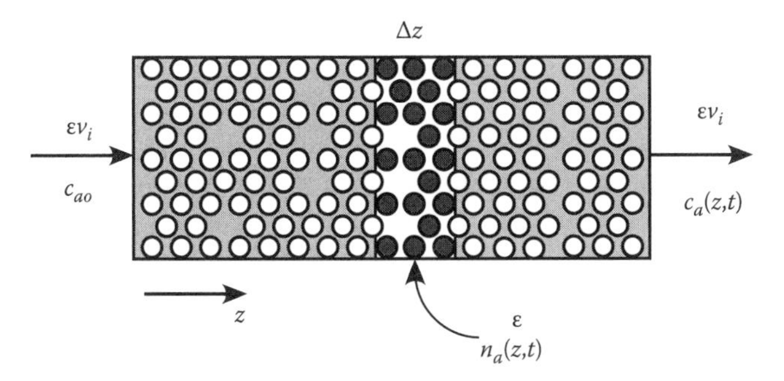

FIGURE P11.27

11.27 Chromatography is a process by which a separation of chemical species is accomplished by selective adsorption on a solid medium. Consider the simple case of a packed column of cross-sectional area A_c and length, L. The void fraction (liquid space) within the column is ε. Fluid containing the adsorbate in a concentration of c_{ao} (moles adsorbate/m^3 fluid) is introduced to the column at a velocity, v_i, characteristic of the fluid velocity in the interstices between adsorbent particles. The concentration of adsorbate on the particles themselves is c_{pa} (moles of adsorbate/m^3 of adsorbent). By performing a balance about a small slice through the column shown in Figure P11.27, show that the differential equation governing the concentration of adsorbate in the fluid is governed by the following equation:

$$\varepsilon v_i \frac{\partial c_a}{\partial z} + \varepsilon \frac{\partial c_a}{\partial t} + (1-\varepsilon)\frac{\partial c_{pa}}{\partial t} = 0$$

where we use the relation

$$(1-\varepsilon)\frac{\partial c_{pa}}{\partial t} = f(c_a, c_{pa}, v_i, \varepsilon, \ldots)$$

to describe the finite rate of transfer of adsorbate from solute to adsorbent. It takes the place of $kA\Delta c$ to describe the mass transfer rate.

11.28 Often in the theory of chromatography, we assume local equilibrium at all points in the column between the adsorbent particles and the adjacent fluid. Under these conditions, we can express c_{pa} as a function of c_a and substitute into the differential equation of the previous problem.

a. Assume $c_{pa} = f(c_a)$ and show that the differential equation describing chromatography can be written as

$$\left[\varepsilon + (1-\varepsilon)\frac{\partial}{\partial t} f(c_a)\right]\frac{\partial c_a}{\partial t} + \varepsilon v_i \frac{\partial c_a}{\partial z} = 0$$

b. The equation in part (a) can be solved using the method of characteristics.* We solve this equation simultaneously with the definition of the total derivative

$$dc_a = \left(\frac{\partial c_a}{\partial z}\right)dz + \left(\frac{\partial c_a}{\partial t}\right)dt$$

* Information on the method of characteristics can be found in [10].

to obtain $\partial c_a/\partial z$ and $\partial c_a/\partial t$. Solve these equations and show that there are certain characteristic velocities at which a particular concentration moves through the bed. (Hint: Solve the equations via matrices and calculate the determinant to get the velocity. $dc_a = 0$ within the velocity wave.)

c. How does this wave velocity depend upon the adsorption isotherm and, in particular, the Langmuir isotherm given by

$$c_{pa} = c_{po}\frac{Kc_a}{1+Kc_a} = f(c_a)$$

REFERENCES

1. Wylie, C.R., *Advanced Engineering Mathematics*, 4th edn., McGraw-Hill, New York (1975).
2. Bird, R.B., Stewart, W.E., and Lightfoot, E.N., *Transport Phenomena*, John Wiley & Sons, New York (1960).
3. Hinze, J.O., *Turbulence*, 2nd edn., McGraw-Hill, New York (1975).
4. Jeffreys, H., The draining of a vertical plate, *Proc. Cambridge Phil. Soc.* **26**, 204 (1930).
5. Liu, A.H., Wayner, P.C., and Plawsky, J.L., Image scanning ellipsometry for measuring the transient, film thickness profiles of draining liquids, *Phys. Fluids A* **6**, 1963 (1994).
6. Smith, J.M., Van Ness, H.C., and Abbot, M.M., *Introduction to Chemical Engineering Thermodynamics*, 6th edn., McGraw-Hill, New York, p. 44 (2001).
7. Geankoplis, C.J., *Transport Processes and Seperation Process Principles*, 4th edn., Prentice Hall, Upper Saddle River, NJ, p. 670 (2003).
8. Van de Vusse, J.G. and Voetter, H., Optimum pressure and concentration gradients in tubular reactors, *Chem. Eng. Sci.*, **14**, 17 (1960).
9. Locke, B.R. and Carbonell, R.G., A theoretical and experimental study of counteracting chromatographic electrophoresis, *Sep. Purif. Methods*, **18**, 1–64 (1989).
10. Lin, C.C. and Siegel, L.A., *Mathematics Applied to Deterministic Problems in the Natural Sciences*, MacMillan Publishing Co., New York (1974).

12 Convective Transport on a Flat Plate (Laminar Boundary Layers)

12.1 INTRODUCTION

We have encountered convective transport at many points in this text, have derived the convective transport equations in microscopic and macroscopic forms, and have solved several convective transport problems. In all instances, we have either not needed a convective transport coefficient or if we needed one, it has been provided without an explanation of how it was obtained or why it has a particular magnitude. The purpose of this chapter is to show how we can use the convective transport equations we derived in Chapter 10 to obtain explicit relations for the transport coefficients.

Our goal is to derive theoretical expressions for the three primary transport coefficients, the friction factor, C_f; the heat transfer coefficient, h; and the mass transfer coefficient, k_c, in laminar flow situations. We know that these coefficients will depend upon

$$C_f = f\,(\text{fluid properties, flow field})$$
$$h = f\,(\text{fluid properties, flow field, temperature field})$$
$$k_c = f\,(\text{fluid properties, flow field, concentration field})$$

To a lesser extent, we will also be concerned with the ionic mass transfer coefficient, $k_\pm$. This coefficient will depend upon

$$k_\pm = f\,(\text{fluid properties, flow field, concentration/charge field, electric field})$$

12.2 CONVECTIVE TRANSPORT COEFFICIENTS, C_f, h, k_c, AND $k_\pm$

We begin by looking at the simplest possible flow geometry, the flat plate. Though the least complicated case, it is of extreme technological importance. From fluid flow over bird wings, airplane wings, aquatic fins, and ski jumpers, to heat transfer from the walls of buildings, stegosaurus plates, and the underbelly of the space shuttle, to mass transfer from fish gills, thin-film adsorbers, and tree leaves, the flat plate represents the first line of attack in nearly all convective transport situations. It is only when the flat plate approximation fails that we are forced to consider more complicated geometries.

The prototypical flat plate transport situation is shown in Figure 12.1. Far away from the leading edge of the plate ($x = 0$), the fluid has a uniform velocity, v_∞; uniform temperature, T_∞; and uniform concentration of species a, $c_{a\infty}$. The plate is assumed immobile, is anchored in space, and may be exposed to an electric field. As fluid flows over the plate, it must obey the no-slip condition at the plate surface, and so it experiences a frictional force that slows it down in a region close to the surface. This flow retardation leads to a transfer of momentum between the free stream, where the fluid is

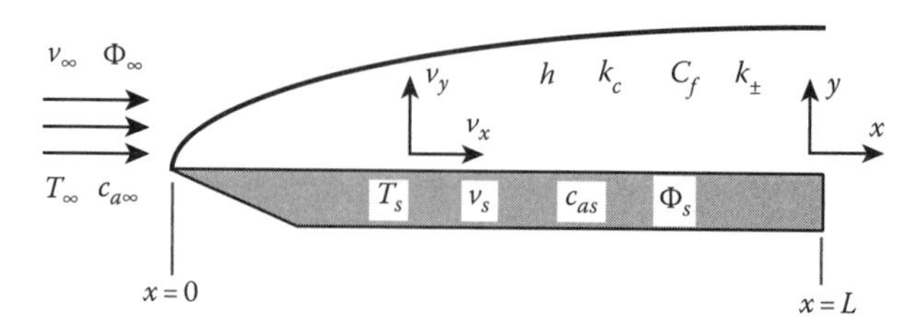

FIGURE 12.1 Convective flow over a flat plate. Fluid enters at a uniform velocity.

still flowing with a uniform velocity, v_∞, and the surface, where the fluid is stationary. Fluid at the surface of the plate will take on values for temperature and concentration characteristic of the plate, and so heat and mass transfer may also occur between the plate surface and the free stream.

Since the fluid experiences a frictional force close to the surface of the plate, it reacts in one of two ways. The fluid molecules can be compressed into a smaller volume and so increase the local pressure, or if the fluid is incompressible (our usual assumption), it begins to flow perpendicular to the plate with velocity, v_y. Those fluid molecules moving perpendicular to the surface are carried down the plate where they merge with faster-moving fluid. The region where the frictional force is felt is called the *boundary layer*, and in this region, the flow field is at least 2-D. As the fluid flows perpendicular to the plate and gets carried down the plate by faster-moving fluid further away, it alters the flow field in the neighborhood of the plate. At steady state, the fluid cannot accumulate, and since the mass flow rate across the plate must be constant, the boundary layer region must increase in size with distance down the plate. As the boundary layer region enlarges, the frictional force is distributed over a larger region and so the stress on the fluid decreases. Thus, the rate of momentum transfer normal to the plate correspondingly decreases. A similar growth process expands the regions of changing temperature and concentration along the plate. These regions are called the *thermal and concentration boundary layers*, and they grow along the length of the plate too. As a general rule, we can assume that all transport perpendicular to the plate occurs via diffusive mechanisms in the boundary layer. As the growth of the boundary layer spreads the driving force over a larger and larger region, the rate of this diffusive-driven transport decreases.

We know from earlier chapters that the heat and mass flux to or from the plate can be defined in terms of a heat transfer coefficient, h, and a mass transfer coefficient, k_c:

$$q'' = h(T_s - T_\infty) \qquad h,\ \mathrm{W/m^2\ K} \tag{12.1}$$

$$N_a = k_c(c_{as} - c_{a\infty}) \qquad k_c,\ \mathrm{m/s} \tag{12.2}$$

The momentum transfer is similarly characterized by the friction factor, C_f, defined by

$$\tau_{xy} = \frac{C_f}{2}\left(\rho v_s^2 - \rho v_\infty^2\right) \qquad C_f,\ \text{dimensionless} \tag{12.3}$$

Charge transfer may take two forms. If an electric field is not present, the charge moves via convective mass transfer, and Equation 12.2 is appropriate to define the ionic mass transfer coefficient. In the presence of an electric field, the flux of charge, J, is driven primarily by drift within the field, and the charge transfer coefficient, $k_\pm$, is defined in terms of a voltage driving force:

$$J = k_\pm(\Phi_s - \Phi_\infty) \qquad k_\pm,\ \mathrm{A/m^2\ V} \tag{12.4}$$

We refer to h, k_c, $k_\pm$, and C_f as *local* transport coefficients since they may vary over the surface of the plate.

To obtain the overall heat flow, mass flow, current, or force on the plate surface, we need to integrate Equations 12.1 through 12.4 over the surface:

$$q = \int_{A_s} q'' dA_s = (T_s - T_\infty) \int_{A_s} h dA_s \tag{12.5}$$

$$\dot{M}_a = \int_{A_s} N_a dA_s = (c_{as} - c_{a\infty}) \int_{A_s} k_c \, dA_s \tag{12.6}$$

$$F = \int_{A_s} \tau_{xy} dA_s = \frac{1}{2}\left(\rho v_s^2 - \rho v_\infty^2\right) \int_{A_s} C_f \, dA_s \tag{12.7}$$

$$I = \int_{A_s} J \, dA_s = (\Phi_s - \Phi_\infty) \int_{A_s} k_\pm \, dA_s \tag{12.8}$$

We can use this integration procedure to define average values for the friction factor, the heat transfer coefficient, the mass transfer coefficient, or the ionic mass transfer coefficient. These average coefficients correspond to the transport coefficient we would normally measure in the course of a simple experiment:

$$\bar{h} = \frac{1}{A_s} \int h \, dA_s \tag{12.9}$$

$$\bar{k}_c = \frac{1}{A_s} \int k_c \, dA_s \tag{12.10}$$

$$\bar{C}_f = \frac{1}{A_s} \int C_f \, dA_s \tag{12.11}$$

$$\bar{k}_\pm = \frac{1}{A_s} \int k_\pm \, dA_s \tag{12.12}$$

Equations 12.5 through 12.8 can be rewritten in terms of these average coefficients:

$$q = \bar{h} A_s (T_s - T_\infty) \tag{12.13}$$

$$\dot{M}_a = \bar{k}_c A_s (c_{as} - c_{a\infty}) \tag{12.14}$$

$$F = \frac{1}{2} \bar{C}_f A_s \left(\rho v_s^2 - \rho v_\infty^2\right) \tag{12.15}$$

$$I = \bar{k}_c A_s (\Phi_s - \Phi_\infty) \tag{12.16}$$

Example 12.1 Local versus Average Friction Factors

Experimental results on a rough surfaced, flat plate, of length, L, and width, W, show that the local friction factor, C_f, fits the relation

$$C_f(x) = \alpha x^{-1/8}$$

where

- α is a coefficient with units ($m^{1/8}$)
- x is the distance from the leading edge of the plate

We would like to determine the relationship between the average friction factor for a plate of length x and the local value.

Using the definition for the average friction factor in (12.11), we find

$$\bar{C}_f = \frac{1}{L}\int_0^L C_f \, dx$$

assuming the plate extends a uniform depth into the page. Substituting our relationship for $C_f(x)$ into the definition and evaluating the integral, we find

$$\bar{C}_f = \frac{1}{L}\int_0^L \alpha x^{-1/8} dx = \frac{8}{7}\alpha L^{-1/8} = \frac{8}{7}C_f(L)$$

12.3 BOUNDARY LAYER DEFINITIONS

In convective transport, all the action takes place in a very small region near the surface of an object or phase called the boundary layer. The hydrodynamic boundary layer, for example, is shown in Figure 12.2. Within the hydrodynamic boundary layer, viscous forces predominate and momentum is transported perpendicular to the plate by diffusion. Outside the boundary layer, inertial effects predominate and momentum is transported by bulk fluid motion alone.

The boundary layer is not a well-defined region of space. We cannot see it easily nor can we truly tell where it ends and the free steam begins. Thus, we define the thickness of the boundary layer as the distance between the plate's surface and the point where the velocity in the boundary layer is 99% of the free stream velocity:

$$\delta_h = f(x) \qquad \text{where } v_x(\delta_h) = 0.99 v_\infty \tag{12.17}$$

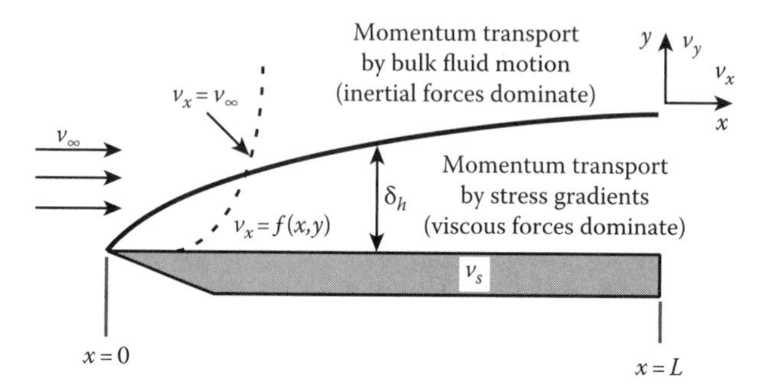

FIGURE 12.2 The hydrodynamic boundary layer on a flat plate.

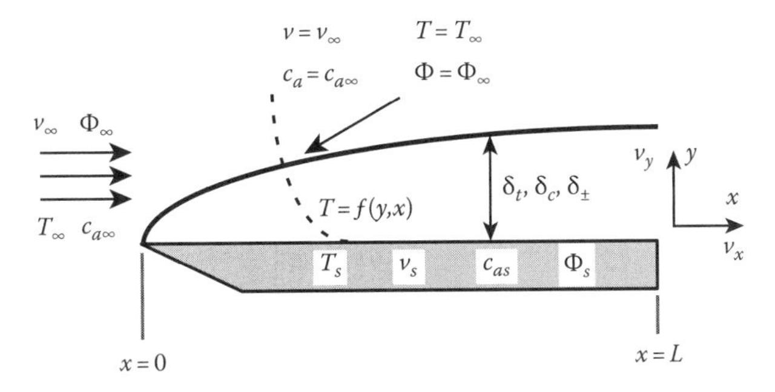

FIGURE 12.3 Thermal and concentration boundary layers developing in flow over a flat plate.

By analogy to the hydrodynamic boundary layer, we may also have thermal, charge, and concentration boundary layers (Figure 12.3). The thermal, charge, and concentration boundary layer thicknesses are defined in exactly the same way as we defined the hydrodynamic boundary layer thickness:

$$T(\delta_T) = 0.99T_\infty \tag{12.18}$$

$$c_a(\delta_c) = 0.99c_{a\infty} \tag{12.19}$$

$$\Phi(\delta_\pm) = 0.99\Phi_\infty \tag{12.20}$$

All transport within the boundary layer is dominated by diffusive processes. Therefore, we can determine the stress, heat flux, and mass or molar flux at the surface of the plate ($y = 0$) using Newton's, Fourier's, Fick's, or Ohm's laws:

$$\tau_{xy} = -\mu \left.\frac{\partial v}{\partial y}\right|_{y=0} \tag{12.21}$$

$$q'' = -k_f \left.\frac{\partial T}{\partial y}\right|_{y=0} \tag{12.22}$$

$$N_a = -D_{ab} \left.\frac{\partial c_a}{\partial y}\right|_{y=0} \tag{12.23}$$

$$J = -\sigma_a \left.\frac{\partial \Phi}{\partial y}\right|_{y=0} \qquad \sigma_a = \mu_{ea} z_{ea} \mathcal{F}_a c_a \tag{12.24}$$

Combining these relationships with those in Equations 12.1 through 12.4 provides the fundamental definitions for C_f, h, k_c, and $k_\pm$ that we will use throughout this text. One should remember that as the boundary layers grow, the surface gradients decrease (the same potential difference is spread across a thicker and thicker region). Therefore, the friction factor and other transport coefficients will start out large and steadily decrease in magnitude as one moves along the plate:

$$C_f = \frac{-2\nu \left.\frac{\partial v}{\partial y}\right|_{y=0}}{v_s^2 - v_\infty^2} \tag{12.25}$$

$$h = \frac{-k_f \left.\dfrac{\partial T}{\partial y}\right|_{y=0}}{T_s - T_\infty} \tag{12.26}$$

$$k_c = \frac{-D_a \left.\dfrac{\partial c_a}{\partial y}\right|_{y=0}}{c_{as} - c_{a\infty}} \tag{12.27}$$

$$k_\pm = \frac{-\sigma_a \left.\dfrac{\partial \Phi}{\partial y}\right|_{y=0}}{\Phi_s - \Phi_\infty} \tag{12.28}$$

All the transport coefficients depend strongly upon the flow conditions. If the plate is long enough, at some point, the character of the boundary layer changes. Initially, the boundary layer is a smooth entity that slowly increases in size. Once it reaches a certain size, a transformation abruptly occurs and the boundary layer becomes more irregular and thicker. In this region, the boundary layer has become a *turbulent* boundary layer, and it grows at a different rate from the laminar layer. The process is shown schematically in Figure 12.4.

The turbulent boundary layer is much thicker than its laminar cousin, but the gradients near the surface of the plate are much steeper due to irregular fluid motion. Information transfer via mixing of large-scale fluid pockets called eddies (large-scale fluid pockets) is more efficient than the molecular transfer of information in a laminar flow situation. One generally separates the turbulent boundary layer into three conceptual regions, the *laminar sublayer*, the *transition region*, and the *turbulent core*. These regions, like the boundary layer thickness itself, do not truly exist, but provide a convenient visualization tool. Transport within the laminar sublayer is envisioned to occur via diffusion, and since the laminar sublayer is so thin, the velocity gradient is high and momentum transport is increased. In the transition and turbulent core regions, transport is also high and efficient due to the mixing of eddies that are now known to be generated at the solid surface due to the high level of vorticity there. These steeper gradients and the mixing that occurs result in much higher rates of heat, mass, and momentum transfer. Since the driving forces across the boundary layer do not change, the transport coefficients increase in turbulent flow to reflect the increase in transport.

We must know the flow field near the solid surface to derive more detailed relationships for the transport coefficients. The Navier–Stokes equations and the species continuity and energy equations we derived in Chapter 10 enable us to attempt this problem.

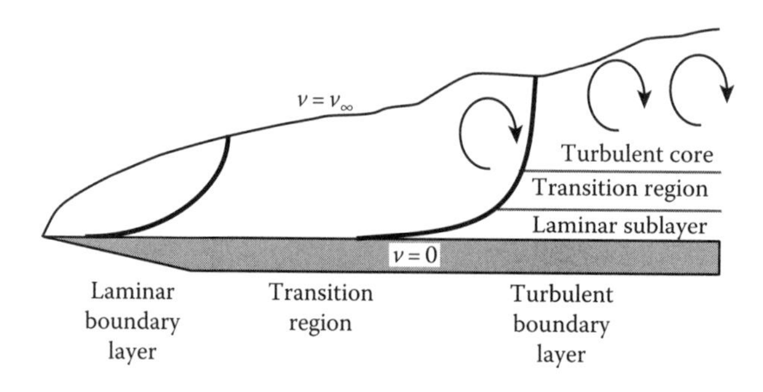

FIGURE 12.4 Laminar and turbulent boundary layers on a flat plate.

Example 12.2 Surface Heat Flux and Boundary Layer Temperature Profile

Experimental measurements of fluid flow over a heated flat plate have shown that the temperature profile within the boundary layer can be represented by the following function:

$$\frac{T - T_s}{T_\infty - T_s} = 1 - \left(0.7 + \frac{v_\infty y}{\nu}\right)^{1/7}$$

If $T_\infty = 400°C$, $T_s = 300°C$, $k_f = 0.037$ W/m K, and $v_\infty/\nu = 5500\ m^{-1}$, what is the surface heat flux and the heat transfer coefficient?

We use the definitions in Equations 12.22 and 12.26 to get this information. First, we must determine the temperature gradient at the surface:

$$\left.\frac{\partial T}{\partial y}\right|_{y=0} = -\frac{1}{7}(T_\infty - T_s)\left(\frac{v_\infty}{\nu}\right)\left(0.7 + \frac{v_\infty y}{\nu}\right)^{-6/7} = -106{,}669\ °C/m$$

The surface heat flux can now be calculated from Equation 12.22:

$$q'' = -k_f \left.\frac{\partial T}{\partial y}\right|_{y=0} = 3946.8\ W/m^2$$

The heat transfer coefficient is obtained from Equation 12.26:

$$h = \frac{-k_f \left.\frac{\partial T}{\partial y}\right|_{y=0}}{T_\infty - T_s} = 39.5\ W/m^2\,K$$

12.4 DERIVATION OF BOUNDARY LAYER EQUATIONS

The 3-D, transient forms of the momentum, energy, charge, and species continuity equations are difficult to solve, and even numerical simulations can only be applied to well-characterized systems. Well-characterized systems exist only in the laboratory and rarely represent what a practicing engineer encounters on the job. For the practicing engineer, a numerical solution to a particular problem may not be necessary or cost effective. Simplifying the conservation equations and determining the relevant physical parameters controlling the system is often a more useful approach and one that will be illustrated later. Such a solution also provides the intuition necessary to judge the validity of a numerical solution.

Our first simplification is to assume steady-state operation. Since steady-state boundary layers have had time to develop to their full potential, we expect them to be thickest and hence the transport coefficients to be smallest. Restricting our analysis to the steady state will provide us with a lower bound for the transport coefficients. The second simplification is to assume incompressible flow. This may be a drastic simplification for gases, but even gases can be approximated as incompressible first (for low Mach number), and corrections applied later.

Most material properties are weak functions of pressure, temperature, and composition, and so we will assume constant properties throughout the system. Engineering transport situations occur often in gases and liquids that behave as Newtonian fluids (air and water to be specific), so we will assume Newtonian behavior in this analysis. Corrections for non-Newtonian behavior can be applied at a later date. Finally, the boundary layer is essentially a 2-D beast. Adding a third dimension will lead to a more accurate result but will vastly complicate the analysis. In most instances, the added accuracy is far outweighed by the amount of work required to achieve it. That can be left for computer analysis. We will restrict our analysis to 2-D systems. Summarizing

1. Steady state operation – $\partial/\partial t = 0$
2. Incompressible flow – $\rho = \text{constant}$
3. Constant properties – $k_f, \mu, D_{ab}, \rho, C_p, \mu_e$
4. Newtonian fluid – $\tau_{xy} = -\mu \dfrac{\partial v_y}{\partial x} \quad \tau_{yx} = -\mu \dfrac{\partial v_x}{\partial y}$
5. 2-D flow – $\partial/\partial z = 0 \quad v_z = 0$

and applying these simplifications to the conservation equations lead to the following set of equations:

Continuity

$$\frac{\partial v_x}{\partial x} + \frac{\partial v_y}{\partial y} = 0 \tag{12.29}$$

Momentum

$$v_x \frac{\partial v_x}{\partial x} + v_y \frac{\partial v_x}{\partial y} = -\frac{1}{\rho}\frac{\partial P}{\partial x} + \nu\left(\frac{\partial^2 v_x}{\partial x^2} + \frac{\partial^2 v_x}{\partial y^2}\right) \tag{12.30}$$

$$v_x \frac{\partial v_y}{\partial x} + v_y \frac{\partial v_y}{\partial y} = -\frac{1}{\rho}\frac{\partial P}{\partial y} + \nu\left(\frac{\partial^2 v_y}{\partial x^2} + \frac{\partial^2 v_y}{\partial y^2}\right) \tag{12.31}$$

Energy

$$v_x \frac{\partial T}{\partial x} + v_y \frac{\partial T}{\partial y} = \alpha\left(\frac{\partial^2 T}{\partial x^2} + \frac{\partial^2 T}{\partial y^2}\right) + \frac{\dot{q}}{\rho C_p} \tag{12.32}$$

Continuity of a

$$v_x \frac{\partial c_a}{\partial x} + v_y \frac{\partial c_a}{\partial y} = D_{ab}\left(\frac{\partial^2 c_a}{\partial x^2} + \frac{\partial^2 c_a}{\partial y^2}\right) + R_a \tag{12.33}$$

Charge continuity of a

$$v_x \frac{\partial c_a}{\partial x} + v_y \frac{\partial c_a}{\partial y} = D_{ab}\left(\frac{\partial^2 c_a}{\partial x^2} + \frac{\partial^2 c_a}{\partial y^2}\right) + \mu_{ea}\left[\frac{\partial}{\partial x}\left(c_a \frac{\partial \Phi}{\partial x}\right) + \frac{\partial}{\partial y}\left(c_a \frac{\partial \Phi}{\partial y}\right)\right] + R_{ea} \tag{12.34}$$

Unfortunately, even with the simplifications we made so far, these equations are still too difficult to solve analytically. We must find additional simplifications and eliminate those terms that contribute very little to the dynamics of the process. Simplification is facilitated by putting the equations in dimensionless form using the following variable substitutions:

$$x^* = \frac{x}{L} \quad y^* = \frac{y}{L} \quad L\text{, length scale (plate length)}$$

$$u^* = \frac{v_x}{v_\infty} \qquad v^* = \frac{v_y}{v_\infty}$$ v_∞, free stream velocity

$$T^* = \frac{T - T_s}{T_\infty - T_s}$$ T_s, wall temperature T_∞, free stream temperature

$$c^* = \frac{c_a - c_{as}}{c_{a\infty} - c_{as}}$$ c_{as}, wall concentration $c_{a\infty}$, free stream concentration

$$\Phi^* = \frac{\Phi - \Phi_s}{\Phi_\infty - \Phi_s}$$ Φ_s, surface voltage Φ_∞, free stream voltage

$$P^* = \frac{P}{\rho v_\infty^2}$$ ρv_∞^2, kinetic energy per unit volume

There is no natural scaling factor for the pressure, so we use the kinetic energy per unit volume of the free stream fluid instead. The transformed equations become

Continuity

$$\frac{\partial u^*}{\partial x^*} + \frac{\partial v^*}{\partial y^*} = 0 \tag{12.35}$$

Momentum

$$u^* \frac{\partial u^*}{\partial x^*} + v^* \frac{\partial u^*}{\partial y^*} = -\frac{\partial P^*}{\partial x^*} + \frac{1}{\mathrm{Re}_\infty}\left(\frac{\partial^2 u^*}{\partial x^{*2}} + \frac{\partial^2 u^*}{\partial y^{*2}}\right) \tag{12.36}$$

$$u^* \frac{\partial v^*}{\partial x^*} + v^* \frac{\partial v^*}{\partial y^*} = -\frac{\partial P^*}{\partial y^*} + \frac{1}{\mathrm{Re}_\infty}\left(\frac{\partial^2 v^*}{\partial x^{*2}} + \frac{\partial^2 v^*}{\partial y^{*2}}\right) \tag{12.37}$$

Energy

$$u^* \frac{\partial T^*}{\partial x^*} + v^* \frac{\partial T^*}{\partial y^*} = \frac{1}{\mathrm{Re}_\infty \mathrm{Pr}}\left(\frac{\partial^2 T^*}{\partial x^{*2}} + \frac{\partial^2 T^*}{\partial y^{*2}}\right) + \frac{\dot{q}L}{\rho C_p v_\infty (T_\infty - T_s)} \tag{12.38}$$

Continuity of a

$$u^* \frac{\partial c^*}{\partial x^*} + v^* \frac{\partial c^*}{\partial y^*} = \frac{1}{\mathrm{Re}_\infty \mathrm{Sc}}\left(\frac{\partial^2 c^*}{\partial x^{*2}} + \frac{\partial^2 c^*}{\partial y^{*2}}\right) + \frac{L}{v_\infty} R_a \tag{12.39}$$

Charge continuity of a

$$u^* \frac{\partial c^*}{\partial x^*} + v^* \frac{\partial c^*}{\partial y^*} = \frac{1}{\mathrm{Re}_\infty \mathrm{Sc}}\left(\frac{\partial^2 c^*}{\partial x^{*2}} + \frac{\partial^2 c^*}{\partial y^{*2}}\right) + \frac{1}{\mathrm{Re}_\infty E_f}\left[\frac{\partial}{\partial x^*}\left(c^* \frac{\partial \Phi^*}{\partial x^*}\right) + \frac{\partial}{\partial y^*}\left(c^* \frac{\partial \Phi^*}{\partial y^*}\right)\right] + \frac{L}{v_\infty} R_{ea} \tag{12.40}$$

Several dimensionless groups appear naturally in these equations. These groupings are important because they show how the key variables of the system interact to govern overall behavior. The Reynolds number, Re_∞, describes the characteristics of the flow field. It represents a ratio of inertial to viscous forces in the fluid. As we increase the free stream velocity or the length scale of the system, the inertial forces increase. If the inertial forces become high enough, the flow will transition from laminar to turbulent:

$$\mathrm{Re}_\infty = \frac{v_\infty L}{\nu} \text{ – Reynolds number – } \frac{\text{Inertial forces}}{\text{Viscous forces}} \tag{12.41}$$

The Prandtl number, Pr, is a measure of how fast thermal energy (heat) is transported through the fluid versus how fast momentum is transported. We will see that it also plays an important role in determining the relative thicknesses of the hydrodynamic and thermal boundary layers:

$$\mathrm{Pr} = \frac{\nu}{\alpha} \text{ – Prandtl number – } \frac{\text{Momentum diffusivity}}{\text{Thermal diffusivity}} \tag{12.42}$$

The Schmidt number, Sc, is the mass transfer analog of the Prandtl number and describes the ratio of momentum to mass diffusivity. It plays an important role in determining the relative thicknesses of the hydrodynamic and concentration boundary layers:

$$\mathrm{Sc} = \frac{\nu}{D_{ab}} \text{ – Schmidt number – } \frac{\text{Momentum diffusivity}}{\text{Mass diffusivity}} \tag{12.43}$$

The unnamed quantity hereafter referred to as the drift number, E_f, is a measure of how fast momentum is transported through the fluid by diffusion versus how fast charge can be transported via drift in the electric field. μ_{ea} is the mobility of the charge carrier within the electric field:

$$E_f = \frac{\nu}{\mu_{ea}(\Phi_\infty - \Phi_s)} \text{ – Drift number – } \frac{\text{Momentum diffusivity}}{\text{Drift diffusivity}} \tag{12.44}$$

The Reynolds, Prandtl, and Schmidt numbers, and to a lesser extent the Drift number, are the primary dimensionless quantities that concern us. If we have generation present in the system, we also have two other dimensionless groups that may arise. The first we may call a thermal generation number. It represents a ratio of heat flux present in the system due to heat generation to the heat flux due to convective transport. The generation number will vary in form depending on the exact form for $\dot{q}$:

$$G_n = \frac{\dot{q}L}{\rho C_p v_\infty (T_\infty - T_s)} - \frac{\text{Generated heat flux}}{\text{Convective heat flux}} \tag{12.45}$$

The final dimensionless group is the Damkohler number, Da. It is a measure of the characteristic chemical reaction time to the characteristic convection time. For a first-order chemical reaction (the type we will consider here), the Damkohler is defined by

$$\mathrm{Da} = \frac{k'' L}{v_\infty} - \frac{\text{Convection time}}{\text{Reaction time}} \tag{12.46}$$

All solutions to the conservation equations will depend upon these dimensionless groups. Even if we had no analytical solutions, we could still develop experimental relationships to describe convective heat, mass, and momentum transport coefficients because they exhibit a functional dependence on various combinations of these dimensionless groups.

The dimensionless groups we derived have yielded some insight into the problem, but we have yet to determine which terms in Equations 12.35 through 12.40 are important. We can use the fact that the boundary layer is very thin to obtain an order-of-magnitude estimate of the size of each term in the equations. There are two methods for deriving this estimate. We can use a formal perturbation analysis or we can use an approximate *order-of-magnitude* analysis. The perturbation analysis is more rigorous and is presented in advanced texts on boundary layer theory [1]. The order-of-magnitude analysis has been used historically and is a bit easier to follow [2].

We start the order-of-magnitude analysis with a few assumptions:

1. The boundary layer thickness is very small – $O(\delta_h)$ in size. All length and velocity scales in the direction perpendicular to the surface are this order of magnitude in size:

$$y^* \approx dy^* \approx v^* \approx dv^* \approx O(\delta_h)$$

2. The velocity in the x-direction (along the plate) is much greater than the velocity in the y-direction (perpendicular to the plate):

$$u^* \gg v^*$$

3. All length and velocity scales in a direction parallel to the surface will be large $\sim O(1)$ in size. The pressure will also be of this size:

$$u^* \approx du^* \approx x^* \approx dx^* \approx P^* \approx O(1)$$

Referring to the assumptions previously, we begin the analysis with the continuity equation. We judge the order of magnitude of the numerator and denominator in each partial derivative:

Continuity

$$\begin{array}{ccc} \dfrac{\partial u^*}{\partial x^*} & + \dfrac{\partial v^*}{\partial y^*} & = 0 \\ \dfrac{1}{1} & \dfrac{\delta_h}{\delta_h} & \end{array} \tag{12.47}$$

All terms in the continuity equation are of equal overall magnitude so all are important to keep. Applying our analysis to the momentum equations gives

Momentum

$$\begin{array}{cccccc} u^*\dfrac{\partial u^*}{\partial x^*} & +\, v^*\dfrac{\partial u^*}{\partial y^*} & = -\dfrac{\partial P^*}{\partial x^*} & + \dfrac{1}{\mathrm{Re}_\infty}\Bigg(& \dfrac{\partial^2 u^*}{\partial x^{*2}} & + \dfrac{\partial^2 u^*}{\partial y^{*2}}\Bigg) \\ 1\dfrac{1}{1} & \delta_h \dfrac{1}{\delta_h} & \dfrac{1}{1} & \delta_h^2 & \dfrac{1}{1} & \dfrac{1}{\delta_h^2} \end{array} \tag{12.48}$$

All terms in the x-momentum equation are of $O(1)$ in size with the exception of the second derivative of u^* with respect to x^*. This term is small by comparison with the second derivative of u^* with respect to y^* and so will be neglected. Most of the change in u^* will be occurring in a direction perpendicular to the plate. (u^* varies from 0 at the surface to 1 at the edge of the boundary layer. By contrast, u^* is essentially constant along the plate.)

Our analysis shows that in order for the remaining viscous term $\partial^2 u^*/\partial y^{*2}$ to be of comparable size to the inertial terms in Equation 12.48, the Reynolds number must be on the order of $1/\delta_h^2$ in size. This result is important because it agrees with experimental evidence that indicates as the Reynolds number increases, the boundary layer thickness decreases. Turning our attention to the y-momentum equation, we find that

$$u^*\frac{\partial v^*}{\partial x^*} + v^*\frac{\partial v^*}{\partial y^*} = -\frac{\partial P^*}{\partial y^*} + \frac{1}{\mathrm{Re}_\infty}\left(\frac{\partial^2 v^*}{\partial x^{*2}} + \frac{\partial^2 v^*}{\partial y^{*2}}\right) \tag{12.49}$$

$$1 \quad \frac{\delta_h}{1} \qquad \delta_h\frac{\delta_h}{\delta_h} \qquad ? \qquad \delta_h^2 \quad \frac{\delta_h}{1} \quad \frac{\delta_h}{\delta_h^2}$$

All terms are of $O(\delta_h)$ in size with the exception of the change in v^* with x^*, $\partial^2 v^*/\partial x^{*2}$. This term is much smaller than the corresponding change in v^* with y^*, so we will neglect it for the same reason we neglected the change in u^* with respect to x^* in Equation 12.48. The only remaining term we are unsure of is the change in pressure. We can calculate the change in pressure over the boundary layer by an integral of the form

$$\Delta P \approx \int_0^{\delta_h/L} \frac{\partial P^*}{\partial y^*} dy^* \tag{12.50}$$

If we assume that the pressure drop across the boundary layer is due solely to the weight of the fluid there, then we can say

$$\Delta P \approx \rho g \delta_h \tag{12.51}$$

and like other terms in Equation 12.49, it is of order δ_h in size. Since the whole of the y-momentum equation is only of order δ_h in size, we will neglect it with respect to the x-momentum equation. This is not a bad approximation since $v_y = 0$ at $y = 0$ and also at $y = \delta_h$. Thus, there is a limit on how large v_y can be.

One consequence of neglecting the y-momentum equation is never being able to solve for the pressure distribution in the fluid. We generally use the continuity equation in conjunction with the momentum equations to solve for the pressure and velocity fields, but when we neglect one momentum equation, we must use the continuity equation to obtain the other velocity component. This means we will either have to prescribe the pressure, $P(x)$, or we will have to assume that it is constant over the system of study. A constant value for the pressure implies that v_∞ is constant in the free stream.

Attempting to determine the relative magnitudes of terms in the energy equation, we find

Energy

$$u^*\frac{\partial T^*}{\partial x^*} + v^*\frac{\partial T^*}{\partial y^*} = \frac{1}{\mathrm{Re}_\infty \mathrm{Pr}}\left(\frac{\partial^2 T^*}{\partial x^{*2}} + \frac{\partial^2 T^*}{\partial y^{*2}}\right) \tag{12.52}$$

$$1\frac{1}{1} \qquad \delta_T\frac{1}{\delta_T} \qquad \delta_T^2 \qquad \frac{1}{1} \qquad \frac{1}{\delta_T^2}$$

Again, all the terms in this equation are large when compared to the second derivative of T^* with respect to x^*. Therefore, we will neglect this term and assume that the major changes in T^* will occur perpendicular to the plate. The $\mathrm{Re}_\infty\mathrm{Pr}$ combination is of order $1/\delta_T^2$ indicating that the *Prandtl* number is generally much smaller than the Reynolds number. We neglect the generation term since its magnitude depends upon many unknown factors.

The species continuity equation follows directly from the energy equation analysis:

Continuity of a

$$u^*\frac{\partial c^*}{\partial x^*}+v^*\frac{\partial c^*}{\partial y^*}=\frac{1}{\mathrm{Re}_\infty\mathrm{Sc}}\left(\frac{\partial^2 c^*}{\partial x^{*2}}+\frac{\partial^2 c^*}{\partial y^{*2}}\right) \tag{12.53}$$

$$1\frac{1}{1}\qquad \delta_c\frac{1}{\delta_c}\qquad \delta_c^2\qquad \frac{1}{1}\qquad \frac{1}{\delta_c^2}$$

We neglect the second derivative of c^* with respect to x^* since changes in that direction are small. Like the $\mathrm{Re}_\infty\mathrm{Pr}$ combination, the $\mathrm{Re}_\infty\mathrm{Sc}$ combination also shows that the *Schmidt* number is generally much smaller than the Reynolds number. We leave the generation term until we know the reaction kinetics.

The charge continuity equation is analyzed along the same lines since it is related to the mass transfer equation:

Charge continuity of a

$$u^*\frac{\partial c^*}{\partial x^*}+v^*\frac{\partial c^*}{\partial y^*}=\frac{1}{\mathrm{Re}_\infty\mathrm{Sc}}\left(\frac{\partial^2 c^*}{\partial x^{*2}}+\frac{\partial^2 c^*}{\partial y^{*2}}\right)+\frac{1}{\mathrm{Re}_\infty E_f}\left[\frac{\partial}{\partial x^*}\left(c^*\frac{\partial \Phi^*}{\partial x^*}\right)+\frac{\partial}{\partial y^*}\left(c^*\frac{\partial \Phi^*}{\partial y^*}\right)\right] \tag{12.54}$$

$$1\frac{1}{1}\qquad \delta_\pm\frac{1}{\delta_\pm}\qquad \delta_\pm^2\qquad \frac{1}{1}\qquad \frac{1}{\delta_\pm^2}\qquad \delta_\pm^2\qquad \frac{1}{1}\qquad \frac{1}{\delta_\pm^2}$$

Within the drift term and the diffusion term, the variation perpendicular to the plate is much higher than that parallel to the plate. Thus, we neglect changes with respect to x^* in both terms. The generation terms require further information to evaluate.

The simplifications we have made lead to the *boundary layer equations*:

Continuity

$$\frac{\partial u^*}{\partial x^*}+\frac{\partial v^*}{\partial y^*}=0 \tag{12.55}$$

Momentum

$$u^*\frac{\partial u^*}{\partial x^*}+v^*\frac{\partial u^*}{\partial y^*}=-\frac{\partial P^*}{\partial x^*}+\frac{1}{\mathrm{Re}_\infty}\left(\frac{\partial^2 u^*}{\partial y^{*2}}\right) \tag{12.56}$$

Energy

$$u^*\frac{\partial T^*}{\partial x^*}+v^*\frac{\partial T^*}{\partial y^*}=\frac{1}{\mathrm{Re}_\infty\mathrm{Pr}}\left(\frac{\partial^2 T^*}{\partial y^{*2}}\right) \tag{12.57}$$

Continuity of a

$$u^*\frac{\partial c^*}{\partial x^*}+v^*\frac{\partial c^*}{\partial y^*}=\frac{1}{\text{Re}_\infty\text{Sc}}\left(\frac{\partial^2 c^*}{\partial y^{*2}}\right) \tag{12.58}$$

Charge continuity of a

$$u^*\frac{\partial c^*}{\partial x^*}+v^*\frac{\partial c^*}{\partial y^*}=\frac{1}{\text{Re}_\infty\text{Sc}}\left(\frac{\partial^2 c^*}{\partial y^{*2}}\right)+\frac{1}{\text{Re}_\infty E_f}\left[\frac{\partial}{\partial y^*}\left(c^*\frac{\partial \Phi^*}{\partial y^*}\right)\right] \tag{12.59}$$

The dimensionless form of the boundary layer equations tells us that the hydrodynamic boundary layer thickness will be a function of its primary dimensionless group, the *Reynolds* number. The thermal boundary layer thickness will be a function of the *Reynolds* number and the *Prandtl* number. The concentration boundary layer thickness will be a function of the *Reynolds* number and the *Schmidt* number. The charge boundary layer will generally be a function of the *Reynolds, Schmidt, and Drift* numbers.

The analysis of the transport equations in terms of their dimensionless form can be extended to include the friction factor, the heat transfer coefficient, and the mass transfer coefficient. We use the dimensionless variables we defined earlier to redefine the gradients in the definition of the transport coefficients:

$$C_f=-\frac{2}{\text{Re}_\infty}\frac{\partial u^*}{\partial y^*}\bigg|_{y^*=0} \tag{12.60}$$

$$h=-\frac{k_f}{L}\frac{\partial T^*}{\partial y^*}\bigg|_{y^*=0} \tag{12.61}$$

$$k_c=-\frac{D_{ab}}{L}\frac{\partial c^*}{\partial y^*}\bigg|_{y^*=0} \tag{12.62}$$

The charge transfer coefficient is a bit more complicated. If charge motion is dominated by diffusion alone, we can use a mass transfer coefficient like the one defined in Equation 12.62. If charge motion is governed primarily by drift in the electric field, we define the ionic mass transfer coefficient of the solution using

$$k_\pm=-\frac{\sigma_a}{L}\frac{\partial \Phi^*}{\partial y^*}\bigg|_{y^*=0} \tag{12.63}$$

Equations 12.60 through 12.63 can be used to define four dimensionless transport coefficients: the friction number for momentum transfer—N_f, the *Nusselt* number for heat transfer—Nu, the *Sherwood* number for mass transfer—Sh, and the conduction number for charge transfer (by drift)—$N_\pm$:

$$N_f=\frac{C_f\text{Re}_\infty}{2}=-\frac{\partial u^*}{\partial y^*}\bigg|_{y^*=0}-f(x,\text{Re}_\infty) \tag{12.64}$$

$$\mathrm{Nu} = \frac{hL}{k} = -\left.\frac{\partial T^*}{\partial y^*}\right|_{y^*=0} - f(x, \mathrm{Re}_\infty, \mathrm{Pr}) \tag{12.65}$$

$$\mathrm{Sh} = \frac{k_c L}{D_{ab}} = -\left.\frac{\partial c^*}{\partial y^*}\right|_{y^*=0} - f(x, \mathrm{Re}_\infty, \mathrm{Sc}) \tag{12.66}$$

$$\mathrm{N}_\pm = \frac{k_\pm L}{\sigma_a} = -\left.\frac{\partial \Phi^*}{\partial y^*}\right|_{y^*=0} - f(x, \mathrm{Re}_\infty, \mathrm{Sc}, E_f) \tag{12.67}$$

Putting the definitions for the friction factor, heat transfer coefficient, and mass transfer coefficient in dimensionless form has given us more engineering *intuition* about the transport processes. If we had no analytical solutions for these coefficients, we could still develop experimental relationships (correlations) for the convective transport coefficients that would be a function of the dimensionless groups shown in Equations 12.64 through 12.67:

$$C_f = K(\mathrm{Re}_\infty)^a \tag{12.68}$$

$$h = K(\mathrm{Re}_\infty)^a(\mathrm{Pr})^b \tag{12.69}$$

$$k_c = K(\mathrm{Re}_\infty)^a(\mathrm{Sc})^b \tag{12.70}$$

$$k_\pm = K(\mathrm{Re}_\infty)^a(\mathrm{Sc})^b(E_f)^c \tag{12.71}$$

The dimensionless groups in Equations 12.64 through 12.67 represent dimensionless gradients at the wall surface. The friction number is the dimensionless velocity gradient, the Nusselt number is the dimensionless temperature gradient, the Sherwood number is the dimensionless concentration gradient, and the conduction number is the dimensionless voltage gradient. If these gradients were related to one another, we could draw analogies between the different dimensionless numbers and, hence, the transport coefficients.

12.5 TRANSPORT ANALOGIES

In most transport processes, we are interested in computing four quantities:

$\mathrm{N_f}$	–	Friction number	–	To determine stresses, pressure drops, and power requirements for pumps or compressors
Nu	–	Nusselt number	–	To determine the heat transfer coefficient and heat transferred between the surface and the fluid
Sh	–	Sherwood number	–	To determine the mass transfer coefficient and mass transferred between phases
$\mathrm{N}_\pm$	–	Conduction number	–	To determine the charge transfer coefficient and the charge transferred between phases

Often these quantities are related through the structure of their boundary layers.

12.5.1 Reynolds Analogy

Let us look again at the momentum, energy, and species continuity boundary layer equations. For simplicity, we will assume no generation terms are present:

Continuity

$$\frac{\partial u^*}{\partial x^*}+\frac{\partial v^*}{\partial y^*}=0 \tag{12.72}$$

Momentum

$$u^*\frac{\partial u^*}{\partial x^*}+v^*\frac{\partial u^*}{\partial y^*}=\frac{1}{\mathrm{Re}_\infty}\left(\frac{\partial^2 u^*}{\partial y^{*2}}\right) \tag{12.73}$$

Energy

$$u^*\frac{\partial T^*}{\partial x^*}+v^*\frac{\partial T^*}{\partial y^*}=\frac{1}{\mathrm{Re}_\infty \,\mathrm{Pr}}\left(\frac{\partial^2 T^*}{\partial y^{*2}}\right) \tag{12.74}$$

Continuity of a

$$u^*\frac{\partial c^*}{\partial x^*}+v^*\frac{\partial c^*}{\partial y^*}=\frac{1}{\mathrm{Re}_\infty \mathrm{Sc}}\left(\frac{\partial^2 c^*}{\partial y^{*2}}\right) \tag{12.75}$$

Charge continuity of a

$$u^*\frac{\partial c^*}{\partial x^*}+v^*\frac{\partial c^*}{\partial y^*}=\frac{1}{\mathrm{Re}_\infty \mathrm{Sc}}\left(\frac{\partial^2 c^*}{\partial y^{*2}}\right)+\frac{1}{\mathrm{Re}_\infty E_f}\left[\left(c^*\frac{\partial \Phi^*}{\partial y^*}\right)\right] \tag{12.76}$$

Notice that if the Prandtl number and Schmidt number are nearly 1, Equations 12.73 through 12.75 are exactly analogous in form. If we have no applied electric field, Equation 12.76 would fit into this analogy as well. Under these special conditions, the dimensionless velocity, temperature, and concentration gradients at the wall will be identical:

$$\frac{\partial u^*}{\partial y^*}=\frac{\partial T^*}{\partial y^*}=\frac{\partial c^*}{\partial y^*} \tag{12.77}$$

With the gradients being identical, the dimensionless transport coefficients should also be identical. This is the heart of the *Reynolds analogy* [3]. It enables us to calculate one transport coefficient by knowing another. Generally, the transport coefficient we know is the friction factor and we want to calculate the heat and mass transfer coefficients:

$$\mathrm{N_f}=\mathrm{Nu}=\mathrm{Sh} \quad \text{Reynolds analogy} \tag{12.78}$$

The presence of an electric field makes the situation much more complicated. Not only must we have $\mathrm{Sc}\approx 1$ but we must also have $E_f \approx 1$ and the special situation of

$$\frac{\partial}{\partial y^*}\left[c^*\frac{\partial \Phi^*}{\partial y^*}\right]\approx\frac{\partial^2 c^*}{\partial y^{*2}} \tag{12.79}$$

Under these conditions, we can extend the Reynolds analogy to include the conduction number.

12.5.2 Chilton–Colburn Analogy and Others

The Reynolds analogy holds over a rather narrow range and is most useful for gases and certain liquids (water or ethanol) where the Prandtl and Schmidt numbers are close to 1. However, the power of the analogy is so great that investigators have long sought for ways to extend its applicability. One of the most successful extensions is the Chilton–Colburn analogy developed in 1933 [4,5]. This is a semiempirical extension derived from a solution to the boundary layer equations. It increases the applicability of the Reynolds analogy to Prandtl and Schmidt numbers between 0.6 and 60:

$$N_f = \mathrm{NuPr}^{-1/3} = \mathrm{ShSc}^{-1/3} \quad \text{Chilton – Colburn analogy} \tag{12.80}$$

Other, more complicated analogies have been developed as boundary layer theory has progressed. They are also primarily empirical and can be quite accurate over a wide range of Prandtl and Schmidt numbers. One of the most accurate is the Friend–Metzner analogy [6] that takes the form

$$\mathrm{Nu} = N_f\left[\frac{\mathrm{Pr}(\mu_\infty/\mu_s)^{0.14}}{1.2+(11.8)(\mathrm{Pr}-1)\mathrm{Pr}^{-1/3}\sqrt{C_f/2}}\right] \tag{12.81}$$

for heat transfer and

$$\mathrm{Sh} = N_f\left[\frac{\mathrm{Sc}}{1.2+(11.8)(\mathrm{Sc}-1)\mathrm{Sc}^{-1/3}\sqrt{C_f/2}}\right] \tag{12.82}$$

for mass transfer.

It is important to remember that no analogy is very good at high heat transfer or high mass transfer rates. In these instances, the boundary layers become distorted because they interact with one another. The temperature and concentration gradients alter the physical properties of the fluid to such a degree that the self-similar relationship between the hydrodynamic and the other boundary layers breaks down, destroying the usefulness of an analogy.

Example 12.3 Chilton–Colburn Analogy

Experiments conducted using water flowing over a stationary flat plate have shown that for a plate 0.2 m × 2 m, the force required to flow the fluid past the plate is 20 N. If the free stream temperature and velocity are 25°C and 1.0 m/s, respectively, and the kinematic viscosity and Schmidt number are 4.3 × 10−6 m²/s and 50, respectively, what are the fiction factor, C_f, and mass transfer coefficient, k_c? The density of water at 25°C is 1000 kg/m³.

We need to use Equation 12.15 relating the friction factor to the force exerted on the plate to determine the friction factor:

$$\overline{C_f} = -\frac{F}{\frac{1}{2}A_s\left(\rho v_s^2 - \rho v_\infty^2\right)}$$

The minus sign assures that C_f will be positive. The surface area of the plate exposed to the fluid is

$$A_s = 2(0.2)(2) = 0.8 \text{ m}^2$$

and includes both sides of the plate. The friction factor can be found directly from the definition:

$$\overline{C_f} = \frac{-20}{-\frac{1}{2}(0.8)(1000)(1)^2} = 0.05$$

To determine the mass transfer coefficient, we use the Chilton–Colburn analogy:

$$N_f = \frac{\overline{C_f}\text{Re}_\infty}{2} = \text{ShSc}^{-1/3}$$

Solving for the mass transfer coefficient, we have

$$\overline{k_c} = \frac{\overline{C_f}\text{Re}_\infty\text{Sc}^{1/3}}{2}\left(\frac{D_{ab}}{L}\right)$$

The diffusivity we obtain from the Schmidt number and the kinematic viscosity. Using a plate length of 2 m, the mass transfer coefficient is

$$\overline{k_c} = \frac{(0.05)\frac{(1.0)(2)}{(4.3\times10^{-6})}(50)^{1/3}}{2}\left(\frac{4.3\times10^{-6}}{2(50)}\right) = 1.84\times10^{-3}\ \frac{\text{m}}{\text{s}}$$

Out of curiosity, we can also use the Friend–Metzner correlation:

$$\overline{k_c} = \frac{\overline{C_f}\text{Re}_\infty D_{ab}}{2L}\left[\frac{\text{Sc}}{1.2+(11.8)(\text{Sc}-1)\text{Sc}^{-1/3}\sqrt{\frac{\overline{C_f}}{2}}}\right]$$

$$\overline{k_c} = \frac{(0.05)(4.65\times10^5)(4.3\times10^{-6}/50)}{(2)2}\times\left[\frac{50}{1.2+(11.8)(50-1)(50)^{-1/3}\sqrt{\frac{0.05}{2}}}\right] = 9.61\times10^{-4}\ \frac{\text{m}}{\text{s}}$$

Substituting our values for the friction factor, Reynolds number, and Schmidt number, we find that both quantities differ by about a factor of 2. In general, the Friend–Metzner correlation is more accurate, and so its value would be more appropriate to use in a design.

12.6 HYDRODYNAMIC BOUNDARY LAYERS

We now begin the process of obtaining analytical solutions to the boundary layer equations and calculating boundary layer thicknesses, heat transfer coefficients, friction factors, and mass transfer coefficients. We may again wonder why we study the flat plate since the geometry is so simplified. One important reason is because we can obtain an analytical solution for our transport coefficients. The expressions we obtain for these coefficients are likely to reappear slightly altered, in the analysis of flow past other geometries. The second important reason for discussing the flat plate lies in the geometrical relationship between the boundary layer thickness and the other physical scales of the problem. The boundary layer thickness is exceedingly small in most cases, much smaller than the dimensions of the plate. In this regard, many geometries can be approximated as a flat plate: hypersonic transport wings, flow over a fin, air flow across the heated wall of a house or window, and flow over curved objects whose curvature is small. Even flow near the wall of a pipe or tube can be approximated first as a flat plate. As an engineer who may one day need a very quick, order-of-magnitude analysis to meet a pressing problem, the flat plate correlations we will develop will be your first tool.

12.6.1 Forced Convection, Laminar Flow

Our first target will be the hydrodynamic boundary layer since hydrodynamics will influence all the other transport operations. We consider the flat plate of Figure 12.5. Fluid flows in the free stream at a constant velocity, v_∞, implying that we have no pressure force acting to accelerate the fluid across the surface of the plate ($dP/dx \sim 0$). The surface of the plate remains stationary, $v_s = 0$, and the plate will be insoluble in the liquid. The hydrodynamic boundary layer equations in their original dimensional form are

Continuity

$$\frac{\partial u}{\partial x}+\frac{\partial v}{\partial y}=0 \tag{12.83}$$

Momentum

$$u\frac{\partial u}{\partial x}+v\frac{\partial u}{\partial y}=\nu\left(\frac{\partial^2 u}{\partial y^2}\right) \tag{12.84}$$

We plan to solve these equations by a partial numerical technique called an integral analysis. In this technique, we will integrate the boundary layer equations over the extent of the boundary layer. A key feature of this analysis is the approximation that the velocity profile, $u(x,y)$, is self-similar in x. No matter where we slice through the flat plate, the shape of the boundary layer velocity profile will be relatively the same. It will only be stretched to cover the entire boundary layer. The integration procedure and our similarity assumption will transform the partial differential equations for u and v into a single ordinary differential equation for the boundary layer thickness, δ_h. A final determination of δ_h and C_f will hinge on our determination of $u(y)$. We will develop a polynomial approximation for u whose constants are determined using the boundary conditions for the problem.

The boundary conditions for flow over a flat plate are located at the surface of the plate and at the edge of the boundary layer. At the surface, we know that the velocities, u and v, must be zero because the plate is stationary and the plate material is insoluble:

$$y=0 \quad u=0 \quad v=0 \tag{12.85}$$

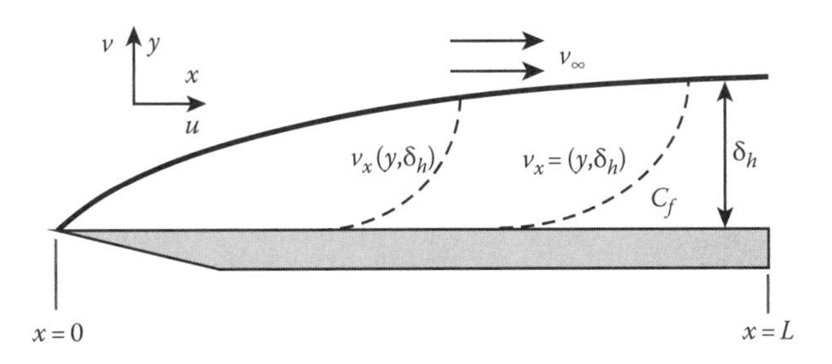

FIGURE 12.5 The hydrodynamic boundary layer on a flat plate.

Evaluating the x-momentum equation at the surface in light of Equation 12.85 shows that

$$y = 0 \quad \frac{\partial^2 u}{\partial y^2} = 0 \tag{12.86}$$

Since both u and v are at $y = 0$ there, the whole left-hand side of Equation 12.84 vanishes. At the edge of the boundary layer, the fluid velocity must equal the free stream velocity:

$$y = \delta_h \qquad u = v_\infty \tag{12.87}$$

The boundary layer velocity and free stream velocity must merge seamlessly since the boundary layer is a *ghost* entity. This requires that the slope of the velocity profile at the edge of the boundary layer vanish:

$$y = \delta_h \quad \frac{\partial u}{\partial y} = 0 \tag{12.88}$$

These four boundary conditions let us represent u as a cubic function of y/δ_h:

$$u = v_\infty f\left(\frac{y}{\delta_h}\right) = v_\infty\left[a_o + a_1\left(\frac{y}{\delta_h}\right) + a_2\left(\frac{y}{\delta_h}\right)^2 + a_3\left(\frac{y}{\delta_h}\right)^3\right] \tag{12.89}$$

We now develop an approximate solution to Equations 12.83 and 12.84 by setting up an integral equation in u with δ_h as a limit. We first eliminate v from the momentum equation by integrating the continuity equation with respect to y:

$$\int_0^y \frac{\partial u}{\partial x} dy + \int_0^y \frac{\partial v}{\partial y} dy = 0 \tag{12.90}$$

Evaluating the second integral at both limits gives

$$v(x,y) - \cancelto{0}{v(x,0)} = -\int_0^y \frac{\partial u}{\partial x} dy \tag{12.91}$$

Since the plate is stationary, $v(x,0) = 0$ and Equation 12.91 becomes

$$v = -\int_0^y \frac{\partial u}{\partial x} dy \tag{12.92}$$

Now we substitute for v in the momentum equation.

$$u\frac{\partial u}{\partial x} - \frac{\partial u}{\partial y}\int_0^y \frac{\partial u}{\partial x} dy = \nu\frac{\partial^2 u}{\partial y^2} \tag{12.93}$$

We multiply the momentum equation through by −1 and integrate it over the boundary layer thickness. Remember, we know exactly what lies beyond the boundary layer in the free stream:

$$\underbrace{-\int_0^{\delta_h}\left(u\frac{\partial u}{\partial x}\right)dy}_{\text{I}}+\underbrace{\int_0^{\delta_h}\frac{\partial u}{\partial y}\left[\int_0^{y}\frac{\partial u}{\partial x}dy\right]dy}_{\text{II}}=\underbrace{-\nu\int_0^{\delta_h}\left(\frac{\partial^2 u}{\partial y^2}\right)dy}_{\text{III}} \tag{12.94}$$

The integral of the second derivative, term III, is the slope evaluated at the limits of integration:

$$\nu\int_0^{\delta_h}\left(\frac{\partial^2 u}{\partial y^2}\right)dy=\nu\left[\cancelto{0}{\left.\frac{\partial u}{\partial y}\right|_{y=\delta_h}}-\left.\frac{\partial u}{\partial y}\right|_{y=0}\right]=-\nu\left.\frac{\partial u}{\partial y}\right|_{y=0} \tag{12.95}$$

Term II is more difficult to integrate. We must use integration by parts $\left[\int_0^y X\,dY=XY\big|_0^y-\int_0^y Y\,dX\right]$. Choosing

$$X=\int_0^y\frac{\partial u}{\partial x}dy\quad Y=u\quad dX=\frac{\partial u}{\partial x}dy\quad dY=\frac{\partial u}{\partial y}dy \tag{12.96}$$

the integration by parts procedure with the boundary conditions on u yields

$$\int_0^{\delta_h}\frac{\partial u}{\partial y}\left[\int_0^{y}\frac{\partial u}{\partial x}dy\right]dy=v_\infty\int_0^{\delta_h}\frac{\partial u}{\partial x}dy-\int_0^{\delta_h}\left(u\frac{\partial u}{\partial x}\right)dy \tag{12.97}$$

Notice that the second integral in Equation 12.97 is the same term as term I in the integrated momentum equation. We can combine Equation 12.94 with Equations 12.95 and 12.97 to give

$$v_\infty\int_0^{\delta_h}\frac{\partial u}{\partial x}dy-2\int_0^{\delta_h}\left(u\frac{\partial u}{\partial x}\right)dy=\nu\left.\frac{\partial u}{\partial y}\right|_{y=0} \tag{12.98}$$

If we fold $2u$ and v_∞ into the derivatives of Equation 12.98, we can combine these two integrals and rewrite the momentum equation as

$$\int_0^{\delta_h}\frac{\partial}{\partial x}\{u(v_\infty-u)\}dy=\nu\left.\frac{\partial u}{\partial y}\right|_{y=0} \tag{12.99}$$

At this point, it is useful to put the momentum equation in dimensionless form. We define a dimensionless velocity, $u^*=u/v_\infty$; exchange the order of integration and differentiation; and rewrite Equation 12.99 as

$$\frac{\partial}{\partial x}\int_0^{\delta_h}\{u^*(1-u^*)\}dy=\frac{\nu}{v_\infty}\left.\frac{\partial u^*}{\partial y}\right|_{y=0} \tag{12.100}$$

The position variable, y, is made dimensionless with respect to the boundary layer thickness, δ_h. The boundary layer thickness is only a function of position along the plate, $\delta_h = f(x)$, and so can be removed from the integral. With $y^* = y/\delta_h$, we find

$$\frac{d}{dx}\delta_h \int_0^1 \{u^*(1-u^*)\}dy^* = \frac{\nu}{v_\infty \delta_h} \left.\frac{\partial u^*}{\partial y^*}\right|_{y^*=0} \tag{12.101}$$

We have succeeded in transforming the partial differential equation, the momentum equation, into an ordinary differential equation for the boundary layer thickness. Both the definite integral and the evaluated derivative in Equation 12.101 are constants:

$$\int_0^1 \{u^*(1-u^*)\}dy^* = A \qquad \left.\frac{\partial u^*}{\partial y^*}\right|_{y^*=0} = B \tag{12.102}$$

Substituting our as yet unknown constants into Equation 12.101 and grouping all the δ_h to one side, we find that

$$\delta_h \frac{d\delta_h}{dx} = \frac{1}{2}\frac{d}{dx}\left(\delta_h^2\right) = \frac{\nu}{v_\infty}\cdot\frac{B}{A} \tag{12.103}$$

Separating and integrating the equation yield an expression for the hydrodynamic boundary layer thickness:

$$\delta_h = \sqrt{\frac{\nu x}{v_\infty}\cdot\frac{2B}{A}} + K \tag{12.104}$$

Since the boundary layer begins at $x = 0$, the *leading edge* of the plate, the boundary condition must be $\delta_h = 0$ at $x = 0$. Thus, the constant in Equation 12.104 vanishes, $K = 0$. We can now rewrite the expression for the boundary layer thickness, grouping terms and incorporating the local Reynolds number into the definition:

$$\delta_h = \sqrt{\frac{x^2}{\mathrm{Re}_x}\cdot\frac{2B}{A}} = \frac{x}{\sqrt{\mathrm{Re}_x}}\sqrt{\frac{2B}{A}} \qquad \mathrm{Re}_x = \frac{v_\infty x}{\nu} \tag{12.105}$$

Notice that the boundary layer grows as $x^{1/2}$ and that it is proportional to $\mathrm{Re}_x^{-1/2}$. To evaluate this expression further, we need an expression for u^* so we can calculate the constants, A and B. We already defined this expression in Equation 12.89. In dimensionless form, we have

$$u^* = a_o + a_1 y^* + a_2 y^{*2} + a_3 y^{*3} \tag{12.106}$$

The constants are determined by applying the boundary conditions, Equations 12.85 through 12.88:

$$y^* = 0 \qquad a_o = 0 \tag{12.107}$$

$$y^* = 0 \qquad 2a_2 = 0 \tag{12.108}$$

$$y^* = 1 \qquad a_o + a_1 + a_2 + a_3 = 1 \tag{12.109}$$

$$y^* = 1 \qquad a_1 + 2a_2 + 3a_3 = 0 \tag{12.110}$$

The resulting velocity profile is

$$u^* = \frac{3}{2}y^* - \frac{1}{2}y^{*3} \tag{12.111}$$

Substituting the velocity profile into the terms of Equation 12.102 gives

$$A = \frac{39}{280} \qquad B = \frac{3}{2}$$

and so the boundary layer thickness becomes

$$\delta_h = \frac{4.64x}{\sqrt{\mathrm{Re}_x}} \tag{12.112}$$

The exact solution of the boundary layer equations [7,8] (Appendix D) shows that our constant, 4.64, is just shy of the more accurate value of 5.

Normally, we are more interested in the friction factor than the boundary layer thickness. We can use the definition of the friction factor and dimensionless variables u^* and y^* to show

$$\frac{C_f}{2} = -\frac{\mu}{\rho v_\infty^2} \cdot \left.\frac{\partial u}{\partial y}\right|_{y=0} = -\frac{\mu}{\rho v_\infty \delta_h} \cdot \left.\frac{\partial u^*}{\partial y^*}\right|_{y^*=0} \tag{12.113}$$

The friction factor is *inversely* proportional to the boundary layer thickness. An inspection of the definition of the friction factor and a look at Figure 12.5 shows why this should be so. The friction factor is a measure of the velocity gradient at the surface of the plate. As the boundary layer gets thinner, the change in velocity occurs over a smaller and smaller distance, and so the velocity gradient is higher.

Substituting the boundary layer thickness and velocity gradient into Equation 12.35 yields

$$C_f = \frac{3}{4.64}\frac{1}{\sqrt{\mathrm{Re}_x}} = 0.646\mathrm{Re}_x^{-1/2} \tag{12.114}$$

We can use numerical methods to simulate a boundary layer flow and see if the simulations can reproduce the general features we have discovered from the analytical solution. Two of those features are that the boundary layer grows as the square root of distance down the plate and that the boundary layer velocity profile is self-similar in x. The power of simulation is that we can handle situations such as non-Newtonian fluids that can't be handled so easily analytically. Figures 12.6 and 12.7 show the boundary layer thickness profile for a Carreau fluid whose viscosity obeys

$$\mu = \mu_\infty + (\mu_o - \mu_\infty)\left[1 + (\lambda\dot{\gamma})^2\right]^{(n-1)/2}$$
$$= 0.01 + (0.01 - 0.001)[1 + (2\dot{\gamma})^2] \tag{12.115}$$

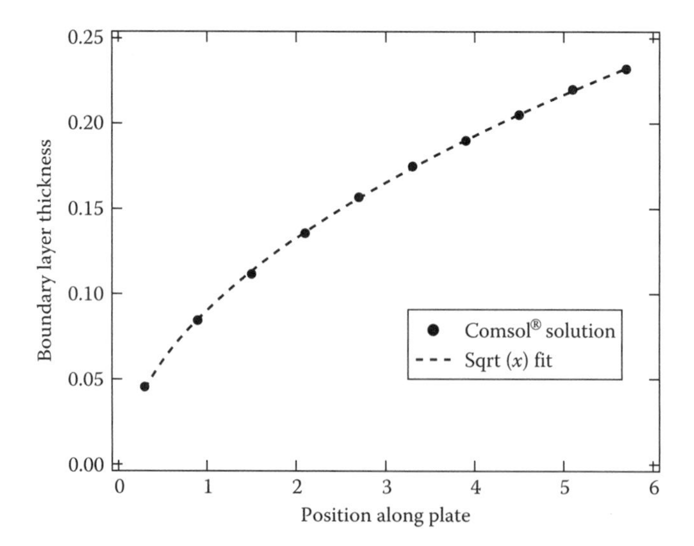

FIGURE 12.6 Boundary layer thickness for a Carreau fluid. Though the fluid does not exhibit Newtonian behavior, the boundary layer still grows with the square root of distance down the plate, and so a Newtonian approximation is good enough (Comsol® module nonNewtBL).

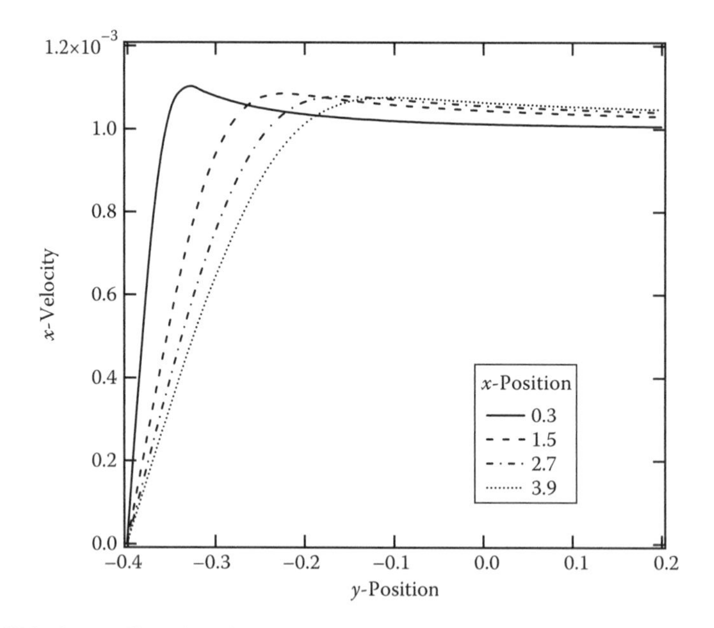

FIGURE 12.7 Velocity profile (u/v_∞) through the boundary layer of a Carreau fluid. Self-similarity still holds along the plate, though the y-velocity generated due to the no-slip condition at the surface of the plate translates into a band of fluid on the opposite side of the boundary layer that moves slightly faster than the free stream velocity.

Example 12.4 Frictional Force Exerted on a Flat Surface

A vulture is riding the thermal air currents wafting up from a canyon. The vulture's speed is on the order of 5 m/s. Its wings can be approximated as flat plates, 1 m long and 0.4 m wide. The air temperature is 25°C. Assuming the wings are perfectly smooth, estimate the average friction factor over the wing surface and the frictional force exerted on each wing.

Before we begin our calculations, we need to determine the physical properties of the air and to define the average friction factor. At 25°C, the air properties are

$$\rho = 1.164\ \text{kg/m}^3, \quad \mu = 184.6 \times 10^{-7}\ \text{N s/m}^2, \quad \nu = 15.89 \times 10^{-6}\ \text{m}^2\text{/s}$$

The average value of the friction factor is given by Equation 12.11:

$$\overline{C_f} = \frac{1}{L}\int_0^L C_f dx = \frac{1}{L}\int_0^L \frac{3}{4.64}\frac{1}{\sqrt{v_\infty x/\nu}} dx = \frac{1.292}{\sqrt{\frac{v_\infty L}{\nu}}}$$

We begin by calculating the Reynolds number based on plate length:

$$\text{Re}_L = \frac{v_\infty L}{\nu} = \frac{5(0.4)}{15.89 \times 10^{-6}} = 125{,}865$$

The transition to turbulent flow over a flat plate occurs at a higher Reynolds number, as we will see in Chapter 14. Therefore, we have laminar flow and our expression for the friction factor will be accurate. The average friction factor is

$$\overline{C_f} = 1.292(\text{Re}_L)^{-1/2} = 3.64 \times 10^{-3}$$

The frictional force exerted on both sides of the wing is

$$F = 2A_s\overline{C_f}\left[\frac{1}{2}\rho v_\infty^2\right] = 2(0.4)(1)(3.64 \times 10^{-3})\left(\frac{1.164}{2}\right)(5)^2 = 4.24 \times 10^{-2}\ \text{N}$$

If the wing were smooth and frictional forces were the only forces acting to retard flight, the frictional forces exerted on the wing would be very small, and it would take little energy to remain aloft. Real bird wings are not perfectly smooth and thus have higher friction factors to contend with. Though this increases the frictional force at low velocities, the benefits of a rougher surface are felt at higher speeds where delayed separation of the boundary layer, less wake, and a smoother overall flight occur as we will see in Chapter 13.

12.6.2 Magnetohydrodynamic Flow

If our fluid happens to be a conductor, such as a molten metal, or an aqueous, ionic solution perhaps with metal particles, then externally imposed electric and magnetic fields can be used to influence the flow [12]. These fields and their influence on flow systems are readily observable in the hot gases of stars, in the auroras on Earth, and in engineering applications such as particle accelerators, plasma vapor deposition systems, power generation systems, propulsion systems (the most famous being the Hunt for Red October by Tom Clancy), and NMR imagers.

An electrical conductor moving through a magnetic field generates an electromotive force (emf) that is proportional to the speed of the conductor and the magnetic field strength. This emf acts to oppose the motion of the conductor. If we replace the conductor by a conducting fluid flowing through the magnetic field, the fluid sets up an emf that acts as a body force to oppose its motion. In a system where we are only applying a magnetic field, B_h, to the fluid, the force is given by

$$\vec{\mathbf{F}} = \vec{J} \times \vec{\mathbf{B}}_\mathbf{h} \tag{12.116}$$

where

$\vec{\mathbf{B}}_\mathbf{h}$ is the magnetic field strength
$\vec{J}$ is the current density (flux)

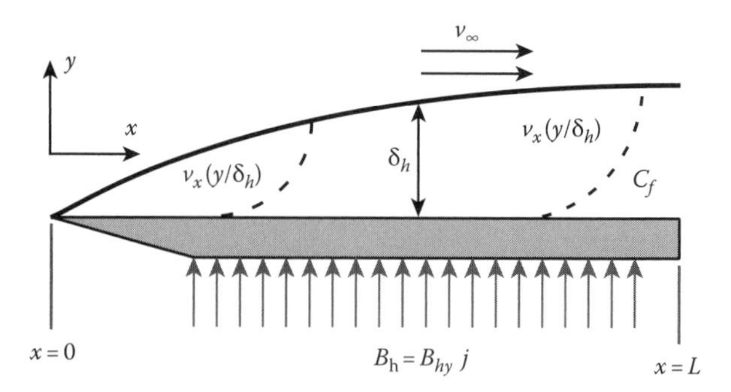

FIGURE 12.8 Flow over a flat plate under the influence of an external magnetic field.

The current generated in the conducting fluid is

$$\vec{J} = \sigma(\vec{\mathbf{v}} \times \vec{\mathbf{B}}_{\mathbf{h}}) \qquad \vec{\mathbf{v}} = u\vec{\mathbf{i}} + v\vec{\mathbf{j}} \tag{12.117}$$

Consider the boundary layer system shown in Figure 12.8. Fluid flows over the flat plate and a magnetic field is imposed across the plate. We would like to determine how the magnetic field affects the fluid and how it changes the friction factor. The boundary layer equations must now include the body force term (generation) arising from the imposed magnetic field. We assume a constant free stream velocity, a constant magnetic field, and zero pressure gradient:

Continuity

$$\frac{\partial u}{\partial x} + \frac{\partial v}{\partial y} = 0 \tag{12.118}$$

Momentum

$$u\frac{\partial u}{\partial x} + v\frac{\partial u}{\partial y} = \nu\frac{\partial^2 u}{\partial y^2} + \frac{1}{\rho}(\vec{J} \times \vec{\mathbf{B}}_{\mathbf{h}})_x \tag{12.119}$$

The body force term is that component of the force acting in the x-direction and retarding the flow. The current carried by the fluid is given in Equation 12.117, and we can substitute that into the momentum equation to rewrite the body force in terms of the velocity and simplify the momentum equation:

$$(\vec{J} \times \vec{\mathbf{B}}_{\mathbf{h}}) = \sigma[(\vec{\mathbf{v}} \times \vec{\mathbf{B}}_{\mathbf{h}}) \times \vec{\mathbf{B}}_{\mathbf{h}}] = -\sigma u B_{hy}^2 \vec{\mathbf{i}} \tag{12.120}$$

$$u\frac{\partial u}{\partial x} + v\frac{\partial u}{\partial y} = \nu\frac{\partial^2 u}{\partial y^2} - \frac{\sigma}{\rho}B_{hy}^2 u \tag{12.121}$$

We apply the same integral procedure we used to solve for the previous hydrodynamic boundary layer thickness in Section 12.6.1. The boundary conditions for this problem are exactly the same as if the fluid were not exposed to the magnetic field. We still must insure no slip at the plate surface and a smooth transition into the free stream. We must also include the condition saying that the curvature is zero at the plate surface since both u and v vanish there:

$$y = 0 \qquad u = 0 \qquad v = 0 \qquad \frac{\partial^2 u}{\partial y^2} = 0 \tag{12.122}$$

$$y = \delta_h \qquad u = v_\infty \qquad \frac{\partial u}{\partial y} = 0 \tag{12.123}$$

Integrating over the boundary layer thickness, multiplying through by −1, and letting $u^* = u/v_\infty$ and $y^* = y/\delta_h$ gives the following differential equation for the boundary layer thickness:

$$\frac{d}{dx}\delta_h \int_0^1 \{u^*(1-u^*)\}dy = \left(\frac{\nu}{v_\infty \delta_h}\right)\frac{\partial u^*}{\partial y^*}\bigg|_{y^*=0} + \left(\frac{\sigma B_{hy}^2}{\rho v_\infty}\right)\delta_h \int_0^1 u^* dy^* \tag{12.124}$$

Since the boundary conditions for this problem are the same as those for the problem we just solved, we can use the same cubic velocity profile for u^* given in Equation 12.111. Evaluating the integrals and derivative in Equation 12.124 shows that

$$\int_0^1 \{u^*(1-u^*)\}dy^* = \frac{39}{280} \qquad \frac{\partial u^*}{\partial y^*}\bigg|_{y^*=0} = \frac{3}{2} \qquad \int_0^1 u^* dy^* = \frac{5}{8} \tag{12.125}$$

Substituting into the differential equation and multiplying through by δ_h gives

$$\left[\frac{39}{560}\right]\frac{d}{dx}\left(\delta_h^2\right) - \left[\frac{5\sigma B_{hy}^2}{8\rho v_\infty}\right]\delta_h^2 - \frac{3\nu}{2v_\infty} = 0 \tag{12.126}$$

Equation 12.126 is a first-order differential equation in δ_h^2 that we can easily solve by using an integrating factor [9] and the knowledge that the boundary layer starts at $x = 0$. The solution is

$$\delta_h^2 = \frac{12\mu}{5\sigma B_{hy}^2}\left\{\exp\left[\frac{350}{39}\left(\frac{\sigma B_{hy}^2}{\rho v_\infty}\right)x\right] - 1\right\} \tag{12.127}$$

If we define a new dimensionless parameter, $\mathcal{N}_{Bx} = \left(\sigma B_{hy}^2/\rho v_\infty\right)x$, we can rewrite Equation 12.127 to show how the boundary layer thickness depends upon the dimensionless groups:

$$\delta_h = 1.55x\left[\exp\left(\frac{350}{39}\mathcal{N}_{Bx}\right) - 1\right]^{1/2} \mathcal{N}_{Bx}^{-1/2}\mathrm{Re}_x^{-1/2} \tag{12.128}$$

Notice that the boundary layer thickness is still inversely proportional to $\mathrm{Re}_x^{-1/2}$ but now the action of the magnetic field is to increase the boundary layer thickness. As either the conductivity of the fluid or the magnetic field strength increases, the boundary layer thickness increases. The field acts as if there were a pressure drop opposing the flow. The velocity parallel to the plate is hindered forcing fluid to flow perpendicular to the plate and thereby increasing the boundary layer thickness. If we were to reverse the polarity of the field, the force would act in the flow direction, and we would decrease the boundary layer thickness.

The friction factor decreases as the field increases:

$$\frac{C_f}{2} = -\left(\frac{\mu}{\rho v_\infty \delta_h}\right)\frac{\partial u^*}{\partial y^*}\bigg|_{y^*=0}$$

$$= 0.97\left[\exp\left(\frac{350}{39}\mathcal{N}_{Bx}\right) - 1\right]^{-1/2} \mathcal{N}_{Bx}^{1/2}\mathrm{Re}_x^{-1/2} \tag{12.129}$$

The use of magnetic fields to stabilize flow is quite common in separation equipment where very small metallic particles are added to the field to render it conductive. Circulation patterns can be controlled by judicious application of the field, and this greatly affects mass transfer properties [10,11].

12.7 THERMAL BOUNDARY LAYERS

If we heat the surface of the plate, we create a thermal boundary layer that starts thin and grows along with the hydrodynamic boundary layer. The characteristics of the thermal boundary layer are governed by the relationship between the temperature and the flow fields. In this section, we will solve the energy equation in the boundary layer using the integral analysis procedure. The solution yields an expression for the thermal boundary layer thickness and the heat transfer coefficient. The boundary layer energy equation is

$$u\frac{\partial T}{\partial x} + v\frac{\partial T}{\partial y} = \alpha\frac{\partial^2 T}{\partial y^2} \tag{12.130}$$

We will consider the simplest case first. We assume the hydrodynamic boundary layer can grow independently of the thermal boundary layer. This requires all material properties to be independent of temperature. Most often, the growth of the hydrodynamic boundary layer depends somewhat upon the thermal boundary layer since the density of most fluids decreases with increasing temperature. The coupling of boundary layers is the hallmark of natural or free convection, our second thermal boundary layer topic.

12.7.1 Forced Convection, Laminar Flow

We assume the plate is stationary in space and that we have laminar flow. These restrictions allow us to use the results of the previous sections for the hydrodynamic boundary layer thickness. The boundary conditions for this problem are virtually identical to the boundary conditions used to solve for the hydrodynamic boundary layer thickness. We know the temperature at the surface of the plate and at the edge of the boundary layer:

$$y = 0 \qquad T = T_s \tag{12.131}$$

$$y = \delta_T \qquad T = T_\infty \tag{12.132}$$

At the edge of the boundary layer, the temperature must smoothly become T_∞:

$$y = \delta_T \qquad \frac{\partial T}{\partial y} = 0 \tag{12.133}$$

At the plate's surface, $u = v = 0$. Evaluating Equation 12.130 there shows that

$$y = 0 \qquad \frac{\partial^2 T}{\partial y^2} = 0 \tag{12.134}$$

We remove v using the continuity equation:

$$v = -\int_0^y \frac{\partial u}{\partial x}\, dy \tag{12.135}$$

Substituting into the energy equation gives

$$u\frac{\partial T}{\partial x}-\frac{\partial T}{\partial y}\int_0^y\frac{\partial u}{\partial x}dy=\alpha\frac{\partial^2 T}{\partial y^2} \qquad (12.136)$$

Now we integrate Equation 12.136 over the extent of the thermal boundary layer from $y = 0$ to $y = \delta_T$ and multiply by −1:

$$\underbrace{-\int_0^{\delta_T}\left(u\frac{\partial T}{\partial x}\right)dy}_{\mathrm{I}}+\underbrace{\int_0^{\delta_T}\left(\frac{\partial T}{\partial y}\int_0^y\frac{\partial u}{\partial x}dy\right)dy}_{\mathrm{II}}=\underbrace{-\alpha\int_0^{\delta_T}\frac{\partial^2 T}{\partial y^2}dy}_{\mathrm{III}} \qquad (12.137)$$

Term III of Equation 12.137 can be integrated easily to give

$$-\alpha\int_0^{\delta_T}\frac{\partial^2 T}{\partial y^2}dy=\alpha\left.\frac{\partial T}{\partial y}\right|_{y=0} \qquad (12.138)$$

Integrating term II by parts gives

$$\int_0^{\delta_T}\left(\frac{\partial T}{\partial y}\int_0^y\frac{\partial u}{\partial x}dy\right)dy=\int_0^{\delta_T}(T_\infty-T)\frac{\partial u}{\partial x}dy-\int_0^0(T_s-T)\frac{\partial u}{\partial x}dy \qquad (12.139)$$

Combining Equations 12.138 and 12.139 with the energy equation leaves us with a fairly simple equation to solve:

$$\int_0^{\delta_T}\frac{\partial}{\partial x}[u(T_\infty-T)]dy=\alpha\left.\frac{\partial T}{\partial y}\right|_{y=0} \qquad (12.140)$$

Unlike the hydrodynamic solution, there are three different cases to consider when solving the energy equation (Figure 12.9). All these cases revolve around the magnitude of the Prandtl number for the fluid. If the Prandtl number is much less than unity, $\mathrm{Pr} \ll 1$, then the diffusion of momentum in the hydrodynamic boundary layer is much slower than the diffusion of energy in the thermal boundary layer. This implies that the hydrodynamic boundary layer thickness will be much less than the thermal boundary layer thickness, $\delta_h \ll \delta_T$. In such circumstances, the fluid velocity inside the thermal boundary layer will be essentially constant at the free stream velocity, v_∞. Substituting this approximation into Equation 12.140 lets us simplify the equation considerably:

$$v_\infty\int_0^{\delta_T}\frac{\partial}{\partial x}(T_\infty-T)dy=\alpha\left.\frac{\partial T}{\partial y}\right|_{y=0} \qquad \mathrm{Pr}\ll 1 \qquad (12.141)$$

If the Prandtl number is of order unity, $\mathrm{Pr} \approx 1$, then the diffusion of momentum and energy occur at the same rate. This indicates that the boundary layers are the same size, $\delta_h \approx \delta_T$, and so the energy

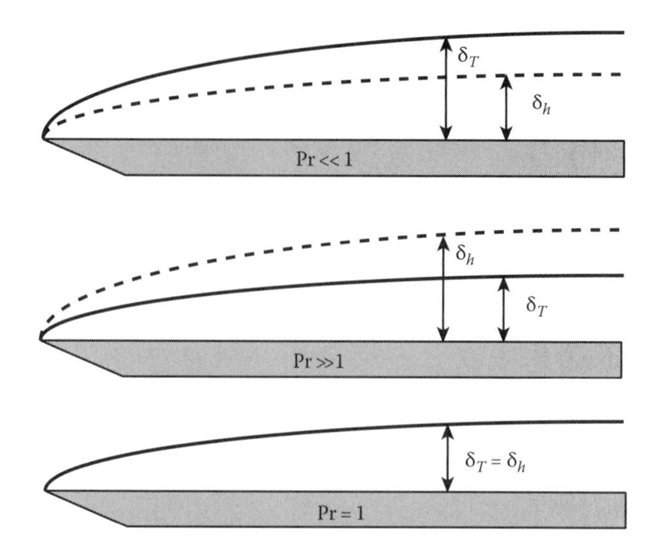

FIGURE 12.9 Relationship of thermal and hydrodynamic boundary layers as a function of the Prandtl number.

equation and momentum equation are of the same form. The Reynolds analogy holds in this situation and we can say

$$\frac{C_f}{2} = \text{St} = \frac{\text{Nu}}{\text{Re}_x\text{Pr}} \qquad \text{St} - \text{Stanton number} \tag{12.142}$$

Using the value for the friction factor we derived in Section 12.6, we can substitute into Equation 12.142 and obtain a relationship for the dimensionless heat transfer coefficient, the Nusselt number, Nu:

$$\text{Nu} = 0.332\text{Re}_x^{1/2} \qquad \text{Pr} = 1 \tag{12.143}$$

If the Prandtl number is much larger than unity, $\text{Pr} \gg 1$, then the diffusion of momentum through the hydrodynamic boundary layer is much faster than the diffusion of energy through the thermal boundary layer. This implies that $\delta_h \gg \delta_T$, and we must integrate Equation 12.140 over the entire thermal boundary layer. To assist in solving the energy equation, we define a new, dimensionless temperature variable:

$$\theta = \frac{T - T_s}{T_\infty - T_s} \tag{12.144}$$

The energy equation, in terms of θ, becomes

$$\int_0^{\delta_T} \frac{\partial}{\partial x}[u(1-\theta)]dy = \alpha \left.\frac{\partial \theta}{\partial y}\right|_{y=0} \qquad \text{Pr} \gg 1 \tag{12.145}$$

We now need to define a temperature distribution in the boundary layer to evaluate the integral and derivative in Equation 12.145. We use another polynomial approximation, similar to the one we employed to define u^*:

$$\theta = b_o + b_1\left(\frac{y}{\delta_T}\right) + b_2\left(\frac{y}{\delta_T}\right)^2 + b_3\left(\frac{y}{\delta_T}\right)^3 \tag{12.146}$$

We have four boundary conditions to evaluate the constants in Equation 12.146. Since these boundary conditions are identical to the hydrodynamic case, we obtain the same cubic profile:

$$\theta = \frac{3}{2}\left(\frac{y}{\delta_T}\right) - \frac{1}{2}\left(\frac{y}{\delta_T}\right)^3 \tag{12.147}$$

$$u^* = \frac{u}{v_\infty} = \frac{3}{2}\left(\frac{y}{\delta_h}\right) - \frac{1}{2}\left(\frac{y}{\delta_h}\right)^3 \tag{12.148}$$

Substituting both of these relationships into the energy equation yields

$$v_\infty \int_0^{\delta_T} \frac{\partial}{\partial x}\left\{\left[\frac{3}{2}\left(\frac{y}{\delta_h}\right) - \frac{1}{2}\left(\frac{y}{\delta_h}\right)^3\right]\left[\frac{3}{2}\left(\frac{y}{\delta_T}\right) - \frac{1}{2}\left(\frac{y}{\delta_T}\right)^3\right]\right\} dy = \alpha \left.\frac{\partial \theta}{\partial y}\right|_{y=0} \tag{12.149}$$

To interchange the order of integration and differentiation in (12.149) (without using Leibnitz's rule) and to more effectively combine the velocity and temperature polynomials, we define a boundary layer ratio, $\Delta_T = \delta_T/\delta_h$, and new coordinate, $y^* = y/\delta_T$. The integral now extends from 0 to 1, and we can interchange the order of integration and differentiation directly:

$$v_\infty \frac{d}{dx} \int_0^1 \Delta_T \delta_h \left[1 - \frac{3}{2} y^* + \frac{1}{2} y^{*3}\right]\left[\frac{3}{2}\Delta_T y^* - \frac{3}{2}\Delta_T^3 y^{*3}\right] dy^* = \frac{\alpha}{\delta_h \Delta_T} \left.\frac{\partial \theta}{\partial y^*}\right|_{y^*=0} \tag{12.150}$$

Evaluating the integral and derivative in Equation 12.150 yields a differential equation for the boundary layer thickness ratio, Δ_T:

$$v_\infty \frac{d}{dx}\left[\delta_h\left(\frac{3}{20}\Delta_T^2 - \frac{3}{280}\Delta_T^4\right)\right] = \frac{3\alpha}{2\delta_h \Delta_T} \tag{12.151}$$

Since the Prandtl number is much larger than unity, $\Delta_T \ll 1$, and we can neglect the Δ_T^4 term with respect to Δ_T^2. Dividing through by v_∞ and multiplying through by $\delta_h \Delta_T$ gives

$$\delta_h \Delta_T \frac{d}{dx}\left[\delta_h \Delta_T^2\right] = \frac{10\alpha}{v_\infty} \tag{12.152}$$

Differentiating by x using the chain rule to separate δ_h and Δ_T gives

$$\delta_h \Delta_T^3 \frac{d\delta_h}{dx} + 2\delta_h^2 \Delta_T^2 \frac{d\Delta_T}{dx} = \frac{10\alpha}{v_\infty} \tag{12.153}$$

From the hydrodynamic boundary layer solution, we know that

$$\delta_h^2 = \frac{280}{13}\frac{\nu x}{v_\infty} \qquad \delta_h d\delta_h = \frac{140}{13}\frac{\nu}{v_\infty} dx \tag{12.154}$$

and so we can substitute this into Equation 12.153 to remove any reference to δ_h. The result is a first-order differential equation for Δ_T^3:

$$\Delta_T^3 + 4x\Delta_T^2 \frac{d\Delta_T}{dx} = \Delta_T^3 + \frac{4}{3}x\frac{d\Delta_T^3}{dx} = \frac{13\alpha}{14\nu} \tag{12.155}$$

Solving for Δ_T^3 by separating and integrating, we have

$$\Delta_T^3 = c_o x^{-3/4} + \frac{13\alpha}{14\nu} = c_o x^{-3/4} + \frac{13}{14}\left(\frac{1}{\mathrm{Pr}}\right) \tag{12.156}$$

If we assume that the entire plate is heated, the boundary layer thickness ratio approaches infinity at the leading edge. This is a direct result of the approximations we used to solve the equations. We can get rid of this inconsistency if we assume that only a portion of the plate is heated and that the heating starts at a position, x_o, from the leading edge. Thus, $\Delta_T = 0$ at $x = x_o$ and the solution is

$$\Delta_T = \left(\frac{13}{14}\right)^{1/3}\left[1 - \left(\frac{x_o}{x}\right)^{3/4}\right]^{1/3} \mathrm{Pr}^{-1/3} \tag{12.157}$$

To see what happens when the entire plate is heated, we take the limit as $x_o \to 0$:

$$\Delta_T = \left(\frac{13}{14}\right)^{1/3} \mathrm{Pr}^{-1/3} \tag{12.158}$$

We can now calculate the thermal boundary layer thickness and, from there, get the heat transfer coefficient. Substituting in for $\delta_h = 4.64x\mathrm{Re}_x^{-1/2}$ in Equation 12.157 gives

$$\delta_T = 4.53\mathrm{Pr}^{-1/3}\mathrm{Re}_x^{-1/2}x\left[1 - \left(\frac{x_o}{x}\right)^{3/4}\right]^{1/3} \tag{12.159}$$

Our definition for the heat transfer coefficient balanced heat conducted through the surface of the plate against heat taken away by the fluid. Using the temperature profile we derived for this problem, we find the heat transfer coefficient to be inversely proportional to the thermal boundary layer thickness:

$$h = \frac{-k_f\left(\frac{\partial T}{\partial y}\right)_{y=0}}{T_s - T_\infty} = \frac{k_f}{\delta_T}\frac{\partial \theta}{\partial y^*}\bigg|_{y^*=0} = \frac{3}{2}\frac{k_f}{\delta_T} \tag{12.160}$$

Just like the friction factor, as the boundary layer thickness decreases, the heat transfer coefficient increases. Substituting the expression for δ_T into Equation 12.160 gives the local value for the heat transfer coefficient:

$$h_x = 0.332\mathrm{Pr}^{1/3}\mathrm{Re}_x^{1/2}k_f\frac{[1 - (x_o/x)^{3/4}]^{-1/3}}{x} \tag{12.161}$$

The Nusselt number becomes

$$\text{Nu}_x = \frac{h_x x}{k_f} = 0.332\text{Re}_x^{1/2}\text{Pr}^{1/3}\left[1-\left(\frac{x_o}{x}\right)^{3/4}\right]^{-1/3} \tag{12.162}$$

The important point to remember about laminar flow and fluids with Pr > 1 is that the Nusselt number grows with $\text{Re}_x^{1/2}$ and with $\text{Pr}^{1/3}$.

The value we have calculated is the local Nusselt number at a distance x from the leading edge of the plate. Often, we are more interested in the average Nusselt number, so we can determine the average value of the heat transfer coefficient and the total heat exchanged with the plate. For the case where $x_o = 0$, we can use the definition of the average to show that

$$\overline{\text{Nu}} = \frac{\int_0^L \text{Nu}_x dx}{\int_0^L dx} = 2\text{Nu}_x(x = L) \tag{12.163}$$

Having solved the most difficult case where Pr > 1, we can now return to the case where Pr ≪ 1. If Pr ≪ 1, then the boundary layer integral equation we must solve is

$$v_\infty \int_0^{\delta_T} \frac{\partial}{\partial x}(T_\infty - T)dy = \alpha \left.\frac{\partial T}{\partial y}\right|_{y=0} \tag{12.164}$$

Rearranging the equation in terms of θ, plugging in the polynomial approximation for the temperature profile, and defining $y^* = y/\delta_T$ lets us exchange the order of integration and differentiation and rewrite Equation 12.164 as

$$v_\infty \frac{d}{dx}\delta_T \int_0^1 \left[1 - \frac{3}{2}y^* + \frac{1}{2}y^{*3}\right]dy^* = \frac{\alpha}{\delta_T}\left.\frac{\partial \theta}{\partial y^*}\right|_{y^*=0} \tag{12.165}$$

Evaluating the integral and derivative leads to a first-order differential equation for δ_T,

$$\delta_T \frac{d\delta_T}{dx} = \frac{4\alpha}{v_\infty} \tag{12.166}$$

whose solution shows that the boundary layer thickness now depends on $x^{1/2}$, just as the friction factor does:

$$\delta_T = \sqrt{\frac{8\alpha}{v_\infty}x} + K = x\sqrt{\frac{8\text{Pr}}{\text{Re}_x}} + K \tag{12.167}$$

For a plate heated along its entire length, $\delta_T = 0$ at $x = 0$ and so $K = 0$. Using this value for the boundary layer thickness in our definition of the heat transfer coefficient shows that heat transfer coefficient and Nusselt number are proportional to $\text{Re}_x^{1/2}$ and $\text{Pr}^{1/2}$:

$$\text{Nu}_x = \frac{hx}{k_f} = 0.53\text{Re}_x^{1/2}\text{Pr}^{1/2} \tag{12.168}$$

Notice that the dependence on the Prandtl number is much greater than before since the thermal boundary layer thickness is much larger than the hydrodynamic boundary layer. This is a direct consequence of the increased thermal conduction through the boundary layer.

Numerical simulation allows us to look at some more complicated situations yet still see how well the approximations we used in developing the analytical solution hold. As an example, we can look at transpirational cooling, a technique widely used to keep items exposed to high temperatures, such as turbine blades or the shells of a hypersonic aircraft, cool. The process involves injecting cooler fluid through the surface and into the boundary layer. The injection affects both the hydrodynamic and thermal boundary layers. Figure 12.10 shows the results of such a simulation. The free stream temperature was set to 500°C, while the injected fluid (about 1% of the free stream velocity) was at 25°C. The thickness of the thermal and hydrodynamic boundary layers are shown as well as fits assuming both grow with square root of distance down the plate. The Prandtl number for this situation is about 3.25. The difference between the thermal and hydrodynamic boundary layer thicknesses is larger than what we would normally assume given the Prandtl number due to the transpiration effect. We can see this more clearly in Figure 12.11, where the boundary layer thickness ratio is presented. Notice that the ratio is highest at the leading edge of the plate where the effect of the constant injection of fluid is greatest. The ratio exponentially decays to a constant value as the curve fit in Figure 12.11 shows.

Example 12.5 Power Dissipation in a Microprocessor

A current generation microprocessor is built on a rectangular die, 2 cm (L) × 1 cm (W) × 0.35 cm (H). It generates heat at a rate of 5 × 10^6 W/m^3. The chip is to be cooled by air at 20°C. If the chip's surface temperature cannot exceed 105°C, how fast must we force air across its surface? Which way should we orient the chip, relative to the air flow, to obtain the maximum heat transfer rate?

The solution to the boundary layer equations shows that for maximum heat transfer, we want to have the thinnest boundary layer over the surface of the chip. To minimize the boundary layer thickness and maximize the heat transfer coefficient, we want to orient the short dimension of the chip (1 cm) parallel to the flow. (What happens to the amount of power necessary to force the air across the chip in both cases? Suppose we had a constraint on the energy we could supply to the air?)

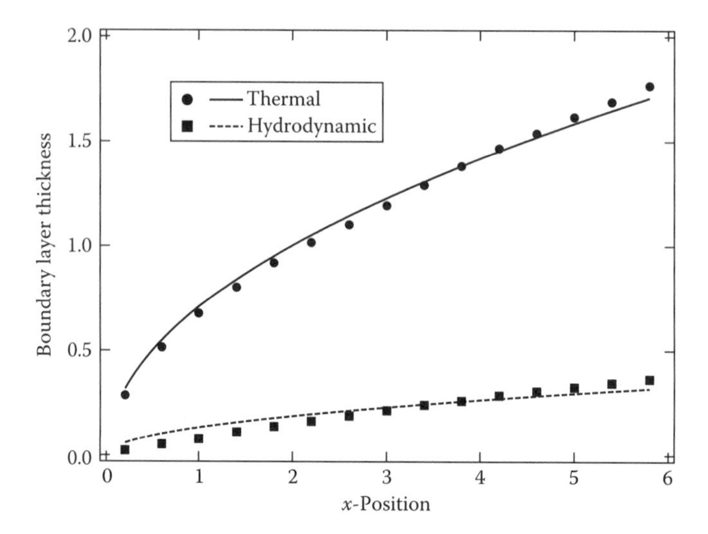

FIGURE 12.10 Simulation of transpirational cooling on a flat plate. Thermal and hydrodynamic boundary layer thicknesses are shown.

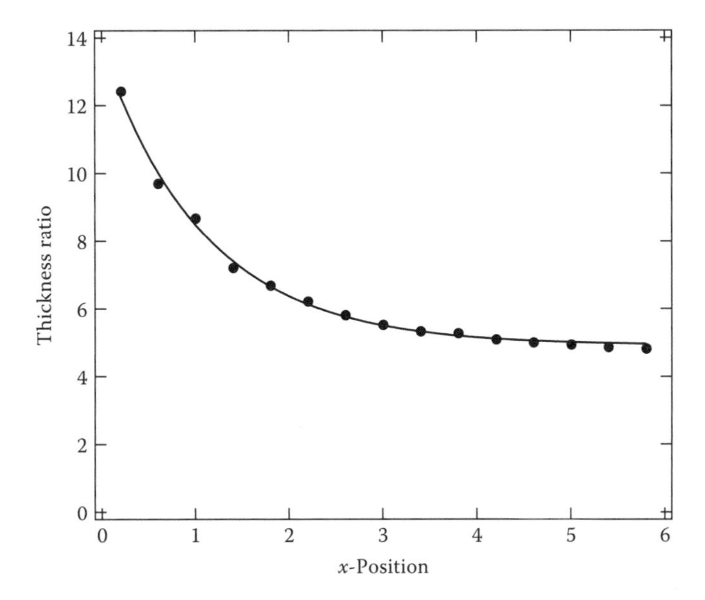

FIGURE 12.11 Thermal versus hydrodynamic boundary layer thickness ratio as a function of position along the surface of the plate. The effect of transpiration at the plate surface is most evident at the leading edge.

At a steady state, an energy balance about the chip tells us that whatever heat we generate, we must remove by convection. Writing the balance in terms of the average heat transfer coefficient, $\bar{h}$, gives

$$Gen = Out$$

$$\dot{q}V = \bar{h}A_s(T_s - T_\infty)$$

Setting the surface temperature to its maximum value, we can solve for the value of the heat transfer coefficient required:

$$\bar{h} = \frac{\dot{q}V}{A_s(T_s - T_\infty)} = 206\ \frac{\text{W}}{\text{m}^2\,\text{K}}$$

We now determine the velocity of air required by using the correlation for the heat transfer coefficient. At this point, we have no idea whether the flow will be laminar or turbulent, but we can assume laminar flow and then check to see if our assumption is correct. Air properties are generally evaluated at the film temperature, $T_f = (T_s + T_\infty)/2 = 62.5$ °C:

$$\nu = 17 \times 10^{-6}\ \text{m}^2/\text{s} \quad \text{Pr} = 0.705 \quad k_f = 0.0273\ \text{W/m K}$$

The Prandtl number here is very close to 1 as it is for most gases. We could, in theory, use the Reynolds analogy. Since Pr < 1, the Reynolds analogy may overestimate the Nusselt number and so Equation 12.162 may be "just right":

$$\overline{\text{Nu}} = \frac{\bar{h}L}{k_f} = 0.664\text{Pr}^{1/3}\text{Re}_L^{1/2} = 0.664\text{Pr}^{1/3}\left(\frac{v_\infty L}{\nu}\right)^{1/2}$$

Solving for v_∞ gives

$$v_\infty = \frac{\nu}{L}\left[\frac{\bar{h}L}{0.664k_f\,\text{Pr}^{1/3}}\right]^2 = 27.7\ \frac{\text{m}}{\text{s}}$$

We should check to insure that we indeed have laminar flow and that our analysis procedure was accurate. Calculating the Reynolds number based on the overall length of the plate (maximum Re) shows that the flow is still laminar. However, the fan could be noisy:

$$\mathrm{Re}_L = \frac{v_\infty L}{\nu} = 16{,}290 \quad \text{Laminar flow}$$

We have used an average value for the heat transfer coefficient to determine the velocity of the air. However, the heat transfer coefficient will vary over the surface of the chip. If the material composing the packaging of the circuit has a very low thermal conductivity, the variation in temperature across the surface may become a problem. Will this solution apply under those conditions? What would need to be done if it didn't?

12.7.2 Free Convection on a Vertical Plate

Previously, we assumed that we could treat the developing hydrodynamic and thermal boundary layers separately. This presumed that the fluid properties were independent of temperature. The density of most fluids decreases with increasing temperature, and this density variation sets up a buoyant force within the fluid that augments or may be totally responsible for the flow.

We are going to develop a boundary layer approximation to describe the flow past a heated vertical plate. The flow will be totally driven by density differences between the fluid in the free stream and the fluid close to the vertical plate. The situation is shown in Figure 12.12. We account for the density differences in the continuity and momentum equations using the Boussinesq approximation first outlined in Chapter 5. The Boussinesq approximation couples the momentum and energy equations by concentrating all the changes in density as a function of temperature in the gravitational body force term of the momentum equation. In all other terms, we assume the fluid is incompressible. The approximation works best for small temperature differences.

The boundary layer equations for natural convection are

Continuity

$$\frac{\partial u}{\partial x} + \frac{\partial v}{\partial y} = 0 \tag{12.169}$$

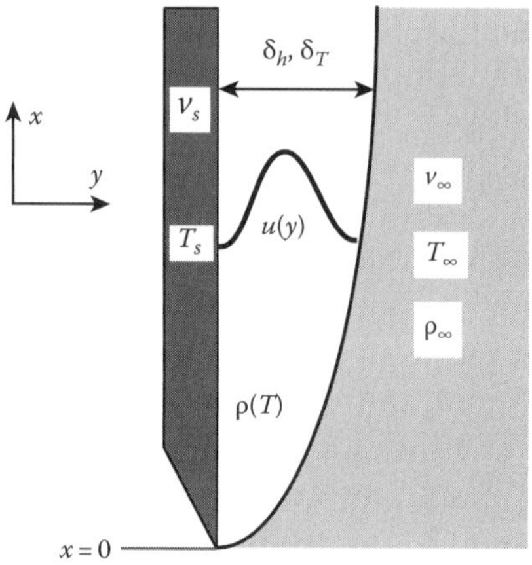

FIGURE 12.12 Natural convection on a vertical plate.

Momentum

$$\rho\left(u\frac{\partial u}{\partial x}+v\frac{\partial u}{\partial y}\right)=-\frac{\partial P}{\partial x}+\mu\frac{\partial^2 u}{\partial y^2}-\rho g \tag{12.170}$$

Energy

$$u\frac{\partial T}{\partial x}+v\frac{\partial T}{\partial y}=\alpha\frac{\partial^2 T}{\partial y^2} \tag{12.171}$$

Any pressure differences perpendicular to the plate can be neglected because the boundary layer is thin. The pressure gradient in the boundary layer is due solely to the weight of the fluid and mirrors the pressure gradient in the free stream. A good approximation is given by

$$\frac{\partial P}{\partial x}=-\rho_\infty g \tag{12.172}$$

The density is a linear function of the temperature for small temperature differences, and the change in density is related to the coefficient of thermal expansion, β, a thermodynamic property:

$$\rho_\infty-\rho=\beta\rho_\infty(T-T_\infty) \qquad \beta=-\frac{1}{\rho_\infty}\left(\frac{\rho_\infty-\rho}{T_\infty-T}\right) \tag{12.173}$$

We substitute Equations 12.172 and 12.173 into the momentum equation (12.170):

$$u\frac{\partial u}{\partial x}+v\frac{\partial u}{\partial y}=\beta g(T-T_\infty)+\nu\frac{\partial^2 u}{\partial y^2} \tag{12.174}$$

The boundary conditions for this problem assume no slip at the wall and a smooth transition into the quiescent free stream for the velocity and temperature profiles. In addition, both u and v vanish at the plate surface and at the edge of the boundary layer. To reflect this, we evaluate the momentum and energy equations there and determine we must have zero curvature:

$$y=0 \qquad u=v=0 \qquad T=T_s \tag{12.175}$$

$$y=0 \qquad \nu\frac{\partial^2 u}{\partial y^2}=-\beta g(T_s-T_\infty) \qquad \frac{\partial^2 T}{\partial y^2}=0 \tag{12.176}$$

$$x=0 \qquad u=v=0 \qquad T=T_\infty \tag{12.177}$$

$$y=\delta_h,\delta_T \qquad u=v=0 \qquad T=T_\infty \qquad \frac{\partial u}{\partial y}=\frac{\partial T}{\partial y}=0 \qquad \frac{\partial^2 u}{\partial y^2}=\frac{\partial^2 T}{\partial y^2}=0 \tag{12.178}$$

We integrate the continuity equation to remove v and then integrate the momentum equation over the hydrodynamic boundary layer thickness while multiplying through by −1:

$$\underbrace{-\int_0^{\delta_h}\left(u\frac{\partial u}{\partial x}\right)dy}_{\text{I}}+\underbrace{\int_0^{\delta_h}\frac{\partial u}{\partial y}\left[\int_0^{y}\frac{\partial u}{\partial x}dy\right]dy}_{\text{II}}=\underbrace{-\int_0^{\delta_h}\beta g(T-T_\infty)dy}_{\text{III}}\underbrace{-\nu\int_0^{\delta_h}\frac{\partial^2 u}{\partial y^2}dy}_{\text{IV}} \tag{12.179}$$

We evaluate term IV exactly as we did in the previous sections. Term II is integrated by parts, but since $v_\infty = 0$, only one term remains. We are left with

$$-2\int_0^{\delta_h}\left(u\frac{\partial u}{\partial x}\right)dy = -\int_0^{\delta_h}\beta g(T-T_\infty)dy + \nu\left.\frac{\partial u}{\partial y}\right|_{y=0} \tag{12.180}$$

We perform the same integration procedure on the energy equation:

$$-\int_0^{\delta_T}\frac{\partial}{\partial x}[u(T-T_\infty)]dy = \alpha\left.\frac{\partial T}{\partial y}\right|_{y=0} \tag{12.181}$$

Since the boundary layer equations are more complicated, we will restrict our analysis to fluids whose Prandtl number is one. This restriction is not severe since the growth of the hydrodynamic and thermal boundary layers must occur at a similar rate in a density-driven flow.

The remaining integrals and derivatives can be simplified if we change variables. Since we have no reference velocity, we assume a characteristic velocity, u_x, that is a function of position along the plate. Letting

$$\upsilon = \frac{u}{u_x} \qquad u_x \text{ is a function of } x \text{ to be determined} \tag{12.182}$$

$$\eta = \frac{y}{\delta} \qquad \delta \text{ is the thermal and hydrodynamic boundary layer thickness} \tag{12.183}$$

$$\theta = \frac{T-T_\infty}{T_s-T_\infty} \tag{12.184}$$

the integral forms of the momentum and energy equations become

$$-\frac{d}{dx}\delta u_x^2\int_0^1\upsilon^2 d\eta = -\beta g\delta(T_s-T_\infty)\int_0^1\theta\, d\eta + \left(\frac{\nu u_x}{\delta}\right)\left.\frac{\partial\upsilon}{\partial\eta}\right|_{\eta=0} \tag{12.185}$$

$$-\frac{d}{dx}\delta u_x\int_0^1\upsilon\theta\, d\eta = \frac{\alpha}{\delta}\left.\frac{\partial\theta}{\partial\eta}\right|_{\eta=0} \tag{12.186}$$

We develop approximate expressions for υ and θ as functions of η by fitting polynomials to the boundary conditions of the problem. With five boundary conditions, we can use fourth-order polynomials:

$$\begin{aligned}\upsilon &= a_o + a_1\eta + a_2\eta^2 + a_3\eta^3 + a_4\eta^4\\ \theta &= b_o + b_1\eta + b_2\eta^2 + b_3\eta^3 + b_4\eta^4\end{aligned} \tag{12.187}$$

The boundary conditions using the new variables are

$$\begin{aligned}&\eta = 0 \qquad \theta = 1 \qquad \upsilon = 0\\ &\frac{\partial^2\theta}{\partial\eta^2} = 0 \qquad \frac{\partial^2\upsilon}{\partial\eta^2} = \frac{\beta g\delta^2}{\nu u_x}(T_s-T_\infty)\end{aligned} \tag{12.188}$$

$$\eta = 1 \qquad \theta = 0 \qquad \upsilon = 0$$
$$\frac{\partial\theta}{\partial\eta} = \frac{\partial\upsilon}{\partial\eta} = \frac{\partial^2\theta}{\partial\eta^2} = \frac{\partial^2\upsilon}{\partial\eta^2} = 0 \tag{12.189}$$

Applying the boundary conditions shows the polynomials of (12.187) to be

$$\theta = 1 - 2\eta + 2\eta^3 - \eta^4$$
$$\upsilon = \frac{\beta g \delta^2}{6\nu u_x}(T_s - T_\infty)(1-\eta)^3\eta \tag{12.190}$$

Now, the factor appearing in our representation for the velocity

$$\frac{\beta g \delta^2}{6\nu u_x}(T_s - T_\infty) = \gamma \tag{12.191}$$

is a constant and, as such, tells us that $\delta \propto \sqrt{u_x}$. We can fold γ into the definition for u_x and simplify the representation for υ:

$$\upsilon = \eta(1-\eta)^3 \tag{12.192}$$

With the temperature and velocity profiles completed, we substitute those into the boundary layer equations and evaluate the integrals. This leads to two differential equations describing the boundary layer thickness:

Momentum

$$-\frac{1}{252}\frac{d}{dx}\left(u_x^2\delta\right) = -\frac{3}{10}\beta g(T_s - T_\infty)\delta + \frac{\nu u_x}{\delta} \tag{12.193}$$

Energy

$$-\frac{11}{504}\frac{d}{dx}(u_x\delta) = -\frac{2\alpha}{\delta} \tag{12.194}$$

From the previous discussion, we know that u_x is proportional to δ^2. We can use this knowledge to our advantage by inserting this relationship into the energy equation. This separates the pair of differential equations so we can solve for the boundary layer thickness alone. Starting with Equation 12.194, we have

$$-\frac{11}{504}\frac{d}{dx}(\delta^3) = -\frac{2\alpha}{\delta} \qquad \text{or} \qquad \frac{d}{dx}(\delta^4) \propto K \tag{12.195}$$

Separating and integrating equation (12.195) shows that $\delta \propto x^{1/4}$. To complete the solution, we use this information and the fact that $u_x \propto \delta^2$ and state

$$\delta = C_1 x^{1/4} \qquad u_x = C_2 x^{1/2} \tag{12.196}$$

Substituting terms from Equation 12.196 into the original differential equations (12.193) and (12.194) leaves us with two simultaneous, nonlinear algebraic equations for C_1 and C_2:

$$-\frac{5}{1008}C_1^2C_2^2 + \frac{3\beta g(T_s - T_\infty)}{10}C_1^2 - C_2\nu = 0$$
$$\frac{33}{4032}C_1^2C_2 - \alpha = 0 \qquad (12.197)$$

The solution for C_1 and C_2 is readily obtained:

$$C_1 = 4.49\mathrm{Pr}^{1/2}\left(\frac{20}{33} + \mathrm{Pr}\right)^{1/4}[\beta g(T_s - T_\infty)]^{-1/4}$$
$$C_2 = 6.05\left(\frac{20}{33} + \mathrm{Pr}\right)^{-1/2}[\beta g(T_s - T_\infty)]^{1/2} \qquad (12.198)$$

and so the boundary layer thickness can be written in terms of two important dimensionless groups, the *Prandtl* and *Grashof* numbers. The Grashof number is a ratio of buoyant to viscous forces in the flow:

$$\frac{\delta}{x} = 4.49\mathrm{Pr}^{-1/2}\left(\frac{20}{33} + \mathrm{Pr}\right)^{1/4}\mathrm{Gr}_x^{-1/4} \qquad (12.199)$$

$$\mathrm{Gr} = \frac{\beta g(T_s - T_\infty)x^3}{\nu^2} \qquad \frac{\text{Buoyant forces}}{\text{Viscous forces}} \qquad (12.200)$$

To determine the heat transfer coefficient, we perform an energy balance about the plate's surface equating heat flow through the plate by conduction to heat flow into the fluid by convection:

$$q_s'' = -\left(\frac{k_f(T_s - T_\infty)}{\delta}\right)\frac{d\theta}{d\eta}\bigg|_{\eta=0} = h(T_s - T_\infty) \qquad (12.201)$$

Using the dimensionless temperature distribution, in Equation 12.190 and the definition of the Nusselt number, we find

$$\mathrm{Nu}_x = \frac{hx}{k_f} = 0.445\left(\frac{20}{33} + \mathrm{Pr}\right)^{-1/4}\mathrm{Pr}^{1/2}\mathrm{Gr}_x^{1/4} \qquad (12.202)$$

Notice that the heat transfer coefficient decays much more slowly in a system governed by free convection than it does in a system governed by forced convection. This is a direct result of the slow growth of the boundary layer along the surface of the plate. In free convection $\delta \propto x^{1/4}$, whereas in forced convection $\delta \propto x^{1/2}$. Figure 12.13 shows the natural convection thermal boundary layer. This picture is a shadowgraph depicting lines of constant temperature. The closer the lines are spaced, the higher the temperature gradient [13].

FIGURE 12.13 Laminar free convection on a vertical plate. Shadowgraph image shows thermal boundary layer and lines of constant temperature. (Photograph courtesy of Eckert and Soehngen. Holman, J.P., Gartrell, H.E., and Soehngen, E.E., *J. Heat Transfer, Ser. C.*, 80, 1960. With permission.)

We can define an average value of the heat transfer coefficient using our averaging technique:

$$\bar{h} = \frac{1}{L}\int_0^L h(x)dx = \frac{4}{3}h(L) \tag{12.203}$$

$$\overline{\text{Nu}} = 0.59\left(\frac{20}{33} + \text{Pr}\right)^{-1/4} \text{Pr}^{-1/4}\text{Ra}^{1/4} \qquad \text{Ra} = \text{GrPr} \tag{12.204}$$

The *Rayleigh* number, Ra, is a grouping of the Grashof and Prandtl number. It is used in free convection systems as an analog of the Reynolds number. One defines the transition from laminar to turbulent free convection when Ra $\approx 10^9$. Most correlations will be written in terms of the Rayleigh number in honor of Lord Rayleigh who first calculated the critical value of this dimensionless parameter signifying the onset of natural convection [14,15].

Example 12.6 Optimal Fin Spacing

We are designing a high-output car stereo system. The power transistors driving the speakers need to dissipate a great amount of heat and are mounted onto large, finned heat sinks. Being a top-of-the-line model, the heat sinks are made of copper and, due to their high thermal conductivity, are at a uniform temperature of 65°C at maximum volume. Each fin is 10 cm high, 1 mm thick, and 5 cm deep. To take full advantage of free convection and achieve maximum power dissipation, we would like to calculate the optimal spacing of these fins. The maximum air temperature in summer will be 27°C and the air properties at this temperature are

$$\rho = 1.1614\ \text{kg/m}^3 \qquad k = 0.0263\ \text{W/m K} \qquad \text{Pr} = 0.707$$

$$\nu = 15.89 \times 10^{-6}\ \text{m}^2/\text{s} \qquad \alpha = 22.5 \times 10^{-6}\ \text{m}^2/\text{s}$$

To obtain the maximum amount of heat transfer from the fins, we want to keep the heat transfer coefficient as high as possible, keep the driving force as high as possible, and pack as many fins as possible into the smallest space. Our knowledge of boundary layer theory tells us that the boundary layer will grow along the length of the fin and if we space the fins too close together, the boundary layers will overlap and we will not have the maximum driving force for heat transfer. If we space the fins too far apart, we do not have enough surface area to maximize the overall heat transfer. The intuitive optimum occurs when the boundary layers just touch at the end of the fins (Figures 12.14 and 12.15).

We can use the solution to the boundary layer equations just derived to calculate the boundary layer thickness at the end of the fins and hence the optimum fin spacing. The Grashof number at maximum length and maximum free stream temperature is

$$\text{Gr} = \frac{g\beta(T_s - T_\infty)x^3}{\nu^2} = \frac{\left(9.8\dfrac{\text{m}}{\text{s}^2}\right)\left(\dfrac{1}{319\text{K}}\right)(338-300\text{K})(0.1\,\text{m})^3}{\left(15.89\times 10^{-6}\dfrac{\text{m}^2}{\text{s}}\right)^2} = 4.623\times 10^6$$

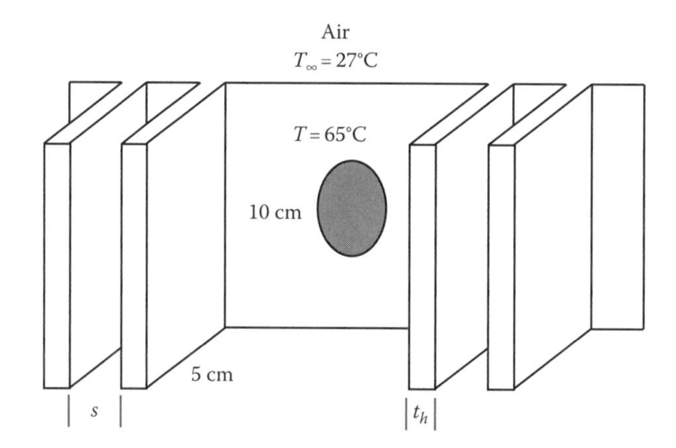

FIGURE 12.14 Fin spacing for natural convection.

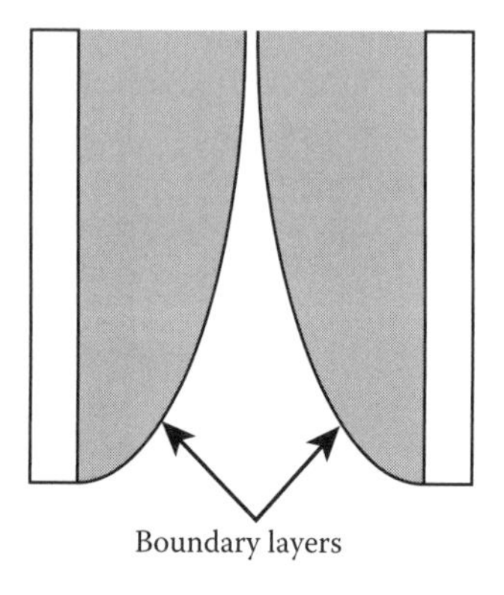

FIGURE 12.15 Optimal boundary layer thickness for natural convection.

The boundary layer thickness at the end of the heat sink is

$$\delta = 4.49\text{Pr}^{1/2}\left(\frac{20}{33}+\text{Pr}\right)^{1/4}\text{Gr}^{-1/4}x$$

$$= 4.49(0.707)^{1/2}\left(\frac{20}{33}+0.707\right)^{1/4}(4.623\times10^{6})^{-1/4}(0.1) = 0.0087\text{ m}$$

Therefore, the optimum separation is twice the film thickness plus the thickness of the fin:

$$\Delta_{opt} = 2\delta + t_h = 2(0.0087) + 0.001 = 0.0184\text{ m}$$

12.7.3 Film Condensation on a Vertical Plate

Our last example of a thermal and hydrodynamic boundary layer system looks at film condensation on a vertical plate. In this system, we have a cooled plate in contact with a vapor phase at the vapor's saturation temperature. On the surface of the plate, the vapor begins to condense and forms a thin liquid film. As the film grows, it begins to flow down the plate under the influence of a gravitational body force. The situation is shown in Figure 12.16.

There is a mechanistic similarity between laminar film condensation and free convection. Regardless of direction, a gravitational force drives the flow. We would expect that the growth of the boundary layer, or the condensing film, would follow the same scaling rules as in free convection. The heat transfer coefficient and Nusselt number should also have the characteristic behavior observed in free convection.

There are some differences in the growth of the boundary layer here because we have a two-phase system. In natural convection, the edges of the hydrodynamic and thermal boundary layers signified the end of any changes in velocity and temperature. Here we actually have two hydrodynamic boundary layers and one thermal boundary layer. The liquid hydrodynamic boundary layer, signified by the film thickness, does not represent the end of velocity changes. A vapor hydrodynamic boundary layer is attached to it and exhibits a velocity gradient because the fluid drags the vapor

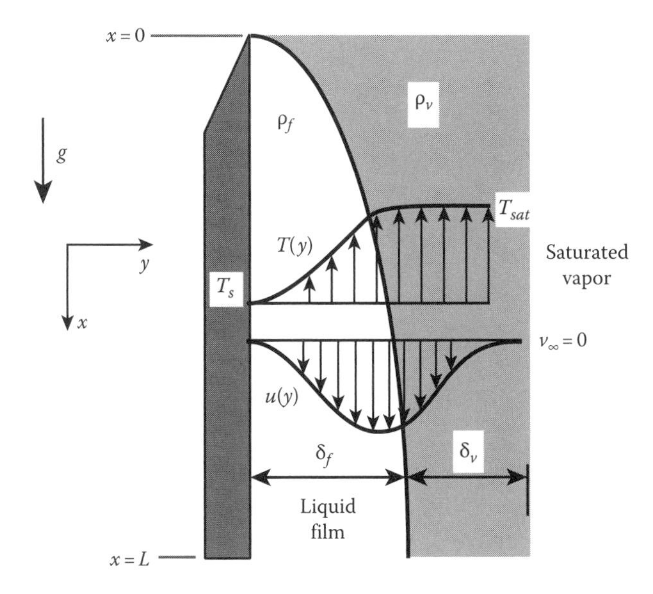

FIGURE 12.16 Laminar film condensation on a vertical plate.

along with it. The thermal boundary exists solely within the liquid since the edge of the boundary layer represents liquid at a temperature of T_{sat}, the same temperature as the surrounding vapor.

To fully analyze this system, we must write the boundary layer equations for each fluid. Borrowing from the discussion of free convection, we assume that the pressure gradient in the x-direction is due solely to the weight of the fluid, evaluated at T_{sat}. Within the liquid, we have

$$\frac{\partial u_f}{\partial x}+\frac{\partial v_f}{\partial y}=0 \tag{12.205}$$

$$u_f\frac{\partial u_f}{\partial x}+v_f\frac{\partial u_f}{\partial y}=\frac{\mu_f}{\rho_f}\frac{\partial^2 u_f}{\partial y^2}+g\left(\frac{\rho_{f\infty}-\rho_f}{\rho_f}\right) \tag{12.206}$$

$$u_f\frac{\partial T_f}{\partial x}+v_f\frac{\partial T_f}{\partial y}=\alpha_f\frac{\partial^2 T_f}{\partial y^2} \tag{12.207}$$

and within the vapor phase,

$$\frac{\partial u_v}{\partial x}+\frac{\partial v_v}{\partial y}=0 \tag{12.208}$$

$$u_v\frac{\partial u_v}{\partial x}+v_v\frac{\partial u_v}{\partial y}=\frac{\mu_v}{\rho_v}\frac{\partial^2 u_v}{\partial y^2}+g\left(\frac{\rho_{v\infty}-\rho_v}{\rho_v}\right) \tag{12.209}$$

$$u_v\frac{\partial T_v}{\partial x}+v_v\frac{\partial T_v}{\partial y}=\alpha_v\frac{\partial^2 T_v}{\partial y^2} \tag{12.210}$$

The boundary conditions for these equations specify what occurs at the plate surface, at the interface between the two boundary layers, and at the interface between the vapor boundary layer and the free stream:

$$y=0 \qquad u_f=v_f=0 \qquad T_f=T_s \tag{12.211}$$

$$y=\delta_f \qquad u_f=u_v \qquad v_f=v_v \qquad T_f=T_v \tag{12.212}$$

$$y=\delta_f \qquad -\mu_f\frac{\partial u_f}{\partial y}=-\mu_v\frac{\partial u_v}{\partial y} \qquad -k_f\frac{\partial T_f}{\partial y}=-k_v\frac{\partial T_v}{\partial y} \tag{12.213}$$

$$y=\delta_f+\delta_v \qquad u_v=v_v=0 \qquad T_v=T_{sat} \tag{12.214}$$

The formal solution to this set of equations was determined by Koh et al. [16]. We will not repeat that solution here but instead will follow an approximate solution first proposed by Nusselt [17].

In the Nusselt's approximation, he assumed that the velocities were very small, and so all the convective terms (acceleration terms) in the momentum equations and the energy equations could be neglected. This means the continuity equation was satisfied identically within the liquid and the vapor and that the flow and temperature fields in each phase were fully developed. The momentum equation in the liquid for fully developed flow becomes

$$\frac{d^2u_f}{dy^2} = g\left(\frac{\rho_f - \rho_v}{\mu_f}\right) \tag{12.215}$$

and the energy equation is

$$\frac{d^2T_f}{dy^2} = 0 \tag{12.216}$$

Since the viscosity of the vapor is much smaller than the viscosity of the liquid, the shear stress at the vapor–liquid interface is virtually zero. Nusselt used this approximation to eliminate the vapor momentum equation by specifying that the velocity gradient was zero at the edge of the condensate film:

$$y = \delta_f \qquad \frac{\partial u_f}{\partial y} = 0 \tag{12.217}$$

This condition is equivalent to stating that the vapor velocity is a constant. Since the velocity is zero in the free stream, the vapor velocity is approximated as being zero at the edge of the condensate film too.

We can easily integrate the momentum equation in the liquid to obtain

$$u_f = \frac{g(\rho_f - \rho_v)y^2}{2\mu_f} + a_1 y + a_2 \tag{12.218}$$

The constants of integration are evaluated using the no-slip condition at the plate surface and the no stress boundary condition at the edge of the condensate film. The final parabolic velocity profile is

$$u_f = \frac{g(\rho_f - \rho_v)\delta_f^2}{2\mu_f}\left[\left(\frac{y}{\delta_f}\right)^2 - 2\left(\frac{y}{\delta_f}\right)\right] \tag{12.219}$$

We use this profile to calculate the mass flow rate of condensate in the same way we calculated the flow rate of fluid draining from the walls of a vessel in Chapter 11. Assuming the plate width is unity, we integrate over the velocity profile to find

$$\dot{m}(x) = \int_0^{\delta_f} \rho_f u_f dy = \frac{g\rho_f(\rho_f - \rho_v)\delta_f^2}{2\mu_f}\int_0^{\delta_f}\left[\left(\frac{y}{\delta_f}\right)^2 - 2\left(\frac{y}{\delta_f}\right)\right]dy$$

$$= \frac{g\rho_f(\rho_f - \rho_v)\delta_f^3}{3\mu_f} \tag{12.220}$$

The mass flow rate is a function of position along the plate because the film thickness, δ_f, changes along the plate.

The energy equation (12.216) can be integrated using boundary conditions that specify the surface temperature of the plate and the temperature at the film–vapor interface. Here, we make the approximation that $T_f = T_{sat}$ when $y = \delta_f$:

$$T_f = T_s + (T_{sat} - T_s)\left(\frac{y}{\delta_f}\right) \tag{12.221}$$

The temperature profile enables us to calculate the heat flow through the plate's surface and the heat transfer coefficient. Applying Fourier's law at the plate's surface and equating it to Newton's law there, we have

$$q'' = k_f \left.\frac{\partial T}{\partial y}\right|_{y=0} = h(T_{sat} - T_s) = \frac{k_f(T_{sat} - T_s)}{\delta_f} \tag{12.222}$$

so $h = k_f/\delta_f$, and h is akin to a conduction-like resistance.

Finally, we must evaluate the film thickness to obtain the velocity profile, mass flow rate, and heat transfer coefficient. We can use a form of macroscopic mass and energy balance to relate the film thickness to the heat transferred during condensation. The differential elements we write the balances about are shown in Figure 12.17. The mass balance states

$$\begin{array}{ccccccc} In & - & Out & + & Gen & = & Acc \\ \dot{m}(x) + \dot{m}_c\Delta x & - & \dot{m}(x+\Delta x) & + & 0 & = & 0 \end{array} \tag{12.223}$$

or in differential form using Equation 12.220

$$\dot{m}_c = \frac{d\dot{m}}{dx} = \left[\frac{g\rho_f(\rho_f - \rho_v)}{3\mu_f}\right]\delta_f^2 \frac{d\delta_f}{dx} \tag{12.224}$$

The energy balance states

$$\begin{array}{ccccc} In & - & Out & = & Acc \\ \rho_f C_{pf} T(x)\dot{m}(x) + \dot{m}_c h_{fg}\Delta x & - & \rho_f C_{pf} T(x+\Delta x)\dot{m}(x+\Delta x) - q''\Delta x & = & 0 \end{array} \tag{12.225}$$

or in differential form

$$\rho_f C_{pf} \frac{d}{dx}(\dot{m}T) = \dot{m}_c h_{fg} - q'' \tag{12.226}$$

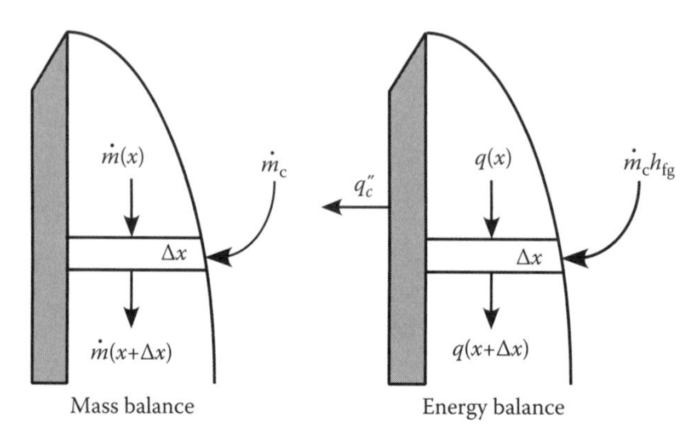

FIGURE 12.17 Macroscopic mass and energy balances.

The release of latent heat of vaporization during condensation is generally large compared to any sensible heat changes that occur in the film due to its change in temperature from T_s to T_{sat}. Therefore, as a first approximation, we neglect the sensible heat term in Equation 12.226 on the left-hand side and rewrite the equation as

$$q'' = \dot{m}_c h_{fg} \tag{12.227}$$

Substituting for $\dot{m}_c$ using Equation 12.224 and for q'' using Equation 12.222 leaves a first-order differential equation for δ_f^4:

$$\frac{d}{dx}\left(\delta_f^4\right) = \frac{k_f \mu_f (T_{sat} - T_s)}{g\rho_f(\rho_f - \rho_v)h_{fg}} \tag{12.228}$$

We integrate this equation realizing that the boundary layer or film begins at the leading edge of the plate ($\delta_f = 0$ at $x = 0$):

$$\delta_f = \left[\frac{4k_f \mu_f (T_{sat} - T_s)}{g\rho_f(\rho_f - \rho_v)h_{fg}} x\right]^{1/4} \tag{12.229}$$

Notice that the boundary layer thickness increases as $x^{1/4}$ just as it did in free convection. The heat transfer coefficient is also similar in behavior to free convection. Using our definition for h, from Equation 12.222, gives

$$h = \left[\frac{g\rho_f(\rho_f - \rho_v)h_{fg}k_f^3}{4\mu_f(T_{sat} - T_s)x}\right]^{1/4} \tag{12.230}$$

$$\text{Nu}_x = \frac{hx}{k_f} = \left[\frac{g\rho_f(\rho_f - \rho_v)h_{fg}x^3}{4\mu_f k_f(T_{sat} - T_s)}\right]^{1/4} \tag{12.231}$$

We can rewrite the Nusselt number as a series of dimensionless groups:

$$\text{Nu}_x = (\text{Ja})^{-1/4}(\text{Gr}_c)^{1/4}(\text{Pr})^{1/4} \tag{12.232}$$

The first dimensionless group is the Jakob number defined as

$$\text{Ja} = \frac{C_p(T_s - T_{sat})}{h_{fg}} - \frac{\text{Sensible heat absorbed}}{\text{Latent heat absorbed}} \tag{12.233}$$

and represents a ratio of the sensible heat absorbed by the condensing surface to the latent heat released by the vapor upon condensing per unit mass. The second dimensionless group is often termed the Grashof number for condensation and is defined as

$$\text{Gr}_c = \frac{g\left(1 - \dfrac{\rho_v}{\rho_f}\right)L^3}{\nu^2} - \frac{\text{Buoyant forces}}{\text{Viscous forces}} \tag{12.234}$$

The magnitude of this number reflects the effect of buoyancy-induced fluid motion on the heat transfer.

12.8 MASS TRANSFER BOUNDARY LAYERS

In the past sections, we have concentrated on the hydrodynamic and thermal boundary layers. We now want to consider the concentration boundary layer and those aspects peculiar to mass transfer. Before we begin to highlight differences between mass transfer and the other transport phenomena, it pays to consider the similarities. Looking at Equations 12.32 and 12.33, one can immediately see that if we neglect the generation terms, the equations are nearly identical in form and so the heat transfer and mass transfer coefficients should be similar.

We can follow the same procedure for solving the species continuity equation that we used in solving the energy equation (Section 12.7). We postulate a boundary layer thickness δ_c for the transport of species a and a boundary layer thickness ratio, Δ_c, which is analogous to the thermal boundary layer ratio, Δ_T, introduced in the last section. The solution of the species continuity equation

$$u\frac{\partial c_a}{\partial x}+v\frac{\partial c_a}{\partial y}=D_{ab}\frac{\partial^2 c_a}{\partial y^2} \tag{12.235}$$

follows the exact same integration procedure we used for the energy equation. Even the boundary conditions are of the same form. Here, we need only assume that the mass flux of species a from the surface is very small so $v \approx 0$ there:

$$y=0 \qquad c_a=c_{as} \qquad u=v=0 \qquad \frac{\partial^2 c_a}{\partial y^2}=\frac{\partial^2 u}{\partial y^2}=0 \tag{12.236}$$

$$y=\delta_c \qquad c_a=c_{a\infty} \qquad u=v_\infty \qquad \frac{\partial c_a}{\partial y}=\frac{\partial u}{\partial y}=0 \tag{12.237}$$

Since the solution for the concentration boundary layer follows so directly from the solution to the thermal boundary layer, we can use the Nusselt number correlations we developed in the previous section to estimate the mass transfer coefficient. All we need to do is replace the Nusselt number with its mass transfer counterpart, the Sherwood number, and replace the Prandtl number with its mass transfer counterpart, the Schmidt number. Thus, for laminar flow, we have

$$\text{Sh}_x=\frac{k_{cx}x}{D_{ab}}=0.332(\text{Re}_x)^{1/2}(\text{Sc})^{1/3} \qquad \text{Sc}\gg 1 \tag{12.238}$$

$$\text{Sh}_x=\frac{k_{cx}x}{D_{ab}}=0.53(\text{Re}_x)^{1/2}(\text{Sc})^{1/2} \qquad \text{Sc}\ll 1 \tag{12.239}$$

Diffusivities are nearly always smaller than kinematic viscosities, especially for fluids (see Chapter 3), and so the Schmidt number is normally much larger than 1. Thus, Equation 12.239 is of very limited utility.

Example 12.7 Formation of Black Ice

Some of the most dangerous road conditions exist on clear fall or spring nights after snow has thawed on the roadway. Though the air temperature may be above freezing, the water may freeze and form a thin layer of invisible ice, called black ice. We consider a roadway with a patch of water on its surface, 0.5 m wide (Figure 12.18). The wind speed is 3 m/s, and the flow is laminar.

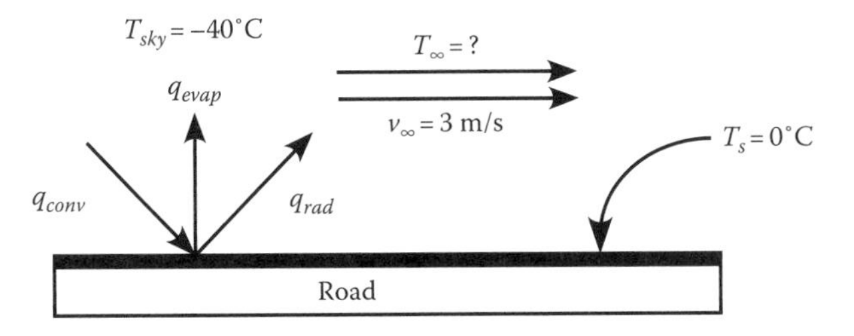

FIGURE 12.18 Black ice formation on a road surface.

The relative humidity of the ambient air is essentially 0. The roadway is asphalt and exposed to a clear night sky whose effective temperature is −40°C. Assuming no water evaporates from the surface and that conduction from the ground is negligible, at what air temperature will water just freeze? What happens if we allow for evaporation?

We can begin to analyze this system as a case of flow over a flat plate. Drawing a simplified diagram of the situation, we see that there are three primary mechanisms we need to consider for heat transfer, evaporation from the surface, radiation, and convection. Performing an energy balance about the water, whatever heat is transferred from the air to the water is lost by evaporation or radiation:

$$In \quad - Out = Acc$$
$$q_{conv} - q_{evap} - q_{rad} = 0$$

We use Newton's law of cooling to describe the convective heat transfer, q_{conv}, and the Stefan–Boltzmann law for q_{rad}, the radiative heat transfer:

$$q_{conv} = \bar{h}A_s(T_s - T_\infty) \qquad q_{rad} = \sigma^r \varepsilon^r \mathcal{F}_{12} A_s\left(T_s^4 - T_{sky}^4\right)$$

The heat removed by evaporation is tied to the mass transfer rate, $\dot{m}$, via the heat of vaporization, h_{fg}:

$$q_{evap} = \dot{m}h_{fg} = N_a A_s M_{wa} h_{fg}$$

We can use the analog of Newton's law for mass transfer to tie the mass flux of water to the mass transfer coefficient:

$$q_{evap} = \bar{k}_c A_s(c_{as} - c_{a\infty})M_{wa}h_{fg}$$

Combining into the overall energy balance, we have

$$\bar{h}A_s(T_\infty - T_s) - \sigma^r \varepsilon^r \mathcal{F}_{12} A_s\left(T_s^4 - T_{sky}^4\right) - \bar{k}_c A_s(c_{as} - c_{a\infty})M_{wa}h_{fg} = 0$$

The area for transport is the same in all cases so it can be removed. The sky envelops the roadway and so the view factor, $\mathcal{F}_{12}$, between the road and sky is 1. We need to calculate the heat transfer coefficient and mass transfer coefficient, and for that, we need the properties of air and water. Since we do not know the ambient air temperature, we can base our properties initially only on the surface temperature, T_s. Later we can adjust properties if necessary:

$\nu_{air} = 13.8 \times 10^{-6}$ m²/s $\quad k_{air} = 24.3 \times 10^{-3}$ W/m K $\quad$ Pr = 0.714

$D_{ab} = 0.26 \times 10^{-7}$ m²/s $\quad h_{fg} = 2502 \times 10^3$ J/kg $\quad P^{sat} = 0.00611$ bar

$\varepsilon^r = 0.85$

Considering the heat transfer coefficient first, we need an average value, and so we use the correlation developed in Section 12.7.1:

$$\overline{\text{Nu}_L} = 0.664\text{Re}_L^{1/2}\text{Pr}^{1/3}$$

The Reynolds number for the flow is

$$\text{Re}_L = \frac{v_\infty L}{\nu_{air}} = \frac{3(0.5)}{13.8\times10^{-6}} = 108{,}696$$

and so we are in the laminar regime. The Nusselt number and heat transfer coefficient are

$$\overline{\text{Nu}_L} = 0.664(108{,}696)^{1/2}(0.714)^{1/3} = 195.7$$

$$\bar{h} = \frac{\overline{\text{Nu}_L} k_f}{L} = \frac{195.7(0.0243)}{0.5} = 9.51\ \text{W/m}^2\,\text{K}$$

Having solved for the heat transfer coefficient, we can deal with the first part of the problem and calculate the required air temperature:

$$\bar{h}(T_\infty - T_s) = \sigma^r\varepsilon^r\left(T_s^4 - T_{sky}^4\right)$$

Substituting in for the temperatures and constants shows that the air temperature can be quite high and yet black ice can form on the surface:

$$T_\infty = 273 + \frac{5.67\times10^{-8}(0.85)}{9.51}[(273)^4 - (233)^4] = 286.2\ \text{K}$$

If we consider evaporation, we can remove even more energy from the water and raise the air temperature for forming black ice further. Using the correlation for the Sherwood number, Equation 12.238, and realizing by analogy that $\overline{\text{Sh}_L} = 2\text{Sh}_x(L)$, we can estimate a mass transfer coefficient:

$$\overline{\text{Sh}_L} = 0.664\text{Re}_L^{1/2}\,\text{Sc}^{1/3} = 0.664(108{,}696)^{1/2}\left(\frac{13.8\times10^{-6}}{0.26\times10^{-7}}\right)^{1/3} = 1772$$

$$\overline{k_c} = \frac{\overline{\text{Sh}_L} D_{ab}}{L} = \frac{1772(0.26\times10^{-7})}{0.5} = 9.22\times10^{-5}\ \text{m/s}$$

The evaporation rate, in kg/s, can now be found using the molecular weight of water:

$$\frac{\dot{m}}{A_s} = k_c M_{wa}\left(\frac{P^{sat}}{RT} - \frac{P_\infty}{RT}\right) = 9.22\times10^{-5}(18)\left[\frac{0.006(1000)}{(8.31\times10^{-2})273} - 0\right]$$

$$= 4.39\times10^{-4}\ \text{kg/m}^2\,\text{s}$$

Substituting into the energy balance and solving for T_∞ yield

$$T_\infty = \frac{\sigma^r\varepsilon^r}{h}\left(T_s^4 - T_{sky}^4\right) + \frac{\dot{m}h_{fg}}{hA_s} + T_s$$

$$= \frac{5.67\times10^{-8}(0.85)}{14.62}(273^4 - 233^4) + \frac{4.56\times10^{-4}(2502\times10^3)}{14.62} + 273 = 359.6\ \text{K}$$

and indeed evaporation does increase the air temperature that can be sustained while still allowing the water to freeze. These temperatures are a too bit high for common sense. On one hand, you can see why evaporative cooling is so effective. On the other hand, these fluxes show how important conduction from the ground would be since actual air temperatures need to be much lower for the ice to form.

12.8.1 Effects of Mass Transfer Rate

The previous solution to the species continuity equation assumes we have a sparingly soluble substrate. If the mass transfer rate is higher, we need to reconsider that solution because the velocity of material perpendicular to the plate will not be zero at the plate's surface. In this section, we will show that under conditions where the mass transfer rate is higher, the mass transfer rate and its direction affect both the hydrodynamic and concentration boundary layers. For simplicity, we will assume that the Schmidt number is nearly one and that $\partial P/\partial x$ is still nearly 0.

The continuity, momentum, and species continuity boundary layer equations are

Continuity

$$\frac{\partial u}{\partial x}+\frac{\partial v}{\partial y}=0 \tag{12.240}$$

Momentum

$$u\frac{\partial u}{\partial x}+v\frac{\partial u}{\partial y}=\nu\frac{\partial^2 u}{\partial y^2} \tag{12.241}$$

Continuity of a

$$u\frac{\partial c_a}{\partial x}+v\frac{\partial c_a}{\partial y}=D_{ab}\frac{\partial^2 c_a}{\partial y^2} \tag{12.242}$$

The boundary conditions for this problem reflect the fact that the mass transfer rate is higher. At the free stream, we have the familiar conditions we have used before:

$$y=\delta_h=\delta_c=\delta \qquad u=v_\infty \qquad c_a=c_{a\infty} \qquad \frac{\partial u}{\partial y}=\frac{\partial c_a}{\partial y}=0 \tag{12.243}$$

At the surface of the plate, we have no slip, and we know that the concentration in the boundary layer must be in equilibrium with the concentration at the surface of the plate. Finally, the flux of liquid at the plate's surface must be zero since the plate is impermeable:

$$y=0 \qquad u=0 \qquad c_a=K_{ea}c_{as} \qquad N_{bo}=0 \tag{12.244}$$

Notice that $v\neq 0$ and so we have fewer boundary conditions than we had in the previous boundary layer problems. The polynomial approximations for the velocity and concentration profiles will be a bit cruder.

We still assume that the velocity and concentration profiles are self-similar in x, allowing us to write u and c_a as functions of y/δ:

$$u = v_\infty f_1\left(\frac{y}{\delta}\right) \qquad c_a = c_{a\infty} f_2\left(\frac{y}{\delta}\right) \tag{12.245}$$

f_1 and f_2 are arbitrary functions and the boundary layer thickness, δ, is a function only of x.

12.8.1.1 Hydrodynamic Boundary Layer Solution

The first task in solving the momentum equation is again to eliminate the velocity, v, using the continuity equation:

$$v = v(x,0) - \int_0^y \frac{\partial u}{\partial x} dy \tag{12.246}$$

Since the mass transfer rate is high, the velocity, $v(x,0)$, is not zero. We can evaluate $v(x,0)$ in terms of the molar flux of a at the surface (the dissolution rate). The total mass flux of species from the surface of the plate is given in terms of the molar fluxes of the two species a and b:

$$\rho v(x,0) = M_{wa} N_{ao} + M_{wb} N_{bo} \tag{12.247}$$

Since the plate is impermeable to b, N_{bo} is zero and the $v(x,0)$ is

$$v(x,0) = \frac{M_{wa} N_{ao}}{\rho} \tag{12.248}$$

Substituting into the integrated continuity equation, we have

$$v = \frac{M_{wa} N_{ao}}{\rho} - \int_0^y \frac{\partial u}{\partial x} dy \tag{12.249}$$

We evaluate N_{ao} using Fick's Law at the surface:

$$N_{ao} = -c_t D_{ab} \left.\frac{\partial x_a}{\partial y}\right|_{y=0} + x_{ao}(N_{ao} + N_{bo}) \tag{12.250}$$

Knowing that N_{bo} vanishes at the surface, we solve for N_{ao} and substitute it into the continuity equation:

$$v = -\left[\frac{M_{wa}}{\rho} \frac{c_t D_{ab}}{(1 - x_{ao})}\right] \left.\frac{\partial x_a}{\partial y}\right|_{y=0} - \int_0^y \frac{\partial u}{\partial x} dy \tag{12.251}$$

Assuming the total concentration, c_t, is constant and substituting into the momentum equation yield

$$u\frac{\partial u}{\partial x} - \frac{\partial u}{\partial y}\left\{\left[\frac{M_{wa} D_{ab}}{\rho(1 - x_{ao})}\right] \left.\frac{\partial c_a}{\partial y}\right|_{y=0}\right\} - \frac{\partial u}{\partial y} \int_0^y \frac{\partial u}{\partial x} dy = \nu \frac{\partial^2 u}{\partial y^2} \tag{12.252}$$

Now we multiply the momentum equation through by −1 and integrate it over the boundary layer thickness, δ:

$$\underbrace{-\int_0^\delta u\frac{\partial u}{\partial x}dy}_{\text{I}}+\underbrace{\left\{\left[\frac{M_{wa}D_{ab}}{\rho(1-x_{ao})}\right]\frac{\partial c_a}{\partial y}\bigg|_{y=0}\right\}\int_0^\delta\frac{\partial u}{\partial y}dy}_{\text{II}}+\underbrace{\int_0^\delta\frac{\partial u}{\partial y}\left[\int_0^y\frac{\partial u}{\partial x}dy\right]dy}_{\text{III}}=\underbrace{-\nu\int_0^\delta\frac{\partial^2 u}{\partial y^2}dy}_{\text{IV}} \quad (12.253)$$

We have evaluated terms III and IV in previous boundary layer equation solutions using integration by parts and direct integration. Integrating term II gives

$$\left\{\left[\frac{M_{wa}D_{ab}}{\rho(1-x_{ao})}\right]\frac{\partial c_a}{\partial y}\bigg|_{y=0}\right\}\int_0^y\frac{\partial u}{\partial y}dy=\left\{\left[\frac{M_{wa}D_{ab}}{\rho(1-x_{ao})}\right]\frac{\partial c_a}{\partial y}\bigg|_{y=0}\right\}(v_\infty-0) \quad (12.254)$$

where we have used the boundary conditions on u to evaluate the integral at its limits. Combining all the previous integrals gives

$$-2\int_0^\delta u\frac{\partial u}{\partial x}dy+v_\infty\int_0^\delta\frac{\partial u}{\partial x}dy=-v_\infty\left\{\left[\frac{M_{wa}D_{ab}}{\rho(1-x_{ao})}\right]\frac{\partial c_a}{\partial y}\bigg|_{y=0}\right\}+\nu\frac{\partial u}{\partial y}\bigg|_{y=0} \quad (12.255)$$

We can now interchange the order of integration and differentiation on the left-hand side and rewrite our momentum equation as

$$\frac{d}{dx}\int_0^\delta[u(v_\infty-u)]dy=-v_\infty\left\{\left[\frac{M_{wa}D_{ab}}{\rho(1-x_{ao})}\right]\frac{\partial c_a}{\partial y}\bigg|_{y=0}\right\}+\nu\frac{\partial u}{\partial y}\bigg|_{y=0} \quad (12.256)$$

This is the basic differential equation for the boundary layer thickness. All the definite integrals and evaluated derivatives are constants.

We divide both sides by v_∞^2, define a dimensionless velocity $u^*=u/v_\infty$, and rewrite Equation 12.256 as

$$\frac{d}{dx}\int_0^\delta[u^*(1-u^*)]dy=-\left\{\left[\frac{M_{wa}D_{ab}}{\rho v_\infty(1-x_{ao})}\right]\frac{\partial c_a}{\partial y}\bigg|_{y=0}\right\}+\frac{\nu}{v_\infty}\frac{\partial u^*}{\partial y}\bigg|_{y=0} \quad (12.257)$$

To complete the representation, we define a dimensionless y-variable $y^*=y/\delta$ and dimensionless concentration $c^*=(c_a-K_{eq}c_{as})/(c_{a\infty}-K_{eq}c_{as})$ so that

$$\frac{d\delta}{dx}\int_0^1[u^*(1-u^*)]dy^*=-\left\{\left[\frac{M_{wa}D_{ab}(c_{a\infty}-K_{eq}c_{as})}{\rho v_\infty\delta(1-x_{ao})}\right]\frac{\partial c^*}{\partial y^*}\bigg|_{y^*=0}\right\}+\frac{\nu}{v_\infty\delta}\frac{\partial u^*}{\partial y^*}\bigg|_{y^*=0} \quad (12.258)$$

Though we have assumed that the concentration and hydrodynamic boundary layers obey the Reynolds analogy, the profiles are not independent. The direction of mass transfer and the magnitude of mass transfer affect both boundary layers.

To evaluate the integrals and derivatives in Equation 12.258, we make use of our similarity hypothesis. With three boundary conditions for the velocity and concentration, we can use quadratic representations. The dimensionless variables we have chosen lead to identical polynomial representations for the concentration and velocity profiles:

$$u^* = a + by^* + cy^{*2} = 2y^* - y^{*2}$$
$$c^* = d + ey^* + fy^{*2} = 2y^* - y^{*2} \tag{12.259}$$

Evaluating the integral and derivatives in Equation 12.258, we obtain the following constants:

$$\int_0^1 [u^*(1-u^*)]\,dy^* = \frac{2}{15} \qquad \left.\frac{\partial u^*}{\partial y^*}\right|_{y^*=0} = \left.\frac{\partial c^*}{\partial y^*}\right|_{y^*=0} = 2 \tag{12.260}$$

Now we substitute back into Equation 12.258 and rearrange the equation to collect all the δ to one side:

$$\frac{2}{15}\delta\frac{\partial \delta}{\partial x} = \frac{2\nu}{v_\infty} - \left[\frac{2M_{wa}D_{ab}(c_{a\infty} - K_{eq}c_{as})}{v_\infty\rho(1-x_{ao})}\right] \tag{12.261}$$

We end up with a first-order differential equation in δ^2 that we can integrate directly to obtain the film thickness:

$$\delta = \sqrt{30\frac{\nu x}{v_\infty} - 30\left[\frac{M_{wa}D_{ab}(c_{a\infty} - K_{eq}c_{as})x}{v_\infty\rho(1-x_{ao})}\right] + C_o} \tag{12.262}$$

Since the boundary layer thickness begins when the fluid contacts the plate, $\delta = 0$ at $x = 0$, and the constant $C_o = 0$. We rewrite the boundary layer thickness in terms of familiar dimensionless numbers and so make a little more sense of the parameters controlling it:

$$\delta = \frac{5.48x}{\sqrt{\text{Re}_x}}\sqrt{1 - \frac{M_{wa}(c_{a\infty} - K_{eq}c_{as})}{\rho(1-x_{ao})}\text{Sc}^{-1}} \tag{12.263}$$

Equation 12.263 demonstrates that the mass flux from the surface not only affects the concentration boundary layer but also affects the hydrodynamic boundary layer. Both the *magnitude* and the *direction* of the mass flux are important. Depending upon which direction the mass is flowing, we can actually use mass transfer to increase or decrease the hydrodynamic boundary layer thickness. This can have profound effects on mass transfer equipment. In liquid–liquid and liquid–gas contactors, for example, coalescence of the dispersed phased and hence the size distribution of that phase has long been recognized to depend upon the direction of mass transfer [18]. If one assumes that coalescence can only occur when the drops get within a hydrodynamic boundary layer or two of one another, then the direction and magnitude of mass transfer can affect the amount of agitation required to bring the particles together and hence affect the coalescence rate.

Normally, we are more interested in the friction factor rather than the boundary layer thickness. Earlier we showed that the friction factor was inversely proportional to the boundary layer thickness. It can be given by

$$C_f = -\left(\frac{2\mu}{\rho v_\infty \delta}\right)\left.\frac{\partial u^*}{\partial y^*}\right|_{y^*=0} = \frac{0.73}{\sqrt{\text{Re}_x}\sqrt{1 - \frac{M_{wa}(c_{a\infty} - K_{eq}c_{as})}{\rho(1-x_{ao})}\text{Sc}^{-1}}} \tag{12.264}$$

To determine the mass transfer coefficient, we return to the Reynolds analogy. In the Reynolds analogy, the Schmidt number is one and the Sherwood number is

$$\mathrm{Sh} = \frac{C_f \mathrm{Re}_x}{2} = \frac{0.37\sqrt{\mathrm{Re}_x}}{\sqrt{1 - \dfrac{M_{wa}(c_{a\infty} - K_{eq}c_{as})}{\rho(1 - x_{ao})}\mathrm{Sc}^{-1}}} \tag{12.265}$$

and we can see that the direction and rate of mass transfer also affects the mass transfer coefficient as well.

12.8.2 Effect of Chemical Reaction [19,20]

We have seen how high mass transfer rates can affect friction factors as well as mass transfer coefficients. Often, mass transfer is accompanied by a chemical reaction and that reaction can affect the mass transfer coefficient depending on whether the reaction acts a source of component, a, or as a sink. In this section, we will consider the limiting case of an incompressible fluid flowing over a sparingly soluble substrate with Sc ~ 1. A first-order or pseudo-first-order chemical reaction occurs in the liquid, and the reaction neither liberates nor absorbs heat from the surroundings.

We must solve the continuity, momentum, and species continuity equations. The species continuity equation now includes a generation term:

Continuity of species a

$$u\frac{\partial c_a}{\partial x} + v\frac{\partial c_a}{\partial y} = D_{ab}\frac{\partial^2 c_a}{\partial y^2} - k'' c_a \tag{12.266}$$

The boundary conditions for this problem are virtually identical to the conditions we used in Section 12.7. The only difference is at the plate's surface where due to the chemical reaction, $\partial^2 c_a/\partial y^2 \neq 0$.

$$y = \delta_h, \delta_c \qquad u = v_\infty \qquad c_a = 0 \qquad \frac{\partial u}{\partial y} = \frac{\partial c_a}{\partial y} = 0 \tag{12.267}$$

$$y = 0 \qquad u = 0 \qquad c_a = c_{as} \qquad \frac{\partial^2 u}{\partial y^2} = 0 \tag{12.268}$$

We already solved the hydrodynamic portion of this problem in Section 12.6, so we need only consider the mass transfer case. To solve the species continuity equation, we eliminate the velocity, v, using the continuity equation:

$$u\frac{\partial c_a}{\partial x} - \frac{\partial c_a}{\partial y}\int_0^y \frac{\partial u}{\partial x}dy = D_{ab}\frac{\partial^2 c_a}{\partial y^2} - k'' c_a \tag{12.269}$$

Now we integrate over the extent of the concentration boundary layer, $y = 0$ to $y = \delta_c$, and multiply through by −1:

$$\underbrace{-\int_0^{\delta_c} u\frac{\partial c_a}{\partial x}dy}_{\mathrm{I}} + \underbrace{\int_0^{\delta_c}\frac{\partial c_a}{\partial y}\left[\int_0^y \frac{\partial u}{\partial x}dy\right]dy}_{\mathrm{II}} = \underbrace{-D_{ab}\int_0^{\delta_c}\frac{\partial^2 c_a}{\partial y^2}dy}_{\mathrm{III}} + \underbrace{\int_0^{\delta_c} k'' c_a dy}_{\mathrm{IV}} \tag{12.270}$$

Again, we integrate each term individually. Combining back into the species continuity equation, we have

$$\int_0^{\delta_c} \frac{\partial}{\partial x}[u(c_{a\infty} - c_a)]dy = D_{ab}\left.\frac{\partial c_a}{\partial y}\right|_{y=0} + k''\int_0^{\delta_c} c_a dy \tag{12.271}$$

The definite integrals and derivative are all constants. The polynomial approximations for c_a and u using the conditions in (12.267) and (12.268) are

$$c_a = c_{as}\left[1 - 2\left(\frac{y}{\delta_c}\right) + \left(\frac{y}{\delta_c}\right)^2\right] \qquad u = v_\infty\left[\frac{3}{2}\left(\frac{y}{\delta_h}\right) - \frac{1}{2}\left(\frac{y}{\delta_h}\right)^3\right] \tag{12.272}$$

Substituting into the differential equation and defining $y^* = y/\delta_c$ and $\Delta_c = \delta_c/\delta_h$ gives

$$-\frac{d\delta_c}{dx}\int_0^1 [1 - 2y^* + y^{*2}]\left[\frac{3}{2}(y^*\Delta_c) - \frac{1}{2}(y^*\Delta_c)^3\right]dy^* = -\frac{2D_{ab}}{v_\infty \delta_c} + \frac{k''\delta_c}{v_\infty}\int_0^1 \left[1 - 2y^* + y^{*2}\right]dy^* \tag{12.273}$$

For simplicity, we have evaluated the concentration gradient at the surface. Evaluating the integrals next and substituting them into Equation 12.273 lead to a differential equation in terms of the hydrodynamic boundary layer thickness, δ_h, and the boundary layer thickness ratio, Δ_c. We solve this, as we did for the heat transfer boundary layer thickness. In mass transfer, Schmidt numbers are generally high so there is only one case to consider:

$$-\frac{d}{dx}\left\{(\delta_h\Delta_c)\left[\frac{\Delta_c}{8} - \frac{\Delta_c^3}{120}\right]\right\} = -2\frac{D_{ab}}{v_\infty \delta_h \Delta_c} + \frac{k''\delta_h\Delta_c}{3v_\infty} \tag{12.274}$$

We can neglect the Δ_c^3 term with respect to the Δ_c term if we assume $\text{Sc} \gg 1$. For most liquids, this is indeed the case, and so multiplying through by $\delta_h\Delta_c$, we have

$$\delta_h\Delta_c\frac{d}{dx}\left(\delta_h\Delta_c^2\right) = \frac{2}{3}\delta_h^2\frac{d\Delta_c^3}{dx} + \Delta_c^3\delta_h\frac{d\delta_h}{dx} = \frac{16D_{ab}}{v_\infty} - \frac{8k''\delta_h^2\Delta_c^2}{3v_\infty} \tag{12.275}$$

Substituting in to eliminate δ_h, letting $\xi = x/L$, and substituting in Equation 12.275 puts the entire equation in dimensionless form and allows us to recover the Damkohler number:

$$\xi\frac{d\Delta_c^3}{d\xi} + \frac{3}{4}\Delta_c^3 + 4\text{Da}\xi\Delta_c^2 + \frac{24}{25}\text{Sc}^{-1} = 0 \tag{12.276}$$

Unfortunately, the presence of even a simple, first-order chemical reaction leaves us with a differential equation that has no simple analytical solution. We can get some insight into the solution by looking at limiting cases. If we have no reaction so that the Damkohler number is 0, we have a first-order differential equation for Δ_c^3 that we can separate and integrate to give

$$\left(\frac{32}{25}\right)\text{Sc}^{-1} - \Delta_c^3 = \frac{a_o}{\xi^{3/4}} \tag{12.277}$$

Since the boundary layer ratio, Δ_c cannot approach ∞ as $\xi \rightarrow 0$, the constant of integration, $a_o = 0$, and we are left with

$$\Delta_c = \left(\frac{32}{25}\right)^{1/3} \mathrm{Sc}^{-1/3} \tag{12.278}$$

If the reaction proceeds slowly so that Da $\ll$ 1, then not much reaction will occur during the characteristic convection time, L/v_∞. We would expect that the boundary layer thickness would not be much different from that predicted by Equation 12.278. An approximate solution to the full equation in this instance can be obtained if we try a polynomial approximation for Δ_c of the form

$$\Delta_c = \left(\frac{32}{25}\right)^{1/3} \mathrm{Sc}^{-1/3}[1 + a_1\xi + a_2\xi^2 + a_3\xi^3 \ldots] \tag{12.279}$$

Substituting into the differential equation, we group like powers of ξ, set each group to 0, and solve the resulting simultaneous equation set for a_1, a_2, a_3, etc. The solution is easy to obtain using a symbolic manipulation program like *Maple or Mathematica*. The expansion should hold for small ξ:

$$\Delta_c = \left(\frac{32}{25}\right)^{1/3} \mathrm{Sc}^{-1/3}[1 - 0.7(\mathrm{DaSc}^{1/3})\xi + 0.13(\mathrm{DaSc}^{1/3})^2\xi^2 + 0.54(\mathrm{DaSc}^{1/3})^3\xi^3] \tag{12.280}$$

Our approximate solution shows that a_1 is negative and indicates that the reaction tends to diminish the boundary layer and increase the flux of material off the surface (i.e., enhance mass transfer to the fluid). As the Damkohler number increases, the boundary layer gets thinner and thinner. The use of chemical reactions to enhance transport is a powerful concept exploited in many separation processes such as reactive distillation or reactive absorption [21]. Biological organisms exploit this phenomenon to power their active transport systems [22,23].

The actual concentration boundary layer thickness and mass transfer coefficient can be found by substituting in for the hydrodynamic boundary layer thickness, δ_h:

$$\delta_c = 5.43x\mathrm{Re}_x^{-1/2}\,\mathrm{Sc}^{-1/3}[1 - 0.7(\mathrm{DaSc}^{1/3})\xi + 0.13(\mathrm{DaSc}^{1/3})^2\xi^2 + 0.54(\mathrm{DaSc}^{1/3})^3\xi^3] \tag{12.281}$$

$$k_c = -\frac{D_{ab}}{\delta_c}\frac{\partial c^*}{\partial y^*}\bigg|_{y^*=0} = \frac{0.37\mathrm{Re}_x^{1/2}\,\mathrm{Sc}^{1/3}D_{ab}}{x[1 - 0.7(\mathrm{DaSc}^{1/3})\xi + 0.13(\mathrm{DaSc}^{1/3})^2\xi^2 + 0.54(\mathrm{DaSc}^{1/3})^3\xi^3]} \tag{12.282}$$

and we can see directly how the coefficient increases as the Damkohler number increases.

The analytical solution allows us to only get a first-order approximation to the boundary layer thickness and mass transfer coefficient. Simulation allows us to handle much more complicated situations such as having a surface reaction that is higher order than or more complicated than the first-order reaction we just considered. Figure 12.19 shows the boundary layer thickness as a function of the surface reaction rate for a second-order surface reaction $\left(-r_s = k_s''c_a^2\right)$. The behavior of the boundary layer changes greatly as the reaction rate changes and is counterintuitive. As the reaction rate constant increases, the boundary layer grows in a manner closer to what we predict from the analytical solution. An increasing surface reaction rate depletes reactant at the surface, and if the reaction were instantaneous, the surface concentration would be zero, a constant over the entire surface. That situation corresponds to what we solved analytically. Lower reaction rates mean a changing concentration on the surface and hence a change in the way the boundary layer thickness grows.

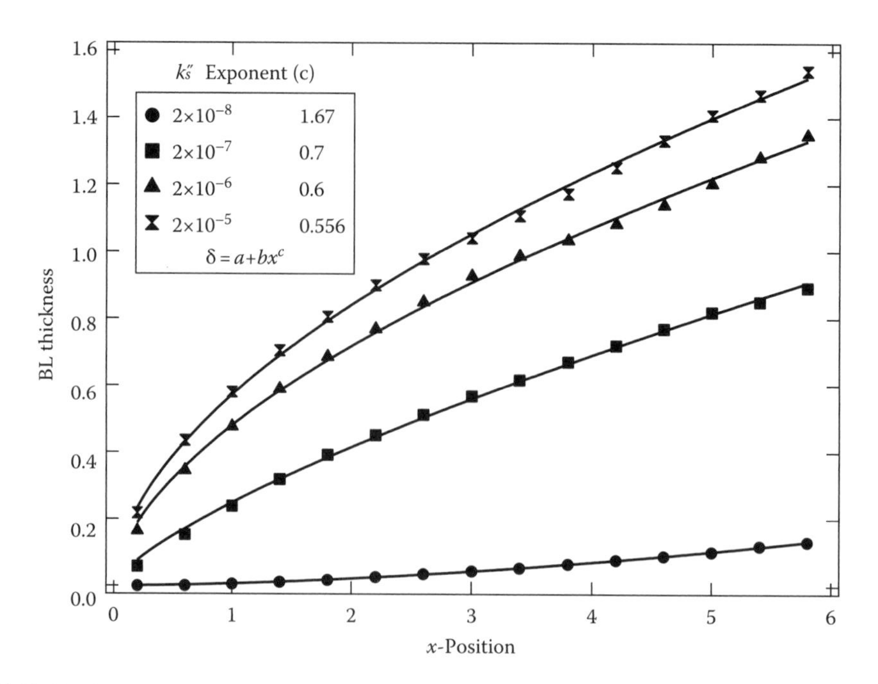

FIGURE 12.19 Boundary layer thickness profile as a function of surface reaction rate.

A similar alteration in behavior can be seen in Figure 12.20 where the local mass transfer coefficient is plotted for the case of fast and slow reaction. Notice the difference in magnitude and behavior for the two cases. The slow reaction flies in the face of one of our tenets that as the boundary layer thickness decreases, the mass transfer coefficient must increase. Here, the concentration gradient at the surface is controlling and overwhelms the effect of boundary layer thickness.

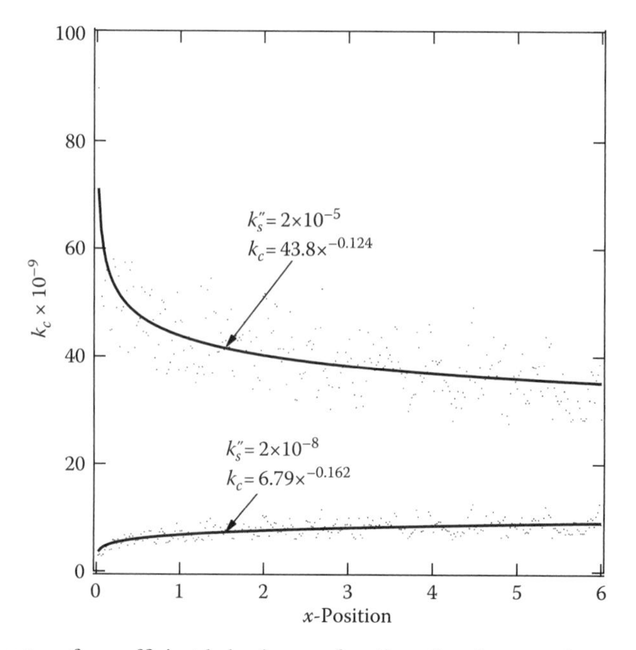

FIGURE 12.20 Mass transfer coefficient behavior as a function of surface reaction rate.

12.9 SIMPLIFIED IONIC BOUNDARY LAYERS

Our final discussion of laminar forced convection mass transfer and concentration boundary layers will consider the effect of imposing an electric field across the boundary layer to assist the motion of ions in the solution. This type of approach to augmenting mass transfer is becoming increasingly important in separations of proteins and peptides where separations can be enhanced by taking advantage of the different charge densities of each species. In this discussion, we will assume a very dilute solution, a uniform temperature across the boundary layer, and a constant applied electric field.

When we charge a surface, the ions that accumulate on that surface attract ions of opposite charge to preserve charge neutrality [24]. This structure is called a double layer and it may greatly affect the transport. We will neglect the formation of a double layer for simplicity. The species continuity equation takes the form

Continuity of a

$$u\frac{\partial c_a}{\partial x}+v\frac{\partial c_a}{\partial y}=D_{ab}\left(\frac{\partial^2 c_a}{\partial y^2}+\beta_e\frac{\partial \Phi}{\partial y}\frac{\partial c_a}{\partial y}\right) \qquad \beta_e=\frac{z_a\mathcal{F}_a}{RT} \tag{12.283}$$

Gauss's law is required to calculate the electric field.

Gauss's law (assuming no charge accumulation, i.e., electroneutrality)

$$\frac{\partial^2 \Phi}{\partial y^2}=0 \tag{12.284}$$

Gauss's law can be integrated directly to show that the potential drop through the boundary layer is linear; hence, $\partial\Phi/\partial y$ is a constant and we can fold it into β_e.

The boundary conditions for this problem are similar to the reaction problem we just finished. We need only add the conditions that specify the voltage in the free stream and at the plate's surface:

$$y=\delta_h,\delta_\pm \qquad u=v_\infty \qquad c_a=0 \qquad \Phi=\Phi_\infty \qquad \frac{\partial u}{\partial y}=\frac{\partial c_a}{\partial y}=0 \tag{12.285}$$

$$y=0 \qquad u=0 \qquad c_a=c_{as} \qquad \Phi=\Phi_s \qquad \frac{\partial^2 u}{\partial y^2}=0 \tag{12.286}$$

We have already solved the hydrodynamic problem and can use the exact solutions from Appendix D. To solve the concentration problem, we eliminate v using the continuity equation:

$$u\frac{\partial c_a}{\partial x}-\frac{\partial c_a}{\partial y}\int_0^y\frac{\partial u}{\partial x}dy=D_{ab}\left(\frac{\partial^2 c_a}{\partial y^2}+\beta'_e\frac{\partial c_a}{\partial y}\right) \tag{12.287}$$

Now we integrate over the extent of the ionic boundary layer, $y=0$ to $y=\delta_\pm$, and multiply by -1:

$$\underbrace{-\int_0^{\delta_\pm}u\frac{\partial c_a}{\partial x}dy}_{\text{I}}+\underbrace{\int_0^{\delta_\pm}\frac{\partial c_a}{\partial y}\left[\int_0^y\frac{\partial u}{\partial x}dy\right]dy}_{\text{II}}=-D_{ab}\left[\underbrace{\int_0^{\delta_\pm}\frac{\partial^2 c_a}{\partial y^2}dy}_{\text{III}}+\underbrace{\beta'_e\int_0^{\delta_\pm}\frac{\partial c_a}{\partial y}dy}_{\text{IV}}\right] \tag{12.288}$$

Integrating and combining all the terms back into the species continuity equation gives

$$\int_0^{\delta_\pm} \frac{\partial}{\partial x}[u(c_{a\infty} - c_a)]\,dy = D_{ab}\left.\frac{\partial c_a}{\partial y}\right|_{y=0} + D_{ab}\beta_e'(c_{as} - c_{a\infty}) \tag{12.289}$$

To evaluate the integral and derivative, we need the polynomial approximations for u and c_a. It helps to define new concentration and velocity variables, c^* and u^*, as

$$c^* = \frac{c_{a\infty} - c_a}{c_{a\infty} - c_{as}} \qquad u^* = \frac{u}{v_\infty} \tag{12.290}$$

Substituting into the differential equation and interchanging the order of differentiation and integration gives

$$v_\infty \frac{\partial}{\partial x}\int_0^{\delta_\pm} u^* c^*\,dy = -D_{ab}\left.\frac{\partial c^*}{\partial y}\right|_{y=0} - D_{ab}\beta_e' \tag{12.291}$$

Based on the boundary conditions we used in this problem, we find we can use the same polynomial approximations for u^* and c^* that we used in Section 12.8.2. Substituting into the differential equation and defining $y^* = y/\delta_\pm$ and $\Delta_\pm = \delta_\pm/\delta_h$ gives

$$\frac{d}{dx}(\delta_h\Delta_\pm)\int_0^1 \left[1 - 2y^* + y^{*2}\right]\left[\frac{3}{2}(y^*\Delta_\pm) - \frac{1}{2}(y^*\Delta_\pm)^3\right]dy^* = \frac{2D_{ab}}{v_\infty \delta_h \Delta_\pm} - \frac{\beta_e' D_{ab}}{v_\infty} \tag{12.292}$$

Evaluating the integral in Equation 12.292 and substituting into the differential equation gives

$$\frac{d}{dx}\left\{(\delta_h\Delta_\pm)\left[\frac{\Delta_\pm}{8} - \frac{\Delta_\pm^3}{120}\right]\right\} = \frac{2D_{ab}}{v_\infty \delta_h \Delta_\pm} - \frac{\beta_e' D_{ab}}{v_\infty} \tag{12.293}$$

We can neglect the $\Delta_\pm^3$ term with respect to the $\Delta_\pm$ term if Sc $\gg$ 1. Multiplying Equation 12.293 through by $\delta_h\Delta_\pm$ and using the chain rule will lead to a first-order differential equation for $\Delta_\pm^3$:

$$\xi\frac{d\Delta_\pm^3}{d\xi} + \frac{3}{4}\Delta_\pm^3 + \frac{12}{5}\mathrm{Sc}^{-1}\beta_e'\sqrt{\frac{L\xi}{\mathrm{Re}_L}}\Delta_\pm = \frac{24}{25}\mathrm{Sc}^{-1} \tag{12.294}$$

If the electric field is negligible, $\beta_e' = 0$, we can show based on our discussion in Section 12.8.2 that the boundary layer thickness ratio is a function of $\mathrm{Sc}^{-1/3}$:

$$\Delta_\pm = \left(\frac{32}{25}\right)^{1/3}\mathrm{Sc}^{-1/3} \tag{12.295}$$

An approximate solution to the full equation can be obtained if we try a polynomial approximation for $\Delta_\pm$. We must form a polynomial in $\xi^{1/2}$ to properly include the electric field term in the solution:

$$\Delta_\pm = \left(\frac{32}{25}\right)^{1/3}\mathrm{Sc}^{-1/3}[1 + a_1\xi^{1/2} + a_2\xi + a_3\xi^{3/2}\ldots] \tag{12.296}$$

Substituting into the differential equation, grouping like terms in powers of ξ, and solving for the constants, a_1, a_2, and a_3 gives

$$\Delta_{\pm} = \left(\frac{32}{25}\right)^{1/3} \mathrm{Sc}^{-1/3}\Big[1 - 0.54(\beta'_e L^{1/2}\mathrm{Re}_L^{-1/2}\mathrm{Sc}^{-1/3})\xi^{1/2} - 0.08\left(\beta'_e L^{1/2}\mathrm{Re}_L^{-1/2}\right)\xi$$

$$-0.01\left(\beta'_e L^{1/2}\mathrm{Re}_L^{-1/2}\right)\xi^{3/2}\Big] \tag{12.297}$$

The approximate solution shows that if the electric field is set up to attract ions to the plate, we can decrease the boundary layer thickness and increase the mass transfer to the surface. If the field is set up to repel ions from the surface, we can increase the boundary layer thickness and decrease the mass transfer. The solution also shows how the Schmidt number, the field strength, and the Reynolds number interact to control the boundary layer. High Reynolds numbers and high Schmidt numbers tend to reduce the effect of the applied field. At higher Schmidt number, we are reducing the diffusivity of the ions and hence reducing their mobility in the field.

12.10 SUMMARY

In this chapter, we introduced the boundary layer equations and showed how we can derive these equations from the full momentum, energy, and continuity equations by suitable approximation. We introduced the concept of the boundary layer and discussed how the thickness of the boundary layer controls the rate of transport. We also derived expressions for the transport coefficients in terms of a friction factor, and a heat transfer or mass transfer coefficient. We discussed two important analogies, the Reynolds and the Chilton–Colburn analogies that are very powerful. These analogies allow one to estimate the value of a transport coefficient knowing the value of any other transport coefficient. Most often one is estimating the heat or mass transfer coefficient from a knowledge of the friction factor.

We solved the boundary layer equations for laminar flow over a flat plate and derived expressions for the boundary layer thicknesses and the transport coefficients. In the next two chapters, we will extend our analysis to geometries other than the flat plate and also extend our flow range to include turbulent flows.

PROBLEMS

HYDRODYNAMIC BOUNDARY LAYERS

12.1 The boundary layer analysis performed in Section 12.6.1 assumed that the fluid was flowing over a stationary plate. However, there is no reason why the fluid cannot be quiescent while the plate moves within it. Consider the situation where the plate is moving from right to left at a speed of v_∞.

a. What are the boundary conditions for the velocities associated with this situation?
b. Using the procedure in Section 12.6.1, derive a formula for the boundary layer thickness.
c. Evaluate the velocity profile using the boundary conditions of part (a) and derive the local boundary layer thickness as a function of Reynolds number.
d. For air at 40°C, a plate that is 100 cm long, and a plate velocity of 1 m/s, how does the boundary layer thickness for this situation compare with the boundary layer thickness for a similar situation but where the plate is stationary?

12.2 In Chapter 9, we introduced the concept of lift in conjunction with potential flow about a sphere. Platelike objects such as your hand outside a moving car window also experience lifting force depending upon your hands "angle of attack" with respect to the direction of air motion. A positive angle of attack provides an upward force while a negative angle provides a downward force. We define a lift coefficient in the same manner we derive the drag coefficient as

$$C_L = \frac{F_L}{\frac{1}{2}\rho v_\infty^2 A_p}$$

For a flat plate, it turns out the lift coefficient is related to the angle of attack by

$$C_L = 2\pi\sin(\alpha) \qquad \alpha\text{, angle of attack(radians)}$$

A falcon has a wingspan of about 80 cm with a wing area of about 0.11 m^2. If the falcon has a mass of 0.6 kg and likes to glide at an angle of attack of about 6°, at what speed must the bird fly so that the lift force just balances out its weight? Assume standard temperature and pressure.

12.3 Based on pure aerodynamics of the type mentioned in Problem 12.2, bumblebees are not supposed to be able to fly. Assume that the average bumblebee has a mass of about 0.9 g, a wingspan of 1.75 cm, and a wing area of about 1.3 cm^2. Such a bee can fly at about 10 m/s.

a. What sort of lift coefficient would be needed to support such a bee?
b. Given the flat plate expression for the lift coefficient, what kind of angle of attack is required?

The actual wing motion and what gives bumblebees their flying prowess are discussed in [25].

12.4 You are evaluating a laboratory report from a senior chemical engineering student. The lab involves measuring the local friction factor over a flat plate. The student states that a plot of C_{fx} versus $x^{1/2}$ should be a straight line if the flow is laminar. Is he/she correct? Why?

12.5 The momentum boundary layer equations were solved to obtain expressions for the velocity profile. Use this information to calculate the following:

a. The velocity potential
b. The stream function
c. The vorticity. How does the vorticity change as one moves farther from the plate's surface?

12.6 To control the drag force or the rate of heat transfer to a surface, a technique called transpirational cooling is employed. The situation is shown later. The surface of the substrate to be cooled is perforated, and positive or negative pressure is applied to induce flow through the surface (Figure P12.6).

a. Using our integral method of analysis, show that the hydrodynamic boundary layer thickness is given by

$$v_o > 0 \qquad x = \frac{7v_\infty}{15v_0}\delta_h - \frac{7\nu}{10v_\infty v_o^2}\ln\left[1 + \frac{2v_o\delta_h}{3\nu}\right]$$

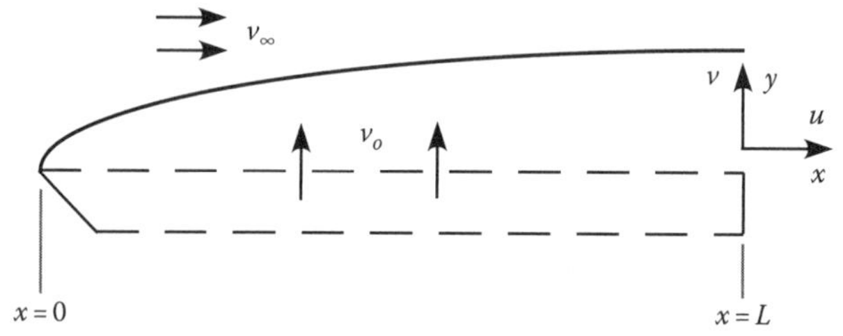

FIGURE P12.6

$$v_o < 0 \qquad x = -\frac{7v_\infty}{15v_0}\delta_h - \frac{7\nu}{10v_\infty v_o^2}\ln\left[1 - \frac{2v_o\delta_h}{3\nu}\right]$$

b. Plot the boundary layer thickness as a function of position along the plate for

$$v_\infty = 1 \text{ m/s}, \qquad v_o = \pm 0.001 \text{ m/s}, \qquad T_s = T_\infty = 300 \text{ K}, \qquad L = 20 \text{ cm}$$

c. What are the implications of your result with regard to
(1) Controlling the heat transfer coefficient?
(2) Controlling the friction factor?

12.7 A sailboard is gliding across a lake at a speed of 20 mph (9 m/s). The sailboard is 3 m long, is 0.75 m wide, and represents a smooth, flat surface. Assume laminar flow.
a. What is the average friction factor for the sailboard?
b. What is the average shear stress on the sailboard?
c. How much power must the wind provide to propel the sailboard?
(Hint: Power = Force × Distance/Time)

12.8 Consider the case of flow over a flat plate where we have a constant pressure drop across the plate ($\partial P/\partial x = \Delta P/L$; $\partial P/\partial y = 0$). Rework the integral boundary layer solution of Section 12.6.1 and determine the following:
a. The boundary layer thickness along the surface
b. The friction factor

12.9 Using a linear velocity profile for the velocity within the boundary layer, use our boundary integral analysis to derive a new expression for the friction number, N_f. How does N_f depend upon the Reynolds number? Assuming your fluid has a Schmidt number > 1, derive an expression for the Sherwood number based on your results for N_f.

Thermal Boundary Layers

12.10 A flat plate 100 cm long and 150 cm wide is held at a temperature of 20°C. The plate is immersed in an air stream at 40°C and 1 atm. The free stream velocity of the air is 5 m/s. Determine the following:
a. The shear stress on the surface of the plate at $x = 50$ cm from the leading edge
b. The heat flux on the surface of the plate at $x = 50$ cm from the leading edge
c. The total drag force on the one side of the plate exposed to the fluid
d. The total heat flow to the one side of the plate from the fluid

12.11 A finned tube is to be used for a heat transfer operation that involves engine oil. The tube and fins are made of copper with the fins being 4 cm on a side and 1 mm thick. The oil flows over them at a velocity of 1 m/s and is at a nominal temperature of 300 K. The tube and fins are at a uniform temperature of 360 K. Determine the minimum fin spacing for this scenario that allows us to use the maximum number of fins per unit length of tube.

12.12 A new fractal surface is being developed for heat transfer by Crinkle, Inc. It is projected to revolutionize the home heating industry. As an agent of Industrial Espionage Ltd., you have stolen some preliminary experimental data. The data indicate that for a plate, 0.25 m long, with air as the heat transfer medium flowing at 5 m/s, the temperature profile in the air was measured to be

$$T(x = L) = T_s - 1.0\times10^5 y + 1.5\times10^5 y^3 \quad \text{plate surface at } y = 0$$

a. What fractional improvement in the heat transfer coefficient (referenced to a conventional, smooth flat plate) was achieved by this new design?
b. Does the hydrodynamic boundary layer lie above or within the thermal boundary layer?

12.13 It is a sunny winter day. In your haste to get to an exam, you accidentally lock *Rover* in the car. The interior dimensions are $L = 2$ m, area = 2.5 m^2, and volume = 4 m^3. Solar radiation provides a heat flux of 116.5 W/m^2 through the moonroof and windows (area = 2.5 m^2). A breeze ($v_\infty = 5$ m/s, $T_\infty = 25$°C) blows over the car. From your final fluids class project, you know that the friction factor for flow over the car obeys

$$\frac{\overline{C_f}}{2} = 0.05\,\mathrm{Re}_L^{-0.35}$$

Assuming only air inside the car ($v_{air} = 0$; $T(t = 0) = 25$°C), the fact that Rover can withstand a temperature of 55°C and that all the net energy from the sun gets transferred to the air inside the car,

a. What is the heat transfer coefficient?
b. How fast does the temperature initially rise or fall in the car?
c. Will Rover expire before you have finished the exam? Why?

12.14 The temperature profile in the boundary layer for air flowing over a heated surface has been found to obey

$$\frac{T - T_s}{T_\infty - T_s} = 1 - \exp\left[-\mathrm{Pr}\left(\frac{v_\infty y}{\nu}\right)\right]$$

where
y is the distance normal to the surface
Pr = 0.7 is the Prandtl number

If $T_\infty = 400$ K, $T_s = 300$ K, and $v_\infty/\nu = 5500$ m^{-1}, what is the surface heat flux?

12.15 A thin, 0.1 m thick plate of copper is brought into contact with a flowing water stream as shown in Figure P12.15. What is the temperature at the center of the plate 10 min later? ($T_\infty = 288$ K, $T_i = 373$ K, $v_\infty = 1$ m/s, $L = 1$ m, $W = 1$ m).

12.16 Atmospheric air is in parallel flow ($v_\infty = 15$ m/s, $T_\infty = 15$°C) over a flat heater surface that is to be maintained at 140°C. The heater surface area is 0.25 m^2, and the airflow is known to induce a drag force of 0.25 N on the heater. What is the electrical power needed to maintain the prescribed surface temperature? What is the heat transfer coefficient?

12.17 For flow over a flat plate with an extremely rough surface, convection heat transfer effects are known to be correlated by the expression

$$\mathrm{Nu}_x = 0.04\,\mathrm{Re}_x^{0.9}\,\mathrm{Pr}^{-1/3}$$

For airflow at 50 m/s, what is the surface shear stress at $x = 1$ m from the leading edge of the plate? Assume air at 300 K.

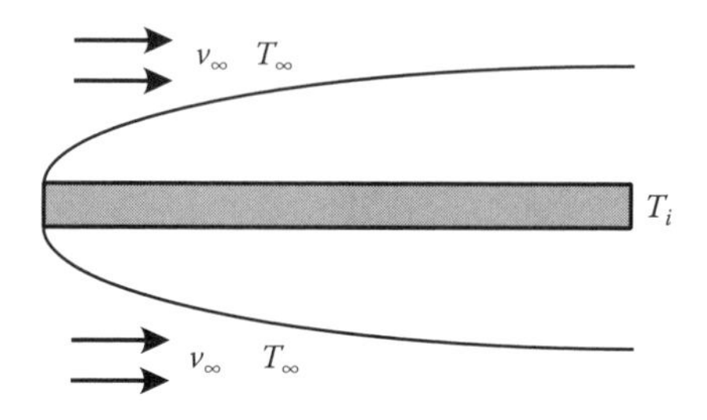

FIGURE P12.15

12.18 Consider flow over a flat plate where the surface of the plate has a temperature distribution. Here, we are solving the situation for the unheated starting length, but once the plate is heated, the steady-state temperature distribution obeys

$$T_s = T_\infty + bx$$

The heat flux for an arbitrary wall temperature distribution is given by

$$q''_s = \int_0^x h(x_o, x)\left(\frac{dT_s}{dx_o}\right)dx_o + \sum_{i=1}^{n} h(x_{oi}, x)\Delta T_{si}$$

where ΔT_{si} represents any abrupt jumps in surface temperature. The idea here is that we break the calculation up into a series of unheated starting length problems and sum them up to get the final heat flux.

a. Calculate the heat flux for the linear temperature ramp. You will encounter an integral of the form

$$\int_0^x Z^{m-1}(1-Z)^{n-1}dZ = \beta_x(m,n)$$

which is the definition of the incomplete beta function. Tabulated values can be found in [26] or in [27].

b. Determine an expression for the average Nusselt number. This expression need not be in closed form or totally evaluated.

12.19 Building on the previous problem allows us to look at predicting the temperature distribution for an arbitrary heat flux specification at the surface. Here, we can solve for the temperature distribution providing the heat transfer coefficient can be written in the following form:

$$h(x_o, x) = f(x)\left(x^a - x_o^a\right)^{-g}$$

Fortunately, our solution for the unheated starting length fits into this form and leads to the following expression for the temperature profile:

$$T_s(x) - T_\infty = \frac{0.623}{k_f}\mathrm{Pr}^{-1/3}\mathrm{Re}_x^{-1/2}\int_0^x \left[1-\left(\frac{x_o}{x}\right)^{3/4}\right]^{-2/3} q_s''(x_o)\,dx_o$$

Assuming a constant surface heat flux, determine an expression for the Nusselt number. How much of an increase do we obtain in the heat transfer coefficient over the constant surface temperature case?

12.20 You are taking a shower in your normally unheated bathroom. Its winter and you notice an extreme amount of condensation on the walls. Given the following information, calculate the following:

a. The velocity profile in the condensed film
b. The boundary layer thickness at the wall/floor interface
c. The average heat flux through the film
d. The mass flow rate of water flowing from the wall onto the floor
e. How much water actually clings to the wall

$$T_s = 10°\text{C} \quad T_{sat} = 40°\text{C}$$

Wall dimensions: 2 m (high) × 1 m (wide) (Paschal, Prosser, and Wan).

12.21 Flow along a flat plate is occurring in a situation where both forced convection and natural convection happen. Assuming natural convection is important when the heat transfer coefficient is > 10% of that for forced convection, determine a general criterion for when we must consider natural convection in our system.

Concentration/Mass Transfer Boundary Layers

12.22 Equation 12.265 indicates that the magnitude and direction of mass transfer can have a large effect on the Sherwood number. It is interesting to explore this a bit. Consider two situations. In the first, we have a saturated sugar/water solution that is being transported to the surface of a plate. A reaction at the surface consumes all the sugar that reaches it. In the second scenario, we have a flat plate made of sugar that is dissolving into water. At a temperature of 25°C, the mole fraction of sugar in water at its solubility limit is 9.65×10^{-2}. The molecular weight of sugar is 342.3 kg/kg-mol. The diffusivity of sugar in water is 0.52×10^{-9} m^2/s. Estimate the maximum difference in the Sherwood numbers for the two cases.

12.23 Wettability gradients are present in many biological systems and have been used to do interesting things like making water run uphill [28]. One way to form such a surface on glass is to immerse the plate in a slowly flowing solution of a silane coupling agent that reacts with OH on the glass surface. The reaction can be nearly instantaneous so that the coupling agent concentration at the surface is essentially 0. We are interested in forming such a surface using $(CH_3)_3SiCl$ in an inert solvent that has the physical properties of water at 25°C. If the free stream concentration of TMCS is 1×10^{-4} kg mol/m^3, the free stream velocity of the fluid is 1 m/s, the plate is 0.01 m long with an area of 2×10^{-4} m^2, and the diffusivity of TMCS can be predicted using the Polson equation (Chapter 3 problems).

a. How long would we have to expose the plate to the fluid to insure that 10% of the available OH on the surface was converted at $x = L$? (There are roughly 2 OH/nm^2 on a glass surface.)

b. How much of the surface OH was converted over the entire plate?

c. What is the surface OH concentration at $x = L/2$?

12.24 A fluid flows along a flat, horizontal plate that is slightly soluble in the liquid. At a distance, $x = 10$ cm, from the leading edge of the plate, the concentration boundary layer thickness, $\delta_c = 10$ mm, and the local value of the mass transfer coefficient, $k_{cx} = 75$ mol/m^2 s. Please determine the numerical values for δ_c at $x = 25$ cm and for the average mass transfer coefficient, $\bar{k}_c$, over the entire length of the plate (25 cm). Assume forced convection and laminar flow with Sc > 1.

12.25 A puddle of water 1 m in diameter and 1 cm deep sits on your driveway. The air temperature is 20°C and a slight breeze is blowing at 4 m/s. The relative humidity is a comfortable 35%. You may assume the water is at the same temperature as the air.

a. What is the driving force for mass transfer?

b. What is the mass transfer coefficient?

c. Under these conditions, how long will it take for the puddle to disappear?

12.26 Consider the situation where we have a flat plate dissolving into the boundary layer. To assist the dissolution process, we treat the fluid so that it contains a reactive component that complexes with the plate material. The system can be described by a first-order chemical kinetics. Using the results from Section 12.8.2, determine the following:

a. An expression for the average mass transfer coefficient

b. An expression for the enhancement factor that arises due to the presence of the reaction

12.27 A fluid containing a solute, a, in dilute solution flows over a flat plate. The solute adsorbs on the plate at a rate characterized by

$$-r_a = \frac{k''c_a}{1+Kc_a}$$

Repeat the boundary layer analysis of Section 12.8.2 to calculate the following:

a. The concentration profile of solute
b. The mass transfer coefficient, k_c
c. The concentration boundary layer thickness.

12.28 A large plate, 0.25 m in diameter is coated with a 1 mm layer of zeolite catalytic material. A gas containing a noxious compound is slowly flowing over the catalytic material. Experiments have shown that the intrinsic, pseudo-first-order rate constant for the destruction of the compound is 1×10^{-6} s^{-1} and the effective diffusivity of the compound in the zeolite is 1×10^{-12} m^2/s. The overall effectiveness factor (Chapter 8) for catalyst was found to be 0.5, and the gas can be approximated as air at 300 K.

a. Estimate at what velocity the fluid is flowing over the surface of the catalyst?
b. What is the maximum effectiveness factor that could be achieved in this system?

REFERENCES

1. Van dyke, M., *Perturbation Methods in Fluid Mechanics*, Academic Press, New York (1964).
2. Schlichting, H., *Boundary Layer Theory*, 7th edn., McGraw-Hill, New York (1979).
3. Reynolds, O., On the extent and action of the heating surface for steam boilers, *Proc. Manchester Lit. Phil. Soc.*, **14**, 7 (1874).
4. Colburn, A.P., A method of correlating forced convection heat transfer data and comparison with fluid friction, *Trans. AIChE* **29**, 174 (1933).
5. Chilton, T.H. and Colburn, A.P., Mass transfer (absorption) coefficients. Prediction from data on heat transfer and fluid friction, *Ind. Eng. Chem.* **26**, 1183 (1934).
6. Friend, W.L. and Metzner, A.B. Turbulent heat transfer inside tubes and the analogy among heat, mass, and momentum transfer, *AIChE J.* **4**, 393 (1958).
7. von Karmán, T., Über laminare und turbulente Reibung, *ZAMM* **1**, 233 (1921).
8. Blasius, H., Grenzschichten in Flüssigkeiten mit Kleiner Reibung, *Z. Math. Phys.* **56**, 1 (1908).
9. Boyce, W.E. and DiPrima, R.C., *Elementary Differential Equations and Boundary Value Problems*, 2nd edn., John Wiley & Sons, New York (1969).
10. Sutton, G.W. and Sherman, A., *Engineering Magnetohydrodynamics*, McGraw-Hill, New York (1965).
11. Romig, M., The influence of electric and magnetic fields on heat transfer to electrically conducting fluids, *Adv. Heat Transfer* **1**, 268 (1964).
12. Rosenzweig, R.E., Fluidization hydrodynamic stabilization with a magnetic field, *Science* **204**, 57 (1979).
13. Holman, J.P., Gartrell, H.E., and Soehngen, E.E. An interferometric method of studying boundary layer oscillation, *J. Heat Transfer, Ser. C.* **80** (1960).
14. Sparrow, E.M. and Gregg, J.L., Laminar free convection from a vertical flat plate, *Trans. ASME* **78**, 435 (1956).
15. Lord R., On convective currents in a horizontal layer of fluid when the higher temperature is on the underside, *Phil. Mag.* **32**, 529 (1916).
16. Koh, J.C.Y., Sparrow, E.M., and Hartnett, J.P., The two-phase boundary layer in laminar film condensation, *Int. J. Heat and Mass Transfer*, **2**, 69 (1961).
17. Nusselt, W., Die Oberflachenkondensation des Wasserdampfes, *Z. Ver. Deut. Ing.*, **60**, 541 (1916).
18. Tavlarides, L.L. and Tsouris, C., Mass transfer effects on droplet phenomena and extraction column hydrodynamics revisited, *Chem. Eng. Sci.* **48**, 1503 (1993).
19. Bird, R.B., Stewart, W.E., and Lightfoot, E.N., *Transport Phenomena*, J. Wiley & Sons, New York (1960).
20. Chambre, P.L. and Young, J.D., On the diffusion of a chemically reactive species in a laminar boundary layer flow, *Phys. Fluids* **1**, 48 (1958).

21. Westerterp, K.R., van Dierendonck, L.L., and de Kraa, J.A. Interfacial areas in agitated gas-liquid contactors, *Chem. Eng. Sci.* **18**, 157 (1963).
22. Stein, W.D. *The Movement of Molecules across Cell Membranes*, Academic Press, New York (1967).
23. Smith, K.A., Meldon, J.H., and Colton, C.K., An analysis of carrier facilitated transport, **19**, 102 (1973).
24. Prentice, G., *Electrochemical Engineering Principles*, Prentice Hall, Englewood Cliffs, NJ (1991).
25. Bomphrey, R.J., Taylor, G.K., and Thomas, A.L.R. Smoke visualization of free-flying bumblebees indicates independent leading-edge vortices on each wing pair, *Experiments in Fluids*, **46**, 811–821 (2009).
26. Kays, W.M., Crawford, M.E., and Weigand, B., *Convective Heat and Mass Transfer*, 4th edn., McGraw-Hill, New York (2005) pp. 162 and 475.
27. Baxter, D.C. and W.C. Reynolds, *J. Aero. Sci.*, **25**, 1958.
28. Chaudury, M.K. and Whitesides, G.M., How to make water run uphill, *Science* **256**, 1539–1541 (1992).

13 Convective Transport
Systems with Curvature

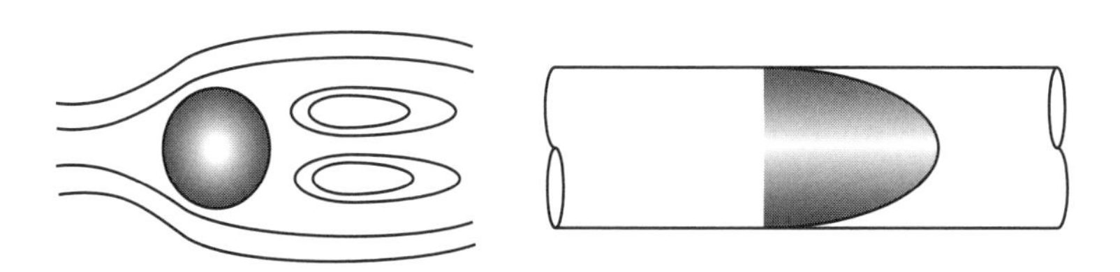

13.1 INTRODUCTION

The flat plate boundary layer analysis we introduced in the last chapter represents the first line of attack in trying to determine values for the convective transport coefficients. A great many systems can be regarded as flat plates if their curvature is sufficiently small. Unfortunately, there are an equal number of technologically important systems where the flat plate analysis fails. Flow over highly curved blunt objects such as cylinders and spheres and flow in tubes and conduits are two of the most important examples. The failures arise because the boundary layer cannot grow in the same, unbounded manner characteristic of the boundary layer on a flat plate. In flow over curved objects, the boundary layer separates from the solid surface and forms a wake where the flow changes direction and recirculates back toward the object. In addition, for flow within a tube, the size of the boundary layer is restricted to the radius of the tube. In this chapter we will discuss boundary layers in these situations and try to determine theoretical values for the convective transport coefficients.

13.2 FLOW OVER CYLINDERS

One of the most important heat transfer systems involves flow over a cylindrical surface. This configuration is used in many heat exchange applications ranging from shell-and-tube heat exchangers to boilers and condensers or the evaporation of sweat from your arms or legs. Mass transfer analogs include tubular membrane systems, extraction columns involving tubular packing (Raschig rings and the like), absorption of nutrients through the villi in your intestine, and optical waveguide manufacture using the outside vapor deposition process.

The flow over a cylindrical object is much more complicated than flow over a flat plate. Most of that complexity arises because at some point, the solid surface begins to curve away from the mean flow direction of the fluid. Fluid flowing around that curve finds it increasingly difficult to stay in *communication* with the solid surface. Centrifugal force thrusts the fluid outward, away from the object. If the fluid is flowing very slowly, frictional forces enable it to negotiate the curve. As the fluid flows faster, the centrifugal force increases. Eventually, this force overcomes the viscous and the intermolecular forces keeping the fluid following the shape of the object. At this point, we say the *boundary layer separates*. Part of the fluid flies off of the cylinder, and the remaining fluid goes into a spin behind the surface. The situation is akin to driving a car around a curve. If one drives slowly, there is no problem negotiating the curve. The faster one drives the more force one must exert on the road to keep the car in the direction of the curve. If one increases one's speed, still further one can either be thrown off the curve or one can overcome the friction of the tires on the road, loose traction, and begin to spin out behind the curve.

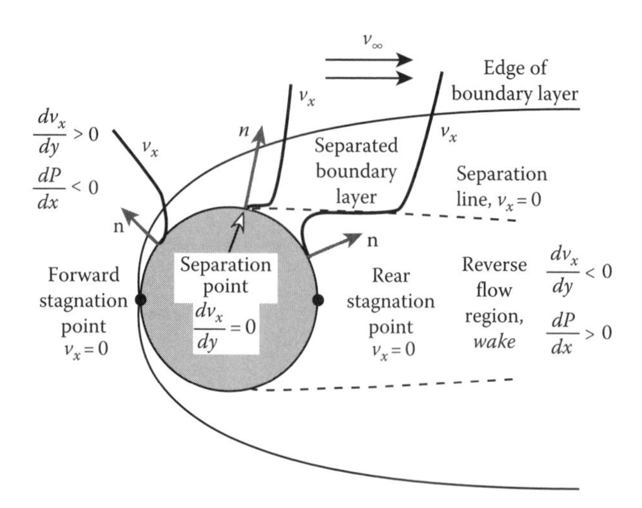

FIGURE 13.1 Boundary layer flow over a solid body of revolution. There are two stagnation points. The forward point is located at the leading edge, and the rear point is located on the opposite side of the object.

The fluid flow situation is shown in Figure 13.1. At the *leading edge* of the surface we have a point along the axis of the flow where the velocity is zero. This is referred to as the *stagnation point*, and it is here where the velocity abruptly changes directions. The stagnation point represents the point of highest pressure, and we can use the mechanical energy balance to evaluate the pressure:

$$P_{st} = P_\infty + \frac{\rho v_\infty^2}{2} \quad \text{Stagnation point pressure} \tag{13.1}$$

The boundary layer is thinnest at the stagnation point. As fluid begins to change direction and flow around the sphere, the pressure drop is negative, the fluid is forced away from the surface, and the boundary layer thickness increases. The effect of the solid surface is to generate vorticity in the fluid at the interface. Eventually this vorticity is enough to overcome the forces keeping the fluid motion uniform. The boundary layer separates from the surface, and rotation cells form to dissipate the vorticity in the wake behind the object. The point at which separation first occurs is characterized by $dv_x/dy = 0$ and $v_x = 0$. The separation and cell formation process is shown in Figure 13.2. Figure 13.3 shows a Comsol® simulation of the situation in Figure 13.2 for a Reynolds number of 26.

(a)

(b)

FIGURE 13.2 Flow past a horizontal cylinder. Note the separation of boundary layer and wake formation. (a) $\text{Re}_d = 1.54$; (b) $\text{Re}_d = 26$. (Photograph courtesy of Taneda, S., *J. Phys. Soc. Jap.*, 11, 302, 1956. With permission).

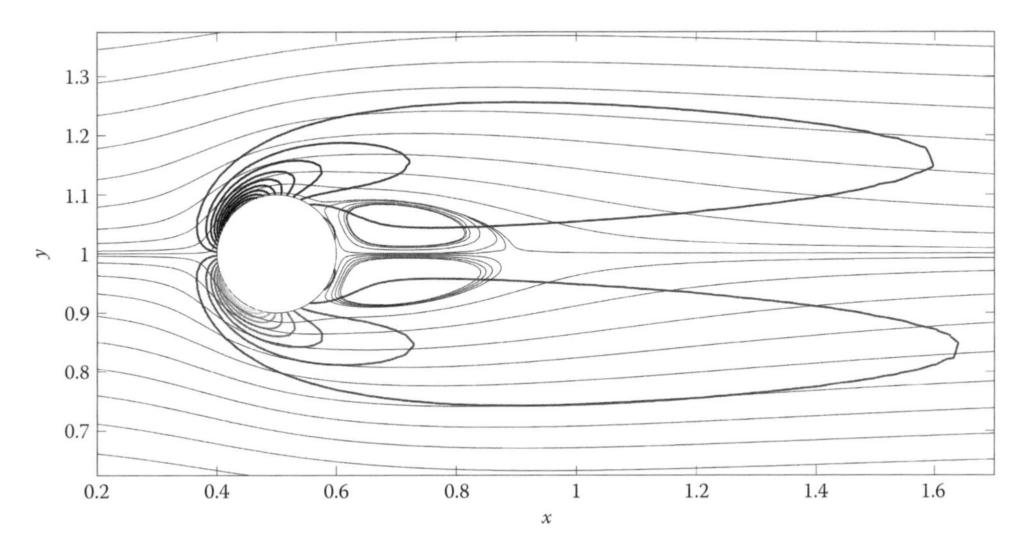

FIGURE 13.3 Comsol® simulation of flow about a cylinder at $Re_d = 26$. Flow streamlines and lines of constant vorticity (grayscale lobes) are plotted. Vorticity lines terminate on the cylinder surface. Comsol® module: Cylinder.

The simulation does a fairly good job of reproducing the experimental data, at least qualitatively, and also allows us to show some other features. The figure shows that the vorticity is generated on the leading edge of the cylinder at essentially two specific locations and that the vorticity dissipates as one moves further and further from the cylinder. One can also use the simulation to pinpoint the location of boundary layer separation by plotting contours where $v_x = 0$ and $dv_x/dy = 0$. In Figure 13.3, this lies at about the point where vorticity lobes and the lobes representing the wake behind the cylinder meet. This location is about 110° from the location of the forward stagnation point.

We can use the boundary layer equations derived in Chapter 12 to describe this flow. Modern numerical integration techniques can find solutions to the differential equations for moderately fast flows, but as the free stream velocity increases, the solutions become more and more difficult and time consuming to generate. Semi-analytic solutions to the equations can also be found using a technique first developed by Blasius [2] and later refined by Heimenz [3] and Howarth [4]. The technique involves representing the potential flow field (see Chapter 9) about the cylinder in terms of a power series in x. Here, x represents the distance from the stagnation point along a particular streamline, and $v_{x,2i-1}$ are known functions that depend only upon the shape of the body:

$$\phi(x) = \sum_{i=1}^{\infty} v_{x,2i-1} x^{2i-1} \quad \text{Velocity potential} \tag{13.2}$$

The continuity equation is solved using the stream function approach (see Chapter 10). In light of our representation for the velocity potential, we choose a similar function for the stream function in which the coefficients, f_i, are functions of y:

$$\psi(x, y) = \sum_{i=1}^{\infty} v_{x,2i-1} x^{2i-1} f_{2i-1}(y) \quad \text{Stream function} \tag{13.3}$$

The stream function can now be used to get the velocity in the x (v_x) and y (v_y) directions.

Once ϕ and ψ are known, we can substitute into the momentum equation, collect like terms, and obtain a set of nonlinear ordinary differential equations to solve for the f functions. The exact solution procedure is given in Schlichting [5] along with tabulated values for the functions.

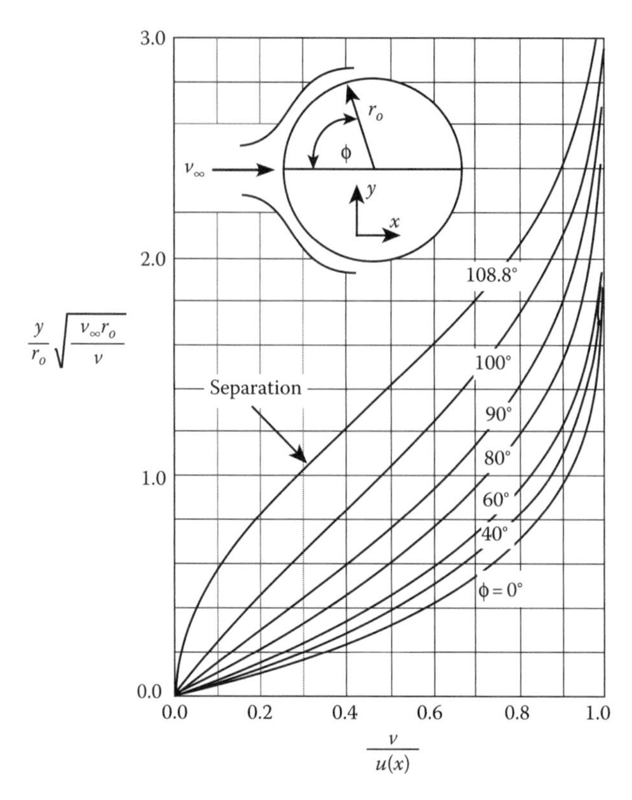

FIGURE 13.4 Velocity profile about a single cylinder in crossflow. Semi-analytic solution obtained from a Blasius series expansion. Laminar flow results in boundary layer separation at 108.8°. (Adapted from Schlichting, H., *Boundary Layer Theory*, 7th edn., McGraw-Hill, New York, 1979. With permission.)

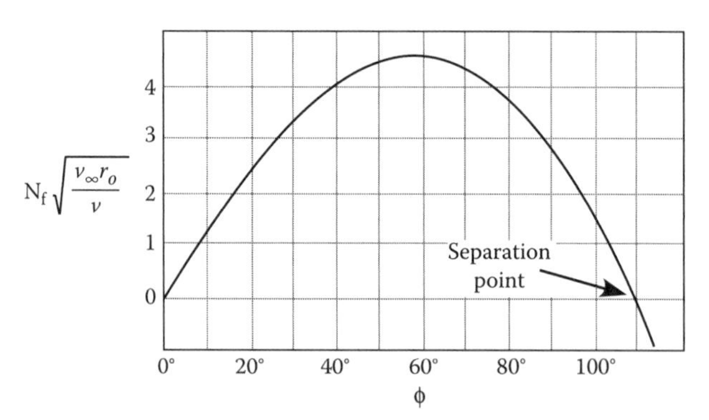

FIGURE 13.5 Friction number behavior for flow about a single cylinder. Results obtained by the Blasius series expansion. Shear stress is zero at separation. (Adapted from Schlichting, H., *Boundary Layer Theory*, 7th edn., McGraw-Hill, New York, 1979. With permission.)

Figures 13.4 and 13.5 show the velocity distribution about a single cylinder and the friction number N_f for flow about that cylinder.

The best way to determine the friction factor and the heat and mass transfer coefficients for flow about a cylinder is by experimentation. An experimental friction factor or drag coefficient plot for single cylinders is shown in Figure 13.6. One can always use these plots with the Chilton–Colburn, Friend–Metzner, or other analogies to obtain an approximate mass or heat transfer coefficient. These plots give average values for the friction factor and hence average values for the heat or mass transfer coefficient. More accurate experimental data also exist. Figure 13.7 shows the variation in

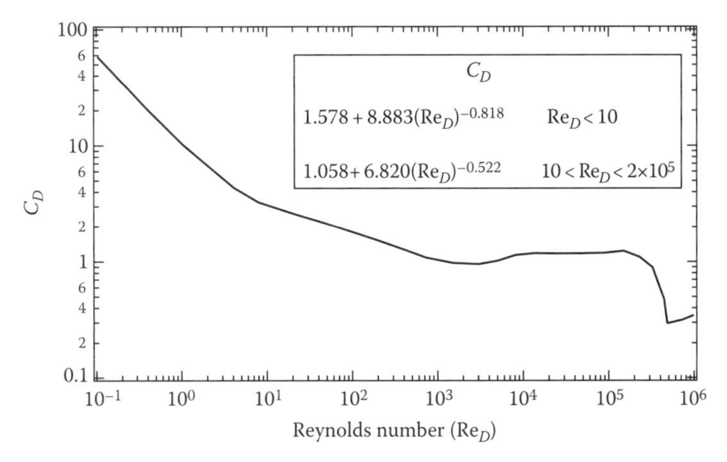

FIGURE 13.6 Friction factor/drag coefficient for flow across a single cylinder. (Adapted from Schlichting, H., *Boundary Layer Theory*, 7th edn., McGraw-Hill, New York, 1979. With permission.)

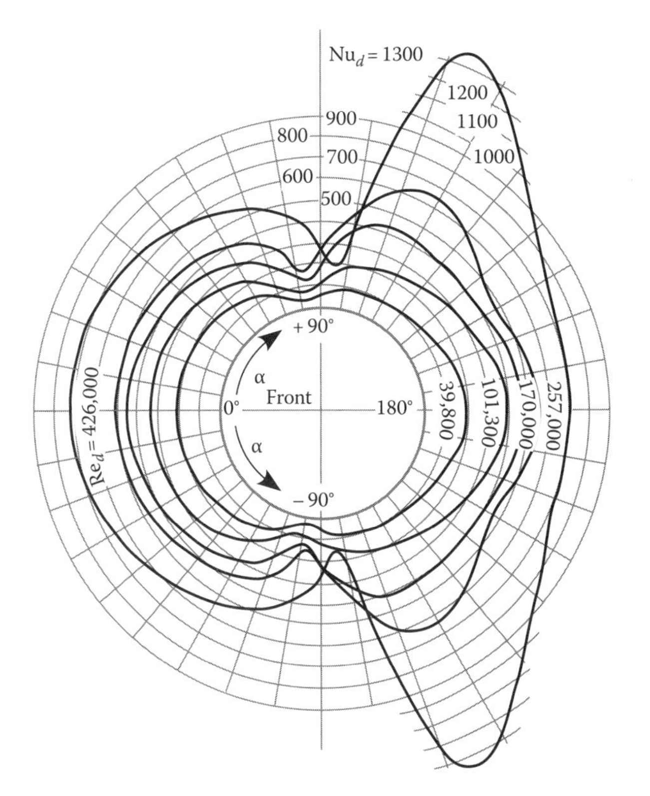

FIGURE 13.7 Variation of the Nusselt number for flow across a cylinder. (Adapted from Schmidt, E. and Wenner, K., *Forsch. Gebeite Ingenieurw.*, 12, 65, 1941. With permission.)

the local Nusselt number as a function of position from the stagnation point for flow about a cylinder. Notice that the coefficient is lowest at the point of separation and highest beyond separation in the wake where significant fluid mixing occurs. Figure 13.8 shows interferograms illustrating the thermal boundary layer about the cylinder. Each dark and light band represents a region of constant temperature. The density of bands represents the temperature gradient, and the location of the first or outer band represents the edge of the boundary layer. Notice the density of bands in the developing boundary layer and the relative uniformity of temperature inside the wake region.

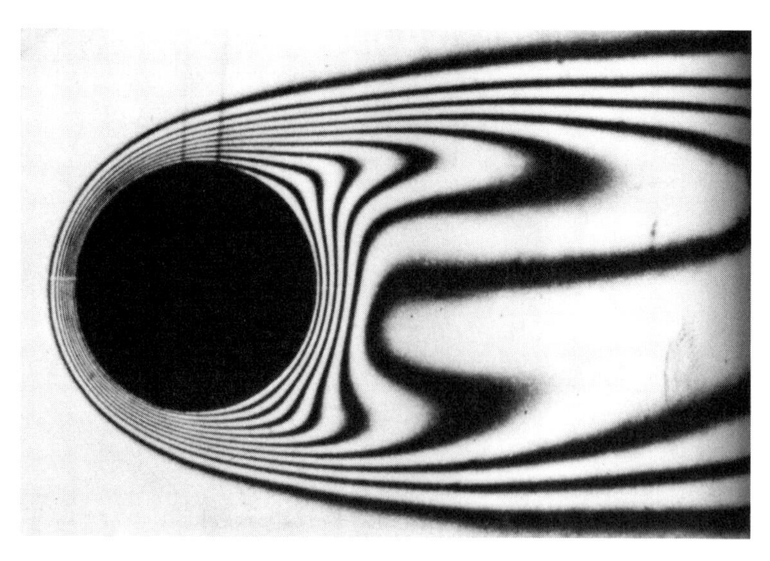

FIGURE 13.8 Boundary layer about a cylinder. Density of light and dark bands is proportional to gradients within the boundary layer. (From Eckert, E.R.G. and Soehngen, E.E., *Trans. ASME,* 74, 343, 1952. With permission.)

If one does not want to use graphs or analogies with friction factors to determine the heat or mass transfer coefficient, one could use any one of a number of correlations that have been developed. One such heat transfer correlation for the single cylinder was developed by Churchill and Bernstein [8]:

$$\overline{\mathrm{Nu}_d} = \frac{\bar{h}d}{k_f} = 0.3 + \frac{0.62\mathrm{Re}_d^{1/2}\mathrm{Pr}^{1/3}}{[1+(0.4/\mathrm{Pr})^{2/3}]^{1/4}}\left[1+\left(\frac{\mathrm{Re}_d}{282{,}000}\right)^{5/8}\right]^{4/5} \tag{13.4}$$

The correlation applies whenever $\mathrm{Re}_d\mathrm{Pr} > 2$ and all properties are evaluated at the film temperature, $T_f = (T_s + T_\infty)/2$. Another popular correlation is the Zhukauskas correlation [9]:

$$\overline{\mathrm{Nu}_d} = C\mathrm{Re}_d^m \mathrm{Pr}^n \left(\frac{\mathrm{Pr}}{\mathrm{Pr}_s}\right)^{1/4} \tag{13.5}$$

Values for C and m are given in Table 13.1. If $\mathrm{Pr} < 10$, $n = 0.37$; if $\mathrm{Pr} > 10$, $n = 0.36$. All properties are evaluated at the free stream temperature except for Pr_s, which is evaluated at the surface temperature.

TABLE 13.1
Constants for Use in Equation 13.5

Re_d	C	m
1–40	0.75	0.4
40–10^3	0.51	0.5
10^3–2×10^5	0.26	0.6
2×10^5–10^6	0.076	0.7

Source: Zhukauskas, A., Heat transfer from tubes in crossflow, in Hartnett, J.P. and Irvine, T.F. Jr., eds., *Advances in Heat Transfer*, Vol. 8, Academic Press, New York, 1972.

One can use the correlations in Equations 13.4 and 13.5 for mass transfer coefficients if one replaces the Prandtl number with the Schmidt number and the Nusselt number with the Sherwood number. Such an approach is shown in the following, for example. One must remember to adjust the range specifications as well to account for the Schmidt number dependence:

$$\overline{\text{Sh}_d} = \frac{\bar{k}_c d}{D_{ab}} = 0.3 + \frac{0.62\text{Re}_d^{1/2}\text{Sc}^{1/3}}{\left[1+\left(\dfrac{0.4}{\text{Sc}}\right)^{2/3}\right]^{1/4}}\left[1+\left(\frac{\text{Re}_d}{282{,}000}\right)^{5/8}\right]^{4/5} \quad (13.6)$$

$$\overline{\text{Sh}_d} = C\text{Re}_d^m\text{Sc}^n\left(\frac{\text{Sc}}{\text{Sc}_s}\right)^{1/4} \quad (13.7)$$

The boundary layers in natural convection are quite different from forced convection. In Chapter 12 we showed that the boundary layer for natural convection grows very slowly and that the transport coefficients are much lower. We can see this in Figure 13.9 that shows the natural convection boundary layer around a single tube. Heat transfer correlations applicable for this situation are based on the Rayleigh number and have been given by Morgan [11]:

$$\overline{\text{Nu}_d} = C(\text{Ra}_d)^n \quad (13.8)$$

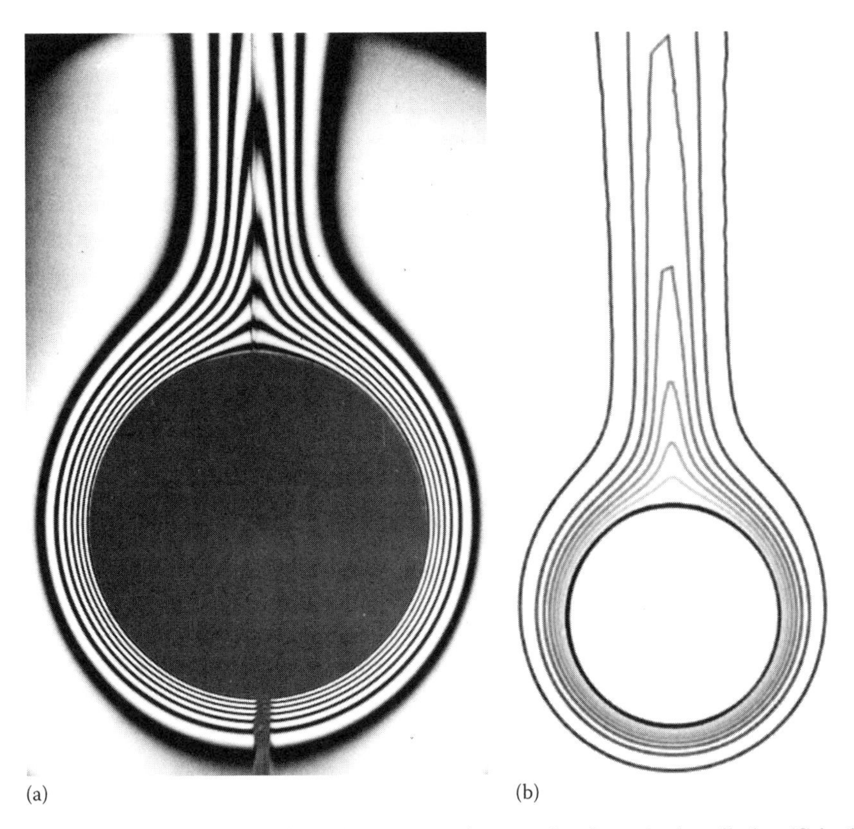

FIGURE 13.9 (a) Thermal boundary layer due to natural convection for a single cylinder. (Grigull, U. and Hauf, W., *Proc. 3rd. Int. Heat Trans. Conf.*, 2, 182, 1966. With permission.) (b) Comsol® simulation at a Rayleigh number of 106,000. The contours represent lines of constant temperature as in part (a).

TABLE 13.2
Constants for Use in Equation 13.8

Ra_d	C	n
10^{-10}–10^{-2}	0.675	0.058
10^{-2}–10^{2}	1.02	0.148
10^{2}–10^{4}	0.850	0.188
10^{4}–10^{7}	0.480	0.250
10^{7}–10^{12}	0.125	0.333

Source: Morgan, V.T., The overall convective heat transfer from smooth circular cylinders, in Irvine, T.F. Jr. and Hartnett, J.P. eds., *Advances in Heat Transfer*, Vol. 11, Academic Press, New York, 1975.

where the constants, C and n, are functions of the Rayleigh number, Ra, and are shown in Table 13.2. All properties are evaluated at the film temperature.

Another widely used correlation for natural convection is from Churchill and Chu [12]. Properties are evaluated at the film temperature:

$$\overline{\mathrm{Nu}}_d = \left\{ 0.6 + \frac{0.387\mathrm{Ra}_d^{1/6}}{\left[1.0 + \left(\dfrac{0.559}{\mathrm{Pr}}\right)^{9/16}\right]^{8/27}} \right\}^2 \qquad 10^{-5} < \mathrm{Ra}_d < 10^{12} \tag{13.9}$$

The corresponding mass transfer correlations can be obtained by replacing the Nu_d with the Sh_d and the Pr number by the Sc number. The mass transfer Rayleigh number in these equations is defined as

$$\mathrm{Ra}_d = \frac{\Delta\rho\gamma_x g d^3}{\rho\nu^2} \tag{13.10}$$

where $\gamma_x = (1/\rho)(\partial\rho/\partial x_i)_{P,T}$ is the coefficient of compositional expansion.

Example 13.1 Effectiveness of a Down Parka

We all know that goose down is an effective insulator, but to design a parka, we need to know how effective that insulation is when protecting someone during sub-zero weather. The average US human can be approximated as a cylinder 0.3 m in diameter and 1.8 m high with an average surface temperature of 24°C. The outside air temperature is –20°C, and the wind speed is 10 m/s. We must calculate the steady-state heat loss from the person with and without a goose down parka (0.075 m thick) assuming laminar flow about the parka and the head covered. The properties of the goose down and properties of the air are given in the following:

$$k_{down} = 0.05 \text{ W/m K} \qquad k_{air} = 0.0235 \text{ W/m K}$$

$$\mathrm{Pr} = 0.72 \qquad \nu = 11.42 \times 10^{-6} \text{ m}^2/\text{s}$$

The heat flow from the person can be given by the expression for convective cooling:

$$q = \bar{h}A(T - T_\infty) = \bar{h}_{sides}A_{sides}(T - T_\infty) + \bar{h}_{top}A_{top}(T - T_\infty)$$

Here, we have included the fact that the cylinder has a top that looks like a flat plate. We can use information from Chapter 12 as well as this chapter to estimate the heat transfer coefficients. Beginning with the flat plate, the first task is to calculate the Reynolds number and evaluate the flow field. Here, the length of the plate is the diameter of the cylinder end:

$$\text{Re}_L = \frac{v_\infty L}{\nu} = \frac{10(0.3)}{11.42\times10^{-6}} = 2.63\times10^5$$

This is laminar flow, as we assumed, so we can use a correlation for the average heat transfer coefficient:

$$\overline{\text{Nu}_L} = \frac{\bar{h}L}{k_f} = 0.664(\text{Re}_L)^{1/2}\text{Pr}^{1/3} = 0.664(2.63\times10^5)^{1/2}(0.72)^{1/3} = 305.0$$

$$\bar{h}_{top} = 23.9\ \frac{\text{W}}{\text{m}^2\,\text{K}}$$

Turning our attention to the side, we can use Equation 13.4:

$$\text{Re}_d = \frac{v_\infty d}{\nu} = \frac{10(0.3)}{11.42\times10^{-6}} = 2.63\times10^5$$

As we will see in the next chapter, this is still laminar flow (barely):

$$\overline{\text{Nu}_d} = \frac{\bar{h}d}{k_f} = 0.3 + \frac{0.62(2.63\times10^5)^{1/2}(0.72)^{1/3}}{\left[1+\left(\frac{0.4}{0.72}\right)^{2/3}\right]^{1/4}}\left[1+\left(\frac{2.63\times10^5}{282{,}000}\right)^{5/8}\right]^{4/5} = 428.9$$

$$\bar{h}_{side} = 33.6\ \frac{\text{W}}{\text{m}^2\,\text{K}}$$

$$\begin{aligned} q &= \pi(0.3)(1.8)(33.6)[24-(-20)] + \pi(0.15)^2(23.9)[24-(-20)] \\ &= 2582\ \text{W} \end{aligned}$$

If we add insulation to the sides, we must recalculate the heat transfer coefficient:

$$\text{Re}_d = \frac{v_\infty d}{\nu} = \frac{10(0.45)}{11.42\times10^{-6}} = 3.94\times10^5$$

$$\overline{\text{Nu}_d} = \frac{\bar{h}d}{k_f} = 0.3 + \frac{0.62(3.94\times10^5)^{1/2}(0.72)^{1/3}}{\left[1+\left(\frac{0.4}{0.72}\right)^{2/3}\right]^{1/4}}\left[1+\left(\frac{3.94\times10^5}{282{,}000}\right)^{5/8}\right]^{4/5} = 583.2$$

$$\bar{h}_{side} = 45.7\ \frac{\text{W}}{\text{m}^2\,\text{K}}$$

We must include the series resistances for the sides and top. The heat flow is

$$q = \frac{(T - T_\infty)}{\dfrac{1}{\bar{h}_{sides}A_{sides}} + \dfrac{\ln(r_o/r_i)}{2\pi k_{down}L}} + \frac{(T - T_\infty)}{\dfrac{1}{\bar{h}_{top}A_{top}} + \dfrac{t_h}{k_{down}A_{top}}}$$

$$q = \frac{44}{\dfrac{1}{\pi(0.3)(1.8)(45.7)} + \dfrac{\ln(0.225/0.15)}{2\pi(0.05)(1.8)}} + \frac{(44)}{\dfrac{4}{\pi(0.3)^2(23.9)} + \dfrac{0.075(4)}{\pi(0.05)(0.3)^2}} = 62.3 \text{ W}$$

The effectiveness of the down can be defined by

$$eff = \frac{q_{no,down}}{q_{down}} = \frac{2582}{62.3} = 41.4$$

and shows how much the insulation would be needed.

Example 13.2 Dissolution of an Aspirin Caplet

Consider an aspirin caplet dissolving in the stomach as shown in Figure 13.10. In the caplet, the aspirin is present in a concentration of $c_{as} = 8$ mol/L. The partition coefficient for dissolution, $K_{eq} = 0.2$.

In the stomach, the aspirin is diluted to 0.025 mol/L. The initial size of the caplet is 3 mm in diameter by 15 mm long. Experiments indicate that under moderate conditions, the stomach provides enough flow for $Re_d \approx 1000$. If the diffusivity of aspirin in water is $D_{as} = 2.5 \times 10^{-10}$ m²/s and the Schmidt number is 100, how long will it be before the caplet has dissolved?

The caplet is in the form of a fairly long cylinder. Since $L > d_p$, we assume that all transport is occurring from the side so that the length of the caplet remains constant while the diameter decreases over time. Using the balance equation, we have

$$\begin{aligned} &In - \quad Out \quad + Gen = \ Acc \\ &0 - \bar{k}_c A_s (K_{eq}c_{as} - c_{a\infty}) + \ 0 \ = \frac{d}{dt}(Vc_{as}) \end{aligned}$$

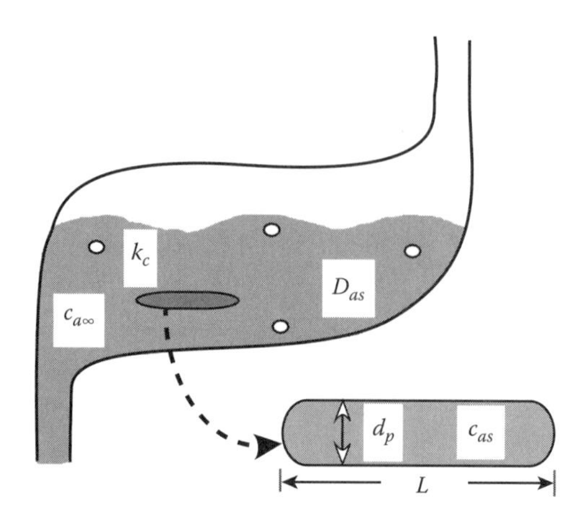

FIGURE 13.10 An aspirin caplet dissolving in the stomach.

At the caplet's surface, c_{as} is constant. Therefore, we are looking at the change in volume of the caplet as a function of time. This volume is given by $A_c \cdot L$ where L is the length of the caplet and $A_c = \pi d_p^2/4$. The differential equation now follows the change in diameter or radius as a function of time:

$$2\pi r L \frac{dr}{dt} = -2\pi r L \bar{k}_c \left(\frac{K_{eq} c_{as} - c_{a\infty}}{c_{as}} \right)$$

Solving this equation subject to the initial condition that

$$t = 0 \qquad r = r_o = \frac{d_p}{2}$$

$$r = r_o - \bar{k}_c \left(\frac{K_{eq} c_{as} - c_{a\infty}}{c_{as}} \right) t$$

We solve for the dissolution time once we know the mass transfer coefficient. Using Equation 13.6 to approximate the mass transfer coefficient gives the time required to dissolve the caplet:

$$\overline{\mathrm{Sh}_d} = \frac{2\bar{k}_c r_o}{D_{as}} = 0.3 + \frac{0.62 \mathrm{Re}_d^{1/2} \mathrm{Sc}^{1/3}}{\left[1 + \left(\frac{0.4}{\mathrm{Sc}}\right)^{2/3}\right]^{1/4}} \left[1 + \left(\frac{\mathrm{Re}_d}{282{,}000}\right)^{5/8}\right]^{4/5}$$

$$\bar{k}_c = 7.74 \times 10^{-6} \ \frac{\mathrm{m}}{\mathrm{s}}$$

$$t = \frac{r_o}{\bar{k}_c \left(\frac{K_{eq} c_{as} - c_{a\infty}}{c_{as}} \right)} = 984 \ \mathrm{s}$$

13.3 FLOW OVER SPHERES

Flow over spherical objects has much in common with flow over cylinders. However, in the spherical case, the surface curves away from the bulk flow in two dimensions rather than in one. The results of this curvature are much the same as we saw in the cylindrical case. The boundary layer is thinnest at the stagnation point (rather than at the stagnation line as in the cylindrical case), and transport coefficients decrease in value as one moves toward the separation point. Beyond the point of separation, the boundary layer detaches, a wake forms, and the transport coefficients begin to rise again. Figure 13.11 shows laminar flow over a sphere at a Reynolds number of 118. As in flow over a cylinder, the wake enlarges as the Reynolds number increases, and mixing in the wake tends to decrease gradients there. The decrease in gradients in the wake corresponds to the increase in transport coefficients there. Note that this behavior is at odds with our discussion of the flat plate boundary layer. The reason the transport coefficients are high in the region between the separation point and the wake is due to the large amount of vorticity in the fluid there. This provides an extra driving force for heat or mass transport.

We can use a Blasius series approach to calculate the boundary layer and flow over a sphere [5]. Figure 13.12 shows the velocity distribution obtained. Experimental measurements still provide the best source of transport coefficients for a *real* spherical object and especially for a group of

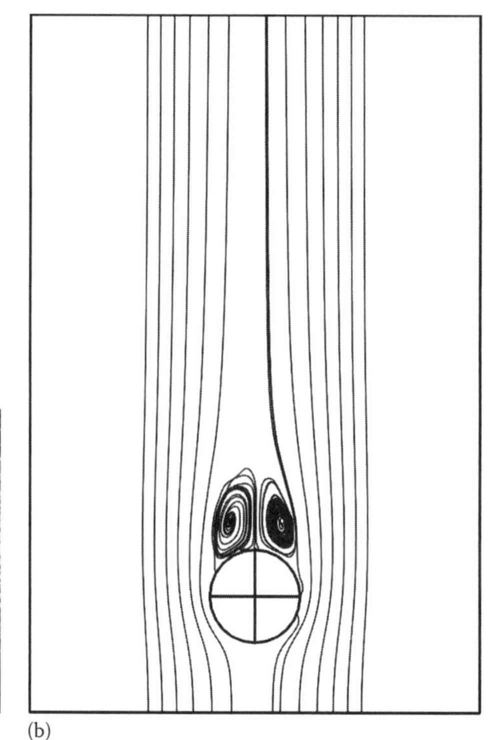

(a) (b)

FIGURE 13.11 (a) Boundary layer in flow over a single sphere at $\text{Re}_d = 118$. Note the qualitative similarity with flow over the cylinder (photograph courtesy of Taneda, S., *J. Phys. Soc. Jap.*, 11, 302, 1956. With permission; Taneda, S., *J. Phys. Soc. Jap.*, 11, 1104, 1956). (b) A 3-D Comsol® simulation (y–z plane shown) from module sphere.

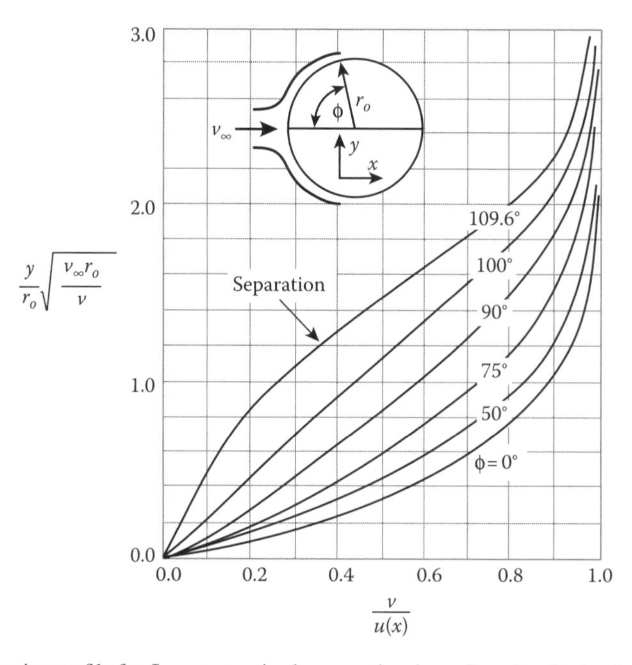

FIGURE 13.12 Velocity profile for flow past a single, smooth sphere. Results obtained using a Blasius series expansion. (Adapted from Schlichting, H., *Boundary Layer Theory*, 7th edn., McGraw-Hill, New York, 1979. With permission.)

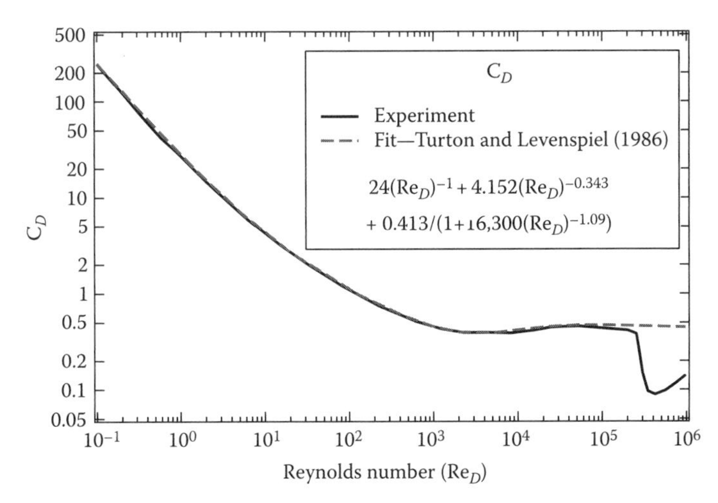

FIGURE 13.13 Friction factor for flow over a single sphere. (Adapted from Schlichting, H., *Boundary Layer Theory*, 7th edn., McGraw-Hill, New York, 1979. With permission; Turton, R. and Levenspiel, O., *Powder Technol.*, 47, 83, 1986.)

spherical objects. One can use any of the *analogies* in conjunction with a friction factor plot like that shown in Figure 13.13 to estimate a heat or mass transfer coefficient. The results will necessarily be approximate since friction factors are defined based on projected areas while we generally calculate transport rates based on the overall surface area in contact with the fluid.

Alternatively, a number of correlations have been developed to account for effects not considered by the analogy–friction factor route. Remember the theoretical descriptions discussed thus far imply that the heat or mass transfer does not affect the hydrodynamic boundary layer. Since our calculations for the flat plate show that natural convection or mass transfer can affect the hydrodynamic boundary layer, the theoretical correlations only apply over a rather limited range. Considering the heat transfer case, we have, for forced convection, the following correlation from Whitaker [14]:

$$\overline{\mathrm{Nu}_d} = 2.0 + \left(0.4\mathrm{Re}_d^{1/2} + 0.06\mathrm{Re}_d^{2/3}\right)\mathrm{Pr}^{0.4}\left(\frac{\mu}{\mu_s}\right)^{1/4} \tag{13.11}$$

that applies over the range

$$\begin{bmatrix} 0.71 < \mathrm{Pr} < 380 \\ 3.5 < \mathrm{Re}_d < 7.6\times 10^4 \\ 1.0 < (\mu/\mu_s) < 3.2 \end{bmatrix}$$

All properties are evaluated at the free stream temperature except for μ_s, which is evaluated at the surface temperature. In many instances, the spherical particle is a freely falling fluid drop and so oscillates in shape. The correlation in Equation 13.12 by Yao and Schrock [15] accounts for this oscillation:

$$\overline{\mathrm{Nu}_d} = 2.0 + 0.6\,\mathrm{Re}_d^{1/2}\mathrm{Pr}^{1/3}\left[25\left(\frac{x}{d}\right)^{-0.7}\right] \tag{13.12}$$

Here, x represents the distance traveled from the starting point. Properties are evaluated at the film temperature, and the Reynolds and Prandtl number ranges are the same as for the Whitaker correlation, Equation 13.11.

For natural convection, the correlation by Churchill [16] is often used:

$$\overline{\mathrm{Nu}_d} = 2.0 + \frac{0.589\,\mathrm{Ra}_d^{1/4}}{\left[1.0 + \left(\frac{0.469}{\mathrm{Pr}}\right)^{9/16}\right]^{4/9}} \qquad \begin{bmatrix} \mathrm{Pr} \ge 0.7 \\ \mathrm{Ra}_d \le 10^{11} \end{bmatrix} \tag{13.13}$$

Properties are evaluated at the film temperature. All these correlations indicate that the lower limit for the Nusselt number is 2. This corresponds to stagnant conditions, a situation difficult to attain in practice. We can derive this result theoretically by looking at a sphere in a quiescent liquid.

The sphere of radius, r_o, is immersed in a liquid of thermal conductivity, k_f. Far away from the sphere, the fluid temperature is T_∞. The surface of the sphere is at a temperature of T_s, and all heat transfer is by conduction. We place a control volume in the fluid as shown in Figure 13.14 and write a differential balance about the volume:

$$\begin{array}{ccccccc} In & - & Out & + & Gen & = & Acc \\ q(r) & - & q(r+\Delta r) & + & 0 & = & 0 \end{array} \tag{13.14}$$

Dividing by Δr, taking the limit, and inserting Fourier's law for the fluid gives

$$\frac{d}{dr}\left(r^2 \frac{dT}{dr}\right) = 0 \tag{13.15}$$

This equation is subject to the boundary conditions that specify constant temperature at the sphere's surface and at a point far away:

$$r = r_o \qquad T = T_s \qquad r = r_\infty \qquad T = T_\infty \tag{13.16}$$

Solving and using the boundary conditions to determine the integration constants leads to the following temperature profile in the liquid:

$$T = T_s + \left[\frac{T_s - T_\infty}{\left(\frac{1}{r_\infty} - \frac{1}{r_o}\right)}\right]\left(\frac{1}{r_o} - \frac{1}{r}\right) \tag{13.17}$$

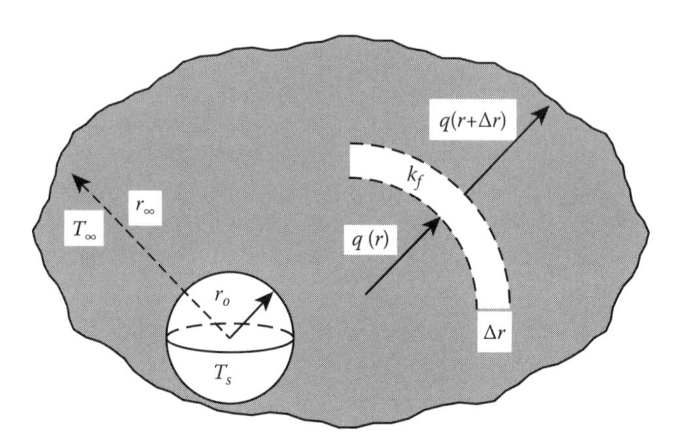

FIGURE 13.14 Schematic for determining the Nusselt or Sherwood numbers when a sphere is immersed in a stagnant liquid.

Using the definition of the heat transfer coefficient and substituting in the temperature profile leads to

$$\bar{h} = \frac{-k_f \left.\dfrac{dT}{dr}\right|_{r=r_o}}{T_s - T_\infty} = -\left[\frac{1}{\left(\dfrac{1}{r_\infty} - \dfrac{1}{r_o}\right)}\right]\frac{k_f}{r_o^2} \tag{13.18}$$

If we let $r_\infty \to \infty$ representing the sphere in an infinite medium, then we can show that the Nusselt number and heat transfer coefficients are constant:

$$\bar{h} \to \frac{k_f}{r_o} = \frac{2k_f}{d} \qquad \text{so: } \overline{\mathrm{Nu}_d} = 2 \tag{13.19}$$

If one tries to treat a cylinder in a quiescent fluid using the previous analysis, the resulting differential equation cannot be solved to give a constant Nusselt number. Thus, the 0.3 factor in Equation 13.4 is an experimentally determined quantity. If we look at a similar mass transfer case for the sphere, we can show that $\overline{\mathrm{Sh}_d} = 2$ under stagnant conditions as well. Replacing the Prandtl number with the Schmidt number and the Nusselt number with the Sherwood number in Equations 13.11 through 13.13 leads to the corresponding mass transfer correlations. Using the mass transfer Rayleigh number leads to a mass transfer coefficient correlation for natural convection based on Churchill's heat transfer correlation, Equation 13.9.

Example 13.3 Evaporation of a Raindrop

Often atmospheric conditions are just right so that raindrops fall from the clouds but evaporate before ever reaching the ground. Consider a 2 mm raindrop falling from a height of 3000 m. The temperature of the drop is 17°C and that of the air is 27°C. We neglect any variation in air temperature with altitude. The relative humidity of the air is 35%. Will the drop reach the ground? What size will it be if it does?

This is a pretty complicated problem that at best must be solved in an iterative fashion. Our first task is to determine the physical properties of the air and the drop. We need the kinematic viscosity of the air, the diffusivity of water through the air, the Schmidt number for the air, and the vapor pressure of water at 27°C and 17°C so that we can determine the driving force for mass transfer. We also need the density of the liquid water:

$\rho_{air} = 1.16\ \mathrm{kg/m^3}$ $\quad\rho_w = 997\ \mathrm{kg/m^3}$ $\quad P_{vap}(17) = 1917\ \mathrm{N/m^2}$

$P_{vap}(27) = 3531\ \mathrm{N/m^2}$ $\quad\nu_{air} = 15.89 \times 10^{-6}\ \mathrm{m^2/s}$ $\quad\mathrm{Sc} = 0.61$

$D_w = 2.54 \times 10^{-5}\ \mathrm{m^2/s}$

We begin by determining how long it will take the drop to fall 3000 m. If the drop falls steadily, at its terminal velocity, the gravitational force must be balanced by the buoyant and drag forces. For this problem, we replace Stokes' drag force expression by an equivalent one involving an experimentally determined drag coefficient that we obtain from Figure 13.13 or correlation therein.

The area used to calculate the drag force is the *frontal area*. The easiest way to envision the shape of the frontal area is to place the object adjacent to a wall and illuminate it. The area of the object's shadow provides the 2-D projection equivalent to the frontal area. Thus, for a sphere, the frontal area would be the largest circle we could get by cutting it, and for a cube the frontal area would be a square. Likewise, for a paraboloid, it would be an ellipse whose dimensions were determined by whether the major or minor axis was oriented parallel to the wall:

$$\underset{\text{Gravitation}}{\rho_w g\left(\frac{4}{3}\pi r_o^3\right)} = \underset{\text{Buoyancy}}{\rho_f g\left(\frac{4}{3}\pi r_o^3\right)} + \underset{\text{Drag}}{\pi r_o^2 C_D\left(\frac{\rho_f v^2}{2}\right)}$$

Solving for the terminal velocity shows that it is proportional to the square root of the radius of the drop:

$$v = \sqrt{\left(\frac{\rho_d - \rho_f}{\rho_f}\right)\left(\frac{8g}{3C_D}\right)r_o} \qquad C_D = \frac{24}{\text{Re}_D} + 4.152\text{Re}_D^{-0.343} + \frac{0.413}{\left[1 + 16{,}300\text{Re}_D^{-1.09}\right]}$$

As evaporation occurs, the terminal velocity will decrease. Looking at Figure 13.13, we also see that the drag coefficient changes as a function of particle size. Some quick Reynolds number calculations will show that for a 2 mm drop, the terminal velocity will be near 3 m/s. The drag coefficient for this entire range hovers around 0.62. We can start our calculations off using this value and then iterate to reach a more refined solution. For a 2 mm drop, the terminal velocity is

$$v = 3.06 \text{ m/s}$$

The mass transfer coefficient for this process can be found using the mass transfer analog of Equation 13.12:

$$\overline{\text{Sh}_d} = 2.0 + \left(0.4\text{Re}_d^{1/2} + 0.06\text{Re}_d^{2/3}\right)\text{Sc}^{0.4}\left(\frac{\mu}{\mu_s}\right)^{1/4}$$

For our 2 mm drop, the Reynolds number is

$$\text{Re}_d = \frac{2vr_o}{\nu_{air}} = 385.1$$

Assuming negligible viscosity change over the range of temperatures encountered, the average mass transfer coefficient is

$$\begin{aligned}\overline{k_c} &= \frac{2.54\times10^{-5}}{2\times10^{-3}}[2.0 + (0.4(385.1)^{1/2} + 0.06(385.1)^{2/3})(0.61)^{0.4}] \\ &= 0.136 \text{ m/s}\end{aligned}$$

To determine the drop size as a function of time, we use a mass balance:

$$\begin{array}{ccccccc} In & - & Out & + & Gen & = & Acc \\ 0 & - & \overline{k_c}A_s(c_{vap} - c_{air}) & + & 0 & = & \dfrac{d}{dt}(Vc_w) \end{array}$$

The concentration of water inside the drop remains constant at

$$c_w = \frac{\rho_w}{M_{ww}} = 55.39 \text{ kg-mol/m}^3$$

The driving force for mass transfer is the difference in concentration between water in the boundary layer at the drop's surface, c_{vap}, and water in the free stream, c_{air}. The former is just the concentration of water in saturated air at the drop's temperature, and the latter is the concentration of water in air at a relative humidity of 35% and a temperature of 27°C. Assuming ideal gases, we have

$$c_{vap} = \frac{P_{vap}(17)}{RT} = 7.95 \times 10^{-4} \text{ kg-mol/m}^3$$

$$c_{air} = 0.35\frac{P_{vap}(27)}{RT} = 4.96 \times 10^{-4} \text{ kg-mol/m}^3$$

Rearranging the balance equation, we have

$$\frac{dr}{dt} = -\overline{k_c}\left(\frac{c_{vap} - c_{air}}{c_w}\right)$$

Integrating gives

$$r = r_i - \overline{k_c}\left(\frac{c_{vap} - c_{air}}{c_w}\right)t$$

The time required to fall 3000 m at the drop's terminal velocity is 980.4 s. The drop radius at that time is

$$r = 1\times10^{-3} - 0.140\left(\frac{7.95\times10^{-4} - 4.96\times10^{-4}}{55.39}\right)(980.4) = 2.58\times10^{-4} \text{ m}$$

and the drop survives the fall though it loses most of its volume. Given the drop size we have just calculated, we can compute an average size over the course of its fall and solve the problem until we converge. Drops not much smaller than a couple of millimeters will end up evaporating under these conditions before hitting the ground. Our analysis here was a bit simplified. Are there any complications that we have failed to account for in this model? What could we do to correct that?

Example 13.4 Fuel Efficiency versus Speed

During the oil crisis of the 1970s, speed limits in the United States were reduced to 55 miles per hour. Considerable fuel savings were projected. If we assume that at highway cruising speeds, the main use of fuel is to overcome aerodynamic drag, we can evaluate this assumption.

In the spirit of the section, we will take an old-style Volkswagen Beetle as the example. The Beetle can be assumed to be a sphere with frontal area of 1.8 m^2, an equivalent radius of 0.378 m, and a drag coefficient of 0.485. The drag force was shown to be

$$F = \frac{1}{2}C_D A_f v^2$$

To determine the power required to overcome this force, we must multiply the force by the speed at which we travel, or

$$P = F * v = \frac{1}{2}C_D A_f v^3$$

Now we can estimate the power needed for cruising (Figures 13.15 and 13.16).

Of course this calculation assumes that the drag coefficient remains constant over the entire speed range. As our experimental data for spheres show, this is not the case. We can modify the standard drag curve to account for this. Power is proportional to Re^3.

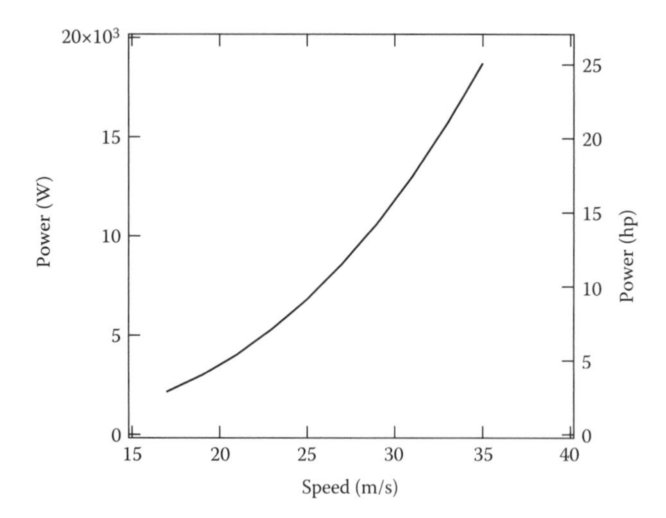

FIGURE 13.15 Horsepower required to overcome drag for a cruising VW Beetle.

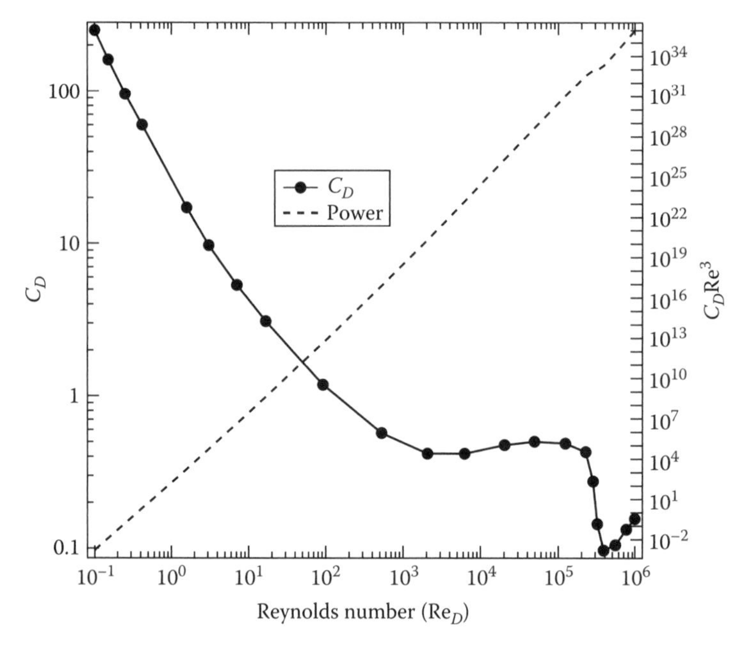

FIGURE 13.16 Power versus the Reynolds number taking the variation in an objects drag coefficient into account. To save power and fuel slower is always better.

More complicated shapes can also be analyzed numerically. Figure 13.17 shows a 2-D simulation for flow across the surface of an Aerobee® flying ring. The figures show the flow streamlines as well as the pressure distribution about the object. Several features are important. The forward stagnation point changes as the angle of attack (angle at which the flow passes over the object) changes with the stagnation point dropping closer to the tip of the leading edge and the angle increases. As the stagnation point drops, notice how the region of low pressure moves from the front of the ring toward the back and how much that low pressure region blooms. This indicates that the flight of the ring will depend critically on the angle at which one throws it, and the abrupt change in pressure distribution is one of the reasons for the exceedingly long flight times of the ring. The Comsol® module, Aerobee, can be used to analyze the situation further and to calculate the viscous and total forces acting on the surface of the Aerobee®.

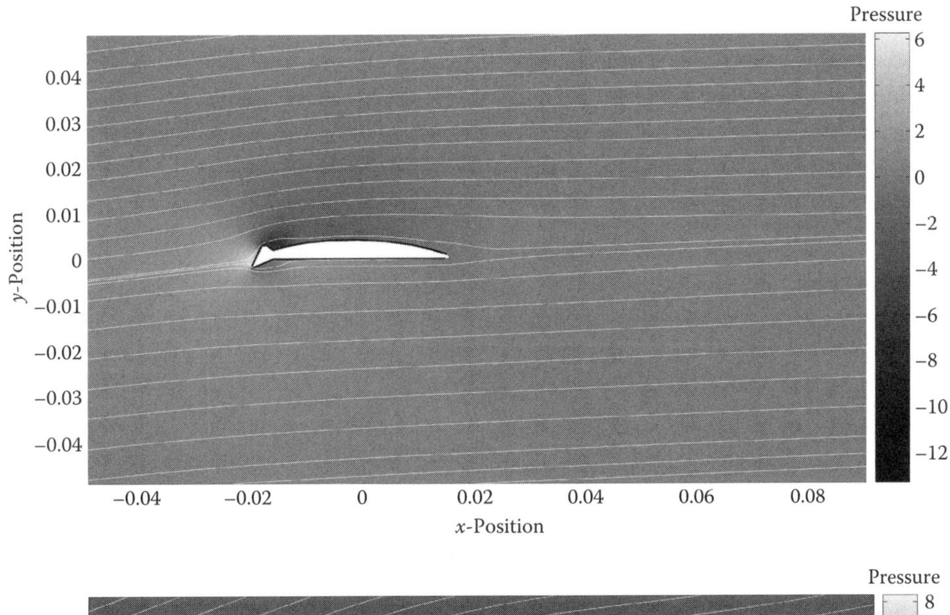

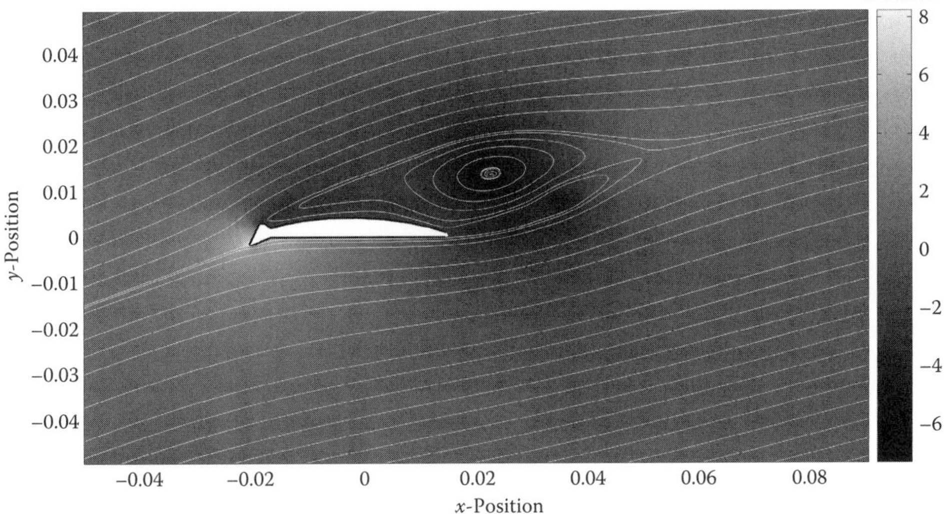

FIGURE 13.17 Pressure distribution and flow streamlines about an Aerobee® flying ring as a function of the throwing angle. The throwing angle is simulated by changing the direction of the flow over the ring. The top figure is at a low angle while the bottom is at a high angle.

Example 13.5 Error between Correlation and Analogy

For oddly shaped objects like the Aerobee®, it would be nice to be able to use friction factor or drag coefficient information to obtain a heat or mass transfer coefficient. However, the difference in reference area may lead to errors since heat or mass transfer depends on the surface area while the drag force depends on the frontal area. Let's consider a simple example to illustrate. Assume we have a 2 cm glass sphere suspended in air. We are interested in calculating the initial heat transfer coefficient when the sphere is at 130°C, the air at 25°C, and the air is flowing over the sphere at a velocity of 20 m/s.

Air properties (350 K)

$$\rho = 0.995\,\frac{\text{kg}}{\text{m}^3} \qquad \mu = 20.82\times10^{-6}\,\frac{\text{N s}}{\text{m}^2} \qquad k = 30.0\times10^{-3}\,\frac{\text{W}}{\text{m K}} \qquad \text{Pr} = 0.700$$

The Reynolds number for this situation is

$$\mathrm{Re}_d = \frac{\rho v_\infty d}{\mu} = \frac{0.995(0.02)(20)}{20.82\times 10^{-6}} = 19120$$

We first start out with the drag coefficient and Chilton–Colburn analogy route. The drag coefficient is

$$\begin{aligned} C_D &= \frac{24}{\mathrm{Re}_d} + 4.152\mathrm{Re}_d^{-0.343} + \frac{0.413}{\left[1+16300\mathrm{Re}_d^{-1.09}\right]} \\ &= \frac{24}{19120} + 4.152(19120)^{-0.343} + \frac{0.413}{[1+16300(19120)^{-1.09}]} = 0.448 \end{aligned}$$

The Nusselt number according to the analogy is

$$N_f \approx \frac{C_D \mathrm{Re}_d}{4} = \overline{\mathrm{Nu}_d}\mathrm{Pr}^{-1/3}$$

$$\overline{\mathrm{Nu}_d} = \frac{C_D \mathrm{Re}_d \mathrm{Pr}^{1/3}}{4} = \frac{0.448(19120)(0.70)^{1/3}}{4} = 1902$$

The product of the heat transfer coefficient and frontal area is

$$\bar{h} = \overline{\mathrm{Nu}_d}\frac{k}{d} = 1902\left(\frac{0.03}{0.02}\right) = 2852\frac{\mathrm{W}}{\mathrm{m}^2\,\mathrm{K}}$$

$$\bar{h}A_f = \frac{\pi d^2}{4}h = 0.90\frac{\mathrm{W}}{\mathrm{K}}$$

If we use the Whitaker correlation and the surface area, we find

$$\begin{aligned} \overline{\mathrm{Nu}_d} &= 2.0 + \left(0.4\mathrm{Re}_d^{1/2} + 0.06\mathrm{Re}_d^{2/3}\right)\mathrm{Pr}^{0.4}\left(\frac{\mu}{\mu_s}\right)^{1/4} \\ &\approx 2.0 + (0.4(19120)^{1/2} + 0.06(19120)^{2/3})(0.7)^{0.4} = 87.2 \end{aligned}$$

$$\bar{h} = \frac{\overline{\mathrm{Nu}_d}k}{d} = 87.2\left(\frac{0.03}{0.02}\right) = 130.7\frac{\mathrm{W}}{\mathrm{m}^2\,\mathrm{K}}$$

$$\bar{h}A_s = \pi d^2 h = 0.164\frac{\mathrm{W}}{\mathrm{K}}$$

The difference is about a factor of 6, and so while the friction factor route can be used in a pinch, it overestimates the dependence on the Reynolds number. Even when the flow velocity approaches 0, the analogy will still be about a factor of 1.5 off.

13.4 VELOCITY PROFILES IN TUBES

Many, if not most, transport operations occur as fluid flows through tubes of one type or another. For instance, a chemical plant or a power plant consists almost entirely of tubes of different size and configuration. Cylindrical symmetry is inherent in the boilers, turbines, heat exchangers, mass transfer equipment, and reaction equipment. Knowing how a transport operation occurs in such

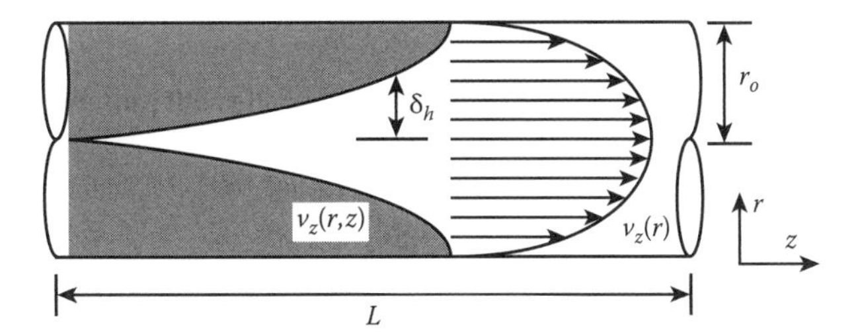

FIGURE 13.18 Developing boundary layers for flow in a tube.

configurations is vital to the proper design and operation of the facility. On the biological side, the flow of sap in trees, blood in arteries and organs, and water in plant roots are all examples of flow in tubes. The proper operation of a complicated organism or the design of an artificial organ requires a knowledge of the transport coefficients in these geometries. In this and subsequent sections, we will consider laminar flow in tubes of all sizes and show how the friction factor, mass transfer coefficient, and heat transfer coefficient depend upon the flow field, the temperature field, and the concentration field.

The boundary layer in the confined space of a tube is much different from that found on a flat plate or over a cylinder or sphere. The situation is shown in Figure 13.18. Suppose we have an initially quiescent fluid and we impose a pressure gradient along the axis of the tube. The force will cause the fluid to begin to flow. Fluid immediately adjacent to the wall is retarded in its motion (no-slip condition), but fluid in the center is free to move in response to the force. The result is the development of a boundary layer. This boundary layer is small initially but grows until it spans the tube. Due to the symmetry of the tube, the boundary layer cannot grow in thickness beyond the tube radius, and so once it reaches that size, it remains the same for all time. We call this ultimate condition, *fully developed flow.* Since the boundary layer cannot increase further, we have a constant friction factor, and correspondingly, a constant heat and mass transfer coefficient. We can perform a simple analysis and see, mathematically, how the flow develops.

Assume that the fluid is incompressible and has constant properties. If there is a uniform pressure gradient along the tube, we can approximate the flow and say that v_r and v_θ are negligible. The assumption of angular symmetry is good, but as we all know, the boundary layer is 2-D and so v_r cannot actually be zero while the boundary layer is developing. However, using these assumptions, we can reduce the momentum and continuity equations to a single equation for the axial velocity, $v_z = f(r,t)$, and approximate how the boundary layer might develop:

$$\rho \frac{\partial v_z}{\partial t} = -\frac{dP}{dz} + \frac{\mu}{r}\frac{\partial}{\partial r}\left(r\frac{\partial v_z}{\partial r}\right) \tag{13.20}$$

To correspond with the scenario in Figure 13.18, we will assume that at $t = 0$, we have the fluid moving uniformly along and then we put a very long pipe in the stream. Once the fluid encounters the tube, it must obey the no-slip condition at the wall, and the velocity profile must be symmetric about the centerline of the tube. Therefore, our initial and boundary conditions are

$$t = 0 \qquad v_z = v_\infty \tag{13.21}$$

$$r = 0 \qquad \frac{\partial v_z}{\partial r} = 0 \tag{13.22}$$

$$r = r_o \qquad v_z = 0 \tag{13.23}$$

Before attempting to solve the equation, we can simplify it by introducing the following dimensionless variables:

$$u = \frac{4\mu v_z}{-\left(dP/dz\right) r_o^2} = \frac{v_z}{v_{max}} \qquad \xi = \frac{r}{r_o} \qquad \tau = \frac{\mu t}{\rho r_o^2} \tag{13.24}$$

The velocity was made dimensionless by dividing through by the maximum velocity at steady state. Substituting the new variables into the differential equation gives

$$\frac{\partial u}{\partial \tau} = 4 + \frac{1}{\xi}\frac{\partial}{\partial \xi}\left(\xi \frac{\partial u}{\partial \xi}\right) \tag{13.25}$$

We must solve this dimensionless equation subject to the transformed boundary conditions:

$$\tau = 0 \qquad u = u_m \tag{13.26}$$

$$\xi = 0 \qquad \frac{\partial u}{\partial \xi} = 0 \tag{13.27}$$

$$\xi = 0 \qquad u = 0 \tag{13.28}$$

At some time in the future, the system will reach a steady state, and from past experience and calculations (see Chapter 5), the velocity profile then must be parabolic. Since Equation 13.25 is linear in u, the solution for all time can be written as a sum (or difference) of steady state and transient/short time solutions:

$$u(\xi, \tau) = u_t(\xi, \tau) + u_\infty(\xi) \tag{13.29}$$

The steady-state solution is governed by the following differential equation (when $\partial u/\partial \tau = 0$):

$$0 = 4 + \frac{1}{\xi}\frac{\partial}{\partial \xi}\left(\xi \frac{\partial u}{\partial \xi}\right) \tag{13.30}$$

This equation can be separated and integrated to yield

$$u_\infty = 1 - \xi^2 \tag{13.31}$$

where we have used the boundary conditions, Equations 13.27 and 13.28, to evaluate the integration constants.

Now we substitute the steady-state solution into Equation 13.29 and then substitute (13.29) back into the differential equation (13.25), to obtain a differential equation describing the transient part of the solution:

$$\frac{\partial u_t}{\partial \tau} = \frac{1}{\xi}\frac{\partial}{\partial \xi}\left(\xi \frac{\partial u_t}{\partial \xi}\right) \tag{13.32}$$

Since the differential equation involves only the short-term solution, the initial condition is changed to

$$\tau = 0 \qquad u_t = u_m - u_\infty \tag{13.33}$$

This causes no complication since the solution for u_t at $\tau = 0$ insures that $u = u_m$ at that time.

Equation 13.32 is suitable for a solution using separation of variables. We define two new functions E and T that depend only on ξ and τ, respectively:

$$u_t = E(\xi)T(\tau) \tag{13.34}$$

Substituting this definition into the differential equation and dividing through by ET gives

$$\frac{1}{T}\frac{dT}{d\tau} = \frac{1}{E}\frac{1}{\xi}\frac{d}{d\xi}\left(\xi\frac{dE}{d\xi}\right) \tag{13.35}$$

Now we can separate the partial differential equation into two ordinary differential equations:

$$\frac{dT}{d\tau} + \lambda^2 T = 0 \tag{13.36}$$

$$\frac{1}{\xi}\frac{d}{d\xi}\left(\xi\frac{dE}{d\xi}\right) + \lambda^2 E = 0 \tag{13.37}$$

Equation 13.36 is just an exponential decay:

$$T = a_o \exp(-\lambda^2\tau) \tag{13.38}$$

The second differential equation (13.37) has a solution given in terms of the Bessel functions:

$$E = a_1 J_o(\lambda\xi) + a_2 Y_o(\lambda\xi) \tag{13.39}$$

where

J_o is the Bessel function of the first kind
Y_o is the Bessel function of the second kind

Graphs of Y_o and J_o were shown in Chapter 6. For $\partial u_\tau/\partial\xi = 0$ at $\xi = 0$, $\partial E/\partial\xi = 0$ must also hold there. The Y_o Bessel function cannot be a solution since its derivative (Y_1) approaches $-\infty$ at $\xi = 0$. Therefore, $a_2 = 0$. The boundary condition at $\xi = 1$ states that E must be zero there. J_o is an oscillating function so there are an infinite number of possible solutions where $E = 0$. Thus, we need to have $J_0(\lambda) = 0$ to satisfy this condition. The first few roots of J_o were also given back in Chapter 6. Reconstituting u_t, we have

$$u_t(\xi,\tau) = a_n \exp\left(-\lambda_n^2\tau\right) J_o(\lambda_n\xi) \tag{13.40}$$

We determine a_n from the initial condition. We can satisfy it only by superimposing an infinite number of solutions like Equation 13.40. Substituting $\tau = 0$ into Equation 13.40 gives

$$u_m - (1-\xi^2) = \sum_{n=1}^{\infty} a_n J_o(\lambda_n\xi) \tag{13.41}$$

To extract the a_n, we multiply Equation 13.41 by $\xi J_o(\lambda_m \xi)d\xi$ and integrate from 0 to 1:

$$\int_0^1 (u_m - 1 + \xi^2) J_o(\lambda_m \xi)\xi\, d\xi = \sum_{n=1}^{\infty} a_n \int_0^1 J_o(\lambda_m \xi) J_o(\lambda_n \xi)\xi\, d\xi \tag{13.42}$$

The only term on the right-hand side that contributes to the sum occurs when $m = n$. The integrals can be evaluated and used to extract the a_n:

$$a_n = \frac{2}{\lambda_n^3}\left[\frac{u_m \lambda_n^2 J_1(\lambda_n) + 2\lambda_n J_0(\lambda_n) - 4J_1(\lambda_n)}{J_o^2(\lambda_n) + J_1^2(\lambda_n)}\right] \tag{13.43}$$

Since $J_0(\lambda_n) = 0$, the final solution for u now becomes

$$u = (1-\xi^2) + 2\sum_{n=1}^{\infty} \frac{1}{\lambda_n^3}\left[\frac{u_m \lambda_n^2 - 4}{J_1(\lambda_n)}\right] J_o(\lambda_n \xi)\exp\left(-\lambda_n^2 \tau\right) \tag{13.44}$$

In Chapter 12, we defined a friction number that was the dimensionless velocity gradient at the wall. For the tube flow case, this becomes

$$N_f = \frac{C_f \operatorname{Re}_d}{4} = -\frac{du}{d\xi}\bigg|_{\xi=1} = -\frac{d}{d\xi}\left[(1-\xi^2) + 2\sum_{n=1}^{\infty} \frac{1}{\lambda_n^3}\left[\frac{u_m \lambda_n^2 - 4}{J_1(\lambda_n)}\right] J_0(\lambda_n \xi)\exp\left(-\lambda_n^2 \tau\right)\right]_{\xi=1} \tag{13.45}$$

Using the property of the Bessel functions that

$$\frac{d}{dx}(J_o(x)) = -J_1(x) \tag{13.46}$$

we have

$$N_f = 2 + 2\sum_{n=1}^{\infty}\left[\frac{u_m \lambda_n^2 - 4}{\lambda_n^2}\right]\exp\left(-\lambda_n^2 \tau\right) \tag{13.47}$$

A graph of v_z/v_{max} for various times is shown in Figure 13.19a. Notice the very flat profile that exists at short times. This profile rapidly transforms itself to the parabolic profile as the boundary layer grows. The maximum centerline velocity is within a percent or two of the parabolic maximum when τ approaches 0.5. The friction number behavior can be seen in Figure 13.19b. One can easily see that the friction number is largest at short times and decreases rapidly as the fully developed profile matures. It reaches its asymptotic value of 2 by the time τ approaches 0.5.

13.5 HEAT AND MASS TRANSFER APPLICATIONS

Previously, we discussed how the velocity profile and friction factor develop when we immerse a tube in a uniform flowing fluid. We now want to investigate heat and mass transfer in a tube and discuss how the thermal and concentration boundary layers grow and how the heat and mass transfer coefficients behave. We begin with an analysis of steady, fully developed laminar flow situations and move to the more complicated case where the flow is fully developed yet the thermal or concentration boundary layers are developing. There are two important cases to consider: constant

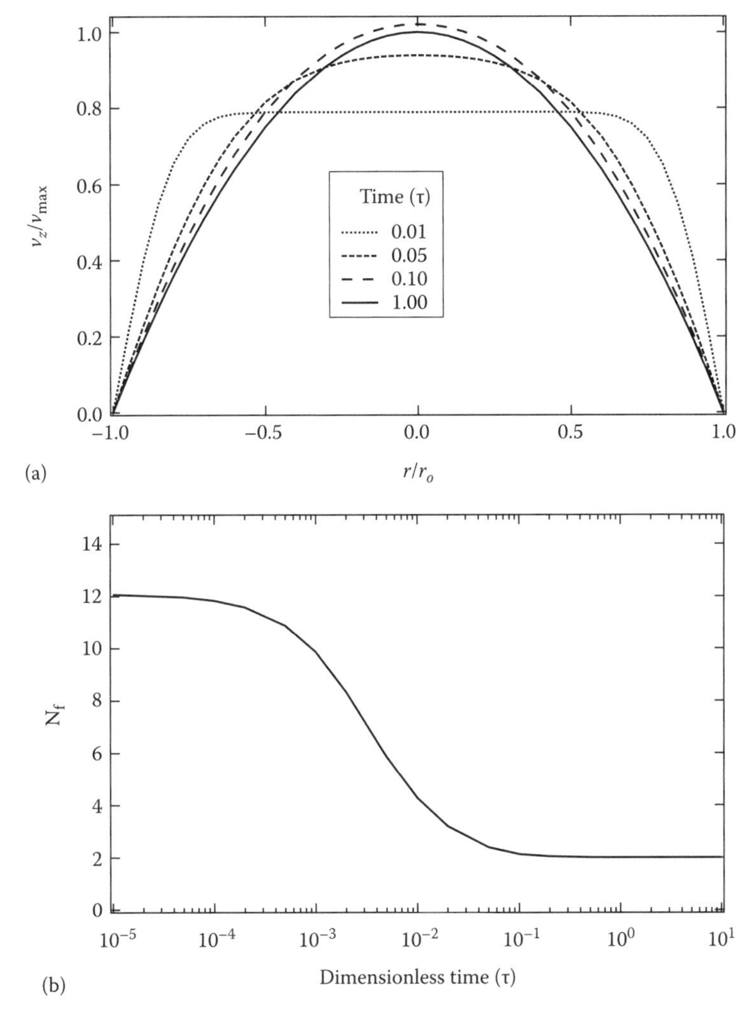

FIGURE 13.19 (a) Developing velocity profiles in a tube. (b) Friction number as a function of time. In both figures, $u_m = 0.75$. The small oscillation in profile in part (a) is due to the finite approximation used for the series in Equation 13.44.

temperature or concentration at the tube wall and constant heat or mass flux at the wall. In all cases, we will assume constant properties and dilute solutions.

We need to determine the friction factor, the heat transfer coefficient, and the mass transfer coefficient. All the coefficients are defined in terms of mean or mixing-cup properties of the fluid in the tube: the mean velocity, the mean temperature, and the mean concentration. The coefficients and mean quantities are defined in Equations 13.48–13.50.

$$f = -\frac{(dP/dz)d}{2\rho\bar{v}_z^{\,2}} \tag{13.48}$$

$$q'' = h(T_s - \bar{T}) \tag{13.49}$$

$$N_a = k_c(c_{as} - \bar{c}_a) \tag{13.50}$$

For tubes of circular cross section,

$$\bar{v}_z = \frac{\displaystyle\int_{A_c} \rho v_z \, dA_c}{\displaystyle\int_{A_c} \rho \, dA_c} = \frac{\displaystyle\int_0^{2\pi} \rho \, d\theta \int_0^{r_o} \frac{\Delta P r_o^2}{4\mu L}\left[1-\left(\frac{r}{r_o}\right)^2\right] r \, dr}{\displaystyle\int_0^{2\pi} \rho \, d\theta \int_0^{r_o} r \, dr} \tag{13.51}$$

$$\bar{T}(z) = \frac{\displaystyle\int_{A_c} \rho C_p v_z T \, dA_c}{\displaystyle\int_{A_c} \rho C_p v_z \, dA_c} = \frac{\displaystyle\int_0^{2\pi} d\theta \int_0^{r_o} \frac{\Delta P r_o^2}{4\mu L}\left[1-\left(\frac{r}{r_o}\right)^2\right] \rho C_p T r \, dr}{\displaystyle\int_0^{2\pi} d\theta \int_0^{r_o} \frac{\Delta P r_o^2}{4\mu L}\left[1-\left(\frac{r}{r_o}\right)^2\right] \rho C_p r \, dr} \tag{13.52}$$

$$\bar{c}_a(z) = \frac{\displaystyle\int_{A_c} v_z c_a dA_c}{\displaystyle\int_{A_c} v_z dA_c} = \frac{\displaystyle\int_0^{2\pi} d\theta \int_0^{r_o} \frac{\Delta P r_o^2}{4\mu L}\left[1-\left(\frac{r}{r_o}\right)^2\right] c_a r \, dr}{\displaystyle\int_0^{2\pi} d\theta \int_0^{r_o} \frac{\Delta P r_o^2}{4\mu L}\left[1-\left(\frac{r}{r_o}\right)^2\right] r \, dr} \tag{13.53}$$

Now we can consider what happens as the fluid flows down a heated tube. Initially the fluid starts out at a uniform temperature of T_o. As the fluid encounters the heated wall, a thermal boundary layer develops. The boundary layer grows as we look along the length of the tube until it spans the entire cross section. If the tube wall is at a constant temperature, then the temperature gradients in the radial and axial directions steadily decrease and the parabolic temperature profile collapses to a uniform temperature. If we have a constant heat flux at the wall, then the temperature gradients need never decrease. The tube wall temperature steadily increases instead. In this scenario, the parabolic profile never changes shape and the gradients remain constant. Both situations are shown in Figure 13.20.

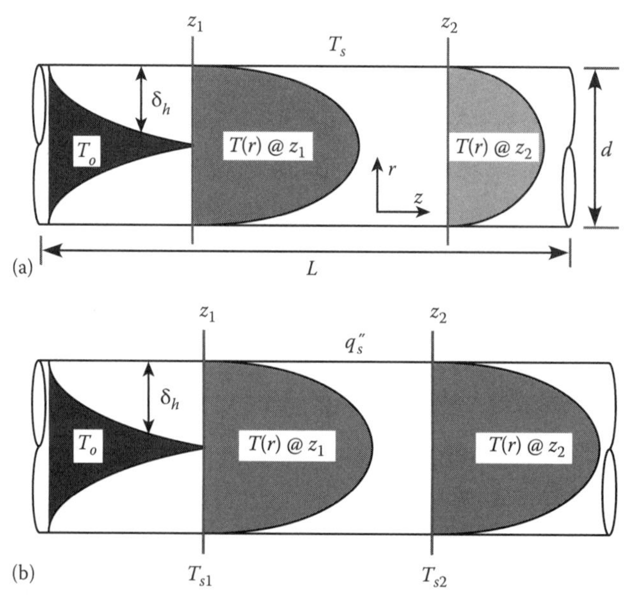

FIGURE 13.20 Developing temperature profiles in internal flow. (a) Constant tube wall temperature. (b) Constant tube wall heat flux.

One of the problems in dealing with internal flow is that there is no free stream temperature to act as an anchor. Each segment of the fluid in the tube must experience an increase in temperature as the fluid flows down along the axis. This means that the mean temperature we calculated in Equation 13.52 must be a function of axial position, z. (Is the mean velocity also a function of z? Why?) A similar conclusion is drawn for mass transfer. The mean concentration must also be a function of axial position. Once the profiles are fully developed, the velocity, temperature, and concentration profiles cannot change their basic geometrical form. This means there is some combination of the temperature or concentration variables that will remain unchanged at any position along the tube. Since the mean temperature must accurately reflect changes occurring in the full temperature profile, any difference between the wall temperature and the actual temperature in the fluid must be mimicked by a similar difference between the wall temperature and mean temperature. This lets us define a new temperature variable using this information that should not be a function of axial position for fully developed conditions:

$$\theta = \frac{T_s(z) - T(r,z)}{T_s(z) - \bar{T}(z)} = \text{Constant} \neq f(z) \tag{13.54}$$

Under fully developed conditions θ is a constant, and we can write

$$\frac{\partial}{\partial z}\left[\frac{T_s(z) - T(r,z)}{T_s(z) - \bar{T}(z)}\right] = 0 \tag{13.55}$$

A similar variable exists for mass transfer and this variable also obeys Equation 13.55:

$$\chi = \frac{c_{as}(z) - c_a(r,z)}{c_{as}(z) - \bar{c}_a(z)} = \text{Constant} \neq f(z) \tag{13.56}$$

$$\frac{\partial}{\partial z}\left[\frac{c_{as}(z) - c_a(r,z)}{c_{as}(z) - \bar{c}_a(z)}\right] = 0 \tag{13.57}$$

We expect that the heat transfer coefficient or the mass transfer coefficient will be constant once fully developed conditions are reached. Remember, we already showed that the friction number and the friction factor were constant. We can use our new temperature and concentration variables to show this conjecture to be true. We take the derivatives of θ and χ with respect to r and evaluate them at the tube wall, r_o:

$$\left.\frac{\partial}{\partial r}\left[\frac{T_s - T}{T_s - \bar{T}}\right]\right|_{r=r_o} = \frac{-\partial T/\partial r\big|_{r=r_o}}{T_s - \bar{T}} = \frac{\bar{h}}{k_f} \tag{13.58}$$

$$\left.\frac{\partial}{\partial r}\left[\frac{c_{as} - c_a}{c_{as} - \bar{c}_a}\right]\right|_{r=r_o} = \frac{-\partial c_a/\partial r\big|_{r=r_o}}{c_{as} - \bar{c}_a} = \frac{\bar{k}_c}{D_{ab}} \tag{13.59}$$

These derivatives should look familiar as two-thirds of the Nusselt number or Sherwood number. All we need to do is add the length scale, the diameter of the tube. Since neither θ nor χ is a function of z, this exercise has shown that neither $\bar{h}$ nor $\bar{k}_c$ is a function of z under fully developed conditions.

If we restrict ourselves to the heat transfer case for the moment, we can expand the derivative in Equation 13.55 and see how T, $\bar{T}$, and T_s behave under conditions where T_s or the heat flux, q''_s, is held constant. Solving for $\partial T/\partial z$, we find

$$\frac{\partial T}{\partial z} = \frac{dT_s}{dz} - \left(\frac{T_s - T}{T_s - \bar{T}}\right)\frac{dT_s}{dz} + \left(\frac{T_s - T}{T_s - \bar{T}}\right)\frac{d\bar{T}}{dz} \quad (13.60)$$

When the surface temperature is constant, $\partial T_s/\partial z = 0$, Equation 13.60 reduces to

$$\frac{\partial T}{\partial z} = \left(\frac{T_s - T}{T_s - \bar{T}}\right)\frac{d\bar{T}}{dz} \quad (13.61)$$

and $\partial T/\partial z$ depends on radial position within the tube. This dependence is necessary to flatten out the parabolic profile for long tube lengths.

When the heat flux at the surface is constant, we use Newton's law, Equation 13.49, to show that

$$\frac{dT_s}{dz} = \frac{d\bar{T}}{dz} \quad (13.62)$$

Equation 13.60 now reduces to

$$\frac{\partial T}{\partial z} = \frac{dT_s}{dz} = \frac{d\bar{T}}{dz} \quad (13.63)$$

Equations 13.61 and 13.63 indicate that $\bar{T}$ and T asymptotically approach T_s when the surface temperature is held constant and that $\bar{T}$, T, and T_s are parallel to one another when the surface heat flux is held constant. The analysis also applies to mass transfer. Thus, $\bar{c}_a$ and c_a asymptotically approach c_{as} for constant surface concentration, and $\bar{c}_a$, c_a, and c_{as} are parallel to one another for constant surface mass flux.

13.5.1 Temperature/Concentration Profiles and Heat/Mass Transfer Coefficients

We are now in a position to solve the energy equation for fully developed flow and to determine the temperature profile within the fluid. We will also calculate the heat transfer coefficient.

13.5.1.1 Steady State: Constant Surface Heat Flux

The constant heat flux case can be achieved in a fired heater, nuclear reactor, or solar collector, where the tube is exposed to a uniform source of thermal radiation. We start with the steady-state energy equation in cylindrical coordinates for a fluid with constant properties and negligible viscous dissipation:

$$k\left(\frac{1}{r}\frac{\partial}{\partial r}\left(r\frac{\partial T}{\partial r}\right) + \frac{1}{r^2}\frac{\partial^2 T}{\partial \theta^2} + \frac{\partial^2 T}{\partial z^2}\right) = \rho C_p\left(v_r\frac{\partial T}{\partial r} + \frac{v_\theta}{r}\frac{\partial T}{\partial \theta} + v_z\frac{\partial T}{\partial z}\right) \quad (13.64)$$

Under fully developed conditions $v_r = v_\theta = 0$, and we have symmetry in the θ direction. We can neglect conduction in the axial direction since convective transport of heat in that direction is so much larger ($\partial^2 T/\partial z^2 < \rho C_p\, \partial T/\partial z$). We can check this assumption by calculating the Peclet number ($\mathrm{Pe} = \mathrm{Re}_d\mathrm{Pr}$ or $\mathrm{Re}_d\mathrm{Sc}$). If $\mathrm{Pe} \gg 1$ then convection dominates in the axial direction. With these simplifications in mind, and substituting in for the velocity profile in terms of the mean velocity $\bar{v}_z$, the temperature in the fluid must obey

$$\frac{1}{r}\frac{\partial}{\partial r}\left(r\frac{\partial T}{\partial r}\right)=\frac{2\overline{v}_z}{\alpha}\left[1-\left(\frac{r}{r_o}\right)^2\right]\frac{\partial T}{\partial z} \tag{13.65}$$

If we have a constant heat flux, we can replace $\partial T/\partial z$ by $d\overline{T}/dz$ in accord with Equation 13.63:

$$\frac{1}{r}\frac{\partial}{\partial r}\left(r\frac{\partial T}{\partial r}\right)=\frac{2\overline{v}_z}{\alpha}\left[1-\left(\frac{r}{r_o}\right)^2\right]\frac{d\overline{T}}{dz} \tag{13.66}$$

This equation is readily separated and integrated since $d\overline{T}/dz$ is a constant and does not depend upon r:

$$T(r,z)=\frac{2\overline{v}_z}{\alpha}\left(\frac{d\overline{T}}{dz}\right)\left[\frac{r^2}{4}-\frac{r^4}{16r_o^2}\right]+a_1\ln r+a_2 \tag{13.67}$$

The first boundary condition for the problem must insure that the temperature remains finite at the center of the tube. Therefore, $a_1 = 0$. We cannot use a flux condition as our second boundary condition because we must fix a temperature to generate a unique solution. To evaluate a_2, we must specify a temperature somewhere, that is, at the surface of the tube, $T_s(z)$:

$$T(r,z)=T_s(z)-\frac{2\overline{v}_z r_o^2}{\alpha}\left(\frac{d\overline{T}}{dz}\right)\left[\frac{3}{16}+\frac{1}{16}\left(\frac{r}{r_o}\right)^4-\frac{1}{4}\left(\frac{r}{r_o}\right)^2\right] \tag{13.68}$$

We can now substitute the temperature and velocity profiles into Equation 13.52 and integrate to get the mean temperature:

$$\overline{T}(z)=T_s(z)-\frac{11}{48}\left(\frac{\overline{v}_z r_o^2}{\alpha}\right)\left(\frac{d\overline{T}}{dz}\right) \tag{13.69}$$

Knowing $\overline{T}$, we can now derive an expression for the heat transfer coefficient. First, we must perform an energy balance about a differential element of the fluid. This is shown schematically in Figure 13.21:

$$\begin{array}{ccccc} In & - & Out & = & Acc \\ \dot{m}C_p\overline{T}(z) & - & \dot{m}C_p\overline{T}(z+\Delta z)-\overline{h}(2\pi r_o\Delta z)(T_s-\overline{T}) & = & 0 \end{array} \tag{13.70}$$

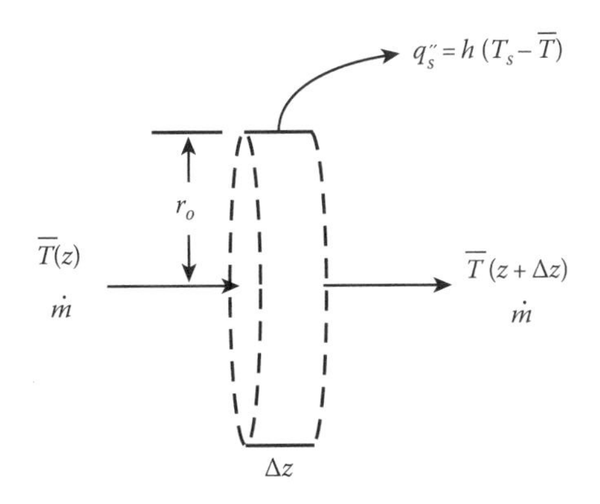

FIGURE 13.21 Energy balance about a small fluid element within the tube.

Dividing through by Δz and taking the limit as $\Delta z \to 0$ shows that

$$\left(\frac{\dot{m}C_p}{2\pi r_o}\right)\frac{d\bar{T}}{dz} = \left(\frac{\rho\bar{v}_z r_o C_p}{2}\right)\frac{d\bar{T}}{dz} = \bar{h}(T_s - \bar{T}) = q_s'' \tag{13.71}$$

Substituting for $\bar{T}$ using our temperature profile,

$$\left(\frac{\rho\bar{v}_z r_o C_p}{2}\right)\frac{d\bar{T}}{dz} = \bar{h}\left[\frac{11}{48}\left(\frac{\bar{v}_z r_o^2}{\alpha}\right)\left(\frac{d\bar{T}}{dz}\right)\right] \tag{13.72}$$

lets us solve for the heat transfer coefficient, $\bar{h}$, and the Nusselt number, $\overline{\mathrm{Nu}}_d$:

$$\overline{\mathrm{Nu}}_d = \frac{\bar{h}d}{k_f} = \frac{48}{11} = 4.36 \tag{13.73}$$

As expected, both the heat transfer coefficient and the Nusselt number are constants. A similar derivation using the species mass balance leads to a constant Sherwood number and mass transfer coefficient:

$$\overline{\mathrm{Sh}}_d = \frac{\bar{k}_c d}{D_{ab}} = 4.36 \tag{13.74}$$

13.5.1.2 Steady State: Constant Surface Temperature

The easiest way to achieve a constant tube wall temperature is to boil or condense a fluid on the surface. As before we begin with the energy equation (13.65), and substitute for $\partial T/\partial z$ using Equation 13.61 to include the mean temperature:

$$\frac{1}{r}\frac{\partial}{\partial r}\left(r\frac{\partial T}{\partial r}\right) = \frac{2\bar{v}_z}{\alpha}\left[1 - \left(\frac{r}{r_o}\right)^2\right]\left[\frac{T_s - T}{T_s - \bar{T}}\right]\frac{\partial \bar{T}}{\partial z} \tag{13.75}$$

Since $\bar{T}$ is found by integrating the temperature profile over the cross section of the tube, Equation 13.75 cannot be solved explicitly. The solution requires a trial and error procedure (see Kays [18]), but the result is similar to the constant heat flux case. Remember once the thermal boundary layer spans the tube, the heat transfer coefficient must be constant and so the Nusselt number must be constant. Since the surface temperature is fixed, and $\bar{T}$ approaches it asymptotically, we cannot sustain as high a temperature gradient as we can in the constant heat flux case. Thus, the Nusselt number is lower:

$$\overline{\mathrm{Nu}}_d = 3.66 \tag{13.76}$$

Similarly, the Sherwood number for mass transfer is

$$\overline{\mathrm{Sh}}_d = 3.66 \tag{13.77}$$

13.5.1.3 Developing Profiles

When we considered flow over a flat plate, we saw that the transport coefficients decreased as the thickness of the boundary layer increased. The same process occurs in tube flow. Consider a tube with a fully developed velocity profile. At some point down the length of the tube, we begin to heat its surface or dissolve some material from the surface. The thermal or concentration boundary layers begin to develop, and we would like to know how the corresponding heat or mass transfer coefficients behave prior to fully developed conditions. We can look at the mass transfer case assuming dilute solutions and extrapolate our results back to the heat transfer case.

We begin by simplifying the species continuity equation in cylindrical coordinates, substituting in for the velocity profile to obtain

$$\frac{1}{r}\frac{\partial}{\partial r}\left(r\frac{\partial c_a}{\partial r}\right)=\frac{2\bar{v}_z}{D_{ab}}\left[1-\left(\frac{r}{r_o}\right)^2\right]\frac{\partial c_a}{\partial z} \tag{13.78}$$

As always, it helps to put things in dimensionless form, and so we use the following variable transformations for r, z, and c_a:

$$\xi=\frac{r}{r_o}\qquad \zeta=\frac{z}{L}\qquad \chi=\frac{c_{as}-c_a}{c_{as}} \tag{13.79}$$

Substituting into Equation 13.78 gives

$$\frac{1}{\xi}\frac{\partial}{\partial \xi}\left(\xi\frac{\partial \chi}{\partial \xi}\right)=\frac{2\bar{v}_z r_o^2}{D_{ab}L}[1-\xi^2]\left(\frac{\partial \chi}{\partial \zeta}\right) \tag{13.80}$$

This equation is subject to boundary conditions specifying the concentration at the wall and symmetry at the tube centerline. We also need to specify the concentration at the inlet:

$$\xi=0\qquad \chi=1 \tag{13.81}$$

$$\xi=0\qquad \frac{\partial \chi}{\partial \xi}=0 \tag{13.82}$$

$$\xi=1\qquad \chi=0 \tag{13.83}$$

The dimensionless quantity

$$\frac{\bar{v}_z r_o^2}{D_{ab}L}=\frac{1}{4}\mathrm{Re}_d\,\mathrm{Sc}\left(\frac{d}{L}\right)=\frac{1}{4}G_{zm}\quad \text{the Graetz number} \tag{13.84}$$

depends upon the Reynolds number, the Schmidt number, and the ratio of tube diameter to tube length. It is called the Graetz number after the man who first solved Equation 13.80. We will see that the value of the Graetz number is a measure of how close the concentration boundary layer is to being fully developed. The analogous Graetz number for heat transfer is defined as

$$\frac{\bar{v}_z r_o^2}{\alpha L}=\frac{1}{4}\mathrm{Re}_d\mathrm{Pr}\left(\frac{d}{L}\right)=\frac{1}{4}G_z \tag{13.85}$$

Equation 13.80 can be solved using separation of variables combined with numerical integration or a series solution technique. We first split the concentration into two functions:

$$\chi=Z(\zeta)E(\xi) \tag{13.86}$$

Substituting into the differential equation and separating the two resulting ordinary differential equations gives

$$\frac{dZ}{d\zeta}+\frac{\lambda^2 G_{zm}}{2}Z=0 \tag{13.87}$$

$$\frac{1}{\xi}\frac{d}{d\xi}\left(\xi\frac{dE}{d\xi}\right)+\lambda^2(1-\xi^2)E=0 \tag{13.88}$$

The first equation for Z is easily integrated:

$$Z=a_o\exp\left(-\frac{\lambda^2 G_{zm}}{2}\zeta\right) \tag{13.89}$$

This solution shows that the profile at the entrance to the mass transfer region will undergo an exponential decay to the fully developed concentration profile.

The second differential equation, (13.88), cannot be solved analytically but has been solved numerically using a series solution. A single one of these series functions cannot satisfy the initial condition so we superimpose a number of partial solutions to generate an acceptable final solution. The first three of the series, or Graetz functions, are shown in Figure 13.22. Notice that these functions bear a passing resemblance to the Bessel functions although their range is limited to between 0 and 1. The full axial and radial concentration profiles for the flow are shown in Figure 13.23. Here, it is clear how the initially uniform profile develops into a parabolic profile at some distance down the length of the tube. From Figure 13.23, we can judge this distance, or entrance length, to be roughly the point where $G_{zm}^{-1}=0.05$ or G_{zm} = 20. If the Graetz number is larger than 20, we have what is called an *entry length* problem. In this region, there is a boundary layer, and because the driving force for transport occurs over a smaller region, heat and mass transfer coefficients are higher here.

If the Graetz number is smaller than 20, we can use the results of the last section for fully developed profiles to find the heat or mass transfer coefficients. The solutions for a variety of constant flux and constant surface composition boundary conditions are shown in Figure 13.24.

Schmidt numbers for normal fluids are generally high so entrance length problems are the rule rather than the exception. This has led to the development of a variety of correlations explicitly

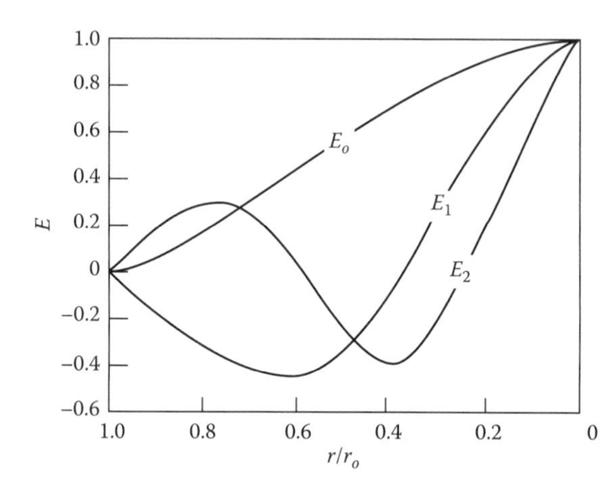

FIGURE 13.22 First three functions in the full series solutions to Equation 13.88. (Adapted from Nusselt, W., *Z. Ver. Deut. Ing.*, 54, 1154, 1910. With permission.)

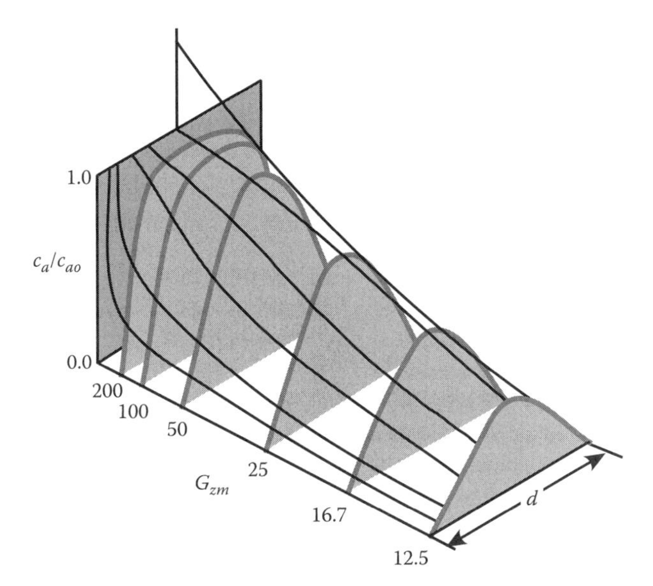

FIGURE 13.23 Developing concentration boundary layer in tube flow. (Adapted from Prandtl, L., *Führer Durch die Strömungslehre*, 7th edn., Vieweg, Brunswick, 1969. With permission.)

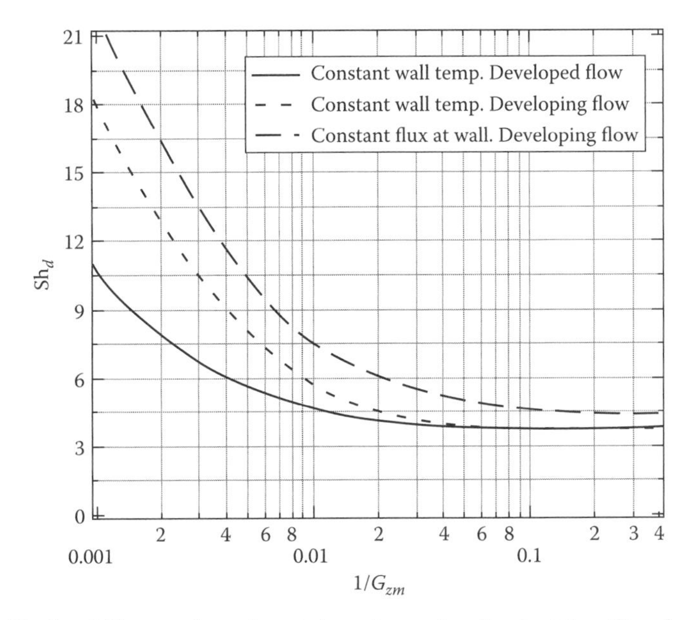

FIGURE 13.24 The local Sherwood number at the entrance length of a tube. (Based on Nusselt number data from Nusselt, W., *Z. Ver. Deut. Ing.*, 54, 1154, 1910; Kays, W.M., *Convective Heat and Mass Transfer*, McGraw-Hill, New York, 1966. With permission.) Sc = 0.7.

designed to deal with such situations. These correlations were originally developed for heat transfer, but we show the mass transfer analog. Properties should be evaluated at the arithmetic averaged mean concentration between the inlet and outlet of the tube $\bar{\bar{c}}_a = (\bar{c}_{ai} + \bar{c}_{ao})/2$.

The first correlation is from Hausen [21], and it assumes a fully developed flow field, but the thermal boundary layer only begins to develop at some point downstream:

$$\overline{\mathrm{Sh}_d} = 3.66 + \frac{0.0668 G_{zm}}{1 + 0.04[G_{zm}]^{2/3}} \tag{13.90}$$

The second correlation assumes combined hydrodynamic and thermal entry lengths and was originally developed by Sieder and Tate [22]:

$$\overline{\mathrm{Sh}}_d = 1.86[G_{zm}]^{1/3}\left(\frac{\mu}{\mu_s}\right)^{0.14} \quad \begin{bmatrix} 0.48 < \mathrm{Sc} < 16{,}700 \\ 0.0044 < \left(\dfrac{\mu}{\mu_s}\right) < 9.75 \end{bmatrix} \tag{13.91}$$

Whitaker [14] has shown that this correlation is valid for the Sherwood numbers in excess of 3.7 or so. Below that, fully developed conditions exist over most of the tube, and little error is incurred in using Equation 13.77.

The corresponding heat transfer correlations can be obtained from the equations presented here by replacing concentration by temperature, the Schmidt number by the Prandtl number, and the Sherwood number by the Nusselt number. You should write out the governing equations for yourself to convince yourself that this is the case.

Example 13.6 Oxygen Uptake by Artificial Blood

A test bed for an artificial blood consists of a small, 2 mm I.D. membrane capillary, 10 m long through which the artificial blood flows. The surface of the membrane is surrounded by air that has a partial pressure of oxygen of 0.21 atm at 298 K. There is a partition coefficient across the membrane such that K_{eq} = 2.15 (kg O_2 in air/kg O_2 after membrane). Assuming the artificial blood enters the capillary free of oxygen at a flow rate of 2×10^{-3} kg/s and that the flow field is fully developed, how much oxygen is absorbed over the length of the test bed? Some physical properties of the artificial blood are

$$D_{Ob} = 2.0 \times 10^{-9}\ \mathrm{m^2/s} \qquad \mu = 1.35 \times 10^{-3}\ \mathrm{Ns/m^2} \qquad \rho = 1100\ \mathrm{kg/m^3} \qquad \mathrm{Sc} = 614$$

We first need to determine the type of problem involved. Since we have a constant partial pressure of oxygen outside the membrane and a constant partition coefficient, we can safely assume we have a constant surface concentration problem. Next, we calculate the Reynolds number to determine the type of flow (Figure 13.25):

$$\mathrm{Re}_d = \frac{4\dot{m}}{\pi d\mu} = \frac{4(0.002)}{\pi(0.002)(1.35\times10^{-3})} = 943 \quad \text{laminar flow}$$

Now we need to determine whether we have fully developed conditions in the capillary. For this we need to calculate the Graetz number for mass transfer:

$$G_{zm} = \mathrm{Re}_d\mathrm{Sc}\left(\frac{d}{L}\right) = (943)(675)\frac{0.002}{10} = 127.3$$

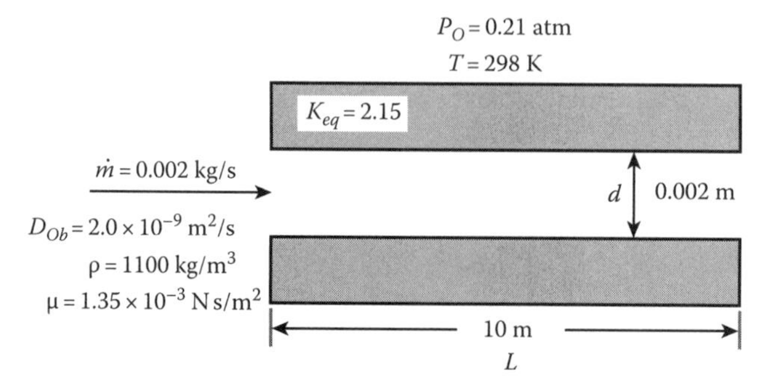

FIGURE 13.25 Oxygen substitute flowing through a capillary membrane.

Since the Graetz number is larger than 20, we do not have fully developed mass transfer conditions and must use an entry length correlation for the mass transfer coefficient:

$$\overline{\text{Sh}_d} = 3.66 + \frac{0.0668(d/L)\text{Re}_d\text{Sc}}{1+0.04[(d/L)\text{Re}_d\text{Sc}]^{2/3}}$$

$$= 3.66 + \frac{0.0668\left(\frac{0.002}{10}\right)(943)(675)}{1+0.04\left[\left(\frac{0.002}{10}\right)(943)(675)\right]^{2/3}} = 7.89$$

$$\bar{k}_c = \frac{7.89(2\times10^{-9})}{0.002} = 7.89\times10^{-6}\ \text{m/s}$$

To determine the total amount of mass transferred, we perform a mass balance about a small point within our capillary (Figure 13.26):

$$\begin{array}{ccccc} In & - & Out & = & Acc \\ \bar{v}_z A_c \bar{c}_O(z) & - & -\bar{v}_z A_c \bar{c}_O(z+\Delta z)\,\pi d \bar{k}_c \Delta z(\bar{c}_O(z)-c_w) & = & 0 \end{array}$$

Dividing through by Δz and taking the limit as $\Delta z \to 0$ yields a first-order differential equation:

$$\frac{d\bar{c}_O}{dz} - \left(\frac{\pi d\bar{k}_c}{\bar{v}_z A_c}\right)(\bar{c}_O - c_w) = 0$$

If we define $\Delta c = \bar{c}_O - c_w$, we can separate and integrate this equation obtaining

$$\frac{\Delta c_o}{\Delta c_i} = \frac{\bar{c}_O(z) - c_W}{\bar{c}_{Oi} - c_W} = \exp\left[-\left(\frac{\pi d\bar{k}_c}{\bar{v}_z A_c}\right)z\right]$$

The total amount of mass transferred is

$$m_o = \bar{v}_z A_c M_{wO}(\bar{c}_O(z) - \bar{c}_{Oi})$$

$$= \bar{v}_z A_c M_{wO}(\bar{c}_{Oi} - c_w)\left[\exp\left[-\left(\frac{\pi d\bar{k}_c}{\bar{v}_z A_c}\right)z\right] - 1\right]$$

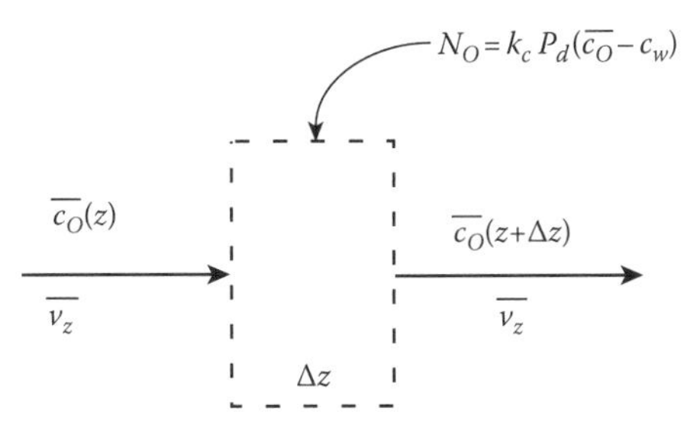

FIGURE 13.26 Control volume for blood substitute mass balance.

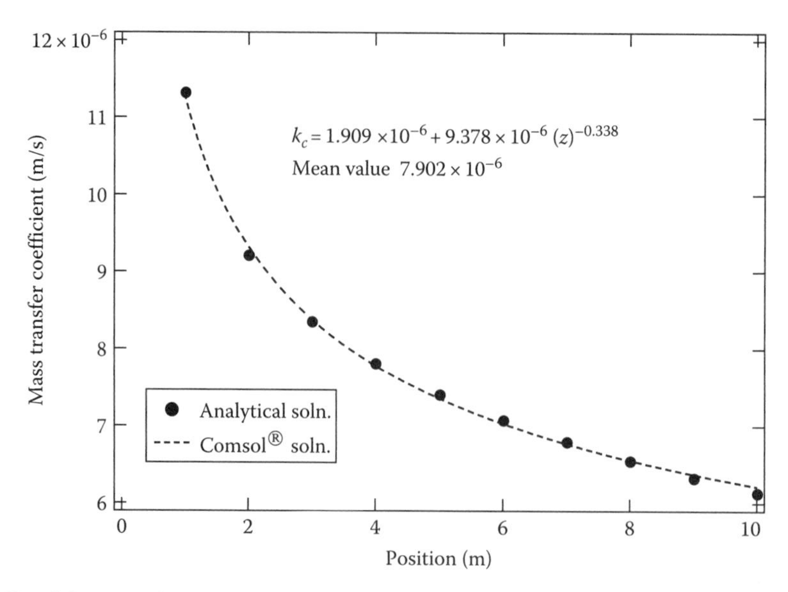

FIGURE 13.27 Mass transfer coefficient as calculated by Comsol® module artificial blood. The analytical and numerical average mass transfer coefficients are very close, and the mass transfer coefficient decreases with axial position to about one-third the power.

The wall concentration, c_w, is found from the partition coefficient

$$c_w = \frac{P_O/RT}{K_{eq}} = \frac{\dfrac{0.21(1.01\times10^5)}{8.314(298)}}{2.15} = 3.98\ \text{mol/m}^3$$

$$\dot{m}_o = 0.579(\pi(0.001)^2)(32)(0-3.98)\times\left\{\exp\left[-\left(\frac{\pi(0.002)(7.63\times10^{-6})(10)}{0.579(\pi(0.001)^2)}\right)\right]-1\right\} = 5.37\times10^{-5}\ \text{kg/s}$$

at $x = 10$ m. The actual mass transfer coefficient as a function of position within the tube can be calculated via numerical simulation. The results are shown in Figure 13.27. As shown, the analytical solution does a very good job.

Example 13.7 Steam Flow through a Duct

A 60 mm diameter, thin-walled copper tube shown in Figure 13.28 is covered with a 30 mm thick layer of urethane foam insulation ($k_i = 0.026$ W/m K). The tube is carrying superheated steam at atmospheric pressure and 165°C. The flow rate of the steam is 5 kg/h. The room air temperature is 20°C, and this provides an outside heat transfer coefficient, $h_o = 10$ W/m² K. At what point along the tube will the steam start condensing? When will its mean temperature reach 100°C?

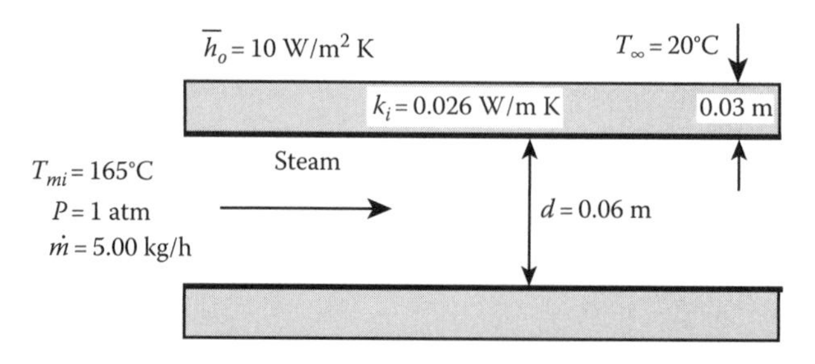

FIGURE 13.28 Steam flowing through and condensing on the surface of a duct.

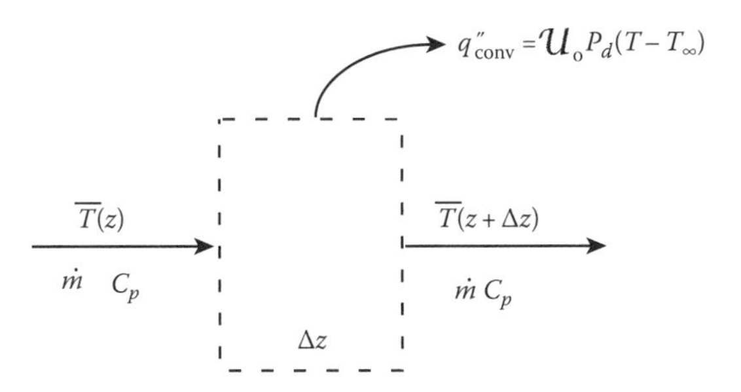

FIGURE 13.29 Control volume for energy balance.

To tackle this problem, we must first determine what type of situation we have: constant surface heat flux or constant surface temperature. We really cannot say we have a constant surface heat flux, but we can assume a constant surface temperature if we let that temperature be T_∞ and use an overall heat transfer coefficient to describe the resistances within the tube, the insulation, and the external flow about the tube. If we take a slice through the tube, we can set up a differential energy balance and determine how the mean temperature of the flow varies downstream.

In Figure 13.29, $\mathcal{U}$ is the overall heat transfer coefficient and serves the same function as the overall resistance of a cylindrical composite wall. We saw in Chapter 4 that such a resistance for two convective and one conduction resistance could be written as

$$\frac{1}{\mathcal{U}_o A_{so}} = \frac{1}{\bar{h}_i A_{si}} + \frac{\ln(r_o/r_i)}{2\pi k_i L} + \frac{1}{\bar{h}_o A_{so}}$$

We now write down our energy balance (no generation or accumulation) (Figures 13.28 and 13.29):

$$\begin{array}{lll} In & - \quad Out & = Acc \\ \dot{m}C_p\overline{T}(z) & - \ \dot{m}C_p\overline{T}(z+\Delta z) - 2\pi r_o \mathcal{U}_o \Delta z(\overline{T} - T_\infty) & = 0 \end{array}$$

Dividing through by Δz and taking the limit as $\Delta z \to 0$ yields a first-order differential equation:

$$-\frac{d\overline{T}}{dz} - \left(\frac{2\pi r_o \mathcal{U}_o}{\dot{m}C_p}\right)(\overline{T} - T_\infty) = 0$$

If we define $\Delta T = \overline{T} - T_\infty$, we can separate and integrate this equation obtaining

$$\frac{\Delta T_o}{\Delta T_i} = \frac{\overline{T}(z) - T_\infty}{\overline{T}_i - T_\infty} = \exp\left[-\left(\frac{2\pi r_o \mathcal{U}_o}{\dot{m}C_p}\right)z\right]$$

We are now in a position to solve the problem. First, we must determine an average temperature for the steam. Here, we might as well assume the steam leaves at 100°C so our average temperature is 132.5°C. Steam properties at this temperature are

$\rho_s = 0.876$ kg/m³ $\quad C_{ps} = 2080$ J/kg K $\quad \mu_s = 12.49 \times 10^{-6}$ N s/m²

$k_s = 0.0258$ W/m K $\quad \Pr_s = 1.004$

The Reynolds number for the flow is

$$\text{Re}_d = \frac{4\dot{m}}{\pi d\mu} = \frac{4\left(\frac{5}{3600}\right)}{\pi(0.06)(12.49\times10^{-6})} = 2360 \quad \text{(laminar flow, barely)}$$

Assuming fully developed profiles, the Nusselt number for laminar flow with constant surface temperature is 3.66. Thus, the inside heat transfer coefficient is

$$\bar{h}_i = \frac{\overline{\text{Nu}_d} k_s}{d} = \frac{3.66(0.0258)}{0.06} = 1.574\ \text{W/m}^2\,\text{K}$$

and the overall heat transfer coefficient per unit length of pipe is

$$\frac{1}{\pi d_o \mathcal{U}_o} = \frac{1}{\pi d_i \bar{h}_i} + \frac{\ln(r_o/r_i)}{2\pi k_i} + \frac{1}{\pi d_o \bar{h}_o} = 22.1\frac{\text{mK}}{\text{W}}$$

The easiest question to answer first is when does the mean temperature reach 100°C. Substituting the value for $\mathcal{U}_o$ into the temperature profile equation and inverting to solve for z gives

$$z = -\frac{\ln\left[\dfrac{\bar{T} - T_\infty}{\bar{T}_i - T_\infty}\right]}{\left(\dfrac{\pi d_o \mathcal{U}_o}{\dot{m} C_p}\right)} = -\frac{\ln\left[\dfrac{100-20}{165-20}\right]}{\left[\dfrac{1/22.1}{\left(\dfrac{5}{3600}\right)2080}\right]} = 37.97\ \text{m}$$

To determine when the surface of the pipe reaches 100°C, we must realize that at a steady state, the amount of heat flowing from the steam to the inner pipe surface is the same as the amount of heat flowing from the steam to the environment at T_∞. Thus, we have

$$q' = \pi d_o \mathcal{U}_o(\bar{T}(z_l) - T_\infty) = \pi d_i \bar{h}_i(\bar{T}(z_l) - T_s(z_l))$$

We solve for $\bar{T}(z_l)$ when $\bar{T}_s(z_l) = 100°\text{C}$:

$$\bar{T}(z_l) = \frac{\pi d_o \mathcal{U}_o T_\infty - \pi d_i \bar{h}_i T_s(z_l)}{\pi d_o \mathcal{U}_o - \pi d_i \bar{h}_i} = 114.4°\text{C}$$

Now we go back to the profile equation and recalculate the axial location where this occurs:

$$z_l = -\frac{\ln\left[\dfrac{114.4-20}{165-20}\right]}{\left[\dfrac{1/22.1}{\left(\dfrac{5}{3600}\right)2080}\right]} = 27.4\ \text{m}$$

The steam starts condensing on the walls about 27 m from the start of the pipe and starts condensing throughout the tube 38 m or so from the inlet.

13.6 TAYLOR DISPERSION [23,24]

We are going to discuss what happens when we have mass transfer occurring in a tube but where the solute is either present in the fluid initially or introduced into the flow stream and is being redistributed under the action of the flow field. This scenario is quite common and has many applications such as measuring the speed of flows, measuring blood and nutrient flow for pharmacological studies, analyzing toxic waste spills in waterways or the atmosphere, and performing tracer experiments for diagnostic purposes in chemical plants.

Consider a system where we have fully developed laminar flow in a tube and want to measure the speed of the flow. Usually we do a *tracer* experiment. We inject an amount of solute, the tracer, into the tube at the inlet or at a particular position and then time how long it takes to make it to a recording station some point further downstream. Experiments have shown that the solute slug does not transform into a paraboloid even though we know that the velocity profile in the tube is parabolic and that the centerline velocity is fully twice that of the mean flow. The slug is observed to move as a unit with the mean speed of the flow and can therefore be used to determine the speed of the stream. This is a remarkable result since it appears to say that the clear fluid in the center of the tube overtakes the slug of solute and then passes through it as if it were a ghost. A second striking result is that the solute slug seems to spread out symmetrically from a point that moves with the mean speed of the flow, even though the flow field is not symmetric in that direction. If we were to inject solute into the tube for a period of 0.01 s, for example, by the time the solute reached our next monitoring station, we might record its passage over a period of 0.1 s or more. The extent of the spread depends upon the mean speed of the flow, the physical properties of the fluid and the solute, and the dimensions of the pipe. The spreading out of the solute by the action of the flow field is termed *dispersion*.

13.6.1 Effect of Molecular Diffusion on Dispersion

Dispersion can occur via convection alone, but that situation is truly applicable only in cases where there is no molecular diffusion. Such a case may be approximated only when the fluid is extraordinarily viscous or the solute is so large that its diffusivity is extremely small. Generally, we have molecular diffusion whenever we perceive the effects of dispersion, and dispersion can be used to measure the diffusivity of a solute.

Our description of dispersion assumes that we have fully developed laminar flow as in Figure 13.30. The species continuity equation for the tracer is

$$D_{ab}\left(\frac{\partial^2 c_a}{\partial r^2}+\frac{1}{r}\frac{\partial c_a}{\partial r}+\frac{\partial^2 c_a}{\partial z^2}\right)=\frac{\partial c_a}{\partial t}+v_{\max}\left(1-\frac{r^2}{r_o^2}\right)\frac{\partial c_a}{\partial z} \tag{13.92}$$

where

D_{ab} is the diffusion coefficient
r_o is the radius of the tube
$v_{\max}$ is the maximum velocity of the fluid at the centerline of the tube

We are assuming that the diffusion coefficient, D_{ab}, is independent of the solute concentration. We also assume dilute solutions or equimolar counterdiffusion and symmetry in the θ-direction.

In fully developed flow at moderate flow rates, the transport of solute down the axis of the tube by convection normally overwhelms any transport in that direction by diffusion. Therefore, we can

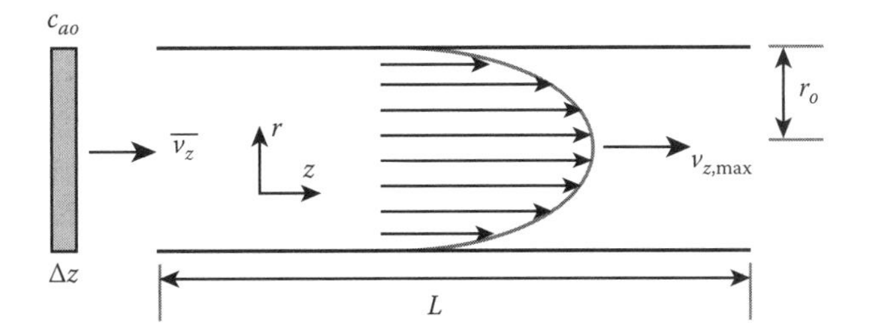

FIGURE 13.30 Geometry for dispersion considering molecular diffusion.

neglect the $\partial^2 c_a/\partial z^2$ term with respect to the diffusion terms in the radial direction. If we define the dimensionless variables,

$$\xi = \frac{r}{r_o} \qquad y = \frac{z}{L} \qquad \chi = \frac{c_a}{c_{ao}} \qquad \tau = \frac{v_{\max} t}{L} \tag{13.93}$$

we transform the species continuity equation to

$$\left(\frac{\partial^2 \chi}{\partial \xi^2} + \frac{1}{\xi}\frac{\partial \chi}{\partial \xi}\right) = \left[\frac{r_o^2 v_{\max}}{D_{ab} L}\right]\frac{\partial \chi}{\partial \tau} + \left[\frac{r_o^2 v_{\max}}{D_{ab} L}\right](1-\xi^2)\frac{\partial \chi}{\partial y} \tag{13.94}$$

with the boundary conditions

$$\xi = 1 \qquad \frac{\partial \chi}{\partial \xi} = 0 \tag{13.95}$$

$$\xi = 0 \qquad \frac{\partial \chi}{\partial \xi} = 0 \tag{13.96}$$

$$y = \infty \qquad \chi = 0 \tag{13.97}$$

The primary dimensionless group in Equation 13.94 is a ratio of the timescale for diffusion to the timescale for convection:

$$\beta_D = \frac{r_o^2 v_{\max}}{D_{ab} L} = \frac{\left(r_o^2 / D_{ab}\right)}{(L/v_{\max})} - \frac{\text{Diffusion time}}{\text{Convection time}} \tag{13.98}$$

There are two situations to consider. If β_D is very large, then diffusion is not important and dispersion is driven solely by convective effects, that is, the velocity profile in the tube causes the tracer or solute to be spread out differentially across the tube radius. If β_D is small, then diffusion tends to wipe out any radial changes in the concentration of tracer. To determine when diffusion is important, we look at the species continuity equation and eliminate the convective component:

$$\left(\frac{\partial^2 \chi}{\partial \xi^2} + \frac{1}{\xi}\frac{\partial \chi}{\partial \xi}\right) = \left[\frac{r_o^2 v_{\max}}{D_{ab} L}\right]\frac{\partial \chi}{\partial \tau} \tag{13.99}$$

The general solution for the concentration, χ, is found by separation of variables:

$$\chi = e^{-\alpha\tau} J_o\left(\sqrt{\beta_D \alpha}\,\xi\right) \tag{13.100}$$

where
 J_o is the Bessel function of zero order
 α is a constant of integration

Applying the boundary condition (13.95), we find that α is given by the solution of

$$J_1\left(\sqrt{\beta_D \alpha}\right) = 0 \tag{13.101}$$

This equation has multiple roots, but the first root that satisfies the equality is

$$\sqrt{\beta_D \alpha} = 3.8 \quad (13.102)$$

This solution represents the time it takes for any radial variation in concentration to decay to $1/e$ of its original value. For diffusion to be important,

$$\beta_D \ll (3.8)^2 = 14.44 \quad (13.103)$$

Notice that β_D plays the same role as the Graetz number did for fully developed flow. In fact, β_D can be written as $\mathrm{Re}_{max}\,\mathrm{Sc}(r_o/L)$.

Diffusion controlled response—Let us look further into the application of Equation 13.94 to the tracer problem and consider the case where diffusion controls the process. It is convenient to change coordinates so that we look at a coordinate system that moves with the average speed of the flow, $v_{max}/2$. This way we can see what happens to the tracer without convection blurring our understanding. We define a new variable, ζ:

$$\zeta = y - \left(\frac{v_{max}}{2L}\right)t = y - \frac{\tau}{2} \quad (13.104)$$

that links time and space and transforms Equation 13.94 to

$$\left(\frac{\partial^2 \chi}{\partial \xi^2} + \frac{1}{\xi}\frac{\partial \chi}{\partial \xi}\right) = \beta_D\left(\frac{1}{2} - \xi^2\right)\frac{\partial \chi}{\partial \zeta} \quad (13.105)$$

Since the mean velocity across a plane for which ξ is a constant is zero (every such plane moves at the mean speed of the flow), the mass transfer across such planes depends only on the radial variation of χ, or alternatively, on diffusion. If χ were independent of z and we had diffusion control, then any radial variation in χ would be quickly damped. Since diffusion is so rapid and controls the process, we can treat $\partial\chi/\partial\zeta$ as being a constant as well. With these restrictions, the governing differential equation becomes

$$\left(\frac{\partial^2 \chi}{\partial \xi^2} + \frac{1}{\xi}\frac{\partial \chi}{\partial \xi}\right) = \frac{1}{\xi}\frac{\partial}{\partial \xi}\left(\xi\frac{\partial \chi}{\partial \xi}\right) = \beta_D\left(\frac{1}{2} - \xi^2\right)\overline{\frac{\partial \chi}{\partial \zeta}} \quad (13.106)$$

Equation 13.106 is reminiscent of the constant surface flux problem we solved in Section 13.5.1. It is easy to separate and integrate yielding a solution that looks like

$$\chi = \chi_{\zeta 0} + \frac{\beta_D}{8}\left(\xi^2 - \frac{1}{2}\xi^4\right)\overline{\frac{\partial \chi}{\partial \zeta}} \quad (13.107)$$

where $\chi_{\zeta 0}$ is the value of χ at $\zeta = 0$. The mass flux across the pipe section at any plane ζ is given by integrating the product of the velocity (relative to the mean speed of the flow) and concentration over the cross-sectional area of the pipe:

$$N_a = -2v_{max}\int_0^1 \chi\left(\frac{1}{2} - \xi^2\right)\xi\,d\xi \quad (13.108)$$

Plugging in for χ from (13.107) and for β_D and integrating gives

$$N_a = -\left(\frac{r_o^2 v_{\max}^2}{192 D_{ab}}\right)\overline{\frac{\partial \chi_\zeta}{\partial \zeta}} \tag{13.109}$$

Since we have assumed that diffusion is so rapid, any radial variations in concentration are negligible with respect to horizontal variations. Thus, the axial change in χ_ζ is mimicked by a corresponding change in the mean concentration, $\overline{\chi}$, analogous to the constant surface flux case we considered in a previous section:

$$\overline{\frac{\partial \chi_\zeta}{\partial \zeta}} \approx \frac{\partial \overline{\chi}}{\partial \zeta} \tag{13.110}$$

$$N_a = -\left(\frac{r_o^2 v_{\max}^2}{192 D_{ab}}\right)\frac{\partial \overline{\chi}}{\partial \zeta} \tag{13.111}$$

Equation 13.111 states that the mean concentration of solute, $\overline{\chi}$, is dispersed relative to a plane that moves at the mean speed of the flow, $v_{\max}/2$, exactly as though it were undergoing a diffusive process that obeys Fick's law. The new *diffusion-type coefficient*, k_D, is given by

$$k_D = \frac{r_o^2 v_{\max}^2}{192 D_{ab}} \tag{13.112}$$

The fact that no material is lost in the process is expressed by the continuity equation for $\overline{\chi}$:

$$\frac{\partial N_a}{\partial \zeta} = -\frac{\partial \overline{\chi}}{\partial t} \tag{13.113}$$

Substituting in for N_a from Equation 13.111 gives the equation describing longitudinal dispersion. k_D is normally referred to as the *dispersion coefficient*:

$$k_D \frac{\partial^2 \overline{\chi}}{\partial \zeta^2} = \frac{\partial \overline{\chi}}{\partial t} \tag{13.114}$$

This equation can now be used to determine the diffusion coefficient from tracer experiments.

Equations 13.111 and 13.114 are remarkable results. Why should this behavior occur, and why should it obey Fick's law? To answer this we must consider what occurs in fully developed flow. Remember that as fluid enters the tube from a state where it was initially in plug flow, boundary layers develop at the tube wall. These boundary layers grow until they meet at the center of the tube. At that point, we have fully developed laminar flow. In the boundary layers, the dominant transport mechanism across the boundary layer is diffusion, not convection. Since the tube is one large boundary layer once fully developed flow exists, transport in the radial direction would be primarily by diffusion and any smearing out of solute should be perceived as occurring by a diffusion-based mechanism. The situation is shown in Figure 13.31. At short times, convection dominates and the slug of fluid is distorted into a parabolic shape. At this point, we have regions of higher concentration spread out radially within the tube. The radial concentration gradient is eliminated by diffusion so that the parabolic shape becomes dispersed throughout a larger area. This process repeats itself indefinitely so that the whole slug seems to move with the mean speed of the flow.

Our dispersion relation also asserts that dispersion is less if the diffusivity is high. This is counterintuitive since one would expect more rapid diffusion to decrease gradients faster. Dispersion is

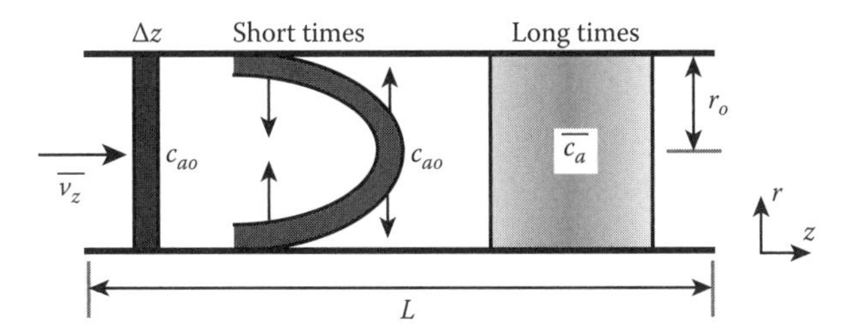

FIGURE 13.31 Process of molecular diffusion smearing out distortions caused by convective dispersion.

slower because we have convective distortion first, and then diffusion acts to remove the distortion. If diffusion is very slow, then convection has time to distort the slug. If diffusion is very rapid, then diffusion removes any gradient first, and convection never has time to distort the slug and cause dispersion to occur.

Though we have considered the process from a mass transfer point of view, can this analysis be applied to heat transfer? What would happen if the tracer were a slug of hot or cold fluid? How would a new dispersion constant be defined, and would dispersion be as potent a force in heat transfer as it is in mass transfer? Why?

Example 13.8 Maximum Frequency of Pulses

We are in a chemical plant and want to perform a tracer experiment to evaluate a new tubular reactor we have just installed. We are concerned about dispersion since the flow in the tubes is laminar. To obtain the most reliable measurements possible, we want to run many tracer experiments, but we need to run them in the shortest time possible. Given a tube length of 150 m, a tube diameter of 10 cm, a diffusivity of tracer of 1×10^{-9} m²/s, and a mass flow rate of 150 kg/h of water, what is the maximum frequency of tracer injection we can tolerate? The tracer is present in a concentration of 2 kg-mol/m³ when injected into the reactor, and we monitor the concentration of tracer at the reactor exit.

A picture of the system is shown in Figure 13.32. At a mass flow rate of 150 kg/h, the average velocity of the water is

$$\overline{v}_z = \frac{\left(150\dfrac{\text{kg}}{\text{h}}\right)\left(\dfrac{1\,\text{h}}{3600\,\text{s}}\right)\left(\dfrac{1\,\text{m}^3}{1000\,\text{kg}}\right)}{\pi(0.05\text{ m})^2} = 0.0053\,\text{m/s}$$

and the time required for the pulse to reach the reactor outlet is

$$t = \frac{150\text{ m}}{0.0053\text{ m/s}} = 28{,}270\text{ s} = 7.85\text{ h}$$

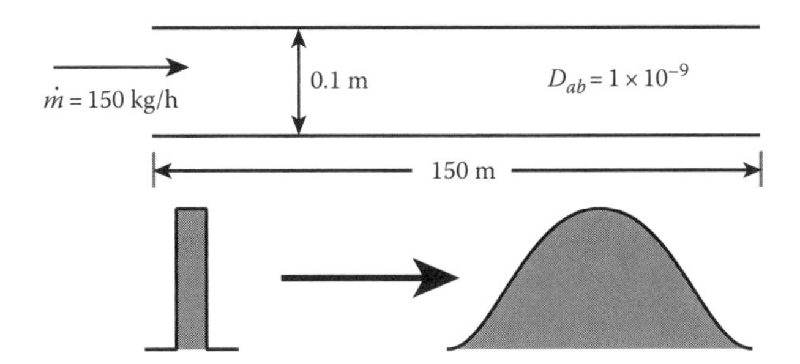

FIGURE 13.32 Reactor geometry for the tracer experiment.

We can solve Equation 13.114 using methods outlined in Chapter 6 to show that the pulse spreads out from its initial slug to form a Gaussian-shaped pulse (Figure 13.32). In dimensional form, the solution to Equation 13.114, assuming an initial mass of solute m_a, is

$$\bar{c}_a = \frac{m_a}{\pi d^2 M_{wa}\sqrt{4\pi k_D t}} \exp\left[-\frac{(z-\bar{v}_z t)^2}{4k_D t}\right]$$

In this shape we know that the average position of the slug is at $\bar{v}_z t$ while the standard deviation or spread of the pulse is $\sigma = \sqrt{2k_D t}$. To distinguish one pulse from another easily, we must time them so that the concentration at the start of the tube has decayed to about $1/e^5$ or less than its starting value. This will insure that the overlap of pulses is no more than 1% or so of the total signal.

Using the data given in the problem, the dispersion coefficient is

$$k_D = \frac{r_o^2 \bar{v}_z^2}{192 D_{ab}} = \frac{(0.05)^2(0.0053)^2}{192(1\times 10^{-9})} = 0.366\ \text{m}^2/\text{s}$$

The time required to reduce the concentration at the inlet of the tube to the required value is given by the solution to

$$\frac{1}{e^5} = \frac{1}{\pi(0.01)^2\sqrt{4\pi(0.366)t}} \exp\left[-\frac{(0-(0.0053)t)^2}{4(0.366)t}\right] \quad t = 36.2\,\text{h}$$

This result indicates that we have to flush the reactor nearly four times between the introduction of each pulse. If we are willing to be a bit less accurate and allow the concentration to drop to only $1/e^3$, the time between pulses would be 14.1 h, and we would have had to flush the reactor only one time between pulses.

Example 13.9 Dispersion between Infinite Flat Plates

Prove for two infinite flat plates, one moving relative to the other with velocity, v_o, that the dispersion coefficient is defined as

$$k_D = \frac{r_o^2 v_o^2}{120 D_{ab}}$$

The system is shown in Figure 13.33.

We can look at this problem in a way parallel to our discussion of dispersion in a tube. The continuity equation is

$$D_{ab}\frac{\partial^2 c_a}{\partial r^2} = \frac{\partial c_a}{\partial t} + v_o\left(\frac{r}{r_o}\right)\frac{\partial c_a}{\partial z}$$

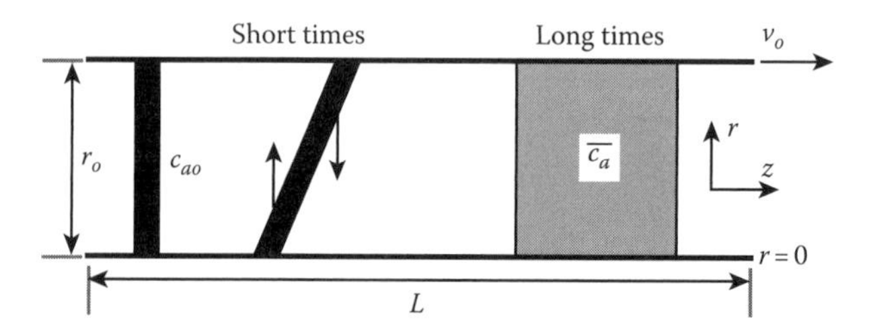

FIGURE 13.33 Dispersion between flat plates.

where we have a linear velocity profile between the plates. We define the dimensionless variables,

$$\xi = \frac{r}{r_o} \qquad y = \frac{z}{L} \qquad \chi = \frac{c_a}{c_{ao}} \qquad \tau = \frac{v_o t}{L}$$

that transforms the species continuity equation to

$$\frac{\partial^2 \chi}{\partial \xi^2} = \beta_D \frac{\partial \chi}{\partial \tau} + \beta_D \xi \frac{\partial \chi}{\partial y}$$

This equation is subject to the boundary conditions:

$$\xi = 1 \qquad \frac{\partial \chi}{\partial \xi} = 0 \qquad \xi = \xi_o \qquad \chi = 0$$

$$y = 0 \qquad \chi = 0$$

We rewrite the equation to follow along with the mean speed ($v_o/2$) of the flow. Letting

$$\zeta = y - \left(\frac{v_o}{2}\right) t = y - \frac{\tau}{2}$$

as before leaves us with a simple differential equation to solve:

$$\frac{\partial^2 \chi}{\partial \xi^2} = \beta_D \left(\xi - \frac{1}{2}\right) \frac{\partial \chi}{\partial \zeta} = \beta_D \left(\xi - \frac{1}{2}\right) \frac{\partial \bar{\chi}}{\partial \zeta}$$

Integrating twice and applying our boundary conditions shows that the concentration profile is

$$\chi = \chi_o + \beta_D \left(\frac{\xi^3}{6} - \frac{\xi^2}{4}\right) \frac{\partial \bar{\chi}}{\partial \zeta}$$

We calculate the flux by integrating the concentration–velocity product over the cross-sectional area of the plates. Here, we perform the calculation per unit width of the plates:

$$N_a = v_o \int_0^1 \left[\chi_o + \beta_D \left(\frac{\xi^3}{6} - \frac{\xi^2}{4}\right) \frac{\partial \bar{\chi}}{\partial \zeta}\right] \left[\xi - \frac{1}{2}\right] d\xi$$

The integration leads to

$$N_a = -\frac{v_o \beta_D}{120} \frac{\partial \bar{\chi}}{\partial \zeta}$$

Equating to Fick's law leads to a dispersion coefficient given by

$$k_D = \frac{r_o^2 v_o^2}{120 D_{ab}}$$

This situation is quite similar to the tube case but applies to a different set of physical phenomena. While the case of flow in tubes is useful for analyzing the performance of tubular chemical reactors, blood flow, or column-type absorption equipment, the case between the parallel plates is more suitable for looking at situations like the dispersion of vehicles on a highway. In such an instance, the velocity profile is not symmetric; the fast lanes lie to one side. This dispersion mechanism is also useful for looking at the spacing of traffic lights or toll booths. Again, the fastest cars are generally skewed in one direction, and dispersion will give an idea about optimal spacing of lights. Finally, such a system is also useful for looking at traffic on heavily traveled freeways. Anyone driving in Los Angeles has encountered the traffic signal approach for allowing cars onto the freeway. These cars will disperse themselves into the flow, and so our analysis is useful for deciding on the timing required to let the cars in smoothly and efficiently.

Dispersion is a very important consideration when one deals with flow through porous media. Such flows are found in oil recovery, chromatography, filtration, and packed bed reactors and in assessing groundwater circulation and contamination issues, such as may occur during hydrofracking operations. The Comsol® module, dispersion, allows one to investigate various aspects of dispersion in porous media by solving the Brinkman equations [25]:

$$\vec{\nabla}\cdot\vec{\mathbf{v}}=0 \qquad \text{Continuity}$$

$$-\left(\frac{\varepsilon_p}{\rho}\right)\vec{\nabla}P+\left(\frac{\mu}{\rho}\right)\nabla^2\vec{\mathbf{v}}+\frac{\varepsilon_p\vec{\mathbf{g}}}{\rho}=\frac{\partial\vec{\mathbf{v}}}{\partial t}+\vec{\nabla}\cdot\left(\frac{\mu\varepsilon_p}{\kappa}\vec{\mathbf{v}}\right)\vec{\mathbf{v}} \qquad \text{Momentum} \tag{13.115}$$

Here

ε_p is the *porosity* of the material

κ is the *permeability* of the porous material

For the simple case of packed spheres, such as assumed in the Comsol® module, the permeability and porosity can be related via

$$\kappa=\frac{\varepsilon_p^3 d_p^2}{180(1-\varepsilon_p)^2} \tag{13.116}$$

where d_p is the diameter of the spheres making up the porous material.

Figure 13.34 shows some results from the simulation looking at the permeation of water through the porous material. In this simulation, a tube, 0.1 m in diameter and 10 m long, was

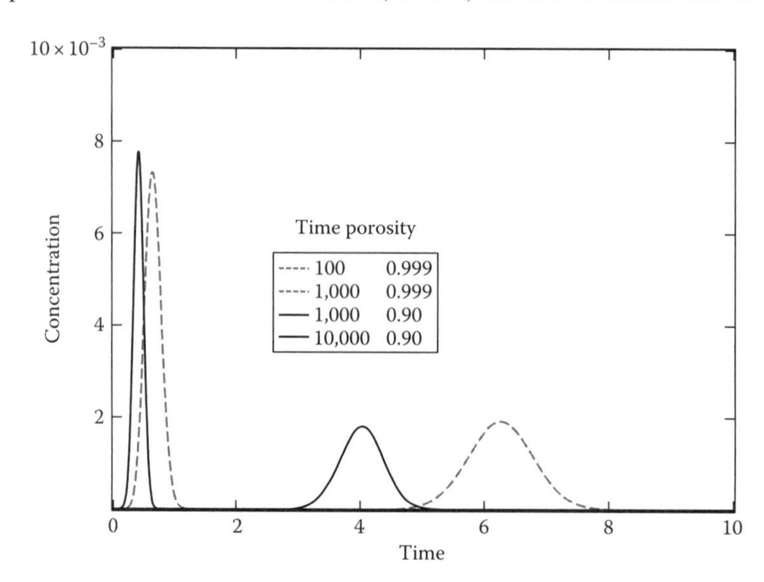

FIGURE 13.34 Dispersion in a packed tube. The effect of increasing porosity is seen as a broadening of the tracer pulse and a decrease in the peak concentration of tracer. Notice how the width of the tracer pulse increases with time and how diffusion blurs the effect of porosity at longer times. Comsol® module dispersion.

filled with the porous material. The concentration of tracer was plotted along the axis of the tube at two different times, 100 and 1000 s after introduction. The simulation reveals that the effects of dispersion are highest for the least porous material. As the porosity of the material decreases, the velocity profile in the radial direction becomes more uniform. The more uniform the velocity front, the less effect dispersion has. One can also see this in the module by altering the inlet velocity and even the radius of the tube to make the velocity profile more dramatic. In real systems, having a porosity distribution, so that fluid can channel along particular path, leads to problems with dispersion.

13.7 SUMMARY

We investigated boundary layer flows about objects that have significant curvature. We showed that this curvature affects the boundary layer development and the transport coefficients. The reason for these effects is that the boundary layer cannot grow indefinitely, as it does on a flat plate. In flow over cylinders or spheres, the boundary layer separates from the surface and forms a wake behind the object. In flow within a tube or conduit, the boundary layer can grow only as large as the tube radius permits. Since the flow fields and geometries are so complicated, experimental methods of determining the transport coefficients are preferred over theoretical solutions.

Our solutions in this and the previous chapter all presumed that we had laminar flow. In the next chapter, we will extend our analysis to turbulent flow and see how the random nature of turbulent motion affects the boundary layers near solid objects and the values of the transport coefficients.

PROBLEMS

General Questions

13.1 Consider fully developed transport in a circular tube. Explain why the Nusselt number, Sherwood number, and friction factor are constants (in the axial direction). Would you expect the same behavior if the tube had a square cross section? Why?

13.2 Consider the composite tube shown in Figure P13.2. Which segment of the tube should have the highest heat transfer coefficient, h? The highest friction factor, C_f? Why?

13.3 Heat transfer coefficients about a sphere or cylinder are highest at the forward stagnation point and decrease as one moves about the object toward the boundary layer separation point. Why?

External Flow: Momentum Transfer

13.4 A student wishes to perform an experiment on terminal velocities. He suspends a spherical-tipped caplet above a vat of water and then drops it end-on and then transversely into the water noting the terminal velocity (see Figure P13.4). The density of the caplet is 1500 kg/m^3. It has a diameter of 4 mm and a length of 25 mm. Estimate the terminal velocities the student measures if the experiment takes place at 25°C.

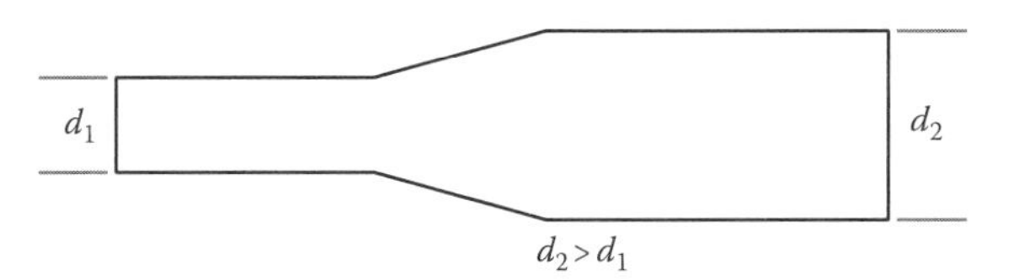

FIGURE P13.2

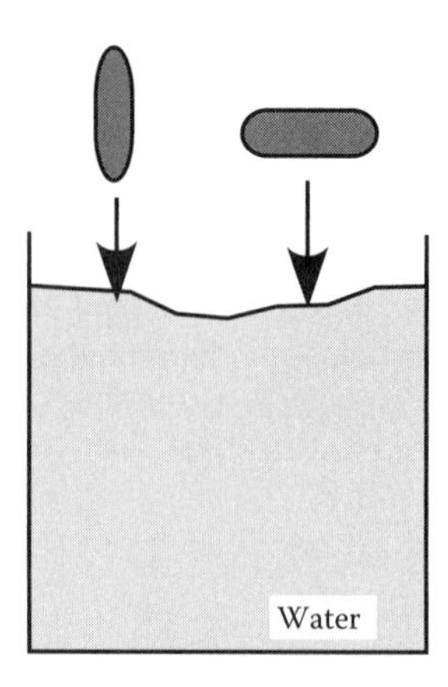

FIGURE P13.4

13.5 In this era of "sustainability," driving a car has become a serious issue. One proposed solution is the introduction of hybrid cars. For example, the fuel economy for Toyota Camry and Honda Civic vehicles is as follows:

Camry XLE 4cyl auto	21/31 mpg
Camry Hybrid	33/34 mpg
Civic Sedan 4cyl auto	26/34 mpg
Civic Hybrid	40/45 mpg

During the last oil crisis, speed limits were reduced to 55 mph in an effort to promote fuel economy. Those limits have now been relaxed to an average of 65 mph on interstate highways.

Hybrid vehicles though marvels of technological innovation are much more complicated machines, cost more initially, and potentially could cost more to repair and maintain such as when large battery packs need to be replaced. The question then is whether it is better to keep our current driving practices and switch to hybrids or whether we can get just as much benefit and lower our environmental impact just by reducing speed.

The Toyota Camry has a drag coefficient on the order of 0.38 and a frontal area of 1.94 m^2. Calculate the drag force on a Camry traveling 55 and 65 mph. Assuming that most of one's fuel consumption goes into overcoming drag, how much could one save by reducing speed back to 55 mph? How does that compare with the energy savings of just switching to a hybrid but keeping the speed at 65 mph?

13.6 A viscous fluid flows slowly by gravity down a 2 cm galvanized iron pipe. The pressures at the higher and lower locations are 120 and 130 kPa, respectively. The horizontal distance between the two locations is 6 m, and the pipe has a slope of 2 m rise per 10 m of run (horizontal distance). For a fluid with a kinematic viscosity of 4×10^{-6} m^2/s and a density of 880 kg/m^3

a. Determine the flow rate (in m^3/s).

b. Is the flow laminar or turbulent?

c. What is the friction factor and head loss for flow through this pipe?

13.7 Tumbleweeds move by rolling across the ground driven by the drag force they experience when exposed to a breeze. Such a concept has been proposed for developing planetary exploration vehicles to be used perhaps, on Mars. We are interested in calculating the drag force required to just get the tumbleweed to move. There is no sliding friction involved, but due to deformation of the tumbleweed as it contact the ground, the normal force acting on the tumbleweed is displaced a bit relative to the center of mass axis as shown in Figure P13.7. The coefficient of static friction, μ_s, is 0.15 [26].

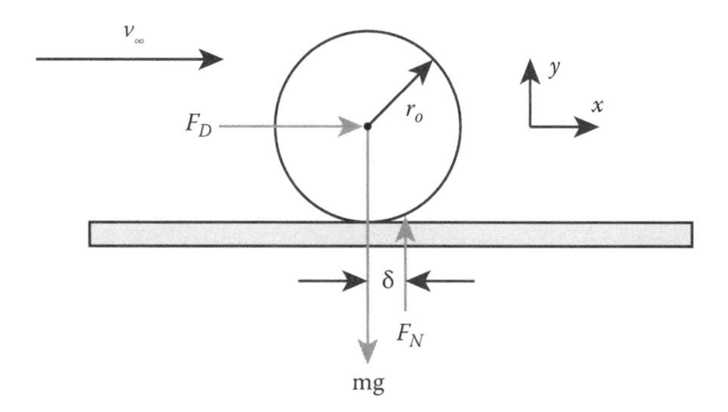

FIGURE P13.7

a. Perform a moment analysis about the center point and estimate the drag force, F_D, required to offset the normal force, F_N, and just get the tumbleweed rolling.
b. What wind speed would be needed?
c. The Martian atmosphere is mostly CO_2 with a density of about 0.016 kg/m^3, an average wind speed of about 3 m/s, and a kinematic viscosity of about 8×10^{-4} m^2/s. If we assume a roving sensor that weights about 3 kg, a drag coefficient that is nearly 1.0, and a normal force displacement of $0.1r_o$, how large would such a tumbleweed have to be to move about?

13.8 The bacterium, *Thiovulum majus*, a species that metabolizes sulfur, is about 18 μm in diameter and can swim, if provoked at speeds of up to 600 μm/s. We would like to look at some of the consequences of that behavior. We assume that *T. majus* moves through a solution that is very much like water at 40°C but with a specific gravity of 1.1 and a viscosity 1.5 times that of water:
a. Given the speed and size of *T. majus*, what is the Reynolds number associated with its motion?
b. What is the drag force experienced by the cell?
c. How much power would the cell have to generate to move at its top speed?
d. How much work does the cell do to travel 1 cm at top speed?
e. If a single molecule of glucose yields 30 molecules of adenosine triphosphate (ATP) to power the cell and metabolism of a molecule of ATP generates roughly 20 k_bT of energy, how much glucose is used to perform the work required in (d)? Assume the temperature is 40°C [27].

13.9 A simple way of measuring the drag force is shown in Figure P13.9. An object, in this case a sphere of weight *W*, and diameter, *d*, is suspended using a relatively weightless wire of length, *L*. A fluid having a free stream velocity of v_∞ flows past the sphere:
a. Develop an expression relating the drag coefficient, C_D; the velocity, v_∞; and the size and weight of the object.
b. Using your expression, determine the drag coefficient for a sphere that is 10 cm in diameter, weighing 2 N, exposed to airstream at 300 K flowing at 20 m/s, and suspended by a wire 50 cm long if the angle the sphere makes with the vertical is 26.5°.

13.10 Hail is formed when ice particles from the upper layers of a thunderstorm fall through the rain, pick up a layer of water, and then get caught in the updraft of the storm, which freezes the accumulated water. This process repeats itself until the weight of the hailstone is too large to be supported by the aerodynamic drag generated by the updraft. Estimate the velocity, v_∞, of the updraft required to make a hailstone 3 cm in diameter. Assume air properties that exist at about 7000 m where the temperature is –30.5°C:

$$\rho_{air} = 0.59\ \frac{\text{kg}}{\text{m}^3} \qquad \mu_{air} = 1.56\times 10^{-5}\ \frac{\text{N s}}{\text{m}^2} \qquad g = 9.785\ \frac{\text{m}}{\text{s}^2} \qquad \rho_{ice} = 948\ \frac{\text{kg}}{\text{m}^3}$$

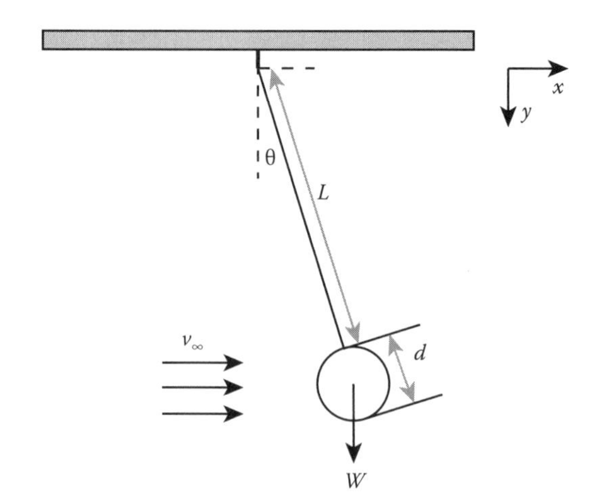

FIGURE P13.9

External Flow: Heat and Mass Transfer

13.11 Air at 40°C flows over a long, 25 mm diameter cylinder with an embedded electrical heater. Measurements of the effect of the free stream velocity, v_∞, and on the power per unit length, P', required to maintain the cylinder surface temperature at 300°C yield the following results:

v_∞ (m/s)	1	2	4	8	12
P' (W/m)	450	658	983	1507	1963

a. Determine the convection coefficient for each of the foregoing test conditions. Plot them.

b. For the corresponding Reynolds number range, determine suitable constants C and m for use with the empirical correlation:

$$\overline{\mathrm{Nu}_d} = C\,\mathrm{Re}_d^m\,\mathrm{Pr}^{1/3}$$

13.12 A very large fermenter is supplied with oxygen sparged in at the bottom to run an aerobic fermentation. Proper control of the device requires that we be able to estimate the percent oxygen absorption. To carry out the estimate, we assume that the vessel contains oxygen free water at 25°C and that the oxygen is supplied as bubbles 4 mm in diameter. The depth of the water is 10 m, and the top of the fermenter is at atmospheric pressure. We may assume that the bubbles are widely separated. How much oxygen is transferred from each bubble assuming the bubble size remains constant? Henry's law constant for oxygen in water is 4.38×10^{-4}.

13.13 A crystal is growing from a supersaturated solution at a rate of 0.2 μm/s. The solution contains 4 mol/L of solute, and saturated conditions have been measured to be 3.95 mol/L at the conditions of the test. The crystal is 1 mm in diameter and the fluid flows past the crystal at a velocity of 0.5 m/s. The viscosity of the fluid is 5×10^{-3} N s/m^2 and its density is 1100 kg/m^3. The solid has a density of 1500 kg/m^3 and its molecular weight is 150. The diffusion coefficient for the solute in the liquid is 4×10^{-10} m^2/s. How fast could the crystal grow in this solution? Is there some resistance to forming the actual crystal, and what is its magnitude?

13.14 Deposition onto the crystal in the previous problem also involves the release of latent heat of fusion in the amount of 6 kJ/mol. The thermal conductivity and heat capacity of the solution can be taken to be that of water while the thermal conductivity and heat capacity of the solid

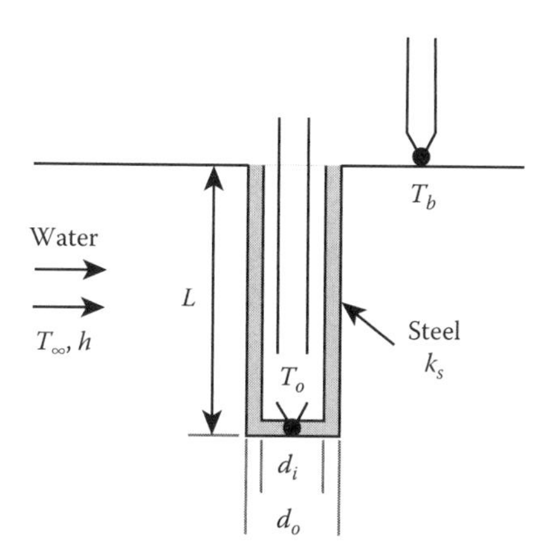

FIGURE P13.15

is 2 W/m K and 850 J/kg K. Assuming a uniform temperature inside the crystal and a stagnant solution, estimate that temperature as a function of time assuming the uniform growth rate of the previous problem. At what size does the assumption of uniform internal temperature fail if the starting temperature at 1 mm is 35°C and the solution temperature is held constant at 25°C?

13.15 The temperature of a liquid stream in a heat exchanger is determined by thermocouples as shown in Figure P13.15. One thermocouple is attached to the tube wall, and it measures a temperature of 300 K. The second thermocouple is soldered to the wall of a steel thermocouple well, and it records a temperature of 350 K. The thermocouple well is 0.05 m long with an inner and outer diameter of 5 and 12.5 mm, respectively. The water is flowing at a rate of 1 m/min. Determine the temperature of the water, T_∞, at the point where the thermocouple is attached to the well.

13.16 A new spray painting system is being evaluated for the auto industry. The paint is delivered via an atomizer that produces 10 μm particles and propels them toward the surface at a velocity of 10 m/s. The particles are a dilute suspension of pigment agents in a solvent and for modeling purposes can be assumed to be pure solvent. To form a good coating, the particles must arrive at the surface with 75% by volume of the solvent remaining. The free stream concentration of the solvent is effectively zero, and the solvent has a specific gravity of 0.85 and a molecular weight of 100. The whole system operates at atmospheric pressure, and at the temperature of deposition, the vapor pressure of the solvent is 250 mmHg. The diffusivity of the solvent in air was measured to be 1×10^{-9} m^2/s, and the Schmidt number for the solvent is 500. How far away from the surface can the painting nozzle be located?

13.17 Spherical polymer pellets impregnated with hormones or other proteins are being investigated as implants to deliver therapeutic agents directly to an affected area. In general, when the agent is released from the polymer matrix, it diffuses from the matrix and also is decomposed by enzymes in the body. Assuming a first-order decomposition reaction, $k''c_a$, a diffusion coefficient, D_{ab}, and a dilute solution of a in the body

a. Derive the steady-state differential equation representing the concentration of a in the body in the neighborhood of the implant.

b. Using the boundary conditions, $r = r_o$, $c_a = c_{ao}$, $r = r_\infty$, $c_a = 0$, solve the differential equation for the concentration profile. You may find it useful to use the variable transformation, $\chi = rc_a$.

c. Show that when $r_\infty \to \infty$, the Sherwood number becomes

$$\text{Sh}_d = 2 + \sqrt{\frac{4k''r_o^2}{D_{ab}}}$$

13.18 A small copper sphere, 10 mm in diameter, is initially at a temperature of 227°C. It must be cooled off to 100°C before it can be packaged and so is dropped into a vat of water whose temperature is held at $T_\infty = 27$°C. The sphere falls at its terminal velocity, u_∞:

a. What is the value of u_∞?
b. What is the heat transfer coefficient for the sphere, h?
c. Is the lumped systems analysis approach valid? Why?
d. How long does it take for the sphere to reach 100°C?
e. What is the total amount of heat transferred from the sphere once it reaches 100°C?

Internal Flow: Heat Transfer

13.19 An insulated tube carries hot water. The tube is 2 cm in diameter (inside) and 100 m long, and the flow rate of fluid is 0.5 kg/min. The tube is made of copper and the wall thickness is 0.2 cm. The insulation is 3 cm thick and its thermal conductivity is 0.035 W/m K. The outer layer of the insulation is held at a constant temperature of 25°C. The water enters the tube at 85°C. What is the water temperature at the outlet?

13.20 Water, initially at 25°C, is slowly flowing through a 1 cm diameter tube whose wall temperature remains constant at −10°C. The water velocity is 0.1 m/s.

a. Assuming fully developed profiles, how far from the inlet of the tube will the mean temperature of the water reach its freezing point?
b. Once the mean temperature has reached its freezing point, how much ice can we form per unit length of tube?

13.21 The entry length problem discussed in Section 13.5.1 can be analyzed using the boundary layer Equation 12.84 for a flat plate. The only caveat is that the thermal boundary layer must be small compared to the tube radius. When this assumption is valid, the velocity $v_y = 0$ for developed flow, and the velocity v_x can be approximated by a linear increase with wall distance, with the rate of that increase dictated by the Poiseuille profile. This is the basis of an analysis by J. Leveque, which resulted in the following equation:

$$\overline{\mathrm{Nu}_d} = 1.16 Gz^{1/3}$$

Starting from the boundary layer equation, describe this result as J. Leveque did in 1928. (Hint: see falling film heat transfer example in Chapter 10).

13.22 Consider fully developed flow inside the annulus shown in Figure P13.22. Both velocity and temperature profiles are fully developed. The heat fluxes in from the two walls are different but are constant along the length of the annulus:

a. Determine the velocity profile inside the annulus for a constant pressure drop ($\Delta P/L$) and fluid of viscosity, μ.

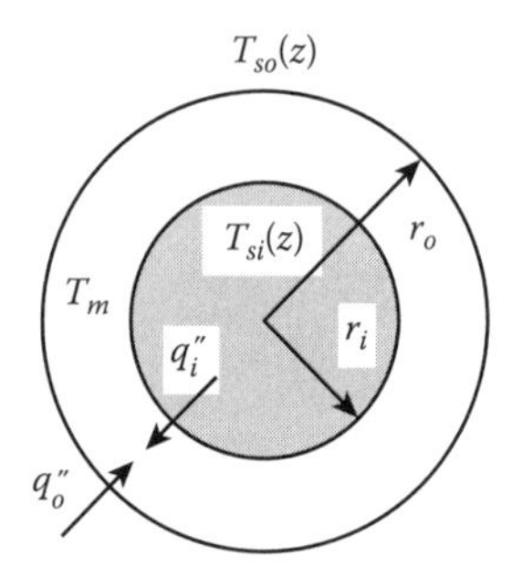

FIGURE P13.22

b. Using the velocity profile from part (a), solve the energy equation using boundary conditions based on the heat flux and surface temperature of the inner cylinder. Repeat using the heat flux and surface temperature for the outer cylinder.
c. Following the procedure outlined in Section 13.5.1, determine a general expression for the Nusselt number and a particular value when $q_o'' = q_i''$ and $r_i = r_o/2$.

13.23 Engine oil at a rate of 0.05 kg/s flows through a 3 mm diameter tube that is 30 m long. The oil has an inlet temperature of 40°C while the tube wall temperature is maintained at 100°C by steam condensing on its outer surface:
a. Estimate the average heat transfer coefficient for internal flow of the oil.
b. Determine the outlet temperature of the oil.

Internal Flow: Mass Transfer

13.24 Water containing 0.1 M benzoic acid flows at 0.1 cm/s through a 1 cm diameter rigid tube of cellulose acetate. The walls of the tube are 0.01 cm thick and are permeable to small electrolytes like the acid. Solutes within the tube wall diffuse as though moving through water. The tube is immersed in a large, well-stirred water bath containing negligible acid. After 50 cm of tubing, what fraction of the 0.1 M benzoic acid solution has been removed? The diffusivity of benzoic acid in water is 1×10^{-9} m²/s.

13.25 Waste gases at a flow rate of 15 kg/min leave a plant through a chimney stack 120 m high. The outer diameter of the chimney is 1.5 m and the walls are made of brick, 0.15 m thick. The gases also contain particulates that deposit on the chimney walls and must be removed periodically. One approach is to use a hot gas steam of mostly oxygen (180°C) to burn it off. The rate of reaction is first order in oxygen concentration:

$$-r_p = k''c_O \; k'' = 5.5 \times 10^{-3}\ \text{s}^{-1}$$

The flow of gas is high enough that the heat generated by the reaction is easily dissipated. The oxygen concentration within the flow during the burnout period is also effectively constant at 25 mol/m³:
a. Calculate the temperature of the waste gases leaving the chimney when there are no deposits on the walls.
b. If the outlet temperature is determined to be 145°C, how thick are the particulate deposits on the wall.
c. For the conditions in part (b), determine the time required to remove 90% of the deposit. You may neglect any variation in diameter when calculating the values of the transport coefficients:

Properties:

Gases: $k = 0.0312$ W/m K $\nu = 2.6 \times 10^{-5}$ m²/s

Pr = 0.68 $\rho = 0.73$ kg/m³ $C_p = 1007$ J/kg K

Particulate

Deposit: $k = 0.047$ W/m K $\rho = 3.33 \times 10^4$ mol/m³

Atmospheric

Conditions: $T_\infty = 20$°C $v_\infty = 25$ km/h

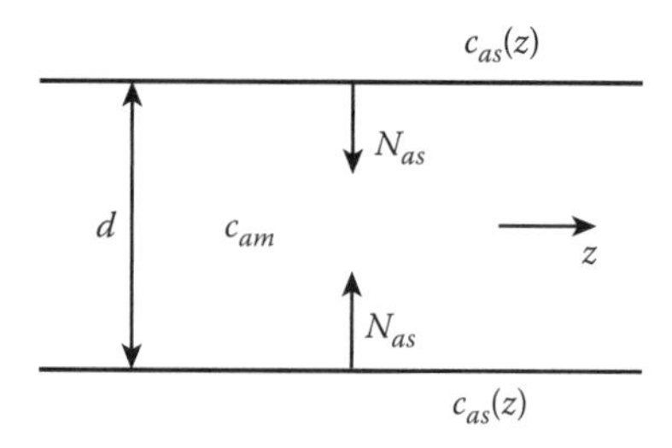

FIGURE P13.27

13.26 We are interested in determining the Sherwood number in fully developed flow between parallel plates. The plates are formed from membranes that provide a constant mass flux into the fluid along their length. Derive the differential equation describing this situation and determine the Sherwood number. How does the Sherwood number compare with the corresponding case for flow in a tube.

13.27 Consider the same situation as in Problem 13.26, but in this case one membrane provides a constant flux into the fluid while the other provides a flux of the same magnitude out of the fluid as shown in Figure P13.27. What is the Sherwood number for each surface and how does that compare to the annulus case (Problem 13.22)?

13.28 Consider the case of axial dispersion in steady, fully developed flow between stationary, parallel plates. The plates are of length, L, and the separation distance between the plates is d. Following the analysis of Section 13.6,

a. Develop a criterion to decide when diffusion is important.

b. By transforming the coordinate system to one that moves with the mean speed of the flow, solve the differential equation and determine a Fick's law–type expression for the axial flux.

c. What is the dispersion constant for this situation?

13.29 The dispersion model is often used to analyze the performance of non-ideal plug flow reactors. In essence, the plug flow reactor model is augmented by the addition of a diffusion-like term involving the second derivative of concentration or mole fraction as a function of axial position and a dispersion constant. Consider such a case without chemical reaction. The differential equation is

$$k_D \frac{\partial^2 c_a}{\partial z^2} - v_m \frac{\partial c_a}{\partial z} = \frac{\partial c_a}{\partial t}$$

a. Why do we not have an explicit expression for the changes in c_a as a function of radial position?

b. Why do we not need to consider $v = f(r)$?

c. Before entering the vessel, we have a vat containing a uniform concentration of reactant, c_{ao}. At the entrance, dispersion and convection occur. At the reactor exit, we assume a perfectly mixed solution with no concentration gradients. Initially, the reactor contains no a. Show, based on flux balances, that the boundary conditions are

$$z = 0 \qquad c_{ao} = -\frac{k_D}{v_m}\frac{\partial c_a}{\partial z} + c_a(z = 0) \qquad z = L \quad \frac{\partial c_a}{\partial z} = 0$$

These are the famous Danckwerts boundary conditions [28]

d. Solve the equation in the steady state using the Danckwerts boundary conditions and compare with a solution specifying that $c_a = c_{ao}$ at $z = 0$.

13.30 Consider the dispersion model for the chemical reactor in the presence of a first-order chemical reaction, $-r_a = k''c_a$.

a. Write the steady-state model equation in dimensionless form. What two dimensionless groups govern the behavior of the model and what do they mean?

b. Using the Danckwerts boundary conditions, solve the steady-state equation for the concentration profile.

c. What happens to the model and the concentration profile if k_D is very large? Very small?

REFERENCES

1. Taneda, S., Experimental investigation of the wakes behind cylinders and plates at low Reynolds numbers, *J. Phys. Soc. Jap.*, **11**, 302 (1956).
2. Blasius, H., Grenzschichten in Flussigkeiten mit keiner Reibung, *Z. Math. u. Phys.*, **56**, 1–37 (1908).
3. Heimenz, K., Die Grenzschicht an einem in den gleichformigen Flkussigkeitsstrom eingetrauchten geraden Kreiszylinder, *Dingl. Polytechn. J.*, **326**, 321 (1911).
4. Howarth, L., On the calculation of steady flow in the boundary layer near the surface of a cylinder in a stream, *Proc. Royal Soc. London A*, **164**, 547–579 (1938).
5. Schlichting, H., *Boundary Layer Theory*, 7th edn., McGraw-Hill, New York (1979).
6. Schmidt, E. and Wenner, K., Wärmeabgabe über den Umfang eines Angeblasenen Geheitzen Zylinders, *Forsch. Gebeite Ingenieurw.*, **12**, 65 (1941).
7. Eckert, E.R.G. and Soehngen, E., Interferometric studies on the stability and transition to turbulence of a free convection boundary layer, *Proceedings of General Discussion, Heat Transfer ASME-IME*, London, U.K. (1951).
8. Churchill, S.W. and Bernstein, M., A correlating equation for forced convection from gases and liquids to a circular cylinder in crossflow, *J. Heat Transfer*, **99**, 300–306 (1977).
9. Zhukauskas, A., Heat transfer from tubes in crossflow, in J.P. Hartnett and T.F. Irvine, Jr., eds., *Advances in Heat Transfer*, Vol. 8, Academic Press, New York (1972).
10. Grigull, U. and Hauf, W., Natural convection in horizontal cylindrical annuli (Tests for measuring heat-transfer coefficients in horizontal annulus filled with gas and visualization of flow), *International Heat Transfer Conference*, 3rd, Chicago, IL (1966).
11. Morgan, V.T., The overall convective heat transfer from smooth circular cylinders, in T.F. Irvine, Jr. and J.P. Hartnett eds., *Advances in Heat Transfer*, Vol. 11, Academic Press, New York (1975).
12. Churchill, S.W. and Chu, H.H.S., Correlating equations for laminar and turbulent free convection from a horizontal cylinder, *Int. J. Heat and Mass Transfer*, **18**, 1049 (1975).
13. Taneda, S., Experimental investigation of the wake behind a sphere at low Reynolds numbers, *J. Phys. Soc. Jap.*, **11**, 1104 (1956).
14. Whitaker, S., Forced convection heat transfer correlations for flow in pipes, past flat plates, single cylinders, single spheres, and flow in packed beds and tube bundles, *AIChE J.*, **18**, 361 (1972).
15. Yao, S.C. and Schrock, V.E., ASME Publication 75-WA/HT-37, *Winter Annual Meeting*, Houston, TX (1975).
16. Churchill, S.W., Free convection around immersed bodies, in E.U. Schlunder, ed. *Heat Exchanger Design Handbook*, Section 2.5.7, Hemisphere Publishing Corp., New York (1983).
17. Turton, R. and Levenspiel, O., A short note on the drag correlation for spheres, *Powder Technol.,* **47**, 83–86 (1986).
18. Kays, W.M., *Convective Heat and Mass Transfer*, McGraw-Hill, New York (1966).
19. Nusselt, W., Die Aghängigkeit der Wärmeübergangszahl von der Rohrlänge, *Z. Ver. Deut. Ing.*, **54**, 1154 (1910).
20. Prandtl, L., *Führer Durch die Strömungslehre*, 7th edn., Vieweg, Brunswick (1969).
21. Hausen, H., Darstellung des Wärmeüberganges in Rohren durch verallgemeinerte Potenzbeziehungen, *Z. Ver. Deut. Ing.*, **4**, 91 (1943).
22. Sieder, E.N. and Tate, G.E., Heat transfer and pressure drop of liquids in tubes, *Ind. Eng. Chem.*, **28**, 1429 (1936).
23. Taylor, G., Dispersion of soluble matter in solvent flowing slowly through a tube, *Proc. Royal Soc. A*, **219**, 186 (1953).
24. Taylor, G., Conditions under which dispersion of a solute in a stream of solvent can be used to measure molecular diffusion, *Proc. Royal Soc. A*, **223**, 446 (1954).

25. Kaviany, M., *Principles of Heat Transfer in Porous Media*, 2nd edn., Springer Verlag, New York (1995).
26. Janes, D.M., The Mars Ball: A Prototype Martian Rover (AAS 87–272). The Case for Mars III, Part II— Volume 75, AAS Science and Technology Series, ed. Carol R. Stoker, Univelt, Inc., San Diego, CA, 1989, pp. 569–574.
27. Wirsen, C.O. and Jannasch, H.W., Physiological and morphological observations on *Thiovulum* sp., *J. Bacteriol.*, **136**, 765–774 (1978).
28. Danckwerts, P.V., Continuous flow systems: Distribution of residence times, *Chem. Eng. Sci.* **2**, 1 (1953).

14 Turbulent Boundary Layers

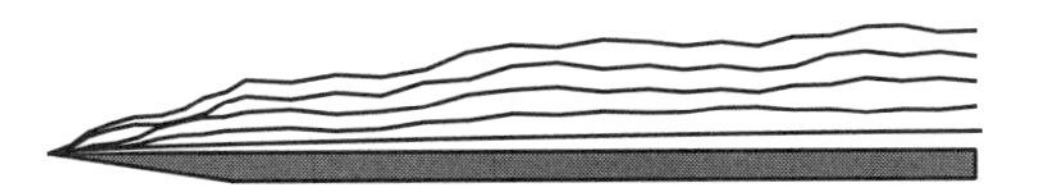

14.1 INTRODUCTION

Our consideration of convective transport has so far been limited to cases where the flow field is laminar. Most transport operations involve not laminar but turbulent flow. The reason behind this is simple: transport coefficients in turbulent flow are often an order of magnitude or more larger than the corresponding coefficients for laminar flow. This difference can be traced to a difference in the structure of the laminar and turbulent boundary layers. Unfortunately, we still cannot solve turbulence equations analytically, and numerical solutions are even more difficult to obtain than solutions for laminar flow over spheres and cylinders. Thus, our knowledge of the turbulent boundary layer's structure is still largely experimental. In this chapter, we will discuss how we represent turbulent flow, the difference between turbulent and laminar boundary layers, and how this difference leads to changes in the values of the transport coefficients.

The first breakthrough in turbulence occurred when a direct correlation between the onset of turbulence and the value of a dimensionless group, the Reynolds number, was made. Table 14.1 shows the critical values of the Reynolds number indicating the transition from laminar to turbulent flow for various geometries. Since the Reynolds number represents a ratio of inertial to viscous forces in the fluid, higher Reynolds numbers mean that the inertial forces driving the flow far outweigh the viscous forces. The reliance on inertial forces for transpoı‸ leads to much more efficient processes.

One can think of the difference in laminar and turbulent flow in terms of information transfer. We've all played the game where we sit around a table and pass a message to the person next to us. He or she passes it to his neighbor and so on down the line until the message returns back to us. This type of information transfer is analogous to laminar flow. In laminar flow, momentum diffuses between a surface and the free stream through the boundary layer. In the game, our information diffuses around the table through each person. The information takes quite a long time to travel from origin to destination and some information is always lost in the process. Thus, the message a person receives directly across the table is not quite the same message that was sent. If we were to talk to that person directly, we would insure the integrity of the message because we send it as a packet directly from the source to the receiver. This is analogous to turbulent flow where eddies transfer momentum information from point to point in the flow field. Thus, the transfer of momentum from the boundary layer in turbulent flow is faster and the integrity of that message is better. One can also make an analogy with the way the Internet works. Laminar flow would be like sending information bit by bit in a continuous stream. Turbulent flow is analogous to the way the Internet really works, sending packets of information in bursts from one point to another. In between laminar and turbulent flow regimes, we have transitional flow where the system oscillates between periods of turbulence followed by periods of laminar flow. The analogy would be where our voice could not carry far enough to reach the other side of the table and we would have to resort to intermediate steps to get our message across.

TABLE 14.1
Critical Values of the Reynolds Number for Transition between Laminar and Turbulent Flow

Geometry	Reynolds Number	Critical Value
Flat plate	$\frac{v_\infty L}{\nu}$	5×10^5
Cylinder in crossflow	$\frac{v_\infty d}{\nu}$	2×10^5
Sphere	$\frac{v_\infty d}{\nu}$	2×10^5
Tube flow	$\frac{\bar{v}_z d}{\nu}$	2000–4000

14.2 TURBULENT BOUNDARY LAYER STRUCTURE

Turbulent boundary layers have a similar structure regardless of the geometry involved. Figure 14.1 shows the turbulent boundary layer on a flat plate. The generation of eddies (swirls) and the chaotic structure of the boundary layer is evident. What is not evident is the structure of the boundary layer close to the solid surface. This structure is seen in Figure 14.2 and schematically in Figure 14.3. Here, we can clearly see what appear to be three distinct regions. There is a region close to the surface that appears to resemble the smooth, laminar boundary layer we have seen previously and has been referred to historically, but inaccurately, as the *laminar sublayer*. Right above that region is a *transition region*, where the boundary layer appears to become unstable and eddies are apparent. Above the transition region is the full turbulent boundary layer or *turbulent core region*. The actual transition point on the flat plate will vary over time but usually occurs when the Reynolds number reaches about 5×10^5.

The laminar sublayer, or the region close to the solid surface, controls the transport of momentum in turbulent flows. We now know that intermittent instabilities and jets exist in this region; gradients in momentum, heat, or mass are very high; and most transport is still dominated by diffusion rather than eddies. Once eddies are formed however, the transport of momentum is fast and efficient. The eddies reduce any momentum gradient that exists within the flow. In Chapter 12, we learned that the boundary layer thickness decreases with increasing Reynolds number, and as the boundary layer thickness decreases, the transport coefficients, which are inversely proportional to the boundary layer thickness, increase. Laminar sublayers in a turbulent boundary layer are

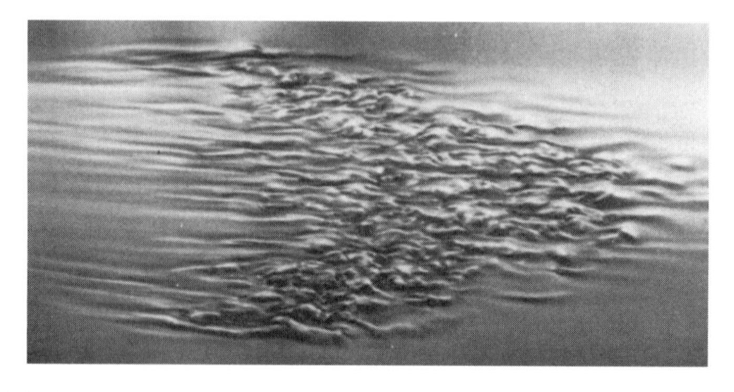

FIGURE 14.1 Turbulent boundary on a flat plate. (From Cantwell, B. et al., *J. Fluid Mech.*, 87, 641, 1978. With permission.)

FIGURE 14.2 Structure of the boundary layer at the plate's surface. (From Cantwell, B. et al., *J. Fluid Mech.*, 87, 641, 1978. With permission.)

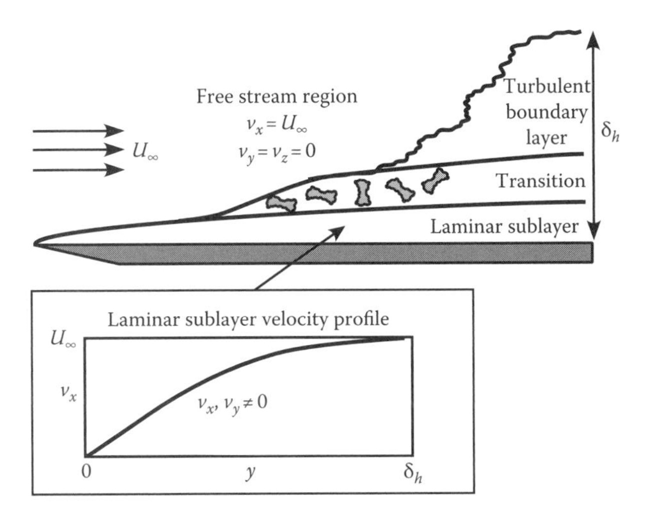

FIGURE 14.3 Schematic of the boundary layer structure.

much thinner than their corresponding regions in laminar flow. It is the thinness of the laminar sublayer and the generation of eddies that lead to the high transport coefficients associated with turbulent flow.

The thinness of the laminar sublayer also leads to important consequences when dealing with flow about submerged objects and flow inside ducts. Figure 14.4 shows laminar and turbulent boundary layers about a convex surface. Though the turbulent boundary layer is thicker, overall, than the laminar boundary layer, the laminar sublayer of the turbulent boundary layer is much thinner. This allows the fluid to *feel* the surface more and lets the flow *hug* the surface for a longer period. Consequently, separation of the turbulent boundary layer does not occur until about 140° from the forward stagnation point.

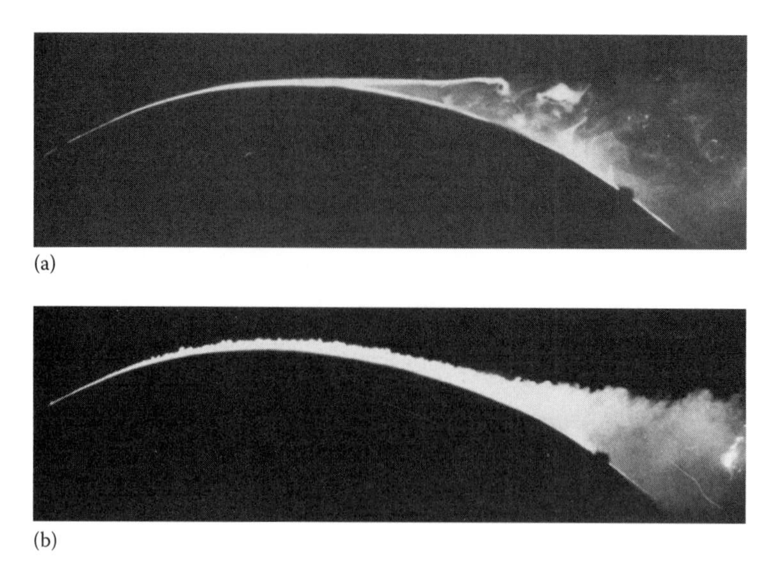

FIGURE 14.4 (a) Laminar and (b) turbulent boundary layers about a convex surface. (From Head, M.R., in *Flow Visualization II*, W. Merzkirch ed., Hemisphere Press, Washington, DC, pp. 399–403. With permission.)

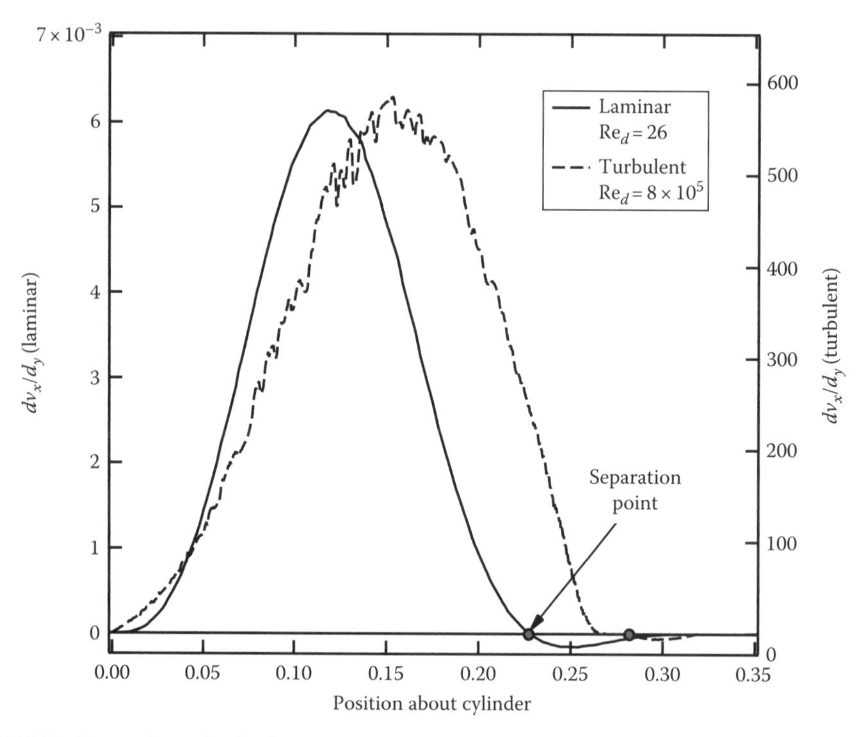

FIGURE 14.5 Separation point for flow over a cylinder calculated using Comsol® modules CylinderLam and CylinderTurb. Turbulent flow delays separation of the boundary layer.

We can see this behavior from simulations using computational fluid dynamics codes such as Comsol®. The software has two built-in turbulence models that, while approximate, represent the gross features of the flow well enough for most purposes. Figure 14.5 shows a plot of dv_x/dy as a function of position about a cylinder. Both laminar and turbulent flows are represented. The point at which $dv_x/dy = 0$ is related to the point where the boundary layer separates from the solid surface.

The figure shows three important differences between laminar and turbulent flows. The first is the location of the peak. The peak for the turbulent case is shifted further toward the back of the cylinder that in this case is 0.2 m in diameter. The second point is the width of the curve itself. The width or spread of the turbulent curve is larger than that for the laminar case. The last, and most important, is the approximate location of the separation point that is greatly delayed in the turbulent case. In this, the simulation agrees at least qualitatively with experimental observations.

14.3 TRANSPORT EQUATIONS IN TURBULENT FLOW

The mathematical representation of turbulent flows has been a subject occupying the minds of the foremost physicists, mathematicians, and engineers since the late nineteenth century. The Navier–Stokes equations are general relationships that apply to all fluids regardless of their state. Therefore, they apply under turbulent conditions as well as laminar. Difficulties arise in applying the Navier–Stokes equations to turbulent flow because (1) there are no real symmetry conditions that can be imposed on turbulent flow so the flows are always 3-D, (2) turbulent flows are inherently time-dependent, and (3) the time and length scales under which the flow is changing are extremely small compared to laminar flows. The first two conditions render an analytical solution to the Navier–Stokes equations for turbulent flow highly unlikely, and the third condition makes numerical solutions of those equations extraordinarily time consuming even for the fastest computers.

The structure of turbulent flow is best described by this little poem by L. F. Richardson, where each whorl can be considered to be an eddy [3]:

Big whorls have little whorls
Which feed on their velocity.
Little whorls have smaller whorls
And so on till viscosity.

The poem implies that there are many scales to a turbulent flow. The largest scale is on the order of the size of the overall flow, that is, the length of a flat plate, the diameter of a pipe or reaction vessel, or the size of a sphere or cylinder. This largest scale describes the overall motion of the fluid. Within this motion, there are large parcels of fluid, called eddies, moving about in a seemingly random fashion. If we were to take a still picture with a high-powered lens, we would see that there are also smaller eddies that interact and feed off the large eddies. As the Reynolds number is increased, the average size of the eddies decreases and more and more smaller eddies feed off of fewer larger ones. What is actually occurring here is an energy transfer process or energy cascade. The largest scale of the flow gets its energy from imposed forces such as a body force from gravity or a surface force due to an applied pressure. This large scale transmits the energy to other parts of the fluid by breaking down into smaller and smaller eddies. These eddies are parasites of the large scale, sucking off their momentum. This process of larger eddies fueling smaller ones cannot continue indefinitely. Eventually, the eddy size becomes small enough that its energy can be dissipated as heat through the action of viscous dissipation (internal fluid friction). It is the smallest eddies that interact with the boundary layer and lose their energy as heat. Thus, turbulence is an inherently dissipative process. If we shut off the body and surface forces, the flow will decay and the fluid will eventually become motionless. Figure 14.6 shows the turbulence dissipation rate for flow behind a cylinder at a Reynolds number of 800,000. The dissipation rate in this plot is a function of radial position passing through a point on the surface of the cylinder near where boundary layer separation occurs. The rate peaks near the surface of the cylinder and exponentially decays as the distance from the cylinder increases.

A.N. Kolmogorov developed a universal equilibrium theory describing the structure of the smallest scales of turbulence. He suggested that the parameters governing the smallest scales, where the

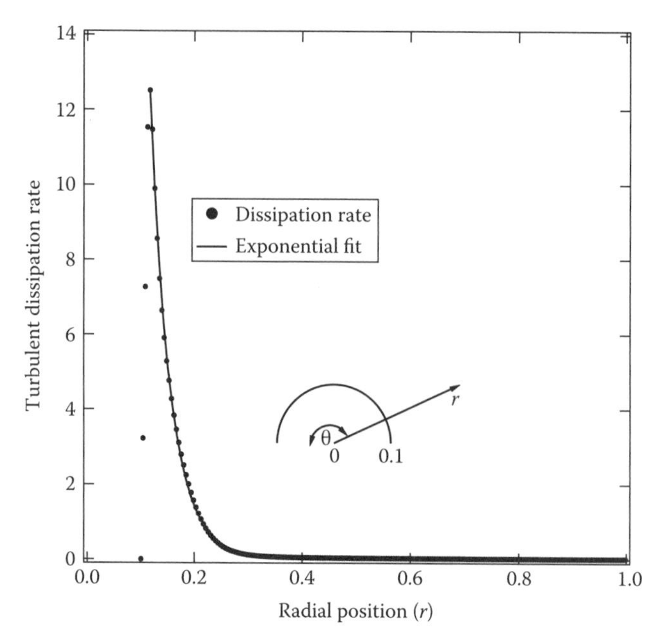

FIGURE 14.6 Turbulent dissipation rate behind the cylinder (Comsol® module CylinderTurb).

turbulent energy finally gets dissipated as heat, include the dissipation rate per unit mass, ε (m^2/s^3), and the kinematic viscosity, ν (m^2/s). These parameters can be used to generate what we now refer to as the Kolmogorov length (l_K), time (t_K), and velocity (v_K) scales:

$$l_K = \left(\frac{\nu^3}{\varepsilon}\right)^{1/4} \qquad t_K = \left(\frac{\nu}{\varepsilon}\right)^{1/2} \qquad v_K = (\nu\varepsilon)^{1/4} \tag{14.1}$$

If we fashion a microscale Reynolds number from these scales, we find

$$\mathrm{Re}_K = \frac{l_K v_K}{\nu} = 1 \tag{14.2}$$

indicating that at the microscale, the flow is dominated by viscous forces. Figure 14.7 shows the Kolmogorov length scale for flow behind the cylinder whose dissipation rate was given in Figure 14.6. The scale grows very rapidly near the surface of the cylinder and then much more slowly in the turbulent core region. As dissipation becomes more important, near the boundary layer, the scale decreases.

To simplify the representation of turbulent flow, we can think of the instantaneous turbulent velocity in terms of an average velocity, $\bar{v}$, and a fluctuating component, v'. In most instances, we do not care about the fluctuating components but would like to be able to determine the particulars about the average flow. We define the instantaneous velocity in terms of the average and fluctuating components:

$$v_i = \bar{v}_i + v_i' \tag{14.3}$$

If we substitute this definition into the Navier–Stokes equations, we derive a new representation for the turbulent flow field in terms of the average flow and fluctuating components. Reynolds was

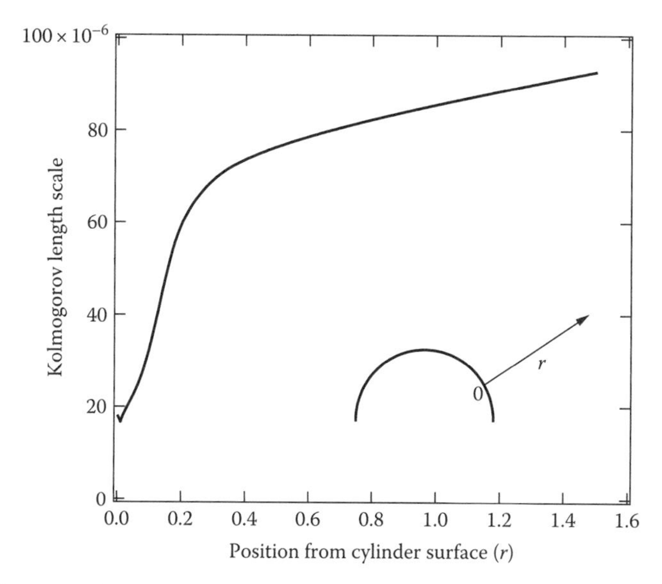

FIGURE 14.7 Kolmogorov length scale for flow across a cylinder. The length scale was calculated using the dissipation rate provided by the Comsol® module CylinderTurb.

the first to consider this approach around the turn of the twentieth century [4]. The averaging process used in defining these new turbulence equations has a few rules that must be observed:

a. The time average of a fluctuating component, v_i', must be 0:

$$\overline{v'} = 0 \tag{14.4}$$

b. Quantities that have already been averaged may be considered as constants during subsequent averaging:

$$\overline{\overline{v}} = \overline{v} \qquad \overline{\overline{v_i v_j}} = \overline{v_i}\,\overline{v_j} \tag{14.5}$$

c. Averaging is distributive across components:

$$\overline{v_i + v_j} = \overline{v_i} + \overline{v_j} \tag{14.6}$$

d. Derivatives of quantities obey the averaging law:

$$\overline{\frac{\partial v_i}{\partial x}} = \frac{\partial \overline{v_i}}{\partial x} \tag{14.7}$$

One consequence of these *Reynolds* rules involves averaging the product of two instantaneous quantities or fluctuating components:

$$\begin{aligned} \overline{v_i v_j} &= \overline{\left(\overline{v_i} + v_i'\right)\left(\overline{v_j} + v_j'\right)} = \overline{\overline{v_i}\,\overline{v_j}} + \overline{\overline{v_i} v_j'} + \overline{\overline{v_j} v_i'} + \overline{v_i' v_j'} \\ &= \overline{v_i}\,\overline{v_j} + \overline{v_i' v_j'} \end{aligned} \tag{14.8}$$

Here, we get a meaningful term containing the product of the averages of the two velocity components, but we also uncover a second term, the average of the product of the two fluctuating components.

Though the average of each fluctuating component is zero, the average of the product is not zero because the fluctuating components do not have to be in phase with one another at all times. These additional terms will complicate the transport equations, but we will see what they represent once we have applied the averaging procedure to the equations.

14.3.1 Continuity and Momentum Equations

We generate the average turbulent equations by taking the Navier–Stokes equations for an incompressible fluid with constant properties, substituting for the instantaneous velocity components and applying Reynolds rules of averaging to each term in the equation. The results are

Continuity

$$\frac{\partial \bar{v}_x}{\partial x}+\frac{\partial \bar{v}_y}{\partial y}+\frac{\partial \bar{v}_z}{\partial z}=0 \tag{14.9}$$

Momentum

$$\frac{\partial \bar{v}_x}{\partial t}+\bar{v}_x\frac{\partial \bar{v}_x}{\partial x}+\bar{v}_y\frac{\partial \bar{v}_x}{\partial y}+\bar{v}_z\frac{\partial \bar{v}_x}{\partial z}=-\frac{1}{\rho}\frac{\partial \bar{P}}{\partial x}+\nu\left(\frac{\partial^2 \bar{v}_x}{\partial x^2}+\frac{\partial^2 \bar{v}_x}{\partial y^2}+\frac{\partial^2 \bar{v}_x}{\partial z^2}\right)$$
$$-\left(\frac{\partial}{\partial x}\left(\overline{v'_x v'_x}\right)+\frac{\partial}{\partial y}\left(\overline{v'_y v'_x}\right)+\frac{\partial}{\partial z}\left(\overline{v'_z v'_x}\right)\right) \tag{14.10}$$

$$\frac{\partial \bar{v}_y}{\partial t}+\bar{v}_x\frac{\partial \bar{v}_y}{\partial x}+\bar{v}_y\frac{\partial \bar{v}_y}{\partial y}+\bar{v}_z\frac{\partial \bar{v}_y}{\partial z}=-\frac{1}{\rho}\frac{\partial \bar{P}}{\partial y}+\nu\left(\frac{\partial^2 \bar{v}_y}{\partial x^2}+\frac{\partial^2 \bar{v}_y}{\partial y^2}+\frac{\partial^2 \bar{v}_y}{\partial z^2}\right)$$
$$-\left(\frac{\partial}{\partial x}\left(\overline{v'_x v'_y}\right)+\frac{\partial}{\partial y}\left(\overline{v'_y v'_y}\right)+\frac{\partial}{\partial z}\left(\overline{v'_z v'_y}\right)\right) \tag{14.11}$$

$$\frac{\partial \bar{v}_z}{\partial t}+\bar{v}_x\frac{\partial \bar{v}_z}{\partial x}+\bar{v}_y\frac{\partial \bar{v}_z}{\partial y}+\bar{v}_z\frac{\partial \bar{v}_z}{\partial z}=-\frac{1}{\rho}\frac{\partial \bar{P}}{\partial z}+\nu\left(\frac{\partial^2 \bar{v}_z}{\partial x^2}+\frac{\partial^2 \bar{v}_z}{\partial y^2}+\frac{\partial^2 \bar{v}_z}{\partial z^2}\right)$$
$$-\left(\frac{\partial}{\partial x}\left(\overline{v'_x v'_z}\right)+\frac{\partial}{\partial y}\left(\overline{v'_y v'_z}\right)+\frac{\partial}{\partial z}\left(\overline{v'_z v'_z}\right)\right) \tag{14.12}$$

The average quantities of the double fluctuating components are referred to as the *Reynolds stresses* and they contain all the information about the turbulence. There are nine components of the Reynolds stresses forming a second-order tensor identical in form to the second-order stress tensor based on the average velocity gradients.

Reynolds stresses

$$\begin{vmatrix} \rho\left(\overline{v'_x v'_x}\right) & \rho\left(\overline{v'_y v'_x}\right) & \rho\left(\overline{v'_z v'_x}\right) \\ \rho\left(\overline{v'_x v'_y}\right) & \rho\left(\overline{v'_y v'_y}\right) & \rho\left(\overline{v'_z v'_y}\right) \\ \rho\left(\overline{v'_x v'_z}\right) & \rho\left(\overline{v'_y v'_z}\right) & \rho\left(\overline{v'_z v'_z}\right) \end{vmatrix} - \text{Stress (N/m}^2\text{)} \tag{14.13}$$

The fact that these fluctuating components have the dimensions of a stress will become very important when we attempt to represent them in terms of the mean components of the velocity. The Reynolds stresses account for the added energy needed to maintain turbulent conditions and for the extra heat dissipated as a result of eddy formation.

There are three normal components, $\rho\left(\overline{v_i'v_i'}\right)$, and six tangential components $\rho\left(\overline{v_i'v_j'}\right)$ of the Reynolds stresses. In general, all the components are different, but in certain circumstances, we may be able to invoke symmetry approximations. We refer to the special case when all the normal Reynolds stress components are equal, and all the tangential components are zero as *isotropic turbulence*. Isotropic turbulence has spherical symmetry, no gradients in the mean velocities, and no global average shear stress. No matter how closely you look at the structure of isotropic turbulence, it appears the same in all directions. Isotropic turbulence can be approximated by flowing the fluid past a fine-meshed screen. Such screens are often used in wind tunnels. Since there is no gradient in the mean velocity, a boundary layer flow cannot exhibit isotropic turbulence, and so isotropic turbulence is normally seen in expanding jets such as those produced in a wind tunnel.

We refer to *homogeneous turbulence* when all the Reynolds stress components are equal and the tangential components exist. Homogeneous turbulence allows for gradients in the mean velocity and can be used to describe flows with some symmetry restrictions. We can approximate turbulent boundary layer flows as homogeneous though in reality, most boundary layer flows are more complicated in structure.

14.3.2 Energy and Species Continuity Equations

We treat the energy and the species continuity equations in exactly the same way we treated the continuity and momentum equations. Here, we have average and fluctuating components for the velocity, $\overline{v} + v'$, the temperature, $\overline{T} + T'$, and the concentration or mole fraction of species $\overline{c}_a + c_a'$. Substituting these definitions into the energy and species continuity equations and using the Reynolds averaging procedure on each term yields

Energy

$$\frac{\partial \overline{T}}{\partial t} + \overline{v}_x \frac{\partial \overline{T}}{\partial x} + \overline{v}_y \frac{\partial \overline{T}}{\partial y} + \overline{v}_z \frac{\partial \overline{T}}{\partial z}$$

$$= \alpha\left(\frac{\partial^2 \overline{T}}{\partial x^2} + \frac{\partial^2 \overline{T}}{\partial y^2} + \frac{\partial^2 \overline{T}}{\partial z^2}\right) - \left(\frac{\partial}{\partial x}\left(\overline{v_x'T'}\right) + \frac{\partial}{\partial y}\left(\overline{v_y'T'}\right) + \frac{\partial}{\partial z}\left(\overline{v_z'T'}\right)\right) \tag{14.14}$$

Reynolds heat fluxes

$$\rho C_p\left(\overline{v_x'T'}\right) \quad \rho C_p\left(\overline{v_y'T'}\right) \quad \rho C_p\left(\overline{v_z'T'}\right) - \text{heat flux W/m}^2 \tag{14.15}$$

Species continuity

$$\frac{\partial \overline{c}_a}{\partial t} + \overline{v}_x \frac{\partial \overline{c}_a}{\partial x} + \overline{v}_y \frac{\partial \overline{c}_a}{\partial y} + \overline{v}_z \frac{\partial \overline{c}_a}{\partial z}$$

$$= D_{ab}\left(\frac{\partial^2 \overline{c}_a}{\partial x^2} + \frac{\partial^2 \overline{c}_a}{\partial y^2} + \frac{\partial^2 \overline{c}_a}{\partial z^2}\right) - \left(\frac{\partial}{\partial x}\left(\overline{v_x'c_a'}\right) + \frac{\partial}{\partial y}\left(\overline{v_y'c_a'}\right) + \frac{\partial}{\partial z}\left(\overline{v_z'c_a'}\right)\right) \tag{14.16}$$

Reynolds mass fluxes

$$\left(\overline{v_x'c_a'}\right) \quad \left(\overline{v_y'c_a'}\right) \quad \left(\overline{v_z'c_a'}\right) \quad \text{Mass flux mol/m}^2\text{ s} \tag{14.17}$$

A first-order chemical reaction generation term would be written as

$$R_a = -k''\,\overline{c}_a \tag{14.18}$$

A second-order reaction would contain two components: the product of average concentrations and the product of fluctuating components $\left(\overline{c}_a\overline{c}_b + \overline{c'_a c'_b}\right)$. The double correlations in the transport equations also contain all the information needed for a description of turbulent transport. The concept of isotropic turbulence can be carried over into the energy or species continuity equations. If isotropic turbulent conditions exist, then all the Reynolds mass fluxes or heat fluxes would be equal.

14.4 REPRESENTING THE REYNOLDS FLUX COMPONENTS

Fundamentally, the individual double averages or double correlations of the fluctuating components represent separate variables in the balance equations. While we have obtained some valuable insights into turbulence by the averaging procedure, we have also introduced more variables into the equations. The number of variables now exceeds the number of equations so we have a *closure* problem. We can manipulate the Navier–Stokes equations to derive additional equations governing the new variables we generated, but every new equation generated introduces another set of unknown variables. Therefore, some further approximations must be made. Several attempts at solving this closure problem have been made in the past, and some of the most notable of these will be discussed in the succeeding text.

14.4.1 Boussinesq Theory [5]

The Reynolds stress terms have the dimensions of a stress, and it is therefore logical to assume that we can relate them to a viscosity and to velocity gradients as was done for fluid stress in laminar flow. Boussinesq used just this sort of logic in one of the first attempts to model turbulence. This led to a concept called the eddy viscosity, μ^e, which is a function of position and direction. In general, the eddy viscosity is not constant, but assuming it is time independent is a useful analogy. The eddy viscosity components in the x-direction are written as

$$\rho\left(\overline{v'_x v'_x}\right) = -\mu^e_{xx}\frac{\partial \overline{v}_x}{\partial x} \tag{14.19}$$

$$\rho\left(\overline{v'_y v'_x}\right) = -\mu^e_{yx}\frac{\partial \overline{v}_x}{\partial y} \tag{14.20}$$

$$\rho\left(\overline{v'_z v'_x}\right) = -\mu^e_{zx}\frac{\partial \overline{v}_x}{\partial z} \tag{14.21}$$

The eddy viscosity is a tensor quantity, so in general, we may have up to nine eddy viscosity components to consider:

$$\vec{\mu}^{\mathbf{e}} = \begin{bmatrix} \mu^e_{xx} & \mu^e_{yx} & \mu^e_{zx} \\ \mu^e_{xy} & \mu^e_{yy} & \mu^e_{zy} \\ \mu^e_{xz} & \mu^e_{yz} & \mu^e_{zz} \end{bmatrix} = \underbrace{\begin{bmatrix} \mu^e & 0 & 0 \\ 0 & \mu^e & 0 \\ 0 & 0 & \mu^e \end{bmatrix}}_{\text{Isotropic turbulence}} = \underbrace{\begin{bmatrix} \mu^e & \mu^e & \mu^e \\ \mu^e & \mu^e & \mu^e \\ \mu^e & \mu^e & \mu^e \end{bmatrix}}_{\text{Homogeneous turbulence}} \tag{14.22}$$

All the Boussinesq scheme does is to replace one unknown, the Reynolds stress, with another, the eddy viscosity. However, the eddy viscosity is a concept that is a bit simpler to understand and something that can be measured easily. With current laser velocimetry equipment, it is possible to measure the mean and fluctuating components of all the velocities and thereby generate experimental data to extract the eddy viscosity.

We can extend the concept of the eddy viscosity to the Reynolds heat and mass fluxes. We obtain analogous quantities termed the *eddy thermal diffusivity* and the *eddy mass diffusivity*. Each has three components:

$$\left(\overline{v_x' T'}\right) = -\alpha_x^e \frac{\partial \overline{T}}{\partial x} \tag{14.23}$$

$$\left(\overline{v_y' T'}\right) = -\alpha_y^e \frac{\partial \overline{T}}{\partial y} \tag{14.24}$$

$$\left(\overline{v_z' T'}\right) = -\alpha_z^e \frac{\partial \overline{T}}{\partial z} \tag{14.25}$$

$$\left(\overline{v_x' c_a'}\right) = -D_{abx}^e \frac{\partial \overline{c}_a}{\partial x} \tag{14.26}$$

$$\left(\overline{v_y' c_a'}\right) = -D_{aby}^e \frac{\partial \overline{c}_a}{\partial y} \tag{14.27}$$

$$\left(\overline{v_z' c_a'}\right) = -D_{abz}^e \frac{\partial \overline{c}_a}{\partial z} \tag{14.28}$$

Now we can substitute these expressions into the balance equations and group together similar terms to obtain a set of simplified equations representing turbulent flow. These equations will have the same form as those we used to represent transport processes in laminar flow. For homogeneous turbulence, we have the following set of six equations:

Continuity

$$\frac{\partial \overline{v}_x}{\partial x} + \frac{\partial \overline{v}_y}{\partial y} + \frac{\partial \overline{v}_z}{\partial z} = 0 \tag{14.29}$$

Momentum

$$\frac{\partial \overline{v}_x}{\partial t} + \overline{v}_x \frac{\partial \overline{v}_x}{\partial x} + \overline{v}_y \frac{\partial \overline{v}_x}{\partial y} + \overline{v}_z \frac{\partial \overline{v}_x}{\partial z} = -\frac{1}{\rho}\frac{\partial \overline{P}}{\partial x} + (\nu + \nu^e)\left(\frac{\partial^2 \overline{v}_x}{\partial x^2} + \frac{\partial^2 \overline{v}_x}{\partial y^2} + \frac{\partial^2 \overline{v}_x}{\partial z^2}\right) \tag{14.30}$$

$$\frac{\partial \overline{v}_y}{\partial t} + \overline{v}_x \frac{\partial \overline{v}_y}{\partial x} + \overline{v}_y \frac{\partial \overline{v}_y}{\partial y} + \overline{v}_z \frac{\partial \overline{v}_y}{\partial z} = -\frac{1}{\rho}\frac{\partial \overline{P}}{\partial y} + (\nu + \nu^e)\left(\frac{\partial^2 \overline{v}_y}{\partial x^2} + \frac{\partial^2 \overline{v}_y}{\partial y^2} + \frac{\partial^2 \overline{v}_y}{\partial z^2}\right) \tag{14.31}$$

$$\frac{\partial \overline{v}_z}{\partial t} + \overline{v}_x \frac{\partial \overline{v}_z}{\partial x} + \overline{v}_y \frac{\partial \overline{v}_z}{\partial y} + \overline{v}_z \frac{\partial \overline{v}_z}{\partial z} = -\frac{1}{\rho}\frac{\partial \overline{P}}{\partial z} + (\nu + \nu^e)\left(\frac{\partial^2 \overline{v}_z}{\partial x^2} + \frac{\partial^2 \overline{v}_z}{\partial y^2} + \frac{\partial^2 \overline{v}_z}{\partial z^2}\right) \tag{14.32}$$

Energy

$$\frac{\partial \overline{T}}{\partial t}+\overline{v}_x\frac{\partial \overline{T}}{\partial x}+\overline{v}_y\frac{\partial \overline{T}}{\partial y}+\overline{v}_z\frac{\partial \overline{T}}{\partial z}=(\alpha+\alpha^e)\left(\frac{\partial^2 \overline{T}}{\partial x^2}+\frac{\partial^2 \overline{T}}{\partial y^2}+\frac{\partial^2 \overline{T}}{\partial z^2}\right) \tag{14.33}$$

Species continuity

$$\frac{\partial \overline{c}_a}{\partial t}+\overline{v}_x\frac{\partial \overline{c}_a}{\partial x}+\overline{v}_y\frac{\partial \overline{c}_a}{\partial y}+\overline{v}_z\frac{\partial \overline{c}_a}{\partial z}=\left(D_{ab}+D_{ab}^e\right)\left(\frac{\partial^2 \overline{c}_a}{\partial x^2}+\frac{\partial^2 \overline{c}_a}{\partial y^2}+\frac{\partial^2 \overline{c}_a}{\partial z^2}\right) \tag{14.34}$$

14.4.2 Prandtl Mixing Length

The concept of eddy viscosity replaced one unknown, the Reynolds stress, with another, the eddy viscosity. Though a useful simplification, the eddy viscosity concept does not provide a complete description of turbulence. We still need to determine how the eddy viscosity will depend on the primary variables of the system.

Early researchers formed a mechanistic model of turbulence. The kinetic theory of gases was well developed at the time people were seriously beginning to tackle the problem of turbulence, and borrowing from that theory, Prandtl expressed the Reynolds stress in terms of the mean velocity by defining a length characteristic of the scale of the turbulence [6]. He called this the mixing length, ℓ, and defined it as the length of the path a mass of fluid or eddy would travel before losing its individuality by mixing or coalescing with its neighbors. This mixing length was analogous to the mean free path in kinetic theory.

Consider a pipe with flow in the radial, r, and axial, z, directions as shown in Figure 14.8. A mass of fluid is assumed to have an average axial velocity, $\overline{v}_z$, in the pipe. There is also a velocity perpendicular to the mean flow, $\overline{v}_r$, which displaces the mass of fluid, or eddy, a distance, ℓ. The change in axial velocity of the mass of fluid is the difference between the velocity at the eddy's starting point and its new position given by $\ell(d\overline{v}_z/dr)$. This change is the definition of the fluctuating velocity component, v_z'. Therefore,

$$v_z'=\ell\frac{d\overline{v}_z}{dr} \tag{14.35}$$

The fluctuation at right angles to this, v_r', must be of the same order of magnitude as v_z' because momentum must be conserved and the time average of the fluctuating components must vanish. v_r' is often taken to be equal to v_z' for convenience:

$$v_z'=\ell\frac{d\overline{v}_z}{dr} \tag{14.36}$$

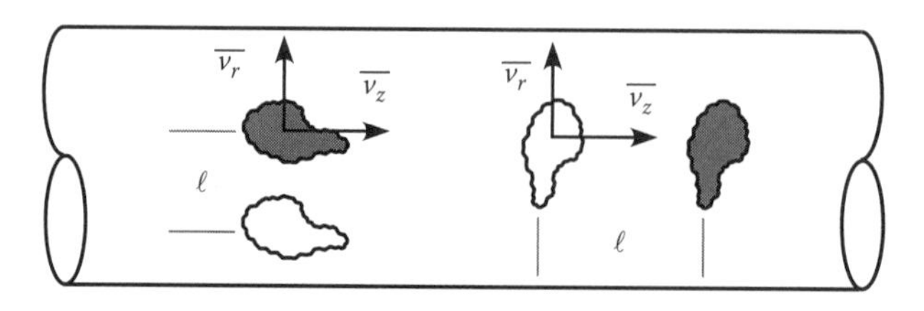

FIGURE 14.8 Mixing lengths in a tube.

We can now use these fluctuating velocity components to write the eddy stress as

$$\rho\overline{v_r'v_z'} = \rho\ell^2\left(\frac{d\bar{v}_z}{dr}\right)^2 \tag{14.37}$$

We continue by considering the Reynolds averaged Navier–Stokes equations in cylindrical coordinates. Soon, we will neglect curvature effects so that the final solution for the turbulent velocity profile within the boundary layer will apply for all boundary layers regardless of their geometry. The boundary layer assumptions are (1) steady-state operation, (2) incompressible flow, (3) constant properties, (4) a Newtonian fluid, and (5) a 2-D flow field. To these five assumptions, we add several more restrictions. We assume we have a fully developed velocity field so that the only mean flow component that exists is $\bar{v}_z(r)$. We keep some semblance of the 2-D boundary layer by including two fluctuating velocity components, v_z' and v_r'. This provides us with the Reynolds stress we need for turbulent conditions. Finally, we assume a constant pressure drop along the tube, $\Delta P/L$. The only momentum equation we need is the z-component equation and that simplifies to

$$0 = -\frac{1}{\rho}\frac{\partial P}{\partial z} + \frac{1}{r}\frac{\partial}{\partial r}\left(r\frac{\partial \bar{v}_z}{\partial r}\right) + \frac{1}{r}\frac{\partial}{\partial r}\left(r\overline{v_r'v_z'}\right) \tag{14.38}$$

Integrating once over r and evaluating from $r = 0$ to the wall ($r = r_o$) gives an expression for the averaged shear stress, $\bar{\tau}_{rz}$, at the wall:

$$\bar{\tau}_{rz} = \bar{\tau}_w = \rho\left(\frac{r}{r_o}\right)(v^*)^2 = -\mu\left(\frac{d\bar{v}_z}{dr}\right) + \rho\overline{v_r'v_z'} \tag{14.39}$$

v^* is called the *friction velocity* and is defined in terms of the stress at the pipe wall. If we divide the stress at the pipe wall by $\rho v^{*2}/2$, we obtain the friction number N_f. Thus, the friction velocity can be found by solving for v^* when the friction number $N_f = 1$ and $r = r_o$:

$$(v^*)^2 = \frac{\bar{\tau}_w}{\rho} = \frac{\frac{r_o}{2}\left[-\frac{\Delta P}{L}\right]}{\rho} \quad \text{Friction velocity} \tag{14.40}$$

We can combine Equation 14.39 with the mixing length expression for the Reynolds stress, Equation 14.37, to obtain a new equation where the only unknown is the mixing length, ℓ:

$$\bar{\tau}_w = \rho\left(\frac{r}{r_o}\right)(v^*)^2 = \underbrace{-\mu\left(\frac{d\bar{v}_z}{dr}\right)}_{\text{Laminar}} + \underbrace{\rho l^2\left(\frac{d\bar{v}_z}{dr}\right)^2}_{\text{Turbulent}} \tag{14.41}$$

Notice that we have separated the equation for the stress into a laminar, or mean flow component, and a turbulent, or fluctuating flow component.

Prandtl solved Equation 14.41 for fully developed turbulent flow in a tube [6]. Before he or she could solve the equation, he or she needed to make some additional assumptions. They were the following:

a. In the region where Equation 14.41 approximates turbulent flow, the turbulent effects are much larger than the viscous effects. Thus, the viscous term $\mu(d\bar{v}_z/dr)$ can be neglected there.
b. In the turbulent region, the stress within the fluid, $\bar{\tau}_{rz}$, can be taken as equivalent to $\bar{\tau}_w$. In other words, we can set $r = r_o$ on the left-hand side of Equation 14.39 or 14.41.

Both assumptions are extreme and assumption (b) is aphysical since we know the stress must decrease from the wall to the free stream. The mixing length concept is considered to be oversimplified and not able to truly describe the flow field. However, it still remains a useful description; if only for approximations. The assumptions greatly simplify Equation 14.41 and lead to

$$v^* = \ell\left(\frac{d\bar{v}_z}{dr}\right) \tag{14.42}$$

c. The final assumption is that the mixing length is proportional to the distance from the wall:

$$\ell = \kappa_T y \qquad y = r_o - r \tag{14.43}$$

Since the boundary layer is very thin and an eddy can be no larger than its distance from the wall, this is not a bad assumption. Notice that by defining the mixing length in terms of the distance from the wall, we have eliminated the last vestige of curvature from this representation.

We now change variables in Equation 14.42 to write it in terms of y and then integrate over y to obtain

$$\frac{\bar{v}_z}{v^*} = \frac{1}{\kappa_T}\ln y + C \tag{14.44}$$

The velocity in the boundary layer is predicted to be a logarithmic function of distance from the wall. κ_T and C are constants that can be determined from experimental measurements of the velocity profile.

The velocity distribution is usually written in the following form:

$$U^+ = \frac{1}{\kappa_T}\ln y^+ + B \tag{14.45}$$

where the dimensionless velocity, U^+, and the dimensionless distance, y^+, the latter a kind of Reynolds number, are defined by

$$U^+ = \frac{\bar{v}_z}{v^*} \qquad y^+ = \left(\frac{v^*}{\nu}\right) y = \left(\frac{v^*\rho}{\mu}\right) y \tag{14.46}$$

In a region very close to the wall (the laminar sublayer of the boundary layer), Prandtl assumed that the fluid motion was greatly influenced by the wall through viscous forces. In that region, no fully formed eddies exist and so the Reynolds stress term in Equation 14.39 could be neglected. Assuming $r = r_o$ in the region near the wall, we obtain another differential equation for the velocity profile:

$$\rho(v^*)^2 = \mu\frac{d\bar{v}_z}{dy} \tag{14.47}$$

Equation 14.47 can be integrated with the boundary condition that the velocity is zero at the wall to give

$$\rho(v^*)^2 = \mu\frac{\bar{v}_z}{y} \tag{14.48}$$

which can be rearranged to give

$$U^{+} = y^{+} \tag{14.49}$$

Thus, near the wall, we have a linear velocity profile, and far from the wall, but still within the boundary layer, we have a logarithmic profile.

We have seen that the turbulent boundary layer can be conceptually separated into three zones. The laminar sublayer is the region where we have the linear velocity profile. Experiments have shown that the extent of this region is $0 \le y^{+} \le 5$. The transition zone follows the logarithmic velocity profile and experiments have shown that the extent of this region is $5 < y^{+} < 30$. Finally, the turbulent core region also has a logarithmic profile and extends out beyond $y^{+} \ge 30$.

The velocity distribution developed for boundary layer flow in pipes based on Prandtl's Equations 14.45 and 14.49 is known as the *universal velocity profile*. Summarizing the results gives

Laminar sublayer

$$U^{+} = y^{+} \qquad y^{+} \le 5 \tag{14.50}$$

Transition zone

$$U^{+} = 5.0 \ln y^{+} - 3.05 \qquad 5 < y^{+} < 30 \tag{14.51}$$

Turbulent core

$$U^{+} = 2.5 \ln y^{+} + 5.5 \qquad 30 \le y^{+} \tag{14.52}$$

Figure 14.9 shows the velocity distribution. Agreement between the universal velocity profile and experimental data is generally good even out toward the center region of the pipe. Though the

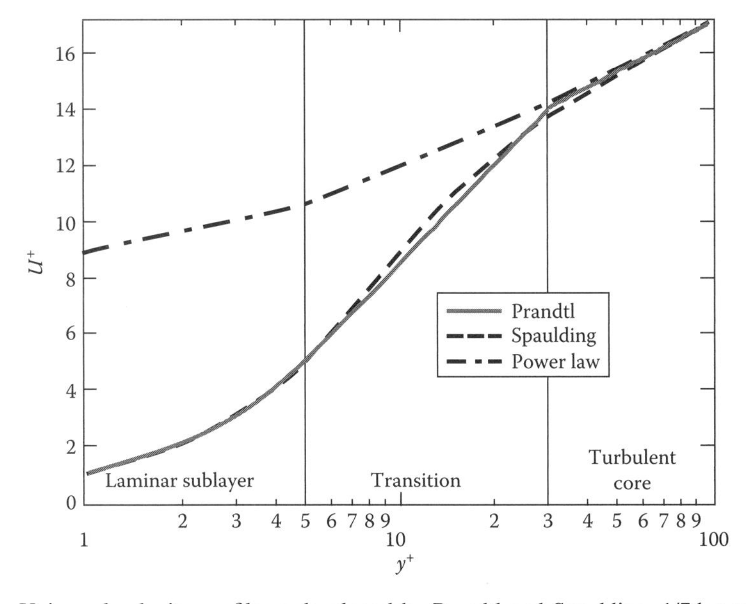

FIGURE 14.9 Universal velocity profile as developed by Prandtl and Spaulding. 1/7th power law profile added for comparison.

mixing length is an aphysical assumption, the results of the model are quite good and have been used for some time as a description for turbulent transport. The only serious difficulty is the break in the slope at $y^+ \approx 30$. Experimental data do not correspond with this behavior (as they shouldn't), and much effort has gone into trying to resolve this dilemma by developing alternatives to the mixing length approach.

14.4.3 Von Karman Similarity Assumption [7]

Prandtl's mixing length seemed a rather arbitrary concept. Looking at flows near the walls of flat plates and the walls of tubes, investigators noticed that in the transition and turbulent core regions of the boundary layer, the shear stress was relatively constant, regardless of the roughness of the wall, its geometry, etc. This led Theodore von Karman to suggest a similarity relationship for the flow patterns in the outer regions of the boundary layer. This relationship is similar to the similarity relationship we invoked for the velocity profile in the laminar boundary layer. Like Prandtl, von Karman assumed that viscosity was important only in the region close to the wall. He or she then assumed that the local flow pattern in the transition and core regions was statistically similar in the neighborhood of every point. The ratio between successive orders of derivative terms has the dimensions of length and von Karman sets this ratio proportional to the mixing length. κ_T is a universal constant:

$$\frac{(d\overline{v}_z/dr)_1}{(d^2\overline{v}_z/dr^2)_1} = \frac{(d\overline{v}_z/dr)_2}{(d^2\overline{v}_z/dr^2)_2} = \frac{\ell}{\kappa_T} \tag{14.53}$$

Solving for the mixing length, substituting into Equation 14.41, and again neglecting the laminar contribution yields a new differential equation for the velocity profile:

$$\rho\left(\frac{r}{r_o}\right)(v^*)^2 = \rho\kappa_T^2\frac{(d\overline{v}_z/dr)^4}{(d^2\overline{v}_z/dr^2)^2} \tag{14.54}$$

Rearranging,

$$\frac{d^2\overline{v}_z}{dr^2} - \frac{\kappa_T}{v^*}\sqrt{\frac{r}{r_o}}\left(\frac{d\overline{v}_z}{dr}\right)^2 = 0 \tag{14.55}$$

Letting $V = d\overline{v}_z/dr$ converts Equation 14.55 to a first-order differential equation that we can solve for the velocity profile. Using the boundary condition that $\overline{v}_z = \overline{v}_{z,\max}$ when $r = r_o$ (now the center of the tube), we derive von Karman's *velocity defect law*:

$$\frac{\overline{v}_{z,\max} - \overline{v}_z}{v^*} = -\frac{1}{\kappa_T}\left[\ln\left(1-\sqrt{\frac{r}{r_o}}\right)+\sqrt{\frac{r}{r_o}}\right] \tag{14.56}$$

Converting to U^+ and y^+ notation,

$$U^+_{\max} - U^+ = -\frac{1}{\kappa_T}\left[\ln\left(1-\sqrt{1-\left(\frac{\nu}{v^*r_o}\right)y^+}\right)+\sqrt{1-\left(\frac{\nu}{v^*r_o}\right)y^+}\right] \tag{14.57}$$

Von Karman's universal velocity profile also only applies in the transition and turbulent core regions of the boundary layer. To obtain a fit to the laminar sublayer, he or she had to assume that there were no Reynolds stress components in that region and so the velocity profile was again $U^+ = y^+$ there. Using measurements of velocity profiles in tubes by Nikuradse and others, von Karman's universal constant was determined to be $\kappa_T \approx 0.36$.

14.4.4 Other Assorted Descriptions

Though von Karman's analysis provided a more theoretical foundation for the mixing length, it failed to produce an improvement in the universal velocity profile. It also suffered from the problem that it assumed no velocity fluctuations in the laminar sublayer. We now know that those fluctuations exist, but they are greatly damped by viscous effects. In true chaotic fashion, microscopically small fluctuations in the laminar sublayer can grow tremendously and lead to large turbulent fluctuations in the core of the boundary layer. Modern measurements have shown that this is indeed the case, and eddies are generated in the layer near the wall by fluctuations that grow exponentially.

Others have attempted to improve upon the mixing length concept by including the fact that the eddy viscosity fluctuates in the laminar sublayer but that it is damped approximately exponentially as one approaches the wall. This has led to a series of empirically based *correlations* for the mixing length or eddy viscosity. One of the most widely used was from Van Driest [8] who assumed the following form for the mixing length:

$$\ell = \left(\frac{\kappa_T \nu}{v^*}\right) y^+ \left[1 - \exp\left(-\frac{y^+}{A}\right)\right] \tag{14.58}$$

Substituting into Equation 14.42 and integrating leads to the velocity profile shown in Equation 14.59 that eliminates the kink in the universal velocity profile at $y^+ = 30$. Van Driest assumed that $\kappa_T = 0.4$ and $A = 27$ in this equation:

$$U^+ = 2\int_0^{y^+} \frac{dy^+}{1 + \left\{1 + 4\kappa_T^2 y^{+2}\left[1 - \exp\left(-\frac{y^+}{A}\right)\right]^2\right\}^{1/2}} \tag{14.59}$$

Spaulding [9] developed a useful engineering approximation for the velocity profile in the boundary layer. He or she wrote the profile in an inverse form that reduces to the linear profile in the laminar sublayer and the logarithmic profile in the turbulent core region:

$$y^+ = U^+ + C\left[\exp(\kappa_T U^+) - 1 - (\kappa_T U^+) - \frac{(\kappa_T U^+)^2}{2!} - \frac{(\kappa_T U^+)^3}{3!} - \frac{(\kappa_T U^+)^4}{4!}\right] \tag{14.60}$$

Here, $\kappa_T = 0.4$ and $C = 0.1108$.

The simplest description of the velocity profile is the 1/7th power law also introduced by Prandtl:

$$U^+ = \frac{\overline{v_{z,\max}}}{v^*}\left(\frac{\nu}{r_o v^*}\right)^{1/7} (y^+)^{1/7} \approx 8.8(y^+)^{1/7} \tag{14.61}$$

Despite the fact that there is no Reynolds number dependence in Equation 14.61, the profile is surprisingly accurate. All the flow dependence is included in $\overline{v}_{z,\max}$. Since it contains no reference to the curvature of the tube, this profile works as well for flat plates as it does for tubes.

Instead of using a mixing-length-type approach to approximate the Reynolds stress, Heng, Chan, and Churchill [10] developed a correlation of the form

$$\overline{v_r' v_z'}^{+} = -\rho \frac{\overline{v_r' v_z'}}{\bar{\tau}_{rz}} = \left(\underbrace{\left[0.7 \left(\frac{y^+}{10} \right)^3 \right]^{-8/7}}_{\text{Wall region}} + \underbrace{\left| \exp\left\{ -\frac{2.294}{y^+} \right\} - \frac{2.294}{a^+} \left(1 + \frac{6.95 y^+}{a^+} \right) \right|^{-8/7}}_{\text{Turbulent core}} \right)^{-7/8} \quad (14.62)$$

based on experiments and numerical simulations by Churchill and Zajic [11] where a^+ and y^+ are defined by

$$a^+ = \frac{r_0 (\rho \bar{\tau}_w)^{1/2}}{\mu} = \frac{r_0 v^*}{\nu} \qquad y^+ = \frac{y (\rho \bar{\tau}_w)^{1/2}}{\mu} = \frac{y v^*}{\nu} \quad (14.63)$$

Equation 14.62 can best be described as the fraction of shear stress that can be attributed to turbulent fluctuations.

The universal velocity profile as developed by Prandtl and later by Spaulding is shown in Figure 14.9. These profiles fit the experimental data fairly well. Notice the abrupt change in slope of the Prandtl profile versus the smooth transition of Spaulding's empirical fit. The 1/7th power law profile is also shown for comparison. Though the power law profile fails miserably in the transition and laminar sublayer regions, it is surprisingly accurate in the turbulent core and can be used to describe the whole velocity profile inside a tube. Equation 14.62 is also accurate but, since it depends upon a^+, does not fit easily onto the same plot.

14.4.5 Numerical Simulation

Numerical simulation has proven to be an extremely useful tool for handling turbulent flows and is used routinely in industrial situations for simulating flows in process equipment such as pumps and compressors, flows through process piping, flows about automobiles and planes, and for weather prediction. Since the Navier–Stokes equations apply, direct numerical simulation of turbulence has also been done. Direct simulation is possible only in limited situations since for any realistic, engineering application, a direct simulation would require more computing power and time than is currently possible.

Figures 14.10 through 14.12 show some of the power of numerical simulation and how well it can reproduce some of the gross features of a turbulent flow. Most numerical models have problems in the wall region so that their predictions of transport coefficients are not always accurate. To combat that tendency, models include *wall functions* based on the universal velocity profile. The usefulness of such a wall function is obvious from Figure 14.10 where the predicted velocity profile follows a $U^+ = y^+$ profile even though the turbulence model does not obey the no-slip condition at the wall.

The numerical models, of which there are many, all begin with the Reynolds averaged Navier–Stokes equations and attempt to model the Reynolds fluxes using a combination of additional balance equations and assumptions. Reviews of turbulence simulation can be found in [12,13]. Two of the most popular models, the k-ε and k-ω models used in Figure 14.11, are from a class of models called two-equation models. They are a good compromise between model complexity and simulation accuracy. Both models start by adding an equation to predict the turbulent kinetic energy, k_e, defined by $k_e = \overline{v_x' v_x'} + \overline{v_y' v_y'} + \overline{v_z' v_z'}$. The k-ε model adds an equation for the dissipation rate,

$$\varepsilon = \nu \left[\overline{\left(\frac{\partial v_x'}{\partial x} \right)^2} + \overline{\left(\frac{\partial v_x'}{\partial y} \right)^2} + \overline{\left(\frac{\partial v_x'}{\partial z} \right)^2} + \overline{\left(\frac{\partial v_y'}{\partial x} \right)^2} + \overline{\left(\frac{\partial v_y'}{\partial y} \right)^2} + \overline{\left(\frac{\partial v_y'}{\partial z} \right)^2} + \overline{\left(\frac{\partial v_z'}{\partial x} \right)^2} + \overline{\left(\frac{\partial v_z'}{\partial y} \right)^2} + \overline{\left(\frac{\partial v_z'}{\partial z} \right)^2} \right] \quad (14.64)$$

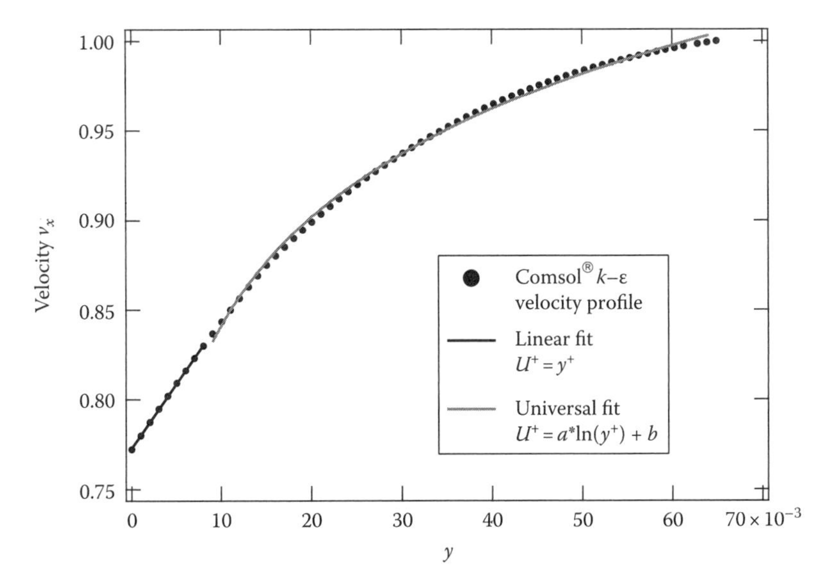

FIGURE 14.10 Turbulence velocity profile prediction for a flat plate versus the universal velocity profile. Only two regions of universal velocity profile were fit to the simulated result. Comsol® module (TurbulentPlate) based on a k-ε approximation.

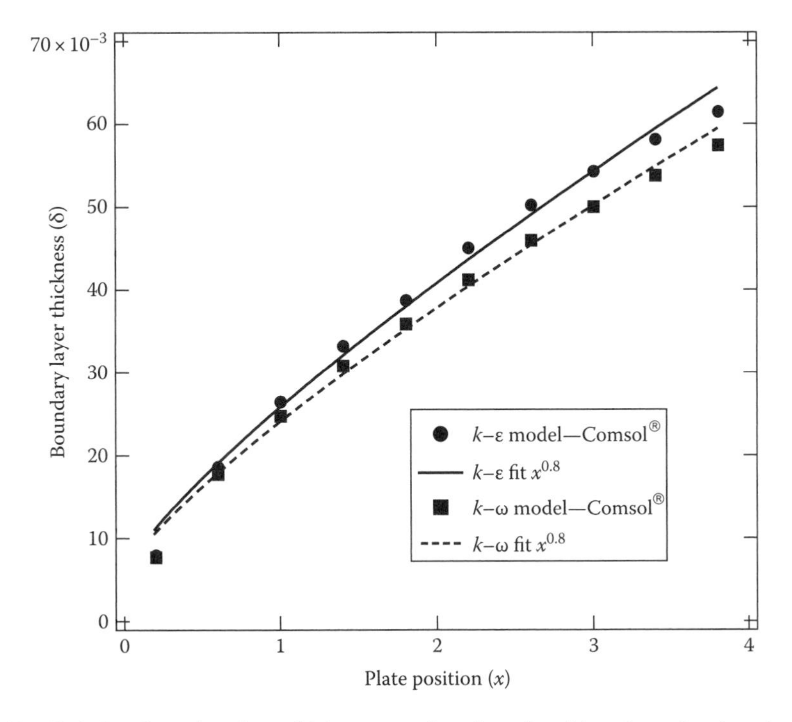

FIGURE 14.11 Turbulent boundary layer thickness as a function of position along the plate. k-ε and k-ω fits using Comsol® module TurbBL. Simulations reproduce the 0.8 power dependence seen from experimental data.

while the k-ω equation replaces the equation for the dissipation rate by an equation for $\omega \propto \varepsilon/k_e$. These equations can be derived from the Reynolds averaged turbulence equations we introduced previously. However, the process is quite a bit more tedious than the processes we used to derive the mechanical energy and vorticity equations from the Navier–Stokes equations in Chapter 10. The derivations are best left to texts on turbulence [14,15]. The full equations are not used by either

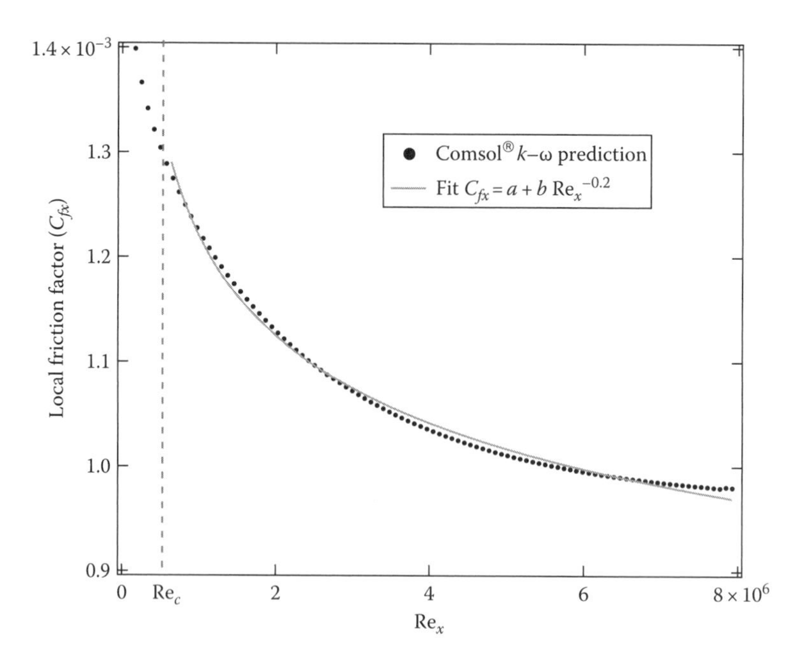

FIGURE 14.12 Local turbulent friction factor on a flat plat. k-ω simulation using Comsol® module TurbBL. Simulation follows experimental Reynolds number dependence fairly well.

the k_e-ε or k_e-ω models. Instead, these models simplify their equations using a kind of Boussinesq/mixing length process where the turbulent kinematic viscosity and double velocity correlation are defined as

$$\nu_t = \frac{c_\mu k_e^2}{\varepsilon} \qquad \text{or} \qquad \nu_t = \frac{k_e}{\omega}\left(\frac{0.15 + k_e/\nu\omega}{6.0 + k_e/\nu\omega}\right) \tag{14.65}$$

$$-\overline{v_i' v_j'} = \nu_t\left(\frac{\partial \bar{v}_i}{\partial x_j} + \frac{\partial \bar{v}_j}{\partial x_i}\right) - \frac{2}{3}k_e\delta_{ij} \qquad \delta_{ij} = \begin{cases} 0 & i \neq j \\ 1 & i = j \end{cases} \tag{14.66}$$

The model equations for k_e, ε, and ω are

$$\frac{Dk_e}{Dt} = \vec{\nabla}\cdot\left[\left(\nu + \frac{\nu_t}{\sigma_k}\right)\vec{\nabla}k_e\right] + \nu_t\left\{2\left[\left(\frac{\partial \bar{v}_x}{\partial x}\right)^2 + \left(\frac{\partial \bar{v}_y}{\partial y}\right)^2 + \left(\frac{\partial \bar{v}_z}{\partial z}\right)^2\right]\right\}$$
$$+\nu_t\left[\left(\frac{\partial \bar{v}_x}{\partial y} + \frac{\partial \bar{v}_y}{\partial x}\right)^2 + \left(\frac{\partial \bar{v}_x}{\partial z} + \frac{\partial \bar{v}_z}{\partial x}\right)^2 + \left(\frac{\partial \bar{v}_z}{\partial y} + \frac{\partial \bar{v}_y}{\partial z}\right)^2\right] - \varepsilon \tag{14.67}$$

$$\frac{D\varepsilon}{Dt} = \vec{\nabla}\cdot\left[\left(\nu + \frac{\nu_t}{\sigma_\varepsilon}\right)\vec{\nabla}\varepsilon\right] + C_{1\varepsilon}\nu_t\left\{2\left[\left(\frac{\partial \bar{v}_x}{\partial x}\right)^2 + \left(\frac{\partial \bar{v}_y}{\partial y}\right)^2 + \left(\frac{\partial \bar{v}_z}{\partial z}\right)^2\right]\right\}\frac{\varepsilon}{k_e}$$
$$+C_{1\varepsilon}\nu_t\left[\left(\frac{\partial \bar{v}_x}{\partial y} + \frac{\partial \bar{v}_y}{\partial x}\right)^2 + \left(\frac{\partial \bar{v}_x}{\partial z} + \frac{\partial \bar{v}_z}{\partial x}\right)^2 + \left(\frac{\partial \bar{v}_z}{\partial y} + \frac{\partial \bar{v}_y}{\partial z}\right)^2\right]\frac{\varepsilon}{k_e} - C_{2\varepsilon}\frac{\varepsilon^2}{k_e} \tag{14.68}$$

$$\frac{D\omega}{Dt}=\vec{\nabla}\cdot\left[\left(\nu+\frac{\nu_t}{\sigma_\omega}\right)\vec{\nabla}\omega\right]+C_{1\omega}\left\{2\left[\left(\frac{\partial\bar{v}_x}{\partial x}\right)^2+\left(\frac{\partial\bar{v}_y}{\partial y}\right)^2+\left(\frac{\partial\bar{v}_z}{\partial z}\right)^2\right]\right\}$$

$$+C_{1\omega}\left[\left(\frac{\partial\bar{v}_x}{\partial y}+\frac{\partial\bar{v}_y}{\partial x}\right)^2+\left(\frac{\partial\bar{v}_x}{\partial z}+\frac{\partial\bar{v}_z}{\partial x}\right)^2+\left(\frac{\partial\bar{v}_z}{\partial y}+\frac{\partial\bar{v}_y}{\partial z}\right)^2\right]-C_{2\omega}\omega^2 \quad (14.69)$$

$$C_\mu=0.09 \qquad C_{1\varepsilon}=1.44 \qquad C_{2\varepsilon}=1.92 \quad (14.70)$$

$$\sigma_k=1.0 \qquad \sigma_\varepsilon=1.3 \qquad \sigma_\omega=2.0 \quad (14.71)$$

$$C_{1\omega}=\frac{5}{9}\left(\frac{6.0+k_e/\nu\omega}{0.15+k_e/\nu\omega}\right)\left(\frac{0.27+k_e/\nu\omega}{2.7+k_e/\nu\omega}\right) \quad (14.72)$$

$$C_{2\omega}=0.09\left[\frac{5/18+(k_e/8\nu\omega)^4}{1.0+k_e/8\nu\omega^4}\right] \quad (14.73)$$

Figure 14.11 shows the boundary layer thickness predictions for the k-ε and k-ω models. Experimental data indicate that the boundary layer thickness should increase as a function of $x^{4/5}$. Both models seem to reproduce the dependence fairly well, and Figure 14.12 shows that the k-ω model reproduces the friction factor dependence on Reynolds number equally well except in the region where the Reynolds number has not yet reached the turbulent transition and near the exit of the plate.

14.5 FRICTION FACTORS AND OTHER TRANSPORT COEFFICIENTS

Once we have the velocity profile at the wall, we are in a position to calculate the friction factor and the other transport coefficients. The equations we developed for the universal velocity profiles had no effect of curvature. They are applicable to any geometry as long as the curvature is much smaller than the boundary layer thickness.

14.5.1 Friction Factors

The friction factor is defined as a nondimensional ratio of the shear stress at a surface, τ_s, to the kinetic energy of the flow. For flow in pipes and tubes, the kinetic energy is related to the mean velocity of the flow, $\rho(\bar{v}_{zm})^2/2$. Here, $\bar{v}_{zm}$ represents an average over time and space. The *Fanning friction factor* based on this definition is

$$C_f=\frac{\tau_s}{(1/2)\rho(\bar{v}_{zm})^2}=\frac{\rho(v^*)^2}{(1/2)\rho(\bar{v}_{zm})^2}=\frac{-\mu((d\bar{v}_z)/(dr))+\rho\overline{v_r'v_z'}}{(1/2)\rho(\bar{v}_{zm})^2} \quad \text{Fanning friction factor} \quad (14.74)$$

We generally relate this to the pressure drop through the tube:

$$C_f=\frac{(\partial P/\partial z)}{(1/2)\rho(\bar{v}_{zm})^2}=\frac{(2r_o(-\Delta P)/4L)}{(1/2)\rho(\bar{v}_{zm})^2} \quad (14.75)$$

For laminar flow, we can integrate over the parabolic velocity profile, obtain $\overline{v}_{zm}$, and substitute into Equation 14.75 to obtain the Fanning friction factor:

$$C_f = \frac{16}{Re_d} \qquad Re_d = \frac{\overline{v}_{zm} d}{\nu} \tag{14.76}$$

The *Moody friction factor*, f_m, is defined as four times the Fanning friction factor:

$$f_m = 4C_f = \frac{64}{Re_d} \quad \text{Moody friction factor} \tag{14.77}$$

Most friction factor plots for tubes, like that by Moody [16], shown in Figure 14.13 do not identify the friction factor except for this laminar equation printed on the plot. The friction factor plot also contains information about the tube wall roughness. This is important for engineers given the wide variety of pipe and conduit materials in use. The conduit that is used for sewage or storm runoff is much different than that used for drinking water or process plant piping. Consequently, the flow behavior in these systems is vastly different. Turbulent eddies arise from disturbances and high stresses within the laminar sublayer. As the surface of the tube gets rougher, it causes flow disturbances within the laminar sublayer and increases the shear stress at the wall. These factors trip up laminar flow and produce turbulence at lower and lower Reynolds numbers. Thus, we see the onset of the relatively constant turbulent friction factor at lower and lower Reynolds numbers for rougher and rougher tubes. The increased stress at the rough surface results in higher values of the friction factor and correspondingly higher values of the heat or mass transfer coefficient.

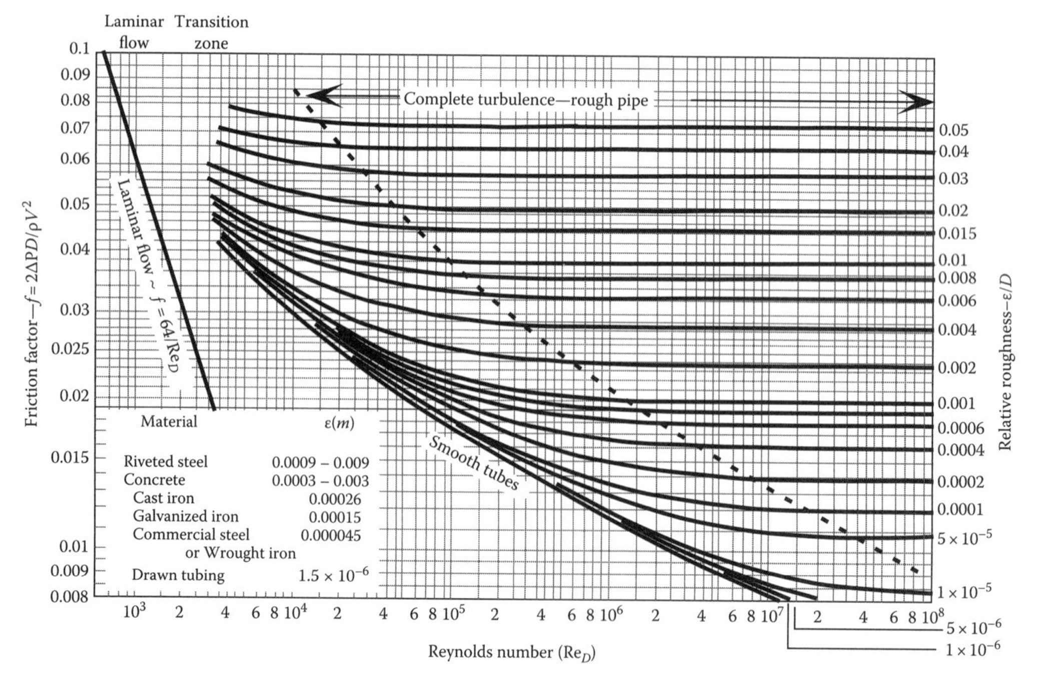

FIGURE 14.13 Friction factor plot for flow in a pipe. (From Moody, L.F., *Trans. ASME*, 66, 671, 1944. With permission.)

Equation 14.74 expresses the friction factor in terms of the friction velocity, v^*. Combining Equations 14.74 and 14.75, we have

$$v^* = \bar{v}_{zm}\sqrt{\frac{C_f}{2}} = \left[\frac{(r_o/2)(\Delta P/L)}{\rho}\right]^{1/2} \tag{14.78}$$

We can predict this friction factor from any one of the velocity profiles we presented previously. Compared with experimental data, all of the velocity profile equations have drawbacks and considerable errors may result if they are used blindly. To derive an expression for the friction factor, we integrate across the tube's cross section and calculate the mean velocity. The simplest case to explore uses the 1/7th power law profile. Expressing Equation 14.61 as

$$\frac{\bar{v}_z}{v^*} = 8.8\left(\frac{(r_o - r)v^*}{\nu}\right)^{1/7} \tag{14.79}$$

we can show that the average velocity is

$$\frac{\bar{v}_{zm}}{v^*} = 7.19\left(\frac{r_o v^*}{\nu}\right)^{1/7} \tag{14.80}$$

Rearranging Equation 14.78 and using Equation 14.80 to eliminate v^* leads to

$$C_f = 2\left(\frac{v^*}{\bar{v}_{zm}}\right)^2 = 0.08\left(\frac{2r_o\bar{v}_{zm}}{\nu}\right)^{-1/4} = 0.08\mathrm{Re}_d^{-1/4} \tag{14.81}$$

which is the Blasius equation [17] for turbulent flow in smooth tubes. It generally underpredicts the friction factor. It is useful for $\mathrm{Re}_d < 1 \times 10^5$. The Moody diagram is essentially just a plot of the earlier Colebrook equation [18]:

$$\frac{1}{\sqrt{f}} = -2.0\log\left(\frac{\varepsilon/d}{3.7} + \frac{2.51}{\mathrm{Re}_d\sqrt{f}}\right) \tag{14.82}$$

This equation is difficult to solve for f because it is implicit. An explicit alternative was developed by Haaland [19]:

$$\frac{1}{\sqrt{f}} \approx -1.8\log\left[\left(\frac{\varepsilon/d}{3.7}\right)^{1.11} + \frac{6.9}{\mathrm{Re}_d}\right] \tag{14.83}$$

Since these early equations, many other representations have been developed [20]. One of the more accurate representations for the friction factor comes from integrating Equation 14.62 and combining it with Equation 14.74. Expressed in terms of the Fanning friction factor, an excellent approximation to the exact integral is

$$\left(\frac{2}{C_f}\right)^{1/2} = 3.2 - 454\left[\frac{\left(2/C_f\right)^{1/2}}{\mathrm{Re}_d}\right] + 10{,}000\left[\frac{\left(2/C_f\right)^{1/2}}{\mathrm{Re}_d}\right]^2 + 2.294\ln\left[\frac{\mathrm{Re}_d}{2\left(2/C_f\right)^{1/2}}\right] \tag{14.84}$$

Example 14.1 Turbulent Boundary Layer Thickness on a Flat Plate

We can use the integral boundary layer analysis from Chapter 12 coupled with the 1/7th power law representation and the Blasius equation for the stress at the wall (14.81) to obtain the boundary layer thickness in turbulent flow over a smooth, flat, plate. Recalling the integral analysis from Chapter 12, the momentum equation was integrated over the boundary layer thickness to yield

$$\int_0^{\delta_h} \frac{\partial}{\partial x}\{u(v_\infty - u)\}dy = \nu \frac{\partial u}{\partial y}\bigg|_{y=0}$$

We defined a dimensionless velocity, $u^* = u/v_\infty$, and dimensionless y coordinate $y^* = y/\delta_h$ and rewrote the equation as

$$\frac{d}{dx}\delta_h \int_0^1 \{u^*(1-u^*)\}\, dy^* = \left(\frac{\nu}{v_\infty \delta_h}\right)\frac{\partial u^*}{\partial y^*}\bigg|_{y^*=0} = \frac{\tau_s}{\rho v_\infty^2}$$

The right-hand side of this equation is just the shear stress at the plate's surface divided by the kinetic energy per unit volume of the bulk flow. We can use the 1/7th power law expressed in terms of the dimensionless variables to describe the velocity profile:

$$u^* = y^{*1/7}$$

Evaluating the integral on the left-hand side of the momentum equation gives

$$\int_0^1 \{u^*(1-u^*)\}\, dy^* = \frac{7}{72}$$

while the shear stress at the surface can be calculated from the Blasius relationship:

$$\frac{\tau_s}{\rho v_\infty^2} = 0.02\left(\frac{\nu}{v_\infty \delta_h}\right)^{1/4}$$

Substituting back into the momentum equation gives

$$\left(\frac{7}{72}\right)\frac{d\delta_h}{dx} = 0.02\left(\frac{\nu}{v_\infty \delta_h}\right)^{1/4}$$

Upon integrating and setting $\delta_h = 0$ at the leading edge of the plate, we have

$$\delta_h = 0.37x\left(\frac{\nu}{v_\infty x}\right)^{1/5} = 0.37x\mathrm{Re}_x^{-1/5}$$

The turbulent boundary layer on a flat plate grows with the 4/5th power of distance down the plate and is inversely proportional to the Reynolds number to the 1/5th power. We can compare this to the laminar case where the boundary layer grew as $x^{1/2}$ and depended on the square root of the Reynolds number. Thus, the turbulent boundary layer grows faster but is less affected by changes in the Reynolds number.

We define the friction factor by

$$\frac{C_f}{2} = \frac{\tau_s}{\rho v_\infty^2} = \left(\frac{7}{72}\right)\frac{d\delta_h}{dx}$$

In this case, this is actually an average friction factor since the shear stress given by the Blasius equation is an average value. Substituting in for the boundary layer thickness gives

$$\bar{C}_f = 0.0575\mathrm{Re}_L^{-1/5} \qquad \bar{N}_f = \frac{\bar{C}_f \mathrm{Re}_L}{2} = 0.029\mathrm{Re}_L^{4/5}$$

Example 14.2 Friction Factor in terms of the Universal Velocity Profile

A more reliable estimate of the friction factor and friction number can be obtained by using the universal velocity profile. Here, we are again looking at the velocity profile in the turbulent core region. Forming a velocity defect expression $\left(U_{\max}^+ - U^+\right)$ from the universal velocity profile Equation 14.45, we then integrate over the tube cross section and obtain an expression for the average velocity:

$$\bar{U}^+ = U_{\max}^+ - \frac{\int_0^{y_{\max}^+} \left[\frac{1}{\kappa_T}\ln\left(\frac{y_{\max}^+}{y^+}\right)\right] y^+ dy^+}{\int_0^{y_{\max}^+} y^+ dy^+} = U_{\max}^+ - \frac{3}{2\kappa_T}$$

Now we substitute into Equation 14.74, expanding $y_{\max}^+$ in terms of v^*, ν, and r_o, and substituting in for $U_{\max}^+$:

$$\sqrt{\frac{2}{C_f}} = \bar{U}^+ = \frac{1}{\kappa_T}\ln\left[\frac{r_o v^*}{\nu}\right] - \frac{3}{2\kappa_T} + B$$

Finally, we replace r_o with the tube diameter, d, and v^* with its representation in terms of C_f and $\bar{v}_{zm}$. This allows us to rewrite the equation for the friction factor in terms of the Reynolds number:

$$\sqrt{\frac{2}{C_f}} = \frac{1}{\kappa_T}\ln\left[\mathrm{Re}_d\sqrt{\frac{C_f}{8}}\right] - \frac{3}{2\kappa_T} + B$$

A number of investigators have used experimental results to determine B and κ_T. Their best results yield the following correlation:

$$\sqrt{\frac{1}{C_f}} = 4.0\log_{10}\left[\mathrm{Re}_d\sqrt{C_f}\right] - 0.4$$

This is called the Prandtl/von Karman universal resistance law for smooth tubes and is valid to Re_d of up to 5×10^6. It is still one of the best correlations available for the friction factor.

Friction factors for flat plates, tube flow, flow around spheres, and cylinders have all been measured experimentally. These measurements provide the best values for these coefficients since they account for variations in surface roughness and cover a range of Reynolds numbers that the correlations do not. This is especially important in the transition region between laminar and turbulent flow. While most of the theoretical expressions for friction factors we presented apply in the limit as $\mathrm{Re} \to \infty$, little is known theoretically about the transition region and so experimental values are all we really have if this range is where one needs to operate. Generally, processes and products are designed to operate either in the laminar or fully turbulent regime.

Example 14.3 Drag on a Giant Manta Ray

A giant manta ray can be nearly 9 m in width and 4 m in length with a volume of 0.95 m^3. They weigh upward of 1350 kg. Great white sharks and killer whales are occasional predators, and when in danger, the rays can swim at speeds up to 12 km/h and use that speed to ram their attacker. Assuming the manta can be viewed as a flat plate and is gliding at its maximum speed, answer the following questions:

a. What is the average friction factor for the ray?
b. What is the average shear stress on its surface?
c. How much power must it generate to travel through the water at this speed?

$$(\text{Hint: Power} = \text{Force} \times \text{Distance/Time})$$

d. Using the flat plate lift coefficient from Chapter 12, $C_L = 2\pi \sin(\alpha)$, how fast must the manta swim at an angle of attack of 5° to support its weight?

The physical properties for water are $\rho = 997.5$ kg/m^3 and $\mu = 9.8 \times 10^{-4}$ N s/m^2.

a. The first job is to determine the Reynolds number for the ray:

$$\text{Re}_L = \frac{v_\infty L}{\nu} = \frac{(12{,}000/3{,}600)4}{9.8\times10^{-4}/997.5} = 1.36\times10^7$$

so it is clearly in the turbulent regime. The friction number is given by

$$\bar{N}_f = \frac{\bar{C}_f \text{Re}_L}{2} = 0.029\text{Re}_L^{4/5} = 0.029(1.36\times10^7)^{4/5} = 1.474\times10^4$$

The friction factor is

$$\bar{C}_f = \frac{2(1.474\times10^4)}{1.36\times10^7} = 2.17\times10^{-3}$$

b. The average shear stress is found from the friction factor:

$$\tau_s = \bar{C}_f\left[\frac{1}{2}\rho v_\infty^2\right] = \frac{2.17\times10^{-3}(997.5)(3.33)^2}{2} = 12.0\ \text{N/m}^2$$

c. The power required to move the ray is determined by multiplying the force required to move the ray by its velocity:

$$\text{Power} = F^* v_\infty = \tau_s A_s v_\infty = 12.0[(2)(9)(4)](3.33) = 2885\ \text{W}$$

so it is no wonder that rays eat upward of 30–60 kg of plankton a day.

d. The ray needs to swim to keep from sinking. The net downward force on the ray is

$$F = W - \rho g V = 1350(9.8) - (0.95)(9.8)997.5 = 3943\ \text{N}$$

This is balanced by the lift force and so the swimming speed should be

$$F = \bar{C}_L A_s \left(\frac{1}{2}\rho v_\infty^2\right) \qquad v_\infty = \sqrt{\frac{2F}{\bar{C}_L A_s \rho}} = 0.63\ \frac{\text{m}}{\text{s}}$$

and we are still well into the turbulent regime (Re = 2.57×10^6).

Example 14.4 Pumping and Piping Revisited

Let's revisit a problem we encountered in Chapter 11. A pump is transferring water from tank A to tank B at a rate of 0.02 m^3/s. The tanks are both open to the atmosphere and tank A is 3 m above ground level. Tank B is 20 m above ground level. The pump is located at ground level, and the cast iron discharge pipe ($\varepsilon = 2.6 \times 10^{-4}$) that enters tank B is 23 m above ground level at its highest point. All the piping is 5.08 cm in diameter, the tanks are 6 m in diameter, and the total length of piping is 75 m. The situation is shown in Figure 14.14. If we include friction losses in the piping, what size pump will we need? If the pump system is 70% efficient, what size motor would we need to drive the pump?

The first step is to apply the continuity equation, but in this case, since what goes in goes out, the continuity equation is trivial. The work the pump must do on the fluid is determined by applying the mechanical energy balance between the two tanks. As a rule, the pump work goes on the left-hand side of the equation (supplier), while the lost work goes on the right-hand side (consumer):

$$\frac{P_1}{\rho g} + \frac{\bar{v}_1^2}{2g} + z_1 + \frac{W_p}{\rho g} = \frac{P_2}{\rho g} + \frac{\bar{v}_2^2}{2g} + z_2 + h_L$$

Substituting what we know from the problem about the uniform piping diameter ($\bar{v}_1 = \bar{v}_2$) and the pressures ($P_1 = P_2 = 0$), we find

$$z_1 + \frac{W_p}{\rho g} = z_2 + h_L$$

$$W_p = \rho g(z_2 - z_1) + \rho g h_L$$

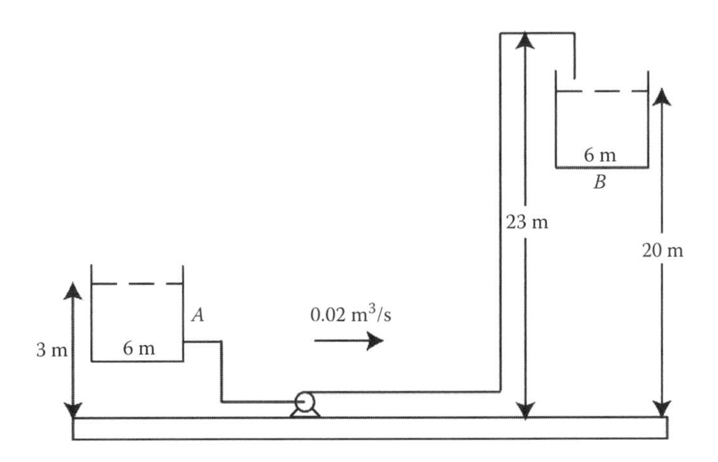

FIGURE 14.14 Pumping and piping between two tanks.

The head loss is related to the friction factor for the flow. Specifically,

$$h_L = \left(f\frac{L}{D}\right)\frac{\bar{v}^2}{2g}$$

We can use the Moody chart, the Colebrook equation, or the Haaland relation to determine the friction factor. Choosing the simplest route, we have

$$\mathrm{Re}_d = \frac{\rho\bar{v}d}{\mu} = \frac{(1000\ \mathrm{kg/m^3})\left(\dfrac{0.02}{\dfrac{\pi}{4}(0.0508)^2}\ \mathrm{m/s}\right)(0.0508\ \mathrm{m})}{0.001\ \mathrm{N\ s/m^2}} = 501{,}275$$

$$\frac{1}{\sqrt{f}} = -1.8\log\left[\left(\frac{2.6\times10^{-4}/0.0508}{3.7}\right)^{1.11} + \frac{6.9}{501{,}275}\right]$$

$$\to f = 0.0308$$

The head loss is

$$h_L = \left(f\frac{L}{d}\right)\frac{\bar{v}^2}{2g} = \frac{0.0308\left(\dfrac{75}{0.0508}\right)\left(\dfrac{0.02(4)}{\pi(0.0508)^2}\right)}{2(9.8)} = 22.9\ \mathrm{m}$$

The work required is

$$W_p = \rho g(z_2 - z_1) + \rho g h_L = 1000(9.8)(23 - 3 + 22.9) = 420{,}420$$

The power required to push all that fluid is equal to the work required multiplied by the volumetric flow rate of the fluid:

$$\text{Power} = W_p^*\dot{V} = 420{,}420\ (0.02) = 8408\ \mathrm{W}$$

At 70% efficient, the motor would have to supply 11.2/0.7 ≈ 16 hp.

14.5.2 Heat Transfer Coefficients

The heat flux in turbulent flow is the product of two components: a laminar-like contribution and a turbulent contribution. The first occurs in the laminar sublayer and is given by the product of thermal conductivity and gradient in the mean temperature. The turbulent component is eddy driven and involves the eddy diffusivity, or eddy thermal conductivity, and the Reynolds heat flux component. We write this contribution in a manner similar to the stress Equation 14.39:

$$\overline{q_T''} = -k_f\frac{\partial\bar{T}}{\partial r} - \rho C_p\frac{\partial}{\partial r}\left(\overline{v_r'T'}\right) \tag{14.85}$$

Our mental picture of turbulent flow involves the transport of momentum by eddies forming from disturbances generated in the laminar sublayer. Energy in turbulent flow must also be transported by eddies. It seems reasonable to assume that the eddies transporting momentum also transport energy simultaneously. Thus, we can invoke a Reynolds analogy of sorts to the turbulent component and state that energy is transported at roughly the same rate as momentum. We assert that the eddy momentum diffusivity and eddy thermal diffusivity are the same. This allows us to use the friction factor models we developed in the previous section to describe the turbulent energy contribution. Thus, we have Prandtl's mixing length model

$$\overline{q''_T} = -k_f \frac{\partial \overline{T}}{\partial r} - \rho C_p \kappa_T^2 r^2 \left(\frac{\partial \overline{v}_z}{\partial r} \right) \frac{\partial \overline{T}}{\partial r} \tag{14.86}$$

von Karman's similarity model

$$\overline{q''_T} = -k_f \frac{\partial \overline{T}}{\partial r} - \rho C_p \kappa_T^2 \left[\frac{(\partial \overline{v}_z / \partial r)^2}{(\partial^2 \overline{v}_z / \partial r^2)} \right] \frac{\partial \overline{T}}{\partial r} \tag{14.87}$$

and van Driest's empirical relation

$$\overline{q''_T} = -k_f \frac{\partial \overline{T}}{\partial r} - \rho C_p \kappa_T^2 r^2 \left[1 - \exp\left(-\frac{v^* r}{\nu A} \right) \right]^2 \frac{\partial \overline{T}}{\partial r} \tag{14.88}$$

Though we have assumed that the turbulent, or eddy Prandtl number, is essentially 1, experiments have shown that the eddy Prandtl number follows the behavior of the normal Prandtl number, and so the Reynolds analogy does not strictly hold. Churchill and Zajic [11] use this relationship to develop a new semitheoretical correlation based on the same approach they used to develop Equations 14.62 and 14.84. Their turbulent heat flux is expressed as

$$\frac{\overline{q''_T}}{\overline{q''_w}} = \left(\frac{\Pr}{\Pr_T} \right) \left[\frac{1}{1 - \overline{v'_r v'_z}^+} \right] \frac{dT^+}{dy^+} \qquad \frac{\Pr_T}{\Pr} = \frac{1 - \left(\rho C_p \overline{T'v'} / \overline{q''_T} \right)}{1 - \left(\rho \overline{v'_r v'_z} / \overline{\tau}_{rz} \right)} \tag{14.89}$$

Based on Equation 14.89, they developed an expression for the Nusselt number of the form

$$\overline{\mathrm{Nu}_d} = \frac{1}{\left(\frac{\Pr_T}{\Pr} \right) \frac{1}{\mathrm{Nu}_{di}} + \left[1 - \left(\frac{\Pr_T}{\Pr} \right)^{2/3} \right] \frac{1}{\mathrm{Nu}_{d\infty}}} \tag{14.90}$$

$$\Pr_T = 0.85 + \frac{0.015}{\Pr} \tag{14.91}$$

$$\mathrm{Nu}_{di} \cong \frac{\mathrm{Re}_d (C_f/2)}{1 + 145(2/C_f)^{-5/4}} \qquad \mathrm{Nu}_{d\infty} = 0.07343 \,\mathrm{Re}_d \left(\frac{\Pr}{\Pr_T} \right)^{1/3} \left(\frac{C_f}{2} \right)^{1/2} \tag{14.92}$$

This correlation works well; however, most calculations of turbulent heat transfer still rely on experimental determinations of the heat transfer coefficient.

Example 14.5 Turbulent Heat Transfer Coefficient Based on Mixing Length

We can use Prandtl's mixing length theories to derive an expression for the heat transfer coefficient and hence the Nusselt number in turbulent flow. We define the heat transfer coefficient as

$$\overline{h} = \frac{\overline{q''_T}}{T_s - T_\infty} = \frac{\rho C_p \kappa_T^2 r^2 \left(\frac{\partial \overline{v}_z}{\partial r} \right) \frac{\partial \overline{T}}{\partial r}}{T_s - T_\infty}$$

We have neglected the laminar portion of the heat flux because for high-enough Reynolds numbers, it is insignificant. To finish the derivation, we must invoke the Reynolds analogy and say that the velocity and temperature profiles are virtually identical in the boundary layer. This lets us substitute in the logarithmic velocity profile in the numerator:

$$\bar{h} = \frac{\rho C_p v^{*2} \kappa_T^2 r_o^2 \left[\frac{\partial}{\partial r}\left(\frac{1}{\kappa_T} \ln\left(\frac{v^* r}{\nu} \right) + C \right)\bigg|_{r=r_o} \right]^2}{T_s - T_\infty} = \frac{\rho C_p v^{*2}}{T_s - T_\infty}$$

Since v^* can be related to the friction factor and the friction factor is a function of the Reynolds number, we can obtain the average heat transfer coefficient in terms of the friction factor:

$$\bar{h} = \frac{\frac{1}{2}\rho C_p \bar{C}_f (\bar{v}_{zm})^2}{T_s - T_\infty} \qquad \overline{\mathrm{Nu}_d} = \frac{\frac{1}{2}\alpha d \bar{C}_f (\bar{v}_{zm})^2}{T_s - T_\infty}$$

Empirical correlations

Empirical correlations for the heat transfer coefficient have been developed for every possible geometry. These correlations are usually accurate to within 10%–20%. Of course, one can always take friction factor data and use any one of the analogies presented in Chapters 12 and 13 to obtain a heat transfer coefficient from a corresponding friction factor, as long as one is aware of the approximation involved.

Flat plate

Flow over a flat plate usually starts out laminar and then undergoes a transition to turbulent when the Reynolds number reaches about 500,000. Experimental values for the local friction factor on a flat plate show that

$$C_f = 0.059\,\mathrm{Re}_x^{-1/5} \qquad 5\times 10^5 < \mathrm{Re}_x < 1\times 10^7 \tag{14.93}$$

If we have a transition across the plate, then we integrate the values for the local coefficients over the length of the plate:

$$\bar{C}_f = \frac{1}{L}\left[\int_0^{x_c} C_{f,lam}\,dx + \int_{x_c}^{L} C_{f,turb}\,dx \right] = \frac{0.074}{\mathrm{Re}_L^{1/5}} - \frac{1742}{\mathrm{Re}_L} \tag{14.94}$$

We can now apply the Chilton–Colburn analogy to obtain a heat transfer coefficient. Using the average value for the friction factor earlier, we have

$$\overline{\mathrm{Nu}_L} = \frac{\bar{h}L}{k_f} = \frac{\bar{C}_f}{2}\mathrm{Re}_L\,\mathrm{Pr}^{1/3} = 0.037\,\mathrm{Re}_L^{4/5}\mathrm{Pr}^{1/3} - 871\,\mathrm{Pr}^{1/3} \tag{14.95}$$

In many instances, the temperature difference between the plate and free stream is high enough that changes in the fluid's physical properties become important. In such instances, the correlation provided by Whitaker [21] is better since it accounts for changes in the viscosity of the fluid:

$$\overline{\mathrm{Nu}_L} = 0.036\,\mathrm{Pr}^{0.43}\left[\mathrm{Re}_L^{4/5} - 9200\right]\left(\frac{\mu_\infty}{\mu_s}\right)^{1/4} \tag{14.96}$$

$$0.7 < \mathrm{Pr} < 380$$
$$2\times10^5 < \mathrm{Re}_L < 5.5\times10^6$$
$$0.26 < \mu_\infty/\mu_s < 3.5$$

In these correlations, all properties are evaluated at the free stream temperature with the exception of the viscosity in the viscosity correction term.

Cylinders and spheres

For flow across cylinders and spheres, we can use the friction factor plots of Chapter 13 coupled with the Chilton–Colburn or other analogy to obtain a value for the heat transfer coefficient. Care must be taken since this approach can have large error. We can also use any of the correlations presented in Chapter 13 since for these geometries, those correlations span the entire range of the data.

Flow in pipes and tubes

We can use a friction factor/analogy approach to obtain heat transfer coefficients in a pinch, but one must be aware of potential error. The only real correlations developed rely on the Blasius expression for the friction factor in smooth tubes coupled with a Chilton–Colburn analogy approach. The Dittus–Boelter [22] relation was one of the first truly useful correlations of its type:

$$\overline{\mathrm{Nu}_d} = \frac{\bar{h}d}{k_f} = 0.023\,\mathrm{Re}_d^{4/5}\,\mathrm{Pr}^n \qquad \begin{array}{c} 0.6 < \mathrm{Pr} < 100 \\ \mathrm{Re}_d > 2000 \\ n = \begin{cases} 0.4\ \text{heating of fluid} \\ 0.3\ \text{cooling of fluid} \end{cases} \end{array} \tag{14.97}$$

The Sieder–Tate [23] relation takes into account changes in the fluid's physical properties:

$$\overline{\mathrm{Nu}_d} = 0.037\,\mathrm{Re}_d^{4/5}\,\mathrm{Pr}^{1/3}\left(\frac{\mu}{\mu_s}\right)^{0.14} \qquad \begin{array}{c} 0.6 < \mathrm{Pr} < 100 \\ \mathrm{Re}_d > 2000 \end{array} \tag{14.98}$$

The Petukhov [24] relation, which is based somewhat on the universal velocity profile relation for the friction factor, can account for tube roughness:

$$\overline{\mathrm{Nu}_d} = \frac{(f/8)\mathrm{Re}_d\mathrm{Pr}}{1.07 + 12.7(f/8)^{1/2}(\mathrm{Pr}^{2/3} - 1)}\left(\frac{\mu}{\mu_s}\right)^n \tag{14.99}$$

$$n = \begin{cases} 0.11\ \text{heating} \\ 0.09\ \text{cooling} \end{cases} \qquad \begin{array}{c} 10^4 < \mathrm{Re}_d < 5\times10^6 \\ 0.6 < \mathrm{Pr} < 2000 \\ 0.08 < \mu/\mu_s < 40 \end{array}$$

Entrance lengths in turbulent flow are not the problem that they are in laminar flow. The good mixing provided by the turbulent eddies ensures that entrance effects are eliminated 10–60 pipe diameters from the inlet.

Example 14.6 Car Trouble

It's a sunny winter day. In your haste to get to a transport exam, you accidentally lock "Rover" in the car. The interior dimensions of the car are $L = 2$ m, $A_s = 2.5$ m^2, and volume = 4 m^3. Solar radiation provides a heat flux of 580 W/m^2 through the roof and windows (the area mentioned earlier). A breeze ($v_\infty = 5$ m/s, $T_\infty = 25$°C) blows over the car. Assuming only air inside the car ($v_{air} = 0$ m/s; $T(t = 0) = 25$°C) under uniform conditions, the following physical properties, the fact that Rover can withstand a temperature of 55°C, and all the net energy from the sun being transferred to the air inside the car:

$$\nu_{air} = 15.60 \times 10^{-6} \text{ m}^2/\text{s} \qquad \alpha_{air} = 0.222 \times 10^{-4} \text{ m}^2/\text{s} \qquad \text{Pr} = 0.707$$

$$k_{air} = 0.026 \text{ W/m K} \qquad \rho_{air} = 1.18 \text{ kg/m}^3 \qquad C_{pair} = 1005 \text{ J/kg K}$$

a. What is the heat transfer coefficient from the car surface to the outside?
b. How fast does the temperature initially rise or fall in the car?
c. Will Rover expire before you have finished the exam? Why?

a. The heat transfer coefficient we find using one of the correlations. First, we calculate the Reynolds number basing our definition on the fact that the surface of the car can be considered to be a flat plate of length, L:

$$\text{Re}_L = \frac{5(2)}{15.6 \times 10^{-6}} = 6.41 \times 10^5$$

Since we are just over the transition Reynolds number, we probably have mixed flow conditions over the top of the car. Thus, we use Equation 14.95 to evaluate the Nusselt number and the heat transfer coefficient:

$$\overline{\text{Nu}_L} = 0.037\text{Re}_L^{4/5}\text{Pr}^{1/3} - 871\text{Pr}^{1/3} = 1457$$

$$\bar{h} = \frac{\overline{\text{Nu}_L}k_f}{L} = \frac{1457(0.026)}{2} = 18.9 \text{ W/m}^2\text{ K}$$

b. To evaluate part (b), we need to use a kind of macroscopic analysis (lumped capacitance). It involves assuming that the air in the car is at a uniform temperature. Since "Rover" cannot sit still inside a car, this is probably a pretty good assumption. We perform an energy balance on the air in the car:

$$\begin{aligned} &\textit{In} \quad - \quad \textit{Out} \quad = \quad \textit{Acc} \\ &q''A_s - \bar{h}A_s(T - T_\infty) = \rho V C_p \frac{dT}{dt} \end{aligned}$$

Letting $\theta = T - T_\infty$, we have

$$q''A_s - \bar{h}A_s\theta = \rho V C_p \frac{d\theta}{dt}$$

Evaluating at $t = 0$, we have $T = T_\infty$ and so

$$\frac{d\theta}{dt} = \frac{q''A_s}{\rho V C_p} = \frac{580(2.5)}{(1.18)(4)(1005)} = 0.31\text{ °C/s}$$

Notice that convection is not important to the initial temperature rise.

c. To find out if Rover will expire, we integrate the balance equation with the boundary condition that at $t = 0$; $\theta = 0$:

$$\frac{d\theta}{q''A_s - \bar{h}A_s\theta} = \frac{dt}{\rho V C_p}$$

$$\frac{\bar{h}A_s}{\rho V C_p} t = \ln\left[\frac{q''A_s}{q''A_s - \bar{h}A_s\theta}\right]$$

Plugging in for the final temperature of 55°C, we find the time to be

$$t = \frac{(1.18)(4)(1005)}{(18.9)(2.5)} \ln\left[\frac{580}{580 - (18.9)(55 - 25)}\right] = 4479 \text{ s}$$

Rover lives but gets mighty uncomfortable.

Example 14.7 External and Internal Tube Flow

Water flows at 0.20 kg/s through a thick-walled tube. The inside diameter of the tube is 25 mm in diameter and the tube is 5 m long. The outside diameter of the tube is 35 mm and the tube material has a thermal conductivity, $k_w = 2$ W/m K. The water enters at a temperature of 25°C, and hot air flows across the tube at a velocity, $v_\infty = 150$ m/s, and a temperature, $T_\infty = 250$°C. What is your estimate of the outlet temperature of the water?

In this problem, it is clear we have convection on the outside and inside of the pipe and so have to use an overall heat transfer coefficient to handle the problem. Our first task is to calculate the inside and outside heat transfer coefficients. Here, we assume properties at temperatures we are sure of, the inlet temperature of the water and the free stream temperature of the air. Later, after we have done the first round of calculations, we can adjust the temperatures and properties:

$$\mu_{water} = 855 \times 10^{-6} \frac{\text{N s}}{\text{m}^2} \qquad Pr_{water} = 5.83 \qquad k_{water} = 0.613 \frac{\text{W}}{\text{m K}} \qquad C_{p,water} = 4179 \frac{\text{J}}{\text{kg K}}$$

$$\nu_{air} = 4.2 \times 10^{-5} \frac{\text{m}^2}{\text{s}} \qquad Pr_{air} = 0.684 \qquad k_{air} = 0.042 \frac{\text{W}}{\text{m K}}$$

$$Re_{d,i} = \frac{4\dot{m}}{\pi d \mu_{water}} = \frac{4(0.2)}{\pi(0.025)(855 \times 10^{-6})} = 11{,}910$$

$$Re_{d,o} = \frac{v_\infty d}{\nu_{air}} = \frac{150(0.035)}{(4.2 \times 10^{-5})} = 125{,}000$$

The Nusselt numbers for the internal and external flow are

$$\overline{Nu}_i = 0.023 Re_d^{4/5} Pr_{water}^{1/3} = 0.023(11{,}910)^{4/5}(5.83)^{1/3} = 75.45$$

$$\overline{Nu}_o = 0.3 + \frac{0.62 Re_d^{1/2} Pr_{air}^{1/3}}{[1 + (0.4/Pr_{air})^{2/3}]^{1/4}} \left[1 + \left(\frac{Re_d}{282{,}000}\right)^{5/8}\right]^{4/5}$$

$$= 0.3 + \frac{0.62(125{,}000)^{1/2}(0.684)^{1/3}}{[1 + 0.4/0.684)^{2/3}]^{1/4}} \left[1 + \left(\frac{125{,}000}{282{,}000}\right)^{5/8}\right]^{4/5} = 246.8$$

The heat transfer coefficients become

$$\bar{h}_i = \frac{\overline{\text{Nu}}_i k_{water}}{d_i} = \frac{(75.45)0.613}{0.025} = 1850 \frac{\text{W}}{\text{m}^2\,\text{K}}$$

$$\bar{h}_o = \frac{\overline{\text{Nu}}_o k_{air}}{d_o} = \frac{(246.8)0.0423}{0.035} = 298.3 \frac{\text{W}}{\text{m}^2\,\text{K}}$$

We can now calculate the overall heat transfer coefficient:

$$\frac{1}{\bar{U}_o A_o} = \frac{1}{\bar{h}_i A_i} + \frac{\ln(d_o/d_i)}{2\pi k_{pipe} L} + \frac{1}{\bar{h}_o A_o}$$

$$= \frac{1}{(1850)\pi(0.025)5} + \frac{\ln(35/25)}{2\pi(2)5} + \frac{1}{(298.3)\pi(0.035)5} = 0.0128 \frac{\text{K}}{\text{W}}$$

The heat exchanged to the fluid we calculate using an approach from Chapter 13. There we showed

$$\bar{T}_{out} = T_\infty + (\bar{T}_{in} - T_\infty)\exp\left[-\left(\frac{\bar{U}_o A_o}{\dot{m} C_{p,water}}\right)\right]$$

$$= 250 + (25 - 250)\exp\left[-\left(\frac{77.95}{0.2(4179)}\right)\right] = 45.0\ ^\circ\text{C}$$

14.5.3 Mass Transfer Coefficients

The mass flux in turbulent flow can also be represented by the product of two components: a laminar-like contribution and a turbulent contribution. The first occurs in the laminar sublayer and is given by Fick's law. The turbulent component is eddy driven and involves the eddy diffusivity and the Reynolds mass flux component. For lack of a better representation, we write this contribution in Fick's law format too:

$$\bar{N}_a = -D_{ab}\frac{\partial \bar{c}_a}{\partial r} - \overline{v'_r c'_a} = -\left(D_{ab} + D^e_{ab}\right)\frac{\partial \bar{c}_a}{\partial r} \tag{14.100}$$

We generally invoke the Reynolds analogy and assume the turbulent Schmidt number is also one, but mass transfer is a bit more complicated than heat transfer and so this analogy does not always work as well as it does in heat transfer. In Chapter 12, we showed that for high mass transfer rates, the mass transfer affects the velocity profile. Such effects are common in turbulent flow since the mass transfer rates are so high. Thus, we use the models and analogies we discussed in previous sections but must realize that these analogies may be in considerable error.

We have Prandtl's mixing length model

$$\bar{N}_a = -D_{ab}\frac{\partial \bar{c}_a}{\partial r} - \kappa_T^2 r^2 \left(\frac{\partial \bar{v}_z}{\partial r}\right)\left(\frac{\partial \bar{c}_a}{\partial r}\right) \tag{14.101}$$

von Karman's similarity model

$$\bar{N}_a = -D_{ab}\frac{\partial \bar{c}_a}{\partial r} - \kappa_T^2 \left[\frac{\left(\frac{\partial \bar{v}_z}{\partial r}\right)^2}{\left(\frac{\partial^2 \bar{v}_z}{\partial r^2}\right)} \right] \left(\frac{\partial \bar{c}_a}{\partial r}\right) \tag{14.102}$$

and van Driest's empirical relation

$$\bar{N}_a = -D_{ab}\frac{\partial \bar{c}_a}{\partial r} - \kappa_T^2 r^2 \left[1 - \exp\left(-\frac{v^* r}{\nu A}\right)\right]^2 \left(\frac{\partial \bar{c}_a}{\partial r}\right) \tag{14.103}$$

Since the eddy Schmidt number is rarely 1 due to the high mass transfer rates and large Sc number, these expressions for the flux are not often used. They do not offer reliable values for the mass transfer coefficients. This is one reason why we rely on experimental determinations of the mass transfer coefficient instead. To start with, we can use the relationships for the heat transfer coefficients we developed earlier and convert them to mass transfer coefficients. All we need to do is replace the Nusselt number by the Sherwood number and the Prandtl number by the Schmidt number. Thus, the Churchill–Zajic approach for flow in tubes becomes

$$\overline{\mathrm{Sh}_d} = \frac{1}{\left(\frac{\mathrm{Sc}_T}{\mathrm{Sc}}\right)\frac{1}{\mathrm{Sh}_{di}} + \left[1 - \left(\frac{\mathrm{Sc}_T}{\mathrm{Sc}}\right)^{2/3}\right]\frac{1}{\mathrm{Sh}_{d\infty}}} \tag{14.104}$$

with

$$\mathrm{Sc}_T = 0.85 + \frac{0.015}{\mathrm{Sc}} \tag{14.105}$$

and

$$\mathrm{Sh}_{di} \cong \frac{\mathrm{Re}_d C_f/2}{1 + 145(2/C_f)^{-5/4}} \qquad \mathrm{Sh}_{d\infty} = 0.07343\,\mathrm{Re}_d \left(\frac{\mathrm{Sc}}{\mathrm{Sc}_T}\right)^{1/3} \left(\frac{C_f}{2}\right)^{1/2} \tag{14.106}$$

The Blasius-type correlation works well for smooth flat plate geometries:

$$\overline{\mathrm{Sh}_L} = \frac{\bar{k}_c L}{D_{ab}} = \frac{\bar{C}_f}{2}\mathrm{Re}_L\,\mathrm{Sc}^{1/3} = 0.037\,\mathrm{Re}_L^{4/5}\,\mathrm{Sc}^{1/3} - 871\,\mathrm{Sc}^{1/3} \tag{14.107}$$

If flow over a cylinder or sphere is involved, any of the mass transfer correlations given in Chapter 13 can be used since they cover the entire flow range. For flow in smooth tubes, we have a modified Dittus–Boelter type of relation:

$$\overline{\mathrm{Sh}_d} = 0.023\,\mathrm{Re}_d^{4/5}\,\mathrm{Sc}^{1/3} \tag{14.108}$$

Example 14.8 Mass Transfer Rate Independent of Orientation

Two identical flat plates are exposed to parallel flow. Each plate is of dimension $L \times 2L$, but one is oriented with its shorter side parallel to the flow, while the other is oriented with its longer side parallel to the flow. We are interested in determining under what conditions the mass transfer rate is independent of the plate orientation (Figure 14.15).

To have the total mass transfer rates equal, we would need the mass transfer coefficients equal, $\overline{k_{cL}} = \overline{k_{c2L}}$. This can only occur if we have different flow rates and different flow regimes across each plate. We can see this more clearly if we plot the mass transfer coefficient as a function of Reynolds number.

For very short plates in laminar flow, the boundary layer is thin and the mass transfer coefficient is high. As the plate length grows, the mass transfer coefficient decreases reaching a minimum at the critical Reynolds number. When the changeover to turbulent flow occurs, the mass transfer coefficient rises, but as the turbulent boundary layer also grows with increasing plate length, the mass transfer coefficient must decrease again. Thus, we have two possibilities for having the same mass transfer rate on the plates.

In the first case, we can have laminar flow on the shorter plate and mixed flow on the longer plate. The second case is mixed flow over both plates (Figure 14.16). If we consider the correlations from Chapters 12 and 14, we can conclude that for laminar flow,

$$\overline{k_{cL}} = 0.664\frac{D_{ab}}{L}\text{Re}_L^{1/2}\,\text{Sc}^{1/3}$$

and for mixed flow,

$$\overline{k_{cL}} = \frac{D_{ab}}{L}\left[0.037\text{Re}_L^{4/5} - 871\right]\text{Sc}^{1/3}$$

Now considering the first case we have

$$0.664\frac{D_{ab}}{L}\text{Re}_L^{1/2}\,\text{Sc}^{1/3} = \frac{D_{ab}}{2L}\left[0.037\text{Re}_{2L}^{4/5} - 871\right]\text{Sc}^{1/3}$$

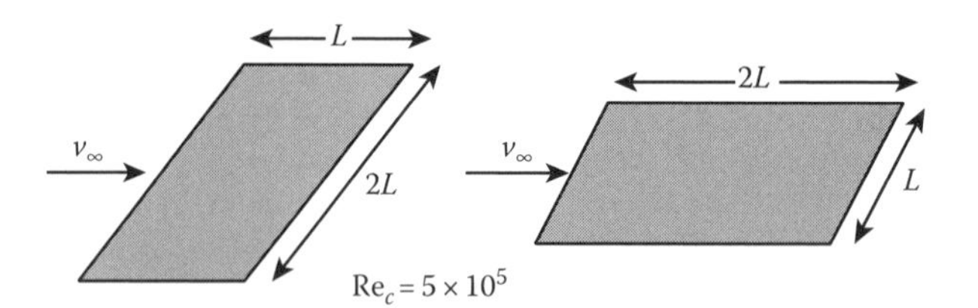

FIGURE 14.15 Laminar–turbulent flow over a flat plate in two orientations.

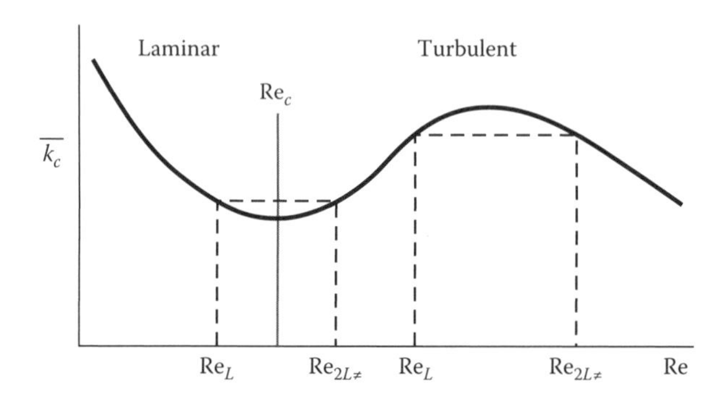

FIGURE 14.16 Mass transfer coefficient versus Reynolds number for a flat plate.

Solving for Re_L that must be $<5 \times 10^5$, we find

$$Re_L = 3.2 \times 10^5 \qquad Re_{2L} = 6.4 \times 10^5$$

The second case requires that we solve

$$\frac{D_{ab}}{L}\left[0.037Re_L^{4/5} - 871\right]Sc^{1/3} = \frac{D_{ab}}{2L}\left[0.037Re_{2L}^{4/5} - 871\right]Sc^{1/3}$$

Again solving for the Reynolds number we have

$$Re_L = 1.5 \times 10^6 \qquad Re_{2L} = 3.0 \times 10^6$$

Finally, we can ask the question, which orientation is better? Clearly both are the same from a mass transfer point of view, but what is the power required to push the fluid across the surface? Is one orientation better than the other?

For this question, we must calculate the friction factor on the surface. Since the friction factor and mass transfer coefficient are both related to the boundary layer thickness, we do not expect there to be any difference. Transport of all types should be the same unless we have other mechanisms in operation such as a chemical reaction at the surface or within the boundary layer.

For $Re < 5 \times 10^5$, the friction factor is defined as

$$\overline{C_f} = 1.328 Re_L^{-1/2}$$

For $Re > 5 \times 10^5$ and mixed flow conditions, the friction factor becomes

$$\overline{C_f} = 0.074 Re_L^{-1/5} - 1742 Re_L^{-1}$$

Considering the first case, we find

$$\overline{C_{fL}} = 0.00235 \qquad \overline{C_{f2L}} = 0.00238$$

and there is not much of a difference. In the second case,

$$\overline{C_{fL}} = 0.0031 \qquad \overline{C_{f2L}} = 0.0032$$

and again there is not much of a difference, just as we expect.

14.6 SUMMARY

We investigated transport processes under turbulent conditions and showed how turbulent velocity profiles influence the rate of momentum, heat, and mass transfer. Turbulent boundary layers were shown to be composed of three regions: the laminar sublayer dominated by diffusive forces, the transition or generation region where turbulent eddies are formed, and the turbulent core region where transport occurs primarily by the turbulent eddies. By developing some simple models for the turbulent core, we were able to develop a universal velocity profile that was instrumental in correlating the transport coefficients and relating them to the velocity field.

Representations of the momentum, energy, and species continuity equations were developed based on Reynolds averaging procedures. The result of this averaging led to the Reynolds fluxes that are quantities describing the effect of turbulence on the transport.

We considered experimentally determined correlations for the transport coefficients that fully account for the effects of turbulent flow and that are much more reliable than any of the theoretical expressions we developed. Turbulence is still a very active area of research and models describing it are still in their infancy. Interested students should consult the texts and papers referenced in this chapter as guides to start understanding what occurs in this field.

PROBLEMS

Momentum Transfer and Drag

14.1 Water at 350 K is flowing through a pipe at a flow rate of 0.3 kg/s. The pipe is 10 mm in diameter and has roughness (ε) of 0.001:

a. What is the friction factor?

b. Estimate the value of the heat transfer coefficient.

14.2 A fluid flows along a flat, horizontal plate that is heated along its entire length. At a distance, $x = 10$ cm, from the leading edge of the plate, the thermal boundary layer thickness, $\delta_t = 10$ mm, and the local value of the heat transfer coefficient, $h_x = 75$ W/m^2 K. Please determine the numerical values for δ_t at $x = 25$ cm and for the average heat transfer coefficient $\bar{h}$ over the entire length of the plate (25 cm). Assume forced convection, turbulent flow, and Pr > 1.

14.3 Estimate the power required to overcome drag for a car traveling at 100 km/h. Typical car dimensions are 3.5 × 1.7 × 1.5 m, but you may assume the car behaves as a cylinder of diameter 1.7 m and length 3.5 m placed in crossflow.

14.4 The universal velocity profile in the turbulent core of smooth tubes was written as

$$U^+ = 2.5 \ln y^+ + 5.5$$

This relationship holds when the *friction* Reynolds number is unity:

$$\mathrm{Re}_f = \frac{v^* \varepsilon}{\nu} = 1 \qquad \varepsilon - \text{roughness } (m)$$

Experiments have shown that for very rough tubes, the turbulent core velocity profile is better represented by

$$U^+ = 2.5 \ln y^+ + 8.5$$

Following the example in the text, derive a Prandtl/von Karman *universal* resistance law for C_f for rough tubes.

14.5 Consider a rotating disk suspended in an infinite fluid. One remarkable result for laminar flow situations was that the boundary layer thickness and hence transport coefficients were uniform across the surface of the disk. Using the 1/7th power law velocity profile, it has been shown that the boundary layer thickness in turbulent flow is a function of radial position:

$$\delta_h = 0.526 r \left(\frac{\nu}{\omega r^2} \right)^{1/5} \qquad \mathrm{Re} = \frac{\omega r_o^2}{\nu}$$

The analog of the friction factor, the average torque coefficient, $\overline{C_T}$, is defined by

$$\overline{C_T} = \frac{2T}{\frac{1}{2} \rho \omega^2 r_o^5}$$

where r_o is the radius of the disk and ω is the rotation rate. In turbulent flow,

$$\overline{C_T} = 0.146\,\mathrm{Re}^{-1/5} \quad \text{(1/7th power law)}$$

$$\frac{1}{\sqrt{\overline{C_T}}} = 1.97\log_{10}\left(\mathrm{Re}\sqrt{\overline{C_T}}\right) + 0.03 \quad \text{(universal profile)}$$

a. For a disk 10 cm in diameter, immersed in water at 25°C, and spinning at 3000 rpm, what is the boundary layer thickness at the edge of the plate, and what is the total torque required to spin the disk?
b. The shear stress at the surface of the disk is proportional to the boundary layer thickness, that is,

$$\tau_w \approx \rho r\omega^2\delta_h$$

Using the Chilton–Colburn analogy, determine the functional form of a correlation for the heat and mass transfer coefficient in turbulent flow.

14.6 When pumping a fluid, the pressure at the entrance to the pump must never drop below the saturation pressure of the fluid. If the pressure does drop below the saturation pressure, cavitation (the forming of vapor bubbles) occurs that can damage the pump impeller. Consider the system shown in Figure P14.6 constructed of commercial steel pipe. For water at 10°C, determine the maximum possible flow rate without cavitation occurring (in m^3/s). The vapor pressure of water is given by

$$\log_{10} P(\mathrm{mmHg}) = 8.07131 - \frac{1730.63}{233.426 + T(°\mathrm{C})}$$

14.7 A sailboard is gliding across a lake at a speed of 20 mph (9 m/s). The sailboard is 3 m long and 0.75 m wide and represents a smooth, flat surface. Using the physical properties for water listed as follows

$$\rho = 997.5\ \mathrm{kg/m^3} \qquad \mu = 9.8\times10^{-4}\ \mathrm{N\ s/m^2} \qquad C_p = 4179\ \mathrm{J/kg\ K}$$

$$k = 0.604\ \mathrm{W/m\ K} \qquad \mathrm{Pr} = 5.85$$

a. What is the *average* friction factor for the sailboard?

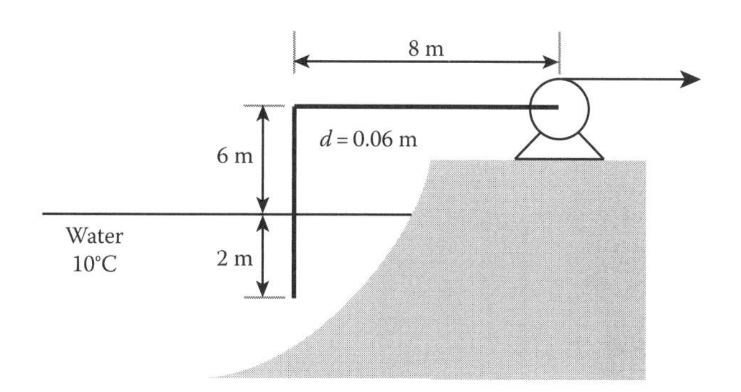

FIGURE P14.6

b. What is the *average* shear stress on the sailboard?
c. How much power must the wind provide to propel the sailboard?

$$\text{(Hint: Power} = \text{Force} \times \text{Distance/Time)}$$

14.8 Golf balls have had dimples in them for well over a century after it was first discovered that dimpled balls could be driven farther. Now we know that the dimples trip up the boundary layer and establish turbulence sooner. As a result, they reduce drag. If the drag coefficient on a golf ball can be adequately represented by

$$C_D = 0.253 + 0.236 \exp\left(\frac{\text{Re}_D - 37430}{6685}\right)$$

compare the drag force on a 3 cm golf ball with that on a similar, but smooth, ping pong ball assuming both are flying through the air at 300 K and 15 m/s.

14.9 Compare predictions of the Prandtl/von Karman universal resistance law and the Heng, Chan, and Churchill/Zajic relation for the friction factor in smooth tube flow, equation 14.84:
a. Plot the friction factor predictions for Reynolds numbers between 10^4 and 10^7.
b. What is the difference in power required to pump 1 m^3/s of water at 25°C through a 40 cm diameter pipe, 1 km long, based on your calculations in part (a)?

14.10 In the late 1940s, it was discovered that the addition of polymers that form random coiled structures in a fluid would reduce the power required to pump that fluid through a tube or to push an object through the fluid. The phenomenon of drag reduction has become very important and is used in the Alaska pipeline. An excellent review of the subject is given by Virk [25]. Drag reduction depends upon the concentration of polymer in the system, but Virk found that the effect asymptotes and that the minimum friction factor is given by

$$\frac{1}{\sqrt{f}} = 19 \log_{10}\left[\text{Re}_d \sqrt{f}\right] - 32.4$$

a. Compare predictions for the friction factor at maximum drag reduction with the Prandtl/von Karman universal resistance law. Plot the difference for Reynolds numbers between 10^4 and 10^7.
b. Compare the power required to flow oil through the Alaska pipeline at a nominal flow rate of 600,000 barrels per day with and without maximum drag reduction. The pipeline is 1287 km long and 122 cm in diameter and the temperature of the oil is 40°C. The specific gravity of the oil is 0.877 and its viscosity is 8.77×10^{-3} N s/m^2.

14.11 Consider a heat exchanger that has 1000 2.5 cm diameter smooth tubes in parallel, each 6 m long. The total water flow of 1 m^3/s at 10°C flows through the tubes. Neglecting entrance and exit losses, determine
a. The pressure drop (in kPa)
b. The pumping power required (in kW)
c. The pumping power for the same flow rate if solid deposits from the water build up on the inner surface of the pipe with a thickness of 1 mm and an equivalent roughness of 0.4 mm

14.12 In the text, we derived a formula for the friction factor based on the universal velocity profile. In essence, it relates $\sqrt{2/C_f} = \bar{U}^+$. The universal velocity profile has the kink at $y^+ = 30$ that makes it a bit aphysical. Though we cannot determine a closed-form, analytical solution, use the van Driest formula for U^+, Equation 14.59, to determine a similar relationship. Plot $\sqrt{2/C_f}$ *vs* y^+. Hint: The van Driest requires numerical integration. Expanding the denominator in a Taylor series about $y^+ = 0$ gives $f(y^+) = 2 + \left(2\kappa_T^2/a^2\right)(y^+)^4$.

Heat and Mass Transfer

14.13 For flow over a flat plate with an extremely rough surface, convection heat transfer effects are known to be correlated by the expression

$$Nu_x = 0.04\,Re_x^{0.9}\,Pr^{1/3}$$

For airflow at 50 m/s, what is the surface shear stress at $x = 1$ m from the leading edge of the plate? Assume air at 300 K.

14.14 Water, initially at 37°C, is flowing through a *cast iron* pipe 43 mm in diameter at a mass flow rate of 10.1 kg/s. The tube is heated by condensing steam (saturated at 1 atm) on its surface:

a. Is the flow laminar or turbulent?
b. Assuming the tube is 15 m long, what is the pressure drop?
c. What is your *best* estimate of the heat transfer coefficient?

14.15 A very thin, lead (Pb), wire of diameter (4×10^{-5} m) and length (1 m) is placed in an air stream (1 atm, $T_\infty = 300$ K) having a flow velocity, v_∞, of 50 m/s. The air is in crossflow over the wire:

$$\text{Melting point of Pb, 600 K} \qquad k_{Pb} = 35.3\ \frac{\text{W}}{\text{mK}}$$

$$\text{Resistance of Pb wire, } 164\ \Omega$$

a. How much heat can be dissipated by the wire? Ignore any temperature gradients in the wire; that is, assume the wire is isothermal.
b. Calculate the maximum electrical current that can be passed through the wire.
c. Show, numerically, whether the assumption of no *radial* temperature gradient in the wire is valid (part (a)).

14.16 An insulated, thin-walled tube, 60 m long and 25 mm inside diameter, is to deliver 65°C hot water to a house. The water enters the tube at 70°C and at a flow rate of 0.25 kg/s. The ambient air temperature in the house is 15°C and the air flows over the tube at a velocity of 15 m/s. The insulation thickness is 25 mm:

a. What are the inside and outside convective heat transfer coefficients, h_i and h_o?
b. What is the magnitude of the overall resistance to heat transfer, 1/UA?
c. What is the rate of heat loss per unit length of tubing at the entrance?
d. Do we have enough to meet the specification? Prove your answer.

14.17 Water flows at 0.20 kg/s through a thick-walled tube. The inside diameter of the tube is 25 mm in diameter and the tube is 5 m long. The outside diameter of the tube is 35 mm and the tube material has a thermal conductivity, $k_w = 2$ W/m K. The water enters at a temperature of 25°C and hot air flows across the tube at a velocity, $v_\infty = 150$ m/s, and a temperature, $T_\infty = 250$°C. What is your estimate of the outlet temperature of the water?

14.18 A spherical, steel, rivet ($r = 5$ mm), initially at a temperature of 250°C, falls from a tall building (100 m) through ambient air (300 K). The rivet's velocity and the heat transfer coefficient are given by

$$v = gt \qquad g\text{—gravitational acceleration} \qquad t\text{—time}$$

$$\overline{Nu_d} = 50(1 + v^{2/3})$$

Assuming the lumped capacitance formulation is a valid approximation for the rivet, what is the temperature of the rivet when it hits the ground 4.5 s later?

14.19 Water at 10°C is in crossflow over the tubes of a shell-and-tube heat exchanger. The mass flow rate is 1 kg/s (6 m/s average). Hot water at 98°C flows through the tubes at a mass flow rate of 0.5 kg/s. The tubes are constructed of stainless steel tubing with a 2.54 cm inside diameter and 3 cm outside diameter, 5 m long. Determine a value for the overall heat transfer coefficient considering convection resistances on the inside and outside of the tube and a conduction resistance through the tube wall.

14.20 A tube of radius $r_0 = 0.2$ m is being dissolved by passing a reactive fluid in the interior. The reactant in the fluid is in great excess, and so we can express the enhancement as a first-order chemical reaction with reaction rate constant, $k'' = 1 \times 10^{-6}$ s^{-1}. The liquid enters the tube with zero concentration of wall material and at a velocity of 0.3 m/s. Liquid density and viscosity barely change throughout the tube, being essentially that of water at 25°C. The solubility of the wall material in the liquid is $c_{aw} = 10$ kg-mol/m^3 and the diffusivity of solute in the liquid is $D_{aw} = 3.5 \times 10^{-10}$ m^2/s:

a. Derive a model describing the average concentration of solute within the liquid as a function of position along the tube.
b. Determine a value for the mass transfer coefficient.
c. What is the outlet concentration of the solute if the tube is 3 m long?

14.21 Repeat the *External and Internal Tube Flow* example using the Churchill/Zajic expression for $\overline{\text{Nu}_d}$. How much does the outlet temperature change compared with the solution in the example?

14.22 Drag reduction is generally done by injecting the polymer into the center of the tube with the fluid. However, then one needs to wait for the polymer to move toward the wall. Injecting through the tube wall should be more efficient. In such circumstances, the friction factor becomes a function of polymer concentration given by

$$\frac{1}{\sqrt{f}} = (4.0 + \delta)\log_{10} \text{Re}_d \sqrt{f} - 0.4 - \delta \log_{10}\left[dW^* \sqrt{2}\right]$$

with

$$R_g W^* = \Omega_L \qquad \frac{\delta}{\sqrt{c}} = 0.67$$

In this situation, we will be using polyethylene oxide (PEO) as the drag reducing agent. The target concentration in parts per million by weight, $c = 300$ wppm. The molecular weight of the PEO is 0.57×10^6 and its radius of gyration, $R_g = 81$ nm. $\Omega_L = 7.1 \times 10^{-3}$. Use the Stokes–Einstein formulation to determine the diffusivity:

a. Based on the targeted, constant wall concentration, what is the friction factor for flow in this system, and how does it compare to the friction factor in the absence of polymer? Assume a 4 cm pipe diameter, a flow rate of 1 kg/s, and a solution having the properties of water at 25°C.
b. If the fluid enters a 10 km pipe, 4 cm in diameter with zero polymer content and at a flow rate of 1 kg/s, what would be the average polymer concentration in the pipe at the exit? Assume the Churchill/Zajic expression for the Sherwood number, Equation 14.104, is appropriate for determining the inside mass transfer coefficient.

REFERENCES

1. Cantwell, B., Coles, D., and Dimotakis, P., Structure and entrainment in the plane of symmetry of a turbulent spot, *J. Fluid Mech.*, **87**, 641 (1978).
2. Hoyt, J.W., Taylor, J.J., and Merzkirch, W., *Flow Visualization—II*, Hemisphere, New York (1982), p. 683.
3. Richardson, L.F., *Weather Prediction by Numerical Process*, Cambridge University Press, Cambridge, U.K. (1922).

4. Reynolds, O., On the dynamical theory of incompressible viscous fluids and the determination of the criterion, *Phil. Trans. Royal Soc. (Lond.)*, **A186**, 123 (1985).
5. Boussinesq, T.V., Théorie de l'écoulement Tourbillonant, *Mem. Pres. Par. Div. Sav. Paris*, **23**, 46 (1877).
6. Prandtl, L., Über die Ausgebildete Turbulenz, *Z. für Angewardte Math. und Mechanik*, **5**, 136 (1925).
7. Von Karman, T. Turbulence and skin friction, *J. Aeronautical Soc.*, **1**, 1 (1934).
8. Van Driest, E.R., On turbulent flow near a wall, *J. Aeronautical Soc.*, **23**, 1007 (1956).
9. Spalding, D.B., A single formula for the "Law of the Wall", *Trans. ASME J. Appl. Mech.*, **28e**, 455 (1961).
10. Heng, L., Chan, C., and Churchill, S.W., Essentially exact characteristics of turbulent convection in a round tube, *Chem. Eng. J.*, **71**, 163–173 (1998).
11. Churchill, S.W. and Zajic, S.C., Prediction of fully developed turbulent convection with minimal explicit empiricism, *AIChE J.*, **48**, 927–940 (2002).
12. Moin, P. and Mahesh, K., Direct numerical simulation: A tool in turbulence research, *Annu. Rev. Fluid Mech.*, **30**, 539–578 (1998).
13. Piomelli, U. and Balaras, E., Wall-layer models for large-eddy simulation, *Annu. Rev. Fluid Mech.*, **34**, 349–374 (2002).
14. Hinze, J.O., *Turbulence*, 2nd edn., McGraw-Hill, New York (1975).
15. Tennekes, H. and Lumley, J.L., *A First Course in Turbulence,* MIT Press, Cambridge, MA (1972).
16. Moody, L.F., Friction factors for pipe flow, *Trans. ASME*, **66**, 671 (1944).
17. Blasius, H., Das Ähnlichkeitsgesetz bei Reibungsvorgängen in Flüssigkeiten, *Forsch. Arb. Ing.-Wes*, **131**, 1 (1913).
18. Colebrook, C.F., Turbulent flow in pipes, with particular reference to the transition region between smooth and rough pipe laws, *J. Inst. Civil Eng. (Lond.)*, **11**, 133–156 (1939).
19. Haaland, S.E., Simple and explicit formulas for the friction factor in turbulent flow, *J. Fluids Eng. (ASME)*, **105**, 89–90 (1983).
20. Taylor, J.B., Carrano, A.L., and Kandlikar, S.G., Characterization of the effect of surface roughness and texture on fluid flow—Past, present, and future, *Int. J. Therm. Sci.*, **45**, 962–968 (2006).
21. Whitaker, S., Forced convection heat transfer correlation for flow in pipes, past flat plates, single cylinders, single spheres, and for flow in packed beds and tube bundles, *AIChE J.* **18**, 361 (1972).
22. Dittus, F.W. and Boelter, L.M.K., Heat transfer in automobile radiators of the tubular type, *Univ. Calif. (Berkeley) Pub. Eng.* **V2**, 443 (1930).
23. Sieder, E.N. and Tate, C.E., Heat transfer and pressure drop of liquids in tubes, *Ind. Eng. Chem.*, **28**, 1427 (1936).
24. Petukhov, B.S., Heat transfer and friction in turbulent pipe flow with variable physical properties, in *Advances in Heat Transfer*, J.P. Hartnett and Irvine, T.F., eds., Academic Press, New York, pp. 504–564 (1970).
25. Virk, P.S., Drag reduction fundamentals, *AIChE J.* **21**, 625–656 (1975).

15 Radiative Transport

15.1 INTRODUCTION

Radiation is one of the three major transport mechanisms. The transport of energy by radiation provides virtually all the energy needed to sustain life on this planet. It drives the weather, feeds the plants we eat, distills the water we drink, provides illumination, and even fuels the vibrations we hear as sound. Eventually, radiation will supplant fossil fuels as our primary energy source whether we choose to harvest it via photovoltaic, solar thermal, wind, hydroelectric, or tidal power. Even the fossil fuels we rely on now represent a form of geological radiation storage.

As engineers, we will be concerned primarily with energy transport via radiation but we can also transfer momentum to a lesser extent [1,2]. The latter is the force behind laser trapping of atoms, laser tweezers for manipulating cells or colloids, and solar sails for powering spacecraft. Our focus will primarily be on heat transfer by radiation, but it is important to realize that electromagnetic energy and potential energy (gravity) are also transported via radiation. In fact, we cannot divorce heat transfer from the transfer of electromagnetic energy since the heat we feel transported by radiation is merely a manifestation of the interaction of that electromagnetic radiation with matter. In this chapter, we will discuss heat transport via radiation from a fundamental viewpoint and then from a more useful, engineering approach.

15.2 PRELIMINARY DEFINITIONS

Heat transferred by radiation is referred to as thermal radiation. Thermal radiation is just electromagnetic radiation in the wavelength range of [3]

$$10^{-7}\ (\mathrm{m}) < \lambda < 10^{-4}\ (\mathrm{m}) \tag{15.1}$$

Thermal radiation is often called infrared (IR) radiation, but the wavelength range for what we normally refer to as thermal radiation encompasses the visible (VIS) and part of the ultraviolet (UV) regions of the spectrum too (Figure 15.1). In a vacuum, thermal radiation, like other forms of electromagnetic radiation, travels as a wave at the speed of light, $c \approx 3 \times 10^8$ m/s. Thus, transport via radiation is conservative. This idea makes sense because there is no way that energy could be transported through the vast distances of the universe if there were even the slightest amount of dissipation. The wavelength and frequency of the radiation are related to one another by [3]

$$c = \lambda\omega \tag{15.2}$$

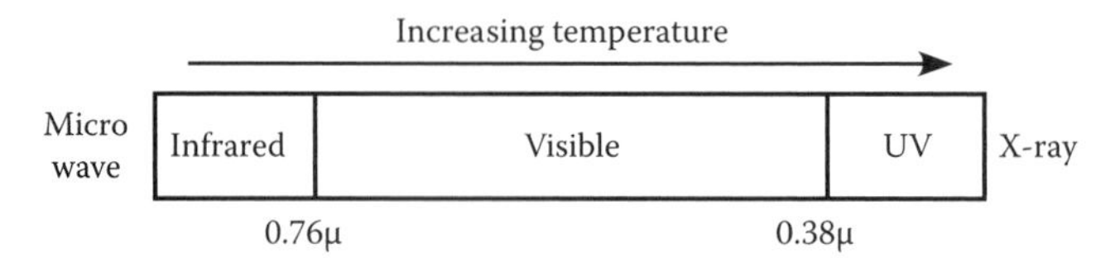

FIGURE 15.1 Color of a body as a function of temperature.

When an object is heated, some of its constituent molecules or atoms are raised to excited states. The heating can be by fire, bombardment by other particles (nuclear decay), radiation (microwaves, radio waves, light, etc.), or electrical excitation. The excited molecules or atoms want to return to their ground state and do so by releasing energy in the form of electromagnetic waves. The release of energy is called *emission* and it comes in discrete packets called photons [3]. The energy of a photon is related to its frequency or, alternatively, its wavelength

$$E_p = \hbar\omega = \frac{\hbar c}{\lambda} \qquad \hbar = \text{Planck's constant} = 6.626 \times 10^{-34}\ \text{J s} \tag{15.3}$$

and increases as the wavelength of the radiation decreases. Recalling thermodynamics, we also expect the energy to increase as the temperature increases. *Therefore, as the temperature of an emitting body increases, the wavelength of the emitted radiation should decrease. We should observe a color change when we change the temperature.*

The energy transferred to a substance per photon is determined by the amount of energy in each photon and by the absorptive properties of the substance:

$$E_p = \frac{\hbar c}{\eta^c \lambda} \qquad \eta^c\text{—complex refractive index of the solid} \tag{15.4}$$

Each electron, atom, or molecule, within the substance, acts as an oscillator, absorbing energy at a specific resonant frequency. In a typical substance, even a pure one, there are many types of oscillators leading to a characteristic absorption spectrum (Figure 15.2). The absorbance, as a function of the wavelength of the incident radiation, can be described using a complex representation for the refractive index:

$$\eta^c(\omega) = \eta_{re}(\omega) + i\eta_{im}(\omega) \tag{15.5}$$

The imaginary component of the refractive index, η_{im}, is related to the absorbance of the material. The absorption of VIS and IR radiation affects molecular bonds causing an increase in the

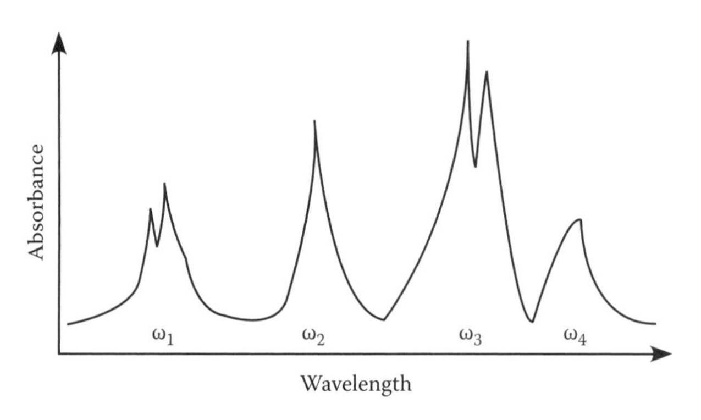

FIGURE 15.2 Absorption spectrum of a compound showing the wavelength dependence.

frequency and amplitude of their vibration. The relaxation of these bonds from their excited state, back to rest, generates heat. Absorption in the UV region may actually break bonds and strip off electrons because each photon contains much more energy and can excite bonds and electrons to vibrate at amplitudes that drive them from their normal bounds. In general, as the wavelength of incident radiation decreases, the absorption grows stronger and the radiation penetrates less into the substance. If we take your body, for example, absorbing UV radiation damages your skin and causes sunburn. Microwave radiation would pass through your skin (little water) to heat internal water molecules and cause damage deep inside your body. Finally, radio waves would pass through your body virtually untouched. UV radiation is ideal for sterilization, and radio waves peer through the sands of the Sahara to reveal ancient rivers or the clouds of Venus to reveal the surface topography. Any radiation that is absorbed is eventually dissipated as heat, and so all of the electromagnetic spectrum can be viewed in that sense as thermal radiation. However, the sources outside of what we normally refer to as thermal radiation represent a very minor component.

15.3 MAXWELL'S EQUATIONS AND HEAT TRANSFER

Radiative heat transfer can be described by Maxwell's equations [3]. The interaction of the radiation with matter is controlled by three physical properties of the material: its conductivity, its permittivity, and its permeability. Most of the interaction is governed by the first two of the three. We can see how thermal energy is exchanged via radiation by considering two cases. The first case will consider the transport of a plane wave in a conducting medium and the second case will consider the transport of a plane wave through an absorbing, dielectric medium. We will show that the two cases are equivalent in that they both lead to the generation of heat within the medium.

Maxwell's equations for a conducting medium are as follows [4]:

$$\vec{\nabla}\cdot\vec{\mathbf{D}}_{\mathbf{e}} = 0 \tag{15.6}$$

$$\vec{\nabla}\cdot\vec{\mathbf{B}}_{\mathbf{h}} = 0 \tag{15.7}$$

$$\vec{\nabla}\times\vec{\mathcal{E}} = -\frac{\partial\vec{\mathbf{B}}_{\mathbf{h}}}{\partial t} \tag{15.8}$$

$$\vec{\nabla}\times\vec{\mathcal{H}} = \vec{\mathcal{J}} + \frac{\partial\vec{\mathbf{D}}_{\mathbf{e}}}{\partial t} \tag{15.9}$$

with constitutive relations

$$\vec{\mathcal{J}} = \sigma\vec{\mathcal{E}} \qquad \vec{\mathbf{D}}_{\mathbf{e}} = \varepsilon\vec{\mathcal{E}} \qquad \vec{\mathbf{B}}_{\mathbf{h}} = \mu_h\vec{\mathcal{H}} \tag{15.10}$$

For simplicity, let's assume that the physical properties of the material—the conductivity, σ; the permittivity, ε; and the permeability, μ_h—are real and constant. We can generate a wave equation for the electric field, $\vec{\mathcal{E}}$, and the magnetic field, $\vec{\mathcal{H}}$, by taking the curl ($\vec{\nabla}\times$) of Equations 15.8 and 15.9 and then substituting the constitutive Equations 15.10 in for $\vec{\mathcal{J}}$, $\vec{\mathbf{D}}_e$, and $\vec{\mathbf{B}}_h$. For the conducting medium, we obtain the following *telegrapher's* equations:

$$\nabla^2\vec{\mathcal{E}} = \mu_h\sigma\frac{\partial\vec{\mathcal{E}}}{\partial t} + \mu_h\varepsilon\frac{\partial^2\vec{\mathcal{E}}}{\partial t^2} \tag{15.11}$$

$$\nabla^2\vec{\mathcal{H}} = \mu_h\sigma\frac{\partial\vec{\mathcal{H}}}{\partial t} + \mu_h\varepsilon\frac{\partial^2\vec{\mathcal{H}}}{\partial t^2} \tag{15.12}$$

If the material is nonconducting, $\sigma = 0$, and the first-order time derivative disappears. If the material is absorbing, then the dielectric constant, ε, is no longer a real number. It is a complex quantity that we write as

$$\varepsilon = \varepsilon_{re} + i\varepsilon_{im} \tag{15.13}$$

The wave equations for the absorbing dielectric can be written to accentuate the imaginary contribution of the index of refraction:

$$\nabla^2 \vec{\mathcal{E}} = \mu_h \varepsilon_{re} \frac{\partial^2 \vec{\mathcal{E}}}{\partial t^2} + i\mu_h \varepsilon_{im} \frac{\partial^2 \vec{\mathcal{E}}}{\partial t^2} \tag{15.14}$$

$$\nabla^2 \vec{\mathcal{H}} = \mu_h \varepsilon_{re} \frac{\partial^2 \vec{\mathcal{H}}}{\partial t^2} + i\mu_h \varepsilon_{im} \frac{\partial^2 \vec{\mathcal{H}}}{\partial t^2} \tag{15.15}$$

Plane waves are generally represented using complex exponential notation. This takes advantage of our knowledge of how the waves propagate in space and time and allows us to focus on how the amplitude and phase vary upon propagation and the interaction with matter:

$$\vec{\mathcal{E}} = E(\vec{\mathbf{r}})\exp[i(\omega t - \vec{\mathbf{k}} \cdot \vec{\mathbf{r}} + \phi)] \tag{15.16}$$

$$\vec{\mathcal{H}} = H(\vec{\mathbf{r}})\exp[i(\omega t - \vec{\mathbf{k}} \cdot \vec{\mathbf{r}} + \phi)] \tag{15.17}$$

where

$E(\vec{\mathbf{r}})$ and $H(\vec{\mathbf{r}})$ are amplitude functions that depend upon the position vector, $\vec{\mathbf{r}}$
ω is the frequency of the radiation
ϕ is an arbitrary phase shift
$\vec{\mathbf{k}}$ is the wave vector whose magnitude is defined by

$$\left|\vec{\mathbf{k}}\right| = \omega\sqrt{\mu_h \varepsilon} = \frac{\left|\eta^c\right|\omega}{c} \tag{15.18}$$

Substituting Equation 15.16 into Equation 15.11 and Equation 15.14 yields two differential equations describing how the amplitude of the electric field vector varies throughout space:

$$\nabla^2 E + \omega^2 \mu_h \left(\varepsilon - i\frac{\sigma}{\omega}\right) E = 0 \quad \text{Conducting medium} \tag{15.19}$$

$$\nabla^2 E + \omega^2 \mu_h (\varepsilon_{re} - i\varepsilon_{im}) E = 0 \quad \text{Absorbing medium} \tag{15.20}$$

Notice that the absorbing medium behaves as if it has some finite conductivity. We can conclude that the complex component of the refractive index serves the same function as the conductivity; both will lead to losses in energy that end up as heat production. Formally, we can associate complex dielectric constants with conducting media such as metals, and as can be seen, the complex dielectric constant has a frequency dependence associated with it [4]. ε_{im} may have any number of poles at various frequencies ω_i:

$$\varepsilon_{im} = \frac{\sigma}{\prod_i (\omega - \omega_i)} \tag{15.21}$$

In general, we do not deal with complex dielectric constants, but instead we speak of a complex index of refraction written as

$$\eta^c = \eta_{re} + i\eta_{im} = \eta_{re}(1 - i\kappa) \tag{15.22}$$

where

κ is called the extinction coefficient

$\eta_{re}\kappa$ is called the absorption coefficient

The magnitude of the wave vector $\vec{\mathbf{k}}$ can be rewritten as

$$\left|\vec{\mathbf{k}}\right| = \left|\eta^c \eta^{c*}\right| \frac{\omega}{c} = \left(\frac{\omega \eta_{re}}{c}\right)\sqrt{1+\kappa^2} \tag{15.23}$$

where $\eta^{c*} = \eta_{re} - i\eta_{im}$ and is the complex conjugate of η^c. We can write the wave vector as a complex quantity $\vec{\mathbf{k}} = \vec{\mathbf{k}}_{re} + i\vec{\mathbf{k}}_{im}$ as well and can see the effect of this complex wave vector by substituting the complex representation for $\vec{\mathbf{k}}$ into Equation 15.16, the definition of the plane wave:

$$\vec{\mathcal{E}} = E(\vec{\mathbf{r}})\exp[i(\omega t - \vec{\mathbf{k}}_{\mathbf{re}} \cdot \vec{\mathbf{r}} + \phi)]\exp[-\vec{\mathbf{k}}_{\mathbf{im}} \cdot \vec{\mathbf{r}}] \tag{15.24}$$

We recover a decaying exponential term, $\exp[-\vec{\mathbf{k}}_{\mathbf{im}} \cdot \vec{\mathbf{r}}]$, that shows that the effect of the complex dielectric constant or finite conductivity of the material is to reduce the amplitude of the wave as it moves through the medium.

The material can do a number of things with this absorbed energy. If it is a conductor with finite conductivity, the absorbed radiation becomes a current that gets dissipated through Joule heating of the material. If the material is a dielectric, part of the absorbed energy is usually reradiated either in the VIS region of the spectrum as fluorescence or in the IR region of the spectrum at a wavelength characteristic of the temperature of the body. The portion of the absorbed radiation not reradiated is converted to heat via dissipation. We can see this a bit more easily if we consider a 1-D case with the wave parallel to the z-axis. Replacing $\vec{\mathbf{k}}$ by its magnitude, Equation 15.23, we find that the electric field can be represented by

$$E = \underbrace{E_o \exp\left[-\left(\frac{\omega \eta_{re}\kappa}{c}\right)z\right]}_{\text{Absorption loss}} \exp\left[i\left(\omega t - \left(\frac{\omega \eta_{re}}{c}\right)z\right)\right] \tag{15.25}$$

The first exponential term describes the absorption loss, while the second term describes the wave-like propagation. The depth, $z = d_s$, where $z = c/(\omega\eta_{re}\kappa)$ is called the *skin depth*. Table 15.1 shows how the skin depth varies with the wavelength of the incident radiation for copper and for dry Earth. The relatively large skin depth for dry Earth is what enables radar imaging to see the rivers that one flowed in the Sahara. Those rivers are now buried beneath many meters of sand dune.

TABLE 15.1
Skin Depth as a Function of Wavelength

Wavelength λ_o (m)	Copper d_s (m)	Dry Earth d_s (m)
10^{-7}	6.2×10^{-10}	1.5×10^{-3}
10^{-5}	6.2×10^{-9}	1.5×10^{-2}
10^{-1}	6.2×10^{-7}	1.5
10^{3}	6.2×10^{-5}	1.5×10^{2}

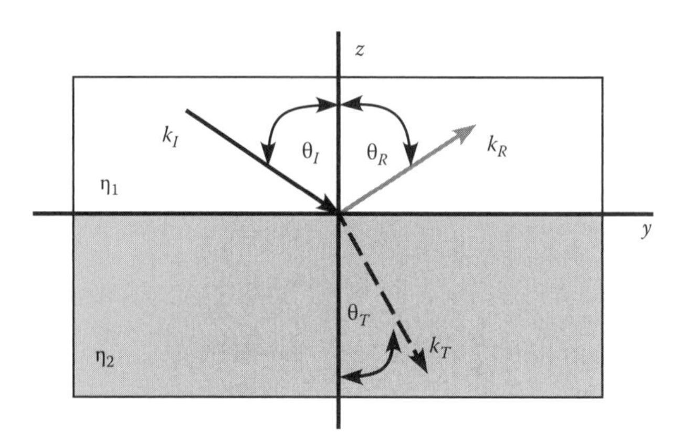

FIGURE 15.3 Electromagnetic radiation incident on the surface of an absorbing medium.

We can use the description for how the radiation propagates through conducting or absorbing media to look at what happens when a plane wave is incident upon a medium that has a complex index of refraction. In Figure 15.3, region 1 is a dielectric medium having a real index of refraction η_{re1} and region 2 contains a material that has a complex index of refraction, η_{re2}.

In Chapter 7, we analyzed reflection and refraction at a plane interface and derived expressions that told us how much of the incident energy was reflected from the interface and how much was transmitted through the interface. The presence of an absorbing medium does not influence the results we obtained. In fact, we can use all the equations for reflectivity and transmittivity we derived if we replace the real index of refraction with the complex index of refraction. Thus, we have for normal incidence

$$\eta^c = \frac{\eta_1^c}{\eta_2^c} \tag{15.26}$$

$$\mathcal{R}_e = \frac{(\eta^c - 1)\left(\eta^{c*} - 1\right)}{(\eta^c + 1)\left(\eta^{c*} + 1\right)} = \frac{(\eta_{re} - 1 + i\eta_{re}\kappa)(\eta_{re} - 1 - i\eta_{re}\kappa)}{(\eta_{re} + 1 + i\eta_{re}\kappa)(\eta_{re} + 1 - i\eta_{re}\kappa)}$$

$$= \frac{(\eta_{re} - 1)^2 + (\eta_{re}\kappa)^2}{(\eta_{re} + 1)^2 + (\eta_{re}\kappa)^2} = 1 - \frac{4\eta_{re}}{(\eta_{re} + 1)^2 + (\eta_{re}\kappa)^2} \tag{15.27}$$

The effect of absorption is to increase the reflectivity over a material having the same refractive index, η_{re}, and an extinction coefficient of 0. Notice that if the index of refraction were purely imaginary, $\eta^c = i\eta_{re}\kappa$, then the reflectivity $\mathcal{R}_e = 1$ and the material would be a perfect reflector. That is why good conductors that have large imaginary components to their refractive indices are also such good reflectors.

We could solve Maxwell's equations and obtain solutions for radiative heat transfer problems from first principles knowing only the geometry of the system and the optical, electronic, and physical properties of the materials. This would be an extraordinarily tedious and cumbersome process for any real system and often the required physical properties are unknown. Since we cannot solve the problem in its entirety, we must make simplifying assumptions. This is the engineering approach for dealing with radiation problems.

15.4 ENERGY FLUXES IN RADIATIVE SYSTEMS [5–7]

Before we can begin to discuss radiative heat transfer, we must first become familiar with the definitions and jargon commonly used to describe it. Fortunately, many of these terms are used by other disciplines more familiar to the layman (photography, astronomy, lighting contractors), and so they are easily recognized.

15.4.1 Incident Radiation: Irradiation

Consider the simple, planar solid surface shown in Figure 15.4. Radiation is incident on its surface. We refer to the total flux of radiant energy striking the surface as the irradiation, $G^r(\lambda)$ (W/m^2-m). The irradiation is only a function of wavelength since it represents an average flux of radiation over the surface. Often the irradiation on a surface is too crude a measure of how the surface interacts with the source of the radiation. For example, the amount of radiation incident on a surface may vary widely from point to point or the illuminating object may not radiate uniformly. To account for these differences and more accurately model radiative transport to all parts of a surface, we define a new function, the intensity of the incident radiation, $I_I(\lambda, \theta, \phi)$. The intensity is a function of direction and wavelength. The irradiation is related to the intensity via the following integral:

$$G^r(\lambda) = \int_0^{2\pi}\int_0^{\pi/2} I_I(\lambda,\theta,\phi)\cos\theta\sin\theta\, d\theta\, d\phi \tag{15.28}$$

The total irradiation on a surface is determined by integrating the irradiation over all wavelengths:

$$G^r = \int_0^{\infty} G^r(\lambda)d\lambda \quad \text{Total irradiation (W/m}^2\text{)} \tag{15.29}$$

Calculating the irradiation from the intensity is an exercise in solid geometry that can best be viewed using Figure 15.5. To visualize the summation, we represent the intensity as rays of electromagnetic

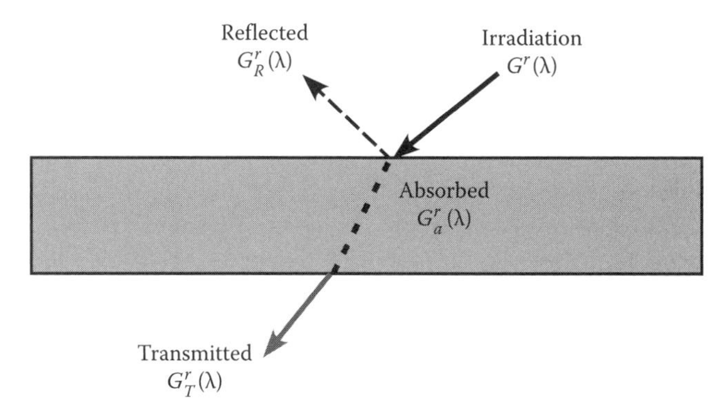

FIGURE 15.4 Light incident on a real object.

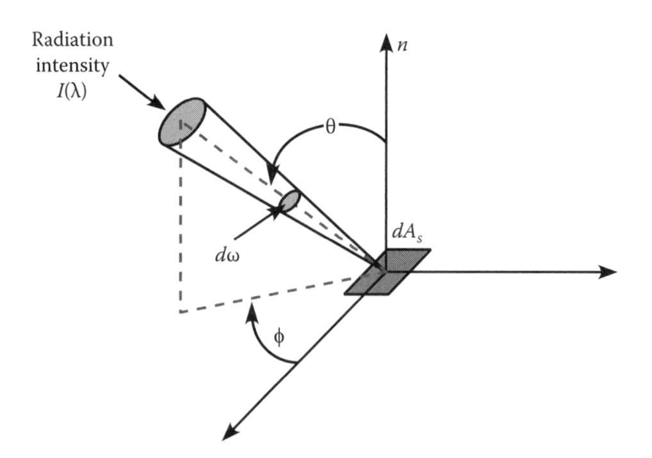

FIGURE 15.5 Directional radiation incident on a surface.

radiation. Equation 15.28 represents the summation of all possible rays emitted from the radiation source that strike the surface of area, dA_s. Figure 15.5 shows a cone or pencil of those rays, subtending a solid angle, $d\omega$, that is incident on the center point of the surface.

If the intensity of radiation is termed *diffuse*, then the intensity is not a function of position and the irradiation and intensity are related via

$$G^r(\lambda) = \pi I_l(\lambda) \tag{15.30}$$

Once the irradiation is determined, it has three paths to follow. The radiation can be transmitted, reflected, or absorbed, all simultaneously. We define the following quantities as measures of these processes:

$$G^r(\lambda) = \underbrace{\rho^r(\lambda)G^r(\lambda)}_{\text{Reflected}} + \underbrace{\alpha^r(\lambda)G^r(\lambda)}_{\text{Absorbed}} + \underbrace{\tau^r(\lambda)G^r(\lambda)}_{\text{Transmitted}} \tag{15.31}$$

where

$\rho^r(\lambda)$ represents the *reflectivity*
$\alpha^r(\lambda)$ represents the *absorptivity*
$\tau^r(\lambda)$ represents the *transmittivity* of the material making up the surface

These quantities are generally functions of wavelength and may be functions of position on a surface. If they are independent of wavelength and position, we refer to them as *total, hemispherical* properties to signify that they are averages over the entire surface. The reflectivity, transmittivity, and absorptivity must sum to one to insure conservation of energy:

$$\rho^r + \alpha^r + \tau^r = 1 \qquad 0 \le \rho^r, \alpha^r, \tau^r \le 1 \tag{15.32}$$

We can obtain values for the reflectivity, absorptivity, and transmittivity of the material by solving Maxwell's equations for a given wavelength using the procedures outlined in the last section. Since the solution will give values as a function of wavelength and position, we then average over all possible incident angles and all wavelengths of interest to get the total hemispherical properties. Fortunately, it is relatively easy to measure reflectivities and absorptivities using optical instruments such as ellipsometers and spectrophotometers, so most data are tabulated. The solution to Maxwell's equations also presumes we know the exact composition of the material and the topography of the surface, a tall order. Some special cases are shown in Table 15.2.

TABLE 15.2
Representative Optical Properties of Common Materials

Material	Property	Comment
Gases, glass	$\rho^r = 0$	Glass – $\rho^r \approx 0.04$
Polished metals	$\alpha^r, \tau^r = 0$	$\rho^r \geq 0.98$
Optically opaque solids	$\tau^r, \rho^r = 0$	$\alpha^r \approx 1.0$
Blackbody (ideal model)	$\alpha^r = 1$	Absorbs all

Example 15.1 Solar Irradiation at the Earth's Surface

Solar radiation incident on the Earth's surface may be divided into two components. The direct component consists of parallel rays incident at a fixed angle corresponding to the angle of the Sun normal to the surface. The diffuse component comes from radiation scattered by the atmosphere and is uniform no matter the angle of the Sun (Figure 15.6).

Let's consider a clear sky where the direct radiation is incident at an angle of 45° with a total flux of $q''_{dir} = 1000\ \text{W/m}^2$. The total intensity of the diffuse radiation is $I_{diff} = 70\ \text{W/m}^2 \cdot \text{sr}$. We would like to calculate the total irradiation on the Earth's surface.

The total irradiation is the sum of the diffuse and direct components:

$$G^r = G^r_{dir} + G^r_{diff}$$

The diffuse component is equal to the intensity of the diffuse radiation multiplied by the solid angle of the exposed surface, in this case π, representing the entire surroundings above the surface:

$$G^r_{diff} = I_{diff}\pi = 70\pi\ \text{W/m}^2$$

To obtain the direct component, we need the vertical component of the direct radiation. This is the vertical projection of the direct radiation onto the surface:

$$G^r_{dir} = q''_{dir}\cos(\theta) = q''_{dir}\sin(\pi/2 - \theta) = 500\sqrt{2}\ \text{W/m}^2$$

zenith angle   horizon angle

$$G^r = 70\pi + 500\sqrt{2} = 927.02\ \text{W/m}^2$$

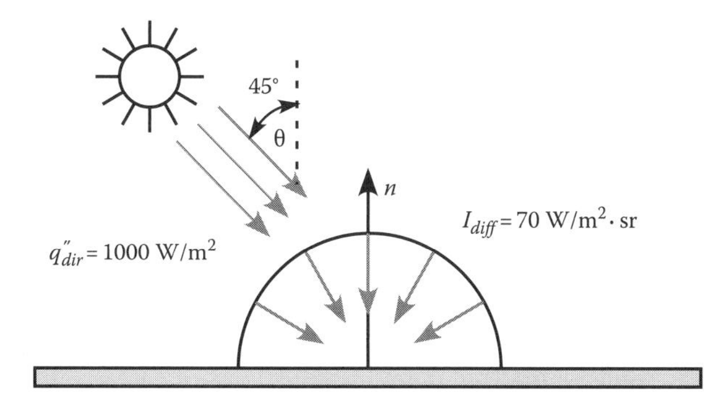

FIGURE 15.6 Diffuse and direct solar radiation incident on the Earth's surface.

15.4.2 Emitted Radiation: Radiosity

All objects have some form of radiation leaving them. Either they emit some as a function of their body temperature, or they reflect or transmit some. We refer to the radiation leaving an object as its *radiosity*, J^r (W/m^2). The radiosity is composed of two parts: radiation emitted from the object as a result of its temperature and all radiation not absorbed by the object:

$$\underset{\substack{\text{Emissive power}\\ \text{due to radiation at}\\ \text{body's temperature}}}{J^r(\lambda) = E^r(\lambda)} + \underset{\substack{\text{Non absorbed}\\ \text{radiation}\\ (\text{if } \tau^r = 0,\ 1-\alpha^r = \rho^r)}}{[1-\alpha^r(\lambda)]G^r(\lambda)} \tag{15.33}$$

The *emissive power*, E^r (W/m^2), is the amount of radiation a body emits due to its temperature. At any temperature above absolute zero, some of the atoms of a substance are promoted to an excited state by thermal motion. They then relax back down to their ground state by emitting radiation. This radiation is the emitted power. The higher the temperature, the more power emitted and the shorter the average wavelength of emission. The radiosity and emissive power are generally both functions of wavelength. To obtain the total radiosity or total emissive power, we must integrate over wavelength, like we did in Equation 15.29.

For an ideal absorber, one that absorbs everything incident upon it, $\alpha^r = 1.0$, and so

$$J^r(\lambda) = E^r(\lambda) \tag{15.34}$$

We can define an intensity of emitted radiation, I_E, and use equations similar to Equations 15.28 and 15.29 to define the radiosity:

$$J^r(\lambda) = \int_0^{2\pi}\int_0^{\pi/2} I_E(\lambda,\theta,\phi)\cos\theta\sin\theta\, d\theta\, d\phi \tag{15.35}$$

$$J^r = \int_0^{\infty} J^r(\lambda)d\lambda = \int_0^{\infty} E^r(\lambda)d\lambda + \int_0^{\infty}[1-\alpha^r(\lambda)]G(\lambda)d\lambda \tag{15.36}$$

15.5 BLACKBODY [6,8]

We cannot measure the dielectric and conductive properties of all materials for all wavelengths nor can we easily solve Maxwell's equations to define the radiative exchange, so we develop various idealizations that make the calculations simpler. The most restrictive idealization is the *blackbody*. The blackbody is a model of a perfect, total absorber. There are no real blackbodies in existence although blackbody absorptive performance can be approached in the laboratory and in nature over discrete regions of the electromagnetic spectrum. Some examples of blackbody-like absorbers are as follows:

1. Carbon nanotube forest—VIS region absorber.
2. Stack of razor blades—IR/VIS region absorber when viewed edge on.
3. Black hole—absorbs all of EM spectrum.
4. Sphere with a pinhole—absorbs UV/VIS/IR. Radiation cannot escape.

The nanotube forest, the hollow sphere, and the razor blade stack work by providing so many internal reflections that any radiation that enters eventually gets absorbed. The hollow sphere (cavity radiator) is used as a radiation power meter for measuring the emitted radiation from many objects such as laser beams and light-emitting diodes. The shell of the sphere is filled with fluid, and the entering beam is trapped within the sphere giving up its energy to raise the temperature of the fluid. The power of the beam is thus related to the temperature rise. The black hole traps radiation by gravitation. Nanotube forests and other similar coatings such as lamp black are used in optical devices like cameras and for camouflage.

A blackbody is not only a perfect absorber, but it also emits more energy at a given temperature and wavelength than any other surface. We specify the blackbody as a diffuse emitter so that the emitted power is independent of the direction from which the body is observed. Examples of blackbody emitters are as follows:

1. Tungsten filament—blackbody at 2800 K
2. The Sun—blackbody at 5800 K
3. Black holes—blackbody at x/gamma ray wavelengths
4. Cavity radiator—blackbody in IR and VIS (same as the sphere with a pinhole)

We can use the results from Section 15.3 to get some optical information about a blackbody. Suppose we have a planar blackbody interface and consider radiation at normal incidence only. We know that the reflectivity of that interface must be zero because the absorptivity of the blackbody must be 1.0. If we set $\mathcal{R}_e = 0$ in Equation 15.27, we find

$$1 = \frac{4\eta_{re}}{(\eta_{re}+1)^2 + (\eta_{re}\kappa)^2} \tag{15.37}$$

Solving gives a relationship between the refractive index and the extinction coefficient:

$$\eta_{re} = \frac{1}{1+\kappa^2} \tag{15.38}$$

A blackbody has an extinction coefficient of infinity, so Equation 15.38 really means the magnitude of its refractive index must approach zero. This indicates that there is no practical way of discovering a blackbody material based on its optical properties alone or of seeing one optically. Some other mechanism for trapping the radiation must be present. In the vacuum of space, a black hole uses gravity to warp the space–time, confine the incident radiation, and so render an effective refractive index of zero.

15.5.1 Emissive Power of a Blackbody

The blackbody is the simplest emitter of radiation envisioned, so it is important to be able to accurately determine how much power is emitted as a function of temperature and wavelength. The breakdown of the classical theory for predicting this power was one of the main driving forces that lead to the development of the quantum theory. At the turn of the twentieth century, Rayleigh and Jeans used classical electromagnetic theory to calculate the energy density of a blackbody formed using the pinhole and hollow sphere approach discussed in the previous section. Inside the blackbody, they assumed a collection of standing electromagnetic waves. The number of waves present within the blackbody was proportional to the frequency of the waves. The higher the frequency, the more standing waves could fit within a given space. The theory assumed a Boltzmann distribution of energies for the waves so that the probability of waves having an energy E_p was solely dependent on the temperature:

$$\mathcal{P}(E_p) = \frac{\exp(-(E_p/k_bT))}{k_bT} \tag{15.39}$$

The average energy of each wave was found by using the definition of the mean:

$$\overline{E_p} = \frac{\int_0^\infty E_p\mathcal{P}(E_p)\,dE_p}{\int_0^\infty \mathcal{P}(E_p)\,dE_p} = k_bT \tag{15.40}$$

This result is called the *law of equipartition of energies* and states that each wave in the enclosure has the same average energy. Note that this theory neglects the fact that radiation at a higher frequency has a higher energy. Using the equipartition law, the *Rayleigh–Jeans* formula for blackbody radiation was developed:

$$\rho_\lambda(\lambda,T)\,d\lambda = \left(\frac{8\pi ck_bT}{\lambda^4}\right)d\lambda \quad \text{Rayleigh–Jeans cavity radiator} \tag{15.41}$$

where

ρ_λ represents the energy density in J/m³/m
c is the speed of light
k_b is Boltzmann's constant

Notice that if we integrate the Rayleigh–Jeans law over all wavelengths, we predict an infinite energy density regardless of the temperature. This is not observed experimentally and is not allowed by thermodynamics. The experimental observations indicate that $\rho_\lambda \to 0$ when $\lambda \to 0$. This paradox between theory and experiment is termed the *UV catastrophe* since Equation 15.41 predicts $\rho_\lambda \to \infty$ when $\lambda \to 0$. To correct this problem, we need some way to cut off the energy contribution from the short wavelength end of the spectrum so that those waves do not contribute to the energy spectrum. The number of waves of short wavelength has to decrease faster than their energy increases.

Max Planck found he could provide such a cutoff if he postulated that the equipartition law was violated. Instead of allowing a continuous spectrum of energies, Planck assumed that the energy must come in discrete intervals and that those intervals were linearly related to the wavelength of the radiation. In essence, what Planck was saying was that instead of having the number of standing waves in the box proportional to their frequency, only standing waves of certain frequencies could exist in the box. Thus, E_p was discrete and given by

$$E_p = 0, \Delta E,\ 2\Delta E,\ 3\Delta E\ldots \qquad \Delta E = \frac{hc}{\lambda} \tag{15.42}$$

In this discretization process, all energies (frequencies) are not represented and so we cannot use the definition in Equation 15.40 to define the mean energy. The evaluation of the mean energy must be calculated using a discrete sum in steps of ΔE:

$$\overline{E_p} = \frac{\sum_{n=0}^\infty E_p\mathcal{P}(E_p)}{\sum_{n=0}^\infty \mathcal{P}(E_p)} = \frac{\sum_{n=0}^\infty n\Delta E\mathcal{P}(n\Delta E)}{\sum_{n=0}^\infty \mathcal{P}(n\Delta E)} = \frac{\hbar c/\lambda}{\exp(\hbar c/\lambda k_bT)-1} \tag{15.43}$$

The result of this sum is that when $\Delta E \ll k_bT$, we recover the classical law of equipartition of energies. When $\Delta E > k_bT$, the average energy is much less than k_bT because there are no waves with energies below k_bT represented in the sum and those waves with energies above k_bT are in short supply. Since a large ΔE occurs at short wavelengths, we throw away a large part of the area under the curve and so have the cutoff we need to avoid overrepresenting the contribution of high-energy waves that leads to the UV catastrophe.

Retracing Rayleigh's and Jeans' calculations with the new postulates led Planck to the following energy density distribution [8]:

$$\rho_\lambda(T)\,d\lambda = \frac{8\pi c\hbar}{\lambda^5[\exp(\hbar c/\lambda k_bT)-1]}\,d\lambda \quad \left(\frac{\text{J}}{\text{m}^4}\right) \tag{15.44}$$

Notice that now as the wavelength approaches zero, the energy density also approaches zero ($\exp(1/\lambda) \gg \lambda^5$) and so the total energy of the system is always finite. The classical energy density and the Planck density are shown in Figure 15.7.

The *Planck distribution*, $\rho_\lambda(T)$, satisfies a number of criteria:

$$\begin{array}{ll}
(1)\ \lambda \to 0 & \rho_\lambda(\lambda,T) \to 0 \\
(2)\ \lambda \to \infty & \rho_\lambda(\lambda,T) \to 0 \\
(3)\ T \to 0 & \rho_\lambda(\lambda,T) \to 0 \\
(4)\ T \to \infty & \rho_\lambda(\lambda,T) \to \infty
\end{array}$$

The first criterion states that no material at a finite temperature can emit an infinite amount of energy. The second states that waves with infinite wavelength contain no energy. The third states that as the temperature approaches absolute zero, all molecules and atoms are in their ground state and so cannot radiate any energy. Finally, the last criterion states that as the temperature approaches infinity, all the atoms and molecules are in excited states and must be radiating a tremendous amount of energy as they cycle back to their ground state.

To determine the energy flux out of the hollow sphere blackbody at a given temperature, we multiply the energy density function, Equation 15.44, by the velocity, $c/4$. This factor accounts for the average velocity of the photons out of the cavity ($c/2$) and the fact that only half the total energy density is attributable to photons moving in a particular direction out of the cavity. We then integrate over all wavelengths:

$$E_{bb}^r = \int_0^\infty \frac{2\pi c^2\hbar}{\lambda^5\left[\exp\left(\dfrac{\hbar c}{\lambda k_b T}\right)-1\right]}\,d\lambda = \left(\frac{2}{15}\frac{\pi^5 k_b^4}{c^2 h^3}\right)T^4 \tag{15.45}$$

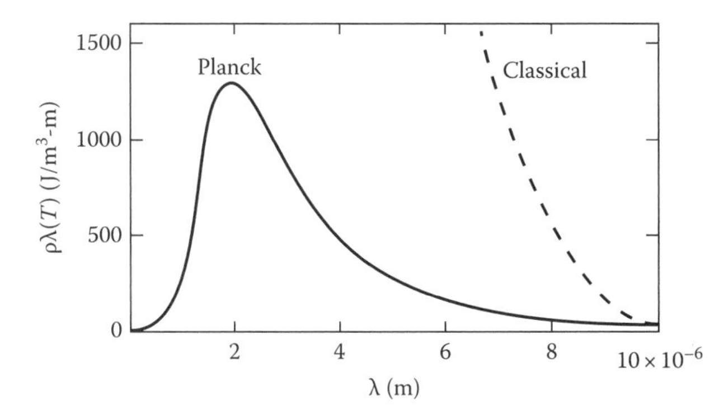

FIGURE 15.7 Planck and classical energy spectra for a blackbody at 1500 K. The classical and Planck spectra agree at $\lambda \to \infty$.

The grouping of constants in parenthesis in Equation 15.45 is what we refer to as the *Stefan–Boltzmann constant*, σ^r, which we have introduced before in Chapter 4. Its value is

$$\sigma^r = \frac{2}{15}\left(\frac{\pi^5 k_b^4}{c^2 h^3}\right) = 5.669\times 10^{-8}\ \frac{\text{W}}{\text{m}^2\text{K}^4} \quad \text{Stefan–Boltzmann constant} \tag{15.46}$$

Example 15.2 Integration and Differentiation of Planck's Law

To integrate Equation 15.45, we introduce a new variable into the integral and use the chain rule:

$$x = \frac{\hbar c}{\lambda k_b T} \qquad E_{bb}^r = -\frac{2\pi k_b^4 T^4}{\hbar^3 c^2}\int_0^\infty \frac{x^3}{\exp(x)-1}dx$$

We can now integrate to obtain the Stefan–Boltzmann radiation law:

$$E_{bb}^r = \frac{2\pi k_b^4 T^4}{\hbar^3 c^2}\left(\frac{\pi^4}{15}\right) = \sigma^r T^4$$

For a given temperature, $\rho_\lambda(\lambda,T)$ exhibits a maximum at λ_{max}. Partly due to this maximum, we observe a hot object to be at a specific color. For example, most objects will begin to appear red at a temperature of 798 K. Substituting x into Equation 15.44, we find

$$\rho_\lambda(T) = \frac{8\pi c^6 \hbar^6}{k_b^5 T^5 x^5 (e^x - 1)}$$

Taking the derivative with respect to x and setting the resulting equation to zero yields

$$5e^{-x} - x - 5 = 0$$

Solving for x gives us the wavelength of maximum emission:

$$x = 4.9651 \quad \text{or} \quad \lambda_{max}T = 2.884\times 10^{-3}\ \text{m K} \quad \text{Wien's law (1893)}$$

This result was first derived by Wien using a thermodynamic argument.

Often we need to know how much energy is emitted within a certain wavelength range. We can easily determine this by evaluating Equation 15.45 from 0 to some arbitrary wavelength, λ. This integration is shown in Appendix E and Figure 15.8. Since Equation 15.45 can be written as a function of λT and is more universally applicable in that form, Appendix E lists the energy fraction emitted as a function of λT. We define the fraction of energy between 0 and λT as

$$\mathcal{F}_{0\to\lambda T} = \frac{2\pi \hbar c^2}{\sigma^r T^4}\int_0^\lambda \frac{d\lambda}{\lambda^5[\exp(\hbar c/k_b \lambda T) - 1]} \tag{15.47}$$

Using the definition for the Stefan–Boltzmann constant, Equation 15.46, we can transform this integral into

$$\mathcal{F}_{0\to\lambda T} = \frac{15}{\pi^4}\int_x^\infty \left(\frac{x^3}{e^x - 1}\right)dx = 1 - \frac{15}{\pi^4}\int_0^x \left(\frac{x^3}{e^x - 1}\right)dx \qquad x = \frac{\hbar c}{\lambda k_b T} \tag{15.48}$$

The integral in Equation 15.48 needs to be evaluated numerically, but if we use the approximation,

$$\frac{1}{1-e^{\varepsilon}} = 1 + e^{\varepsilon} + e^{2\varepsilon} + \cdots \qquad \varepsilon \le 1 \tag{15.49}$$

we can integrate by parts and obtain the following approximation [6]:

$$F_{0\to\lambda T} = \frac{15}{\pi^4}\sum_{n=1}^{\infty}\left[\frac{e^{-nx}}{n}\left(x^3 + \frac{3x^2}{n} + \frac{6x}{n^2} + \frac{6}{n^3}\right)\right] \tag{15.50}$$

The series converges rapidly and is what is plotted in Figure 15.8 and listed in Appendix E.

The Stefan–Boltzmann law, Equation 15.45, was discovered much earlier than Planck's distribution. It was first put forth as an empirical observation by Stefan in 1879 and then derived from thermodynamics and electromagnetics by Boltzmann in 1884. We can derive the law by considering blackbody radiation within an enclosure at equilibrium. Figure 15.9 shows a container of volume, V, held at a constant temperature, T. The surface of the piston is a perfect reflector.

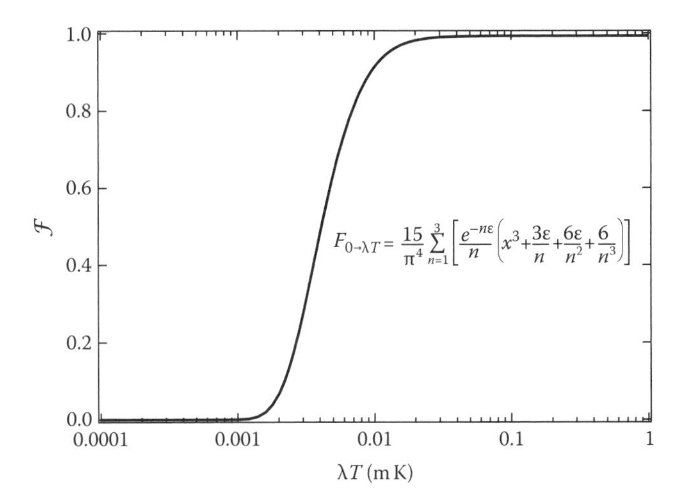

FIGURE 15.8 Fraction of radiation between $\lambda T = 0$ and $\lambda T = \infty$.

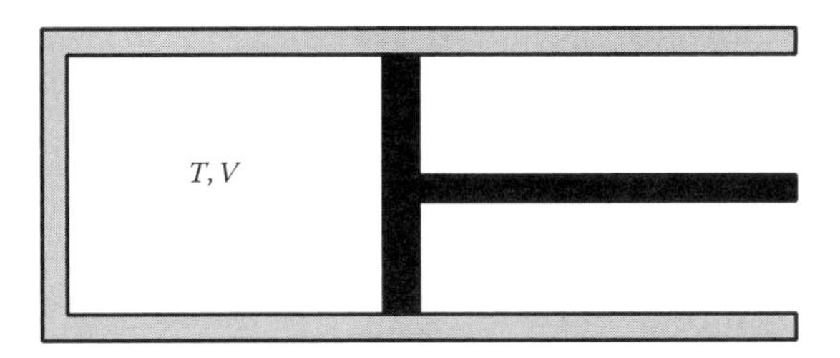

FIGURE 15.9 Blackbody radiation in an enclosure.

From electromagnetic theory (Maxwell 1865), we know that radiation hitting an object imparts a force to it so that inside the cylinder we should have a pressure due to light striking all the walls. This pressure is very small but given enough surface area can be used to propel a spacecraft (using a solar sail). At the Earth's surface, the radiation pressure from the Sun is about 4×10^{-6} N/m^2, so solar sail craft works by passing close to the Sun before heading to the outer planets. We also know that since light moves with a finite velocity, there must be a finite amount of kinetic energy contained per unit volume in a radiation field. The radiation pressure, P^r, and the radiation density, $\rho_{\lambda bb}$, that describe the light force and energy are given by

$$\rho_{\lambda bb} = \frac{4E^r_{bb}}{c} \qquad P^r = \frac{\rho_{\lambda bb}}{3} \tag{15.51}$$

The total energy in the enclosure is found by multiplying the radiation density by the volume of the container:

$$\rho_{\lambda bb,t} = \rho_{\lambda bb} V \tag{15.52}$$

Now from thermodynamics, we know that

$$\left(\frac{\partial \rho_{\lambda bb,t}}{\partial V}\right)_T = T\left(\frac{\partial P^r}{\partial T}\right)_V - P^r \tag{15.53}$$

but for a blackbody,

$$\left(\frac{\partial \rho_{\lambda bb,t}}{\partial V}\right)_T = \rho_{\lambda bb} \qquad \text{and} \qquad \left(\frac{\partial P^r}{\partial T}\right)_V = \frac{1}{3}\frac{d\rho_{\lambda bb}}{dT} \tag{15.54}$$

so

$$\rho_{\lambda bb} = \frac{T}{3}\left(\frac{d\rho_{\lambda bb}}{dT}\right) - \frac{\rho_{\lambda bb}}{3} \tag{15.55}$$

Rearranging gives

$$\frac{d\rho_{\lambda bb}}{4\rho_{\lambda bb}} = \frac{dT}{T} \tag{15.56}$$

We can integrate Equation 15.56 to yield the following Stefan–Boltzmann equation:

$$\rho_{\lambda bb} = k_b T^4 \qquad E^r_{bb} = \sigma^r T^4 \qquad \sigma^r\text{—undetermined constant} \tag{15.57}$$

Since radiation has an energy density associated with it and also a pressure, it can exert a force or a stress and can transfer momentum. This ability to transfer momentum has gained engineering and science fiction importance in connection with space flight where radiation pressure has been proposed as a means of propelling spacecraft [9]. One of the most interesting applications of radiative momentum transfer lies in the field of spectroscopy. Here, laser light is used to cool and trap atoms [1,2].

Example 15.3 Temperature on the Surface of the Moon

The Sun radiates as a blackbody at a temperature of 5800 K. Assuming the surface of the Moon to be black, estimate its temperature. The diameters of the Moon and Sun are 3.5×10^6 m and 1.39×10^9 m, respectively, and the distance between the Moon and Sun is 1.5×10^{11} m.

The first task is to perform an energy balance about the Moon. The situation is shown in Figure 15.10:

$$In - Out = 0$$

$$q_{in} - q_{out} = 0$$

$$A_p G_m^r - A_s E_{bb}^r(T_m) = 0$$

A_p is the area blocked by the presence of the Moon. It is equal to the cross-sectional area taken at the Moon's equator. A_s is the total surface area of the Moon. We need this because the Moon will emit over its entire surface, while it will receive radiation only from that part of the surface that blocks radiation from the Sun. Thus, we have

$$\left(\frac{\pi d_m^2}{4}\right) G_m^r - \pi d_m^2 \left(\sigma^r T_m^4\right) = 0$$

The radiation from the Sun is emitted from a surface of area of πd_s^2. By the time that radiation reaches the Moon, it has spread out to cover an area of

$$\pi\left(r_o - \frac{d_s}{2} - \frac{d_m}{2}\right)^2$$

We can equate the amount of energy contained in the two areas to obtain the irradiation experienced at the Moon's location:

$$\underbrace{\pi d_s^2 G_s^r}_{\text{Sun surface}} = \underbrace{\pi\left(r_o - \frac{d_s}{2} - \frac{d_m}{2}\right)^2 G_m^r}_{\text{Moon surface}}$$

The irradiation at the Sun's surface is just its emissive power and so

$$G_s^r = \sigma^r T_s^4$$

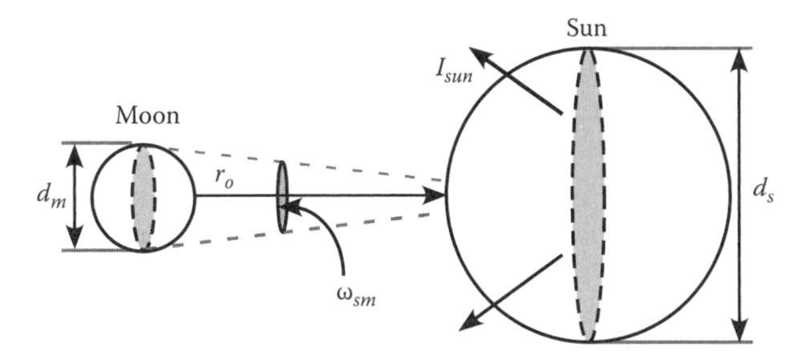

FIGURE 15.10 Radiative exchange between the Sun and the Moon.

Substituting into the previous energy balance allows us to solve for G_m^r:

$$G_m^r = \frac{\pi d_s^2 \sigma^r T_s^4}{\pi(r_o - (d_s/2) - (d_m/2))^2}$$

Finally, substituting back into the energy balance about the Moon gives us the temperature, T_m. Of course, the Moon's surface temperature is quite a bit different from this since the Moon does not radiate as a blackbody at a single temperature. The dark side is much colder than the average temperature we calculated, and the side exposed to the Sun is higher. The surface also reflects some of the incident light and we have not accounted for that either:

$$T_m = \left(\frac{G_m^r}{4\sigma^r}\right)^{1/4} = \left[\frac{d_s^2 T_s^4}{4(r_o - (d_s/2) - (d_m/2))^2}\right]^{1/4} = 395\,\text{K}$$

15.6 GRAYBODY [6]

Most real surfaces do not behave as blackbodies. They offer up some resistance to absorbing or emitting radiation. Figure 15.11 shows the energy flux emitted by a blackbody at a given temperature and a corresponding real body at that temperature. The real body emits much less energy than the blackbody. It also has a much more complicated emission pattern since it is not a smooth function of wavelength. Still, the most interesting feature to note is that the wavelength of maximum emission is pretty close to that of the blackbody at the same temperature. We can use this feature to our advantage and smooth out the real body's response to obtain our next level of approximation, the *graybody*.

The graybody uses the same type of energy spectrum as the blackbody but scales the blackbody's output by a multiplicative constant termed the *emissivity*, ε^r. The total radiative power of the graybody is then

$$E^r = \varepsilon^r \sigma^r T^4 \tag{15.58}$$

The power of the graybody approximation lies in the way we relate the energy given off by real bodies to that given off by an equivalent graybody. Real bodies have very complicated emissivities that are defined spectroscopically and that vary with wavelength and position. In essence, the radiation spectrum of the real body is measured and then compared, point by point, to the

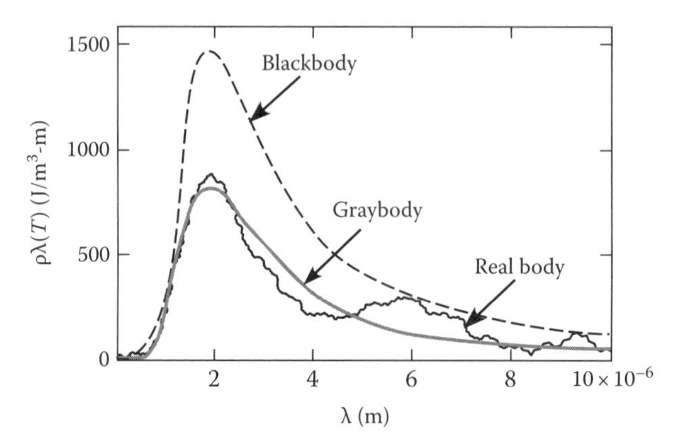

FIGURE 15.11 Comparison of ideal versus real radiators.

blackbody spectrum at the same temperature. Thus, if we use intensities of radiation, we can define the emissivity as

$$\varepsilon^r(\lambda,\theta,\phi,T) = \frac{I_E(\lambda,\theta,\phi,T)}{I_{bb}(\lambda,T)} \tag{15.59}$$

or for engineering calculations using emissive powers that only depend upon wavelength and temperature, we have

$$\varepsilon^r(\lambda,T) = \frac{E^r(\lambda,T)}{E^r_{bb}(\lambda,T)} \tag{15.60}$$

We approximate the real body as a graybody by insuring that the overall energy given off by each body at a given temperature is the same. Thus, we define a *total, hemispheric emissivity* as an average over all wavelengths:

$$\varepsilon^r(T) = \frac{\int_0^\infty \varepsilon^r(\lambda,T)E^r_{bb}(\lambda,T)\,d\lambda}{E^r_{bb}(T)} \tag{15.61}$$

We can make several generalizations about emissivities as defined by Equation 15.61:

a. The emissivity of metal surfaces and other highly reflecting surfaces is small and can be made to be <0.02 by fine polishing.
b. The presence of oxide layers on the surface of a metal surface (rust) can significantly increase the surface emissivity. For instance, polished aluminum has an emissivity of 0.04 (300 K), whereas anodized aluminum has an emissivity closer to 0.8 (300 K). This is why anodized pans cook better.
c. The emissivity of dielectrics is generally large, >0.6 and often >0.8.
d. The emissivity of conducting surfaces increases with temperature, but the emissivity of dielectrics may increase (Teflon) or may decrease (Al_2O_3) with increasing temperature, depending upon the material.

The graybody approximation also extends to those elements of radiation associated with incident radiation. Thus, we can define *total hemispherical absorptivities, reflectivities, and transmittivities* to calculate the overall amount of radiation absorbed, reflected, or transmitted by a body:

$$\alpha^r(T) = \frac{\int_0^\infty \alpha^r(\lambda,T)G^r(\lambda)\,d\lambda}{\int_0^\infty G^r(\lambda)\,d\lambda} \tag{15.62}$$

$$\rho^r(T) = \frac{\int_0^\infty \rho^r(\lambda,T)G^r(\lambda)\,d\lambda}{\int_0^\infty G^r(\lambda)\,d\lambda} \tag{15.63}$$

$$\tau^r(T) = \frac{\int_0^\infty \tau^r(\lambda,T)G^r(\lambda)\,d\lambda}{\int_0^\infty G^r(\lambda)\,d\lambda} \tag{15.64}$$

15.6.1 Kirchhoff's Law

In the last section, we saw that highly reflecting surfaces have low emissivities. It seems logical to suggest that highly absorbing materials would also have high emissivities since they approximate the absorbing properties of a blackbody. This relationship between the emissivity and the other surface properties can be extended and quantified under certain circumstances.

Consider the situation shown in Figure 15.12. Three bodies are exchanging radiation between themselves and an enclosure. All surfaces are at a temperature of T_s. Since the enclosure is large and isolated from the surroundings, the radiation emitted from it to the small bodies inside is that of a blackbody at a temperature of T_s. Thus, all the internal surfaces are irradiated with a radiative power $G^r = E^r_{bb}(T_s)$. If we perform an energy balance about the individual bodies within the enclosure, we find

$$\begin{aligned} In \quad &- \quad Out \quad = 0 \\ \alpha^r_1 G^r A_1 \; &- \; E^r_1(T_s)A_1 \; = 0 \end{aligned} \tag{15.65}$$

$$\alpha^r_2 G^r A_2 - E^r_2(T_s)A_2 = 0 \tag{15.66}$$

$$\alpha^r_3 G^r A_3 - E^r_3(T_s)A_3 = 0 \tag{15.67}$$

Since $G^r = E^r_{bb}(T_s)$, we can substitute into Equations 15.65 through 15.67, solve for $E^r_{bb}(T_s)$, and find

$$\frac{E^r_1(T_s)}{\alpha^r_1} = \frac{E^r_2(T_s)}{\alpha^r_2} = \frac{E^r_3(T_s)}{\alpha^r_3} = E^r_{bb}(T_s) \tag{15.68}$$

This is a statement of *Kirchhoff's law* and since $\alpha^r_i < 1$, no surface can emit more energy than a blackbody. If we substitute for the emissive power using the total hemispherical emissivity (*gray-body approximation*), $E^r_i(T_s) = \varepsilon^r_i E^r_{bb}(T_s)$, we can derive an alternate form of Kirchhoff's law:

$$\frac{\varepsilon^r_1}{\alpha^r_1} = \frac{\varepsilon^r_2}{\alpha^r_2} = \frac{\varepsilon^r_3}{\alpha^r_3} = 1 \tag{15.69}$$

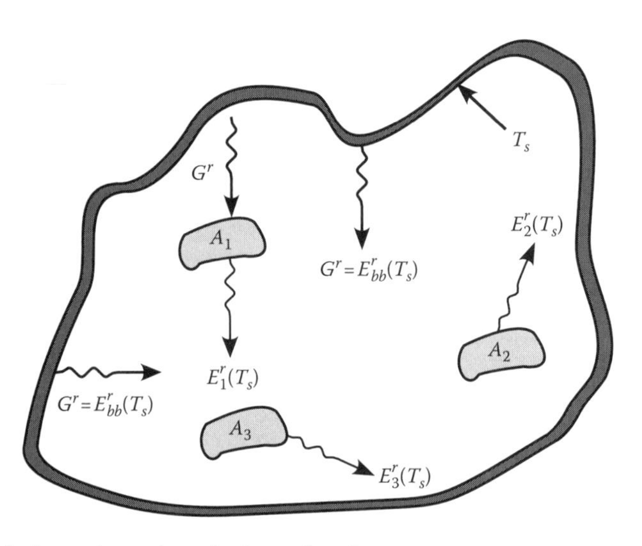

FIGURE 15.12 Radiative exchange in an isothermal enclosure.

Thus, for any surface in the enclosure, the total hemispherical emissivity and the total absorptivity are equal: $\varepsilon^r = \alpha^r$. This statement of Kirchhoff's law has some severe restrictions imposed upon it among them: equilibrium between all surfaces, blackbody cavity radiation from the enclosure surface, and diffuse surfaces.

The previous derivation has been repeated for spectral (wavelength dependent) conditions and leads to a less restrictive form of Kirchhoff's law where the surfaces need only be diffuse radiators:

$$\varepsilon_i^r(\lambda) = \alpha_i^r(\lambda) \tag{15.70}$$

A form of Kirchhoff's law for which there are no restrictions involves spectral, directional properties:

$$\varepsilon_i^r(\lambda,\theta) = \alpha_i^r(\lambda,\theta) \tag{15.71}$$

Equation 15.71 is always applicable because $\varepsilon^r(\lambda, \theta)$ and $\alpha^r(\lambda, \theta)$ are inherent properties of the surface and are independent of the spectral and directional distributions of the emitted or incident radiation. If the irradiation or the surface is diffuse, then there is no directional dependence and Equation 15.70 is applicable.

Example 15.4 Glass Annealing

A glass with the spectral radiative properties given in Figure 15.13 is being annealed in a large oven to relieve the stresses set up when the glass was initially cooled. The walls of the oven are lined with a diffuse, gray, refractory material having an emissivity of 0.8. The walls of the oven are maintained at $T_w = 1600°C$. The glass is being annealed at a temperature of 500°C. At these conditions

a. What are the total reflectivity, total transmittivity, and total emissivity of the glass?
b. What is the net radiative heat flux to the glass?

a. The total reflectivity is obtained using Equation 15.63. Since the furnace walls represent so much more area than the glass surface and are isolated from the surroundings, we can say $G^r(\lambda) \approx E_{bb}^r(\lambda, T_w)$. Substituting into the definition and realizing that the lower integral is just $\sigma^r T_w^4$, we have

$$\rho^r = \frac{\int_0^\infty \rho^r(\lambda) E_{bb}^r(\lambda) d\lambda}{\sigma^r T_w^4} = \frac{1}{\sigma^r T_w^4}\left[0.1\int_0^{1.65} E_{bb}^r(\lambda) d\lambda + 0.5\int_{1.65}^{\infty} E_{bb}^r(\lambda) d\lambda\right]$$

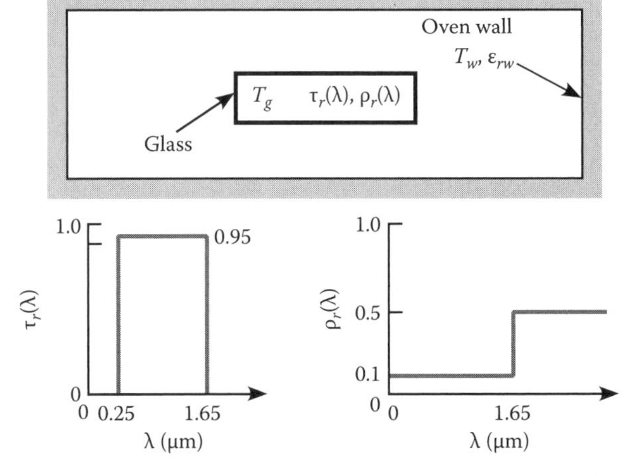

FIGURE 15.13 Glass annealing in a furnace. Glass spectral properties are also given.

The integrals can be evaluated using Appendix E:

$$\rho^r = 0.1\mathcal{F}(0 \to 1.65T_w) + 0.5\mathcal{F}(1.65T_w \to \infty)$$

$$= 0.1\mathcal{F}(0 \to 1.65T_w) + 0.5\mathcal{F}(0 \to \infty) - 0.5\mathcal{F}(0 \to 1.65T_w) = 0.383$$

We perform the same type of manipulation for the total transmittivity:

$$\tau^r = \frac{\int_0^\infty \tau^r(\lambda)G^r(\lambda)d\lambda}{\sigma^r T_w^4}$$

$$= \frac{1}{\sigma^r T_w^4}\left[0.0\int_0^{0.25} E_{bb}^r(\lambda)d\lambda + 0.95\int_{0.25}^{1.65} E_{bb}^r(\lambda)d\lambda + 0.0\int_{1.65}^{\infty} E_{bb}^r(\lambda)d\lambda\right]$$

$$= 0.95[\mathcal{F}(0 \to 1.65T_w) - \mathcal{F}(0 \to 0.25T_w)] = 0.279$$

The total absorptivity can be calculated by difference and is

$$\alpha^r = 1 - \rho^r - \tau^r = 0.338$$

We can calculate the total emissivity of the glass by first figuring out where we lie in the spectral range. At 773 K, the total energy emitted at wavelengths less than 1.65 μm is negligible:

$$\lambda T_g = 1.65(773) = 1275.5 \qquad \mathcal{F}(0 \to 1275.5) \approx 0.0037$$

Thus, all the emitted energy is at wavelengths longer than 1.65 μm. For those wavelengths, $\tau^r(\lambda) = 0$, and $\rho^r(\lambda) = 0.5$. Thus, $\alpha^r(\lambda) = 0.5$ and using Kirchhoff's law, we can say

$$\varepsilon^r(\lambda) = \alpha^r(\lambda) = 0.5 \quad \text{so}$$

$$\varepsilon^r = 0.5 \quad \text{since } \varepsilon^r(\lambda) \text{ is constant for } \lambda > 1.65\,\mu\text{m}$$

b. To calculate the net radiative heat flux to the glass, we must perform an energy balance about the glass. Referring to the figure, we see that

$$(q''_{net})_{rad} = (\dot{E}''_{in})_{rad} - (\dot{E}''_{out})_{rad} = \alpha^r G_w^r(T_w) - \varepsilon^r E_{bb}^r(T_g)$$

Evaluating the individual terms and remembering that the glass has two surfaces exposed to the walls, we find

$$(q''_{net})_{rad} = 2\sigma^r[0.338(1873)^4 - 0.5(773)^4] = 451 \text{ kW/m}^2$$

If this were the only energy exchange, the glass would heat up quite rapidly. For steady-state annealing, we must have an additional heat loss and that loss comes from convection to the furnace air that can be exhausted from the furnace.

15.7 VIEW FACTORS [7]

Radiation travels through space in straight lines. Straight-line travel is also a reasonable approximation over relatively short distances in the atmosphere, at least for the engineering applications considered here. Since the radiation has a direction associated with it, to calculate the radiative exchange between two surfaces, we must be able to calculate how much energy emitted by each surface actually collides, or is intercepted by, the other surface. This involves projecting the surface of each emitting object onto the surface of the corresponding object receiving the radiation. It is a geometrical exercise. The quantity that we calculate from the projection is called the *view factor.*

15.7.1 View Factor Integral

The view factor F_{ij} is defined as the fraction of radiation leaving surface i that is intercepted by surface j. To develop a general expression for F_{ij}, we use the two arbitrarily oriented surfaces shown in Figure 15.14. We define two differential patches: dA_{si} on surface A_{si} and dA_{sj} on surface A_{sj}. The two regions are connected by a line of length, r_o. This line makes an angle of θ_i with respect to the normal, $\vec{\mathbf{n}}_i$, to surface i and an angle of θ_j with respect to the normal, $\vec{\mathbf{n}}_j$, to surface j.

The rate at which radiation leaves dA_{si} and is intercepted by dA_{sj} may be expressed as

$$dq_{i\to j} = I_{Ei}\cos\theta_i\, dA_{si}\, d\omega_{ji} \tag{15.72}$$

Here, I_{Ei} is the intensity of radiation leaving surface i and $d\omega_{ji}$ is the solid angle subtended by area dA_{sj} when viewed from dA_{si}. This solid angle is defined by

$$d\omega_{ji} = \frac{\cos\theta_j\, dA_{sj}}{r_o^2} \tag{15.73}$$

To define a view factor for the surface A_{si}, we assume the surface reflects and emits diffusely so $I_{Ei} = J_i^r/\pi$. Substituting these relations into Equation 15.72, we have

$$dq_{i\to j} = J_i^r\left(\frac{\cos\theta_i\cos\theta_j}{\pi r_o^2}\right)dA_{si}\,dA_{sj} \tag{15.74}$$

The total rate of energy exchange between A_{si} and A_{sj} is now found by integrating over both surfaces:

$$q_{i\to j} = J_i^r\int_{A_{sj}}\int_{A_{si}}\left(\frac{\cos\theta_i\cos\theta_j}{\pi r_o^2}\right)dA_{si}\,dA_{sj} \tag{15.75}$$

Notice that we assumed the radiosity, J_i^r, was uniform over surface i. We need not make that simplification, but the integral becomes more complicated as we'll see in the following discussion.

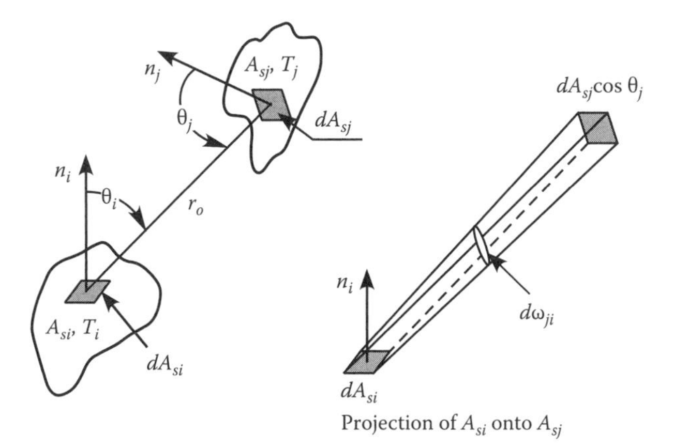

FIGURE 15.14 View factor associated with radiative exchange between differential components of surfaces A_{si} and A_{sj}.

Using the definition of the view factor as the fraction of radiation leaving A_{si} that is intercepted by A_{sj}, we define it as

$$F_{ij} = \frac{q_{i \to j}}{J_i^r A_{si}} = \frac{1}{A_{si}} \int_{A_{sj}} \int_{A_{si}} \left(\frac{\cos\theta_i \cos\theta_j}{\pi r_o^2} \right) dA_{si} dA_{sj} \tag{15.76}$$

We can perform a similar derivation to obtain F_{ji}, the fraction radiation leaving surface j that is intercepted by surface i:

$$F_{ji} = \frac{q_{j \to i}}{J_j^r A_{sj}} = \frac{1}{A_{sj}} \int_{A_{sj}} \int_{A_{si}} \left(\frac{\cos\theta_i \cos\theta_j}{\pi r_o^2} \right) dA_{si} dA_{sj} \tag{15.77}$$

The important points to note about using Equations 15.76 and 15.77 are that the equations are applicable only to surfaces that are *diffuse* emitters or reflectors and to surfaces whose radiosity is uniform over the surface. If these conditions are not met, the view factor varies from point to point over the surface, and one must use directional intensities and ray tracing computer packages to obtain the radiative exchange. It is interesting to note that the same packages used to render objects for animated movies are based on the same technology used to calculate radiative exchange under conditions where the surfaces are not diffuse.

15.7.2 Relations between View Factors

There are two important relations between view factors that are commonly used when analyzing radiative exchange. The first relation follows directly from Equations 15.76 and 15.77. Since the integrals are identical,

$$A_{si} F_{ij} = A_{sj} F_{ji} = \int_{A_{sj}} \int_{A_{si}} \left(\frac{\cos\theta_i \cos\theta_j}{\pi r_o^2} \right) dA_{si} dA_{sj} \qquad \text{Reciprocity} \tag{15.78}$$

This is termed the *reciprocity relation* and is valid for any pair of view factors. It is useful in determining a second view factor if the first view factor and the surface areas are known.

The second important relation concerns view factors within an enclosure. Referring to Figure 15.15, we choose surface 1 as a reference. The total amount of energy that surface 1 emits to all other surfaces is

$$q_1 = A_{s1} J_1^r \tag{15.79}$$

and this is split up among all the other surfaces that surface 1 *sees*:

$$q_1 = \sum_{j=1}^{n} q_{ij} = \sum_{j=1}^{n} A_{s1} F_{1j} J_1^r \tag{15.80}$$

If surface 1 is diffuse, then we can remove the area and radiosity from Equation 15.80 and equate 15.79 with (15.80) to find that for any surface, i,

$$\sum_{j=1}^{n} F_{ij} = 1 \qquad \text{Summation rule} \tag{15.81}$$

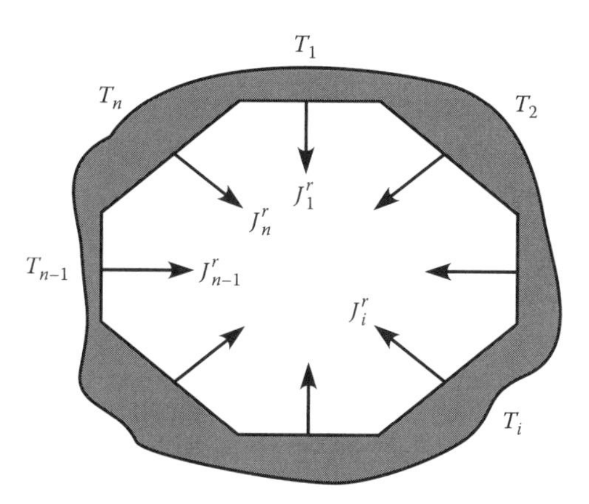

FIGURE 15.15 Radiative exchange within an enclosure whose surfaces are all held at different temperatures.

This relationship is termed the *summation rule* and it applies strictly to enclosures. It is just a statement of conservation of energy within an enclosure. One can always use this rule by making a hypothetical enclosure. Whether the hypothetical enclosure makes the determination of the view factors easier depends on the geometry of the enclosure and whether one can easily evaluate the view factors for the hypothetical walls one built. The use of these rules will be shown in the following examples.

Example 15.5 View Factor Calculations

Determine the following view factors given the geometries in Figure 15.16:

a. View factors F_{12} and F_{21}.
b. View factor for the channel to the surroundings.
c. View factors F_{12} and F_{21}.

We begin with case (a). Radiation from the sphere leaves normal to its surface. Thus, all points on the sphere below its equator, where the normal points toward the plane, represent radiation that would be intercepted by the plane. Fully half of the radiation leaving the sphere would be intercepted by the plane. The view factor F_{12} must be 1/2. Since the normal to the sphere's surface always points away from the surface, $F_{11} = 0$. The same is true for the plane so $F_{22} = 0$. We can obtain F_{21} using the reciprocity theorem:

$$A_{s1}F_{12} = A_{s2}F_{21}$$

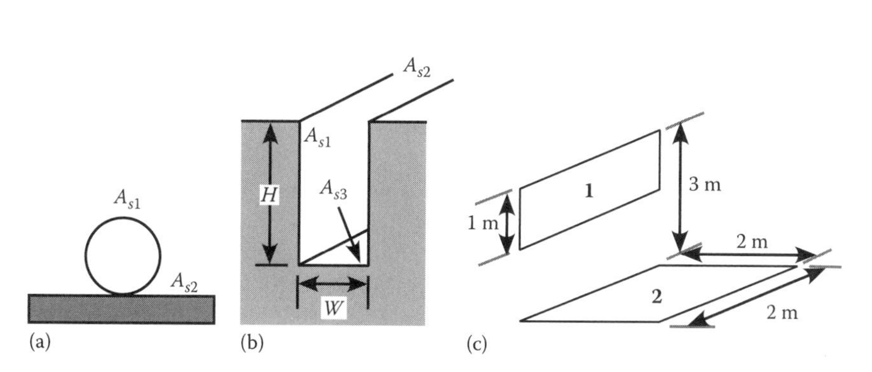

FIGURE 15.16 Surfaces exchanging radiation. (a) Sphere lying atop an infinite plane. (b) Long, rectangular duct. (c) Two opposing walls.

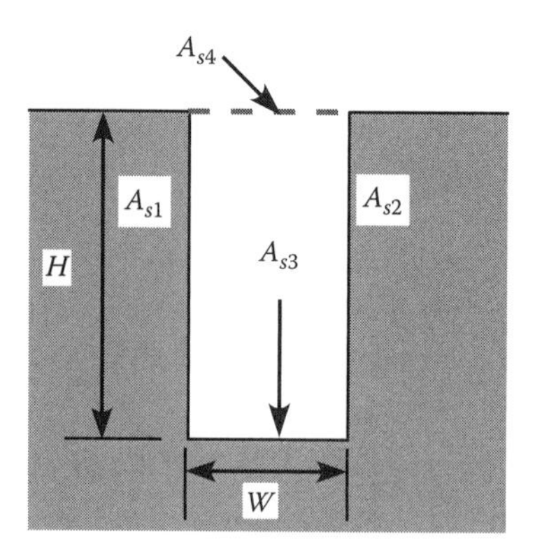

FIGURE 15.17 Case (b) showing the hypothetical surface forming an enclosure.

Since $A_{s2} = \infty$, $F_{21} = 0$.

Case (b) is most easily solved if we use a common trick. We place a hypothetical wall, A_{s4}, on top of the duct to form an enclosure. The situation is shown in Figure 15.17. Wall A_{s4} effectively represents the surroundings since any radiation leaving the groove must pass through the wall. We can set up two surfaces, wall 4, and the composite system, walls 1, 2, and 3. From reciprocity, we have

$$A_{s123}F_{(123),4} = A_{s4}F_{4,(123)}$$

From the summation rule, we know that

$$F_{4,(123)} + F_{44} = 1$$

Substituting for the respective areas will allow us to solve for the view factors:

$$A_{s4} = W \cdot L \qquad A_{s123} = (H + H + W)L$$

Solving, knowing that $F_{44} = 0$ because the normal to the surface always points away from the surface, gives

$$F_{4,(123)} = 1 \qquad F_{(123),4} = \frac{W}{H + H + W}$$

It is important to analyze the situation carefully to try to find the path of least resistance. Here, we used the summation rule on wall 4 because we realized that $F_{44} = 0$. We could not say for sure what $F_{(123),(123)}$ would be at first glance. Now we know that it must be

$$F_{(123),4} + F_{(123),(123)} = 1 \qquad F_{(123),(123)} = \frac{2H}{H + H + W}$$

To work on case (c), we must use the graph presented in Figure 15.18b or the formulae in Table 15.3, and we must also resort to the trick we used previously. Here, we insert a hypothetical wall that connects surfaces 1 and 2 as shown in Figure 15.19. Performing an energy balance about the radiation leaving wall 2 that is intercepted by wall 1 and wall 3, we know that

$$q_{2\to1,3} = q_{2\to1} + q_{2\to3}$$

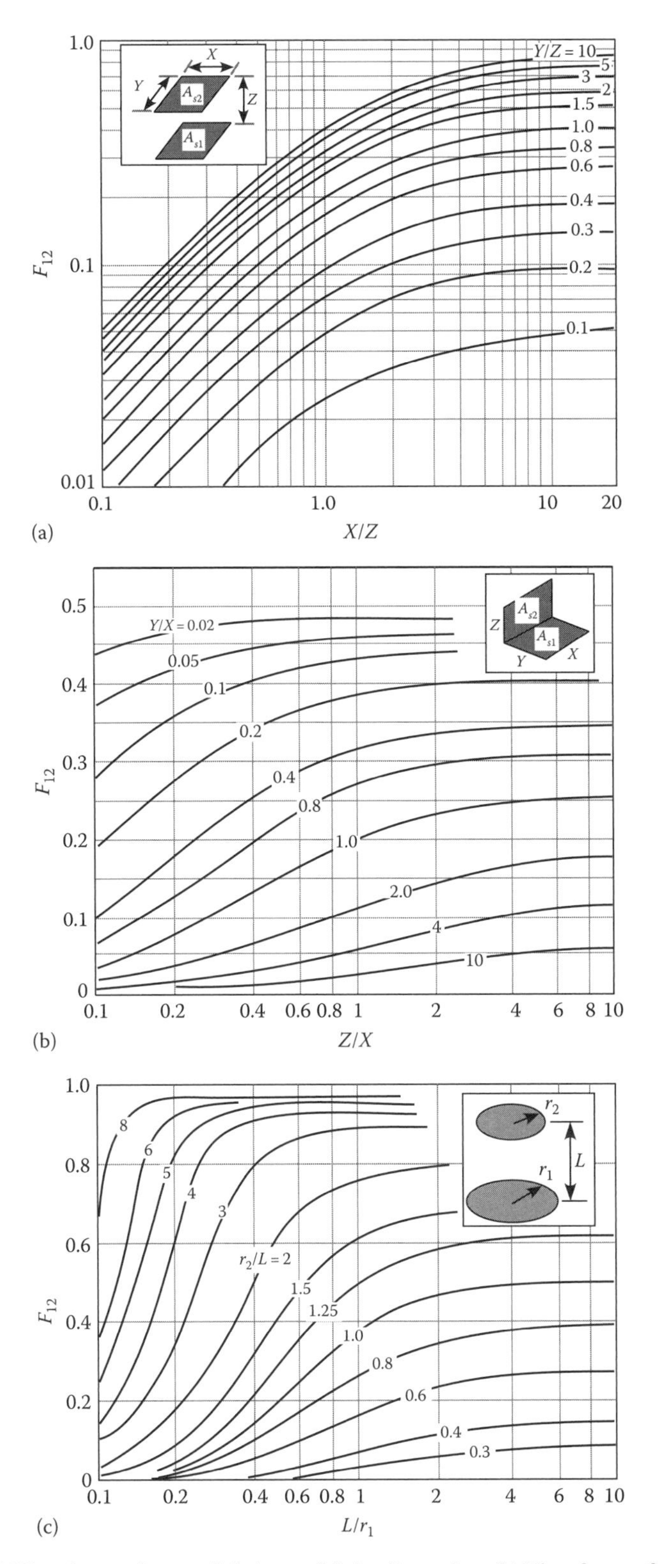

FIGURE 15.18 (a) View factors for parallel plates of finite dimension. (b) View factors for attached, perpendicular walls. (c) View factors for parallel disks of finite dimension. (Adapted from Incropera, F.P. and Dewitt, D.P., *Introduction to Heat Transfer,* 4th edn., John Wiley & Sons, New York, 1998. With permission.)

TABLE 15.3
View Factors for Selected 2-D and 3-D Geometrical Configurations

Geometry **View Factor**

2-D Systems

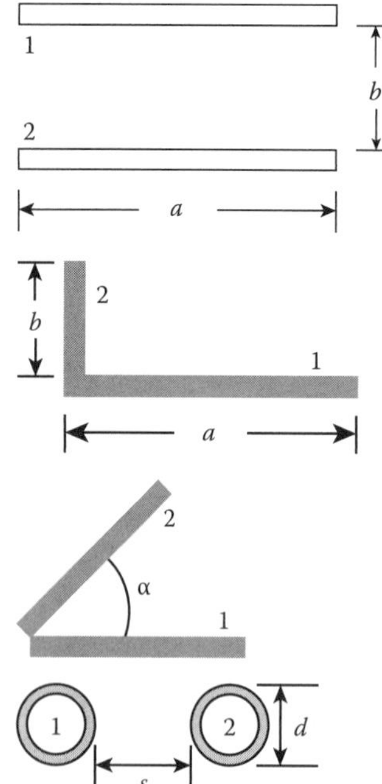

$$F_{12} = F_{21} = \left[1 + \left(\frac{b}{a}\right)^2\right]^{1/2} - \left(\frac{b}{a}\right)$$

$$F_{12} = \frac{1}{2}\left\{1 + \left(\frac{b}{a}\right) - \left[1 + \left(\frac{b}{a}\right)^2\right]^{1/2}\right\}$$

$$F_{12} = F_{21} = 1 - \sin\left(\frac{\alpha}{2}\right)$$

$$\theta = 1 + (s/d)$$

$$F_{12} = F_{21} = \frac{1}{\pi}[(\theta^2 - 1)^{1/2} + \sin^{-1}(1/\theta) - \theta]$$

3-D Systems

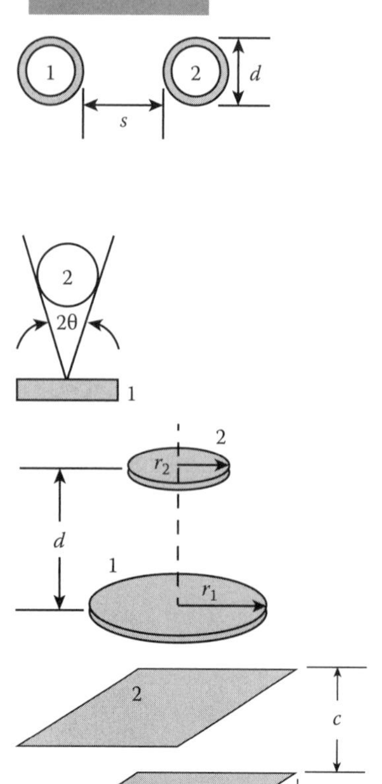

$$F_{12} = \sin^2\theta$$

$$L_1 = \frac{r_1}{d} \qquad L_2 = \frac{r_2}{d} \qquad \theta = 1 + \frac{1 + L_2^2}{L_1^2}$$

$$F_{12} = \frac{1}{2}\left\{\theta - \left[\theta^2 - 4\left(\frac{L_2}{L_1}\right)^2\right]^{1/2}\right\}$$

$$X = \frac{c}{b} \qquad Y = \frac{a}{c}$$

$$F_{12} = \frac{2}{\pi XY}\left\{\ln\left[\frac{(1+X^2)(1+Y^2)}{1+X^2+Y^2}\right]^{1/2} - X\tan^{-1}(X) - Y\tan^{-1}(Y)\right.$$

$$\left. + X(1+Y^2)^{1/2}\tan^{-1}\left[\frac{X}{(1+Y^2)^{1/2}}\right] + Y(1+X^2)^{1/2}\tan^{-1}\left[\frac{Y}{(1+X^2)^{1/2}}\right]\right\}$$

TABLE 15.3 (continued)
View Factors for Selected 2-D and 3-D Geometrical Configurations

Geometry	View Factor

3-D Systems

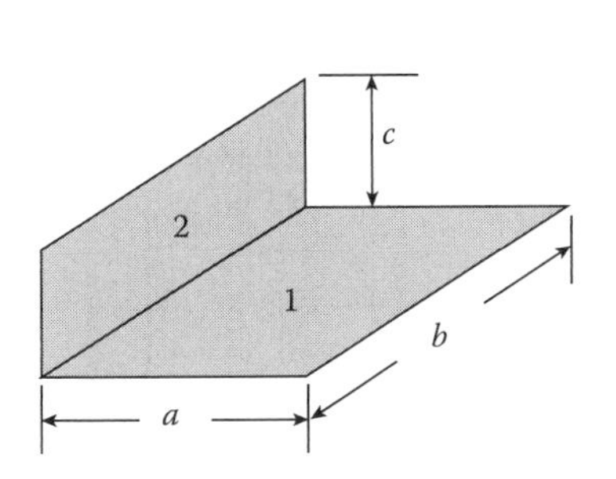

$$X = \frac{a}{b} \qquad Y = \frac{c}{b}$$

$$F_{12} = \frac{1}{\pi X}\left\{X \tan^{-1}\left(\frac{1}{X}\right) + Y \tan^{-1}\left(\frac{1}{Y}\right) - (X^2+Y^2)^{1/2}\tan^{-1}\left(\frac{1}{(X^2+Y^2)^{1/2}}\right)\right.$$

$$\left. + \frac{1}{4}\ln\left[\left(\frac{(1+X^2)(1+Y^2)}{1+X^2+Y^2}\right)\left(\frac{X^2(1+X^2+Y^2)}{(1+X^2)(X^2+Y^2)}\right)^{X^2}\left(\frac{Y^2(1+X^2+Y^2)}{(1+Y^2)(X^2+Y^2)}\right)^{Y^2}\right]\right\}$$

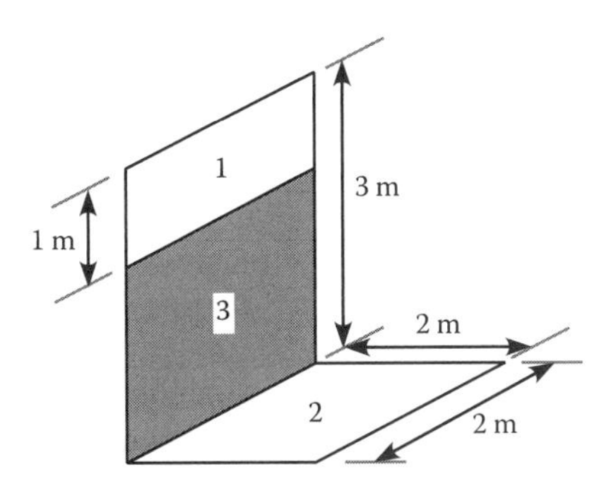

FIGURE 15.19 Hypothetical wall inserted between surfaces 1 and 2.

must hold. Since we are interested in F_{21}, we need to evaluate the three terms. Letting walls one and three have the same temperature, we can factor out the temperature differences and the area of wall 2 to obtain

$$F_{2,(13)} = F_{21} + F_{23}$$

From Figure 15.18b, we can easily evaluate $F_{2,(13)}$ and F_{23}:

$$F_{2,(13)} = 0.22 \qquad F_{23} = 0.2$$

This leaves

$$F_{21} = 0.02 \qquad F_{12} = 0.04 \quad \text{by reciprocity}$$

15.8 RADIATIVE ENERGY EXCHANGE

In the previous sections, we discussed the properties of surfaces, how to calculate those properties, how to calculate the amount of energy emitted by a surface, and how to calculate the amount of radiation emitted by one surface that is intercepted by another when no view factors were involved. We can now calculate the radiative exchange between surfaces.

15.8.1 Radiative Heat Transfer between Blackbodies

In general, a surface can emit or reflect radiation (most surfaces dealt with in engineering situations are opaque), and the receiving surface can absorb or reflect some of that radiation. This makes the calculation of radiative exchange for real surfaces very complicated. Fortunately, life is much simpler if the surfaces exchanging radiation are blackbodies. All radiation leaving a blackbody is emitted and all radiation incident on a blackbody is absorbed. Let's consider what happens between two arbitrary blackbodies.

Figure 15.20 shows two arbitrary blackbodies exchanging radiation. We define $q_{i\to j}$ as the rate at which radiation leaves surface i and is intercepted by surface j. From the previous section, we have

$$q_{i\to j} = J_i^r A_{si} F_{ij} \qquad q_{j\to i} = J_j^r A_{sj} F_{ji} \tag{15.82}$$

Since the radiosity for a blackbody is just its emissive power (no reflection), we can substitute for J_i^r and J_j^r in Equations 15.82:

$$q_{i\to j} = E_{bbi}^r A_{si} F_{ij} = \sigma^r T_i^4 A_{si} F_{ij} \tag{15.83}$$

$$q_{j\to i} = E_{bbj}^r A_{sj} F_{ji} = \sigma^r T_j^4 A_{sj} F_{ji} \tag{15.84}$$

The net radiative exchange between surfaces i and j is the difference between the rate at which radiation leaves surface i and is intercepted by surface j and the rate at which radiation leaves surface j and is intercepted by surface i:

$$q_{ij} = q_{i\to j} - q_{j\to i} \tag{15.85}$$

Using the reciprocity relation for the view factors lets us write the net exchange in terms of temperatures:

$$q_{ij} = A_{si} F_{ij} \sigma^r \left(T_i^4 - T_j^4\right) \tag{15.86}$$

Note that for the two-body system, $q_{ij} = -q_{ji}$. Equation 15.86 requires the use of absolute temperatures and the sign convention is the same as we have used throughout the text. If q_{ij} is positive, we have net heat transfer from surface i to surface j and so $T_i > T_j$.

The derivation leading up to Equation 15.86 can be followed to determine the net radiative exchange between surface i and all other surfaces in an enclosure. For n surfaces, we have n

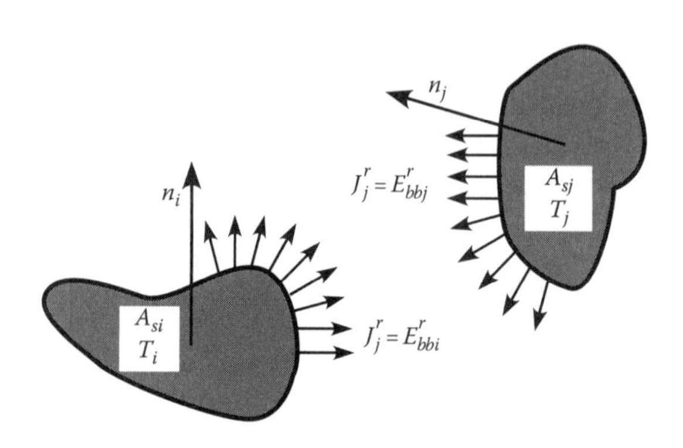

FIGURE 15.20 Radiative exchange between two blackbodies.

equations like Equation 15.86 and the total energy transferred from surface i is just the sum of those n equations:

$$q_i = \sum_{j=1}^{n} q_{ij} = \sum_{j=1}^{n} A_{si} F_{ij} \sigma^r \left(T_i^4 - T_j^4\right) \quad (15.87)$$

15.8.2 Radiative Heat Transfer between Graybodies

The results of the previous section are useful to a point, but the assumption of blackbody behavior is often too restrictive. We can generally achieve a better approximation to radiative transfer by considering the surfaces to be gray. Each surface is assumed to be opaque ($\tau^r = 0$), diffuse, and isothermal and to be characterized by a uniform radiosity and irradiation. The usual circumstance is one in which all the surface temperatures are known and we are interested in determining the heat fluxes associated with each surface. We will confine ourselves to situations within enclosures where we can make use of Kirchhoff's laws.

Figure 15.21 shows two diffuse, gray, surfaces exchanging energy. The net rate at which radiation leaves surface i, q_i, reflects the net effect of radiative interactions at the surface and is the rate at which energy must be transferred to keep the surface at a constant temperature, T_i. We define q_i by a steady-state energy balance, treating q_i as an *In* term. This makes q_i positive if the object radiates more energy than it receives:

$$\begin{aligned} In \quad &- \quad Out = 0 \\ q_i + A_{si} G_i^r \quad &- \quad A_{si} J_i^r = 0 \end{aligned} \quad (15.88)$$

The radiosity for surface i is composed of its emissive power and the fraction of incident radiation that is reflected:

$$J_i^r = E_i^r + \rho_i^r G_i^r \quad (15.89)$$

The emissive power can be written in terms of the emissive power of a blackbody, E_{bbi}^r, and the emissivity of the surface, ε_i^r. Using the fact that $\rho_i^r = 1 - \alpha_i^r$ and substituting in Kirchhoff's law, $\alpha_i^r = \varepsilon_i^r$, we can rewrite Equation 15.89 in terms of the emissivity and emissive power of the blackbody at T_i:

$$J_i^r = \varepsilon_i^r E_{bbi}^r + \left(1 - \varepsilon_i^r\right) G_i^r \quad (15.90)$$

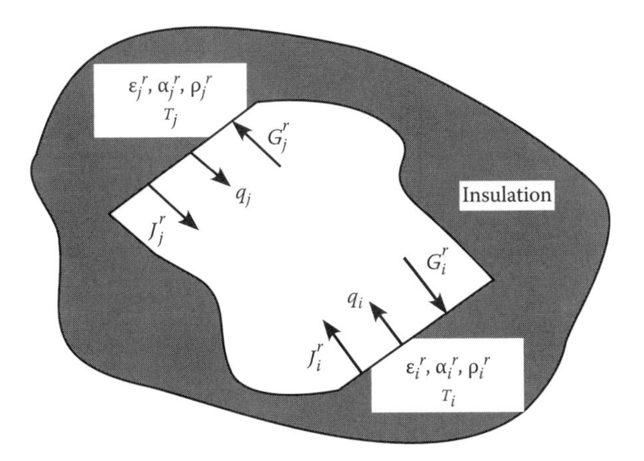

FIGURE 15.21 Radiative exchange between graybodies in an enclosure.

Solving for G_i^r in Equation 15.90 and then substituting for G_i^r in Equation 15.88 lets us represent q_i as

$$q_i = A_{si}\left(J_i^r - \frac{J_i^r - \varepsilon_i^r E_{bbi}^r}{1-\varepsilon_i^r}\right) = \frac{E_{bbi}^r - J_i^r}{\left(\left(1-\varepsilon_i^r\right)/\varepsilon_i^r\right)A_{si}} = \frac{\text{Driving force}}{\text{Resistance}} \tag{15.91}$$

Equation 15.91 represents the *net radiative heat transfer* from surface *i*. We write this flow in terms of a driving force and a resistance. The driving force is the difference between the emissive power of a blackbody at T_i and the actual amount of radiation leaving the surface, the radiosity, J_i^r. The resistance, which is called a *surface resistance*, describes the difficulty the surface has in approaching the performance of a blackbody. Note that the surface resistance is zero for a blackbody since $\varepsilon_i^r = 1$ in that case.

Having defined the net radiative heat transfer from the surface, we are in a position to describe radiative transfer between surfaces. To evaluate J_i^r in Equation 15.91, we need to determine the irradiation experienced by surface *i*. This irradiation is a result of the radiosity of all the other surfaces in the enclosure and the fraction of that radiation that surface *i* intercepts. Thus,

$$A_{si}G_i^r = \sum_{j=1}^{n} A_{sj}F_{ji}J_j^r = \sum_{j=1}^{n} A_{si}F_{ij}J_j^r \tag{15.92}$$

Removing the A_{si} from Equation 15.92 using reciprocity and substituting for G_i^r in Equation 15.88, we find

$$q_i = A_{si}\left(J_i^r - \sum_{j=1}^{n} F_{ij}J_j^r\right) \tag{15.93}$$

Using the summation rule, we substitute for J_i^r and then factor out the view factors, F_{ij}:

$$q_i = A_{si}\left(\sum_{j=1}^{n} F_{ij}J_i^r - \sum_{j=1}^{n} F_{ij}J_j^r\right) = \sum_{j=1}^{n} A_{si}F_{ij}\left(J_i^r - J_j^r\right) \tag{15.94}$$

The last sum in Equation 15.94 is a sum over all the net radiative transfer components between surface *i* and the other surfaces in the enclosure. We can also write this in terms of a sum of driving forces and resistances:

$$q_i = \sum_{j=1}^{n} q_{ij} = \sum_{j=1}^{n} \frac{J_i^r - J_j^r}{1/A_{si}F_{ij}} = \frac{\text{Driving force}}{\text{Resistance}} \tag{15.95}$$

For radiative exchange between surfaces, the driving force is the difference between radiosities, and the resistance is just the geometrical function that describes how much of the radiation coming from surface *j* is intercepted by surface *i*. We refer to this as a *geometrical resistance*. All radiation problems can be written as a combination of geometrical and surface resistances. For steady-state problems, we can use the resistance diagram approach of Chapter 5 to help set up and solve problems.

Example 15.6 Radiative Transport from a Lightbulb

At steady state, a 100 W, soft white, incandescent lightbulb has a surface temperature of 155°C when the ambient room temperature is 25°C. If the bulb is modeled as a 75 mm diameter diffuse, gray, sphere, with a surface emissivity of 0.82, what is the radiant heat transferred from the bulb surface to the room surroundings?

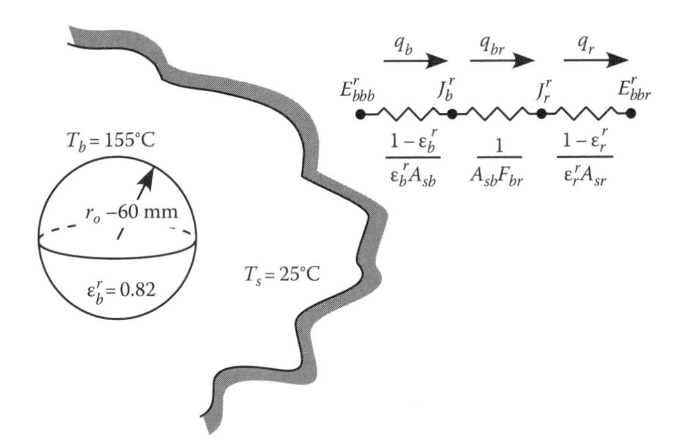

FIGURE 15.22 Lightbulb radiating into a room.

The first task in solving this problem is to draw a picture and then come up with a resistance diagram for the situation. Here, there are three resistances: a surface resistance for the bulb since the bulb does not radiate as a blackbody, a similar resistance for the walls of the room, and a geometrical resistance from the bulb to the room. The overall situation and the resistance diagram are shown in Figure 15.22. For every surface that is not a blackbody, we must include a surface resistance. There is one geometrical resistance for every two bodies exchanging radiation. The nodes in the diagram represent blackbody emissive powers or radiosities. The geometrical resistances are connected via radiosity nodes. Surface resistances connect blackbody emissive power nodes and corresponding radiosity nodes. For every resistance, there is an associated current representing the heat flow through that resistance. These currents are also shown in the diagram.

In this problem, we are given no information regarding the emissivity of the room. Thus, we must assume something, and a blackbody is the simplest. One reason why this assumption will be valid concerns the area of the room. Notice that the surface resistance of the room is inversely proportional to the radiating area. Since the room area is so large, its surface resistance is very small and it may as well be a blackbody. The geometrical resistance between the bulb and room is simple to evaluate. Since the room entirely encloses the bulb, the view factor, $F_{br} = 1$. The heat flow between the bulb and room is just the overall driving force, divided by the sum of resistances:

$$q_{b\to r} = \frac{\sigma^r\left(T_b^4 - T_r^4\right)}{\dfrac{1}{A_{sb}F_{br}} + \dfrac{1-\varepsilon^r_b}{\varepsilon^r_b A_{sb}}} = \frac{5.67\times10^{-8}(428^4 - 298^4)}{\dfrac{1}{\pi(0.075)^2} + \dfrac{1-0.82}{0.82(\pi(0.075)^2)}} = 21.1\,\text{W}$$

Fully one-fifth of the power consumed by the lightbulb is lost as heat via radiation. Another 40%–70% will be lost via convection to the atmosphere and conduction through the base leaving only about 10%–20% of the power available as light. Soft white bulbs have a phosphor coating on the inside. A clear bulb would lose even more heat because the filament is at ~2900 K.

15.8.3 Radiation Shields

One of the most important heat transfer devices is the radiation shield. This is a low-emissivity, high-reflectivity, material used to reduce the net radiative exchange between two surfaces. One common radiation shield is the silver windshield screen used in cars on a sunny day. This material reflects sunlight coming in through the windshield and reduces the net radiative transfer between the Sun and the interior of the car. Radiation shields are also important in the construction of Airstream® trailers, in the spacecraft and space suits worn by astronauts, and in simple things like

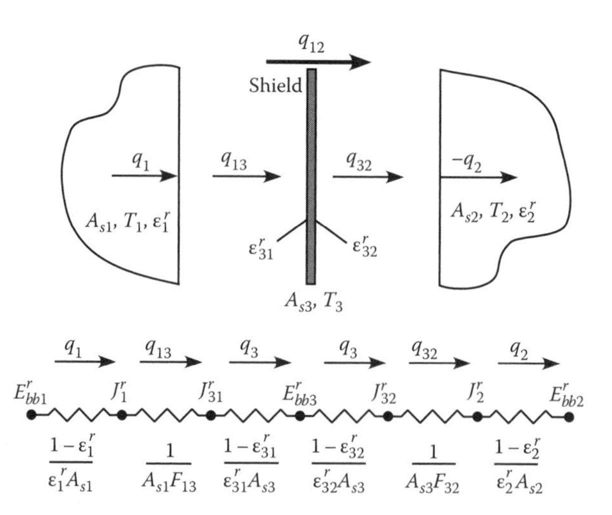

FIGURE 15.23 Radiation shield between two large, parallel surfaces.

a toaster oven where the shields keep the outer surface of the oven from getting too hot. Radiation shield technology is now being incorporated into home attic insulation and camping stoves and has always been important in the design of microwave ovens and thermos bottles. The protein melanin provides for UV radiation shielding in a number of animals and fungi.

Figure 15.23 shows a typical radiation shield system along with the resistance network for the system. All the surfaces are assumed to be very large and parallel to one another. If the surfaces are large relative to the spacing between them, we can view them as a pseudo-enclosure since very little radiation will escape from the ends.

Once we have the resistance network constructed, it is a simple matter to calculate the heat flow. The driving force is always the difference between blackbody emissive powers, in this case E_{bb1}^r and E_{bb2}^r. We then need to sum the resistances. In this case, all the areas are the same $A_{s1} = A_{s3} = A_{s2}$ and all the resistances are in series, so

$$q_{12} = \frac{\sigma^r\left(T_1^4 - T_2^4\right)}{\dfrac{1-\varepsilon_1^r}{\varepsilon_1^r A_{s1}} + \dfrac{1}{A_{s1}F_{13}} + \dfrac{1-\varepsilon_{31}^r}{\varepsilon_{31}^r A_{s3}} + \dfrac{1-\varepsilon_{32}^r}{\varepsilon_{32}^r A_{s3}} + \dfrac{1}{A_{s3}F_{32}} + \dfrac{1-\varepsilon_2^r}{\varepsilon_2^r A_{s2}}} \tag{15.96}$$

Notice that the orientation of the radiation shield does not affect the heat flow. It does however affect the temperature of the radiation shield. You can convince yourself of this by using the results of Equation 15.96 to calculate the temperature of the radiation shield for the two different orientations. In general, we want to have the lowest emissivity side facing the highest temperature surface to insure the lowest radiation shield temperature.

Example 15.7 Automobile Catalytic Converter

A typical automobile catalytic converter catalyst operates at about $T_c = 850°C$. The surface temperature of the converter may easily exceed 500°C. This high temperature may be one cause of the high incidence of asphalt and concrete failures at intersections. The repeated heating and cooling while cars are stopped at traffic lights results in the deterioration of the paving. To alleviate the problem, auto manufacturers are assessing the value of radiation shields. For a typical catalytic converter area of 0.0625 m^2 and distance from the road of 0.25 m, what would the reduction in heat flux be if a radiation shield of emissivity, $\varepsilon_s^r = 0.1$, were installed between the converter and ground? Assume $T_g = 25°C$, $\varepsilon_g^r = 0.9$ (ground), and $\varepsilon_c^r = 0.8$ (converter).

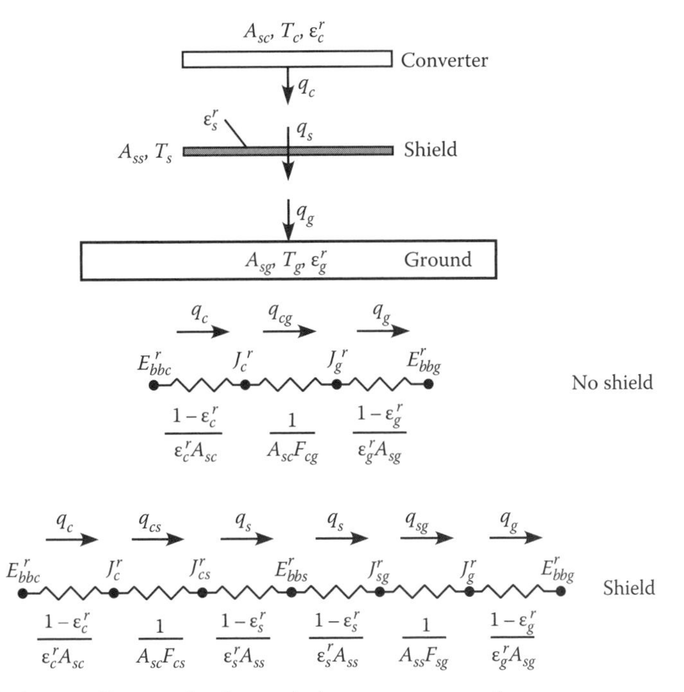

FIGURE 15.24 Resistance diagrams for the catalytic converter example.

We compare the overall resistances with and without the radiation shield. The resistance diagrams are shown in Figure 15.24. The driving forces for both cases are identical and the radiation shield reduces the heat flux by a factor of

$$f = \frac{\frac{1-\varepsilon_c^r}{\varepsilon_c^r A_{sc}} + \frac{1}{A_{sc}F_{cs}} + \frac{1-\varepsilon_s^r}{\varepsilon_s^r A_{ss}} + \frac{1-\varepsilon_s^r}{\varepsilon_s^r A_{ss}} + \frac{1}{A_{ss}F_{sg}} + \frac{1-\varepsilon_g^r}{\varepsilon_g^r A_{sg}}}{\frac{1-\varepsilon_c^r}{\varepsilon_c^r A_{sc}} + \frac{1}{A_{sc}F_{cg}} + \frac{1-\varepsilon_g^r}{\varepsilon_g^r A_{sg}}}$$

In this case, the area of the ground is effectively infinite and so its resistance to heat transfer is zero. Assuming the shield covers the entire converter, we have

$$f = \frac{\frac{1-0.8}{0.8(0.0625)} + \frac{1}{(0.0625)1} + 2\left(\frac{1-0.1}{0.1(0.0625)}\right) + \frac{1}{(0.0625)1} + 0}{\frac{1-0.8}{0.8(0.0625)} + \frac{1}{(0.0625)1} + 0} = 16.2$$

and so can reduce the heat transfer to the asphalt by a factor of over 16. The key issue would be trying to keep the radiation shield's emissivity at a value of 0.1 through road salt, abrasion, and every other detrimental agent the underside of a car comes in contact with.

Example 15.8 Radiation Shield about a Cryogenic Duct

A diffuse, gray radiation shield of 60 cm diameter and emissivities of $\varepsilon_{si}^r = 0.01$ and $\varepsilon_{so}^r = 0.1$ on the inner and outer surfaces, respectively, is placed about a long tube transporting liquid oxygen to a rocket. The tube surface is also diffuse and gray with an emissivity of $\varepsilon_t^r = 0.85$ and diameter of 20 cm. The region between the tube and the shield is evacuated. The exterior surface of the shield is exposed to a large room whose walls are at 17°C and the shield experiences convection

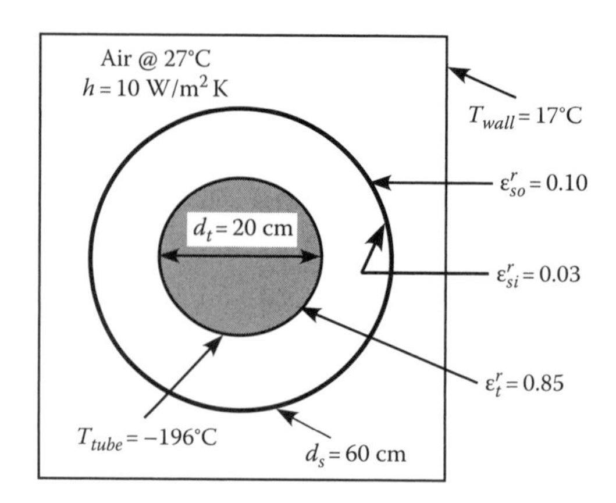

FIGURE 15.25 Radiation shield surrounding a duct carrying a hot process stream.

with the surrounding air at 27°C with h = 10 W/m² K. The system is shown in Figure 15.25. Determine the steady-state operating temperature of the shield if the tube is maintained at −196°C. You may assume the duct is 10 m long.

This problem is an example of multimode heat transfer incorporating radiation. The first task is to assemble the resistance diagram for this situation. Considering only radiation, we have a classic radiation shield resistance diagram. At the outer surface of the shield, we also have convection that sets up a resistance in parallel with the radiation. The resistance network is given in Figure 15.26. Notice that the split with convection occurs at E^r_{bbs} the blackbody emissive power of the shield. This is the appropriate point since E^r_{bbs} is defined solely in terms of the temperature, so we have a reference temperature point as a node for the convective resistance. Splitting off at a radiosity node is meaningless since the surface resistance of the shield affects the split of heat flow between radiation and convection. High surface resistances would force more heat to flow through the convective arm and vice versa.

All the view factors in the diagram are one since the process tube is completely surrounded by the shield and the shield is surrounded by the wall. In addition, A_w, is very large so the room effectively behaves as a blackbody at a temperature of 17°C. It is not convenient here to use an overall driving force–resistance concept since the final two driving forces are different functions of the temperature. However, we can split the problem in two considering the heat flow through the purely radiative component and the heat flow from the surface of the shield to the room via convection. At a steady state, those two flows must be equal. We apply conservation of energy at the shield, a form of Kirchhoff's current law:

$$q_t = q_w + q_{conv}$$

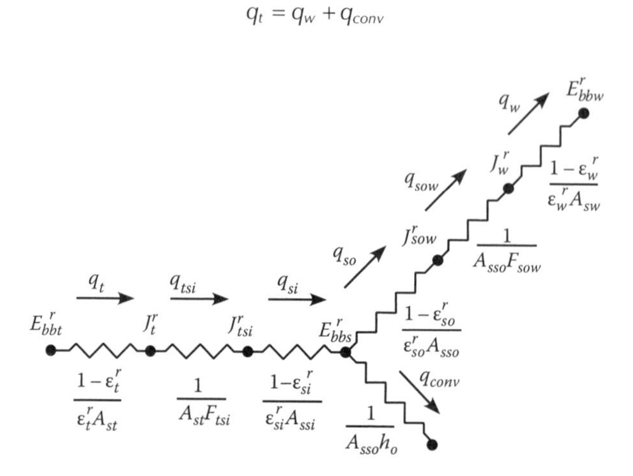

FIGURE 15.26 Resistance network for the process stream problem.

From the driving force/resistance concept, we can obtain q_{si} which is the same as q_t:

$$q_t = \frac{\sigma^r\left(T_t^4 - T_s^4\right)}{\dfrac{1-\varepsilon_t^r}{\varepsilon_t^r A_{st}} + \dfrac{1}{A_{st}F_{tsi}} + \dfrac{1-\varepsilon_{si}^r}{\varepsilon_{si}^r A_{ssi}}} = 2.97\times10^{-8}\left[(77)^4 - T_s^4\right]$$

q_w is evaluated in the same way keeping in mind the great area of the wall:

$$q_w = \frac{\sigma^r\left(T_s^4 - T_w^4\right)}{\dfrac{1-\varepsilon_{so}^r}{\varepsilon_{so}^r A_{sso}} + \dfrac{1}{A_{sso}F_{sow}} + \dfrac{1-\varepsilon_w^r}{\varepsilon_w^r A_{sw}}}$$

$$= 1.07\times10^{-7}\left[T_s^4 - (290)^4\right]$$

q_{conv} is evaluated based on the heat transfer coefficient given:

$$q_{conv} = \bar{h}A_{sso}(T_s - T_{air}) = 10(18.85)(T_s - 300)$$

Equating the heat flows and solving for T_s gives

$$T_s \approx 298.3\text{ K} \qquad q_t = -234.1\text{ W}$$

15.8.4 Radiative Heat Transfer in Three-Surface Enclosures

The last topic we will discuss in radiation is the three-surface enclosure. This is a common situation and extrapolation to more than three surfaces is straightforward. The situation is shown in Figure 15.27. There are three separate surfaces all held at different temperatures. The objective of the problem is to solve for the three radiosities and then obtain the corresponding heat flows from each surface.

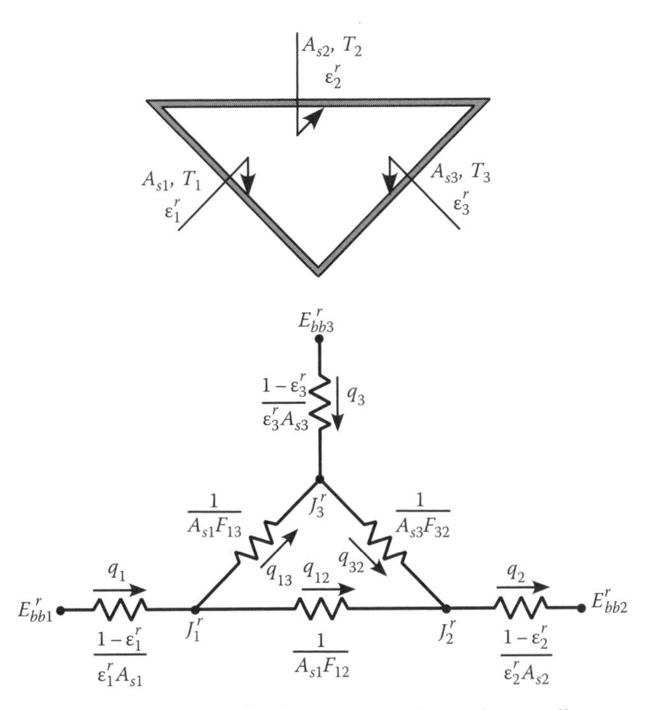

FIGURE 15.27 A three-surface enclosure with the accompanying resistance diagram.

The three-surface enclosure introduces the concept of matrix solutions to radiation problems since we will develop a set of algebraic equations for the unknowns, J_1^r, J_2^r, and J_3^r. The resistance diagram is a clue for how to proceed. We borrow Kirchhoff's current law from electric circuit theory and apply it to the radiosity nodes. The application of Kirchhoff's current law is really just a steady-state energy balance applied at each node. In essence we know that at steady state, whatever heat flows into the node must also flow out. In the resistance diagram, we set up arbitrary directions for the heat flows. As long as we follow the arrow conventions in the diagram, the mathematics will correct us and give us the correct heat flow directions. The energy balances at the nodes are as follows:

$$\begin{aligned} &In - Out = 0 \\ &q_1 - q_{12} - q_{13} = 0 \qquad J_1^r \end{aligned} \tag{15.97}$$

$$\begin{aligned} &In - Out = 0 \\ &q_{12} + q_{32} - q_2 = 0 \qquad J_2^r \end{aligned} \tag{15.98}$$

$$\begin{aligned} &In - Out = 0 \\ &q_{13} + q_3 - q_{32} = 0 \qquad J_3^r \end{aligned} \tag{15.99}$$

We now fill in each of the heat flows using the driving force–resistance concept and the direction arrows in Figure 15.27:

$$\frac{E_{bb1}^r - J_1^r}{\dfrac{1-\varepsilon_1^r}{\varepsilon_1^r A_{s1}}} - \frac{J_1^r - J_2^r}{\dfrac{1}{A_{s1}F_{12}}} - \frac{J_1^r - J_3^r}{\dfrac{1}{A_{s1}F_{13}}} = 0 \qquad J_1^r \tag{15.100}$$

$$\frac{J_1^r - J_2^r}{\dfrac{1}{A_{s1}F_{12}}} + \frac{J_3^r - J_2^r}{\dfrac{1}{A_{s3}F_{32}}} - \frac{J_2^r - E_{bb2}^r}{\dfrac{1-\varepsilon_2^r}{\varepsilon_2^r A_{s2}}} = 0 \qquad J_2^r \tag{15.101}$$

$$\frac{J_1^r - J_3^r}{\dfrac{1}{A_{s1}F_{13}}} + \frac{E_{bb3}^r - J_3^r}{\dfrac{1-\varepsilon_3^r}{\varepsilon_3^r A_{s3}}} - \frac{J_3^r - J_2^r}{\dfrac{1}{A_{s3}F_{32}}} = 0 \qquad J_3^r \tag{15.102}$$

The three previous equations can be rearranged and placed in matrix form to solve for the vector of unknowns, J_1^r, J_2^r, and J_3^r. The form looks like

$$\begin{bmatrix} K_{11} & K_{12} & K_{13} \\ K_{21} & K_{22} & K_{23} \\ K_{31} & K_{32} & K_{33} \end{bmatrix} \begin{bmatrix} J_1^r \\ J_2^r \\ J_3^r \end{bmatrix} = \begin{bmatrix} F_1 \\ F_2 \\ F_3 \end{bmatrix} \tag{15.103}$$

where

K is the coefficient matrix composed solely of surface and geometrical resistances

F is the force vector that is a function of the surface and geometrical resistances, but also contains all the information about the blackbody emissive powers

This formulation can be easily extended to any number of surfaces and solved with many symbolic or numeric manipulation programs. Once the J_i^r are determined, the equation set (15.100 through 15.102) can be reentered to determine any of the heat flows between the surfaces.

Example 15.9 Insulated or Reradiating Surface

One of the most common examples of three-surface radiation involves the case where one surface is insulated and so reradiates into the enclosure. Figure 15.28 shows the schematic situation and the associated resistance network. Determine an expression for the heat flows q_1 and q_2.

Since the reradiating wall is insulated, $q_3 = 0$. The driving force–resistance formulation for the leg between J_3^r and E_{bb3}^r allows us to show that $J_3^r = E_{bb3}^r$ for the reradiating surface:

$$q_3 = 0 = \frac{E_{bb3}^r - J_3^r}{\left(\left(1-\varepsilon_3^r\right)/\varepsilon_3^r A_{s3}\right)}$$

$$J_3^r = E_{bb3}^r = \sigma^r T_3^4$$

Thus, if we can calculate J_3^r, we can readily obtain the temperature of the reradiating surface.

Applying an energy balance about all three radiosity nodes gives

$$\frac{E_{bb1}^r - J_1^r}{\dfrac{1-\varepsilon_1^r}{\varepsilon_1^r A_{s1}}} - \frac{J_1^r - J_2^r}{\dfrac{1}{A_{s1}F_{12}}} - \frac{J_1^r - J_3^r}{\dfrac{1}{A_{s1}F_{13}}} = 0 \qquad J_1$$

$$\frac{J_1^r - J_2^r}{\dfrac{1}{A_{s1}F_{12}}} + \frac{J_3^r - J_2^r}{\dfrac{1}{A_{s3}F_{32}}} - \frac{J_2^r - E_{bb2}^r}{\dfrac{1-\varepsilon_2^r}{\varepsilon_2^r A_{s2}}} = 0 \qquad J_2$$

$$\frac{J_1^r - J_3^r}{\dfrac{1}{A_{s1}F_{13}}} + 0 - \frac{J_3^r - J_2^r}{\dfrac{1}{A_{s3}F_{32}}} = 0 \qquad J_3$$

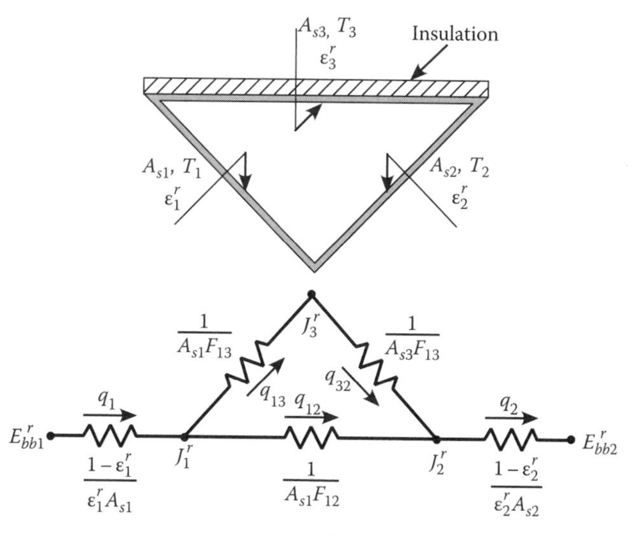

FIGURE 15.28 A three-body enclosure with one reradiating wall.

Notice that we have no surface resistance for wall 3 and so the emissivity of that wall will not enter into any of the resulting calculations. We can now solve for the radiosities using matrix methods, or we can solve for what we are really interested in, q_1 and q_2. With $q_3 = 0$, the resistance network is a simple series–parallel combination that we can get the equivalent resistance for:

$$\mathcal{R}_{eq} = \frac{1-\varepsilon_1^r}{\varepsilon_1^r A_{s1}} + \frac{1}{A_{s1}F_{12} + \left(1 \Big/ \left(\dfrac{1}{A_{s1}F_{13}} + \dfrac{1}{A_{s2}F_{23}}\right)\right)} + \frac{1-\varepsilon_2^r}{\varepsilon_2^r A_{s2}}$$

Solving for q_1 using the driving force between surface 1 and 2 gives

$$q_1 = -q_2 = \frac{E_{bb1}^r - E_{bb2}^r}{\mathcal{R}_{eq}}$$

Example 15.10 Conduction and Radiation in a Sierpinski Carpet

Finite element techniques can be used to solve complex, coupled radiation problems that would prove to be intractable analytically. An example of one of these problems is conduction and radiation in a Sierpinski carpet. A Sierpinski carpet is fractal structure. One kind of Sierpinski carpet is formed by taking a square and repeatedly removing smaller squares that are 1/3 the size of the square in the prior generation. The structure formed by repeating this four times is shown in Figure 15.29. This kind of fractal structure can be viewed as a model porous material.

The Comsol® module Sierpinski considers radiation and conduction through a glass-based Sierpinski carpet with an applied heat flux on one surface and radiation to the ambient environment at a temperature $T_{amb} = 298$ K. At these temperatures, glass is opaque to thermal radiation and the glass has an emissivity approaching 0.9. Figure 15.29 shows the temperature contours throughout the material. Notice how the temperature snakes around the porous regions. This can be seen more clearly in Figure 15.30 where two temperature profiles have been plotted: one near the large central pore and the second near one of the next-generation pores. While the

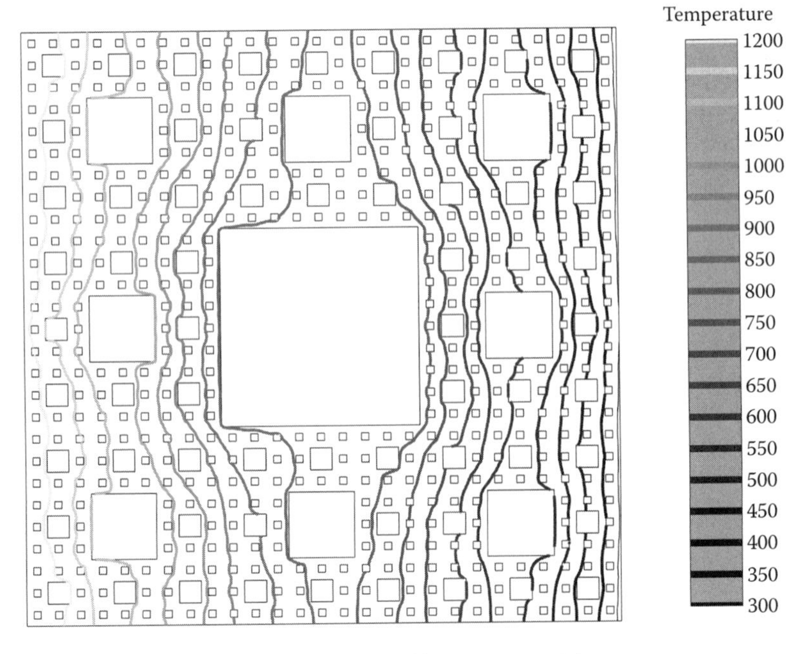

FIGURE 15.29 Temperature contours in the Sierpinski porous material.

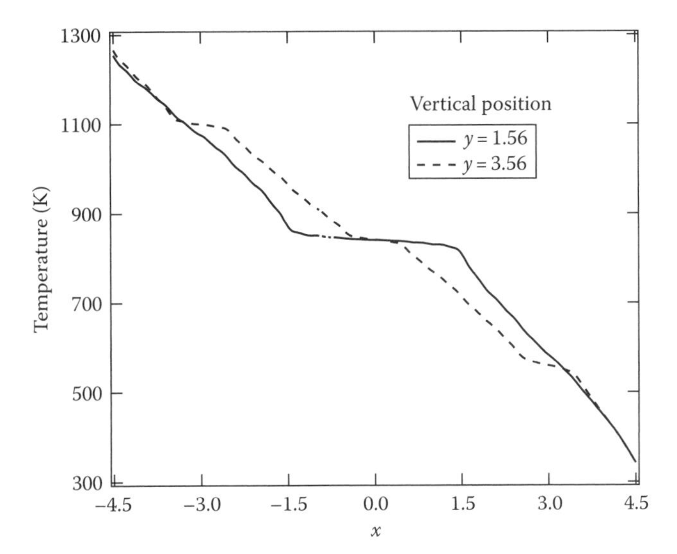

FIGURE 15.30 Temperature profile across the Sierpinski porous material at two different locations showing the effect of periodic radiating elements.

temperature gradient varies throughout the material, depending on the local porosity, the overall temperature difference between the two ends is relatively independent of position. You will be able to delve into the variation in effective thermal conductivity as a function of local porosity for the carpet in the homework at the end of this chapter.

15.9 SUMMARY

We have seen how radiative heat transfer is governed by the propagation of electromagnetic energy and the interaction of that energy with matter. In most instances, a complete description of radiative heat transfer starting from the basic equation governing the interaction of electromagnetic waves with matter is intractable. Thus, we resorted to simplified descriptions of the irradiation of a surface, the energy emitted by a surface, and the properties of the surface. Two basic surfaces were introduced: the ideal blackbody and the more realistic graybody. Radiative exchange between these surfaces was discussed in terms of the surface properties and the geometrical orientation of the surfaces. These factors gave rise to surface and geometrical resistances that we could use in a sum-of-resistance context to describe the radiative exchange.

PROBLEMS

GENERAL

15.1 Explain why the net heat exchange between two blackbodies at temperatures T_1 and T_2, $q_{12}(BB)$, is greater than the net heat exchange between two graybodies, $q_{12}(GB)$, at the same temperatures.

15.2 A diffuse surface has the following spectral emissivity:

$$\varepsilon(\lambda) = \begin{cases} 0.5 & \lambda < 1.5\ \mu\text{m} \\ 0.75 & 1.5\ \mu\text{m} < \lambda < 2.5\ \mu\text{m} \\ 0.25 & \lambda > 2.5\ \mu\text{m} \end{cases}$$

Sketch, *qualitatively*, the behavior of the *total emissivity* as a function of temperature.

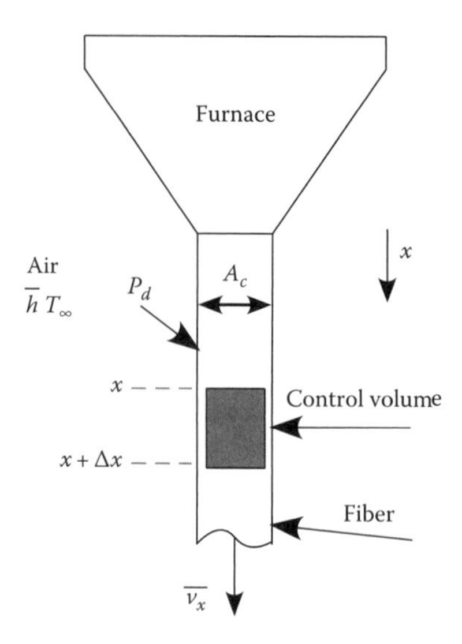

FIGURE P15.3

15.3 Optical fibers are made by drawing the hot glass from a furnace as shown in Figure P15.3. Heat, generated in the furnace, is transferred by conduction along the drawn glass fiber, by convection to the atmosphere at an ambient temperature, T_∞, through a heat transfer coefficient, $\bar{h}$, and by radiation to the ambient air. The fiber is pulled at a constant velocity, $\overline{v_x}$. The fiber temperature is $T(x)$, and the cross-sectional area and perimeter are A_c, and P_d, respectively. Assuming the fiber may be treated as a fin (1-D conduction) under steady-state conditions, start with an energy balance on a control volume (fixed in space) of length Δx, and show that the following equation for $T(x)$ results (conductivity k, density ρ, and heat capacity C_p are known. The air and fiber can be assumed to be black bodies):

$$\frac{d^2T}{dx^2} - \frac{\rho C_p \overline{v_x}}{k}\frac{dT}{dx} - \frac{\bar{h}P_d}{kA_c}(T - T_\infty) - \frac{\sigma^r P_d}{kA_c}\left(T^4 - T_\infty^4\right) = 0$$

15.4 A recent mission to Venus used radar to penetrate the clouds and map the surface. A similar process has been used to penetrate the sands of the Sahara and see the long dead rivers that once used to flow there. Explain why this technique works. Would it work if the surface were highly metallic?

15.5 Spacecraft must be cooled via radiative mechanisms and one means of doing this is radiation fins. In many instances, these fins are heated rods that protrude from the spacecraft. Assume that the rod is of length, L; cross-sectional area, A_c; and perimeter, P_d. Its base temperature is T_b, and the rod has a thermal conductivity and emissivity, k and ε^r, respectively. Set up the differential equation describing the temperature profile within the rod assuming that space is a blackbody at T_s.

Radiation Properties and Blackbody Radiation

15.6 The emissivity of a body is generally a function of temperature and this makes calculation of the heat exchange difficult. Show that if the emissivity varies linearly with temperature, we can recover the form of the two-body exchange law,

$$Q_{12} = (\varepsilon_1)_{ref}\,\sigma^r A_1\left(T_1^4 - T_2^4\right)$$

provided the emissivity is evaluated at a reference temperature defined by

$$T_{ref} = \frac{T_1^5 - T_2^5}{T_1^4 - T_2^4}$$

15.7 Two approximations to Planck's law are useful in the extreme low and high limits of λT.

a. Show that in the limit where $(C_1/\lambda T) \gg 1$, Planck's spectral distribution reduces to the following form:

$$E_{b\lambda}(\lambda, T) \approx \frac{C_o}{\lambda^5} \exp\left(-\frac{C_1}{\lambda T}\right) \quad \text{Wien's law}$$

Compare this result to Planck's distribution and determine when the error between the two is less than 1%.

b. Show that in the limit $(C_1/\lambda T) \ll 1$, Planck's distribution law reduces to

$$E_{b\lambda}(\lambda, T) \approx C_2 \frac{T}{\lambda^4} \quad \text{Rayleigh–Jeans law}$$

Compare with Planck's distribution and determine when the two are in error by less than 1%.

15.8 An opaque, diffuse surface has a spectral reflectivity that varies with wavelength as shown in Figure P15.8. The surface is held at 850 K and one side is exposed to thermal radiation, while the other side is insulated. The irradiation is wavelength dependent as shown in Figure P15.8.

a. What are the total absorptivity and total emissivity of the surface?

b. What is the net radiative heat flux to the surface?

15.9 What fraction of the Sun's energy lies in the following spectral range?

a. $0 < \lambda < 2\ \mu m$

b. $2 < 10\ \mu m$

c. $2 < \lambda < 4\ \mu m$

Assume the Sun radiates as a blackbody at a temperature of 5800 K.

15.10 Contractors in the southwest generally coat roofs with light materials designed to keep the interior of homes cool. It has been suggested that the same practice be followed in the north. We would like to evaluate this prospect. Consider two diffuse, opaque, coatings with spectral absorptivities shown in Figure P15.10. Which one results in a lower roof temperature? Which coating is preferred for winter use? Sketch the spectral distribution of $\alpha(\lambda)$ that would be

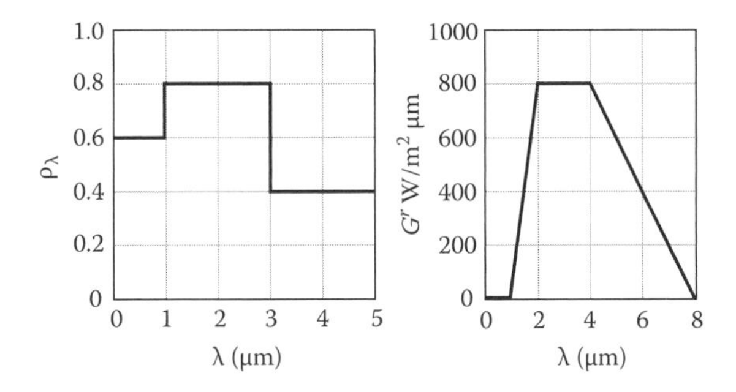

FIGURE P15.8

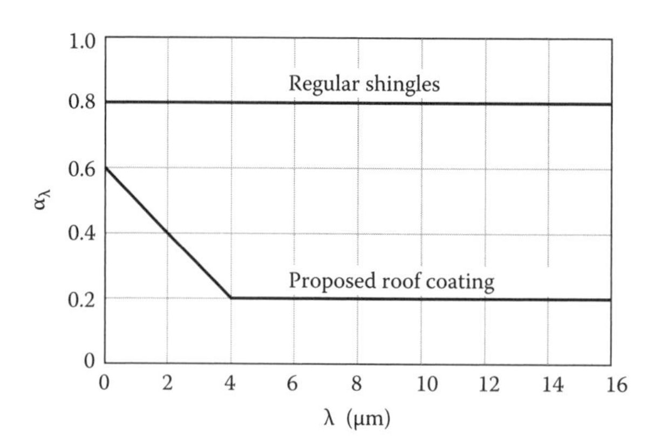

FIGURE P15.10

ideal for winter use? Is there a best compromise for the two seasons? What does its spectral absorptivity look like? Use data for the Sun radiating as a blackbody at 5800 K and deal on a 1 m² basis. You may assume the house interior is at 298 K.

15.11 People who apply facial makeup are often vexed by their difference in appearance under daylight, incandescent, and fluorescent lighting conditions. Assuming daylight acts like illumination from a blackbody at 5800 K, incandescent as illumination from a blackbody at 2900 K, and fluorescent lighting as illumination from a blackbody at 4000 K, make some calculations and comment on how a Caucasian complexion might change under the different conditions. The emissivity of skin is about 0.95. You may have to look for data on skin reflectivity as a function of wavelength.

15.12 Planck's radiation law assumes a body radiating into a vacuum. If the body is radiating within another medium, the wavelength of the emitted radiation changes because the speed of light in the medium changes even if the frequency of the radiation remains the same. This leads to changes in Wien's displacement law and the power output of a blackbody of the following form:

$$E_{bb}^{r} = \int_{0}^{\infty} \frac{2\pi c_m^2 \hbar}{\lambda_m^5 \left[\exp\left(\dfrac{\hbar c_m}{\lambda_m k_b T}\right) - 1\right]} d\lambda_m \qquad \lambda_{\max} T = \frac{2.884 \times 10^{-3}}{\eta}$$

a. Integrate the previous power density function to show that in the medium of refractive index, η, the blackbody emissive power is

$$E_{bb}^{r} = \eta^2 \sigma^r T^4$$

Here, we know that in the medium, the wavelength and velocity are related to their vacuum wavelength and velocity via

$$\lambda_m = \frac{\lambda}{\eta} \qquad c_m = \frac{c}{\eta}$$

b. If we have an object in a dense medium of refractive index 1.5 that is radiating at a temperature of 3000 K, determine the blackbody emissive power and the wavelength of maximum emission.

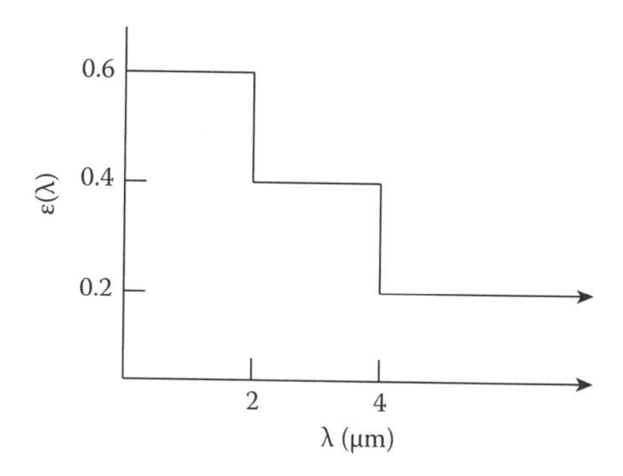

FIGURE P15.13

15.13 The spectral emissivity for a material is shown in Figure P15.13 and does not depend much on the temperature.

a. What is the total hemispherical absorptivity for the material if it is exposed to radiation from a blackbody source at 1500 K?

b. What is the total hemispherical absorptivity if the material is exposed to radiation from a graybody source at 1550 K with an emissivity, $\varepsilon = 0.75$?

c. What is the total hemispherical absorptivity if the material is exposed to radiation from an identical material acting as the source at 1500 K?

View Factors

15.14 Find the view factors for the following enclosures:

a. Long, hemispherical, duct of radius, r_o

b. A paraboloid ($z = 1 - x^2 - y^2$) bounded at the bottom by the plane $z = 0$

15.15 Find the view factor F_{12} for the rectangles shown in Figure P15.15.

Radiative Exchange

15.16 A diffuse, gray radiation shield 90 mm in diameter with emissivities of $\varepsilon_{2,i}$ and $\varepsilon_{2,o}$, on its inner and outer surfaces, respectively, is concentric with a long tube transporting a hot process fluid. The tube surface is diffuse and gray with an emissivity of ε_t and a diameter of 50 mm. The area between the tube and the radiation shield is evacuated. The outer surface

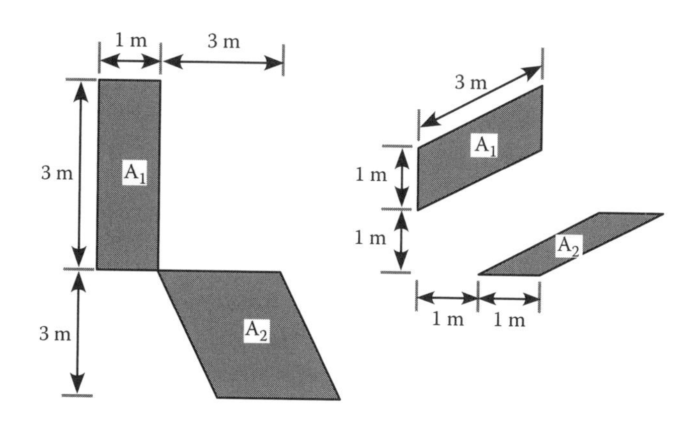

FIGURE P15.15

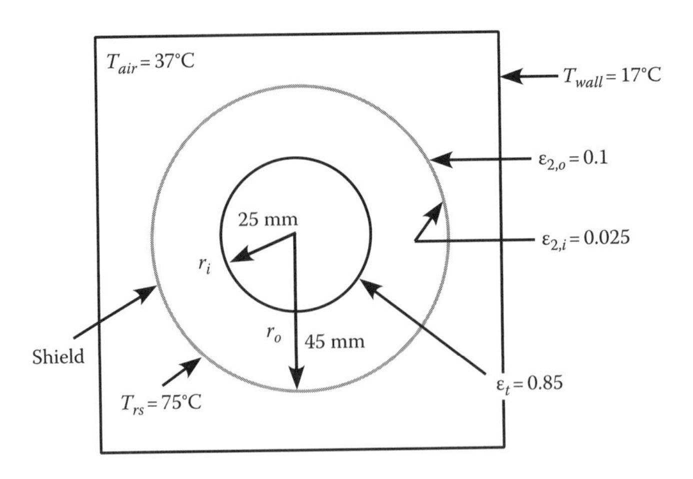

FIGURE P15.16

of the shield is exposed to a very large room whose walls are at 17°C and the shield experiences convection with the surrounding air at 37°C and with a heat transfer coefficient $h = 10$ W/m² K. Determine the steady-state operating temperature of the inner tube surface if the shield is held at 75°C (Figure P15.16).

15.17 Pizza's in large restaurants are often made in belt furnaces. The pizza is placed on a translating belt that carries it through the oven for a specified length of time. Assume a semicylindrical oven, diameter 0.75 m, radiating from a hemispherical element like a blackbody at a temperature of 2000 K. The pizza, 1 cm thick and 0.45 m in diameter, is cooked when its centerline temperature reaches at least 225°C. If the oven is 1.5 m long, how fast must the belt move to insure a correct baking time? You may neglect convection in this analysis. The emissivity of the pizzas may be assumed to be 0.95, and its other thermal properties may be approximated as those of cake batter with a heat capacity of 2500 J/kg K. (You may have to approximate the view factor between the pizza and oven.)

15.18 As an engineer for a heating and cooling company, you must design the refrigeration for a new ice skating rink. As a first crack, you assume a rink 35 m in diameter covered by a dome 45 m in diameter. The rink is kept at a temperature of 0°C, while the dome is at 15°C. If both behave as blackbody radiators, what is the net heat transfer between the dome and rink and the amount of refrigeration required?

15.19 Cryogenic gases are often stored in double-walled, spherical containers. The space between the two walls is evacuated so that all heat transfer between the walls occurs via radiation. Assume liquid oxygen is being stored in the inner container (10 m in diameter) at a temperature of 95 K. The outer shell of the container is held at 280 K, on average, and is 15 m in diameter. If the walls of the container are opaque, diffuse, gray surfaces with emissivities of 0.05,

a. What is the net radiative exchange between the walls of the vessel?
b. How much oxygen will be lost via evaporation? ($\Delta H_{vap} = 6.82$ kJ/mol)
c. Why store the material in spherical containers?

15.20 A toaster oven is fabricated from aluminum that has been oxidized over time (properties akin to anodized aluminum). The cylindrical heater element (0.5 cm in diameter) is located 5 mm from the bottom surface. This element radiates like a blackbody at 1000 K. The toaster oven sits 3 cm from a wood countertop. What would be the reduction in heat flux to the countertop if a radiation shield of emissivity 0.05 were placed 3 mm from the heater element between the element and the outer surface of the oven?

15.21 A new radiant heating system is installed in the floor of a room 4 m long by 3 m wide. The floor is held at a temperature of 35°C and the ceiling, located 2.75 m from the floor, is well insulated. The walls leak heat and have a surface temperature of 17°C. If all the surfaces have an emissivity of 0.9, how much heat is transferred via radiation from the floor?

15.22 Consider the previous room problem (Problem 15.21). Of the four walls in the room, two are external walls with surface temperatures of 17°C. The other two are internal walls whose temperatures are at 20°C.

a. Set up the equations describing radiative heat transfer between the walls, floor, and ceiling.
b. What are the view factors?
c. Solve the set of equations to obtain the heat flows between the walls, floor, and ceiling.
d. How much energy must be supplied to the floor?

15.23 Derive an expression for the net radiation exchange between two uniform concentric spheres. You may assume each sphere is a diffuse graybody with emissivities ε_1 and ε_2. The outer sphere is insulated (Figure P15.23).
If $\varepsilon_1 = 0.5$, $\varepsilon_2 = 0.8$, $T_1 = 500$ K, $T_2 = 1000$ K, $r_1 = 0.5$ m, and $r_2 = 1.5$ m, what is the net rate of heat exchange between the two objects?

15.24 The frustrum of a cone has its base heated as shown in Figure P15.24. The top of the cone is held at 600 K and the bottom at 1500 K, and its sides are insulated.

a. What is the net rate of heat transfer between surfaces 1 and 3?
b. What is the temperature of surface 2?

15.25 Two hollow rods of diameter 0.04 m and length of 1 m run parallel to one another separated by a distance of 0.25 m. The rods can be assumed to be diffuse radiators with an emissivity of 0.8.

a. If rod 1 is held at a temperature of 1800 K and rod 2 at 600 K, what is the net rate of heat transfer between the two?
b. If rod 1 is held at a temperature of 1800 K and rod 2 is insulated on the inside, what is the temperature of rod 2?
c. If rod 1 is held at a temperature of 1800 K and rod 2 cannot exceed a temperature of 500 K, how far apart must the rods be spaced if the maximum rate of heat exchange between the rods cannot exceed 100 W?

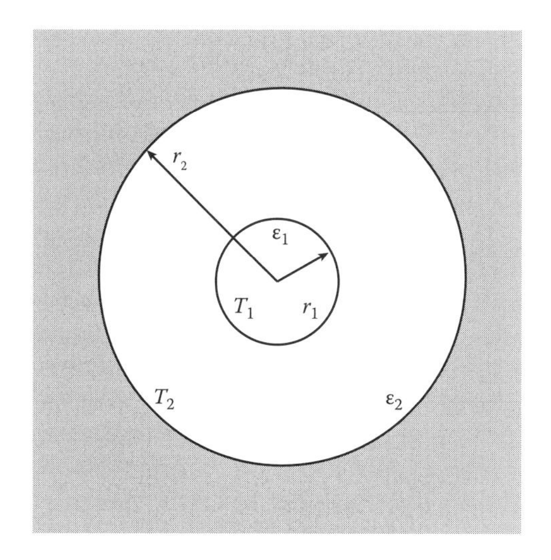

FIGURE P15.23

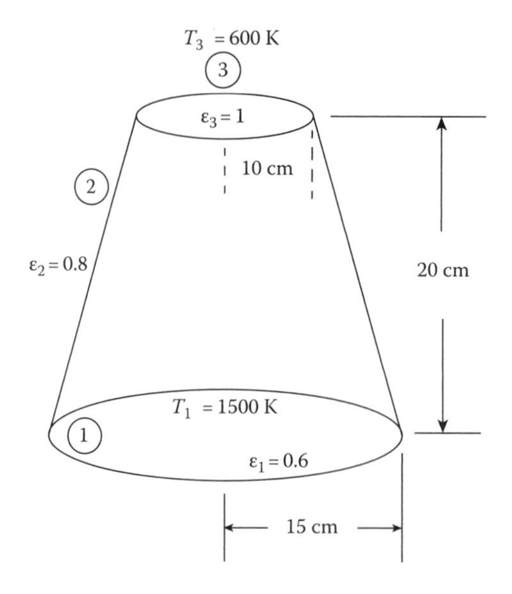

FIGURE P15.24

15.26 A copper ball bearing of radius $r_0 = 0.01$ m is exposed to a small heat source that operates at a temperature of 2000 K and has an emissivity of 0.75. The situation is shown in Figure P15.26 with the copper sphere being 0.05 m away from the surface. The copper needs to be heated for tempering and must reach a temperature of 500 K from room temperature.

a. Derive the differential equation governing the temperature of the sphere. You may assume the emissivity of the copper does not depend on temperature.

b. How long does it take for the copper to reach its tempering temperature? You may want to solve the problem using what is called a radiation heat transfer coefficient

$$q_{net} = \frac{\sigma^r\left(T_1^4 - T_2^4\right)}{\sum R} = \underbrace{\frac{\sigma^r\left(T_1^2 + T_2^2\right)(T_1 + T_2)}{\sum R}}_{h_{rad}A}(T_1 - T_2)$$

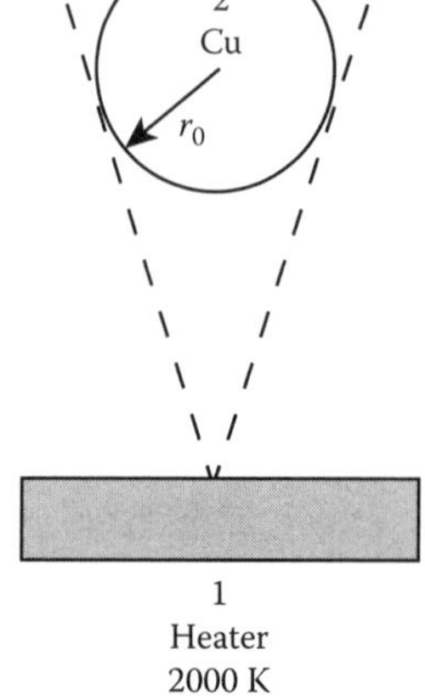

FIGURE P15.26

15.27 You are on a quest to develop the perfect meringue topping for a pie. The surface needs to become a golden brown, a process that results from caramelization of the sugars in the meringue. To form such a meringue, you bake the pie to set the meringue and, as the final step, place it under a broiler element for a short time. A pie is roughly 20 cm in diameter and we can approximate the pie as a hemisphere exposed to a small heating element of surface area 2×10^{-5} m^2. The problem is quite complicated since the rate of heat transfer to the pie needs to be controlled to avoid burning the meringue.

a. If the heating element can deliver 5 W at 1800 K with an emissivity of 0.9 to the meringue surface and the meringue cannot be exposed to a temperature of more than 500 K, how far must the pie be from the heating element? Meringue, being porous, is a very good insulator and so will reradiate into space as a blackbody.

b. If the caramelization reaction is essentially a zero-order reaction obeying

$$r_c\left(\frac{g}{s}\right) = k''(T_{surface} - 422) \qquad k'' = 1\times10^{-4}\,\frac{1}{\text{Ks}}$$

and we need to react 2.5 g of meringue to adequately produce the desired color, how long must the broiler be on?

15.28 Most cooks know that you must tent a large bird during roasting to prevent the skin from burning. Assume we have a turkey in a 475 K oven. The oven surface area is 2 m^2. To keep the oven at that temperature, the heating elements and walls are at about 750 K with surface emissivities of 0.8. Assuming the turkey can be approximated as a hemisphere, 30 cm in diameter, with an emissivity of 0.75, that is totally enclosed by the oven:

a. Compare the rate of heat transfer between the turkey and the oven in the absence of a radiation shield and when an aluminum radiation shield is used with surface emissivities of 0.05 on the polished side and 0.4 on the rough side.

b. When one tents a bird, the radiation shield generally touches the bird skin. If that happens, we would like to have the temperature of the radiation shield as low as possible. Which side of the shield should be pointed toward the oven walls?

15.29 Repeat the "Temperature of the Moon's Surface" example for a more realistic scenario. The Moon has an emissivity of about 0.94 and an albedo or reflectivity of 0.12. Perform a new energy balance and determine an alternative temperature for the Moon.

REFERENCES

1. Phillips, W.D. and Metcalf, H.J., Cooling and trapping atoms. *Sci. Am.*, **256**, 50–56 (March 1987).
2. Meystre, P. and Stenholm, S., Introduction to feature on the mechanical effect of light, *JOSA B*, **2**(11), 1706 (1985). http://dx.doi.org/10.1364/JOSAB.2.001706.
3. Eisberg, R. and Resnick, R., *Quantum Physics of Atoms, Molecules, Solids, Nuclei, and Particles,* 2nd edn., John Wiley & Sons, New York (1985).
4. Guenther, R.D., *Modern Optics*, John Wiley & Sons, New York (1990).
5. Eckert, E.R.G. and Drake, R.M. Jr., *Analysis of Heat and Mass Transfer*, McGraw-Hill, New York (1972).
6. Howell, J.R., Siegel, R., and Menguc, M.P., *Thermal Radiation Heat Transfer,* 5th edn., CRC Press, Boca Raton, FL (2010).
7. Hottel, H.C. and Sarofim, A.F., *Radiative Transfer*, McGraw-Hill, New York (1967).
8. Planck, M., *The Theory of Heat Radiation*, Dover Publications, New York (1979).
9. Leipold, M., Eiden, M., Garner, C.E., Herbeck, L., Kassing, D., Niederstadt, T., Krüger, T. et al., Solar sail technology development and demonstration, *Acta Astronautica*, **52**, 317–326 (2003).
10. Incropera, F.P. and Dewitt, D.P., *Introduction to Heat Transfer,* 4th edn., John Wiley & Sons, New York (1998).

Nomenclature

a acceleration (m/s^2)
A area (m^2)
a_a activity of species a
A_c cross-sectional area (m^2)
a_g acceleration (m/s^2)
A_i interfacial area (m^2)
A_m area of moving wetted surfaces (m^2)
A_p projected or permeable area (m^2)
A_s surface area (m^2)
A_t total area (m^2)
a_v specific surface area (m^2/m^3)
b channel width (m)
B_h magnetic displacement (Wb/m^2)
Bi Biot number
Bi_m Biot number for mass transfer
C capacitance (F)
c speed of light (m/s)
Ç convective transport coefficient
c_a concentration of a; may be composed of average, $\bar{c}_a$, and fluctuating, c'_a components (kmol/m^3)
c_{as} concentration of a on the plate's surface (kmol/m^3)
$c_{a\infty}$ free stream concentration (kmol/m^3)
c_b concentration of b (kmol/m^3)
c_{bb} concentration at fin base (kmol/m^3)
$c_{b\infty}$ free stream concentration of b (kmol/m^3)
C_D drag coefficient/friction factor
c_+ concentration of positively charges species (kmol/m^3)
c_e concentration of electronic charge carriers (C/m^3)
$c_{\pm\infty}$ external/free stream charged species concentration (C/m^3)
C_f Fanning friction factor
$\overline{C_f}$ average Fanning friction factor
$\overline{C_D}$ average drag coefficient
c_h hormone concentration (kmol/m^3)
C_L lift coefficient
c_λ wave velocity (m/s)
$c_{\lambda g}$ speed of sound in the gas (m/s)
$c_{\lambda l}$ speed of sound in the liquid (m/s)
c_n concentration of negative charge carriers or electrons (kmol/m^3)
C_p heat capacity at constant pressure (J/kg K)
c_p concentration of particles or holes (kmol/m^3)
C_{pl} heat capacity of liquid (J/kg K)
C_{ps} heat capacity of solid (J/kg K)
c_t total concentration (kmol/m^3)
C_v heat capacity at constant volume (J/kg K)
D diffusivity (m^2/s)
d diameter (m)

D_+ diffusivity of positively charged species (m^2/s)
D_- diffusivity of negative ions (m^2/s)
Da Damkohler number
D_{aa*} tracer or self-diffusivity (m^2/s)
D_{ab} diffusivity of a through b (m^2/s)
d_{ab} mean diameter (m)
D_{ap} diffusivity of a through a particle (m^2/s)
D_{as} diffusivity of a through s (m^2/s)
D_{bs} diffusivity of b through s (m^2/s)
D_e electric displacement (C/m^2)
D^e_{ab} eddy diffusivity (m^2/s)
D_{eff} effective diffusivity (m^2/s)
D_g gravitational flux density ($kg^2/N\ m\ s^2$)
D_{hs} diffusivity of hormone (m^2/s)
D_{ij} binary diffusivity of i through j (m^2/s)
$\mathcal{D}_{ij}$ multicomponent diffusivity of i through j (m^2/s)
D_{mw} diffusivity of cations in water (m^2/s)
d_p particle diameter (m)
D_p diffusivity of particles (m^2/s)
D_s diffusivity in a solid (m^2/s)
d_s skin depth (m)
$\mathcal{D}^T$ thermal diffusivity ($W\ m^2/kmol$)
D_{xw} diffusivity of anions in water (m^2/s)
$D_\pm$ diffusivity of charge carriers (m^2/s)
$\mathcal{E}$ electric field (V/m)
E energy (J)
e charge on an electron (C)
E_a activation energy (J/kmol)
E_c energy transferred due to mass diffusion through permeable walls (J)
E_f drift number
E_μ lost work (J)
$\mathcal{E}_o$ reference electric field strength (V/m)
E_p photon energy (J)
E^r emissive power (W/m^2); may be a function of wavelength
E^r_b emissive power of a blackbody (W/m^2); may be a function of wavelength
E_{tot} total energy of the system (J)
F Force (N)
$F_\perp$ normal force (N)
f Moody friction factor
f fin number
f^c fugacity coefficient
F_{12} view factor between surfaces 1 and 2
$\mathcal{F}a$ Faraday's constant (C/kmol)
F_c force resulting from diffusion of mass through permeable walls (N)
F_h magnetic field force (N)
F_{ij} view factor between surfaces i and j
F_μ viscous forces (N)
f_m Moody friction factor
Fo Fourier number
$F_\parallel$ tangential force (N)
$\mathcal{G}$ gas flow rate (kg/s)

G Gibb's energy (J)
g gravitational acceleration (m/s^2)
G_g universal constant of gravitation (N m^2/kg^2)
$\overline{G}_i$ partial molar Gibbs energy (J)
Gn generation number
Gr Grashof number
G^r irradiation on surface (W/m^2)
G_a^r absorbed component of irradiation (W/m^2)
Gr_{ab} Grashof number for mass transfer
Gr_c condensation Grashof number
G_R^r reflected component of irradiation (W/m^2)
G_T^r transmitted component of irradiation (W/m^2)
Gz Graetz number for heat transfer
Gz_m Graetz number for mass transfer
H enthalpy (J)
H_a Henry's law constant (Pa)
$\mathcal{H}$ magnetic field (Henry)
$\hbar$ Planck's constant (J s)
h heat transfer coefficient (W/m^2 K); height (m)
h_{fc} heat transfer coefficient, forced convection (W/m^2 K)
h_{fg} heat of phase change (J/kg)
h_L head loss (m)
h_{vap} heat of vaporization (J/kg)
$\overline{H}_i$ partial molar enthalpy (J)
h^r radiative heat transfer coefficient (W/m^2 K)
h_x local value of the heat transfer coefficient (W/m^2 K)
$\overline{h}$ average heat transfer coefficient (W/m^2 K)
$\mathcal{I}$ current (A)
I_{bb} intensity of emission of a blackbody (W/m^2 sr)
$\mathcal{I}_C$ current through the capacitor (A)
I_E intensity of emission (W/m^2 sr)
I_I incident intensity (W/m^2 sr)
$\mathcal{I}_R$ current through the resistor (A)
$\mathcal{J}$ current density (C/m^2 s)
J molar flux (kmol/m^2 s)
$\mathcal{J}_+$ flux of positive ions (A/m^2)
$\mathcal{J}_-$ flux of negative ions (A/m^2)
Ja Jakob number
j_i^c mass flux of species i relative to the molar average velocity (kg/m^2 s)
J_i molar flux of species i relative to the molar average velocity (kmol/m^2 s)
j_i mass flux of species i relative to the mass average velocity (kg/m^2 s)
J_p flux of particles
J^r radiosity (W/m^2)
J_b^r radiosity of a blackbody (W/m^2)
J_i^ρ molar flux of species i relative to the mass average velocity (kmol/m^2 s)
K circulation (s^{-1})
k'' first order reaction rate constant (s^{-1})
k thermal conductivity (W/m K)
$\mathbf{k}$ wave vector (m^{-1}); may be complex $\mathbf{k} = \mathbf{k}_{re} + i\mathbf{k}_{im}$ (Chapter 7)
$K.E.$ kinetic energy (J)
k_b Boltzmann's constant (J/K)

K_c overall mass transfer coefficient (m/s)
k_c mass transfer coefficient (m/s)
k^c thermal conductivity of conductor (W/m K)
k_{cg} mass transfer coefficient (gas phase) (m/s)
k_{cl} mass transfer coefficient (liquid phase) (m/s)
k_{cx} mass transfer coefficient (x-phase) (m/s)
k_{cy} mass transfer coefficient (y-phase) (m/s)
$\overline{k_c}$ average mass transfer coefficient (m/s)
k_D dispersion coefficient (m^2/s)
k_e electronic contribution to the thermal conductivity (W/m K)
k_{eff} effective thermal conductivity (W/m K)
K_{eq} partition or equilibrium constant
k_f thermal conductivity of the fluid (W/m K)
k_g'' growth constant (s^{-1})
k_i thermal conductivity of insulation (W/m K)
k_l thermal conductivity of liquid (W/m K)
k_μ convective momentum transfer coefficient (kg/m^2 s)
K_m Michaelis constant ($kmol/m^3$)
k_p pressure conductivity (W m/N)
k_{perm} permeability (m^2)
k_{ph} phonon contribution to the thermal conductivity (W/m K)
k_s thermal conductivity of solid or steel (W/m K)
k_s'' surface area based version of k'' (m/s)
k_{sp} spring constant (N/m)
k^T thermal diffusion ratio
$k_\pm$ charge transfer coefficient (m/s)
$\overline{k_\pm}$ average charge transfer coefficient (m/s)
L length (m)
$\mathcal{L}$ solvent flow rate (kg/s)
ℓ mixing or diffusion length (m)
L_f lift force (N)
l_k Kolmogorov length scale (m)
$\mathcal{L}o$ Lorenz constant ($W\ \Omega/K^2$)
L_{tot} total momentum (kg m/s)
Lw Lewis number
m mass (kg)
m_a total amount of a transferred between phases (kg)
$M_{a,tot}$ total moles of a in the system (kmol)
$m_{a,tot}$ total mass of a in the system (kg)
m_{ab} reduced mass (kg)
m_e mass of an electron (kg)
m_k partition coefficient
M_{tot} total momentum of the system (kg m/s)
m_{tot} total mass of the system (kg)
M_w molecular weight (kg/kmol)
$\dot{M}$ molar flow rate (mol/s)
$\dot{m}$ mass flow rate (kg/s)
$\dot{m}_c$ mass flow rate of cold side fluid (kg/s)
$\dot{m}_d$ mass flow via diffusion through permeable walls (kg/s)
$\dot{m}_h$ mass flow rate of hot side fluid (kg/s)

$\dot{m}_i$ rate of ice formation (kg/s)
$\mathcal{N}$ number of particles
N mass flux (kmol/m^2 s)
$\mathcal{N}_+$ total number of positive charge carriers (holes)
$\mathcal{N}_-$ total number of negative charge carriers (electrons)
N_a flux of a particles (kmol/m^2 s)
n_a mass flux of a (kg/m^2 s)
N_{ao} reference flux of a (kmol/m^2 s)
$\mathcal{N}_{av}$ Avogadro's number
$N_{a\infty}$ free stream molar flux (kmol/m^2 s)
N_b molar flux of b (kmol/m^2 s)
n_b mass flux of b (kg/m^2 s)
$\mathcal{N}_c$ number of particles yet to collide with another particle
N_f friction number
$\mathcal{N}_n$ number of moles
Nu Nusselt number
$\overline{\text{Nu}}$ average Nusselt number
Nu_d Nusselt number based on diameter
Nu_L Nusselt number based on length
Nu_x local value of the Nusselt number
$\text{N}_\pm$ conduction number
P pressure (N/m^2)
$P.E.$ potential energy (J)
$\mathcal{P}_c$ collision probability
P_d perimeter (m)
Pe Peclet number
$P_{i,sat}$ partial pressure at saturation conditions (N/m^2)
$P_{i,\infty}$ free stream partial pressure (N/m^2)
P^r radiation pressure (N/m^2)
Pr Prandtl number
Pr_s Prandtl number evaluated at surface conditions
P_{st} pressure at the stagnation point (N/m^2)
$\mathcal{P}_v$ probability that a particle has a velocity between v and $v + dv$
P_i^v vapor pressure of species i (N/m^2)
$\dot{p}$ momentum generation (kg/m^2 s)
Q heat (J)
$\mathcal{Q}$ charge (C)
q heat flow (W)
$\dot{q}$ heat source (W/m^3)
q'' heat flux (W/m^2)
q_a charge transfer due to diffusion of a through permeable wall (C/s)
$Q_{a,tot}$ total charge of a in the system (C)
q_{ae} charge transfer of a due to an applied electric field (C/s)
$q_{i\to j}$ heat transferred from surface i to surface j (W)
q_{ij} net heat transferred between surfaces i and j (W)
q_m heat flow due to mass diffusion through system boundaries (W)
Q heat (J)
$Q^\pm$ total charge (C)
Q_o maximum heat transferred (J)
q_r radiative heat flow (W)

q''_s heat flux at the plate's surface (W/m^2)
R gas constant (J/kmol K)
$\mathcal{R}$ resistance (m^2 K/W)
r_+ electrical neutralization reaction rate (C/m^3 s)
R_a reaction/recombination rate of "a" (kmol/m^3 s)
Ra Rayleigh number (GrPr)
r_a reaction rate of a (kg/m^3 s)
Ra_d Rayleigh number based on diameter
$\mathcal{R}_c$ convection resistance (m^2 K/W)
$\mathcal{R}_d$ resistance from dielectric (m^2 K/W)
Re Reynold's number
$\mathcal{R}_e$ electrical resistance (Ohms)
Re_d Reynolds number based on diameter
$\mathcal{R}_{ef}$ reflectivity
$\mathcal{R}_{efn}$ reflectivity at normal incidence
Re_L Reynolds number based on length
Re_x local value of the Reynold's number
r_h hydrodynamic radius (m)
$\mathcal{R}_h$ reluctance (Henry)
$\mathcal{R}_I$ internal resistance (m^2 K/W)
r_i inner radius (m)
r_o reference radius (m)
r_p particle radius (m)
$\mathcal{R}_T$ thermal resistance (K/W)
$\mathcal{R}_t$ total resistance (K/W)
$R_{x,y}$ reaction force (N)
$\dot{R}_a$ mass generation rate (kmol/m^3 s)
$r_\pm$ radius of an ion (m)
S entropy (J/K)
s surface recombination velocity (m/s), scalar
Sc Schmidt number
Sc_s Schmidt number evaluated at surface conditions
Sh Sherwood number
$\overline{\mathrm{Sh}}_L$ average Sherwood number
Sh_d Sherwood number based on diameter
Sh_x local value of the Sherwood number
S_k convective charge transfer coefficient (C/m^2)
St Stanton number
T temperature; may be composed of average, $\overline{T}$, and fluctuating, T' components (°C, K)
t time (s)
T_b base temperature (°C, K)
T_c cold side temperature (°C, K)
T_f liquid temperature (°C or K)
t_h thickness (m)
T_h hot side temperature (°C, K)
t_K Kolmogorov time scale (s)
$\mathcal{T}_{rans}$ transmittivity
T_s surface temperature (°C, K)
T_{sat} saturation temperature (°C or K)
T_∞ free stream temperature (°C or K)

U	internal energy (J)
$\mathcal{U}$	overall heat transfer coefficient (W/m^2 K)
u	velocity parallel to the plate (m/s)
U^+	dimensionless universal velocity
u_f	liquid velocity parallel to the plate (m/s)
u_v	vapor velocity parallel to the plate (m/s)
V	volume (m^3); voltage (V)
v	velocity; may be composed of average, $\bar{v}$ and fluctuating, v' components (m/s)
v^*	friction velocity (m/s)
v_a	velocity of system boundaries (m/s)
$V_{applied}$	applied voltage (V)
V_b	molar volume of liquid at its normal boiling point (m^3/kmol)
V_c	molar volume of gas at its critical point (m^3/kmol)
v^c	molar average velocity (m/s)
v_d	drift velocity of a charge carrier (m/s)
V_f	molar volume of solid at its freezing point (m^3/kmol)
v_f	liquid velocity perpendicular to the plate (m/s)
$\overline{V_i}$	partial molar volume of species i (m^3/kmol)
v_K	Kolmogorov velocity scale (m/s)
V_m	molar volume of the liquid (m^3/kmol)
$\mathcal{V}_\mu$	viscous dissipation function (W/m^3)
v_{mass}	mass average velocity (m/s)
v_{Mole}	molar average velocity (m/s)
v_M	molar velocity (m/s)
v_n	velocity of fluid normal to surface (m/s)
V_p	pore/particle volume (m^3)
v_θ	angular velocity (m/s)
V_r	volume ratio
v^ρ	mass average velocity (m/s)
v_s	plate velocity (m/s)
V_t	total or tank volume (m^3)
v_t	transient component of velocity (m/s)
$V_{thermal}$	thermal voltage (Chapter 5) (V)
v_v	vapor velocity perpendicular to the plate (m/s)
V_x	variance
v_{zm}	mixing cup velocity in a tube (m/s)
$v_{z,max}$	maximum axial velocity (m/s)
$\dot{v}$, $\dot{V}$	volumetric flow rate (m^3/s)
v_∞	free stream velocity (m/s)
$\dot{v}_s$	volumetric flow rate of side feed per unit length (m^3/ms)
W	Work (J)
w_a	mass fraction of a
W_μ	work done by viscous forces (J)
W_p	work done by machinery on the fluid (J)
W_{rev}	reversible work interaction (J)
x_a	mole fraction of a
x_h	mole fraction of holes
x_p	mole fraction of particles
Y	Young's modulus (N/m^2)
y^+	dimensionless distance from wall
y_a	mole fraction of a in the gas phase

$z_{\pm i}$ valence of charge carrier i
α thermal diffusivity (m^2/s)
α_{ab} Seebeck thermoelectric coefficient (V/K)
α_d damping coefficient
α^e eddy thermal diffusivity (m^2/s)
α_f thermal diffusivity of liquid (m^2/s)
α^r absorptivity; may be a function of wavelength
α_s thermal diffusivity of solid (m^2/s)
α_T Seebeck coefficient (V/K)
α_v thermal diffusivity of vapor (m^2/s)
β coefficient of thermal expansion (K^{-1})
β_D Graetz number for dispersion
Δ_c concentration boundary layer thickness ratio δ_c/δ_h
δ_c concentration boundary layer thickness (m)
$\Delta_{\chi lm}$ log-mean mole fraction difference
$\delta_{d(x)}$ Dirac delta function
$\Delta\Phi$ electrical potential difference (V)
Δ_f film thickness (m)
δ_f boundary layer thickness of liquid (m)
$\Delta G^\dagger$ activation energy for liquid jumps (J/kmol)
ΔH enthalpy change (J/kmol or J/kg)
δ_h hydrodynamic boundary layer thickness (m)
ΔH_f heat of fusion (J/kmol or J/kg)
ΔH_m heat of mixing (J/kmol or J/kg)
ΔH_r heat of reaction (J/kmol or J/kg)
δ_λ wave height (m)
δ_μ hydrodynamic penetration depth (m)
$\delta_\pm$ ionic boundary layer thickness (m)
ΔP pressure drop (N/m^2)
ΔT thermal boundary layer thickness ratio δ_t/δ_h
δ_T thermal penetration depth (m)
δ_t thermal boundary layer thickness (m)
ΔT_{lm} log-mean temperature difference (°C or K)
δ_v boundary layer thickness of vapor (m)
δ_x concentration penetration depth (m)
$\Delta\chi_{lm}$ log mean concentration difference (kmol/m^3)
$\delta_\pm$ ionic boundary layer thickness (m)
ε relative permittivity; may be complex $\varepsilon = \varepsilon_{re} + i\varepsilon_{im}$ (F/m); turbulent dissipation rate per unit mass (Chapter 14) (m^2/s^3)
ε_f Lennard-Jones parameter (J)
ε_g gravitational permittivity (kg^2/N m^2)
ε_o permittivity of a vacuum (F/m)
ε_p porosity
ε_r emissivity; may be a function of wavelength
ϕ particle volume fraction; velocity potential function (m^2/s)
Φ electrical potential (V)
Φ_g gravitational potential (m^2/s^2)
ϕ_p Thiele modulus
Φ_s surface electrical potential (V)
Φ_u potential energy (J)
Φ_∞ external/free stream voltage (V)

γ strain (m/m); heat capacity ratio (Chapter 11)
γ_{ij} interfacial tension (N/m)
$\dot{\gamma}$ shear rate (1/s)
η refractive index
η^c complex refractive index; $\eta^c = \eta_{re} + i\eta_{im}$
η^{c*} complex conjugate of complex refractive index
η_f fin efficiency
η_g overall transport efficiency
η_p transport efficiency
η_v villi efficiency
ϑ_i state variable (temperature, velocity, mole fraction, etc.)
ϑ_∞ free stream value of the state variable
κ absorption coefficient; permeability (Chapter 13) (m^2)
κ_μ coefficient of bulk viscosity (N s/m^2)
κ_T universal constant
Λ flow of the transport quantity (heat, mass, etc.)
λ wavelength (m)
Λ'' flux of transport quantity
λ_a characteristic length (A_c/A_s)
λ_f mean free path (m)
λ_v characteristic length (V/A_s)
μ viscosity of the fluid (N s/m^2)
μ^c chemical potential (J/particle or J/kmol)
μ_e mobility (m^2/s V)
μ^e eddy viscosity (N s/m^2)
μ_h permeability (H/m)
μ_{ho} permeability of a vacuum (H/m)
μ_o^c reference chemical potential (J/particle or J/kmol)
μ_s viscosity evaluated at surface conditions (N s/m^2)
μ_∞ viscosity at free stream temperature (N s/m^2)
ν kinematic viscosity (m^2/s)
ν^e eddy kinematic viscosity (m^2/s)
ν_f kinematic viscosity of liquid (m^2/s)
ν_i stochiometric number for species i
ν_j frequency of diffusional jumps (s^{-1})
ν_v kinematic viscosity of vapor (m^2/s)
π_T Peltier thermoelectric coefficient (V)
ρ density (kg/m^3)
$\overline{\rho}$ mean density (kg/m^3)
ρ_a density of a (kg/m^3)
ρ_b density of b (kg/m^3)
ρ_f density of liquid (kg/m^3)
ρ_λ spectral radiation density (J/m^3-m)
$\rho_{\lambda b}$ radiation density of a blackbody (J/m^3-m)
ρ^r reflectivity; may be a function of wavelength
ρ_r resistivity (Ohms-m)
ρ_s density of solid (kg/m^3)
ρ_v density of vapor (kg/m^3)
ρ_∞ fluid density in the free stream (kg/m^3)
$\rho^\pm$ charge density (C/m^3)
σ conductivity (Siemens), surface tension (N/m)

σ_c collision cross-section (m)
Σ_f transport property of the fluid
σ_f Lennard-Jones parameter
$\sigma_\perp$ normal stress (N/m^2)
σ_m molar conductivity (S/kmol)
σ^r Stefan–Boltzmann constant ($W/m^2\ K^4$)
Σ_w transport property of the wall material
τ stress (N/m^2)
τ_e Thompson thermoelectric coefficient (V/K)
τ_o yield stress (N/m^2)
τ^r transmitivity; may be a function of wavelength
τ_s surface shear stress (N/m^2)
τ_t time constant; dimensionless time
τ_w shear stress at the wall (N/m^2)
υ^+ stochiometric coefficient for cations in the solute
υ^- stochiometric coefficient for anions in the solute
ϖ density ratio
ω frequency (s^{-1})
ω_a composite mole fraction of *a* in the gas phase
Ω_c total or overall collision frequency between particles (s^{-1})
ω_c collision frequency per particle; critical frequency (s^{-1})
ω_{cr} critical frequency (s^{-1})
Ω_D Chapman–Enskog diffusion parameter
ω_f frequency of forward jumps (s^{-1})
Ω_μ Chapman–Enskog viscosity parameter
ω_μ vorticity (s^{-1})
ω_o rotation frequency (s^{-1})
ω_r frequency of reverse jumps (s^{-1})
Ω_v rotation rate (rev/s)
Ω_w collision frequency per unit area of wall ($m^2\ s^{-1}$)
ξ reaction coordinate/extent of reaction
ψ stream function (m^2/s)
Ψ_h magnetic potential (Amp-turns)

Appendix A: Vector Mathematics

A vector is a mathematical quantity defined by a magnitude and direction. It is defined in terms of unit vectors along orthogonal axes of a coordinate system. In our general 3-D space, we have three axes to consider. An example of a vector in Cartesian coordinates is the electric field, $\vec{\mathbf{E}}$:

$$\vec{\mathbf{E}} = E_x\vec{\mathbf{i}} + E_y\vec{\mathbf{j}} + E_z\vec{\mathbf{k}} \tag{A.1}$$

The magnitude of $\mathbf{E} = |\mathbf{E}| = \sqrt{E_x^2 + E_y^2 + E_z^2}$.

A.1 ADDITION AND SUBTRACTION

The addition or subtraction of two or more vectors involves separating the individual components, performing the mathematical operation, and reassembling the vector again. Thus,

$$\vec{\mathbf{E}} = E_x\vec{\mathbf{i}} + E_y\vec{\mathbf{j}} + E_z\vec{\mathbf{k}} \qquad \vec{\mathbf{H}} = H_x\vec{\mathbf{i}} + H_y\vec{\mathbf{j}} + H_z\vec{\mathbf{k}} \tag{A.2}$$

$$\vec{\mathbf{E}} + \vec{\mathbf{H}} = (E_x + H_x)\vec{\mathbf{i}} + (E_y + H_y)\vec{\mathbf{j}} + (E_z + H_z)\vec{\mathbf{k}} \tag{A.3}$$

A.2 MULTIPLICATION: THE DOT, CROSS, AND DYAD PRODUCTS

Since vectors have both magnitude and direction, multiplication is not straightforward, unless one is merely multiplying the vector by a scalar. Scalar multiplication proceeds by multiplying each component of the vector by the scalar:

$$s\vec{\mathbf{E}} = sE_x\vec{\mathbf{i}} + sE_y\vec{\mathbf{j}} + sE_z\vec{\mathbf{k}} \tag{A.4}$$

The dot product transforms two vectors into a scalar. It gives the projection of one vector onto another. The dot product is written as

$$\vec{\mathbf{E}} \cdot \vec{\mathbf{H}} = |\vec{\mathbf{E}}||\vec{\mathbf{H}}|\cos\theta \tag{A.5}$$

where θ is the angle between vectors $\vec{\mathbf{E}}$ and $\vec{\mathbf{H}}$. The dot product of vectors perpendicular to one another is zero. In terms of the individual components, the dot product is

$$\vec{\mathbf{E}} \cdot \vec{\mathbf{H}} = E_xH_x + E_yH_y + E_zH_z \tag{A.6}$$

The cross or vector product takes two vectors and forms a third vector perpendicular to the original two. It is written as

$$\vec{\mathbf{E}} \times \vec{\mathbf{H}} = |\vec{\mathbf{E}}||\vec{\mathbf{H}}|\sin\theta\,\vec{\mathbf{n}} \tag{A.7}$$

where $\vec{\mathbf{n}}$ is the unit vector normal to both $\vec{\mathbf{E}}$ and $\vec{\mathbf{H}}$. The easy way to view the multiplication in terms of vector components is using the matrix approach:

$$\vec{\mathbf{E}}\times\vec{\mathbf{H}} = \begin{vmatrix} \vec{\mathbf{i}} & \vec{\mathbf{j}} & \vec{\mathbf{k}} \\ E_x & E_y & E_z \\ H_x & H_y & H_z \end{vmatrix} \tag{A.8}$$

Expanding the matrix multiplication yields

$$\vec{\mathbf{E}}\times\vec{\mathbf{H}} = \left(E_yH_z - E_zH_y\right)\vec{\mathbf{i}} - \left(E_xH_z - E_zH_x\right)\vec{\mathbf{j}} + \left(E_xH_y - E_yH_x\right)\vec{\mathbf{k}} \tag{A.9}$$

The dyad product transforms two vectors into a tensor or matrix. It is a special product written and evaluated as

$$\vec{\mathbf{E}}:\vec{\mathbf{H}} = \begin{vmatrix} E_xH_x & E_xH_y & E_xH_z \\ E_yH_x & E_yH_y & E_yH_z \\ E_zH_x & E_zH_y & E_zH_z \end{vmatrix} \tag{A.10}$$

One can view this procedure as transforming the two vectors into three separate vectors corresponding to the rows of the matrix:

$$\begin{aligned} &E_xH_x\vec{\mathbf{i}} + E_xH_y\vec{\mathbf{j}} + E_xH_z\vec{\mathbf{k}} \\ &E_yH_x\vec{\mathbf{i}} + E_yH_y\vec{\mathbf{j}} + E_yH_z\vec{\mathbf{k}} \\ &E_zH_x\vec{\mathbf{i}} + E_zH_y\vec{\mathbf{j}} + E_zH_z\vec{\mathbf{k}} \end{aligned} \tag{A.11}$$

A.3 DIFFERENTIATION: DIVERGENCE, GRADIENT, CURL, AND LAPLACIAN

There are several different operations involving differentiation that one can perform with or on vectors. We first define a vector operator for differentiating called "del," $\vec{\nabla}$:

$$\vec{\nabla} = \frac{\partial}{\partial x}\vec{\mathbf{i}} + \frac{\partial}{\partial y}\vec{\mathbf{j}} + \frac{\partial}{\partial z}\vec{\mathbf{k}} \tag{A.12a}$$

$\vec{\nabla}$ can operate on vectors or on scalars. When it operates on a scalar function, $V(x,y,z)$, it forms the gradient of that function. The gradient is the vector giving the magnitude and direction of the fastest rate of change of the scalar:

$$\vec{\nabla}V = \frac{\partial V}{\partial x}\vec{\mathbf{i}} + \frac{\partial V}{\partial y}\vec{\mathbf{j}} + \frac{\partial V}{\partial z}\vec{\mathbf{k}} \tag{A.12b}$$

The same operation on a vector is a dyad product and leads to a tensor.

$\vec{\nabla}$ can operate on vectors using either the dot or cross products. The dot product of a vector $\vec{\mathbf{E}}$ with $\vec{\nabla}$ is called the divergence. The divergence is a scalar function that gives the flux flowing toward

(–) or away (+) from a point. If the divergence is zero, there are no sources or sinks within the control volume:

$$\vec{\nabla}\cdot\vec{\mathbf{E}} = \frac{\partial E_x}{\partial x} + \frac{\partial E_y}{\partial y} + \frac{\partial E_z}{\partial z} \tag{A.13}$$

The cross product of $\vec{\nabla}$ with a vector is called the curl. It is a vector that describes the magnitude and direction of circulation or rotation about a point:

$$\vec{\nabla}\times\vec{\mathbf{E}} = \begin{vmatrix} \vec{\mathbf{i}} & \vec{\mathbf{j}} & \vec{\mathbf{k}} \\ \dfrac{\partial}{\partial x} & \dfrac{\partial}{\partial y} & \dfrac{\partial}{\partial z} \\ E_x & E_y & E_z \end{vmatrix}$$

$$= \left(\frac{\partial E_z}{\partial y} - \frac{\partial E_y}{\partial z}\right)\vec{\mathbf{i}} - \left(\frac{\partial E_z}{\partial x} - \frac{\partial E_x}{\partial z}\right)\vec{\mathbf{j}} + \left(\frac{\partial E_y}{\partial x} - \frac{\partial E_x}{\partial y}\right)\vec{\mathbf{k}} \tag{A.14}$$

Finally, we can combine the operations of divergence and gradient to obtain the Laplacian. The Laplacian of a scalar function is

$$\vec{\nabla}\cdot\left(\vec{\nabla}V\right) = \nabla^2 V = \frac{\partial^2 V}{\partial x^2} + \frac{\partial^2 V}{\partial y^2} + \frac{\partial^2 V}{\partial z^2} \tag{A.15}$$

The Laplacian of a vector function results in a vector

$$\nabla^2\vec{\mathbf{E}} = \vec{\mathbf{i}}\nabla^2 E_x + \vec{\mathbf{j}}\nabla^2 E_y + \vec{\mathbf{k}}\nabla^2 E_z \tag{A.16}$$

A.4 OTHER USEFUL RELATIONS

There are a number of other relations between the vector operations that we will find useful:

$$\vec{\mathbf{E}}\cdot\left(\vec{\mathbf{H}}\times\vec{\mathbf{L}}\right) = \vec{\mathbf{H}}\cdot\left(\vec{\mathbf{L}}\times\vec{\mathbf{E}}\right) = \vec{\mathbf{L}}\cdot\left(\vec{\mathbf{E}}\times\vec{\mathbf{H}}\right) \tag{A.17}$$

$$\vec{\mathbf{E}}\times\left(\vec{\mathbf{H}}\times\vec{\mathbf{L}}\right) = \vec{\mathbf{H}}\left(\vec{\mathbf{E}}\cdot\vec{\mathbf{L}}\right) - \vec{\mathbf{L}}\left(\vec{\mathbf{E}}\cdot\vec{\mathbf{H}}\right) \tag{A.18}$$

$$\vec{\nabla}\left(\Phi + \Psi\right) = \vec{\nabla}\Phi + \vec{\nabla}\Psi \tag{A.19}$$

$$\vec{\nabla}\left(\Phi\Psi\right) = \Psi\vec{\nabla}\Phi + \Phi\vec{\nabla}\Psi \tag{A.20}$$

$$\vec{\nabla}\cdot\left(\vec{\mathbf{E}} + \vec{\mathbf{H}}\right) = \vec{\nabla}\cdot\vec{\mathbf{E}} + \vec{\nabla}\cdot\vec{\mathbf{H}} \tag{A.21}$$

$$\vec{\nabla}\cdot\left(\Psi\vec{\mathbf{E}}\right) = \vec{\mathbf{E}}\cdot\left(\vec{\nabla}\Psi\right) + \Psi\left(\vec{\nabla}\cdot\vec{\mathbf{E}}\right) \tag{A.22}$$

$$\vec{\nabla}\times\left(\vec{\mathbf{E}}+\vec{\mathbf{H}}\right)=\vec{\nabla}\times\vec{\mathbf{E}}+\vec{\nabla}\times\vec{\mathbf{H}} \tag{A.23}$$

$$\vec{\nabla}\times\left(\Psi\vec{\mathbf{E}}\right)=\vec{\nabla}\Psi\times\vec{\mathbf{E}}+\Psi\left(\vec{\nabla}\times\vec{\mathbf{E}}\right) \tag{A.24}$$

$$\vec{\nabla}\cdot\left(\vec{\mathbf{E}}\times\vec{\mathbf{H}}\right)=\vec{\mathbf{H}}\cdot\left(\vec{\nabla}\times\vec{\mathbf{E}}\right)-\vec{\mathbf{E}}\cdot\left(\vec{\nabla}\times\vec{\mathbf{H}}\right) \tag{A.25}$$

$$\vec{\nabla}\times\left(\vec{\mathbf{E}}\times\vec{\mathbf{H}}\right)=\vec{\mathbf{E}}\left(\vec{\nabla}\cdot\vec{\mathbf{H}}\right)-\vec{\mathbf{H}}\left(\vec{\nabla}\cdot\vec{\mathbf{E}}\right)+\left(\vec{\mathbf{H}}\cdot\vec{\nabla}\right)\vec{\mathbf{E}}-\left(\vec{\mathbf{E}}\cdot\vec{\nabla}\right)\vec{\mathbf{H}} \tag{A.26}$$

$$\vec{\nabla}\cdot\vec{\nabla}\Psi=\nabla^2\Psi \tag{A.27}$$

$$\vec{\nabla}\cdot\left(\vec{\nabla}\times\vec{\mathbf{E}}\right)=0 \tag{A.28}$$

$$\vec{\nabla}\times\vec{\nabla}\Psi=0 \tag{A.29}$$

$$\vec{\nabla}\times\left(\vec{\nabla}\times\vec{\mathbf{E}}\right)=\vec{\nabla}\left(\vec{\nabla}\cdot\vec{\mathbf{E}}\right)-\nabla^2\vec{\mathbf{E}} \tag{A.30}$$

$$\vec{\nabla}\left(\vec{\mathbf{E}}\cdot\vec{\mathbf{H}}\right)=\left(\vec{\mathbf{E}}\cdot\vec{\nabla}\right)\vec{\mathbf{H}}+\left(\vec{\mathbf{H}}\cdot\vec{\nabla}\right)\vec{\mathbf{E}}+\vec{\mathbf{E}}\times\left(\vec{\nabla}\times\vec{\mathbf{H}}\right)+\vec{\mathbf{H}}\times\left(\vec{\nabla}\times\vec{\mathbf{E}}\right) \tag{A.31}$$

Appendix B: Mathematical Functions

TABLE B.1
Error Function

x	erf(x)	x	erf(x)	x	erf(x)
0.00	0.0000	0.36	0.3893	1.04	0.8587
0.02	0.0226	0.38	0.4090	1.08	0.8733
0.04	0.0451	0.40	0.4284	1.12	0.8868
0.06	0.0676	0.44	0.4662	1.16	0.8991
0.08	0.0901	0.48	0.5028	1.20	0.9103
0.10	0.1125	0.52	0.5379	1.30	0.9340
0.12	0.1348	0.56	0.5716	1.40	0.9523
0.14	0.1570	0.60	0.6039	1.50	0.9661
0.16	0.1790	0.64	0.6346	1.60	0.9764
0.18	0.2009	0.68	0.6638	1.70	0.9838
0.20	0.2227	0.72	0.6914	1.80	0.9891
0.22	0.2443	0.76	0.7175	1.90	0.9928
0.24	0.2657	0.80	0.7421	2.00	0.9953
0.26	0.2869	0.84	0.7651	2.20	0.9981
0.28	0.3079	0.88	0.7867	2.40	0.9993
0.30	0.3286	0.92	0.8068	2.60	0.9998
0.32	0.3491	0.96	0.8254	2.80	0.9999
0.34	0.3694	1.00	0.8427	3.00	1.0000

TABLE B.2
Bessel Functions

x	$J_0(x)$	$J_1(x)$	x	$J_0(x)$	$J_1(x)$
0.0	1.0000	0.0000	1.6	0.4554	0.5699
0.1	0.9975	0.0499	1.7	0.3980	0.5778
0.2	0.9900	0.0995	1.8	0.3400	0.5815
0.3	0.9776	0.1483	1.9	0.2818	0.5812
0.4	0.9604	0.1960	2.0	0.2239	0.5767
0.5	0.9385	0.2423	2.1	0.1666	0.5683
0.6	0.9120	0.2867	2.2	0.1104	0.5560
0.7	0.8812	0.3290	2.3	0.0555	0.5399
0.8	0.8463	0.3688	2.4	0.0025	0.5202
0.9	0.8075	0.4059	2.406	0.0000	
1.0	0.7652	0.4400	2.5	−0.0484	0.4971
1.1	0.7196	0.4709	2.6	−0.0968	0.4708
1.2	0.6711	0.4983	2.7	−0.1424	0.4416
1.3	0.6201	0.5220	2.8	−0.1850	0.4097
1.4	0.5669	0.5419	2.9	−0.2243	0.3754
1.5	0.5118	0.5579	3.0	−0.2601	0.3391

x	$e^xK_0(x)$	$e^xK_1(x)$	x	$e^xK_0(x)$	$e^xK_1(x)$
0.0	∞	∞	5.5	0.5233	0.5690
0.5	1.5241	2.7310	6.0	0.5019	0.5422
1.0	1.1445	1.6362	6.5	0.4828	0.5187
1.5	0.9582	1.2432	7.0	0.4658	0.4981
2.0	0.8416	1.0335	7.5	0.4505	0.4797
2.5	0.7595	0.9002	8.0	0.4366	0.4631
3.0	0.6978	0.8066	8.5	0.4239	0.4482
3.5	0.6490	0.7365	9.0	0.4123	0.4346
4.0	0.6093	0.6816	9.5	0.4016	0.4221
4.5	0.5761	0.6371	10.0	0.3916	0.4108
5.0	0.5478	0.6003			

x	$e^{-x}I_0(x)$	$e^{-x}I_1(x)$	x	$e^{-x}I_0(x)$	$e^{-x}I_1(x)$
0.0	1.0000	0.0000	5.5	0.1745	0.1577
0.5	0.6450	0.1564	6.0	0.1667	0.1521
1.0	0.4658	0.2079	6.5	0.1598	0.1469
1.5	0.3674	0.2190	7.0	0.1537	0.1423
2.0	0.3085	0.2153	7.5	0.1483	0.1380
2.5	0.2700	0.2066	8.0	0.1434	0.1341
3.0	0.2430	0.1968	8.5	0.1390	0.1305
3.5	0.2228	0.1874	9.0	0.1350	0.1272
4.0	0.2070	0.1788	9.5	0.1313	0.1241
4.5	0.1942	0.1710	10.0	0.1278	0.1213
5.0	0.1835	0.1640			

Appendix C: First Eigenvalue for 1-D Transient Conduction with External Convection

TABLE C.1
First Eigenvalue for 1-D Transient Conduction with External Convection

$Bi = \frac{h\lambda_v}{k}$	Geometry		
	Infinite Plane Wall	Infinite Cylinder	Sphere
0.01	0.0998	0.1412	0.1730
0.02	0.140	0.1995	0.2445
0.03	0.1732	0.2439	0.2989
0.04	0.1987	0.2814	0.3450
0.05	0.2217	0.3142	0.3852
0.06	0.2425	0.3438	0.4217
0.07	0.2615	0.3708	0.4550
0.08	0.2791	0.3960	0.4860
0.09	0.2956	0.4195	0.5150
0.10	0.3111	0.4417	0.5423
0.15	0.3779	0.5376	0.6608
0.20	0.4328	0.6170	0.7593
0.25	0.4801	0.6856	0.8448
0.3	0.5218	0.7465	0.9208
0.4	0.5932	0.8516	1.0528
0.5	0.6533	0.9408	1.1656
0.6	0.7051	1.0185	1.2644
0.7	0.7506	1.0873	1.3525
0.8	0.7910	1.1490	1.4320
0.9	0.8274	1.2048	1.5044
1.0	0.8603	1.2558	1.5708
2.0	1.0769	1.5995	2.0288
3.0	1.1925	1.7877	2.2889
4.0	1.2646	1.9081	2.4556
5.0	1.3138	1.9898	2.5704
6.0	1.3496	2.0490	2.6537
7.0	1.3766	2.0937	2.7165
8.0	1.3978	2.1286	2.7654
9.0	1.4149	2.1566	2.8044
10.0	1.4289	2.1795	2.8363
20.0	1.4961	2.2881	2.9857
30.0	1.5202	2.3261	3.0372
40.0	1.5325	2.3455	3.0632
50.0	1.5400	2.3572	3.0788
100.0	1.5552	2.3809	3.1102
∞	$\pi/2$	2.4050	π

Appendix D: Exact Solution to the Boundary Layer Equations

The boundary layer equations are a set of coupled partial differential equations that we solved, approximately, in Chapter 12 using an integral technique. A more rigorous solution can be obtained using a series solution technique or a numerical solution. The series solution was first developed by Blasius in 1908 [1], and the numerical results were obtained by Howarth in 1938 [2].

The physical situation is shown schematically in Figure D.1. We need to solve the continuity and x-momentum equations.

Continuity

$$\frac{\partial v_x}{\partial x} + \frac{\partial v_y}{\partial x} = 0 \tag{D.1}$$

Momentum

$$v_x \frac{\partial v_x}{\partial x} + v_y \frac{\partial v_x}{\partial x} = \nu \frac{\partial^2 v_x}{\partial y^2} \tag{D.2}$$

We can satisfy the continuity equation exactly by using the stream function representation discussed in Chapter 9. Thus, we define

$$v_x = \frac{\partial \psi}{\partial y} \qquad v_y = -\frac{\partial \psi}{\partial x} \tag{D.3}$$

and the momentum equation becomes

$$\frac{\partial \psi}{\partial y}\frac{\partial^2 \psi}{\partial y \partial x} - \frac{\partial \psi}{\partial x}\frac{\partial^2 \psi}{\partial y^2} = \nu \frac{\partial^3 \psi}{\partial y^3} \tag{D.4}$$

We have eliminated one equation in favor of increasing the order of the second.

In the previous boundary layer analysis, we said that the velocity profile would be self-similar along the length of the plate. Thus, we should be able to develop a similarity variable for the problem. Our free parameters are x, y, v_∞, and ν. Letting $\eta = (x)^a (y)^b (v_\infty)^c (\nu)^d$, we can come up with the following similarity variable:

$$\eta = \left(\frac{v_\infty y^2}{\nu x}\right)^{1/2} \tag{D.5}$$

Though the velocity, v_x, along the plate is only a function of this similarity variable, the stream function is not since it must also represent v_y. To get a stream-type function that is just a function of η, we must redefine it to be

$$f = \frac{\psi}{\sqrt{\nu v_\infty x}} \tag{D.6}$$

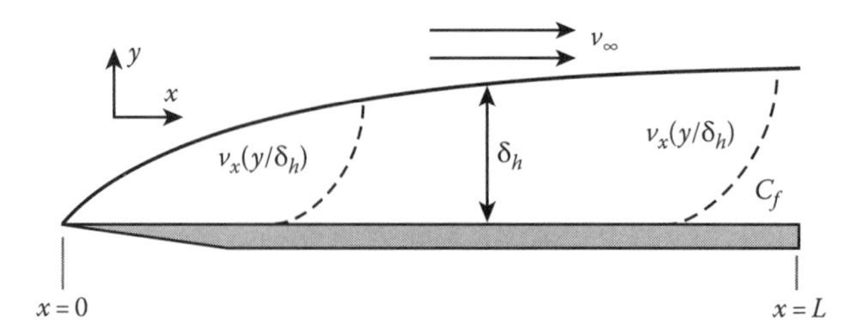

FIGURE D.1 The hydrodynamic boundary layer on a flat plate.

Now we can transform the variables and the required derivatives in Equation D.4:

$$-\frac{\partial \psi}{\partial x} = \frac{1}{2}\left(\frac{\nu v_\infty}{x}\right)^{1/2}\left(\eta \frac{\partial f}{\partial \eta} - f\right) \tag{D.7}$$

$$2\frac{\partial^3 f}{\partial \eta^3} + f\frac{\partial^2 f}{\partial \eta^2} = 0 \tag{D.8}$$

$$\frac{\partial^2 \psi}{\partial y \partial x} = -\frac{v_\infty}{2x}\frac{\partial^2 f}{\partial \eta^2} \tag{D.9}$$

$$\frac{\partial^2 \psi}{\partial y^2} = v_\infty \left(\frac{v_\infty}{\nu x}\right)^{1/2}\frac{\partial^2 f}{\partial \eta^2} \tag{D.10}$$

$$\frac{\partial^3 \psi}{\partial y^3} = \left(\frac{v_\infty^2}{\nu x}\right)\frac{\partial^3 f}{\partial \eta^3} \tag{D.11}$$

Substituting Equations D.7 through D.11 into the momentum equation, (D.4) gives

$$2\frac{\partial^3 f}{\partial \eta^3} + f\frac{\partial^2 f}{\partial \eta^2} = 0 \tag{D.12}$$

The boundary conditions for the problem specify no slip at the surface of the plate, $v_x(x,0) = v_y\,(x,0) = 0$, and unidirectional velocity at the edge of the boundary layer ($v_x\,(x,\infty) = v_\infty$). In terms of the transformed variables, these conditions become

$$\eta = 0 \qquad f = 0 \qquad \frac{df}{d\eta} = 0 \tag{D.13}$$

$$\eta = \infty \qquad \frac{df}{d\eta} = 1 \tag{D.14}$$

The momentum equation can now be solved numerically or via a power series technique. Regardless, the solution for v_x is shown in Figure D.2. The edge of the boundary layer, as we have defined it, is located where $v_x \approx 0.99\ v_\infty$. From Figure D.2, we can see that this occurs at $\eta \approx 5$. If we

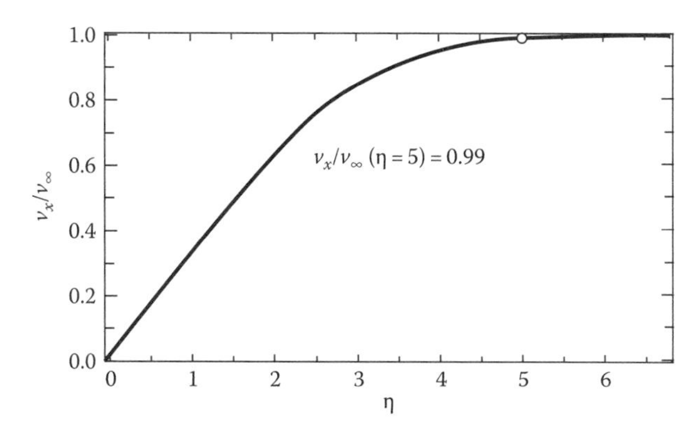

FIGURE D.2 Velocity along the plate surface as a function of position.

transform back to the coordinates of Chapter 12 in an attempt to determine a functional form for the boundary layer thickness, we find that

$$\delta_h = \frac{5x}{\sqrt{\mathrm{Re}_x}} \tag{D.15}$$

which is more precise than the integral analysis formula we derived in Chapter 12.

REFERENCES

1. Blasius, H., Grenzschichten in Flüssigkeiten mit Kleiner Reibung, *Z. Math. Phys.*, **56**, 1 (1908).
2. Howarth, L., On the solution to the laminar boundary layer equations, *Proc. R. Soc. Lond. Ser. A*, **164**, 547 (1938).

Appendix E: Blackbody Emission Functions

TABLE E.1
Fraction of Blackbody Emission in the Wavelength Range ($0 \rightarrow \lambda T$)

λT (m K)	$\mathcal{F}(0 - \lambda T)$	λT (m K)	$\mathcal{F}(0 - \lambda T)$	λT (m K)	$\mathcal{F}(0 - \lambda T)$
0	0	0.004	0.48003	0.0085	0.87382
0.0002	3.158e−27	0.0042	0.51518	0.009	0.88916
0.0004	1.795e−12	0.0044	0.54797	0.0095	0.90212
0.0006	9.070e−08	0.0046	0.57848	0.01	0.91312
0.0008	1.615e−05	0.0048	0.60678	0.0105	0.92251
0.001	0.00032	0.005	0.63299	0.011	0.93056
0.0012	0.00211	0.0052	0.65723	0.0115	0.93749
0.0014	0.00772	0.0054	0.67964	0.012	0.94348
0.0016	0.01958	0.0056	0.70034	0.013	0.95323
0.0018	0.03910	0.0058	0.71947	0.014	0.96069
0.002	0.06637	0.006	0.73714	0.015	0.96647
0.0022	0.10042	0.0062	0.75347	0.016	0.97102
0.0024	0.13969	0.0064	0.76857	0.018	0.97752
0.0026	0.18246	0.0066	0.78254	0.02	0.98179
0.0028	0.22716	0.0068	0.79547	0.025	0.98747
0.003	0.27245	0.007	0.80745	0.03	0.98996
0.0032	0.31728	0.0074	0.82885	0.04	0.99185
0.0034	0.36089	0.0076	0.83841	0.05	0.99247
0.0036	0.40275	0.0078	0.84730	0.075	0.99286
0.0038	0.44253	0.008	0.85556	0.1	0.99294

The entries in the table were evaluated from $\mathcal{F}_{0\rightarrow\varepsilon} = \frac{15}{\pi^4}\sum_{n=1}^{3}\left[\frac{e^{-n\varepsilon}}{n}\left(\varepsilon^3 + \frac{3\varepsilon^2}{n} + \frac{6\varepsilon}{n^2} + \frac{6}{n^3}\right)\right]$

Appendix F: Thermodynamic and Transport Properties of Materials

TABLE F.1
Diffusion Coefficients in Gases

Gas Pair		Diffusion Coefficient ($\times 10^5$ m²/s)	Temperature (K)
Air	Benzene	0.96	298.2
	Butanol	0.87	299.1
	CH_4	1.96	273
	CO_2	1.42	276.2
	Ethanol	1.02	273
	H_2	6.11	273
	H_2O	2.6	298.2
	n-Hexane	0.80	294
H_2	Cyclohexane	3.19	288.6
	D_2	10.24	288.2
	Ethane	5.37	298
	H_2O	9.15	307.1
	He	11.32	298.2
	n-Butane	3.61	287.9
	O_2	6.97	273
N_2	Ethane	1.48	298
	H_2O	2.56	307.5
	He	6.87	298
O_2	Benzene	1.01	311.3
	H_2O	2.82	308.1
	He	7.29	298
	n-Hexane	0.75	288.6

Sources: Hirschfelder, J. et al., *Molecular Theory of Gases and Liquids*, Wiley, New York, 1954; Reid, R.C. et al., *Properties of Gases and Liquids*, 3rd edn., McGraw-Hill, New York, 1977.

TABLE F.2
Diffusion Coefficients in Water at Infinite Dilution

Solute	Diffusion Coefficient ($\times 10^9$ m^2/s)	Temperature (K)
Acetic acid	1.21	298
Acetone	1.16	293
Air	2.00	298
Carbon dioxide	1.92	298
Chlorine	1.25	298
Ethane	1.20	298
Ethanol	0.84	283
Ethylene glycol	1.16	298
Fibrinogen	0.02	298
Helium	6.28	298
Hemoglobin	0.069	298
Hydrogen	4.50	298
Isopropanol	0.87	288
Methane	1.49	293
Methanol	0.84	283
Nitric acid	2.60	298
Nitrogen	1.88	298
Oxygen	2.10	298
Propane	0.97	298
Sucrose	0.523	298
Sulfuric acid	1.73	298
Urea	1.38	298

Sources: Sherwood, T.K. et al., *Mass Transfer*, McGraw-Hill, New York, 1975; Cussler, E.L., *Diffusion: Mass Transfer in Fluid Systems*, Cambridge University Press, New York, 1984.

TABLE F.3
Diffusion Coefficients in Nonaqueous Solvents at Infinite Dilution

Solvent	Solute	Diffusion Coefficient ($\times 10^9$ m²/s)	Temperature (K)
Acetone	Acetic acid	3.31	298
	Benzoic acid	2.64	293
	Water	4.56	298
Benzene	Acetic acid	2.09	298
	Cyclohexane	2.09	298
	Ethanol	2.25	288
	Toluene	1.85	298
Ethanol	Benzene	1.81	298
	Carbon dioxide	3.20	290
Ethanol	Oxygen	2.64	302.6
	Water	1.24	298
n-Hexane	Carbon tetrachloride	3.70	298
	Dodecane	2.73	298
	Propane	4.87	298
	Toluene	4.21	298

Sources: Reid, R.C. et al., *Properties of Gases and Liquids*, 3rd edn., McGraw-Hill, New York, 1977; Sherwood, T.K. et al., *Mass Transfer*, McGraw-Hill, New York, 1975.

TABLE F.4
Diffusion Coefficients in Solids

System	Temperature (°C)	Diffusion Coefficient (m²/s)
Ag–Si	1200	3.83×10^{-14}
Au–Ag	760	3.6×10^{-14}
C–Fe	1100	4.5×10^{-11}
Cu–Si	1200	2.53×10^{-10}
Ge–Si	1200	5.2×10^{-18}
H_2–Fe	100	1.24×10^{-11}
H_2–Si	1200	2.1×10^{-8}
H_2–SiO_2	200	6.5×10^{-14}
Ni–Si	1200	1.72×10^{-12}
O–Si	1200	1.20×10^{-15}
P–Si	1200	2.4×10^{-16}
H_2 in Lexan	25	0.64×10^{-10}
O_2 in Lexan	25	0.021×10^{-10}
H_2 in natural rubber	25	10.2×10^{-10}
O_2 in natural rubber	25	1.58×10^{-10}
H_2 in polystyrene	25	4.36×10^{-10}
O_2 in polystyrene	25	0.11×10^{-10}

Sources: American Society for Metals, *Diffusion*, ASM, Metals Park, OH, 1973; Barrer, R.M., *Diffusion in and through Solids*, MacMillan, New York, 1941; Ghandi, S.K., *VLSI Fabrication Principles*, Wiley-Interscience, New York, 1983.

TABLE F.5
Thermal Properties of Selected Metals and Alloys

Composition	Density ρ (kg/m^3)	Heat Capacity C_p (J/kg K)	Thermal Conductivity k (W/m K)	Thermal Diffusivity α ($\times 10^6$ m^2/s)
Aluminum	2,702	903	237	97.1
Copper				
Pure	8,933	385	401	117
Bronze	8,800	420	52	14
Brass	8,780	355	54	17
Gold	19,300	129	317	127
Iron	7,870	447	80.2	23.1
Carbon steel	7,854	434	60.5	17.7
Stainless steel				
AISI 304	7,900	477	14.9	3.95
AISI 316	8,238	468	13.4	3.48
Lead	11,340	129	35.3	24.1
Platinum				
Pure	21,450	133	71.6	25.1
Silicon	2,330	712	148	89.2
Silver	10,500	235	429	174
Titanium	4,500	522	21.9	9.32
Tungsten	19,300	132	174	68.3

Sources: Touloukian, Y.S. and Ho, C.Y., eds., *Thermophysical Properties of Matter*, Vols. 1–9, Plenum Press, New York, 1972; Touloukian, Y.S. and Ho, C.Y., eds., *Thermophysical Properties of Selected Aerospace Materials, Parts I & II.* Thermophysical and electronic properties Information Analysis Center, CINDAS, Purdue University, West Lafayette, IN, 1976; Ho, C.Y. et al., *J. Phys. Chem. Ref. Data*, 3(Suppl 1), 1974; Desai, P.D. et al., Part I: Thermophysical properties of carbon steels, Part II: Thermophysical properties of low chromium steels, Part III: Thermophysical properties of nickel steels, Part IV: Thermophysical properties of stainless steels, CINDAS Special Report, Purdue University, West Lafayette, IN, September 1976; American Society for Metals, *Metals Handbook, Properties and Selection of Metals*, 8th edn., Vol. 1, ASM, Metals Park, OH, 1961; Hultgren, R. et al., *Selected Values of the Thermodynamic Properties of the Elements*, American Society of Metals, Metals Park, OH, 1973; Hultgren, R. et al., *Selected Values of the Thermodynamic Properties of Binary Alloys*, American Society of Metals, Metals Park, OH, 1973.

TABLE F.6
Thermal Properties of Selected Nonmetals

Composition	Density ρ (kg/m^3)	Heat Capacity C_p (J/kg K)	Thermal Conductivity k (W/m K)	Thermal Diffusivity α ($\times 10^6$ m^2/s)
Aluminum oxide (sapphire)	3970	765	36	11.9
	3970	765	46	15.1
Boron	2500	1105	27.6	9.99
Carbon				
Amorphous	1950		1.6	
IIa diamond	3500	509	2300	1291
Graphite				
‖ to layers	2210	709	1950	
ǀ to layers	2210	709	5.7	
Pyroceram	2600	808	3.98	1.89
Silicon carbide	3160	675	490	230
Quartz				
‖ to c axis	2650	745	10.4	
⊥ to c axis	2650	745	6.21	
Silicon dioxide	2220	745	1.38	0.834
Silicon nitride	2400	691	16	9.65
Sulfur	2070	708	0.206	0.141
Titanium dioxide	4157	710	8.4	2.8

Sources: Touloukian, Y.S. and Ho, C.Y., eds., *Thermophysical Properties of Matter*, Vols. 1–9, Plenum Press, New York, 1972; Touloukian, Y.S. and Ho, C.Y., eds., *Thermophysical Properties of Selected Aerospace Materials, Parts I & II*, Thermophysical and Electronic Properties Information Analysis Center, CINDAS, Purdue University, West Lafayette, IN, 1976; Ho, C.Y. et al., *J. Phys. Chem. Ref. Data*, 3(Suppl 1), 1974; Hultgren, R. et al., *Selected Values of the Thermodynamic Properties of the Elements*, American Society of Metals, Metals Park, OH, 1973.

TABLE F.7
Thermal Properties of Selected Building Materials

Composition	Density ρ (kg/m³)	Heat Capacity C_p (J/kg K)	Thermal Conductivity k (W/m K)
Gypsum/plasterboard	800	1000	0.17
Cement board	1920	1550	0.58
Plywood	545	1215	0.12
Sheathing	290	1300	0.055
Acoustic tile	290	1340	0.058
Hardboard siding	640	1170	0.094
Particleboard	1000	1300	0.17
Hardwoods (oak, maple)	720	1255	0.16
Softwoods (pine, fir)	510	1380	0.12
Cement mortar	1860	780	0.72
Brick, common	1920	835	0.72
Gypsum plaster	1680	1085	0.22
Fiberglass blanket paper faced	16		0.046
	28		0.038
	40		0.035
Fiberglass duct liner coated	32	835	0.038
Polystyrene boards			
Extruded	55	1210	55
Molded beads	16	1210	16
Corkboard	120	1800	0.039
Glass fiber blown in	16	835	0.043
Urethane foam in place	70	1045	0.026
Aluminum foil/glass fiber mat (150 K)			
10–12 layers	40		0.00016
75–150 layers	120		0.000017

Sources: Touloukian, Y.S. and Ho, C.Y., eds., *Thermophysical Properties of Matter*, Vols. 1–9, Plenum Press, New York, 1972; American Society of Heating, Refrigerating and Air Conditioning Engineers, *ASHRAE Handbook of Fundamentals*, ASHRAE, New York, 1994; Mallory, J.F., *Thermal Insulation*, Van Nostrand Reinhold, New York, 1969; Hanley, E.J. et al., The thermal transport properties at normal and elevated temperature of eight representative rocks, *Proceedings of the Seventh Symposium on Thermophysical Properties*, American Society of Mechanical Engineers, New York, 1977; Sweat, V.E., *A Miniature Thermal Conductivity Probe for Foods*, American Society of Mechanical Engineers, Paper 76-HT-60, August 1976; Kothandaraman, C.P. and Subramanyan, S., *Heat and Mass Transfer Data Book*, Halsted Press/Wiley, New York, 1975.

TABLE F.8
Thermal Properties of Selected Other Materials

Composition	Density ρ (kg/m³)	Heat Capacity C_p (J/kg K)	Thermal Conductivity k (W/m K)	Temperature T (K)
Asphalt	2115	920	0.062	300
Bakelite	1300	1465	1.4	300
Fireclay brick (burned 1725 K)	2325	960	1.3	773
Fireclay brick	2645	960	1.5	922
Clay	1460	880	1.3	300
Coal, anthracite	1350	1260	0.26	300
Concrete	2300	880	1.4	300
Cotton	80	1300	0.06	300
Foodstuffs				
Banana	980		0.481	300
Apple, red	840	3350	0.513	300
Cake batter	720	3600	0.223	300
Cake baked	280		0.121	300
Chicken breast			0.489	293
Glass	2500	750	1.4	300
Ice	920	2040	1.88	273
		1945	2.03	253
Leather	998		0.159	300
Paper	930	1340	0.180	300
Rock				
Granite	2630	775	2.79	300
Limestone	2320	810	2.15	300
Sandstone	2150	745	2.90	300
Rubber				
Soft	1100	2010	0.13	300
Hard	1190		0.16	300
Sand	1515	800	0.27	300
Soil	2050	1840	0.52	300
Snow				
Powder	110		0.049	273
Packed	500		0.190	
Teflon	2200		0.35	300
			0.45	400
Tissue, human				
Skin			0.37	300
Adipose (fat)			0.2	300
Muscle			0.41	300
Wood				
Cypress	465		0.097	300
Fir	415	2720	0.11	300
Oak	545	2385	0.17	300
Yellow pine	640	2805	0.15	300
White pine	435		0.11	300

Sources: Touloukian, Y.S. and Ho, C.Y., eds., *Thermophysical Properties of Matter*, Vols. 1–9, Plenum Press, New York, 1972; American Society of Heating, Refrigerating and Air Conditioning Engineers, *ASHRAE Handbook of Fundamentals*, ASHRAE, New York, 1994; Mallory, J.F., *Thermal Insulation*, Van Nostrand Reinhold, New York, 1969; Hanley, E.J. et al., The thermal transport properties at normal and elevated temperature of eight representative rocks, *Proceedings of the Seventh Symposium on Thermophysical Properties*, American Society of Mechanical Engineers, New York, 1977; Sweat, V.E., *A Miniature Thermal Conductivity Probe for Foods*, American Society of Mechanical Engineers, Paper 76-HT-60, August 1976; Kothandaraman, C.P. and Subramanyan, S., *Heat and Mass Transfer Data Book*, Halsted Press/Wiley, New York, 1975.

.operties of Selected Gases at Atmospheric Pressure

..nperature T (K)	Density ρ (kg/m³)	Heat Capacity Cp (J/kg K)	Viscosity $\mu \cdot 10^6$ (N s/m²)	Thermal Conductivity $k \cdot 10^3$ (W/m K)	Prandtl Number Pr
Air					
100	3.5562	1,032	7.11	9.34	0.786
150	2.3364	1,012	10.34	13.8	0.758
200	1.7458	1,007	13.25	18.1	0.737
250	1.3947	1,006	15.96	22.3	0.720
300	1.1614	1,007	18.46	26.3	0.707
350	0.9950	1,009	20.82	30.0	0.700
400	0.8711	1,014	23.01	33.8	0.690
450	0.7740	1,021	25.07	37.3	0.686
500	0.6964	1,030	27.01	40.7	0.684
550	0.6329	1,040	28.84	43.9	0.683
600	0.5804	1,051	30.58	46.9	0.685
650	0.5356	1,063	32.25	49.7	0.690
700	0.4975	1,075	33.88	52.4	0.695
750	0.4643	1,087	35.46	54.9	0.702
800	0.4354	1,099	36.98	57.3	0.709
850	0.4097	1,110	38.43	59.6	0.716
900	0.3868	1,121	39.81	62.0	0.720
950	0.3666	1,131	41.13	64.3	0.723
1000	0.3482	1,141	42.44	66.7	0.726
1100	0.3166	1,159	44.90	71.5	0.728
1200	0.2902	1,175	47.30	76.3	0.728
1300	0.2679	1,189	49.60	82	0.719
1400	0.2488	1,207	53.0	91	0.703
1500	0.2322	1,230	55.7	100	0.685
1600	0.2177	1,248	58.4	106	0.688
1700	0.2049	1,267	61.1	113	0.685
1800	0.1935	1,286	63.7	120	0.683
1900	0.1833	1,307	66.3	128	0.677
2000	0.1741	1,337	68.9	137	0.672
Helium					
100	0.4871	5,193	9.63	73.0	0.686
120	0.4060	5,193	10.7	81.9	0.679
140	0.3481	5,193	11.8	90.7	0.676
180	0.2708	5,193	13.9	107.2	0.673
220	0.2216	5,193	16.0	123.1	0.675
260	0.1875	5,193	18.0	137	0.682
300	0.1625	5,193	19.9	152	0.680
400	0.1219	5,193	24.3	187	0.675
500	0.09754	5,193	28.3	220	0.668
700	0.06969	5,193	35.0	278	0.654
800	0.06023	5,193	38.2	304	0.652
Hydrogen					
100	0.24572	11,229	4.212	66.5	0.712
150	0.16371	12,602	5.595	98.1	0.718
200	0.12270	13,540	6.813	128.2	0.719
250	0.09819	14,059	7.919	156.1	0.713
300	0.08185	14,314	8.963	182	0.706
350	0.07016	14,436	9.954	206	0.697
400	0.06135	14,491	10.86	228	0.690
450	0.05462	14,499	11.78	251	0.682
500	0.04918	14,507	12.64	272	0.675
550	0.04469	14,532	13.48	292	0.668

TABLE F.9 (continued)
Thermal Properties of Selected Gases at Atmospheric Pressure

Temperature T (K)	Density ρ (kg/m³)	Heat Capacity Cp (J/kg K)	Viscosity $\mu \cdot 10^6$ (N s/m²)	Thermal Conductivity $k \cdot 10^3$ (W/m K)	Prandtl Number Pr
600	0.04085	14,537	14.29	315	0.664
700	0.03492	14,574	15.89	351	0.659
800	0.03060	14,675	17.40	384	0.664
Nitrogen					
100	3.4388	1,070	6.88	9.58	0.768
150	2.2594	1,050	10.06	13.9	0.759
200	1.6883	1,043	12.92	18.3	0.736
250	1.3488	1,042	15.49	22.2	0.727
300	1.1233	1,041	17.82	25.9	0.716
350	0.9625	1,042	20.00	29.3	0.711
400	0.8425	1,045	22.04	32.7	0.704
450	0.7485	1,050	23.96	35.8	0.703
500	0.6739	1,056	25.77	38.9	0.700
550	0.6124	1,065	27.47	41.7	0.702
600	0.5615	1,075	29.08	44.6	0.701
700	0.4812	1,098	32.10	49.9	0.706
Oxygen					
100	3.945	962	7.64	9.25	0.796
150	2.585	921	11.48	13.8	0.766
200	1.930	915	14.75	18.3	0.737
250	1.542	915	17.86	22.6	0.723
300	1.284	920	20.72	26.8	0.711
350	1.100	929	23.35	29.6	0.733
400	0.9620	942	25.82	33.0	0.737
450	0.8554	956	28.14	36.3	0.741
500	0.7698	972	30.33	41.2	0.716
550	0.6998	988	32.40	44.1	0.726
600	0.6414	1,003	34.37	47.3	0.729
700	0.5498	1,031	38.08	52.8	0.744
Superheated steam					
380	0.5683	2,060	12.71	24.6	1.06
400	0.5542	2,014	13.44	26.1	1.04
450	0.4902	1,980	15.25	29.9	1.01
500	0.4405	1,985	17.04	33.9	0.998
550	0.4005	1,997	18.84	37.9	0.993
600	0.3652	2,026	20.67	42.2	0.993
650	0.3380	2,056	22.47	46.4	0.996
700	0.3140	2,085	24.26	50.5	1.00
750	0.2931	2,119	26.04	54.9	1.00
800	0.2739	2,152	27.86	59.2	1.01
850	0.2579	2,186	29.69	63.7	1.02

Sources: American Society of Heating, Refrigerating and Air Conditioning Engineers, *ASHRAE Handbook of Fundamentals*, ASHRAE, New York, 1994; Vargaftik, N.B., *Tables of Thermophysical Properties of Liquids and Gases*, 2nd edn., Hemisphere Publishing, New York, 1975; Eckert, E.R.G. and Drake, R.M., *Analysis of Heat and Mass Transfer*, McGraw-Hill, New York, 1972; Incropera, F.P. and DeWitt, D.P., *Introduction to Heat Transfer*, 3rd edn., Wiley, New York, 1994; Geankoplis, C.J., *Transport Processes and Unit Operations*, 3rd edn., Prentice Hall, Englewood Cliffs, NJ, 1993; Liley, P.E., in *Handbook of Heat Transfer Fundamentals*, 2nd edn., Rohsenow, W.M., Hartnett, J.P., and Ganic, E.N., eds., McGraw-Hill, New York, 1985; Bolz, R.E. and Tuve, G.L., eds., *CRC Handbook of Tables for Applied Engineering Science*, 2nd edn., CRC Press, Boca Raton, FL, 1979.

TABLE F.10
Thermal Properties of Selected Saturated Liquids

Temperature T (K)	Density ρ (kg/m^3)	Heat Capacity Cp (J/kg K)	Viscosity $\mu \cdot 10^6$ (N s/m^2)	Thermal Conductivity $k \cdot 10^3$ (W/m K)	Prandtl Number Pr	Thermal Expansion $\beta \cdot 10^3$ (1/K)
Engine oil (new)						
273	899.1	1796	385	147	47,000	0.70
280	895.3	1827	217	144	27,500	0.70
290	890.0	1868	99.9	145	12,900	0.70
300	884.1	1909	48.6	145	6,400	0.70
310	877.9	1951	25.3	145	3,400	0.70
320	871.8	1993	14.1	143	1,965	0.70
330	865.8	2035	8.36	141	1,205	0.70
340	859.9	2076	5.31	139	793	0.70
350	853.9	2118	3.56	138	546	0.70
360	847.8	2161	2.52	138	395	0.70
370	841.8	2206	1.86	137	300	0.70
380	836.0	2250	1.41	136	233	0.70
390	830.6	2294	1.10	135	187	0.70
Ethylene glycol						
273	1130.8	2294	6.51	242	617	0.65
280	1125.8	2323	4.20	244	400	0.65
290	1118.8	2368	2.47	248	236	0.65
300	1114.4	2415	1.57	252	151	0.65
310	1103.7	2460	1.07	255	103	0.65
320	1096.2	2505	0.757	258	73.5	0.65
330	1089.5	2549	0.561	260	55.0	0.65
340	1083.8	2592	0.431	261	42.8	0.65
350	1079.0	2637	0.342	261	34.6	0.65
360	1074.0	2682	0.278	261	28.6	0.65
370	1066.7	2728	0.228	262	23.7	0.65
373	1058.5	2742	0.215	263	22.4	0.65
Mercury						
273	13595	1404	0.1688	8180	0.0290	0.181
300	13529	1393	0.1523	8540	0.0248	0.181
350	13407	1377	0.1309	9180	0.0196	0.181
400	13287	1365	0.1171	9800	0.0163	0.181
450	13167	1357	0.1075	10400	0.0140	0.181
500	13048	1353	0.1007	10950	0.0125	0.182
550	12929	1352	0.0953	11450	0.0112	0.184
600	12809	1355	0.0911	11950	0.0103	0.187
R134a						
250	1367	1286.9	363.25	102.5	4.56	2.22
260	1336.60	1308.60	316.570	97.9	4.23	2.33
270	1304.7	1333.0	277.54	93.4	3.96	2.50

TABLE F.10 (continued)
Thermal Properties of Selected Saturated Liquids

Temperature T (K)	Density ρ (kg/m^3)	Heat Capacity Cp (J/kg K)	Viscosity $\mu \cdot 10^6$ (N s/m^2)	Thermal Conductivity $k \cdot 10^3$ (W/m K)	Prandtl Number Pr	Thermal Expansion $\beta \cdot 10^3$ (1/K)
280	1271.4	1361.1	244.34	89.0	3.74	2.69
290	1236.3	1393.7	215.64	84.6	3.55	2.92
300	1199.3	1432.8	190.46	80.3	3.40	3.20
310	1159.5	1481.1	168.04	76.1	3.27	3.57
320	1116.4	1543.1	147.78	71.8	3.18	4.06
330	1068.8	1627.3	129.20	67.5	3.12	4.76
340	1014.7	1751.3	111.81	63.1	3.10	5.80
350	951.0	1962.0	95.095	58.6	3.18	6.70
Saturated liquid water [22,24]						
273.15	1000	4217	1,750	569	12.99	−68.05
280	1000	4198	1,422	582	10.26	46.04
290	999.0	4184	1,080	598	7.56	174.0
300	997.0	4179	855	613	5.83	276.1
310	993.1	4178	695	628	4.62	361.9
320	989.1	4180	577	640	3.77	436.7
330	984.3	4184	489	650	3.15	504.0
340	979.4	4188	420	660	2.66	566.0
350	973.7	4195	365	668	2.29	624.2
360	967.1	4203	324	674	2.02	697.9
370	960.6	4214	289	679	1.80	728.7
373.15	957.9	4217	279	680	1.76	750.1
380	953.3	4226	260	683	1.61	788
390	945.2	4239	237	686	1.47	841
400	937.2	4256	217	688	1.34	896
410	928.5	4278	200	688	1.24	952
420	919.1	4302	185	688	1.16	1010
430	909.9	4331	173	685	1.09	
440	900.9	4360	162	682	1.04	
450	890.5	4400	152	678	0.99	
460	879.5	4440	143	673	0.95	
470	868.1	4480	136	667	0.92	
480	856.9	4530	129	660	0.89	
490	844.6	4590	124	651	0.87	
500	831.3	4660	118	642	0.86	
510	818.3	4740	113	631	0.85	
520	803.9	4840	108	621	0.84	
530	788.6	4950	104	608	0.85	
540	772.8	5080	101	594	0.86	
550	755.9	5240	97	580	0.87	
560	738.0	5430	94	563	0.90	

(*continued*)

TABLE F.10 (continued)
Thermal Properties of Selected Saturated Liquids

Temperature T (K)	Pressure P (N/m^2)	Density ρ (kg/m^3)	Heat Capacity Cp (J/kg K)	Viscosity $\mu \cdot 10^6$ (N s/m^2)	Thermal Conductivity $k \cdot 10^3$ (W/m K)	Prandtl Number Pr
Saturated water vapor [22,24]						
273.15	0.00611	0.00485	1,854	8.02	18.2	0.815
280	0.00990	0.00767	1,858	8.29	18.6	0.825
290	0.01917	0.01435	1,864	8.69	19.3	0.841
300	0.03531	0.02556	1,872	9.09	19.6	0.857
310	0.06221	0.04361	1,882	9.49	20.4	0.873
320	0.1053	0.07153	1,895	9.89	21.0	0.894
330	0.1719	0.113	1,911	10.29	21.7	0.908
340	0.2713	0.174	1,930	10.69	22.3	0.925
350	0.4163	0.2600	1,954	11.09	23.0	0.942
360	0.6209	0.3781	1,983	11.49	23.7	0.960
370	0.9040	0.5374	2,017	11.89	24.5	0.978
373.15	1.0133	0.5956	2,029	12.02	24.8	0.984
380	1.2869	0.7479	2,057	12.29	25.4	0.999
390	1.794	1.02	2,104	12.69	26.3	1.013
400	2.455	1.37	2,158	13.05	27.2	1.033
410	3.302	1.81	2,221	13.42	28.2	1.054
420	4.370	2.35	2,291	13.79	29.8	1.075
520	37.70	19.0	3,700	17.33	47.5	1.35
530	44.58	22.5	3,960	17.72	50.6	1.39
540	52.38	26.7	4,270	18.1	54.0	1.43
550	61.19	31.5	4,640	18.6	58.3	1.47
560	71.08	37.2	5,090	19.1	63.7	1.52
570	82.16	43.9	5,670	19.7	76.7	1.59
580	94.51	51.8	6,400	20.4	76.7	1.68
590	108.3	61.4	7,350	21.5	84.1	1.84
600	123.5	73.0	8,750	22.7	92.9	2.15
610	137.3	87.0	11,100	24.1	103	2.60
620	159.1	106	15,400	25.9	114	3.46
630	179.7	133	22,100	28.0	130	4.8
640	202.7	175	42,000	32.0	155	9.6
647.3	221.2	221	∞	45.0	238	∞

Sources: American Society of Heating, Refrigerating and Air Conditioning Engineers, *ASHRAE Handbook of Fundamentals*, ASHRAE, New York, 1994; Vargaftik, N.B., *Tables of Thermophysical Properties of Liquids and Gases*, 2nd edn., Hemisphere Publishing, New York, 1975; Eckert, E.R.G. and Drake, R.M., *Analysis of Heat and Mass Transfer*, McGraw-Hill, New York, 1972; Incropera, F.P. and DeWitt, D.P., *Introduction to Heat Transfer*, 3rd edn., Wiley, New York, 1994; Geankoplis, C.J., *Transport Processes and Unit Operations*, 3rd edn., Prentice Hall, Englewood Cliffs, NJ, 1993; Liley, P.E., in *Handbook of Heat Transfer Fundamentals*, 2nd edn., Rohsenow, W.M., Hartnett, J.P., and Ganic, E.N., eds., McGraw-Hill, New York, 1985; Bolz, R.E. and Tuve, G.L., eds., *CRC Handbook of Tables for Applied Engineering Science*, 2nd edn., CRC Press, Boca Raton, FL, 1979; NIST Webbook, http://webbook.nist.gov

TABLE F.11
Emissivity of Selected Metals

Material	100	300	600	1000
Aluminum				
Polished film	0.02	0.04	0.06	
Anodized		0.82		
Copper				
Polished		0.03	0.04	0.04
Oxidized			0.50	0.80
Gold				
Polished	0.01	0.03	0.04	0.06
Nickel				
Polished			0.09	0.14
Oxidized			0.40	0.57
Silver				
Polished			0.03	0.08
Stainless steels				
Polished		*0.17*	*0.19*	*0.30*
Cleaned		*0.22*	*0.24*	*0.35*

Sources: Touloukian, Y.S. and Ho, C.Y., eds., *Thermophysical Properties of Matter*, Vols. 1–9, Plenum Press, New York, 1972; Mallory, J.F., *Thermal Insulation*, Van Nostrand Reinhold, New York, 1969; Gubareff, G.G. et al., *Thermal Radiation Properties Survey*, Minneapolis-Honeywell Regulator Co., Minneapolis, MN, 1960; McAdams, W.H., *Heat Transmission*, 3rd edn., McGraw-Hill, New York, 1954.

Note: Values for the hemispherical emissivity are in plain type. Values for the normal emissivity are in italics.

TABLE F.12
Emissivity of Some Common Materials

Material	Temperature (K)	Emissivity (ε)
Aluminum oxide	1000	*0.55*
Asphalt pavement	300	0.85–0.93
Building materials		
Brick, red	300	0.93–0.96
Gypsum plasterboard	300	0.90–0.92
Wood	300	0.82–0.92
Cloth	300	0.75–0.90
Concrete	300	0.88–0.93
Glass, window	300	0.90–0.95
Ice	273	0.95–0.98
Black paint	300	0.98
White paint, acrylic	300	0.90
Paper, white	300	0.92–0.97
Pyrex	600	*0.80*
Pyroceram	600	*0.78*
Refractories (furnace liners)		
Kaolin brick		*0.57*
Magnesia brick	1000	*0.36*
Alumina brick		*0.33*
Rocks	300	0.88–0.95
Rubber		
Soft, gray	300	*0.86*
Hard, black, rough	300	*0.95*
Sand	300	0.90
Silicon carbide	1000	*0.87*
Skin	300	0.95
Snow	273	0.82–0.90
Soil	300	0.93–0.96
Teflon	300	0.85
Vegetation	300	0.92–0.96
Water	300	0.96

Sources: Touloukian, Y.S. and Ho, C.Y., eds., *Thermophysical Properties of Matter*, Vols. 1–9, Plenum Press, New York, 1972; Mallory, J.F., *Thermal Insulation*, Van Nostrand Reinhold, New York, 1969; Gubareff, G.G. et al., *Thermal Radiation Properties Survey*, Minneapolis-Honeywell Regulator Co., Minneapolis, MN, 1960; McAdams, W.H., *Heat Transmission*, 3rd edn., McGraw-Hill, New York, 1954.

Note: Values for the hemispherical emissivity are in plain type. Values for the normal emissivity are in italics.

TABLE F.13
Lennard-Jones Parameters for Selected Gases

Species	Collision Radius $\sigma_f/2$ (nm)	Interaction Energy (ε/k_B)
Acetone (CH_3COCH_3)	0.230	560.2
Acetylene (CH_2CH_2)	0.202	231.8
Air	0.185	78.6
Ammonia (NH_3)	0.145	558.3
Argon (Ar)	0.177	93.3
Benzene (C_6H_6)	0.267	412.3
Bromine (Br_2)	0.215	507.9
Carbon dioxide (CO_2)	0.197	195.2
Carbon monoxide (CO)	0.185	91.7
Carbon tetrachloride (CCl_4)	0.297	322.7
Chlorine (Cl_2)	0.211	316.0
Chloroform ($CHCl_3$)	0.269	340.2
Cyclohexane (C_6H_{12})	0.309	297.1
Ethane (C_2H_6)	0.222	215.7
Ethanol (CH_3CH_2OH)	0.227	362.6
Ethylene (C_2H_4)	0.208	224.7
Helium (He)	0.113	10.22
Hydrogen (H_2)	0.141	59.7
Hydrogen chloride (HCl)	0.167	344.7
Hydrogen peroxide (H_2O_2)	0.210	289.3
Methane (CH_4)	0.188	148.6
Methanol (CH_3OH)	0.181	481.8
n-Butane (C_4H_{10})	0.234	531.4
n-Hexane (C_6H_{14})	0.297	399.3
Neon (Ne)	0.141	32.8
Pentane (C_5H_{12})	0.284	345

Source: Reid, R.C. et al., *Properties of Gases and Liquids*, 3rd edn., McGraw-Hill, New York, 1977.

COLLISION INTEGRALS [29]

The following expressions for the collision integrals are strictly appropriate for nonpolar molecules only. For polar molecules, a correction involving the dipole moment and properties at the boiling point is required. Rules for this correction and for estimating the viscosity of mixtures can be found in [2,30]:

$$\Omega_{\mu,k} = \frac{1.16145}{\left(k_B T/\varepsilon\right)^{0.14874}} + \frac{0.52487}{\exp\left[0.77320\left(k_B T/\varepsilon\right)\right]} + \frac{2.16178}{\exp\left[2.43787\left(k_B T/\varepsilon\right)\right]}$$

$$\Omega_d = \frac{1.06036}{\left(k_B T/\varepsilon\right)^{0.1561}} + \frac{0.193}{\exp\left[0.47635\left(k_B T/\varepsilon\right)\right]} + \frac{1.03587}{\exp\left[1.52996\left(k_B T/\varepsilon\right)\right]} + \frac{1.76474}{\exp\left[3.89411\left(k_B T/\varepsilon\right)\right]}$$

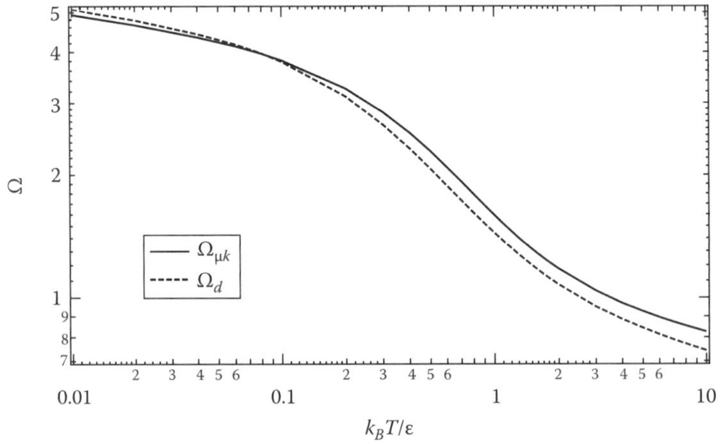

REFERENCES

1. Hirschfelder, J., Curtiss, C.F., and Bird, R.B., *Molecular Theory of Gases and Liquids*, Wiley, New York (1954).
2. Reid, R.C., Sherwood, T.K., and Prausnitz, J.M., *Properties of Gases and Liquids*, 3rd edn., McGraw-Hill, New York (1977).
3. Sherwood, T.K., Pigford, R.L., and Wilke, C.R., *Mass Transfer*, McGraw-Hill, New York (1975).
4. Cussler, E.L., *Diffusion: Mass Transfer in Fluid Systems*, Cambridge University Press, New York (1984).
5. American Society for Metals, *Diffusion*, ASM, Metals Park, OH (1973).
6. Barrer, R.M., *Diffusion in and through Solids*, MacMillan, New York (1941).
7. Ghandi, S.K., *VLSI Fabrication Principles*, Wiley-Interscience, New York (1983).
8. Touloukian, Y.S. and Ho, C.Y., eds., *Thermophysical Properties of Matter*, vols. 1–9, Plenum Press, New York (1972).
9. Touloukian, Y.S. and Ho, C.Y., eds., *Thermophysical Properties of Selected Aerospace Materials, Parts I & II*, Thermophysical and Electronic Properties Information Analysis Center, CINDAS, Purdue University, West Lafayette, IN (1976).
10. Ho, C.Y., Powell, R.W., and Liley, P.E., Thermal conductivity of the elements: A comprehensive review, *J. Phys. Chem. Ref. Data*, 3(Suppl 1), 756–796 (1974).
11. Desai, P.D., Chu, T.K., Bogaard, R.H., Ackermann, M.W., and Ho, C.Y., Part I: Thermophysical properties of carbon steels, Part II: Thermophysical properties of low chromium steels, Part III: Thermophysical properties of nickel steels, Part IV: Thermophysical properties of stainless steels, CINDAS Special Report, Purdue University, West Lafayette, IN (September 1976).

12. American Society for Metals, *Metals Handbook, Properties and Selection of Metals*, 8th edn., Vol. 1, ASM, Metals Park, OH (1961).
13. Hultgren, R., Desai, P.D., Hawkins, D.T., Gleiser, M., and Kelley, K.K., *Selected Values of the Thermodynamic Properties of the Elements*, American Society of Metals, Metals Park, OH (1973).
14. Hultgren, R., Desai, P.D., Hawkins, D.T., Gleiser, M., and Kelley, K.K., *Selected Values of the Thermodynamic Properties of Binary Alloys*, American Society of Metals, Metals Park, OH (1973).
15. American Society of Heating, Refrigerating and Air Conditioning Engineers, *ASHRAE Handbook of Fundamentals*, ASHRAE, New York (1994).
16. Mallory, J.F., *Thermal Insulation*, Van Nostrand Reinhold, New York (1969).
17. Hanley, E.J., Dewitt, D.P., and Taylor, R.E., The thermal transport properties at normal and elevated temperature of eight representative rocks, *Proceedings of the Seventh Symposium on Thermophysical Properties*, American Society of Mechanical Engineers, New York (1977).
18. Sweat, V.E., *A Miniature Thermal Conductivity Probe for Foods*, American Society of Mechanical Engineers, St. Louis, MO, Paper 76-HT-60 (August 1976).
19. Kothandaraman, C.P. and Subramanyan, S., *Heat and Mass Transfer Data Book*, Halsted Press/Wiley, New York (1975).
20. Vargaftik, N.B., *Tables of Thermophysical Properties of Liquids and Gases*, 2nd edn., Hemisphere Publishing, New York (1975).
21. Eckert, E.R.G. and Drake, R.M., *Analysis of Heat and Mass Transfer*, McGraw-Hill, New York (1972).
22. Incropera, F.P. and DeWitt, D.P., *Introduction to Heat Transfer*, 3rd edn., Wiley, New York (1994).
23. Geankoplis, C.J., *Transport Processes and Unit Operations*, 3rd edn., Prentice Hall, Englewood Cliffs, NJ (1993).
24. Liley, P.E., Thermophysical properties in *Handbook of Heat Transfer Fundamentals*, 2nd edn., W.M. Rohsenow, Hartnett, J.P., and Ganic, E.N., eds., McGraw-Hill, New York (1985).
25. Bolz, R.E. and Tuve, G.L., eds., *CRC Handbook of Tables for Applied Engineering Science*, 2nd edn., CRC Press, Boca Raton, FL (1979).
26. NIST Webbook. http://webbook.nist.gov
27. Gubareff, G.G., Janssen, J.E., and Torborg, R.H., *Thermal Radiation Properties Survey*, Minneapolis-Honeywell Regulator Co., Minneapolis, MN (1960).
28. McAdams, W.H., *Heat Transmission*, 3rd edn., McGraw-Hill, New York (1954) (data compiled by H.C. Hottel).
29. Neufeld, P.D., Janzen, A.R., and Aziz, R.A., Empirical equations to calculate 16 of the transport collision integrals $\Omega(l,s)^*$ for the Lennard-Jones (12–6) potential, *J. Chem. Phys.*, 75, 1100 (1972).
30. Wilke, C.R., Viscosity equation for gas mixtures, *J. Chem. Phys.*, 18, 517 (1950).

Appendix G: Comsol® Modules

This appendix contains a listing of the Comsol® modules associated with the text. The modules can be located at http://www/rpi.edu/~plawsky and are listed here organized by chapter. These modules are continually updated and added to so this list is only a subset of what you might find on the site.

Chapter 2

1. Comsol® Module: MatProp
   In many materials of engineering interest, material properties are a function of position. This is especially true for "engineered" materials like bone, teeth, and carbon–carbon composites and even for natural materials such as graphite. The Comsol® module "MatProp" allows you to explore this for graphite. The system consists of a very simple geometry, a square of graphite whose boundaries are set at a given temperature. At the center of the square is a point source of heat of strength, Qsource.

Chapter 4

1. Comsol® Module: BloodShearTest
   A number of blood diseases can be analyzed by looking at blood's rheological properties. Blood can be represented as a Carreau–Yasuda power law fluid where the viscosity is given by

$$\frac{\mu-\mu_\infty}{\mu_o-\mu_\infty}=\left[1+(\lambda\dot{\gamma})^a\right]^{n-1/a}=\left[1+\left(\lambda\frac{dv_x}{dy}\right)^a\right]^{n-1/a} \tag{G.1}$$

$$\mu_\infty=0.0035 \qquad \mu_o=0.16 \qquad \lambda=8.2\,\text{s} \qquad a=0.64 \qquad n=0.313$$

   Assume we are performing a diagnostic experiment shearing blood between two plates. The Comsol® model "BloodShearTest" performs a diagnostic simulation for the full Carreau–Yasuda model shown earlier.
2. Comsol® Module: TwoRegionDiffusion
   One of the major differences between mass transfer and either heat or momentum transfer concerns the boundary conditions at the interface between two media. At the boundary, we must specify two conditions: a condition that links the dependent variable in the two regions and a condition that links the flux of the dependent variable in each region. For heat transfer and momentum transfer, we generally assume continuity of the flux across the boundary and also continuity of the dependent variable. Thus, at the interface, we have only one velocity or one temperature on both sides of the interface. However, in mass transfer, we normally have continuity of the mass flux across the boundary, but most often, the concentration is discontinuous. In one dimension, the boundary conditions are usually of the form

$$\begin{aligned} &x=x_b \qquad c_2=f(c_1) \qquad \text{or} \qquad c_2=K_{eq}c_1 \\ &x=x_b \qquad N_1=N_2 \qquad \text{or} \qquad -D_1\frac{dc_1}{dx}=-D_2\frac{dc_2}{dx} \end{aligned} \tag{G.2}$$

   Comsol® and most finite element programs assume continuity of both flux and dependent variable as their default conditions. To handle conditions like Equation G.2, we need to resort to a trick. This Comsol® module uses an approximation to simulate diffusion across a boundary.

3. Comsol® Module: WallConduction
   This module contains a composite wall geometry composed of a set of American red oak studs and fiberglass batt insulation. The goal of the module is to use the 1-D sum-of-resistance concept to calculate the heat flow through the wall and the temperature at various nodes and then compare that to what the computer predicts in 2-D.

Chapter 5

1. Comsol® Module: *Power Law*
   This module contains a simple parallel plate geometry with which you can see the effects of various power laws on the velocity profile, volumetric flow rate, shear stress distribution, and pressure drop for flow through the channel.
2. Comsol® Module: DiffReact
   Diffusion with simultaneous chemical reaction is a classic problem that can be used to simulate a wide variety of problems ranging from heterogeneous and homogeneous catalysis to the operation of a diode or transistor. Unfortunately, outside of reaction rate expressions that are zero or first order, solutions to the diffusion/reaction problem are impossible to solve analytically to obtain a closed-form solution. Finite element modeling is therefore essential for complicated rate expressions. This module allows you to experiment with changes in rate law.
3. Comsol® Module: *TransistorSS*
   The module solves the steady-state diffusion and reaction problem used to generate Figures 5.20 and 5.21. Actually, the module is set up to solve a more complicated problem involving the movement of positive and negative charges through a variable electric field.

Chapter 6

1. Comsol® Module: Lumps
   When in doubt, an engineer's first response is to use a lumped parameter model to gauge the response of a transient system. The Comsol® module "Lumps" is designed to allow you to investigate when the lumped parameter assumption is valid and how it can be extended to systems that involve generation.
2. Comsol® Module: Cake
   This module is designed to solve the problem of baking a cake similar to the homework problem at the end of the chapter. In fact, the physical property data and dimensions are the same.
3. Comsol® Module: Breakdown
   This module simulates the time to failure of a gate dielectric in an integrated circuit. The structure is a sandwich consisting of a metal anode (Cu), a silicon dioxide dielectric, and a metal cathode (Al). A voltage, V_o, is supplied across the sandwich causing the copper anode to dissolve in the dielectric. The action of the electric field drives the copper toward the cathode where the ions pile up. Once the concentration of copper at the cathode causes the local electric field there to exceed the breakdown field, E_{bd}, the dielectric breaks down and is useless. The module is formed from two equations similar to the diode problem in Chapter 5. One is the continuity equation for the copper ions and the second is Poisson's equation to describe the local electric field.

Chapter 7

1. Comsol® Module: Wave
   This module explores a 1-D tidal wave as it progresses from a shallow channel to a deeper one. The module sets up an initial small wave that progresses over time through the system. The module actually contains three separate regions so that you can change velocities in the different regions and see how the wave changes shape as it propagates.

2. Comsol® Module: Tweeter
   In considering the horn tweeter, we made some simplifications to the model equation to affect an analytic solution. This module allows us to explore the effects of those simplifications as well as more exotic horn shapes. Such solutions cannot be obtained analytically but are musts for contemporary speaker designers to simulate the behavior of complicated systems such as the Bose Wave® radio.

Chapter 8

1. Comsol® Module: PoreDiffusion
   This module looks at diffusion and surface reaction inside the nanopore of a porous dielectric material as it is exposed to a plasma containing a polymerizable agent. In this material, the pores are bottlenecked in shape. We are interested in investigating the transport and growth of polymer material on the walls of the pore. We assume dilute solutions, a first-order chemical reaction on the pore wall surface, negligible convective transport within the pore, and Knudsen diffusion that depends upon the pore diameter.
2. Comsol® Module: PoreEtch
   The module looks at diffusion and surface reaction inside the nanopore of a porous dielectric material as it is exposed to a plasma containing an etchant agent. This is really the sister module to the previous PoreDiff module.

Chapter 9

1. Comsol® Module: Aerocapture
   Aerocapture is an experimental maneuver designed to insert a spacecraft into planetary orbit in an economical way. Instead of using rocket motors for active braking, the spacecraft uses friction with a planet's atmosphere as the braking mechanism. Since spacecraft may be traveling upward of 25 km/s, a great deal of heat is generated during the maneuver, and the delicate instruments within the craft need to be protected by a sophisticated insulation system.

   The "Aerocapture" module is designed to simulate entry into the atmosphere of Titan, Saturn's largest moon. The heat flux to the planet's surface, encountered during entry, the temperature of the moon's atmosphere, and the properties of the insulation materials as a function of temperature are all known and included in the module. The goal of the module is to design an appropriate composite insulation system that protects the internal contents and provides minimum weight and maximum internal volume for the craft. This is clearly a problem that cannot be solved analytically because the material properties and the boundary conditions are all functions of time and temperature. There are two basic materials used: fiberform, a porous amorphous carbon, and aerogel, a highly porous silica. The maximum temperature the aerogel can withstand is 1200 K, while to protect the internal contents of the craft, the bond line temperature at the insulation/aluminum skin interface cannot exceed 250°C.
2. Comsol® Module: CVBRadiation
   The Comsol® module "CVBRadiation" looks at the testing of a miniature heat pipe in space. In this case, the system is being tested in a vacuum chamber on the ground. Two possible scenarios are of interest, one where the heater surface (boundary 1) is held at constant temperature and the other where a constant heat flux is input onto that surface. Since the chamber is evacuated, the only way heat can be transferred to the surroundings is by radiation.
3. Comsol® Module: Skin
   The Comsol® module "Skin" looks at diffusion of a drug through the epidermis and into the dermis layer of the skin. The idea is to evaluate the effectiveness of a transdermal nicotine patch. The diffusivity data for the epidermis and dermis are taken from rat skin models.

4. Comsol® Module: Doublet

   One of the methods we showed for producing the effect of a circular obstruction in the flow field is to use a doublet immersed in a uniform flow. The doublet consists of a source and sink of momentum located in close proximity to one another. The analytical solution to the doublet was shown to be

$$\phi = -\phi_o x - \phi_o L^2 \frac{\cos\theta}{r_o} = -\phi_o L\cos\theta - \phi_o L^2 \frac{\cos\theta}{r_o} \tag{G.3}$$

$$\psi = -\phi_o L\sin\theta - \phi_o L^2 \frac{\sin\theta}{r_o}$$

   where

   ϕ is the velocity potential
   y is the stream function
   $\phi_o L^2$ is the strength of the doublet
   L is the separation between the source and sink
   r_o is the radial distance from the doublet
   θ is the angle from the horizontal

   The Comsol® module solves this for you to compare with the analytical solution.

5. Comsol® Module: Drain

   The module "Drain" mimics the irrotational flow solution combining a uniform flow with a tornado-like flow. The solution to this problem was shown to be

$$\psi = \frac{K}{2\pi}\ln r - \frac{q}{2\pi}\theta + u_o r\sin\theta \tag{G.4}$$

   where

   u_o was the velocity of the plane flow
   K was the circulation around the tornado origin
   q was the strength of the sink at the center of the tornado

   The Comsol® module simulates this system using an initially counterclockwise circulation. The sink is defined in terms of a pressure, while the circulation around the system is defined by defining a small circle around the point sink and defining x and y velocity components on the boundaries of that circle to yield a rotational flow.

Chapter 10

1. Comsol® Module: Sphere

   The Comsol® module "Sphere" is designed to solve the creeping flow problem we solved analytically, numerically. It has the big advantage that it is not limited to creeping flow.

2. Comsol® Module: Curveball

   One of the most interesting phenomena to study exists in sports where the curveball, be it in baseball or soccer, continues to fascinate people. We can learn how such a system behaves using a simulation. The Comsol® module "Curveball" simulates in two dimensions, the action of a rotating sphere immersed in a uniform flow field.

3. Comsol® Module: FluidJet

   Expanding jets occur very often in engineering situations. The Comsol® module "FluidJet" explores one such scenario where a water jet expands into another water stream. The fluid jet is moving relatively quickly with respect to the larger stream.

4. Comsol® Module: Biochip
   The recent development of rapid screening techniques has led to a simultaneous growth in integrated laboratory devices. The Comsol® module "Biochip" builds on a canned Comsol® example to look at diffusion and reaction in a small, integrated module. In this case, we use electrokinetic flow to drive the fluids but also include some packing material that has a catalytic surface. Thus, we can inject a fluid containing a reactant and have that reactant interact with the packing in the integrated column.
5. Comsol® Module: LabChip0-3
   The popularity of lab-on-a-chip systems has brought forth a flaw in their operation, namely, the difficulty in getting two streams, exhibiting laminar flow, to mix. In such instances, the only mixing occurs via diffusion. The Comsol® modules "LabChip0-3" investigate the ways in which mixing can be promoted in these devices. Labchip1 provides no internal surfaces to promote mixing, while LabChip1-3 incorporate various geometries of internals to help aid the mixing.
6. Comsol® Module: Rayleigh
   A classic problem in coupled heat and momentum transport concerns the onset of natural convection in a pool of liquid heated from below. We have all observed this phenomenon, the onset of a cellular flow pattern. This flow pattern, termed Rayleigh–Benard convection, has been studied extensively since it is the process that governs the motion of continents, currents in the ocean, and atmosphere among other phenomena. The onset of natural convection occurs at a critical value of the Rayleigh number, Ra, that is defined as

$$\mathrm{Ra} = \frac{\beta g d^3 \Delta T}{\alpha \nu} \qquad \mathrm{Ra}_c = 1707.8 \quad \text{Closed system} \tag{G.5}$$

   where
   β is the thermal expansion coefficient
   g is the gravitational acceleration
   d is the fluid depth
   ΔT is the temperature difference between hot and cold surfaces
   α is the thermal diffusivity
   ν is the kinematic viscosity

   The Comsol® module allows one to solve this problem in a cavity.
7. Comsol® Module: ReactingWall
   Falling film systems are among the easiest systems to analyze and understand how convection affects mass and heat transfer coefficients. The Comsol® modules "ReactingWall" and "HeatedWall" investigate these phenomena. The nice part about these systems is that we have analytical solutions to the simplest cases.
8. Comsol® Module: HeatedWall
   Falling film systems are among the easiest systems to analyze and understand how convection affects mass and heat transfer coefficients. The Comsol® module "HeatedWall" investigates these phenomena for Newtonian and non-Newtonian fluids.

Chapter 12

1. Comsol® Module: BLVelocity
   This module considers boundary layer flow for a Newtonian fluid. One of the key issues involving the solution is what viscous force is on the solid surface, that is, what the friction factor is. The Comsol® module "BLVelocity" can be used to test these solutions and to see what the effect of the boundary layer approximations is.

2. Comsol® Module: BLnonNewtonian
   Boundary layer theory was developed for Newtonian fluids and also specified that the viscosity would be independent of temperature or composition. Adding non-Newtonian behavior increases the nonlinearity of the problem and the difficulty in obtaining a solution. The Comsol® module "BLnonNewtonian" explores boundary layer flow over a flat plate using non-Newtonian fluids.
3. Comsol® Module: BLTemperature
   Boundary layer theory allowed us to predict the heat transfer coefficient from a knowledge of the thermal and flow properties of a fluid. As we have seen before, the heat transfer coefficient is dependent upon two fundamental dimensionless numbers, the Reynolds number and the Prandtl number. The Comsol® module "BLTemperature" allows us to explore this in more detail.
4. Comsol® Module: BLConcentration
   Boundary layer theory allowed us to predict the mass transfer coefficient from a knowledge of the mass transfer and flow properties of a fluid. As we have seen before, the mass transfer coefficient is very similar to the heat transfer coefficient and is dependent upon two fundamental dimensionless numbers, the Reynolds number and the Schmidt number. The Comsol® module "BLConcentration" allows us to explore mass transfer in more detail.

Chapter 13

1. Comsol® Modules: NACA2412L/NACA2412T/NACA2412time
   The precursor to NASA, NACA (National Advisory Committee for Aeronautics), was instrumental in furthering the development of flight, both military and civilian. As part of their work, they did extensive wind tunnel testing on a variety of standard-shaped airfoils. These data are still available and serve as a benchmark for fluid mechanics simulation programs. We can benchmark Comsol® against these systems too.
   The Comsol® module "NACA2412" was written by students to investigate the properties of the standard airfoil #2412 and to see if Comsol® would reproduce the experimental results.
2. Comsol® Module: Nusselt
   Heat transfer around a cylindrical object is extremely important since most heat exchangers operate using a shell and tubes. The flow field surrounding a cylinder is very complicated, and so the heat transfer coefficient is a function of angular position. The Comsol® module "Nusselt" explores heat transfer about a cylinder in an attempt to see if numerical simulation can represent some features of this figure.
3. Comsol® Modules: Whalefin and Whalefin0
   Recent research shows that humpback whale flipper is more efficient from both a drag and lift standpoint than a conventionally shaped airfoil (Miklosovic et al. *Phys Fluids* 16, L39, 2004). These advantages are being tested in aircraft wings and wind turbine blades. The Comsol® modules "Whalefin" and "Whalefin0" are designed to try and show this numerically in a very simple way. Each looks at a simple, 2-D geometry using a fin or airfoil in water:
   a. Run the two modules and compare the flow fields and the viscous stresses, that is, the drag around each object. Which geometry exhibits the least drag? What aspects of the flow field hint at lowering the stress? Does the drag advantage, if evident, exist for all velocity ranges that you can get a solution for?
   b. Alter the boundary conditions above and below the fin (boundaries 2 and 3) to make them neutral conditions. Then you can alter the input velocity on boundary #1 to simulate different angles of attack. Run the modules for several angles of attack and calculate the lift force on each structure. Is there a lift advantage for either structure?

c. For the runs in part (b), also calculate the total drag force. Then combine parts (b) and (c) to calculate the lift to drag ratio. Is there an advantage for a particular geometry? What does that say about performing acrobatic maneuvers?

Chapter 14

1. Comsol® Module: TurbBL
   Turbulent boundary layers are fundamentally different from their laminar counterparts, and there are still no analytical solutions to describe them or their structure. What we have is an abundance of experimental evidence that shows that the friction factor, heat transfer coefficient, and mass transfer coefficient all depend upon the Reynolds number raised to a power larger than one-half and that the velocity profile obeys the "universal" law. This module allows you to investigate the behavior.
2. Comsol® Module: Truck
   Everyone has experienced the influence of drag on the highway when a truck passes by. The other drafting experience is in automobile or bicycle racing where a following vehicle follows extremely closely behind a leading vehicle to effectively travel in the lead vehicle's wake. Similar phenomena likely exist in the natural world in the formation of geese and schools of fish. The Comsol® module "Truck" looks at one such scenario, drafting in the wake of a semi. The module uses the k-ε turbulence function in Comsol® to calculate the flow field behind a truck in 2-D.
3. Comsol® Module: Turbcylinder
   Turbulent flow about a cylindrical object is quite a bit different than laminar flow. The boundary layer separates at an angle of 140° rather than the laminar angle of 109.9°. Since boundary layer separation is delayed, the drag on the cylinder is also less. The price one pays is that energy dissipation in turbulent flow is higher than in laminar flow. This module investigates turbulent flow behind a cylinder.

Chapter 15

1. Comsol® Module: Radiation Fin
   This model lets you explore radiative exchange using a pair of fins designed for radiation. You can modify the module and geometry to see how radiative exchange affects the axial and lateral temperature profile in the fins.
2. Comsol® Module: Toaster Oven Operation
   It does not take much complication before we can no longer handle view factors and radiation analytically. Even the simplest of items, the toaster oven can be too much to handle. Toaster ovens are essentially boxes that have a set of heating elements at the top and bottom. Most elements are made of Calrod®, a ceramic tube that contains a resistive metal-heating element inside. While toaster ovens are versatile, their few heating elements, if not placed correctly, make horrible toast. Figure P15.15 shows the configuration of a typical oven. It contains four heating elements of emissivity 0.9, and two radiation shields of emissivity 0.05, located behind the heating elements. The heating elements draw 1500 W total power. The sidewalls of the oven have an emissivity of 0.8. The simulation looks to toast a slice of bread whose thermal properties are $k = 0.106$ W/m K, $C_p = 2516$ J/kg K, $\rho = 236$ kg/m^3, $\varepsilon = 0.8$.
3. Comsol® Module: Sierpinski
   The Comsol® module "Sierpinski" is a slightly less complicated version of the carpet in the chapter. That makes it a bit easier to solve and display. The module assumes a constant heat flux into the left-hand face of the carpet shown in Figure P15.16, radiation to the ambient environment at 298 K on the right-hand face, and an insulated top and bottom.

The material properties are set as those of fused silica glass with internal surfaces set to have an emissivity of 0.8.

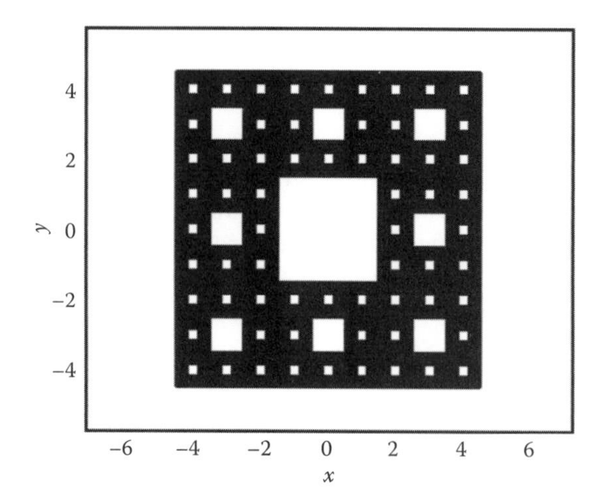

Index

G

H

J

K

L

N

R

S

T

U

V

W